Solar Collectors, Energy Storage, and Materials

Solar Heat Technologies: Fundamentals and Applications
Charles A. Bankston, editor-in-chief

1. *History and Overview of Solar Heat Technologies*
Donald A. Beattie, editor

2. *Solar Resources*
Roland L. Hulstrom, editor

3. *Economic Analysis of Solar Thermal Energy Systems*
Ronald E. West and Frank Kreith, editors

4. *Fundamentals of Building Energy Dynamics*
Bruce Hunn, editor

5. *Solar Collectors, Energy Storage, and Materials*
Francis de Winter, editor

6. *Active Solar Systems*
George Löf, editor

7. *Passive Solar Buildings*
J. Douglass Balcomb and Bruce Wilcox, editors

8. *Passive Cooling*
Jeffrey Cook, editor

9. *Solar Building Architecture*
Bruce Anderson, editor

10. *Fundamentals of Concentrating Systems*
Lorin Vant-Hull, editor

11. *Distributed and Central Receiver Systems*
Lorin Vant-Hull, editor

12. *Implementation of Solar Thermal Technology*
Ronal Larson and Ronald E. West, editors

Solar Collectors, Energy Storage, and Materials

edited by Francis de Winter

The MIT Press
Cambridge, Massachusetts
London, England

This book was set in Times Roman by Asco Trade Typesetting Ltd., Hong Kong and was printed and bound in the United States of America.

Library of Congress Cataloging-in-Publication Data

Solar collectors, energy storage, and materials / edited by Francis de Winter.
   p. cm.—(Solar heat technologies, fundamentals, and applications; 5)
   Includes index
   ISBN 0-262-04104-9
   1. Solar collectors. I. de Winter, Francis. II. Series: Solar heat technologies; 5.
TJ812.S63   1990
621.47—dc20
                                                           90-43142
                                                             CIP

Contents

Series Foreword by Charles A. Bankston   ix

Preface   xi

Acknowledgments   xvii

I   SOLAR COLLECTORS

1   Overview   3
Francis de Winter

2   Collector Concepts and Designs   25
Ari Rabl

3   Optical Theory and Modeling of Solar Collectors   56
Ari Rabl

4   Thermal Theory and Modeling of Solar Collectors   99
Noam Lior

5   Testing and Evaluation of Stationary Collectors   183
Robert D. Dikkers

6   Testing and Evaluation of Tracking Collectors   289
David W. Kearney

7   Optical Research and Development   305
Roland Winston

8   Collector Thermal Research and Development   358
Charles A. Bankston

9   Collector Engineering Research and Development   442
Charles F. Kutscher

10   Solar Pond Research and Development   468
John R. Hull and Carl E. Nielsen

11   Cost Issues and Opportunities   511
John A. Clark

12   Reliability and Durability of Solar Collectors   538
William Freeborne

13   Environmental Degradation of Low-Cost Solar Collectors: Research Issues and Opportunities   579
Fred Loxsom and Eugene Clark

II   ENERGY STORAGE FOR SOLAR SYSTEMS

14   Overview   611
C. J. Swet

15   Storage Concepts and Design   625
C. J. Swet

16   Analytical and Numerical Modeling of Thermal   692
Energy Storage
John R. Hull

17   Testing and Evaluation of Thermal Energy   731
Storage Systems
Robert D. Dikkers

18   Storage Research and Development   763
Allan I. Michaels and C. J. Swet

19   Issues and Opportunities   796
C. J. Swet

III   MATERIALS FOR SOLAR TECHNOLOGIES

20   Overview   831
Stanley W. Moore

21   Materials for Solar Collector Concepts and Designs   874
Richard Silberglitt and Hien K. Le

22   Theory and Modeling of Solar Materials   904
Carl M. Lampert

23   Testing and Evaluation of Solar Materials   933
Hien K. Le and Richard Silberglitt

24   Exposure Testing and Evaluation of Performance   955
Degradation
Thomas E. Anderson

25   Solar Materials Research and Development   974
Richard Silberglitt and Hien K. Le

26   Solar Materials Issues and Opportunities   1040
C. J. Swet

Contributors   1061
Index   1073

Series Foreword

Charles A. Bankston

This series of twelve volumes summarizes research, development, and implementation of solar thermal energy conversion technologies carried out under federal sponsorship during the last eleven years of the National Solar Energy Program. During the period from 1975 to 1986 U.S. Department of Energy's Office of Solar Heat Technologies spent more than $1.1 billion on research development, demonstration, and technology support projects, and the National Technical Information Center added more than 30,000 titles on solar heat technologies to its holdings. So much work was done in such a short period of time that little attention could be paid to the orderly review, evaluation, and archival reporting of the significant results.

It was in response to the concern that the results of the national program might be lost that this documentation project was conceived. It was initiated in 1982 by Frederick H. Morse, director of the Office of Solar Heat Technologies, Department of Energy, who had served as technical coordinator of the 1972 NSF/NASA study "Solar Energy as a National Resource" that helped start the National Solar Energy Program.

The purpose of the project has been to conduct a thorough, objective technical assessment of the findings of the federal program using leading experts from both the public and private sectors, and to document the most significant advances and findings. The resulting volumes are neither handbooks nor textbooks, but benchmark assessments of the state of technology and compendia of important results. There is a historical flavor to many of the chapters, and volume 1 of the series will offer a comprehensive overview of the programs, but the emphasis throughout is on results rather than history.

The goal of the series is to provide both a starting point for the new researcher and a reference tool for the experienced worker. It should also serve the needs of government and private-sector officials who want to see what programs have already been tried and what impact they have had. And it should be a resource for entrepreneurs whose talents lie in translating research results into practical products.

The scope of the series is broad but not universal. It is limited to solar technologies that convert sunlight to heat in order to provide energy for application in the building, industrial, and power sectors. Thus it explicitly excludes photovoltaic and biological energy conversion and such thermally driven processes as wind, hydro, and ocean thermal power. Even with this limitation, though, the series assembles a daunting amount of information. It represents the collective efforts of more than 200 authors and editors. The volumes are logically divided into those dealing with general topics such as the availability, collection, storage, and economic analysis of solar energy and those dealing with applications.

This volume covers the components and materials that make up most solar energy systems. It is devoted primarily to the research, development, and analysis of solar collectors, energy storage, and special solar materials. Mathematical, numerical, and physical modeling of components are emphasized, and numerous laboratory studies of the basic optical, thermal, fluid dynamic, and mechanical behavior of components and materials are reviewed and their important findings reported. The reader will find the general characteristics of virtually every type of collector and energy storage technique discussed, but for the most part they are treated as generic devices or concepts rather than products.

There is a heavy emphasis on the physics, chemistry, and engineering of solar components and materials, and a lighter emphasis on their cost-effectiveness. This reflects the fact that few of the components or special materials covered here are in widespread commercial use. The use of solar components in systems is covered in other volumes of the series. The reader is referred to volumes 6 through 11 which deal with the many applications of solar heat, and to volume 12 in which the efforts made in the 1970s and 1980s to encourage the general use of solar energy and the development of a viable solar industry are described.

Preface

The oil embargo of the early 1970s led to a widespread belief that an energy crisis existed. This produced a period of almost explosive growth in work on energy-related issues. Governments formed departments or secretariats of energy, and they set up bilateral and multilateral agreements and programs. Private research institutes were created. Large amounts of money were allocated. At universities new courses were begun. Research laboratories and companies were founded to investigate and commercialize energy technology.

Before the embargo there had been almost no concern about energy usage or energy efficiency. Alternative energy sources were, by and large, unexplored. Energy was considered to be cheap and virtually inexhaustible by most of the public and by most in government and industry. Homes, appliances, and a myriad of other devices were built as cheaply as possible, with no regard for energy efficiency. Somebody else often paid the energy bills anyway, and furthermore the bills were cheap. Many wasteful energy practices were made advantageous by taxation laws, institutional practices, building codes, or other regulations and customs.

It had of course been known for many years that fossil fuel resources were finite. In 1970 the United States became the first major oil producer to reach the M. King Hubbert oil production peak. M. King Hubbert, a geologist working for the Shell Oil Company, had already predicted in 1956 that the production (the extraction) of a finite mineral resource such as oil would increase at first as the large and easy sources of oil were found and tapped to satisfy an increasing demand, and that later the production would peak and decrease in an inexorable fashion as the large and easy sources slowly were exhausted and the new sources became harder to find, harder to tap, and smaller. Hubbert had accurately predicted in 1956, to the disbelief and chagrin of some, that the peak of U.S. production would occur in 1970.

The U.S. oil production peak came and went in 1970 without much notice. It was not until the 1973 oil embargo that the public became worried about energy supply. Major programs started throughout the world to find and extract more oil, coal, and gas. A serious effort was made to quantify the energy consumption everywhere, and to switch from oil to coal, gas, or other fuels. A major effort started in saving on energy consumption by cutting back on speed limits and thermostat settings, by turning out lights when not needed, and by changing other habits which were felt to be responsible for unnecessary energy consumption. Many people began thinking about geothermal energy, nuclear fusion, and fission. Major programs were initiated in renewable energy technologies including wind power, photovoltaics, wave energy, tidal power, ocean thermal energy, and power towers—technologies that people had only dreamed about earlier. Work intensified on more familiar renewable energy technologies, such as small hydroelectric devices, solar heating and cooling of buildings, and solar heating of water and swimming pools.

Legal and institutional questions were examined. Tax credits were extended to energy conservation and renewable energy devices. Codes and regulations were promulgated that required standards of energy efficiency in buildings, appliances, and other devices. Automobile mileage (efficiency) requirements were established. A better understanding was gained of the institutional and legal factors that favored the use of fossil fuel. The environmental effects of using fossil fuel became more clearly recognized.

These activities have had major impacts on national energy utilization. In renewable energy many changes are already visible with many more on the way. At present, wind and solar thermal power plants supply more than 1% of the electrical power for California. Solar energy is used worldwide for the heating of small swimming pools. Biogas and biomass have grown in importance. There are hundreds of thousands of homes that make use of the sun for passive heating, heating water, and daylighting. Many new homes, even in extremely cold climates, use virtually no energy for space heating. This is due, in part, to the use of passive solar heating but mostly to energy conservation features that have been found to be cost-effective. Photovoltaics has had only a small impact, but growth in the use of photovoltaics is widely considered to be unstoppable.

The major impact, however, has been in energy conservation. The efficiency of new automobiles has increased enormously. Many appliances have been made more efficient, and further improvements are on the way. As mentioned above, some new homes are incredibly efficient when compared with the standards of the past. Heat recovery possibilities have been found in many processes and buildings. The energy efficiency of many processes and industries has been improved. Airliners have been made more efficient. Cogeneration is used with small internal combustion engines *and* in large power plants to utilize heat that was formerly wasted. The new Public Utility Regulatory Policy Act (PURPA) legislation requires the local utility to buy excess power at reasonable prices. Cogeneration combined with PURPA has had such an impact that, on many utility grids, a significant fraction of recently installed power comes from private cogeneration facilities.

The oil embargo and the subsequent increases in fossil fuel prices had serious financial impacts. The energy crisis mentality that followed led to the development of new and improved energy technology, resulting in the use of less energy—much less energy per unit of GNP. As might have been expected, the end result was an "oil glut." After all, the exploration, production, and depletion processes for oil resources are processes involving decades, and in a period of less than one decade, the energy consumption practices of society were changed, suddenly and significantly.

Oil resources are, of course, still finite, but with the development of the oil glut, the costs of oil, coal, gas, and gasoline have gone down. The public interest in energy questions also has diminished. These factors have had an effect in many areas. Large and inefficient cars are making a comeback. In 40 states the speed limit has gone back up to 65 MPH. Budgets in energy conservation R&D, renewable energy R&D, and all other areas of energy R&D have been reduced significantly. Many energy companies have gone bankrupt, many laboratories have closed, and many groups have disbanded. Many people have left the energy field for other professional pursuits.

At some time in the future, a major effort on solar energy applications may be started again. In a few decades the depletion of fossil fuels will inevitably force society to take solar energy seriously. In the near future it is, however, unlikely that a worldwide oil shortage will be the cause of much in renewed interest. The "oil glut" is unlikely to end soon, and oil disruptions are painful to producers and consumer alike. Renewed interest could, however, come as a result of concern over the dwindling U.S. oil reserves, the mounting U.S. foreign debt, and recognition of the effect of oil imports on the U.S. balance of payments. There might be a recognition of the contributions that energy conservation and renewable energy can make to domestic employment. Another cause for renewed interest could be recognition of the contribution renewable sources of energy could have on the reduction of acid rain and air pollution, including the "greenhouse effect" caused by the uncontrolled addition of carbon dioxide and other gases to the atmosphere. These atmospheric pollution effects are becoming better defined, more serious, and more widely known.

Prior to 1970 nearly all solar energy R&D was initiated because individual researchers thought the work useful or interesting and because they had the time and money to get the work done. Much useful work was done that way, and much can be done in the future, but the pace can be greatly accelerated by government support and public concern.

Future work in solar energy must be based on the body of information created in the 1970s and 1980s. To do so requires a coherent overview of the state-of-the-art. These volumes are designed to bridge the gap between the solar energy R&D effort which has been winding down and the solar energy R&D work which is inevitable in the future.

The authors were chosen from among the most active leaders in their areas of professional concentration. It is important to get their interpretation of recent work—*and* of the earlier work on which it was based—before the memories fade and the files are lost. The objective was to prepare a type of annotated bibliography—not a textbook or a handbook but a treasure map for textbook authors or for future researchers in the field.

Other volumes in this series are concerned with solar energy technology development, demonstrations, or implementation programs, or with specific technical topics of various kinds. Some of these are insolation characterization, solar energy economics, or specific applications such as active systems, passive systems (solar architecture), and solar thermal electric applications.

This volume deals with components or subsystems—with heat collectors, thermal storage, and materials. There are other components, such as pumps, valves, and controllers that are not truly solar devices and are not unique to solar systems. It seemed neither necessary nor desirable to include such equipment in this volume.

This volume is divided into three parts: collectors of thermal energy, thermal storage, and materials. At the beginning of each part is an introductory chapter that summarizes and comments on the part's organization and contents. The treatment for each of the three topics has been roughly similar. Each includes results on performance, from both a thermal performance and durability perspective, provides an overview of the criteria and techniques available for hardware design, and documents design concepts that have been shown to be successful as well as some that have failed.

The durability of system components and materials is critical. For a device to be as cost-effective as possible, it should not be overdesigned for either performance or durability. In terrestrial solar energy applications, this means that one may not be able to afford a sturdy device that withstands all the damages of ultraviolet radiation, the environment, and the weather, but may need a device that has been made as inexpensive and as flimsy as one dares. The terrestrial solar energy environment is still not very well understood. Materials can respond in unexpected ways. Many degradation mechanisms are quite complex and difficult to model. It is only when the degradation mechanisms are relatively simple and well understood that one can perform accelerated life tests that can be easily related to expected product behavior. Most solar energy equipment should survive several decades to be economically viable, yet most solar energy R&D started in the mid-1970s, only about one decade ago. We cannot yet have all the answers to the durability questions, and it may be many more years before we do. In the meantime solar energy may have to depend on "safe" materials that are well known but expensive, such as glass and the more durable metals.

The treatment of the costs of components is somewhat uneven in the three parts of the volume. Material cost is a straightforward subject and is not discussed to any great extent. The costs of storage subsystems may not be straightforward, but they are not very controversial. Collector costs are something else. The literature cites many unrealistic cost estimates and projections for collectors and collector concepts. Collector cost projections have been made for collectors as installed on site in a system whose costs are far below the costs encountered in real systems involving commercial hardware that is marketed, sold, and installed following normal industry practices. Therefore it seemed worthwhile to include a thorough review of the typical escalation of costs in the various steps between the purchase of the raw materials needed for a collector's manufacture and the cost of a collector finally installed in a working system.

It can be appreciated that with all the authors, the reviewers, and others listed in the acknowledgments, I have had an enormous amount of help putting this volume together. If after all the help there are still errors or omissions in this volume, it is quite clear that I must take the blame.

Acknowledgments

Many people cooperated in the creation of this series of volumes, and in the preparation of the present volume. Frederick Morse realized a thorough overview of the state-of-the-art in solar thermal energy would be especially valuable at this time, and he made the preparation of these volumes into a significant part of the current DOE SOLAR program. Charles Bankston, with the assistance of Lynda McGovern-Orr, organized and oversaw the whole multivolume preparation effort, and was also involved in the establishment of the outline of this volume and in the choice of authors and reviewers. Oscar Hillig took care of the contractual details for the authors and editors involved. Paul Notari and Charles Berberich headed the SERI involvement, and Nancy Reece was in charge of the editing and the word processing at SERI. At Altas the work would have been impossible without the constant assistance of Elizabeth Clark.

It is quite evident that for many of the authors the work was of enormous importance, and many of the chapters are indeed much more detailed and complete than anything we might have hoped for. In all cases the authorship represented a major undertaking, and this was made clear to the authors at the start. Some who felt unable to accept this responsibility nevertheless were able to serve as reviewers, and the reviewers were able to make significant contributions by pointing out references or programs that had been overlooked at the first-draft stage, or by pointing out nuances or details that had not yet been formally incorporated.

I would like to thank the following reviewers: Tom Anderson, Charles A. Bankston, William A. Beckman, David K. Benson, Paul Berdahl, Karl Böer, Chris Cameron, Robert Cameron, James Clinton, Roger Cole, Francis de Winter, Francis Dudley, Robert French, Gordon Gross, E. Harley, K. G. T. Hollands, Allan Hurley, S. Karaki, S. A. Klein, Jan F. Kreider, Frank Kreith, LBL reviewers, Carlo LaPorta, Carl M. Lampert, Noam Lior, G. O. G. Löf, Harold Lorsch, Allan Michaels, Stanley W. Moore, Dennis J. Morrison, Don Neeper, Ari Rabl, Kent Reid, Stephen L. Sargent, John Schultz, B. O. Seraphin, Lorin Vant Hull, Roland Winston, and Federica Zangrando.

I SOLAR COLLECTORS

1 Overview

Francis de Winter

1.1 Introductory Comments

This volume was prepared as an extended, annotated bibliography in the solar thermal energy collection field, documenting the state-of-the-art in the late 1980s. It covers collectors of solar thermal energy, including salt gradient solar ponds, flat plate collectors, compound parabolic concentrators, and other stationary and tracking collection systems. Collectors that are used for building applications are emphasized since power and industrial applications are considered in other volumes.

Collector technology is closely related to the subjects covered in other volumes of this series. Knowledge of the available solar radiation is essential to the study of solar collectors—*Solar Resources* (volume 2). In the *Economic Analysis of Solar Thermal Energy Systems* (volume 3), collectors are shown to be the single most important cost component. *Active Solar Systems* (volume 6) is entirely based on the use of collectors. Some use of collectors is noted in *Passive Solar Buildings* (volume 7) but little is mentioned in *Passive Cooling* (volume 8). *Solar Architecture and Planning* (volume 9) is often concerned with integrating collectors into a building as elegantly as possible. *Distributed and Central Receiver Systems* (volume 11) deals with high temperature collection systems, as does *Fundamentals of Concentrating Systems* (volume 10). *Implementation of Solar Thermal Technology* (volume 12) also has collection as a principal concern.

Collectors are normally incorporated into a system involving many other components, such as pumps, heat storage systems, plumbing parts needed to complete fluid loops, heat exchangers, controllers, and valves. University texts that include a good coverage of the usage of collectors in complete systems are Kreith and Kreider (1978), Duffie and Beckman (1980), Lunde (1980), and Rabl (1985).

There are collectors of many types intended for different applications. Collectors used on buildings for water or space heating are often flat plate collectors and normally involve little or no concentration. Industrial process heat collection may involve salt gradient solar ponds or concentrating collectors. Thermal power cycles normally use concentration to increase the collection temperature and the conversion efficiency.

My overview of this volume is arranged in three sections. In section 1.2 I have included my personal comments on the literature, research, performance, and technology of thermal energy collection which may be helpful for those who are not familiar with the field. In section 1.3 the organization of the section topics is explained and the content of individual chapters discussed. Section 1.4 gives my comments on possible uses for this volume.

1.2 General Comments

Solar thermal energy collection is a peculiar technical field. People have worked with collectors for centuries, yet much of the technology has never come to full fruition as has been the case with the bicycle, the television, the gas stove, or any one of the many other devices which we use in daily life. The equipment often looks simple, yet producing a cost-effective and durable design requires many talents and much experience.

1.2.1 Comments on the Collector Literature

To appreciate the objectives and the usefulness of the present work on collectors, a review of some unfortunate aspects of the previous work on solar energy and collectors as reported in the literature seems appropriate.

Solar energy has attracted many competent people, but it has also attracted many curious followers. Nowhere is this more evident than in the literature on heat collectors. For each useful paper in the literature, there are many that are frivolous, irrelevant, trivial, or wrong.

For centuries people have studied various concentrating and flat plate collectors.[1] Solar energy collection has been plagued by the deceptive simplicity of the physics. Many people feel qualified to design collectors to heat swimming pools after having experienced hot water coming from a garden hose left lying in the sun. Many who have seen one flat plate collector soon feel they understand all there is to know about them. After playing with a magnifying glass, many feel confident they have a good grasp of solar concentration. This apparent accessibility has led to much trivial work. Many have worked in the field without understanding what they were doing, and without knowing (or caring to know) what others had done before them.

Collector R&D is usually applied R&D tied to a specific collector concept or design. A specific collector concept or design makes commercial sense only if it is potentially cost-effective and durable. Many projects noted in the literature suffer from excessive optimism about costs or durability. Many of the cost claims in the literature are not reliable or consistent. Few of those involved in collector R&D were aware of the true costs of a commercial venture: the costs of fabrication, distribution, R&D, sales, advertising, transportation, installation, service, and warranty fulfillment. Many had heard of the 'markups' involved, believed these markups were primarily connected with profits, and believed they were exorbitant and unnecessary. Just as some have studied collector designs without realizing that they might never be cheap enough, still others have studied designs without realizing that they might not be sturdy or durable enough. Durability and sturdiness can only be demonstrated in the real world—the world of real weather, real installers, real customer neglect, real service personnel, and real warranty fulfillment costs. With many of the authors confused, it is easy to appreciate that the literature can be confusing.

The cost issues reported in the literature on do-it-yourself collector concepts are quite different. Many do-it-yourselfers have built questionable designs that nevertheless have satisfied their builder-owner *and* reduced fossil fuel consumption. One must recognize that the do-it-yourself movement often has curious economic rules. Often labor is not counted as an expense but as a benefit. A labor-intensive device that uses cheap materials does not need a good design or even a long life to be successful in the eyes of the do-it-yourselfer. However, these same designs may have no commercial significance. The same may be true of a collector that is built properly with good materials but is only economical because the labor was not only free but perhaps therapeutic.

The energy crisis and oil embargo in the early 1970s provided a stimulus to solar energy R&D in the United States, just as sputnik some 15 years earlier had stimulated aerospace R&D. The result was an explosive growth in the solar energy literature that paralleled the earlier growth in the aerospace literature. In both cases the quality of the literature was uneven.

A further aerospace analogy is instructive. The space age made radiation heat transfer a very important field because it was necessary for the thermal analysis and design of spacecraft. In the most prestigious peer-reviewed journals of the engineering literature of the early 1960s, one can find a number of the radiation view factors originally published by Lambert (1760) more than two centuries earlier, derived again and published again by people who apparently barely knew (or cared) that Lambert ever existed.

Similarly the solar energy field in the mid-1970s suffered an influx of people who were neither knowledgeable nor thorough. Very good work has been done in the years since. For anyone who wants to find this work, it is necessary to wade through mountains of literature, including an enormous number of items that are simply research reports required by a contract, by a specific job, or by the researcher's ego.

In the 1970s funding sources were numerous, including international, national, and state agencies as well as private foundations. Private industry,

industry groups, and academic institutions provided funds for internal work and for some outside contracts. Rejection of a questionable proposal often did not mean the end of the proposal but simply a search for new funding sources. Many questionable proposals were funded. Much research was nonsearch.

There was (and continues to be) a disappointingly small reaction to the flood of trivial and incorrect presentations, claims, papers, and other publications in the solar energy field. Too few sought to refute misleading claims or to rebut frivolous presentations and publications. There was even a solar energy college "textbook" published that was quite whimsical and misleading. The textbook and its authors faded from the solar energy scene shortly after a critical review of the book was published in the April 1977 issue of *Solar Age*. Many unwarranted and unchallenged projections on solar energy collection and usage had however already found their way into national television coverage, into the *Wall Street Journal, Fortune, National Geographic, Time, Newsweek, Science*, and many other publications. Those who question the need for more serious standards in the solar energy field would do well to review this curious episode.

Valuable contributions to the solar energy collector technology have been produced by people in many different fields, from different organizations and countries. The results, in several languages, have appeared in a wide variety of journals, reports, and conference proceedings. No single, widely accepted publication has emerged, and this makes the search for information more difficult.

If these comments about the literature seem negative, they have pointed out that finding good work on solar energy collectors is even more difficult than finding a needle in a haystack. In the needle in a haystack case, it is difficult to mistake a needle when it is found. One may, however, need to read a good paper repeatedly before recognizing one has found a gem, and it is easy to overlook a good paper altogether.

This volume has been intended to provide a reliable guide to the literature. If this book can help to limit the number of questionable publications and projects in the future, then the effort in preparing it will have been amply justified. The authors of the individual chapters are to be congratulated on their efforts which have been herculean. The solar energy field is indebted to them.

1.2.2 Comments on Previous Collector Research

Prior to the 1970s most work on collectors was performed by individuals and private groups as documented by Butti and Perlin (1980), Jensen (1959), Veltford (1942), and de Winter (1975a).

John Ericsson funded a major solar energy effort in the 1860s and continued to work in solar energy, but he made no lasting contribution to collector technology. In the late nineteenth-century the Smithsonian Institution, first under Samuel Pierpont Langley and then under Charles Greeley Abbot, started a major solar energy effort that continued until the Smithsonian Institution Radiation Biology Laboratory was closed recently. Radiation measurements and plant response to solar energy were the major areas of interest. There was only a modest interest in collectors, and there were no major contributions to the technology.

The most important effort in solar collector technology prior to 1970 was the work on flat plate collectors done by the Godfrey L. Cabot Solar Energy Program at the Massachusetts Institute of Technology. The program was started with a donation of 100 shares (equivalent to about $650,000 at the time) of Cabot Company stock by Dr. Godfrey L. Cabot in 1938; the income from these funds was to be used for the next 50 years on research on the utilization of solar energy by nonliving organisms. Professor Hoyt C. Hottel headed the MIT work. The year before the establishment of the MIT program, a parallel program was established at Harvard University (with similar funding) to investigate the utilization of solar energy by living organisms— basically a biomass program.

The first full year of work on the Godfrey L. Cabot program was devoted to a careful study of the literature in the solar energy field (Hottel 1970s). This has clearly not been done by many of the solar energy groups in the last few years, but it might serve as an example for future programs. Hottel and Woertz (1942) was the first quantitative and solid paper in the flat collector field, laying the groundwork for Whillier and others who followed.

In the late 1950s and early 1960s in Israel, Tabor's work on collectors included pioneering research on selective surfaces, collector analysis, and salt gradient solar ponds. Another notable effort was that of Baum in originating the "power tower" concept in the U.S.S.R. in the late 1950s. The book of Butti and Perlin (1980) describes much of this early work. Most of the thorough work has however occurred in the last 20 years.

Before the oil embargo the National Aeronautics and Space Administration (NASA) had an important program in solar energy. The program involved photovoltaics, selective coatings, and solar power sources such as solar Brayton cycles and solar thermionics. Parabolic dish concentrators were studied in the 1960s, but that work was discontinued by 1970. One reason for this work was the early pessimism about the capabilities of photovoltaics— both in efficiency and radiation resistance. Much of this pessimism proved unjustified.

For many years the National Science Foundation (NSF) sponsored R&D work in solar energy (e.g., the team of Edwards, Gier, Dunkle et al. at UCLA in the early 1960s), just as it sponsored work in other technical fields. In 1972 NSF initiated the RANN (Research Applied to National Needs) program in solar energy. This program was later turned over to the Energy Research and Development Administration (ERDA)—the precursor of the Department of Energy (DOE) program.

In the last two decades research and development has been funded by various national agencies (NSF, NASA, ERDA, DOE, the Department of Housing and Urban Development), by individual states, by the United Nations, the Organization of American States, UNESCO, the International Energy Agency, NATO and other international agencies, and by numerous private groups. Many other nations initiated R&D programs in the mid-1970s (de Winter et al. 1976) that are continuing (Balcomb 1988).

Much of the work covered in subsequent chapters of this volume was sponsored by the DOE Solar Heating and Cooling Program which was actually initiated before the formation of DOE as an ERDA program (ERDA 1976). This program resulted in more than 250 research contracts with university, industry, and private researchers and a corresponding effort at the government national laboratories. It included about 120 projects on solar collectors and solar collector materials. There was, in parallel with the heating and cooling program, an equally large effort in solar thermal technology for power and industry. This program sponsored most of the R&D on concentrating collectors and high temperature materials that was done at the national laboratories. These government programs started in 1976 and are continuing today. The peak funding years for these programs were from 1980 to 1982.

1.2.3 Comments on the Analysis of Collector Performance

It has become almost universal to express the efficiency of flat plate collectors as a linear function of the differences between a collector temperature and the ambient temperature divided by the solar radiation incident on the collector. Most of these linear equations derived from the early MIT work of Hottel and his coworkers.

The Hottel-Whillier (HW) linear model of the flat plate collector has in the past few years received more attention than even Hottel or Whillier probably intended. There is no question about its usefulness, particularly in its extended form with the "heat exchanger factor" (de Winter 1975c). However, it should be recognized that there is more to the analysis of a collector than simply the specification of the parameters F_R, U_L, and $\alpha\tau$, and some incident angle modifier numbers. Radiation and convection losses are nonlinear. U_L varies with temperature. The heat absorption in the glazing affects the performance. The linear model is interesting but only approximate.

Although Hottel, Woertz, and Whillier derived all the linear equations of the HW model, second-order effects were carefully considered even though at the time it was an excruciatingly time-consuming task to do the calculations with slide rules, math tables, and mechanical desktop calculators. The linear equations were only considered to be an approximate description of collector behavior. Now, when the average engineer has easy access to an incredibly powerful and inexpensive computer capability, it is ironic that the simple HW linear equations are used almost exclusively. It may well be that most solar energy practitioners have only read the Bliss (1959) paper in which the linear derivations are summarized, and have never read the Whillier (1953) thesis or the other Hottel, Woertz, or Whillier papers in which the derivations were first published. Those interested in accurate flat plate collector analysis would, however, do well to examine these earlier papers, and to obtain a clear view of recent work on the second-order effects from the material presented in this volume. Present-day computers do not need to artificially restricted to the consideration of straight-line equations.

Tabor, Hottel, Woertz, Whillier, and many others often worked with make-shift and inadequate results on convection heat transfer coefficients in the flat plate collector glazing systems. For some curious reason the results of Juerges (1924) were used for decades by almost everyone in the field for the external convection heat transfer losses. The Juerges values led to the prediction of very low collection efficiencies for unglazed collectors. Standard boundary layer theory led to much lower losses and agreed fairly well with experimental results (de Winter 1975b). Convection heat transfer for flat plate collectors is now understood quite well.

Ordinarily solar collectors should be designed for maximum cost effectiveness, even if this reduces collector efficiency somewhat. Many collectors have been more efficient and less cost-effective than desirable. A careful cost optimization of flat plate collectors with copper fins and tubes was made using realistic costs for materials (de Winter and Lyman 1978) and for soldering, joining, and other fabrication processes. Both flat and tapered fins were examined. The optimum collectors had thinner fin material, wider tube spacing, and lower F_R values than is normally thought to be reasonable for liquid-heating flat plate collectors. This may be due in part to the emphasis on collector efficiency (rather than cost-effectiveness) in reporting test results.

The theory of serpentine flat plate collectors was developed by Abdel-Khalik (1976) and Zhang et al. (1983). These collectors had been used for decades, but it is only recently that their performance can be predicted with ease and accuracy.

The possibilities of solar concentration have long been appreciated. More than 2,000 years ago the point-focusing properties of parabolic dishes were understood (Butti and Perlin 1980), and solar energy was used to light ceremonial fires. Although many scientists have worked on solar concentrators and on furnaces employing concentrating devices, such as the parabolic dish or trough, spheres and other concave reflectors, heliostat arrangements, and the Fresnel and other lenses, a precise understanding of the actual optics is fairly recent, requiring the use of high speed computers. The CPC-compound parabolic concentrator technology represents a completely new technology that has developed in the last two decades.

Until recently it was virtually impossible to analyze or design many concentrators and receiver cavities properly because the optical calculations had to be performed with crude calculation tools. Computers have had an impact. In the 1960s ray-tracing schemes became widespread in thermal radiation heat transfer work, and these allowed equipment to be analyzed and designed accurately and inexpensively using computers to do the ray tracing. Monte Carlo ray-tracing programs were written at the Jet Propulsion Laboratory in the 1960s for solar thermionic systems using parabolic dishes (de Winter 1968a, 1968b). In the 1970s and 1980s many additional programs of this type were written and used. The quality of analytical tools is no longer a limiting factor.

1.2.4 Comments on Collector Technology

The technical performance and reliability of new collector design concepts can often be evaluated quite conclusively in a research setting or in field tests. Costs can be tightened up once the concept looks attractive. The durability of a new ultra-low-cost collector design concept can, however, not be determined in an ivory tower setting; it must be established in the real world. Before it can be considered a commercial success such a design concept must be able to attract venture capital, demonstrate that it can be manufactured reliably, capture a market share, show that it can survive the real world well enough so that warranty costs are not excessive, *and* last for many years after the lapse of the warranty.

Many companies have incurred enormous costs in warranty fulfillment because collectors were marketed that were not sufficiently sturdy. A number of these companies went bankrupt in the process. Trial and error, and the process of natural selection in the marketplace, yielded designs of relatively low cost and of relatively low warranty liability in both plastic and metal collectors. Despite this it is inevitable that there will be those who propose designs based on their belief that weight and cost can be reduced much further without significant decreases In durabillty, and who do not realize that ivory tower programs in drastic cost reduction are doomed to failure.

Solar swimming pool heating, which is not discussed in this volume, has become well established. Without U.S. government support, swimming pool heating has become an almost exclusively solar application (at least for small pools[2]) with a viable industry that has outlasted the expiration of most local tax credits. Highly durable and cost-effective plastic collectors have been developed by a number of companies. Cost-effective do-it-yourself designs are available for copper solar swimming pool heaters. A prototype solar pool heater based on a design manual for swimming pool heating systems (de Winter 1975b) has been in operation in Pasadena, California, since March 1973 with few service requirements (de Winter 1988). Commercial copper and plastic solar swimming pool heaters have been marketed for many years by several companies.

1.3 Discussion of Topics and Chapter Contents

The topic of collectors of solar thermal energy is very complex. First, there are collectors of many types, serving many purposes. Some have become cost-effective, while others are still at the research stage. They range from very low temperature swimming pool heaters to concentrating collectors designed for very high temperatures, and include solar salt-gradient ponds that can serve many uses.

Second, several professional disciplines are involved in the development and study of collectors to address specific problem areas. Many of the problems overlap, while others affect only specific types of collectors.

Finally, material technology is an important issue in all collectors as well as thermal energy storage systems. Only some questions on applications of materials are discussed in this part of the volume on collectors. The general subject of materials for solar applications is treated in depth in the third part.

Considering these complications, this volume was planned as an in-depth study of collectors that can be used by the practitioners of many disciplines in the solution of many problems. Some topics will be examined by several authors from different points of view. Although this plan may result in some duplication, it guards against serious gaps.

1.3.1   Collector Concepts and Designs

Many schemes have been used or proposed for intercepting or gathering solar energy (with or without concentration), converting it to heat, and using it, while minimizing the heat losses and the equipment costs. Many of these schemes have been around for centuries, whereas others are quite new. More than 2,000 years ago solar concentrators were used to light ceremonial fires (Butti and Perlin 1980). The flat plate collector, or hot box, was invented in the eighteenth century (Ackerman 1915; de Winter 1975a). The compound parabolic concentrator is only a few decades old, as is the solar salt gradient pond.

In chapter 2 Ari Rabl describes different types of collectors, involving compound parabolic concentrators or concentrators of other types, or involving a flat plate collector geometry. Many have been invented, and reinvented or revived repeatedly. To compete in the energy market, a collector should involve a cost-optimized design. The right (most cost-effective) materials must be chosen. Operating experience is needed to establish durability. Product experience is needed to confirm durability results, and to establish consumer acceptance and warranty needs and possibilities. All of these steps are critical in the process of bringing something from the stage of design concept to the stage of final design.

The term "final design" is perhaps optimistic. Many products are not ready to be implemented when first marketed. Improvements continue, based on consumer reactions, new materials, product requirements, or market conditions. As an example, the last decade has seen enormous design improvements in the automobile, many years after the introduction of the automobile. A design concept proposed with no design research or hardware experience to back it may however represent little more than wishful thinking.

Ari Rabl was an active member of the Winston group in the Chicago area. Later he worked at SERI for several years before moving to the Center for Energy and Environmental Studies at Princeton. He is currently at the Ecole des Mines in Paris. Rabl has made significant contributions to the technology of compound parabolic concentrators, parabolic trough collectors, and solar ponds. His book *Active Solar Collectors and Their Applications*, published by Oxford University Press in 1985, is a widely used and valuable reference.

1.3.2   Optical Theory and Modeling of Solar Collectors and Thermal Theory and Modeling of Solar Collectors

The development of theoretical relations and the performance of mathematical or numerical modeling are the first steps needed to bring a collector design concept to reality.

Many solar practitioners have an intense distrust of equations and numbers. However, theory and modeling lie at the heart of engineering and technology. The easiest way to distinguish between good engineers and poor ones is to determine how well they can predict the performance and behavior *before* a system or device is built. Such predictions cannot be "inspired guesses." They must be based on a quantitative understanding of the way nature works at the subcomponent level and how subcomponents interact as a system.

It is difficult to build and test any device successfully without some theory and modeling. As an example, the first flight of a modern airliner is preceded by thousands of simulated flights made by analytical models on a computer. The building of the Golden Gate Bridge in San Francisco was not preceded by tests on full-scale hardware prototypes. It was preceded by calculation "tests" on full-scale analytical models, and by the changing of the design until the best Golden Gate Bridge design was established.

"Black box" models that do not include the detailed effects of the individual design parameters are not very useful. One cannot optimize the costs of components and materials without a model that includes explicit descriptions of their influence on performance. If a subcomponent is not understood, then R&D on *that* subcomponent, not on the complete device, may be in order to develop an analytical model.

The Hottel-Whillier model of the flat plate collector is based on physical principles, if one includes not just the first-order but also the second-order (nonlinear) effects. The model was built from the ground up and includes the effects of all the important design parameters. For a particular flat plate design, one or two iterations of analysis, prediction, and tests should result in confirming the model's parameters to within a few percent. This process provides the analytical tools needed to cost-optimize the design concept. Only by comparing the cost-optimized design concept to other concepts or products will a designer be able to determine the merit of a new concept. The cost optimization of any design parameter should not be limited to the examination of the performance at the beginning of the component's life. Components and materials should be sturdy enough so that the mean life to failure is acceptable.

Millions of dollars have been wasted in the last 30 years because solar energy hardware was built without the benefit of careful preliminary work in theory and modeling. This author's first exposure to solar energy engineering was in a program involving such a fiasco. In the mid-1960s solar thermionics seemed to be a competitive spacecraft power supply (compared to photovoltaics) for spacecraft missions approaching the sun. The thermionic diodes on an experimental prototype were supposed to have their emitter surfaces heated to about 2,000°C in the receiver cavity of a parabolic dish concentrator. The heat was indeed available, but the diodes had been placed in a relatively cool part of the cavity and the device simply did not work. Tests and analysis confirmed this (de Winter 1968a, b). The analysis was however, done after the fact. The hardware had been "designed" long before, based not on analysis but on inspired guesswork, and a multimillion dollar program was quietly killed and buried. The Venus–Mercury Flyby Mission was completed successfully with a photovoltaic power supply, protected from high temperatures by tilting the panels when close to the sun. The use of solar thermionics was out of the question. Although the stated reason was that the technology was not sound, the real reason was that the designers had not done a good job. Millions of dollars were wasted for lack of a few months of careful modeling and design spent at the right time in the program.

There are several distinct areas of theory and modeling needed in solar collection work. Before calculating the solar energy absorbed and converted to heat, one must understand the optics involved. After absorption, one must understand heat transfer, thermodynamics, and fluid flow to describe the processes. To ensure long life, one must avoid corrosion problems and provide structural strength to withstand wind loads, fluid pressures, and thermal stress.

This volume includes one chapter on optical theory and modeling and one on thermal theory and modeling. The engineering aspects of theory and modeling were included in a later chapter on engineering R&D. Chapter 3 on optical theory and modeling was prepared by Ari Rabl, whose background and experience was introduced earlier. Noam Lior, author of chapter 4 on thermal theory and modeling, has been at the University of Pennsylvania for a number of years as part of a group that has made significant contributions to the understanding of heat transfer mechanisms in collectors and thermal and thermodynamic interactions of the components of active solar heating and cooling systems.

The chapters of Rabl and Lior are detailed and thorough. In the early 1970s the state-of-the-art in both optical and thermal modeling was primitive, and, as their chapters demonstrate, progress in this technology is impressive. There is some material included in both of these chapters on distributed and central receiver systems, but a more complete treatment on such systems will be found in volume 11 of this series.

1.3.3   Testing and Evaluation of Stationary Collectors and Testing and Evaluation of Tracking Collectors

Once a collector has been designed and constructed, it must be tested to determine its durability and to characterize its performance. It must not only be able to survive the expected hazards of transportation and installation but also the normal conditions of weather and use. It must survive them well, with little degradation, for long periods of time. The collector environment may be low humid areas or dry, high desert conditions in which the ultraviolet radiation environment can be more brutal. It may be required to survive hailstones, freezing weather, rain or snowstorms, or high wind velocities. The collector may operate without cooling for many successive high temperature days, or it may be cooled by water that is very hard or very soft, may contain high concentrations of carbon dioxide or other gases, or may have a pH far from neutral. It may subjected to air pollution, dust, or sandstorms. It may be subjected to consumer neglect or to handling by indolent technicians, but such human factors can not be addressed in formal equipment tests. One can not guarantee a product is *really* foolproof.

Test objectives are clearly different for a preliminary version of an engineering prototype than for a collector that a manufacturer wants to certify for the marketplace. For an engineering prototype there may be one or two pressing questions of interest. In product certifications the test objectives are to protect consumers and the sales volume of the good products already on the market. When solar energy (or other) products get a bad reputation, the whole field suffers the consequences.

Many square miles of flat plate collectors were sold and installed in the 1970s and 1980s. A significant industry developed, and a network of solar energy contractors, installers, and salespeople developed across the nation. Many used high pressure sales tactics. Many overpriced systems were made attractive by tax credits (Baer and Shurcliff 1988). Many collectors or systems failed to live up to expectations. The need to protect against poor products and poor installations was recognized early, and as a result certification criteria and testing standards were developed and supported by government and industry.

Tracking collectors appeal to a more sophisticated market. They are purchased primarily by energy system developers, industry, or utilities. This market has had significant implications on the testing and evaluation of tracking collectors. Interest in certification criteria and testing standards has not been widespread, and the amount of work done in these areas has been correspondingly less. Much of the testing has been done by and for technical people.

For about ten years the National Bureau of Standards (NBS) was the central U.S. agency in the development and establishment of collector-testing procedures and standards, although other groups were involved. There were round-robin testing programs in which many laboratories tested the same hardware. There were professional society committees and subcommittees (primarily of ASHRAE and ASTM) in which the participants contributed to the formulation of testing procedures, standards requirements, and terminology. These standards were developed by consensus and used voluntarily by all participating organizations. Ultimately a number of ASHRAE standards resulted: for flat plate collectors without glazing, for heating liquids (e.g., swimming pool heaters), ASHRAE Standard 96-80; for collectors involving a phase change, ASHRAE Standard 109-86; and for both nontracking and tracking collectors, ASHRAE Standard 93-77, subsequently modified to 93-86.

Collectors involving a significant amount of concentration require special test and reporting methods. The second- (and higher-) order heat losses are too important to ignore, and special work was needed to address these effects. SERI did much of this work which culminated in a modified version of ASHRAE Standard 93-77, and was later adopted by the ASTM as ASTM Test Method E 905-82.

A number of states in the United States, and a number of foreign countries had parallel programs in the development of testing procedures and equipment standards. Many ended up adopting the ASHRAE standards.

Chapter 5 on testing and evaluation of stationary collectors was prepared by Robert Dikkers of the NBS. Chapter 6 on testing and evaluation of tracking collectors was prepared by David W. Kearney, who directed the industrial process heat (IPH) program at SERI.

1.3.4   Collector Optical Research and Development, Collector Thermal Research and Development, and Collector Engineering Research and Development

Chapters 7 through 9 describe the important results of the research and development in these fields. R&D involves the formal process of developing understanding and creating products. Theory and modeling involves the formal process of arranging the understanding systematically and applying the results of R&D. It is clear this topic arrangement involves some overlap, and makes some duplication inevitable. Just as there are people interested and involved in conducting R&D, there are others primarily involved in applying it. There are two different objectives and two different audiences.

Roland Winston, author of chapter 7 on optical R&D, is the inventor of the compound parabolic, or Winston, concentrator and has been involved for many years in developing the theory of nonimaging optics. His work has resulted in an advance of the practical limit of concentration approaching the thermodynamic limit. His development of devices achieving a concentration of 70,000 suns raises the possibility of solar-driven lasers. Winston's direction of a very active group of researchers at the University of Chicago, and his work with Argonne National Laboratory, has educated and inspired many others.

Charles Bankston, who also serves as chief editor of this series, prepared chapter 8 on thermal R&D. He has been active in heat transfer and the thermal sciences for several decades. While at the Los Alamos National Laboratory, he was in charge of collector R&D for the U.S. government. In private practice the last few years he has been active in International Energy Agency (IEA) activities involving collectors, notably in seasonal storage applications, and in the design and operation of large arrays of collectors.

Charles Kutscher prepared chapter 9 on engineering R&D. There is no parallel chapter on theory and modeling, since the engineering aspects of theory and modeling are not unique to solar energy. There are, however, many areas in which engineering R&D has been performed to address solar energy collector concerns, and this is the topic of this chapter. At SERI, Charles Kutscher was the principal author of a major manual for the design of large industrial process heat systems using solar energy; this manual stressed many practical engineering aspects.

In each of these three chapters there is some information on central receiver systems and distributed receiver systems. Additional information appears in volume 11.

1.3.5 Solar Pond Research and Development

Solar salt gradient ponds constitute a different type of solar collector, including not just a collection but also a storage function. Years ago such a pond was found in nature in a Yugoslavian river system. Since then another has been found under a layer of Antarctic ice. The study of salt gradient ponds for solar energy applications began in the 1960s, and the technology has progressed rapidly. Practical applications already exist, and many more are anticipated.

In Israel a low temperature Rankine cycle turbine using thermal energy from a salt gradient pond has been used to produce electricity. The potential of such systems seems promising in Israel, Australia, and perhaps elsewhere.

In the province of Jujuy in northern Argentina, salt gradient ponds are being used on a commercial basis to purify sodium sulphate. A mixture of roughly equal parts of sodium sulphate and sodium chloride, brought from a nearby salt flat, is purified to sodium sulphate of about 99% purity. Previously a large amount of fuel was needed for the process, but now with the pond only a small pump and a small staff of semiskilled laborers are needed. The facilities are being expanded. The plant was developed by the team of Luis Saravia, Gabriela Lesino, and others at the Universidad Nacional de Salta, Argentina (see the paper of Lesino referenced in chapter 10; see also Lesino et al. 1990).

Salt gradient ponds are widely recognized as promising devices for producing industrial process heat. There is little doubt that uses will proliferate in the next few decades, and that the devices can be useful up to fairly high latitudes.

John Hull and Carl Nielsen, the authors of chapter 10 on solar ponds, have long been involved in solar pond R&D—Hull at Argonne National Laboratory, and Nielsen at Ohio State University.

1.3.6   Collector Cost Issues

As mentioned earlier, there are many costs that are not readily apparent to those not involved in a successful solar collector company with an established distribution network. This has caused intense confusion in the R&D work and in the literature. Many cost projections have involved no more than wild guesses or wishful thinking.

Warranty fulfillment costs can be significant for any product. Some gas-fired domestic water heaters are sold in both a five-year warranty "model" and a ten-year warranty "model." The models happen to be identical, except for the price and the final color coat of paint. The price is different to account for the expectable difference in warranty fulfillment cost for the two "models."

There is an intensive interplay between the equipment sturdiness (i.e., cost), the environment faced by the equipment in the real world, warranty fulfillment costs, and equipment success and manufacturer survival. Both underdesign and overdesign call for design changes. The environment can include unexpected factors, is quite random, and may be largely beyond control. Customers, installers, and service personnel can be quite careless. Dust, soot, acid rain, solar ultraviolet radiation, storms, water quality, and other factors can vary widely. This real world is the one and only arena in which successful low-cost equipment can be developed. Ivory tower programs on ultra-low-cost equipment are doomed to failure from the start. Since experience in this area is so costly and so critical to most companies, normally all details are kept secret. As an example, for decades the Hughes Tool Company had a virtual monopoly on oil well drilling toolbits. They never sold the equipment: They leased it and replaced it whenever necessary. Only the Hughes Tool Company knew what worked, what did not, and why. Others could only guess. Such outsider guesses are uncertain and speculative even for mature products.

"Hidden" manufacturing costs, including advertising, breakage, reject, and theft costs, can be significant and must be charged to the product. R&D costs also are significant and can range as high as 30% of sales. Sales commissions add costs, but without them sales may suffer. Maintaining inventories can be costly, yet long delivery times may discourage sales. Management costs, support staff costs, facility, insurance, and utility costs can be significant. For most companies all such costs are confidential, and all one can determine or estimate involve one or more "markup" numbers. Only a small part of such markups involves a net profit.

Chapter 11 on cost issues and opportunities was prepared by John Clark of the University of Michigan. Clark has studied both the solar energy industry and manufacturing processes. In this chapter he discusses the background of the development of the solar industry, with candid comments on the effects of the tax credits, R&D programs, demonstration programs, and the changing times for solar energy. He documents the trend toward simpler systems and the history of some technologies that have not met early expectations. He provides a detailed breakdown for different "cost/price chains" for solar collectors, showing the markups involved in different steps of the process from manufacturing to final installation. He includes many interesting figures on costs, cost trends, design trends, and markup levels for the period between 1975 and 1985.

1.3.7 Reliability and Durability of Solar Collectors

A federal solar heating and cooling demonstration program initiated in 1975 was designed to provide a market for an emerging solar industry, identify effective materials and design concepts, and demonstrate the viability of solar technology to the public.

Although solar-heated water had been popular in California until natural gas became available inexpensively in the 1920s and in Florida until the cost of electricity dropped after World War II (Butti and Perlin 1980), these early solar water-heating designs had been developed before there really was any understanding of collector performance—before the paper of Hottel and Woertz (1942). Many new materials were developed after World War II for which no solar energy equipment experience was available. There was no existing industry at the time of the oil embargo; there were no equipment standards, nor accepted test procedures. There were no experienced designers, manufacturers, installers, or service people. In the 1970s when solar energy suddenly acquired nationwide importance, the reliability and durability of specific material and design concepts had to be established, and many people had to be trained.

William Freeborne's chapter 12 on reliability and durability issues and opportunities is primarily based on the experience of the demonstration programs of HUD, in which he was involved from the inception, and other U.S. agencies and laboratories. In these programs many unacceptable materials and design features were identified. Many of the failures were quite visible and catastrophic. Such types of failures could have been anticipated as the technology was not ready for rapid implementation. Because there had not been time to find them in a laboratory setting, the development problems were found in the field.

The testing and certification programs established today were designed on purpose to prevent many of the problems found in the early demonstration efforts. Badly designed or built equipment simply will not pass the tests. Manufacturers continue to find and address problems with their products. Most of these problems are minor ones since systems now are based on reasonable designs.

In mature and cost-effective hardware, the equipment must be reliable and durable, but not much too reliable and durable. In attempting to achieve low-cost collector designs, care must be taken to see that the threshold of durability below which a warranty program is simply not feasible is not crossed. Conversely, if there are never problems of durability reported in the field, the product may be overdesigned; if so, the safety margins could be reduced, and the product could be made less expensively.

1.3.8   Environmental Degradation of Low-Cost Collectors

Plastic has become the most popular material for unglazed swimming pool heating collectors, which operate at very low temperatures and at pressures that are normally also very low. The collectors sold by the companies that are still in business are very reliable. It was, however, a very painful process—a process taking many years—for collector designs to be developed that were sturdy and durable enough to stand up to the service requirements. In these designs the plastic formulations had to be able to withstand the ultraviolet radiation environment the collectors experienced out in the field for many years, *and* the quality control process had to be stringent enough so as to ensure that the plastic formulation was within specifications and that the collector was fabricated properly. Some companies survived this developmental process; most did not.

A collector used for heating swimming pools is expected to heat water to about 30°C. A collector used for heating domestic hot water (DHW) or for other active solar heating functions is normally expected to deliver heat at about 60°C. The difference between typical *operating* temperatures is hence about 30°C. The difference between the *stagnation* temperatures is likely to be much greater. The swimming pool heating collector is typically unglazed, whereas the higher temperature collector normally with one glazing has a much higher stagnation temperature. Plastic material degradation rates increase with temperature. The classical model for the effect of temperature on reaction rates involves the Arrhenius equation which, for many reactions, may predict a doubling of reaction rates for every increase in temperature of about 20°C. Based on the above, it is likely most collectors will not be as cheap as swimming pool heaters. There is, however, a great need to quantify the degradation.

Many plastics otherwise suitable for high temperature applications have molecular bonds that can be damaged by ultraviolet radiation. These degradation processes are extremely complex, and nearly impossible to model accurately. Different molecular bonds are damaged by different parts of the spectrum. The ultraviolet spectrum and intensity, which vary from place to place, may be much less in a humid location at sea level than at a high and dry location. They also vary with the season, the weather, and the current condition of the ozone layer. Plastics that are exposed to the sun can be protected by incorporating carbon black or other materials that block the sun. It is not easy nor inexpensive to do this properly. Plastics used as a transparent glazing are even more difficult to protect.

Chapter 13 on environmental degradation of low-cost collectors has been prepared by Fred M. Loxsom and Eugene S. Clark of Trinity University in San Antonio, Texas. In the 1970s and 1980s Clark and Loxsom directed work in characterizing the ultraviolet intensity and the spectrum, and in the testing of solar energy equipment under realistic conditions. Trinity was one of eight centers conducting research in the careful measurement of a standardized set of irradiance parameters including ultraviolet as well as direct, diffuse, and global insolation.

1.4   Using This Volume

It has been mentioned several times that much of the previous research and resulting literature have been of little value—and may even have hurt the solar energy field. Just as poor products can scare customers away from related products that are good, so frivolous publications or programs can lead competent researchers to believe there is a lack of seriousness in the solar energy field.

It is hoped this volume will improve the quality of solar energy work in the future and help future researchers avoid needless duplication of earlier work and activities that have been shown to be unnecessary or undesirable. In case this does not happen, it is hoped the volume will help reviewers of proposals or papers identify the projects or publications that have little value. For both researchers and reviewers, it is hoped that this volume will provide many useful references to facilitate their work.

The authors generally have not included in full the results shown in a particular reference but have limited the coverage to a discussion of the results and their significance. Anyone seeking to use the results may need to consult the original reference.

Interesting and useful research will continue on solar collector R&D regardless of societal or government support. Prior to 1970 most of the important solar activities were privately funded and organized, and R&D could continue in that environment. Society may well show increased interest in solar energy in the future. In many other countries solar energy programs continue to be developed (Balcomb 1988). There is little doubt that solar energy will become increasingly important in the decades ahead.

Notes

1. By far the best historical treatise on solar energy is the book by Butti and Perlin (1980), with an earlier overview by Ackerman (1915) and an overview limited to research work by Jensen (1959). The literature on flat plate collectors was covered by Veltford (1942) and by de Winter (1975a).

2. For large pools it tends to be more cost effective to use the waste or cooling heat from a cogeneration system. A 50-100 kW internal combustion engine produces about the right amount of waste heat to heat a typical institutional or municipal pool, and the electrical power can be used for local buildings, floodlights, etc. This approach has become very popular, particularly since the U.S. Federal PURPA (Public Utilities Regulatory Policy Act) legislation requires the local utility to purchase the excess electrical power at a reasonable price.

References

Abdel-Khalik, S. I. 1976. Heat removal factor for a flat-plate solar collector with a serpentine tube. *Solar Energy* 18: 59–64.

Ackerman, A. S. E. 1915. The utilization of solar energy *Annual Report of the Smithsonian Institution*, pp. 141–166.

Baer, S., and Shurcliff, W. 1988. *Subsidizing the Sun—A Collection of Essays and Letters*. Self-published by Steve Baer. Copyrighted in 1988, S. Baer, P. O. Box 1327, Corrales, NM 87048.

Balcomb, J. D. 1988. Report on the 1987 Hamburg ISES Conference to the ASES Board of Directors. *Solar Today* (January/February): 6.

Bliss, R. W., Jr. 1959. The derivation of several plate-efficiency factors' useful in the design of flat-plate solar heat collectors. *Solar Energy* 3, 4: 55–64.

Butti, K., and Perlin, J. 1980. *A Golden Thread: 2500 Years of Solar Architecture and Technology*. Palo Alto: Cheshire Books.

de Winter, F. 1968a. Computer studies of parabolic solar concentrator performance for solar thermionic studies. Jet Propulsion Laboratory, Space Programs Summary 37–49. Vol. 3, February 1968, pp. 99–102.

de Winter, F., and Merrill, O. S. 1968b. Analytical characterization of parabolic solar concentrator performance. Paper presented at the Fourth Annual Meeting of ISES. Palo Alto, CA, October 21–23, 1968.

de Winter, F. 1975a. Solar energy and the flat plate collector—An anotated bibliography. ASHRAE report S-101, New York (now Atlanta), February 1975.

de Winter, F. 1975b. *How to Design and Build a Solar Swimming Pool Heater.* Manual Published and Distributed free of charge by the Copper Development Assoication, P. O. Box 1840, Greenwich, CT 06836.

de Winter, F. 1975c. Heat exchanger penalties in double loop solar water heating systems. *Solar Energy* 17, 6: 335–337.

de Winter, F., and de Winter, J. W. 1976. Description of the solar energy R&D programs of many nations. ERDA Report SAN/1122-76/1. February 1976. Available from NTIS, Springfield, VA.

de Winter, F., and Lyman, W. S. 1978. Optimum collection geometries for copper tube—copper sheet flat plate collectors. *Proceedings of the ISES Congress.* Vol. 2. New Delhi, India, January 16–21, 1978. Elmsford, NY: Pergamon Press, pp. 895–899.

de Winter, F., and Lyman W. S. 1988. CDA's do-it-yourself pool heater doing well after 15 years. *Solor Today* (May/June): 13–19.

Duffie, J. A., and Beckman, W. A. 1980. *Solar Engineering of Thermal Processes.* New York: Wiley.

ERDA 76-144. Interim Report—National Program Plan for Research and Development in Solar Heating and Cooling. Energy Research and Development Agency, Washington, DC, November 1976.

Hottel, H. C., and Woertz, B. B. 1942. The performance of flat-plate solar heat collectors. *Trans. ASME* (February): 91–104.

Hottel, H. C., and Whillier, A. 1955. Evaluation of flat-plate solar collector performance. *Trans. Conference on the Use of Solar Energy—The Scientific Basis.* Vol. 2. Tucson, AZ, October 31–November 1, pp. 74–104.

Hottel, H. C. 1970s. Personal Communication regarding the Godfrey L. Cabot Solar Energy Program at MIT. At the 1988 Annual Meeting of the American Solar Energy Society, held at the MIT Campus during June 18–24, 1988, Prof. Hottel gave an invited lecture with an overview of the Cabot Program after 50 years. The lecture will be published in *Solar Energy.*

Jensen, J. S. 1959. Applied solar energy research. Stanford Research Institute, Menlo Park, CA.

Juerges, W. 1924. Die Warmeubergang an einer Ebenen Wand. *Beihefte zum Gesundheits-Ingenieur* series 1, suppl. 19. This paper gives experimental results for heat transfer coefficients from a 50 cm × 50 cm steam-heated, vertical, sharp-edged plate, with the wind velocity parallel to the surface. It is not clear whether the high surface temperature produced a significant *natural* convection component. The 50-cm length is less than that of a typical collector, so heat loss predications can be expected to be too high. For nearly four decades, nobody seemed to notice this correlation was inappropriate and that more reliable values could easily be calculated using boundary layer theory.

Kreith, F., and Kreider, J. A. 1978. *Principles of Solar Engineering.* McGraw-Hill., New York.

Lambert, J. H. 1760. *Photometria.* See *Ostwald's Klassiker der Exakten Wissenschaften.* Nos. 31–33 for a German translation of the 1760 work, published in Leipzig in 1892.

Lesino, G., Saravia, L., and Galli, D. 1990. Industrial production of sodium sulfate using solar ponds. *Solar Energy* 45, 4: 215–219. (This issue of *Solar Energy* is entirely on solar ponds, with John Hull as guest editor.)

Lunde, P. J. 1980. *Solar Thermal Engineering—Space Heating and Hot Water Systems.* New York: Wiley.

Rabl, A. 1985. *Active Solar Collectors and Their Applications.* Oxford: Oxford University Press.

Veltford, T. E. 1942. Solar water heaters. Copper and Brass Research Assoication (internal report), New York, March 3, 1942.

Whillier, A. 1953. Solar energy collection and its utilization for house heating. Sc.D. thesis in Mechanical Engineering. MIT.

Zhang, H. F., and Lavan, Z. 1983. Thermal performance of a serpentine absorber plate. *Progress in Solar Energy. Proc. Annual Meeting of the American Solar Energy Society*, Vol. 6, pp. 471–476.

2 Collector Concepts and Designs

Ari Rabl

2.1 Introduction

Over the years practically every conceivable combination of abgorber, cover, reflector, and container seems to have been proposed as a solar collector, and sometimes even built and tested.[1] It is unlikely that any significant concept has been overlooked, especially after the vigorous, worldwide research efforts of the past decade. This chapter surveys the most important concepts and designs, with some indication as to which of them hold promise of being practical. The presentation is arranged according to concentration ratio and tracking requirement, beginning with the flat plate. The discussion is summarized in tables 2.1 to 2.3, which include some information on costs. For collectors in commercial production, cost and performance data can be found in the *Solar Products Specifications Guide*, published annually by *Solar Age** (1983). Typical operating temperatures for practical applications are indicated in tables 2.1 to 2.3. The optimal operating temperature depends on many factors, especially optical efficiency, heat loss coefficient, and collector cost.

For any thermal collector the efficiency η decreases as the operating temperature is increased. This can be described by

$$\eta = \eta_0 - \frac{U(T_{\text{abs}} - T_{\text{amb}})}{I}, \tag{1}$$

where η_0 = optical efficiency (also known as $\tau - \alpha$ product in the flat plate literature), T_{abs} = absorber temperature, T_{amb} = ambient temperature, U = collector heat loss coefficient (in W/m² °C), and I = solar irradiance on collector aperture (in W/m²). This equation is based on the temperature of the absorber surface. The heat loss coefficient U depends on temperature, but in practice a constant value often gives a good approximation, especially for low temperature collectors. At higher temperatures U increases, but if the range of operating temperatures is fairly limited, one can still use a linear equation of the form in equation (1) if one modifies the parameters η_0 and U according to the procedure described by Cooper and Dunkle (1981). The linearized equation is convenient because it allows the efficiency to be plotted against a single variable $\Delta T/I$ rather than separate variables ΔT and I.

In practice, equation (1) is not quite suitable because one is concerned with the temperature of the heat transfer fluid, not the absorber surface temperature. In most collectors a significant performance penalty results from the

---

* *Solar Age* is now no longer published.

Table 2.1
Nontracking collectors

Collector type	Approximate maximum operating temperature, °F (°C)	Cost in 1985 $, $/ft² ($/m²)		Comments
		Now	Goal	
Shallow solar pond	104 to 140 (40 to 60)	19.4 (208)[a] (complete system, including storage for one day)		Plastic covers may need to be replaced every 5 years or so for good performance
Deep solar pond (salt gradient)	104 to 194 (40 to 90)	3.6 to 7.2 (39 to 78)[b] (includes storage)		Collector and long-term storage in one unit; for seasonal storage, depth should be about 3 m; low cost, but low efficiency (10%–20%)
Flat plate	104 to 176 (40 to 80)	12.1 to 18.1 (130 to 195)[c]		Best-known and most developed of all collector types
Nonevacuated CPC fixed-tilt, or summer-to-winter tilt adjustment	176 to 248 (80 to 120)	18.1 (195)[c]		
Evacuated tubes (with reflector enhancement including CPC)	212 to 392 (100 to 200)	24.2 to 36.3 (260 to 390)[c]	12.1 (130)	Many opportunities for cost reduction by mass production and for performance improvements through R&D

Source: Rabl (1985).
a. Personal communication, Solar Energy Group, Lawrence Livermore Laboratory, Livermore, CA (1980).
b. A 29.5-ft (3-m)-deep salt gradient solar pond was built in Miamisburg, Ohio, for $492/ft² ($45/m²) in 1978. Solar ponds without long-term storage or in locations with free salt should cost considerably less.
c. Based on *Solar Age* (1979, *Solar Products Specifications Guide*).

Table 2.2
One-axis tracking collectors

Collector type	Approximate maximum operating temperature °F (°C)	Cost in 1985 $, $/ft² ($/m²) Now	Goal	Comments
Inflated cylindrical reflector	284 (140)	5.9 to 8.4 (65 to 91)[a]	—	Does not need continuous tracking but requires weekly tilt adjustments; plastic cover may need to be replaced every 5 years or so
Parabolic trough	572 (300)	18.1 to 30.2 (195 to 325)[b]	13.2 (110)[c]	Continuous accurate tracking; sensitive to dirt
Line-focus Fresnel reflector	482 (250)			The most cost-effective design for process steam has not been built yet; may combine advantages of parabolic trough and of central receiver for temperatures below 250°C
Fixed line-focus reflector with tracking receiver	482 (250)			Problems with dirt accumulation on reflectors and large cosine losses

Source: Rabl (1985).
a. Personal communication, Solar Energy Group, Lawrence Livermore Laboratory, Livermore, CA (1980).
b. Based on *Solar Age* (1980, *Solar Products Specifications Guide*).
c. Cost goals for 1990 (in 1985) from *Solar Thermal Program Multiyear Plan*. Golden, CO: Solar Energy Research Institute. Consistent cost goals have not been developed for all collector types.

thermal resistance between absorber surface and heat transfer fluid because the absorber must be hotter than the fluid. This effect can be analyzed most conveniently in terms of a multiplicative factor for the collector efficiency. When the collector temperature is specified as temperature of the heat transfer fluid, the efficiency is given by

$$\eta = F' \left[\eta_0 - \frac{U(T_m - T_{amb})}{\dot{I}} \right], \tag{2}$$

where T_m = mean fluid temperature.

The factor F' is the plate efficiency factor, and numerical values range typically from about 0.8 to 0.9 for air collectors, from 0.9 to 0.95 for liquid collectors, and from 0.95 to 0.99 for evacuated collectors. This means, for example, that an air collector delivers about 10% to 20% less energy at a given temperature than it would with perfect heat transfer between absorber and air. F' does not vary much with operating conditions, and in practice, it is usually treated as a constant. In fact, it is not even measured separately in

Table 2.3
Two-axis tracking collectors

Collector type	Approximate maximum operating temperature, °F (°C)	Cost in 1985 $, $/ft² ($/m²)		Comments
		Now	Goal	
Parabolic dish or point-focus Fresnel lens	2,700 (1,500) (possibly more)	121 (1,303)[a]	15.7 (169)[b]	Good if energy can be used directly in focal zone (e.g., photovoltaics or solar thermal power); otherwise, transporting heat to point of use is problematic
Central receivers	1,800 (1,000) (possibly more)	60.5 (651)[c] 24.2 (260)[d] + tower[e]	12.1 (130)[b] + tower[e]	Optical transport of energy
Fixed-hemispherical reflector, tracking receiver	752 (400)			Problems with heat transport to point of use, and with dirt accumulation on reflector

Source: Rabl (1985).
a *Solar Age* (1979, *Solar Products Specifications Guide*).
b. Cost goals for 1990 (in 1985) from *Solar Thermal Program Multiyear Plan*, Golden, CO: Solar Energy Research Institute. Consistent cost goals have not been developed for all collector types.
c. Average cost of heliostats for Barstow solar power plant.
d. Incremental cost of heliostats for Barstow after tooling costs, etc., have been paid. (P. Eicker, Sandia Livermore Laboratories, personal communication, 1980.)
e. Cost of tower is estimated to be approximately 10% of heliostat cost.

standard collector tests (e.g., see ASHRAE 1977; Jenkins and Hill 1980; BSE 1978); rather, only the combinations $(F'\eta_0)$ and $(F'U)$ are determined.

In most applications the inlet temperature T_{in} is given, not the mean fluid tempeature $T_m = (T_{in} + T_{out})/2$. However, the basic form of the efficiency equation remains valid even if the collector temperature is specified as T_{in}, provided one introduces another multiplicative factor, the heat removal factor F_R. This factor depends on the flow rate-specific heat product $(\dot{m}c)$ of the heat transfer fluid and is given by

$$F_R = \frac{\dot{m}c}{AU}\left[1 - \exp\left(-\frac{AUF'}{\dot{m}c}\right)\right]. \tag{3}$$

It allows the efficiency to be written in the form

$$\eta = F_R\left[\eta_0 - \frac{U(T_{in} - T_{amb})}{I}\right], \tag{4}$$

where T_{in} = fluid inlet temperature.

This equation is known as the Hottel-Whillier-Bliss equation. It is less exact than equation (1) because for its derivation (e.g., see Rabl 1985) a constant value has been assumed for U. But to the extent that equation (1) can be linearized (Cooper and Dunkle 1981), the Hottel-Whillier-Bliss equation holds for any collector type,[2] even though it was originally used only for flat plates.

2.2   Flat Plate Collectors

The oldest and best-developed collector is the flat plate, shown in figures 2.1 and 2.2 for water and for air, respectively, as heat transfer fluid. Flat plates are discussed in many articles and books (e.g., Löf 1980; Kreider and Kreith 1980). The earlier literature on flat plates has been surveyed in an annotated bibliography published by de Winter (1975).

The back and sides of a flat plate collector should be covered with adequate insulation. Recommendations for the optimal thickness of this insulation have been published in a classic paper by Tabor (1958a), a summary of which can also been found in Rabl (1985, ch. 9). Frontal heat losses can be reduced by

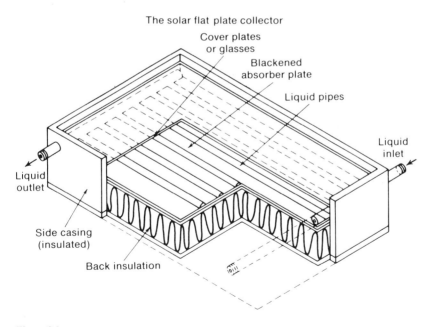

Figure 2.1
Liquid flat plate collector. Source: Lunde (1980).

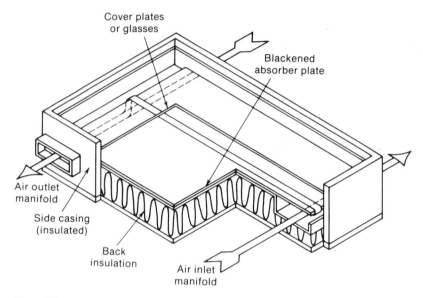

Figure 2.2
Air flat plate collector. Source: Lunde (1980).

means of additional covers. Of course the extra covers also reduce the transmitted solar radiation, and one must weigh optical versus thermal losses. In practice, few collectors have been built with more than two glazings. With the introduction of selective coatings of high quality and sufficiently low cost, the motivation for multiple covers is disappearing. For swimming pool heating in mild climates, the operating temperatures may be so low that simple unglazed collectors may be the most cost-effective.

There are many design variations, especially in the absorber design. In most liquid collectors the heat transfer fluid flows in tubes that are attached to the absorber plate, either in serpentine or parallel arrangements. In some liquid collectors, and in many air collectors, the fluid flows between two plates that are sealed at the edges, covering the entire absorber area: This arrangement yields good heat transfer between the absorber surface and the heat transfer fluid, but in liquid collectors it tends to make the heat capacity large. Also the plates need to be bonded together at interior support points to prevent bulging when the absorber is exposed to the relatively high pressure of the water mains (typically several times atmospheric pressure). In some air collectors the air flows directly between absorber surface and cover (Hollands and Shewen 1981). An interesting variation of the air collector is the transpired absorber, shown in figure 2.3, where the air is drawn through a porous absorber. The

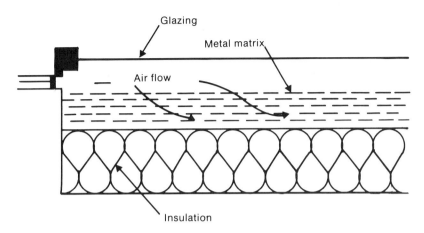

Figure 2.3
Transpired flat plate collector. Source: Löf (1980).

thermal efficiency of a transpired collector can be very good, but there is a penalty for increased pumping power (Rhee and Edwards 1983).

The heat transfer between absorber surface and heat transfer fluid can be improved by adding internal fins or turbulence generators, but then the pumping power increases. With liquid collectors there is another solution that can actually raise the plate efficiency factor F' to values slightly above unity. This solution is the black liquid collector in which the absorber is transparent and the fluid itself is black (Minardi and Chuang 1975). Implementing this concept has been impeded by the difficulties of finding a liquid with excellent stability and low cost. Furthermore this absorber design is not compatible with selective absorber coatings. The only way to achieve selectivity in black liquid collectors is with heat mirror coatings (coatings that are transparent to solar radiation but reflective for low temperature thermal radiation). The black liquid collector is unlikely to be competitive given the heat mirrors that have become available until now.

The most common absorber design consists of tubes bonded to the absorber plate. Copper is a favorite choice for the absorber material because it is quite durable and has high thermal conductivity. But since it is expensive, one would like to minimize its mass. This involves a trade-off between material cost and performance: As the thickness of the copper plate is reduced, the plate efficiency factor F' also decreases. A profile that is tapered between the tubes optimizes the utilization of materials (Kovarik 1978; de Winter and Lyman 1978).

One absorber variation for which extravagant claims were made is the so-called trickle collector. Here the absorber consists simply of open channels aligned vertically and painted black. The sun warms the water as it trickles down the channels. The heat losses are large because some of the water evaporates and then condenses on the cover, acting like a heat pipe from absorber to ambient. Furthermore the water droplets on the cover strongly absorb the near-infrared portion of the solar spectrum. The cost advantage, if any, is not sufficient to compensate for the low efficiency of the trickle collector.

Many attempts have been made to reduce the heat loss of flat plate collectors. Selective absorber coatings—coatings with high absorptivity for solar radiation and low emissivity for low temperature thermal radiation—can be very effective. While low emissivity is particularly desirable for evacuated collectors, in flat plates convection begins to dominate once the emissivity is reduced below about 0.3. For that reason, low-cost selective paints look attractive for flat plates although their emissivity is approximately 0.3, which is much higher than the values of approximately 0.1 achievable with the more expensive coatings such as black chrome.

Evacuation is another approach for combating heat loss. However, in the flat plate configuration, ordinary glass is not strong enough to withstand the resulting air pressure; the natural geometry for an evacuated collector is the tube, which is discussed below. Some investigators have tried convex acrylic covers as an alternative, but problems with durability and leak protection seem insurmountable. Another approach is to use a tightly packed array of plain evacuated tubes as cover for an ordinary flat plate (Jones and Shaw 1980; Alben and Hardcastle 1981; Herrick 1983). With uncoated glass the heat loss reduction is not dramatic because radiative transfer within the tubes remains high. But, if the interior of the tubes is coated with a heat mirror, the low solar transmissivity of the coatings available results in low optical efficiency. The version tested by Herrick (1983) combined uncoated evacuated tubes with a transpired absorber, and it achieved impressive performance for an air collector.

Several researchers (e.g., Kenna 1983) have propoged using a thick sheet of acrylic as the collector cover, a design sometimes called a "thermal trap" (a misleading name since all solar collectors are thermal traps). An acrylic sheet several centimeters thick does indeed possess the desirable qualities of a transparent insulator, but its cost is prohibitive if the thickness is enough to make the insulation effective.

Frontal heat losses can also be reduced by honeycombs, which inhibit both convection and radiation (e.g., Hollands 1965; Buchberg and Edwards 1976; Meyer et al. 1978; Felland and Edwards 1978; Hollands et al. 1978). Honeycombs can allow for very low frontal heat loss coefficients. But honeycombs also reduce the optical efficiency, especially at large incidence angles. Glass is the most durable honeycomb material, but the associated fabrication costs are prohibitive unless the honeycombs are produced in very large quantities.

Many attempts have been made to reduce the cost of flat plate collectors. An obvious first step is to select low-cost materials. Even waste materials such as broken glass and aluminum cans have been tried. A tinkerer with imagination can convert almost any scrap into a contraption that will become warm under the sun. But waste materials of sufficient quality tend to become costly after collection, cleanup, and recycling.

For residential applications two other ideas have been advocated for cost reduction: integration of the collector into the roof or wall of a house, and construction of the collector at the site. Kohler et al. (1978) investigated these two ideas in combination and claimed significant cost reduction. If the collector can do double duty as part of the building envelope, appreciable savings are indeed possible. Site construction, on the other hand, is problematic: The interior of the collector is likely to get dirty, and quality control is more difficult than in a factory. Site construction looks appealing only in special circumstances where good workers are willing to work for low wages. Even the construction of houses is likely to shift more to factories where automation is easier and productivity can be much higher. This shift has already happened in Sweden (Schipper 1984).

An interesting variation of the integrated collector is the thermic diode collector (Buckley 1978). Each module contains a collector on the outside of the building and a storage layer on the inside of the building. Collector and storage are separated by insulation. The heat transfer fluid is water, which flows by natural convection. A check valve prevents reverse flow. Each module is designed to replace a section of the wall of a building and acts as a self-contained passive solar heater.

Recently a promising low-cost, lightweight collector with a plastic absorber was developed by Andrews and Wilhelm (1980). It is suitable for applications where the water quality inside the collector is not critical, for instance, in indirect domestic hot water systems (with a heat exchanger) and in direct industrial process heat systems. Another low-cost, flat plate design has been reported by Atkinson and Caesar (1983). The principal challenge with plastic collectors is durability. Long-term testing is required before manufacturers can sell such collectors with confidence. But in view of the enormous potential for cost reduction, it is essential to continue the research and development of plastic collectors.

2.3 Solar Ponds

An important variation of the flat plate collector is the so-called shallow solar pond (Clark and Dickinson 1980). It consists of a shallow, horizontal water bag, a few centimeters deep, covered by one or two plastic films and air layers for insulation, as shown in figure 2.4. The pond is filled in the morning and drained into a storage tank in the evening. Because of its horizontal aperture and high heat capacity, the shallow solar pond is best suited for sunny locations not too far from the equator.

A totally different concept is the deep solar pond, shown in figure 2.5, also known as a nonconvecting solar pond (Nielsen 1980; Tabor 1981). A thick layer [about 3.3 ft (1 m)] of nonconvecting water provides thermal insulation. Convection is prevented by adding salt to establish a concentration gradient, with the saltier water at the bottom. The saltier water is heavy enough to stay at the bottom even when warmed by the sun. One meter of nonconvecting water offers as much thermal resistance as 2 in. (5 cm) of styrofoam, but it transmits much of the incident solar radiation.

The concentration gradient can be maintained quite easily by washing the surface with fresh water while reinjecting concentrated brine at the bottom. The annual salt consumption is approximately 10% of the salt inventory. Beneath the nonconvecting layer is a convecting layer of salt water for thermal storage and heat extraction; its thickness is 0.7 to 6.6 ft (0.2 to 2 m), depending on the desired amount of storage. Figure 2.5 also indicates a convective zone of approximately 0.3 to 1.3 ft (0.1 to 0.4 m) at the surface. Such a convective surface layer is undesirable, but it seems to be very difficult to avoid in practice. Solar ponds are unique because they combine collector and storage into a

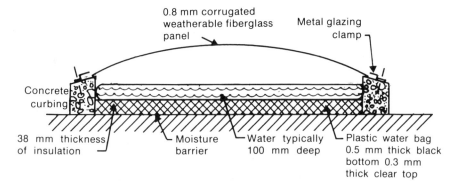

Figure 2.4
Shallow solar pond. Source: Clark and Dickinson (1980).

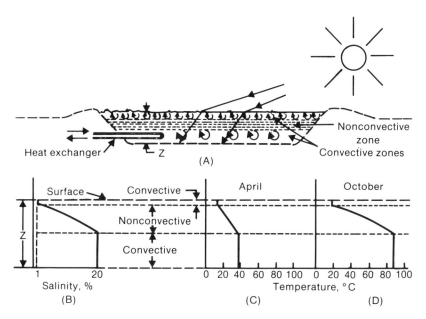

Figure 2.5
Deep solar pond. Source: Nielsen (1980).

single low-cost element, and they are one of the few concepts suitable for annual heat storage. A convenient starting point for a literature search on solar ponds is the November 1983 issue of the ASME *Journal of Solar Energy Engineering* (ASME 1983), which is devoted to solar ponds.

Several nonconvecting solar ponds have been built in Israel (Tabor 1981), Portugal (Collares-Pereira et al. 1982a), the United States, and elsewhere. Results for ponds in the United States have been reported by Nielsen (1982), Jones and Meyer (1982), Wittenberg and Harris (1980), and Zangrando and Bryant (1978). The largest ponds have been built at the Dead Sea, with the goal of generating electric power (Assaf 1978). A 5-MW$_e$ power plant has been operating there since 1984, supplied by heat from two solar ponds with total area of 2,700,000 ft^3 (250,000 m^2) [one pond has 432,000 ft^2 (40,000 m^2), the other, 2,268,000 ft^2 (210,000 m^2)] (Bronicki et al. 1984).

2.4   Evacuated Collectors

Evacuation is very effective for reducing the heat loss of a collector. The natural configuration for an evacuated collector is the glass tube. There are many possible designs, and quite a few manufacturers are selling evacuated tubular collectors. For a survey of evacuated collectors, see Graham (1979). All of the collectors have selective coatings as the absorber because with a nonselective absorber, radiative losses would remain large. Eliminating convection alone would not be effective. If one goes to the trouble of evacuating, one should certainly add a selective coating.

Several designs of evacuated collectors are shown in figure 2.6. In figure 2.6a the absorber is metal, and glass-to-metal seals are needed to get the heat transfer fluid into and out of the tube. Inlet and outlet are shown at the same tube end. They could be at opposite ends, provided one includes bellows to accommodate the differential thermal expansion of the absorber tube and the glass envelope. The heat pipe offers another way of extracting the heat from the absorber; it achieves excellent heat removal efficiency. Heat pipe evacuated tubular collectors have been built and tested in Europe (Mahdjuri 1979; Bloem et al. 1981) and in the United States (Ortabasi and Buehl 1980), including an all-glass collector with glass heat pipe (Ribot and McConnell 1983).

The need for glass-to-metal seals can be avoided if one uses the Dewar flask Thermos™ bottle) design. Here the absorber is a glass tube, sealed to the glass envelope at one end; at the other end both tubes are closed as shqwn in figures 2.6b and c. The selective absorber coating ig applied to the outside (vacuum side) of the inner tube. There are several methods of extracting the heat from the absorber surface. The collector in figure 2.6b places a feeder tube concentrically inside the absorber tube. The heat transfer fluid enters through the feeder tube. The feeder tube is open at the other end and allows the heat transfer fluid to return through the annulus between the feeder and absorber tube. In the design of figure 2.6c a copper sheet is placed inside the absorber tube, touching the glass to permit flow of heat from glass to copper. The heat is then collected from the copper sheet by a hairpin duct bonded to the sheet.

The heat extraction methods of these designs seem to involve significant performance penalties. In figures 2.6a and b there is a thermal short circuit between incoming and outgoing fluid. Compared to a collector without such a short circuit, this design lowers the outlet temperature for a given average temperature. Hence for a given outlet temperature the average absorber temperature and the heat losses are increased by the short circuit (Window 1983). Fortunately the heat loss coefficient of an evacuated collector is low enough that the associated performance penalty is acceptably small. In figure 2.6c the thermal resistance between absorber surface and fluid seems large, since in practice the copper sheet will not make perfect contact with the glass. This performance penalty is also quite small. The resistance between absorber

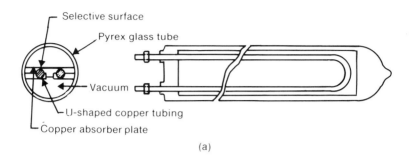

(a)

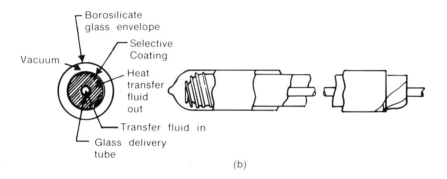

(b)

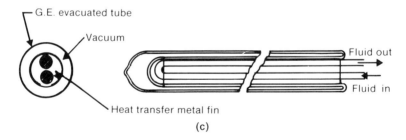

(c)

Figure 2.6
Several designs for evacuated tubular collector: (a) metal absorber plate; (b) glass Dewar flask with inner feeder tube; (c) glass Dewar flask with copper sheet and hairpin loop. Source: Graham (1979).

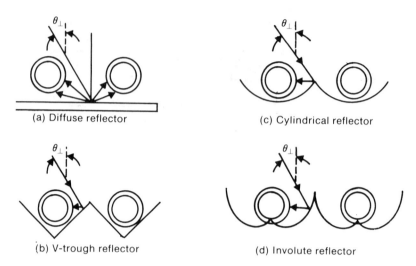

Figure 2.7
Several reflector shapes for tubular collectors. Source: O'Gallagher et al. (1980).

surface and ambient is so large that the resistance between absorber and fluid does not matter much. For the same reason evacuated collectors perform quite well as air collectors, despite the poor heat transfer characteristics of air.

Evacuated tubular collectors are hermetically sealed and contain getters to absorb any molecules that outgas into the vacuum. The tubes are expected to have a maintenance-free lifetime of approximately 20 years, not an unreasonable expectation in view of experience with vacuum tubes for radio and television. Evacuated tubular collectors are usually deployed with fixed tilt. Many designs include some kind of reflector enhancement, as shown in figure 2.7. They are well suited for 212° to 302°F (100° to 150°C). A design with a shaped glass tube [a compound parabolic concentrator (CPC) reflector] has been demonstrated to operate at 392°F (200°C) above ambient with 50% efficiency (O'Gallagher et al. 1982).

2.5   Nontracking Concentrators

Heat losses are approximately proportional to absorber area. By concentrating the radiation incident on the aperture onto a smaller absorber, one can reduce the heat loss per collector aperture area. Optical concentration must inevitably reduce the field of view of the collector (Winston 1970; Rabl 1976). Thus the higher the concentration, the lower the acceptance for diffuse insolation. Also the higher the concentration, the closer one must track the motion of the sun.

Very low concentration ratios,[3] generally 1.0 to 1.3 averaged over the year, can be achieved by flat side reflectors (Seitel 1975; Grassie and Sheridan 1977; Chiam 1981). Grimmer et al. (1978) reported measured solar radiation data on tilted surfaces augmented by flat side reflectors. For concentration ratios up to about 2, the need for tracking or tilt adjustments can be avoided by using a compound parabolic concentrator (CPC) (Winston 1974), also known as a nonimaging concentrator. The CPC concept was discovered independently in three places in 1966 (Baranov and Melnikov 1966; Hinterberger and Winston 1966; Ploke 1966).

The CPC is well suited for evacuated tubes, and most of the evacuated collectors sold in the United States today use CPC reflectors. There are many embodiments of the CPC design, depending on absorber shape and other considerations. The CPC of figure 2.8a is appropriate for flat, one-sided absorbers. Figure 2.8b shows the cross section of a cusplike CPC reflector for a tubular receiver (Winston and Hinterberger 1975). In figure 2.8c the evacuated glass tube itself is shaped like a CPC; since the reflector is protected from the environment, a high reflectance silver coating can be used, and the optical efficiency is excellent (O'Gallagher et al. 1982). In the design of practical CPC collectors a problem arises from the optical requirement to have the reflector go all the way to the absorber. There are several possible modifications of the basic CPC that allow for a gap between reflector and absorber. The gap design of Winston (1980) is the preferred choice for tubes because it does not incur any optical loss.

Sometimes the V-trough reflector has been used. It is a classic reflector design for low concentration ratios (e.g., see Tabor 1958b; Tabor, 1966; Hollands 1971). In a sense the design is a straight-line approximation of the CPC with a flat, one-sided absorber, and, for a specified field of view, it does not achieve as high a concentration as the CPC. The higher the desired concentration ratio, the greater the difference between a V-trough and a CPC. For example, if one allows summer-to-winter tilt adjustment, then a CPC can achieve threefold concentration, whereas the V-trough is limited to a factor of about two. As an intermediate step between the simple V-trough and a CPC, one can use a segmented V-trough (Olvera and Bannerot 1981; Kwan and Bannerot 1984). Selcuk (1979) designed and tested a reversible (summer/ winter) V-trough with an evacuated tube absorber. The simplicity of straight reflectors is especially appealing in small-scale production. In mass production the detailed shape of the reflector has less influence on the collector cost, and the CPC is likely to be preferred over the V-trough. The advantage of the CPC lies not only in the higher concentration ratio but also in its ability to

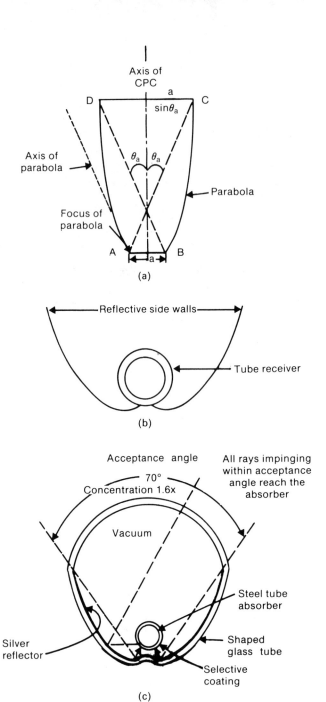

Figure 2.8
Compound parabolic concentrator: (a) CPC for flat one-sided absorber (source: Rabl 1985); (b) CPC for tubular absorber (source: Rabl 1985); (c) evacuated tube, shaped as CPC (source: O'Gallagher et al. 1982).

match any absorber shape; the V-trough is suitable only for flat, one-sided absorbers.

Because of its proportions, the CPC is generally not used as a single, large reflector for ordinary flat plate collectors. Rather, several small CPC troughs are assembled to form collector modules of reasonable depth. Several non-evacuated CPC collectors have been built and tested with concentration ratios between 3 and 6 (Rabl, O'Gallagher, and Winston 1980), and as stationary collectors with concentrations of 1.6 (Collares-Pereira 1982b). The latter surpasses the performance of flat plates from about 140° to 212°F (60° to 100°C), and its production cost is lower than for a flat plate (it is in commercial production in Portugal). A basic difficulty with this design is the potential for heat leaks through the reflector, caused by the optical requirement that the reflector should reach close to the absorber. Choice of materials is critical for this design. Not only must the reflectivity be high, but the thermal conductivity should be low. If heat transfer problems are not carefully considered, one may find disappointing collector performance in designs that use reflector troughs with nonevacuated absorbers, as many investigators have learned the hard way. A study of the relevant heat transfer mechanisms can be found in Iyican, White, and Bayazi (1981).

When research and development of CPC solar collectors started in 1974, emphasis was placed on attaining high operating temperatures for power generation. This emphasis forced the early designs toward concentration ratios of 5 to 10; in retrospect this range was too high to be practical for the CPC. The reason lies in the shape: At high concentrations the ratio of reflector area over aperture area becomes unfavorable, and for reasonable absorber tubes the size becomes too bulky. Recent optimization studies show that the optimal concentration ratio for fixed CPC collectors is very close to unity (the precise value depending on cost ratios and other factors) for both evacuated (Mills, Bassett, and Derrick 1984) and nonevacuated collectors (Gordon 1986). The best solar applications of the CPC lie either at very low concentrations (1 to at most 3) or as second-stage concentrators for focusing systems, as discussed below.

2.6   Tracking Concentrators

While CPCs are most suitable for low concentration ratios (for fixed collectors and for collectors with seasonal tilt adjustments), focusing optics are preferable for the design of tracking collectors. Collectors with high concentration require accurate and continuous tracking, and they are quite sensitive to dirt

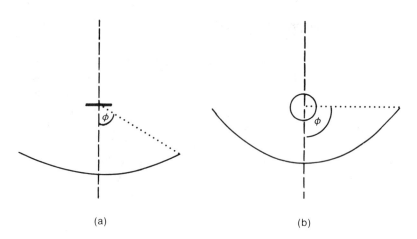

Figure 2.9
Collector with parabolic reflector: (a) flat receiver; (b) cylindrical or spherical receiver.

accumulation on both the reflector and the receiver. Cleaning every two weeks may be necessary in dirty environments (Freese 1978).

The parabolic trough, figure 2.9, is the current favorite for 300° to 572°F (150° to 300°C), and there are several manufacturers (Dudley and Workhoven 1979; Sandia 1980). The design depends on the shape of the absorber, as suggested in figures 2.9a and 2.9b: The optimal rim angle ϕ is approximately 45 deg for a flat absorber, and approximately 90 deg for a round absorber. Another candidate for this temperature range is the linear Fresnel reflector. It is the linear equivalent of the central receiver shown in figure 2.10 and is discussed below. One linear Fresnel collector with reflector slats of approximately 7.9 in. (0.2 m) width has been commercially produced.

To a certain extent the shape of a parabola can be approximated by a circle, if one compensates for the resulting aberrations by making the receiver large enough. This concept was first tested by Tabor and Zeimer (1962). A more recent design of this type is shown in figure 2.11 with a typical ray diagram (Gerich 1978, 1979). The cylindrical reflector shape could be manufactured at a very low cost because it can be obtained simply by inflating a plastic cylinder, the bottom half of which is metallized. The design in figure 2.11 has been developed for operating temperatures up to about 284°F (140°C); it does not require continuous tracking but only weekly tilt adjustments if it is deployed with an east–west axis.

With the concentrators shown in figures 2.9 to 2.11, tracking is accomplished by moving the entire reflector. An interesting alternative is a collector where the reflector is fixed and only the receiver moves. With the arrangement

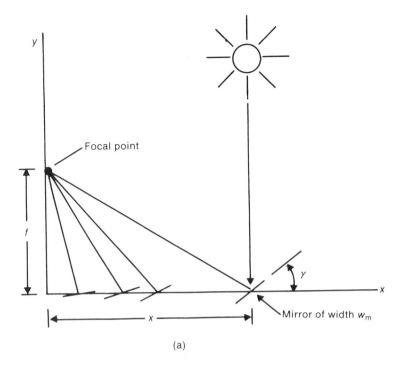

(a)

(b)

Figure 2.10
Fresnel reflector: (a) schematic diagram; (b) central receiver solar power plant at Barstow, CA.
Source: Sandia National Laboratories.

Ari Rabl

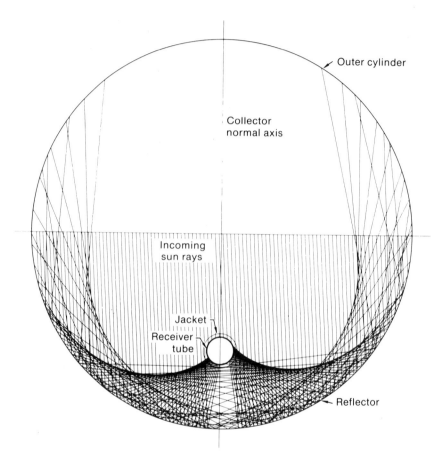

Figure 2.11
Collector with circular cylindrical reflector. Source: Gerich (1979).

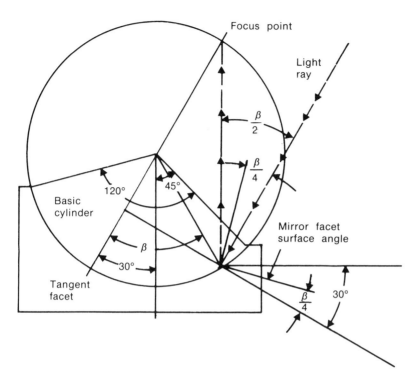

Figure 2.12
Collector with fixed reflector slats and tracking receiver. Source: Eggers et al. (1979).

of reflector slats shown in figure 2.12, concentration ratios comparable to the parabolic trough can be achieved (Eggers et al. 1979). The main problem posed by a fixed reflector is dirt accumulation on the mirror. A tracking reflector can be turned upside down whenever the sun is not shining, but a fixed reflector collects dirt at all times. Dirt accumulation is particularly severe in the early morning when dew condenses on dust particles in the atmosphere and then settles on exposed reflector surfaces (Berg 1978). Furthermore the fixed reflector (or at least a large part of it) lacks the proper tilt to be cleaned by rain (Freese 1978).

Another optical design with fixed reflector and tracking receiver is the hemispherical reflector, shown in figure 2.13. The receiver is aligned radially, with its axis pointing toward the sun (Kreider 1975; Clausing 1976). Test data for this collector have been reported by Fructer, Grossman, and Kreith (1982).

The suitability of lenses for solar energy is somewhat limited. Ordinary lenses are awkward for large apertures (more than a few centimeters in diameter) because their mass per aperture area increases linearly with the

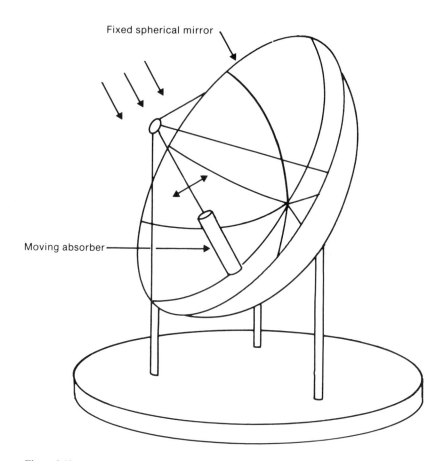

46

Ari Rabl

Figure 2.13
Hemispherical reflector with tracking receiver. Source: Lunde (1980).

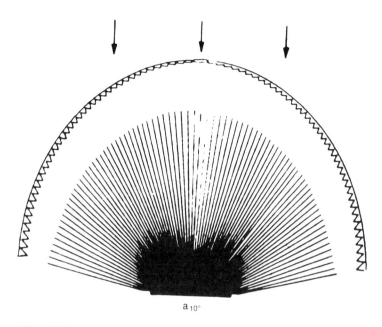

Figure 2.14
Fresnel lens. Source: Kritchman et al. (1979).

aperture width. Water-filled lenses have on occasion been suggested, but the absorption losses are excessive, to say nothing of the cost of the containment and support structure. For thermal applications only the Fresnel lens appears practical.

Line-focus Fresnel lenses, shown in figure 2.14, can achieve concentration ratios comparable to parabolic troughs as long as the aperture is normal to the sun (O'Neill 1979; Kritchman, Friesem, and Yekutieli, 1979). But unlike a line-focus reflector, a linear Fresnel lens suffers from off-axis aberrations. These aberrations become severe at incidence angles beyond about 20 deg. Thus a linear Fresnel lens requires tracking either about the polar axis (north–south axis with tilt equal to latitude) or about two axes, which increases the cost, especially in thermal applications.

In practice, the upper range of operating temperatures for line-focus collectors seems to be 570° to 750°F (300° to 400°C). For significantly higher temperatures, for power generation and some industrial processes, one needs point-focus concentrators. The principal types are the parabolic dish, the point-focus Fresnel lens, and the central receiver. The last, also known as the power tower, is shown in figure 2.10. The basic idea dates back to antiquity, although the claim that Archimedes set the Roman fleet afire by lining up polished shields along the shore is controversial. Modern pioneers of the power tower are Baum (1957) and his colleagues in Russia and Francia (1968) in Italy. In the United States the subject has been under investigation since the early 1970s (Vant-Hull and Hildebrandt 1976).

Several central receivers have been built in the United States and abroad. In the United States a 5-MW_t central receiver test facility has been operating at Sandia National Laboratories in Albuquerque, New Mexico, since 1977 (Holmes 1981). The largest central receiver to date is the solar power plant in Barstow, California, rated at 10 MW_e, and operating since 1981 (Sunworld 1981; Bartel and Skarna 1983). This plant produces steam at about 930°F (500°C). The upper limit of practical operating temperatures for point-focus solar collectors is much higher, in the range of 1,800° to 3,600°F (1,000° to 2,000°C). The efficiency at high temperatures may be improved by using second-stage concentrators of the CPC type at the receiver, as indicated in figure 2.15 (Kritchman 1981). More recent articles on the central receiver can be found in the February 1984 issue of the ASME's *Journal of Solar Energy Engineering*, which is devoted entirely to this topic.

Selecting the best collector type for a given application depends on many factors. The radiation available to the aperture depends on the concentration ratio and the tracking mode. Collectors differ widely in optical and thermal efficiency. A particularly important consideration is the cost of delivering heat from the collectors to the point of use. This involves not only capital costs but also heat losses and pumping power. Collecting heat from a field of individual

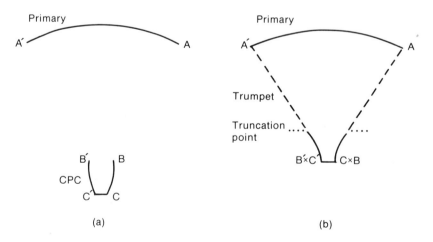

Figure 2.15
Second-stage concentrators for focusing parabola: (a) CPC; (b) trumpet. Source: adapted from Kritchman (1981).

point-focus collectors is costly. Piping runs are long, and there are high steady-state heat losses, aggravated further by transient losses from the night cooldown. This problem is less serious in an array of parabolic troughs because the absorber tube can act as a transport pipe if the trough modules are arranged and connected appropriately. The central receiver minimizes heat collection costs by transporting the energy optically rather than thermally.

The difficulties of transporting hot fluids increase with temperature, and the central receiver is likely to be the best choice above about 570°F (300°C) if central heat collection is required, as shown in a study by Iannucci (1980). On the other hand, point-focus parabolas and Fresnel lenses are excellent if the solar energy can be utilized directly at the focus of each collector. This is the case for photovoltaic cells and for Stirling engines that can be placed at the focus. Always normal to the sun, parabolic dishes collect more energy per aperture area than heliostats, for which the incidence angle varies with time. On the other hand, heliostats are simpler and less expensive.

Early concentrator designs tended to be rather expensive. But many attempts have been made to reduce costs. Of particular interest are pressure-stabilized plastic or metal films. One possibility is a thin reflector film stretched over a circular frame, and slightly depressurized on the nonreflective side to approximate a focusing mirror (Brantley 1977). The aberrations are negligible as long as the f-number is fairly high (small rim angle ϕ in figure 2.9). Another possibility is to enclose a heliostat or an entire collector under a transparent dome. This has been tested for heliostats by Boeing (1976). Although the pumping power for pressurization is negligible, the materials requirements are severe. The cover must have high specular transmissivity and long life when exposed to full sunlight, not to mention mechanical strength. The great attraction of the transparent dome is that it is very lightweight and allows a low-cost design for the concentrator.

Many other low-cost concentrators have been proposed. Low concentration ratios with crude tracking and large error tolerances have often been suggested as ways to reduce costs (e.g., see Pierce 1978). Obviously high accuracy increases cost, but it is not clear how large the incremental cost of high accuracy will be for mass-produced solar concentrators. The difference between low accuracy (approximately 10 deg) and medium accuracy (approximately 1 deg) may well turn out to be small, in which case the motivation for the crude concentrator approach disappears. For example, the first generation of tracking controllers left much to be desired, but with current technology, tracking accuracy comparable to the angular diameter of the sun seems to be no problem (Gee 1983).

Among other cost-cutting attempts one could mention a clever fabrication technique for manufacturing flat, point-focus Fresnel reflectors of moderate size (a few meters in diameter), invented by Steenblick and reported by Walton (1980). For parabolic troughs a promising technique is based on applying a bending moment to the edges of an elastic reflector sheet (McCormick 1981). Conical reflectors are another possibility (Kurzweg 1980).

2.7 Conclusion

It is difficult to predict which of the many possible collector types will be practical for future use. Often the choice of a collector is determined by the availability of materials. For example, the comparison of nonevacuated CPC and flat plate is based on the cost and reflectivity of reflector surfaces relative to the cost of absorber materials. A change in the respective materials technologies or costs can shift the balance. Tremendous progress may be possible in the fields of selective coatings, heat mirrors, antireflective coatings, cover materials, and reflector surfaces (Butler and Claassen 1980). New windows with high thermal resistance now in research (Neeper and McFarland 1982), such as aerogel glass and evacuated flat borosilicate glass panels, may also be suitable as covers for more efficient collectors. New schemes for system design and integration may be developed; for example, two-phase transport of heat by evaporation in the collector and condensation at the point of use (e.g., see May and Murphy 1983; Hedstrom 1984). Finally, with the fully automated factories of the future, the production processes and costs may be quite different from what we know today.

Notes

1. A fascinating account of the early history of solar energy can be found in the book *The Golden Thread* by K. Butti and J. Perlin, Van Nostrand-Reinhold, New York, 1980. A summary, by the same authors, has also been published as chapter 1 of the *Solar Energy Handbook* (Kreider and Kreith 1980).

2. With the exception of collectors that have significant thermal short circuit between inlet and outlet, as occurs in certain evacuated tubes (Window 1983).

3. Geometric concentration ratio, defined as ratio of aperture area over absorber surface area.

References

Alben, R., and K. Hardcastle. 1981. Theoretical and experimental study of an air collector with an evacuated tube cover. *ASME J. Solar Energy Eng.* 103: 251.

Andrews, J. W., and W. G. Wilhelm. 1980. *Thin-Film Flat-Plate Solar Collectors for Low-Cost Manufacture and Installation.* BNL 51124. Upton, NY: Brookhaven National Laboratory.

ASHRAE. 1977. *Standard 93–77, Collector Test Procedure*. Atlanta, GA: American Society of Heating, Refrigeration and Air Conditioning Engineers.

ASME J. Solar Energy Eng. 1983. 105: 339–384.

ASME J. Solar Energy Eng. 1984. 106: 22–105.

Assaf, G. 1978. The Dead Sea: Scheme for a solar lake. *Solar Energy* 18: 294.

Atkinson, B., and R. Caesar. 1983. The Volkspanel Model T. *Solar Age* (April): 33.

Baranov, V. K., and G. K. Melnikov. 1966. *Sov. J. Opt. Tech*. 33: 408.

Bartel, J. J., and P. E. Skarna. 1983. *Overview of the Construction and Start-up of the 10 MW_e Solar Thermal Central Receiver Pilot Plant*. SAND83-8021. Albuquerque, NM: Sandia National Laboratories.

Baum, V. A., R. R. Aparissi, and B. A. Garf. 1957. High power solar installations. *Solar Energy* 1: 6.

Berg, R. S. 1978. *Heliostat Dust Buildup and Cleaning Studies*. SAND 78-0035. Albuquerque, NM: Sandia Laboratories.

Bloem, H., V. C. de Grijs, and R. L. C. de Vaan. 1981. Evacuated tubular collectors with two-phase heat transfer into the system. In *Solar World Forum, Proceedings of the International Solar Energy Society Congress*. Vol. 1. Brighton, England, p. 176.

Boeing Corp. 1976. *Central Receiver Solar Thermal Power System*. SAN-111-76-2, Seattle, WA: Boeing.

Brantley, L. W., Jr. 1977. A pressure stabilized solar collector. Presented at ERDA Conference on Concentrating Solar Collectors. Georgia Instiute of Technology, Atlanta, GA.

Bronicki, L., B. Doron, A. Raviv, and H. Tabor. 1984. *Progress in Solar Ponds in Israel*. Yavne, Israel: Ormat Turbines, Ltd.

BSE. 1978. *BSE Guidelines and Directions for Determining the Utilizability of Solar Collectors: A Solar Collector Efficiency Test*. Bundesverband fuer Solarenergie, Kruppstrasse 5, 4300 Essen 1, Federal Republic of Germany.

Buchberg, H., and K. K. Edwards. 1976. Design consideration for solar collection with cylindrical glass honeycombs. *Solar Energy* 18: 193.

Buckley, S., 1978. Thermic diode solar panels for space heating. *Solar Energy* 20: 495.

Butti, K., and J. Perlin. 1980, *The Golden Thread*. New York: Van Nostrand-Reinhold.

Butler, B. L., and R. S. Claassen. 1980. Survey of solar materials. *ASME J. Solar Energy Eng*. 102: 175.

Chiam, H. F., 1981. Plannar concentrators for flat plate solar collectors. *Solar Energy* 26: 503.

Clark, A. F., and W. C. Dickinson. 1980. Shallow solar ponds. In W. C. Dickinson and P. N. Cheremisinoff (eds.), *Solar Energy Technology Handbook*. New York: Marcel Dekker, p. 377.

Clausing, A. M. 1976. The performance of a stationary reflector/tracking absorber solar concentrator. Presented at ISES Solar Energy Conference. Vol. 2. Winnipeg, Canada, p. 304.

Collares-Pereira, M., A. Joyce, and L. Valle. 1982a. A salt gradient solar pond for greenhouse heating applications. *Progress in Solar Energy* 5, 1: 221.

Collares-Pereira, M. 1982b. Nonevacuated CPC type 1.6 × concentrator for applications up to 100°C—Description and performance. *Progress in Solar Energy* 5, 1: 287.

Cooper, P. I., and R. V. Dunkle. 1981. A non-linear flat plate collector model. *Solar Energy* 26: 133.

de Winter, F. 1975. *Solar Energy and the Flat Plate: An Annotated Bibliography*. ASHRAE Report S-101. Atlanta, GA: American Society of Heating, Refrigeration and Air Conditioning Engineers.

de Winter, F., and Lyman, W. S. 1978. Optimum collection geometries for copper tube—copper sheet flat plate collectors. *Proceedings of the ISES Congress*. Vol. 2. New Delhi, India, January 16–21, 1978. Elmsford, NY: Pergamon Press, pp. 895–899.

Dickinson, W. C., and P. N. Cheremisinoff (eds.). 1980. *Solar Energy Technology Handbook*. New York: Marcel Dekker.

Dudley, V. E., and R. M. Workhoven. 1979. *Summary Report: Concerning Solar Collector Test Results Collector Module Test Facility*. SAND-78-0977. Albuquerque, NM: Sandia National Laboratories.

Eggers, G. H., A. J. Housman, F. L. Openshaw, S. L. Russell, Jr. 1979. Solar collector field subsystem program on the fixed mirror solar concentrator. Final Report. GA-A14209 (Rev.). San Diego, CA: General Atomic.

Felland, J. R., and D. K. Edwards. 1978. Solar and infrared radition properties of parallel-plate honeycomb. J. Energy 2: 309.

Francia, G. 1968. Pilot plants of solar steam generating stations. *Solar Energy* 12: 51.

Freese, J. M. 1978. *Effects of Outdoor Exposure on the Solar Reflectance Properties of Silvered Glass Mirrors*. SAND 78-1649. Albuquerque, NM: Sandia National Laboratories.

Fructer, E., G. Grossman, and F. Kreith. 1982. An experimental investigation of a stationary reflector/tracking absorber solar collector at intermediate temperature. *ASME J. Solar Energy Eng*. 104: 340.

Gee, R. C. 1983. *An Experimental Performance Evaluation of Line Focus Sun Trackers*. SERI/TR-632-646. Golden, CO: Solar Energy Research Institute.

Gerich, J. W. 1978. An inflated cylindrical solar concentrator. In *Proc. 1978 Annual Meeting Am. Section of ISES*. Denver, CO, p. 889.

Gerich, J. W. 1979. A nontracking inflated cylindrical solar concentrator. Presented at the International Solar Energy Society Congress. Atlanta, GA.

Gordon, J. M. 1986. Low concentration CPC's for low-temperature solar energy applications. *ASME J. Solar Energy Eng*. 108: 49.

Graham, B. J. 1979. *A Survey and Evaluation of Current Design of Evacuated Collectors*. Contract No. DE-AC04-78CS05350. Final Report. Annapolis, MD: Trident Engineering Associates, Inc.

Grassie, S. L., and N. R. Sheridan. 1977. The use of planar reflectors for increasing the energy yield of flat plate collectors. *Solar Energy* 19: 663.

Grimmer, D. P., K. G. Zinn, K. C. Herr, and B. E. Wood. 1978. Augmented solar energy collection using different types of planar reflectives surfaces; theoretical caluclations and experimental results. *Solar Energy* 21: 497.

Hedstrom, J. C. 1984. *Performance of an Active/Passive Hybrid Solar System Utilizing Vapor Transport*. LA-UR-84-2003, Los Alamos, NM: Los Alamos National Laboratory.

Herrick, C. S. 1983. An air-cooled solar collector using all-cylindrical elements in a low-loss body. *Solar Energy* 30: 217.

Hinterberger, H., and R. Winston. 1966. *Rev. Sci. Inst*. 37: 1094.

Hollands, K. G. T. 1965. Honeycomb devices in flat plate solar collectors. *Solar Energy* 9: 159.

Hollands, K. G. T. 1971. A concentrator for thin-film solar cells. *Solar Energy* 13: 149.

Holland, K. G. T., K. N. Marshall, and R. K. Webel. 1978. An approximate equation for predicting the solar transmittance of transparent honeycombs. *Solar Energy* 21: 231.

Hollands, K. G. T., and E. C. Shewen. 1981. Optimization of flow passage geometry for air-heating plate type solar collectors. *ASME J. Solar Energy Eng*. 103: 323.

Holmes, J. T. 1981. Heliostat operation at the central receiver test facility. In *Proc. STTF Testing for Long-Term Systems Performance Workshop*. Albuquerque, NM, p. 179.

Iannucci, J. J. 1980. Thermal energy centralization networks: design, cost and performance for dish, trough and central receiver systems. In *Proc. of AS of ISES Annual Meetings*. Vol. 3.1. Phoenix, AZ, p. 44.

Iyican, L., L. C. White, and Y. Bayazi, 1981. Natural convection and sidewall losses in trapezoidal groove collectors. *ASME J. Solar Energy Eng.* 103: 167.

Jenkins, J. P., and J. E. Hill. 1980. A comparison of test results for flat plate water heating solar collectors using the BSE and ASHRAE procedures. *ASME J. Solar Energy Eng.* 102: 2.

Jones, D. E., and L. E. Shaw. 1980. Development of a high performance air heater through use of an evacuated tube cover design. In *Proc. of AS of ISES Annual Meeting*. Phoenix, AZ. Vol. 3.1, p. 433.

Jones, G. F., and K. A. Meyer. 1982. Description and initial operation of the Los Alamos National Laboratory salt gradient solar pond. *Int. Solar Pond Letters* 1, 1: 2.

Kenna, J. P. 1983. The thermal trap solar collector. *Solar Energy* 31: 335.

Kohler, J. T., et al. 1978. Evaluation of six desgins for a site-fabricated, bulding-integrated air heater. In *Proc. 1978 Annual Meeting AS of ISES* 2, 1: 239.

Kovarik, M. 1978. Optimal distribution of heat conducting material in the finned pipe solar energy collector. *Solar Energy* 21: 477.

Kreider, J. F. 1975. Thermal performance analysis of the SRTA solar concentrator. *J. Heat Transfer* 97 (3).

Kreider, J. F., and F. Kreith, eds. 1980. *Solar Energy Handbook*. New York: McCraw-Hill.

Kritchman, E. M. 1981. Nonimaging second-stage elements —a brief comparison. *Appl. Opt.* 20: 3824.

Kritchman, E. M., A. A. Friesem, and G. Yekutieli. 1979. Efficient lens for solar concentration. *Solar Energy* 22: 119.

Kurzweg, U. H. 1980. Maximum solar flux concentration achievable with axicon collectors. *Solar Energy* 25: 221.

Kwan, B. M., and R. B. Bannerot. 1984. Improved optical design of nontracking concentrators. *ASMS J. Solar Energy Eng.* 106: 271.

Löf, G. O. G. 1980. Flat plate and nonconcentrating collectors. In W. C. Dickinson and P. N. Cheremisinoff (eds.), *Solar Energy Technology Handbook*. New York: Marcel Dekker.

Lunde, P. J. 1980. *Solar Thermal Engineering*. New York: Wiley.

Mahdjuri, F. 1979. Evacuated heat pipe solar collectors. *Energy Conversion* 9: 85.

May, E. K., and L. M. Murphy. 1983. Performance benefits of the direct generation of steam in line-focus solar collectors. *ASME J. Solar Energy Eng.* 105: 126.

McCormick, P. G. 1981. Optical evaluation of cylindrical elastic concentrators. *Solar Energy* 26: 519.

Meyer, B. A., M. M. El-Wakil, and J. W. Mitchell. 1978. Natural convection heat transfer in small and moderate aspect ratio enclosures—an application to flat-plate collectors. In F. Kreith, R. Boehm, J. Mitchell, and R. Bannerot (eds.), *Thermal Storage and Heat Transfer in Solar Energy Systems*. New York: ASME.

Mills, D. R., I. M. Bassett, and G. H. Derrick, 1984. Relative cost-effectiveness of CPC reflector designs suitable for evacuated absorber tube solar collectors. Sydney, Australia: School of Physics, University of Sydney.

Minardi, J. E., and H. H. Chuang. 1975. Performance of a black liquid flat plate solar collector. *Solar Energy* 17: 179.

Neeper, D. A., and R. D. McFarland, 1982. *Some Potential Benefits of Fundamental Research for the Passive Solar Heating and Cooling of Buildings*. LA-9425-MS. Los Alamos, NM: Los Alamos National Laboratory.

Nielsen, C. A. 1980. Nonconvective salt-gradient solar ponds. In W. C. Dickinson and P. N. Cheremisinoff (eds.), *Solar Energy Technology Handbook*. New York: Marcel Dekker, ch. 11.

Nielsen, C. A. 1982. Work in progress at the Ohio State University. *International Solar Pond Letters* 1, 1: 2.

O'Gallagher, J. J., A. Rabl, R. Winston, and W. McIntire. 1980. Absorption enhancement in solar collectors by multiple reflections. *Solar Energy* 24: 323.

O'Gallagher, J. J., K. Snail, R. Winston, C. Peak, and J. D. Garrison. 1982. A new evacuated CPC collector tube. *Solar Energy* 29: 575.

Olvera, A., and R. B. Bannerot. 1981. Design, evaluation and testing of a moderately concentrating nontracking solar energy collector. *ASME J. Solar Energy Eng.* 103: 34.

O'Neill, M. J. 1979. A unique new Fresnel lens solar concentrator. Presented at International Solar Energy Congress, Atlanta, GA. Also U.S. patent No. 4,069,812. 1978. Solar concentrator and energy collection system.

Ortabasi, U., and W. M. Buehl. 1980. An internal cusp reflector for an evacuated tubular heat pipe solar thermal collector. *Solar Energy* 25: 67.

Pierce, N. T. 1978. Efficient low-cost concentrating solar collectors. *Solar Energy* 19: 395.

Ploke, M. 1966. Lichtfuehrungseinrichtungen mit starker Konzentrationswirkung. *Optik* 25: 31.

Rabl, A. 1976. Comparison of solar concentrators. *Solar Energy* 18: 93.

Rabl, A. 1985. *Active Solar Collectors and Their Applications*. Oxford: Oxford University Press.

Rabl, A., J. O'Gallagher, and R. Winston. 1980. Design and test of nonevacuated solar collectors with compound parabolic concentrators. *Solar Energy* 25: 335.

Rhee, S. J., and D. K. Edwards. 1983. Comparison of test results for flat plate, transpired flat plate, corrugated and transpired corrugated solar air heaters. *ASME J. Solar Energy Eng.* 105: 231.

Ribot, J., and R. D. McConnell. 1983. Testing and analysis of heat pipe solar collector. *ASME J. Solar Energy Eng.* 105: 440.

Sandia. 1980. *Line Focus Solar Thermal Energy Technology Development*. FY 1979 Annual Report. SAND 80-0865, Albuquerque, NM: Sandia National Laboratories.

Schipper, L. 1984. Lecture on the Swedish housing industry. American Council on Energy Efficient Economy, Santa Cruz, CA. August.

Seitel, S. C. 1975. Collector performance enhancement with flat reflectors. *Solar Energy* 17: 291.

Selcuk, M. K. 1979. Analysis, development and testing of a fixed tilt solar collector employing reversible vee-trough reflectors and vacuum tube receivers. *Solar Energy* 22: 413.

Solar Products Specification Guide. Published annually by *Solar Age*.

Sunworld. 1981. Issues 3 and 4 of *Sunworld*, vol. 5, are devoted to solar thermal power generation, and the central receiver is described in issue 4.

Tabor, H. 1958a. Radiation, convection and conduction coefficients in Solar collectors. *Bulletin of the Research Council of Israel*. Vol. 6C, p. 155.

Tabor, H. 1958b. Stationary mirror system for solar collectors. *Solar Energy* 2: 27.

Tabor, H. 1966. Mirror boosters for solar collectors. *Solar Energy* 10: 111.

Tabor, H. 1981. Solar ponds. *Solar Energy* 27: 181.

Tabor, H., and H. Zeimer. 1962. Low cost focusing collector for solar power units. *Solar Energy* 6: 55.

Vant-Hull, L. L., and A. F. Hildebrandt. 1976. Solar thermal power system based on optical transmission. *Solar Energy* 18: 31.

Walton, J. D. 1980. Development of the spiral fresnel concentrator. Presented at International Symposium on Solar Thermal Power and Energy Systems, 15–20 June, Marseilles, France.

Window, B. 1983. Heat extraction from single ended glass absorber tubes. *Solar Energy* 31: 159.

Winston, R. 1970. Light collection within the framework of geometrical optics. J. Optical Soc. America 60: 245.

Winston, R. 1974. Solar concentrators of a novel design. *Solar Energy* 16: 89.

Winston, R. 1980. Cavity enhancement by controlled directional scattering. *Appl. Opt.* 19: 195.

Winston, R., and H. Hinterberger. 1975. Principles of cylindrical concentrators for solar energy. *Solar Energy* 17: 255.

Wittenberg, L. J., and M. J. Harris. 1980. The Miamisburg salt gradient solar pond. In *Nonconvective Solar Pond Workshop Proceedings.* Desert Research Institute, University of Nevada System, July 30 and 31.

Zangrando, F., and H. C. Bryant. 1978. A salt gradient solar pond. *Solar Age* 3: 21.

3 Optical Theory and Modeling of Solar Collectors

Ari Rabl

3.1 Basic Principles

3.1.1 Optical Efficiency

The goal of the optical analysis of a solar collector is to calculate the optical efficiency η_0 defined as the fraction of the solar radiation incident on the collector aperture that is absorbed by the absorber. The term collector designates the complete assembly, consisting of concentrator (e.g., reflector or lens) and receiver (i.e., absorber plus associated enclosure). How complicated this calculation is depends on the collector type and the desired accuracy. For many collectors it is easy to derive approximate formulas for the optical efficiency. For example, in a flat plate collector with a single cover of transmissivity τ and an absorber of absorptivity α, the optical efficiency can usually be approximated by $\tau\alpha$.

For greater accuracy one may want to include multiple reflections (between the surfaces of the cover and between cover and absorber), variation with incidence angle, and perhaps even polarization effects. Some people use the notation $(\tau\alpha)$ and the name transmittance-absorptance product to indicate that these effects have been included.

If τ and α depend on wavelength, then separate calculations may be necessary for different portions of the solar spectrum. In concentrating collectors, one may have to calculate ray diagrams, and the calculation can become complicated. This chapter provides the basic tools for the optical analysis. No spectral dependence of material properties is explicitly indicated, and in most cases a calculation with properties averaged over the solar spectrum is adequate. Only when two or more materials in one collector display strong spectral variation will a breakdown into separate spectral regions be necessary. However, even in that case the same formulas apply with different parameters for different spectral regions and with a summation over the entire solar spectrum.

The basic factors that enter into the optical analysis are the transmissivity τ of the cover(s), if any; the reflectivity ρ of the reflector(s), if any; the absorptivity α of the absorber; and the intercept factor γ. The latter is defined as the fraction of the rays incident on the collector aperture that reach the absorber.

In flat plate collectors the intercept factor is usually assumed to be unity; it may be slightly less than unity at nonnormal incidence because of shading by the collector enclosure. In most concentrating collectors the intercept factor is 0.9 to 1.0, and its calculation can be complicated. Once the intercept factor has been determined, one has a good approximation for the optical efficiency in the form

$$\eta_0 = \gamma\rho\tau\alpha. \tag{1}$$

The optical efficiency varies with incidence angle. To account for that, one defines an incidence angle modifier $K(\theta)$ such that the optical efficiency at incidence angle θ is

$$\eta_0(\theta) = \eta_0 K(\theta), \tag{2}$$

where η_0, without argument, indicates the value at normal incidence. The variation of $K(\theta)$ is caused by the change in transmissivity (given by Fresnel's equations) and absorptivity; in addition there may be geometrical effects. In most collectors the optical efficiency is fairly constant up to about 45 deg and falls off rapidly beyond about 70 deg (see figures 3.4 and 3.5 below).

The optical efficiency is one of the three key parameters in the Hottel-Whillier-Bliss equation that describe the performance of the collector. The other two parameters of the Hottel-Whillier-Bliss equation characterize thermal behavior in terms of heat loss coefficient and heat removal efficiency.

The geometrical optics of solar concentrators is determined by Snell's laws for reflection (for mirrors) and for refraction (for lenses). To trace the passage of rays through concentrating collectors, it is convenient to use vector algebra.

The Law of Reflection   In terms of the three unit vectors in figure 3.1,

$\hat{i} =$ direction of incident ray,

$\hat{n} =$ direction of normal of reflector surface,

$\hat{r} =$ direction of reflected ray,

all three of which point away from the surface, the law of specular reflection states that

1. The angle of incidence is equal to the angle of reflection,

$$\hat{i}\cdot\hat{n} = \hat{r}\cdot\hat{n}. \tag{3}$$

2. $\hat{i}$, $\hat{n}$, and $\hat{r}$ lie in the same plane,

$$(\hat{i} \times \hat{r})\cdot\hat{n} = 0. \tag{4}$$

Given any two of these vectors, the third one is uniquely determined by equations (3) and (4), apart from trivial minus signs. If $\hat{i}$ and $\hat{r}$ are specified, $\hat{n}$ must be a linear combination $a\hat{i} + b\hat{r}$ because of coplanarity, and the coeffi-

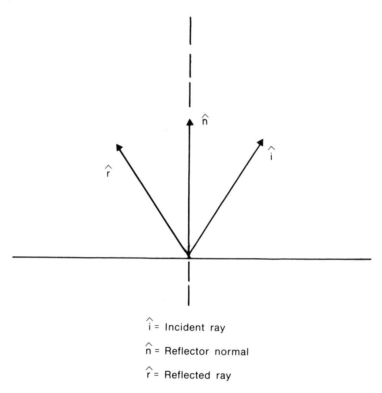

$\hat{i}$ = Incident ray

$\hat{n}$ = Reflector normal

$\hat{r}$ = Reflected ray

Figure 3.1
Unit vectors for the law of specular reflection.

cients a and b are fixed by equation (3) and by normalization. The result is

$$\hat{n} = (\hat{i} + \hat{r})(2 + 2\hat{i}\cdot\hat{r})^{-1/2}. \tag{5}$$

On the other hand, with $\hat{i}$ and $\hat{n}$ given, a similar argument shows that $\hat{r}$ is

$$\hat{r} = -\hat{i} + 2(\hat{i}\cdot\hat{n})\hat{n}. \tag{6}$$

Snell's Law of Refraction   Snell's law of refraction states that the incident ray $\hat{i}$, the normal to the surface $\hat{n}$, and the refracted (transmitted) ray $\hat{t}$ lie in a plane and that the angle of incidence $\theta_i = \arccos(\hat{i}\cdot\hat{n})$ and the angle of refraction $\theta_t = \arccos(\hat{t}\cdot\hat{n})$ satisfy

$$n_i \sin\theta_i = n_t \sin\theta_t, \tag{7}$$

where n_i and n_t are the indices of refraction of the two media. In vector notation, coplanarity of $\hat{i}$, $\hat{n}$, and $\hat{t}$ implies that $\hat{t}$ must be a linear combination of $\hat{i}$ and $\hat{n}$:

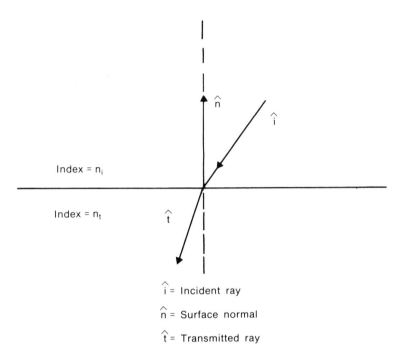

$\hat{i}$ = Incident ray

$\hat{n}$ = Surface normal

$\hat{t}$ = Transmitted ray

Figure 3.2
Unit vectors for law of refraction.

$$\hat{t} = a\hat{i} + b\hat{n}. \tag{8}$$

The coefficients a and b are uniquely determined (apart from trivial minus signs) by equation (7) and by normalization. One can readily verify that for the vectors indicated in figure 3.2 the refracted ray $\hat{t}$ is related to $\hat{i}$, $\hat{n}$, and refractive indices by

$$\hat{t} = \frac{n_i}{n_t}[\hat{i} - \hat{n}(\hat{i} \cdot \hat{n} + \sqrt{(\hat{i} \cdot \hat{n})^2 + (n_t/n_i)^2 - 1})]. \tag{9}$$

Fresnel Equations   In addition to the direction of a refracted ray, one needs to know how much radiation is reflected and how much is transmitted. The reflection coefficients r (the ratio of reflected radiation over incident radiation) are different for the parallel and perpendicular components of polarization. Parallel and perpendicular refer to the plane spanned by the incident direction and the surface normal, as indicated by the subscripts $\parallel$ and $\perp$. For radiation passing from a medium with refractive index n_i to a medium with index n_t the reflection coefficients are given by

$$r_\parallel = \frac{\tan^2(\theta_t - \theta_i)}{\tan^2(\theta_t + \theta_i)} \tag{10}$$

for the parallel component,

$$r_\perp = \frac{\sin^2(\theta_t - \theta_i)}{\sin^2(\theta_t + \theta_i)} \tag{11}$$

for the perpendicular component, and

$$r = \frac{I_r}{I_i} = \frac{1}{2}(r_\parallel + r_\perp) \tag{12}$$

for unpolarized radiation. θ_i and θ_t are the angles of incidence and refraction, related by Snell's law, equation (7).

For radiation at normal incidence, θ_i and θ_t vanish and equation (12) yields, in the limit $\theta_i \to 0$, the reflection coefficient

$$r = \left.\frac{I_r}{I_i}\right|_{\theta_i=0} = \left(\frac{n-1}{n+1}\right)^2, \tag{13}$$

where

$$n = \frac{n_t}{n_i} \tag{14}$$

is the relative index of refraction. Most glazing materials have an index near 1.5, and the corresponding reflection coefficient at normal incidence is

$$r = \left(\frac{0.5}{2.5}\right)^2 = 0.04. \tag{15}$$

At grazing incidence, r approaches 1. For intermediate angles $r_\perp$ increases monotonically as θ_i increases, whereas $r_\parallel$ first decreases until it vanishes at the Brewster angle, where $\theta_i + \theta_t = \pi/2$, and then increases to unity at $\theta_i = \pi/2$.

Direct solar radiation is unpolarized, and the partial polarization of diffuse sky radiation has negligible effects. Hence one can consider incident solar radiation to be unpolarized. Nonetheless, there could be polarization effects when calculating transmission of solar radiation through multiple covers. The difference between reflection and transmission coefficients for different polarizations implies that at nonzero incidence anlges the transmitted radiation is partially polarized even if the incident radiation was not.

Multiple Reflection   In many applications radiation is to be transmitted through a slab of material; in that case there are two parallel interfaces, each with reflection losses. As a first approximation for small reflection coefficients r (at each surface), one can approximate the transmission through the slab by

$$\tau \approx 1 - 2r, \tag{16}$$

if there is no absorption in the slab.

A more accurate calculation takes into account all the multiple reflections as shown in figure 3.3 for a nonabsorbing slab. The calculation is done separately for each component of polarization, keeping track of the fact that $r_{\parallel}$ and $r_{\perp}$ are functions of θ_i. At the first interface a fraction $(1 - r_{\perp})$ of the perpendicular component is transmitted into the slab. At the second interface $(1 - r_{\perp})^2$ is transmitted, while $(1 - r_{\perp})r_{\perp}$ is reflected back up. Of the latter portion, some will be reflected back down at the first interface, thus repeating the pattern.

Contributions from successive round trips correspond to successive terms in a geometric series, and the sum of all transmitted terms is

$$\tau_{\perp} = (1 - r_{\perp})^2 \sum_{n=0}^{\infty} r_{\perp}^{2n} = \frac{(1 - r_{\perp})^2}{(1 - r_{\perp}^2)} = \left(\frac{1 - r_{\perp}}{1 + r_{\perp}}\right). \tag{17}$$

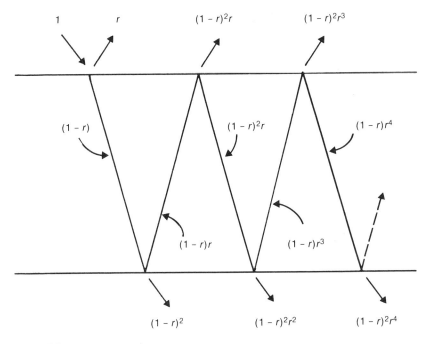

Figure 3.3
Transmission of radiation through nonabsorbing slab of material. Source: Duffie and Beckman (1980).

The same formula holds for the other polarization component. Hence the transmittance of unpolarized radiation is given by the average transmittance for the two components:

$$\tau = \frac{1}{2}\left(\frac{1 - r_\parallel}{1 + r_\parallel} + \frac{1 - r_\perp}{1 + r_\perp}\right). \tag{18}$$

For the above example, a medium with $n = 1.5$ at normal incidence, we find

$$\tau = \frac{1 - 0.04}{1 + 0.04} = 0.923, \tag{19}$$

only slightly larger than the naive result 0.920 of equation (16). Of course the effect of multiple reflection becomes more important if the reflection coefficients are larger (as indeed they are at nonnormal incidence), or if there is more than one cover (Edwards 1977). For a system with N identical covers, Duffie and Beckman (1980) give the result

$$\tau = \frac{1}{2}\left[\frac{1 - r_\perp}{1 + (2N - 1)r_\perp} + \frac{1 - r_\parallel}{1 + (2N - 1)r_\parallel}\right], \tag{20}$$

again for the case where nothing is absorbed in the material. This equation has been plotted in figure 3.4 versus incidence angle for 1, 2, 3, and 4 covers of glass with $n = 1.526$. One sees that the transmittance remains almost constant to 40 deg and then begins to drop, slowly at first, then rapidly beyond 60 deg. At 90 deg the transmittance vanishes.

Absorption in the Cover   Absorption of radiation in a partially transparent medium is described by Bouguer's law. The fraction α of radiation that is absorbed over a path length L in a medium of extinction coefficient K is given by

$$1 - \alpha = \exp(-KL). \tag{21}$$

For a slab of thickness d the path length is

$$L = \frac{d}{\cos\theta_t}, \tag{22}$$

where θ_t is the angle between the surface normal and the direction of the light ray in the slab.

Bouguer's law holds for each wavelength λ, but if K varies with λ, one must replace equation (21) by a superposition of exponentials, for example, as for solar pond calculations. For glass it is safe to assume a single extinction coefficient, which ranges from $4\ \mathrm{m}^{-1}$ for "water white" glass to approximately

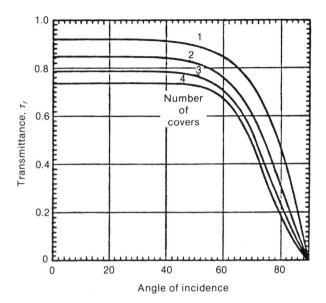

Figure 3.4
Transmittance of 1, 2, 3, and 4 nonabsorbing covers having an index of refraction of 1.526. Source: Duffie and Beckman (1980).

32 m^{-1} for ordinary window glass (which appears greenish at the edge). Thus a $d = 5$-mm thick pane of ordinary window glass absorbs $\alpha = 1 - \exp(-32 \times 0.005) = 15\%$ of the solar radiation at normal incidence.

It is straightforward to modify figure 3.3 and the transmission equations (17) and (18) for absorption in the slab. The result is

$$\tau_{\parallel} = \frac{(1 - \alpha)(1 - r_{\parallel})^2}{1 - r_{\parallel}^2(1 - \alpha)^2} \tag{23}$$

for the parallel component of polarization, with α given by equation (21). The analogous formula holds for the other polarization component, and the transmission coefficient for unpolarized radiation is the average of $\tau_{\parallel}$ and $\tau_{\perp}$. For good materials the approximation

$$\tau \simeq 1 - \alpha - 2r \tag{24}$$

will be adequate. Numerical results for the transmission of cover systems with absorption are shown in figure 3.5 for a slab with a refractive index of 1.526 and various absorption coefficients as indicated by the values of Kd. The total reflectance $\rho_{\parallel}$ of a slab with absorption is obtained in a similar manner:

64

Ari Rabl

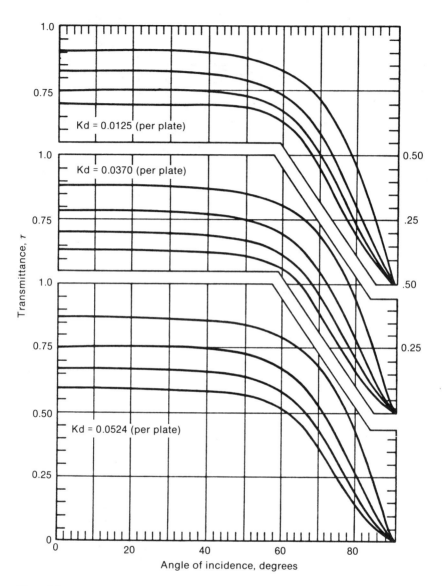

Figure 3.5
Transmittance of radiation through 1, 2, 3, and 4, covers with index of refraction of 1.526 and various absorption coefficients. Source: Duffie and Beckman (1980).

$$\rho_\parallel = r_\parallel [1 + (1 - \alpha)\tau_\parallel], \tag{25}$$

with the analogous formula for the $\rho_\perp$. Note that ρ includes multiple reflections between the slab surfaces, whereas r is the reflection coefficient for a single surface.

Absorption at the Absorber   Ultimately the solar radiation is of course to be absorbed by the absorber. Since no absorber is perfectly black, one has to consider multiple reflections between cover system and absorber plate, if one wants to be exact. The summation of multiple reflections leads again to a geometric series for the fraction of incident solar radiation that is absorbed by the absorber. This fraction is frequently called the effective $\tau\alpha$ product and designated by $(\tau\alpha)$:

$$(\tau\alpha) = \tau\alpha \sum_{n=0}^{\infty} [(1 - \alpha)\rho_d]^n$$
$$= \frac{\tau\alpha}{1 - (1 - \alpha)\rho_d}, \tag{26}$$

where α is now the absorptance of the absorber surface. For the cover reflectance we take ρ_d, the diffuse reflectance, because most solar absorbers are diffuse reflectors. As shown by Duffie and Beckman (1980), ρ_d can be approximated by evaluating equation (25) at an angle for incidence $\theta_i \approx 60°$. In practice the difference between $(\tau\alpha)$ and the simple product of τ and α is likely to be very small. Even for selective coatings α will usually be at least 0.9, while the cover reflectance ρ_d is about 0.15. For such a case the difference between $(\tau\alpha)$ and $\tau\alpha$ is at most 1.5%.

3.1.2   Concentration Ratio and Acceptance Angle

For a discussion of concentrating collectors, one must define two important characteristics: the concentration ratio and the acceptance angle. The geometrical concentration C is the ratio of collector aperture area A and absorber surface area A_{abs}:

$$C = \frac{A}{A_{abs}}. \tag{27}$$

This quantity depends only on the geometry. Sometimes a flux concentration ratio has been used, defined as intensity ratio at aperture and absorber; it depends on absorption effects in addition to geometry. In this chapter the geometrical concentration ratio is used exclusively.

The acceptance angle or field of view is defined as the angular range over which all or almost all of the rays incident on the collector aperture reach the absorber, without moving all or part of the collector. The acceptance angle is one of the more important characteristics of a solar concentrator because it determines the tracking requirement. By considering phase space conservation (Winston 1970) or reciprocity relations for radiation shape factors (Rabl 1976a), it can be seen that the second law of thermodynamics imposes an upper limit on the concentration ratio achievable by any optical system with nonzero acceptance angle; this is sometimes called the thermodynamic (or ideal) limit of concentration.

There must be a connection between geometrical optics and the second law of thermodynamics because, if solar radiation could be concentrated onto an arbitrarily small receiver, then the receiver temperature could exceed the surface temperature of the sun. This would obviously be a violation of the second law, which states that heat cannot flow from a cold surface to a hot surface without an external source of work. The maximum possible concentration for a given acceptance half angle θ_a is

$$C_{\text{ideal, 2D}} = \frac{1}{\sin \theta_a} \tag{28a}$$

for a two-dimensional (troughlike, line-focus) concentrator, and

$$C_{\text{ideal, 3D}} = \left(\frac{1}{\sin \theta_a}\right)^2 \tag{28b}$$

for three-dimensional types (cones, dishes, pyramids, point-focus). Since the angular radius of the sun is $\theta_s \simeq 1/4$ deg $\simeq 5$ mrad, this implies a limit of 200 for collectors with single-axis tracking and 40,000 for dual-axis tracking.

However, the concentration achievable in practical systems is reduced by the following factors:

1. Most conventional concentrators, focusing types in particular, are based on optical designs that fall short of the thermodynamic limit by a factor of 2 to 4.

2. Receiver misalignment, tracking errors, and errors in mirror surface and contour necessitate design acceptance angles that are considerably larger than the angular diameter of the sun.

3. No lens or mirror material is perfectly specular; therefore the acceptance angle must be enlarged further.

4. Because of atmospheric scattering, a significant portion of the solar radiation may come from directions other than the solar disc itself.

The concentration ratio achievable with nontracking collectors is determined by the magnitude of the angular motion of the sun during the day and the year (Rabl 1976a). The highest practical concentration ratio for fixed collectors is about 2. This value can be increased to about 3 if a summer-to-winter adjustment of the collector tilt is made. With daily tilt adjustments, the concentration limit for nontracking collectors is about 10. These limits can be reached only by nonfocusing collectors of the CPC (compound parabolic concentrator) type (Welford and Winston 1978).

Since nontracking solar concentrators must have a large acceptance angle, they can collect a significant amount of diffuse radiation. A precise calculation of this effect would require detailed information about the angular distribution of diffuse sky radiation. Since few data on this distribution are available, one usually assumes that the hemispherical (or total) irradiance I_h incident on the aperture is the sum of the beam (or direct) component I_b (multiplied by the cosine of the incidence angle θ) and an isotropic background of diffuse irradiance I_d; that is,

$$I_h = I_b \cos \theta + I_d. \tag{29}$$

There is a simple proof based on radiation shape factors that a concentrator of concentration C accepts a fraction $1/C$ of isotropic radiation (Rabl 1976a). Hence the irradiance within the acceptance angle of a solar collector can be approximated by

$$I = I_b \cos \theta + \frac{I_d}{C}. \tag{30}$$

Because of the predominance of near forward scattering in the atmosphere, the sky radiation tends to be centered around the sun; therefore the isotropic model may underestimate the actual acceptance for diffuse radiation slightly, by at most 1% according to data of Reed (1976). A model for the anisotropy of sky radiation can be found in Hay (1979a, b).

The intercept factor and optical efficiency depend on the specification of the incident radiation. Several different conventions are possible. For consistency with standardized collector test results, the following is recommended: for flat plates and for collectors with concentration close to unity, the efficiency should be based on hemispherical irradiance; for collectors with high concentration, it should be based on the beam irradiance. For collectors with intermediate concentration, insolation can be specified according to equation

(30). The chosen convention and the diffuse radiation during the test should be stated clearly to permit conversion to a different insolation specification.

3.2 Nontracking Collectors

The principal nontracking collector types are flat plates, flat plates enhanced by side reflectors or V-troughs, tubular collectors, and compound parabolic concentrators (CPC).

3.2.1 Flat Plate Collectors

The optical analysis of most flat plate collectors is relatively straightforward, as discussed previously. Once the transmissivity of the cover has been calculated as a function of angle (as displayed in figures 3.4 and 3.5), then even the incidence angle modifier can be determined easily.

A typical incidence angle modifier for a flat plate collector is shown in figure 3.6. When the incidence angle modifier for a flat plate collector is plotted versus $1/\cos(\theta)$ instead of θ, it can usually be approximated by a linear fit. Since $K(0) = 1$, this fit can be written in the form

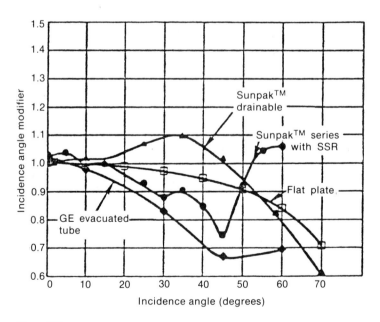

Figure 3.6
Incidence angle modifiers for flat plate and several evacuated tubular collectors. (incidence angle in direction perpendicular to tubes).

$$K(\theta) = 1 + b\left[\frac{1}{\cos(\theta)} - 1\right],\tag{31}$$

where b is a constant that depends on collector design and materials.

There are many variations of the basic flat plate collector design; for example, flat plate collectors with honeycombs or with corrugated covers. Calculating the optical efficiency and the incidence angle modifier for these collectors can be complicated (e.g., see Hollands et al. 1979; Felland and Edwards 1978; Symons 1982).

3.2.2 Compound Parabolic Concentrators

Concentrators that reach the thermodynamic limit of concentration [equations (28a) or (b)] have been called ideal concentrators because of their optical properties. In the solar energy literature, names such as compound parabolic concentrators (CPCs) and nonimaging concentrators have also been used. We

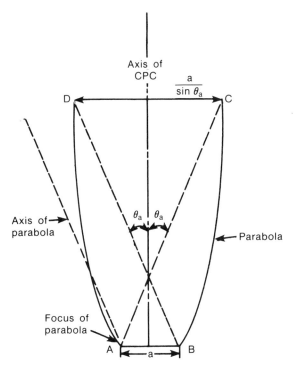

Figure 3.7
Compound parabolic concentrator (CPC). Aperture width $= a/\sin\theta_a$, absorber width $= a$, concentration ratio $C = 1/\sin\theta_a$. Axis and focus for parabolic section of the right-hand portion of the reflector are shown.

will refer to all concentrators of this class as CPCs, even though some of them are not even parabolic.

Ideal concentrators are a surprisingly recent discovery. The first example of a CPC, shown in figure 3.7, was found independently in the United States by Hinterberger and Winston (1966), in Germany by Ploke (1966), and in the U.S.S.R. by Baranov and Melnikov (1966). This CPC consists of parabolic reflectors that funnel the radiation from aperture to absorber. The right and left halves belong to different parabolas, as expressed by the name compound parabolic concentrator. The axis of the right branch, for example, makes an angle θ_a with the collector midplane, and its focus is at A. At the end points C and D, the slope is parallel to the collector midplane.

Tracing a few sample rays reveals that this device has the following angular acceptance characteristic: All rays incident on the aperture within the acceptance angle, that is, with $\theta < \theta_a$, will reach the absorber, whereas all rays with $\theta > \theta_a$ will bounce back and forth between the reflector sides and eventually reemerge through the aperture. This property, plotted schematically by the solid line in figure 3.8, can be shown to imply that the concentration is equal to the thermodynamic limit.

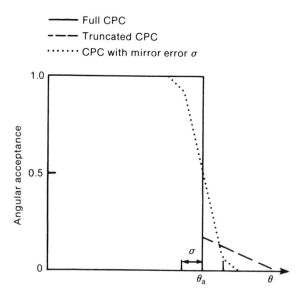

Figure 3.8
Fraction of the radiation incident on aperture at angel θ that reaches absorber, for CPC with acceptance half-angle θ_a, assuming reflectivity $\rho = 1$. *Solid line*: untruncated CPC with perfect reflectors; *dashed line*: truncated CPC with perfect reflectors; *dotted line*: untruncated CPC with surface errors σ.

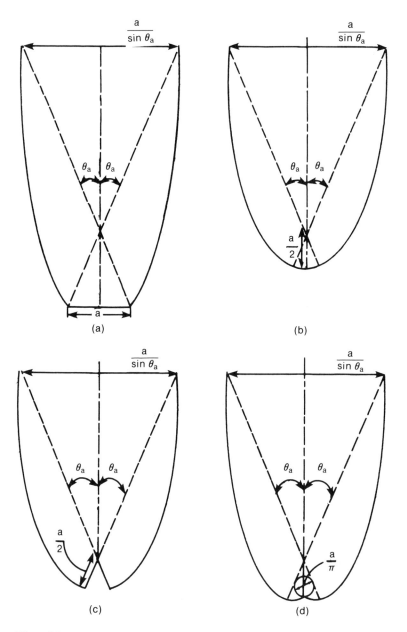

Figure 3.9
CPCs with different absorber shapes. All have the same absorber perimeter a and acceptance half-angle θ_a.

Subsequent to the discovery of the basic CPC (figure 3.7), several generalizations relevant to special applications have been described. These generalizations concern the following:

1. The use of arbitrary receiver shapes such as fins and tubes (see figure 3.9) (Winston and Hinterberger 1975; Rabl 1976c).

2. The restriction of exit angles at the receiver to values less than 90 deg (Rabl and Winston 1976). This is important because some receivers have poor absorptance at large angles of incidence.

3. The asymmetric orientation of source and aperture (for the design of collectors with seasonally varying outputs) (Rabl 1976a; Mills 1978).

4. The matching of a CPC to a finite source of radiation (useful as second-stage concentrators to collect radiation from a first stage that is a finite distance away) (Rabl and Winston 1976).

The design goal is to concentrate radiation maximally, subject to any of the subsidiary conditions (l) through (4) that may have been specified.

The design of two-dimensional concentrators is determined uniquely by the extreme rays (rays coming from the edge of the source). The concentrator must be designed in such a way that extreme rays from the source are transformed into extreme rays at the absorber; that is, they must either hit the edge of the absorber or reach the absorber tangentially.

Explicit solutions and equations for the reflector shapes can be found in the individual references for each CPC type. In three dimensions (geometry of cones and pyramids) the design is, in general, overdetermined, but for flat receivers an excellent compromise is achieved by choosing a surface of revolution whose cross section has been determined by the two-dimensional solution (Welford and Winston 1978).

As for the choice between different CPC types, the configurations with fin or tube absorbers, figures 3.9b and 3.9d, are preferable for most solar applications. Not only is the absorber material used more efficiently than in other designs, but heat loss through the back is low. Also tubes are natural for thermal collectors because they are needed to carry the heat transfer fluid. In most cases these advantages will more than compensate for the slightly higher optical losses (the average number of reflections for the tube or fin configurations is approximately 0.5 higher than for the configuration with a flat, one-sided absorber).

In their optical properties, all CPC types are exactly or nearly alike. Above all, the same relation, equation (28) exists between their concentration and angular acceptance, with the sharp cutoff implied in figure 3.8. The flux distribution at the absorber depends on the angle of incidence and on absorber shape, and in general, it can be determined only by detailed ray tracing.

However, the following important statement can be made about all CPCs, without any need for ray tracing: If the radiation incident on the aperture is spread uniformly over the entire acceptance angle, then it will be isotropic when it reaches the absorber. This consideration of uniform illumination is important because it yields a simple and reliable estimate of the average performance of a CPC solar collector. When beam insolation is incident at certain angles, hot spots of high flux concentration (up to about 40 for the 2D case) may appear on the absorber. In a real collector there will be slope errors caused by imperfections in the optical figure and some nonspecularity introduced by the reflecting material. These effects tend to reduce the actual hot spots of flux concentration to perhaps a factor equal to ten times the average illumination. This may still be high enough to pose a problem in particular configurations and needs to be taken into account in the collector design.

CPCs have large reflector areas. Fortunately, this disadvantage can be alleviated by truncation; that is the top portion of a CPC can be cut off with minimal loss in concentration since it does not intercept much radiation. Detailed graphs showing the effects of truncation on height, reflector area, number of reflections, and concentration ratio have been published by Rabl (1976a, b), McIntire (1979), and Carvalho et al. (1985). For the configuration of figure 3.9a, the relation between reflector area, acceptance half angle, and concentration ratio is shown in figure 3.10.

To analyze the optical performance of a CPC, it is convenient to think in terms of the average number of reflectiong $\langle n \rangle$. (The number of reflections varies both with angle of incidence and with point of incidence on the aperture.) A good estimate of the fraction τ_{CPC} of radiation transmitted through a CPC is given by the simple formula

$$\tau_{\text{CPC}} = \rho^{\langle n \rangle}. \tag{32}$$

The validity of this approximation has been demonstrated by Rabl (1977). This reference also shows how $\langle n \rangle$ can be calculated in closed form for a large number of concentrators and radiation passages. Numerical results (Carvalho et al. 1985) for the tubular configuration of figure 3.9d are displayed in figure 3.11 for several values of the design acceptance half-angle θ_a; for each θ_a curve shows how $\langle n \rangle$ varies with the actual concentration ratio as the CPC is truncated.

Regarding sensitivity to mirror surface errors, the analysis is equally simple for all CPCs because their geometry implies that all rays incident near the cutoff angle (i.e., with $\theta \lesssim \theta_a$) undergo exactly one reflection on the way to the

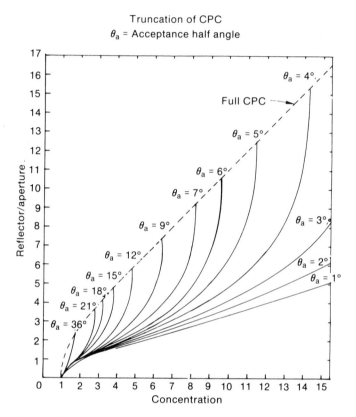

Figure 3.10
Reflector/aperture area ratio for full and truncated CPCs with flat, one-sided absorbers (configuration of figure 3.9a). Source: Rabl (1976b).

absorber. In most practical applications the acceptance half-angle θ_a will be larger than 5 deg, and the mirror surface error σ will usually be small compared to θ_a. If the error distribution is rectangular, all of the rays with $\theta < \theta_a - \sigma$, and none of the rays with $\theta > \theta_a + \sigma$ are accepted; between these two angles, some rays are accepted and some are rejected. Qualitatively this feature holds true for other error distributions as well. The resulting angular acceptance is shown schematically by the dotted line in figure 3.8. For an analysis of the effect on collector performance, the reader is referred to Carvalho et al. (1985) and Mills, Bassett, and Derrick (1984). Further details on optical and geometric properties of CPCs can be found in Welford and Winston (1978) and Rabl (1985).

An important design problem for thermal CPC collectors arises from the need for a gap between reflector and absorber. The design principles for CPC

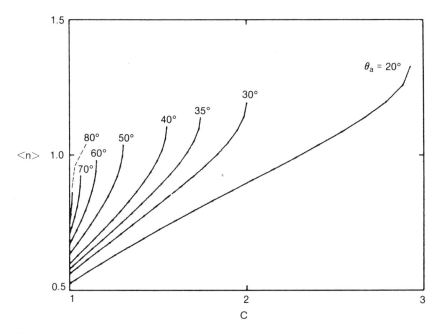

Figure 3.11
Average number of reflections $\langle n \rangle$ for CPC with tubular absorber, averaged over all rays within field of view, as function of concentration ratio C, for $\theta_a = 20, 30, 35, 40, 50, 60,$ and 70 deg. For $\theta_a = 90$ deg and $C = 1$, $\langle n \rangle = \pi/4$. Source: Carvalho et al. (1985).

reflectors demand that the reflector extend all the way to the absorber. However, this is not possible with evacuated tubes because of the finite thickness of the glass and because of the finite spacing between absorber surface and glass envelope. A gap is also needed in nonevacuated CPC collectors if the reflector has a high conductivity; otherwise, the reflector can act as a cooling fin for the absorber.

Several modifications of the basic CPC design are possible to deal with this gap problem. One could, for example, leave the reflector intact and reduce the absorber size—or, leave the absorber as it is and truncate the adjacent reflector. These and several other solutions have been investigated, and the following can be recommended as the best, in terms of minimizing the optical loss.

For flat absorbers, the optical loss of the gap is minimized if the reflector is truncated adjacent to the absorber, as sketched in figure 3.12. As shown in Rabl, Goodman, and Winston (1979), the optical loss for uniform illumination within the acceptance angle can be calculated in closed form using radiation shape factors. This is an excellent approximation for the yearly average loss

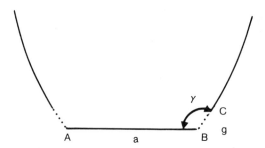

Figure 3.12
Truncation of CPC to accommodate gap between flat, one-sided absorber and reflector. $a = \overline{AB}$, $g = \overline{BC}$.

in actual operation. For small gaps in the configuration of figure 3.12, the fraction of the incident radiation that is lost is in the range of $0.2g/a$ to $0.5g/a$ (depending on CPC shape) where $g = \overline{BC}$ is the gap width and $a = \overline{AB}$ is the absorber width.

For tubular absorbers one could also truncate the reflector, as suggested in figure 3.14d. The associated gap loss has been calculated by Rabl, Goodman, and Winston (1979) and turns out to be quite small. But with tubular absorbers one can do even better by avoiding gap losses altogether. This possibility was suggested by McIntire (1981) and explored systematically by Winston (1980). This design replaces the cusp by a grooved cavity as shown in figure 3.13. The number and the dimensions of the grooves depend on the gap width.

With a single groove the gap can be as large as 0.27 times the tube radius. A design with two grooves permits a lossless solution up to about $0.4r$. With a larger number of grooves the lossless design can be extended up to a maximum gap equal to the radius. The lossless design necessarily entails a certain sacrifice of concentration ratio. For example, the design of figure 3.13 achieves only a concentration of 1.0 even though the thermodynamic limit for this acceptance angle is $1/\sin 60$ deg $= 1.15$. In most solar applications high optical efficiency is more important than the attainment of the highest possible concentration ratio; hence one will usually choose the lossless solution.

It turns out that using the grooved cavity as a means of suppressing gap loss has the following beneficial effect, which leads to a compromise design: As one increases the gap beyond the strict gap-loss limit, the optical loss goes up only quadratically with the size of the gap. As a result even a single V-groove cavity can accommodate gaps up to about 40% of the radius while keeping the optical loss to $< 1\%$. Consequently the single V-cavity is likely to be the most practical CPC design for a tubular absorber when even a substantial gap between the absorber and reflector is required (Winston 1983).

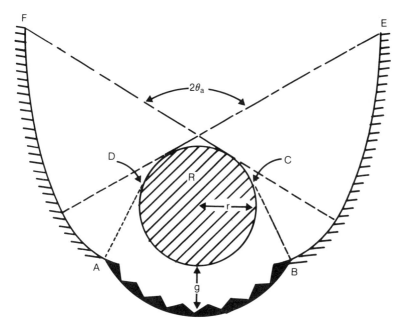

Figure 3.13
Modification of CPC for tubular absorber with finite gap g and zero optical loss. For the example chosen, the gap is $g = 0.8r$, the groove angle $2\psi = 118$ deg the acceptance half-angle $\theta_a = 60$ deg and the concentration ratio $C = 1$. Source: Winston (1980).

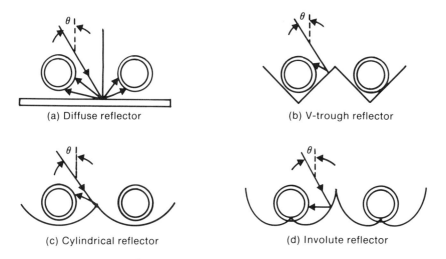

Figure 3.14
Several reflector configurations for evacuated tubes: (a) diffuse reflector; (b) V-trough reflector; (c) circular reflector; (d) involute reflector (CPC). Source: O'Gallagher et al. (1980).

There is an alternative design that maximizes the flux on the absorber by designing the CPC profile for a virtual "ice cream cone" shaped absorber for which there is no gap. For the real absorber this design does have a gap loss. However, it has the offsetting advantage of producing a geometric concentration ratio somewhat larger than $C_{ideal} = 1/\sin(\theta_a)$. There is no contradiction with equation (4a) (the thermodynamic limit) because this design has some optical loss. In fact the optical loss is precisely equal to the increase in concentration beyond the "thermodynamic limit." For small gaps this may be the preferred design, the choice being a subject for a detailed optimization that must include consideration of such factors as optical efficiency, thermal performance, and cost.

Some new types of concentrators discovered recently reach or closely approach the thermodynamic limit, even though they are quite different from the CPC. Winston and Welford (1979) formulated a generalized design principle that yields not only the CPC but also a novel type of second-stage concentrator, called trumpet because of its shape. The trumpet is well-suited as a second- stage concentrator for point-focus collectors.

3.2.3   Reflectors for Evacuated Tubes

By and large, the cost per unit area of reflectors is lower than the cost of evacuated tubes. For that reason most of the evacuated collectors sold today use some kind of reflector enhancement. Because of imperfect reflectivity, the use of a reflector incurs some optical loss. At the same time heat loss is reduced because there is less absorber area per aperture area. Thus, the choice of a reflector involves trade-offs between optical performance, thermal performance, and cost. In general, one prefers a large acceptance angle for nontracking collectors to minimize the need for tilt adjustments.

The choice of the reflector depends on the shape of the absorber. Figure 3.14 shows several reflector arrangements that have been used with tubular absorbers. The diffuse reflector in figure 3.14a is just a plain white surface behind the tubes. It has the lowest cost and the lowest performance since much of the reflected radiation misses the tubes. The V-groove and circular cylindrical reflectors, shown in figures 3.14b and 3.14c, are easy to fabricate, and in small quantities their cost can be significantly lower than the cost of CPCs. The V-trough reflectors are well matched to flat absorbers, but with tubular absorbers they do not utilize the back of the tubes very well. The best optical performance is achieved with a CPC reflector, as shown in figure 3.14d.

It is worth pointing out that, in some cases, one may want to design for a geometric concentration ratio below unity. Some absorber coatings of evacuated tubes have a rather low absorptivity. Together, with reflections off the glass envelope, these tubes may reflect 20% to 30% of the incident radiation. If the reflected rays have another chance to contact a tube, the overall effective absorptivity of the collector is improved. This kind of absorption enhancement is possible if the geometric concentration ratio is less than unity.

In effect the absorber tubes plus their reflectors can act like a radiation cavity whose absorptivity exceeds the absorptivity of the absorber surface itself. This question has been studied by O'Gallagher et al. (1980), and the involute (CPC with concentration less than unity) was shown to give the greatest possible absorption enhancement. For a detailed optical analysis of evacuated tubes with reflectors, see the articles by McIntire (1980) and Window and Basset (1981).

3.2.4 V-Troughs

The V-trough is a classic concentrator design for applications with a large acceptance angle (e.g., see Tabor 1958; Hollands 1971). To analyze the multiple reflections in V-troughs, the method of images is convenient. In principle, one could reach the thermodynamic limit with a V-trough if it were made infinitely deep and narrow and with perfect reflectors. But, in practice, the reflection losses would be prohibitive. The higher the desired concentration, the greater is the relative advantage of the CPC over the V-trough.

The upper limit of concentration for a practical V-trough is about 3 (as nontracking collector with daily tilt adjustments) compared to 10 for CPC. With summer-to-winter adjustments only, the V-trough is limited to concentration values below 2, compared to 3 for a CPC. For a completely fixed collector a V-trough gives almost no concentration, whereas a CPC can yield almost a factor of 2. Furthermore the V-trough is limited to flat, one-sided absorbers and can achieve very little useful concentration for tubular absorbers. An interesting asymmetric V-trough that requires summer–winter adjustment has been described by Selcuk (1979).

3.2.5 Side Reflectors

Side reflectors can be a useful means of increasing the output of a flat plate collector. In situations where a suitable surface is available next to a collector, it can be cost-effective to add a reflector, such as a sheet of anodized aluminum. For example, if a tilted collector is mounted on a flat horizontal roof, one could place a reflector in front. In large installations several rows of collectors can be mounted one behind another. Reflectors can be placed behind the collector rows, creating a sawtooth pattern of alternating collector and reflector surfaces. If a separate support structure is needed for a side reflector, then the cost may be too high to be practical.

Flat plates with a side reflector tend to produce a nonuniform output over the course of a year, just like an extreme asymmetric (or one-sided) CPC (Rabl 1976a) to which it is in some sense a straight-line approximation. Whether such nonuniform output is desirable depends on the load distribution. Depending on the geometry, during part of the year the reflector may cast a shadow on the collector and actually reduce the performance.

There is one case, however, in which a reflector is guaranteed to improve the collector output. A single tilted collector always receives some radiation from the ground during daylight hours; hence a horizontal reflector in front of the collector can only help. Whether it is cost-effective is another question. As for specularity, a specular reflector is certainly preferable because a diffuse side reflector offers minimal benefit, as shown by Grassie and Sheridan (1977). To evaluate the benefit of side reflectors, look at system performance, not just the optical enhancement. Since the heat loss is the same with or without a reflector, the increase of the collector output can be significantly higher than the optical enhancement factor.

Apart from these simple facts it is difficult to make general statements about side reflectors. A detailed evaluation may be needed in each case since the benefit of side reflectors depends on a large number of variables: not only the geometric/optical parameters but also the system configuration and load. Even the optical analysis of the reflector/collector combination is complicated, especially if both reflector and collector are short enough to make edge effects significant. For more specific results, in particular for optical performance, see the detailed studies in McDaniels et al. (1975), Seitel (1975), Larson (1980), and Chiam (1981).

3.3 Tracking Collectors

In principle any collector, even a flat plate, could be made to track the sun, but in practice one will usually resort to tracking only with collectors of high concentration ratio. Hence the acceptance angle of tracking collectors is relatively small, and care must be exercised in the optical design. Since even direct solar radiation is not perfectly collimated but comes from a range of directions with further spread caused by optical errors, the receiver must exceed a minimum size if it is to intercept most of the radiation incident on a given collector aperture.

The choice of the optimal absorber size involves a compromise between optical and thermal performance. If the absorber is too large, most of the incident solar radiation will be intercepted, but the heat loss is excessive. On the other hand, a small absorber has low heat loss but misses too much of the available solar radiation. The optimal concentration ratio is found by equating the incremental heat loss with the incremental optical loss. A proper analysis of these effects needs to take into account the angular distribution of direct solar radiation and optical errors.

3.3.1 Image Spread Caused by Finite Width of Sun and Optical Errors

When viewed from the earth, the sun appears as a disk of angular radius

$$\Delta_s = 4.7 \, \text{mrad}. \tag{33}$$

Seasonal variations in Δ_s, caused by the eccentricity of the earth's orbit, are so small ($\pm 1.7\%$) that they can be neglected. One need not worry about the detailed brightness distribution of the solar disk for most applications. Only collectors with high concentration are sensitive to details of the angular brightness distribution of the sun. A nominal standard solar brightness distribution is listed in table 3.1 as $B(\theta)$ (in W/m^2-sterad) versus angular distance θ from the center of the sun; it is also plotted in figure 3.15 in dimensionless form. This distribution (Rabl and Bendt 1982) is the simple average of all solar and circumsolar data obtained by the Lawrence Berkeley Laboratory circumsolar telescope (Grether, Hunt, and Wahlig 1977).

The brightness distribution of table 3.1 implies a particular radiation level: 682.6 W/m^2 for the irradiance from the solar disk and 25.6 W/m^2 for the circumsolar irradiance, with a circumsolar ratio of 3.6%, which is typical of fairly clear skies. However, this distribution really has a more general interpretation. Scaled up or down according to actual insolation levels, it can be used as standard sun shape for the calculation of the intercept factor, yielding the most accurate results for the long-term average performance. Circumsolar effects for a particular climate are treated by assigning a separate weighting factor to the circumsolar region of the distribution, as described by Rabl and Bendt (1982).

The distribution in figure 3.15 is seen to be fairly flat over the center of the sun, but it decreases near the edge; this phenomenon is known as limb darkening. At the edge the brightness drops sharply by several orders of magnitude. The region beyond 4.7 mrad and out to 50 mrad is called the circumsolar region. The limb darkening of the solar disk is caused by absorption and scattering in the photosphere of the sun and the atmosphere of the earth. If the sun radiated isotropically and if there were no narrow angle scattering in the atmosphere, then the brightness of the solar disk would be perfectly uniform.

Table 3.1
Nominal standard solar brightness distribution

θ(mrad)[a]	$B(\theta)$(W/m^2 sterad)[b]	
0.218	13,631,252	
0.654	13,561,148	
1.091	13,441,701	
1.527	13,263,133	
1.963	13,012,331	
2.400	12,668,456	Solar disk
2.836	12,185,726	
3.272	11,443,798	
3.709	10,061,047	
4.145	7,002,494	
4.581	1,730,196	
5.018	144,818	
5.454	63,325	
5.890	51,116	
6.327	43,669	
6.763	36,973	
7.199	31,935	
7.636	27,979	
8.072	24,767	
8.508	21,830	
9.381	17,444	
10.690	13,084	
11.999	10,194	
13.308	8,177	
14.617	6,716	
15.926	5,634	
17.235	4,807	
18.544	4,147	Circumsolar region
19.853	3,633	
21.162	3,213	
22.471	2,867	
23.780	2,579	
25.089	2,337	
26.398	2,133	
27.707	1,958	
29.016	1,805	
30.325	1,672	
31.634	1,550	
32.943	1,444	
34.252	1,353	
35.561	1,268	
36.870	1,194	
38.179	1,127	
39.488	1,070	
40.797	1,018	
42.106	969	

Break in size of angular interval (at 8.508)

Table 3.1 (continued)

θ(mrad)[a]	$B(\theta)$(W/m^2 sterad)[b]	
43.415	923	
44.724	882	
46.033	845	
47.342	811	
48.651	777	Circumsolar region
49.960	747	
51.269	722	
52.578	700	
53.887	681	
55.196	665	

Source: Rabl and Bendt (1982).
a. Solar radius 0.275 deg (disk plus resolution of instrument).
b. Brightness at angle θ from center of sun.

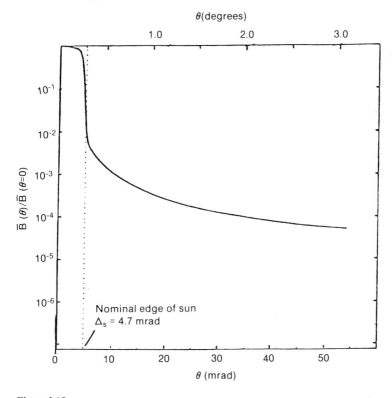

Figure 3.15
Nominal standard brightness distribution of solar and circumsolar region of table 3.1, plotted in dimensionless form. Source: Rabl and Bendt (1982).

Snell's law of reflection is an idealization based on perfectly smooth surfaces. Real surfaces tend to have irregularities, from microscopic roughness to macroscopic undulations. Furthermore real reflector surfaces will not conform exactly to the design shape, and tracking collectors may have alignment errors. All these effects contribute to enlarging the focal zone of a solar concentrator. For solar energy calculations, a statistical analysis of these effects is adequate since one is only interested in overall or averaged performance features, quite apart from the fact that it would be practically impossible to measure and monitor the reflector surface to its last detail.

When discussing the features of reflector surfaces, it is conceptually helpful to distinguish three scales, as illustrated in figure 3.16. The dotted line shows the design shape, typically a parabola. The solid line shows the actual reflector surface with its waviness. The dashed line averages the undulations. The difference between the design shape (dotted line) and the average shape (dashed line) represents large-scale optical errors caused by gravity, wind, stress, or manufacturing errors. The waviness, with typical wavelengths on the order of centimeters to decimeters, represents medium-scale errors. Finally, there is the small-scale, microscopic surface roughness that causes scattering away from the specular direction.

This classification is not always distinct, and in some cases there could be a continuous transition from small- to large-scale errors. Nonetheless, it seems to be useful for solar concentrators, and it corresponds to different measuring

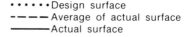

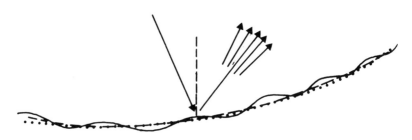

Figure 3.16
Schematic sketch of large-scale, medium-scale, and microscale errors of reflector surface. Microroughness of surface causes scattering of reflected radiation over a range of angles.

methods. Small-scale errors are a property of the reflector surface itself, whereas medium- and large-scale errors are due to the substrate and the support structure. Thus small-scale errors are determined by measuring the spread of a collimated beam of light after it has been reflected from a small sample of the material. Medium- and large-scale errors can be determined by mapping the contour of the reflector using mechanical or laser ray tracing. In addition to these errors of the reflector surface, there are tracking errors; that is gross pointing errors of the entire collector.

Each of these optical errors can be characterized by a statistical distribution function. Only angular variations need to be considered. Since image preservation is irrelevant, displacements in position need to be considered only to the extent that they cause deviations in angle. Ideally, the distributions should be derived from measurements. From the available data it appears that the surface error distributions can usually be approximated by a Gaussian or normal distribution (Butler and Pettit 1977).

Fortunately in many cases the details of the distributions do not matter even if they are not Gaussian. This follows from the central limit theorem of statistics, which says that the convolution of a large number of independent distributions approaches a Gaussian one, even if the individual distributions are not Gaussian (see Cramer 1947 or Adams 1974). Thus only the standard deviations of the individual distributions are needed, provided they have zero mean as they usually do in solar applications. Because of the central limit theorem, one can often approximate the brightness distribution of the sun by a single Gaussian distribution, with the exception of collectors with very high concentrations ($C \gtrsim 2000$) and very small errors.

The lack of specularity of solar reflector materials has been investigated extensively by Pettit (1977). Pettit found that for many reflector materials the scattering of radiation about the specular direction can be described by a Gaussian distribution

$$R(\theta) \propto \rho \exp\left(-\frac{\theta^2}{2\sigma^2}\right), \tag{34}$$

where θ is the angular deviation of a particular ray from the specular direction, $R(\theta)$ is the intensity of radiation reflected into the direction θ, ρ is the hemispherical reflectance and is also known as total reflectance, and σ is the width of the distribution. Some materials—for example, Alzak, a rolled and polished aluminum sheet—exhibit different surface roughnesses in different directions. For these materials Pettit found it necessary to fit the data with a super-position of two different Gaussian distributions.

The rms width σ_{optical} for the total optical error is obtained by adding the squares of the individual widths:

$$\sigma_{\text{optical}}^2 = 4\sigma_{\text{contour}}^2 + \sigma_{\text{specular}}^2 + \sigma_{\text{displacement}}^2 + \sigma_{\text{tracking}}^2. \tag{35}$$

σ_{contour} is multiplied by 2 because of Snell's law; in Fresnel reflectors, σ_{tracking} must also be multiplied by 2. The displacement term accounts for misalignment between concentrator and absorber. The total rms width is obtained by adding the rms width of the sun according to

$$\sigma^2 = \sigma_{\text{optical}}^2 + \sigma_{\text{sun}}^2. \tag{36}$$

3.3.2 Parabolic Reflectors

One of the best-known solar concentrators is the parabolic reflector; it can be built either as a trough or as a dish. The absorber can take a variety of shapes, the most common being round (cylindrical or spherical) or flat. The parabola is the unique reflector shape that focuses a collimated beam of radiation into a single point. Perfect focusing is possible only for rays that are incident parallel to the optical (symmetry) axis of the parabola. A collimated beam coming from other directions will not only miss the focus, but because of off-axis aberrations of the parabola, it will not even converge into a single point.

If the absorber is to intercept most or all rays, it must be sufficiently large. The example of the parabolic trough reflector with a cylindrical absorber tube in figure 3.17 illustrates this problem for focusing collectors. The absorber tube is placed concentrically around the focal line. If the ray with the largest deviation (θ_a) is to reach the absorber just barely, as shown by the dashed line in figure 3.17, then the concentration must be

$$C_{\text{cyl,abs}} = \frac{D}{\pi d} = \frac{\sin \phi}{\pi \sin \theta_a} = \frac{\sin \phi}{\pi} C_{\text{ideal}}, \tag{37}$$

where ϕ is the rim angle $\sphericalangle$ AOB. The maximum occurs for $\phi = 90$ deg and falls a factor π short of the ideal limit. This is typical of all single-stage focusing concentrators; that is, they reach only one-fourth to one-half of the thermodynamic concentration limit. Equation (37) determines how large the absorber must be to intercept all rays incident within the acceptance angle (i.e., within $\pm \theta_a$ from the optical axis). But this is not the complete answer since the distribution of incident rays does not have a sharp cutoff. Also, in a non-ideal concentrator, some rays from outside the nominal acceptance angle are accepted. An accurate analysis becomes complicated.

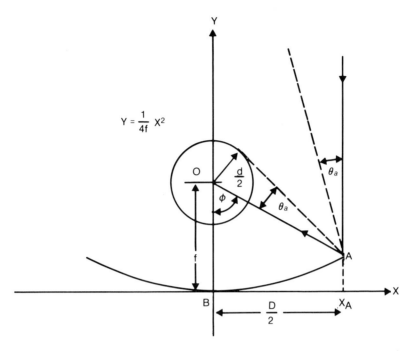

$$Y = \frac{1}{4f} X^2$$

Figure 3.17
Focusing parabola with aperture width D, absorber radius $d/2$, and acceptance half-angle θ_a.

As a simple shortcut one frequently takes the rule of choosing θ_a equal to twice the rms width of the radiation distribution incident on the absorber, as given by equation (37). This rule corresponds to intercepting 95% for line focus (and 84% for point focus) of the radiation for a Gaussian distribution, and it seems to be a reasonable compromise for line focus collectors.

A more accurate optimization of the concentration ratio involves weighing incremental optical gains against incremental thermal losses. Usually this problem has been treated with computer ray trace programs (Treadwell 1976; Biggs and Vittitoe 1979). Several approaches can be taken to treat the statistics of sun shape and optical errors. The Monte Carlo technique traces a large number of individual rays from source to absorber, assigning at each optical interface appropriate probabilities to different propagation directions. The Monte Carlo technique is very flexible, but the actual calculation can be slow for complicated configurations. Cone optics is another approach. Here only the ray for the central direction is traced explicitly, but each such ray is surrounded by a cone of radiation according to the distribution of sun shape and optical errors. A cone optics program can be quite fast if the convolution

(a)

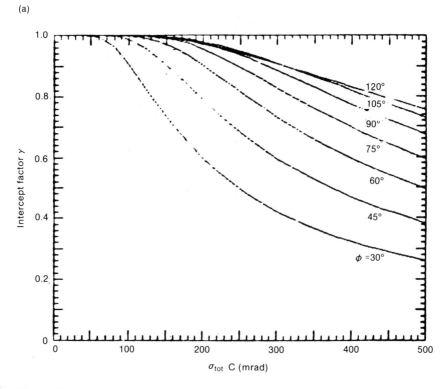

Figure 3.18
Intercept factor γ versus σ_{tot} C for parabolic trough with different rim angles φ (Gaussian approximation) for (a) a cylindrical receiver and (b) a flat receiver. Source: Bendt et al. (1979).

of the distribution functions can be carried out beforehand in closed analytical form.

Recently analytical or semianalytical solutions have been published by Rabl, Bendt, and Gaul (1982) for parabolic troughs, and by Bendt and Rabl (1981) for parabolic dishes. The solution is particularly simple if one can make the approximation that the distribution of radiation at the absorber is a Gaussian of width σ. Then the intercept factor γ depends only on rim angle ϕ, on absorber shape, and on the quantity (σC) for troughs or $(\sigma^2 C)$ for dishes. Thus it can readily be presented in graphic form. Figure 3.18 shows γ for a parabolic trough, with a cylindrical receiver and a flat, one-sided receiver; for the parabolic dish with a flat, one-sided receiver, γ is shown in figure 3.19. Bendt et al. (1979) show how three-dimensional effects need to be accounted for if a parabolic trough is analyzed in two dimensions.

(b)

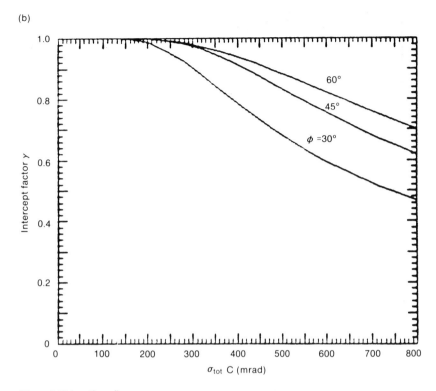

Figure 3.18 (continued)

3.3.3 Fresnel Reflectors and Central Receivers

The smooth optical surface of a reflector or lens can be replaced by individual segments, an arrangement called a Fresnel reflector or Fresnel lens. There is a wide variety of different possible designs. For example, the mirror facets could be fixed on the aperture, tracking the sun as a monolithic unit, or they could be movable, following the sun individually. For solar applications the latter mode is most interesting because it permits the design of very large collectors. Moving a monolithic reflector much larger than 393 in. (10 m) in diameter is impractical because of excessive stresses from wind and gravity. A better alternative is to move individual reflector segments. Of particular importance is the point-focus Fresnel reflector; it forms the basis of the central receiver (ASME 1984). This collector consists of a receiver located on top of a tower surrounded by a field of two-axis tracking mirrors called heliostats.

The optical analysis of central receivers is complicated. Since the location but not the orientation of the heliostats is fixed, the angle of incidence of the

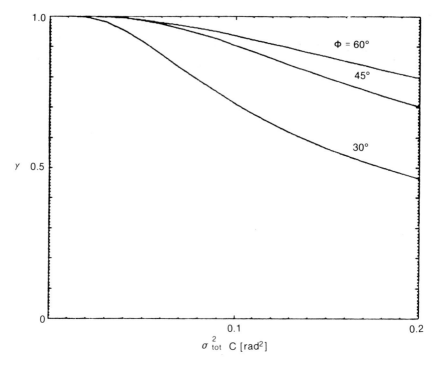

Figure 3.19
Intercept factor γ for parabolic dish with flat receiver geometric concentration C, and rim angle φ, if source plus errors are Gaussian with width σ_{tot}. Source: Bendt and Rabl (1981).

sun on the heliostats varies with time. This necessitates careful analysis of shading and blocking. Shading occurs if direct sunlight fails to reach a helio-stat because the heliostat is in the shade of another heliostat; blocking occurs if light reflected by a heliostat fails to reach the absorber because it is inter-cepted by the back of another heliostat. In the interest of efficient mirror utilization, the heliostats should be spaced far enough apart to avoid excessive shading and blocking. Too large a separation is also undesirable because then the radiation level at the receiver would be too low to be practical. To optimize the spacing, one needs to take into account the distribution of incidence angles as well as the relative cost of heliostats and receiver. The spacing is usually reported in terms of the ground cover ratio ψ, defined as ratio of heliostat area and ground area. For the heliostats of a central receiver, optimal values of ψ are in the range of 0.2 to 0.6. The spacing decreases toward the edge of the heliostat field.

A further complication is caused by the varying incidence angles of the sun. The size and shape of the solar "images" from the individual reflectors change with time of day and year. This effect is particularly complicated if the individual mirrors have some curvature. If the number of mirrors is large, then the gain in concentration achievable by curving the reflectors is small. When the number of mirrors exceeds on the order of 100 for linear or 10,000 for point-focus Fresnel systems, one gets diminishing returns from the extra expense of curving the mirrors and may be better off with flat mirrors. On the other hand, when the number of mirrors is small, the concentration can be increased significantly by curving the mirrors. However, then the off-axis aberrations become more severe.

In view of the complexity of the problem, it is not surprising that no closed form analytical formulas are available for calculating the optical performance of central receivers. Only some partial solutions have been attempted. For example, Riaz (1976) has derived some limiting equations for the effects of shading and blocking. Various fast algorithms have been developed for carrying out the computations (Vant-Hull 1976), but the complete analysis seems to require a computer program.

Several different types of computer programs have been developed for the optical analysis of central receivers. One can distinguish between programs for calculating the performance of a specified design and programs for design optimization. HELIOS (Biggs and Vittitoe 1979), MIRVAL (Leary and Hankins 1977), UH-NS, and UH-IH (Lipps and Vant-Hull 1980a, b, c) are examples of the former. HELIOS is based on numerical integration of the distribution functions that describe the sunshape and the optical errors. MIRVAL uses Monte Carlo techniques. These programs are accurate and flexible, and they can take into account many details such as field layout, heliostat shape, error distribution, circumsolar radiation, and atmospheric attenuation. They require as input the definition of the entire system. The process of using these programs is slow and not well suited for optimization.

By contrast, programs like DELSOL (Dellin and Fish 1979) and UH-RC (Laurence and Lipps 1980) can optimize the design. For example, DELSOL I finds tower height, receiver size, and heliostat number, given the heliostat locations. UH-RC and DELSOL II can optimize the heliostat spacings and the field boundary; by iterating with UH-RC or DELSOL II, one can optimize the system (Lipps and Vant-Hull 1980a).

For further information on the modeling of central receivers, the reader is referred to the work of Biggs and Vittitoe (1979) and of Vant-Hull and his coworkers (Vant-Hull 1976; Walzel, Lipps, and Vant-Hull 1977; Lipps and Vant-Hull 1978; Lipps and Vant-Hull 1980b).

A particularly interesting result of these calculations is shown in table 3.2 (Eicker 1979). The table shows the field efficiency (essentially the incidence

Table 3.2
Field efficiency for two central receiver designs

Small central receiver (2.4 MW$_t$) ϕ_s = azimuth (deg; south = 0) Elevation = 90 deg − θ_z (deg, horizon = 0)	0	30	60	75	90	110	130
5	0.384	0.404	0.366	0.330	0.300	0.240	0.212
15	0.701	0.687	0.576	0.495	0.429	0.367	0.315
25	0.789	0.771	0.662	0.584	0.521	0.445	0.391
45	0.814	0.814	0.757	0.708	0.661	0.603	0.544
65	0.811	0.806	0.754	0.753	0.724	0.689	0.642
89.5	0.723	0.729	0.748	0.726	0.730	0.736	0.736
Large central receiver (195 MW$_t$) ϕ_s = azimuth (deg; south = 0) Elevation = 90 deg − θ_z (deg, horizon = 0)	0	30	60	75	90	110	130
5	0.216	0.215	0.206	0.204	0.199	0.194	0.192
15	0.446	0.448	0.425	0.423	0.405	0.392	0.385
25	0.560	0.558	0.537	0.522	0.516	0.498	0.491
45	0.719	0.640	0.626	0.618	0.605	0.594	0.599
65	0.684	0.670	0.671	0.668	0.660	0.655	0.641
89.5	0.683	0.683	0.686	0.672	0.682	0.687	0.681

Source: Eicker (1979).
Note: Includes losses resulting from the effects of cosine, tower shadowing, blocking and shading, atmospheric attenuation, intercept factor, and a value $\rho = 0.90$ for the heliostat reflectivity.

angle modifier) $\rho K(\theta_z, \phi_s)$ for particular central receiver designs. For any values of the solar zenith angle θ_z and the solar azimuth angle ϕ_s, the product of beam irradiance and $\rho K(\theta_z, \phi_s)$ yields the radiation intercepted by the receiver.

The field efficiency as listed in table 3.2 includes (and is limited to) the effects of incidence angle cosine, shading, blocking, mirror reflectivity ρ, tower shadowing, atmospheric attenuation between mirror and receiver, and intercept factor (loss of rays that miss the receiver).

The atmospheric attenuation depends on the distance from mirror to receiver, and is approximately 10%/km. (For an accurate model of this attenuation, the reader is referred to Pitman and Vant-Hull 1982.) The reflectivity of the mirrors is assumed to be $\rho = 0.90$. The absorptivity of the receiver is not included in these tables.

The need for different tables 3.2a and 3.2b arises from the difference in mirror field designs for small and large systems. For small systems at intermediate latitudes, the heliostats will be deployed only to the north of the tower, whereas in large systems, the mirror field will surround the tower (albeit with more mirrors to the north than to the south). This design difference reflects different economies of scale in tower and heliostat costs.

The top of table 3.2 is for a small field of 2.4 MW_t capacity; the bottom, for a large field of 195 MW_t capacity. Both are optimized for a geographic latitude of 35 deg and for a specified ratio of tower and heliostat costs. The reader should understand that different designs, as appropriate for different latitudes, cost ratios, or system sizes, may have different incidence angle modifiers.

3.3.4   Fresnel Lenses

An ordinary lens is impractical in most solar applications because it would have to be too thick. The mass per aperture ratio is proportional to aperture width, and for widths larger than a few centimeters, the mass and weight become excessive. Optically the principal difference between an ordinary lens and a Fresnel lens is that in a Fresnel lens the optical path length is different for different rays, whereas in a normal lens the optical path from object to image is the same for all rays. This difference is not important for solar energy since imaging and coherence properties are irrelevant.

Starting with the extreme ray principle for an ideal concentrator, Kritchman, Friesem, and Yekutieli (1979) discovered an elegant lens design that maximizes concentration. An example of such a lens with an acceptance half angle $\theta_a = 10$ deg is shown in figure 3.20. Included in this figure are the ray paths for two incident directions, $\theta = 0$ deg and $\theta = 10$ deg. In keeping with the extreme ray design principle, the rays coming from $\theta = \theta_a = 10$ deg all contact the edge of the absorber. The geometrical concentration of these lenses is at least 75% of the thermodynamic limit.

To understand why this lens falls short of the thermodynamic limit, despite the extreme ray principle used for its design, note the following: On the absorber side of the lens each prism has two faces; one can be called active, the other inactive. The radiation is supposed to pass entirely through the active face, while the inactive face only serves to link up to the neighboring prism. However, it is impossible to position the inactive face in such a way that it does not intercept at least some rays that are intended for the receiver. For instance, if the inactive face is chosen parallel to the rays that come from the central direction, then some rays from other directions within the acceptance angle will intercept it. This feature is related to the fact that there is a discontinuity in optical path length from one prism facet to another. Since the extreme ray condition implies imaging for the extreme rays, it cannot be entirely consistent with Fresnel optics.

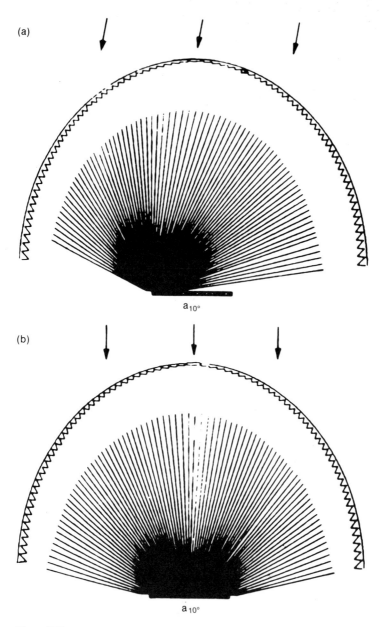

Figure 3.20
Shape of Fresnel lens with maximal concentration for acceptance half-angle of 10 deg. Source:
Kritchman, Friesem, and Yekutieli (1979).

Like the CPC these lenses can be truncated with minimal loss in concentration; that is, the outer portions of the lens do not intercept much radiation but require a large amount of material per aperture area. In practice, it will probably be reasonable to truncate the lens to about half of the full depth; for the lens shown in figure 3.20, such a truncation reduces the concentration from 0.75/sin 10 deg to 0.67/sin 10 deg. If designed with an acceptance angle equal to the solar size (i.e., $\theta_a = \theta_s = 5$ mrad), then such a lens reaches a geometric concentration ratio of 0.745/sin $\theta_a = 149$ in line-focus geometry. Truncated to half of its full depth, its concentration is still equal to 133, compared with the thermodynamic limit of 200 for this acceptance angle.

Compared with other Fresnel lenses, the design of Kritchman, Friesem, and Yekutieli has the following advantages:

1. It reaches the highest geometric concentration possible with a Fresnel lens.

2. Reflection losses are minimal because the angle between incident ray and first surface is close to the angle between exiting ray and prism surface.

3. The curved shape adds mechanical strength to the lens.

4. The loss of performance for nonzero elevation angles (i.e., when the sun is not in plane of the paper of figure 3.20) is less severe than for other linear Fresnel lenses. Moreover it is less sensitive to chromatic aberrations.

A similar design with comparable performance has been developed by O'Neill (1979). It starts by requiring equality of incident and exit angles at each prism, which minimizes reflection losses.

References

Adams, W. J. 1974. *The Life and Times of the Central Limit Theorem.* New York: Kaedman.

ASME. 1984. *ASME J. of Solar Energy Eng.* 106 (February): 22–103.

Baranov, V. K., and G. K. Melnikov. 1966. *Soviet J. of Optical Tech.* 33: 408.

Bendt, P., A. Rabl, H. W. Gaul, and K. A. Reed. 1979. *Optical Analysis and Optimization of Line Focus Solar Collectors.* SERI/TR-36-092. Golden, CO: Solar Energy Research Institute.

Bendt, P., and A. Rabl. 1981. Optical analysis of point focus parabolic radiation concentrators. *Applied Optics* 20: 674.

Biggs, F., and C. N. Vittitoe. 1979. *The Helios Model for the Optical Behavior of Reflecting Solar Concentrators.* SAND76-0347. Albuquerque, NM: Sandia National Laboratories.

Butler, B. L., and R. B. Pettit. 1977. Optical evaluation techniques for reflecting solar concentrators. *Optics Applied in Solar Energy Conversion.* Bellingham, WA: SPIE—The International Society for Optical Engineering, vol. 114, p. 43.

Carvalho, M. J., M. Collares-Pereira, J. M. Gordon, and A. Rabl. 1985. Truncation of CPC solar collectors and its effect on energy collection. *Solar Energy* 35: 393.

Chiam, H. F. 1981. Planar concentrators for flat plate solar collectors. *Solar Energy* 26: 503.

Cramer, H. 1947. *Mathemetical Methods of Statistics*. Princeton, NJ: Princeton University Press.

Dellin, T. A., and M. J. Fish. 1979. *A User's Manual for DELSOL—A Computer Code for Calculating the Optical Performance, Field Layout, and Optimal System Design for Solar Central Receiver Plants*. SAND79-8215. Albuquerque, NM: Sandia National Laboratories.

Duffie, J. A., and W. A. Beckman. 1980. *Solar Eng. of Thermal Processes*. New York: Wiley Interscience.

Edwards, D. K. 1977. Solar absorption by each element in an absorber-coverglass array. *Solar Energy* 19: 401.

Eicker, P. J., 1979. Letter of 18 April 1979 to J. Thornton, Solar Energy Research Institute, Livermore, CA: Sandia Livermore Laboratories.

Felland, J. R., and D. K. Edwards. 1978. Solar and infrared radiation properties of parallel-plate honeycomb. *J. Energy* 2: 309.

Grassie, S. L., and N. R. Sheridan. 1977. The use of planar reflectors for increasing the energy yield of flat plate collectors. *Solar Energy* 19: 663.

Grether, D. F., A. Hunt, and M. Wahlig. 1977. *Techniques for Measuring Circumsolar Radiation*. LBL-8345. Berkeley, CA: Lawrence Berkeley Laboratory.

Hay, J. E. 1979a. Calculation of monthly mean solar radiation for horizontal and inclined surfaces. *Solar Energy* 23: 301.

Hay, J. E. 1979b. *Study of Shortwave Radiation on Non-horizontal Surfaces*. Canadian Climate Center Report 79-12. Downsview, Ontario: Atmospheric Environment Service.

Hinterberger, H., and R. Winston. 1966. *Rev. Sci. Instr.* 37: 1094.

Hollands, K. G. T. 1971. A concentrator for thin-film solar cells. *Solar Energy* 13: 149.

Hollands, K. G. T., G. D. Raithby, B. B. Russel, and R. G. Wilkinson. 1979. *Methods for Reducing Heat Losses from Flat Plate Solar Collectors*. Final Report, Ontario, Canada: University of Waterloo.

Kritchman, E. M., A. A. Friesem, and G. Yekutieli. 1979. Efficient lens for solar concentration. *Solar Energy* 22: 119.

Larson, D. C. 1980. Optimization of flat plate collector—Flat mirror system. *Solar Energy* 24: 203.

Laurence, C. L., and F. W. Lipps. 1980. *User's Manual for the University of Houston Computer Code RC: Cellwise Optimization for the Solar Receiver Project*. Houston, TX: Energy Lab., University of Houston. DOE/SF/10763-T7, available through National Technical Information Service, 5285 Port Royal Road, Springfield, VA 22161.

Leary, P., and J. Hankins. 1977. *A User's Guide for MIRVAL Computer Code for Comparing Designs of Heliostat Receiver Optics for Central Receiver Solar Power Plants*. SAND77-8280. Livermore, CA: Sandia National Laboratories.

Lipps, F. W., and L. L. Vant-Hull. 1978. A cell-wise method for the optimization of large central receiver systems. *Solar Energy* 20: 505.

Lipps, F. W., and L. L. Vant-Hull. 1980a. *User Manual for the University of Houston Solar Central Receiver, Cellwise Performance Model: NS*. Houston, TX: Energy Lab., University of Houston. DOE/SF/10763-T9, available through National Technical Information Service, 5285 Port Royal Road, Springfield, VA 22161.

Lipps, F. W., and L. L. Vant-Hull. 1980b. *Notes on Collector Field Optimization: Extensions of the Basic Theory*. Houston, TX: Energy Lab., University of Houston. DOE/SF/10763-T6, available through National Technical Information Service, 5285 Port Royal Road, Springfield, VA 22161.

Lipps, F. W., and L. L. Vant-Hull. 1980c. *Programmer's Manual for the University of Houston Computer Code RCELL: Cellwise Optimization for the Central Receiver Project*. Houston, TX: Energy Lab., University of Houston. DOE/SF/10763-T5, available through National Technical Information Service, 5285 Port Royal Road, Springfield, VA 22161.

McDaniels, D. K., D. H. Lowdes, H. Mathew, J. Reynolds, and R. Gray. 1975. Enhanced solar energy collection using reflector solar-thermal collector combinations. *Solar Energy* 17: 277.

McIntire, W. R. 1979. Truncation of nonimaging cusp concentrators. *Solar Energy* 23:351.

McIntire, W. R. 1980. Stationary concentrators for tubular evacuated receivers: optimization and comparison of reflector designs. In *Proc. 1980 Annual Meeting AS Int. Solar Energy Soc.*, Phoenix, AZ, p. 505.

McIntire, W. R. 1981. New reflector design which avoids losses through gaps between tubular absorbers and reflectors. *Solar Energy* 25: 215.

Mills, D. R. 1978. The place of extreme asymmetrical non-focusing concentrators in solar energy utilization. *Solar Energy* 21: 431.

Mills, D. R., I. M. Bassett, and G. H. Derrick, 1984. *Relative Cost-Effectiveness of CPC Reflector Designs Suitable for Evacuated Absorber Tube Solar Collectors.* Sydney, Australia: School of Physics, University of Sydney.

O'Gallagher, J. J., A. Rabl, R. Winston, and W. McIntire. 1980. Absorption enhancement in solar collectors by multiple reflections. *Solar Energy* 24: 323.

O'Neill, M. J. 1979. A unique new Fresnel lens solar concentrator. International Solar Energy Congress, Atlanta, GA. Also, U.S. Patent No. 4,069,812. Solar concentrator and energy collection system (1978).

Pettit, R. B. 1977. Characterization of the reflected beam profile of solar mirror materials. *Solar Energy* 19: 733.

Pitman, C. L., and L. L. Vant-Hull. 1982. Atmospheric transmission model for a solar beam propagating between a heliostat and a receiver. In *Proc. ASES Congress, Progress in Solar Energy.* Houston, TX, p. 1247.

Ploke, M. 1966. Lichtfuehrungseinrichtungen mit starker Konzentrationswirkung. *Optik* 25: 31.

Rabl, A. 1976a. Comparison of solar concentrators. *Solar Energy* 18: 93.

Rabl, A. 1976b. Optical and thermal properties of compound parabolic concentrators. *Solar Energy* 18: 497.

Rabl, A. 1976c. Solar concentrators with maximal concentration for cylindrical absorbers. *Applied Optics* 15: 1871.

Rabl, A., and R. Winston. 1976. Ideal concentrators for finite sources and restricted exit angles. *Applied Optics* 15: 2880.

Rabl, A. 1977. Radiation transfer through specular passages. *Int. J. Heat Mass Transfer* 20: 323.

Rabl, A., N. B. Goodman, and R. Winston. 1979. Practical design considerations for CPC solar collectors. *Solar Energy* 22: 373.

Rabl, A., and P. Bendt. 1982. Effect of circumsolar radiation on performance of focussing collectors. *ASME J. Solar Energy Eng.* 104: 237.

Rabl, A., P. Bendt. and H. W. Gaul. 1982. Optimizatinn of parabolic trough collectors. *Solar Energy* 29: 407.

Rabl, A. 1985. *Active Solar Collectors and Their Applications.* Oxford: Oxford University Press.

Reed, K. 1976. Instrumentation for measuring direct and diffuse insolation in testing thermal collectors. *Optics in Solar Energy Utilization II.* Bellingham, WA: SPIE—The International Society for Optical Engineering, Vol. 85.

Riaz, M. 1976. A theory of concentrators of solar energy on a central receiver for electric power generation. *J. Eng. for Power* 98: 375.

Seitel, S. C. 1975. Collector performance enhancement with flat reflectors. *Solar Energy* 17: 291.

Selcuk, M. K. 1979. Analysis, development and testing of collector employing reversible V-trough reflectorg and vacuum tube receivers. *Solar Energy* 22: 413.

Symons, J. G. 1982. The solar transmittance of some convection suppression devices for solar energy applications. *ASME J. Solar Energy Eng.* 104: 251.

Tabor, H. 1958. Stationary mirror systems for solar collectors. *Solar Energy* 2: 27.

Treadwell, G. W. 1976. *Design Considerations for Parabolic Cylindrical Solar Collectors.* SAND76-0082. Albuquerque, NM: Sandia National Laboratories.

Vant-Hull, L. L. 1976. An educated ray trace approach to solar tower optics. In *Proc. Society of Photo-Optical Instrumentation Engineers* 85: 111.

Walzel, M. D., F. W. Lipps, and L. L. Vant-Hull. 1977. A solar flux density calculation for a solar tower concentrator using a two-dimensional hermite function expansion. *Solar Energy* 19: 239.

Welford, W. T., and R. Winston. 1978. *The Optics of Nonomaging Concentrators.* New York: Academic Press.

Window, B., and I. M. Bassett, 1981. Optical collection efficiencies of tubular solar collectors with specular reflectors. *Solar Energy* 26: 341.

Winston, R. 1970. Light collection within the framework of geometrical optics. *J. Opt. Soc. Am.* 60: 245.

Winston, R. 1978. Ideal flux concentrators with reflector gaps. *Appl. Opt.* 17: 1668.

Winston, R. 1980. Cavity enhancement by controlled directional scattering. *Appl. Opt.* 19: 195.

Winston, R. 1983. Compound parabolic concentrator with cavity for tubular absorber. U.S. letters patent 4,387,961.

Winston, R., and H. Hinterberger. 1975. Principles of cylindrical concentrators for solar energy. *Solar Energy* 17: 255.

Winston, R., and W. T. Welford. 1979. Geometrical vector flux and some new nonimaging concentrators. *J. Opt. Soc. Am.* 69: 532.

4 Thermal Theory and Modeling of Solar Collectors

Noam Lior

This chapter reviews the advances made in thermal theory and modeling of nonconcentrating and concentrating solar collectors. It attempts not to emphasize thermal *research*, which is covered in chapter 8 of this book, nor optical aspects (in coatings that affect radiative transfer or in concentrating collectors), which are covered in chapter 7. An effort is made to describe the current, least speculative understanding of the thermal theory of solar collectors and of the state-of-the-art modeling. Each major section ends with a "progress summary" that highlights the progress made since the early 1970s. According to the editor's dictate, this chapter is not to serve as a source of all equations needed for thermal modeling, but it should be an up-to-date resource for the relevant references and progress accomplished. The missing equations and further detail should be sought in the references quoted, where they can be found in their most original form, unspoiled by any errant middleman.

Based on the work of solar collector theory pioneers such as Hottel (see Hottel and Woertz 1942; Hottel and Erway 1963), Tabor (1955, 1958), Whillier (1953, 1964a, b), Bliss (1959), and others, the state-of-the-art in the 1960s was summarized by Hottel and Erway in the book edited by Zarem and Erway (1963), by Whillier (1967) in an ASHRAE book, and in the first edition of the book *Solar Energy Thermal Processes* by Duffie and Beckman (1974). These publications serve as the datum level to which later progress is compared in this chapter.

The rapid progress made in the mid- and late 1970s was summarized in an updated (1977) version of the ASHRAE book (Liu and Jordan 1977), the booklet by Edwards (1977), the book by Kreith and Kreider (1978), the 1980 edition of the book by Duffie and Beckman, a number of review chapters by Löf (1980), Rabl (1980, 1981), and Kreith and Kreider (1981) in the volumes edited by Dickinson and Cheremisinoff and by Kreider and Kreith, the book by Garg (1982), and several other books, such as by Howell et al. (1982) and Rabl (1985). The number of books and comprehensive reviews on this topic diminished precipitously (alongside with the solar energy research budget) in the mid-1980s, and newer information is primarily contained in journals and conference publications.

This chapter begins by describing, in section 4.1, the overall thermal balance in the collector. The first section then deals in more detail with its various components, such as the collector exterior, the window system, enclosed insulating spaces, and the absorber and the working fluid. It addresses the effect of transient conditions, performance sensitivity to design parameters, and the differences between the modeling of a single collector and of collector arrays. At the end of the chapter a state-of-the-art modeling method is outlined.

4.1   The Thermal Energy Balance for Solar Collectors

4.1.1   General Description of Solar Collectors

Types of Nonconcentrating Solar Collectors   *Flat plate solar collectors* utilize a flat absorber plate to convert the electromagnetic energy of solar radiation to heat. To reduce heat losses to the ambient, the insolated front of the absorber is usually separated from the ambient by a "window" that allows transmittance of solar radiation to the absorber but impedes heat losses from the absorber to the ambient. This is usually accomplished by one or more panes of glass or plastic transparent to radiation in the solar (shortwave) spectrum. These panes are mounted parallel to the absorber, with small air gaps between them. The uninsulated back and sides of the collector are insulated by conventional, opaque insulation. Useful heat is taken away from the absorber by putting a working fluid in contact with it: either directly by flow over or below the absorber (as typically done in air-heating collectors) or by flow-through conduits (often simply tubes) in good thermal contact with the absorber. Manifolding devices are used to distribute the fluid properly over the absorber area at the inlet to the collector and to collect the heated fluid into the outlet conduit. Figures 4.1 and 4.2 show typical configurations of liquid-heating and air-heating flat plate solar collectors, respectively.

To eliminate the resistance to heat transfer associated with the conduit for the heated fluid, a number of flat plate collector designs in which the solar radiation is absorbed directly into a flowing layer of the heated fluid have been designed and tested. The fluid is made more absorbent to radiation by adding solid suspensions or dyes that absorb solar energy. The "black fluid" in the designs considered so far either flows as a thin film on the absorber plate or passes through transparent tubing. The film or tubing is typically contained in a conventional solar collector box separated by one or more transparent panes from the environment.

Cylindrical solar collector are usually constructed in the shape shown in figure 4.3. This design allows the use of a relatively thin glass window for containing an insulating vacuum between the absorber and this exterior glass window. Evacuation of the gap between the absorber and the window reduce both convective and conductive heat losses significantly, and reduce them altogether at a perfect vacuum. Many designs have been proposed, built, and tested (see Graham 1979). One design by Corning Glass (figure 4.3a) has

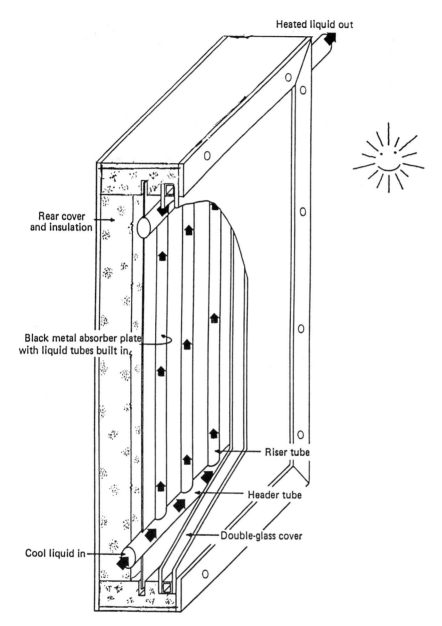

Heated liquid out

Rear cover
and insulation

Black metal absorber plate
with liquid tubes built in

Riser tube

Header tube

Double-glass cover

Cool liquid in

Figure 4.1
Typical liquid-heating flat plate solar collector.

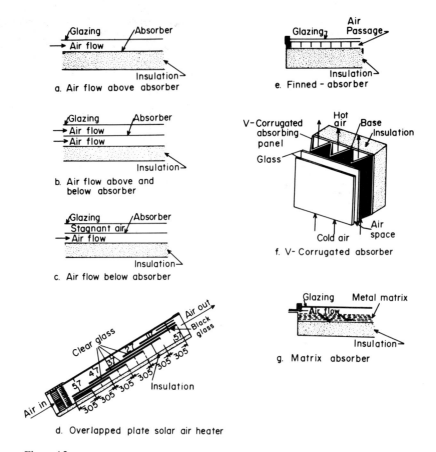

Figure 4.2
Air-heating solar collector configurations: (a) airflow above absorber; (b) airflow above and below absorber; (c) airflow below absorber; (d) overlapped plate solar air heater; (e) finned absorber; (f) V-corrugated absorber; (g) matrix absorber.

simply a flat plate absorber in the middle of the cylinder attached to a U-tube conducting the fluid. In another by Phillips (figure 4.3b) the U-tube is not attached to an absorber. In the Owens-Illinois design (figure 4.3c) the fluid is fed in through a central small tube and is returned through a concentric annulus formed between that tube and a larger cylindrical absorber. Figure 4.3d shows a design in which a Thermos bottle design is employed for the glass vacuum envelope and a copper U-tube is attached to a copper sleeve simply inserted into the center.

Types of Concentrating Solar Collector Receivers   In a flat plate collector the heat loss area is equal to the insolation collection area. The primary advantage of the concentrating solar collector is that the heat loss area (that of the receiver) is smaller (up to several thousand times) than the insolation collection area (that of the reflector). This allows higher efficiency for a given useful output temperature, or a higher temperature for the same efficiency as attained by a flat plate solar collector. The various methods of concentrating solar radiation will not be discussed here because they are not within the scope of this chapter (some discussion of concentrators appears in chapters 2, 3, and 7). In general, a concentrator would be considered in the thermal design only if it has some thermal effect on the receiver such as in producing a radiative exchange surface or in confining or channeling convection. For thermal design it is usually sufficient to know the radiative flux incident on the receiver.

At the lower energy collection temperature levels (e.g., $< 200°C$) the receiver configurations are similar to the flat plate collectors shown in figures 4.1 and 4.2 and to the evacuated cylindrical collectors shown in figure 4.3. In some designs using trough (line-focusing) concentrators, the receiver is simply a blackened tube conducting the heated fluid. The tube is often insulated from the ambient by a larger concentric glass tube, with a thermally insulating annular air gap between the two tubes. The glass tube can also be coated with an antireflective coating. Cavity receivers and external receivers are most often used for the high temperatures typically obtained from point-focusing concentrators (dish or distributed mirrors focusing on a central tower; see Hildebrandt and Dasgupta 1980; Kreith and Wang 1986; Lior 1986).

Cavity receivers consist of an insulated enclosure with an optical aperture just large enough to admit the concentrated solar radiation beam, which contains on its interior the heated fluid conduits (see McKinnon et al. 1965; Grilikhes and Obtemperanskii 1969; Kugath et al. 1979; Strumpf et al. 1982; Davis 1982; Wu et al. 1983; Borgese et al 1984; Harris and Lenz 1985). Its geometry has the objective of maximizing the absorption of the incident radiation, minimizing heat losses (radiant and convective) from the cavity to

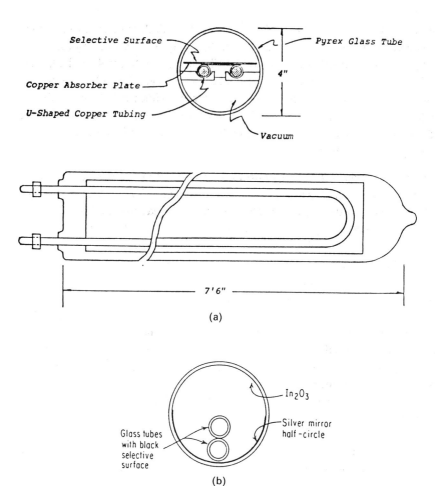

Figure 4.3
Evacuated cylindrical solar collectors: (a) Corning Cortec™ evacuated tube collector; (b) Philips (Aachen) GmbH with In_2O_3 infrared reflector; (c) Owens-Illinios SUNPAK™ evacuated tube collector; (d) General Electric TC-100 SOLARTRON™ evacuated tube collector.

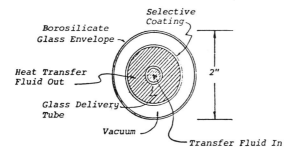

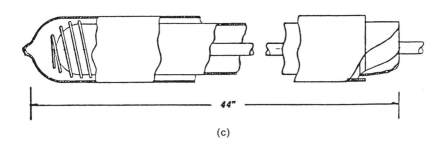

(c)

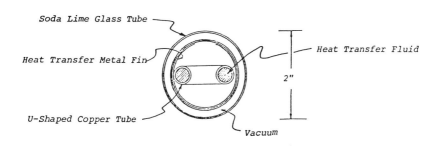

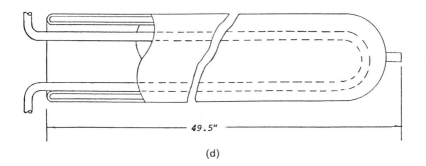

(d)

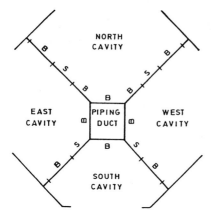

B= BOILER TUBE PANEL. S=SUPERHEATER TUBE PANEL

a. "Quad" central tower cavity receiver:
   steam boiler and superheater (Wu et al., 1983)

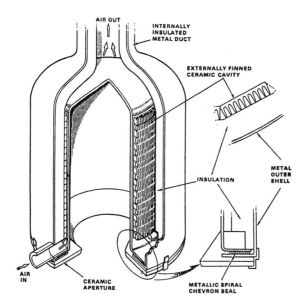

b. Ceramic finned-shell air heating receiver
   for parabolic-dish concentrator (Strumpf et al., 1982)

Figure 4.4
Cavity receivers: (a) "quad" central tower cavity receiver—steam boiler and superheater (Wu et al. 1983); (b) ceramic finned-shell air heating receiver for parabolic dish concentrator (Strumpf et al. 1982); (c) central tower molten salt receiver (Martin-Marietta).

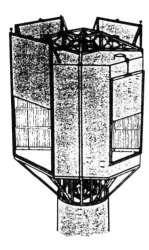

c. Central tower molten salt receiver
   (Martin-Marietta)

Figure 4.4 (continued)

the ambient, accommodating the heat exchangers that transfer the radiant energy to the heated fluid, and distributing the radiant flux over the surface of the heat exchangers properly. Typical cavity receiver configurations are shown in figure 4.4.

External receivers are similar to nonconcentrating collectors in that they expose the heated fluid conduits (surface treated to increase absorptance of the radiation) directly to the radiative flux without the radiation-trapping benefits of the cavity (see Dunn and Vafaie 1981; Chiang 1982; Durant et al. 1982; Wu et al. 1983; Yeh and Wiener 1984). They consequently have lower efficiencies than cavity receivers but are smaller and more economical to manufacture. In a design study such a receiver could be considered for use with the central tower solar concentrator power scheme, as shown in figure 4.5.

To reduce resistance to heat flow and to eliminate the need for exposing the fluid conduits to high radiative fluxes and temperatures, a number of receiver designs have been proposed, and partially tested, in which the solar radiation is absorbed directly into either a falling liquid (molten salt) film (figure 4.6; see Bohn 1985; Lewandowski 1985) or into free falling solid particles (figure 4.7; see Hruby 1985; Hunt et al. 1985).

More detail on various collector designs and configurations is given in chapter 2 of this volume.

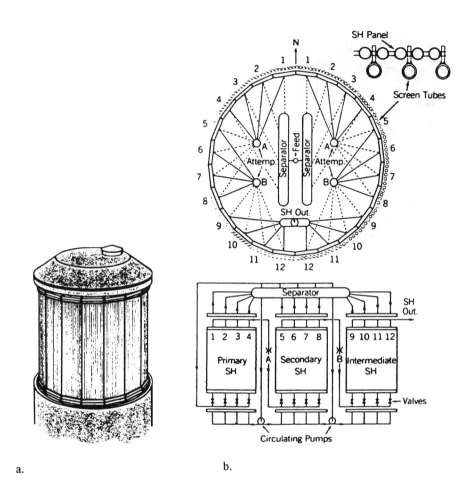

Figure 4.5
External receivers: (a) sodium (General Electric); (b) steam (Durrant et al. 1982).

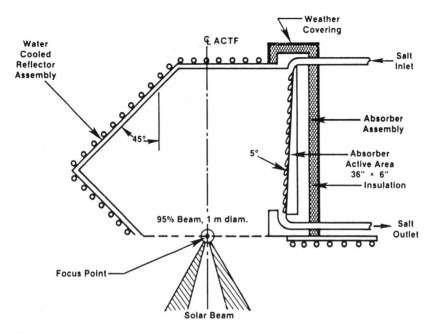

Figure 4.6
Direct absorption molten-salt receiver. Source: Kreith and Anderson (1985).

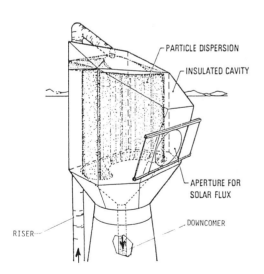

Figure 4.7
Solid particle cavity receiver. Source: Hruby (1985).

4.1.2 The Thermal Energy Balance

The description of the thermal theory and modeling of solar collectors will start here with the energy conservation equation. Each component will be identified and then decomposed into its heat transfer rate subcomponents. The progress made in the ability to evaluate these components will be described in the subsequent sections 4.2 through 4.11.

In its general form the conservation law states that the rate of energy input (E_i) is equal to the sum of the rates of useful heat obtained from the collector (q_u), energy losses from the collector (E_l), and heat storage in the collector (q_s):[1]

$$E_i = q_u + E_l + q_s. \tag{1}$$

In many cases the thermal capacity of the collector is small, or the transients are moderate. In these cases the last term drops out. In steady-state analyses the last term is zero, and the rest of the terms are assumed to be quasi-invariant with time.

The *radiative* energy input E_R is composed of the components of the beam insolation I_b and diffuse insolation I_d (I_b and I_d are measured on a horizontal surface at the same location and time) that are normal to the collector surface (see figure 4.8), multiplied by the collector window area (A_w):

$$E_R = (R_b I_b + R_d I_d) A_w = R I A_w, \tag{2}$$

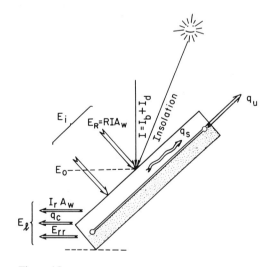

Figure 4.8
Overall energy balance (external envelope control volume) of a solar collector. E_0 is any nonsolar heat input.

where R_b, R_d, and R are the ratios of the beam, diffuse, and total radiation on the collector surface to that on the horizontal surface,[2] respectively;

$$R_b = \frac{I_{bc}}{I_b},$$ (3)

$$R_d = \frac{I_{dc}}{I_d},$$ (4)

$$R = \frac{I_c}{I},$$ (5)

where the subscript c indicates the collector surface, and

$$I = I_b + I_d.$$ (6)

The early work by Liu and Jordan (1963) and many subsequent analyses have assumed that the diffuse radiation component is directionally isotropic. Progress made since then indicates that circumsolar diffuse radiation can be as much as an order of magnitude higher than the diffuse radiation from directions farthest from the sun and that the distribution of diffuse radiation also depends significantly on the composition of the atmosphere and the cloud cover. This consideration is very important in areas where the diffuse component is a significant fraction of the total solar radiation. Consequently R_d at a given geographic location will not only be a function of the tilt of the collector but will also depend on the solar angle. With a clear sky, it is often assumed that $R_b = R_d = R$.

From the available data it appears that the diffuse component has a spectral distribution similar to that of the total solar radiation, with possibly a very slight shift to the shorter wavelengths. This allows the assumption in thermal modeling that both of the radiation components have the same spectral distribution.

Any further discussion of the directional and spectral dependence of insolation is beyond the scope of this chapter and is treated in more detail in volume 2 and in chapter 3 of this book. The preceding comments have been made primarily to indicate that the energy input term is a function of direction and wavelength; this fact influences the thermal phenomena in the collector.

As the radiation q_R incident on the collector strikes its exterior surface, it proceeds to interact with the collector components through one or more reflective, absorptive, and transmissive (the latter if a transmissive component is in its way) processes. Absorption of radiation into any of the collector components is a process that converts the electromagnetic energy of radiation into heat. Reflected and transmitted radiation are still in the original form of electromagnetic energy. But since all practical reflection and transmission processes are accompanied by some amount of absorption, the amount of the radiant energy of a beam after reflection or transmission is always smaller than the beam's radiant energy before such interactions have occurred. The control volume drawn around the external envelope of the collector (Figure 4.8) indicates the energy fluxes crossing that envelope, which must satisfy conservation regardless of the processes occurring inside the control volume. It shows the fraction lost due to reflection from exterior and interior surfaces, I_r, the fraction reradiated from the exterior and interior surfaces to the ambient, E_{rr}, the fraction convected to the cooler ambient, q_c, and the useful heat picked up by the heated fluid, q_u. The energy difference remaining, q_s, is the rate of energy storage in the collector structure (with nonsolar energy inputs ignored):

$$I_c A_w - I_r A_w - E_{rr} - q_c - q_u = q_s, \tag{7}$$

but

$$I_c - I_r = (1 - \rho)I_c = \alpha I_c \tag{8}$$

(since none of the incident radiation is usually transmitted through the entire collector), where ρ is the reflectance of the collector assembly and α the absorptance of the collector assembly.

Using the notation of equation (1) in equation (7) the total energy loss from the collector can be obtained:

$$E_l = I_r A_w + E_{rr} + q_c. \tag{9}$$

Using equation (8), the reflection loss is separated from the other terms on the right-hand side of equation (9) to yield the following equation for the useful heat:

$$q_u = \alpha I_c A_w - (E_{rr} + q_c) - q_s. \tag{10}$$

The second term on the right-hand side of equation (10) represents the heat losses due to reradiation to the ambient (from the exterior surface of the collector), as well as to the fraction of radiation emitted from the interior surfaces and subsequently transmitted into the ambient, and due to convection from the collector exterior surface to the ambient. To solve equation (10) for q_u, all the terms on the right-hand side must be expressed as known functions of given quantities. Thus for a given collector component configuration,

$$\alpha = \alpha \text{ (direction, spectrum, temperature)},$$

$I_c = I_c$ (direction, spectrum),

$E_{rr} = E_{rr}$ (temperature, and emittance, transmittance, and reflectance at this spectrum and temperature of all participating components of the collector; temperature of the ambient),

$q_c = q_c$ (temperature of the exterior surface of the collector, temperature of the ambient, thermophysical transport properties),

$q_s = \partial(\rho V c_p T)/\partial t = q_s$ (temperature, density, and volume of each of the collector components, including the working fluid; time).

It is important to note that a solution of the real problem described by equation (10) requires the determination of complete radiative transfer—namely, transmission, reflection, and absorption—and of all of the temperatures in the entire collector structure.

Electrical circuit analogies have been used to aid in clarifying the energy balance as well as the internal energy transfer. An early analogy of this type that includes the consideration of thermal capacity of the collector elements was suggested for flat plate collectors by Parmelee (1955) and used in the transient analysis of such collectors by Kamminga (1985). These analogies were also used for the description of steady-state energy balances and heat transfer by Duffie and Beckman (1974, 1980), Kreith and Kreider (1981), and many others.

Superimposed on a sketch of a two-plate solar collector, a fairly comprehensive circuit analogy is shown in figure 4.9. The incident insolation I is diminished in passage through each of the transmissive covers by reflection from each of the cover/air interfaces and by absorption in the cover. The amount I_2 arrives at the absorber surface, and a fraction of it is absorbed and converted into heat. The remainder is reflected. Not noted on the schematic for reasons of clarity is the fact that all reflected radiation, as well as radiation emitted by the collector components, undergoes further transmissive, reflective, and absorptive processes as it interacts with collector components its path. Further detail on this is given in subsection 4.2.1. Thermal resistances wired in parallel indicate that both radiative and convective, or conductive, transport are present. The exterior surfaces of the collector exchange radiation with the objects "visible" to these surfaces. Typically the sky-exposed side of the collector will exchange radiation with the sky, at the "sky temperature" T_s, while other parts of the collector surface may exchange radiation with the surrounding ground, obstructions such as trees, buildings, or other collectors, and possibly the sky too. An equivalent temperature of this part of the environment can be formulated, shown in figure 4.9 as T_b. The capacitors C_{g1},

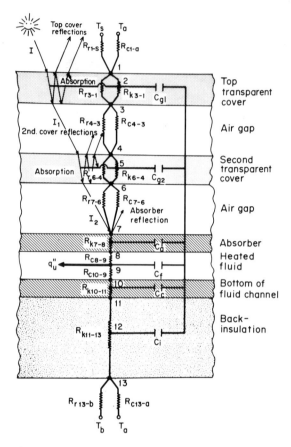

Figure 4.9
Heat transfer electric circuit analog for double-glazed solar collectors.

C_{g2}, C_a, C_c, C_f, and C_i represent the thermal capacitance of the two ᵥ plates, the absorber and the fluid channels bottom, the working fluid, and ᵢ. back insulation, respectively. q_u'' is the useful heat flux (electric current in the analog circuit) taken away by the working fluid. In operation I, T_a, T_s, T_b, R_{c1-a} (through the wind velocity), and q_u (through working fluid flow rate and inlet temperature) are (or have) transient inputs.

Although figure 4.9 may be applied to the entire collector, with area-averaged parameters, one must recognize that practically all of the resistances and fluxes are location dependent. Consequently such a schematic may be used also to describe transport in a differential element of the collector and integrated over the area. Two other aspects not shown in figure 4.9 are (1) the effect of the collector sides in partially shading the absorber from the incident radiation, on convection and radiation in their vicinity, and in additional heat losses through the sides, and (2) the existence and influence of the fluid distribution manifolds (headers).

It is often assumed that the collectors operate in quasi-steady state, that is, that the conditions remain steady for given time periods (e.g., for an hour) with step-updates in the forcing parameters after each such period. Other often-used assumptions are that all internal emission may be ignored (since most of it gets absorbed in the internal components anyway), that there is one-dimensional heat flow through the covers and the back insulation, that the conditions over the collector are uniform, that there is negligible temperature drop through a cover plate, and that heat losses from the collector surfaces are all to the same heat sink at temperature T_a. An analog circuit diagram of this simplified system is shown in figure 4.10a. Substitution of equivalent resistances, and computation of the insolation striking the absorber, with a further assumption that solar radiation does not generate heat as it is attenuated in passage through the cover plates, results in the most simplified (and most-often-used) analog circuit of the solar collector, depicted in figure 4.10b. This reduces equation (10) to the well-known energy balance expression

$$q_u = IR(\alpha\tau)_{ef} A_w - U_p(T_p - T_a) = I_c(\alpha\tau)_{ef} A_w - U_p \Delta T_{pa}, \tag{11}$$

where $(\alpha\tau)_{ef}$ is the overall effective product of absorptance and transmittance of the collector and T_p the absorber temperature. Comparison of equation (11) with (10) shows that in (11) the energy storage term was eliminated (steady state), that the window cover plates act only to attenuate the incoming solar radiation without their thermal participation, and that the convective and radiative heat losses were combined into a Newtonian overall heat loss coefficient U_p multiplied by the first-order temperature difference between

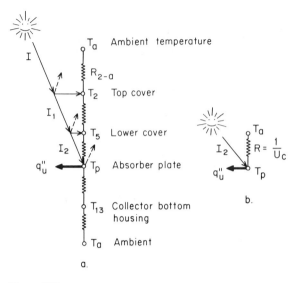

Figure 4.10
Simplified heat transfer circuit analogs for solar collectors: (a) first simplification; (b) most simplified.

the *absorber plate* and the ambient T_{pa}. Despite the seemingly simple form of equation (11), it should be noted that the theoretical determination of $(\alpha\tau)_{ef}$ and U_p requires the analysis (albeit somewhat simplified) of radiative-convective heat transfer within the collector.

The useful heat from the collector, q_u, amounts to the increase in energy of the heated fluid as it passes through the collector. This heat is usually acquired either by transfer from the hotter walls of the fluid conduits or by direct absorption of solar energy into the fluid (or solid particles, in the case of one of the receivers considered).

The temperature of the absorber plate, T_p, is not convenient for practical analysis and design of collectors since it is usually unknown. The known or desired temperature is usually the temperature of the heated fluid. In collectors where the heated fluid does not absorb radiation directly, T_p is expressed in terms of the heated fluid temperature by analyzing the heat transfer between the absorber/fluid-conduit system and the fluid. If the fluid is absorbing radiation directly, this absorption is added to the heat transfer analysis. In both cases, as will be explained further in section 4.3 a relationship is found between the absorber and fluid temperatures, completing the thermal design process.

At this point it should be noted that the efficiency of a collector, η, is defined as

$$\eta = \frac{q_u}{I_c A_w}.$$

As mentioned above, I_c is the solar radiation normal to the surface of the collector before any possible transmission, reflection, and absorption with any of the collector components occur.

Using equation (10), the efficiency can be expressed as

$$\eta = \alpha - \left[\frac{(E_{rr} + q_c) + q_s}{I_c A_w} \right]. \tag{13}$$

Using the simplified energy balance [equation (11)] gives

$$\eta = (\alpha\tau)_{ef} - \left(\frac{U_p \Delta T_{pa}}{I_c A_w} \right), \tag{14}$$

the most frequently used expression for collector efficiency.

4.2 Heat Transfer in the Collector Window System

4.2.1 Radiative Transfer through the Transparent Covers

Radiative transfer in a system containing an absorber separated from the ambient by one or more partially transparent plates consists of multiple transmissions, reflections, absorptions, and emissions. Two such plates are depicted in figure 4.11. Because of the nature of the radiation and the properties of the related materials that these phenomena depend on the wavelength and the direction, temperature, and various surface properties. Although most texts indicate only the multiple reflections, absorptions, and transmissions that a directional beam undergoes in the solar collector window system, it is worth noting that emitted (infrared) radiation undergoes similar processes too. The relationship with temperature couples the radiative problem with other heat transfer processes in the system, such as convection and the ubiquitious conduction. Strictly speaking, the formulation of this problem is difficult, and it would involve a set of integro-differential equations that are hard to solve. It is no surprise therefore that earlier formulations and solutions (starting perhaps with Stokes 1862) assumed that radiative transfer could be decoupled from other heat transfer, that the fully spectral models could be represented by two-band models (shortwave in the solar spectrum, and infrared), that glassplates that are opaque to the infrared band were used, and often that the absorption of the shortwave band in the plates would be negligible (i.e., that imperfect transmittance was due only to reflections). The state of the art

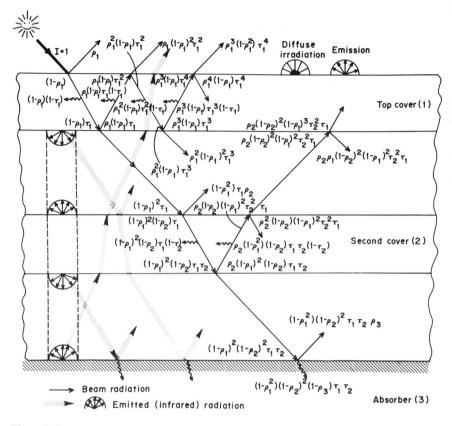

Figure 4.11
Radiative transfer in the window system of a double-glazed solar collector.

around 1970 is described by the earlier review by Dietz (1963) and by the work of Mitalas and Stephenson (1962), Stephenson (1965), and Whillier (1953, 1963), Whillier's review (1967), and the book by Duffie and Beckman (1974).

A review of the radiative characteristics of a single semitransparent plate was performed by Viskanta and Anderson (1975). Siegel (1973) showed that the "net radiatlon method" was much easier to employ for finding the overall transmittance of a stack of parallel semitransparent plates than the older ray-tracing technique. This method was employed by Shurcliff (1973) for the case of nonabsorbing covers. Sharafi and Mukminova (1975) and Viskanta and Taylor (1976) obtained solutions for multilayer systems with varying optical properties, such as with short-wavelength radiation antireflective and infrared reflective coatings. Wijeysundra (1975) included the determination of the radiation absorbed in each cover by solving a system of N equations, where

N is the number of the collector plates (plus an additional cover plate for the absorber). Edwards (1977) proposed and successfully used the "embedding technique," which is computationally simpler, for the solution of the same problem. Viskanta et al. (1978) presented a general analysis (using the net radiation method) to predict the spectral directional radiation characteristics of single and multiple plates of semitransparent material, with parallel optically smooth surfaces that can also be coated with one or more thin-film materials to achieve desired spectral selectivity. Scattering inside the glass was assumed to be negligible, and absorption in the plates was taken into consideration. Elsayed (1984) extended the last two studies by considering both the perpendicular and parallel components of the unpolarized incident radiation throughout the plate stack, rather than use the average plate properties for both parallel and perpendicular components as done in the previous studies. He found that the differences are small for one or two covers, but they increase with the number of covers, to about 25% for six covers. Morris et al. (1976) analyzed radiative transfer through thin-walled glass honeycombs used for convection suppression in collectors and found that high shortwave transmittance and low emittance can be obtained with this design. A good review of radiative behavior of windows, of window stacks relative to solar collector applications, and of window coatings was given in the book by Siegel and Howell (1981, pp. 718–747).

The earlier analyses assumed that the cover plates were opaque to infrared radiation, and they ignored infrared interactions (emission, absorption, reflection, and transmission) between the semitransparent plates of the stack. This could lead to serious errors, especially for plates that are significantly transmissive in the infrared. Lior et al. (1977) developed a computer program for the analysis of solar collectors and systems (SOLSYS) that included the infrared emissive, absorptive, and reflective interactions in the stack, but they still assumed the plates to be opaque in the infrared spectrum. More recent work by Hassan (1979), Hollands and Wright (1983), and Edwards and Rhee (1981) includes all of the infrared interactions in the analysis of the combined radiant/convective transfer in solar collectors. The analysis by Hollands and Wright (1983) allows the use of different radiative properties on each side of the plate and makes the assumption that each cover is radiantly grey in the wavelength range of interest (3–30 μm). Edwards and Rhee (1981) have made one step further in allowing nongrey behavior. The additional rigor included in the last two studies results in more cumbersome expressions for the radiant transfer, but efficient computer algorithms have been developed for their solution.

Collectors with cylindrical glass envelopes interact with radiation in a different way from those with flat covers because of the difference in geometry. For example, the transmittance of a cylindrical envelope is smaller than that of a flat plate for normally incident insolation because the solar rays strike the cylindrical envelope at incidence angles of 0–90 deg while the flat plate incident angle is uniformly 90 deg. Perhaps the earliest analysis of this configuration (for evacuated single-cover cylindrical collectors) was performed by Felske (1979) who considered a configuration with an absorber plate spanning the diameter of the glass tube and running along its length (as in figure 4.3a). He determined the transmittance of the cylindrical cover in terms of the fraction of solar radiation striking the absorber but neglected absorption in the cover, change of ray angle due to refraction, and internal reflections, and did not analyze the absorber–cover radiant interactions. This work was extended by Garg et al. (1983), with similar simplifications to collectors with two concentric covers.

Saltiel and Sokolov (1982) used three-dimensional ray tracing (the algorithm, however, was not presented in the paper) to perform an optical and thermal analysis of a cylindrical evacuated collector, with a cylindrical semitransparent absorber placed eccentrically inside. Their analysis is significantly more comprehensive than the two described above in that they refrain from the above-listed simplifications. Still, they assume that the inner and outer cylinders are opaque in the infrared range and that the thermal radiation is at a single wavelength. A ray-tracing technique using the Monte Carlo method was developed as a computer program for analyzing cylindrical double-wall evacuated solar collectors by Window and coworkers (Window and Zybert 1981; Window and Bassett 1981; Chow et al. 1984).

Cellular structures, such as honeycomb panels, have been considered for placement between the absorber and the cover plate to suppress convection and reduce reradiative losses. Sparrow et al. (1972), Tien and Yuen (1975), Felland and Edwards (1978), and Symons (1982) studied the radiative transfer in such structures, and presented results that could be used in thermal design.

"Thermal traps," in which the solar radiation is converted into heat by absorption in a semitransparent solid or stagnant fluid, were proposed and analyzed in detail by Cobble and coworkers (Cobble 1964a, b; Safdari 1966; Pellette et al. 1968; Lumsdaine 1970), and the results were validated experimentally (Cobble et al. 1966; Lumsdaine 1969). More recently both experimental and theoretical studies were performed by Abdelrahman et al. (1979) and Arai, Hasatani, and coworkers (see Arai et al. 1980, 1984; Bando et al. 1986) on volume heat trap solar collectors by using absorption in semitransparent fluids, in most cases containing in suspension absorbing fine particles. Absorption coefficients and transient temperature profiles were obtained by a simultaneous solution of the radiative-conductive problem, using a multiband model for the radiative properties.

Progress Summary   *The primary progress made was in (1) relaxing most of the simplifying assumptions–for example, by including the transmission of infrared radiation through the covers and infrared radiation exchange among the covers–by allowing spectral dependence of the properties, and by being able to analyze the effects of optical coatings on the semitransparent covers and thus extending the ability to consider a variety of materials and techniques for improving the transmittance of insolation and reducing the heat and radiative losses; (2) developing efficient algorithms for the solution of these more complicated problems; (3) formulating and solving the radiative problem of cylindrical evacuated collectors; (4) proposing, and advancing the state of knowledge on, "volume heat trap" collectors consisting of fine-particle suspensions in fluids.*

4.2.2   Natural Convection

The Basic Phenomena   In many types of collectors and receivers air is contained between the hotter absorber and the cooler ambient. The temperature difference causes natural convection of the air, creating greater heat losses from the absorber to the ambient than would have taken place if the air were stagnant and the heat loss were by conduction only. In solar collectors the insulating air is typically confined in an enclosure formed by the sides of the collector, one or two of the window panes, or the absorber plate . If convection suppression partitions (or honeycomb structures) are used in the collector, the convection occurs in the volume enclosed by these partitions and by the window panes or absorber. Some cavity receivers in concentrating collectors have an open solar radiation entance aperture, and because of this opening on one or more walls, the natural convection enclosure is incomplete.

Relatively little was known about natural convection in solar collectors at the early 1970s. Many still used the dimensional empirical equation proposed by Hottel and Woertz (1942) and Whillier (1953) for the heat transfer coefficient h_{12}, between parallel plates 1 and 2, at temperatures T_1 and T_2 respectively:

$$h_{12} = c(T_1 - T_2)^{1/4} \equiv c\Delta T_{12}^{1/4}. \tag{15}$$

Tabor (1958) summarized and consolidated a number of empirical results on natural convection between flat parallel plates that existed at that time (Mull and Reiher 1930; Robinson and Powlitch 1954); some of these correlations already included the effects of tilt and Grashof (or Rayleigh) number (Fishenden and Saunders 1950; de Graaf and Van der Held 1953, 1954),

although they were in some conflict with each other. In addition these correlations have been obtained for configurations and boundary conditions that
did not represent solar collectors well and were based on a relatively small
number of data points. These equations were used through the early 1970s
(see Duffie and Beckman 1974).

The relevance to the improvement of solar collector efficiency, the interest
in the development and evaluation of methods for the suppression of convection, the serious paucity of necessary knowledge of the phenomena, and
perhaps also the intrinsic interest in the scientific fundamentals of natural
convection have combined to motivate a very large amount of research in that
field over the last 15 years or so. This has resulted in significant improvements
in understanding and in the ability to make quantitative predictions over a
very large range of parameters. Some of that progress relative to solar energy
applications has been summarized by Buchberg et al. (1976), Lior et al. (1983),
Kreith and Anderson (1985), Hoogendorn (1985), Wang and Kreith (1986),
and Lior (1986).

The state-of-the-art of natural convection in enclosures at the beginning
and mid-1970s was reviewed by Ostrach (1972) and Catton (1978). Much work
has been done till that time on the development of empirical heat transfer
correlations and on the prediction of onset of instability in horizontal fluid
layers heated from below, and analytical/numerical methods have been established for the solution of laminar two-dimensional natural convection problems in horizontal and inclined layers of infinite span. Aided by the advent of
more powerful computers and motivated in part by solar energy applications,
the development of effective computer programs for the analysis of three-
dimensional laminar natural convection in rectangular, cylindrical, and spherical enclosures with arbitrary boundary conditions, and their experimental
confirmation, was begun in the mid-1970s. This is exhibited in the publications
of the research team of Ozoe, Churchill, Lior, and coworkers (see Ozoe et al.
1976, 1977a, b, 1978, 1979, 1981a, 1983a, 1985a; Chao et al. 1981, 1983a, b),
Chan and Banerjee (1979a, b), and in the reviews by Ostrach (1982) and Lior
et al. (1983). Some of the advances resulting from this work are highlighted
below.

It is now understood more clearly that real natural convection flows in
enclosures are not two-dimensional and that a velocity component parallel
to the familiar roll-cell axis is also present. The flow thus resembles a double
helix, with fluid particles moving along both the circumference of the roll cell
and the direction of its principal axis—from the walls into the enclosure, up
to a certain distance, and then back toward the walls. This third flow velocity
component is due both to the drag at the end walls and to thermal gradients

generated at these walls because of the diminished rate of circulation. The significance of the three-dimensionality of the flow becomes even more pronounced when the enclosure is tilted or when partial partitions are inserted. It was determined that steady laminar natural convection in horizontal enclosures with a bottom temperature higher than the top is characterized by a train or roll cells with their axis parallel to the short side of the enclosure, an observation that also served to justify the many two-dimensional analyses of the phenomenon. This is no longer correct when the box is tilted: *Inclination*, about its longer side causes the axis of the roll cells to become oblique to the side, and at some critical angle all the roll cells form one large cell that has an axis perpendicular to the short side of the box. Tilting the box along its shorter side gradually merges the parallel roll cells into one large circulating cell, with its axis still parallel to the short side of the box. As the box is tilted from the horizontal position, the Nusselt number is first seen to decrease gradually to a minimum that coincides with the transition from one convective pattern to another, reaches a maximum at a higher angle of inclination (~ 60 to 90 deg), and then diminishes monotonically as the angle is increased to 180 deg. Both the minimum and maximum of the Nusselt number occur at slightly higher values for boxes inclined about the long side than for those inclined about the shorter side, but the values of Nu are about the same in both cases and similar to those of figure 4.12.

Three-dimensional calculations produce lower Nu values than two-dimensional ones, for the same case, principally because the three-dimensional calculations account for the slowdown and redirection of the circulation by the solid ends.

Computer time and memory are major stumbling blocks in the computation of natural convection in enclosures of practical size. The research team of Ozoe, Churchill, Lior, and coworkers (Ozoe et al. 1982, 1983b) used the observation that the convection occurs in roll cells confined to almost-fixed volumes in the enclosure to develop a new computational method that would reduce computer time and memory. In this method a number of typical (according to boundary conditions) cells are computed individually, and the solutions are then patched together to produce overall heat transfer coefficients for the enclosure. The results obtained for both horizontal and inclined enclosures were quite encouraging.

The fact that the minimal Nu number occurred at angles of inclination at which the roll cells were lined up with the longest axis, or the one along which the motion is most tortuous, indicated that the manipulation of roll-cell orientation by such means as *internal partial baffles* may result in the reduction of Nu. Numerical and experimental studies by Chao et al. (1983a, b) for a

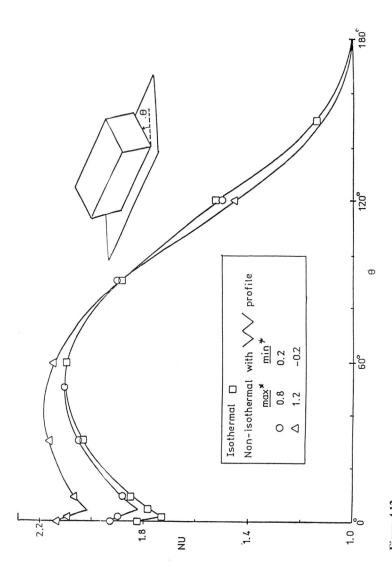

Figure 4.12
Mean natural convection Nusselt number across and enclosure (Ra = 6,000, Pr = 10) as a function of the inclination angle θ. The hotter surface has a double-sawtooth temperature distribution as shown. Source: Chao et al. (1981).

$2 \times 1 \times 1$ box with a partial fin perpendicular to the long side of the enclosure and attached to it, and for Ra up to 2×10^4, have shown that the fin produces a 30% reduction in Nu (at Ra $= 6 \times 10^3$) for all angles of inclination as compared to the finless box and that the dependence on angle of inclination is similar to that in figure 4.12. Similar effects of partial partitions were shown by Lin and Bejan (1983) for $10^9 \leq$ Ra $\leq 10^{10}$ and by Nansteel and Greif (1984) for $10^{10} \leq$ Ra $\leq 10^{11}$.

It is well-known that natural convection phenomena are quite *sensitive to the boundary conditions*. Three-dimensional computations, which in some cases were validated in by experiments (see Chao et al. 1981; Ozoe et al. 1983c), have shown that the Nusselt number decreases as the temperature of the hot plate becomes more uniform: For example, figure 4.12 shows the results of computations for a double-sawtooth temperature distribution on the lower surface (to simulate the effect of cooling pipes along the absorber), where Nu is up to about 20% higher for the nonuniform boundary condition investigated.

Although it is obvious that *radiative exchange among the enclosure surfaces* in a solar collector may have an important role in the determination of the boundary conditions and consequently on the natural convection, accelerated progress on the understanding of the interactions between radiation and natural convection was made only in the last few years (see reviews by Viskanta 1982 and Yang and Lloyd 1985). Except for a few special cases. such as receivers containing solid suspensions, the gas in solar collectors is non-participating. This fact simplifies the analysis significantly. Analytical studies by Edwards and Sun (1971) and Hatfield and Edwards (1982) have determined that radiation increases the critical Rayleigh number for the onset of convection because it tends to smooth out temperature differences on the surfaces. In considering both wall conductance and radiation, Kim and Viskanta (1983, 1984) pointed out that Nu is not representative of the overall heat transfer across the enclosure and that, for their conditions, radiant exchange tended to reduce the intensity of natural convection.

Natural convection in solar collector or receiver enclosures is often *turbulent*. Due both to uncertainties in the modeling of turbulence in this case and to numerical obstacles, solutions of turbulent natural convection in enclosures are still in their infancy and are mostly for two-dimensional models (see Fraikin et al. 1980; Farouk and Guceri 1982; Markatos and Pericloeous 1984; Ozoe et al. 1985b). Hjertager and Magnussen (1977), and more recently Ozoe et al. (1986), developed and solved a three-dimensional model for the turbulent case. The models use the $k - \varepsilon$ formulation that was originally developed for forced convection, and there is thus still uncertainty about the values of the coefficients to be used for turbulent natural convection (for an analysis of sensitivity to model coefficients, see Ozoe et al. 1985b).

More detail on natural convection in enclosures is available in the recent reviews by Hoogendorn (1986) and de Vahl Davis (1986). A summary of results specific to typical solar collector configurations is given below.

Natural Convection between Parallel Plates (Large Aspect-Ratio Enclosures)
The interest in solar collectors prompted the rexamination of older correlations on natural convection between parallel plates and the development of improved and more appropriate empirical correlations and theoretical anaylses, which also include effects of inclination (see Hollands and Konicek 1965; Arnold et al. 1976; Hollands et al. 1976).

Elsherbiny et al. (1982) conducted a comprehensive experimental investigation of the heat transfer in air-filled, high aspect ratio enclosures with isothermal walls and produced results that at this time are recommended for use in thermal design of collectors. Their experiments covered the ranges $10^2 \leq \mathrm{Ra_L} \leq 2 \times 10^7$, $5 \leq H/L \leq 110$ (H is the length along the inclined side of the collector and L the distance between the two plates) and $0 \leq \phi \leq$ 90 deg, where ϕ is the angle of the enclosure axis with respect to the horizontal. They found the transition from the conduction to convection regimes in vertical enclosures to be a strong function of aspect ratio when $H/L < 40$. The recommended heat transfer correlations for vertical layers and enclosures are as follows:

For vertical layers ($\phi = 90$ deg),

$$\mathrm{Nu_1} = 0.0605\,\mathrm{Ra_L^{1/3}}, \tag{16}$$

$$\mathrm{Nu_2} = \left[1 + \left\{\frac{0.104\mathrm{Ra_L^{0.293}}}{1 + (6310/\mathrm{Ra_L})^{1.36}}\right\}^3\right]^{1/3}, \tag{17}$$

$$\mathrm{Nu_3} = 0.242(\mathrm{Ra_L}/AR)^{0.272}, \tag{18}$$

$$\mathrm{Nu_L} = [\mathrm{Nu_1}, \mathrm{Nu_2}, \mathrm{Nu_3}]\mathrm{max} = \mathrm{Nu_{90}}. \tag{19}$$

For inclined layers ($\phi = 60$ deg),

$$\mathrm{Nu_1} = \left[1 + \left\{\frac{0.0936\mathrm{Ra_L^{0.313}}}{1 + G}\right\}^7\right]^{1/7}, \tag{20}$$

$$G = \frac{0.5}{[1 + (\mathrm{Ra_L}/3160)^{20.6}]^{0.1}}, \tag{21}$$

$$\text{Nu}_2 = \left(0.104 + \frac{0.175}{\text{AR}}\right)\text{Ra}_L^{0.283}, \tag{22}$$

$$\text{Nu}_L = [\text{Nu}_1, \text{Nu}_2]\text{max} = \text{Nu}_{60}. \tag{23}$$

The notation used in equation (19) and (23) indicates that the correlation with the maximum value should be used.

For tilt angles between 60 deg and 90 deg, Elsherbiny et al. (1982) suggest a linear interpolation between the limiting correlations given above:

$$\text{Nu}_\phi = \frac{(90\deg - \phi)\text{Nu}_{60} + (\phi - 60\deg)\text{Nu}_{90}}{30\deg}. \tag{24}$$

For tilt angles between 0 and 75 deg, Hollands et al. (1976) recommend the correlation

$$\overline{\text{Nu}}_L = 1 + 1.44\left(1 - \frac{1708}{\text{Ra}\cos\phi}\right)^{\cdot}\left[1 - \frac{1708(\sin 1.8\phi)^{1/6}}{\text{Ra}\cos\phi}\right]$$

$$+ \left[\left(\frac{\text{Ra}\cos\phi}{5830}\right)^{1/3} - 1\right]^{\cdot}, \tag{25}$$

where L is the distance between the plates at temperature T_1 and T_2, respectively, and the Rayleigh number Ra is given by

$$\text{Ra} = \frac{2g(T_1 - T_2)L^3}{v^2(T_1 + T_2)}\text{Pr} \tag{26}$$

and

$$[X]^{\cdot} \equiv \frac{|X| + X}{2}.$$

It should be noted that when $\text{Ra} < 1708/\cos\phi$, the Nusselt number in equation (25) is exactly equal to unity. Since by definition

$$q_c = A\overline{h}(T_1 - T_2) = A\overline{\text{Nu}}\frac{k}{L}(T_1 - T_2), \tag{27}$$

the condition $\overline{\text{Nu}} = 1$ implies that the heat transfer is by pure conduction.

Natural convection in cylindrical enclosures pertians primarily to partially evacuated collectors, line-focus concentrator receivers, and CPC collectors. Kuehn and Goldstein (1976a, b, 1978) developed a comprehensive correlation for the Nusselt number for natural convection between horizontal concentric (1976a, b, 1978) and eccentric (1978) cylinders at constant (but different) temperatures, as a function of the Prandtl and Rayleiyh numbers, for $10^2 <$ Ra $< 10^{10}$ and $10^{-2} <$ Pr $< 10^3$. Kuehn and Goldstein (1980) also presented a simplified correlation for Pr $= 0.71$ and laminar flow. Addressing natural convection between eccentric cylinders both numerically and experimentally, Lee et al. (1984) considered $10^2 <$ Ra $< 10^6$, diameter ratios of 1.25 to 5, and eccentricity ratios of up to ± 0.9 for air (Pr $= 0.71$). They illustrated the internal convective flow patterns and isotherms and have provided the overall heat transfer coefficient as a function of Ra.

Compound parabolic concentrators (CPC) are similar to line-focusing collectors. One of the differences is in a design proposing to enclose the entire concentrator trough, containing the receiver, with a front cover. This forms an irregular, noncircular cylinder. Natural convection in this device, oriented vertically, was computed for $2 \times 10^3 <$ Ra $< 1.3 \times 10^6$ by Abdel-Khalik et al. (1978). Meyer et al. (1980) determined the heat losses from a trough collector as a function of Ra and tilt angle. Some experimental results, summarized by Kreith and Anderson (1985), have also been obtained by a number of researchers.

Iyican et al. (1981) developed a correlation for natural convection and convective/conductive heat transfer in trapezoidal groove collectors.

Suppression of Natural Convection and Its Consequences   As early as 1929 (see Veinberg 1959) honeycomb structures were placed in the air space between the absorber and the window cover in solar collectors to suppress natural convection and to reduce reradiated energy transfer to the ambient, and thus to reduce heat losses. It is obvious that such structures must not impede significantly the absorption of solar radiation by the absorber and that they should add minimal heat losses by conduction through their walls. Early work on this concept was done by Francia (1962) and by a number of researchers in France (see Perot et al. 1967). Perhaps the first analysis that considers both convection and radiation in such structures was performed by Hollands (1965), who used a rather simplified model. Tabor (1969) pointed out that the structures should be transparent to solar radiation and preferably opaque to infrared. To minimize cost, weight, and heat conduction, he recommended very thin materials that should also withstand the temperatures and the degradation due to insolation. Excluding glass from most applications because of cost and weight, he recommended the development of appropriate plastic materials.

Hollands (1973) performed experlments on natural convection in horizontal honeycomb panels (vertical cell axis), and he determined that the critical Rayleigh number for the onset of convection lies between those determined for perfectly conducting and perfectly insulating side walls. Charters and Peterson (1972) brought up the important question of the critical conditions for inducing convection in such a honeycomb cell and commented that the fluid inside inclined honeycomb structures is always unstable, and thus no significant suppression effect can be expected in that configuration. It is worthwhile noting that (1) realistic boundary conditions, which exhibit non-uniform temperatures on the hot and cold walls and/or on the side walls, make the enclosed fluid always unstable, and there is essentially no critical Rayleigh number (i.e., $Ra_{cr} = 0$ always) even for horizontal plates (vertical cell axis), and (2) the larger surface-to-volume ratio associated with "convection-suppression devices" nevertheless produces less vigorous convection and reduces heat losses due to convection; at low Ra numbers experiments indicated that the overall heat transfer coefficient is very close to conduction only.

Among the many types of convection-suppression devices developed and tested for solar collectors, successful results were obtained with 0.2-mm wall glass tubes (see Buchberg and Edwards 1976; McMurrin et al. 1977), and 0.076-mm-thick Lexan and Mylar (Marshall et al. 1976). "Bubble-sheet" packaging material was indicated to have potential if made of proper plastic materials, and aluminized plastic honeycombs, with their axis tilted toward the sun, that use reflection rather than transmission for channeling the solar energy to the absorber and for reducing reradiation were tested with reasonable success (Lior and Saunders 1973). Experiments by Symons and Gani (1980) have shown that flat collectors with a single antireflection etched low-iron glass cover, a convection-suppression device, and selective black absorber, performed up to 110°C better than commercially available cylindrical evacuated collectors configured with a specular reflector in the back. The advantage is realized primarily because the $(\alpha\tau)_{ef}$ of the cylindrical collector was significantly lower because of both the shape of the outer cover and its higher surface reflectivity.

For an inclined square honeycomb the Nusselt number depends on the Rayleigh number, the inclination (angle ϕ between the plates and the horizontal), and the aspect ratio of the honeycomb $AR = L/D$ (=depth/width). For the range $0 < Ra < 600 \cdot AR^4$, $30 < \phi < 90$ deg, and $AR = 3, 4$, and 5, the Nusselt number for air is given by Cane et al. (1977) in the form

$$\overline{Nu} = \frac{\bar{h}L}{k} = 1 + 0.89\cos(\phi - 60 \text{ deg})\left(\frac{Ra}{2420AR^4}\right)(2.88 - 1.64\sin\phi). \qquad (28)$$

This relation may also be used for hexagonal honeycombs if D is replaced by the hydraulic diameter. For engineering design the honeycomb should be chosen to give a Nusselt number of 1.2 according to Hollands et al. (1976).

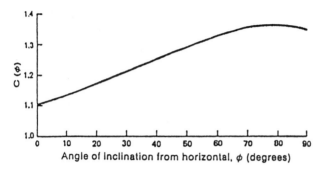

Figure 4.13
Function $C(\phi)$ for use in equation (29). Source: Hollands et al. (1976).

The optimum geometry is found from

$$AR = C(\phi)\left(1 + \frac{200}{T_m}\right)^{1/2}\left(\frac{100}{T_m}\right)(T_1 - T_2)^{1/4}L^{3/4} \tag{29}$$

if L is in centimeters and T in Kelvin. For air at atmospheric pressure and moderate temperatures, $280\,K < T_m < 370\,K$. The function $C(\phi)$ is plotted in figure 4.13. Empirical correlations for Ra_{cr} and Nu for inclined *rectangular-celled* diathermaneous honeycombs were further developed by Smart et al. (1980). They also concluded that square honeycombs are superior in most cases to rectangular ones.

Slats (rectangular enclosure with very large planar aspect ratios) placed along the east–west axis of the collector were also considered for suppressing natural convection. Meyer et al. (1979) found in small-scale laboratory experiments that convection was reduced for aspect ratios (distance between slats/depth of slats) below 0.5 and that convection heat transfer actually increased above the values observed for enclosures without stats for large aspect ratios, with a maximum occurring for aspect ratios of 1 to 2. Experiments with solar collectors using thin glass slats by Guthrie and Charters (1982) have confirmed that the slats improve the collector efficiency for normally incident insolation (by about 40% at 100°C) but that solar transmittance is reduced significantly for other solar incidence angles.

It was determined that small gaps between the honeycomb panel and the absorber and top glass cover do not affect the convection suppression capacity of the panel (Edwards et al. 1976). Demonstrating the strong coupling that exists between radiative transfer and the heat conduction in honeycombs, Hollands et al. (1984) have shown that an analysis that decouples the modes may severely underpredict the real heat transfer rates across the honeycomb panel. The radiation tends to increase the temperature gradients in the gas at the hot plate. An acceptably simplified analytical method that considers the coupled problem has also been presented. In addition Hollands and Iynkaran (1985) observed that conduction through the air layer next to the hot plate raises the temperature of the honeycomb panel and thereby also raises the reradiated energy loss from the panel. The use of such a panel thus tends to diminish the improvement one may expect from having a selectively coated (low emissivity) absorber in the collector. To take best advantage both of the reduction of radiative losses by the use of low emittance coatings and of the reduction of convective losses by using cellular convection suppression structures, they recommended and tested a configuration in which the honeycomb was separated by a 10-mm air gap from the hot plate, thus reducing the coupling between conduction to and radiation from the honeycomb panel. Such a collector was built and tected, and its improved performance was demonstrated (Symons and Peck 1984).

Progress Summary   *Motivated in large part by solar energy applications, researchers in the past decade have made enormous progress in understanding natural convection in enclosures of various tilt and aspect ratios and in the ability to predict it. Their methods were both experimental and numerical. The new numerical techniques they developed allow the solution of three-dimensional problems for moderate aspect ratios and laminar flows. This solution has begun to be extended to turbulent flows, using the k − ε model. To day the effect of convection-suppression devices is reasonably well understood and predictable, and much progress has also been made in understanding the coupling between conduction and radiation in such devices and enclosures.*

4.3   Heat Transfer in the Absorber and the Heated Fluid

4.3.1   The Plate Tube Absorber

The absorber of many liquid-heating flat plate collectors consists of a number of parallel tubes (risers) connected at the inlet to a distribution manifold and at the outlet to a collection manifold, and attached with good thermal contact to the absorber plate. This is shown in figure 4.1, and in better detail in figure 4.14. Many cross-sectional shapes of the tubes, and methods for their attachment to the absorber plate, have been used.

   The solar energy converted to heat at the absorber surface S is conducted through the absorber and the tube walls into the fluid that is flowing through

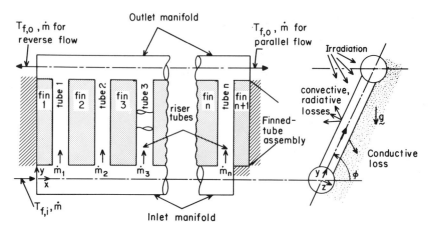

Figure 4.14
Plate-tube absorber manifold assembly geometry.

the tubes and that is thus being heated. The conventional approach for determining the heat gain of the fluid was established by the Hottel-Whillier (HW) model, in which the plate-tube assembly is modeled as the well-known fin-tube problem (figure 4.15). In that model it is assumed that for each differential element of length along the tube, the temperature distribution in the plate is one-dimensional, in the x-direction perpendicular to the tubes, and any gradients in the y-direction along the tubes and z-direction perpendicular plate are assumed to be zero. It also assume that the temperature profile is symmetric about the tube, with the tubes's centerline at the minimum temperature point and the fin's centerline (at $w/2$) at the maximum temperature. Performing an energy balance with this model for a unit length of the fin-tube (see Duffie and Beckman 1980), it is easy to show that the amount of heat collected by the fin q'_{fin} is

$$q'_{\text{fin}} = (w - d)\eta_{\text{f}}[S - U_{\text{p}}(T_{\text{b}} - T_{\text{a}})], \qquad (30)$$

where η_{f} is the fin's efficiency, S is the radiative flux absorbed per unit of absorber, U_{p} is the overall heat transfer coefficient from the absorber to the ambient, and T_{b} is the fin's base temperature (at the junction with the tube). Although typically the absorber plate (fin) is of uniform thickness, fins thicker at the root are known to have higher efficiency. Kovarik (1978) developed a method to determine the optimal profile of the fin in solar collector, based on an objective function of minimum cost per unit of heat output.

Since the tube also collects energy directly over its projected area d, the total useful energy gain per unit length of tube is

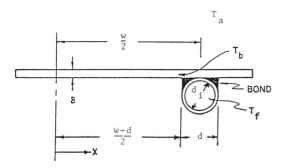

Figure 4.15
The basic fin-tube schematic.

$$q'_u = [(w - d)\eta_f + d][S - U_p(T_b - T_a)]. \tag{31}$$

That useful heat gain is transferred to the fluid, and the model expresses this in terms of the sum of the convective resistance between the fluid and the tube wall, and the contact resistance due to the bond between the tube and the plate (fin)

$$q'_u = \frac{T_b - T_a}{(1/h_{f,i}\pi d_i) + (1/C_b)}, \tag{32}$$

where $h_{f,i}$ is the heat transfer coefficient between the tube wall and the fluid, d_i is the internal diameter of the tube, and C_b is the bond conductance (detailed discussion of the effect of C_b can be found in Whillier 1964b and Whillier and Saluja 1965).

Another energy balance is made along the tube, equating the q'_u with the increase of enthalpy of the heated fluid for that length, which is expressed by the equation

$$\dot{m}c_p\left(\frac{dT_f}{dy}\right) - nwF'[S - U_p(T_f - T_a)] = 0, \tag{33}$$

where $\dot{m}$ is the overall mass flow rate through the collector, n is the number of parallel tubes, c_p is the specific heat of the fluid, T_f is the local temperature of the fluid, and F' is the "collector efficiency factor." The collector efficiency factor is indentified from equations (30) through (33) as

$$F' = \frac{1/U_p}{w\{[1/U_p(d + (w - d)\eta_f)] + (1/C_b) + (1/\pi d_i h_{f,i})\}}. \tag{34}$$

It is usually assumed at this point that U_p and F' are independent of position,

and equation (33) is integrated to yield an expression for the temperature rise
of the fluid as a function of distance along the tube and the other parameters
(which are assumed to be constant):

$$\frac{T_f(y) - T_a - (S/U_p)}{T_{f,i} - T_a - (S/U_p)} = e^{-(U_p n w F' y / \dot{m} c_p)}, \tag{35}$$

where $T_{f,i}$ is the fluid temperature at the inlet to the tube.

Equations (33) and (35), together with the assumption of constancy of the
heat transfer coefficients and F' along the tube (y-direction), indicate that the
plate temperature must also vary exponentially with y.

Collector efficiency can now be expressed in the conventional way, using the
"collector heat removal factor" F_R, which is defined as

$$F_R = \frac{\dot{m} c_p}{A_w U_p} [1 - e^{-(A_w U_p F' / \dot{m} c_p)}], \tag{36}$$

and the collector efficiency [see also equations (10) through (12)]

$$\eta = \frac{q_u}{A_w I_c} = F_R \left[(\alpha \tau)_{ef} - \frac{U_p(T_{f,i} - T_a)}{I_c} \right]. \tag{37}$$

Abdel-Khalik (1976) developed an analytical model of an absorber that has a
serpentine tube bonded to the plate and solved it for two segments of the
serpentine. He concluded that a general equation can be projected for the
calculation of the heat removal factor F_R for any number of segments, with
small error.

Zhang and Lavan (1985) extended the solution to four segments of the
serpentine and that extrapolation from the two-segment solution can lead to
much larger errors than predicted by Abdel-Khalik (1976). Either of these
solutions calculates the heat transfer only for the straight parts of the serpentine, ignoring its U-bend portions, and assumes essentially one-dimensional
heat transfer in the plate.

Equations (30) through (35) serve not only to show the conventional method
of calculating heat transfer through the absorber but also to highlight the
critical assumptions made in the model. In addition to the simplifications
mentioned already, several others stand out: (1) U_p is actually a complicated
combination of convective and radiative heat transfer between the absorber
and the ambient, as discussed in sections 4.1 and 4.2; to say the least, it is not
constant, and it depends on the temperature, in a nonlinear manner at that.
(2) $h_{f,i}$ is not constant either; it depends strongly on the location along the
tube if the internal flow is developing, on the temperature because the convec-

tion is often mixed (forced and natural), and because of the properties. (3) The temperature might not be uniform through the thickness of the fin or the tube, especially if low conductivity materials are used. (4) The plate's temperature field perpendicular to the tube may not be symmetrical because of the effects of unequal flow though the parallel tubes and edge effects. (5) The temperature T_b at the base of the fin might not be equal to the temperature at the interior diameter of the tube nor might the latter be constant around the internal circumference of the tube (obviously the top is heated and the bottom is not). To be able to understand collector operation better and to design more efficient and economical units, research during the last decade has examined many of these simplifications, and the results are summarized below.

Rao et al. (1977) solved anlytically the two-dimensional fin-tube problem, still assuming constant heat transfer coefficients, and concluded that the HW model results are accurate enough for the design of conventional flat plate solar collectors but not for collector design optimization. Chiou (1979, 1980) solved the two-dimensional problem numerically and also found excellent agreement with the HW model. Taking an analytical approach to the heat conduction problem in the absorber, Phillips (1979) noted that heat is actually conducted along the absorber in a direction counter to the flow of the heated fluid, reducing the amount of heat transferred to the fluid and thus collector efficiency. He found that the HW model therefore predicts efficiencies that are too high, by up to about 30% in the range of the parameters he considered.

Typically it is assumed that the flow of the heated fluid through each of the parallel risers is the same, and this is indeed desirable: The studies by Chao et al. (1981) and Jones and Lior (Jones 1981; Jones and Lior 1987) indicate that a uniform temperature absorber produces a higher efficiency collector than one with a nonuniform temperature. At the same time it can not be taken for granted that the flow is distributed equally through the risers: The dual manifold system must be designed to meet that objective. Few studies of flow distribution in such collector manifold systems were made. The earliest proposed for solar collectors was by Dunkle and Davey (1970) who have established, and solved analytically, a highly simplified flow model with a continuous slit (instead of discrete risers) distribution and collection inertia-dominated manifolds. Bajura and Jones (1976) have treated the inertia-dominated dual-manifold system with discrete risers. Jones and Lior (1978), Jones (1981; see also Jones and Lior 1987), Menuchin et al. (1981), and later Hoffman and Flannery (1985) included both inertial and frictional effects in the analysis of dual-manifold systems. Apart from having established a method to compute flow distribution in such a manifold system, it was suggested that essentially uniform water flow distributions are obtained if the riser-to-manifold tube diameter ratio d_r/d_m is $\frac{1}{4}$ or less in the range of flows pertinent to collectors.

Jones and Lior (Jones 1981; also in Jones and Lior 1987) examined the effects of flow maldistribution on collector efficiency using a three-dimensional conjugate model of an unglazed collector (see subsection 4.3.3), and they found negligible influence ($<2\%$) in the range of $\frac{1}{4} < d_r/d_m < \frac{3}{4}$ and pertinent flat plate liquid-heating collector parameters. Solving a two-dimensional collector model numerically and examining arbitrarily imposed flow maldistributions (in most cases of much larger magnitude than those found by Jones and Lior), Chiou (1982) found reductions of 2%–20% in collector efficiency due to flow maldistribution.

Progress Summary *The primary progress in this well-trodden area was in advancing from the one-dimensional absorber plate–tube models to two-dimensional ones, in the ability to determine flow distribution among the risers in a more correct way, and in examining the effect of maldistribution on absorber heat transfer. It was confirmed that collector effjciency improves somewhat as the absorber temperature becomes more uniform.*

4.3.2   Convection to the Heated Fluid

Much is known about convective heat transfer in conduits (see Shah and London 1978; Kreith and Kreider 1978), and when judiciously applied, this available information can be directly used in the thermal analysis and design of solar collectors.

Three aspects related to the proper choice of the convection correlation or to the formulation of the analytical/numerical problem that may be important in solar collectors are flow and thermal development, nonuniformity of the boundary conditions on the interior surface of the conduit tube), and buoyancy effects on the convection (existence of mixed convection).

Usually the flow rate in conduits of liquid-heating collectors is very low, and the flow is consequently laminar and possibly developing for a fair fraction of the tube length. This should be examined by applying one of the conventional criteria for flow development before a decision is made on the correlation or analytical method to be used. It should, however, be noted that most of these criteria were established for constant temperature or heat flux boundary conditions, and without consideration of buoyancy effects—that is, for conditions that do not represent the situation in collectors exactly. For example, Lior et al. (1983) computed that in mixed convection in a vertical tube with linearly increasing wall temperature the Nusselt number is 28%–40% higher than that for the constant wall temperature case, and that flow development is different too. Morcos and Abou-Ellail (1983) examined numerically buoyancy effects in the entrance region of an inclined collector composed of parallel rectangular channels with realistic boundary conditions, and found Nusselt numbers up to 300% higher (at $Ra = 10^5$) than those predicted without the inclusion of buoyancy effects.

Cheng and Hong (1972) analyzed numerically the case of mixed laminar convection in uniformly heated tubes of various inclination and determined that both friction factor and Nusselt number increase significantly with the inclination angle. Baker (1967) observed that augmented mixed convection would occur in solar collector tubes due to variations in the circumferential temperature and that the heat transfer coefficients (for $370 < Re < 2,700$ laminar flow in horizontal tubes) were about 10% higher than those for tubes with circumferentially uniform temperature. Such augmentation becomes particularly important when the heating is from below. In turbulent horizontal flow, on the other hand, circumferentially nonuniform heating appears to increase the thermal development length but to have either essentially no effect on the average Nusselt number (see Black and Sparrow 1967; Schmidt and Sparrow 1978; Knowles and Sparrow 1979) or a small opposite one, seen to reduce it by up to 20% for heating from below (Tan and Charters 1970). The reduction in heat transfer due to buoyancy effects may result from the tendency to relaminarize the flow. This is also consistent with the general conclusion of several of these researchers that the highest local convective coefficients are encountered at the least heated circumferential positions, and vice versa.

For mixed convection in a tube attached to an absorber plate, with conditions typical to flat plate solar collectors, Sparrow and Krowech (1977) and Jones (1981) concluded from analysis that circumferential variations in the thermal conditions of the tube can be neglected.

As seen from equations (36) and (37), increasing the overall flow rate through the collector improves its efficiency and the rate of heat collection. At the same time the efficiency improvement is an asymptotic function of the flow rate; increases of flow rate require more energy and capital equipment investment in pumping, and they increase the operating pressure in the collector and balance of system. Hewitt et al. (1978) and Hewitt and Griggs (1979) proposed a method, based on economic optimization, for determining the optimal flow rate through liquid- and air-heating collectors (but they have not considered possible implications of the pressure increase in the system, which is required for increasing the flow). Optimal control strategy of mass flow rates in flat plate solar collectors was determined for several combinations of objective functions and system models by Kovarik and Lesse (1976) and Winn and Winn (1981).

Progress Summary   *Much work was done and important progress was made in providing information on convection with flow inside inclined tubes, with circumferentially varying thermal boundary conditions, flow development, and buoyancy effects. Ways to determine collector flow rate based on economic considerations were established.*

4.3.3   Formulation and Solution of the Overall (Conjugate) Thermal/Flow Behavior of the Collector

Many years of experience have shown that the Hottel-Whillier thermal model for flat plate solar collectors is adequate for designing conventional collectors and for estimating their performance; experience has also shown that overall collector efficiency is not too sensitive to most of the design parameters when perturbed around a base-case conventional design system definition (see also section 4.10). Realizing, however, that numerous extreme simplifications are inherent in this model (many of which are described in this chapter), it is clear that collector optimization and the development of new collector designs require more rigorous thermal/fluid models. To avoid the need for specifying approximate or arbitrary boundary conditions for each thermal subproblem of the overall collector problem—namely, the radiative and thermal transfer in the window/absorber system, natural convection in the window system, conduction in the absorber, flow distribution in the collector, and convection to the heated fluid—and thus to avoid adding to the solution error, the conjugate heat problem describing the entire collector should be solved. In the conjugate solution all of the subproblems are solved simultaneously, with one serving automatically as the boundary condition for the other.

A comprehensive three-dimensional computer program and effective solution technique were developed for this purpose by Jones (1981; see also Jones and Lior 1987). The specific example solved was a 1.5 × 2 to 6 ft unglazed solar collector with a dual-manifold system containing four risers. The problem was divided into three subproblems: (1) dual-manifold system hydrodynamics, (2) radiant-conductive finned-tube heat transfer, and (3) riser tube fluid dynamics and heat transfer. Each subproblem was solved numerically, and the resulting system of equations was solved simultaneously using an iterative scheme. The solution is refined with each cycle of the iteration since any one subproblem is solved subject to boundary conditions that result from the most recent solutions of the remaining two.

The subproblem models and solutions are fairly general but oriented to solar collector conditions. For example, the solution to the mixed convection heat transfer and fluid mechanics problem in inclined collector risers was bounded by the one providing the lowest Nu, occurring in the case of horizontal tubes without buoyancy effects, and by the one providing an upper limit for Nu, corresponding to vertical tubes with buoyancy effects included. Flow development was included in the analysis, and the effects of buoyancy on heat transfer and flow distribution were established. Notably, in comparison to the case where buoyancy is neglected, the effects of buoyancy on flow in the this system indicate a maximal (1) 5% increase in riser Nusselt number, (2) 24% decrease in flow maldistribution fraction, and (3) 38% reduction in overall dual-manifold pressure drop.

Solution of the conjugate problem was found to provide remarkable insight into the behavior of the collector and gave quantitative relationships among the different components and the operating conditions. It also confirmed the fact that the efficiency of reasonably well-designed conventional flat plate collectors can be predicted by the HW model to within 3% of the value obtained from the significantly more elaborate conjugate model.

Morcos and Abou-Ellail (1983) developed a partially conjugate numerical model of an inclined solar collector with parallel rectangular flow channels in which they considered mixed developing laminar convection in the channels, with circumferentially nonuniform thermal conditions on the channel walls; the latter conditions were determined through simultaneous solution of the channel wall conduction problem. They found that entry length is reduced as the Rayleigh number increases and that buoyancy serves to increase the Nusselt numbers over those predicted with buoyancy neglected.

As an alternative to the solution of the complicated conjugate problem once it was realized that the thermal behavior of the collector is a nonlinear function of the temperature, attempts were made to improve on the conventional HW linear relationship between efficiency and temperature difference by developing and examining nonlinear relationships. Cooper and Dunkle (1981) developed a nonlinear model in which three nondimensional groups were added, but they determined that little improvement was obtained over the conventional linear model in characterizing the daily performance. Phillips (1982) also developed a nonlinear model, with coefficients determined empirically, and found that it correlated simulated collector data better than the linear model. Comparison with collector test data was, however, not made. It should be noted, in summary, that it is already common in collector testing to express the efficiency in terms of a quadratic polynomial in $(\Delta T/I)$.

Progress Summary   *For thorough analysis of the thermal-fluid behavior of collectors and collector components, a conjugate flow distribution and heat transfer model was developed, and a solution technique was developed and used* *successfully. The model does not yet include natural convection in the insulating spaces as one of the subproblems. Nonlinear lumped-system collector models were developed and shown to represent collector efficiency a little better than the linear HW model.*

4.3.4 Plate Absorbers with Flow over Their Entire Surface

To reduce sealing complexity and thus the cost of the collector, liquid-heating collectors typically channel the liquid flow through discrete tubular conduits attached to the absorber plate, as described in subsection 4.3.1. This reduced flow area also increases the velocity of the liquid, and consequently the heat transfer coefficients between the tubes and the liquid. Because of the relatively low heat transfer coefficients between solids and gases and because leakage of air is somewhat more tolerable than leakage of liquids, the heated air in air-heating solar collectors is exposed typically to the entire area of the absorber and is not channeled through a number of discrete small conduits (figure 4.2). The thermal analysis (see Whillier 1964a and a more recent detailed review by Gupta 1982) is very similar in principle to that for plate-tube absorbers (subsections 4.3.1 and 4.3.2), with the main difference in the fact that absorber-to-fluid heat transfer coefficients are calculated for plates instead of tubes.

Going beyond the conventionally used lumped-system energy balance solutions, Liu and Sparrow (1980) investigated numerically the effects of radiative transfer in a shallow and wide rectangular airlflow channel with black walls and a configuration consisting of the top plate heated and the bottom insulated, as is used in many solar collectors in which the air to be heated flows below the absorber (see figure 4.2b). They found that radiative transfer from the hotter to the cooler plate is significant in the developed laminar flow region in that it allows the lower plate to transfer by convection up to 40% of the heat to the air while reducing the temperature of the upper plate and consequently its losses to the ambient. A theoretical model of a similar configuration for turbulent flow was developed and validated experimentally by Diaz and Suryanarayana (1981), but they did not consider radiative exchange. The importance of the large heat transfer coefficients at the channel entry was confirmed. Diab et al. (1980) performed a two-dimensional analysis of a number of air-heating solar collector configurations by applying a nodal formulation. Apart from the value of the technique for this purpose, an important contribution of this study was to indicate, once again (see subsection 4.3.3), that the conventional single-module lumped system analysis predicts efficiencies that are within 5% of a more elaborate (here six-module) analysis. An integral solution for the single rectangular-channel solar collector, with fully developed flow in the channel, was obtained by Grossman et al. (1977), using a second-degree polynomial for the temperature profile, and by Naidu and Agarwal (1981) who used a more correct, fourth-order, polynomial.

To improve the thermal performance of air-heating collectors, a number of designs that increase the contact area between the air and the absorber, or the heat transfer coefficients, or both, have been proposed and analyzed. The effects of such improvements on collector cost and needed pumping power were examined and constrained to acceptable levels in the designs that were recommended for development and production. Many of these designs involve a rather complex and different flow geometry. It would not be practical to present general results, but some of the main studies will be mentioned for reference.

An analysis of the *overlapped glass plate* air heater (figure 4.2d) proposed by Miller (1943) and developed by Löf et al. (1961) was formulated and performed by Selçuk (1971). Christopher and Pearson (1982) produced an analysis of convective heat transfer (by solving the two-dimensional conti-nuity, momentum, and energy equations) in an air heater with *overlapped opaque louvers*.

Analyses of air heaters with *finned absorbers* (figure 4.2e) were performed by Bevill and Brandt (1968), Cole-Appel and Haberstroh (1976), Cole-Appel et al. (1977), and Youngblood and Mattox (1978), among others. Fin dimen-sions, spacing, and surface radiative properties were some of the parameters considered for improving the collector performance. *V-corrugated* absorbers (figure 4.2f), with air flowing in the V-channels below the absorber, were analyzed by Hollands (1963) and Shewen and Hollands (1981), and compar-isons were made with rectangular-channel flow passages under flat plate absorbers. Short flow passages were recommended to take advantage of the high heat transfer coefficients in the developing flow, and the V-corrugated design was found to produce a more efficient collector as compared to the flat plate absorber design for the same pressure drop and airflow rate.

Passage of the air through an absorber formed of a *porous matrix* (figure 4.2g), as proposed by Bliss (1955), increases the contact area significantly, and possibly also the heat transfer coefficients. Chiou et al. (1965) laid the founda-tion for the design and analysis of such collectors and determined friction fac-tors for various slit-and-expanded aluminum foil matrices. Swartman and Ogunlade (1966), Beckman (1968), Lalude and Buchberg (1971), Buchberg et al. (1971) (the latter two studies proposed a honeycomb convection suppres-sion device on top of the porous bed), and Lansing et al. (1979) proposed and laid the ground for the analysis of *packed and porous bed absorbers. Jet impingement* solar air heaters in which the heated air impinges via many small jets upon the back of the absorber have been studied by Honeywell (1977).

Progress summary   *Some progress was made in the understanding of heat transfer in ducts with asymmetric boundary conditions, in advancing to two-dimensional models, and in better understanding of ways to enhance heat transfer between the fluid and the absorber.*

4.3.5   Heat Transfer to the Liquid in Solar Collectors for Boiling Liquids

Many applications require the generation of steam or vapor by solar energy, for example, for driving prime movers. Even if there is no need for steam or vapor, the high heat transfer coefficients associated with boiling in tubes have appeal for the improvement of collector efficiency. At the same time flow boiling in tubes is likely to introduce higher pressure drop than single phase flow and heat transfer, and the flow is apt to be less stable.

A few theoretical and experimental studies have been made on this subject. Experiments with boiling acetone and petroleum ether (Soin et al. 1979), fluorocarbon refrigerants such as R-11 and R-114 (Downing and Waldin 1980; Al-Tamimi and Clark 1983), butane (Bol and Lang 1978), and boiling water in a the tubular receiver of a line-focus parabolic trough solar concentrator (Hurtado and Kast 1984) have demonstrated both feasibility and improved heat transfer coefficients. All three of the reported analyses, for the plate-tube collector by Al-Tamimi and Clark (1983) and Abramzon et al. (1983) and for the cylindrical receiver of a line-focusing collector by May and Murphy (1983), essentially use the Hottel-Whillier equation and consider two regimes along the tube: first the subcooled regime along which the liquid is rising in temperature but not boiling yet, and the performance is evaluated with the conventional single-phase coefficients, followed by the boiling regime, downstream of the section at which saturation temperature was attained. The calculation incorporates the determination of the distance along the tube at which transition from the subcooled single-phase regime to the boiling two-phase regime occurs. The analytical approaches are fairly similar, differing only in the fact that Abramzon et al. (1983), and May and Murphy (1983) perform energy balances on small elements along the tube and integrate the equations along the same path, whereas Al-Tamimi and Clark (1983) use the available HW results, which already have been integrated for each of the two regimes based on the HW approach. Consequently Al-Tamimi and Clark (1983) proposed the same equation for the boiling collector as (37),

$$\eta_{\mathrm{B}} = F_{\mathrm{BR}}\left\{(\alpha\tau)_{\mathrm{ef}} - \left[\frac{U_{\mathrm{p}}(T_{\mathrm{i}} - T_{\mathrm{a}})}{I}\right]\right\}, \tag{38}$$

in this case with the heat removal factor F_{BR} expressed as

$$F_{BR} = F_R F_B, \tag{39}$$

where F_R is the HW heat removal factor evaluated for single-phase conditions and the same flow rate used in the boiling collector [see equation (37)], and F_B is the correction to the heat removal factor to account for boiling:

$$F_B = \frac{1 - \exp(-az^*)}{1 - \exp(-a)} + \frac{(1 - z^*)\exp(-az^*)}{F_R/F_B'}, \tag{40}$$

where

$$a \equiv \frac{U_p F'}{(\dot{m}/A_w)c_{pl}}, \tag{41}$$

$$z^* \equiv \frac{L_{nb}}{L}, \tag{42}$$

L_{nb} is the distance along tube needed for the liquid to rise to the saturation temperature (determined from an energy balance on the liquid), L is the overall length of the tube, and F_B' is F' [equation (34)] for the boiling part of the tube. The boiling (internal) heat transfer coefficient was determined usually from the correlations by Chen (1966) and Bennet and Chen (1980).

Abramzon et al. (1983) further concluded that collectors with internal boiling come within 2%–3% of attaining "ideal" efficiency, that is, the efficiency that could be attained for internal heat transfer coefficients and Reynolds number approaching infinity.

Deanda and Faust (1981, AiResearch Mfg. Co.) have designed and developed an insulated, cylindrical coiled tube boiler that is mounted at the focal plane of a parabolic solar reflector. It was designed to perform as once-through boiler, with or without reheat, for generating steam for a Rankine cycle.

Progress Summary *Almost all of the work in this area was done during the last decade. As expected, practically ideal internal heat transfer coefficients (from tke collector efficiency standpoint) can be obtained if boiling is allowed inside the collector tubes. Experiments with a few fluids demonstrated this fact By adapting the HW model to this problem, a reasonable first step was made in the ability to predict boiling collector efficiency. More work is needed, primarily in determining possible adverse effects of increased pressure drop and flow instability and in determining boiling heat transfer and pressure drop for the boundary conditions specific to solar collector systems.*

4.3.6   Collectors with Free Flow and/or Direct Absorption of Radiation in the Heated Fluid

Improvements in heat transfer and in collector capital costs might be realized if the conduits for the heated fluid could be eliminated or simplified: Instead of irradiating the solid conduit (absorber) and then transferring the heat via conduction through the wall, and convection, to the heated fluid, the flowing fluid could be exposed to the solar radiation directly. If the fluid layer is highly absorptive to solar radiation (this is affected both by the opacity of the fluid and by the thickness of the layer), all of the radiation would be converted to heat as it passes through the fluid. Otherwise, the design would ensure that most of the radiation transmitted through the fluid layer be absorbed in the solid substrate on which the fluid is flowing. As the temperature of the substrate thus rises above that of the fluid, heat would be transferred from the substrate to the fluid by convection and absorption of the substrate radiosity. Several such configurations were evaluated, and the primary interest at present is in evaluating the applicability of these concepts to high temperature central-solar-tower-type receivers.

Free flowing collectors with at least some absorption in the substrate, such as the Thomason "Solaris Trickle Collector," were evaluated and analyzed by Beard et al. (1977, 1978), and the analysis was advanced a little by Vaxman and Sokolov (1985). The Solaris collector exhibited a problem that needs to be addressed in all free flow collectors that heat a liquid with a free surface: evaporation from the free surface with subsequent condensation on the inside of the transparent cover plate. This poses several difficulties, the primary being an effective transfer of heat from the liquid to the ambient, and reduction of cover transmittance due to condensation, both resulting in low collector efficiency. The use of liquids that have a very low vapor pressure at the operating temperature can alleviate this problem (see Beard et al. 1977).

A much more rigorous analytical approach that takes into consideration detailed (spectral, directional) radiative interactions in the liquid/substrate system was developed and validated by Arai et al. (1980), Hasatani et al. (1982), Bando et al. (1986), Wang and Copeland (1984), and Webb and Viskanta (1985). The latter two studies have also included the film fluid mechanics in the analysis. Webb and Viskanta (1985) also discovered that there exists a critical fluid layer opacity that yields the optimal radiation collection efficiency: Fluid layers that are less optically thick than this critical value are too thin to adequately absorb and transport the incident radiation; layers whose opacities are greater than the critical thickness absorb too strongly near the surface, and subsequent emission reduces their performance.

Fully absorbing 'black liquid' collector studies include those by Minardi and Chuang (1975), Trentelman and Wojciechowski (1977), Landstrom et al. (1978, 1980), Samanó and Fernandez (1983), and Janke (1983). Efficiencies 10%–15% higher than those of conventional plate-tube absorber collectors were observed experimentally. Landstrom et al. (1978, 1980) were also engaged in the development of a low-cost commercial collector made primarily from acrylic material. Black dyes dissolved or suspended in the heated liquid were used to produce the "black liquid." Stability problems were encountered, and an acceptable and proven "black liquid" needs yet to be found. A complete set of radiative properties of india ink suspensions of different concentration were measured as a function of wavelength by Wagner et al. (1980).

A radiation-absorbing fluid can also be produced by suspending very fine (often in the submicron size) absorbing solid particles in a flowing gas. This concept has been explored, both analytically and experimentally, for a "volume heat trap" collector by Arai et al. (1984) and recommended by Hunt (1978) for use with air as the working medium in a solar central receiver. Analysis and experiments by Hunt et al. (1983, 1985) have indicated that the gas assumes almost immediately the particle (magnetite was used) temperature if the particles are small enough (about 0.1 μm or less), and that very small concentrations of solids ($\approx 10^{-3}$ kg/m^3) are needed to achieve effective heating of the gas. It is noteworthy that the thermal and radiative exchange between the particles and the gas cannot be calculated by continuum theories because the particle size is of the order of the mean free molecular path of the gas (Yuen et al. 1986).

Another approach was taken by Sandia Laboratory (Hruby 1985) in which larger particles (sand-size, hundreds μm) fall in front of the receiver aperture and are thereby irradiated and heated. Rather than just serve to heat the air, these particles serve as the heat transfer medium and are also used as the medium for heat storage. The particle materials proposed were sintered bauxite for temperatures up to 1,000°C and doped fused zircon for higher temperatures. Analysis and experiments were being conducted.

Progress Summary   *"Black liquid" collectors have been proposed and briefly tested, but little or no activity is evident in this area now. Promising work is proceeding on direct absorption of radiation in fine particles suspended (or falling) in air.*

4.3.7   Shading Effects

Whenever the solar rays are not normal to the collector, the collector side walls shade a part of the absorber. Considering typical flat plate collector configurations, the use of a "shading factor" s of about 0.97 was recommended by which to multiply the conventional collector efficiency prediction equation. In this way the thermal analysis of the collector and the prediction of its efficiency can be made, assuming that no shading occurs, and the final efficiency is simply multiplied by this shading factor.

The use of such a constant factor is imprecise, in that the factor depends on collector configuration, internal detail, orientation, location, and the diffuse fraction of insolation. It is obvious that deeper collectors will have a larger part of their area shaded by the side walls and will thus have a lower shading factor. This is of particular interest with the use of convection-suppression devices in front of the absorber. Furthermore, as Lior et al. (1977) remarked, the shaded areas still accept diffuse insolation, and both the shaded and unshaded parts of the absorber will also intercept some radiation through reflections and reradiations from interior surfaces of the side walls and from window panels. The amount of radiation absorbed in this way of course depends on the configuration and on radiative properties. In their analysis of partially shaded collector arrays, the shaded areas were computed as a function of time (and the solar incidence angle), and it was assumed that these areas accept only diffuse radiation, not the radiosity coming from other internal surfaces. Nahar and Garg (1980) developed the equation for the unshaded area fraction $(1 - s)$ of an equator-facing collector:

$$(1 - s) = 1 - \left(\frac{x_0 x + D_y - x_y}{x_0 D} \right), \tag{43}$$

where x_0 is the collector length, D is the collector width, and

$$x = d \tan \theta_t \sin \gamma_t, \tag{44}$$

$$y = d \tan \theta_t \cos \gamma_t, \tag{45}$$

where d is the depth of collector, θ_t the angle of incidence on the tilted equator-facing collector, and γ_t the azimuth angle of the tilted collector.

They also proposed a simplistic correction for diffuse radiation reaching the absorber due to reflections from the interior surfaces of the side walls. This correction may have the same order of magnitude of error as the conventional assumption of a constant shading factor. It should also be noted (Lior et al. 1977) that a shading factor like this, which simply uses the fractional unshaded area to multiply the overall equation for collector efficiency, is inherently in error (up to 500%, as shown in the computations performed by these authors) since it indiscriminately multiplies both the energy input and energy loss terms in the equation: Shaded areas indeed may collect less energy, but they continue to lose it. This fact was included in the computer program SOLSYS developed by Lior et al. (1977).

Shading may also occur due to various objects between the sun and the collector, including other collectors (e.g., in an array configuration with more than one row of collectors). Computation of the position, shape, and size of shaded areas based on the geometry of the obstructions and of the target area (e.g., the collector's absorber surface) and on the position of the sun is well understood (see. U.S. Post Office 1969; DOE-2 1981), although new and more effective techniques are being developed (see Budin and Budin 1982; Sassi et al. 1983). Jones and Burkhart (1981) developed analytically an extension of the Liu and Jordan (1961) model for insolation incident on a collector, for mutual shading by parallel rows of solar collectors, and pointed to an error in a previous analysis by Appelbaum and Bany (1979).

Progress Summary   *The extent and effect of collector shading can now be predicted correctly. This also allows optimal spacing of collector arrays where the collector mounting area is constrained, by allowing partial shading during some periods while producing more heat overall than could be obtained from fewer rows (which are placed farther apart and are thus always unshaded).*

4.4   Heat Transfer through the Back and Side

4.4.1   Collector Insulation

As shown figure 4.9, the heat loss through the back and sides of the collector is governed by the resistance due to heat conduction through the insulation R_{k11-13}, which is also usually predominant, in series with the parallel combination of the convective resistance R_{c13-a} and the radiative resistance R_{r13-b} between the exterior surface of the collector back and sides and the ambient. It is common practice to neglect the resistances to the ambient relative to the one through the thermal insulation and to assume that the temperature difference driving both back and edge losses is the same: T_{11} (or T_7) $- T_a$. The calculation of heat loss is thus straightforward (see Gilleland 1980).

Although the above-described approach is often adequate for design purposes, the oversimplified model used does not allow precise optimization of insulation and does not address adequately collector designs and installations for which the simplifications do not apply. One obvious error, as Tabor (1958) also pointed out, is in the description of the edge losses due to heat conduction through the side insulation. The absorber plate is located perpendicular to the edge insulation and is not parallel to it, and the heat flow from the absorber

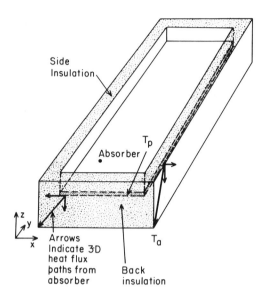

Figure 4.16
Three-dimensional conductive heat transfer from absorber through collector side- and back-insulation.

to the ambient is two-dimensional (three-dimensional in the corner regions), as shown in figure 4.16. Tabor (1958) computed correction factors to account for this and recommended that a good starting point for design is to specify the same thickness for the side insulation as selected for the back insulation.

Another error may arise due to the neglection of the convective and radiative exchange with the ambient. This may become important in areas where no wind is present at the collector and back edge (for the convective resistance) or where radiative exchange may become significant. The latter can occur if (1) the outer surface of the back and edges is at high temperature, either due to high absorber temperatures or smaller amount of insulation, (2) the temperature of the ambient surfaces (or sky) is relatively low, and (3) the temperature of the ambient surfaces and/or the albedo is high, thus actually acting to add energy to the collector through its back and sides.

Satcunanathan and Gandhidasan (1981) pointed out that for small angles of inclination, such as those used in low latitudes, natural convection in enclosures that are hot at the top and cold at the bottom is very small, and they have therefore recommended that the solid back insulation could be replaced with an air gap, preferably including a plate placed parallel to the absorber that would serve as radiation shield between the absorber and the back cover. Their experiments indicated that a collector with such an air gap insulation performed perhaps even a little better than a collector with fibrous back insulation.

Jones and Lior (1979) developed an insulation design procedure for solar heating systems and presented optimal insulation thickness selection graphs based on a present-value life-cycle cost analysis.

4.4.2 Double-Exposure Collectors

Going an important step beyond the idea of eliminating the solid insulation from the back of the collectors (see the discussion in subsection 4.4.1), Souka (1965) recommended construction of a collector that is glazed on both sides, with the exposure of the back of the absorber to insolation reflected from a mirror placed behind the collector. This almost doubles the amount of solar energy incident on the same collector. His experiments, as well as those by others (see Savery et al. 1976; Savery and Larson 1978) indicated significant improvement in total energy collected. Souka and Safwat (1966, 1969) have also developed a simplified theoretical thermal model of such collectors and have made recommendations for the optimal orientation of the collector and of the mirrors.

Boosting of the solar radiation incident on the front of conventional single-exposure flat plate collectors with flat mirror reflectors has been the subject of a number of studies, beginning with Shuman's work on a solar water-pumping system in Philadelphia in 1911 (see Tabor 1966; McDaniels et al. 1975; Grassie and Sheridan 1977; Baker et al. 1978; Kaehn et al. 1978; Larson 1980a, b). The results were encouraging, and the techniques used for the optical analysis can be also adapted to the optimization of double-exposure collector-mirror systems.

Progress Summary *Almost half of the collector surface area is located at its back. The back can be just insulated to reduce heat losses, or it can be designed to even add to the heat input to the collector, such as is done in double-exposure collectors. Although three-dimensional conduction analysis in the back and side insulation would provide more precise information for insulation optimization than the currently used one-dimensional calculations, enough was already known from the practical standpoint, so not much progress was needed in thermal theory and modeling in this area, and indeed little was done. The replacement of solid insulation in the back with a suppressed-convection air space warrants an economical feasibility study.*

Exposure of the back of the collector to solar radiation reflected from mirrors has conclusively shown a marked improvement in the thermal performance of such a double-exposure collector, but it was not made clear yet whether the additional costs associated with this installation, and the need to periodically adjust the mirror position and to maintain the mirror surface, can be justified.

Heat Transfer in Partially Evacuated Enclosures

It was obvious at least from the beginning of the century, when Emmet (1911) of the General Electric Company patented a tubular evacuated solar collector module, that vacuum between the absorber and the cover would reduce heat losses due to both convection and conduction and thus would improve efficiency significantly. Renewed interest was expressed following the work by Speyer (1965), who built and successfully tested several variations of a tubular evacuated solar collector, and Blum (Blum et al. 1973; Eaton and Blum 1975), who proposed, built, and tested flat plate evacuated collectors. The 1970s saw vigorous development of many types of evacuated solar collectors (see. Graham 1979), as shown in figure 4.3, and they also attained some market acceptance.

Aspects related to radiative transfer through the cylindrical cover to the absorber were discussed in subsection 4.2.1. Heat transfer in the absorber and the heated fluid were included in section 4.3. Overall performance analysis is described later in section 4.11.

Practically all of the analyses of evacuated solar collectors have assumed the existence of a perfect vacuum, that is, the absence of any conduction or convection in the evacuated space. This simplifies the analysis relative to that needed for non vacuated collectors and is correct for collectors that have been evacuated sufficiently to make these modes of heat transfer small enough to be negligible. For example, a vacuum of about 10^{-2} mmHg (absolute, or 10^{-5} Torr) is needed to reduce the conductivity of air to 1% of its value at atmospheric pressure, and this is indeed the vacuum that has been used in most of the collector designs. The costs of manufacturing evacuated collectors are, however, somehow proportional to the vacuum level that needs to be attained and maintained during the life of the collector. It is therefore of interest to determine overall heat transfer as a function of the absolute pressure (i.e., degree of vacuum) in the enclosure and also to understand the factors that may increase the pressure during the life of the collector, such as joint leakage, volatilization of internal components, and penetration of gases from the adjoining spaces (e.g., the ambient and the working fluid) through the vacuum enclosure (usually glass) into the evacuated space.

Lou and Shih (1972a, b), Thomas (1973, 1979), and Wideman and Thomas (1980) studied conduction heat transfer in rarefied (partial vacuum) between parallel plates and concentric cylinders and spheres and developed recommended equations.

Glasses are permeable to helium, which is present in the atmosphere at a partial pressure of about 4×10^{-3} Torr. This pressure is usually higher than that used in evacuated collectors (10^{-5} Torr), and if it penetrates into the collector and comes to its equilibrium pressure, it will cause a reduction in collector efficiency. Thomas (1981) presented a method for calculating the helium penetration rates and the consequent conduction heat flux. He concluded that penetration time constants of about 50 years can be expected if the glass envelope is kept close to ambient temperatures but that this would be reduced to a matter of a few months if the glass was operated at temperatures over 200°C. If helium came to its atmospheric equilibrium pressure in the collector, the conductive heat flux could rise from about 1% of the radiative flux, at 10^{-4} Torr, to about 25%.

If the correlations expressing the Nusselt number as a function of the Rayleigh (or Grashof) number just after convection onset are also correct for high vacuums, it is easy to determine the absolute pressure at which natural convection can be suppressed, as Eaton and Blum (1975) have done. They suggested that convection would be suppressed at pressures less than about 7 Torr for conditions typical to flat plate solar collectors operating at absorber temperatures up to 175°C. They also confirmed this experimentally in a qualitative way. It is important to note that convection can thus be suppressed even if the higher vacuums needed to suppress conduction have not been attained (or have slightly deteriorated in time due to leakage).

Progress Summary *Major progress was made in evacuated solar collectors. Such collectors existed only in concept at the turn of the 1970 decade, and yet they have begun competing effectively for a share of the market at the end of that decade. Understanding of the thermal theory of present-generation collectors of this type, and the ability to predict their behavior, became in that short time at least as good as that for conventional flat plate collectors. The primary aspects in which progress needs to be made, apart from the ever-present need for cost reduction, are in improving the energy collection rate $[(\alpha\tau)_{ef}]$ and perhaps in the understanding of heat transfer the in partial vacuum in such collector configurations.*

4.6   Heat Transfer from the Collector Exterior

Unless the collector is well shielded from wind, the convective thermal resistance $R_{c1\text{-}a}$ between its exterior and the ambient (figure 4.9) is dominated by forced convection due to wind flow over the collector (otherwise, the resistance is associated with natural convection). It is a quaint fact that the only way to calculate the heat transfer coefficient h_w due to forced convection over flat collector plates till fairly recently was by using the dimensional empirical correlation developed by Jurges in 1924 (see McAdams 1954) for flow parallel to a plate:

$$h_w = 5.7 + 3.8V, \qquad (46)$$

where V is the wind velocity, with all units in SI. The lack of characteristic length and the independence from properties and inclinations limit the validity of this correlation severely. The Russian literature shows the use of dimensionless correlations in the laminar flow regime, which also take angles of attack and yaw into account (Avezov et al. 1973a, b; Avezov and Vakhidov 1973).

Before development of improved predictive equations for convective wind effects is attempted, it must be realized that (1) wind varies with time in speed, direction, and turbulence, and its mean velocity changes with height, (2) wind arriving at the collector is affected by the topography upstream of the collector and surrounding it (see Kind et al. 1983; Kind and Kitaljevich 1985; Lee 1987), (3) sharp differences in wind speed and direction were found even on the face of each collector (Oliphant 1979), (4) free stream turbulence of the wind, in part generated by upstream and surrounding obstacles, has an important effect on heat transfer and can explain the difference between wind tunnel results with low free stream turbulence and results obtained in the natural environment, where the free stream turbulence may be 20% (based on the local velocity) and the convective heat transfer coefficient twofold higher (Test et al. 1981; Francey and Papaioannou 1985; Lee 1987), and (5) the shape of the leading edge of the collector affects convection at its surface downstream through such phenomena as separation, reattachment, and redevelopment (see Ota and Kon 1979; Test and Lessman 1980).

Sparrow and coworkers (Sparrow and Tien 1977; Sparrow et al. 1979, 1982) conducted a series of experiments in a wind tunnel, to develop correlations for convective heat transfer, using the naphtalene sublimation technique as analog to heat transfer. Square and rectangular plates of about 2 to 5 in. size were placed at different angles of attack and yaw, and both windward and leeward plate configurations were investigated for $2 \times 10^4 < \text{Re} < 10^5$ (laminar flow). They have developed a correlation for windward orientations:

$$j = 0.86\text{Re}_L^{-1/2}, \qquad (47)$$

where j is the Colburn j-factor $= \text{Nu}/(\text{RePr})$, and

$$Re_L = \frac{V(4A/C)}{v}, \tag{48}$$

where A is the area of the plate and C the length of its perimeter. They found practically no effect of angles of attack or yaw. With the collector leeward (wind blowing at its back) they found that for Reynolds numbers below 6×10^4 windward-face plates exhibit heat transfer coefficients about 10% higher than leeward-face ones, but this was reversed as Re exceeded 6×10^4: at Re $= 10^5$ the windward-face coefficient became 15% lower than the leeward-face one. They also determined that adding coplanar plates at the edges of the collector moves the highly convecting edge zones to these passive edge plates, and the heat loss can be reduced by up to about 10%.

The above correlation, as well as experimental results obtained by Kind et al. (1983) and Kind and Kitaljevich (1985) obtained in highly turbulent non-uniform flows generated in a wind tunnel, give heat transfer coefficients that may be as much as four-times lower than the Jurges correlation. Onur and Hewitt (1980) made convective heat transfer experiments with 6 in. models under a free jet and obtained results about 10% lower than those of Sparrow and coworkers. Kind and Kitaljevich (1985) also found that heat transfer coefficients for solar collectors mounted at an angle on a flat horizontal roof are 50% higher than those for collectors mounted flush with an inclined roof.

Truncellito et al. (1987) obtained numerical solutions for turbulent forced convection over a plate with an angle of attack for Reynolds numbers up to 3×10^5. They found that the Nusselt number increases slightly with the angle of attack and that the j-factor is the same for Re $\approx 3 \times 10^4$ as that predicted by equation (47), but it is increasingly larger as Re increases. Correlations of Nu as a function of Re, Pr, and the angle of attack were provided.

Lior and Segall (1986) have done experimental studies in a wind tunnel to determine convective heat transfer coefficients on a solar collector array composed of three parallel rows, all facing the wind, with variable spacing and inclination, and $4.8 \times 10^4 <$ Re $< 8.5 \times 10^5$. For the upstream plate the Colburn j-factor was found to be slightly higher than that predicted by equation (47). It was up to about 40% higher for the second plate, due to effects of the wake generated by the plate upstream, but only up to about 30% higher for the third plate, due to the flow pattern. It was a weak function of inclination and spacing for the two downstream plates.

Forced convection heat transfer information for external flow around cylindrical collectors can be found in the book by Zukauskas (1985) which deals exclusively with heat transfer from cylinders exposed to external flow.

Progress Summary   *Although important progress was made during the last decade in the understanding of forced convection over inclined plates for Reynolds numbers below 10^5, and in advancing beyond the limitations of the Jurges correlation, we have only begun to understand and to try to predict heat transfer due to wind in the natural environment and for realistic collector/surround geometries. The extension of the past work to larger Reynolds numbers, attainment of better agreement between results of different investigators, and the accounting for real geometries and the natural environment are needed.*

4.7   Transient Effects

The transient nature of solar radiation, ambient temperature, wind, and the heat load indicates that there may be merit in investigating the transient behavior of collectors, instead of using steady-state models such as the HW one. This is particularly important if the collectors have a large time constant, if frequent and strong variations in insolation occur in a region, or if the collector performance has a rapid influence on the ultimate heat load, on other parts of the solar system, or on system controllers. Klein et al. (1974) compared the zero-thermal-capacitance HW model with one-node (all thermal capacity lumped into one term) and multinode (transient energy balances for each component solved simultaneously) models that include the thermal storage term in the energy equation [equation (1)] for conventional flat plate collector parameters. They determined that the collector responded to step changes of the meteorological variables within a fraction of an hour and that therefore the zero-capacitance (steady-state) model is adequate when hourly (or longer-period) meteorological data are used. In other words, they recommended that transient effects need not be considered in performance modeling of conventional collectors. Wijeysundera (1976, 1978) developed a detailed transient model for an air-heating collector, and his results essentially concurred with those by Klein et al. (1974): A two-node model gave accurate results for collectors with up to three cover plates, a single-node model was satisfactory for collectors with one cover plate, but even a steady-state model was adequate if only hourly meteorological data were used.

Siebers and Viskanta (1978), de Ron (1980), Saito et al. (1984), and Kamminga (1985) modeled flat plate collectors by a set of at least three coupled equations, one each describing the transient energy balance in the fluid, each of the cover plates, and absorber. Whereas de Ron (1980) and Kamminga (1985) ignored heat conduction in the absorber and glass and thus ended up with these two equations being first-order ordinary differential, Siebers and

Viskanta (1978) and Saito et al. (1984) in addition consider conduction in the flow direction and thus have two (or three for two cover plates) second-order partial differential equations and one first-order partial differential equation. The latter approach produced excellent agreement with experimental data. Both de Ron (1980) and Kamminga (1985), however, introduced simplified linear models that can represent collector behavior well without the need to use the rigorous, complex models. Other researchers have pointed out that apart from the already-recognized importance of the heat capacity of the fluid, the heat transfer coefficient between the tubes and the heated fluid is an important parameter that should not be ignored in transient analysis and that the single-node model (see Klein et al. 1974) does not describe the transient behavior well. It was found that the transient temperature of the heated fluid in collectors in which the absorber is well-insulated from the cover plate (as it usually is) can be predicted well from models that use only the transient energy balance equations for the fluid and the absorber and ignore the transient terms in the equation for the cover plate.

Interested in investigating transient performance of evacuated tubular collectors, which have a large time constant, Mather (1982) applied a similar analysis to that of de Ron (1980) and obtained excellent agreement with experimental data. A similar model for the same purpose was developed later by Bansal and Sharma (1984). Morrison and Ranatunga (1980) developed a transient model for thermosyphon collectors and verified it experimentally.

Edwards and Rhee (1981) proposed a useful correction in the experimental determination of instantaneous efficiency of solar collectors that uses the time constant of the collectors determined by separate experiment (with no insolation).

In closing, it should be noted that good understanding of the theory of the transient behavior of collectors can also lead to the development of techniques for the rapid experimental determination of the parameters that characterize the collector and its performance. This could be used in collector performance testing and diagnostics and possibly in collector and component R&D.

Progress Summary   *As found in the early 1970s, steady-state models are adequate for describing the energy collection performance of conventional flat plate collectors, when hourly (or longer-period) meteorological data are used. Transient modeling is, however, necessary for collectors with large time constants (e.g., many of the evacuated-tube collectors) or when the transient combination of weather, insolation, load, and system operation are such as to require it. Very good transient models with that capability have been developed and verified during the last decade. Good understanding of the theory of the transient* *behavior of collectors can lead to the development of experimental techniques for rapid diagnostics and evaluation of collector performance.*

4.8 Thermal Design of Solar Concentrator Receivers

Receivers are essentially solar collectors too, and their thermal theory would therefore be reviewed here. On the other hand, the few that exist have been custom designed for the specific system in which they operate, and there isn't nearly as much information on their thermal modeling and optimization as available for solar collectors. The review of this subject would be rather brief.

From the viewpoint of thermal theory and modeling, solar concentrator receivers have many obvious similarities to conventional nonconcentrating solar collectors, but they also differ from them in several aspects:

1. For high concentration ratios it is often not practical to transfer the maximally focused radiant flux [about 250 Btu/ft^2 s (2.8 MW/m^2) as it enters or strikes the receiver, an order of magnitude greater than used in conventional fuel-fired boilers] into the working fluid even with the highest practical conductive/convective heat transfer coefficients between the exterior wall of the fluid conduit and the fluid itself, especially if the highest fluid outlet temperature is desired. An attempt to apply that flux to fluid conduits may result in poor efficiency and in hot spots that can damage the receiver. A typical remedy is to redistribute the radiant flux over a larger heat exchanger area inside the receiver once the beam entered it, while keeping the inlet aperture small to reduce radiative and convective losses.

2. Radiant energy exchange becomes dominant and requires much more precise calculation.

3. Due to the larger temperature differences between the receiver and the ambient, and in some cases due to the larger characteristic dimensions, the Grashof number reaches up to 10^{14}, and natural convection becomes highly turbulent and much more vigorous. For open cavities in wind, the forced convection Reynolds number at the same time may reach 10^7. This requires both theoretical and empirical heat transfer information, which is still quite scarce (see Abrams 1983; Siebers and Kraabel 1984).

4. The large temperature differences incurred require the consideration of the temperature dependence of the radiative and convective properties of the materials.

5. In contrast with nononcentrating collectors, the diffuse component of solar radiation usually needs not be considered in the thermal analysis.

A plate-tube design (with tubes spaced very closely, often touching each other) is commonly used for receivers. Tube flow patterns are determined by considerations of heat transfer, thermal stress, heated fluid quality (fraction of vapor), and cost (see Sobin et al. 1974). The plate-tube system gains energy from the solar flux and in cavity receivers also from irradiation and reflections from other surfaces that it views, and it loses heat by reradiation, natural convection, and possibly forced convection if exposed to wind. The useful heat is gained by the working fluid, which may exit in the same phase in which it entered, or phase change may occur during passage. The latter may be in a boiler, in which a subcooled liquid may first be brought to saturation temperature and then change phase into steam. Finally, the generated steam may be superheated before it exits the receiver.

Cavity Receivers (figure 4.4)   Cavity receivers are designed to minimize radiative losses by absorbing as much as possible of the incoming radiation into internal walls that do not view the opening. Due to the large temperature differences between the interior walls of the cavity and the ambient, and often the large size of the receiver, natural convection in the cavity can be vigorous, and it would carry some of the heat from the walls to the cavity aperture. That aperture is often open to the ambient because of the high temperatures that such windows may be exposed to, and it also reduces transmittance losses in the window.

Rozkov (1977) studied natural convection in cylindrical receivers. Humphrey and Jacobs (1981) developed a numerical model for predicting laminar flow and temperature fields in a small open cubical cavity subject to external wind, and they calculated the heat loss from the cavity. Since many receivers would usually incur turbulent convection, the model needs to be developed further and compared with experiments. Clausing (1983) developed a simplified engineering model to determine heat losses from such a cavity receiver, and it was found that his predictions were in close agreement with the experimental data of McMordie (1984) and Mirenayat (1981). He also developed a semiempirical correlation for the heat transfer coefficients between the inner surface of the cavity and the air for different surface angles. Later (Clausing et al. 1986) he determined from experiments that the area of the aperture and its location have a major influence on the overall Nusselt number for heat loss. Natural convection from a 2.2-m electrically heated cubical cavity was measured by Kraabel (1981). The aperture was vertical, comprising one side of the cube. The Grashof number in the experiments was varied between 9.4×10^{10} and 1.2×10^{12}. The experimental results were used to develop a correlation for the Nusselt number, and the correlation was reported to fit well also data obtained by Mirenayat (1981) down to $Gr = 5 \times 10^7$. An interesting result from Kraabel's correlation was that for surface temperatures above 420 K, the Nusselt number for the cavity was larger than that for a single vertical plate, and thus the total heat loss from a cavity can exceed that calculated from the sum of losses from the individual walls.

Wind effects on cavity receiver heat transfer have not yet been studied experimentally, and the theory is also in its infancy.

Radiative exchange inside cavity receivers, and conclusions for the improvement of their performance, have been calculated by McKinnon et al. (1965) and Grilikhes and Obtemperanskii (1969). In one cavity receiver, the receiver consisted of a conically wound tube, with the base (opening) of the cone acting as the aperture for the incoming concentrated solar flux (Rice et al. 1981). Rice et al. performed a thermal analysis on a rather simplified model of this receiver and determined its performance. As expected, smaller cone angles for the same size of the cone opening (base) provided higher receiver efficiency because of the reduction of the radiative loss. Convection inside the cone was treated in an oversimplified way, and the conclusions must be regarded as being primarily qualitative.

A detailed design of a 550-MW$_t$ output quad-cavity receiver (figure 4.4) for generating and superheating steam in a solar central power plant has been presented by Wu et al. (1983). In comparing this design to an external receiver for similar duty, the authors found that the losses for this cavity receiver were about half of those for the external receiver. It is noteworthy, though, that not enough is known yet about losses due to combined natural convection and wind effects from cavity receivers; they may be larger than estimated by these authors (see Hildebrandt and Dasgupta 1980). Harris and Lenz (1985) analyzed thermal performance of concentrator/cavity receiver systems and made recommendations for optimization.

A cylindrical cavity receiver with an aperture parallel to the axis of the cylinder (figure 4.17), which is suitable for line-focusing concentrators, was recommended and analyzed by Boyd et al. (1976), and a version of it was analyzed, built, and tested successfully in a parabolic-trough concentrator (Barra and Franceschi 1982), with further recommendations for optimization. CPC collectors may also be regarded in some sense as cavity receivers. Thermal analyses of heat transfer and natural convection in a CPC receiver were performed numerically by Abdel-Khalik et al. (1978), and thermal analysis for an evacuated glass-jacketed CPC receiver was performed by Thodos (1976).

It is important to note that there is a trade-off in cavity receiver designs between having a large receiver aperture, associated with a low concentration ratio, to capture all of the incoming energy, and having a small aperture (high concentration ratio) to minimize radiation and convection losses (see

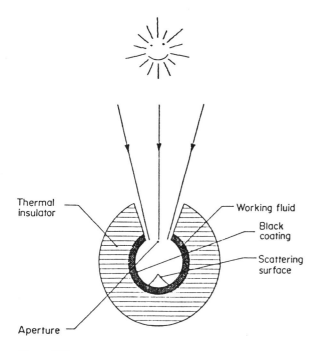

Figure 4.17
Cross section of the cylindrical cavity receiver. Source: Boyd et al. (1976).

Pons 1980). This fact couples the thermal analysis with the design of the concentrator.

Internal radiative exchange and heat transfer through the working fluid conduit walls into the fluid become quite complicated when optimized designs are sought, and the solution of such problems must be done numerically. A finite difference code for simulating heat transfer in cylindrical solar receivers, named "HEAP," was developed at the Jet Propulsion Laboratory (Lansing 1979). A program to analyze cavity radiation exchange ("CREAM") was developed at the University of Houston (Lipps 1983). Other programs have been developed by the Aerospace Corporation and modified by JPL (El Gabalawi et al. 1978) and by the Pacific Northwest Laboratory (Bird 1978), by the Solar Energy Research Institute (Finegold and Herlevich 1980), and by Sandia National Laboratories (MIRVAL: Leary and Hankins 1979; HELIOS: Vittitoe and Biggs 1981). Some of these programs also apply to external receivers.

External Receivers (figure 4.5)   The external receiver of Solar 1, the 10-MW solar thermal power plant at Barstow, California, is a once-through boiler of

7-m diameter and 12.5-m height. The average temperature of the exterior of the receiver is 600°C, and the wind velocities perpendicular to the receiver are 0 to 25 m/s. These conditions result in Reynolds numbers (based on diameter) of 0 to 10^8 and Grashof and Rayleigh numbers of 10^{12} to 10^{14}.

Siebers et al. (1982, 1983) conducted experiments on a 3 × 3 m electrically heated plate in a wind tunnel and developed correlations for natural, mixed, and forced convection in the laminar and turbulent regimes. Mixed convection was found to exist in the range $0.7 < Gr/Re^2 < 10.0$. A correlation was also developed to indicate the transition from laminar to turbulent flow. It should be noted that natural convection on this receiver is always turbulent, and it can be neglected relative to the wind-induced forced convection only when the wind velocity exceeds about 3 m/s. Since wind velocities vary between 0 and 25 m/s, there are many periods in which natural convection is the dominant convection mode. Mixed convection for these configurations was studied by Afshari and Ferziger (1983).

Details of the thermal analysis and overall design of an external cylindrical receiver to generate and superheat steam for a 100-MW_e solar central power plant (figure 4.5) has been presented by Yeh and Wiener (1984) and Durrant et al. (1982). The importance of heat losses due to wind and their effect on overall efficiency were determined as a result of the analysis: For example, when wind velocity increased from 0 to 13 m/s, the receiver efficiency decreased from about 91% to 86% (at an ambient temperature of 25°C). A general purpose computer program for the design of solar thermal central external receiver plants, DELSOL2, was developed by Sandia National Laboratories (Dellin et al. 1981).

External receivers, usually in the form of a tube coiled helically around a cylindrical mandrel, are also used with the fixed-mirror distributed-focus concentrator in which the concentrator is a fixed hemispherical dish and the receiver tracks the focal point. The tube carrying the heated fluid is exposed to a highly asymmetric radiative flux. Dunn and Vafaie (1981) have developed empirical correlations for the Nusselt number for both single and boiling water/steam flow inside the coil.

As stated earlier, external receivers in the form of a simple tube conducting the heated fluid are often used with line-focusing concentrators. The tube's centerline is placed at the focal line, and in many designs it is placed concentrically inside another (transparent) tube to insulate it by the annular air (or vacuum) gap formed thereby. Analyses of such receivers were done by Barra et al. (1978a, b), Deanda and Faust (1981), Harrison (1982), and May and Murphy (1983), and reviewed by Chiang (1982), and a numerical analysis of the effects of natural convection in the annular air gap on heat losses was performed by Ratzel et al. (1979).

Dirct absorption receivers, where the solar flux is absorbed directly in the working fluid, such as in powders, particles, and molten salt films, are in the research stage, and their thermal principles have been reviewed in section 4.3.6.

Progress Summary *Much progress has been realized in the thermal analysis, design, and construction of both small and large solar receivers. Information about wind effects and natural convection, separately and combined, on heat losses in large receivers is still inadequate. Interesting concepts of receivers that employ direct absorption of the solar flux into powder/air suspensions, solid particles, and falling liquid films are being explored.*

4.9 Solar Collector Arrays

It has been recognized in both theory and practice that arrays composed of a large number of collectors do not perform in a way that could be predicted by the simple addition of the performances of the individual collectors that compose them. Large arrays typically performed poorer, in some installations at half the efficiency predicted from single collector data. The reasons for the difference include increased heat and pressure losses from array piping, flow maldistribution among the collectors, increased heat losses due to wind because of the influence of the array on the wind at its location, much larger thermal capacitance, and sometimes mutual shading of collectors.

Supported by the USDOE, Lior and coworkers at the University of Pennsylvania developed a computer simulation program to address these array design and performance problems (Menuchin et al. 1981), and they have started the experimental study of wind effects on the thermal performance of collector arrays because no data in this area were available (Lior and Segall 1986, described in further detail in section 4.6). The problems associated with large array performance have recently brought about a workshop of the International Energy Agency (Bankston 1984) in which comprehensive information on operating, design, and research aspects of large arrays was presented and discussed. General agreement was obtained among the workshop attendees that many systems operated at half of the predicted efficiency, or worse.

The objective of the work at the University of Pennsylvania was to optimize design of collector arrays by determining the best configuration of collector rows as a function of available collector mounting area, interrow spacing, row orientation, collector inclination and height, wind effects, and system cost.

Two comprehensive computer programs were developed for the simulation and optimization: SOLRAY, which computes the hydrodynamics of flow in arbitrarily piped (any parallel series combination) solar collector arrays, with the inclusion of both inertia and friction effects (based on the work by Jones and Lior 1978), and the University of Pennsylvania program SOLSYS (Edelman et al. 1977) which computes the thermal efficiency of the arrays by using as input the individual flow rates computed by SOLRAY. SOLSYS can also compute the thermal performance of partially shaded collectors. From the information provided by SOLSYS, SOLRAY then determines overall energy efficiencies and costs of different solar collector array configurations, to find the optimum. Heat losses from interconnecting pipes could be calculated from another program developed by Jones and Lior (1979).

For the size and piping of typical flat plate solar collectors, it was found (Menuchin et al. 1981) that the flow in the parallel flow dual-manifold system is friction dominated; that is, the flow rate is lowest in the inner collectors. Since parallel flow dual-manifold systems provide more uniform flow distribution than reverse flow ones, only parallel flow was considered in the optimization. The maldistribution increases with the spacing between the parallel piped collectors and with their number. It also increases as the collector pressure drop is decreased and as the ratio between the diameter of the tube connecting the collector to the manifold and the diameter of the manifold is increased. A life-cycle present-value cost analysis (for 1978 conditions, and includes also both energy and capital costs of pumping) showed that the cost has a minimum for a given number of collectors piped in parallel, which for the total number of collectors considered (48, 96, 192, and 288), consisted of 16 to 24 collectors. Other information useful to array design was also provided.

Using a simple HW model for collector performance and considering a single parallel piped row, Culham and Sauer (1984) suggested that flow imbalance in arrays would have little effect as long as the flow in the collector does not fall below 35% of the recommended design value. Mansfield and Eden (1978) made recommendations on the ways to use thermography for determining malfunctions or flow maldistribution in collector arrays.

Array piping and configuration is of great importance in distributed concentrator solar thermal heat and power systems, particularly because of the high temperature and the greater spacing between the units. A number of computer programs have been developed for economic optimization of the piping of such systems (see Barnhart 1979; Fujita et al. 1982).

It was recognized (Eck et al. 1984) that differences between array performance and single-collector performance predictions may well arise also from the fact that the standard test conditions for collectors (e.g., that by *ASHRAE*

Standard 93-77) are not representative of operating conditions. For example, the standard for testing collectors specifies solar radiation that is often higher than seen by the operating collectors, and wind speeds and incidence angles that are lower. McCumber and Weston (1979) have indeed demonstrated that if the highly scattered field data for arrays is filtered from points measured at conditions beyond those required by the ASHRAE test standard, it clusters well around the straight-line HW equation. The standard also specifies only "instantaneous" rather than all-day efficiency, and requires quasi-steady operating conditions. It is therefore worthwhile to explore the development and use of standard performance characterization methods that give more realistic long-term operating results.

Progress Summary   *The fact that large collector arrays perform worse than predicted by single-collector test data was foreseen and has indeed materialized in most such installations. The basis for computer programs that could predict array behavior correctly or optimize the design has been developed, and the programs should be enhanced as needed and made available to designers. A basis for knowledge of wind effects on heat losses from arrays has also been established (for Re $< 8.5 \times 10^5$, parallel flow), but the work should be expanded to address the entire spectrum of wind conditions, and validated with full-scale systems in the natural environment. Flat plate collector performance standards, such as ASHRAE 93-77, were found to produce performance expectations that cannot be met in field operation of large arrays and should be reviewed and modified to produce more realistic results.*

4.10   Collector Performance Sensitivity to Design Parameters

Parametric studies to determine collector performance sensitivity to its components (number of covers, their thickness, tube dimensions, tube-to-fluid heat transfer coefficient, plate spacings, emittance of cover plates, emittance of absorber plate, absorptance of absorber plate, thermal conductivity of cover plates, thermal conductivity and thickness of the absorber plate, and thermal resistance of the insulation), to operating factors (type of fluid and its inlet temperature and flow rate), and to meteorological variables (insolation, ambient temperature, wind velocity, and sky temperature) have been performed by a number of researchers (Test 1976; Wolf et al. 1981; Arafa et al. 1978), using more detailed models than HW. The highlights of the sensitivity to the most influential parameters are presented in this section.

For a single-glazed water-heating collector, increase of the tube/fluid heat transfer coefficient up to about 300 kJ/m^2 hr K improves efficiency significantly and has little effect thereafter. Similarly increasing flow rate (at constant tube/fluid heat transfer coefficient) up to 20 kg/m^2 hr improves efficiency significantly and has little effect thereafter. The (conductivity × thickness) product of the absorber has an important effect: Its reduction from 0.5 W/K to 0.001 W/K typically reduces the efficiency by at least one-third. The efficiency increases strongly and linearly with absorber plate absorptance and decreases with emittance. Lowering the emittance of the cover plate from 0.95 to 0.6 improves the efficiency by at least 10%. The effect of insulation (fiberglass) thickness depends on the ambient temperature, but little is gained beyond 5 cm.

Insolation, ambient temperature and fluid inlet temperature have an important influence on collector performance, as expected, and wind has a moderate effect.

For a double-glazed collector the main difference is that there is weaker dependence on absorber emittance. The number of glazings has a major effect, but it depends strongly on the temperature difference between the absorber and the ambient.

Studies of the effect of the type of fluid on collector efficiency were conducted by Youngblood et al. (1979) who used four different fluids: water, ethylene-glycol/water solution (50% by weight), a silicone based heat transfer fluid (Syltherm 444), and a synthetic hydrocarbon (Therminol 44). Each of the fluids was tested in four types of commercial flat plate collectors. It was found that relative to water the other liquids reduced the collector efficiency by (1) 2%–4% for the ethylene-glycol/water solution, (2) 3%–10% for Syltherm 444, and (3) 3.5% for Therminol 44.

Progress Summary *The advanced models developed for collector analysis allow good quantitative sensitivity analysis for the purposes of collector research, development, and design.*

4.11   Summary: A Thermal Modeling Guide and Theory Outlook

Conventional flat plate collector design and performance determination can still be performed quite successfully by using the Hottel-Whillier lumped model, but with up-to-date values for the required coefficients as described in the preceding sections. The most recent and probably most accurate correlation for the overall top heat loss coefficient is the one by Garg and Datta (1984). More detailed information, and somewhat better modeling, can be obtained by performing individual energy balances for each cover and solving the equations simultaneously, but with convective coefficients that are now much more accurate and possibly with a more precise description of the radiant exchange. For collector R&D, new configurations, and materials, including coatings for the modification of radiative properties, two- (or even three-) dimensional models of the Navier-Stokes and energy equations need to be solved as a conjugate problem if the most accurate representation is needed.

From the thermal modeling standpoint, evacuated collectors are in one respect easier to model since no conduction or convection between the absorber and cover exists (if adequately high vacuum can be assumed). In another respect they are somewhat more difficult to model due to the radiative transfer in the cylindrical geometry, which is often internally asymmetric (see Saltiel and Sokolov 1982; Garg et al. 1983). Bhowmik and Mullick (1985) developed an expression for the overall top loss coefficient for such collectors, for use in an HW lumped model. Several more detailed thermal models have been developed (Mather 1982; Behrendorff and Tanner 1982; Rahman et al. 1984; Banzal and Sharma 1984; the first and last ones transient) to serve as a good start for more precise modeling.

Although thermal modeling of collectors was in the past restricted to steady state, good transient models have been developed in the last decade to be used where appropriate.

Receiver analysis is typical done by using three-dimensional finite difference or element programs, and probably the only uncertainty is in the effects of natural, forced, and mixed convection, especially in large cavity receivers.

Notes

1. One may note that the energy input is primarily in the form of thermal radiation (electromagnetic wave energy), and the losses are partially radiative and partially thermal. The useful energy output is assumed to be only heat, and so is the energy storage term.

2. Radiation on the horizontal surface serves as the basis because solar radiation is usually measured and reported in the horizontal plane.

References

Abdel-Khalik, S. I. 1976. Heat removal factor for a flat-plate solar collector with a serpentine tube. *Solar Energy* 18: 59–64.

Abdel-Khalik, S. I., H.-W. Li, and K. R. Randall. 1978. Natural convection in compound parabolic concentrators—A finite-element solution. *J. Heat Transfer* 100, 2: 199–204.

Abdelrahman, M. A., P. Fumeaux, and P. Suter. 1979. Study of solid-gas suspensions used for direct absorption of concentrated solar radiation. *Solar Energy* 22: 45–61.

Abrams, M. 1983. The status of research on convective losses from solar central receivers. Sandia Report 83-8224. Sandia National Laboratories, Livermore, CA.

Abramzon, B., I. Yaron, and I. Borde. 1983. An analysis of a flat-plate solar collector with internal boiling. *ASME J. Solar Energy Eng.* 105: 454–460.

Afshari, B., and J. H. Ferziger. 1983. Computation of orthogonal mixed convection heat transfer. *Proc. ASME-JSME Thermal Engineering Joint Conference.*, vol. 3, pp. 169–173.

Al-Arabi, M., and M. M. El-Rafael. 1978. Heat transfer by natural convection from corrugated plates to air. *Int. J. Heat Mass Transfer* 21: 357–359.

Al-Tamimi, A. I., and J. A. Clark. 1983. Thermal analysis of a solar collector containing a boiling fluid. *Progress in Solar Energy*, vol. 6, ASES Annual Meeting, pp. 319–324.

Ambrosone, G., A. Andretta, F. Bloisi, S. Catalanotti, V. Cuomo, V. Silvestrini, and L. Viçari. 1980. Long-term performance of flat-plate solar collectors. *Appl. Energy* 7, 1-3: 119–128.

Appelbaum, J., and J. Bany. 1979. Shadow effects of adjacent solar collectors in large scale systems. *Solar Energy* 23: 497–507.

Arafa, A., N. Fisch, and E. Hahne. 1978. A parametric investigation on flat-plate solar collectors. *Sun, Mankind's Future Source of Energy*. Proc. 1978 ISES Congress, New Delhi, India. Elmsford, NY: Pergamon, pp. 917–923.

Arai, N., Y. Itaya, and M. Hasatani. 1984. Development of a "volume heat-type" type solar collector using a fine-particle semitransparent liquid suspsension (FPSS) as a heat vehicle and heat storage medium. *Solar Energy* 32, 2: 49–56.

Arai, N., Y. Kato, M. Hasatani, and S. Sugiyama. 1980. Unsteady heat transfer in an optically thick semitransparent liquid layer heated by radiant heat source. *Heat Transfer—Japanese Research* 9: 22–31.

Arnold, J. N., I. Catton, and D. K. Edwards. 1975. Experimental investigation of natural convection in inclined rectangular regions of differing aspect ratios. ASME paper no. 75-HT-62. AIChE-ASME, Heat Transfer Conference, San Francisco, CA, August 11–13.

Arnold, J. N., D. K. Edwards, and I. Catton. 1977. Effect of tilt and horizontal aspect ratio on natural convection in rectangular honeycomb solar collectors. *J. Heat Transfer* 99: 120–122.

Avezov, R., A. Akhmadaliev, and N. A. Kakhdarov. 1973a. Effect of the angle of attack on the efficiency and heat Transfer of the glass cover of a solar installation under laminar-flow conditions. *J. Appl. Solar Energy* 9, 5: 45–48.

Avezov, R., A. Azizov, S. Khatamov, and G. Sharipov. 1973b. Experimental study of the aerodynamic resistance of a shingle heat-strong unit. *J. Appl. Solar Energy* 9, 5: 49–52.

Avezov, R. R., and A. T. Vakhidov. 1973. Choice of controlling dimension in the oblique flow past the glass surface of a solar installation. *J. Appl. Solar Energy* 9, 6: 90–91.

Bajura, R. A., and E. H. Jones. 1976. Flow distribution manifolds. *J. Fluids Eng.* 98: 654–666.

Baker, L. H. 1967. Film heat-transfer coefficients in solar collector tubes at low Reynolds numbers. *Solar Energy* 11: 78–85.

Baker, S. H., D. K. McDaniels, H. D. Kaehn, and D. H. Lowndes. 1978. Time integrated calculation of the insolation collected by a reflector-collector system. *Solar Energy* 20: 415–417.

Bando, T., M. Nishimura, M. Kuraishi, T. Kasuga, and M. Hasatani. 1986. Effect of optical depth on outdoor performance of "volume heat trap" type solar collector. *Heat Transfer—Japanese Researach* 15: 57–71.

Bankston, C. A., comp. 1984. *Proc. International Energy Agency Workshop on the Design and Performance of Large Solar Thermal Collector Arrays*. San Diego, CA, June 10–14. Solar Energy Research Institute report SERI/SP-271-2664.

Bansal, N. K., and A. K. Sharma. 1984. Transit theory of a tubular solar energy collector. *Solar Energy* 32: 67–74.

Barnhart, J. S. 1980. ETRANS: An energy transport system optimization code for distributed networks of solar collectors. Pacific Northwest Laboratory report PNL-3327.

Barra, O., M. Conti, E. Santamata, R. Scarmozzion, and R. Visentin. 1977. Shadows' effect in a large scale solar plant. *Solar Energy* 19: 759–762.

Barra, O. A., M. Conti, L. Correra, and R. Visentin. 1978a. Thermal regimes in a primary fluid heated by solar energy in a linear collector. *Il Nuovo Cimento* 1C: 185–195.

Barra, O. A., M. Conti, L. Correra, and R. Visentin. 1978b. Transient temperature variations in the primary network of a solar plant. *Il Nuovo Cimento* 1C: 167–184.

Barra, O. A., and L. Franceschi. 1982. The parabolic trough plants using black body receivers: Experimental and theoretical analyses. *Solar Energy* 28: 163–171.

Beard, J. T., F. A. Iachetta, L. U. Lilleleht, F. L. Huckstep, and W. B. May. 1978. Design and operation influences on thermal performance of "Solaris" solar collector. *ASME J. Eng. for Power* 100: 497–502.

Beard, J. T., F. A. Iachetta, R. F. Messer, F. L. Huckstep, and W. B. May. 1977. Performance and analysis of an open fluid-film solar collector. *Proc. AS/ISES Ann. Meet.* 1: 126–129.

Beckman, W. A. 1968. Radiation and convection heat transfer in a porous bed. *ASME J. Eng. for Power* 90: 51–54.

Behrendorff, M. J., and R. I. Tanner. 1982. Transient performance of evacuated tubular solar collectors. *J. Solar Energy Eng.* 104: 326–332.

Bennet, D. L., and J. C. Chen. 1980. Forced convective boiling in vertical tubes for saturated pure components and binary mixtures. *AIChE J.* 26: 454–461.

Bevill, B., and H. Brandt. 1968. A solar energy collector for heating air. *Solar Energy* 12: 19–36.

Bhowmik, N. C., and S. C. Mullick. 1985. Calculating of tubular absorber heat loss factor. *Solar Energy* 35: 219–225.

Bird, S. 1978. Modification of the JPL solar thermal simulation code for use in the PNL small solar thermal power system analysis. Battelle Pacific Northwest Laboratory, Richland, WA.

Black, A. W., and E. M. Sparrow. 1967. Experiments on turbulent heat transfer in a tube with circumferentially varying thermal boundary conditions. *J. Heat Transfer* 89: 258–268.

Blanpied, M., G. Clark, and J. S. Cummings. 1982. Effect of tilt angle on infrared radition received on tilted surfaces. *Annual Meeting American Solar Energy Society*, vol. 5, pt. 2 of 3, pp. 673–678.

Bliss, R. W., Jr. 1959. The derivations of several "plate-efficiency factors" useful in the design of flat-plate solar heat collectors. *Solar Energy* 3, 4: 55–64.

Bliss, R. W., Jr. 1955. Multiple gauge flat plate solar air heaters. *Proc. World Symp. on Applied Solar Energy*. Phoenix, AZ, pp. 151–158.

Blum, H. A., J. M. Estes, and E. E. Kerlin. 1973. Design and feasibility of flat plate solar collectors to operate at 100–150°C. Paper no. E18. International Solar Energy Society Conference, Paris.

Bohn, M. S. 1985. Direct absorption receiver. *Proc. Solar Thermal Research Program Annual Conference*. SERI/CP-251-2680, DE85002938, pp. 89–96.

Bol, K., and M. Lang. 1978. Two-phase working fluid in solar collectors. AIAA paper N354.

Borgese, D., G. Dinelli, J. J. Faure, J. Gretz, and G. Schober. 1984. Eurelios, the 1-MW(el) heliostatic power plant of the European community program. *ASME J. Solar Energy Eng.* 106: 66–77.

Boyd, D. A., R. Gajewski, and R. Swift. 1976. A cylindrical blackbody solar energy receiver. *Solar Energy* 18: 395–401.

Buchberg, H., I. Catton, and D. K. Edwards. 1976. Natural convection in enclosed spaces—A review of applications to solar energy collection. *J. Heat Transfer* 98, 2: 182–188.

Buchberg, H., and D. K. Edwards. 1976. Design considerations for solar collectors with cylindrical glass honeycomb. *Solar Energy* 18: 182–188.

Buchberg, H., O. A. Lalude, and D. K. Edwards. 1971. Performance characteristics of rectangular honeycomb solar—Thermal converters. *Solar Energy* 13: 193–221.

Budin, R., and L. Budin. 1982. A mathematical model of shading calculation. *Solar Energy* 29, 4: 339–349.

Cane, R. L. D., K. G. T. Hollands, G. D. Raithby, and T. E. Unny. 1977. Free convection heat transfer across inclined honeycomb panels. *J. Heat Transfer* 99: 86–91.

Catton, I. 1978. Natural convection in enclosures. *6th International Heat Transfer Conference.* Montreal, Canada, pp. 13–30.

Chan, A. M. C., and S. Banerjee. 1979a. Three-dimensional numerical analysis of transient natural convection in rectangular enclosures. *J. Heat Transfer* 101: 114–119.

Chan, A. M. C., and S. Banerjee. 1979b. A numerical study of three-dimensional roll cells within rigid boundaries. *J. Heat Transfer* 101: 233–237.

Chao, P., H. Ozoe, and S. W. Churchill. 1981. The effect of a non-uniform surface temperature on laminar natural convection in a rectangular enclosure, *Chem. Eng. Commun.* 9: 245–254.

Chao, P. K., H. Ozoe, N. Lior, and S. W. Churchill. 1983a. The effect of partial baffles on natural convection in an inclined rectangular enclosure. Part I: Experimental observations. *Chem. Eng. Fund.* 2, 2: 39–49.

Chao, P. K., H. Ozoe, N. Lior, and S. W. Churchill. 1983b. The effect of partial baffles on natural convection in an inclined rectangular enclosure. Part II: Numerical solution. *Chem. Eng. Fund.* 2, 2: 38–49.

Charters, W. W. S., and L. F. Peterson. 1972. Free convection suppression using honeycomb cellular materials. *Solar Energy*, 13: 353–361.

Chen, J, C. 1966. Correlation for boiling heat transfer to saturated fluids in convective flow. *Ind. Eng. Chem. Process Design Develop.* 5: 322–329.

Cheng, K. C., and S. W. Hong. 1972. Effect of tube inclination on laminar convection in uniformly heated tubes for flat-plate solar collectors. *Solar Energy* 13: 363–371.

Chiang, C. J. 1982. Thermal receiver designs for line-focusing solar collectors. Sandia National Laboratories report. SAND81-1862.

Chiou, J. P. 1982. The effect of nonuniform fluid flow distribution on the thermal performance of solar collector. *Solar Energy* 29, 6: 487–502.

Chiou, J. P. 1980. The effect of longitudinal heat conduction on the thermal performance of the flat plate solar collector. ASME paper 80-C2/Sol-5.

Chiou, J. P. 1979. Noniterative solution of heat transfer equation of fluid flow in solar collector. ASME paper 79-WA/Sol-24, New York, 6 pp.

Chiou, J. P., M. M. El-Wakil, and J. A. Duffie. 1965. A slit-and expanded aluminum-foil, matrix solar collector. *Solar Energy* 9, 2: 73–80.

Chow, S. P., G. L. Harding, B. Window, and K. J. Cathro. 1984. Effect of collector components on the collection efficiency of tubular evacuated collectors with diffuse reflectors. *Solar Energy* 32: 251–262.

Christopher, D. M., and J. T. Pearson. 1982. Convectice heat transfer in a louvered air-heating solar collector. ASME paper 82-HT-41, 9 pp.

Clausing, A. M. 1983. Convective losses from cavity solar receivers—Comparisons between analytical predictions and experimental results. *J. Solar Energy Eng.* 105: 29–33.

Clausing, A. M., L. D. Lister, and J. M. Waldvogel. 1986. An experimental investigation of convective losses from cavity solar receivers. *ASES Ann. Meet.*, pp. 248–251.

Cobble, M. H. 1964a. Irradiation into transparent solids and the thermal trap effect. *Franklin Institute J.* 278: 383–393.

Cobble, M. H. 1964b. Heating a fluid by solar radiation. *Solar Energy* 8, 2: 65–68.

Cobble, M. H., P. C. Fang, and E. Lumsdaine. 1966. Verification of the theory of the thermal trap. *Franklin Institute J*. 282: 102–107.

Cole-Appel, B. E., and R. D. Haberstroh. 1976. Performance of air-cooled flat plate collectors. *Proc. ISES/SESC Joint Meeting "Sharing the Sun"* 2: 94–106.

Cole-Appel, B. E., G. O. G. Löf, and L. E. Shaw. 1977. Performance of air-cooled flat plate collectors. *Proc. 1977 AS/ISES Ann. Meet.*, pp, 2.6–2.10.

Cooper, P. I., and R. V. Dunkle. 1981. A Non-linear flat-plate collector model. *Solar Energy* 26: 133–140.

Culham, R., and P. Sauer. 1984. The effects of unbalanced flow on the thermal performance of collector arrays. *ASME J. Solar Energy Eng*, vol. 106, pp. 165–170.

Davis, S. B. 1982. Ceramic high temperature receiver. *Proc. Parabolic Dish Solar Thermal Power Annual Program Review*. Report DOE/JPL-1060-52, pp. 247–256.

Deanda, L. E., and M. Faust. 1981. Steam rankine solar receiver, phase 2. Final Technical Report. Report NASA-CR-168656; NAS 1 26-168656, 95 pp.

Dellin, T. A., M. J. Fish, and C. L. Yang. 1981. User's manual for DELSOL2: A computer code for calculating the optical performance and optimal system design for solar-thermal central-receiver plants. Sandia National Laboratories report SAND-81-8237, 76 pp.

de Graaf, J. G. A., and E. F. M. Van der Held. 1954. The relation between the heat transfer and the convection phenomena in enclosed plane air layers. *Appl. Sci. Res*. A4: 460–461.

de Graaf, J. G. A., and E. F. M. Van der Held. 1953. The relation between the heat transfer and the convection phenomena in enclosed plain air layers. *Appl. Sci. Res*. A3: 393–409.

de Ron, A. J. 1980. Dynamic modelling and verification of a flat-plate solar collectors. *Solar Energy* 24, 2: 117–128.

de Vahl Davis, G. 1986. Finite difference methods for natural and mixed convection in enclosures. *Heat Transfer 1986, Proc. Eighth International Heat Transfer Conf.*, Washington, DC: Hemisphere, pp. 101–109.

de Winter, F. 1975. Solar energy and the flat plate collector—An anotated bibliography. ASHRAE report S-101, New York (now Atlanta). February 1975.

Diab, M. R., J. T. Pearson, and R. Viskanta. 1980. A two-dimensional analysis of flat plate air-heating solar collectors. ASME paper no. 80-HT-117. *Joint ASME/AIChE Nat. Heat Transfer Conf.*, Orlando, FL, 9 pp.

Diaz, L. A., and N. V. Suryanarayana. 1981. Temperature variation in the absorber plate of an air heating flat plate collector. *J. Solar Energy Engineering* 103, 2: 153–157.

Dietz, A. G. H. 1963. Diathermaneous materials and properties of surfaces. In A. M. Zarem and D. D. Erway, eds., *Introduction to the Utilization of Solar Energy*. New York: McGraw-Hill, ch. 4.

DOE-2 Reference Manual Version 2.1A. 1981. Los Alamos Scientific Laboratory report LA-7689-M.

Downing, R. C., and V. H. Waldin. 1980. Phase-change in solar hot water heating using R-11 and R-114. *Trans. ASHRAE* 86, pt. 1: 848–856.

Duffie, J. A., and W. A. Beckman. 1980. In *Solar Engineering of Thermal Processes*. New York: Wiley, ch. 8.

Duffie, J. A., and W. A. Beckman. 1974. In *Solar Energy Thermal Processes*, New York: Wiley, chs. 4–8.

Dunkle, R. V., and E. T. Davey. 1970. Flow distribution in solar absorber banks. *Proc. ISES Conf.*, Melbourne, Australia.

Dunn, J. R., and F. N. Vafaie. 1981. Forced convection heat transfer for two-phase helical flow in a solar receiver. ASME, paper 81-WA/HT-13. Winter Annual Meeting, 9 pp.

Durrant, O. W., T. J. Capozzi, and R. H. Best. 1982. The development and design of steam/water central solar receivers for commerical application. *J. Solar Energy Eng.* 104: 173–181.

Eaton, C. B., and H. A. Blum. 1975. The use of moderate vacuum environment as a means of increasing the collection efficiencies and operating temperatures of flat plate solar collectors. *Solar Energy* 17: 151–158.

Eck, T. F., T. L. Logee, and S. M. Rossi. 1984. Performance of collectors in the national solar data network. *Proc. International Energy Agency Workshop on the Design and Performance of Large Solar Thermal Collector Arrays*, C. A. Bankston, comp. Solar Energy Research Institute report SERI/SP-271-2664. San Diego, CA, pp. 116–135.

Edwards, D. K. 1977. *Solar Collector Design*. Philadelphia: Franklin Institute Press.

Edwards, D. K. 1977. Solar absorption by each element in an absorber-coverglass array. *Solar Energy* 19, 4: 401–402.

Edwards, D. K., J. N. Arnold, and I. Catton. 1976. End-clearance effects on rectangular-honeycomb solar collectors. *Soler Energy* 18: 253–257.

Edwards, D. K., J. N. Arnold, and P. S. Wu. 1979. Correlations for natural convection through high L/D rectangular cells. *J. Heat Transfer* 101: 741–743.

Edwards, D. K., snd S. J. Rhee. 1981. Experimental correction of instantaneous collector efficiency for transient heating or cooling. *Solar Energy* 26: 267–270.

Edwards, D. K., and W. M. Sun. 1971. Effect of wall radiation on thermal instability in a vertical cylinder. *Int. J. Heat Mass Transfer* 14: 15–18.

El Gabalawi, N., G. Hill, J. Bowyer, and M. Slonski. 1978. A modularized computer simulation program for solar thermal power plants. Jet Propulsion Laboratory report JPL-5102-80.

Elsayed, M. M. 1984. Calculation of radiation properties of a stack of partially transparent plates with and absorber plate. *ASME J. Solar Energy Eng.* 106: 370–372.

Elsherbiny, S. M., K. G. T. Hollands, and G. D. Raithby. 1980. Free convection across inclined air layers with one surface V-corrugated. *J. Heat Transfer* 100: 410–415.

Elsherbiny, S. M., G. D. Rathby, and K. G. T. Hollands. 1982. Heat transfer by natural convection across vertical and inclined air layers. *J. Heat Transfer* 104: 96–102.

Emmet, W. L. R. 1911 (January 3). Apparatus for utilizing solar heat. U.S. patent no. 980,505.

Farouk, B., and S. I. Guceri. 1982. Laminar and turbulent natural convection in the annulus between horizontal concentric cylinders. *J. Heat Transfer* 104: 631–636.

Felland, J. R., and D. K. Edwards. 1978. Solar and infrared radiation properties of parallel plate honeycomb. *AIAA J. Energy* 2, 5: 309–317.

Felske, J. D. 1979. Analysis of an evacuated cylindrical solar collector. *Solar Energy* 21: 567–570.

Finegold, J. G., and F. Ann Herlevich. 1980. BALDR-1: A solar thermal system simulation. Solar Energy Research Institute report SERI/TP-351-464, 6 pp.

Fishenden, M., and O. A. Saunders. 1950. *Introduction to Heat Transfer*. Oxford: Oxford University Press.

Fraikin, M. P., T. J. Portier, and C. J. Fraikin. 1980. Application of a k-8 turbulence model to an enclosed buoyancy-driven recirculating flow. ASME paper 80-HT-68.

Francey, J. L., and J. Papaioannou. 1985. Wind related heat losses of a flat plate collector. *Solar Energy* 35: 15–19.

Francia, G. 1962. A new collector of solar radiant energy-theory and experimental verification—Calculation of the efficiencies. SAE paper no. 594B. National Aerospace Engineering and Manufacturing Meeting, Los Angeles, CA, October 8–12, 19 pp.

Fujita, T., W. Revere, J. Biddle, H. Awaya, and D. DeFranco. 1982. Design options for low cost thermal transport piping networks for point focusng parabolic dish solar thermal systems. *Proc. ASES Ann. Meet.*, pp, 339–343.

Garg, H. P. 1982. *Treatise on Solar Energy: Fundamentals of Solar Energy*. Vol. 1. New York: Wiley.

Garg, H. P., R. A. Shukla, R. C. Agnihotri, and S. Chakravertty, and Indrajit. 1983. Study of transmittance in an evacuated flat plate collector with double cylindrical glass cover. *J. Energy Convers. Mgmt.* 23, 1: 33–36.

Garg, H. P., V. K. Sharma, and B. Bandyopadhyay. 1982. Transient analysis of a solar air heater of the second kind. *Energy Convers. Mgmt.* 22: 47–53.

Garg, H. P., and Datta, G. 1984. The top loss calculation for flat plate solar collectors. *Solar Energy* 32, 1: 141–143.

Gilleland, F. W. 1980. Insulation for solar collectors important to heat loss control. *Solar Eng.*: 20–21.

Graham, B. J. 1979. A survey and evaluation of current design of evacuated collectors. Final technical report ALO/5350-1 from Trident Engineering Associates to the USDOE, 178 pp.

Grassie, S. L., and N. R. Sheridan. 1977. The use of planar reflectors for increasing the energy yield of flat-plate collectors. *Solar Energy* 19: 663–668.

Grilikhes, V. A., and F. V. Obtemperanskii. 1969. Analysis of radiative heat exchange processes in cylindrical cavity-type collectors of solar power plants. *Geliotekhnika* 5: 40–48.

Guthrie, K. I., and W. W. S. Charters. 1982. An evaluation of a transverse slatted flate plate collector. *Solar Energy* 28: 89–97.

Harris, J. A., and T. G. Lenz. 1985. Thermal performance of solar concentrator/cavity receiver systems. *Solar Energy* 34: 135–142.

Harrison, T. D. 1982. Program for predicting thermal performance based on test data of low- to medium-temperature line-focusing, concentrating solar collectors. Sandia National Laboratories report SAND-82-0092, 20 pp.

Hasatani, M., N. Arai, Y. Bando, and H. Nakamura. 1982. Collection of thermal radiation by a semitransparent fluid layer flowing in an open channel. *Heat Transfer-Japanese Research* 11: 17–30.

Hassan, K. 1979. Heat transfer through collector glass covers. *Proc. ISES Silver Jubilee Congress*, K. Boer and B. Glenn, eds. Elmsford, NY: Pergamon, pp. 312–316.

Hatfield, D. W., and D. K. Edwards. 1982. Effects of wall radiation and conduction on the stability of a fluid in a finite slot heated from below. *Int. J. Heat Mass Transfer* 25: 1363–1376.

Hewitt, H. C., and E. I. Griggs. 1979. Method to determine and optimal air-flow rate in solar collectors. *J. Energy* 3, 6.

Hewitt, H. C., B. K. Parekh, and G. L. Askew. 1978. Test for determining a best flow rate through solar collectors. *J. Energy* 2, 6: 342–345.

Hildebrandt, A. F., and S. Dasgupta. 1980. Survey of Power Tower Technology. *ASME J. Solar Energy Eng.* 102: 91–104.

Hjertager, B. H., and B. F. Magnussen. 1977. Numerical prediction of three-dimensional turbulent buoyant flow in a ventilated room. In D. B. Spalding and N. Afgan, eds., *Heat Transfer and Turbulent Buoyant Convection*, vol. 2. Washington, DC: Hemisphere, pp. 429–442.

Hoffman, E. W., and V. C. Flannery. 1985. Investigation of flow distribution in solar collector arrays. *Proc. ISES Conf.*, Montreal, Canada, pp. 1008–1012.

Hollands, K. G. T., and K. Iynkaran. 1985. Proposal for a compound-honeycomb collector. *Solar Energy* 34: 309–316.

Hollands, K. G. T., G. D. Raithby, F. B. Russell, and R. G. Wilkinson. 1984. Coupled radiative and conductive heat transfer across honeycomb panels and through single cells. *Int. J. Heat Mass Transfer* 27: 2119–2131.

Hollands, K. G. T. 1973. Natural convection in horizontal thin-walled honeycomb panels. *J. Heat Transfer* 95: 439–444.

Hollands, K. G. T. 1965. Honeycomb devices in flat-plate solar collector. *Solar Energy* 9, 3: 159–164.

Hollands, K. G. T., and L. Konicek. 1965. Experimental study of the stability of differentially heated inclined air layers. *J. Heat Transfer* 87: 1467–1476.

Hollands, K. G. T. 1963. Directional selectivity, emittance, and absorptance properties of vee corrugated specular surfaces. *Solar Energy* 7, 3: 108–116.

Hollands, K. G. T., and E. C. Shewen. 1981. Optimization of flow passage geometry for air-heating, plate-type solar collectors. *J. Solar Energy Engg.* 103, 4: 323–330.

Hollands, K. G. T., T. E. Unny, G. D. Raithby, and L. Konicek. 1976. Free convection heat transfer across inclined air layers. *J. Heat Transfer* 98, 2: 189–193.

Hollands, K. G. T., and J. L. Wright. 1983. Heat loss coefficients and effective products for flat-plate collectors with diathermanous covers. *Solar Energy* 30, 3: 211–216.

Honeywell. 1977. Jet impingement solar air heater. Report on USDOE contract EY-76-C-02-2929 "Low Cost Solar Air Heater."

Hoogendorn, C. J. 1986. Natural Convection in Enclosures. *Heat Transfer 1986, Proc. 8th International Heat Transfer Conference*, vol. 1. Washington, DC: Hemisphere, pp. 111–120.

Hoogendron, C. J. 1985. Natural convection supression in solar collectors. In S. Kakac, W. Aung, and R. Viskanta, eds., *Natural Convection Fundamentals and Applications*. Washington, DC: Hemisphere, pp. 940–960.

Hottel, H. C., and D. D. Erway. 1963. Collection of solar energy. In A. M. Zarem and D. D. Erway, eds., *Introduction to the Utilization of Solar Energy*, New York: McGraw-Hill, ch. 5.

Hottel, M. C., and B. B. Woertz. 1942. Performance of flat plate solar-heat collectors. *Trans. ASME* 64: 91–104.

Howell, J., B. R. Bannerot, and G. C. Vilet. 1982. *Solar-Thermal Energy Systems: Analysis and Design*. New York: McGraw-Hill, chs. 4–5.

Hruby, J. M. 1985. Solid particle receivers: a status report. *Proc. Solar Thermal Research Program Annual Conference*. SERI/CP-251-2680, DE85002938, pp. 79–87.

Humphrey, J. A. C., and E. W. Jacobs. 1981. Free-forced laminar flow convective heat transfer from a square cavity in a channel with variable inclination. *Int. J. Heat Mass Transfer* 24: 1589–1597.

Hunt, A. J. 1978. Small particle heat exchangers. Lawrence Berkeley Laboratory report LBL-7841.

Hunt, A. J., and C. T. Brown. 1983. Solar test results of an advanced direct absorption high temperature gas receiver (SPHER). *Proc. 8th Biennial Congress of ISES*, vol. 2. Perth, Australia, pp. 959–963 (also Report LBL-16497).

Hunt, A. J., Ayer, and P. Hull. 1985. Direct radiant heating of particle suspensions. *Proc. Solar Thermal Research Program Annual Conference*. SERI/CP-251-2680, DE85002938, pp, 139–148.

Hurtado, P., and M. Kast. 1984. Experimental study of direct in-situ generation of steam in a line-focus solar collector. Report DOE/SF/11946-T1 (DE85000497), 71 pp.

Iyican, L., L. C. Witte, and Y. Bayazitoglu. 1981. Natural convection and sidewall losses in trapezoidal groove collectors. *J. Solar Energy Eng.* 103: 167–172.

Janke, S. H. 1983. A three-dimensional model for determining the solar radiation absorbed in a transparent circular cylinder containing a blackened fluid. *Progress in Solar Energy*, vol. 6. American Solar Energy Society Annual Meeting, pp. 459–464.

Jones. G. F. 1981. Heat transfer and flow distribution within radiant-convective finned-tube manifold assemblies. Ph.D. dissertation. Dept. Mech. Engrg. & Appl. Mech., University of Pennsylvania, Philadelphia.

Jones, G. F., and N. Lior. 1979. Optimal insulation for solar heating system pipes and tanks. *Energy*, 4: 593–621.

Jones, G. F., and N. Lior. 1978. Isothermal flow distribution in solar collectors and collector manifolds. *Proc. Annual Meeting of the AS/ISES*, vol. 2.1, Denver, co, pp. 362–372.

Jones, G. F., and N. Lior. 1987. Conjugate heat transfer and flow distribution in an assembly of manifolded finned tubes. ASME Paper.

Jones, R. E., and J. F. Burkhart. 1981. Shading effects of collector rows tilted toward the equator. *Solar Energy* 26: 563–565.

Kaehn, H. D., M. Geyer, D. Fong, F. Vignola, and D. K. McDaniels. 1978. Experimental evaluation of the reflector-collector system. *Proc. AS/ISES Ann. Meet.*, vol. 2.1, pp. 654–659.

Kamminga, W. 1985. The approximate temperature within a flat-plate solar collector under transient conditions. *Int. J. Heat Mass Transfer* 28: 433–440.

Kim, D. M., and R. Viskanta. 1984. Effect of wall heat conductance and radiation on natural convection in a rectangluar cavity. *Num. Heat Transfer* 7: 449–470.

Kim, D. M., and R. Viskanta. 1985. Effect of wall heat conduction on natural convection in a square enclosure. *J. Heat Transfer* 107: 139–146.

Kind, R. J., D. H. Gladstone, and A. D. Moizer. 1983. Convective heat losses from flat-plate solar collectors in turbulent winds. *J. Solar Energy Eng.* 105: 80–85.

Kind, R. J., and D. Kitaljevich. 1985. Wind-induced heat losses from solar collector arrays on flat-roofed buildings. *ASME J. Solar Energy Eng.* 107: 335–342.

Klein, S. A., J. A. Duffie, and W. A. Beckman. 1973. Transient considerations of flat-plate solar collectors. *ASME J. Eng. for Power* 96A: 109–113.

Knowles, G. R., and E. M. Sparrow. 1979. Local and average heat transfer characteristics for turbulent airflow in an asymmetrically heated tube. *J. Heat Transfer* 101: 635–641.

Kovarik, M. 1978. Optimal distribution of heat conducting material in the finned pipe solar energy collector. *Solar Energy* 21: 477–484.

Kovarik, M., and P. F. Lesse. 1976. Optimal control of flow in low temperature solar heat collectors. *Solar Energy* 18: 431–435.

Kraabel, J. S. 1983. An experimental investigation of the natural convection from a side-facing cubical cavity. *ASME-JSME Thermal Engineering Joint Conference*, vol. 1, pp. 299–306.

Kreith, F., and J. F. Kreider. 1981. Nonconcentrating solar collectors. In J. F. Kreider and F. Kreith eds., *Solar Energy Handbook*. New York: McGraw-Hill.

Kreith, F., and J. F. Kreider. 1978. *Principles of Solar Engineering*. Washington DC: Hemisphere.

Kreith, F., and R. Anderson. 1985. Natural convection in solar systems. In S. Kakac, W. Aung and R. Viskanta, eds., *Natural Convection: Fundamentals and Applications*. Washington, DC: Hemisphere, pp. 881–939.

Kuehn, T. H., and R. J. Goldstein. 1980. A parametric study of Prandtl number and diameter ratio effects on natural convection heat transfer in horizontal cylindrical annuli. *J. Heat Transfer* 102: 768–770.

Kuehn, T. H., and R. J. Goldstein. 1978. An experimental study of natural convection heat transfer in concentric and eccentric horizontal cylinders. *J. Heat Transfer* 100: 635–640.

Kuehn, T. H., and R. J. Goldstein. 1976. Correlating equations for natural convection heat transfer between horizontal circular cylinders. *Int. J. Heat Mass Transfer* 19: 1127–1134.

Kuehn, T. H., and R. J. Goldstein. 1976. An experimental and theoretical study of natural convection in the annulus between horizontal concentric cylinders. *J. Fluid Mech.* 74: 695–719.

Kugath, D. A., G. Drenker, and A. A. Koenig. 1979. Design and development of a paraboloidal dish solar collector for intermediate temperature service. *Proc. ISES Silver Jubilee*, vol. 2, Atlanta, GA, pp. 449–453.

Lalude, O., and H. Buchberg. 1971. Design and application of honeycomb porous-bed solar-air heaters. *Solar Energy* 13: 223–242.

Landstrom, D. K., G. H. Stickford, Jr., S. G. Talbert, and R. E. Hess. 1978. Development of a low-temperature, low-cost solar collector using a black-liquid concept. *Proceedings of the 1978 Annual Meeting, American Section of the International Solar Energy Society, Inc.*, vol. 2.1, Denver, CO, pp. 228–233.

Landstrom, D. K., S. G. Talbert, and V. D. McGinniss. 1980. Development of a low-temperature, low-cost black liquid solar collector phase II. *Proc. Ann. DOE Active Solar Heating and Cooling Contractors' Review Meeting.*

Lansing, F. L. 1979. HEAP: Heat energy analysis program—A computer model simulating solar receivers. Report DOE/JPL-1060-13 (JPL publication 79-3).

Lansing, F. L., V. Clarke, and R. Reynolds. 1979. A high performance porous flat plate solar collector. *Energy* 4: 685–694.

Larson, D. C. 1980a. Optimization of flat-plate collector-flat mirror systems. *Solar Energy* 24: 203–207.

Larson, D. C. 1986b. Concentration ratios for flat plate solar collectors with adjustable flat mirrors. *AIAA J. Energy* 4: 170–175.

Leary, P. L., and J. D. Hankins. 1979. A user's guide for MIRVAL—A computer code for comparing designs of heliostat-receiver optpics for central receiver solar power plants. Sandia National Laboratories Report SAND-77-8280.

Lee, B. E. 1987. Wind flow over roof-mounted flat plate collectors. *Solar Energy* 38: 335–340.

Lee, T. S., N. E. Wijeysundera, and K. S. Yeo. 1984. Free convection fluid motion and heat transfer in horizontal concentric and eccentric cylindrical collector systems. In D. Y. Lowani, ed., *Solar Engineering 1984*. ASME, New York, pp. 194–200.

Lewandowski, A. A. 1985. Direct absorption receiver system studies. *Proc. Solar Thermal Research Program Annual Conference.* SERI/CP-251-2680, DE85002938, pp. 107–117.

Lin, N. N., and A. Bejan. 1983. Natural convection in a partially divided enclosure. *Int. J. Heat Mass Transfer* 26: 1867–1878.

Lior, N. 1986. Natural convection in high technology applications. In W.-J. Yang and Y. Mori, eds., *Heat Transfer in High Technology and Power Engineering*. Washington, DC: Hemisphere, pp. 573–596.

Lior, N., J. O'Leary, and D. Edelman. 1977. Optimized spacing between rows of solar collectors. *Proc. Annual Meeting, American Section of the International Solar Energy Society*, vol. 1. Orlando, FL, pp. 3-15–3-20.

Lior, N., H. Ozoe, P. Chao, G. F. Jones, and S. W. Churchill. 1983. Heat transfer considerations in the use of new energy resource. In T. Mizushina and W.-J. Yang, eds., *Heat Transfer in Energy Problems*. New York: Hemisphere/Springer-Verlag, pp. 175–188.

Lior, and N., and A. P. Saunders. 1973. Solar collector performance studies. University of Pennsylvania report NSF/RANN/SE/GI27976/TR73/1 to NSF/RANN, 114 pp.

Lior, N., and R. N. Segall. 1986. Heat transfer with air flow across an array of plates with angle of attack. AIAA paper 86-1365. AIAA/ASME 4th Thermophysics and Heat Transfer Conference, Boston, MA.

Lipps, F. W. 1983. Geometric configuration factors for polygonal zones using Nusselt's unit sphere. *Solar Energy* 30: 413–419.

Liu, C. H., and E. M. Sparrow. 1980. Convective-radiative interaction in a parallel plate channel——Application to air-operated solar collectors. *Int. J. Heat Mass Transfer* 23: 1137–1146.

Liu, B. Y. H., and R. C. Jordan. 1965. Performance and evaluation of concentrating solar collectors for power generation. *Trans. ASME J. Eng. for Power* 87: 1–7.

Liu, B. Y. H., and J. C. Jordan. 1963. A rational procedure for predicting the long-term average performance of flat-plate solar-energy collectors. *Solar Energy* 7, 2: 53–74.

Liu, B. Y. H., and R. C. Jordan. 1961. Daily insolation on surfaces tilted toward the equator. *ASHRAE J.* 3: 53–59.

Löf, G. O. G., M. M. El Wakil, and J. A. Duffie. 1961. The performance of Colorado solar house. *United Nations Conference on New Sources of Energy*, Rome.

Löf, G. O. G. 1980. Flat-plate and nonconcentrating collectors. In W. C. Dickinson and P. N. Cheremisinoff, eds., *Solar Energy Technology Handbook*, part A. New York: Marcel Dekker, ch. 9.

Lou, Y. S., and T. K. Shih. 1972a. Nonlinear heat conduction in rarefied gases. *AIAA J.* 10: 1149–1150.

Lou, Y. S., and T. K. Shih. 1972b. Nonlinear heat conduction in rarefied gases confined between concentric cylinders and spheres. *Phys. Fluids* 15: 785–788.

Lumsdaine, E. 1969. Solar Heating of a fluid through a semi-transparent plate: Theory and experiment. *Solar Energy* 12: 457–467.

Lumsdaine, E. 1970. Transient solution and criteria for achieving maximum fluid temperature in solar energy applications. *Solar Energy* 13: 3–19.

Mansfield, R. G., and A. Eden. 1978. The application of thermography to large arrays of solar energy collectors. *Solar Energy* 21: 533–537.

Marcus, S. L. 1983. An approximate method for calculating the heat flux through a solar collector honeycomb. *Solar Energy* 30, 2: 127–131.

Markatos, N. C., and K. A. Pericloeous. 1984. Laminar and turbulent natural convection in an enclosed cavity. *Int J. Heat Mass Transfer* 27: 755–772.

Marshall, K. N., R. K. Wedel, and R. E. Dammann. 1976. Development of plastic honeycomb flat-plate solar collectors. Final report SAN/1081-76/1 to the USDOE. Lockheed Missile Co. LMSC/D462879, 178 pp.

Mather, G. R., Jr. 1982. Transient response of solar collectors. *ASME J. Solar Energy Eng.* 104: 165–172.

May, E. K., and L. M. Murphy. 1983. Performance benefits of the direct generation of steam in line-focus solar collectors. *J. Solar Energy Eng.* 105, 2: 126–133.

McAdams, W. H. 1954. *Heat Transmission*. 3d ed. New York: McGraw-Hill.

McCumber, W. H., and M. W. Weston. 1979. The analysis and comparison of actual to predicted collector array performance. *IBM J. R&D* 23: 239–358.

McDaniels, D. K., D. H. Lowndes, H. Mathew, J. Reynolds, and R. Gray. 1975. Enhanced solar energy collection using reflector-solar thermal collector combinations. *Solar Energy* 17: 277–283.

McKinnon, R. A., A. D. Turrin, and G. L. Schrenk. 1965. Cavity receiver temperature analysis. Paper no. 65–470. AIAA Second Annual Meeting, 24 pp.

McMordie, R. K. 1984. Convection heat loss from a cavity receiver. *J. Solar Energy Eng.* 106, 1: 98–100.

McMurrin, J. C., N. A. Djordjevic, and H. Buchberg. 1977. Performance measurements of a cylindrical glass honeycomb solar collector compared with predictions. *J. Heat Transfer* 99C, 2: 169–173.

Menuchin, Y., S. Bassler, G. F. Jones, and N. Lior. 1981. Optimal flow configuration in solar collector arrays. *Proc. of the 1981 Annual Meeting of the AS/ISES*, vol. 4.1, Philadelphia, PA, pp. 616–620.

Meyer, B. A., M. M. El-Wakil, and J. W. Mitchell. 1979. Natural convection heat transfer in small and moderate aspect ratio enclosures—An application to flat plate collectors. *J. Heat Transfer* 101: 655–659.

Meyer, B. A., J. M. Mitchell, and M. M. El-Wakil. 1980. Convective heat transfer in trough and CPC collectors. *Proc. Ann Meet. AS/ISES*, vol. 3.1, pp. 437–440.

Miller, K. W. 1943 (July). Solar heat trap. Memorandum to the Office to Production, Research and Development, WPB.

Minardi, J. E., and H. N. Chuang. 1975. Performance of a "black" liquid flat-plate solar collector. *Solar Energy* 17: 179–183.

Mirenayat, J. 1981. Experimental study of heat loss through natural convection from an isothermal cubic open cavity. SANDIA Report 81-8014. Sandia National Laboratories, Livermore, CA, pp. 165–174.

Mitalas, G. P., and D. G. Stephenson. 1962. Absorption and transmission of thermal radiation by single and double glazed windows. Research paper no. 173. Division of Building Research of National Research Council, Ottawa, Canada.

Morcos, S. M., and M. M. M. About-Ellail. 1983. Buoyancy effects in the entrance region of an inclined multirectangular-channel solar collector. *J. Solar Energy and Eng.* 105, 2: 157–162.

Morrison, G. L., and D. B. J. Ranatunga. 1980. Transient response of thermosyphon solar collectors. *Solar Energy* 24: 55–61.

Mull, W., and H. Reiher. 1930. Der Warmeschutz von Luftschichten. *Beiheft zum Gesund. Ing. Beiheft*, vol. 28, no. 1, 26 pp.

Nahar, N, M., and H. P. Garg. 1980. Free convection and shading due to gap spacing between an absorber plate and the cover glazing in solar energy flat-plate collectors. *Applied Energy*, 7, 1-3: 129–145.

Naidu, M. G., and J. P. Agarwal. 1981. A theoretical study of heat Transfer in a flat-plate solar collector. *Solar Energy*, 26, 4: 313–323.

Nansteel, M., and R. Greif. 1984. An investigation of natural convection in enclosures with two- and three-dimensional partitions. *Int. J. Heat Mass Transfer* 27: 561–571.

Oliphant, M. V. 1979. Measurement of wind speed distributions across a solar collectors. *Solar Energy* 24: 403–405.

Onur, N., and J. C. Hewitt, Jr. 1980. A study of wind effects on collector performance, ASME paper 80-C2/Sol-4, 6 pp.

Ostrach, S. 1982. Natural convection heat transfer in cavities and cells. *7th Int. Heat Transfer Conf.*, vol. 1. Munich, pp. 365–379.

Ostrach, S. 1972. Natural convection in enclosures. In J. P. Hartnett and T. F. Irvine, Jr., eds., *Advances in Heat Transfer*, vol. 8. New York: Academic Press, ch. 12.

Ota, T., and N. Kon. 1979. Heat transfer in the separated and reattched flow over blunt flat plates—Effects of nose shape. *Int. J. Heat Mass Transfer* 22: 197–206.

Ozoe, H., P. K.-B Chao, S. W. Churchill, and N. Lior. 1983. Laminar natural convection in an inclined rectangular box with the lower surface half heating and half insulated, *ASME J. Heat Transfer* 105: 425–442.

Ozoe, H., K. Fujii, N. Lior, and S. W. Chruchill. 1982. A theoretically based correlation for natural convection in horizontal rectangular enclosures heated from below with arbitrary aspect ratios. Paper no. NC23. *7th International Heat Transfer Conference*, vol. 2. Munich. Washington, DC: Hemisphere, pp. 257–262.

Ozoe, H., K. Fujii, N. Lior, and S. W. Churchill. 1983b. Long rolls generated by natural convection in an inclined rectangular enclosure. *Int. J. Heat Mass Transfer* 26, 10: 1427–1438.

Ozoe, H., K. Fujii, T. Shibata, H. Kuriyama, and S. W. Churchill. 1985a. Three-dimensional numerical analysis of natural connvectional in a spherical annulus. *Numerical Heat Transfer* 8: 383–406.

Ozoe, H., A. Mouri, M. Hiramitsu, S. W. Churchill, and N. Lior. 1986. Numerical calculation of three-dimensional turbulent natural convection in a cubical enclosure using a two-equation model for turbulence. *J. Heat Transfer* 108: 806–813.

Ozoe, H., A. Mouri, M. Ohmuro, S. W. Churchill, and N. Lior. 1985b. Numerical calculations of laminar and turbulent natural convection of water in rectangular channels heated and cooled isothermally on the opposing vertical walls. *Int. J. Heat Mass Transfer* 28, 1: 125–138.

Ozoe, H., M. Ohmuro, A. Mouri, S. Mishima, H. Sayama, and S. W. Churchill. 1983a. Laser-doppler measurements of the velocity along a heated vertical wall of a rectangular enclosure. *J. Heat Transfer* 105: 782–783.

Ozoe, H., T. Okamoto, and H. Sayama, and S. W. Churchill. 1978. Natural convection in doubly inclined rectangular boxes. *Proc. 6th International Heat Transfer Conference*, vol. 2. Washington, DC: Hemisphere, pp. 293–298.

Ozoe, H., H. Sayama, and S. W. Churchill. 1977a. Natural convection patterns in along inclined rectangular box heated from below. Part I: Three dimensional photography. *Int. J. Heat Mass Transfer* 20: 123–129.

Ozoe, H., H. Sayama, and S. W. Churchill. 1977b. Natural convection patterns in along inclined rectangular box heated from below. Part II: Three dimensional numerical results. *Int. J. Heat Mass Transfer* 20: 131–139.

Ozoe, H., T. Shibata, and S. W. Churchill. 1981a. Natural convection in an inclined circular cylindrical annulus heated and cooled on its end plates. *Int. J. Heat Mass Transfer* 24: 727–737.

Ozoe, H., K. Yamamoto, and S. W. Churchill. 1979. Three-dimensional numerical analysis of natural convection in an inclined channel with a square cross section. *AIChE J.* 25: 709–716.

Ozoe, H., K. Yamamoto, S. W. Churchill, and H. Sayama. 1976. Three-dimensional, numerical analysis of laminar natural convection in a confined fliuid heated from below. *J. Heat Transfer* 98C: 202–207.

Parmelee, G. V. 1955. Application of thermal circuit analysis to collector design. *Trans. Int. Conf. on the Use of Solar Energy: The Scientific Basic*, vol. 2, pt. 1, Sec. A. Tucson, AZ, pp. 115–119.

Pellette, P., M. Cobble, and P. Smith. 1968. Honeycomb thermal trap. *Solar Energy* 12: 263–265.

Perrot, M., P. Gallet, G. Peri, J. Desautel. and M. Touchaies. 1967. Les structures cellulaires antirayonnantes et leurs applications industrielles. *Solar Energy* 11, 1: 34–38.

Phillips, W. F. 1982. A simplified nonlinear model for solar collectors. *Solar Energy* 29, 1: 77–82.

Phillips, W. F. 1979. The effects of axial conduction on collector heat removal factor. *Solar Energy* 23: 187–191.

Pons, R. L. 1980. Optimization of a point-focusing distributed receiver solar thermal electric system. *ASME J. Solar Energy Eng.* 102: 272–280.

Rabl, A. 1985. *Active Solar Collectors and Their Applications*. Oxford: Oxford University Press.

Rabl, A. 1981. Yearly average performance of the principal solar collector types. *Solar Energy* 27, 3: 215–234.

Rabl, A. 1981. Intermediate concentrating solar collectors. In J. F. Kreider and F. Kreith, eds., *Solar Energy Technology Handbook*. New York: McGraw-Hill, ch. 8.

Rabl, A. 1980. Concentrating collectors. In W. C. Dickinson and P. N. Cheremisinoff, eds., *Solar Energy Technology Handbook*, pt. A. New York: Marcel Dekker, ch. 10.

Rao, P. P., J. E. Francis, and T. J. Love, Jr. 1977. Two-dimensional analysis of a flat-plate solar collector. *J. Energy* 1, 5: 324–328.

Ratzel, A. C., C. E. Hickox, and D. K. Gartling. 1979. Techniques for reducing thermal conduction and natural convection heat losses in annular receiver geometries. *J. Heat Transfer* 101: 108–113.

Rice, M. P., M. F. Modest, D. N. Borton, and W. E. Rogers. 1981. Heat transfer analysis of receivers for a solar concentrating collectors. *Chem. Engineering Comm.*, 8, 4-6: 353–365.

Robinson, H. E., and F. J. Powlitch. 1954. Thermal insulation values of airspaces. U. S. Housing and Home Finance Agency, Division of Housing, research paper no. 32. April.

Rozkov, I. A. 1977. Study of the characterictics of convective heat transfer in cylindrical solar receivers. *Geliotekhnika* 2: 56–63.

Safdari, Y. B. 1966. Radiation heating through transparent and opaque walls. *Solar Energy* 10: 53–58.

Saito, A., Y. Utaka, T. Tsuchio, and K. Katayama. 1984. Transient response of flat plate solar collector for periodic solar intensity variation. *Solar Energy* 32, 1: 17–23.

Saltiel, C. J., and M. Sokolov. 1982. Thermal and optical analysis of an evacuated circular cylindrical concentrating collector. *Solar Energy* 29: 391–396.

Samano, A., and A. Fernandez. 1983. Simulation of a solar evacuated collector with black fluid. *Progress in Solar Energy*, vol. 6. American Solar Energy Society Annual Meeting, pp. 377–380.

Sassi, G. 1983. Some notes on shadow and blockage effect. *Solar Energy* 31, 3: 331–333.

Satcunanathan, S., and P. Gandhidasan. 1981. The role of rear insulation in the performance of flat plate solar collectors. *ASME J. Solar Energy and Engineering* 103: 282–284.

Savery, C. W., P. M. Anderson, and D. C. Larson. 1976. Double-exposure collectors with mirrors for solar-heating systems, ASME paper no. 76-WA/HT-16, 8 pp.

Savery, C. W., and D. C. Larson. 1978. Performance of a double-exposure solar collector. *Proc. AS/ISES Ann. Meet.*, vol. 2.1, pp. 649–653.

Schmidt, R. R., and E. M. Sparrow. 1978. Turbulent flow of water in a tube with circumferentially nonuniform heating, with or without buoyancy. *J. Heat Transfer* 100: 403–409.

Seitel, S. C. 1975. Collector performance enhancement with flat reflectors. *Solar Energy* 17: 291–295.

Selcuk, M. K. 1971. Thermal and economic analysis of the overlapped-glass plate solar-air heater. *Solar Energy* 13: 165–191.

Selcuk, M. K. 1964. Flat-plate solar collector performance at high temperatures. *Solar Energy* 8, 2: 57–62.

Shah, R. K., and A. L. London. 1978. *Laminar Flow Forced Convection in Ducts.* New York: Academic Press.

Sharafi, A. Sh., and A. G. Mukminova. 1975. Procedure for computing the reflectively, absorptivity, and transmission coefficient for radiant energy in multillayer systems with varying optical properties. *Geliotekhnika* 11: 2: 50–54.

Shoemaker, M. J. 1961. Notes on a solar collector with unique air permeable media. *Solar Energy* 5, 4: 138–141.

Shouman, A. R., and I. A. Tag. 1981. Method of testing solar collectors for determination of heat transfer parameters. Solar Energy Division of the ASME, Winter Annual Meeting, Washington, DC, November 15–20, pp. 1–5.

Shurcliff, W. A. 1974. Transmittance and reflection loss of multi-plate planar window of a solar-radiation collector: Formulas and tabulation of results for the case $n = 1.5$. *Solar Energy* 16: 149–154.

Siebers, D. L., and J. S. Kraabel. 1984. Estimating convective energy losses from solar central receivers. *Proc. International Energy Agency Workshop on the Design and Performance of Large*

Solar Thermal Collector Arrays, C. A. Bankston, comp. Solar Energy Research Institute report SERI/SP-271-2664. San Diego, CA, pp. 401–440.

Siebers, D. L., R. L. Moffat, and R. G. Schwindt. 1983. Experimental, variable properties natural convection from a large vertical flat surface. *ASME-JSME Thermal Engineering Joint Conference*, vol. 3, pp. 269–275.

Siebers, D. L., R. G. Schwindt, and R. J. Moffat. 1982. Experimental mixed convection from a large vertical plate in a horizontal flow. *Proc. 7th International Heat Transfer Conference*. Paper MC 13.

Siebers, D. L., and R. Viskanta. 1979. Thermal analysis of some flat-plate solar collector designs for improving performance. *J. Energy* 3: 8–15.

Siebers, D. L., and R. Viskanta. 1978. Some aspects of the transient response of a flat plate solar energy collector. *Energy Conversion* 18: 135–139.

Siegel, R. 1973. Net radiation method for transmission through partially transparent plates. *Solar Energy* 15: 273–276.

Siegel, R., and J. R. Howell. 1981. *Thermal Radiation Heat Transfer*. 2d ed. New York: McGraw-Hill.

Smart, D. R., K. G. T. Hollands, and G. D. Raithby. 1980. Free convection heat transfer across rectangular-called diathermaneous honeycomb. *J. Heat Transfer* 102: 75–80.

Sobin, A., W. Wagner, and C. R. Easton. 1974. Central collector solar energy receivers. *Proc. AS/ISES Ann. Meet.*, 44 pp.

Soin, R. S., K. Sangameswar Rao. D. P. Rao, and K. S. Rao. 1979. Performance of flat plate solar collector with fluid undergoing phase change. *Solar Energy* 23: 69–73.

Souka, A. F. 1965. Double exposure flat-plate collector. *Solar Energy* 9, 3: 117–118.

Souka, A. F., and H. H. Safwat. 1969. Theoretical evaluation of the performance of a double exposure flat-plate collector using a single reflector. *Solar Energy* 12: 347–352.

Souka, A. F., and H. H. Safwat. 1966. Determination of the optimum orientation for the double exposure flat-plate collector and its reflectors. *Solar Energy* 10: 170–174.

Sparrow, E. M., W. J. Bifano, and J. A. Healy. 1972. Efficiencies of honeycomb absorbers of solar radiation. *J. Spacecraft* 9, 2: 67–68.

Sparrow, E. M., and R. J. Krowech. 1977. Circumferential variations of bore heat flux and outside surface temperature for a solar collector tube. *J. Heat Transfer* 99, 3: 360–366.

Sparrow, E. M., J. S. Nelson, and W. Q. Tao. 1982. Effect of leeward orientation, adiabtaic framing surfaces, and eaves on solar-collector-related heat transfer coefficients. *Solar Energy* 29: 33–41.

Sparrow, E. M., J. W. Ramsey, and E. A. Mass. 1979. Effect of finite width on heat transfer and fluid flow about an inclined rectangular plate. *J. Heat Transfer* 101: 199–204.

Sparrow, E. M. and K. K. Tien. 1977. Forced convection heat transfer at an inclined and yawed square plate—Applications to solar collectors. *J. Heat Transfer* 99, 4: 507–512.

Speyer, E. 1965. Solar energy collection with evacuated tubes. *Trans. ASME J. Eng. for Power*, pp. 270–276.

Stephenson, D. G. 1965. Equations for solar heat gain through windows. *Solar Energy* 9, 2: 81–86.

Stokes, G. G. 1982. On the intensity of the light reflected from or transmitted through a pile of plates. *Proc. Roy. Soc. (London)* 11: 545–556.

Strumpf, H. J., D. M. Kotchick, and M. G. Coombs. 1982. High temperature ceramic heat exchanger element for a solar thermal receiver. *Parabolic Dish Solar Thermal Power Annual Program Review Proceedings*. Report DOE/JPL-1060-52, pp. 233–246. Reprinted in *J. Solar Energy Eng.* 104: 305–309.

Swartman, R. K., and O. Ogunlade. 1966. An investigation on packed-bed collectors. *Solar Energy* 10, 3: 106–110.

Symons, J. G. 1982. The solar transmittance of some convection suppression devices for solar energy applications: An exerimental study. *ASME J. Solar Energy Eng.* 104: 251–256.

Symons, J. G., and R. Gani. 1079. Thermal performance predictions and sensitivity analysis for high temperature flat-plate solar collectors. *Solar Energy* 24: 407–410.

Symons, J. G., and M. K. Peck. 1984. An overview of the CSIRO project on adavnced flat-plate solar collectors. *Proc. ISES Biennial Meeting, Solar World Congress*, vol. 2. Perth, Australia, 1983, New York: Pergamon, pp. 748–752.

Tabor, H. 1969. Cellular insulation (honeycombs). *Solar Energy* 12: 549–552.

Tabor, H. 1966. Mirror boosters for solar collectors. *Solar Energy* 10: 111–118.

Tabor, H. 1958. Radiation, convection and conduction coefficients in solar collectors. *Bull. Res, Counc. of Israel* 6C: 155–176.

Tabor, H. 1955. Solar Energy collector design with special reference to selective radition. *Bull. Res. Counc. of Israel* 5C: 5–27.

Tan, H. M., and W. W. S. Charters. 1970. An experimental investigation of forced-convective heat transfer for fully-developed turbulent flow in a rectangular duct with asymmetric heating. *Solar Energy* 13: 121–125.

Test, F. L. 1976. Parametric study of flat plate solar collectors. *Energy Conversion* 16: 23–33.

Test, F. L., and R. C. Lessman. 1980. An experimental study of heat transfer during forced convection over a rectangular body. *J. Heat Transfer* 102: 146–151.

Test, F. L., R. C. Lessman, and A. Johary. 1981. Heat transfer during wind flow over rectangular bodies in the natural environment. *J. Heat Transfer* 103: 262–267.

Thodos, G. 1976. Predicted heat transfer performance of an evacuated glass-jacketed CPC receiver: Countercurrent flow design. Argonne National Laboratory Report ANL-76-67, 37 pp.

Thomas, J. R. 1981. Helium penetration in evacuated solar collectors: Theory and the effect on their performacne. *ASME J. Solar Energy Eng.* 103: 175–177.

Thomas, J. R., Jr. 1979. Heat conduction through partial vacuum. U.S.D.O.E. Report ALO/5367-1.

Thomas, J. R., Jr., T. S. Chang, and C. E. Siewert. 1973. Heat transfer between parallel plates with arbitrary surface accommodation. *Phys. Fluids* 16: 2116–2120.

Tien, C. L., and W. W. Yuen. 1975. Radiaton characteristics of honeycomb solar collectors. *Int. J. Heat Mass Transfer* 18: 1409–1413.

Trentelman, J., and P. H. Wojciechowski. 1977. Performance analysis of a black liquid absorbing collector (BLAC). *Proc. AS/ISES Ann. Meet.* 1: 1-21–1-25.

Truncellito, N., H. Yeh, and N. Lior. 1987. Numerical solutions of turbulent convection over a flat plate with angle of attack. *J. Heat Transfer* 109: 238–242. Longer version also published as ASME paper 83-WA/HT-3, 1983.

U.S. Post Office. 1969. Computer program for analysis of energy utilization in post facilities. Report on Project No. 67138.

Umarov, G. Ya., R. R. Avezov, F. Soatov, and K. Babakulov. 1975. Determination of the entrance coefficient for solar rays transmitted by a glazed cover of a hot-box type solar installation. *Geliotekhnika* 11, 3-4: 70–73.

Vaxman, M., and M. Sokolov. 1985. Analysis of a free-flow solar collector. *Solar Energy* 35: 287–290.

Veinberg, V. B. 1959. *Optics in Equipment for the Utilisation of Solar Energy.* State Publishing House of Defense Ministry, Moscow. Trans. by U.S. Department of Army Intelligence Translation 44787, or USAEC Translation AEC-tr-4471.

Viskanta, R. 1982. Radiation heat transfer: interaction with conduction and convection and approximate methods in radiation. *7th Int. Heat Transfer Conf.*, vol. 1. Munich, pp. 103–121.

Viskanta, R., and E. E. Anderson. 1975. Heat transfer in semitransparent solids. In *Advances in Heat Transfer*, vol. 11. New York: Academic Press, pp. 317–441.

Viskanta, R., D. L. Siebers, and R. P. Taylor. 1978. Radiation charactericstics of multiple-plate glass system. *Int. J. Heat Mass Transfer* 21: 815–818.

Vittitoe, C. N., and F. Biggs. 1981. User's guide to HELIOS: A computer program for modeling the optical behavior of reflecting solar concentrators. Part I. Introduction and code input. Sandia National Laboratory Report SAND-81-1180, 64 pp.

Wagner, T. R., W. G. Houf, and F. P. Incropera. 1980. Radiative property measurements for india ink suspensions of varying concentrations. *Solar Energy* 25, 6: 549–554.

Wang, K. Y., and F. Kreith. 1986. Heat transfer research for high temperature solar energy systems. In W.-J. Yang and Y. Mori, eds., *Heat Transfer in High Technology and Power Engineering*. Washington, DC: Hemisphere, pp. 541–560.

Wang, Y. Y., and R. J. Copeland. 1984. Heat transfer in a solar radiation absorbing molten salt film flowing over an insulated substrate. ASME Paper 84-WA/Sol-22.

Webb, B. W., and R.. Viskanta. 1985. Analysis of heat transfer and solar radiation absorption in an irradiated thin, falling molten salt film. *ASME J. Solar Energy Eng.* 107: 113–119.

Whillier, A. 1953. Solar energy collection and its use in house heating. Sc.D. dissertation. Department of Mechanical Engineering, MIT.

Whillier, A. 1963. Plastic covers for solar collectors. *Solar Energy* 7, 3: 148–151.

Whillier, A. 1964a. Performance of black-painted solar air heaters of conventional design. *Solar Energy* 8, 1: 31–37.

Whillier, A. 1964b. Thermal resistance of the tube-plate bond in solar heat collectors. *Solar Energy* 8, 3: 95–98.

Whillier, A., and G. Saluja. 1965. Effects of materials and construction details on the thermal performance of solar water heaters. *Solar Energy* 9, 1: 21–26.

Whillier, A. 1967. Design factors influencing collector performance. In R. C. Jordan, ed., *Low Temperature Engineering Application of Solar Energy*. New York: ASHRAE, ch. 3.

Whillier, A. 1977. Prediction of performance of solar collectors. In R. C. Jordan and B. Y. H. Liu, eds., *Applications of Solar Energy for Heating and Cooling of Buildings*. NY: ASHRAE GRP 170, ch. 8.

Wideman, D. C., and J. R. Thomas, Jr. 1980. Heat conduction in rarefied gases between concentric cylinders: Arbitrary accommodation on both surfaces. *Proc. 1980 Heat Transfer and Fl. Mech. Inst.*, Gerstein and Choudhury, eds., Stanford: Standford University Press, pp.186–195.

Wijeysundera, N. E. 1978. Comparison of transient heat transfer models for flat plate collectors. *Solar Energy* 21, 6: 517–521.

Wijeysundera, N. E. 1976. Response time of solar collectors. *Solar Energy* 21: 517–521.

Wijeysundera, N. E. 1975. A net radiation method for the transmittance and absorptivity of a series of parallel regions. *Solar Energy* 17: 75–77.

Window, B., and I. M. Bassett. 1981. Optical collection efficiencies of arrays of tubular collectors with specular reflectors. *Solar Energy* 26: 341–346.

Window, B., and J. Zybert. 1981. Optical collection efficiencies of arrays of tubular collectors with diffuse collectors. *Solar Energy* 26: 325–331.

Winn, R. C., and C. B. Winn. 1981. Optimal control of mass flow rates in flat plate solar collectors. *ASME J. Solar Energy Eng.* 103: 113–120.

Wolf, D., A. I. Kudish, and A. N. Sembira. 1981. Dynamic simulation and parametric sensitivity studies on a flat-plate solar collector. *Energy* 6, 4: 333–351.

Wu, S. F., T. V. Narayana, and D. N. Gorman. 1983. Conceptual design of an advanced water/steam receiver for a solar thermal central power system. *J. Solar Energy Engineering* 105: 34–41.

Yang, K. T., and J. R. Lloyd. 1985. Natural convection-radiation interaction in enclosures. In S. Kakac, W. Aung, and R. Viskanta, eds., *Natural Convection Fundamentals and Applications.* Washington, DC: Hemisphere, pp. 381–410.

Yeh, L. T., and M. Wiener. 1984. Thermal analysis of a solar advanced water/steam receiver. *ASME J. Solar Energy Engineering* 106: 44–49.

Youngblood, W. W., and P. I. Mattox. 1978. Evaluation of heat transfer enhancement in air heating collectors. *Proc. 3rd Ann. Solar Heating and Cooling Cotractors' Meeting,* pp. 106–107.

Youngblood, W. W., W. Schultz, and R. Barber. 1979. Solar collector fluid parameter study. National Bureau of Standards Report NBS-GCR 79-184.

Yuen, W. W., F. J. Miller, and A. J. Hunt. 1986. Heat transfer characteristics of small particle/gas mixtures. *Int. Comm. Heat Mass Transfer* 13: 145–154.

Zhang, H.-F., and Z. Lavan. 1983. Thermal performance of a serpentine absorber plate. *Progress in Solar Energy,* vol. 6. Proc. ASES Annual Meeting, pp. 471–476.

Zukauskas, A. 1985. *Heat Transfer of a Cylinder in Crossflow.* Washington, DC: Hemisphere.

5 Testing and Evaluation of Stationary Collectors

Robert D. Dikkers

Many reference sources report on the development of test procedures and evaluation methods for stationary collectors. A list of these is presented at the end of this chapter. The methods for determining thermal performance are necessary to assume that design guidelines will be satisfied when component specifications are selected. The operation and performance after installation requires that durability and reliability be sustained over service lifetimes for installation and usage under a wide range of environmental parameters. The following summary review presents the requirements for thermal performance, durability/reliability, and miscellaneous performance testing.

5.1 Thermal Performance

5.1.1 Basic Equations Governing Thermal Performance

The relationship that governs the thermal performance of a solar collector under steady-state conditions is based upon the difference between the energy absorbed and energy losses, which equals the energy conveyed by the transfer fluid. This relationship in equation-form, often referred to as the Hottel-Whillier equation, is (Duffie and Beckman 1980)

$$\frac{\dot{Q}_u}{A_a} = F_R I_t(\tau\alpha)_e - F_R U_L(t_{f,i} - t_a) = \frac{\dot{m}c_p}{A_a}(t_{f,e} - t_{f,i}), \tag{1}$$

where

$\dot{Q}_u$ = rate of heat loss from the collector (W),

A_a = aperture area of the collector (m^2),

F_R = collector heat removal factor,

I_t = total solar energy incident on the plane of the collector per unit time per unit arc,

$(\tau\alpha)_e$ = effective transmittance-absorptance product for the collector,

U_L = heat transfer loss coefficient for the collector (W/m^2 $^\circ$C)

$t_{f,i}$ = temperature of the heat transfer fluid entering the collector ($^\circ$C),

t_a = ambient air temperature ($^\circ$C),

$\dot{m}$ = mass flow rate of the heat transfer fluid through the collector (kg/s m^2),

c_p = specific heat of the transfer fluid (J/kg s),

$t_{f,e}$ = temperature of the heat transfer fluid leaving the collector ($^\circ$C).

If the solar collector efficiency is defined by

$$\eta_a = \frac{\dot{Q}_u/A_a}{I_t}, \tag{2}$$

then the efficiency from equation (1) can be expressed by

$$\eta_a = F_R I_t(\tau\alpha)_e - F_R U_L \frac{(t_{f,i} - t_a)}{I_t} \tag{3}$$

or

$$\eta_a = \frac{\dot{m}c_p(t_{f,i} - t_a)}{A_a I_t}. \tag{4}$$

Equation (3) indicates that a linear relationship exists for the efficiency as a function of $(t_{f,i} - t_a)/I_t$ if F_R, U_L, and $(\tau\alpha)_e$ are assumed constant. The slope is $-F_R U_L$, and the intercept is $F_R(\tau\alpha)_e$. Efficiency data taken near normal incidence [when $(\tau\alpha)_e = (\tau\alpha)_{e,n} \approx$ constant] for most collectors operating at temperatures $<80°C$ ($175°F$) can be correlated reasonably well with equation (3).

For transient conditions the equation for thermal performance of the collector must be modified to account for the absorbed energy that heats the collector and its components. The rate of energy storage of the collector for transient conditions is (Duffie and Beckman 1980; Klein, Duffie, and Beckman 1974)

$$\frac{C_A}{A_a}\frac{dt_f}{d\theta} = F_R I_t(\tau\alpha)_e - F_R U_L(t_{f,i} - t_a) - \frac{\dot{m}c_p}{A_a}(t_{f,e} - t_{f,i}), \tag{5}$$

where

C_A = effective heat capacity of the collector, its components, and the transfer fluid in the collector (J/°C),

$\dot{t}_f$ = average temperature of the heat transfer fluid in the collector (°C),

θ = time, incident angle between the direct solar beam and the outward-drawn normal to the plane of the collector.

The solution of equation (5) for the exit temperature of the transfer fluid $t_{f,e}$ as a function of time θ follows from the following assumptions:

1. The exit temperature of the transfer fluid is related to the average temperature by

$$\frac{dt_f}{d\theta} = K \frac{dt_{f,e}}{d\theta},\tag{6}$$

where (Simon 1976)

$$K = \frac{\dot{m}c_p}{F'U_L A_a}\left(\frac{F'}{F_R} - 1\right)\tag{7}$$

and F' is the collector efficiency factor.

2. I_t, $(\tau\alpha)_e$, U_L, t_a, $\dot{m}$, c_p, and $t_{f,i}$ are all constant for the period covered by the transient solution.

The solution to equation (5) is

$$\frac{F_R I_t(\tau\alpha)_e - F_R U_L(t_{f,i} - t_a) - (\dot{m}c_p/A_a)(t_{f,e} - t_{f,i})}{F_R I_t(\tau\alpha)_e - F_R U_L(t_{f,i} - t_a) - (\dot{m}c_p/A_a)(t_{f,e,\text{initial}} - t_{f,i})} = \exp\frac{\dot{m}c_p}{KC_A}\theta.\tag{8}$$

The system time constant described by an equation of the form of equation (8) is obtained from the quantity in the exponential. The physical interpretation of the time constant is that it is the time required for the left-hand side of equation (8) to decrease from 1.0 to 0.368 (where $0.368 = 1/e$). This concept should be maintained to be consistent with the accepted interpretation of time constant. This has not always been practical for solar collectors. Simon (1976) published the value of "time constant" for nine different water-heating collectors using a 0%–99% change time. Wijeysundera (1976) computed the "response time" of typical one-, two-, and three-cover air heaters using the 0%–99% change time.

The effective transmittance-absorptance product $(\tau\alpha)_e$, of a flat plate solar collector can be described (Souka and Safwat 1966) by

$$(\tau\alpha)_e = K_{\alpha\tau}(\tau\alpha)_{e,n},\tag{9}$$

where $K_{\alpha\tau}$ is a function of incident angle only. The relationship was derived by assuming that the optical properties of the cover plate(s) were the only factors affecting the change in absorbed solar radiation as incident angle changes. In testing solar collectors, a similar relationship was assumed (Simon and Buyco 1976) in the form

$$F_R(\tau\alpha)_e = K_{\alpha\tau}F_R(\tau\alpha)_{e,n}.\tag{10}$$

Determination of $K_{\alpha\tau}$ as a function of its independent variables was made from a test procedure developed for the particular collector being evaluated. The incident angle modifier (IAM) $K_{\alpha\tau}$ simply describes how the optical efficiency of the collector changes as a function of incident angle and other independent variables. It represents an essential correction factor that is applied to the efficiency curve in determining the all-day performance of most stationary and single-axis tracking collectors.

For any incident angle, the collector thermal efficiency is defined by

$$\eta_a = F_R K_{\alpha\tau}(\tau\alpha)_{e,n} - \frac{F_R U_L(t_{f,i} - t_a)}{I_t}. \tag{11}$$

$K_{\alpha\tau}$ influences only the intercept; it does not affect the slope of the efficiency curve. Figure 5.1 shows how $K_{\alpha\tau}$ varies with the incident angle for four collectors. For many flat plate collectors, $K_{\alpha\tau}$ can be correlated with equations (12) and (13), as is shown in figure 5.2:

$$K_{\alpha\tau} = 1 - b_0 \left[\left(\frac{1}{\cos\theta} \right) - 1 \right], \tag{12}$$

where b_0 is a dimensionless constant.

For all collectors $K_{\alpha\tau}$ may not be symmetrical with the normal aperture plane. Mather and Beekley (1976) suggested that the optical characteristics of these collectors can be described using a biaxial IAM. For parabolic trough

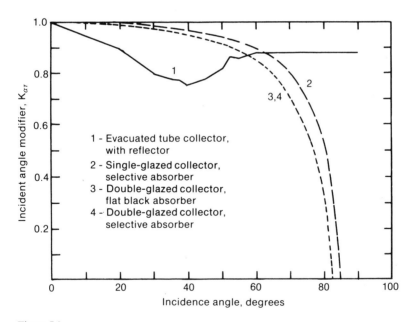

Figure 5.1
Incident angle modifier as a function of the incident angle for four stationary solar collectors.
Source: Hill, Wood, and Reed (1985).

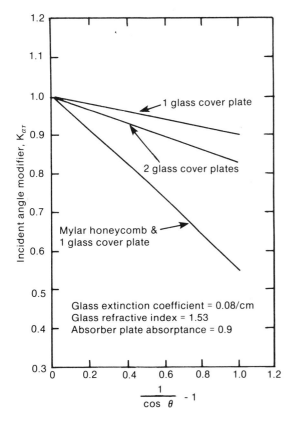

Figure 5.2
Incident angle modifier for three flat plate solar collectors with nonselective coatings on the absorber. Source: ASHRAE (1977).

concentrating collectors and evacuated tubular collectors, there are two mutually perpendicular directions of symmetry: one parallel to the longitudinal absorber axis and the other perpendicular to the axis. Following (Mather and Beekley 1976), $K_{\alpha\tau}$ can be expressed as the product of two separate IAM values, K_1 and K_2, each determined in the mutually perpendicular directions of symmetry so that

$$K_{\alpha\tau} = K_1 K_2. \tag{13}$$

The Solar Rating and Certification Corporation (SRCC) standard for rating solar collectors (SRCC 1981) adopted that concept.

5.1.2   Testing Solar Collectors under Clear-Sky, Full-Irradiance Conditions

ASHRAE standards   The American Society of Heating, Refrigerating, and Air-Conditioning Engineers Standards 93-77 (ASHRAE 1977b) and 96-80 (ASHRAE 1980) provide test methods for determining the thermal performance of solar collectors, which heat fluids for use in thermal systems. Their provisions will be briefly reviewed, and details can be found in the standards or in studies of applications of the test methods from reports listed in the references. A new standard, ASHRAE 93-86 which is a revision of ASHRAE 93-77, is discussed at the end of this section.

ASHRAE Standard 93-77 applies only to collectors in which fluid enters through a single inlet and leaves through a single outlet, or to collectors with more than a single inlet/outlet in which the piping can be connected to effectively provide a single inlet and outlet. The standard does not apply to collectors in which the thermal storage unit is an integral part so that the collector and storage processes cannot be separated for the purpose of making measurements. The method is applicable only for determining steady-state efficiency and not transient responses. Provisions are included for conducting tests outdoors under natural solar irradiation and indoors under simulated solar irradiation.

The standard provides for either liquid or gas transfer fluids, but not a mixture of the two phases. For liquid transfer fluids, a closed-loop testing configuration is shown in figure 5.3. It further specifies mounting precautions, average irradiation, collector orientation, and wind velocity measurements appropriate for specific tests. The standard specifies that outdoor tests are to be conducted only on days in which the 15-minute integrated average solar radiation is at least 630 W/m^2 (200 Btu/h ft^2). Further restrictions are given for incident angle and the range of ambient temperature for the test points in the efficiency curve. The preconditioning consists of stagnation heat in a

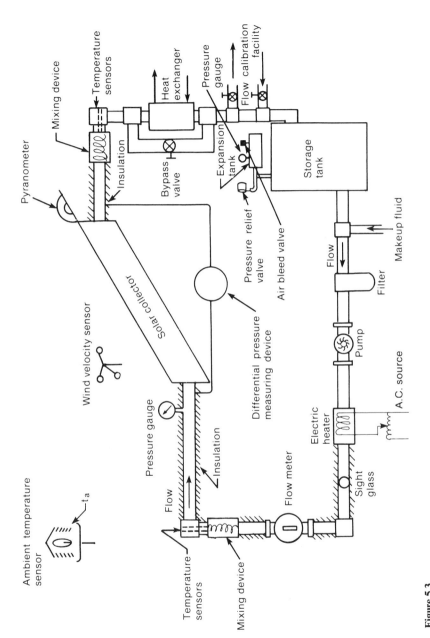

Figure 5.3
Closed-looped testing configurations for a solar collector when the transfer fluid is a liquid. Source: ASHRAE (1977).

nonoperational mode under dry conditions for three days with a cumulative mean incident solar radiation of not less than 17,000 kJ/m² day (1,500 Btu/ft² day).

The standard details the instrumentation for the pyranometer and its location for flow metering in air systems. Instrumentation is required for measuring total shortwave radiation (pyranometer), direct normal irradiation (pyrheliometer), temperature and temperature differences, liquid flow, airflow, pressure, time and mass, and wind velocity. Parameters to be measured during the testing procedure include temperature, temperature differences, pressure drop, and airflow. The standard provides three test configurations for testing liquid solar collectors (i.e., one closed-loop and two open-loop configurations). Allowable tolerances and levels of accuracy are given for all required instrumentation and measurements.

The test procedures require experimental measurement of the rate of incident solar radiation onto the solar collector, as well as the rate of energy addition as it passes through the solar collector. From a combination of values of incident radiation, ambient temperature, and inlet fluid temperature, values of instantaneous efficiency can be obtained. In addition tests are specified for determining the time response characteristics of the collector and the way in which steady-state thermal efficiency varies with the incident angle between the direct beam irradiation and the collector. The testing procedure provides for a sequence of tests. First is the experimental determination of the collector time constant (by one of two methods). The time constant is defined as the time required for the fluid leaving a solar collector to attain 63.2% of its steady-state value following a step change in irradiation or inlet fluid temperature. Second, a series of thermal efficiency tests are conducted to determine the governing efficiency curve. Finally, to permit the prediction of the solar collector performance under a wide range of conditions and/or time of day with varying angles of incidence, the standard requires the determination of the collector incident angle modifier, using either one of two experimental methods. This latter procedure is not necessary for flat plate collectors whose angular response characteristics are known.

Performance equations are given for calculating collector thermal efficiency, collector time constant (for evaluating transient behavior), and collector incident angle modifier. For determining the efficiency curve, at least four different values of inlet fluid temperature must be used, and for each inlet fluid temperature, at least four data "points" must be taken, two preceding and two symmetrically following solar noon. Each data "point" represents an integrated value where the period is the time constant or 5 minutes, whichever is larger.

Although the scope of ASHRAE Standard 93-77 does not specifically exclude unglazed flat plate collectors operating with a liquid as the transfer fluid, such collectors have special performance characteristics when used in low temperature applications such as swimming pool heaters or heat pumps. Unglazed collectors have a greater sensitivity to environmental conditions and to operating flow rates. Also the liquid heat transfer fluid may have a temperature lower than the ambient air. These characteristics need to be considered in developing test procedures for unglazed flat plate solar collectors. Accordingly, a new standard, ASHRAE Standard 96-80, Methods of Testing to Determine the Thermal Performance of Unglazed Flat-Plate Liquid-Type Solar Collectors, was promulgated. This standard closely follows ASHRAE Standard 93-77, with the following exceptions: It contains tighter requirements on collector test loop stability, it requires greater instrumentation accuracy, it restricts allowable environmental variables during the tests, it requires that during the tests the wind does not exceed 1.3 m/s (3 mph), and it specifies that the range of ambient temperatures for all reported test points making up the efficiency curve must be less than 10°C (50°F).

Further differences between ASHRAE 96-80 and 93-77 are that ASHRAE 96-80 applies only to flat plate nonconcentrating collectors using liquid heat transfer media, whereas ASHRAE 93-77 applies to both liquid and air collectors, as well as concentrating and flat plate collectors. These differences in scope made it possible to omit the requirement for determining the collector time constant from ASHRAE 96-80.

Recognizing the need to update ASHRAE Standard 93-77 because of major changes in the technology of testing solar collectors, ASHRAE Standard Project Committee (93-77R) began work in 1980 to prepare a revised standard. As a result of their efforts, a new ASHRAE (1986a) Standard 93-86 was recently approved for publication. Although the new standard maintains the original methodology of ASHRAE Standard 93-77, it contains several significant changes that were made to reduce the experimental uncertainties. (These uncertainties will be discussed later in this section). The following briefly summarize some of the revisions:

1. For unglazed solar collectors, testing is to be in accordance with ASHRAE Standard 96-80. For collectors in which the heat transfer fluid changes phase, testing is to be conducted using another recently approved standard, ASHRAE Standard 109-86 (1986b).

2. The heat transfer fluid used in the collector during testing is required to be the same fluid as recommended by the collector manufacturer and to have a

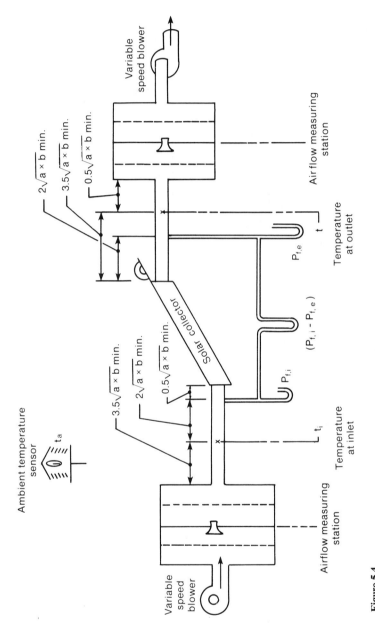

Figure 5.4
Testing configuration for solar collectors where heat transfer fluid is air. Source: ASHRAE (1986a).

known temperature dependence for its density and specific heat over the temperature range of the fluid during the test period.

3. The three-day preconditioning specified in ASHRAE Standard 93-77 has been deleted.

4. Specifications for instrumentation have been upgraded, especially those for solar irradiance measurements.

5. When a solar simulator is used in lieu of outdoor testing, the simulator shall provide a spectral distribution as defined by the standard air mass 1.5 solar spectrum (ASTM Standard E-892) (ASTM 1982).

6. Test conditions and procedures for air-heating collectors have been revised to minimize the effects of air leakage. The recommended test configuration for collectors using air as a transfer fluid is shown in figure 5.4.

7. Only one experimental method—same as Method No. 1 in ASHRAE 93-77—is provided for the determination of the collector time constant.

Limitations of the Assumed Collector Model   The determination of the performance of a solar collector as given by equations (3) and (5) assumes that the heat loss coefficient U_L is a constant. That term of the equations involving the loss coefficient represents the rate of energy loss from the collector by conduction, convection, and radiation. Those losses, however, depend on linear and nonlinear temperature differences between the working fluid and the surroundings. The losses are physically associated with the operating temperature of the collector (i.e., the plate temperature or fluid temperature), the tilt angle of the collector (effect on the free convection loss component), and the ambient weather conditions (i.e., environmental temperature and wind velocity and direction).

Outdoor test conditions for collectors can only be established within prescribed ranges. The variability of ambient weather with very low wind conditions, settings for other tilt angles, and different times of test or different test site locations cause variations in results obtained for the same collector. To account for other than a constant U_L in testing collectors and correlation of test results, various options have been used. The large majority of techniques (Hill, Wood, and Reed 1985) were developed for flat plate collectors that operate at relatively low temperatures for solar domestic water heating and space heat applications. An empirical relationship developed for the loss coefficient includes four major variables (in addition to material properties) that affect the coefficient: tilt angle, ambient temperature, mean plate temperature, and wind heat transfer coefficient at the outer surface of the exterior glazing. (The heat loss coefficient U_L differs from the top loss coefficient only by the amount of losses occurring through the edges and back of the collector, which are generally small.) Because, in general, both radiation and free convection occur in the heat loss process, the top loss coefficient is a function of both the average of the mean plate and ambient temperatures and the difference between them.

In "A Generalized Method for Testing All Classes of Solar Collectors, Part II—Evaluation of Collector Thermal Constants," Proctor (1982a) proposes to account for convection and radiation by the relation

$$F'U_{L} = (C_{8} + C_{9}v)\frac{\overline{t_{f}} - t_{e}}{I_{t}} + (C_{10} + C_{11}v)\frac{(\overline{t_{f}} - t_{e})z}{I_{t}}$$

$$+ (C_{12} + C_{13}\overline{t_{f}})\frac{[\overline{T_{f}}^{4} - \overline{T_{s}}^{4}]}{I_{t}}, \tag{14}$$

where

C_{8}–C_{22} = constants,

t_{e} = the effective environmental temperature (°C),

v = the wind speed (m/s),

$\overline{T_{f}}$ = average temperature in the heat transfer fluid (K),

$\overline{T_{s}}$ = average sink temperature for radiation loss (K).

In ASHRAE Standards 93-77 and 96-1980 the possibility that test data when plotted will not result in a linear relationship as indicated by equation (3) (and hence variable U_{L}) is acknowledged by an allowance to use a second-order fit to the data with the independent variable being $(t_{f,i} - t_{a})/I_{t}$.

If a second- or higher-order fit is to be used with the test data to account for variations in U_{L}, there are two choices to properly correlate the data (Hill, Wood, and Reed 1985) The first is to use a correlation that is consistent with the assumed dependence of U_{L} on independent variables. For example, in accordance with the Australian Standard (Standards Association of Australia 1982) the data are correlated by

$$\eta_{a} = C_{17} + C_{18}\frac{(\overline{t_{f}} - t_{e})}{I_{t}} + C_{19}\frac{(\overline{t_{f}} - t_{e})^{2}}{I_{t}}. \tag{15}$$

A second choice is to continue to use the kind of plot that has become accepted through the use of ASHRAE Standard 93-77 $(t_{f,i} - t_{a})/I_{t}$ as the independent variable but to plot the data on several curves to account for the correct dependence of U_{L}. For example, for low temperature applications of flat plate collectors, equation (15) could be written

$$\eta_a = C_{17} + C_{18}\frac{(\dot{t}_f - t_e)}{I_t} + I_t C_{19}\frac{(\dot{t}_f - t_e)^2}{I_t},\tag{16}$$

and the type of plot called for in the ASHRAE standards used, provided separate curves are given for different levels of irradiance.

For the higher operating temperatures radiation heat loss becomes significant. A correlation that has been used (Solar Energy Industries Association 1979) to account for that is

$$\eta_a = C_{20} + C_{21}\frac{(t_{f,i} - t_a)}{I_t} + C_{22}\sigma\frac{(T_{f,i}^4 - T_a^4)}{I_t}.\tag{17}$$

In consideration of the performance of high temperature flat-plate collectors that have relatively low heat loss coefficients, the loss by free convection was recognized as very low and U_L is primarily a function of the average fluid temperature. As a result of the analysis, which was similar to the one above, the "standard plot" was recommended (Gani, Proctor, and Symons 1981), providing different curves for different values of fluid temperature

It should be noted that commonly used design procedures for predicting the performance of solar energy systems (Beckman, Klein, and Duffie 1977; Klein and Beckman 1979) require a linear model for the solar collector as indicated by equations (3) and (5). Multiple correlation parameters should be used only with great care (see Hill, Wood, and Reed 1985, figs. 7–12). Consequently, if the results of a collector test warrant the use of a higher-order "fit" as discussed earlier, care must be taken in "linearizing" the results before using them in the system's models (see Gani, Proctor, and Symons 1981; Cooper and Dunkle 1981; Proctor 1982b).

Evaluation of Results for Outdoor Tests   Four major solar collector programs involving the testing of one or more collectors at several test facilities were conducted to determine the comparability of thermal performance data. In a round-robin test program, which was undertaken prior to ASHRAE Standard 93-77 adoption, the NBS interim test procedure was specified. Some testing facilities, however, did use the more restrictive requirements set forth in the final standard (which was adopted during the course of the round-robin). The results of the round-robin were reported for the two commercially available collectors (Streed et al. 1978; Streed et al. 1979). Collector 1 had two tempered glass cover plates over an aluminum absorber with integral flow passages and a nonselective black coating backed with glass-fiber insulation and contained a sheet-metal enclosure. Collector 2 had a single, tempered glass cover over a stitch-welded and pressure-expanded absorber made from two steel sheets.

The absorber was coated with a selective surface and backed with glass-fiber insulation.

Each participant in the test series reported the test conditions for each data point and also plotted collector efficiency as follows:

η_a versus $\dfrac{[(t_{f,e} + t_{f,i})/2] - t_a}{I}$.

The values of $F'(\tau\alpha)_e$ and collector heat removal factor $F'U_L$ were determined by a first-order least square fit to all measured data points. Mean and standard deviation values were collected for these two parameters from each laboratory. This statistical analysis indicated important deviations, particularly for collector 2. The coefficient of variation (standard deviation expressed as a percent of the mean value) for the y-intercept of efficiency curve for collectors 1 and 2 were 7.7% and 4.7%, respectively, and for the slopes assuming a linear plot were 15.8% and 24.9%, respectively.

Determination of the effects of environmental conditions at the different test sites on the results was carried out by a detailed computerized thermal model of each of the two collectors. Changes in efficiency were computed that would result from a change in test conditions.

Figure 5.5 shows the efficiency of collector 2 as reported by ten test sites reporting complete data. The efficiency values as reported (figure 5.5a) show considerable scatter. Theoretical efficiencies, from the analytical model, for the actual test conditions reported are shown in figure 5.5b for the allowable scatter in the absence of experimental error. Results of correcting the measured efficiencies of a set of "reference" test conditions (chosen to be the mean of all actual test conditions to minimize to overall adjustment) are shown in figure 5.5c. The mean value of the square of the distance from the points to the curve ("mean square") was used in the analysis to quantify the closeness of correlations. A mean square value of 17.5 in figure 5.5a was reduced to 13.5 in figure 5.5c. Figure 5.5b shows the scatter theoretically justified in figure 5.5a by the variations in the environmental test conditions, a mean square value of only 3.7. Consequently the scatter remaining in figure 5.5c was caused by reasons other than the different test environments.

Figure 5.6 shows the results of collector 2 for the subset of tests, which reportedly met the requirements of ASHRAE Standard 93-77. The mean square value was substantially reduced to 4.6 in figure 5.6a, and figure 5.6b had a mean square value of 3.7. Applying the correction procedure, the mean square value was only slightly reduced to a value of 4.0 in figure 5.6c. The substantial improvement in results when considering those tests, which reportedly met the requirements of Standard 93-77, was probably due as much to

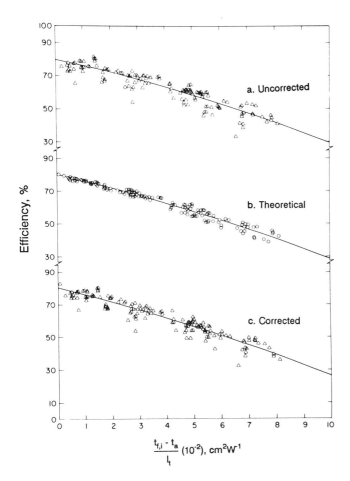

Figure 5.5
Collector 2 efficiency curves based on measurements from 10 test sites (106 values) in the NBS round-robin program. Source: Streed et al. (1979).

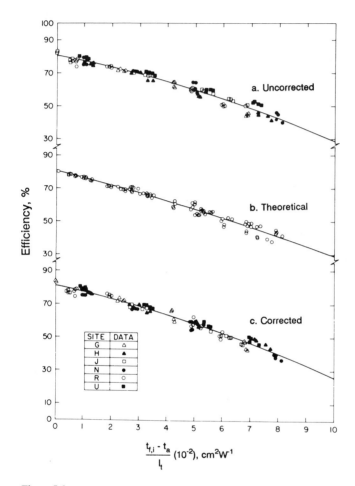

Figure 5.6
Collector 2 efficiency curves based on measurements from 6 test sites (96 values) meeting
ASHRAE Standard 93-77 requirements in the NBS round-robin program. Source: Streed et al.
(1979).

the improved experimental procedures being used by those participants as to the more restrictive requirements.

The data of figure 5.6 showed that test results from individual sites exhibited relatively small scatter. Further analysis (Streed et al. 1978; Streed et al. 1979) indicated that a more probable error band based on random occurrence of the instrumentation inaccuracy fell within the spread of data. Thus systematic differences from facility to facility resulted in data scatter. Less than one-third of the scatter could be attributed to different test conditions.

There were two possible sources of systematic error: systematic facility error and systematic instrument error. Systematic facility error sources included

1. Heat transfer between the test apparatus and the collector.
2. Reduced heat losses to the ambient environment caused by shielding of the collector support stand.
3. Test apparatus not at steady-state or "quasi-steady-state" condition during test.
4. Change in specific heat of the transfer fluid.
5. Inexperienced technicians conducting the tests.

Systematic instrument error sources included

1. Conduction errors in thermocouple installations.
2. Pyranometer calibration error.
3. Flowmeter calibration error.

ASHRAE Standard 93-77 was modified to contain a requirement for using only a first-class pyranometer or pyrheliometer as classified by the World Meteorological Organization (WMO).

A round-robin program similar to the one described above was initiated as part of the International Energy Agency tasks on solar heating and cooling, and tests completed in 1977 and 1978 (Talarek 1979). Two liquid-heating flat plate collectors (one of which was identical to collector 2 in the NBS round-robin program) were distributed to 16 laboratories in 12 different countries throughout the Northern Hemisphere. The reported data fell within a band bracketed by theoretical efficiency curves established by assuming that extremes of allowable environmental test conditions occurred simultaneously with allowable measurement errors. No prominent grouping of data emerged around the mean values; therefore systematic differences between test facilities were suspected.

An NBS project investigated accelerated and real-time test methods for predicting collector durability and reliability (Waksman et al. 1981). Representative, commercially available flat plate collectors and their materials were exposed outdoors in four different U.S. climatic regions.

Tests consisted of efficiency and incident angle modifier tests on eight collector types at each of the four sites, with two to four collectors being tested at each site. The reproducibility of results between test sites and the repeatability of results within a single test are reported in Streed and Waksman (1981a, b). The average coefficient of variation for the y-intercept for all collector types evaluated was 2.4%, and for the slope, 8.4%. Those results showed improvement over the results of the first NBS round-robin program (4.7%–7.7% in y-intercept and 15.8%–24.9% in slope) and indicated advances in experimental procedures for collector testing over the intervening two to three years. Within a test site the average coefficient of variation in the y-intercept and slope reduced to 2.1% and 5.9%, respectively.

The incident angle modifier (IAM) results were reported in Thomas et al. (1982). The concept of the incident angle modifier was originally conceived as accounting for the change in material radiative properties of the glazing system and absorber of a flat plate collector and consequently only a function of θ (Souka and Safwat 1966). The data test from the durability/reliability program (Thomas et al. 1982) showed that the effects of shading the absorber by the ends and sides of the four collectors considered was the same order of importance as the change in transmitting properties of the cover absorber assembly as the incident angle changed. Consequently flat plate collectors may exhibit a bidirectional angular response.

Figure 5.7 shows the results for the four collector types analyzed (Thomas et al. 1982). The IAM calculated from theory would lie within the area bracketed by the dotted lines depending on the orientation of the collector (effects of shading mentioned earlier). The mean IAM curve based on all measurements (figure 5.7) corresponded closely to curves 8, 7, 5, and 6 for collectors A, B, D, and E, respectively. The relative agreement between calculated and measured results was about the same for the collectors with flat black absorbers (B and E) and those with selective absorbers (A and D). The probable uncertainties in results due only to the uncertainties in measurements made were calculated and are shown in figure 5.8 as the dashed lines. (Assuming the uncertainties in the individual measurements were equally probable, this is the square root of sum of the squares of the individual uncertainties; see Kline and McClintock 1953).

The large uncertainty in test results for the IAM of conventional flat plate collectors was found to have only a minor effect on the uncertainty in predicted

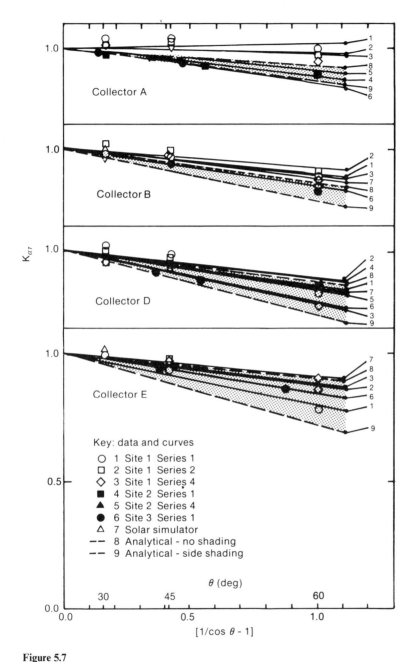

Figure 5.7
Incident angle modifier data, correlations, and calculated curves for four collectors. Source:
Thomas et al. (1982).

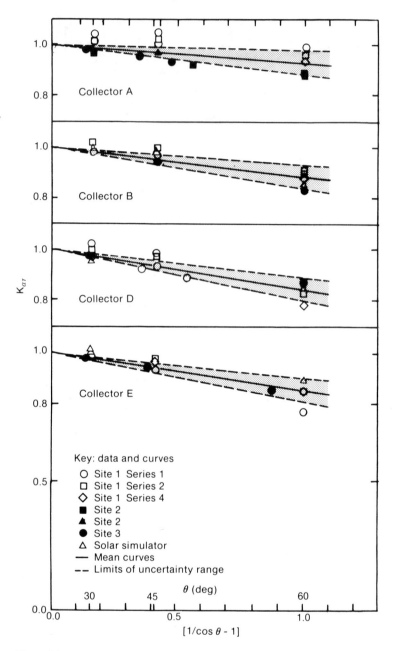

Figure 5.8
Correlations and uncertainty ranges for incident angle modifiers based on measurements for four collectors. Source: Thomas et al. (1982).

seasonal performance of solar space heating and domestic hot water systems with relatively efficient collectors and met a reasonable load. For example, the uncertainty in calculated seasonal solar fraction for a combined system in Madison, Wisconsin, was only 0.05 as a result of the uncertainty for collectors D and E indicated in figure 5.7.

Thomas et. al (1982) concluded that for flat plate collectors, the uncertainty in determining the IAM experimentally is the same order of magnitude as the efficiency reduction from the normal to the maximum incident angle, and therefore a simplified calculation procedure was recommended in lieu of the experimental procedure. Also it was recommended that the experimental procedure be used for collectors with a strong thermal performance dependence on incident angle (i.e., with complex geometries on concentrating features). However, the IAM for these collectors is expected to depend strongly on the scattered fraction of solar radiation, which is not accounted for in the present procedure.

The most recent collector program was the DOE Solar Collector Testing Program to provide uniform and comparable thermal performance test data on a significant portion of the solar collectors commercially available in the United States in the late 1970s. It was a joint effort of the federal government and the U.S. solar industry to accelerate the development and implementation of a nationally recognized certification, rating, and labeling program for solar collectors. The efficiency range of nine flat plate collector types that were evaluated in the test program is summarized in table 5.1 (Kirkpatrick 1983).

Testing Concentrating Collectors   Testing procedures required to test concentrating collectors were developed primarily for flat plate collectors and can be applied if the individual tests are adapted to fit the particular type of concentrator under consideration.

Closed- and open-loop test configurations for collectors operating at temperatures below approximately 100°C (212°F) and at only moderate pressures are shown in figure 5.3 and Hill, Wood, and Reed (1985). A facility to allow testing to 315°C (600°F) and 0.51 MPa (75 psig) is described in Harrison, Dworzok, and Folkner (1977). Because of the high temperatures and pressures, heat transfer oils are frequently used as the working fluid. Reference heat sources or calorimeters are utilized to determine the collector thermal output (Jenkins 1979; Reed and Allen 1978; Collares-Pereira et al. 1981).

It is unrealistic for tracking collectors with high concentration ratios to be exposed to preconditioning tests (e.g., the three-day test specified in ASHRAE Standard 93-77) because of the extremely high temperatures likely to occur.

Table 5.1
Collector test performance summary

Units/group tested	Heat transfer fluid	Number of glazings	Absorber coating	Thermal performance equations (first-order, English units)
6	Air	1	Flat-black	max. $\eta_g = 0.590 - 1.210P$ med. $\eta_g = \begin{cases} 0.538 - 1.127P \\ 0.438 - 1.512P \end{cases}$ min. $\eta_g = 0.331 - 0.759P$
12	Air	2	Flat-black	max. $\eta_g = 0.624 - 1.129P$ med. $\eta_g = 0.552 - 0.995P$ min. $\eta_g = 0.428 - 0.906P$
1	Air	3	Flat-black	med. $\eta_g = 0.394 - 0.983P$
3	Air	1	Selective	max. $\eta_g = 0.606 - 0.919P$ med. $\eta_g = 0.549 - 0.569P$ min. $\eta_g = 0.476 - 1.125P$
1	Liquid	0	Flat-black	med. $\eta_g = 0.672 - 2.483P$
43	Liquid	1	Flat-black	max. $\eta_g = 0.792 - 1.355P$ med. $\eta_g = 0.654 - 1.217P$ min. $\eta_g = 0.472 - 0.925P$
19	Liquid	2	Flat-black	max. $\eta_g = 0.691 - 0.894P$ med. $\eta_g = 0.619 - 0.833P$ min. $\eta_g = 0.421 - 0.604P$
26	Liquid	1	Selective	max. $\eta_g = 0.762 - 0.624P$ med. $\eta_g = 0.675 - 0.973P$ min. $\eta_g = 0.512 - 0.645P$
6	Liquid	2	Selective	max. $\eta_g = 0.686 - 0.874P$ med. $\eta_g = \begin{cases} 0.665 - 0.896P \\ 0.571 - 0.759P \end{cases}$ min. $\eta_g = 0.492 - 0.591P$

Source: Kirkpatrick (1983).
Note: $P = (t_{f,i} - t_a)/I_t$.

The normal control strategy with such collectors is to defocus the concentrator whenever the absorber temperature reaches a preset temperature.

The time constant test can be applied as described earlier. The time constant test for a single-axis, tracking, linear, parabolic, trough-concentrating collector was performed (Wood, Fiore, and Christopherson 1979) by maintaining the inlet temperature to the collector while in the defocused position and suddenly moving it to the focused position, and vice versa. The time constant was found to be 1.4 minutes and 1.8 minutes, respectively. The full range of tests specified in ASHRAE Standard 93-77 were also reported.

Tests of five different concentrating solar collectors having geometric concentration ratios up to 67:1 were reported (Dudley and Workhoven 1978), and the results are shown in figure 5.9 (the average fluid temperature appears in the axis of the plot rather than the inlet fluid temperature normally used). The tests in Standard 93-77 were modified to meet the specific characteristics of the collectors. The results are in agreement with Wood, Fiore, and Christopherson (1979). The tests (Wood, Fiore, and Christopherson 1979; Dudley and Workhoven 1978) were conducted at times close to solar noon

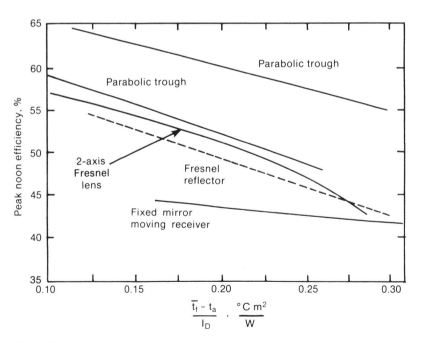

Figure 5.9
Comparison of peak solar noon efficiency for five concentrating solar collectors. Source: Dudley and Workhoven (1978).

with the collectors in the tracking mode. The optical efficiency of a simple flat plate collector is the same at an incident angle of 30 deg at 0 deg, whereas at those two incident angles it can be quite different for a concentrator. Therefore the near-normal incidence efficiency tests should be conducted well within the "acceptance angle" of the collector, and not just anywhere between 0 to 30 deg.

ASHRAE Standard 93-77 requires that both a pyrheliometer and pyranometer be used for concentrators so that both the total incident radiation and the direct component can be determined. Two efficiency curves are then required, one that uses the global incident radiation to calculate efficiency and a second that uses only the direct component (see Dudley and Workhoven 1978).

Using a single curve to describe the performance is even less appropriate for concentrating collectors than for flat plate collectors because of the change in heat loss coefficient with operating conditions. Following the recommendation of Gani, Proctor, and Symons (1981), the heat loss from high temperature collectors is predominately a function of operating temperature; therefore curves for different operating temperatures provide a more complete characterization of the collectors' performance.

The most significant deviation from the specifications of a procedure (e.g., ASHRAE Standard 93-77) to adequately test concentrators is required to conduct the incident angle modifier test. For concentrators the incident angle modifier test should be conducted based on the collector design and intended use, whereas for flat plate collectors the incident angle modifier test can be conducted for specific incident angles of 30, 45, and 60 deg.

An analysis of the characteristics of the single-axis tracking linear concentrator tests (Wood, Fiore, and Christopherson 1979) led to the conclusion that the incident angle modifier could be described by

$$K_{\alpha\tau} = K_0 K_E K_{A_P},\tag{18}$$

where

K_0 = optical loss coefficient to account for losses caused by off-normal reflection or refraction from the concentrator and incident angle effects for the receiver,

K_E = end loss coefficient to account for the fact that part of the receiver is not illuminated during off-normal operation,

K_{A_P} = $\cos\theta$ to account for the change in projected aperture area normal to the direct solar beam.

Since the collector had an east–west axis and tracked north–south, tests were conducted in accordance with ASHRAE Standard 93-77 for various incident angles in the east–west direction measured relative to the outward drawn normal. The results are shown in figure 5.10. Since the last two terms in equation (18) could be calculated knowing the incident angle θ and the geometry of the collector, the optical loss coefficient was calculated from the test results and is shown in figure 5.11 with the other two coefficients. Note that the end losses and projected aperture losses were greater than the optical losses for incident angles up to 45 deg. Beyond 45 deg, the optical losses were dominant.

If the concentrator is a two-axis tracking collector, the incident angle modifier should be identically one except for tracking error. At present there is no procedure incorporated in any of the specific standard testing procedures for determining tracking error.

Collector Testing with Solar Simulators   As noted previously, ASHRAE Standard 93-77 contains provisions for conducting tests indoors under simulated solar irradiation as well as outdoors under natural solar irradiation. These provisions are also included in ASHRAE Standard 96-1980 for unglazed collectors and ASHRAE Standard 93-1986 (a revision of ASHRAE Standard 93-77).

Although higher cost factors could represent a major drawback to the widespread use of solar simulators, the inherent advantages that solar simulators have over outdoor testing, such as time saving, stability of test conditions, reproducibility, and accuracy, has resulted in a significant growth in the use of solar simulators in the United States, Canada, and various European countries (Aranovich and Gillet 1982).

It is not possible to duplicate natural outdoor conditions exactly therefore some differences will exist between test results obtained outdoors and indoors in a simulator. These differences may be attributed to the following factors: the ratio of diffuse to direct radiation, the degree of collimation of the light source, the sky temperature seen by the collector, and differences in the spectra of the solar simulator and the sun (Nielson 1984). To evaluate the performance of solar simulators, a number of studies have been conducted in which collector thermal performance was determined with a solar simulator under natural outdoor conditions. Indoor test results compared very well with the outdoor test results in these studies (Nielson 1984; Dokos 1982; Gillet and Moon 1982).

5.1.3   Testing Solar Collectors under Zero-Irradiance Conditions

A German Bundesverband Solarenergie (BSE) working group recommended in May 1978 a standard test procedure for solar collectors that is different

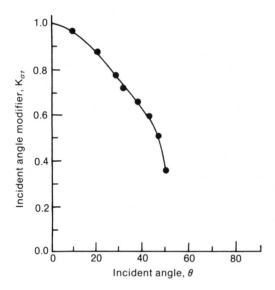

Figure 5.10
ASHRAE Standard 93-77 incident angle modifier test results for a concentrator. Source: Wood,
Fiore, and Christopherson (1979).

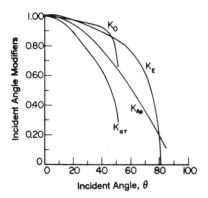

Figure 5.11
Loss coefficients that make up the incident angle modifier for a concentrator. Source: Wood,
Fiore, and Christopherson (1979).

from the other standard procedures proposed and adopted. The BSE procedure requires the collector optical efficiency and thermal loss characteristics to be determined independently through outdoor and indoor tests, respectively; other procedures generally require complete testing under full solar irradiance. The optical efficiency is determined during outdoor testing in which the operating conditions are regulated so that the collector experiences negligible heat loss. The thermal loss characteristics are determined under indoor laboratory conditions with zero solar irradiance while circulating the working fluid through the collector over a range of operating temperatures above ambient air temperature. The two separately determined properties are then used to generate the collector near-normal-incidence instantaneous efficiency curve as a function of various operating conditions.

Before BSE published its results, research on the technique was completed by Symons (1976), Christie (1976), Smith and Weiss (1977), and Whillier (1965). NBS evaluated the experimental procedure by testing five single- and double-glazed flat plate water-heating collectors and two unglazed flat plate water-heating collectors (Jenkins and Hill 1980a, b; Jenkins and Reed 1982; Jenkins 1982) using both the BSE and ASHRAE procedures. The studies showed that the difference in efficiency between the two procedures was less than the experimental uncertainty of the efficiency determined in accordance with the ASHRAE procedures, that the repeatability of test results was improved by using the BSE procedure because outdoor testing was kept to a minimum and indoor testing was performed under controlled laboratory conditions, and that the total test time associated with the BSE procedure was reduced by minimizing the dependence of testing upon favorable outdoor weather conditions. The procedure was found to be particularly advantageous for the unglazed collectors because their efficiency is very sensitive to environmental conditions and varied over a whole range of wind speeds and directions.

Despite the experiences and recent experimental evaluations, researchers have been reluctant to adopt the concept of an indoor heat loss test because of differences between collector thermal losses determined indoors and those experienced outdoors. Losses determined indoors will normally be lower than those experienced outdoors because the environmental and operating conditions are generally more favorable, and for the same environmental and operating conditions, the fundamental heat loss mechanisms are different. Consequently the collector efficiencies will be higher when calculated based on indoor heat loss tests. However, the NBS experimental studies (Jenkins and Hill 1980a, b; Jenkins and Reed 1982; Jenkins 1982) have shown how the environmental and operating conditions can be controlled during the indoor laboratory tests to closely match those occurring outdoors. The differences in heat loss likely to occur between indoor and outdoor tests have been analytically studied (Svendson 1978; Gillet 1981; Jenkins and Bushby n.d.). In addition recent analytical studies have shown that collector differences can be predicted and then used to correct the results after the mixed indoor/outdoor tests are completed.

Figure 5.12 conceptually shows the difference in efficiency, $\Delta\eta$, that occurs assuming environmental and operating conditions are identical. Figure 5.12 is a typical result from an outdoor test under full-irradiance conditions. Average fluid temperature $\bar{t}_f$ was used in the x-axis instead of inlet fluid temperature $t_{f,i}$ for convenience of analysis in Jenkins and Bushby (n.d.). An identical outdoor test is conducted to determine the efficiency curve y-intercept η_0 when the mixed indoor/outdoor test procedure is used as shown in figure 5.12. The collector thermal losses are determined in the indoor heat loss test as shown in figure 5.12 and then combined with η_0 to generate an efficiency curve for a selected irradiance level as shown in figure 5.12b. If the results of the mixed indoor/outdoor tests are plotted in a manner consistent with the results of the outdoor tests under full-irradiance conditions (figure 5.12), the curves in figure 5.12 are obtained. Note that the curve intercepts are equal because of the identical procedures used to determine η_0. The magnitude of $\Delta\eta$ is directly related to differences between characteristic slope of the two efficiency curves $\Delta(F'U_L)$. However, because the collector efficiency factor F' is essentially constant for a specific collector and set of operating conditions, $\Delta\eta$ fundamentally depends on the difference between the collector loss coefficients ΔU_L for the same operating conditions.

As shown in figure 5.13, the collector heat transfer loss coefficient U_L can be represented as a linear function of the absorber plate temperature $\bar{t}_p$. The slope and magnitude of the U_L versus $\bar{t}_p$ curve is unique for a specific collector and set of environmental conditions. From the figure, ΔU_L, and consequently $\Delta\eta$, depend on factors that contribute to differences between the mean absorber plate temperature $\Delta\bar{t}_p$ during indoor testing $(\bar{t}_{p,i})$ and outdoor testing $(\bar{t}_{p,o})$ for the same fluid temperature. The fluid temperature is the reference temperature in characterizing the collector efficiency. The primary factor that contributes to $\Delta\bar{t}_p$ is inverse fluid-absorber temperature profiles between indoor and outdoor testing.

Reverse modes of heat transport between indoor and outdoor testing are responsible for the inverse fluid-absorber temperatures profiles. During outdoor testing, the useful thermal energy is transported from the absorber to the working fluid. As a result the mean absorber plate temperature $\bar{t}_{p,o}$ is

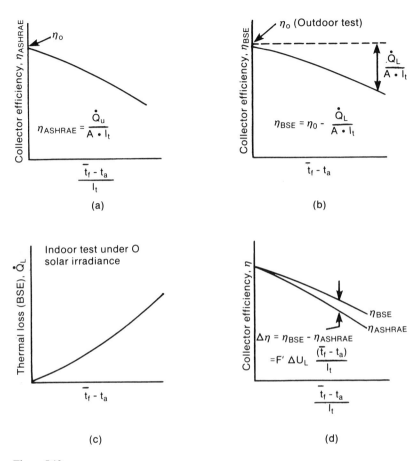

Figure 5.12
(a) ASHRAE collector efficiency curve; (b) BSE collector efficiency curve. (c) BSE collector thermal losses; (d) comparison of BSE and ASHRAE efficiency curves. Source: Hill, Wood, and Reed (1985).

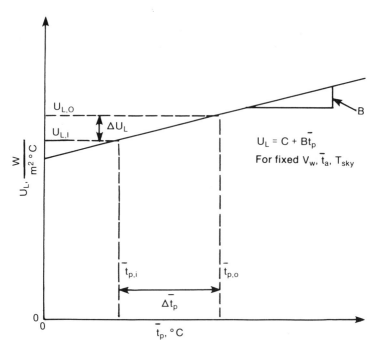

Figure 5.13
Collector loss coefficient U_L as a function of the mean absorber plate temperature $\bar{t}_p$.

higher than the mean working fluid temperature $\bar{t}_f$. Conversely, during indoor testing under zero irradiance conditions, the thermal losses from the working fluid are transported to the absorber and then from the collector into the environment. Consequently the mean absorber plate temperature $\bar{t}_{p,i}$ is less than $\bar{t}_f$. Because of these reverse modes of heat transport, the absorber plate temperatures are reversed such that $\bar{t}_{p,0} \geq \bar{t}_f \geq \bar{t}_{p,i}$ for a given value of t_f. Therefore the magnitude of Δt_p depends on the absorber-fluid thermal resistance and the magnitude of heat transfer occurring in the collector.

Jenkins and Bushby (n.d.) developed an equation for $\Delta\eta$ containing four terms and verified its accuracy by modeling the thermal performance of selected single- and double-glazed flat plate collectors operating both outdoors and indoors. The equation predicted $\Delta\eta$ to within ± 2.5 efficiency points and was valid for any type of collector whose performance could be expressed in terms of the Hottel-Whillier equation. More important, Jenkins and Bushby developed a short procedure that calculated all the terms in the equation for $\Delta\eta$ from the results of the mixed indoor/outdoor tests.

5.1.4   Considerations in Testing Air Collectors

Air Leakage   Air leakage in air heaters is a troublesome problem that can affect the actual performance of an installed system as well as test results. The array should be tested so that the air leakage occurring during the tests will indicate leakage that will occur in an actual installation as a result of the collector design and recommended installation practice. Measurements should be made in such a way as to determine the true output of the collector array. During testing, the test loop can be sealed sufficiently well with duct tape, caulking, etc.

The results of an analysis of air leakage in air-heating collectors reported in Close and Yusoff (1978) assumed a constant leakage rate along the length of the collector. The effect on the efficiency measurements was determined for all combinations of operating the collector under negative pressure (air leaking in) and positive pressure (air leaking out), and measuring the airflow rate upstream of the collector or downstream of the collector.

The results are shown in figure 5.14. The abscissa is the leakage rate divided by the measured flow rate and the ordinate is the ratio of the actual collector efficiency to the measured efficiency. For three of the testing configurations, the discrepancy between measured and actual efficiency is a direct function of the leakage rate. For the case where the collector is operating under negative pressure (air leaking in), the difference in efficiency depends on the difference in temperature between the ambient air and the entering airstream to the collector. Therefore the error in efficiency for this case is larger for the data points at the higher inlet fluid temperatures relative to ambient. Also note that when the air leaks in, the actual efficiency is larger than indicated by the measurements but just the opposite is true when air leaks out. If the airflow measurement is upstream of the collector, the collector is heating a larger quantity of air than is indicated by the measurements. When the flow rate is measured upstream and air leaks out of the collector, the quantity of useful heated air is less than indicated by the measurements. However, if air leaks out and the airflow measurement is made downstream of the collector, the quantity of useful heated air is precisely what is measured, and as a result there is no difference between actual and measured collector efficiency. Such data would produce a horizontal straight line with an ordinate value of 1 in figure 5.14.

The implications of this analysis for testing air-heating collectors are

1. If the collector is normally operated under positive pressure, it should be tested while operating under positive pressure, and the airflow rate should be measured downstream of the collector.

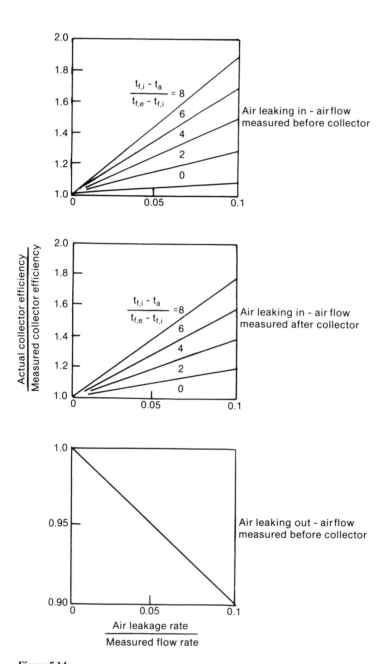

Figure 5.14
Relationship between actual efficiency and measured efficiency as a function of air leakage rate for air-heating solar collectors. Source: Close and Yusoff (1978).

2. If the collector is normally operated under negative pressure, it should be tested while operating under negative pressure, and the airflow rate should be measured both upstream and downstream of the collector to quantify the leakage rate. An estimate of the actual collector efficiency could be made by the user of the collector with that data provided, even though the "correction" might not be made and published as part of the test results.

Predicting Collector Array Performance from Tests on Single Modules   As indicated by equation (1), the useful heat output of the collector is directly proportional to the collector heat removal factor F_R. If the collector test results are to be used without correction to predict the performance of the collector in an actual field installation, then the value of F_R that occurs during the test should be approximately the same as the value that will occur in the field. The expression for F_R is

$$F_R = \frac{\dot{m}c_p}{A_a U_L} \left\{ 1 - \exp - \left[\frac{A_a U_L F'}{\dot{m}c_p} \right] \right\}. \tag{19}$$

Therefore the conditions during testing and in actual field operating should be such that $\dot{m}/A_a$, c_p, U_L, and F' are approximately the same in both cases. For liquid-heating collectors this is readily accomplished.[1] The flow rate per unit collector area, $\dot{m}/A_a$, the type of fluid and hence specific heat, c_p, and the collector tilt angle, temperature, incident solar radiation, and ambient conditions are such that the heat loss coefficient U_L is approximately the same. In addition the collector efficiency factor F' is primarily a function of the geometry of the absorber (materials, thickness, distance between tubes, etc.) and is the same for the tested and installed collectors at any flow rate.

The situation is slightly different for air heaters. Although $\dot{m}/A_a$, c_p, and U_L can be made the same in a similar fashion as with the liquid-heating collectors, ensuring the same value of F' is more difficult. For air heaters F' is primarily a function of the convection heat transfer coefficient between the absorber and the airstream. Since the airflow is nearly always in the turbulent flow range (to maximize the heat transfer), the heat transfer coefficient is determined primarily by the value of the Reynolds number in the collector.

Quite often air collectors are designed to be installed with at least two modules in the series. A procedure has been developed so that the test results on a single module can be corrected to predict the performance of the collectors after installation in the array (Hill, Jenkins, and Jones 1979; Oonk, Jones, and Cole-Appel 1979). The single module must be tested at a flow rate that will result in the same flow velocity (and hence Reynolds number) that will occur in the array of collector modules in series. The collector output for the single module is then multiplied by a correction factor to obtain the output for the array. The correction procedure was experimentally verified by tests on one and then two modules of a flat plate air heater connected in series (Hill, Jenkins, and Jones 1979).

5.1.5 Calculating All-Day Collector Performance

Calculation Including Diffuse Solar Irradiance   The standard test methods described previously were developed to provide sufficient data for calculating the useful daily energy collected by the solar collector. This is essential for design purposes, and since the performance of collectors can vary widely, it becomes necessary to consider their daily output when making a comparison between any two of them.

The ASHRAE Standard 93-77 and 96-1980 data can be used to calculate the collector all-day performance as illustrated in figure 5.15. Excellent agreement between calculated and measured all-day performance for clear-sky conditions has been obtained (Wood, Fiore, and Christopherson 1979; Zerlaut, Dokos, and Heiskell 1977; Beach 1979; Hill et al. 1979). The agreement in thermal energy collected was typically $\pm 5\%$ for a wide range of collectors. The calculated and measured all-day performance of a flat plate collector was compared under various sky conditions (Farber, Dixon, and Wix 1981); even for cloudy days, the agreement was $\pm 6\%$.

The all-day performance of a collector is a priori dependent on the intensity-weighted directionality of the incoming solar irradiance. Consequently on a clear day the majority of solar irradiance strikes the aperture plane of the collector at incident angles defined by the sun's path through the sky (beam component). On partially diffuse or diffuse days, a large portion of the incident solar irradiance may come at incident angles significantly different from those experienced on a clear day.

The ASHRAE standard test methods are prescribed for clear skies with low levels of diffuse solar irradiance. The previous equations do not explicitly consider the response of the collector to diffuse solar irradiance; however, the required test data does include both the hemispherical and direct normal irradiance from which the global diffuse irradiance determined. Equation (11) can be modified to explicitly consider both the beam and diffuse components of the solar irradiance[2] as

$$\eta_a = \frac{A_a}{A_g} \frac{F_R(\tau\alpha)_{e,n}(K_{\alpha\tau} I_{dN} \cos\theta + K_d I_d)}{I_t} - \frac{A_a}{A_g} \frac{F_R U_L(t_{f,i} - t_a)}{I_t}, \tag{20}$$

where

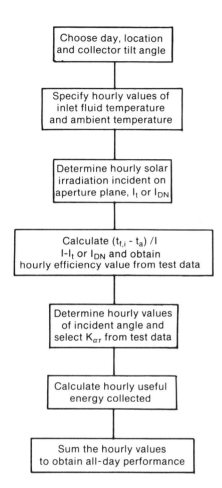

Figure 5.15
Calculating all-day useful energy collected using ASHRAE Standard 93-77 data.

A_g = gross area of the collector (m^2),

I_d = direct component of solar energy incident on the plane of the collector per unit time per unit area (W/m^2),

I_t = total solar energy incident on the plane of the collector per unit time per unit area (W/m^2),

K_d = diffuse incident angle modifier (Simon and Buyco 1976) such that

$$K_d = \frac{\int_0^{2\pi} I_d(\omega) K_\alpha \tau(\omega) \cos \theta d\omega}{\int_0^{2\pi} I_d(\omega) \cos \theta d\omega}. \tag{21}$$

For the case of isotropically distributed diffuse irradiance and for collectors whose $K_{\alpha\tau}$ can be described by equation (12), K_d becomes

$$K_d = (1 - b_0). \tag{22}$$

Note that $(1 - b_0)$ is the value of $K_{\alpha\tau}$ at an incident angle of 60 deg. This means that beam irradiance at an incident angle of 60 deg has the same transmittance-absorptance as all of the isotropic diffuse irradiance (i.e., the effective incident angle for isotropic diffuse irradiance at 60 deg). Brandemuehl and Beckman (1980) calculated the effective beam angle for solar collectors that "see" both the sky and ground.

SRCC Rating Calculation Methods   The Solar Rating and Certification Corporation (SRCC) developed a standard for calculating the all-day thermal performance of solar collectors using ASHRAE (1981) standard test data. The rating itself is a calculated set of numbers representing the all-day energy output of the solar collector under prescribed rating conditions. SRCC included a diffuse solar irradiance term very similar to that in equation (20), which assumes isotropic diffuse irradiance.

SRCC uses two kinds of incident angle modifiers. First, for most flat plate collectors $K_{\alpha\tau}$ is assumed to be symmetrical with respect to the normal direction from the aperture plane (i.e, $K_{\alpha\tau}$ is a function of the magnitude of θ only and not the direction of the incident rays). The diffuse incident angle modifier is given by equation (21) which, for isotropic diffuse irradiation, becomes

$$K_d = \int_0^{\pi/2} K_{\alpha\tau}(\theta) \sin 2\theta d\theta. \tag{23}$$

This integral can be evaluated either numerically or analytically, if an expression for $K_{\alpha\tau}(\theta)$ is available.

Second, the SRCC method considers the response of the collectors when $K_{\alpha\tau}$ is a function of both the angle of incidence and the direction of the solar irradiance. SRCC uses the biaxial incident angle modifier described by equation (13) (Mather and Beekley 1976). Two separate incident angle modifiers are experimentally determined as a function of the off-altitude angle and the off-azimuth angle. Mather (1980) suggested that the angle Ψ rather than the angle Γ was better to use in the correlation for evacuated tubes; hence

$$K_{\alpha\tau} = K_1(\Omega)K_2(\Psi), \tag{24}$$

where

K_1 = off-azimuth incidence angle modifier,

K_2 = off-altitude incidence angle modifier,

Ω = off-azimuth angle,

Ψ = off-altitude angle.

It was indicated that $K_2(\psi)$ was a better measure of the Fresnel reflection effects that occur when the solar rays are not perpendicular to the tube axis. Using ψ allows equation (21) to be simplified to

$$K_{\mathrm{d}} = \frac{4}{\pi} \int_0^{\pi/2} K_1(\Omega) - \cos\Omega\, d\Omega \int_0^{\pi/2} K_2(\Psi)\cos^2(\Psi)\, d\Gamma. \tag{25}$$

Effects of Diffuse Irradiance on Calculations To validate the SRCC method for calculating the all-day energy collection using the diffuse incident angle modifier, the Solar Energy Research Institute funded DSET Laboratories, Inc. to measure the performance of flat plate and stationary concentrating calculators and to compare the calculations using ASHRAE 93-77 test data and several models with all-day measurements. The following summary is from Putman, Evans, and Wood (1982).

An extensive literature survey showed that the effect of diffuse irradiance on the thermal performance of solar collectors could be taken into account for the majority of commercially available solar collectors as long as their directional dependence could be established. The major obstacle was the inherent inability to characterize the solar irradiance in a meaningful way.

The thermal performance of three generic collector types—flat plate, single-glazed, nonselective absorber; flat plate, single-glazed, selective absorber; and evacuated tube, stationary, concentrating reflector—were determined in accordance with ASHRAE Standard 93-77, including extensive tests to determine the incident angle modifiers. Subsequently the all-day, optical performance of each collector was determined (i.e., the collector inlet fluid temperature was maintained at ambient temperature during each test day) under three distinctly different sky conditions of clear, partly cloudy, and overcast. The collectors were oriented due south at a 34-deg tilt angle. The directional intensity of the incoming diffuse radiation during the Standard 93-77 and all-day tests was measured with a Directional Diffuse Radiation Flux Mapper, an instrument designed specifically for this project. These data were then used to compare the predicted performance using the various techniques to the measured all-day performance. The uniqueness of this research was its ability to measure the actual direction and intensity of the radiation during testing and thereby to evaluate unproven sky and collector models.

The measured sky radiation for the overcast sky was isotropic (i.e, uniformly distributed over the sky). Similar data for the partly cloudy day and the clear-day data showed that the diffuse radiation was anisotropic. It was concluded that the clear-sky diffuse radiation could be modeled as a pseudo-solar beam and a constant or isotropic background. Because of the random distribution of discrete clouds, there is no generalized model for diffuse irradiance for a partly cloudy day.

The optical characteristics of the evacuated-tube collector were not symmetrical when comparing off-altitude and off-azimuth incident angle modifier data. However, the optical characteristics of the other two collectors were found to be symmetrical.

Measurements of the time-dependent variables were based upon averages for each 15-minute interval over each of the three different test days. Using these measurements, the all-day performance was calculated using five different calculation techniques. Each of the calculation techniques treated the diffuse solar irradiance differently. The five methods were as follows:

1. The *BEST* method used the beam irradiance as measured with the Directional Diffuse Radiation Flux Mapper and the measured incident angle modifier for each view angle defined by the flux mapper. The BEST calculations were the only ones that included the actual measured distribution of the sky diffuse radiation. Four different methods were used to calculate the incident angle modifier. In BEST 1 and BEST 2 a biaxial incident angle modifier was defined as the product of the incident angle modifiers from two mutually perpendicular directions. A full two-dimensional mapping of the incident angle modifier was completed for the evacuated tube collector, and the all-day calculations were designated the BEST 4 method. The BEST 3 method applied to collectors with axisymmetric properties such as the flat-plate collectors. In BEST 3 the incident angle modifier depended only on the incident angle and not on the off-altitude and off-azimuth angles.

2. The *SRCC-ASYM* method used the beam irradiance and assumed the measured diffuse irradiance to be isotropic. The incident angle modifiers were biaxial and defined as the product of the off-altitude and off-azimuth incident angle modifiers as in BEST 1 and BEST 2. The all-day calculations were designated as SRCC-ASYM 1 and SRCC-ASYM 2. The SRCC-ASYM 1 method followed the SRCC method for "anisotropic" collectors.

3. The *SRCC-SYM* method used the beam irradiance and assumed the measured diffuse irradiance to be isotropic. The incident angle modifiers were symmetrical with respect to the normal direction from the collector aperture plane. This method was the same as the SRCC "isotropic" collector method.

4. The *SYMFIT* method used the beam irradiance and assumed isotropic diffuse irradiance. The incident angle modifier was determined from a single correlation for all the measured incident angle data [i.e., equation (12)]. The effective incident angle for the diffuse irradiance was taken as 60 deg.

5. The *TOT* method treated the global irradiance as beam irradiance with all incident angle modifier data "fit" to get b_0 in equation (12).

All of these methods gave surprisingly similar results for all-day performance. In fact, treating the global irradiance as if it were all beam irradiance, an assumption that greatly simplifies the calculations gave very good results for all collectors on clear, partly cloudy, and even totally cloudy (diffuse) days. The agreement between the measured all-day performance and the SRCC-SYM or SRCC-ASYM methods was improved by substituting an anisotropic diffuse sky model for the isotropic diffuse model. Using a biaxial incident angle modifier as done in the SRCC method appears to be reasonable. The results of Putman, Evans, and Wood (1982) are summarized in table 5.2.

It is interesting to note that the all-day collector optical efficiencies for each collector were essentially the same for each of the three solar days. This further suggests that the different diffuse irradiance distributions had little effect on the all-day collector performance.

5.2 Durability/Reliability Performance

5.2.1 Test Program Overview

Environmental exposure testing at both the collector component and material levels is required to ensure satisfactory long-term performance of solar collectors. Such tests should be capable of predicting the effects of long- term exposure in a relatively short time and at a reasonable cost. Under a DOE-sponsored program initiated in 1977, NBS undertook research to evaluate

Table 5.2
Percent difference between calculated and measured all-day performance for three collectors using various calculations methods

	Best				SRCC				TOT	Measured all-day efficiency
	1	2	3	4	ASYM1	ASYM2	SYM	SYMFIT		
Collector 1										
Clear day	4.8	4.8	4.5		3.5	3.0	2.6	1.3	3.9	53.8
Partly cloudy day	5.1	5.1	4.5		4.0	2.5	2.1	0.2	5.3	53.0
Diffuse day	0.9	0.9	1.1		0	−2.6	−2.4	−4.4	3.9	54.0
Collector 2										
Clear day	−1.0	−1.0	−0.7		−2.1	−2.6	−2.5	−2.9	−1.0	68.0
Partly cloudy day	−1.0	−1.0	−1.5		−2.2	−3.9	−4.0	−4.9	−0.3	67.3
Diffuse day	−3.4	−3.4	−2.8		−5.2	−7.9	−7.0	−8.2	0.7	67.2
Collector 3										
Clear day	3.4	3.4	4.6	2.7	1.6	1.4	3.0	6.2	7.0	43.9
Partly cloudy day	3.5	3.5	3.5	0.3	2.8	0.9	1.9	5.6	5.3	43.1
Diffuse day	0.2	0.2	1.2	−0.2	0	−3.2	−0.9	2.6	0.7	43.1

Source: Putman, Evans, and Wood (1982).

test and exposure procedures, which were intended to determine the influence of environmental exposure parameters that could affect the degradation of solar collectors and their materials. The procedures were also intended, to the extent possible, to provide a correlation between changes that occur at the material and collector component levels. The data obtained through their use was expected to provide more meaningful reliability/durability tests for solar collectors and their materials. A detailed description of the research is provided in Waksman, Streed, and Seiler (1981) and is summarized in the following sections.

Collector Tests   Table 5.3 summarizes outdoor exposure tests of solar collectors. The four test sites selected represent the extreme and median U.S. climatological conditions, as follows:

EXTREME CLIMATOLOGICAL CONDITIONS

1. Site 1

a. Hot, dry

b. High solar radiation

c. High UV radiation

d. Rural, desert environment

The hot, dry condition can be found in the southwestern states of Arizona, Nevada, and New Mexico. DSET Laboratories, Inc. in Phoenix, Arizona, was selected as the test site.

2. Site 2

a. Hot, humid

b. High solar radiation

c. Low to moderate UV radiation

d. Coastal, salt-air environment

The hot, humid condition can be found in Florida and along the Gulf Coast in Alabama, Mississippi, Louisiana, and Texas. The Florida Solar Energy Center in Cape Canaveral, Florida, was selected as the test site.

MEDIAN CLIMATOLOGICAL CONDITIONS

3. Site 3

a. Moderate temperature, dry

b. High solar radiation

Table 5.3
Summary description of field test series on solar collectors

Test series	Collector performance measurement	Conditions for weathering exposure	Properties of test series[c]
Series 1	Initial measurement in accordance with ASHRAE 93-77	Each collector preconditioned for each weathering exposure by purging with dry air to remove the remaining heat transfer fluid	1. Observation of collector performance and other characteristics for various weathering terms
Dry stagnation	Performance re-test after 3, 15, 30, 60, 120, and 240 days of exposure[a]		2. Provide data for comparing solar simulator performance with field data
			3. Provide data for comparing collector performance (with and without 3-day preexposure)
Series 2	Initial measurement in accordance with ASHRAE 93-77	Collectors same as in series 1, except that tests per NBSIR 78-1305A will be performed during series 2 test collectors only	1. Observation of effects of no-flow stagnation on collector performance and other characteristics
No-flow stagnation	Performance retests same as in series 1		2. Observations of effects of thermal shock tests representing (a) filling a hot collector with cool heat transfer medium and (b) summer rain on a hot collector
			3. Observation of static pressure leakage after 30 and 120 days of exposure
Series 3	Performance of test collectors measured in accordance with ASHRAE	During weathering exposure heat transfer flow rate maintained at 25% of operational flow rate for liquid	1. Observation of effects of normal operation on collector performance and other characteristics
Controlled flow	Performance retests same as in series 1		

Table 5.3 (continued)

Series 4	Initial measurement same as in series 1	Preconditioning and weathering exposures same as in series 1 except that a reflector[b] will be used on each collector during each day of weathering exposure;[a] solar radiation measurements required both with and without reflector	1. Observation of effects of dry stagnation on collector performance and other characteristics with solar radiation amplified by a reflector
Dry stagnation with augmentation reflectors	Performance retests same as in Series 1		2. Obtaining temperature history within collectors for most severe exposure conditions

Source: Waksman, Streed, and Seiler (1981).
a. Individual days with solar radiation of 17,000 kJ/m^2 day or greater as measured in the plane of the collector aperture without the influence of a reflector.
b. The reflector is described in Waksman et al. (1981).
c. All series include provision of data for comparing test series, test sites (climatic regions), collectors designs), etc.

c. Moderate UV radiation

d. Urban environment

The moderate, dry condition can primarily be found in parts of California. The Lockheed Research Laboratory in Palo Alto, California, was selected as the test site.

4. Site 4

a. Moderate temperature, humid

b. Moderate solar radiation

c. Moderate to low UV radiation

d. Suburban environment

The moderate, humid condition can be found in the Pacific Northwest, Mid-Atlantic, and Mid-South regions of the United States. The NBS test facility in Gaithersburg, Maryland, served as the test site.

All four test series, each having a sample of the eight collector types, were conducted at sites 1 and 2, and series 1 and 2 tests were also conducted at sites 3 and 4.

Eight types of liquid-heating, flat plate solar collectors were selected for use in the test program. The designs chosen were representative of commonly used materials and types of construction. All collectors from each manufacturer were from the same production lot. The collector cover and absorber materials, their pertinent optical properties, and average collector areas (gross and aperture) are listed in table 5.4. The absorber materials optical property data are based on measurements of at least ten samples taken from an actual absorber of each collector type prior to aging. The solar transmittance of the glass materials was obtained using ASTM Standard E 424, Method B (ASTM 1971a). The solar transmittance of nonglass cover materials and the solar absorptance values were obtained from spectral measurements with an integrating sphere (ASTM Standard E 424, Method A 1971a). Emittance was measured using a portable instrument employing a thermopile and infrared reflectance technique, in accordance with ASTM Standard E 408, Method A (1971b).

A description of each collector and its pertinent material properties are listed in appendix B of NBS Technical Note 1140 (Streed and Waksman 1981a).

Thermal performance measurements were made on the collector in each test series. The ASHRAE Standard 93-77 thermal performance test procedure was used in this program both as a full test (four temperature, four data points at each temperature) and as a three-point performance retest (three tempera-

Table 5.4
Test collector specimen description

Collector code	Cover material Outer	Inner	Solar transmittance	Absorber material Material	Solar absorptance	Emittance	Average Area[a] Gross m²	Aperture m²
A	Water white glass	—	0.90	Black nickel	0.87	0.13	2.150	1.831
B	Low iron glass	Low iron glass	0.88	Black velvet paint	0.97	0.96	1.732	1.602
C	Plate glass	Thin-film heat trap	0.86	Black velvet paint	0.98	0.92	2.589	1.924
D	Etched glass	Etched glass	0.96	Black chrome	0.97	0.07	1.655	1.402
E	FRP[b] (type 1)	—	0.85	Lacquer primer	0.95	0.87	1.892	1.720
F	Water white glass	—	0.90	Copper oxide	0.96	0.75	1.922	1.769
G	FRP[b] (type II)	FEP[c] film	0.84[b] 0.96[c]	Porcelain enamel	0.93	0.86	2.563	2.188
H	Polyester[d] film	FEP[c] film	0.85[d] 0.96[c]	Siliconized polyester paint	0.95	0.89	2.916	2.641

Source: Waksman, Thomas, and Streed (1984).
a. Average of values reported by four test sites.
b. Glass-fiber-reinforced plastic.
c. Fluorinated (ethylene propylene) copolymer.
d. Poly(ethylene terephthalate).

tures, four data points at each temperature). The retest was conducted to demonstrate the magnitude of changes in the intercept and slope of the efficiency curve as a function of environmental exposure time. The measurements were required to be spread over the range of $0.02°$ to $0.07°C\ m^2/W$.

Data collected in addition to collector thermal performance included the following:

1. Key environmental parameters—high and low daily ambient temperatures, total and diffuse solar radiation, peak hourly solar radiation, wind velocity, precipitation, and visual sky and weather conditions on a daily basis.

2. Maximum daily collector absorber plate temperature (for all collectors).

3. Daily profiles of absorber plate temperature (for collectors D and H), ambient temperature, irradiance, and wind velocity at all four outdoor exposure sites for a period of one year.

4. Visual observations of changes in collector appearance during outdoor exposure and disassembled collectors following the completion of outdoor exposure and thermal performance testing.

5. Optical property measurements on coupon samples of polymeric cover materials and absorber materials taken from disassembled exposed and unexposed collectors and microstructural studies on specimens showing degradation.

Collector Materials Tests   Coupon specimens of cover plate and absorber materials were subjected to several different types of laboratory and outdoor environmental exposure test conditions. Specimens consisted of samples taken from the eight types of full-size collectors used in the test program and of several additional materials of interest. Changes in the optical properties of these materials were measured as a function of exposure time. Microstructural studies were conducted on materials showing visible degradation.

Outdoor exposure conditions at the materials coupon specimen level "real-time" exposure in simulated collectors and exposure to concentrated radiation in machines, described in ASTM E 838 (1981a). Indoor laboratory tests conducted included exposure to temperature, combined temperature and humidity, combined temperature and radiation, and thermal cycling (absorber materials only). Additional exposure tests of materials level were conducted in xenon arc and tungsten lamp solar simulators. The outdoor "real-time" materials level exposures were conducted concurrently with those on full-scale collectors at the four outdoor exposure test sites.

The exposure conditions used for the cover and absorber materials are summarized in tables 5.5 and 5.6, respectively. The exposure conditions are

Table 5.5
Exposure tests for cover materials

Exposure conditions	Value of range	Exposure time
Temperature (indoor)	70°C (158°F) 90°C (194°F) 125°C (257°F)	500, 1,000, and 2,000 hours
Temperature and humidity (indoor)	70°C (158°F) and 95% RH 90°C (194°F) and 95% RH	500, 1,000, and 2,000 hours
Temperature and radiation (indoor)	Xenon arc weathering machine 70°C (158°F) 90°C (194°F)	500, 1,000, and 2,000 hours
Solar simulator	Tungsten Xenon simulators with irradiance of ~ 950 W/m^2 and ~ 70°C (158°F)	30, 60, and 120 cycles[a]
"Real-time" outdoor	1 sun at ~ 60°C (140°F)	80, 160, 240, and 480 days[b]
Accelerated outdoor	~ 6 suns at ~ 70°C (158°F)	6, 12, and 24 equivalent months[c]

Source: Waksman, Thomas, and Streed (1984).
a. Each cycle consists of 5 h irradiation and 1 h cooling.
b. Days having a minimum radiant exposure of 17,000 kJ/m^2.
c. One equivalent month equals 6.625×10^8 J/m^2 (15,835 Langleys).

Table 5.6
Exposure tests for absorber materials

Exposure conditions	Value of range	Exposure time
Temperature (indoor)	150°C (302°F) 175°C (347°F)	1,000 and 2,000 hours
Temperature and humidity (indoor)	90°C (194°F) and 95% RH	1,000 and 2,000 hours
Thermal cycling (indoor)	$-10°-175$°C (23°-115°F)	5, 15, and 30 cycles
Temperature and radiation (indoor)	Xenon arc weathering machine at 90°C (194°F)	1,000 and 2,000 hours
Solar simulator	Tungsten Xenon simulators with irradiance of ~ 950 W/m^2 and ~ 130°C (266°F)	30, 60, and 120 cycles[a]
"Real-time" outdoor	1 sun at ~ 140°C (~ 284°F) and ~ 160°C	80, 160, 240, and 480 days[b]
Accelerated outdoor	~ 6 suns at ~ 150°C (~ 302°F)	6, 12, and 24 equivalent months[c]

Source: Waksman, Thomas, and Streed (1984).
a. Each cycle consists of 5 h irradiation and 1 h cooling.
b. Days having a minimum radiant exposure of 17,000 kJ/m^2.
c. One equivalent month equals 6.625×10^8 J/m^2 (15,835 Langleys).

intended to simulate a broad range of environmental stress conditions. Primary emphasis was placed on exposure to temperature, solar radiation, and moisture. Other degradation factors such as hail, pollutants, and dust are localized in nature and were assessed via their occurrence at the "real-time" outdoor exposure test sites participating in the program.

Test specimens were obtained from the cover and absorber, as listed in tables 5.7 and 5.8. Material specimens having code letters A through H were cut from solar collectors of the same type and batch as those exposed as full-size collectors. The apparatus in which test specimens were mounted for outdoor exposure are shown in Waksman, Streed, and Seiler (1981) and Waksman, Thomas, and Streed (1984).

Data collected include integrated solar transmittance and absorptance per ASTM E 424-71, Method A (1971a); emittance per ASTM E 408-71, Method A (1971b); normal and hemispherical spectral transmittance curves measured on spectrophotometers with and without integrating spheres; visual observation of changes; microscopic evaluations using both optical and scanning electron microscopes; key environmental parameters, the same as for solar collector exposure; and test apparatus exposure temperatures on a daily basis.

Table 5.7
Cover test materials

Code	Solar cover material	Transmittance (controls)[c]
E	FRP[d] type Ia	0.85
G	FRP[d] type II	0.84
H[a]	PET[e]/FEP[f] (outer/inner)	0.85/0.96
J	Polycarbonate	0.88
K	Poly(vinyl fluoride)	0.89
L	FRP[d] type Ib	0.84
M	FRP[d] type III	0.78
N	Poly(methyl methacrylate)	0.90
O[b]	Glass[g]/poly(vinyl fluoride) (outer/inner)	0.86/0.89

a. Code letters E, G, and H indicate materials coupon specimens cut from solar collectors E, G and H. Codes J, K, L, M, N, O tested at the materials level only.
b. Materials exposed as a combination in the cover miniboxes and in the accelerated exposure cover miniboxes. Materials exposed individually in all other tests. Glass and FEP materials were not exposed individually.
c. These properties depend on the formulation and manufacturing processes used. Other products within a generic class of materials may have significantly different properties.
d. Glass-fiber-reinforced plastic.
e. Poly(ethylene terephthalate).
f. Fluorinated (ethylene propylene) copolymer.
g. Ordinary plate glass.

Table 5.8
Absorber test materials

Absorber material			Optical properties[b]	
Code[a]	Coating	Substrate	Absorptance[c]	Emittance[c]
A	Black nickel	Steel	0.87	0.13
B	Flat black paint	Copper	0.98	0.92
D	Black chrome	Steel (nickel flashed)	0.97	0.07
E	Flat black paint	Copper	0.95	0.87
F	Copper oxide	Copper	0.96	0.75
G	Black porcelain enamel	Steel	0.93	0.86
H	Black siliconized polyester	Aluminum	0.95	0.89
I	Black chrome	Stainless steel	0.88	0.19
J	Black chrome	Aluminum	0.98	0.14
L	Lead oxide	Copper	0.99	0.29
M	Oxide anodized	Aluminum	0.94	0.10
N	Oxide conversion coating	Aluminum	0.93	0.51
P	Black chrome	Copper (nickel flashed)	0.96	0.08

Source: Waksman, Thomas, and Streed (1984).
a. Code letters *A* through *H* indicate materials coupon specimens cut from solar collectors A through H. Codes *I* through *P* tested at the materials level only.
b. These properties depend on the formulations and manufacturing processes used. Other products within a generic class of materials may have significantly different properties.
c. Average values based on a minimum of ten test specimens.

5.2.2   Test Results and Discussion

Collector Thermal Performance Dependence on Material Properties   A mathematical model for collector thermal performance was used to calculate expected changes in efficiency curve parameters as a result of arbitrary and measured changes in several key material properties. The results are shown for changes in absorber plate emittance and absorptivity, cover normal beam transmittance, and conductivity of thermal insulation. The results calculated for typical changes in measured properties are then compared with measured changes of the collector efficiency curve parameters.

Collectors D and H were selected for the theoretical investigation. The efficiency curve parameters and material properties for these two collectors showed significant changes after exposure. Collector D has two glass covers and a selective absorber. Tests on actual samples of the absorber plate showed that the emittance changed appreciably during the exposure period for some of the collectors exposed. Collector H has an outer cover of poly(ethylene terephthalate) (PET) and an inner fluorinated (ethylene propylene) copolymer (FEP) cover with a flat plate absorber. Tests showed that significant changes had occurred in the solar transmittance of the two cover systems primarily because of outgassing deposits.

The mathematical model used to calculate the thermal performance is based on the Hottel-Whillier-Bliss analysis with an extension to account for the serpentine flow configuration of collector H. A detailed description of the mathematical model is given in Thomas (1980b). Table 5.9 shows the base case parameters and dimensions required by the analytical model for the two collectors. Tables 5.10 and 5.11 show calculated efficiencies for the base case and arbitrary changes in properties. The calculations shown are for changing only one property, with others held at the base case value for a range of $[t_i - t_a]/G$. In order to compare with measured results, a linear curve was fit to the calculated efficiency values shown. The abscissa values were selected in accordance with ASHRAE Standard 93-77 (1977b) as approximately 0.1, 0.3, 0.5, and 0.7 of $[t_i - t_a]/G$ at stagnation conditions. The F_R ($\tau\alpha$) and $F_R U_L$ values shown in the tables are then the intercept and negative slope of the correlating curves. The mean residual errors for the curve fit based on these four sets of values are also included in the table. In the tables τ_{sb} designates the solar beam transmittance of the inner cover (1) and outer cover (2).

Several observations follow from examining the effect of $F_R U_L$ ($\tau\alpha$) and $F_R U_L$ of changes in the four materials. Changes by 0.01 W/m °C in the thermal conductivity from the base-case value have a strong effect on the slope parameter but a small effect on the intercept parameter. Although conductivities were not measured before and after exposure, this property could change as a result of compaction, moisture entering fibrous insulation (collector D), thermal damage, or moisture entry into open pores of organic foam insulation (collector H).

An increase in the absorber plate emittance would be expected to increase the loss coefficient U_L and have an insignificant effect on the optical parameter ($\tau\alpha$) as shown since this property controls long wavelength radiation from the absorber. The increase in ε_p by 0.1 increases $F_R U_L$ much more for collector D than for collector H since the former has a selective surface with relatively low base-case value. Actual tests of collector D absorber plate samples showed in some cases that the plate emittance changed by approximately this amount in going from an initial value of 0.07 to 0.17 after exposure. The emittance of test samples from collector H decreased slightly after exposure from 0.89 to 0.86.

A decrease of 0.1 in the solar absorptance of the plate decreases the intercept parameters of both collectors by about 8% and has a negligible effect on the efficiency curve slopes. Tests on absorber samples from both collectors showed

Table 5.9
Base case collector parameter for collecting thermal performance dependence on material properties

Dimension or property	Units	Collector D	Collector H
Absorber			
Flow configuration	—	Parallel	Serpentine
Effective length	m	1.726	2.237
Effective width	m	0.813	1.130
Flow tubes			
Number	—	10	8
O.D.	mm	8.1	15.88
Hydraulic diameter	mm	4.93	12.70
Wetted perimeter	mm	15.49	39.90
Thickness	mm	0.90	1.778
Thermal conductivity	W/m°C	45.00	200.0
Emittance	—	0.07	0.89
Solar absorptance	—	0.97	0.96
Cover assembly			
Number of covers	—	2	2
Air space: under cover 1/under cover 2	mm	37.0/25.0	25.4/12.7
Infrared emittance: cover 1/cover 2	—	0.84/0.84	0.33/0.76
Infrared transmittance: cover 1/cover 2	—	0.02/0.02	0.60/0.12
Index of refraction: cover 1/cover 2	—	1.30/130	1.33/1.68
Extinction coefficient: cover 1/cover 2	mm	0.0021/0.0021	0.0275/0.1874
Thickness: cover 1/cover 2	mm	3.18/3.18	0.0254/0.1778
Insulation			
Thickness			
Back	mm	88.9	25.4
Edge	mm	25.4	25.4
Conductivity			
Back	W/m°C	0.04	0.02
Edge	W/m°C	0.04	0.02
Aperture area	m^2	1.39	2.64
Gross area	m^2	1.67	2.93

Source: Waksman, Thomas, and Streed (1984).

Table 5.10
Effects of material property changes on $F_R(ta)$ and $F_R U_L$ for collector D

$\Delta t/G$ ($^\circ$C m^2/W)[a]	Calculated efficiency (aperture area basis, %)						
	Base case	k + 0.01 (W/m $^\circ$C)	k − 0.01 (W/m $^\circ$C)	e_p + 0.10	a_p − 0.10	t_{sb1} − 0.10 (inner)	t_{sb} − 0.10 (both)
0.02	71.2	69.3	73.2	70.3	64.5	66.0	58.9
0.05	58.3	55.1	61.6	56.6	51.6	53.3	46.3
0.08	45.8	41.3	50.4	43.0	38.9	40.7	33.6
0.12	26.5	20.2	33.0	21.7	19.6	21.4	14.4
$F_R(ta)$	0.806	0.796	0.816	0.806	0.738	0.754	0.683
$F_R U_L$ (W/m^2 $^\circ$C)	4.458	4.902	4.006	4.847	4.470	4.450	4.445
$\sigma_\eta \times 100$ (%)[b]	0.564	0.583	0.552	0.749	0.539	0.597	0.6032

Source: Waksman, Thomas, and Streed (1984).
a. Base case stagnation Wt/G rise = 0.17°C m^2/W.
b. σ_η is the mean residual error of the linear correlating curve.

the solar absorptance essentially unchanged from the base-case value after exposure.

The effect of F_R ($\tau\alpha$) and $F_R U_L$ of changes in the solar transmittance of the cover system is consistent with expectations. The decrease in transmittance decreased F_R ($\tau\alpha$) proportionately but decreased $F_R U_L$ very little. The calculations are based on the assumption that the cover reflectance is unchanged; a decrease in transmittance is assumed to increase the cover absorptance by the same amount. Consequently the covers are somewhat warmer than in the base case and slightly reduce heat loss from the absorber. Although the glass cover material of collector D will not degrade with exposure, the antireflective etching can lose its effectiveness because of build-up of outgassing condensation or dust. The cover system of collector H is more susceptible to degradation from exposure. Table 5.11 shows the effects of arbitrary decreases in solar beam transmittance of 0.10 from the base case value. The last four columns show additional results calculated for actual measured changes in transmittance for the two covers. Two samples of the outer cover, an ultraviolet-resistant PET film, experienced a decrease in transmittance from 0.85 to 0.84 and 0.81.

Samples of the inner cover, FEP film, showed apparent changes from 0.96 to 0.81 and 0.79 after 480 days exposure. The transmittance changes in the inner cover resulted from an outgassing deposit on the inner surface, which could be removed by washing. The results for the four degraded transmittance combinations are shown in table 5.11.

Table 5.11
Effects on material property changes on $F_R(\tau\alpha)$ and $F_R U_L$ for collector H

$\Delta t/G$ (°C m²/W)[a]	Calculated efficiency (aperture area basis, %)											
	Base case	$k + 0.01$ (W/m °C)	$k - 0.01$ (W/m °C)	$\varepsilon_p + 0.10$	$\alpha_p - 0.10$	$\tau_{sb1} - 0.10$ (inner)	$\tau_{sb2} - 0.10$ (outer)	$\tau_{sb} - 0.10$ (both)	$\tau_{sb1/2} =$ 0.81/0.81	$\tau_{sb1/2} =$ 0.81/0.84	$\tau_{sb1/2} =$ 0.79/0.81	$\tau_{sb1/2} =$ 0.79/0.84
0.01	66.8	65.6	67.9	66.3	60.9	63.4	59.8	57.0	58.9	60.7	58.1	59.9
0.03	54.6	52.3	56.9	53.7	48.8	51.3	47.9	45.2	47.2	49.0	46.4	48.2
0.05	40.6	37.1	44.0	39.1	34.8	37.5	34.2	31.4	33.5	35.3	32.9	34.6
0.07	25.1	20.5	29.7	23.1	19.4	22.1	18.8	16.2	18.3	20.0	17.7	19.4
$F_R(\tau\alpha)$	0.746	0.740	0.751	0.744	0.687	0.711	0.675	0.647	0.665	0.684	0.657	0.675
$F_R U_L$ (W/m² °C)	6.953	7.527	6.370	7.217	6.929	6.880	6.834	6.812	6.766	6.784	6.730	6.746
$\sigma_\eta \times 100$ (%)	0.824	0.814	0.832	0.858	0.830	0.830	0.876	0.855	0.888	0.892	0.878	0.882

a. Base case stagnation $\Delta t/G$ rate = 0.12°C m²/W.

Comparing calculations with the measured changes in the efficiency curve parameters, the observed variations in many cases for $F_R(\tau\alpha)$ are of approximately the same order. In many other cases, particularly for $F_R U_L$, the measured changes are not consistent with those expected solely from changes in collector materials properties. Data scatter in the collector efficiency measurements most likely obscured the observation of these changes.

Absorber Stagnation Temperature Results The absorber plate is considered to be one of the components of the collector most susceptible to damage because of the magnitude and range of temperatures experienced during operational and nonoperational periods. A comparison of the absorber plate temperatures for collector D when exposed at sites 1, 2, and 4 and to the xenon and tungsten simulators is shown in figure 5.16. The higher peak temperature obtained in the tungsten simulator is attributed to the deviation in spectral distribution from the sun (larger infrared portion) and the higher environmental temperature. The higher peak temperature obtained in the xenon simulator is partially the result of the higher peak flux of 1,100 W/m² as

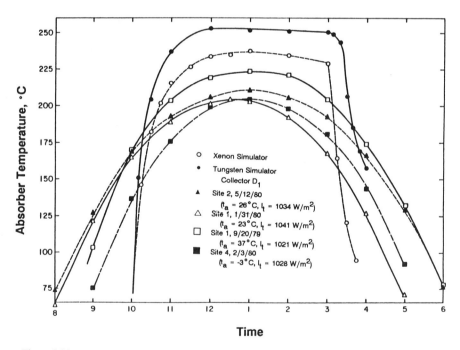

Figure 5.16
Comparison of absorber stagnation temperature profiles using solar irradiance simulators and outdoor exposure, collector D.

compared to about 1,000 W/m^2 experienced outdoors. The sharper rise and fall of the temperature in the simulator results from the abrupt turn-on and shutoff of the simulator lamps.

An analytical and experimental investigation was carried out to evaluate an alternative to the energy output method for measuring thermal degradation of materials used in flat plate collectors. The proposed method is based on measuring the temperature of the absorber under a stagnation condition before and after prolonged exposure (Birnbreier 1978). For a given solar irradiance level, the measured absorber stagnation temperature depends on cover transmittance, solar absorptance and infrared emittance of the absorber, and the collector loss coefficient. The method, test procedures, and results for applying the stagnation temperature methods under nearly steady-state conditions are discussed in detail in Dawson, Thomas, and Waksman (1982, 1983a).

The investigations showed that the proposed method is as sensitive to small changes in collector material properties as the currently used method based on measuring the energy output. Figure 5.17 shows the sensitivity of the new method to a range of property changes for typical collectors. Property changes on the order of 0.1 in plate absorptance, cover solar transmittance, and plate

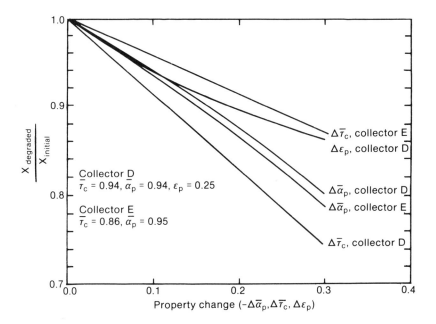

Figure 5.17
Sensitivity of normalized absorber stagnation temperature to changes in material properties.

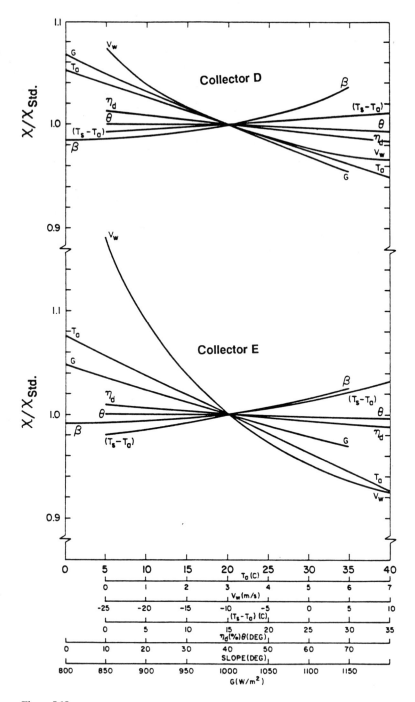

Figure 5.18
Effect of environmental test conditions on normalized absorber stagnation measurements.

emittance would be detectable using the new method with the exception of cover transmittance for collector E, which would be only marginally detectable. Figure 5.18 shows the variation of normalized stagnation temperature rise as a function of environmental parameters for collectors D and E. Clearly wind speed, ambient temperature, and solar irradiance have the largest effect on measured stagnation temperature. These parameters, however, primarily affect U_L rather than $(\tau\alpha)$.

A method for reconciling an averaging method to account for short-term transients in the solar irradiance profile as well as long-term variations in daily solar radiation is presented in Thomas, Dawson, and Waksman (1983). This method is based on measuring the absorber temperature continuously over a period of several days along with the total daily solar irradiation. The absorber temperature rise above ambient is then integrated to determine a daily value. The measurements provide data for a graph of the integrated absorber temperature parameter versus total daily irradiation in the solar collector aperture. A comparison of two such graphs, based on data obtained over a period of several days before and after a prolonged durability exposure, shows if the thermal properties of the collector have changed significantly. It was shown that the effect of short-term transients in the daily irradiance profile is in-

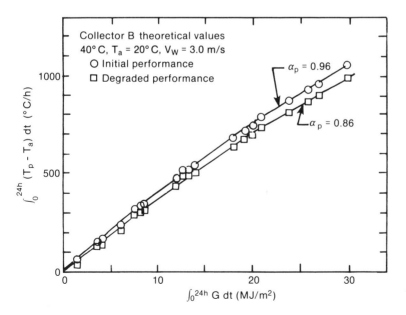

Figure 5.19
Sensitivity of the all-day integration method to a 0.10 change in plate absorptance for collector B.

significant. Figure 5.19 shows expected results from using all-day integrated parameters for stagnation temperature and solar irradiation for a typical change in plate absorptivity.

The all-day integration method was shown to be a viable approach for detecting change in material properties that has advantages over test methods based on steady-state measurements of either absorber stagnation temperature or collector energy output. The principal limitation of both stagnation temperature methods, however, is the strong effect of other environmental conditions, particularly wind speed, on the test results. Neither the temperature nor energy output measurement method can directly identify the particular material property that changed in a collector. With limitations on wind speed and for relatively clear days, the preliminary investigation suggests that stagnation temperature rise is a reproducible collector performance parameter that is at least as sensitive as the ASHRAE Standard 93-77 method and much less expensive to measure.

5.2.3 Conclusions and Recommendations

This section summarizes significant observations and recommendations resulting from the test program (Waksman, Thomas, and Streed 1984). The findings are primarily concerned with an evaluation of test methods for determining the thermal performance and durability of solar collectors and their materials.

Cover Materials

1. Outdoor exposure at sites having a combination of high prevailing humidity and high solar radiation generally produced more severe changes in polymeric cover materials than the other test sites. It is essential that outdoor exposure testing include the combinations of environmental exposure parameters that will occur in normal use.

2. Outdoor "real-time" exposure on cover miniboxes for 480 days ($>17,000$ kJ/m^2 day) is required for many polymeric materials to induce degradation that is detectable without sophisticated analysis. This is equivalent to two or more calendar years. Changes such as microcracking and embrittlement were not readily observable in shorter periods of time for many materials.

3. Accelerated outdoor exposure of polymeric cover materials for 120 calendar days produced changes similar to those occurring in 480 days of "real-time" exposure ($>17,000$ kJ/m^2 day). The cover specimens were mounted on the accelerated exposure cover miniboxes and exposed to concentrated sunlight (~ 6 suns).

4. Outdoor exposure of cover materials at elevated temperatures representative of both operational and stagnation conditions is needed to assess the durability of polymeric covers of solar collectors. Exposure to stagnation temperatures resulted in the loss of ultraviolet radiation screening additives from several of the materials studied.

5. Currently recognized indoor laboratory exposure tests are not capable of reproducing many of the changes observed for polymeric materials exposed outdoors. The indoor laboratory exposure testing, which used test procedures similar to those in ASTM E 765 (1980b), did not duplicate the extensive microcracking observed outdoors. Other testing for 1,200 hours using a xenon arc weathering machine and an intermittent water spray, as specified in ASTM E 765, was also unsuccessful in duplicating the microcracking.

6. Heat stability testing of cover materials at a temperature of 90°C (194°F) in addition to the temperatures currently specified in ASTM E 765 (1980a) is desirable. Since the covers of single-glazed flat plate solar collectors can reach 90°C (194°F) under stagnation conditions with zero wind speed and solar radiation levels of 1,000 W/m^2, it is desirable to test covers for such collectors at this temperature.

7. Accelerated aging of cover materials using xenon arc weathering machines should be performed at temperatures representative of stagnation conditions. Another consideration is the loss or degradation of ultraviolet radiation screening additives at elevated temperatures.

8. Exposure testing of polymeric glazings using elevated temperature and humidity conditions produced changes considerably different from those caused by outdoor exposure of samples of the same materials. A need exists for the development of an aging test that accounts for the synergistic effects of moisture, temperature, and sunlight. Cyclic moisture exposure to cause shrinking and swelling of the polymers should be a part of this test.

9. Normal and hemispherical spectral transmittance curves measured in the ultraviolet visible spectral region are more sensitive indicators of cover materials degradation than integrated spectral transmittance values determined in accordance with ASTM E 424 (1971a). More emphasis should be placed on the analysis of spectral curves.

10. Normal ultraviolet visible spectral transmittance measurements have greater sensitivity to chemical and/or physical changes in cover materials than hemispherical transmittance spectra. The normal measurements are a sensitive indicator of changes in light scattering.

11. There is a need for data on the infrared spectral properties of organic materials (e.g., emittance and diffuse transmittance). These data are needed for the mathematical modeling of collector designs and the calculation of exposure temperatures.

Absorber Materials

1. Outdoor exposure of small-scale absorber samples mounted on the absorber of a simulated collector is an effective method for determining the thermal stability of large numbers of samples under stagnation conditions; however, it is not a valid test for determining the stability in the presence of moisture of these materials when used in solar collectors. Observations in the test program showed moisture penetration in full-size solar collectors appears to be a common occurrence. This moisture penetration resulted in corrosion and appearance changes not observed with small-scale specimens.

2. Both the simulated solar collectors and the accelerated exposure miniboxes used in the test program are useful for exposing small absorber test specimens to stagnation temperatures. It is not clear that exposure to concentrated solar radiation accelerated the photolytic degradation of absorber materials.

3. Indoor laboratory exposure of absorber test specimens to temperature characteristics of stagnating solar collectors is an effective method for determining their thermal stability.

4. Exposure testing of absorber materials with continuous elevated temperature and humidity exposure conditions produced changes that are considerably more severe than those produced by the outdoor exposure of materials of the same composition in full-size solar collectors and exposure test boxes. The moisture exposure conditions, specifically ASTM E 744 (1980c) 90°C (194°F) and 95% relative humidity for 30 days, need to be reexamined; those conditions may be unduly severe. Materials that are severely degraded as a result of exposure using the test condition still may be capable of providing adequate service as absorbers in flat plate solar collectors.

5. Exposure to xenon arc radiation at elevated temperatures produced significant optical property changes in several of the selective absorber materials studied. Exposure testing with xenon arc radiation should be performed at temperatures representative of solar collector operating conditions. Testing should be performed with the glazings to be used in the actual collector installed between the light source and the absorber surface, or with the worst-case configuration possible. The light source should be filtered to match the solar spectrum.

6. The thermal cycling test most closely simulated the types of corrosion and other changes that were observed in full-size solar collectors. In this test coupon specimens were removed from a changer at $-10°C$ ($14°F$) and allowed to equilibrate at room temperature before being placed in an oven at $177°C$ ($351°F$). Because of the apparent importance of condensed moisture, the humidity needs to be controlled during the equilibration process.

7. Research needs to be conducted to define the extent to which the absorber materials in operational solar collectors are exposed to moisture. Factors such as wetting time, absorber temperature, diurnal breathing, and environmental exposure conditions need to be considered in this research. The presence of porosity in many absorber coatings means that moisture can condense in these pores at humidities lower than 100% relative humidity. It is more likely that moisture would condense out on the absorber at night when it is cool rather than in the daytime.

8. Spectral reflectance curves are a more sensitive indicator of absorber materials degradation than solar absorptance values determined by integrating over a standard solar energy distribution curve. Spectral reflectance changes in the near-infrared region are a sensitive indicator of early degradation in many absorber materials.

9. Absorber materials generally exhibit larger changes in emittance than in integrated solar absorptance for both the outdoor exposure and indoor laboratory tests.

10. The absorber samples tested must duplicate the full-scale collector material in substrate, preparation, and coating application techniques to provide valid test results.

Solar Collectors

1. Neither energy output measurements, based on ASHRAE Standard 93-77, nor stagnation temperature measurement methods are satisfactory for determining changes in material properties observed after outdoor exposure. Collector thermal performance measurements are not sufficiently sensitive and precise under typical test environments to detect the relatively small changes in efficiency curve slope and intercept resulting from changes in material properties caused by 480 days of outdoor exposure ($> 17,000$ kJ/m^2 day).

2. The absorber stagnation temperature methods evaluated are at least as good an indicator of solar collector thermal performance changes, with the exception of bond conductance as the ASHRAE Standard 93-77 test method and are less expensive to implement.

3. The final selection of materials for use in solar collectors should be based both on the results of small-scale materials tests and on an evaluation of the properties of materials samples taken from full-size solar collectors exposed outdoors. Exposure testing of full-size collectors permits evaluation of materials interactions that may not be obvious from small-scale materials tests. Additional consideration needs to changes in the mechanical properties of solar collector materials, especially thin-film glazings.

4. The uncertainty in the measured efficiency curve intercept and slope parameters are as expected considering the inherent precision in the ASHRAE Standard 93-77 test method. The coefficient of variations for the initial baseline tests of the eight collector types, from all test sites, were from 1.5% to 4.0% for the intercept parameter and from 5% to 13% for the slope parameter.

5. For flat plate collectors of conventional design, calculation procedures are capable of giving results at least as good as those obtained with the ASHRAE Standard 93-77 measurement procedure for the incident angle modifier.

6. A comparison of results obtained under natural outdoor test conditions with those obtained using tungsten-halogen and xenon arc simulators showed that higher thermal performance results can be obtained with the solar irradiance simulators for collectors having polymeric covers. The highest results were obtained with the xenon arc simulator.

7. Stagnation exposure testing of flat plate solar collectors intended for use in systems with solar radiation augmentation reflectors should be conducted with these reflectors in place.

8. The peak stagnation temperature for flat plate solar collectors should be measured at a distance approximately one-fourth of the way below the top of the collector and with the collector tilted so that its aperture is normal to the sun.

9. Over a period of approximately three years, pyranometer sensitivity changes, if not calibrated, could give rise to errors comparable to the uncertainty in the measured collector efficiency.

10. There is a need for a water leakage test for flat plate solar collectors similar to the test described in ASTM E 331 (1977b) for exterior windows, curtain walls, and doors.

Several of the collectors used in this test program and others observed in the field showed signs of excessive moisture, which appeared to result from leakage rather than condensation caused by diurnal collector breathing. The presence of moisture and resulting degradation of materials appears to be strongly dependent on the design of collector joints and seams and the quality of workmanship in assembling the collectors.

11. Outgassing deposits were observed on the glazings of virtually all the solar collectors exposed in the test program; however, changes caused by this outgassing could not be discerned by thermal performance measurements.

12. The thermal shock/water spray test did not cause thermal shock problems in any of the solar collectors in these tests.

5.3 Miscellaneous Performance

5.3.1 DOE Collector Testing Program

Under the Department of Energy (DOE) national solar energy program, testing was undertaken in 1977 to establish a data base on the thermal performance and reliability of collectors and to study the development of needed test methods for flat plate solar collectors. The National Bureau of Standards (NBS) prepared a preliminary document delineating provisional solar collector testing procedures in the areas of thermal performance and reliability, durability and environmental exposure, structural integrity, and fire safety. This document, "Provisional Flat Plate Solar Collector Testing Procedures," NBSIR 77-1305 (1977), was later revised and published as NBSIR 78-1305A, 'Provisional Flat Plate Solar Collector Testing Procedures: First Revision' (Waksman et al. 1978).

For the collector testing program, DOE organized the work into the following four categories:

Work area 1. Thermal performance and reliability testing and design verification.

Work area 2. Environmental exposure and durability testing.

Work area 3. Structural testing.

Work area 4. Fire safety testing.

DOE decided to perform the first work area (thermal performance) during 1978. Eight testing laboratories were awarded contracts for this phase, and collectors were procured by DOE from the solar collector manufacturers who desired to participate on a cost-sharing basis with DOE for this thermal performance phase. The results of this testing phase were discussed previously and are summarized and reported by SERI in Kirkpatrick (1983).

For the evaluation of the provisional test methods in work areas 2, 3, and 4, DOE awarded contracts during the summer of 1979 as follows:

Work Area 2—Environmental exposure.
Approved Engineering Test Laboratories (AETL)
University of Florida, Gainesville (U/F)

Work Area 3—Structural.
Approved Engineering Test Laboratories (AETL)
Dayton T. Brown, Inc. (DTB), including air overpressure
Underwriters Laboratories (UL), including air overpressure

Work Area 4—Fire safety.
Underwriters Laboratories (UL)
Approved Engineering Test Laboratories (AETL)

The reported laboratory investigations (Dayton T. Brown, Inc. 1980; Approved Engineering Test Laboratories 1980a, 1980b; University of Florida 1981; Underwriters Laboratories 1981a) were based on the provisional test methods for environmental exposure and structural integrity described in Waksman et al. (1978).

Collectors were selected for each provisional test method work area based on significant collector characteristics that would possibly have an influence on collector performance or reliability in that test area. The collectors were selected from the same list of those that had been submitted by manufacturers to DOE for participation in the collector thermal performance test program (work area 1). The number of different collectors selected for each test area was determined by the resources available to DOE for this provisional test program. Collectors selected for each test area had a wide range of generic characteristics so that the impact, if any, of these characteristics on the validity of the test method could be assessed.

The solar collectors selected for environmental exposure tests are shown in tables 5.12 and 5.13 while those selected for structural (including hail impact) and air overpressure tests are presented in tables 5.14 and 5.15.

5.3.2 Test Results and Discussion

Rain The adequacy of two proposed test methods for determining the resistance of solar collectors to water penetration when subjected to wind-driven rain was evaluated.One proposed rain test utilized the ASTM E331 (1977b) test procedure, which incorporates a pressure differential to simulate the effects of wind-driven rain. The second test was a procedure used by Underwriters Laboratories (1974) to test water penetration into electrical equipment.

Table 5.12
Collectors selected for rain tests

Generic collector characteristics	DOE manufacturer control number	Assigned test laboratory	Number to be tested
Liquid			
Container: metal Cover: glass Sealant: silicone	53L	AETL	1
Container: metal Cover: glass Gasket: EPDM	23L	U/F	1
Container: metal Cover: glass Gasket: silicone	28L	U/F	1
Container: metal Cover: plastic Gasket: silicone	57L	AETL	1
Container: metal Cover: plastic Sealant: rubber	25L	U/F	1
Container: metal Cover: plastic Sealant: butyl	73L	U/F	1
Container: plastic Cover: plastic Sealant: silicone	4L	AETL	1
Container: wood Cover: glass Sealant: silicone	35L	U/F	1
Air			
Container: metal Cover: glass Gasket: EPDM	79A	AETL	1
Container: metal Cover: plastic Sealant: silicone	81A	AETL	1
Total			10

Source: Street, Skoda, and Cattaneo (1985).

Table 5.13
Collectors selected for thermal cycling tests

Generic collector characteristics	DOE manufacturer control number	Assigned test laboratory	Number to be tested
Nonload-bearing absorber			
Materials: compatible (Cu)	46L	AETL	1
Materials: compatible (Al)	16L	U/F	1
Materials: compatible rollbond (Cu)	41L	AETL	1
Materials: compatible rollbond (Al)	9L	U/F	1
Materials: compatible (Al) (AIR)	76A	AETL	1
Materials: not compatible (Al/Cu)	28L	U/F	1
Load-bearing absorber			
Materials: compatible (Cu)	10L	U/F	1
Materials: compatible rollbond (Cu)	43L	AETL	1
Materials: not compatible (Al/Cu)	54L	U/F	1
Total			9

Source: Street, Skoda, and Cattaneo (1985).

This method was considered as an alternative in NBSIR 78-1305A (Waksman et al. 1978) because it did not include a means for simulating wind-driven rain. The test results and a critique of the test methods, which are described in Street, Skoda, and Cattaneo (1985), are summarized next.

Ten different collectors were rain tested (table 5.12). Each collector was tested on its front and rear face using the ASTM test, followed by comparable tests using the UL test procedure. Collectors with rear faces not designed for outside (above the roof line) exposure were not tested on this face.

Results of the laboratory tests are presented in tables 5.16 and 5.17. Using the ASTM method, water leakage into the collector was observed in nine out of ten collectors tested. Water leakage varied from 0.1 lb to as much as 6.3 lb. Using the UL method, water leakage into the collector was observed in seven out of the nine collectors tested. Water leakage varied from as little as 0.1 lb to as much as 5.6 lb. In general, these data indicate that water leakage through the front (cover plate) side of the collector was greater using the ASTM procedure. By contrast, water leakage through the collector back side using the UL procedure was about the same as or somewhat higher than using the ASTM procedure. There does not appear to be any correlation between the amount of water leakage observed and the generic class of the collector or its cover plate configuration using either test.

For the ASTM method one laboratory sealed the test box framing where it contacted the collector sides. Another laboratory made this interface as tight as possible but did not use gaskets or sealants, which allowed water to blow parallel to the sides of the collector frame. That raised the possibility of water

Table 5.14
Collectors selected for structural tests

Generic collector characteristics				DOE manufacturer control number	Assigned test laboratories	Number to be tested
Fluid type	Cover material	Cover type	Container construction			
Liquid	Glass	Tempered: double	Aluminum extrusion sides Aluminum sheet back	55L	DTB UL AETL	3
Liquid	Glass	Tempered: single	Aluminum sheet sides Aluminum sheet back	53L	DTB UL AETL	3
Liquid	Glass	Tempered: single	Aluminum extrusion sides Aluminum sheet back	110L	DTB UL AETL	3
Liquid	Glass	Plain: single (in sections)	Wood sides Plywood back	54L	DTB UL AETL	3
Liquid	Plastic	FRP: single	Galvanized steel sheet sides Galvanized steel sheet back	108L	DTB UL AETL	3
Liquid	Plastic	FRP over film	Molded plastic sides and back (monolithic)	63L	DTB UL AETL	3
Liquid	Film	Single	Aluminum extrusion sides, open back (insulation only)	16L	DTB UL AETL	3
Air	Glass	Tempered: single	Aluminum extrusion sides open back (insulation only)	87A	DTB UL AETL	3
Air	Plastic	FRP: single	Galvanized steel sheet sides, chipboard back	77A	DTB UL AETL	3
Total						27

Source: Street, Skoda, and Cattaneo (1985).

Table 5.15
Collectors selected for over pressure tests (air type only)

Generic collector features		Cover					
Container					DOE manufacturer	Assigned test	Number to
Sides	Back	Outer	Inner	Sealant	control number	laboratories	be tested
Al I-beam	FRP	FRP	Film	PVC foam tape	76A	DTB UL	2
Sheet steel	Chipboard	FRP	—	Silicone	77A	DTB UL	2
Sheet steel	Sheet steel	Tempered glass sections	Tempered glass	EPDM, silicone	79A	DTB UL	2
Al extrusion	Al sheet	Acrylic	Glass	Silicone	81A	DTB UL	2
Al extrusion	Insulation only	Tempered glass sections	—	Butyl tape, silicone	87A	DTB UL	2
Wood	Insulated installation in wall	Tempered glass sections	Tempered glass	Silicone	91A	DTB UL	2
Al sheet: foam sandwich	Al sheet: foam sandwich	FRP	FRP	Rubber	104A	DTB UL	2
Wood	Fiberboard	Double polycarbonate	—	Rubber, butyl calk	119A	DTB UL	2
Total							16

Source: Street, Skoda, and Cattaneo (1985).

Table 5.16
Summary results of collector rain tests using test method 1 (pressure box—ASTM)

Generic collector characteristics	DOE collector number	Tested at (laboratories)	Surface tested	Test method 1: pressure box (ASTM)				
				Weight before, lb	Weight after, lb	Weight, lb	Percent weight, lb	Water presence after test
Liquid								
Container: metal Cover: glass Sealant: silicone	53L	AETL	Front Back	92.0 92.3	92.7 92.4	0.7 0.1	0.8 0.1	Slight None
Container: metal Cover: glass Gasket: EPDM	23L	U/F	Front Back	115.6 115.6	116.2 115.9	0.6 0.3	0.5 0.3	Medium Slight
Container: metal Cover: glass Gasket: silicone	28L	U/F	Front Back	172.0 172.0	172.4 172.8	0.4 0.8	0.2 0.5	Slight No data
Container: metal Cover: plastic Sealant: silicone	57L	AETL	Front Back (not tested)	83.7	84.0	0.3	0.4	None
Container: metal Cover: plastic Gasket: rubber	25L	U/F	Front Back	91.7 91.7	97.0 92.4	5.3 0.7	5.8 0.8	No data Slight
Container: metal Cover: plastic Tape: butyl	73L	U/F	Front Back	109.6 109.6	111.6 112.3	2.0 2.7	1.8 2.5	Slight Large
Container: plastic Cover: plastic Sealant: silicone	4L	AETL	Front Back (not tested)	149.1	149.1	0	0	None
Container: wood Cover: glass Sealant: silicone	35L	U/F	Front Back (not designed for rain leak test)	109.8	112.6	2.8	2.6	Large

Table 5.16 (continued)

Generic collector characteristics	DOE collector number	Tested at (laboratories)	Surface tested	Test method 1: pressure box (ASTM)				
				Weight before, lb	Weight after, lb	Weight, lb	Percent weight, lb	Water presence after test
Air								
Container: metal Cover: glass Gasket: EPDM	79A	AETL	Front Back (not tested)	204.6	210.9	6.3	3.1	None
Container: metal Cover: plastic Sealant: silicone	81A	AETL	Front Back (not tested)	201.0	204.8	3.8	1.9	Medium

Source: Street, Skoda, and Cattaneo (1985).

Table 5.17
Summary results of collector rain tests using test method 2 (rain stand—UL)

Generic collector characteristics	DOE collector number	Tested at (laboratories)	Surface tested	Test method 2: rain stand—UL				
				Weight before, lb	Weight after, lb	Weight, lb	Percent weight, lb	Water presence after test
Liquid								
Container: metal Cover: glass Sealant: silicone	53L	AETL	Front Back	92.4 92.4	92.4 92.5	0 0.1	0 0.1	None Slight
Container: metal Cover: glass Gasket: EPDM	23L	U/F	Front Back	115.6 115.6	115.6 116.7	0 1.1	0 1.0	No data No data
Container: metal Cover: glass Gasket: silicone	28L	U/F	Front Back	172.0 172.0	172.0 172.0	0 0	0 0	None None
Container: metal Cover: plastic Sealant: silicone	57L	AETL	Front Back	84.3 84.2	84.1 84.6	0.8 0.4	1.0 0.5	No data Slight
Container: metal Cover: plastic Gasket: rubber	25L	U/F	Front Back	91.7 91.7	91.7 91.8	0 0.1	0 0.1	No data No data
Container: metal Cover: plastic Tape: butyl	73L	U/F	Front Back	109.6 109.6	110.4 115.2	0.8 5.6	0.7 5.1	No data No data
Container: plastic Cover: plastic Sealant: silicone	4L	AETL	Front Back	149.1 149.0	149.1 149.0	0 0	0 0	None None
Container: wood Cover: glass Sealant: silicone	35L	U/F	Front Back (not designed for rain leak test)	109.8	110.8	1.0	0.9	No data

Table 5.17 (continued)

Generic collector characteristics	DOE collector number	Tested at (laboratories)	Surface tested	Test method 2: rain stand—UL				
				Weight before, lb	Weight after, lb	Weight, lb	Percent weight, lb	Water presence after test
Air								
Container: metal	79A	AETL	Front	204.8	208.4	3.6	1.8	None
Cover: glass			Back					
Gasket: EPDM			(not tested)					
Container: metal	81A	AETL	Front					
Cover: plastic			(not tested)					
Sealant: silicone			Back					
			(not tested)					

Source: Street, Skoda, and Cattaneo (1985).

penetration at inlet/outlet ports or air vents if the collector were not properly sealed at these points. Also the required air blower pumping power is increased to maintain the proper positive pressure inside the test box. The test procedure (Waksman et al. 1978, sec. 7.5.4) specifies sealing of this interface between collector side. Since differences in interpretation occurred between the two laboratories, it would appear that the test procedure should be rewritten to emphasize that this interface requires a seal.

Nozzles that provided a fine mistlike spray were necessary to obtain a uniform density spray at low flow rates specified in the ASTM method. The small water droplet size that resulted did not appear to simulate real rain characteristics in either size or impact force. Water penetration into the collector depended primarily on the effects of air pressure on the collector joints and mating surfaces (e.g., the coverplate to collector framing junction). Consideration should be given to increasing droplet size and velocity to be more representative of wind-driven rain with a change in the specified water-flow rates. A possible alternative to this test method might be to use a large fan with suitable spray rates and a droplet size of heavy wind-driven rain.

The spray configuration defined by the UL procedure does not fully cover the width of a collector that is more than approximately 4 ft (1.2 m) wide. Many collectors are specified by the manufacturer to be mounted with their long axis horizontal. To test these type collectors, two or more UL spray rigs or a much longer redesigned UL type spray rig would be required to ensure that the water spray covers both ends of the collector. The test method (Waksman et al. 1978, sec. 7.5.4.) does not mention this limitation. The water spray rig as defined by the test procedure generates a spray pattern that does not cover most collectors completely in a fore-and-aft direction, resulting in little collector area being covered by impacting water. The lower collector sections were only wetted by runoff water, and certain parts were not wetted at all.

The spray impact using the UL test method simulates only a moderate rainfall and fails even to approximate wind-driven rain. In addition the spray coverage horizontally and fore and aft of the collector is incomplete. Therefore the UL method is not a sufficiently severe test to adequately determine collector leakage potential. Also there is no simulation of the air pressure that would be exerted by the wind itself on the collector faces. It was suggested that this test method be eliminated from further consideration as a proposed rain test for solar collectors (Street, Skoda, and Cattaneo 1985).

The question still remains as to whether or not the ASTM test procedure really duplicates the condition of strong wind-driven rain that a collector

could potentially be exposed to in a real world situation. Further testing and
research should be conducted with water flow rates that provide droplet size
more closely simulating a heavy wind-driven rain. The change could require
different spray nozzle configurations as well as higher water-flow rates.

Thermal Cycling Thermal cycling tests are intended to determine if solar
collectors intended for use under freezing conditions will perform reliably after
exposure to such conditions. The tests impose stresses caused by differential
thermal expansion and freezing of heat transfer fluid, if any, that would occur
in collectors when they are used in accordance with their manufacturer's
recommendations. A proposed cycling test in which the temperature in the
chamber was cycled from ambient to $-25°F$ $(-32°C)$ at a rate of $40°-60°F$
$(4°-16°C)$ per hour was contained in NBSIR (1977). During a review of the
proposed test, an alternate procedure that subjects the test specimen to
repeated cycles of placement in a chamber at $-25°F$ $(-32°C)$ followed by
removal and recovery to ambient temperature was considered. It is simpler
and less expensive to perform; however, it may lead to condensation problems
inside the collector. Both test procedures were incorporated in Waksman et
al. (1978) for purposes of comparative testing and evaluation.

Nine collectors were subjected to thermal cycling tests (see table 5.13). The
two different test procedures mentioned earlier were used on all collectors.
Summaries of data are presented in tables 5.18 and 5.19. Data in table 5.18 are
from test method no. 1 in which the collectors are allowed to remain in the
environmental chamber while the temperature is varied between $70°F$ $(21°C)$
and $-25°F$ $(-32°C)$ for ten cycles. The data in table 5.19 are from test method
no. 2 where the collectors are put into a $-25°F$ $(-32°C)$ chamber atmosphere
until the low temperature equilibrium is achieved and then removed from the
chamber to an ambient laboratory condition of $70°F$ $(21°C)$ where they remain
until the higher temperature equilibrium is achieved. This in-and-out procedure is then repeated until the same number of temperature cycles used in
Method 1 (ten cycles) is completed. Requirements relating to test specimens
preparation, installation of temperature sensing device on the collector, and
static pressure leakage are identical for both test methods and are described
in Waksman et al. (1978, secs. 7.6.3, 7.12).

The results of both test methods were the same (i.e, "no visible damage or
leakage in any collector"). Increased air leakage in the one air collector tested
was reported. Additional tests are required on a variety of air collector types
of construction that do meet initial air-type leakage requirements to determine
if any failures occur. For the liquid-type collectors tested, the inclusion of a
50/50 mix of water and antifreeze, which prevents freezing down to $-34°F$

Table 5.18
Summary data of thermal cycling tests using test method 1

Generic collector characteristics	DOE collector number	Tested at (laboratories)	Fluid in collector during test[a]	Number of cycles	Average cooling cycle, h	Average heating cycle, h	Average cooling temperature, °F	Average heating temperature, °F	Average cooling rate, °F/h	Average heating rate, °F/h
Nonload-bearing absorber										
Materials: compatible	46L	AETL	50/50	9	2.40	2.00	74	69	31	35
	16L	U/F	50/50	10	2.35	2.33	84	80	36	34
	41L	AETL	50/50	9	2.40	2.00	75	70	31	35
	9L	U/F	50/50	10	2.82	3.28	79	80	28	24
	76A	AETL	Air	9	2.40	2.00	90	86	38	43
Material: not compatible	28L	U/F	Drained	10	2.00	2.15	81	79	41	37
Lord-bearing absorber										
Materials: compatible	10L	U/F	50/50	10	1.63	1.83	82	80	50	44
	43L	AETL	50/50	9	2.40	2.00	79	74	33	37
Material: not compatible	54L	U/F	50/50	10	2.85	2.82	84	81	29	29

Source: Stree, Skoda, and Gattaneo (1985).
a. 50/50 = mixture of 50% antifreeze, 50% water.

Table 5.19
Summary data of thermal cycling tests using test method 2

Generic collector characteristics	DOE collector number	Tested at (laboratories)	Fluid in collector during test[a]	Test method 2						
				Number of cycles	Average cooling cycle, h	Average heating cycle, h	Average cooling temperature, °F	Average heating temperature, °F	Average cooling rate, °F/h	Average heating rate, °F/h
Nonload-bearing absorber										
Materials: compatible	46L	AETL	50/50	10	1.10	1.35	60	84	55	62
	16L	U/F	50/50	10	1.28	1.57	85	81	66	52
	41L	AETL	50/50	10	1.10	1.35	51	77	46	57
	9L	U/F	50/50	10	2.00	2.52	84	80	42	32
	76A	AETL	Air	10	1.10	1.35	58	75	53	56
Material: not compatible	28L	U/F	Drained	10	1.33	1.63	86	80	65	49
Lord-bearing absorber										
Materials: compatible	10L	U/F	50/50	10	0.93	1.90	83	81	89	43
	43L	AETL	50/50	10	1.10	1.35	53	74	48	55
	54L	U/F	50/50	10	2.80	5.42	83	80	30	15
Material: not compatible										

Source: Street, Skoda, and Cattaneo (1985).
a. 50/50 = mixture of 50% antifreeze, 50% water.

(-37°C) eliminates the risk of freezing the heat transfer fluid during testing to -25°F (-32°C). Test method no. 2 appears to be somewhat more severe.

Examination of the measured cooling and heating rates for the collectors achieved during the tests (see tables 5.18 and 5.19) indicates that those rates were much higher for the method no. 2 test compared to those contained in the method no. 1 test. Both methods required a cooling and heating rate of 40°–60°F (4°–16°C) per hour. Only two of the nine collectors in the method no. 1 tests achieved this rate in the cooling mode, but only one collector achieved this rate in the heating mode. In all other cases the heating and cooling rates were less than specified. In the method no. 2 tests, five of the nine collectors were cooled at the prescribed rate while three collectors were cooled at a rate greater than 60°F (16°C) per hour, and one collector was cooled at a rate of less than 40°F (4°C) per hour. In the heating mode, six of the nine collectors achieved the required heating rate, but one was heated at more than 60°F (16°C) per hour and two collectors were heated at less than 40°F (4°C) per hour. Those results indicate that the temperature rate control is more easily achieved in method no. 2. However, since no failures occurred under either method, the significance of temperature rate control appears to be questionable.

The reason for the inability to achieve specific temperature change rates is that each collector has a different thermal capacity, which in turn depends on features of construction. Therefore, as required by the proposed test procedure [i.e., the absorber plate temperature must reach -10°F (-23°C)], collectors with less insulation are subjected to a shorter period of freezing temperatures than collectors constructed with more insulation that have a greater low temperature tolerance. It appears that the well-insulated collector can be penalized for being built to higher insulation standards.

The applications for which the test procedures would have validity would be in testing collectors that are designed to be "freeze tolerant." Those collectors would be filled with plain water and then subjected to cycle of freezing and thawing; they would have to withstand (without leakage) the internal stresses caused by water expanding upon freezing. No such collectors were included in this test program. In addition only one draindown-type collector was tested; after the water was allowed to drain out of it and the collector was thermally cycled, no visual damage was observed.

The test procedures should be modified to require a given period (in hours) during which the specimen collector is to be subjected to the low and high ambient temperatures and not be based on attained absorber plate temperatures. Also the tests should be considered only for collectors that are designed to be 'freeze tolerant' or are of the draindown type. The removal to laboratory ambient method procedure might be useful for air-type collector testing since condensation is introduced, but testing of additional air collectors would have to be completed to verify the utility of such tests.

Structural   Structural test methods are intended to provide the means for evaluating the structural integrity o,f solar collectors under the action of various service loads such as wind and overpressure. The purpose of conducting a series of structural tests was to evaluate the provisional methods contained in Waksman et al. (1978) and to generate typical data on structural behavior of various generic types of flat plate collectors. Specimens were selected from approximately 200 different solar collectors declared available in 1978 by manufacturers participating in the DOE collector testing program. The solar collectors selected for the structural (including hail impact) and air overpressure tests are listed in tables 5.14 and 5.15.

The environmental live-load structural tests performed simulated positive live loads (causing flexure of the collector), negative or combination (i.e., positive plus negative) wind loads, longitudinal (in-plane) loads, and hail (impact) loads. In addition the tests were accompanied by a static pressure leakage test performed on the specimen collector before testing and after release of each of the applied incremental test loads. The leakage test was conducted concurrently with the first two tests and was used to help determine the degree of loading damage to the specimen's fluid transport system.

A fifth test, which simulated accidental internal overpressure (positive or negative) in air-type collectors, was also performed to evaluate the procedure. Eight additionally selected air-type collectors were used as test specimens. The static pressure leakage test was not used in conjunction with this rupture/collapse test for air-type collectors because the performance acceptance test criterion is the absence of rupture (or collapse) and of readily visible joint movement during application of a specified excessive test pressure.

No replicate tests were conducted at a given laboratory and the one collector of each brand selected for use in the environmental live-loads structural tests was scheduled for use in all of them. Replication, therefore, occurred only between laboratories. The longitudinal load test, considered to be the least detrimental to any collector as a whole, was scheduled as the first test on all nine collectors at all three laboratories. Additional information on the sequence of testing, etc., is documented in Street, Skoda, and Cattaneo (1985). This reference also contains a critique of the structural tests.

POSITIVE LIVE LOADS   The purpose of this test is to determine the ability of a flat plate collector to function satisfactorily after being subjected to uniformly distributed loads resulting from snow accumulation or positive wind pressure (whichever predominates) on the cover plate of the collector. Use of the method is intended only with rigid flat plate collectors.

The procedure (Waksman et al. 1978, sec. 7.7) was derived from one of the optional methods ("Transverse Load—Specimen Horizontal") in ASTM Standard Method E72 (1977a) developed for testing wall and floor panels. The positive live loads test results are summarized in table 5.20. As indicated, the required test load for the collectors tested ranged from 82.5 psf to 135 psf. In Waksman et al. (1978) the specified maximum test load that a collector should be able to resist is based on its manufacturer's design load rating for uniform pressure applied to its cover (normal to the exterior surface), increased by a suitable safety factor. In the absence of a load rating, the maximum test load is based on an anticipated extreme loading condition (e.g., maximum wind load for a 25-year mean recurrence interval, likewise increased by the safety factor) that would make possible the collector's use in any U.S. locality. The condition assumes vertical mounting in the wind-driven force of the building with the back not subjected to negative loads. In this test program 135 psf was selected in the absence of a manufacturer's rating. However, failure of the specimen collector by the described maximum load (selected for the conditions it represents) does not mean that a model would not provide satisfactory service under different (lower expected maximum) service loads at other locations.

NEGATIVE AND COMBINATION WIND LOADS The purpose of this test is to determine the ability of a rigid flat plate collector to function satisfactorily after being subjected to uniformly distributed loads resulting from either of two cases. Case 1 applies when the wind has access only to the cover plate surface of the collector and creates a suction (negative pressure) on that surface. Case 2 applies when the wind has access to both surfaces of the collector and creates a combined load of suction on the cover plate surface and uplift (positive pressure) on the back of the collector. The condition necessary for applicability of case 2 is described in Waksman et al. (1978) as the mounting of a collector on an open rack or mounting in such a way that any portion of its back surface is 8 in. (20 cm) or more from the supporting surface.

Negative wind testing and combined (negative plus positive) wind loads testing of flat plate solar collectors are based, in part, on techniques existing in ASTM Standard Method E72 (1977a) applied singly or in combination. When manufacturers indicate the back surface of their installed collector would be protected from wind loads (e.g., when set into a roof surface), case

Table 5.20
Positive live loads test results summary

Tested at (laboratories)	DTB — Required test load (psf)	DTB — Leakage test pressure (psi or in. W.C.)	AETL — Required test load (psf)	AETL — Leakage test pressure (psi or in. W.C.)	UL — Required test load (psf)	UL — Leakage test pressure (psi or in. W.C.)
Data	Maximum load applied (psf); Physical damage by this test	(Leakage) pressure loss; or, air volume transfer before and after load increment (psim/psim/...); or, (cfh/cfh/...)	Maximum load applied (psf); Physical damage by this test	(Leakage) pressure loss; or, air volume transfer before and after load increment (psim/psim/...); or, (cfh/cfh/...)	Maximum load applied (psf); Physical damage by this test	(Leakage) pressure loss; or, air volume transfer before and after load increment (psim/psim/...); or, (cfh/cfh/...)
Collector						
16L	No test Damaged in negative wind loads test		90 90 None seen while loading; popping sounds at 44 psf; cover damage observed after test knockdown	80 −/0/0/0	90 90 Center deflection of 1 inch	80 0/0/0/0
53L	135 135 $\frac{3}{8}$ in. permanently deflected not considered failure	100 0/0/0/0	No test Damaged in combined wind loads test		135 109 (81% of required) Cover shattered; container bowed $1\frac{1}{4}$ in. at center	0.022 0/*/−/ (*excessive)
54L	188 118 (40 ÷ 0.34) 54 (46% of required) Two central glass panes broke	0/0/0	No test Glass damaged in handling after longitudinal loads tests		135 45 (33% of required) Cover shattered	80 0/0

55L	92 92 None	30 0/0/0/0	30 0/0/0/0	No test Damaged in negative wind loads test
63L	60 60 None	25 2.5/2.5/2.5/2.5	No test Damaged in wind loads test	25 135 135 Metal retainer strips over long edges of of cover buckled 0.67/0.53/0.47/0.40
108L	No test Damaged in combined wind loads test		80 135 62.4 (46% of required) Cracking sounds and load drop-off halted test; cover tore at 3 corners –/–/0	80 135 114 (84% of required) Supported brackets buckled; collector buckled; cover tore a 3 edges 0.27/0.33/0.63/11
110L	82.5 82.5 None	75 Not attainable 0/0/0/0 Excessive/–	50 82.5 82.5 None –/0/0/0	50 82.5 82.5 None 0/0/0/0
77A	118 (40 ÷ 0.34) 118 Cover distorted; cover buckled in; back separated from sides ⅛ in.		No test No specimen	No test Damaged in combined loads test
87A	135 135 None	0.5 15/15/15/15	0.5 135 135 None 5.9/6.8/6.3/6.3	0.5 135 135 None visible 170/173/185/195

Source: Street, Skoda, and Cattaneo (1985).

1 (negative load only) was applied; otherwise, case 2 (negative pressure on the glazing and positive uplift pressure on the back surface) was applied.

Clarification on the intent of this test as permitting the use of either vacuum lift cups or vacuum chamber was provided to the laboratories for simulation of negative wind pressure on collector glazing. Negative wind load and combination wind load test results are summarized in table 5.21, and detailed discussions appear in Street, Skoda, and Cattaneo (1985). In these tests the maximum load to be imposed on each surface was 120 psf where there was no design load rating specified by the manufacturer (Waksman et al. 1978).

LONGITUDINAL LOADS   The objective of this test is to determine the ability of the collector's mounting hardware (used to attach or connect the collector to its support structure) to withstand cyclic loads in the plane of the collector. Accordingly the intent is to determine the performance of connector points on the collector under the action of wind gusts or seismic inertial loads. Its use is primarily for those types of connector points (*not the supporting frame*) that are of innovative design and do not lend themselves to a load determination by an engineering analysis.

The test method consists of a technique (i.e., one of two optional methods given in Waksman et al. 1978, sec. 7.9.5) that simulates the in-plane loads. The three test laboratories used the same method (no. 1) in which the collector is mounted parallel to a test bed and loaded by hydraulic jack. One laboratory also conducted leakage tests before and after applying the in-plane loads sequence.

The results of the longitudinal loads test are listed in table 5.22. No permanent physical damage was reported for any of the specimens. The preload force was specified to be equal to the maximum filled weight of the collector. The imposed additional load was required to be 50% of the preload force.

HAIL LOADS   The purpose of this test is to determine the ability of cover plates (glazing) on flat plate collectors to withstand impact forces from hailstones or similar flying objects. The procedure as presented is limited to simulating hailstones with ice balls impelled at velocities that are the results of their free-fall terminal speeds and a driving horizontal wind speed of 45 mph (72 kmh).

The laboratory method for simulating the effects of a falling, wind-driven hailstone on a surface (Waksman et al. 1978, sec. 7.10) evolved from initial work reported by Laurie (1960). Further development of the technique is presented in Cattaneo et al. (1981) and Jenkins and Mathey (1982). The procedure proposed in Waksman et al. (1978, sec. 7.10), which was used by

Table 5.21
Negative or combined wind loads test results summary

Tested at (laboratories)	DTB					AETL					UL				
Data	Required test loads, on cover/back (psf/psf)	Maximum loads applied (psf/psf)	Physical damage by this test	Leakage test pressure (psi or in. W.C.)	(Leakage) pressure loss; or, air volume transfer before and after load increment (psim/psim/...); or, (cfh/cfh/...)	Required test loads, on cover/back (psf/psf)	Maximum loads applied (psf/psf)	Physical damage by this test	Leakage test pressure (psi or in. W.C.)	(Leakage) pressure loss; or, air volume transfer before and after load increment (psim/psim/...); or, (cfh/cfh/...)	Required test loads, on cover/back (psf/psf)	Maximum loads applied (psf/psf)	Physical damage by this test	Leakage test pressure (psi or in. W.C.)	(Leakage) pressure loss; or, air volume transfer before and after load increment (psim/psim/...); or, (cfh/cfh/...)
Collector															
16L	−120/0	−40/0 (33% of required)	Cover pulled out of frame	80	0/0	−120/−120	−(>40/>40) (>33% of required)	None up to load level applied (test not completed: negative load vacuum system leakage)	80	0/0/0	−120/−120	−73/−73 (61% of required)	Cover deflected up 2 in. and separated from frame for 5 ft along side	80	0/0/0
53L	−120/−120 −120/−120	Two clips deformed at 2/3 R_m; repair and retest without damage		100	0/0/0/0	−120/−120	−70.2/63.4 (~56% of required) Cover shattered		80	−/0/0	No test Damaged in positive live loads test				
54L	No test Damaged in positive live loads test					No test Glass damaged in handling after longitudinal loads test					No test Damaged in positive live loads test				

Table 5.21 (continued)

Tested at (laboratories)	DTB					AETL					UL				
Data	Required test loads, on cover/back (psf/psf)	Maximum loads applied (psf/psf)	Physical damage by this test	Leakage test pressure (psi or in. W.C.)	(Leakage) pressure loss; or, air volume transfer before and after load increment (psim/psim/...); or, (cfh/cfh/...)	Required test loads, on cover/back (psf/psf)	Maximum loads applied (psf/psf)	Physical damage by this test	Leakage test pressure (psi or in. W.C.)	(Leakage) pressure loss; or, air volume transfer before and after load increment (psim/psim/...); or, (cfh/cfh/...)	Required test loads, on cover/back (psf/psf)	Maximum loads applied (psf/psf)	Physical damage by this test	Leakage test pressure (psi or in. W.C.)	(Leakage) pressure loss; or, air volume transfer before and after load increment (psim/psim/...); or, (cfh/cfh/...)
Collector															
55L	-120/-120	-120/104 (/87% of required)	Six rivets popped out of back surface	30	0/0/0/0	-120/0	-80/0 (67% of required)	Double glazing bowed up out of framing on one side	30	0/0/0	-120/0	-73/0 (61% of required)	Cover edge retainer strip pulled screws away from container frame	30	0/0.045/0.045
63L	-60/0	-60/0	Cover edge retainer strip distored; not considered a failure	25	2.5/2.5/2.5/2.5	-120/-120	-(<40)/<40	Collector damaged in repeated attempts; cover frame distored; test not completed	—	-/-	-120/120	-62.4/80 (~60% of required)	Bottom surface separated from side of container	25	0.43/0.23/0.21

Collector	Structural load test (−/+; % of required; damage)	Positive live loads test	Impact test
108L	−120/120; −52.5/45.7 (~40% of required); Cover and support strips bowed up 2 in.; cover ripped from end frame	80 · 0/0/0 — No test / Damaged in positive live loads test; No test / Damaged in positive live loads test	—
110L	−40.3/42.2; −40.3/42.2; None; −120/120; −34.3/40 (~30% of required); Not attainable	75 · 0/0/0 — −40/40; −56/40; None; No test; No specimen; 50; 0/0/0 — −40/40; −40/40; None; −120/120; −88/120 (73% of required); Container bowed up 1 in.; cover punctured at screw/washer location; 50; 0/0/0	0.5; 100/95.3/96.7/ >111
77A	Excessive/−; Cover pulled out from edge retainer strip; collector bowed up 1½ in.	0.5 — −120/0; −120/0; Some increase in leakage; −/4.0/4.2/5.9; 0.5	0.5
87A	−120/0; −120/0; None	0.5; 15/15/15/15 — −120/0; −120/0; None visible; 0.5	163/164/163/168; 15/15/15/15

Source: Street, Skoda, and Cattaneo (1985).

Table 5.22
Longitudinal cyclic loads test results summary

Tested at (laboratories) →	DTB					AETL					UL				
Collector (Data ↓)	Preload (lbf)	Maximum loads (lbf)	Loading rate (lbf/s)	Loaded edge	Physical damage by this test	Preload (lbf)	Maximum loads (lbf)	Loading rate (lbf/s)	Loaded edge	Physical damage by this test	Preload (lbf)	Maximum loads (lbf)	Loading rate (lbf/s)	Loaded edge	Physical damage by this test
16L	70	105	20	Long	None	65	105	—	Short	None	64	96	—	Long	None
53L	97	146	25	Long	None	96	144	—	Short	None	94.4	142	—	Long	None
54L	218	327	55	Short	None	158.4	237.5	—	Short	None (glass damaged in handling after test)	184	276	—	Long	None
55L	200	300	50	Short	None	186.5	279	—	Short	None (flexed under load but not permanently)	221	332	—	Short	None
63L	100	150	26	Short	None	102	—	—	Short	None	98	147	—	Short	None

108L	119	Short	118.6	Short	118	Short
	178	None	177.9	None	177	None
	30	Short	—	Short	—	Short
110L	164	Short	164	Short	152	Short
	246	None	246	None (flexed under load but not permanently)	228	None
	41		—		—	
77A	160	Short	no test	Short	145	Short
	240	None (end of container buckled under load but not permanently)	No specimen		217.5	Loaded end flexed under load but not permanently
	40					
87A	105	Short	105	Short	104	Short
	158	None	152	None	152	None
	28		—		—	

Source: Street, Skoda, and Cattaneo (1985).

the laboratories in this test program, is basically the same procedure subsequently approved by ASTM as a standard procedure (ASTM 1981b).

Test specimens were cover plates in complete collectors, and simulated
hailstones were spherically molded ice balls. Launchers were quick-release,
compressed air guns with multiple or interchangeable barrels of corresponding diameters and made of steel or PVC pipe. Propelled ice ball velocities were
measured at all three laboratories by an electronic gate pulse-counter controlled by photoelectric or expandable foil tape trigger circuits placed in the
path of the ice ball at the beginning and end of a fixed distance near the target
cover plate.

The results of the hail impact loads tests are presented in table 5.23. The
test data include size of ice ball used (1), measured velocity (2), measured
velocity deviation percent from the required velocity (3), impact location given
by coordinate measurements from the nearest inside corner of the cover
support frame (4), and observed damage to the cover (5).

One laboratory (DTB) reported a departure from the referenced procedure,
impacting each collector cover with progressively smaller ice balls (instead of
larger ones as specified). Also, in the UL results, it was not certain whether
the reported impact velocities were intended as nominal or measured values;
no deviations were reported. Since UL reported impacting each cover with
all four sizes of ice balls, the dashes in column (S) presumably indicate no
damage. Values for the average diameter and weight of each size hailstone
were submitted by DTB (see table 5.24).

In related DOE-sponsored studies statistical model to assess the risk of hail
damage to flat plate solar collectors has been developed (Cox and Armstrong
1979), and the need for hail protection devices for solar collectors has been
evaluated (Armstrong, Cox, and de Winter 1980).

AIR COLLECTOR RUPTURE AND COLLAPSE   The purpose of these tests is to
determine the ability of air-type solar collectors to meet the pressure requirements (positive or negative) established by Underwriters Laboratories (1981b)
for the air duct systems with which the collectors will be integrated. Generally
it is intended that such a collector be capable of withstanding an internal
pressure for a given period that is a specified multiple of the collector's
operating pressure. The rupture (or collapse) test procedure for air-type solar
collectors (Waksman et al. 1978, sec. 7.11) is based on UL Standard 181
(Underwriters Laboratories 1981b), which specifies that the required test
overpressure for duct sections having joints be 2.5 times the manufacturer's
rated (positive or negative) operating pressure (but not less than 1.25 in., water
gauge), and sets the test period for overpressurizing at one hour.

Table 5.23
Hall impact loads test results summary

Tested at laboratory: DTB

Collector (cover)	(1) Ice ball diameter, in.	(2) Measured velocity, ft/s	(3) Deviation from required velocity, % (83 91 98 112)	(4) Location of impact (x, y) from corner, in.	(5) Physical damage by this size ice ball at this velocity
16L (7 to 8 mil polyester)	No test	Damaged in negative wind loads test			
53L (1/16 in. tempered glass)	1/2	—			
	3/4	—			
	1				
	1 1/4	109	(−3%)	7 × 7	None
54L (5 1/8 in. glass panes)	No test	Damaged in positive loads test			
55L (Double 0.16 to 0.186 in. tempered glass)	1/2	—			
	3/4	—			
	1	81	(−17%)	6 × 6	None
	1 1/4	128	(+14%)	6 × 7	None
63L (0.045 in. FRP over 1 mil polyfilm)	1/2	156	(+88%)	7 × 7	None
	3/4	92	(+1%)	7 × 6	Crazed 1/2 in. diameter
	1	—			
	1 1/4	117	(+4%)	6 × 6	Crazed 1 in. diameter

Tested at laboratory: AETL

Collector (cover)	(1) Ice ball diameter, in.	(2) Measured velocity, ft/s	(3) Deviation from required velocity, % (83 91 98 112)	(4) Location of impact (x, y) from corner, in.	(5) Physical damage by this size ice ball at this velocity
16L (7 to 8 mil polyester)	1/2	84	<(+1%)	6 × 7	None
	3/4	71	(−22%)	5 × 6	Small dimple
	1	97	(−1%)	6 × 6 1/4	Dimple: 1/4 × 1/32 in. deep
	1 1/2	119	(+6%)	6 × 6	Dimple: 1/2 × 0.1 in. deep
53L (1/16 in. tempered glass)	No test	Damaged in combined wind loads test			
54L (5 1/8 in. glass panes)	No test	Glass damaged in handling after longitudinal loads test			
55L (Double 0.16 to 0.186 in. tempered glass)	No test	Damaged in negative wind loads test			
63L (0.045 in. FRP over 1 mil polyfilm)	No test	Damaged in combined wind loads test			

Tested at laboratory: UL

Collector (cover)	(1) Ice ball diameter, in.	(2) Measured velocity, ft/s	(3) Deviation from required velocity, % (83 91 98 112)	(4) Location of impact (x, y) from corner, in.	(5) Physical damage by this size ice ball at this velocity
16L (7 to 8 mil polyester)	1/2	83	—	—	—
	3/4	91	—	—	—
	1	98	Sight indentation		
	1 1/2	112		Indentation: 1/4 × 1/16 in. deep	
53L (1/16 in. tempered glass)	No test	Damaged in positive live loads test			
54L (5 1/8 in. glass panes)	No test	Damaged in positive live loads test			
55L (Double 0.16 to 0.186 in. tempered glass)	1/2	83	—	—	No visible structural damaged to cover
	3/4	93	—	—	
	1	98	—	—	
	1 1/4	112	—	—	
63L (0.045 in. FRP over 1 mil polyfilm)	1/2	83	—	—	
	3/4	91	—	—	
	1	98	—	—	
	1 1/4	112	—	—	Puncture: 2 × 2 1/2 in.

Table 5.23 (continued)

Tested at (laboratories)	DTB					AETL					UL				
Data	(1)	(2)	(3)	(4)	(5)	(1)	(2)	(3)	(4)	(5)	(1)	(2)	(3)	(4)	(5)
Collector (cover)	Ice ball diameter in.	Measured velocity ft/s	Deviation from required velocity, % (83 91 98 112)	Location of impact (x, y) from corner, in.	Physical damage by this size ice ball at this velocity	Ice ball diameter in.	Measured velocity ft/s	Deviation from required velocity, % (83 91 98 112)	Location of impact (x, y) from corner, in.	Physical damage by this size ice ball at this velocity	Ice ball diameter in.	Measured velocity ft/s	Deviation from required velocity, % (83 91 98 112)	Location of impact (x, y) from corner, in.	Physical damage by this size ice ball at this velocity
108L (0.05 in. FRP)	No test	Damaged in combined wind loads test				No test	Damaged in positive loads test				$\frac{1}{2}$	83	—	—	—
											$\frac{3}{4}$	91	—	—	—
											1	98	—	—	—
											$1\frac{1}{2}$	112	—	—	Puncture: 3 in. long
110L ($\frac{1}{16}$ in. tempered glass)	$\frac{1}{2}$	—				$\frac{1}{2}$	82	(−1%)	7 × 7	None	$\frac{1}{2}$	83	—	—	No visible
	$\frac{3}{4}$	—				$\frac{3}{4}$	89	(−2%)	6 × 6	None	$\frac{3}{4}$	91	—	—	structural
	1	52	(−47%)	6 × 6	None	1	100	(+2%)	6 × 7	None	1	98	—	—	cover
	$1\frac{1}{2}$	112	(0)	6 × 6	None	$1\frac{1}{2}$	129	(+15%)	6 × 6	None	$1\frac{1}{2}$	112	—	—	damage cover
77A (0.045 in. FRP)	$\frac{1}{2}$	149	(+80%)	7 × 7	None	No test	No specimen				$\frac{1}{2}$	83	—	—	—
	$\frac{3}{4}$	81	(−11%)	6 × 8	Crazed $\frac{1}{4}$ in. diameter						$\frac{3}{4}$	91	—	—	—
	1	77	(−21%)	6 × 7	Crazed $\frac{3}{4}$ in. diameter						1	98	—	—	—
	$1\frac{1}{2}$	—									$1\frac{1}{2}$	112	—	—	Slight indentation with shatter marks
87A ($\frac{1}{16}$ in. tempered glass)	$\frac{1}{2}$	—				$\frac{1}{2}$	83	(0)	6 × 6	None	$\frac{1}{2}$	83	—	—	No visible
	$\frac{3}{4}$	—				$\frac{3}{4}$	91	(0)	6 × 6	None	$\frac{3}{4}$	91	—	—	structural
	1	—				1	98	(0)	6 × 6	None	1	98	—	—	damage to
	$1\frac{1}{2}$	158	(+41%)	6 × 6	None	$1\frac{1}{2}$	109	(−3%)	6 × 6	None	$1\frac{1}{2}$	112	—	—	cover

Source: Street, Skoda, and Cattaneo (1985).

Table 5.24
Average diameter and weight of "hailstones"

Required diameter, in.	Actual diameter, in.					Average diameter, in.
	1	2	3	4	5	
0.5	0.492	0.491	0.491	0.490	0.499	0.493
0.75	0.781	0.77	0.770	0.778	0.775	0.7748
1.0	0.99	1.0	0.99	1.005	0.995	0.996
1.5	1.451	1.470	1.470	1.450	1.46	1.464

Hailstone diameter, in.	Required weight, g	Actual weight, g					Average weight, g	Average weight, lb
		1	2	3	4	5		
0.5	0.986	0.965	0.873	0.96	0.952	0.991	0.9782	0.00215
0.75	2.258	3.97	3.92	3.93	3.96	3.89	3.932	0.0086
1.0	7.71	8.45	8.45	8.4	8.55	8.45	8.47	0.0186
1.5	26.3	25.9	26.5	26.1	25.8	26.1	26.08	0.0575

Source: Street, Skoda, and Cattaneo (1985).

Results of the rupture or collapse tests for the air-type collectors (see table 5.15 for collector descriptions) are given in table 5.25. None of the collectors tested were physically damaged. Excessive leakage during application of the test overpressure was reported by one laboratory (DTB) as a measured quantity; the other (UL) gave a more general description. It is not clear in the latter case how the dividing line was defined between "excessive leakage" and "no evidence of excessive leakage," other than the inability to develop and maintain the required test overpressure.

The extent of leakage during overpressure in these tests (in relation to the materials of construction and sealants) varied randomly so that no general tendency was indicated. To draw inferences from the observed amounts of overpressure leakage relative to air duct systems in which such collectors will operate, table 5.26 was compiled of specimen collector properties relating to airflow. Entries in the second, third, and fourth columns are based on collector dimensions obtained from data sheets and drawings originally submitted to DOE by the manufacturers. Column 5 contains values of collector airflows to be expected in service; they are based on providing a flow of 2 cfm for each square foot of absorber plate area (column 2) unless more was stipulated by the manufacturers (ASHRAE 1977b). The air velocity values (column 6) were derived (from columns 3 and 5) merely to confirm the "low- pressure-velocity" classification of the collectors according to Sheet Metal and Air Conditioning

Table 5.25
Air-type collector rupture or collapse test results summary

Tested at (Lab)	DTB		UL	
Data	Manufacturers recommended operating pressure (in. water column)		Manufacturers recommended operating pressure (in., water column)	
Collector	Test pressure (in. water column)	Leakage, cfm / Physical damage by this test	Test pressure (in., water column)[a]	Leakage / Physical damage by this test
76A	±1.0 / −1.6	21.8 / None	— / −1.25	No evidence of excessive / None
77A	+0.5 / +1.25	36.2 / No additional	— / +0.8	Excessive[b] / None
79A	+2.0 / +1.25	No measurable leakage / None	— / +1.25	No evidence of excessive / None
81A	+0.5	12.5 / None	—	Excessive[b] (existing 0.25-in. joint gaps) / —
87A	+0.5 / +1.25	5.9 / None	— / +1.25	No evidence of excessive / None
91A	Not known / −1.25	4.0 / None	— / −1.25	No evidence of excessive / None
104A	−0.5 / +1.25	2.8 / None	— / −1.25	No evidence of excessive / None
119A	+2.0 / +2.0	4.0 / None	+0.2	Excessive[b] / None

Source: Street, Skoda, and Cattaneo (1985).
a. The dash indicates that information on manufacturer's recommended operating pressure was not available at time tests were initiated.
b. Repeated inability to maintain required pressure.

Table 5.26
Properties of rupture/collapse test specimen collectors

Air-type collector number	Absorber plate area, ft²	Flow path cross-sectional area, ft²	Collector volume, ft³	Operating airflow, cfm	Operating air velocity, fpm	"SMACNA expected" operating duct leakage, cfm	"U.L.-181 allowable" duct leakage, cfm	DTB-observed leakage at test overpress, cfm	Prorated leakage at operating pressure, cfm
76	30.7	0.688	5.4	61.4	89.2	3.1	1.8	21.8	15.6
77	30.0	0.319	2.5	60.0	188.0	3.0	0.83	36.2	20.0
79	18.5	0.590	3.7	37.0	62.7	1.9	1.2	0	0
81	42.8	0.399	4.7	200.0[a]	501.0	10.0	1.6	12.5	6.9
87	19.7	0.278	3.3	39.4	142.0	2.0	1.1	5.9	3.3
91	17.2	0.802	5.0	51.6[a]	64.3	2.6	1.7	4.0	2.2
104	19.4	1.18	8.1	146.0[a]	124.0	7.3	2.7	2.8	1.5
119	42.3	0.851	12.3	127.0[a]	149.0	6.4	4.1	4.0	2.2

Source: Street, Skoda, and Cattaneo (1985).
a. More than 2 cfm/ft².

Contractors National Association (SMACNA) standards (1976) (i.e., static pressure: 0.5 in., W.G.; velocity: 2,000 fpm and less). That standard indicates ducts with joints and seams (conforming to the standard) requirements and having the corners of their transverse joints sealed (with mastic plus tape, or gasketing) can be expected to have leakage less than 5% of system operating airflow. The values in column 7 are 5% of the corresponding values in column 5. Column 8 shows values (in cfm) of the maximum allowable leakage in one hour (according to UL Standard 181) for air duct specimens 8 ft (2.4 m) long subjected to an air pressure of not less than 0.5 in., W.G. (i.e., a total hourly leakage of 20 times the internal volume of the specimen, column 4). Column 9 repeats the leakage at test overpressure observed from DTB in table 5.25.

The last column (10) in table 5.26 contains estimates of leakage at *operating* pressure that were obtained by prorating from the DTB observed leakages at *test* overpressure (column 9). The calculations were based on the relationship (ASHRAE 1977a)

$$Q = C(\Delta P)^n,$$

where Q is the volume flow rate of air leakage (cfm) for a pressure difference ΔP; C is the flow coefficient, defined as the volume flow rate per unit length of crack or unit area at pressure difference; and n is the flow exponent (between 0.5 and 1.0, and usually near 0.65 for leakage openings). An estimate of the leakage at operating pressure as derived by ratio to the leakage observed at the test overpressure conditions is obtained from

$$Q_{oper} = Q_{overp} \left(\frac{\Delta P_{oper}}{\Delta P_{overp}} \right)^{0.65}.$$

The leakages observed at DTB in these tests (on this basis of comparing column 10 with columns 7 and 8), specimens 76, 77, and 87 clearly gave evidence of excessive air leakage, with 81 and 91 marginal and the remaining three (79, 104, and 119) apparently satisfactory.

Fire   As part of the DOE Collector Testing Program (see subsection 5.3.1), a study was undertaken to investigate the use of ASTM E108 (1980a) (NFPA 256, UL 790) for testing roof-mounted solar energy collectors (Walton and Waksman 1981). The ASTM E 108 test method is commonly referenced in building codes as the procedure for determining fire characteristics of roof coverings. As a possible rating basis Waksman et al. (1978) suggested that solar collectors meet the requirements for a class C rating for intermittent flame, spread of flame, and burning brand tests in ASTM E 108.

The testing was conducted at two different laboratories to evaluate the use of ASTM E 108 as a means for determining the influence of roof-mounted collectors on the fire characteristics of roof coverings; determine modifications in the test procedure that might be required to make it applicable to roof-mounted collectors; and determine the influence that flat plate solar collectors constructed of various materials might have on the fire characteristics of roof coverings. Details are presented in Walton and Waksman (1981). The study did not attempt to rate the collectors tested nor did it address other fire-related features of solar energy systems (i.e., potential self-ignition of collectors, or the fire resistance of the roof assemblies with solar collectors when exposed to interior fires).

The tests were conducted on 11 collectors constructed of materials indicated in table 5.27. The four mounting configurations are shown in figure 5.20, and the test apparatus is shown in figure 5.21. All collectors were tested without heat transfer fluid. The test roofs were wood decks covered with class C asphalt-impregnated organic-felt shingles and were sloped 23 deg (from the horizontal) in most tests. A summary of the burning brand, intermittent flame, and spread of flame tests conducted is given in table 5.28.

For the burning brand tests, class A, B, or C brands were placed on the collector covers, and for collectors mounted on standoffs, the brands were placed beneath the collectors directly on the roof surface. For the intermittent flame test (class C), the test deck (including collector) was subjected to a luminous gas flame—at a temperature of $704°C \pm 28°C$ ($1,300°F \pm 50°F$)—approximately the width of the deck, which uniformly bathed the surface. The flame was applied intermittently for a specified period (1 minute on and 2 minutes off) for three cycles. For the spread of flame test (class C), the same type of flame was applied continuously for 4 minutes. As indicated in table 5.28, the majority of tests conducted were spread-of-flame tests.

The class C brands ignited one collector cover with acrylic glazing and destroyed the collector, but had no effect on either tempered or annealed glass and caused only a small area of blackening on the fiber-reinforced plastic (FRP) covers. Both class A and B brands readily ignited FRP covers, shattered tempered glass glazing, and broke annealed glass, but the absorber plates prevented further fire penetration of the collector.

In the intermittent flame tests conducted on three collectors constructed of FRP, wood, and aluminum, neither the shingles nor the collector were ignited. During the flame spread tests the glazing on all the collectors with FRP covers ignited, and flames traveled rapidly to the end of the collector, usually in less than one minute. The spread of flame under collectors was shown to be primarily a function of separation or standoff distance between the collector

Robert D. Dikkers

Table 5.27
Description of collectors subjected to ASTM E108 tests

Collector code	Materials		Glazing	Insulation	Absorber	Approximate dimensions length × width × depth
	Case sides	Case back				
29L	Aluminum	Aluminum	Glass (tempered)	Foam plastic[b]	Aluminum	102 × 36 × 3 1/2 in. (259 × 127 × 17.7 cm)
32L	Wood	Plywood	FRP[a]	Fiberglass	Aluminum	120 × 48 × 5 in. (305 × 127 × 17.7 cm)
41L	Wood	Plywood	Double glass (tempered)	Urethane foam	Copper	94 × 36 × 3 3/8 in. (239 × 91 × 8.6 cm)
43L	Molded FRP	Molded FRP	FRP	Urethane foam	Copper	96 × 36 × 5 3/4 in. (244 × 91 × 14.6 cm)
45L	Aluminum	Aluminum	FRP	Foam plastic	Aluminum	118 × 37 3/4 × 3 in. (300 × 96 × 7.6 cm)
46L	Aluminum	Hardboard	Glass (tempered)	Isocyanurate foam	Copper	93 × 35 × 3 1/2 in. (236 × 89 × 8.9 cm)
54L	Wood	Plywood	Glass (annealed)	Fiberglass	Aluminum	99 × 49 3/4 × 3 1/2 in. (251 × 126 × 8.9 cm)
103L	Aluminum	Aluminum	Acrylic	Polystyrene (loose fill)	Aluminum	96 × 48 × 7 1/2 in. (244 × 122 × 19.1 cm)
108L	Steel	Steel	FRP	Fiberglass	Steel	86 1/2 × 40 × 3 in. (220 × 102 × 7.6 cm)
126L	Aluminum	Aluminum	Glass (tempered)	Fiberglass	Copper	77 × 35 × 5 in. (196 × 89 × 12.7 cm)
78A	Aluminum	Aluminum	FRP	None	Aluminum	196 3/4 × 24 × 7 1/4 in. (500 × 61 × 18.4 cm)

Source: Walton and Waksman (1981).
a. FRP stands for glass-fiber-reinforced plastic.
b. Rigid foam plastic, type unknown.

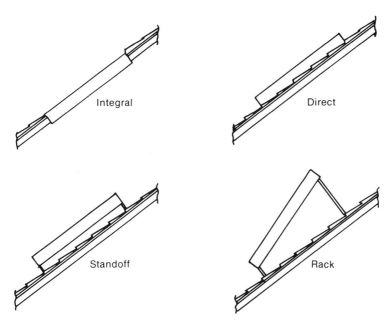

Figure 5.20
Collector mounting configurations. Source: Walton and Waksman (1981).

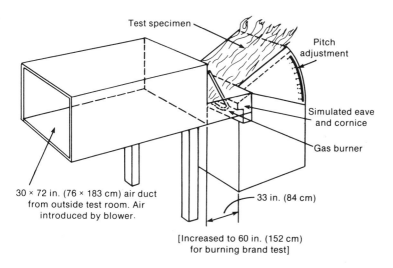

Figure 5.21
Schematic for fire test apparatus. Source: Walton and Waksman (1981).

Table 5.28
Summary of fire test configurations

Collector code	Test[a] type	Test number	Test[b] laboratories	Mounting[c]
—	SF	21	UL	Calibration deck only
—	SF	29	UL	Calibration deck only
—	SF	36	AETL	Calibration deck only
—	SF	48	AETL	Calibration deck only
—	C brand	22	UL	4-in. (10.2-cm) standoff, brand under plywood (0 incline)
29L	SF	35	AETL	2-in. (5.1-cm) standoff
29L	SF	43	AETL	2-in. (5.1-cm) standoff
32L	SF	46	AETL	Direct
32L	SF	47	AETL	1-in. (2.5-cm) standoff
41L	IF	4	UL	4-in. (10.2-cm) standoff
41L	SF	5	UL	4-in. (10.2-cm) standoff
41L	SF	17	UL	4-in. (10.2-cm) standoff
43L	IF	1	UL	4-in. (10.2-cm) standoff
43L	IF	2	UL	Direct
43L	SF	3	UL	Direct
43L	C brand	10	UL	Direct (2 in 12 incline)
43L	A brand	11	UL	Direct (2 in 12 incline)
43L	C brand	12	UL	4-in. (10.2-cm) standoff, brand under (2 in 12 incline)
43L	B brand	13	UL	4-in. (10.2-cm) standoff, brand under (2 in 12 incline)
43L	SF	20	UL	Direct
43L	C brand	23	UL	Direct (0 incline) (no wind)
45L	SF	37	AETL	Direct
45L	SF	38	AETL	2-in. (5.1-cm) standoff
45L	SF	39	AETL	4-in. (10.2-cm) standoff
46L	IF	6	UL	4-in. (10.2-cm) standoff
46L	SF	7	UL	4-in. (10.2-cm) standoff
46L	SF	19	UL	4-in. (10.2-cm) standoff
54L	B brand	31	AETL	Direct
54L	C brand	32	AETL	Direct
54L	SF	45	AETL	3/4-in. (1.9-cm) standoff
78A	B brand	14	UL	No deck (integral) (2 in 12 incline)
78A	C brand	15	UL	No deck (integral) (2 in 12 incline)
78A	A brand	16	UL	No deck (integral) (2 in 12 incline)
78A	SF	18	UL	No deck (integral)
103L	SF	26	UL	No deck (integral
103L	SF	27	UL	6-in. (14.2-cm) standoff
103L	SF	28	UL	2–24-in. (5.1–61.0-cm) rack (4 in 12 deck incline and 22 in 93 collector incline to deck)
103L	C brand	30	UL	Direct (3 in 12 incline)
108L	SF	40	AETL	2-in. (5.1-cm) standoff
108L	SF	41	AETL	Direct
108L	SF	42	AETL	4-in. (10.2-cm) standoff
126L	SF	8	UL	4-in. (10.2-cm) standoff

Table 5.28 (continued)

Collector code	Test[a] type	Test number	Test[b] laboratories	Mounting[c]
126L	A brand	9	UL	Direct
126L	SF	24	UL	4-in. (10.2-cm) standoff
126L	SF	25	UL	4-in. (10.2-cm) standoff
126L	SF	33	AETL	4-in. (10.2-cm) standoff
126L	SF	34	AETL	2-in. (5.1-cm) standoff
126L	SF	44	AETL	9/16-in. (1.4-cm) standoff
126L	SF	49	AETL	1-in. (2.5-cm) standoff
126L	SF	50	AETL	1–9/16-in. (4.0-cm) standoff

Source: Walton and Waksman (1981).
a. Test type: (IF) intermittent flame; (SF) spread of flame; (A brand) class A burning brand; (B brand) class B burning brand; (C brand) class C burning brand.
b. Test laboratory: (AETL) Approved Engineering Test Laboratories; (UL) Underwriters Laboratories.
c. All inclines 5 in 12 except as noted.

and the roof covering. The rate of flame travel did not depend significantly on the material (wood, steel, and aluminum) forming the collector back.

Walton and Waksman (1981) concluded that the burning brand test of ASTM E 108 could be applied with only minor modifications to solar collectors. They also determined that collectors meeting the burning brand test and the spread of flame test also met those for the intermittent flame test. The presence of the collector on the roof and its associated mounting configuration appears to have the most significant impact on the flame spread test. The results indicate that collectors mounted on roofs with a standoff greater than approximately 3.8 cm (1.5 in.) above a class C roof covering will result in flame travel under the collector greater than that allowed by the test criteria, regardless of collector construction. The results also indicate that collectors with combustible glazing could provide a path for rapid flame spread from one area of a roof to another area.

It was recommended that further investigations should be conducted to determine the suitability of small-scale flammability tests such as ASTM D 635 for evaluating the combustibility of glazing materials. (See Braun and Allen 1984 for additional information.) In addition the possibility of fire penetration into the building air duct system from burning collector covers should be investigated.

Nomenclature

A_a	aperture area of the collector, m^2
A_g	gross area of the collector, m^2
b_0	constant used in incident angle modifier equation, dimensionless
$C_8 \ldots C_{22}$	constants in equations (14)–(17)
C_A	effective heat capacity of the collector, its components, and the transfer fluid in the collector, $J/°C$
C_P	specific heat of the transfer fluid, $J/kg\ s$
F'	collector efficiency factor
F_R	collector heat-removal factor
I_d	direct component of solar energy incident upon the plane of the collector per unit time per unit area, W/m^2
I_t	total solar energy incident upon the plane of the collector per unit time per unit area, W/m^2
K	defined by equation (7)
$K_{\alpha\tau}$	incident angle modifier
K_d	diffuse incident angle modifier
K_0	optical loss coefficient
K_E	end loss coefficient
K_{A_P}	$\cos\theta$
K_1	off-azimuth incident angle modifier
K_2	off-altitude incident angle modifier
m	mass flow rate of the heat transfer fluid through the collector, $kg/s\ m^2$
$\dot{Q}_u$	rate of useful energy extracted from the collector, W
t_a	ambient air temperature, $°C$
t_e	effective environmental temperature, $°C$
$\overline{T}_f$	average temperature of the heat transfer fluid in the collector, K
$\overline{t}_f$	average temperature of the heat transfer fluid in the collector, $°C$
$t_{f,e}$	temperature of the heat transfer fluid leaving the collector, $°C$
$t_{f,i}$	temperature of the heat transfer fluid entering the collector, $°C$
$\overline{t}_p$	average temperature of the collector absorber, $°C$

$\dot{t}_{\mathrm{p,i}}$	average temperature of the collector absorber during indoor testing, °C
$\bar{t}_{\mathrm{p,0}}$	average temperature of the collector absorber during outdoor testing, °C
$\bar{T}_{\mathrm{s}}$	average sink temperature for radiation loss, K
U_{L}	heat-transfer loss coefficient for the collector, W/m^2 °C
v	wind speed, m/s
η_{a}	collector efficiency based on collector aperture area
η_{ASHRAE}	collector efficiency determined under full-irradiance conditions in accordance with ASHRAE procedure
η_{BSE}	collector efficiency determined by a combined indoor/outdoor test in accordance with the BSE procedure
Δ_{η}	$\eta_{\mathrm{BSE}} - \eta_{\mathrm{ASHRAE}}$
θ	time, incident angle between the direct solar beam and the outward-drawn normal to the plane of the collector
$(\tau\alpha)_{\mathrm{e}}$	effective transmittance-absorptance product for the collector
$(\tau\alpha)_{\mathrm{en}}$	effective transmittance-absorptance product for the collector at normal incidence.
τ_{sb}	solar beam transmittance of inner and outer covers

Notes

1. However, if a fluid that is different from the fluid used in the actual field installation is used, substantial differences in collector efficiency can result. Youngblood, Schultz, and Barber (1979) and Thomas (1980a) give test results for typical flat plate collectors using water, mixtures of water, and a commercial ethylene glycol-base antifreeze, and a mineral-base heat transfer oil.

2. Equation (11) has also been modified by the inclusion of $A_{\mathrm{a}}/A_{\mathrm{g}}$ since the efficiency as determined in the ASHRAE standards is based on gross collector area.

References

Approved Engineering Test Laboratories. 1980a. *Report of Structural Tests Performed on Solar Collectors.* Test report no. 976-3645-1 Performed for U.S. Department of Energy, George C. Marshall Space Flight Center, Huntsville, AL.

Approved Engineering Test Laboratories. 1980b. *Report of Rain and Thermal Tests Performed on Solar Collectors.* Test report no. 976-3645-2 (and Revision A). Performed for U.S. Department of Energy, George C. Marshall Space Flight Center, Huntsville, AL.

Aranovich, E., and B. Gillet, eds. 1982. In *Proc. Workshop on Solar Simulators.* Ispra, Italy: Joint Research Centre.

Armstrong, P. R., M. Cox, and F. de Winter. 1980. *Need for and Evaluation of Hail Protection Devices for Solar Flat Plate Collectors.* AL0-4291-2, Santa Cruz, CA: Altas Corp.

ASHRAE. 1977a. Infiltration and ventilation. Chapter 21 in *ASHRAE Handbook—1977 Fundamentals.* New York: American Society of Heating, Refrigerating, and Air-Conditioning Engineers.

ASHRAE. 1977b. *Methods of Testing to Determine the Thermal Performance of Solar Collectors.* ASHRAE Standard 93-77. Atlanta, GA: ASHRAE.

ASHRAE. 1980. *Methods of Testing to Determine the Thermal Performance of Unglazed Flat-Plate Liquid-Type Solar Collectors.* ASHRAE Standard 96-1980. Atlanta, GA: ASHRAE.

ASHRAE. 1986a. *Methods of Testing to Determine the Thermal Performance of Solar Collectors.* ASHRAE Standard 93-1986. Atlanta, GA: ASHRAE.

ASHRAE. 1986b. *Methods of Testing to Determine the Thermal Performance of Flat-Plate Solar Collectors Containing a Boiling Liquid.* ASHRAE Standard 109-1986. Atlanta, GA: ASHRAE.

ASTM. 1971a. *Test for Solar Energy Transmittance and Reflectance (Terrestrial) of Sheet Materials.* ASTM E 424-71. Philadelphia, PA: American Society for Testing and Materials.

ASTM. 1971b. *Test for Total Normal Emittance of Surfaces Using Inspection-Meter Techniques.* ASTM E 408-71. Philadelphia, PA: ASTM.

ASTM. 1977a. *Standard Methods of Conducting Strength Tests of Panels for Building Construction,* ASTM Standard E 72-77. Philadelphia, PA: ASTM.

ASTM. 1977b. *Standard Method of Test for Water Penetration of Exterior Windows, Curtain Walls, and Doors by Uniform Static Air Pressure Difference.* ASTM Standard E 331-71. Philadelphia, PA: ASTM.

ASTM. 1980a. *Fire Tests of Roof Coverings.* ASTM E 108-80, Philadelphia, PA: ASTM.

ASTM. 1980b. *Practice for Evaluating of Cover Materials for Flat-Plate Solar Collectors.* ASTM E 765-80. Philadelphia PA: ASTM

ASTM. 1980c. *Practice for Evaluating Solar Absorptive Materials for Thermal Applications.* ASTM E 744-80. Philadelphia, PA: ASTM.

ASTM. 1981a. *Practice for Performing Accelerating Outdoor Weathering Using Concentrated Natural Sunlight.* ASTM E 838-81. Philadelphia, PA: ASTM.

ASTM. 1981b. *Standard Practice for Determining Resistance of Solar Collector Covers to Hail by Impact with Propelled Ice Balls.* ASTM Standard E 822-81. Philadelphia, PA: ASTM.

ASTM. 1982. *Standard for Terrestrial Solar Spectral Irradiance Tables at Air Mass 1.5 for a 37° Tilted Surface.* ASTM Standard E 892-82. Philadelphia, PA: ASTM.

Beach, C. D. 1979. *A Solar Collector Testing Program. Part 2: All-Day Test Results.* Final report for DOE Grant EG-77-G-05-5561, Report FSEC-RD-79-2. Florida Solar Energy Center.

Beckman, W. A., S. A. Klein, and J. A. Duffie. 1977. *Solar Heating Design by the F-Chart Method.* New York: Wiley.

Birnbreier, H. 1978. Durability tests by stagnation temperature measurements. Presented at the International Energy Agency Task III Meeting, Heidelberg, West Germany.

Brandemuehl, M. J., and W. A. Beckman. 1980. Transmission of diffuse radiation through CPC and flat-plate collector glazings. *Solar Energy* 24: 511.

Braun, E., and P. J. Allen, 1984. *Flame Spread on Combustible Solar Collector Glazing Materials.* NBSIR 84-2887. Gaithersburg, MD: National Bureau of Standards.

Cattaneo, L. E., J. R. Harris, T. A. Reinhold, E. Simiu, and C. W. F. Yancy. 1981. *Wind, Earthquake, Snow and Hail Loads on Solar Collectors.* NBSIR 81-2199. Gaithersburg, MD: National Bureau of Standards.

Christie, E. A. 1976. *A Method Measuring the Performance Characteristics of Flat-Plate Solar Collectors.* Report no. TR 6. CSIRO Division of Mechanical Engineering.

Close, D. J., and M. B. Yusoff. 1978. The effects of air leaks on solar air collector behavior. *Solar Energy* 20(6): 459–463.

Collares-Pereira, M., J. Duque, C. Saraiva, and A. Rego-Teixeira. 1981. A calorimeter for solar thermal collector testing. *Solar Energy* 27(6): 581–582.

Cooper, P. I., and R. V. Dunkle, 1981. A non-linear flat-plate collector model. *Solar Energy* 26: 133–140.

Cox, M., and P. R. Armstrong. 1979. *A Statistical Model for Assessing the Risk of Hail Damage to Any Ground Installation.* AL0-4291-1. Santa Cruz, CA: Altas Corp.

Dawson, A. G., III, W. C. Thomas, and D. Waksman. 1982. *Evaluation of Absorber Stagnation Temperature as a Characteristic Performance Parameter of Flat-Plate Solar Collectors.* ASME paper 82-WA/Sol-5. New York: American Society of Mechanical Engineers.

Dawson, A. G., III, W. C. Thomas, and D. Waksman. 1983a. Solar collector durability evaluation by stagnation temperature measurement. *J. Solar Energy Engineering* 105: 259–267.

Dawson, A. G., III, W. C. Thomas, and D. Waksman. 1983b. Testing solar collector materials durability by integrated day-long stagnation temperature measurements. In *Proc. ASME Solar Enerey Division: Fifth Annual Technical Conference*, pp. 301–307.

Dayton T. Brown, Inc. 1980. *Structural Testing of Solar Collectors.* Prepared for U.S. Department of Energy, George C. Marshall Space Flight Center, Huntsville, AL, contract no. AC-01-79-CS-30231.

Dokos, W. 1982. Presentation of some U.S. solar simulator characteristics. In *Proc. Workshop on Solar Simulators.* Ispra, Italy: Joint Research Centre, p. 51.

Dudley, V. E., and R. M. Workhoven. 1978. *Summary Report: Concentrating Solar Collector Test Results, Collector Module Test Facility.* SAND78-0815. Sandia Laboratories.

Duffie, J. A., and W. A. Beckman. 1980. *Solar Engineering of Thermal Processes.* New York: Wiley.

Farber, E. A., R. W. Dixon, and S. D. Wix. 1981. *Investigation of the Accuracy of Predicting Day-Long Solar Collector Performance from the ASHRAE 93-77 Thermal Performance and Angle Modifier Curves.* Unpublished report. Gainesville, FL: University of Florida.

Gani, R., D. Proctor, and J. G. Symons. 1981. Linear efficiency characterization for high temperature flat-plate collectors. *Solar Energy* 26: 271–273.

Gillet, W. B. 1981. The equivalence of outdoor and mixed indoor/outdoor solar collector testing. *Solar Energy* 25: 543–548.

Gillet, W. B., and J. E. Moon. 1982. Results from solar simulators in the CEC collaborative collector testing programme. In *Proc. Workshop on Solar Simulators.* Ispra, Italy: Joint Research Centre, p. 130.

Harrison, T. D., W. D. Dworzok, and C. A. Folkner, Jr. 1977. *Solar Collector Module Test Facility, Instrumentation Fluid Loop Number One.* SAND76-0425. Sandia Laboratories.

Hill, J. E., J. P. Jenkins, and D. E. Jones. 1979. *Experimental Verification of a Standard Test Procedure for Solar Collectors.* NBS Building Science Series 117. Gaithersburg, MD: National Bureau of Standards.

Hill, J. E., B. D. Wood, and K. A. Reed. 1985. Testing solar collectors. In K. W. Böer and J. A. Duffie, eds., *Advances in Solar Energy*, vol. 2. New York: Plenum Press, ch. 7.

Jenkins, D. R., and R. G. Mathey, 1982. *Hail Impact Testing Procedure for Solar Collector Covers.* NBSIR 82-2487. Gaithersburg, MD: National Bureau of Standards.

Jenkins. J. P. 1979. *The Design and Evaluation of a Reference Heat Source to be Used in Conjunction with Solar Collector Testing.* Letter report. Gaithersburg, MD: National Bureau of Standards.

Jenkins, J. P., and K. A. Reed. 1982. *A Comparison of Unglazed Flat-Plate Liquid-Heating Solar Collector Thermal Performance Using the ASHRAE Standard 96-1980 and Modified BSE Procedures.* NBSIR 82-2522. Gaithersberg, MD: National Bureau of Standards.

Jenkins, J. P., and S. T. Bushby. N.d. Differences in determining collector thermal performance between the BSE and ASHRAE collector test procedures. Unpublished paper. Gaithersburg, MD: National Bureau of Standards.

Jenkins, J. P., and J. E. Hill. 1980a. A comparison of test results for flat-plate water-heating solar collectors using the BSE and ASHRAE procedures. *ASME J. Solar Energy Eng.* 102: 2–15.

Jenkins, J. P., and J. E. Hill. 1980b. *Testing Flat-Plate Water-Heating Solar Collectors in Accordance with the BSE and ASHRAE Procedures.* NBSIR 80-2087. Gaithersburg, MD: National Bureau of Standards.

Jenkins, J. P., and K. A. Reed. 1982. *A Comparison of Unglazed Flat-Plate Liquid Solar Collector Thermal Performance Using ASHRAE Standard 96-1980 and Modified BSE Test Procedures.* NBSIR 82-2522. Gaithersburg, MD: National Bureau of Standards.

Kirkpatrick, D. L. 1983. *Flat-Plate Solar Collector Performance Data Base and User's Manual.* SERI/STR-254-1515. Golden, CO: Solar Energy Research Institute.

Klein, S. A., and W. A. Beckman. 1979. A general design method for closed loop solar energy systems. *Solar Energy* 22: 269.

Klein, S. A., J. A. Duffie, and W. A. Beckman. 1974. Transient considerations of flat-plate solar collectors. *ASME J. Eng. Power* 96A: 109.

Kline. S. J., and F. A. McClintock. 1953. Describing uncertainties in single-sample experiments. *Mech. Eng.* (January): 3.

Laurie, J. A. P. 1960. *Hail and Its Effects on Building.* Bulletin 21. Pretoria, South Africa: National Building Research Institute.

Mather, G. R. 1980. ASHRAE 93-77 instantaneous and all-day tests of the Sunpak evacuated-tube collector. *ASME J. Solar Energy Eng.* 102: 294.

Mather, G. R., and D. C. Beekley. 1976. Performance on an evacuated tubular collector using non-imaging reflectors. In *Proc. 1976 International Solar Energy Society Conference*, Winnipeg, Canada, August 15–20, vol. 2, p. 74.

NBSIR. 1977. *Provisional Flat Plate Solar Collectors Testing Procedures.* NBSIR 77-1305. Gaithersburg, MD: National Bureau of Standards.

Nielson, V. H. 1984. *Three Years' Experience with the Solar Simulator at the National Solar Test Facility.* Mississauga, Ontario, Canada: Ontario Research Foundation.

Oonk, R. L., D. E. Jones, and B. E. Cole-Appel. 1979. Calculation of the performance of N collectors in series from test data on a single collector. *Solar Energy* 23: 535–536.

Proctor, D. 1982a. A generalized method for testing all classes of solar collectors. Part II: Evaluation of collector thermal constants. *Solar Energy* 32(3): 387–394.

Proctor, D. 1982b. A generalized method for testing all classes of solar collectors. Part III: Linearized efficiency equations. *Solar Energy* 32(3): 395–399.

Putman, W. J., D. E. Evans, and B. D. Wood. 1982. *The Effect of Different Sky Conditions on the Optical Performance of Flat-Plate and Stationary Concentrating Collectors.* Final report for SERI contract XX-1-1178-1.

Reed, K. A., and J. W. Allen. 1978. Thermal performance testing of solar collectors: The calorimetric ratio technique. In *Proc. 1978 American Section Meeting of the International Solar Energy Society.* pp. 345–346.

SMACNA. 1976. *Low Pressure Duct Construction Standards.* Vienna, VA: Sheet Metal and Air Conditioning Contractors National Association, Inc.

Simon, F. F. 1976. Flat-plate solar-collector performance evaluation with a solar simulator as a basis for collector selection and performance prediction. NASA TM X-71793. *Solar Energy* 18: 451–466.

Simon, F. F., and E. H. Buyco. 1976. Outdoor flat-plate collector performance prediction from solar simulator test data. NASA TM X-7107. In *Proc. 10th AIAA Thermal Physics Conference.* Denver, Colorado, May 27–29.

Smith, C. C., and T. A. Weiss. 1977. Design application of the Hottel-Whillier-Bliss equation. *Solar Energy* 19(2): 109–114.

Solar Energy Industries Association. 1979. *Product Certification Standard 1–79.* Washington, DC: Solar Energy Industries Association.

Souka, A. F., and H. H. Safwat. 1966. Determination of optimum orientations for the double-exposure, flat-plate collector and its reflector. *Solar Energy* 10(4): 170–174.

SRCC. 1981. *Methodology for Determining the Thermal Performance Rating for Solar Collectors.* SRCC Standard RM-l. Washington, DC: Solar Rating and Certification Corp.

Standards Association of Australia. 1982. *Glazed Flat-Plate Solar Collectors with Water as the Heat-Transfer Fluid-Method for Testing Thermal Performance.* Australian Standard 2535-1982. North Sydney, NSW: Standards House.

Streed, E. R., J. E. Hill, W. C. Thomas, A. G. Dawson, and B. D. Wood. 1979. Results and analysis of a round-robin test program for liquid-heating flat-plate solar collectors. *Solar Energy* 22: 235–249.

Streed, E. R., W. C. Thomas, A. G. Dawson, B. D. Wood, and J. E. Hillo. 1978. *Results and Analysis of a Round-Robin Test Program for Liquid-Heating Flat-Plate Solar Collectors.* NBS technical note 975. Gaithersburg, MD: National Bureau of Standards.

Streed, E., and D. Waksman. 1981a. *Uncertainty in Determining Thermal Performance of Liquid-Heating Flat-Plate Solar Collectors.* NBS technical note 1140. Gaithersburg, MD: National Bureau of Standards.

Streed, E. R., and D. Waksman. 1981b. Uncertainty in determining thermal performance of liquid-heating flat-plate solar collectors. *ASME Journal of Solar Energy Engineering* 103: 26–134.

Street, W. G., L. F. Skoda, and L. E. Cattaneo. 1985. *Laboratory Investigations of Provisional Flat Plate Solar Collector Testing Procedures: Rain, Thermal Cycling, and Structural.* NBSIR draft. Gaithersburg, MD: National Bureau of Standards.

Svendson, S. 1978. *Theoretical Investigation of the Methodical Errors of the BSE-Procedure for Testing Solar Collectors.* Report no. 78–27. University of Denmark.

Symons, J. G. 1976. *The Direct Measurement of Heat Loss from Flat-Plate Solar Collectors on an Indoor Testing Facility.* CSIRO Division of Mechanical Engineering report no. TR 7.

Talarek, H. D. 1979. *Task III: Performance Testing of Solar Collectors, Results and Analysis of IEA Round-Robin Testing.* Kernforschungsanlage, Julich, Federal Republic of Germany.

Thomas, W. C. 1980a. *Effects of Test Fluid Composition and Flow Rates on the Thermal Efficiency of Solar Collectors.* NBS GCR 80–254. Gaithersburg, MD: National Bureau of Standards.

Thomas, W. C. 1980b. *Solar Collector Test Procedures: Development of a Method to Refer Measured Efficiencies to Standardized Test Conditions.* Report VPI-E-80.23. Blacksburg, VA: Virginia Polytechnic Institute and State University. Also NBS-GCR-84-459. Gaithersburg, MD: National Bureau of Standards.

Thomas, W. C., A. G. Dawson, D. Waksman, and E. R. Streed. 1982. Incident angle modifiers for flat-plate solar collectors analysis of measurements and calculation procedures. *ASME J. Solar Energy Eng.* 104: 349.

Thomas, W. C., A. G. Dawson, III, and D. Waksman. 1983. Testing solar collector materials durability by integrated day-long stagnation temperature measurements. In *Proc. ASME Solar Energy Division: Fifth Annual Technical Conference,* pp. 301–307.

Underwriters Laboratories, Inc. 1974. *Refrigeration and Air Conditioning Condensing and Compressor Units.* UL Standard No. 303. Northbrook, IL: Underwriters Laboratories, Inc.

Underwriters Laboratories, Inc. 1981a. *Summary Report on Fire Safety and Structural Tests on Solar Collectors*. Prepared for U.S. Department of Energy, George C. Marshall Space Flight Center, Huntsville, AL, contract no. AC-01-79-CS-30060.

Underwriters Laboratories, Inc. 1981b. *Factory-Made Air Duct Materials and Air Duct Connectors*. UL Standard no. 181. Northbrook, IL: Underwriters Laboratories, Inc.

University of Florida. 1981. *Testing of Solar Collector—Rain and Thermal Cycling Tests*, Project report (revised). Department of Energy Contract No. AC-01-79-30232. University of Florida, Solar Energy Laboratory, Gainesville.

Waksman, D., E. R. Streed, T. W. Reichard, and L. E. Cattaneo. 1978. *Provisional Flat Plate Solar Collector Testing Procedures: First Revision*. NBSIR 78-1305A, Supersedes NBSIR 77-1305. Gaithersburg, MD: National Bureau of Standards.

Waksman, D., E. R. Streed, and J. Seiler. 1981. *Solar Collector Durability/Reliability Test Program Plan*. NBS technical note 1136. Gaithersburg, MD: National Bureau of Standards.

Waksman, D., W. C. Thomas, and E. R. Streed. 1984. *NBS Soiar Collector Durability/Reliability Test Program: Final Report*. NBS technical note 1196. Gaithersburg, MD: National Bureau of Standards.

Walton, W. D., and D. Waksman. 1981. *Fire Testing of Roof-Mounted Solar Collectors by ASTM E 108*. NBSIR 81-2344. Gaithersburg, MD: National Bureau of Standard.

Whillier, A. 1965. The thermal performance of solar water heaters. *Solar Energy* 9(1).

Wijeysundera, N. W. 1976. Response time of solar collectors. *Solar Energy* 18: 65–68.

Wood, B. D., P. J. Fiore, and C. R. Christopherson. 1979. Application of ASHRAE Standard 93-77 for testing concentrating collectors for the purpose of pedicting all-day performance. In *Proc. Silver Jubilee Congress of the International Solar Energy Society*, Atlanta, GA, vol. 1, pp. 487–491.

Youngblood, W. W., W. Schultz, and R. Barber. 1979. *Solar Cullector Fluid Parameter Study*. NBS GCR 79-184. Gaithersburg, MD: National Bureau of Standards.

Zerlaut, G. G., W. T. Dokos, and R. F. Heiskell. 1977. The use of ASHRAE Standard 93-77 in predicting all-day performance of flat-plate collectors. In *Proc 1977 Flat-Plate Solar Collector Conference*. Florida Solar Energy Center.

6 Testing and Evaluation of Tracking Collectors

David W. Kearney

6.1 Introduction

The technology for tracking collectors encompasses a wide range of concentrating collectors following the sun in either one or two axes. Single-axis or line-focusing collectors, however, were developed and commercialized more rapidly than two-axis systems such as central receiver or parabolic dish systems. Many technical and institutional factors led to this situation, though certainly a major contributor was the more difficult technical problems associated with the higher temperature operation of the two-axis tracking collectors. Line-focus collectors, on the other hand, were perceived from the beginning as quite appropriate for a broad range of intermediate temperature thermal applications such as heating water, air, and steam for industrial process heat needs and for both organic and steam Rankine cycle electrical generation.

The development and commercial production of concentrating collectors led to the recognition by the Department of Energy (DOE) and industry that systematic and standardized collector test procedures were necessary to evaluate this technology so that potential users could assess collectors with a consistent measure. While the major focus of this work has been line-focus collectors, the approaches and facilities were developed with the expectation that they would be applicable to both line- and point-focus technologies.

The pioneering work in testing for the tracking collectors began in 1973 with Sandia National Laboratories, Albuquerque (SNLA), under DOE funding. By 1975 the SNLA Midtemperature Solar System Test Facility (MSSTF), which included the Collector Module Test Facility and the System Test Facility, was operational. Additional concentrating collector testing capabilities were developed for line-focus collectors at the Solar Energy Research Institute (SERI) in 1979 and, through SNLA, at commercial testing laboratories in 1980. A parabolic dish test facility was constructed under Jet Propulsion Laboratory sponsorship in 1980 but was primarily oriented toward testing of prototypes rather than commercial products.

ASHRAE Test Procedure 93-77 was published in 1977 to provide guidelines for testing both nontracking and tracking solar collectors. Later, in parallel with the development of tracking collector test facilities, many investigators involved in collector development and evaluation worked to develop a rigorous test procedure specifically for tracking collectors under the auspices of the American Society for Testing and Materials (ASTM), resulting in the publication of an ASTM standard test method (ASTM E905) in 1983.

However, since tracking collectors are typically used in large arrays, testing single modules requires understanding system performance. This understanding should encompass systems that include piping and other balance-of-system components as well as non-steady-state conditions. DOE initiated several programs to gain this information. The first involved extensive field tests of numerous large parabolic trough collector systems at industrial sites. The second program, termed the Modular Industrial Solar Retrofit (MISR), was more narrowly defined and aimed specifically at developing and testing advanced parabolic trough collector systems for industrial steam applications.

6.1.1 Development of Test Procedures

Early testing of tracking collectors at SNLA and later at SERI was based on the general techniques in ASHRAE Standard 93-77. This method was not wholly adequate, however, because testing tracking collectors required broad interpretations, analyses, and precise testing techniques not specifically covered in that standard. This incompleteness was recognized by the Solar Energy Committee of the ASTM, which established a subcommittee to develop a series of standards for determining the annual energy output of a specific tracking collector for a specific location, based on an economical test method.[1] The subcommittee operated under the procedures of voluntary consensus, and its membership represented manufacturers, users, and others drawn from industry, government, universities, and test laboratories. SERI developed the initial draft of the standard for the subcommittee. The final product of the subcommittee, which is described in some detail below, was ASTM Standard Test Method E905-82: "Determining Thermal Performance of Tracking Concentrating Solar Collectors" (1983).

There are numerous technical issues specific to the test requirements of concentrating collectors. The main considerations are

1. the effects of the tracking/drive systems and reflector surface accuracy on performance,

2. selection of the proper normalizing factor for the incident solar radiation

3. development of quasi-steady-state test conditions applicable to high concentration ratio collectors,

4. a separate determination of the $m \cdot C_p$ product by calorimetric methods due to the lack of sufficient accuracy in C_p for the high temperature oils to be used in testing,

5. the need to define the requirements for normal-incidence and off-angle incidence testing and analysis,

6. elimination of the need for preconditioning the collector since exposure without flow could cause damage,

7. the need to restrict testing to outdoor, clear-sky conditions due to the difficulties and uncertainties introduced by most existing simulators for large concentrators.

The published standard applies to outdoor testing of one- or two-axis concentrating thermal collectors in which the effects of only direct radiation are important (i.e., the influence of diffuse radiation is negligible). The testing procedures determine the optical response for various incident angles of solar radiation and the thermal performance at various operating temperatures for the condition of near-normal incidence. The method requires attaining quasi-steady-state conditions, measuring environmental parameters, and determining the temperature rise and $m \cdot C_p$ product between the inlet and outlet of the collector. Thermal performance is then determined as the rate of heat gain of the collector relative to the solar power incident on the plane of the collector aperture. The test method provides experimental and calculational procedures to determine the following parameters: response time, incident angle modifiers, near-normal incidence angular range, and rate of heat gain at near-normal incidence angles.

Response time is the time required for a specified change in the temperature rise across the collector after a step change in the solar irradiance. It establishes the time required for quasi-steady-state conditions to exist for valid testing conditions.

The incident angle modifier is measured for single-axis tracking collectors so that the thermal performance at arbitrary angles of incidence can be predicted from the thermal performance measured at near-normal incidence angles. The effects of incidence angle on performance are caused by geometric variations and by changes in optical properties as a function of off-normal irradiation. The incident angle modifier is measured where thermal heat losses are minimized, namely, where the inlet temperature of the heat transfer fluid to the collector is at or near ambient temperature.

If the test apparatus includes a two-axis tracking platform on which the collector to be tested is mounted, then performance tests can be carried out under normal incidence conditions throughout the day. If this is not the case (i.e., the test platform is single axis or a line-focus collector is driven by its own tracking mechanism), then the range over which "near-normal" data can be obtained must be determined. The test standard provides methods for finding the angular range over which the thermal performance decreases no more than 2%.

The effect of angle on testing of near-normal incidence thermal performance is also a critical factor in the ability to maintain a constant level of irradiation during a test, which in turn determines the ability to maintain quasi-steady-state conditions. The test standard lays down tight requirements on the amount of change permitted during a test in inlet temperature, temperature rise through the collector, heat input or $m \cdot C_p$ change, ambient temperature, and irradiation. In addition, the tests must be conducted with a maximum in average wind speed of 4.5 m/s (10 mph) and with at least 630 W/m^2 (214 Btu/h ft^2) direct normal solar irradiance. The near-normal thermal performance testing is to be repeated four times with equally spaced values of inlet fluid temperatures.

The results of a thermal performance test are typically presented in terms of efficiency versus the temperature parameter, namely, where

$$\eta = \frac{\text{thermal heat gain}}{\text{direct radiation in the plane of the collector}},$$

$$\frac{\Delta T}{I} = \frac{T_{\text{inlet}} - T_{\text{ambient}}}{\text{direct radiation in the plane of the collector}}.$$

The test standard does not call for testing at different levels of solar irradiance, but Dudley and Workhoven (1982) and others have found that irradiation level may have a marked influence on collector efficiency. This point is discussed in more detail in chapter 17 by Galowin. Until the test standard is revised to account for this fact, it is recommended in Blackmon and Linskens (1982) and ASTM (1983) that performance tests be conducted over the range of direct radiation levels likely to be encountered by the collector system in question.

6.1.2   Collector Test Facilities

The test facilities developed between 1974 and 1980 to test tracking collectors are listed in table 6.1, with a summary of their scope and capabilities. Though all were designed and constructed before the publication of the ASTM test standard, the basic tenets of ASHRAE Standard 93-77, which are retained in the ASTM standard, were the bases of the design of these test facilities. Moreover a number of the test facility designers also participated in the ASTM Solar Energy subcommittee.

It is important to note that all the facilities listed in table 6.1 other than the Parabolic Dish Test Site were designed to test line-focus tracking collectors. This situation reflects the fact that, at the time of these efforts, line-focus parabolic trough collectors were closer to commercialization and were manu-

Table 6.1
Concentrating collector test facilities

Test facility	Tested
Collector module test facility Sandia National Laboratories Albuquerque, New Mexico	Line-focus collector modules, including three test stations, each served by its own fluid loop and data-acquistion capability and able to handle a collector up to $45\,\text{m}^2$ ($486\,\text{ft}^2$) in aperture area. Different fluids in each loop permitted testing to different temperatures:
	Loop 1 Therminol 66 315°C (600°F) Loop 2 Syltherm 800 425°C (797°F) Loop 3 Water 315°C (600°F)
	Loop 2 also incorporated a rotating (two-axis) test platform, which allowed for any orientation over time desired for a particular test. A weather station collected all necessary meteorological data.
Mid-temperature collector research facility Solar Energy Research Institute Golden, Colorado	Line-focus collector modules including single test loop with two-axis tracking test platform and full data acquisition capabilities. Temperature limitation was 230°C (446°F) with water and 340°C (644°F) with heat transfer fluid. Weather station for meteorological data.
DOE-approved commercial testing laboratories DSET Laboratories, Inc., Phoenix, Arizona Wyle Laboratories, Huntsville, Alabama BDM Corporation, Albuquerque, New Mexico	SNLA established a program in 1989 by which commercial laboratories satisfying certain stringent criteria were approved as qualified to test mid-temperature concentrating collectors. The basic requirements were that collector modules could be tested at temperatures between 100° and 425°C (212°–797°F) using water or heat transfer oil at a flow rate from 1 to 10 gpm (6.3×10^{-5}–$63 \times 10^{-5}\,\text{m}^3/\text{s}$). Two-axis tracking, full data acquisition, and a meteorological station were also required. Added criteria were familiarity with standard test procedures, proven engineering capability, and the ability to achieve established accuracy criteria.
Parabolic dish test site Constructed by Jet Propulsion Lab at Edwards Test Station, California, in 1979. Moved to SNLA in 1984.	Since parabolic dishes have different structural supports and tracking subsystems depending on the specific design, this facility was flexible enough to accommodate the construction of tailored test facilities. Facility permitted testing at temperature levels from 300° to 1,700°C (572°–3,092°F), with full data acquisition and weather station capabilities. Electrical measuring equipment provided for gathering of electrical generation performance of dish-mounted engine generators. The primary mission of this facility has been component development and prototype testing both at JPL and SNLA.

factured by several firms in the United States. Note also that each of these facilities included two-axis test platforms, allowing good control of test conditions and permitting direct normal testing to take place during most of the daylight hours. Although not classified in this chapter as "collector testing facilities," facilities for developmental testing of central receiver/heliostat technologies were operating during this same period at the Georgia Institute of Technology, Sandia National Laboratories, Albuquerque, and the Solar One site at Daggett, California.

6.1.3 Test Results and Evaluation of Data

Under DOE-sponsored testing programs at SNLA and at the DOE-approved commercial testing laboratories, approximately 30 tracking collectors were tested from 1978 to 1982 (Leonard and Klimas 1984). These collectors were usually commercial prototypes or commercially available models. Between 1980 and 1981 tracking collectors were being tested at SERI, entirely in support of the development of the ASTM testing standard (Lewandowski 1983). Since 1982 limited proprietary testing has been carried out at the DOE-approved laboratories, primarily DSET, for commercial development.

The tests in the ASTM procedure include determining thermal efficiency, heat loss, and incident angle modifier. The test procedure used by SNLA for tracking collectors preceded the ASTM standard but encompassed many of the elements recommended in that document. Figure 6.1 shows typical results determined from SNLA tests on three collectors (Harrison, 1980a, b, 1981).

The collector efficiency curves shown in figure 6.1 are subject to the criticism that this method of presentation of test data does not fully describe the performance of a tracking collector at high operating temperatures and at different levels of solar insolation. The performance parameter $\Delta T/I$ implies that heat losses are linearly dependent on ΔT, which is adequate for low temperature collectors such as flat plates where radiation terms can be linearized. For high temperature collectors, however, radiative heat losses become much more important. In 1981 Dudley and Workhoven published a paper in which they argue that level of radiation is a critically important parameter. Figure 6.2 shows their results for a series of tests in which thermal efficiency was measured as a function of temperature with the parameter I (level of direct beam insolation in the plane of the collector) fixed. Experience at SNLA and SERI shows that adding at least one nonlinear heat loss term models the data with acceptable accuracy and reproduces the effects shown in figure 6.2, though the exact form of the additive terms is still open to question.

From first principles one would expect an optical efficiency term that includes the effect of incidence angle and heat loss terms that model convective

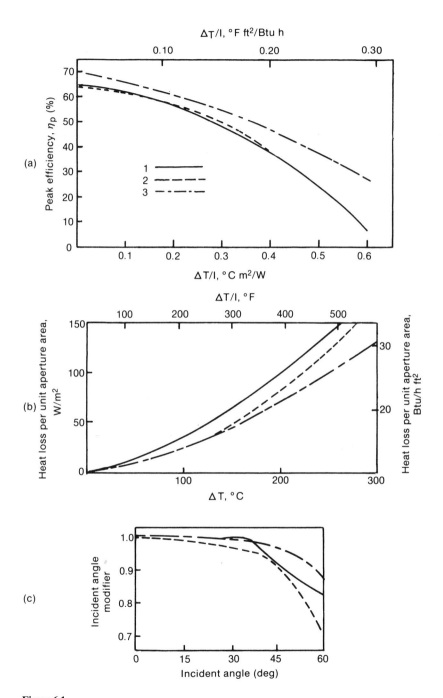

Figure 6.1
Typical collector test results. Source: Harrison (1980a, b, 1981).

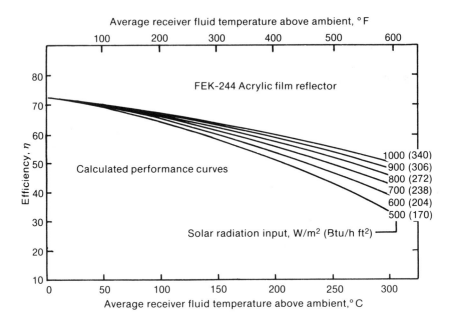

Figure 6.2
Line-focus collector efficiency curves. Source: Dudley and Workhoven (1982).

and radiative heat losses. The driving temperature differences are not identical for these two types of losses, and the effect of temperature is linear for convection but a fourth-power effect for radiation. Exact treatment of these factors does not lend itself to a straightforward test procedure nor to a convenient prediction of collector performance.

In an experiment or system application, a driving temperature difference which is easily measured or known is that between the fluid and the ambient temperature. For line-focus concentrating collector Cameron and Dudley (1984) and Dudley and Workhoven (1982) have found the following form of equation to be useful in a practical sense and to model experimental results well:

$$\eta = \eta_0 K - C_1 K \Delta T - \frac{C_2 \Delta T}{I} - \frac{C_3 (\Delta T)^2}{I},$$

where

η_0 = optical efficiency,

K = incident angle modifier,

ΔT = average fluid temperature (ambient temperature),

I = direct radiation in plane of collector,

C_i = constants determined by test.

Although not precise in a physical sense, the second term in this equation accounts for the influence of insolation level on measured efficiency, the second term models convective/conductive losses, and the final term models radiative losses.

In summary, the physical mechanisms involved in the gain and loss of heat in concentrating systems are complex. At high collector operating temperatures the influences of many of the contributing factors are more significant than at lower temperatures, and hence they must be included in both test procedures and models. Although much progress has been made in establishing test methods and identifying the most important parameters, continued development of test methods, forms of presentation, and testing standards is required for tracking concentrating collectors.

6.2   System and Subsystem Testing

The ultimate concern of the end user of solar technology is the performance of a collector field or array, not an isolated collector alone. In this context efficiency tests on a single collector are a critical but not self-sufficient element. System piping heat losses during operation, the influence of the load on collector operating temperature, and system cool-down losses at night or during weather transients are important additional factors. The design and operation of large collector arrays to maximize performance are discussed in several reports (Kutscher et al. 1982; Bankston 1984; Cameron and Dudley 1985). The following section briefly discusses two DOE-sponsored programs in which system and subsystem testing and performance monitoring of line-focus tracking collectors were very important.

6.2.1   MISR Testing

In 1981 SNLA initiated the Modular Industrial Solar Retrofit (MISR) program for DOE (Cameron and Dudley 1984, 1985). This included the design and subsystem testing of five line-focus collector systems capable of producing industrial process steam up to about 1.3 MPa (about 250 psia). Each system design included an equipment skid and modular solar field array with an aperture area of 175–460 m^2 (1,890–4,968 ft^2) and an outlet temperature of 250°–290°C (482°–554°F). Table 6.2 gives the characteristics of the five systems. The full series of tests carried out on the MISR systems included safety

Table 6.2
MISR system design characteristics

System designer	Collector	Reflector	Receiver O.D., cm (inch)	Delta-temperature string		
				Length, m (ft)	Aperture, m² (ft²)	Flow, L/s (cfm)
Acurex	Acurex 3011	Thin glass laminate	3.5 (1.4)	146 (5,748)	2.0 (21.6)	1.0 (2.1)
BDM	SKI T-700	FEK-244	4.4 (1.7)	219 (8,622)	2.1 (22.7)	1.3 (2.7)
CEI	CEI-Budd	Chem-core glass	3.5 (1.4)	146 (5,748)	3.0 (32.4)	1.0 (2.1)
F W	Suntec Mark V	Sagged glass	4.8 (1.9)	146 (5,748)	3.0 (32.4)	0.7 (1.5)
SKI	SKI T-800	FEK-244	4.4 (1.7)	73 (2,874)	2.4 (25.9)	1.3 (2.7)

System	Number of delta-T strings[a]	Total aperture m² (ft²)	Heat exchanger surface area, m² (ft²)	HTF temperature at heat exchange outlet, °C (°F)[c]	Skid HTF flow rate, L/s (cfm)
Acurex	8	2,484 (26,827)	45/10[b] (484/11)	190 (374)	7.6 (16.1)
BDM	5	2,324 (25,099)	60 (648)	210 (410)	6.3 (13.3)
CEI	8	2,305 (24,894)	45/10[b] (484/11)	190 (374)	7.6 (16.1)
F W	12	2,586 (27,929)	62 (670)	210 (410)	8.3 (17.6)
SKI	14	2,472 (26,698)	79 (853)	210 (410)	16.4[d] (33.9)

Source: Cameron and Dudley (1984).
a. Only a single delta-temperature string was installed for each qualification test system. The remaining strings were simulated with propane-fired heater systems.
b. Figure applies to feedwater preheater.
c. Indicates average skid operating temperature and temperature of HTF returning to collectors. Data measured when generating 1,000 kg/h (2,204.6 lb/h) of 1.7 MPa steam with feedwater temperature of 90°C (194°F).
d. Design flow rate of 17.7 L/s (37.5 cfm) could not be achieved due to pressure drops in test facility equipment.

checks, complete functional tests in automatic operation, thermal performance tests, and life-cycle evaluations. Four MISR systems using heat transfer oil were at SNLA, and one using pressurized water was at SERI.

The performance tests were carried out on a number of collector drive groups connected in series and termed a "delta-temperature string." The equipment skid associated with each test string included pumps, steam generator, expansion tank, and other components. The equipment skid was designed to accommodate a full field of 2,300 m² (24,840 ft²); hence auxiliary propane-fired heater operated in parallel with the solar subsystem to simulate a full-sized solar field. The performance tests consisted of measurements of thermal efficiency as a function of operating temperature, thermal losses of the collector string and equipment skid, incident angle modifier, heat capacities of the collector string and equipment skid, and the parasitic power requirements.

The techniques used in the MISR testing and the handling of the test data follow the SNLA procedures described in section 6.1. In general, the consistent use of these methods resulted in a good correspondence between subsystem tests and single collector tests. The heat capacities of the delta-temperature strings agreed well with calculations based on the mass of metal and fluid in the system. Heat loss, heat capacity, and parasitic power were measured on the equipment skids as well. These results plus incident angle modifier data allowed the prediction of annual performance for each MISR system. Full test results for the MISR program were published in 1985. These data, in addition to the inherent value in terms of the performance of specific hardware, increase our understanding of the relationships between collector and subsystem testing.

6.2.2   Industrial Process Heat Field Tests

Beginning in 1976, DOE funded a series of field tests of large collector systems installed at industrial sites to provide process heat. The field tests involved measuring system performance, reliability, and maintainability. The system design called for thermal performance measurements to be made continuously by data acquisition systems; thus system transients are an inherent part of the data, and comparison to controlled steady-state test data must consider this. While the full program included a number of collector types, many of the systems (and particularly those constructed in the later part of the program) were line-focus tracking collector systems. Table 6.3 from Harley and Stine (1984) gives summary data on the parabolic trough systems that were installed as part of this program.

Table 6.3
Summary of the IPH field test line-focus projects

Location	Process	Industrial partner/contractor	Collectors	Array size, ft² (m²)	Steam conditions	Fluid conditions	Designed annual energy delivery, Btu/ft² yr (kJ/m² yr)
Low temperature steam IPH field tests, 110°–177°C (230°–350°F)							
Pasadena, California	Commercial laundry	Home Cleaning and Laundry/Jacobs-Del Solar Systems	Parabolic trough	6,496 (70,157)	0.86 MPa (125 psia) 171°C (340°F)		249 (2,826)
Sherman, Texas	Gauze bleaching	Johnson & Johnson Acurex Corp.	Parabolic trough	11,520 (124,416)	0.86 MPa (125 psia) 174°C (345°F)		122 (1,385)
Fairfax, Alabama	Fabric drying	West Point Pepperell Honeywell, Inc.	Parabolic trough	8,313 (89,780)	0.58 MPa (85 psia) 158°C (317°F)		132 (1,498)
Intermediate temperature steam IPH field tests, 177°–288°C (350°–550°F)							
Dalton, Georgia	Latex production	Dow Chemical/Foster-Wheeler Dev. Corp.	Parabolic trough	9,930 (107,244)	1.14 MPa (165 psia) 185°C (366°F)		252 (2,860)
San Antonio, Texas	Brewery	Lone Star Brewing Co./Southwest Res. Inst.	Parabolic trough	9,450 (102,060)	0.96 MPa (140 psia) 177°C (351°F)		338[a] (3,836)
Ontario, Oregon	Potato processing	Ore-Ida Co./TRW	Parabolic trough	9,520 (102,816)	2.07 MPa (300 psia) 214°C (417°F)		200 (2,270)
Hobbs, N. Mexico	Oil refinery	Southern Union Co./Monument Solar Corp.	Parabolic trough	10,080 (108,864)	1.27 MPa (185 psia) 193°C (380°F)		351 (3,984)
Intermediate temperature cost-shared IPH field tests, 100°–288°C (212°–550°F)							
San Leandro, California	Pressurized hot water for washing	Caterpillar Tractor/Southwest Res. Inst.	Parabolic trough	50,400 (544,320)		Water 0.21 MPa (30 psia) 113°C (235°F)	269 (3,053)
Haverhill, Ohio	Chemical plant-polystyrene	U.S.S. Chemicals/Columbia Gas/H.A. Williams & Assoc.	Parabolic trough	50,400 (544,320)		Steam 1.03 MPa (150 psia) 227°C (440°F)	143 (1,623)

Source: Harley and Stine (1984).
a. Collector output.

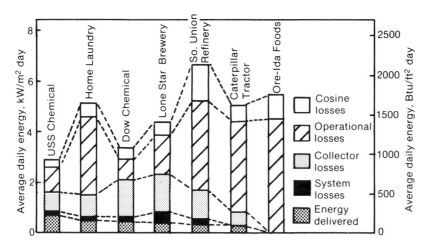

Figure 6.3
IPH field test measured system performance. Source: Harley and Stine (1984).

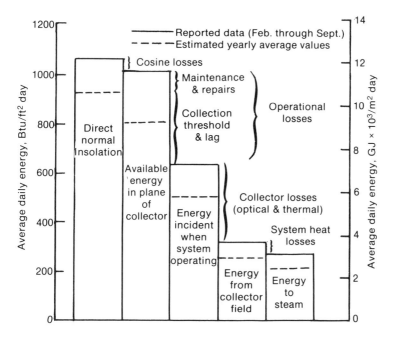

Figure 6.4
Energy losses from the USS Chemicals system. Source: Harley and Stine (1984).

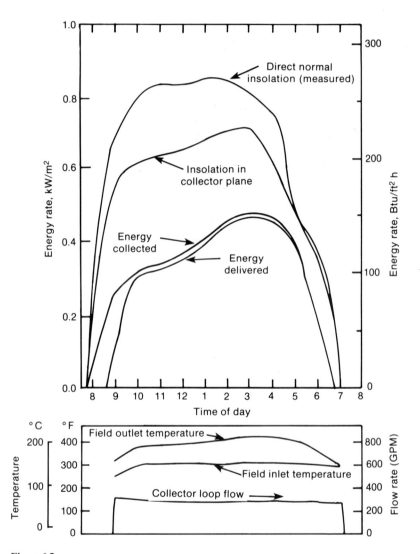

Figure 6.5
Clear-day performance of the USS Chemicals system for September 8, 1983. Source: Harley and Stine (1984).

Figures 6.3 through 6.5 show examples of the test data developed from the IPH field tests. Figure 6.3 gives a summary of the measured energy delivered to the industrial loads by these systems. Starting from a basis of direct normal incident radiation, the cosine losses account for the energy collection loss due to single-axis tracking. The operational losses are quite different in that they are associated with the availability of the system; that is, the losses are associated with nonoperation of the solar energy system or the inability of the industrial process to use the heat that could otherwise have been delivered. In addition this term includes losses associated with thermal lag due to transients and minimum radiation thresholds that prevent performance during low insolation periods. The collector losses include those from both optical and thermal mechanisms. In field tests of this nature, however, other elements such as soiled collectors, tracking system inaccuracies, and degradation of physical properties or collector adjustments also contribute. System losses include piping and insulation heat losses, both in operation and at shutdown; electrical parasitics for pumps, tracking, and control; and field-to-process heat exchanger losses. In the overall energy delivery by these systems, thermal performance of a single collector is a significant but not dominant factor in solar field performance.

Further insight into the measurement of system performance for tracking concentrators can be obtained by a more detailed examination of one of the IPH systems. The solar energy system at the USS Chemicals plant supplied saturated steam at 151°C (302°F) for continuous production of industrial chemicals. Therminol 60 circulated through a field of Solar Kinetics collectors and then was fed to an unfired steam generator. Both the supply and return headers were stepped down in size along their lengths. All valves, pipes, pipe anchors, and other equipment were very well insulated, and the pipe supports eliminated metal-to-metal contact to minimize thermal loss. The results of this concern for detail are evident in the energy performance of this system. Figure 6.4 shows more detailed information on the long-term performance of this system, and figure 6.5 gives the diurnal performance for a single clear day. Figure 6.6 is an excellent example of the test measurements resulting from the IPH field test program.

Valid test measurements are the basis for system performance predictions for these large arrays. Lewandowski (1984) focuses on this subject, giving results for many of the line-focus IPH systems. Particularly good agreement between measured and predicted performance was found for the USS Chemicals system, presumably because its design and construction features allowed the model to characterize its thermal characteristics well.

Note

1. The description here of the background, objectives, and summary of the ASTM method is condensed from the comments of Blackmon and Linskens (1982).

References

ASTM. 1983. Standard Test Method E 905-82 for "Determining Thermal Performance of Tracking Concentrating Solar Collectors." Approved Oct. 29, 1982.

Bankston, C. F., ed. 1984. In *Proc. International Energy Agency Workshop on the Design and Performance of Large Solar Collector Arrays*. San Diego, CA.

Blackmon, J. B., and M. C. Linskens. 1982. *Development of a Consensus Standard for Determining Thermal Performance of High Concentration Ratio Solar Collectors*. SERI/TP-253-1839. Golden, CO: Solar Energy Research Institute. Also in the *Proc. ASME Solar Energy Division 5th Annual Technical Conference*, Orlando, FL, April 1983.

Cameron, C. P., and V. E. Dudley. 1984. Testing of the MISR industrial process heat systems. In *Proc. International Energy Agency Workshop on the Design and Performance of Large Solar Collector Arrays*. San Diego, CA.

Cameron, C. P., and V. E. Dudley. 1985. *Modular Industrial Solar Retrofit Project Final Report*. SAND85-0755. Albuquerque, NM: Sandia National Laboratory.

Dudley, V. E., and R. M. Workhoven. 1982. Variation of collector efficiency and receiver thermal loss as a function of solar irradiance. In *Proc. ASME Solar Energy Division 4th Annual Technical Conference*. Albuquerque, NM. Also see Dudley, V. E. 1982. *Performance Testing of the Solar Kinetics T-700A Solar Collector*. SAND81-0984. Albuquerque, NM: Sandia National Laboratory.

Harley, E. L., and W. B. Stine. 1984. *Solar Industrial Process Heat (IPH) Project Technical Report*. SAND84-1812. Albuquerque, NM: Sandia National Laboratory.

Harrison, T. D. 1980a. *Midtemperature Solar Systems Test Facility Predictions for Thermal Performance of the Solar Kinetics T-700 Solar Collector with FEK-244 Reflector Surface*. SAND80-1964/1. Albuquerque, NM: Sandia National Laboratory.

Harrison, T. D. 1980b. *Midtemperature Solar Systems Test Facility Predictions for Thermal Performance of the Suntec Solar Collector with Heat-Formed Glass Reflector Surface*. SAND80-1964/2. Albuquerque, NM: Sandia National Laboratory.

Harrison, T D. 1981. *Midtemperature Solar Systems Test Facility Predictions for Thermal Performance of the Acurex Solar Collector with FEK 244 Reflector Surface*. SAND80-1964/3. Albuquerque, NM: Sandia National Laboratory.

Kutscher, C. F., R. L. Davenport, D. A. Dougherty, R. C. Gee, T. M. Masterson, and E. K. May. 1982. *Design Approaches for Solar Industrial Process Heat Systems*. SERI/TR-253-1356. Golden, CO: Solar Energy Research Institute.

Leonard, J.A., and C. R. Klimas. 1984. *A Bibliography of Reports of the Sandia Solar Thermal Distributed Receiver Systems Project*. SAND84-0313. Albuquerque, NM: Sandia National Laboratory.

Lewandowski, A. 1983. Testing of prototype Fresnel-lens concentrator for thermal applications. In *Proc. ASME Solar Energy Division 5th Annual Technical Conference*. Orlando, FL.

Lewandowski, A. 1984. Comparison of predicted and reported performance for DOE-sponsored IPH field test experiments. In *Proc. International Energy Agency Workshop on the Design and Performance of Large Solar Collector Arrays*. San Diego, CA.

7 Optical Research and Development

Roland Winston

7.1 Introductory Remarks

It has been said that research and development (R&D) in the United States makes significant strides only in periods of crisis. The perceived "energy crisis" of the 1970s certainly caused significant strides in the optical R&D component of solar energy technology. At the beginning of the 1970s solar energy optics still relied on antiquated, nineteenth-century principles for designing concentrating solar collectors. A decade later a new optical principle, nonimaging optics, had evolved, leading to nontracking and even stationary concentrating solar collectors; sophisticated computer-aided techniques had been applied to tracking collectors; and nonconventional optical elements such as holographic concentrators had been explored.

7.2 Brief Review of Relevant Optical Principles

7.2.1 Geometrical Optics Applied to Solar Conversion

The Concepts of Geometrical Optics Geometrical optics is used as the basic tool in designing almost any optical system, including solar concentrators. We use the intuitive ideas of a ray of light, roughly defined as the path along which radiant energy travels, together with surfaces that reflect or transmit the light. When light is reflected from a smooth surface, it obeys the well-known law of reflection, which states that the incident and reflected rays make equal angles with the normal to the surface and that both rays and the normal lie in one plane. When light is transmitted, the ray direction changes according to the law of refraction, Snell's law. This law states that the sine of the angle between the normal and the incident ray bears a constant ratio to the sine of the angle between the normal and the refracted ray; again, all three directions are coplanar.

In geometrical optics we represent power density across a surface by the density of ray intersections with the surface, and the total power by the number of rays. We take N rays spaced uniformly over the entrance aperture of a concentrator at an angle of incidence θ, as in figure 7.1. If only N' rays emerge through the exit aperture, whose dimensions are determined by the desired concentration ratio, then the power transmission for the angle θ is N'/N. Clearly N must be taken large enough to ensure that a thorough exploration of possible ray paths in the concentrator is made. By assigning weights to

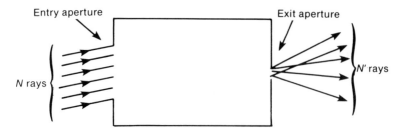

Figure 7.1
Determining the transmission of a concentrator by ray tracing.

individual rays, this procedure is readily generalized to include losses resulting from imperfect reflections and attenuation in the medium.

Ray Tracing   In tracing rays through reflecting and refracting surfaces, it is most convenient to cast the laws of reflection and refraction in vector form. Figure 7.2 shows the geometry with unit vectors $\mathbf{r}$ and $\mathbf{r}''$ along the incident and reflected rays and a unit vector $\mathbf{n}$ along the normal pointing into the reflecting surface. Then the law of reflection is expressed by the vector equation

$$\mathbf{r}'' = \mathbf{r} - 2(n \cdot \mathbf{r})n. \tag{1}$$

Ray tracing through a refracting surface is similar, but first we have to formulate Snell's law vectorially. Figure 7.3 shows the relevant unit vectors. It is similar to figure 7.2 except that $\mathbf{r}'$ is a unit vector along the refracted ray. We denote by n, n' the refractive indexes of the media on either side of the refracting boundary. The law of refraction is usually stated as

$$n' \sin I' = n \sin I. \tag{2}$$

However, the preferred form for ray tracing is

$$n'\mathbf{r}' = n\mathbf{r} + (n'\mathbf{r}' \cdot n - n\mathbf{r} \cdot n)n. \tag{3}$$

To use this equation, we first find $\cos I'$ using equation (2).

If a ray travels from a medium of refractive index n toward a boundary with another of index $n' < n$, then it can be seen from equation (2) that it would be possible to have $\sin I' > 1$. Under this condition the ray is completely reflected at the boundary. This is called total internal reflection and is a useful effect in some concentrator designs. Further details can be found in Welford and Winston (1989).

Some Elementary Properties of Image-Forming Systems   In principle, ray tracing tells us all there is to know about the geometrical optics of a given

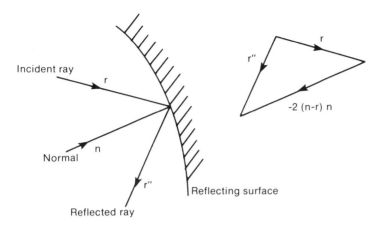

Figure 7.2
Vector formulation of reflection; **r**, **r″**, and **n** are all unit vectors.

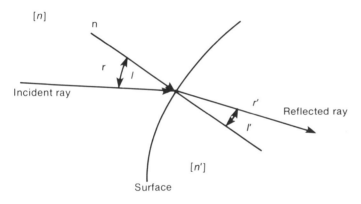

Figure 7.3
Vector formulation of refraction.

system. However, alone it is of little use for inventing new systems having good properties for a given purpose: We need ways of describing the properties of optical systems in terms of general performance. In this subsection we introduce some of these concepts.

Consider first a thin converging lens, which would be used to increase the flux on a solar cell (shown in figure 7.4). If we have rays coming from a point at a great distance to the left, thus substantially parallel as in the figure, they meet approximately at a point F, the focus. The distance from the lens to F is called the *focal length*, denoted by f. If the rays come from an object of finite size at a great distance, the rays from each point on the object converge to a separate focal point and we get an image; this is of course what happens when a burning glass forms an image of the sun or when the lens in a camera forms an image on the film. It is indicated in figure 7.5 where the object subtends the

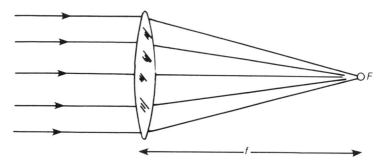

Figure 7.4
A thin converging lens bringing parallel rays to a focus; since the lens is technically "thin," we do not have to specify the exact plane in the lens from which the focal length f is measured.

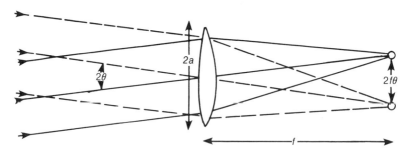

Figure 7.5
An object at infinity has an angular subtense 2θ; a lens of focal length f forms an image of size $2f\theta$.

small angle 2θ; the size of the image is then $2f\theta$. This is easily seen by considering the rays through the center of the lens because these pass through undeviated.

Figure 7.5 implicitly contains one of the fundamental notions used in solar concentrator theory, the concept of light of a certain diameter and angular extent. The diameter is that of the lens, $2a$, and the angular extent is given by 2θ. (The paraxial approximation is implied so that $\theta \approx \sin\theta$.) These two can be combined as a product $\approx \theta a$, a quantity known variously as extent, *étendue*, phase space acceptance, Lagrange invariant, etc. It is in fact an invariant through the optical system, provided that there are no obstructions in the light beam and that we ignore certain losses caused by properties of the materials such as absorption and scattering. For example, at the plane of the image the *étendu* becomes the image height θf multiplied by the convergence angle a/f of the image-forming rays, giving again θa. In discussing three-dimensional systems, the square of this quantity, $(a\theta)^2$, is also called the *étendue*. We can put an aperture of diameter $2f\theta$ at the focus of the lens, as in figure 7.6; then the system will accept only rays within the angular range $\pm\theta$ and inside the diameter $2a$. Now suppose that a flux of radiation $B(W/m^2$ sr) is incident on the lens from the left. The system will actually accept a total flux $(\pi\theta a)^2 W$; thus the *étendue* $(\theta a)^2$ is a measure of the power flow that can pass through the system.

The same discussion shows how the concentration ratio C appears in the context of classical optics; the accepted power $B(\pi\theta a)^2 W$ must flow out of the aperture to the right of the system. Thus the system is acting as a concentrator with concentration ratio $C = (a/f\theta)^2$ for the input semiangle 3. Clearly for a given lens diameter we gain by reducing the focal length as much as possible.

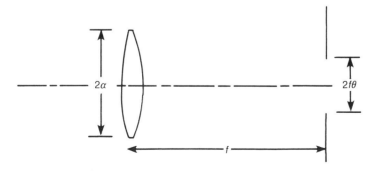

Figure 7.6
An optical system of acceptance, throughput, or *étendue* $a^2\theta^2$.

Aberrations of Image-Forming Systems   According to the simplified picture presented in preceding discussion, we could make a lens system with indefinitely large concentration ratio by simply decreasing the focal length sufficiently. This is, of course, not so partly because of aberrations in the optical system and partly because of the fundamental limit on concentration prescribed by the second law of thermodynamics.

We can explain the concept of aberrations by reference to our example of the thin lens of figure 7.4. We suggested that all the parallel rays shown converged to a single point F after passing through the lens; in fact this is true only in the limiting case when the diameter of the lens is taken as indefinitely small. The theory of optical systems under this condition is called *paraxial* or *Gaussian optics*, and it is a very useful approximation for getting at the principal large-scale properties of image-forming systems. If we take a simple lens with a diameter that is a sizable fraction of the focal length, say, $f/4$, we find that the rays do not all converge to a single image point.

In general, for a convex lens the rays from the outer part of the lens aperture meet the axis closer to the lens than the paraxial rays, an effect known as spherical aberration (see figure 7.7). Even if we could eliminate spherical aberration, we would still find that the rays from an object pointing away from the axis do not form point images; in other words, there would be off-axis aberrations. For details, see the specialized treatments in Welford (1974). Similar considerations apply to imaging-reflecting systems. For example, a paraboloid reflects rays parallel to the axis to a single point (no spherical aberration) but has severe off-axis aberrations (Welford and Winston 1989). In general, spherical aberrations but not off-axis aberrations can be eliminated completely for reflecting and refracting systems. Suppose this has been done for the simple concentrator of figure 7.4. Some rays of the beam at the extreme angle θ will fall outside the defining aperture of diameter $2f\theta$. We can see this more clearly by representing the aberration as a spot diagram; this is a diagram in the image plane with points plotted to represent the intersections of the various rays in the incoming beam. Such a spot diagram for the extreme angle θ might appear as in figure 7.8; the ray through the center of the lens (principal ray) meets the rim of the collecting aperture, which blocks a considerable amount of the flux. Conversely, some flux from beams at an angle greater than θ will be collected.

The graph in figure 7.9 shows the fraction collected at different angles up to the theoretical maximum, θ_{max}. An ideal concentrator would behave according to the solid line, collecting all flux within θ_{max} and none outside.

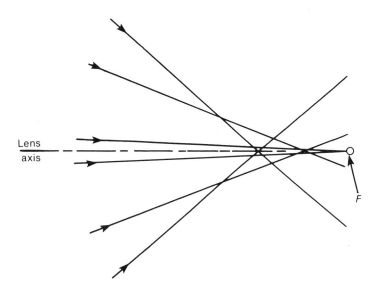

Figure 7.7
Rays near the focus of a lens showing spherical aberration.

Frequently textbooks on geometrical optics give the impression that aberrations are in some sense "small"; this is true in optical systems designed to form reasonably good images, such as camera lenses, but these systems do not operate with large enough convergence angles (a/f in the notation of figure 7.5) to approach the maximum theoretical concentration ratio. A conventional image-forming system under such conditions would have very large aberrations, which would severely depress the concentration ratio. For example, a paraboloidal reflector falls short of the ideal concentration limit by a factor of at least four (Harper et al. 1976). This is one limitation that has led to the development of the new nonimaging solar devices. In fact the emphasis of solar applications on flux concentration as opposed to image-formation has increased understanding of the concentration limits of both image-forming and nonimaging optical systems (Winston and Welford 1982; Bassett and Winston 1984).

Optical Path Length and Fermat's Principle   There is another way of looking at geometrical optics and the performance of optical systems. The speed of light in a medium of refractive index n is c/n, where c is the speed in vacuum, so light travels a distance s in the medium in time ns/c. The quantity ns is called the *optical path length* corresponding to the length s. If a point source O emits light into an optical system, as in figure 7.10, we can trace any number

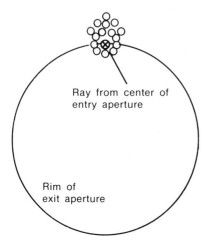

Figure 7.8
A spot diagram for rays from the beam at the maximum entry angle for an image-forming
concentrator. Some rays miss the edge of the exit aperture due to aberrations, and the *étendue* is
thus less than the theoretical maximum.

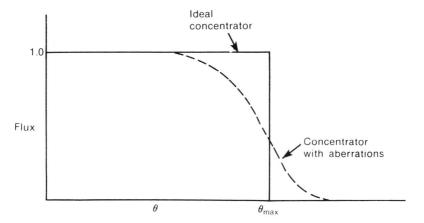

Figure 7.9
A plot of collection efficiency against angle; the ordinate is the proportion of flux entering the
collector aperture at angle θ, which emerges from the exit apertures.

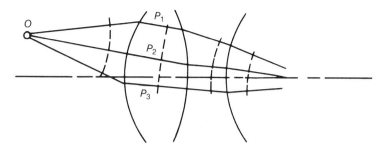

Figure 7.10
Rays and (in broken line) geometrical wave fronts.

of rays through the system and then mark off along these rays points P_1, P_2, etc., that are all at the same optical path length from O. We do this by making the sum of the optical path lengths from O in each medium the same:

$$\sum ns = \text{constant.} \tag{4}$$

These points can be joined together to form a surface called a wave front, and we can construct wave fronts at all distances along the bundle of rays from O. We now introduce a key principle based on the concept of optical path length, which is a way of predicting the path of a ray through an optical medium. Suppose that we have any optical medium, which can have lenses and mirrors and even regions of continuously varying refractive index. We wish to predict the path of a light ray between two points A and B, in this medium (figure 7.11). We can propose an infinite number of paths, of which three are indicated, but unless A and B happen to be object and image (and we assume that they are not), only one or perhaps a small finite number of paths will be physically possible, for example, paths that rays could take according to the laws of geometrical optics. Fermat's principle states that a physically possible ray path is one for which the optical path length along it from A to B is an extremum as compared to neighboring paths. It is possible to derive all of geometrical optics from Fermat's principle; it also leads to the result that the rays are normal to the wave fronts. This in turn tells us that if there is no aberration (i.e., all rays meet at one point), the wavefronts must be portions of spheres; thus the optical path length from object point to image point is the same along all rays. Thus an alternative way of expressing aberrations is in terms of the departure of wave fronts from the ideal spherical shape. This concept will be useful when we discuss the different senses in which an image-forming system can form "perfect" images.

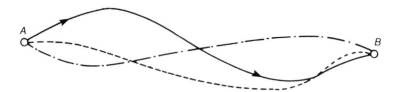

Figure 7.11
Fermat's principle; it is assumed in the diagram that the medium has a continuously varying refractive index; the full-line path has a stationary optical path length from A to B and is therefore a physically possible ray path.

Generalized *Etendue* and the Phase Space Concept   We now introduce a concept that is essential to the development of the optical principles of all solar collectors. We recall that the quantity $(a\theta)^2$ is a measure of the power accepted by the system; here a is the radius of the entrance aperture and, is the semiangle of the beams accepted, and we found that in paraxial approximation for an axisymmetrical system, this is invariant through the optical system. If we are considering a region of refractive index different from unity, say, the inside of a lens or prism, the invariant becomes $(na\theta)^2$. The reason for this can be seen from figure 7.12, which shows a beam at the extreme angle θ entering a plane-parallel plate of glass of refractive index n; inside the glass the angle $\theta' \approx \theta/n$, by the law of refraction, so that the invariant in this region is

$$\text{étendue} = (na\theta)^2. \tag{5}$$

The generalization of the *étendue* to rays at finite angles to the axis has been known for some time (e.g., Winston 1970), but since it has not been used to any extent in classical optical design, it is not described in many classical optics texts. It applies to solar optical systems of any or no symmetry and of any structure, such as refracting, reflecting, or continuously varying refractive index.

Let the system be bounded by homogeneous media of refractive indices n and n' as in figure 7.13, and suppose that we have a ray traced exactly between the points P and P' in the respective input and output media. We wish to consider the effect of small displacements of P and of small changes in direction of the ray segment through P on the emergent ray so that these changes define a beam of rays of a certain cross section and angular extent. To do this, we set up a Cartesian coordinate system $Oxyz$ in the input medium and another, $O'x'y'z'$, in the output medium. The positions of the origins of these coordinate systems and the directions of their axes are quite arbitrary

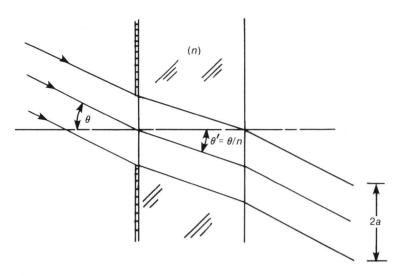

Figure 7.12
Inside a medium of refractive index n the étendue become $n^2 a^2 \theta'^2$.

with respect to each other, to the directions of the ray segments, and of course to the optical system. We specify the input ray segment by the coordinates of $P(x, y, z)$ and the direction cosines of the ray (L, M, N) and similarly for the output segment. We can now represent small displacements of P by increments dx and dy to its x and y coordinates, and we can represent small changes in the direction of the ray by increments dL and dM to the direction cosines for the x and y axes. Thus we have generated a beam of area $dxdy$ and angular extent $dLdM$; this is indicated in figure 7.14 for the y section. Corresponding increments dx', dy', dL', and dM' will occur in the output ray position and direction. Then the invariant quantity turns out to be $n^2 dxdydLdM$, and we have

$$n'^2 \, dx' \, dy' \, dL' \, dM' = n^2 dxdydLdM. \tag{6}$$

Details of the proof of this theorem can be found in Welford and Winston (1978).

The generalized *étendue* is sometimes written in terms of the optical direction cosines $p = nL$, $q = nM$, when it takes the form

$$dxdydpdq, \tag{7}$$

and it can be related to phase space conservation (Winston 1970). We can now use the *étendue* invariant to calculate the theoretical maximum concentration

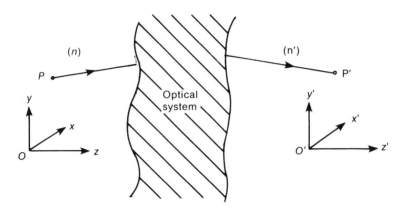

Figure 7.13
The generalized *étendue*.

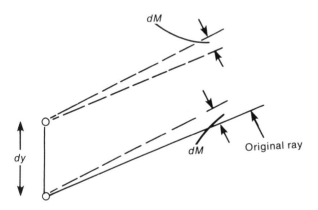

Figure 7.14
The generalized *étendue* in the *y* section.

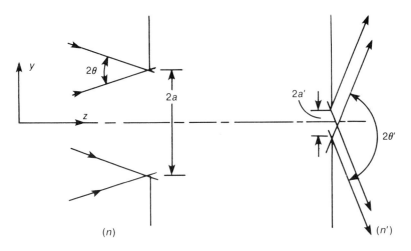

Figure 7.15
The theoretical maximum concentration ratio for a two-dimensional optical system.

ratio of solar collectors. Consider first a two-dimensional concentrator as in figure 7.15; from equation (6) we have

$$n\, dy\, dM = n'\, dy'\, dM' \tag{8}$$

for any ray bundle that traverses the system, and integrating over y and M, we obtain

$$4na \sin \theta + 4n'a' \sin \theta' \tag{9}$$

so that the concentration ratio is

$$\frac{a}{a} = \frac{n' \sin \theta'}{n \sin \theta}. \tag{10}$$

In this result a' is a dimension of the exit aperture large enough to permit any ray that reaches it to pass and θ' is the largest angle of all the emergent rays. Clearly θ' cannot exceed $\pi/2$, so the theoretical maximum concentration ratio is

$$C_{\max} = \frac{n'}{n \sin \theta}. \tag{11}$$

Similarly for the three-dimensional case we can show for an axisymmetrical concentrator the theoretical maximum is

$$C_{\max} = \left(\frac{a}{a'}\right)^2 = \left(\frac{n'}{n \sin \theta}\right)^2, \tag{12}$$

where again θ is the input semiangle. The results in equations (11) and (12) are maximum, not necessarily achievable, values. In practice, if the exit aperture has the diameter given by equation (10), some of the rays within the incident collecting angle and aperture do not pass it. In some systems incident rays are actually turned back by internal reflections and never reach the exit aperture. There are also losses caused by absorption, imperfect reflection, etc. Thus equations (11) and (12) give the theoretical upper bounds on performance of solar collectors in much the same way that the Carnot efficiency gives the upper bound of performance of heat engines; in the same spirit it provides a useful figure of merit for solar concentrators.

Basic Characteristics of Image-Forming Concentrators   The simplest hypothetical image-forming solar concentrator functions as in figure 7.16. The rays are coded to indicate that rays from one direction, from the sun, are focused at one point in the exit aperture (i.e., the concentrator images the sun at the exit aperture). The exit angle θ' must be $\pi/2$ for maximum concentration. Clearly such a concentrator would have to be constructed with glass, acrylic, or some other medium of refractive index greater than unity forming the exit surface, as in figure 7.17, and the angle θ' in the glass, etc., would be such that $\sin \theta' - 1/n$ so that the emergent rays just fill the required $\theta'' = \pi/2$ angle. For typical materials the angle θ' would be ≈ 40 degrees. Figure 7.17 brings out an important point about the optics of such a concentrator. We have labeled the central or principal ray of the two extreme angle beams a and b, respectively, and at the exit end these rays have been drawn normal to the exit face. This would be essential if the concentrator were used with air as the final medium since, if rays a and b were not normal to the exit face, some of the extreme angle rays would be totally internally reflected and thus the concentration ratio would be reduced. In fact the condition that the exit principal rays should be normal to what is in ordinary lens design terms the image plane is not usually fulfilled. Such an optical system, called *telecentric*, needs to be specially designed, and the requirement imposes constraints that would certainly worsen the attainable performance. An alternative configuration for an image-forming concentrator would be as in figure 7.18. The concentrator collects rays over $2\theta_{\max}$ as before, but the internal optics form an image of the entrance aperture at the exit aperture, as indicated by the arrow coding of the rays. In ordinary optics terminology this would be a telescopic or afocal system. We can also regard as ideal a system that brings all incident rays within

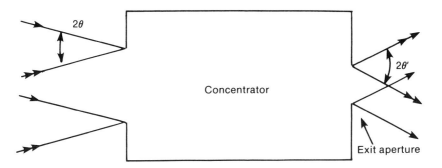

Figure 7.16
An image-forming concentrator; an image of the source at infinity is formed at the exit aperture
of the concentrator.

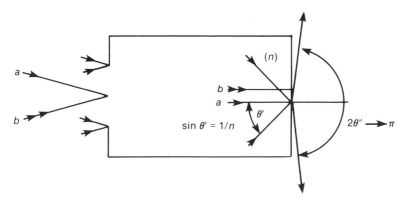

Figure 7.17
In an image-forming concentrator of maximum theoretical concentration ratio, the final medium
in the concentrator would have to have a refractive index n greater than unity; the angle θ' in
this medium would be arcsin $1/n$, giving an angle $\pi/2$ in the air ouside.

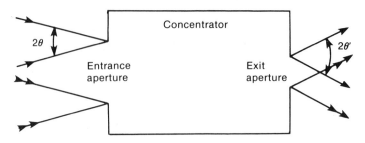

Figure 7.18
An alternative configuration of an image-forming concentrator. The rays collected from an angle
to form an image of the entrance aperture at the exit aperture.

θ_{max} out within θ'_{max} and inside an exit aperture a' given by equation (10) (i.e.,
$a' = na \sin \theta / n' \sin \theta'_{max}$). Such a concentrator will be ideal but will not have
the theoretical maximum concentration.

The concentrator sketched in figures 7.16 and 7.17 must contain something
either of a very large aperture (small f number) photographic objective or of
a high-power microscope objective used in reverse. The speed of a photographic objective is indicated by its f number or aperture ratio; thus an $f/2$
objective has a focal length twice the diameter of its entrance aperture. For a
variety of reasons this description is not suitable for imaging systems in which
the rays form large angles approaching $\pi/2$ with the optic axis. In discussing
the resolving power of such systems, the most useful measure of performance
is the "numerical aperture," or NA, a concept invented by Ernst Abbe in connection with the resolving power of microscopes. Figure 7.19 shows an optical
system with entrance aperture of diameter $2a$. It forms an image of the axial
object point at infinity, and the semiangle of the cone of extreme rays is α'_{max}.
Then the numerical aperture is defined by

$$NA = n' \sin \alpha'_{max},\tag{13}$$

where n' is the refractive index of the medium in the image space. We assume
that all the rays from the object point focus sharply at the image point (i.e.,
there is no spherical aberration). Abbe showed that off-axis object points will
also be sharply imaged if the condition

$$h = n' \sin \alpha' \times \text{constant}.\tag{14}$$

is fulfilled for all the axial rays. In this equation h is the distance of the
incoming ray from the axis and α' is the angle at which that ray meets the axis
in the final medium. Equation (14) is a form of the celebrated Abbe sine

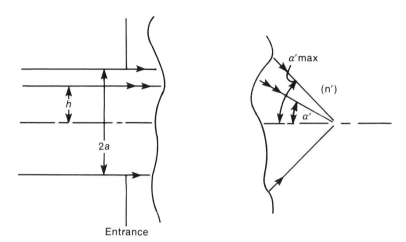

Figure 7.19
The definition of the numerical aperture of an image-forming system. The numerical aperture is $n' \sin \alpha$.

condition for good image formation. It does not ensure perfect image formation for all off-axis object points, but it does ensure that aberrations that grow linearly with off-axis angle are zero. These aberrations are various kinds of *coma*, and the condition of freedom from spherical aberration and coma is called *aplanatism*. Clearly a necessary condition for our image-forming concentrator to have the theoretical maximum concentration or even for it to be ideal, but without the theoretical maximum concentration, is that the image formation should be aplanatic. Unfortunately, this is not a sufficient condition.

The constant in equation (14) has the significance of a focal length. The definition of focal length for optical systems with media of different refractive indices in the object and image spaces is more complicated than for the thin lenses discussed above. In fact it is necessary to define two focal lengths, one for the input space and one for the output space, and their magnitudes are in the ratio of the refractive indices of the two media. In equation (14) the constant is the input side focal length f. From equation (14) we have for the input semiaperture

$$a = f \, \text{NA}, \tag{15}$$

and also from equation (14)

$$a' = a \sin \frac{\theta_{max}}{\text{NA}} \tag{16}$$

so that by substituting from equation (15) into equation (16), we have

$$a' = f \sin \theta_{\max}, \tag{17}$$

where $\theta_{\max}$ is the input semiangle. In an aplanatic system the focal length is a constant, independent of the distance h of the ray from the axis used to define it. Thus equation (17) tells us that in an imaging concentrator with maximum theoretical concentration, the diameter of the exit aperture is proportional to the sine of the input angle. This is true even if the concentrator has an exit numerical aperture less than the theoretical maximum n', provided it is ideal in the sense defined above. In conventional lens optics the result of equation (17) is well known; it is simply another way of saying that the largest aperture aplanatic lens with air as the exit medium is $f/0.5$ since equation (17) tells us that $a' \leq f$. The importance of equation (17) is that it tells us something about one of the shape-imaging aberrations required of the system, namely distortion. A distortion-free lens imaging onto a flat field must obviously have an image height proportional to $\tan \theta$ so that our concentrator lens system is required to have what is usually called barrel distortion; this is illustrated in figure 7.20.

Our picture of a concentrator is gradually taking shape, and we begin to see that certain requirements of conventional imaging can be relaxed. Thus, if we can get a sharp image at the edge of the exit aperture and if the diameter of the exit aperture fulfills the requirements of equations (14) to (17) we do not need perfect image formation for object point at angles smaller than $\theta_{\max}$. For example, the image field could be curved, provided we take the exit aperture in the plane of the circle of image points for the direction $\theta_{\max}$, as in figure 7.21. Also the inner parts of the field can have point-imaging aberrations, provided these were not so large as to spill rays outside the circle of radius a'. An image-forming concentrator need not, in principle, be so difficult to design as an imaging lens since the aberrations need to be corrected only at the edge of the field. This leads us to a valuable principle for nonimaging concentrators, the "edge ray principle." Not only is it unnecessary to have good aberration correction except at the exit rim, but we do not even need point imaging at the rim itself. It is only necessary that all rays entering at the extreme angle $\theta_{\max}$ leave from some point on the rim and that the aberrations inside not push rays outside the rim of the exit aperture.

Image-Forming Mirror Systems   Concave mirrors have, of course, been used for many years as collectors for solar furnaces. However, little was published on angle-transmission curves for such systems before the 1970s. Consider a

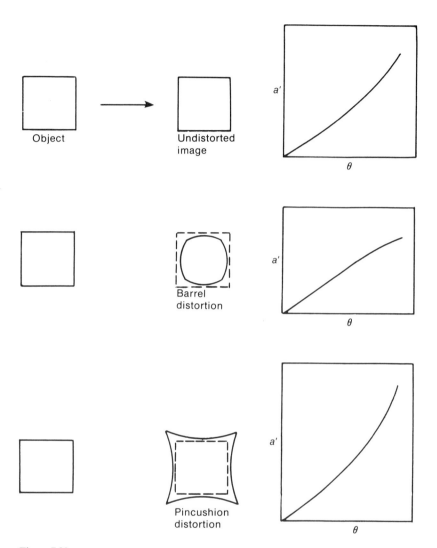

Figure 7.20
Distortion in image-forming systems. The optical systems are assumed to have an axis of rotation.

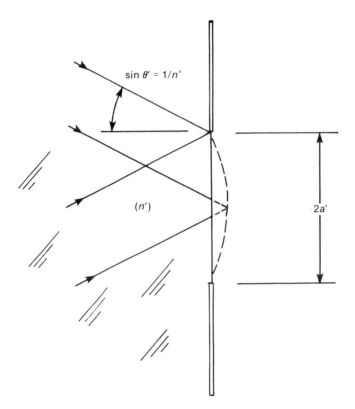

Figure 7.21
A curved image field with a plane exit aperture.

simple paraboloidal mirror as in figure 7.22. As is well known, this mirror
focuses rays parallel to the axis to a point; that is, it has no spherical aberration. However, off-axis rays are badly aberrated. In the meridian section (the
section of the diagram) it is easily shown by ray tracing that the edge rays at
angle θ meet the focal plane farther from the axis than the central ray (under-
corrected coma), so this cannot be an ideal concentrator even for emergent
rays at angles much less than $\pi/2$. An elementary argument (e.g., see Harper et
al. 1976) shows how big the exit aperture must be to collect all the rays in the
meridian section. As in figure 7.23 we draw a circle passing through the ends
of the mirror and the absorber (i.e., the exit aperture). Then by a well-known
property of the circle, if the absorber subtends $4\theta_{max}$ at the center of the circle,
it subtends $2\theta_{max}$ at the ends of the mirror, so the collecting angle is $2\theta_{max}$.
The mirror is not specified to be any particular shape except that it must reflect

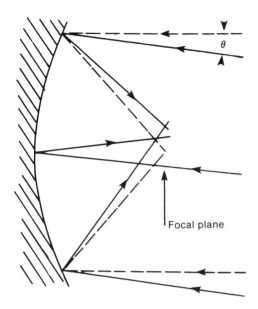

Figure 7.22
Coma of a paraboloidal mirror. The rays of an axial beam are shown in broken line. The outer rays from the oblique beams at angle θ meet the focal plane that is farther from the axis than the central ray of this beam.

all inner rays to the inside of the exit aperture. If the mirror subtends 2θ at the center of the circle, we find

$$\frac{a'}{a} = \sin\frac{2\theta_{max}}{\sin\theta}, \tag{18}$$

and the minimum value of a' is clearly attained when $\theta = \pi/2$. At this point the concentration ratio is (allowing for obstruction by the absorber)

$$\left(\frac{a'}{a}\right)^2 - 1 = \left(\frac{1}{4}\sin^2\theta_{max}\right)\left(\frac{\cos^2 2\theta_{max}}{\cos^2\theta_{max}}\right). \tag{19}$$

This is less than 25% of the theoretical maximum concentration ratio than 50% of the ideal for the emergent angle used. The large loss in concentration is basically because the single concave mirror has large coma; that is, it does not satisfy Abbe's sine condition [equation (14)]. The large amount of coma introduced into the image spreads the necessary size of the exit aperture and so lowers the concentration below the ideal value.

There are image-forming systems that satisfy the Abbe sine condition and have large relative apertures. The prototype of these is the Schmidt camera,

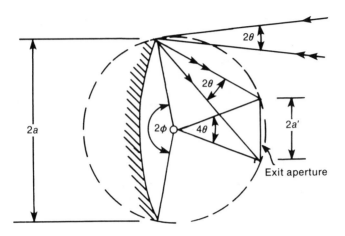

Figure 7.23
Collecting all the rays froma concave mirror.

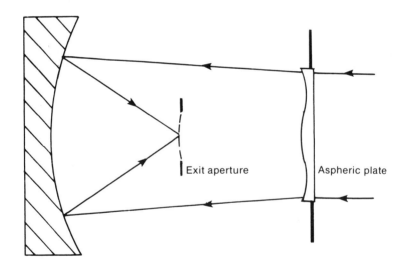

Figure 7.24
The Schmidt camera. This optical system has no spherical aberration or coma so that in principle it could be a good concentrator for small collecting angles. However, there are serious practical objections such as cost and the central obstruction of the aperture.

which has an aspheric plate and a spherical concave mirror, as in figure 7.24. The aspheric plate is at the center of curvature of the mirror, and thus the mirror must be larger than the collecting aperture. Such a system would have the ideal concentration ratio for a restricted exit angle apart from the central obstruction. However, a system of this complexity is clearly not to be considered seriously for solar energy work. As to theoretical possibilities, it is certainly possible to have an ideal concentrator of maximum concentration ratio if we use a spherically symmetric geometry, a continuously variable refractive index, and quite unrealistic materials properties (i.e., refractive index between 1 and 2 and no dispersion; see, e.g., the Luneburg lens in Luneburg 1964). But whatever the theoretical possibilities, practical concentrators based on image-forming designs fall far short of the ideal.

7.2.2 Nonimaging Optics

Introduction   Nonimaging optics depart from the methods of traditional optical design to develop instead techniques for maximizing the collecting power of concentrating elements and systems. Designs that exceed the concentration attainable with focusing techniques by factors of four or more and approach the thermodynamic limit are possible.

The Edge Ray Principal   Nonimaging optics are concerned with the collection and redirection of light (or, more generally, electromagnetic radiation) by means of optical systems that do not use image formation in their design. In nonimaging optics the mighty edifice of aberration theory used in ordinary lens design is dismantled and replaced by a single key idea. According to this idea maximum concentration is achieved by ensuring that rays collected at the extreme angle for which the concentrator is designed are redirected to strike points at the edge of the absorber. A nontrivial example is the compound parabolic concentrator (CPC) invented in 1965 for collecting Cherenkov radiation from large volumes of gas and concentrating it onto the relatively small area of a photomultiplier cathode (Hinterberger and Winston 1966). This task would, according to conventional optical practice, be performed by a lens or mirror image-forming system of high numerical aperture, but much greater concentration has been achieved by a comparatively simple device, the CPC. The key was to abandon the principle of imaging with high numerical aperture and instead to get the collected rays onto as small an area as possible without attempting to produce an image. The skillful application of this principle can ensure that devices with the maximum concentration are obtained, at least in two-dimensional (troughlike) geometry. Actually the first

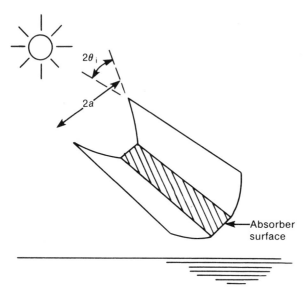

Figure 7.25
A two-dimensional concentrator.

CPC-like devices described in the literature were three dimensional (cone shaped) (Hinterberger and Winston 1966; Baranov and Melnikov 1966; Ploke 1967). It was only with the interest in two-dimensional geometry for solar energy work that this case was explored in detail, with the result that ideal concentrators can now be designed for any two-dimensional convex receiver (Welford and Winston 1989).

Basic Design Principles of Nonimaging Concentrators

GENERALIZING THE TWO-DIMENSIONAL CONCENTRATOR   Two-dimensional or troughlike nonimage-forming concentrators can be made to have the maximum theoretical concentration ratio. Thus if the concentrator (figure 7.25) collects energy in an aperture of width $2a$ and over angle $\pm\theta_i$, all of this energy falls on the absorber surface of width $2a' = 2a\sin\theta$ at angles of incidence ranging from 0 to $\pi/2$. If they are assumed to be infinitely long in the trough direction or, the practical equivalent, terminated with plane end mirrors, they can be regarded as ideal concentrators in the following sense:

1. The *étendue* of the entering beam appears at the exit aperture, since all rays meet the absorber.

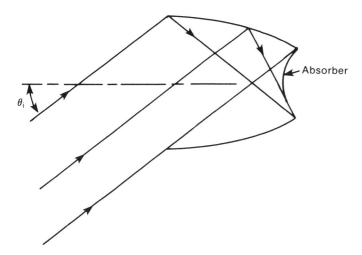

Figure 7.26
A concentrator with a nonplane absorber.

2. A system with a convex absorber as in figure 7.26 can be designed to ensure that rays entering at the extreme angle θ_i meet the absorber tangentially if they do not meet its edge, and here also it is found that all rays meet the absorber and that the relation $2a' - 2a\sin\theta_i$ is satisfied, where now $2a'$ is the arc length around the absorber.

3. Let the source be at a finite distance as in figure 7.27; here PP' is the source, RR' is the absorber, and QQ' is the entry aperture of the concentrator. Thus the *étendue* of the entering beam is

$$H = PQ' + P'Q - P'Q' - PQ, \tag{20}$$

and the concentrator can be designed to ensure that all of this flux is received by the absorber RR'. Also it can be arranged that the arc length $RR' = H/2$, so the system has the maximum theoretical concentration. Equation (20) is equivalent to the celebrated Hottel string formula for radiation shape factors (Hottel 1954). We now set up and outline the solution of the general problem of radiant energy concentration in two dimensions.

Let AB and $A'B'$ in figure 7.28 represent the entry and exit apertures of a concentrator, and let the medium between and on either side of these curves have some distribution of refractive index (in a practical system, one or more lenses might be interposed). The radiant energy source lies to the left of the entry aperture, and $\Sigma\alpha$ and $\Sigma\beta$ are extreme rays in the sense that at any point on AB the incoming rays will all be between the extreme rays. We have thus

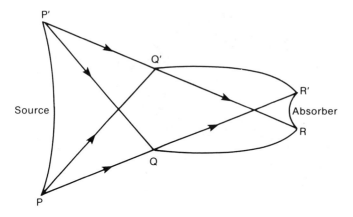

Figure 7.27
A concentrator for a source at a finite distance.

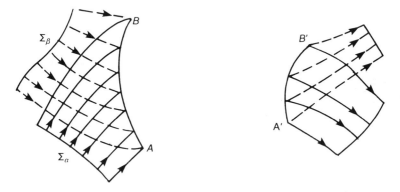

Figure 7.28
Entry and exit surfaces for a general two-dimensional concentrator.

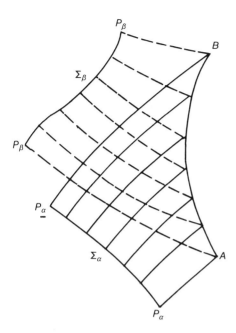

Figure 7.29
Calculating the étendue by means of the Hilbert integral.

defined a beam of a certain *étendue H*, although we have not shown how to calculate it. We can now propose a set of wave fronts emerging from $A'B'$ of which $\Sigma\alpha'$ and $\Sigma\beta'$ are extreme wave fronts, and we can postulate that the beam so defined has the same *étendue H* as the entering beam. For this to be possible, the optical path length along the arc $A'B'$ must be $\geq H/2$. But given that condition, there is still considerable freedom of choice in the pattern of emergent rays.

CALCULATING THE *ETENDUE*   To calculate the *étendue* of a beam in an inhomogeneous medium crossing an aperture defined by a curve AB, as in figure 7.29, we use the Hilbert integral adapted to geometrical optics (see Luneburg 1964 or Born and Wolf 1975). Let P_1 and P_2 be two points somewhere in a pencil of rays (i.e., the rays from a single wave front). Then the Hilbert integral from P_1 to P_2 is

$$I(P_1, P_2) = \int n\mathbf{k}\, ds, \tag{21}$$

where n is the local refractive index, $\mathbf{k}$ is a unit vector along the ray direction, and ds is an element of the path $P_1 P_2$. Thus the Hilbert integral is simply the optical path length along any ray between the wave fronts that pass between P_1 and P_2, and it depends only on the positions of the end points of the path of integration. Returning to figure 7.29, the Hilbert integral from A to B for the rays from Σ_a is

$$I_a(A, B) = \int n \sin \phi \, ds, \tag{22}$$

where ϕ is the angle of incidence of a ray on the line element ds. Thus from the definition of *étendue* we have for the *étendue* of the entering beam

$$H = I\alpha(A, B) - I\beta(A, B). \tag{23}$$

But as remarked above,

$$I\alpha(A, B) = [P\underline{\alpha}B] - [P\alpha A], \tag{24}$$

where $[P\underline{\alpha}B]$ is the optical path length along the ray from $P\underline{\alpha}$ to B; similarly for $[P\alpha A]$, we find for the *étendue*

$$H = [P\underline{\alpha}B] + [P\beta A] - [P\alpha A] - [P\underline{\beta}B]. \tag{25}$$

This is a straightforward generalization of equation (20).

DESIGN OF THE CONCENTRATOR   Following the edge ray principls, we postulate that the optical system between AB and $A'B'$ must be such that one extreme pencil (i.e., from the wave front $\Sigma\alpha$) is exactly imaged into one of the emergent extreme pencils and similarly for the other extreme incident pencil from $\Sigma\beta$. Then the system takes $\Sigma\alpha$ and $\Sigma\beta$ into $\Sigma\alpha'$ and $\Sigma\beta'$, and we wish it to do so without loss of *étendue*. We write down the optical path length from $P\alpha$ to $P\alpha'$ and equate it to that from $P\underline{\alpha}$ to $P\underline{\alpha}'$ and similarly for the other pencil:

$$[P\underline{\alpha}B] + [BA']\alpha + [A'P'\alpha] = [P\alpha A] + [AB']\alpha + [B'P\alpha'], \tag{26}$$

$$[P\beta A] + [AB']\beta + [B'P\beta'] = [P\underline{\beta}B] + [BA']\beta + [A'P'\beta'], \tag{27}$$

where $[BA']\beta$ means the optical path length from B to A' along the ray of the β pencil, and similarly for the other symbols. From these we find

$$\{[P\underline{\alpha}B] + [P\beta A] - [P\alpha A] - [P\underline{\beta}B]\} - \{[A'P\beta']$$
$$+ [B'P\alpha'] - [A'P\underline{\alpha}'] - [B'P\beta']\}$$
$$= [AB']\alpha - [AB']\beta + [BA']\beta - [BA']\alpha. \tag{28}$$

If we compare the left side of this equation with equation (25), we see that it is

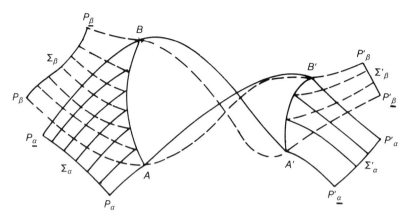

Figure 7.30
Designing the concentrator.

the difference between the *étendues* at the entry and exit apertures. We want this difference to vanish, so we have to make the right side of equation (28) vanish. One simple way to do this is to make the optical system (figure 7.30) such that the α and β ray paths from A to B' coincide exactly and similarly with those from B to A'. We can do this by starting mirrors at A and B in directions that bisect the angles between the incoming α and β rays. We then continue the mirror surfaces in such a way as to make all rays join up with the corresponding α' rays (i.e., we image the α pencil directly into the α' pencil, and similarly for the other mirror surface). This construction then specifies two mirrors connecting A to A' and B to B'. It completes the design and uses up all available degrees of freedom in doing so. This is the design prescription for the generalized CPC trough collector. The design procedure will be successful so long as the reflecting surface is everywhere well defined; this amounts to saying that the extreme rays must not form caustics before reaching the reflecting side walls. If we had not incorporated reflectors, the vanishing of the right side of equation (28) would no longer be built into the method of solution but would appear as an additional constraint. It is possible for refracting systems to also satisfy this constraint (Welford and Winston 1989), but the complicated all-lens designs are probably not practical for solar applications.

ELABORATIONS OF NONIMAGING DESIGNS   Various elaborations of nonimaging designs have been proposed. A number of these have been successfully implemented in solar work and are worth noting. Designs for second-stage applications in the focal plane of mirror or lens concentrators are actually included in the general design prescription already described. This is also true of designs that restrict the angle of incidence on the absorber (Rabl and Winston 1976). Designs that employ total internal reflection in a transparent medium have been applied to photovoltaic work (Winston 1976). An important design consideration that arises in thermal collectors is to allow an insulating gap between the absorber and reflector while maintaining high efficiency (Rabl, Goodman, and Winston 1979). A solution for small gaps was proposed by McIntire (1980) in the form of a W-shaped cavity. This was later generalized to cavities for moderate sized gaps and to V-shaped cavities (Winston 1978). The V-shaped cavity is probably the most practical configuration for thermal collectors.

More recently, designs have been developed that depart from the edge ray principle to construct reflecting surfaces along the mean directions of energy flow from a radiant source. In certain (fortunate) cases the energy flow pattern is undisturbed, and one transforms a virtual radiant source (or absorber) into a smaller, real source, thereby producing a concentrator with no aberrations (Winston and Welford 1979). So far the only case under active investigation is a disk source (e.g., the aperture of a blackbody cavity) for which the matching reflecting surface is hyperboloidal. This "trumpet-shaped" concentrator appears promising in second-stage applications for parabolic dishes (O'Gallagher and Winston 1986).

Optical Tolerances   In the design of image-forming systems there are two kinds of optical tolerances. The first kind is theoretical tolerances for aberrations, and the second is constructional or manufacturing tolerances. It is impossible to design image-forming systems completely free from aberrations for finite aperture and field, thus the need for aberration tolerances. Aberration in the present sense is essentially a concept of geometrical optics. The most stringent system of tolerances uses the approach of reducing the geometrical aberrations until the residual image imperfections are on the scale of diffraction effects. This system (originally designed by Strehle) is described by Born and Wolf (1975). In solar energy applications, considerably larger amounts of aberrations are allowed. Constructional tolerances are concerned with the differences between the system as designed and as made (errors of surface slope, refractive index inhomogeneity, misalignment, etc). Additionally one must consider tolerance caused by tracking error. The alignment and optical distortion of solar concentrators can be characterized experimentally by the method of reverse illumination (Masterson and Gaul 1981) based on the reversibility of light rays in an optical system. One simply places an optical target, often subdivided into colored zones, at the receiver aperture and observes the concentrator at a sufficiently great distance so that the angular subtense of the concentrator's entrance aperture is less than the sun's. The fraction of the concentrator's aperture filled by various zones gives useful information on the optical performance.

Tolerances for Nonimaging Concentrators Aberration tolerances deal with failure to achieve the maximum theoretical concentration. Here the problem is not very profound because the untruncated (two-dimensional) CPC achieves the theoretical limit so that one trades off concentration and truncation (Rabl, 1976b). In three dimensions the CPC falls slightly short of the theoretical limit. However, no degrees of freedom are available for improving performance, so little can be said about aberrational tolerances. The analysis for constructional tolerances is equally simple for all CPCs because the edge ray principle implies that all rays near the cutoff angle θ_{max} undergo exactly one reflection on their way to the absorber. Therefore, for a mirror slope error of Δ, all the rays with $\theta < \theta_{max} - 2\Delta$ and none of the rays with $\theta > \theta_{max} + 2\Delta$ will reach the absorber. Hence the rectangular angular acceptance of figure 7.9 (solid curve) is smeared by $\pm 2\Delta$ (dashed curve).

7.2.4 Holographic Elements for Concentrating Sunlight

To discuss the application of holography to solar energy concentration, it is useful to consider each portion of a hologram as a special kind of diffraction grating that deflects all incident energy in one direction (the first order). Just like a diffraction grating, the classical hologram is made for one wavelength and one direction. However, the solar spectrum is extremely broadband, ranging in wavelength from under 400 nm to well over 1,200 nm. This more than 3:1 range of wavelengths is but one of the factors that make the solar application of holography problematic. In addition, for a stated concentration ratio it is desirable to have the largest possible angular acceptance to reduce tolerance and tracking requirements on the overall system and hence the cost.

However, in this respect holograms have a difficult time competing with low focal ratio conventional optical elements (e.g., f/l Fresnel lenses). Nevertheless, holograms offer exciting possibilities for low-cost replication, various multiplexing schemes, and simplicity of deployment (flat sheets), which has attracted significant interest in their application to solar energy concentration (National Technical Systems 1983).

Broadband Focusing When solar radiation with its broadband spectrum is focused by a diffraction grating, the 3:1 range of wavelengths is converted into a comparable range of deflection angles, which severely defocuses the energy. This limits the achievable concentration to ≤ 10 in point-focus geometry (2-axis tracking) and ≤ 3 in line-focus geometry (1-axis tracking), which is one to two orders of magnitude less than attained by conventional systems. One must therefore stack a number of holograms, each designed for a different wavelength, to cover a substantial portion of the solar spectrum. The stack A, B, C, ..., could actually be encoded into one physical hologram (multiplexing) and might operate as follows: Component A focuses sunlight from 300 nm to 400 nm and transmits the rest, B focuses from 400 nm to 500 nm, transmitting the rest, and so on. For this scheme to work even in principle, each element needs to have square transmission characteristics with sharp cut-on and cutoff. Any overlap between the cutoff of one element and the cut-on of another will crosstalk wherein energy focused by A is defocused by B. This degrades the overall system concentration. When one considers the compound effects from more than ten elements with overlapping characteristics, the degradation can be severe. Just how sharp the characteristics of individual holographic elements can be in practice is a subject of ongoing research; in this respect reflection holograms are much more favorable than transmission. In fact the cross-talk characteristics of transmission holograms probably make them unsuitable for broadband focusing.

Angular Multiplexing   One must state from the outset that naive applications of angular multiplexing (e.g., A focuses light from 0 to 10 degrees and transmits the rest, B focuses from 10 to 20 degrees and transmits the rest, and so on) will not work! If they did, one could concentrate solar radiation by large amounts with stationary holograms; that is, one could "track optically." Unfortunately, some proposals have suggested that this could actually be done (Margarinos and Coleman 1981), beating the second law of thermodynamics in the process (Winston and Welford 1982). What happens in detail is that in addition to deflection of the desired rays, certain conjugate rays are also inevitably deflected, causing defocusing of the beam. Nevertheless, multiplexing of holograms can achieve results not available to nondiffractive systems. For example, different wavelength ranges can be focused onto different locations, and portions of the solar spectrum can be focused while other portions are rejected. This give holograms a flexibility that may be advantageous in specific applications. Whether any multiplexing scheme has sufficient efficiency and usefulness for solar energy applications is a subject of ongoing research.

Comparison with Conventional Optics   The possibility of quasi-thin-film holographic elements that simply emulate conventional optical elements is attractive because of the potential for low-cost fabrication and installation. For example, a flat sheet that could emulate a parabolic dish or trough might be extremely useful. In fact, thin, flexible holograms have been successfully fabricated (National Technical Systems 1984). For solar energy applications it is useful to use fast (i.e., low focal ratio) elements to improve the trade-off between concentration and angular acceptance. However, it is known that fast holographic elements need to be curved (Welford 1973); otherwise, they have aberrations that degrade performance. Alternatively, one might consider holographic emulators of CPCs. Unfortunately, reflecting holograms lose efficiency at large angles of incidence (near Brewster's angle only one plane of polarization is diffracted). But large angles of incidence are characteristic of CPC-like reflectors.

Conclusions The prospect of quasi-thin-film holographic elements is as appealing for solar energy concentrators as, for example, thin-film solar cells are for photovoltaics. However, classical holograms operate efficiently only for a narrow range of wavelengths and directions. Any attempt to use holograms to concentrate the broadband, wide-angle solar spectrum encounters serious technical obstacles; some of these have been outlined above. Without attempting to minimize the problems, which are formidable, it is fair to point out that the potential is also considerable, possibly justifying further research in this area.

7.3 State of Knowledge and Practice before 1973

The state of the art of optics applied to solar energy several decades ago can be found in many textbooks from that period but is succinctly represented in some excellent papers by H. Tabor (e.g., Tabor 1958). Thus one could derive the limits on concentration for various systems useful for solar applications:

$$C \leq \frac{1}{\pi \sin \theta} \quad \text{for parabolic troughs,} \tag{28}$$

$$C \leq \frac{1}{4 \sin^2 \theta} \quad \text{for parabolic dishes,} \tag{29}$$

$$C \leq \frac{1}{2f \sin^2 \theta} \text{for point focus lenses,} \tag{30}$$

and so on. In fact one could confidently expect that nontracking but daily adjusted concentrators ($\theta \approx 7$ deg) were limited to $C \leq 3$, while a totally stationary concentrator ($\theta \approx 30$ deg) would more likely deconcentrate ($C \leq 1$). The involute, a $C \leq 1$ construction, was proposed for solar applications by Trombe and independently by Meinel (Meinel and Meinel 1976). A variety of useful and speculative concentrator designs are catalogued in Meinel's book. Special mention should be made of the circular Fresnel mirror trough, introduced by Russel (1976), which remains fixed while the absorber tracks the sun. In addition the spherical bowl concentrator was proposed for solar work by Stewart (see Kreider and Kreith 1975). Of course spherical mirror with a large rim angle has huge on-axis aberrations and actually produces a line focus that requires a line absorber that tracks the sun. Nevertheless, this concept received substantial attention in the past decade (e.g., the Crosbytown project).

7.4   Review of Optional R&D Projects, 1973 to 1983

7.4.1   Low Concentration/Nontracking

The introduction of the concept of nonimaging optics for solar concentrators can be traced to a system for coupling photomultipliers to Cherenkov counters and other extended sources (Hinterberger and Winston 1966) illustrated in profile in figure 7.31. Essentially, in designing a concentrator as an image-forming system, unnecessarily high standards are being set since the image of the sun is not needed. All that is needed is that the rays collected will all fall, within as small an area as possible, on the proposed absorber surface.

The application of the Cherenkov coupler to nontracking solar collectors was suggested in 1973 (Winston 1974), and a considerable amount of solar applications development was carried out at Argonne National Laboratory. Significantly, it was recognized that the two-dimensional form of the Cherenkov coupler had certain nonobvious advantages for solar work. The three-dimensional form (i.e., with the profile of figure 7.31 rotated about its axis of symmetry) is capable of high concentration ratios but must be guided to face the sun with a precision better than the angular acceptance. A two-dimensional version—looking like a trough with parabolic cylindrical sides appropriate to a flat absorber, as illustrated in figure 7.32, with the long axis aligned in the east–west direction—can achieve moderate concentration (up to ≈ 10) with no diurnal tracking and even useful concentration (up to ≈ 2) with fixed position year long. Modifications of the two-dimensional design were made for various nonplane absorber shapes (Winston and Hinterberger 1975). Several of the most common solutions recommended for thermal applications are shown in figure 7.33 and illustrate the wide acceptance angles possible. Reflecting troughs with these profiles achieve the limit of equation (11), collecting all rays within $\pm\theta$ at the entrance aperture and directing them to the absorber. The solution for a flat absorber (figure 7.31) consists of two tilted

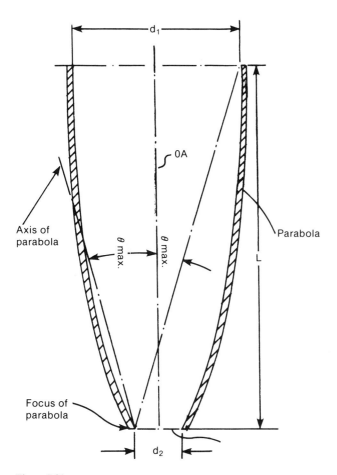

Figure 7.31
Cross-sectional profile of nonimaging cone concentrator. The device collects radiation entering the circular entrance aperture of diameter d within angle of $\pm\theta_{max}$ and directs it to the smaller diameter aperture d_2 with negligible losses and thus closely approaches the thermodynamics limit.

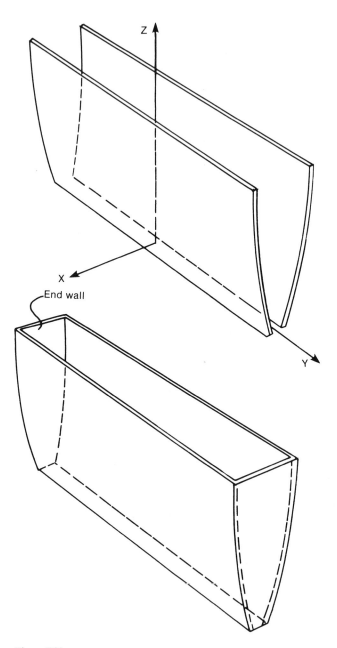

Figure 7.32
A two-dimensional or trough concentrator having the same profile as figure 7.31 aligned with its y axis in the east–west direction has an acceptance field of view well matched to solar motion and can achieve moderate concentration with no diurnal tracking. Vertical reflecting end walls effectively recover shading and end losses.

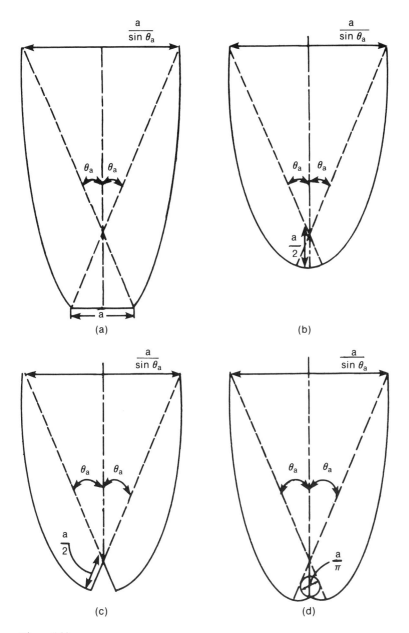

Figure 7.33
Cross-sectional profiles of ideal trough concentrators generalized to absorbers of different shapes. In practice, the reflectors are usually truncated to about half their full height to save reflector material with only negligible loss of concentration.

parabolic segments, each with its axis tilted at an angle θ relative to the collecting aperture normal and its focus at the lower edge of the opposite mirror. For the fin and tube solutions (figure 7.33b and d), the profile consists of an involute of the absorber inside the "shadow lines" defined by the extreme rays tangent to the absorber at $\pm\theta$ and a curve outside the shadow lines whose slope at each point is defined by the condition that a ray parallel to the extreme ray is reflected such that it is just tangent to the absorber. For the fin this corresponds to two tilted parabolas outside the shadow lines and a circular arc segment inside. For the tube the profile is nonparabolic. These all became known as the CPC (compound parabolic concentrator), although for non-plane absorbers the cylinder profile is frequently nonparabolic. As is evident from figure 7.33 the upper portion of the profile is essentially parallel to the axis, hence contributing only marginally to concentration. It follows that there is an economic, and even a performance benefit, to truncation of the reflector. An elegant study of the relevant considerations were performed by Rabl (1976b), and more recently by Carvalho et al. (1985).

7.4.2 Nonevacuated CPCs

Nonevacuated CPCs, although clearly not high temperature collectors, have a measured performance that exceeds that for typical flat plate collectors at temperatures above $\approx 50°C$ ($122°F$) and should have economic advantages as well if properly constructed (Rabl, O'Gallagher, and Winston 1980). The recommended design constraints derived from more than two years of non-evacuated collector development are

1. using one of the "backless" or "wraparound" designs such as the cusp for a tubular absorber or vertical fin design of figure 7.33b and d,

2. using selective coatings on the absorber,

3. avoiding the use of cavity absorbers since they are characterized by high heat losses,

4. minimizing heat loss via conduction through the reflectors. This can be accomplished by using reflectors with negligible thickness compared to the overall dimensions (height, aperture, etc.) of the trough, such as plastics or films, and by thermally decoupling the absorber by a gap maintained by insulating standoffs. In this connection gap-loss suppressing designs previously mentioned are advantageous.

A prototype with $C = 3$ for a vertical fin absorber was built to verify the production design for an experimental array of low temperature CPC collec-

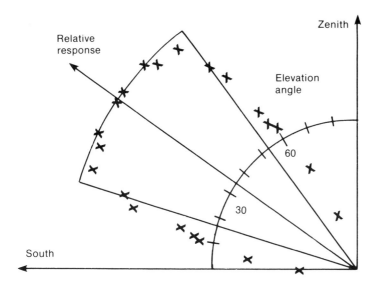

Figure 7.34
The "ideal" and measured angular response function for the 3 × nonevacuated CPC installed at the Navajo school project. The latitude is about 37 deg at this location, so the collector "sees" the sun during the winter only. For year-round collection two tilt adjustments would be necessary.

tors at the Bread Springs School, an elementary school on the Navajo Indian Reservation near Gallup, New Mexico. This array, installed in 1978, is operating at the present time. Since this collector was designed for heating applications only, the acceptance angle was chosen to be ± 18 deg and the collector tilted relative to the zenith by the latitude angle plus 18 deg as shown on figure 7.34. This ensures collection for at least 7 hours per day for the six-month period between the fall and spring equinoxes, with no tilt adjustments. The absorber is a fin (configuration of figure 7.33b) with a center tube; each collector consists of two-trough module 1.4 m² (15.1 ft²) in net area, and the troughs are 0.45 m (1.48 ft) deep. An aluminum sheet 0.5 mm thick is used as the reflector ($p = 0.84$). The outer enclosure was constructed of glass fiber epoxy with water-white glass as a cover. No other insulation was employed. Measured performance yielded an optical efficiency of $\eta = 0.61 \pm 0.03$ with a heat loss coefficient of $U = 2.7 \pm 0.2$ W/m² K (0.475 + 0.35 Btu/ft² hr °F). This should be compared with typical values for flat plates with η from $0.70 - 0.78$ and U from $4.5 - 7$ W/m² K (0.795 − 1.237 Btu/ft² hr °F). Despite lower optical efficiency these CPCs outperform typical flat plate collectors at temperatures ≥ 35°C above ambient.

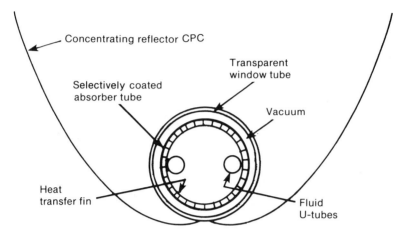

Figure 7.35
Cross-sectional profile of contemporary evacuation tube CPC according to the basic design
developed at Argonne National Laboratories.

7.4.3   Evacuated CPCs

The development of stationary concentrators was strongly influenced by the
emergence of the Dewar-type evacuated absorber. This thermally efficient
device, developed by two major manufacturers, Owens Illinois and the General Electric Co., could, when coupled to CPC reflectors, supply heat at higher
temperatures than flat plate collectors while retaining the latter's advantage of
a fixed mount. For example, with external anodized aluminum reflectors
designed for an acceptance angle of ± 35 deg and truncated to a net geometric concentration of between 1.2 and 1.5, such concentrators attain optical
efficiencies of between 0.52 and 0.62, depending on whether a cover glass
is used. The thermal performance is excellent with values of $c \approx 0.05$ and
$U \approx 0.5$ W/m^2 K (0.088 Btu/ft^2 hr °F). An illustration of the basic design is
shown in figure 7.35. Other versions have opened up the acceptance angle to
± 50 deg with $C \approx 1.2$ to allow polar orientation and are characterized by
$U = 1.3$ W/m^2 K. The trends in improving performance are the use of gap-loss
suppressing designs or extended cusp designs (Winston 1978), higher coatings,
and higher reflectivity mirrors (silver-coated films). Several manufacturers
introduced collectors of these types for applications ranging from heating to
absorption cooling to driving Rankine cycle engines. A number of installations each in excess of 1,000 m^2 (10,760 ft^2) have been deployed. Examples
include a "SOLERAS" in Phoenix driving a Rankine engine (figure 7.36), the
Holiday Inn Virgin Islands array for cooling and desalinization (figure 7.37),

Figure 7.36
Evacuated tube CPC array installed in Phoenix, Arizona, as part of SOLARAS project.

and the Illinois Department of Agriculture Headquarters array for heating and cooling in Springfield (figure 7.38). The concentrators are arranged in panels resembling flat plate collectors. There are two methods for extracting the heat: Either the fluid is simply allowed to contact the glass, being introduced into the Dewar by a feeder tube, or it flows through a closed piping network with metal heat exchange fins actually contacting the inner glass wall (figure 7.35).

7.4.4   An Advanced CPC: The Integrated Concentrator

The optical efficiency of evacuated CPC solar collectors can be significantly improved over that of the commercial versions discussed above by shaping the outer glass envelope of the evacuated tube into the concentrator profile (Garrison 1977). Improved performance results directly from integrating the reflecting surface and vacuum enclosure into a single unit. This concept is the basis of a new evacuated CPC collector tube that has a substantially higher optical efficiency and a significantly lower rate of exposure-induced degrada-

Figure 7.37
Polar-oriented evacuated tube CPCs at Frenchman's Reef in the Virgin Islands.

tion than external reflector versions. These performance gains are a consequence of two obvious advantages of the integrated design:

1. Placing the reflecting surface in a vacuum eliminates degradation of the mirror's reflectance, and thus high quality (silver or aluminum $\rho = 0.91–0.96$) first-surface mirrors can be used instead of anodized aluminum sheet metal or thin-film reflectors ($\rho = 0.80–0.85$) typical of the external reflector designs.

2. The transparent part of the glass vacuum enclosure also functions as an entrance window and thus eliminates the need for an external cover glazing. This increases the optical efficiency by a factor $1/T$, where typical transmittances are $T = -0.88–0.92$.

Research on this collector concept has been under way at the University of Chicago and at Argonne (figure 7.39) (Snail, O'Gallagher, and Winston 1984). A 2-m^2 (22-ft^2) prototype panel has been tested for several years, permitting reliable performance projections. The optical characteristics are shown in

Figure 7.38
Part of a 12,000 ft² (1,115 m²) evacuated tube CPC array deployed on the Illinois Department of Agriculture Building in Springfield, Illinois.

figures 7.40 and 7.41. Figure 7.42 shows that the efficiency of this collector is competitive with flat plates at low temperatures [< 100°C (< 212°F)] and with tracking troughs out to ≥ 200°C (≥ 392°F), offering the exciting possibility of efficient thermal conversion over a wide temperature range by a fixed collector.

7.4.5 Line-Focus/Tracking Collectors

Over the past decade parabolic troughs were brought to a high state of development and successfully transferred to industry. Their excellent performance up to ≥ 300°C (≥ 572°F) depends on accurate tracking and high quality back-silvered glass reflectors. Much of the analysis of the optical properties of troughs was based on computer ray-trace simulations. However, an analytic formulation can lend more insight; such an approach was taken by Rabl and his colleagues at SERI (Bendt et al. 1979; Bendt, Gaul, and Rabl 1980; Bendt and Rabl 1980).

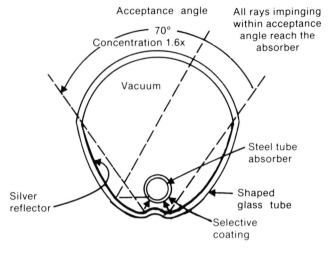

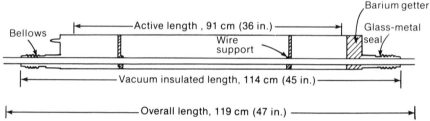

Figure 7.39
Cross-sectional and longitudinal (not to scale) views of the Integrated Stationary Evacuated Concentrator (ISEC) design.

7.4.6   Point-Focus/Tracking Collectors

Point-focus concentrators offer the only means to achieve high temperature, $\geq 400°C$ ($\geq 752°F$), from sunlight. Perhaps the most visibly demonstrated solar thermal technology has been the central receiver concept (e.g., see ASME 1984). This is a special case of a Fresnel reflector whose scale is so large that the mirror segments become heliostats and the point focus is situated on top of a tower. The optical analysis of Fresnel reflectors is complex and most amenable to fairly sophisticated computer modeling. Among the factors the designer must consider are the following: (1) shading and blocking of individual heliostats—shading occurs when direct sunlight fails to reach one heliostat because it is in the shadow of others, and blocking when the reflected energy from one heliostat is intercepted by other heliostats before reaching the receiver; (2) location and spacing strategy of heliostats—this interacts

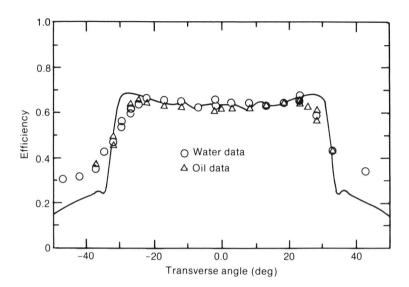

Figure 7.40
Prototype ISEC efficiency versus transverse angle. The solid line is a cubic spline interpolation
to results from a ray trace program run at 1 deg intervals on a single tube.

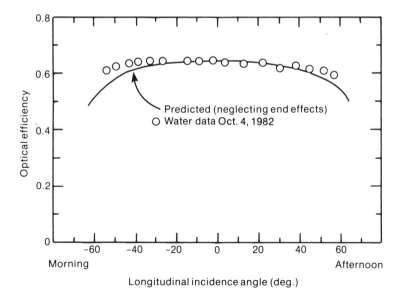

Figure 7.41
Prototype ISEC efficiency versus longitudinal angle. The solid curve is a polynomial fit to data
from a typical east–west collector.

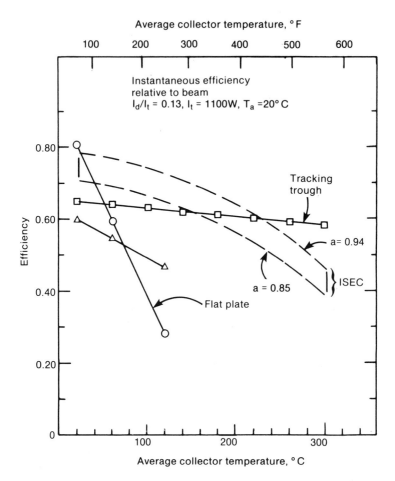

Figure 7.42
Instantaneous performance curves relative to beam used in the long-term energy delivery calculations. The open triangles are for a contemporary evacuated tube collector with an external
reflector CPC and a Dewar receiver.

strongly with shading and blocking losses; and (3) spreading of the reflected beam—spreading can be minimized by compound curving individual heliostats but is ultimately limited by the finite angular subtense of the sun. In addition the size and shape of the "image" of the sun from individual heliostats depends on the angle of incidence, hence on the time of day and year.

When one considers that the number of heliostats is very large (10,000 or more), the case for a sophisticated computer code is persuasive. Such analysis has been carried out primarily by the University of Houston group (Vant-Hull 1976; Lipps and Vant-Hull 1978; see also Riaz 1976). Several computational models have been developed to calculate the optical performance of a central receiver system. The HELIOS program (Biggs and Vittitoe 1979) can model such details as circumsolar radiation, arbitrary heliostat shapes, and second-stage concentrators. However, such programs tend to be too slow for many important application (e.g., optimization studies based on performance with realistic weather data). Therefore computer codes have been developed that make various approximations to speed up computations. Examples are DELSOL (Dellin and Fish 1979) and SCRAM (Bergeron and Chiang 1980). A 10 MW$_e$ central receiver power plant in Barstow, California, has been producing power since 1982. This successful operation of "Solar One" testifies to the validity of the complex calculations.

As for distributed point-focus systems, a frequently investigated design is a paraboloidal mirror, although Fresnel lenses are occasionally considered (O'Neill 1982). As previously noted, the inherent optical aberrations of a paraboloid limit the achievable concentration to less than 1/4 the thermodynamic limit, which is still very high ($1/4 \sin^2 \theta_p$), where θ_p is the half angle of the field of view and must include such effects as sun-size, mirror slope and alignment errors, and tracking tolerance. For example, the high quality test bed concentrator operated by JPL readily achieves concentrations in excess of 2,000. However, commercial versions require a larger budget for errors and tolerance to be practical. For high temperature operation it may therefore be advantageous to deploy a second stage concentrator in the focal plane. Although, in principle, one could gain back the factor 4 in concentration, a practical boost with reasonable rim angle dishes would be closer to a factor ≈ 2. There are at least two choices for the design of the secondary concentrator (Winston and Welford 1980). One could use either a version of the CPC or the "trumpet-shaped" flow-line design. The latter has the advantage of intercepting a smaller fraction of the beam, which results in higher throughput and smaller thermal loading of the secondary. For this reason the flow-line design was selected for tests with the test bed concentrator. The overall concentration

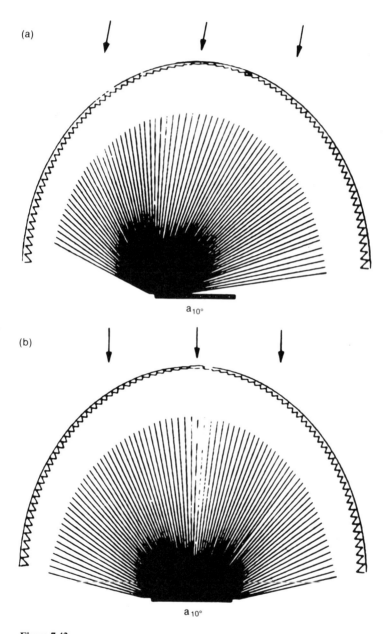

Figure 7.43
A lens of finite thickness and finite grooves, designed for $\theta_0 = 10$ deg; (a) $\theta = 10$ deg; (b) $\theta = 0$ deg.
Source: Kritchman et al. (1979).

was increased from 2200:1 to 4800:1 at a loss of only 4% in throughput (O'Gallagher and Winston 1982).

Significant advances in Fresnel lens design have been made by O'Neill (1979, 1982) and by Kritchman, Friesem, and Yekutieli (1979). In both approaches the Fresnel lens lies on a spherical surface with radius equal to its focal length in order to satisfy the Abbe sine condition [equation (13)], thereby correcting spherical and coma aberrations. Kritchman, Friesem, and Yekutieli (1979) have emphasized that such a lens can come close to the theoretical limit of concentration [equation (12)] by making the lens shape approach a hemisphere as in figure 7.13. O'Neill (1979, 1982) has shown that a Fresnel lens has significantly greater tolerance to surface contour errors than reflecting parabolas. The argument is that a slope error α in, say, the facet of a Fresnel lens, perturbs a ray direction by $\approx (n - 1)\alpha$, where n is the index of refraction (≈ 1.5 for typical lens materials). By contrast, the same slope error in a mirror perturbs the ray direction by 2α, which is worse than the lens by a factor ≈ 4!

7.5 Further Needs and Future Opportunities in Optical R&D

7.5.1 Narrowed Options

After more than a decade of research, a sufficient variety of designs have been investigated to permit some perspective on the more prevalent approaches to solar thermal conversion. An elegant and comprehensive treatment can be found in Rabl (1985). One way to classify these approaches is by their end use temperature. Any such description risks oversimplifying the technology and underemphasizing some useful directions.

1. Very high temperature [$\geq 600°C$ ($\geq 1112°F$)]. Chemical processes, Brayton cycles, and Stirling cycles operate in this range, which exceeds conventional power cycle temperatures. Such temperatures are readily attainable by point-focus concentrators, the principal types being parabolic dishes and central receivers with tracking heliostats. It may be desirable to employ second-stage concentration to boost system performance or simply relax the optical requirements on the primary reflector. Provided the materials problems of maintaining high reflectivity and controlling thermal loading can be solved, it may always be advantageous to trade off the added cost of a secondary concentrator against that of the primary reflector and balance of system.

2. High temperature [$400°-600°C$ ($752°-1112°F$)]. This temperature is suitable for driving Rankine cycle engines. The central receiver is the most visibly demonstrated of the solar thermal technologies and appears the most ap-

propriate for large-scale power applications. The point-focus parabolic dish or some close variant such as segmented mirrors on a paraboloidal contour also appears promising. Again, second-stage concentration may be employed to advantage.

3. Intermediate temperature [250°–400°C (482°–752°F)]. Line-focus parabolic troughs have been successfully developed and transferred to industry. Their high performance depends on durable back-silvered glass reflectors and accurate tracking. Since the geometric concentration is already fairly high, the performance would not be greatly enhanced by either an improved selective coating or vacuum insolation. This then would appear to be the most mature of the solar thermal technologies.

4. Moderate temperature [90°–250°C (194°–482°F)]. Nontracking collectors employing nonimaging concentration are capable of supplying heat in the moderate temperature range. So far only the lower end of this range is adequately represented by the commercial vacuum Dewar collector with external CPC mirrors. However, research versions that integrate the concentrating optics and the receiver into a single glass envelope are efficient over the entire temperature range. Because stationary collectors necessarily have low concentration, their performance is highly sensitive to coating selectivity, vacuum quality, and efficiency of heat extraction. They would benefit from development in these areas.

7.5.2 Suggested Near-Term Goals

From this brief survey, several productive areas for optical R&D can be identified:

1. The point-focus systems for high to very high temperature may incorporate some version of a second-stage concentrator in the focal zone. It would therefore be useful to develop and test optical designs that optimize the combination of primary and secondary with respect to performance and cost.

2. The low concentration stationary collectors supply heat over a temperature range that overlaps with the capabilities of flat plates at the low end of the scale and tracking troughs at the high end. Consequently they address significant requirements in space conditioning and industrial process heat. The development of a manufacturable and efficient collector, possibly an integrated version with a heat pipe, would open up broad areas of applications for solar thermal energy.

3. Promising proposals for optical concentration of solar radiation should be carefully evaluated and, where appropriate, tested for feasibility. An example

is the application of diffractive optics (holography) to concentration of solar radiation. Although the technical problems are formidable, the extent to which this technology can be adapted for solar thermal purposes may be an appropriate subject for research.

References

Baranov, V. K., and G. K. Melnikov. 1966. Study of the illumination characteristics of hollow focons. *Sov. J. Opt. Technol.* 33: 408–411.

Bassett, I. M., and R. Winston. 1984. Limits to concentration in physical optics and wave mechanics. *Optica Acta, 31*, 499–505.

Bendt, P., H. W. Gaul, and A. Rabl. 1980. *Determining the Optical Quality of Focussing Collectors without Laser Ray Tracing.* SERI/TR-333-359. Golden, CO: Solar Energy Research Institute.

Bendt, P., and A. Rabl 1980. *Optical Analysis of Point Focus Parabolic Radiation Concentrators.* SERI/TR-631-336. Golden, CO: Solar Energy Research Institute.

Bendt, P., A. Rabl, H. W. Gaul, and K. A. Reed. 1979. *Optical Analysis and Optimization of Line Focus Solar Collectors.* SERI/TR-34-092. Golden, CO: Solar Energy Research Institute.

Carvalho, M. J., M. Collares-Pereira, J. M. Gordon, and A. Rabl. 1985. Truncation of CPC solar collectors and its effect on energy collection. *Solar Energy* 35: 393–399.

Bergeron, K. D., and C. J. Chiang. 1980. *SCRAM: A Fast Computational Model for the Optical Performacne of Point Focus Solar Central Receiver Systems.* SAND 80-0433. Albuquerque, NM: Sandia Laboratories.

Biggs, F., and C. N. Vittitoe. 1979. *The Helios Model for the Optical Behavior of Reflecting Solar Concentrators.* SAND 76-0347. Albuquerque, NM: Sandia Laboratories.

Born, M., and E. Wolf. 1975. *Principles of Optics.* 5th ed. Oxford: Pergamon.

Dellin, T. A., and M. J. Fish. 1979. *A Users Manual for DELSOL—A Computer Code for Calculating the Optical Performance, Field Layout, and Optimal System Design for Solar Central Receiver Plants.* SAND 79-8215. Albuquerque, NM: Sandia Laboratories.

Garrison, J. D. 1977. An optimally designed thermal solar collector. Paper presented at the *Ann. Meet. Int. Solar Energy Soc.* (Am. Sect.). Orlando, FL, June.

Harper, D. A., R. H. Hildebrand, R. Stiening, and R. Winston. 1976. Heat trap: an optimized far infrared filed optics system. *Appl. Opt.* 15: 53–60.

Hinterberger, H., and R. Winston. 1966. Efficient light coupler for threshold Cherenkov counters. *Review of Scientific Instruments* 37: 1094.

Hottel, H. 1954. Radiant heat transmission. In W. H. McAdmas (ed.), *Heat Transmission* 3d ed. New York: McGraw-Hill.

Jannsen, T., H. Tin, and H. A. Yu. 1983. Wideband angular spectrum aberrations-free holoconcentrators. *Proc. of SPIE*, 441. Bellingham, WA: SPIE—The International Society for Optical Engineering.

Kreider, J. F., and F. Kreith. 1975. *Solar Heating and Cooling.* New York: McGraw-Hill.

Kritchman, E. M., A. A. Friesem, and G. Yekutieli. 1979. Efficient lens for solar concentration. *Solar Energy* 22: 119.

Laser Focus (December 1981), pp. 38–42.

Lipps, F. W., and L. L. Vant-Hull. 1978. A cell-wise method for the optimization of large central receiver systems. *Solar Energy* 20: 35.

Luneburg, R. K. 1964. *Mathematical Theory of Optics.* Berkeley, CA: University of California Press.

Magariños, J. R., and D. J. Coleman. 1981. *J. Opt. Soc. Am.* 71: 1614A. Short abstract in *Opt. News* 7,69 (1981).

Masterson, K., and H. Gaul. 1981. *Optical Characterization Method for Concentrating Reverse Illumination.* SERI/TP-641-1179. Golden, CO: solar Energy Research Institute.

McIntire, W. 1980. New reflector design which avoids losses through gaps between tubular absorber and reflector. *Solar Energy* 25: 215–220.

McIntire, W. R. 1979. Truncation of nonimaging cusp concentrators. *Solar Energy* 23: 351.

Meinel, A. B., and M. P. Meinel. 1976. *Applied Solar Energy.* Reading, MA: Addison-Wesley.

National Technical System. 1984. *Development of Holoconcentrators.*

O'Gallagher, J., and R. Winston. 1986. Test of a "trumpet" secondary concentrator with a paraboloidal dish primary. *Solar Energy* 36: 37–44.

O'Neill, M. J. 1979. A unique new Fresnel lens solar concentrator. *International Solar Energy Congress.* Atlanta, GA.

O'Neill, M. V. Goldberg, and D. Muzzi. 1982. A transmittance-optimized, point focus fresnel lens solar concentrator. Fourth parabolic dish solar thermal power program review by Jet Propulsion Lab for the Department of Energy. JPL-pub 83-2, pp. 209–220.

Ploke, M. 1967. Lichtführungseinrichtungen mit starker Konzentrationswirkung. *Optik* 25: 31–43.

Rabl, A. 1976a. Comparison of solar energy concentrators. *Solar Energy* 18: 93–111.

Rabl, A. 1976b. Optical and thermal properties of compound parabolic concentrators. *Solar Energy* 18: 497–511.

Rabl, A. 1976c. Solar concentrators with maximal concentration for cylindrical absorbers. *Appl. Opt.* 15: 1871–1873.

Rabl, A. 1985. *Active Solar Collectors and Their Applications.* Oxford: Oxford University Press.

Rabl, A., N. B. Goodman, and R. Winston. 1979. Practical design considerations for CPC solar collectors. *Solar Energy* 22: 373–381.

Rabl, A., J. O'Gallahger, and R. Winston. 1980. Design and test of nonevacuated solar collectors with compound parabolic concentrators. *Solar Energy* 25: 335.

Rabl, A., and R. Winston. 1976. Ideal concentrators for finite sources and restricted exit angles. *Appl. Opt.* 15: 2880–2883.

Riaz, M. 1976. A theory of concentrators of solar energy on a central receiver for electric power generation. *J. Engineering for Power* 98: 375.

Russell, J. L. 1976. Principles of the fixed mirror solar concentrator. In *Proc. Society of Photo-optical Instrumentation Engineers* 85: 139–145.

Schuster, J. R., G. H. Eggers, and J. L. Russel, Jr. 1978. Operating experience with the general atomic fixed mirror solar concentrator. In *Proc. Annual Meeting of the American Section of ISES.* Denver, CO, p. 863.

Snail, K. A., J. J. O'Gallagher, and R. Winston. 1984. A stationary evacuated collector with integrated concentrator. *Solar Energy* 33: 441–449.

Tabor, H. 1958. Stationary mirror systems for solar collectors. *Solar Energy* 2: 27.

Vant-Hull, L. L. 1976. An educated ray-trace A to solar tower optics. In *Proc. SPIE, 85.* Bellingham, WA: SPIE—The International Society for Optical Engineering, p. 111.

Welford, W. T. 1973. *Opt. Commun.* 9: 268.

Welford, W. T. 1974. *Aberrations of the Symmetrical Optical System.* New York: Academic Press.

Welford, W. T., and R. Winston. 1979. Two-dimensional nonimaging concentrations with refracting optics. *J. Opt. Soc. Am.* 69: 917–919.

Welford, W. T., and R. Winston. 1989. *High Collection Nonimaging Optics*. New York: Academic Press.

Winston, R. 1970. Light collection within the framework of geometrical optics. *J. Opt. Soc. Am.* 60: 245–247.

Winston, R. 1974. Principles of solar concentrators of a novel desgin. *Solar Energy* 16: 89–95.

Winston, R. 1976. Dielectric compound parabolic concentrators. *Appl. Opt.* 15: 291–292.

Winston, R. 1978. Ideal flux concentrators with reflector gaps. *Appl. Opt.* 17: 1668.

Winston, R. 1980. Cavity enhancement by controlled directional scattering. *Appl. opt.* 19: 195–197.

Winston, R., and H. Hinterberger. 1975. Principles of cylindrical concentrators for solar energy. *Solar Energy* 17: 255.

Winston, R., E. Kritchman, and J. O'Gallagher. 1982. Concentration enhancement of a point focus dish using a nonimaging secondary: A full scale optical experiment. In *Proc. Annual Meeting of the American Solar Energy Soc.* Houston, vol. 1. p. 311.

Winston, R., and W. T. Welford. 1978. Two-dimensional concentrators for inhomogenous media. *J. Opt. Soc. Am.* 68: 289.

Winston, R., and W. T. Welford. 1979. Geometrical vector flux and some new nonimaging concentrators. *J. Opt. Soc. Am.* 69: 532–536.

Winston, R., and W. T. Welford. 1980. Design of nonimaging concentrators as second stages in tandem with image forming first stage concentrators. *Appl. Opt.* 19: 347.

Winston, R., and W. T. Welford. 1982. The efficiency of nonimaging concentrations in the physical optics model. *J. Opt. Soc. Am.* (November). 72, 1564–1566.

8 Collector Thermal Research and Development

Charles A. Bankston

8.1 Introduction

8.1.1 Scope

Thermal research is a broad term used in this chapter to include all the research conducted in order to better understand the energy exchange within solar collectors and between solar collectors and the environment. Three main areas of thermal science are covered: thermodynamics, fluid mechanics, and heat transfer. It was originally intended to include a section on mass transfer, but very little relevant work was found.

The assessment is approached from the technological perspective rather than from the purely research view. The emphasis is on research results that will be of interest to engineers and designers. However, I have also tried to identify key research issues that have been resolved and those that have not, so the review should also be valuable to the new researcher or the researcher interested in collector thermal science for the first time.

Prodigious testing and evaluation of solar collectors has contributed to our overall understanding of the thermal behavior of these devices. This work is not covered here except indirectly as it influenced other research work. The distinction between research and testing and evaluation is based on the purpose and interpretation of the work rather than the way the work was done. Investigations that attempted to answer fundamental questions were included even if they dealt with commercial products.

The review generally covers the last decade, but the starting point is rather indistinct. Most of the work included was performed, or at least reported, since 1975, but prior work is also included in those fields where there have been few major advances.

Finally, this chapter covers thermal research relevant to all collector technologies except solar ponds, which are covered in chapter 10. There is a definite tilt toward research that is more relevant to the collector technologies used in solar heating and cooling of buildings, but some of the basic work related to high temperature line- and point-focus tracking collectors for power and industry is included as well. More detailed research findings pertinent to tracking collectors may be found in volumes 10 and 11.

8.1.2 Background

Prior to 1974 only a small amount of government support was available for solar collector research through the National Science Foundation (NSF), the

Atomic Energy Commission (AEC), and the National Aeronautics and Space Administration (NASA). The earlier work in the field had largely been done by a few pioneering researchers at universities: Hottel and colleagues at the Massachusetts Institute of Technology in the 1950s; Löf at Colorado State University in the 1940s and 1950s; Buchberg, Edwards, and colleagues at the University of California at Los Angeles in the early 1960s; Tabor at Hebrew University in Israel in the 1950s; and a few others.

Substantial government interest began early in the 1970s as concerns about the future of inexpensive oil supplies began to mount. NSF, NASA, and AEC initiated research and development work in solar energy. NSF evolved the first cohesive plan for solar R&D that was implemented in 1974. Shortly thereafter the Energy Research and Development Administration (ERDA) was formed and assumed the responsibility for solar research. One of the first priorities of the ERDA Solar Heating and Cooling Research program was to commission an assessment of the technology. This assessment, which was conducted in 1975 and 1976 by Balcomb and associates at the Los Alamos Scientific Laboratory (LASL) (Balcomb and Perry 1977), covered solar collectors, energy storage, air conditioning and heat pumps, systems and controls, and nonengineering aspects of solar heating and cooling. The assessment concluded that the problem areas in the theory and basic understanding of phenomena of solar collectors included these points:

1. Free convection in collectors that have very rough surfaces (V-corrugations, tube grooves, etc.) is not well understood.

2. Convection is not well understood when the absorber plate is not at a single temperature and the glazing is not at a constant temperature.

3. To optimize collector performance, more information is needed in the design of honeycombs regarding cell shape and wall conduction effects.

4. The theory for gaseous thermal conduction at rarefied pressures in evacuated collectors is not well understood.

5. In the case of transpired air heaters, the combination of free and forced convection is not well understood.

6. Forced convection between plates of various rough shapes is not well understood.

All six fall in the general area of collector thermal research.

The LASL assessment led to the development of the National Program Plan for Research and Development in Solar Heating and Cooling, which was published in 1976 as ERDA 76-144. This document, with its now familiar 11 paths and more than 300 tasks, was the basis for most of the research and development that was subsequently carried out by the Department of Energy. The solar heating and cooling R&D was widely distributed among the national laboratories, universities, and industrial and private researchers, with the major portion of the work conducted by industry.

In addition to the heating and cooling program, ERDA established a large program for solar power. This program was centered in three major national laboratories—Sandia Albuquerque, Sandia Livermore, and the Jet Propulsion Laboratory—but it, too, involved university and private research.

A more comprehensive account of the development of the government solar energy programs is found in volume 1. For purposes of the present chapter, it is sufficient to note that the bulk of collector research work for heating and cooling involved over 100 projects and had a total budget in excess of $20 million. In addition to these projects, the inhouse and contract research of the major solar laboratories, including SERI, has been included in this review. The open literature has been researched as well.

8.2 Thermodynamics

Thermodynamics provides the scientific basis for the testing, analysis, understanding, and application of solar collectors. The laws of thermodynamics determine the work and thermal energy that can flow to and from any thermodynamic system and place constraints upon the states that can be achieved by those systems. Application of the first law of thermodynamics—conservation of energy—is familiar even to the nonscientist. The second law of thermodynamics—which is usually thought of in terms of its many corrollaries, one of which prohibits a decrease in entropy of an isolated system—is another powerful tool for the analysis of applications of solar energy. Some consequences of the application of purely thermodynamic considerations to solar collectors will be reviewed in this section before delving into the detailed energy, mass and momentum transport phenomena that govern the performance of specific devices in sections 8.3 and 8.4.

8.2.1 First Law Analysis

A solar thermal energy conversion device—any solar thermal collector—is easily analyzed by application of the principle of conservation of energy to a simple steady-state open system as depicted in figure 8.1.

$$A_c I = \dot{m}(h_2 - h_1) + Q_L \tag{1}$$

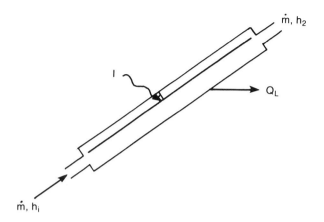

Figure 8.1
Solar collector heat balance.

where A_c is the area of the collector, I is the total incident radiation flux, $\dot{m}$ is the mass flow through the collector, h is the enthalpy of the fluid, and Q_L is the heat loss to the environment from the collector. Equation (1) simply states that the energy received by the collector from the sun must be equal to the energy delivered to the working fluid, the useful output, plus the energy lost to the environment. Since the energy delivered to the working fluid is the desired output and I is the available solar resource, the collector efficiency is defined as

$$\eta_c = \frac{\dot{m}(h_2 - h_1)}{A_c I}. \tag{2}$$

This relation is the basis for the usual measure of thermal performance. When properly and uniformly applied, it permits the comparison of different collectors for specific applications. The simplicity of equation (2), of course, belies the complexity of the efficiency. Collector efficiency cannot be represented as a single number. It depends upon $\dot{m}$, h_2, h_1, and I in rather complicated ways. There is, however, much to be learned from equations (1) and (2) and simple variants. For example, it is common to divide the collector losses into optical losses due to reflection of the incident flux by the collector and the environment. The reflection losses can be represented by an effective reflectivity ρ, and the thermal losses by a loss coefficient U_L. Equation (2) can be expressed as

$$\eta_c = \frac{A_c I - \rho A_c I - U_L(T_c - T_A)}{A_c I} \tag{3}$$

or

$$\eta_c = (1 - \rho) - U_L(T_c - T_A)I. \tag{4}$$

This equation has two useful limits: When $T_c \rightarrow T_A$, the thermal losses vanish so that

$$\eta_c = 1 - \rho = \eta_o \tag{5}$$

which is usually called the optical efficiency η_o. The optical efficiency can be calculated from the optical properties of the glazings, the absorber, and the structural parts of the collector, or measured directly by testing at a very high flow rate so that $T_c \rightarrow T_A$.

The second interesting limit occurs when $\eta_c = 0$. From equations (4) and (5) this yields

$$\eta_c = \eta_o - \frac{U_L(T_c - T_A)}{I}, \tag{6}$$

and for $\eta_c = 0$,

$$T_c = T_s = T_A + \frac{\eta_o I}{U_L}. \tag{7}$$

The $\eta_c = 0$ condition is referred to as the stagnation condition since it is the state assumed by the collector if there is no flow of the working fluid. It also represents a limitation upon the maximum temperature the collector absorber can attain. At this condition no useful energy can be delivered to a load. It is of interest to compute the magnitude of T_s for the ideal condition of no convective or conductive heat losses. The so-called adiabatic plate temperature is the state that would be achieved by a flat plate in radiative equilibrium with the solar input. Figure 8.2 shows the adiabatic plate temperatures for two values of the plate emissivity 1.0 (black) and 0.1.

Figure 8.2 shows that the maximum temperatures that can be attained by flat plate collectors without concentration at normal insolation levels of 1,000 W/m² are between 116° and 378°C depending on the emissivity of the surface. Higher temperatures may be achieved by concentrating the solar flux from a larger area upon the receiver by use of mirrors. The first law places no limit on the temperature that could be attained by focusing the energy from a very large area onto a small spot. As we shall see, the second law does limit the maximum temperature and also provides some valuable insights into the most thermodynamically advantageous ways of using solar energy.

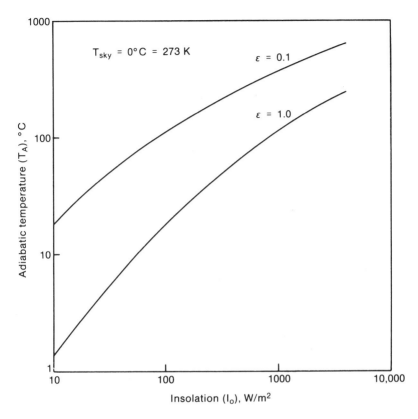

Figure 8.2
Adiabatic plate temperatures for a black absorber ($\alpha = \varepsilon = 1$), and a selective absorber with $\alpha = 1$, $\varepsilon = 0.1$.

8.2.2 Second Law Analysis

The second law of thermodynamics can be stated as "It is impossible to construct an engine that will work in a complete cycle and produce no effect except to raise a weight and exchange heat with a single reservoir" (Keenan 1941). This statement introduces the notion of an engine (heat engine) working in a closed cycle and its interaction with the rest of the universe. The law has many forms and many corollaries. One useful corollary is: "It is impossible to construct an engine to work between two heat reservoirs, each having a fixed and uniform temperature, that will exceed in efficiency a reversible engine working between the same reservoirs." The efficiency of a reversible engine operating between two fixed reservoirs is the well-known Carnot efficiency:

$$\eta = 1 - \frac{T_2}{T_1}. \tag{8}$$

Thus the second law prohibits the construction of a heat engine working between T_2 and T_1 that will exceed the efficiency of a Carnot engine.

A more general development of the second law leads to an expression of the thermodynamic property of a system that determines the capacity of the system to produce work when coupled with a passive environment (e.g., Kennan, Hatsopoulos, and Gyftopsulos 1980). This property is called by various authors available work, available energy, availability, or *exergy*. The form of expression for exergy depends upon the type of system and the way it is coupled to the passive environment or reservoir. The simplest expression of exergy is for a closed system with rigid adiabatic walls. For such a system an expression for exergy is

$$E = U - T^*S - \bar{E}, \tag{9}$$

where E is the exergy, U is the energy, S is the entropy, T^* is the temperature of the reservoir, and $\bar{E}$ is a constant to be evaluated (it turns out to be the Gibbs free energy of the system when it is in equilibrium with the reservoir; Kestin 1980).

Combining this definition with the first and second laws results in an expression for the work W that can be accomplished or required to transform a system state 1 to state 2:

$$W = E_1 - E_2 - T^*\theta, \tag{10}$$

and when the transformation is reversible,

$$W_{\text{rev}} = E_1 - E_2. \tag{11}$$

Thus the difference in exergy between two states of the system is the optimum work that is involved in the change. The function θ is the entropy production or entropy generation term, and it denotes the extent to which a real transformation differs from a reversible one.

The practical significance of equation (10) for solar collectors or solar system analysis is the observation that the optimization of useful work that can be accomplished by the system is achieved by maximizing the exergy collected by the system (collector) and minimizing the entropy production.

Application of Second Law Analyses to Solar Collectors   Second law analyses of solar collectors generally take the approach of optimizing the exergy collected or minimizing the entropy generation of the heat exchange processes within the collector or between the collector and the environment. The appropriate form of the exergy equation, equation (9), or the corresponding form for open (flowing) systems is generally used. Some of the important consequences of this type of analysis can be derived quite simply for an isothermal collector. The isothermal collector may be considered a heat reservoir at temperature T_c. Then, according to the corollary stated above, the maximum power that can be obtained by a heat engine operating between T_c and a reservoir at ambient temperature T_a is

$$P = \dot{W} = \dot{E} - \dot{E}_a = Q_c \left(\frac{1 - T_a}{T_c} \right). \tag{12}$$

Substituting the usual collector efficiency, we find

$$\dot{E}_c = I\eta_c \left(\frac{1 - T_a}{T_c} \right) \tag{13}$$

is the exergy that can be delivered from a collector operating at temperature T_c to the ambient environment. In analogy to equation (2) we can define a second law efficiency:

$$\eta_{II} = \frac{\dot{E}_c}{I} = \eta_c \left(\frac{1 - T_a}{T_c} \right). \tag{14}$$

Scholton (1983, 1985) shows that this expression is approximately correct for nonisothermal collectors cooled by either incompressible fluids or perfect gases so long as $T_0 - T_i$ is not too large.

Figure 8.3 shows the second law efficiencies for several commercial and laboratory collectors at an insolation level of 1,000 W/m^2 and ambient temperatures of 27°C. These curves illustrate the enormous differences between

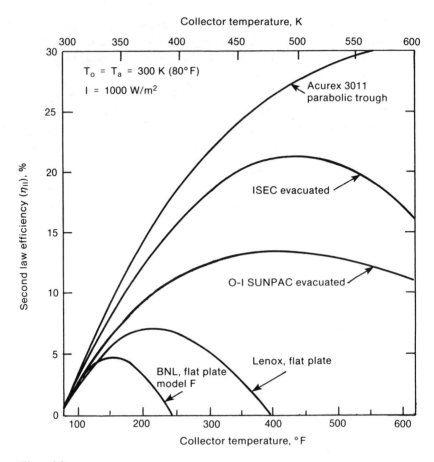

Figure 8.3
Second law efficiencies of some commercial and developmental solar collectors

the exergy output of collectors that have fairly similar first law efficiency curves, and show that both the collector type and the operating temperature must be carefully selected. The same general conclusions were deduced by Howell and Bannerot (1977) by maximizing the power output of a solar plant. The advantage of applying the second law to the analysis of the collector alone is that it also provides both the means to optimize the operating strategy for time-varying insolation (as shown by Bejan 1982) and a basis for improving the design of collectors by minimizing the irreversibilities associated with internal heat exchange (as shown by Bejan, Kearney, and Kreith 1981). An interesting question that remains controversial concerns the maximum exergy of solar radiation (see discussion of the paper by Gribik and Osterle 1986). The obvious result from equation (14) that $\dot{E}_{max} = I(1 - T_a/T_{sun})$ is the upper limit for the efficiency of solar conversion, but more thorough analyses of the radiant exchange between the sun and collector leads to lower efficiencies. Although theoretically interesting, the differences resulting from the various analyses are very small when T_0/T_s is small (e.g., $300/6{,}000 = 0.05$).

Nonisothermal collectors were examined by Bejan, Kearney, and Kreith (1981). These studies deal with the irreversibility associated with entropy production within the collector and at the collector–load heat exchanger. Minimization of these irreversibilities provides the relations for optimization of the working fluid flow rate and other heat exchanger parameters.

Collector control strategies based on maximization of exergy have been developed by Bejan (1982) and Zarea and Mayer (1984). These strategies take the form of variable flow control or operational mode control. It is generally possible to increase the collector exergy output substantially over conventional control schemes (constant flow, constant temperature, etc.).

The second law also prohibits the transfer of heat from a low temperature source to a sink at a higher temperature without the addition of work from an external source. Thus it is impossible to construct a concentrating collector that will produce a temperature greater than the surface temperature of the sun ($\sim 6{,}000$ K). From this thermodynamic limit and geometric optics, it is possible to derive the relation for the maximum concentration ratio. Winston (1970) shows that in three dimensions the maximum concentration ratio is $1/\sin^2 \theta$ and that in two dimensions it is $1/\sin \theta$, where θ is the acceptance angle of the concentrator. A three-dimensional concentrating collector with an acceptance angle of 7×10^{-5}, steradians (equal to the solid angle subtended by the sun's disk) would have to have a concentration ratio of 40,000, and a two-dimensional concentrator would have to have a maximum concentration ratio of 2,000. The maximum temperature for a terrestrial concentrator would be approximately 4,800 K due to irreversibilities in the earth's atmosphere. Hence the second law, unlike the first, places bounds on the range of temperatures that can be achieved and on the extent to which solar energy can be concentrated.

8.2.3   System Considerations

Although the purpose of this chapter is to assess the developments in collector thermal research, we cannot move to a detailed review of heat transfer and fluid mechanics research without commenting on the importance of properly matching the thermodynamic state of the collector with the thermal performance characteristics of the load and other components of the system. Second Law analysis, utilizing exergy flows, as described above, provides a powerful tool both for selecting the proper collector and operating conditions for a particular load and system configuration and for examining and minimizing the irreversibilities that arise within the collector (due to temperature differences, for example) and between components.

These are the considerations that have led to energy transport systems in which the working fluid changes phase in the collector and at the load. Since evaporative and condensing heat transfer coefficients are much larger than single-phase heat transfer coefficients, the temperature differences required at the collector and load heat transfer surfaces are smaller and the entropy production is reduced. Considerable research has been conducted in adapting the heat pipe or closed thermosyphons to the absorber of flat plate (Bienert and Wolf 1976) or evacuated collectors (Ernst 1979, 1981; Ortabasi, Hull, and Schertz 1986). Although the advantages of heat pipe collectors have been well established, and prototypes have been operated successfully in the laboratories, they have not been adopted by commercial manufacturers in this country. However, commercial heat pipe collectors are manufactured in Europe and in Japan by Philips, Sanyo, Hatachi, and others.

Recently, research has been conducted to determine the feasibility of boiling the working fluid in the collector for direct vapor phase transfer to the heat load or to a heat engine. Theoretical studies of Rankine cycle engines coupled directly to collectors through an organic fluid were performed at Honeywell, Barber-Nichols (Batton and Barber 1979), and Los Alamos National Laboratory (Hedstrom 1982). Laboratory investigations of boiling in commercial collectors have been conducted by Thermacore (1981), Argonne National Laboratory, and Sandia National Laboratory-Albuquerque. Of course the central receiver of the experimental Solar One plant at Barstow, California, is a prime example of a boiling collector.

8.3   Fluid Mechanics

With the exception of solar ponds, fluid mechanics research has not attracted a great deal of attention in the solar thermal energy program. The solar thermal technologies are generally concerned only with the fluid dynamic process to the extent that it influences the heat transfer process or the structural or mechanical requirements of the collector. The heat transfer implications of a variety of internal and external fluid fields are reviewed in section 8.4. In this section the studies that have contributed to the understanding of the actual flow processes internal and external to solar collectors will be reviewed briefly. That is, we shall be interested in research for which results are expressed directly in fluid mechanics terms such as velocity distributions, turbulence, streamlines, streaklines, density gradients, etc.

Fluid mechanics processes of interest are divided into internal and external flows. Generally, the flow external to a solar thermal energy collector is required to define the velocity field for solution of the energy equation. This is used to determine external heat losses or to determine the fluid dynamic loading that must be withstood by the collector structures. Internal fluid mechanics are of interest for their heat transfer implications, but pumping power and flow balancing are also important reasons for understanding the mechanics and dynamics of internal flows.

8.3.1   External Flow

Flow patterns, over all types of solar collectors, have been studied as a part of the government solar energy program. Flows over flat plate (stationary) collectors were studied as a part of the solar heating and cooling program. Research at Texas A&M (Chevalier et al. 1979) and the National Bureau of Standards (Cattaneo et al. 1981; Tieleman et al. 1980) focused upon structural requirements and the dynamic response of the collectors to time-varying free stream velocities. Studies by researchers at Tennessee Technological University (Hewitt and Griggs 1979; Hewitt 1980) and the University of Pennsylvania (Lior and Yeh 1980; Segall, Lior, and Yeh 1983) treated both the external fluid dynamics of flows over single and multiple rows of flat plate collectors and the associated heat transfer. The fluid mechanics findings of these studies will be discussed in this section, but the heat transfer finding will be deferred to subsection 8.4.1. Flows over concentrating collectors have been studied extensively in the technology development programs: line-focus collectors, Sandia National Laboratory-Albuquerque; the central receiver, Sandia National Laboratory-Livermore; and the point-focus collectors, Jet Propulsion

Laboratory. Wind loading is a crucial issue in both the design and operation of tracking collectors and has received considerable attention. In addition heat losses from cylindrical- and cavity-type central receivers are highly dependent on the external flow field. Studies and analyses of central receiver heat losses are treated in section 8.4.

Flow over Stationary Collectors   The flow over a single solar collector with a blunt leading edge is depicted in figure 8.4. A separation region characteristically occurs near the leading edge and occupies approximately one-tenth of the collector chord length (i.e., the length of the collector). Small vortices occur near the surface. After reattachment the flow is relatively uniform and parallel to the surface of the plate. The nature of the flow depends on the length of the separation zone (which depends on the bluntness of the leading edge), the free stream velocity, the free stream turbulence, the roughness of the plate, and the distance from the leading edge. Ordinarily a laminar boundary layer will form at the leading edge or at the point of reattachment, and a transition to a turbulent boundary layer will occur when the local Reynolds number reaches a critical value. Surface roughness and free stream turbulence reduce the critical value of the Reynolds number. Most of the phenomena just described have been studied in wind tunnels for decades and are well understood. The pattern of the flows and the characteristic of the associated boundary layers can be found in standard fluid mechanics text and reference books (e.g., Schlichting 1960).

The flow over solar collectors may be influenced by the surroundings in which the collectors are located. Flow over single roof-mounted collectors will be influenced by the base building and by the surrounding buildings, trees, and other obstructions. The obstructed rows of multiple-row collector arrays will be influenced by the separation zones and turbulence produced by the leading rows. The flow patterns produced by such obstructions, and their ultimate effect on heat losses, were studied using wind tunnel models by researchers at both Tennessee Technological University and the University of Pennsylvania.

At Tennessee Technological University, four wind tunnel experiments were performed to determine the wind flow patterns around buildings. The velocity profiles around the buildings and over the collector surfaces were measured. In addition water table experiments were run to give a visual representation of the flow around buildings. Although the existence of the separated zone near the leading edge was confirmed, the effect of building, orientation, and wind direction on the flow field over the collector could not be quantified. In view of the infinite variety of possible configurations, free stream directions,

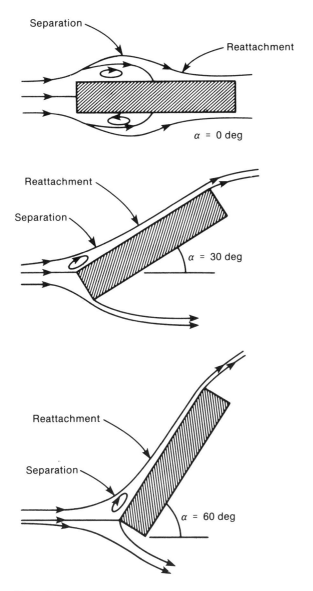

Figure 8.4
Flow patterns around a single inclined rectangular plate for various angles of attack α. Note the regions of flow separation and reattachment. Source: Segall, Lior, and Yeh (1983).

and turbulence, the investigators concluded that it would be impossible to use wind weather data to predict flow patterns. They found that the heat losses could be correlated with the free stream velocity (Hewitt and Griggs 1979), as will be discussed in section 8.4.

The University of Pennsylvania experiment (Segall, Lior, and Yeh 1983) involved a three-row model array which was tested in a low speed wind tunnel. Each plate, representing a row of flat plate collectors, was rectangular, and all three were identical. The interrow spacing, the collector tilt, and the free stream velocity could all be varied. The flow field was visualized by attaching short yarn tufts to the surface and edges of the plates. Photographs taken through the wind tunnel window allowed the determination of attached and separated flow regions and the direction of the flow at the surfaces and edges of the model. All the photographs made in this experiment were made at a single Reynolds number ($Re_L = 1.8 \times 10^5$). Visual study of the flow at six different velocities, three angles of attack (30, 45, and 60 deg), and two row spacings led to the following conclusions:

1. The flow over the unobstructed upstream collector was laminar (in the sense that the yarn tufts were quiescent) for all speeds and angles of attack except for a small ($x/L \approx 0.1$) region of separation near the leading edge. The flow in the separated zone was toward the sides of the plate.

2. The flow at the surface of the second plate was completely separated, turbulent, and three dimensional, with components of the velocity directed from the edges toward the center. The pattern was attributed to vortices shed from the trailing edge and sides of the upstream plate.

3. The flow at the surface of the third plate appeared to be attached but was not as stable as the flow over the first plate. There was some degree of turbulence, and the vortices shed by upstream plates were pronounced (especially at a tilt of 60 deg).

The latter situation—as depicted in figure 8.5—is difficult to explain from theory, but the investigators' hypothesis is that it may be a result of flow around the sides of the second plate being accelerated by the free stream flow over the array.

In addition to the flow field studies, the local and average heat transfer coefficients were measured and correlated, as discussed in section 8.4.

Flow over Tracking Collectors   Two separate studies of wind loading on parabolic trough collectors were conducted as a part of Sandia National Laboratory-Albuquerque's technology development program (Lindsey 1976;

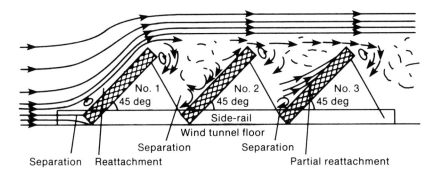

Figure 8.5
Streamlines and flow patterns around an array of three inclined plates at $\alpha = 45$ deg. Note the vortices shed from the top edges of the second and third plates. Source: Segal, Lior, and Yeh (1983).

Peterka, Sinou, and Cermak 1980). The results are summarized in the 1980 paper by Randall, McBride, and Tate. The first study was run in a nonviscid uniform flow of a low speed wind tunnel at Vought Corporation, using single-collector models. The collector geometry, angle of attack, and yaw were varied, and the lift, drag, and pitching moments were measured using force balance techniques. In addition the pressure distribution was measured over the front and rear of the collector and used to compute loads for comparison with the balance measurements. No velocity measurements or flow fields were reported.

The second major study of wind loading on parabolic troughs was made in the low speed wind tunnel at Colorado State University. Single and multi-row models were immersed in a 48-inch thick boundary layer that had been specially tailored to resemble the 1/7 power law believed to represent the atmospheric boundary layer over open terrain. Again, the emphasis was on direct determination of the forces acting upon the collectors and their supports as functions of the collector geometry and orientation. In addition there were studies of the effect of row spacing, the gap between adjacent collectors in the same row, and the effect of wind protection fences. Flow visualization techniques were used to study wind barrier effectiveness. The results of these two series of experiments are summarized by Randall, McBride, and Tate (1980):

1. The forces and moments on parabolic trough collector modules increase monotonically with mounting height above the ground.

2. The peak forces and moments on individual collector modules increase with aspect ratio up to ratios of 10 or greater.

3. Intermodule gaps as narrow as 6% of the aperture between end-to-end

collectors within a row are sufficient to permit collectors to function aero-dynamically as individual modules, effectively nullifying any long-row aspect ratio influence.

4. Collector modules installed within large arrays, even those within the second row of an array, experience an interference effect which provides a significant reduction (50%–65%) of the peak lateral and lift forces originating with the wind.

5. The interference-induced load reduction does not extend to the collector pitching moment, indicating that a pressure distribution change accompanies the interference effect.

6. Appropriate fence or berm configurations can provide reduction of lateral and lift forces in perimeter rows equivalent to the interference effect within collector arrays.

7. A fence or berm height of approximately three-fourths the maximum collector height provides the major fraction of the force reduction achievable.

The emphasis of the central receiver research on external flows has been primarily the determination of mixed convection (forced and natural) heat transfer rates for cylinders and various cavity-shaped receivers. Because of the large dimension and high operating temperatures of central receivers, the effects of buoyancy are far more pronounced than those usually encountered in industrial processes. This has made study of the flow fields very difficult to conduct in other than full-scale models, although wind tunnel tests have been conducted (Clausing 1982) using a cryogenic environment to achieve simultaneously both a high Reynolds number and a high Grashof number. Since the results of these experiments are usually in the form of heat transfer coefficients or Nusselt numbers, they are discussed in section 8.4. Figure 8.6, however, illustrates schematically the complexities of the mixed convection over a short cylindrical receiver.

Wind tunnel testing of central receiver heliostats and parabolic dish receivers has also been conducted. As with the line-focus collectors, the primary objective of this research was the quantification of the aerodynamic loading on the structure of the collector and its foundation. The desire to reduce collector costs by reducing the mass of structural material will assure that interest in the aerodynamics of collector continues. New, strong, lightweight design utilizing stretched membranes or composite materials are evolving. Some of these are designed to withstand high winds by assuming a special stow configuration that reduces the aerodynamic forces. Bubble enclosures for heliostats and plastic "dish covers" may also be used to eliminate or reduce

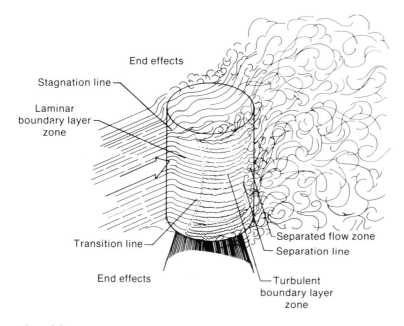

Stagnation line

End effects

Laminar
boundary layer
zone

Transition line

Separated flow zone

Separation line

End effects

Turbulent
boundary layer
zone

Figure 8.6
Schematic of the convection heat transfer zones on a cylindrical, external-type receiver. Source: Siebers and Kraabel (1984)

aerodynamic loads. More detailed results of the studies of external flows over tracking collectors may be found in volumes 10 and 11 of this series.

8.3.2   Internal Flow

The nature of the flow of the working fluid in the solar collector or collector array not only has an important influence upon the heat transfer from the receiver, and therefore, the efficiency of the collector, but also determines the power required to pump the fluid through the collector, how the fluid will divide within the collector flow passages or between collectors, and whether the collector will be susceptible to damage from fouling, erosion, or cavitation. Since most solar collectors are made up of relatively familiar fluid flow elements (pipes, elbows, parallell wall channels, etc.), internal fluid mechanics has not received a great deal of research attention in the solar energy program. This section will review briefly some of the important concepts and engineering data that apply to many solar collector situations.

Channel Flow   The critical Reynolds number for a tube is 2,100 and is slightly higher for parallel plates (2,400). Since the Reynolds number is a function of the dynamic viscosity as well as the mass flow rate, it also depends upon temperature. The viscosity of liquids decreases with temperature while the viscosity of gases increases. This leads to the possibility that the nature of the flow may change within the channel due to heating or cooling. The typical tube and sheet flat plate collector with water as the working fluid is likely to have a tube Reynolds number in the low turbulent range, say $Re < 10^4$. If it is oil cooled, or if the passages are very small, the flow will normally be laminar. Air-cooled collectors are frequently designed to operate with fairly large temperature gains and therefore tend to operate in the laminar turbulent flow transition region ($10^3 < Re < 10^4$).

Concentrating collectors are normally cooled by single-phase liquid flow, but boiling with net steam generation (two-phase, steam-water flow) occurs in central receiver boilers and occasionally in line-focus and CPC collectors (Thermacore 1981). Since the heat flux in a concentrating collector receiver is typically 5 to 1,000 times higher than in flat plate collectors, higher mass flows are required to remove the heat, so the Reynolds numbers are well above critical at design conditions. However, many concentrating collectors use oils as the working fluid, and since the viscosity is quite temperature dependent, the flow may be laminar during start-up or shutdown or when operating at low solar input.

The pressure distribution in the collector is of interest to predict the end-to-end pressure drop for use in system calculations, or to predict the onset of nucleate boiling, cavitation, or condensation. The pressure difference between the entrance and any point in the channel is determined by two processes: acceleration of the fluid and wall friction,

$$\Delta p = \int_a^c \rho u \, du + \int_0^l \rho \frac{f u^2}{2} \, du. \tag{15}$$

For fully developed incompressible flow, the velocity profile is not a function of x so that

$$\Delta p = \int_0^l \rho \frac{f u^2}{2} \, du. \tag{16}$$

The friction factor f is a function of the channel shape, the Reynolds number, and, in the entry region, the distance from the inlet. Laminar flow solutions for fully developed flow in rectangular and cylindrical annular channels are shown in figures 8.7 and 8.8. The length scale for the Reynolds number in the figures is the hydraulic diameter

$$D_h = \frac{4 \text{ cross-sectional area}}{\text{perimeter}}. \tag{17}$$

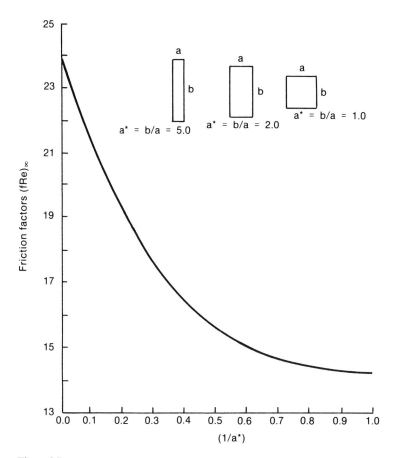

Figure 8.7
Friction coefficients for fully developed laminar flow in rectangular tubes. Source: Kays (1966).

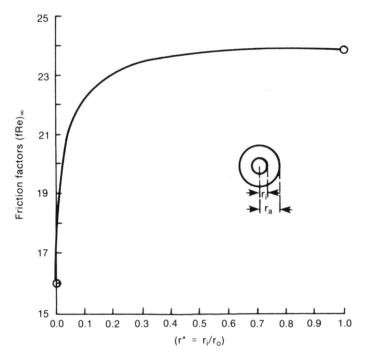

Figure 8.8
Friction coefficients for fully developed laminar flow in circular tube annuli. Source: Kays (1966).

For isosceles triangular channels, the friction factor is given by

$$f\,\mathrm{Re} = 13.33. \tag{18}$$

The local and average friction factors depend upon the length of the hydraulic entry region (generally the distance from a sharp inlet) as shown in figure 8.9. Exact solutions for turbulent flows do not exist. Approximate solutions, based on various models of the turbulent shear stress, are available in the literature, but it is more common to find the friction factor from the well-known Moody diagram, figure 8.10. Unlike the laminar flow, in a turbulent channel flow, most of the change in velocity takes place very near the wall and hence is not influenced by other walls (i.e., the shape of the channel). Figure 8.10 may be applied to practically any shape channel using the hydraulic diameter concept of equation (17).

In the transition region between Reynolds numbers of 2,000 and 4,000, the friction factor increases abruptly. Since the location of the transition usually depends on the details of the channel entry or the roughness of the wall, it is

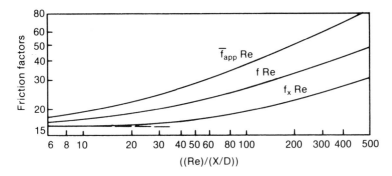

Figure 8.9
Friction coefficients for laminar flow in the hydrodynamic entry length of a circular tube. Source: Kays (1966).

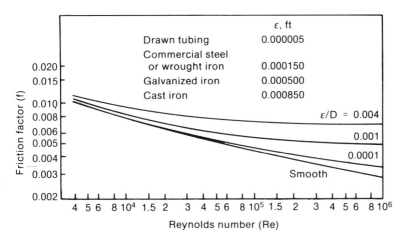

Figure 8.10
Friction coefficients for fully developed turblent flow in smooth and rough wall circular tubes
Source: Kays (1966).

very difficult to determine transition friction factors accurately. The effects of boyancy may also be significant in some collectors, so friction factor may depend upon the rate of heating and the inclination of the channel.

The available literature of fluid mechanics should be adequate to predict both the type of flow and the pressure distribution in most solar collectors. Further the pressure drop in the cooling channels of most collectors is not the most significant loss in the system. Losses in distribution and header pipes, elbows, valves, and flow control orifices are often more important (see ASHRAE Handbooks or Handbook of Fluid Mechanics). Usually the main need for the pressure drop across a collector is to achieve flow balancing in multicollector arrays. Often the array designer uses a measured pressure drop obtained from a standard test (specified flow) of the collector module. However, as the sophistication of control strategy evolves to include flow modulation, the pressure drop as a function of flow rate (i.e., Reynolds number) will assume greater importance.

Flow Distribution   The distribution of flow among a group of parallel channels is an important consideration both in the design of single collectors and in large arrays. Poor flow balancing can result in some channels being deprived of sufficient fluid, so they overheat and operate inefficiently while other channels are overcooled. In some circumstances the distribution or allocation of flow may be unstable in the sense that if it is slightly perturbed so that one channel receives less flow than the rest, the flow impedance of that channel will increase, and the flow will thereby decrease until it is shut off completely. This situation exists in gas-cooled nuclear reactors (Bankston 1965; Bankston, Sibbitt and Skoglund 1966) and in parallel tube boilers in which the entering liquid is subcooled. Ordinarily low temperature solar collectors do not fall into the flow regimes where flow distribution instabilities are a concern. However, water-boiling receivers, such as the Solar One plant at Barstow, may be afflicted with the flow distribution problem known as "boiling disease." In order to avoid such problems, the Solar One receiver was fitted with flow-control orifices at the outlet of each boiler tube.

In single-phase, low temperature collectors, flow distribution problems are usually due to the pressure distribution in the feed and discharge headers, creating unequal pressure differences across different channels or different collectors. Although the basic equation and analysis techniques for properly designing and sizing manifolds and header tubes to achieve acceptable flow distributions have been available for some time, they have been infrequently applied in the design of solar systems.

The University of Pennsylvania conducted a study on the optimization of flat plate collector arrays. Flow distribution is a major issue in the optimization process, and this led the Pennsylvania researchers to formulate the problem carefully and to include the solution procedure in their optimization process (Lior et al. 1977; Lior and Yeh 1980; Lior et al. 1980; Menuchin et al. 1981).

The pressure distributions in the manifolds of the branching system are determined by the combination of inertia forces—due to acceleration and deceleration of the fluid—and friction forces due to the wall shear stress. Conservation of mass and momentum and mechanical energy balances are applied at each branch of the manifold to determine the branch-to-branch pressure differences between manifolds. Note that correct modeling of the manifold requires the analysis of discrete branch joints rather than continuous removal or addition of fluid.

The method has been applied to an individual collector to calculate the flow distribution among the riser tubes and the pressure distribution in the inlet and outlet manifolds (Jones and Lior 1978). The same method has been applied to the analysis of flow distribution among collectors piped in various parallel series arrangements in arrays. In the latter analysis each branch consists of one or more collectors, interconnecting piping with appropriate fittings, and perhaps flow control valves.

Figure 8.11 shows the calculated pressure distribution in the parallel flow (flow in the same direction) inlet and outlet manifolds of a typical seven-tube, water-cooled flat plate collector. The pressure distribution in this rather short manifold is inertia dominated. The resulting flow distribution is shown in figure 8.12 for two total flow rates. At the highest flow rate there is more than a 60% variation in flow between the first and last riser tube.

Figure 8.13 shows the nondimensional flow distribution among collector modules connected between parallel flow manifolds for groups of 8, 16, 24, and 48 collectors for a fixed manifold-to-connector tube diameter ratio and nominal collector flow rate. As the number of collectors in parallel increases, the manifold pressure distribution becomes friction dominated. The pressure in the inlet manifold rises slightly at first and then falls as frictional losses outweigh inertial gains. In the outlet manifold, the inertial and frictional forces add to create a continuously decreasing static pressure. The resulting flow maldistribution becomes increasingly severe as the length of the manifold (number of collectors) increases. Collectors near the inlet and outlet of the manifold are overfed while those near the center are starved. It is theoretically possible for the flow in some collectors to stop entirely or reverse. Reverse flows have, in fact, been observed in experiments in France (Lebru et al. 1984).

The reverse flow manifold configuration in which the direction of flow in the outlet manifold is in the opposite direction of the flow in the inlet manifold

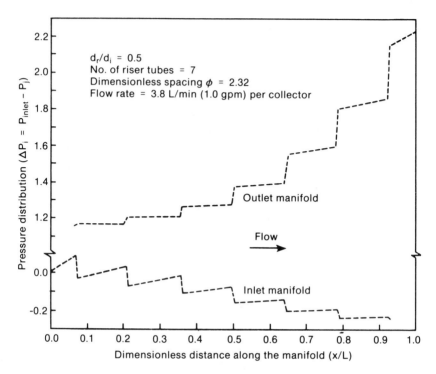

Figure 8.11
Manifold pressure distribution for parallel flow.

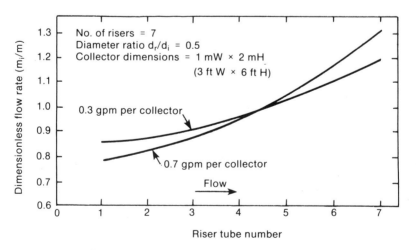

Figure 8.12
Flow distribution among riser tubes in a collector.

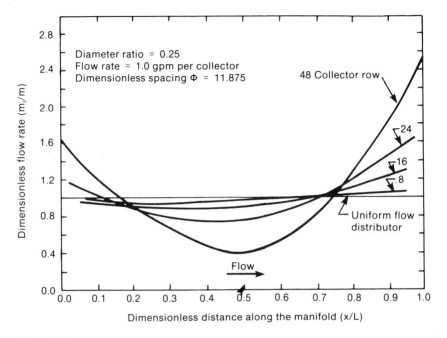

Figure 8.13
Flow distribution in flat plate collector row, parallel flow.

provides improved flow distribution for inertial-dominated manifolds but allows the collectors or risers at the end of the friction-dominated manifolds to be badly starved unless the channel impedance is very high or control valves or orifices are used. It does, however, require the shortest piping and may be considered in an economic optimization.

Figure 8.14 shows the fraction of the intermanifold pressure drop attributed to each component. Note that for this collector, the collector itself contributes only a small fraction. Acceleration and turning losses are larger. The Pennsylvania work, confirmed by other analysts, concludes that

1. maldistribution of flow increases with the number of collectors or risers in parallel,

2. maldistribution increases with the ratio of the connecting tube-to-manifold diameters,

3. maldistribution decreases with the impedance of the collector or connecting piping.

The flow distribution is little affected by the inlet temperature, but the magni-

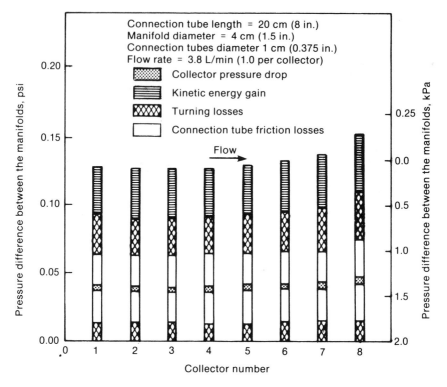

Figure 8.14
Pressure profile in a row of eight collectors.

tude of the pressure drop does vary significantly with temperature and must be considered in the combination of parallel rows in large arrays.

The effect of flow maldistribution, of course, is to reduce the efficiency of the entire array. The magnitude of this effect can be substantial if actual plugging occurs, and even in well-designed arrays, may range from 2% to 20%. The impact of flow maldistribution on the collection efficiency is included in the University of Pennsylvania optimization process. The direct effect of flow maldistribution on array efficiency may be incorporated in a single collector efficiency deterioration factor as shown by Chiou (1982).

Computer codes are available to the designer of large collector arrays for analyzing the flow distribution—ETRANS (Barnhart 1980)—and design studies have been reported by a number of investigators (Morton 1981, 1983; Sharp 1981; Sharp et al. 1983; Kutscher et al. 1982).

8.4   Heat Transfer

The heat transfer processes that occur between the solar collector and its environment and between the solar collector and the system to which it delivers energy determine the efficiency of the collection process and ultimately of the solar energy system. These external heat transfer processes are dependent upon the temperatures and material characteristics of the heat exchange surface of the solar collectors, which in turn depend upon the internal heat transfer processes within the collector. The collector analyst must therefore be able to analyze both internal and external heat transfer processes involving radiation, conduction, and convection modes.

Research has provided methods of analyzing the most common configurations and materials found in a variety of solar collectors. Before reviewing the relevant literature in detail, however, it is informative to examine the general dependence of solar collector efficiency and absorber temperature on the various internal and external heat transfer modes and the accuracy with which they can be estimated.

An analysis following the general assumption and treatment employed in the Hottel-Whillier-Bliss (HWB) model (Whillier 1964) and described by Duffie and Beckman (1980) results in a useful nondimensional representation of the efficiency and temperature. The results of such an analysis (McEligot and Bankston 1980) are below

$$\eta = \frac{1}{\text{NTU}}\left[\alpha\tau - \frac{U}{I}(t_i - t_a)\right]\left(1 - \exp\left\{-\text{NTU}\left(\frac{1}{1 + U/h}\right)\right\}\right) \tag{19}$$

and

$$\frac{t_p - t_i}{t_s - t_a} = \left[1 - \frac{U}{\alpha\tau I}(t_i - t_a)\right]fn\left(\text{NTU}, \frac{U}{I}\right),$$

where

$$fn\left(\text{NTU}, \frac{U}{h}\right) = 1 - \left[1 - \frac{U}{h}\left(\frac{1}{1 + U/h}\right)\right]\exp\left\{-\text{NTU}\left(\frac{1}{1 + U/h}\right)\right\}, \tag{20}$$

$\text{NTU} = UPx/\dot{m}c_p$ is the "number of transfer units" (Kays and London 1964), U is the overall effective external heat exchange coefficient between the absorber and the ambient environment, and h is the internal effective heat transfer coefficient between the absorber and the working fluid.

The internal heat transfer enters only through the ratio h/U. Thus for many liquid-cooled collectors where $h \gg U$, the efficiency and the absorber temperature are virtually independent of h. Figure 8.15 shows the efficiency and nondimensional absorber temperature for $\alpha\tau = 0.8$ and $t_i = t_a$. The importance of both $\dot{m}$, I, U, and h can be easily estimated from these figures with the aid of a few simple calculations. McEligot and Bankston analyed four cases: a typical liquid-cooled flat plate collector, an air-cooled collector, a CPC receiver, and a parabolic trough collector. For the liquid-cooled flat plate $U(t_i - t_a)/\alpha\tau I \simeq 0.15$, NTU $\simeq 0.05$, and $U/h \simeq 0.075$, giving an efficiency of 0.59. For these conditions the efficiency is already near its maximum ($U/h = 0$), and improvements in the internal heat transfer process (i.e., increased h) will have little effect. The external heat exchange coefficient is approximately equally dependent upon the infrared radiation from the glass and the convection due to wind. A strong wind could double the convection coefficient, but this would only reduce the collector efficiency by about 0.09. Thus at typical liquid flat plate collector operating conditions, the designer need not be overly concerned with either internal or external convection processes. For a typical air heating collector, $U/h \sim 1$, NTU ~ 1, and the efficiency is only ~ 0.3. The performance is sensitive to h, U, and $\dot{m}$. Substantial improvements are possible through convection augmentation, and the designer is quite concerned with the internal convection. Much of the heat transfer research that has been conducted as a part of the solar thermal energy program has addressed the prediction and enhancement of the internal heat transfer coefficient, h.

In evacuated receivers and concentrating receivers operating at low temperatures, U/h and NTU are generally both small so that the efficiency is determined primarily by optical considerations. At high temperatures or for bare receivers (e.g., central receiver towers), external convection becomes a crucial issue.

In the following sections we will review the research results that allow the collector designer or analyst to make reliable estimates of the various modes of external and internal heat transfer processes in solar collectors. The reader should be aware that the relevant heat transfer literature is vast. This review can only hope to include work that was done with solar energy application in mind.

8.4.1 External Heat Transfer

Radiation and convection are the dominant modes of external heat transfer between a solar collector and its environment. Conduction, for example, through the back insulation of a flat plate solar collector is generally a straightforward process if the geometry and material properties are known. Low temperature infrared radiation from the external surfaces of the collector to the environment, which was well described in the early literature (e.g.,

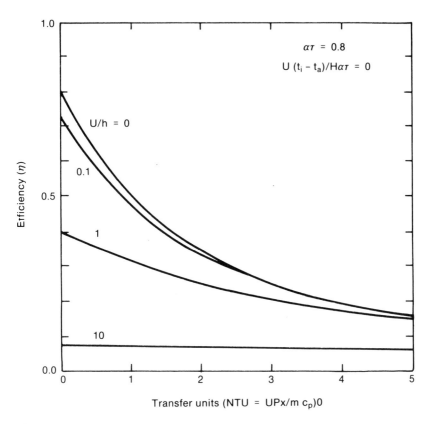

Figure 8.15
Efficiency of typical solar collectors with single glass covers. Source: McEligot and Bankston (1980).

Whillier 1967), will not be covered in this section. The radiant exchange between the sun and the collector is covered in the sections on optics. The convection loss from collectors is one heat transfer phenomenon that has received much research attention.

External Radiaton   Most aspects of external radiation are well understood and already in the literature in standard texts on radiative heat transfer. The reader who is interested in radiation, however, should also consult the section of this volume on materials for further details on the radiative properties of wavelength selective materials, and perhaps volume 8 on natural cooling for more information on radiaton to the sky and surroundings.

External Convection

PLANE SURFACES   Much of the early design and analysis work on flat plate solar collectors employed an equation of the form

$$h = a + bV^n. \tag{21}$$

This correlation was developed from the 1924 measurements of Jürges. It was widely used because it appeared in the most popular heat transfer reference book of the 1950s and 1960s (McAdams 1954) in spite of its two major deficiencies: It contains no length scale, and it applies only to flow parallel to the plate. The experiments were in fact conducted with a vertical 0.5×0.5 m copper plate with airflow at room temperature. Coefficients were provided for air speeds in the ranges 0 to 5 and 5 to 30 m/s.

In 1949 Drake measured laminar flow heat transfer coefficients on an incline plate at angles of attack of 0 to 40 deg. The Reynolds numbers, based on the plate length L, were 3.5 to 8.5×10^5. The experimental results were correlated with the relationship

$$Nu_x = C\left(\frac{x}{L}\right)^n \sqrt{Re_L}, \tag{22}$$

where x is the distance from the leading edge, and n and C are dependent upon the angle of attack. The form of the equation was confirmed theoretically. Many investigators of the 1950s and 1960s studied the heat transfer from flat plates at 0 deg angle of attack and a variety of thermal boundary conditions (e.g., Reynolds, Kays, and Kline 1960). There have also been studies of natural convection from exposed inclined plates (Vliet 1969; Yung and Oetting 1969), but for solar collectors in an outdoor environment, breezes sufficient to dominate the external flow field almost always exist.

In the last decade the U.S. solar program has included at least four specific studies of heat transfer from single and multiple rows of flat plate collectors: University of Minnesota, Tennessee Technological University, American Heliothermal, and the University of Pennsylvania.

At the University of Minnesota, Sparrow and his coworkers studied inclined square and rectangular plates at various angles of attack and yaw in a wind tunnel. They actually measured the sublimation of naphthalene from plates and used the analogy between heat and mass transfer to infer the heat transfer characteristics. For square plates at Reynolds numbers from 2×10^4 to 10^5, Sparrow and Tien (1977) reported the general relationship

$$j = 0.931 \mathrm{Re}_L^{-1/2}, \tag{23}$$

where j is the Colburn j-factor $= \mathrm{St} \mathrm{Pr}^{2/3} = (h/\rho c_p U_\infty) \mathrm{Pr}^{2/3}$.

Within the experimental error of their experiments, this relation appeared to be independent of yaw or angle of attack.

Sparrow, Ramsey, and Mass (1979), working with rectangular plates, proposed the general relationship for independent angle and aspect ratio:

$$j = 0.86 \mathrm{Re}_L^{-1/2}, \tag{24}$$

where the Reynolds number is based on the characteristic length $L = 4A/C$ (where A is the plate area and C is its perimeter). These experimental findings contradict boundary layer theory applied to wedge flows which predicts much greater variation of h with angle of attack (Eckert and Drake 1972). The magnitudes of the average heat transfer coefficient predicted by the relation of equations (23) and (24) for typical solar collector dimensions are much lower than those obtained from equation (21).

The study of external flow over collectors at Tennessee Technological University was also conducted in a small wind tunnel in which a multicollector model was subjected to a variety of angles of attack, yaw angles, and free stream turbulence (representing the effects of surrounding buildings, trees, fences, etc.). The Reynolds number range was 2.5×10^4 to 1.1×10^5, angle of attack was 30 or 45 deg, and the yaw angle was 0 or 180 deg. The results of this study, which were presented by Hewitt and Griggs (1979) and Hewitt and Onur (1980) in the form

$$h = a + b\sqrt{U_\infty} \tag{25}$$

for different free stream conditions, average approximately 10% below the results of Sparrow and Tien. The Tennessee experimental results agreed well with their numerical boundary layer analysis.

A full-scale, outdoor study of convective heat losses from flat plate collectors was conducted by American Heliothermal Corporation of Denver, Colorado, (Skartvedt et al. 1980). In this study 2×6 ft collector panels containing two $6\frac{1}{4} \times 6\frac{1}{4}$ inch heated copper plates were exposed to natural wind conditions at ground and roof levels. Both convective heat transfer coefficients and night sky temperatures were inferred from the data. The results from the roof-mounted plates were correlated by the equation

$$Nu = 0.74Re^{0.555} \, Pr^{0.33}. \tag{26}$$

Over the range of conditions encountered in the experiments, this equation yields Nusselt numbers that are approximately 50% higher than Sparrow and Tien's [equation (23)]. The results from the ground level panels were even higher (93%). The investigator suggests that the higher heat transfer coefficients may be attributed to a greater intensity of small-scale turbulence in the ambient air as compared to wind tunnel tests. This is consistent with the finding of higher heat transfer coefficients for the panels at ground level where one would expect higher turbulence intensities.

Skartvedt et al. (1980) also reports results for quiet periods in which the recorded wind velocity was zero as

$$Nu = 1.48(GrPr)^{0.282}. \tag{27}$$

This result is three to four times the value predicted by relations obtained in laboratory experiments and tends to confirm the earlier assertion that free convection conditions seldom exist in the outdoors.

The most recent results are those from the University of Pennsylvania, Segall, Lior, and Yeh (1983).[1] In this study a three-plate array was tested in a large, low speed wind tunnel. The Reynolds number ranged as high as 7.5×10^5. Six different velocities, three angles of attack (30, 45, and 60 deg), and two interrow spacings were tested. Flow patterns and local heat transfer coefficients were determined. Flow patterns show zones of separation near the leading edge of the first collector, over the entire surface of the second collector, and over much of the third collector. The details of the flow separation, reverse flow regions, and transverse flow are complex and difficult to quantify. The heat transfer results are quantitative but still complex. Segall et al. (1983) present a table of correlations and comparisons for various collector positions, spacings, and free stream velocities. One of the more important of their findings is that the average j-factor for the undisturbed (first) collector is given by

$$j = 0.749Re_L^{-0.485} \tag{28}$$

for $Re_L < 6.4 \times 10^4$. This result is independent of the angle of attack for Reynolds numbers less than 3.5×10^5, but at high Reynolds numbers the results show a pronounced decrease in the magnitude of the slope of the j-factor Reynolds number curve. This effect is most pronounced for a tilt angle of 45 deg where the slope actually becomes positive. Over the range of the experiments, this expression averages 2% lower than the Sparrow and Tien result, equation (23); 27% lower than Jürges, equation (21); and 13% higher than Hewitt, equation (25).

Heat losses from the second and third rows are higher than from the unobstructed plate by about 40% due to the vortices shed from the edges of the upstream plate(s). This result is relatively insensitive to row spacing within the range of the experiments (i.e., 1.4 to 2.9 collector lengths).

From the foregoing review it seems reasonable to conclude that the average heat transfer coefficient for an unobstructed flat plate collector of typical dimensions is nearly independent of angle of attack (tilt) and small yaw angles. For Reynolds numbers less than about 5×10^5, the heat transfer coefficient can be estimated from either equation (23) or (28) with reasonable accuracy. However, if there are obstructions in the form of other collector rows, buildings, or trees to generate vortices or small-scale turbulence, the heat transfer coefficients can be substantially higher, 40% to 50%. This may explain a part of the common observation that collector array performance is lower than expected from single-module test data.

Reliable results are not available for high Reynolds numbers (large modules), sheltered collectors (i.e., wind fences), or flow parallel to long rows or from the rear.

In another study of heat transfer from solar collector glazings at MIT's Lincoln Laboratory, Raghuraman and Kon (1981) found that covers with crisscrossing rectangular grooves could reduce the convective heat transfer coefficient by a factor of four. The ratio of groove spacing to groove depth, w/l, has to be less than 2 for reduction, and the best covers had $w/l \times 1/3$. The transmissivity of the cover is said to be unaffected by the grooves. The practicality of grooved covers was not discussed.

Cylindrical Surfaces and Cavities   Cylindrical surfaces are common elements of many collectors or collector array components. Interconnecting plumbing is the most common example found in distributed systems of all kinds, but the receivers of line-focus tracking collectors and stationary concentrating collectors (CPC type) are usually right circular cylinders, and the external type of absorber of a central receiver system is often cylindrical. The heat transfer literature pertaining to convection from cylindrical objects is vast because so many common engineering systems employ them. The most common means of measuring and quantifying turbulent fluid flows for many years was the hot wire anemometer, and this application alone spawned thousands of publications dealing with some aspect of heat transfer from cylinders. While much of the literature does apply to the components and fluid flow conditions found in typical solar collectors—for example, the piping of collector arrays and the receiver tubes of parabolic troughs—some solar components fall outside the normal range of heat transfer experience. A notable example is the absorber of a central receiver for a power plant application which, because of its size (10–20 m diameter) and its operating temperature (up to 800°C), operates in a regime of Reynolds and Grashof numbers that is unprecedented. Figure 8.16a illustrates the operating regime of typical solar central receivers along with the boundaries of existing experimental results in 1979. Figure 8.16b shows how the map has been filled in by the concerted effort launched in 1979 under the DOE Central Receiver Loss Program (Abrams 1983). Although the number and range of experimental results have grown impressively in the past few years, major uncertainties remain, and the theoretical analysis still lags the experiments (Abrams 1983). Some of the important findings along with the outstanding research issues will be reviewed briefly in this section. More detail on central receiver research is found in volumes 10 and 11.

Unlike the flat plate collectors discussed in the previous section, large central receiver absorbers are often operated in a regime where both free and forced convection effects are important. The literature of mixed convection heat transfer was reviewed by Churchill (1977) who examined the literature of mixed convection in which the buoyant and inertial forces act in the same direction over bodies that can be characterized by a single-length scale (spheres, vertical plates, infinitely long cylinders, etc.). For this class of heat transfer problems, it has been found that the Nusselt number can be correlated by an expression of the form

$$\text{Nu}_{l,mx} = (\text{Nu}_{l,\text{fc}}^a + \text{Nu}_{l,\text{nc}}^a)^{1/a}, \tag{29}$$

where l is the common length scale and the value of the power lies between 2 and 4 (Churchill 1977). Other reviewers have confirmed this "vectorial" addition of Nusselt number with a value of $a = 3$ as the most common recommendation. When the buoyant forces and the inertial forces are opposed, equation (29) with a value of $a = 3$, overpredicts the heat transfer. Under opposing free and forced convection flow conditions, the heat transfer will decrease below either value taken alone. The appropriate exponent under these conditions may be negative (Siebers and Kraabel 1984).

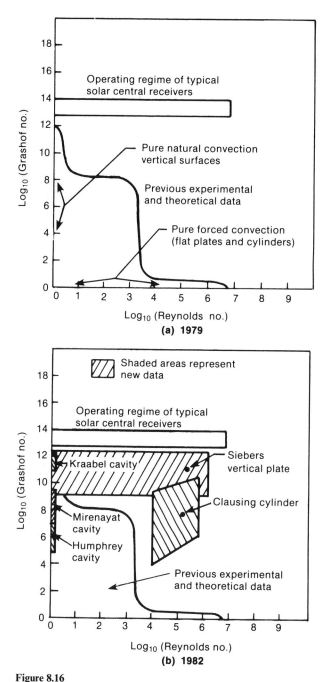

Figure 8.16
Progress in convective loss research over the period 1979–82. The properties in the Reynolds and Grashof numbers are evaluated at the ambient temperature. The boundaries of the regions are approximate. Source: Abrams (1983).

The more common situation in the solar field is the case in which the buoyant and inertial forces interact at some angle (usually orthogonal) and where the appropriate length scale for the forced convection, usually the diameter of a vertical cylinder, is different from the appropriate length scale for the free convection (i.e., the height of the vertical cylinder). Modern experimental investigation of horizontal flow over heated vertical cylinders have been conducted by Clausing et al. (1983), Young and Ulrich (1983), Oosthuizen and Leung (1978), Yao and Chen (1981), and Evans and Plumb (1982, 1983). These experiments are reviewed by Abrams (1983) and by Siebers and Kraabel (1984). Siebers also conducted an experimental investigation of mixed convection heat transfer from large vertical surfaces in horizontal flows (1982, 1983) and found that the mixed convection was best correlated by an equation of the form

$$\bar{h} = (\bar{h}_{fc}^a + \bar{h}_{nc}^a)^{1/a}, \tag{30}$$

where $\bar{h}_{fc}$ is the average heat transfer coefficient that would result from the forced convection alone, and $\bar{h}_{nc}$ is the average heat transfer coefficient that would result from the natural convection acting alone. In computing $\bar{h}_{fc}$, the horizontal dimension is used, and in computing $\bar{h}_{nc}$, the vertical dimension is used. Siebers and Kraabel (1984) argue that results for the heated cylinder experiment should also be correlated in this way, rather than using a single-length scale as practiced by Clausing et al. (1983) and Oosthuizen and Leung (1978). Neither theory nor experimental precision are as yet adequate to settle the question of one length scale or two, and as it turns out there are often much greater uncertainties.

Siebers and Kraabel (1984) have employed the two-length-scale approach, however, in developing a methodology explicitly for cylindrical central receiver absorbers that employs the most recent results available. Their procedure includes estimates of the uncertainties in its application to a tube ribbed absorber. The procedure utilized the rough cylinder forced convection results of Achenbach (1977).

For $k_s/D = 0$ (a smooth cylinder), use

$$\text{Nu}_D = 0.3 + 0.488\text{Re}_D^{0.5}[1.0 + (\text{Re}_D/282000)^{0.625}]^{0.8} \qquad \text{for all Re}_D. \tag{31}$$

For $k_s/D = 75 \times 10^{-5}$, use smooth cylinder correlation, equation (31) when $\text{Re}_D \leq 7.0 \times 10^5$, or

$$\text{Nu}_D = 2.57 \times 10^{-3}\text{Re}_D^{0.98} \qquad \text{when } 7.0 \times 10^5 < \text{Re}_D < 2.2 \times 10^7 \tag{32}$$

or

$\mathrm{Nu}_D = 0.0455\mathrm{Re}_D^{0.81}$ when $\mathrm{Re}_D \geq 2.2 \times 10^7$.

For $k_s/D = 300 \times 10^{-5}$, use smooth cylinder correlation, equation (31) when $\mathrm{Re}_D \leq 1.8 \times 10^5$, or

$\mathrm{Nu}_D = 0.0135\mathrm{Re}_D^{0.89}$ when $1.8 \times 10^5 < \mathrm{Re}_D < 4.0 \times 10^6$

or

$\mathrm{Nu}_D = 0.0455\mathrm{Re}_D^{0.81}$ when $\mathrm{Re}_D \geq 4.0 \times 10^6$.

For $k_s/D = 900 \times 10^{-5}$, use smooth cylinder correlation, equation (31) when $\mathrm{Re}_D \leq 1.0 \times 10^5$, or

$$\mathrm{Nu}_D = 0.0455\mathrm{Re}_D^{0.81} \quad \text{when } \mathrm{Re}_D > 1.0 \times 10^5. \tag{34}$$

The natural convection coefficient is calculated from the turbulent flow relation obtained by Siebers et al. (1982) for smooth plane surfaces,

$$\mathrm{Nu}_H = 0.098\mathrm{Gr}_H^{1/3}\left(\frac{T_w}{T_\infty}\right)^{-0.14}, \tag{35}$$

and adjusted for the extended surface of the ribbed cylinder by

$$h_{nc}(\text{rough}) = \left(\frac{\pi}{2}\right)h_{nc}(\text{smooth}). \tag{36}$$

The result of this calculation applied to the receiver of Solar One at Barstow is shown in figure 8.17. Note that the roughness of the cylinder is quite important and that, for wind speeds above about 2 m/s, the heat transfer is dominated by the forced convection field. In an analysis of the uncertainties of the prediction, Siebers and Kraabel consider both the impact of uncertainty in the values of the parameters in the analysis (wind speed, temperature, etc.), which are combined by the traditional method (Kline and McClintock 1953), and the unknown or indeterminate effects due to the differences between the prototype environment and experiments. Figure 8.18 shows the result of the uncertainty analysis for the Solar One receiver operating at 400°C. Although random uncertainties are likely to lead to a $\pm 35\%$ error band, the unknown effects may result in losses 70% greater than predicted. The greatest unknown is the effect of free stream turbulence in the outdoor environment. Uncertainty in the treatment of roughness and end effects also contributes to the lack of confidence in the result.

Low temperature convective loss measurements from the actual receiver panels of the Solar One Plant at Barstow have been reported by Stoddard and Evans (1984). These results confirm the method of Siebers and Kraabel

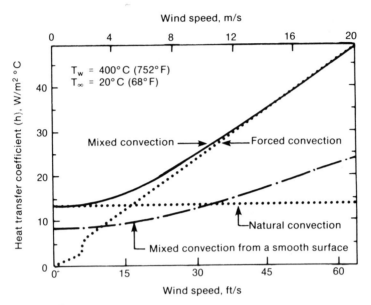

Figure 8.17
The heat transfer coefficient versus wind speed for the Barstow receiver at $T_w = 400°C$. Source: Siebers and Kraabel (1984).

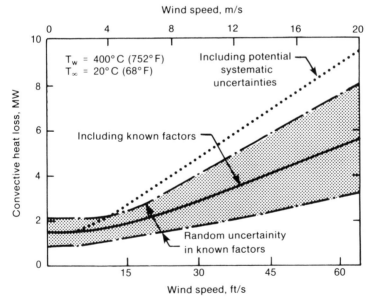

Figure 8.18
The uncertainty in the total convective heat loss versus wind speed from the Barstow receiver at $T_w = 400°C$. Source: Siebers and Kraabel (1984).

within the expected accuracy of the experiment and the analysis. The experiments, however, were at such low temperature differences that the forced convection dominated the flow field. Higher temperature experiments have been carried out recently and the results correlated by Boehm (1987). Siebers and Kraabel also propose a method for dealing with cavity receivers such as the International Energy Agency's (IEA) Small Solar Power Systems Project in Almeria, Spain. A cavity receiver has one or more openings facing the heliostat field on to which the radiation is focussed. Internal panels absorb the radiation. The object is to suppress both wind-driven and natural convection. The geometry of cavities, along with the variability of the wind speed and direction makes the cavity heat transfer and fluid mechanics very complex. The thermally driven flow in the core entrains fluid from the exterior and, in the presence of wind, interacts with the external flow field to produce recirculating and separated flow regions.

The natural convection heat transfer from the interior of open cavities has been studied by Kraabel (1983) and by Mirenayat (1981). The results are correlated by Kraabel in the equation

$$Nu = 0.088 Gr^{1/3} \left(\frac{T_w}{T_\infty}\right)^{0.18} \tag{37}$$

for $10^5 < Gr < 10^{12}$.

Since the Grashof number enters to the $\frac{1}{3}$ power, the heat transfer coefficient predicted by this relation is independent of cavity size and apparently is relatively insensitive to cavity shape. Cavity losses appear to be somewhat sensitive to the area of the aperture. Although the sensitivity is not great—a reduction of aperture area of 67% in one experiment only resulted in a heat transfer reduction of 17% (Siebers and Kraabel 1984). The losses are dependent upon the location of the aperture relative to the cavity. A lip over the floor of the cavity has practically no effect, whereas a lip at the roof of the cavity decreases the heat transfer. The natural convection for a tilted cavity with upper and lower lips may be estimated from the equation

$$h_{nc} = h_{nc,0} \left(\frac{A_1}{A_2}\right)\left(\frac{A_3}{A_1}\right)^{0.63}, \tag{38}$$

where A_1 is the interior area of the cavity, A_2 is A_1 minus the area of the lower lip, and A_3 is the interior area below a horizontal plane passing through the top of the aperture. This equation is reported to be reliable within $\pm 5\%$ for downward tilted cavities with tilt angles less than 30 deg; for tilt angles greater than 30 deg the exponent should be 0.8 (Siebers and Kraabel 1984).

The influence of wind on the cavity has not been well established. Experiments at the Central Receiver Test Facility showed no systematic influence of wind (McMordie 1981), whereas limited data from the IEA cavity at Almeria (Kraabel 1983a, b, 1982) did show an influence. Siebers and Kraabel recommend estimating the additional heat transfer due to wind-driven convection as though the aperture were a smooth surface at the temperature of the interior. For a flat plate

$$\text{Nu}_W = 0.0287\text{Re}_W^{0.8}\text{Pr}^{1/3}. \tag{39}$$

Properties are evaluated at the film temperature [i.e., $(T_c + T_a)/2$]; the length scale is the aperture width, and the velocity is the free stream wind speed.

For cavity receivers the individual heat losses are calculated independently and added. Figure 8.19 shows a comparison of predicted and measured heat transfer based on aperture area for the Almeria cavity. Although the experimental and prediction uncertainties are both large, it appears that the sensitivity to wind speed is larger than predicted.

The current understanding of mixed convection from large, high temperature objects such as cylindrical and cavity solar receivers is far from complete. Prediction methods are available, but the uncertainties are large. Research is needed to determine the effect of atmospheric free stream turbulence, the roughness of tubular ribs, and the flow induced in cavities by unsteady external winds. Similarity will generally require full-scale testing for accurate determination of mixed flow heat transfer. Since full-scale testing is very costly and time-consuming, the challenge for researchers is to devise dependable ways of combining free and forced convection results. The use of numerical flow modeling is essential and needs greater attention (Abrams 1983).

8.4.2 Internal Heat Transfer

In this subsection the research results that pertain to all modes of heat transfer within the solar collector will be reviewed. Internal heat transfer may be characterized as belonging to one of two categories: heat transfer to the working fluid and heat losses to the environment. Heat losses eventually must result in external heat transfer, which was discussed in the preceding subsection. Here, however, we are concerned with the heat exchange between the absorbing surface and the collector envelope for those classes of collectors in which they are not the same. In terms of the analysis introduced earlier in this section, the heat loss terms all go into the overall heat loss coefficient U. The loss coefficient usually includes terms representing convection, radiation, and, sometimes, conduction that are combined in the manner described in chapter 4 (i.e., as series and parallel thermal resistance).

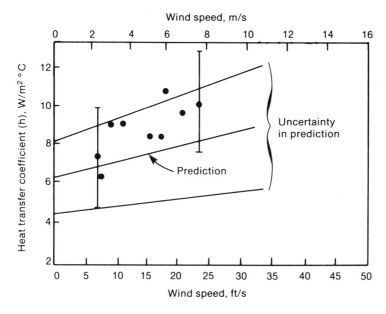

Figure 8.19
Comparison of the convective loss predictions and measurements for the cavity receiver of the International Energy Agency small solar electric program. Source: Siebers and Kraabel (1984).

The heat transfer to the fluid, on the other hand, is normally as direct as possible and therefore is usually represented by a single coefficient, h, that relates the heat transfer rate to the difference between the absorber temperature and the fluid temperature. The heat transfer to the fluid is normally a purely convective process, but depending upon the geometry, fluid and flow rate may involve a variety of fluid mechanical regimes (laminar, turbulent, transitional, entry region, fully developed compressible, incompressible, buoyant, single phase, multiphase, absorbing, transparent, reacting, etc.). The rich variety of flow regimes and physical processes that occur in solar receivers means that the relevant heat transfer literature is very large. Fortunately, as indicated earlier in this section, great accuracy is not often needed for the design of efficient collectors.

Internal Radiation   Radiation may be the dominant mode of heat loss for absorbers that operate at high temperatures or evacuated receivers in which conductive and convective losses are largely eliminated. However, for flat plate collectors at modest operating temperatures, the effective radiative heat transfer coefficient is normally less than the heat transfer coefficient for natural convection between the absorber and the cover glass, and this coefficient will be much smaller if the absorber surface is highly selective (McEligot and Bankston 1980). For glass covers, which are opaque to the infrared radiation from the low temperature absorber surface, the radiation is easily calculated by the methods shown in chapter 4. The situation is more interesting if one or more of the covers is made of a plastic that transmits some long wavelength radiation. Approximate equations for multiple plastic glazing were worked out by Whillier (1964). A more exact analysis of heat transfer across an arbitrary number of parallel flat plate solar collector covers, each of which can be partly transparent to long wavelength thermal radiation, was presented by Hollands and Wright (1983), and similar calculational procedures are used in the work of Buchberg and Edwards and their students, which is discussed later in this section in the subsection on convection. Owing to the simple geometry of flat plate collectors, there has been little additional research interest in the radiative heat transfer in flat plate collectors except for the determination of optical properties of absorbing and transmitting materials.

Although the importance of radiation losses in the overall heat balance increases with the temperature of the absorber, no really new technical issues arise, except of complex geometries, as long as the surfaces involved are opaque solids. Translucent solar ponds have been studied by Tabor and Weinberger (1980) and others, translucent plastic absorbers by Lumsdaine (1969) and others, and two-phase absorbers of a gas with entrained solid or liquid particles by Hunt (1981). When the temperatures are high, such as in the central receiver concept, the absorbing, radiating, and possibly reacting fluid presents a complex heat transfer problem.

Internal Conduction   Like radiation, the thermal conduction within most types of solar collectors has been well understood for a long time. Conduction is important to the performance in two ways: (1) most absorbers rely upon conduction to transport the energy from the absorbing surface to the convecting surface where it is removed by the working fluid—in this context the conduction path should offer minimum resistance; (2) a poorly conducting region usually separates the absorber from the collector envelope—in this context the thermal resistance should be large to minimize heat losses.

Heat conduction in absorbers is important in most flat plate collectors in which the coolant passages are attached to a sheet absorber. The geometry and boundary conditions are relatively simple, and closed-form analytical solutions are possible. Usually the heat conducted to the tube by the absorber is expressed in terms of a "fin" efficiency, as described in chapter 4. A simple solution for a plane, constant thickness fin with a constant heat loss coefficient gives

$$\eta_f = \frac{\tanh[m(W-D)/2]}{m(W-D)/2},$$ (40)

where $m^2 = U_L/\lambda t$, D is the tube diameter, W is the tube spacing, t is the thickness, and λ is the thermal conductivity.

Solutions for common fin geometries and boundary conditions were tabulated by Whillier (1967), and solutions for a variety of conditions are available in most heat transfer texts and solar reference books (by Jakob, Eckert and Drake, Duffie and Beckman, Kreith and Kreider, etc.). When the geometry is more complex or the temperature is high, so that the heat loss coefficient cannot be taken as a constant or the heat flux is nonuniform, as in concentrating collectors, it may be necessary to resort to numerical methods to determine the fraction of the energy absorbed by the receiver that is actually conducted to the fluid interface. Examples of detailed numerical analysis of nonuniformly illuminated absorbers may be found in the work of Bingman (1977) who analyzed the conduction in nonuniformly illuminated receivers of compound parabolic concentrators as part of the CPC development program at Argonne National Laboratory.

One aspect of the heat conduction problem that remains largely empirical is the thermal resistance between separate conductors in intimate contact (i.e., contact resistance). Contact resistances for a variety of methods of attaching cooling tubes to absorber plates was the subject of at least one project in the U.S. solar program. The final report of Mega Engineering's investigation of low-cost collector manufacturing techniques contains results for a variety of bonded and nonbonded tube-sheet joints (Dame 1979, 1980).

There have been a number of attempts to develop low-cost collectors using inexpensive materials—for example, concrete, concrete block (Payne and Brown 1979), ceramics, foam glass (Loth 1979a, b), volcanic scoria (Simpson 1977a, b), and various polymers as the absorber-heat exchanger (Nelson and Muller 1980; Muller et al. 1981; Leffingwell and Schwendeman 1979; Ball and Leffingwell 1979; Wildhelm 1981; Wildhelm 1982). Thermal analysis is the most critical element of the design of such collectors since temperature differences developed by the flow of heat through even very thin sections of some of these materials may result in unacceptable losses or in mechanical failure due to thermal stress.

The second aspect of heat conduction in solar collectors (i.e., heat loss reduction) has received more attention in the government solar program. Much of the research involved the investigation of materials and the development of improved thermal insulation and are covered in chapters 20 through 26. The thermal researcher, however, will want to be aware of the number and type of materials that are available and of their thermal properties. An extensive tabulation of the insulating materials available for low temperature application is contained in the final report of Versar Corporation's study of available thermal insulation materials for solar heating and cooling systems (Cheng and Gift 1980).

Perhaps the largest research and development effort related to internal heat conduction is the work on evacuated tubular collectors. The glass envelope of an evacuated collector is intended to suppress both convection and gas-based conduction. Since the convection is related to the Rayleigh number, which is proportional to the square of the absolute pressure, modest reductions in pressure (e.g., to 0.1 atm) effectively eliminate convection. Gas phase conduction, on the other hand, is independent of pressure in the continuum region, and the Fourier Law applies. As the pressure is reduced into the free molecular flow region (i.e., $< 10^{-4}$ Torr), gas conduction becomes insignificant. Evacuated collectors are generally produced with an initial pressure $< 10^{-5}$ Torr by a combination of evacuated bake out and gettering. However, glasses are not completely impervious to gas molecules. Borosilicate glasses in particular are known to be permeable to helium molecules. Thus, even though the concentration of helium in the atmosphere is very small ($P = 4 \times 10^{-3}$ Torr), over a long period of time the helium pressure within a borosilicate enclosure will build up, and since helium is an excellent conductor compared to the heavier gases, the heat loss may increase substantially above its initial level.

The buildup of helium and its effect on the performance of borosilicate glass collectors was the subject of two substantial government research projects. An experimental investigation by Owen-Illinois (Moan and Beekley 1980, 1982) and a theoretical investigation at Virginia Polytechnical Institute (Thomas 1979a, b, 1980; Wideman and Thomas 1980) provides a satisfactory resolution of this question.

Theoretical methods were developed for calculating the conduction heat losses from concentric cylinders and parallel plates over the full range of Knudsen number applicable to solar collectors. The Knudsen number is the ratio of the molecular mean free path, λ, to the characteristic dimension for conduction (e.g., λ/s for parallel plates separated by a distance s). Since the mean free path of helium at a mean temperature of 100°C and equilibrium pressures (4×10^{-3} Torr) is about 0.05 m, the Knudsen numbers of interest are in the range 2 to 20.

In the free molecule regime ($Kn > 10$), collisions between molecules are relatively rare. At much higher pressures ($Kn < 0.01$) the gas may be treated as a continuum, and conduction is independent of pressure (or Kn). Between these extremes lie two regimes: the transition regime ($0.1 < Kn < 10$) and the temperature jump regime ($0.01 < Kn < 0.1$).

Heat transfer in the free molecule regime is calculated from classical kinetic theory. A variety of solutions are available in standard texts—such as Bird, Steward and Lightfoot, and Kennard.

Heat conduction in a transition regime must be calculated from the Boltzmann equation for the molecular distribution function (which always requires approximation). Thomas (1979) and Wideman and Thomas (1980) developed calculational methods based on several approximations in the literature for both parallel plane surfaces and concentric cylinders. For example, for small T_1/T_2, Wideman and Thomas suggest that

$$q = \frac{Q}{Q_{fm}} = \frac{(R_1/R_2)[(1/\alpha_2) - 1] + (1/\alpha_1)}{(R_1/R_2)[(1/\alpha_2) - 1] + (1/\alpha_1) + (4/15)(R_1/\lambda_1)\ln(R_2/R_1)}, \tag{41}$$

where α_1 and α_2 are the thermal accommodation coefficients for the inner and outer concentric cylinders of radius R_1 and R_2, and Q_{fm} is the free molecule solution found in Kennard (or Thomas 1979a). When T_1/T_2 is greater than about 1.5, the complete nonlinear equation must be solved using the numerical method described by Wideman and Thomas. There is little experimental data for concentric cylinders (except for experiments with fine wires for which $R_1 \ll R_2$) and even less with appreciable T_1/T_2; however, based upon similar analysis and comparison to experiments for parallel plates, Thomas (1980) believes the method is accurate within about 10% over the Knudsen number range of interest.

Figure 8.20 shows a calculation of the conduction heat flux for typical evacuated collector geometry, temperatures, and emissivities. Even at the equilibrium helium pressure (i.e., 4×10^{-3}) the conduction is only about 25% of the radiation flux. Unfortunately, the calculation requires the thermal accommodation coefficient for each of the surfaces. The thermal accommodation coefficient can be thought of as the fraction of molecules striking a surface that acquires the thermal energy corresponding to the wall temperature. Accommodation coefficients are dependent upon the gas and the properties and history of the surface. Since they are known to be sensitive to absorbed surface layers, most of the values found in the scientific literature are for ultraclean surfaces, and for helium they tend to be quite low (0.01 to 0.1).

A few values for helium on normal surfaces are available including a value of

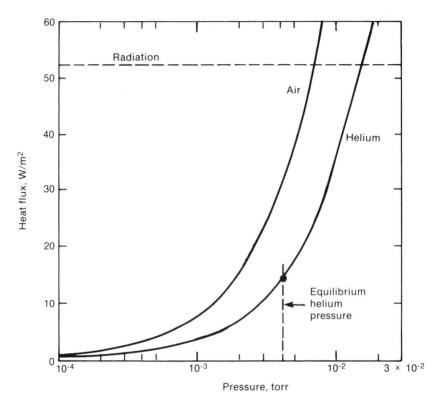

Figure 8.20
Heat conduction in air and helium in an evacuated tube collector. $T_1 = 240°F$, $T_2 = 68°F$, $\varepsilon_1 = 0.06$, $\varepsilon_2 = 0.90$. For air, $\alpha_1 = \alpha_2 = 0.95$; for helium, $\alpha_1 = 0.60$, $\alpha_2 = 0.40$. Source: Thomas (1980).

0.35 for helium on glass, and a value of 0.53 for helium on tungsten. No values were found for typical collector absorber surfaces.

It was hoped that the experiments conducted at Owens-Illinois would produce data from which the effective thermal accommodation coefficients typical of evacuated solar collectors could be determined. Aging and helium backfill experiments were conducted, and the results were analyzed. But the results were so inconsistent with theory that it was not possible to infer anything about the accommodation coefficients. The experimental results of Moan and Beekley did, however, yield useful information regarding the rate of permeation of atmospheric helium into the vacuum. The partial pressure in an evacuated region varies with time according to the relation (Thomas 1979a)

$$P(t) = P_0(1 - e^{-t/\tau}), \tag{42}$$

where P_0 is the helium partial pressure in the atmosphere (4×10^{-3} Torr at standard conditions) and τ is the time constant, given by

$$\tau = \frac{V}{S}\left(\frac{N_0}{RT}\right)\frac{\Delta x}{K},$$

in which N_0 is Avogadro's number; x is the wall thickness of a vessel of volume V, surface area S, and temperature T; and K is the permeation rate constant.

The theoretical temperature dependence of τ is

$$\tau = \frac{C_1}{T}\exp\left(\frac{C_2}{T}\right). \tag{43}$$

These relations were used to interpret the borosilicate glass tube aging results of Moan and Beekley (1982). Figure 8.21 shows the experimental results at tube temperatures from 600° to 800°F (315°–425°C). The curve represents an empirical data correlation using the two highest temperature points to evaluate the constants C_1 and C_2:

$$\tau = \frac{136}{T}\exp\left(\frac{1940}{T}\right), \tag{44}$$

where τ is in years and the temperatures are Rankine. This fit is clearly the most conservative (pessimistic) interpretation of the results. For example, at 700°F (370°C) it predicts a time constant of 0.62 years compared to the mean value from the experiments of one year. Based on equation (44), Thomas estimates that all borosilicate glass evacuated collectors operating at 240°F (116°C) for an average of 8 hours would have a helium pressure of 3×10^{-3} Torr in about 4.5 years. At this pressure the total heat loss would increase by about 20%. The envelope of evacuated collectors with an internal metal absorber generally operates at much lower temperatures and would be less vulnerable to helium penetration. Ordinary soda-lime glass is much less permeable to helium than borosilicate glass, so collectors made from that material could be expected to be virtually unaffected by helium buildup (Thomas 1979).

To summarize the accomplishments regarding helium buildup in evacuated collectors and its effect on heat loss, an accurate method is available for calculating losses in both plane and cylindrical geometries; but reliable values for thermal accommodation coefficients are not available for typical solar absorber surfaces. Applications of the theoretical analysis to typical collector geometries using experimental permeation rates, but assumed thermal accommodation coefficients (believed to be conservative), indicate that the increase

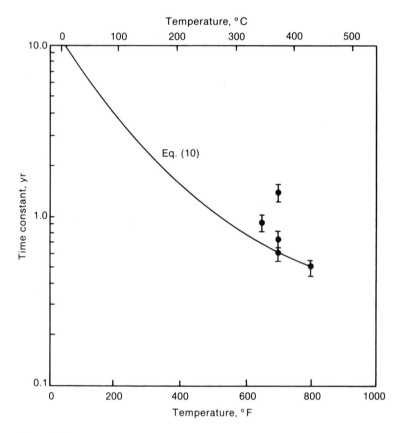

Figure 8.21
Temperature dependence of time constant for helium permeation. Source: Thomas (1980).

in heat losses at equilibrium would be on the order of 20%, and that would take four to five years for borosilicate glass. Reliable measurements of thermal accommodation coefficients for real collector absorber surfaces (selective coatings) would be desirable additions to the understanding of evacuated collector behavior but would require a rather expensive experimental program. The theoretical and experimental results available to date do not indicate that helium penetration is likely to be a life-limiting factor for evacuated collectors.

Internal Convection   Convection within a solar collector includes the heat transfer to the collector working fluid and the heat transfer to closed systems within the collector envelope (e.g., the convection from absorber to cover). In terms of the earlier analysis in this section, the heat transfer to the working fluid is represented by the heat transfer coefficient h, while the heat transfer to internal fluids is a component of the overall heat loss coefficient U. As shown earlier, the collector efficiency and temperature distribution depends on the ratio U/h, $UA/\dot{m}c_p$, and $U(t_i - t_a)/I$. Thus it is usually more important, in terms of prediction or improvement, to control U. For many collector designs, the U/h ratio is inherently small and, as shown in figure 8.15, the efficiency changes little between $U/h = 0.1$ and $U/h = 0$. However, when U and h are of the same order of magnitude, as they often are in air-heating collectors, the performance is quite sensitive to U/h, and substantial improvement may be achieved if the effective h can be increased. Also the loss coefficient is usually made up of a convective part and a radiative part (see chapter 4) and is therefore temperature dependent. Since the temperature of the absorber surface at any point depends on the fluid temperature and h, the loss coefficient also depends upon h. The research relevant to the determination and enhancement of h will be reviewed in the next subsection. Research that pertains primarily to U will be addressed in a subsequent section.

FORCE CONVECTION HEAT TRANSFER COEFFICIENT, h Most liquid heating collectors employ channel geometries and fluid conditions for which the heat transfer characteristics are well known and generally available in the literature—often in standard heat transfer texts. Therefore there has been little heat transfer research explicitly for solar applications. A review of the design literature, however, indicates that solar designers are often unaware of the modern heat transfer literature and frequently use archaic relations or apply relations incorrectly. Therefore a few general remarks about available heat transfer relations may be in order. A good text on convection such as Kays (1966) may be consulted for additional detail.

Heat transfer correlations relating the Nusselt number to the Reynolds number, Prandtl number, and nondimensional position are available for common channel sections, including circles (pipes), rectangles, triangles, and parallel plates; for both laminar and turbulent flow, for various thermal boundary conditions; and hydraulic entry conditions (Kays 1966). In all cases the Nusselt number is hD_h/k where D_h is 4 × flow area/perimeter. Note that the Nusselt numbers depend upon the shape and the type of boundary condition (i.e., constant temperature or constant heat flux, one heated surface or two). For laminar flow the heat transfer in the thermal entry region is generally correlated in terms of the nondimensional length parameter

$$x^+ = \frac{x/r_0}{\text{Re}\,\text{Pr}}.\tag{45}$$

A typical variation of local Nu compared with x^+ for a circular tube with constant surface temperature is shown in figure 8.22. Series solutions and numerical results are available for many boundary conditions. Note that the local Nusselt number approaches the fully developed value within about 1% for $x^+ = 0.10$, but the mean Nusselt number

$$\text{Nu}_m = \frac{1}{x^+} \int_0^{x^+} \text{Nu}_x \, dx^+\tag{46}$$

is still 27% greater than the fully developed limit.

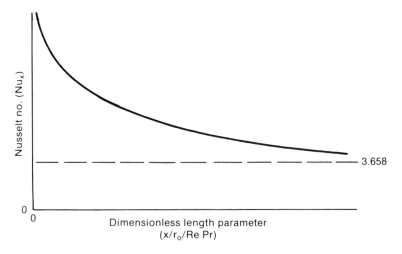

Figure 8.22
Variation of local Nusselt number in the thermal entry region of a tube with constant surface temperature. Source: Kays (1966).

For water at $100°F$ ($Pr \simeq 4.5$) and Reynolds number $= 1,500$, $x^+ = 0.1 \Rightarrow$ $x/D_h = 0.05 \times 4.5 \times 1,500 = 337.5$ or an entry length of 3.4 m for a 1-cm tube. The error resulting from neglecting thermal and hydraulic entry lengths in solar collector heat transfer calculation can be substantial as this example illustrates.

Similar results are available for turbulent flow, although they are more often empirical equations based on experiments. Turbulent flows are more likely to be encountered in concentrating collectors even when the working fluids are oils.

Most of the convective heat transfer research effort for flat plate solar collectors had the objective of improving h for air heating collectors. The Government collector R&D program had at least six projects whose primary or secondary goals were to improve the absorber-to-air heat transfer effectiveness: Honeywell (Rask and Mueller 1977; Rask 1979), UCLA (Buchberg and Edwards 1980), Northrup (Mattox 1979), Purdue (Pearson 1983), Solaron (Cole-Appel et al. 1978), and LANL (Neeper 1979). Rask, at Honeywell, employed and tested an absorber plate that was cooled by an array of normal impinging air jets. Cole-Appel et al. at Solaron investigated various heat transfer augmentation approaches including boundary layer trips and extended surfaces (fins). Buchberg and his colleagues at UCLA investigated a variety of transpired absorbers including porous plates, packed beds, and louvered plates. Collier (Collier and Arnold 1980) and Neeper (1980) at Los Alamos studied transparent and translucent porous absorbers, and Pearson and his coworkers at Purdue investigated corrugated, finned, and louvered absorbers. Mattox, at Northrup, conducted a theoretical and numerical evaluation of the various techniques that have been suggested for augmenting absorber plate heat transfer in air-cooled collectors.

Most of the experimental investigations cited above were conducted with full-scale or model-collector absorber plates that were tested in much the same way as a solar collector. Often the experimental devices embodied a number of efficiency-enhancing modifications. This makes determination of cause and effect rather difficult. All of the investigators found it possible to increase h above that for a smooth flat plate, as one would expect, but there is always a price—either in increased materials and fabrication costs or in increased friction—which results in increased operating cost. It appears, on the basis of most of the evidence, that rather simple back fins or plate louvers may be most cost effective, but there are so many trade-offs involved, that firm conclusions are difficult. We shall review the heat transfer findings without too much regard for their cost implications.

Air-heating collectors attracted considerable attention prior to the establishment of a government research program. De Winter's annotated bibliography on flat plate solar collectors (de Winter 1975) cites the work of George Löf and others (Löf 1946, 1950, 1954; Löf and Nevens 1953; Daniels and Duffie 1955) on the development of a novel air heater in which air is transpired through a staggered stack of partly transparent, partly blackened glass plates. The concept reduces losses to the cover by separating the warmed airflow from the colder air near the cover, and it enhances the convective heat transfer coefficient by restarting the thermal boundary layer on each plate.[2] The fluid mechanics and thermal analysis of the Löf air heater are quite complex (Selcuk 1971) and have not been verified through detailed temperature and velocity measurement.

Other transpired air heaters were studied before 1970. Bliss (1956) used a blackened screen as the absorber, and similar concepts were studied by Whillier (1964), Gupta and Garg (1967), and Chiou et al. (1965). In Australia, Close (1963) conducted early studies of air-heating collectors.

Edwards and Buchberg and their coworkers and students at UCLA began a serious and extended investigation of air-heating collectors in the midsixties (Edwards and Leung 1964) which is still active (Rhee and Edwards 1983) and was partly supported by the government solar programs. The UCLA approach has usually been to combine fundamental theoretical or numerical analysis of the fluid mechanics and heat transfer on specific geometries with the construction and testing of prototype solar collectors in an outdoor environment. Thus the emphasis is on the performance of the collector rather than the internal heat transfer processes, and the most significant accomplishments—in addition to a number of well-trained solar engineers and researchers—has been to establish a performance hierarchy that includes a variety of transpired and solid absorber surfaces, flow arrangements, glazings, and optical characteristics. It appears that porous beds offer the greatest improvement in internal heat transfer (because of the greatly increased surface area) (Galanter 1980), but other transpired absorbers, including slotted and microperforated plates (Rhee and Edwards 1983), achieve nearly the same effectiveness. The performance of porous bed absorber collectors could be adequately predicted using a model that treated the bed (in this case a fibrous matrix) as an isothermal bed with a heat exchanger effectiveness of unity. Performance was also accurately predicted by the more detailed simultaneous conduction, convection, and radiation model of a porous semi-infinite, gray, nonscattering medium originally developed by Weiner and Edwards (1963) and modified by Edwards and Leung (1964), but it was necessary to include an allowance for recirculation of warm air through the porous bed. A recirculation fraction of 0.2 (20% recirculation) was required in order to achieve good agreement with experimental results. The calculated collector efficiency with a recirculation fraction of 0 was about 20% above the data at $[(T_L + T_0)/2 - T_A]/I = 0.08°C$ m^2/W and about 8% high at the $\Delta T = 0$ intercept. The fact that the more complete model predicts a higher efficiency for a medium without circulation than does the isothermal model or the experimental data suggests that there may be even more improvement available if the recirculation is controlled. This outcome would be expected since the more complete optical modeling assures that the top surface of porous absorber operates at a lower temperature than the isothermal model temperature and hence has lower heat losses. Galanter (1980), however, does not suggest any means of eliminating the recirculation.

Absorber plates with slots or microperforations were investigated by Rhee (1980), who also performed theoretical studies and prototype collector testing. Numerical solutions were obtained for four basic sets of boundary conditions which could be combined to represent a variety of uniform thermal boundary conditions. The boundary conditions that apply to flat plate solar collectors are generally those of a semiporous cavity (one side permeable and one side impermeable to flow) with either continuous aspiration (porous beds, screens, etc.) or intermittent aspiration (slots, louvers, etc.). The boundary condition on the impermeable wall can be isothermal, adiabatic, or mixed, and the boundary condition at the permeable wall can be specified heat flux or temperature. Unfortunately, numerical instabilities limited the solution for intermittent wall aspiration to wall Reynolds numbers $v_w D_h/v < 10$ to 15.

Rhee constructed and tested five model solar collectors to validate the semi-porous channel analysis and to determine which configurations possess the greatest technical merit. The experimental absorbers included an impermeable reference panel, a slotted plane panel, a microperforated plane panel, a plane corrugated panel, and a slotted corrugated panel. All models had two glass covers, and the absorber surfaces were Highland Platings black chrome ($\alpha = 0.965$, $\varepsilon = 0.16$). The two transpired absorber models also had a Teflon sheet between the absorber and the inner glazing to confine the entering flow. Careful comparison of the impermeable and the microperforated absorber model test results with the theoretical prediction confirmed the validity of the model within 2% to 3% (of collector efficiency) and established confidence in the experimental findings. However, the test results showed that the slotted transpired plane absorber performed as well as the microperforated absorber and better than the other configurations. In addition the pressure drop through the slotted plate was much lower than the pressure drop through the microperforated plate.

The experiments at UCLA on various transpired absorbers are a bit difficult to summarize because there are so many variables. In addition to the fact that the experiments were conducted over a period of time by different investigators, the nature of the absorber dictated some differences in configuration. Some of the model collectors required Teflon inner glazing, some had selective surfaces, some had additional convection suppression devices, and so on. The reviewer's perceptions of the conclusions that may be drawn from this excellent research are as follows: Porous or transpired absorber plates increase the effective h substantially and improve collector efficiency at operating temperatures by about 20%. The question of which type of porous absorber is best is very difficult to answer even in the context of thermal performance alone (i.e., without considering cost or practicality). Metal absorbers with slots or microperforations can be coated with selective absorber coatings to reduce radiation losses. Porous beds such as wire mesh, fibrous materials, or crushed glass can in effect be made selective because the emitting surfaces may operate at lower temperatures than the average absorber temperature, but recirculation of warmed air through the bed may reduce potential gains. Performance gains for all transpired absorbers are greatest when the entering air is near ambient temperature. When the inlet temperature is much above ambient, an interior glazing is a necessity. If the interior glazing also had a high IR reflectance (i.e., a heat mirror), the nonselective porous bed might be a better performer.

From practical consideration one is tempted to conclude that a slotted or louvered absorber with a selective coating, a single glass cover, and one or two internal Teflon films (depending on operating temperature) is the most attractive collector configuration of those studied at UCLA. In addition, although there are still some outstanding problems, the theoretical modeling capability appears to be adequate for most porous absorbers.

The research conducted at Purdue and supported by the government solar program had the specific objectives of surveying the applicable heat transfer and fluid mechanics information relevant to flat plate air heating collectors and conducting studies and experiments leading to performance improvements (Pearson 1983). Their analysis of the literature led them to concentrate on two types of absorbers for which existing information seemed incomplete: transpired absorbers made of metallic louvers and impermeable absorbers with some form of extended surface. The results of the literature survey are contained in progress reports (Pearson 1979, 1980) and Purdue theses (Christopher 1982; Diab 1981; Shockey 1981). The literature survey and the preliminary theoretical work that followed it led to the conclusions that

1. for collectors having no airflow above the absorber, the back side of the absorber should have an extended surface;

2. for collectors having airflow above the absorber, the absorber should be louvered so that only unheated air flows above the absorber and only heated air flows behind it;

3. porous matrix surfaces offer no advantage over louvered surfaces;

4. corrugated absorbers offer little advantage.

Four arrangements were selected for further analysis and experimentation: a louvered absorber, a back corrugated absorber, a continuous finned absorber, and an interrupted fin absorber. All of these configurations resulted in improved thermal efficiencies, by 12% to 25%, relative to the flat plate with back flow only, at approximately the same thermal and mechanical condition (i.e., $\dot{m}$ and Δp). This improvement in efficiency is a result of a three- to four-fold increase in the effective conductance. Most of the improvement in the aforementioned absorber configurations comes from increases in the convection surface area. The louvered absorber and the interrupted fin absorber also benefit from high $\bar{h}$ as a result of the interruptions of the thermal and hydraulic boundary layers (Pearson 1980; Shockey, Pearson, and DeWitt 1983; Christopher and Pearson 1980).

Once again cost and other practical considerations may govern the choice to a greater extent than the relatively minor differences in performance. The extended surfaces need to be adequate conductors in good contact with the absorber (the Purdue studies indicate thin aluminum or steel may be used, and cement bonding is adequate if the contact surfaces are large). The louvered absorber may be mild steel, and the thickness is determined by structural rather than thermal requirements.

Solaron Corporation, under contract from DOE, also conducted a study of improvements of solar air heaters. The study consisted of a literature review followed by the design, construction, and testing of prototype collectors employing the most promising absorber configurations from the literature (Cole-Appel et al. 1978). The Solaron work excluded any configurations with flow above the absorber. Three configurations were tested, in addition to the flat plate reference collector: a flat plate collector with boundary layer trips, a flat absorber with extended surface on the back, and a dimpled plate absorber. All configurations were tested at the same two flow rates (0.01 m^3/m^2s and 0.02 m^3/m^2s). The efficiency improvement ratio for both the extended surface absorber and the dimpled absorber was 1.11, but the boundary layer trips only improved the efficiency by 3% (the channel Reynolds number was in the transition range). Since the dimpled absorber also increased the pressure drop by a factor of 7, the finned absorber was chosen for cost studies. The Solaron findings seem to be in substantial agreement with the Purdue results.

A theoretical analysis of heat exchange effectiveness was conducted by Northrop Services for DOE (Mattox 1979). Mattox used a transient numerical finite difference code to compute the dynamic and steady-state response of various collector absorber configurations to step changes in solar insolation. Smooth ducts, continuous fin absorbers, and interrupted fin absorbers were studied, and a very substantial advantage was claimed for the interrupted fin. Mattox used the interrupted fin design utilized in the University of Delaware experiment on their "Solar One" house (Kuzay et al. 1971) and the continuous fin design of Solaron (Cole-Appel et al. 1978), but in his evaluation of effective convective conductance, he found much higher values for the interrupted fin geometry than those found by Cole-Appel et al. (1978) or Pearson (1983) (27.9 cf. 5–7 Btu/h ft^2 °F). These high conductances are not supported by the Purdue experiments (Pearson 1983).

Hollands and Shewen (1981) reexamined the optimization of flow passage geometry for air-heating collectors. Hollands argues that the mass flow rate and pressure drop are essentially fixed by system requirements—that is, $\dot{m} = 0.01$ kg/m^{2}s in order to assure an adequate temperature rise for space heating, and $\Delta p \approx 60$ Pa for practical reasons including blower availability, leak control, and flow balancing. For impermeable plates Hollands shows that the fixed pressure drop and mass flow requirements dictate the relation between the channel height and its length so that the designer is only free to select one characteristic dimension. He provides the detailed methodology and equations for finding the dimension of parallel wall channels and V-corrugated channels. Hollands finds that substantial improvements in efficiency (of the order of 10%) are possible with flow passages less than 0.5 m long.

In the same study Hollands reports a gain of 7% to 12% for V-corrugated channels compared to parallel walls, and an increase of about 3% as a result of increasing the emissivity of the back surface of the absorber and the back plate from 0.20 (bare metal) to 0.9 (black paint). However, if only one surface is painted, the increase in efficiency is only about 0.5%.

HEAT LOSS COEFFICIENT U The dominance of the heat loss coefficient in determining the overall performance of solar collectors was illustrated at the start of this section, and subsequently the role of internal radiation was discussed. At the relatively low operating temperatures of flat plate liquid and air-heating collectors, it is usually the natural convection between the absorber surface and the glazing(s) that dominates the heat loss. With the advent of effective selective surfaces for absorbers and IR-reflective coatings for glazing, the convection remains important even at high temperatures. The important role of natural convection in collector performance and the fact that the inclination of the convection zone (because of collector tilt) sets the solar application apart from most areas of traditional scientific and industrial interest in natural convection has resulted in an impressive amount of heat transfer research aimed at understanding and controlling free convection in the air spaces within solar collectors.

The early treatment of the free convection coefficient between a collector absorber and its glazing (Hottel 1954; Whillier 1967) included an empirical relation for the tilt angle dependence of the collector and for the temperature difference between the heat transfer surfaces. Hottel indicated that glazing spacing greater than 0.5 inch (1.3 cm) had little effect on reducing the heat loss. With nonselective absorbers and uncoated glazing, radiation was about as important as convection, so the gap spacing was less critical. Measures for reducing the convective losses were introduced in the 1960s by Francia (1962) who suggested a deep honeycomb cover assembly for high temperature collectors. Honeycombs were also investigated by Hollands (1965), and Perrot et al. (1965) in the 1960s and by UCLA researchers in the early 1970s (Buchberg et al. 1971; Lalude and Buchberg 1971).

The pertinent research for convective heat transfer in inclined enclosures similar to those in flat plate solar collectors was reviewed by Buchberg, Catton, and Edwards (1976). Their review dealt primarily with high-aspect ratio spaces inclined at angles up to 60–70 deg above horizontal. The important aspects of such convective problems are

1. For a horizontal enclosure heated from below by an isothermal surface, the heat transfer is pure conduction for Rayleigh numbers less than the critical Rayleigh number, 1,708, and the Nusselt number is unity (a purely conducting layer with no convection is difficult to obtain in practice). When the Rayleigh number exceeds 1,708, a multicellular convection is established consisting of an integral number of "roll cells" whose axes are parallel to the shorter dimension of the enclosure. The Nusselt number is well represented by the theoretical expression of Malkus and Veronis (Ozoe et al. 1982a):

$$\mathrm{Nu} = 1 + 1.446\left(1 - \frac{1,708}{\mathrm{Ra}}\right)S\{1,708\}$$

$$+ 1.664\left(1 - \frac{17,610}{\mathrm{Ra}}\right)S\{17,610\}, \tag{47}$$

where $S\{1,708\} = 0$ for Ra $< 1,708 := 1$ for Ra $> 1,708$ and $S\{17,610\} = 0$ for Ra $17,610 := 1$ for Ra $> 17,610$.

2. When the surface is tilted at any angle greater than zero, a unicellular base flow is established in which the direction of circulation is up the lower heated surface and down the cooled surface above. The Nusselt number for this base flow is greater than one, but represents a negligible heat loss for aspect ratios in the range 20–200 (Buchberg, Catton, and Edwards 1976). However, as the Rayleigh number increases due to larger spacing or greater temperature differences between the inclined surfaces, the base flow breaks up into a multicellular structure similar to the supercritical flow between horizontal surfaces. It is this flow that significantly augments the heat loss.

3. Clever (1973) showed theoretically that for infinite Prandtl number fluids, the g (acceleration of gravity) in the critical Rayleigh number could be replaced by $g \cos \phi$. Hollands et al. (1975) showed theoretically that the scaling could also be applied to fluids with finite Prandtl number. This confirmed the experimental observation of Hollands and previous investigators. Nusselt number correlations in the supercritical flow regime for inclined spaces were advanced by several researchers (Tabor 1958; Buchberg, Catton, and Edwards 1976; Hollands et al. 1975). According to Buchberg, Catton, and Edwards (1976), all the correlations are adequate for prediction of heat losses and the optimization of convective spaces in solar collectors. The correlation of Hollands employs a single equation:

$$
\mathrm{Nu} = 1 + 1.44 \left[1 - \frac{1{,}708}{\mathrm{Ra} \cos \phi} \right]^{\bullet} \left(1 - \frac{(\sin 1.8\phi)^{1.6} \, 1{,}708}{\mathrm{Ra} \cos \phi} \right)
$$

$$
+ \left[\left(\frac{\mathrm{Ra} \cos \phi}{5{,}830} \right)^{1/3} - 1 \right]^{\bullet}. \tag{48}
$$

Note that the dotted tracked terms are zero when negative—that is, $[x]^{\bullet} = (|x| + x)/2$; ϕ, the tilt angle, is greater than 0; and that all properties are evaluated at the average gap temperature.

4. Optimization of convection spaces in collectors, based on the correlation above, leads to the conclusion that air gaps should be 4–8 cm, which is substantially more than suggested by Hottel. In practice, of course, trade-offs must consider absorber shading and cost.

Buchberg, Catton, and Edwards also reviewed the status of convection suppression structures based upon their own work and that of Francia (1961), Perrot (1967), Hollands (1975), and Charters and Peterson (1972). The basic premise of suppression is to reduce the characteristic dimension of the convection cell below the critical value (i.e., the dimension for which $R_a = R_{a_c}$) Buchberg et al. (1968) suggested that rectangular cells with their long dimension running east–west would give the best optical performance and also suggested the cell width d should be about $\frac{1}{2}$ cm and the thickness of the structure, L, should be three to eight times d for absorbers operating near 100°C. In subsequent works the effect of radiation, wall conduction, and tilt were shown to be significant. Buchberg, Catton, and Edwards presented a design method in their 1976 review that accounts for radiation and conduction. The method leads to a value of d that gives Nu = 1 for horizontal collectors and Nu = 1.25 for tilted ones.

The importance of understanding and controlling the natural convention process in reducing collector heat losses was recognized early in the government solar programs (Balcomb and Perry 1977), and a number of projects related to the measurement, analysis, or suppression of convection were supported. In addition to the heat transfer projects reviewed below, there were at least two materials R&D projects aimed at developing cost-effective honeycomb structures. Lockheed (Marshall 1977; McCargo 1979) worked on innovative methods of fabricating all glass honeycombs, and Battelle Northwest Laboratories (Gordon 1979; Chicalla 1979) worked on the development of polymer structures.

The University of Waterloo, Canada, entered the U.S. solar program in 1974 when Hollands received a research grant from the National Science Foundation for "Studies on Methods for Reducing Heat Losses from Solar Collectors." Subsequent grants and contracts authorized additional work on the effectiveness of various methods of reducing heat losses and the optical characteristics of such devices. The primary emphasis of this program has been the careful experimental determination of the heat transfer between isothermal plates as functions of the intervening convection suppression structures, the orientation of the plates, and the Rayleigh number. The experimental apparatus is housed in a pressure vessel so that the Rayleigh number is varied by controlling the pressure rather than the temperature difference between surfaces—a characteristic that distinguishes the Waterloo equipment from most free convection experiments. It was a very productive program with a large number of publications. A drawing of the aparatus is shown in figure 8.23.

The apparatus was automated so that a programmable calculator controlled the experiment and processed the data. The calculator was programmed to adjust the "guard heater" current appropriately and to increment the pressure to adjust to a new Rayleigh number. A single configuration of plates and spacers (i.e., honeycomb) was run through a series of about ten different Ra by adjusting the pressure from well below 1 atm to about 300 psia. Since the temperature changes resulting from changes in Ra are small, equilibrium was achieved quickly (~ 30 min) and a series could be completed

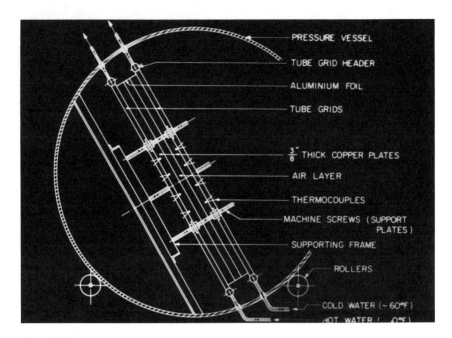

Figure 8.23
Hollands' natural convection aparatus: internal view.

in one day. An example of the experimental results from this apparatus is shown in figure 8.24, in which the heated surface is a V-corrugated (60) plate separated from a plane-cooled plate by a distance of four times the depth of the V-corrugations.

The principal results of the Waterloo research are a large data base of experimental heat transfer results specifically applicable to solar collectors and appropriate correlations and methods for the designer (Hollands et al. 1978, 1979). The experimental results and correlation for inclined plane iso-thermal surfaces of large-aspect ratios have already been cited in equation (48). Additional results have included a comprehensive study of V-corrugated surfaces that lead to the correlation

$$
\mathrm{Nu} = \mathrm{Nu_c} + K\left[1 - \frac{\mathrm{Ra_c}}{\mathrm{Ra}\cos\theta}\right]^{\bullet}\left(1 - \frac{\mathrm{Ra_c}(\sin 1.8\theta)^{1.6}}{\mathrm{Ra}\cos\theta}\right)
$$

$$
+ \mathrm{B}\left[\left(\frac{\mathrm{Ra}\cos\theta}{\mathrm{Ra_t}}\right)^{1/3} - 1\right]^{\bullet} \tag{49}
$$

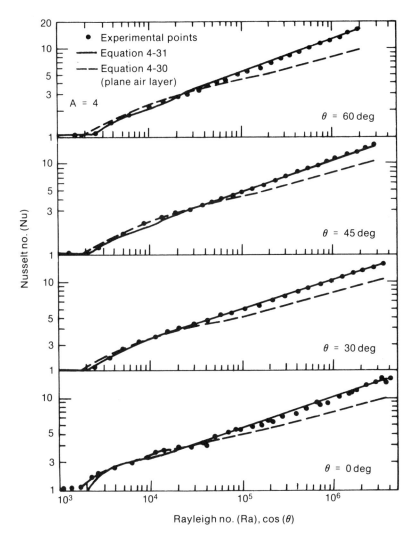

Figure 8.24
Measurements of Nusselt number as a function of $\mathrm{Ra}\cos\theta$ for $A = 4$. Source: Hollands et al. (1979).

where

$$[x]^{\cdot} = \frac{|x| + x}{2},$$

$$\text{Ra}_c = 1{,}708\left[1 + \frac{0.0360}{A} + \frac{2.69}{A^2} - \frac{1.703}{A^3}\right], \tag{50}$$

$$K = \frac{2{,}459}{\text{Ra}_c}\left[1 - \frac{0.195}{A} + \frac{5.97}{A^2} - \frac{4.16}{A^3}\right], \tag{51}$$

$$B = 2.225 - 0.01226\theta + 0.340 \times 10^{-3}\theta^2, \tag{52}$$

$$\text{Ra}_t = 11{,}290\{1 + 0.204\sin[4.50(\theta - 37.8°)]\}. \tag{53}$$

The experiments also showed that corrugation can increase the heat transfer coefficient substantially if the spacing is small, but V-corrugation (and presumably other forms of surface roughness) has little effect on the natural convection if the ratio of the roughness height to the separation distance is less than 0.25 and the Rayleigh number is smaller than about 10^5.

A number of convection-suppression configurations were also studied theoretically and experimentally at Waterloo. One of the major findings was that horizontally slit honeycombs show greater convection-suppression capabilities than square-celled honeycombs using the same amount of material for tilt angles more than 60 deg. For tilt angles less than 45 deg, the horizontal slits are significantly less effective than the square cells. Hollands also analyzed the effect of conduction and radiation on the convective heat transfer in honeycomb cells and found a dependence that had not been hypothesized by previous investigators. Specifically, Hollands' analysis showed that if either the hot or the cold plate emissivity is low, the coupling with the honeycomb wall temperature enhances the convective heat transfer. He also showed that the critical Rayleigh number for tilted honeycombs can be calculated by dividing by cos ϕ, provided the platform aspect ratio is reasonably large (as for horizontal slits).

The studies conducted by the University of Wisconsin are unique among those conducted within the solar program in that interferrometric techniques were used to determine local heat transfer coefficients, and correlations were developed for both local and average Nusselt numbers (El-Wakil and Mitchell 1979).

The results were categorized as belonging to one of three heat transfer regimes: conduction (Ra < Ra$_c$), boundary layer transition, and fully developed. In the conduction regime the Nusselt number is high in the starting corner and low in the finishing corner, and it has a constant value of unity over most of the channel. Likewise the fully developed regime exhibits a high Nusselt number at the low end of the cavity and a reduced Nu at the upper extreme, and it is relatively constant over the central portion. The penetration depth for the starting and departure corners is symmetrical and an increasing function of tilt angle. The Nusselt number at the center of the channel is equal to the average Nusselt number. This is important since it supports the use of Hollands' results, which were measured near the center of a large aspect ratio channel, for smaller aspect ratios.

Figure 8.25 shows the variation of local Nusselt number with nondimensional distance along the cavity for three aspect ratios with a tilt angle of 90 deg (vertical) and a Grashof number of 93,000 for heat transfer in the transition regime.

The transition regime begins at the critical Rayleigh number, $1,708/\cos\phi$, but the upper boundary depends upon both Ra and the aspect ratio. As

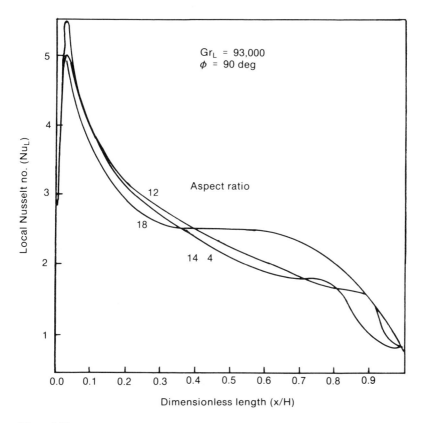

Figure 8.25
Local heat transfer coefficient along the surface of a vertical enclosure for laminar boundary layer regime. Source: Randall et al. (1979).

indicated by figure 8.25, either high Ra or high aspect ratios drive the flow into the fully developed boundary layer regime.

Average Nusselt numbers were obtained by integration of the local values over the channel length. Randall et al. (1979) proposed a correlation for the average Nusselt number of

$$\overline{Nu}_L = 0.118[Gr_L Pr\cos^2(\phi - 45)]^{0.29}, \tag{54}$$

where L is the plate separation and ϕ is the tilt. This equation correlates the Wisconsin data within $\pm 8\%$ for $Pr = 0.7$ (air), $4 \times 10^3 < Gr_L < 3.1 \times 10^5$, and aspect ratios from 9 to 36.

Natural convection research at the University of Pennsylvania and Okayamo University in Japan has been under the collaborative direction of Churchill, Ozoe, and Lior. The work is primarily theoretical and takes advantage of recent advances in computer technology and numerical methods to solve the full Navier-Stokes, energy, and continuity equations in three dimensions. Selected experiments were also performed on small cavities (1 × 2 inch) to verifty the results. The solution method employs the Boussinesq approximations and a transformation of the equations to vorticity, vector velocity potential form from Ozoe et al. (1983). The solution by the finite difference alternating direction, ADI, method is covered in more detail in Ozoe, Yamamoto, and Churchill (1979). The main advantages of the theoretical investigation are that it permits the study of a wide range of geometries and boundary conditions and that the computer-generated graphical results often offer greater insight into the fluid mechanics than is possible in the laboratory. For example, the perspective view of streamlines in figure 8.26, which is taken from Ozoe, Fujii, Lior, and Churchill (1983), shows the evolution of the horizontal roll cell into a boundary layer flow as the cell is tilted from horizontal to vertical. The Rayleigh number is 4,000. The pattern at the end of the roll cell resembles a double helix with fluid particles moving along the direction of the central axis of the roll cell from the end walls into the enclosure, up to a certain point, and from the back toward the end wall. These flows are due to the drag associated with the end wall and the temperature gradients associated with the reduced circulation. The three dimensionality of the flow becomes even more pronounced as the tilt angle is increased. The variation of Nusselt number with tilt angle is shown in figure 8.27. The rather dramatic minima occurs at the point of transition from multiple roll cells to a single boundary layer flow. The decrease in Nusselt number with tilt also occurs in small aspect ratio cells (Lior et al. 1983), but, for the 1 × 1 × 2 cells, the decrease is small and the minima occurs at tilt angles of only a few

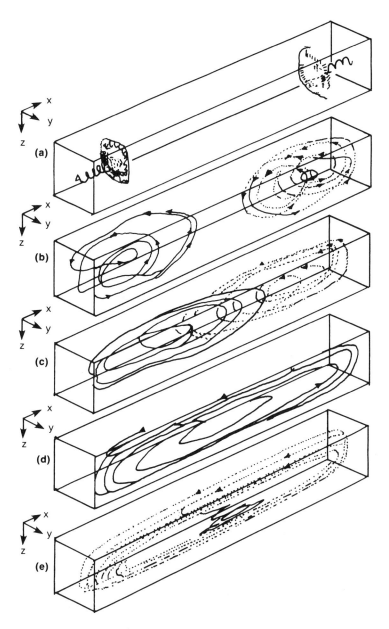

Figure 8.26
Perspective view of streaklines for various inclinations of free-rigid cell with $l = 1$. The Δ-starting points are the same as before. Eye point $(-100, 100, -70)$. (a) $\theta = 0$ rad. (b) $\theta = 20\pi/180$ rad. (c) $\theta = 40\pi/180$ rad. (d) $\theta = 50\pi/180$ rad. (e) $\theta = 90\pi/180$ rad. Source: Ozoe et al. (1983).

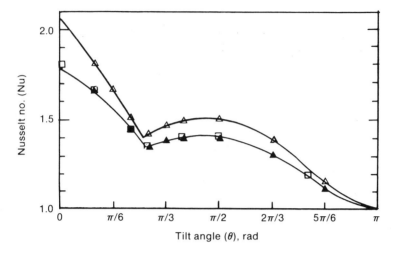

Figure 8.27
Effect of inclination on average Nusselt number of cell: $-\triangle-$free-free boundaries; $l = 1$. $-\blacktriangle-$free-rigid boundaries; $l = 1$. $-\square-$free-rigid boundaries; $l = 1.1$. Source: Ozoe et al. (1983).

degrees. The minima is followed by a substantial increase to a maxima near $\pi/2$ and then decreases to unity at π. The experimental results of Hollands show smaller dependence upon tilt as do the Wisconsin results.

It is well known that most laminar flow heat transfer processes, including natural convection, are quite sensitive to the thermal boundary conditions. Virtually all of the experimental work related to solar collectors has been done with isothermal test surfaces. Since absorber and glazing are seldom isothermal in practice, a major objective of the University of Pennsylvania project was to determine the influence of nonisothermal boundaries on the convention losses, and if possible, to devise ways of utilizing small temperature variations (and perhaps the variation between collector tubes) advantageously. Numerical experiments were performed in which the basic $1 \times 2 \times 1$ convection cell was subjected to sawtooth and double sawtooth temperature distribution on the heating surface. These calculations revealed that the direction of circulation was controlled by the temperature distribution (up flow over the hot spot, down flow at the cold spot) but that the effective Nusselt number was virtually unaffected. However, recent experiments (by Schinkel and Hoogendoorn 1983) has indicated that the constant heat flux hot-wall boundary conditions (along with an isothermal cold wall) can result in an increase in isothermal boundary heat transfer coefficient of 19%. These results were for $10^5 < \text{Ra} < 10^7$ and $20 < \phi < 90$ deg.

The numerical modeling work at the University of Pennsylvania began with studies of laminar flows in small aspect ratio enclosures due to numerical and computer time limitations. Advances in modeling technique as well as computational power have now relaxed some of those restrictions. In recent papers turbulent flows and laminar flows in high aspect ratio enclosures are calculated and compared with experiments (Ozoe et al. 1985, 1982).

The early work at UCLA on natural convection and convection suppression has already been described. Most of the recent work has involved the use of honeycomb glazing over porous absorbers. As previously mentioned, these investigations usually took the form of a thorough theoretical treatment of the solar collector configuration and experiments on a model or prototype collectors. Since the convective heat transfer coefficients are difficult to infer from collector tests, the theoretical methods used in these studies may be of interest to researchers as well as designers. The reports by Rhee (1980), Galanter (1980), and Scott (1980) devote substantial space to presentation of analytical methods applicable to natural convection and honeycomb convection separation devices. These methods, like the recently published procedure by Hollands and Wright (1983), permit accurate determination of both convective and radiative surfaces with real (nongray) optical properties.

The review of natural convection heat transfer in solar collectors has, so far, been confined to planar surfaces. Concentrating collectors generally employ cylindrical (linear concentrators) or cavity (point-focus) receivers that may or may not be protected by transparent covers. Early CPC and V-trough stationary collectors were often designed with no cover surrounding the cylindrical receiver tube (normally the trough is covered). A number of experiments and analyses (Rabl et al. 1980; Woo 1983; Collares Perreira 1981) have shown that heat losses are reduced by a factor of two or more by enclosing the receiver in a cylindrical transparent envelope inside the trough. Thus the majority of linear concentrator receivers are now designed with a cylindrical cover that may or may not be concentric. Even when no cover is used, it may be necessary to approximate the heat transfer problem as a pair of eccentric concentrators.

Heat transfer between concentric and eccentric circular cylinders with air as the convecting media (Pr = 0.7) was investigated by Lee et al. (1984). Results were obtained for diameter ratios from 1.25 to 5.0, eccentricities of -0.9 to $+0.9$, and Rayleigh numbers, Ra (based on average gap dimension), from 10^2 to 10^6. Figure 8.28 shows the isotherms and streamlines for concentric cylinders with diameter ratios of 1.25, 2.6, and 5.0 at Ra of 10^2, 10^4, and 10^6. The

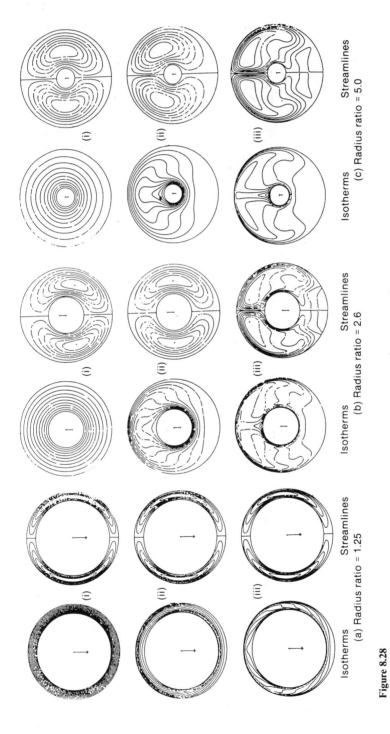

Figure 8.28
Isotherms and streamlines at radius ratios of 1.25, 2.6, and 5.0 for (i) Ra $= 10^2$, (ii) Ra $= 10^4$, and (iii) Ra $= 10^6$. Source: Lee et al. (1984).

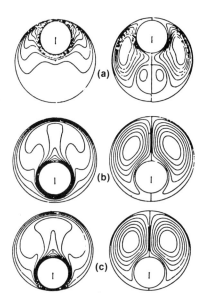

Figure 8.29
Isotherms and streamlines at eccentricity (c) 0.9, for Ra $= 5 \times 10^5$ and radius ratio of 2. Source: Lee et al. (1984).

influence of eccentricity is shown in figure 8.29 in which isotherms and stream-lines are plotted for eccentricities of -0.67, 0.33, and 0.67 with $D_o/D_i = 2.6$ and Ra $= 5 \times 10^5$. The average Nusselt numbers for four eccentricities are shown in figure 8.30.

The range of Rayleigh numbers and Prandtl numbers was greatly enlarged in a series of studies by Kuehn and Goldstein (1976a, b, 1978, 1980). They developed a general correlation for long horizontal concentric cylinders that covers a broad range of diameter ratios, Rayleigh numbers from 10^2 to 10^{10}, and Prandtl numbers from 0.01 to 1,000. The general correlation is quite complex and requires iteration because of the dependence on inner and outer surface temperature. The relation can be greatly simplified for Pr $= 0.71$ and laminar flow which is the normal case in receivers for CPC and parabolic trough solar collectors and is embodied in the equation

$$\text{Nu}_{D_i \text{ conv}} = \frac{2}{[1 + (2/0.4\text{Ra}_{D_i}^{1/4})]/[1 - (2/0.587\text{Ra}_{D_o}^{1/4})]}. \tag{55}$$

Equation (55) or the curves in figure 8.30 are considered satisfactory for typical horizontal troughs or CPC collectors even though they do not account for

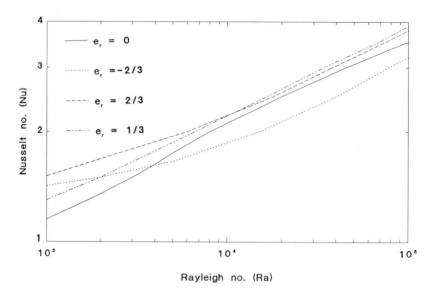

Figure 8.30
Overall heat transfer coefficient versus Rayleigh number at eccentric positions. $r_0/r_i = 5.0$ numerical results. Source: Lee et al. (1984).

the nonuniform temperature of the receivers (owing to nonuniform illumination). CPC and V-trough collectors are often oriented north–south and inclined at fairly steep tilt angles just as flat plate collectors are. Meyer et al. (1980) determined the losses from V-trough collectors as a function of tilt and found that the tilt affects the heat transfer rate in much the same way as in plane enclosures [i.e., by replacing g with $g\cos\phi$; see equation (48)].

Cavity receivers for central receiver solar plants have already been discussed earlier in this section. The small cavity receivers normally found on parabolic dish collectors, however, represent an even more complex situation since the orientation of the cavity with respect to the gravitational field changes continually as the collector tracks the sun. Kugath et al. (1981) studied cavity receivers in connection with the design of the Shenandoah, Georgia, system of parabolic dish collectors, and proposed the following correlation:

$$\overline{\mathrm{Nu}_L} = 0.52P(\phi)K^{1.75}\mathrm{Ra}_L^{0.25},$$

where

$$P(\phi) = \cos^{3.2}\phi \qquad \text{when } 0 \le \phi \le 45 \text{ deg},$$
$$= 0.707\cos^{2.2}\phi \qquad \text{when } 45 \le \phi \le 90 \text{ deg},$$

$$K = \frac{R_{aperture}}{R_{cavity}} \quad \text{when } R_{aperture} \leq R_{cavity},$$

$$K = 1 \quad \text{when } R_{aperture} \geq R_{cavity},$$

$$L = \sqrt{2}R_{cavity},$$

$$\phi = \text{angle of declination.} \tag{56}$$

Their studies also showed that cavity losses are quite sensitive to wind. With a 10-mph wind normal to the cavity aperture, the losses were four times the natural convection value.

Finally, an excellent review of natural convection in all applications of solar thermal systems has recently been completed by Anderson and Kreith (1986) and will soon be published by SERI. That review covers much of the material reviewed here and in addition covers natural convection in buildings.

The depth of understanding of natural convection phenomena in solar collectors has increased considerably as a result of government-supported research in the past decade. There is now a wealth of application and accurate experimental results, adequate theory for most scientific and design purposes, and a capability, in the form of reliable numerical analysis, to investigate three-dimensional flows in considerable detail. This latter capability is important not only in the analysis of complex shapes but can also be very valuable in evaluating the effect of variables that cannot be totally eliminated in experiments, such as temperature gradients and end walls. The goal of developing effective convection-suppression devices has been met only partially. There appears to be an adequate fundamental understanding of the controlling phenomena and the performance trade-offs for practical designs, but cost-effective structures have not evolved. This is partly due to the failure to develop either inexpensive glass or polymer structures and partly a result of the improvement in selective surfaces, which has provided the collector designer with simple and reliable alternatives to complex glazing systems.

There is still a significant improvement in collector efficiency to be gained by the suppression of convection and conduction losses in collector glazing systems, but research attention has now turned to the development of microporous (aerogel) glass materials or evacuated glazing assemblies.

8.5 Theoretical, Mathematical, and Numerical Methods

The material originally intended for this section included the development of second law methods of analysis for collectors, radiation calculations (Monte

Carlo), two- and three-dimensional numerical solutions of the Navier-Stokes equations, numerical solutions of the boundary layer equations, turbulence models, collector and collector array optimization, and perhaps a few more. These topics have been touched briefly in the foregoing material or in other chapters with primary emphasis on the results obtained. Since research methodology advances are almost always published in archival journals and since the important advances are usually recognized and taken up immediately, it does not seem appropriate nor productive to devote the time and space necessary to cover these broad topics in adequate detail. The reader will find most of the important results in the *Journal of Heat Transfer*, the *International Journal of Heat and Mass Transfer*, the *Journal of Fluid Mechanics*, *Physics of Fluids*, *Advances in Heat Transfer*, and *Advances in Fluid Mechanics*, just to name a few.

8.6 Conclusions

The variety of solar collectors is so rich that there always will be configurations and combinations of boundary conditions for which the heat transfer and fluid flow processes are not well known. Thus solar collectors will continue to provide a plethora of research problems for the thermal scientist. The solar engineer, however, should not be dismayed. The present understanding and engineering data for most of the important thermal processes in solar collectors would be adequate for design and analysis if only they were properly applied in practice. This seems to be true of all the internal processes within closed collector envelopes with the possible exception of buildings that function as solar collectors. Thermal conduction, convection, and radiation can be calculated within acceptable levels of uncertainty from known principles and data.

The external processes, such as heat losses from central receivers, are less well understood and probably represent the most promising area for thermal research. As long as the absorbing surfaces of solar collectors are exposed to variable and turbulent atmospheric conditions, analyses of their heat losses will be a challenge for the heat transfer specialist. Thus unglazed solar collectors, whether they are flat plate, solar ponds, CPC troughs, parabolic dish receivers, or central receivers, are likely to be the continuing subject of research for many years. Because of the difficulty of performing field experiments or of replicating atmospheric variations in the laboratory, numerical simulation may be the most appropriate tool for investigation of these complex phenomena.

Nomenclature

a	coefficient
A	area
b	coefficient
C	constant
D	diameter of a cylinder
E	exergy
f	friction factor
g	gravitational constant
Gr	Grashof number, $g\beta(T_\omega - T_\infty)l^3/v^2$
h	enthalpy
h	heat transfer coefficient
H	enthalpy
I	insolation
k	thermal conductivity
l	a general characteristic length
L	length
n	exponent
$\dot{n}$	flow of species
Nu	Nusselt number, hl/k
N_0	Avogadro's number
Pr	Prandtl number, $\mu c_p/k$
p	pressure
q	heat flux
Q	rate of heat transfer
r, R	radius
Ra	Rayleigh number, GrPr
Re	Reynolds number
S	entropy
t	temperature, time
T	temperature
U	heat loss coefficient
u, U	velocity component
U	internal energy
v, V	velocity component
V	volume
W	work
W	width

Greek

α	solar absorption coefficient
α	thermal diffusivity
α	thermal accommodation coefficient
β	coefficient of volumetric expansion
ν	kinematic viscosity
μ	dynamic viscosity
ρ	density
η	efficiency
τ	solar transmission coefficient
θ	entropy production
θ, ϕ	tilt angle
θ	acceptance angle of concentrating collector
λ	thermal conductivity

Subscripts and Superscripts

a, A	ambient
c, C	collector
D	diameter
f	film temperature, $(T_C + T_A)/2$
fc	forced convection
H	height
h	hydraulic
i	inlet
l	general length scale
L	length, or loss
mx	mixed convection
nc	natural convection
o	no temperature difference, inlet condition
s	stagnation (no flow) condition
w	wall or surface
B	free stream or ambient
*	dead state
–	average (bar over symbol)
rev	reversible
II	based on second law

Notes

1. In 1986 the work by Segall, Lior, and Yeh was recorrelated by the authors to span a wider range with a single correlation.

2. The concept was first described by K. W. Miller in 1943 who also held a 1954 patent (see Miller 1954).

References

Abrams, M. 1983. *The Status of Research on Convective Losses from Solar Central Receivers.* SAND83-8224. Livermore, CA: Sandia National Laboratories.

Achenbach, E. 1977. The effect of surface roughness on the heat transfer from a circular cylinder to the cross flow of air. *Int. J. Heat Mass Transfer* 20: 359–369.

Balcomb, J. D., and J. E. Perry, Jr. 1977. *Assessment of Solar Heating and Cooling Technology.* LA-6379-MS. Los Alamos, NM: Los Alamos Scientific Laboratory.

Ball, G. L, III, and J. W. Leffingwell. 1979. Medium temperature air heater base on durable transparent films. *Proc. 3rd Annual Solar Heating and Cooling Research and Development Branch Contractors' Meeting.* Washington, D.C., September 24–27. Washington, DC: U.S. Department of Energy, pp. 93–94.

Bankston, C. A. 1965. Fluid friction heat transfer turbulence and interchannel flow stability in the transition from turbulent to laminar flow in tubes. Dissertation. University of New Mexico.

Bankston, C. A., W. L. Sibbitt, and V. J. Skoglund. 1966. Stability of gas flow distribution among parallel heated channels. Presented at the AIA Second Propulsion Joint Specialist Conference. Colorado Springs, CO. AIA paper no. 66-589.

Barnhart, J. S. 1980. *ETRANS A Energy Transport System Optimization Code for Distributed Network of Solar Collectors.* PNL-3327. Richland, WA: Pacific Northwest Laboratory.

Batton, W. D., and R. E. Barber. 1979. *Solar Powered Rankine Cycle Irrigation Pump.* Final report, DE-AC03-78ET20491 (DOE/SAN). Denver, CO: Barber Nichols Engineering Company.

Bejan, A. 1982. Extraction of exergy from solar collectors under time-varying conditions. *Int. J. Heat Fluid Flow* 3(2): 67–72.

Bejan, A., D. W. Kearney, and F. Kreith. 1981. Second law analysis and synthesis of solar collector systems. *J. Solar Energy Eng.* 103: 23–27.

Bienert, W. B., and D. A. Wolf. 1976. *Heat Pipe Applied to Flat Plate Solar Collectors.* Final report, COO/2604/76/1, E(11-1)-2604. Dynatherm.

Bingman, R. 1977. *Analysis of Design Changes Necessary to Adapt Receiver to Compound Parabolic Reflector.* Final report for Contract 31-109-38-3805.

Bird, R. B., W. E. Sewart, and E. N. Lightfoot. 1960. *Transport Phenomena.* New York: Wiley.

Bliss, R. W., Jr. 1956. Solar house heating, a panel. *Proc. World Symposium of Applied Solar Energy.* Phoenix, AZ, November 1955. Menlo Park, CA, pp. 151–158.

Boehm, R. F. 1987. Review of thermal loss evaluation of solar central receivers. *J. Solar Energy Eng.*

Buchberg, H., D. K. Edwards, and O. A. Lalude. 1968. *Design Consideration for Cellular Solar Collectors.* ASME paper no. 68-WA/SOL-3.

Buchberg, H., I. Catton, and D. K. Edwards. 1976. Natural convection in enclosed spaces— Review of application to solar energy collection. *J. Heat Transfer* 98(2): 182–188.

Buchberg, H., O. A. Lalude, and D. K. Edwards. 1969. Performance characteristics of rectangular honeycomb solar-thermal converters. *Solar Energy* 13: 193–221.

Buchberg, H., and D. K. Edwards. 1980. Transpired solar air heaters. *Proc. Annual DOE Active Solar Heating and Cooling Contractors' Review Meeting.* Incline Village, NV, March 26–28. Washington, DC: U.S. Department of Energy, pp. 9–20, 21.

Buchberg, H. K., O. A. Lalude, and D. K. Edwards. 1971. Performance characteristics of rectangular honeycomb solar-thermal converters. *Solar Energy* 13: 193–211.

Cattaneo, L., J. R. Harris, T. Reinhold, E. Simiu, and C. W. C. Yancey. 1981. *Wind, Earthquake, Snow and Hail Loads on Solar Collectors.* NBSIR 81-2199. Gaithersburg, MD: National Bureau of Standards.

Charters, W. W. S., and L. F. Peterson. 1971. Free convection suppression using honeycomb cellular materials. *Solar Energy* 13: 335–361.

Cheng, B. H., and R. D. Gift. 1980. *Survey and Evaluation of Available Thermal Insulation Materials for Use on Solar Heating and Cooling Systems.* Final report, DE-AC04-78CS35363. Versar, Inc.

Chevalier, H. L., and D. J. Norton. 1979. Structural integrity of solar collectors. Final report, DE-AC04-79AL11774. College Station, TX: Texas A&M University.

Chevalier, H. L., and R. A. Wilke. 1980. Structural integrity of solar collectors. *Proc. Annual DOE Active Solar Heating and Cooling Contractors' Review Meeting.* Incline Village, NV, March 26–28. Washington, DC: U.S. Department of Energy, pp. 9–60, 61.

Chicalla, T. D. 1979. Development of improved cover plates for solar collectors. *Solar Heating and Cooling Research and Development Project Summaries.* DOE/CS-0010. Washington, DC: U.S. Department of Energy, pp. 1–9.

Chiou, J. P., et al. 1965. A slit-and-expanded aluminum-foil matrix solar collector. *Solar Energy* 9(2): 73–80.

Chiou, J. P., 1983. The effect of non-uniform fluid flow distribution on the thermal performance of solar collectors. *Solar Energy* 29(6): 487–502.

Christopher, D. M., and J. T. Pearson. 1980. Parametric analysis of louvered air-heating solar collectors. *ASME* 80-WA/Sol-38.

Churchill, S. W. 1977. Comprehensive correlating equation for laminar, assisting, forced and free convection. *AIChE* 23(1): 10–16.

Clausing, A. M. 1982. Advantages of a cryogenic environment for experimental investigation of convective heat transfer. *Int. J. Heat Mass Transfer* 25: 1255–1257.

Clausing, A. M., K. A. Wagner, and R. J. Skarda. 1983. An experimental investigation of combined convection from a vertical cylinder in a crossflow. *Proc. ASME-JSME Thermal Engineering Joint Conference, vol. 3.* Honolulu, HI, March 20–24, pp. 155–161.

Clever, R. M. 1973. Finite amplitude longitudinal convection rolls in an inclined layer. *J. Heat Transfer, Trans ASME* 95, series C: 407–408.

Close, D. J. 1963. Solar air heaters for low and moderate temperature applications. *Solar Energy* 7(3): 117–124.

Cole-Appel, B. E. 1978. The improvement of solar air collectors—Study and experimental research project. Final report, May 1976–June 1978, ALO-3713-TI. Albuquerque, NM: U.S. Department of Energy, Albuquerque Operations Office.

Cole-Appel, B. E., G. O. P. Löf, L. E. Shaw, and B. B. Fischer. 1979. The improvement of solar air collectors—Study and experimental research project. *Proc. 3rd Annual Solar Heating and Cooling Research and Development Branch Contractors' Meeting.* Washington, D.C., September 24–27, 1978. Washington, DC: U.S. Department of Energy, pp. 131–134.

Collares Pereira, M., J. Dugue, A. Joyce, M. Delgado, G. Serrudo, and A. Rego-Teixeira. 1981. A 3X CPC-type concentrator with tubular receiver and tubular glass envelope to reduce convective losses: Description and performance. *Proc. International Solar Energy Congress*. Solar World Forum, Brighton, 1718–1722.

Collier, R. K., and F. H. Arnold. 1980. Comparison of transpired beds for solar collector applications. *Proc. 1980 Annual Meeting of the American Section of the International Solar Energy Society*, pp. 451–455.

Dame, R. E. 1979. *Final Report Study and Analysis of a Low-Cost Cement Bonded Flat-Plate Solar Collector*. Final report, ALO-41221T1.

Dame, R. E. 1980. Selection and test of high temperature solar collector adhesives. *AS/ISES, Proc. 1980 Annual Meeting*, vol. 3.1. Phoenix, AZ, p. 470.

Daniels, F., and J. A. Duffie. 1955. *Solar Energy Research*. Madison, WI: University of Wisconsin Press.

de Winter, F. 1975. Solar energy and the flat plate collector, an annotated bibliography. ASHRAE report S-101 (also pp. 56–59, *ASHRAE J.*, 1975).

Drake, R. M. 1949. Investigation of the variation of point unit heat transfer coefficient for laminar flow over an inclined flat plate. Transactions of ASME, *J. Appl. Mech.* 71: 1–8.

Duffie, J. A., and W. A. Beckman. 1974. *Solar Engineering of Thermal Processes*. New York: Wiley.

Eckert, E. R. G., and R. M. Drake. 1972. *Analysis of Heat and Mass Transfer*. New York: McGraw-Hill.

Edwards, D. K., and A. T. Leung. 1964. Collection of solar energy in a porous bed. Part IV: *Basic Heat Transfer Studies Related to the Use and Control of Solar Energy*. Final report, NS G20246, report no. 64-14. Los Angeles: UCLA, Department of Engineering, pp. 131–142.

Ernst, D. M. 1979. Cost effective solar collectors using heat pipes. *Proc. 3rd Annual Solar Heating and Cooling Research and Development Branch Contractors' Meeting*, September 24–27. Washington, D.C., CONF-780983, p. 146.

Ernst, D. M. 1981. *Cost Effective Solar Collectors Using Heat Pipes*. Final report, July 1979–August 1981, DOE/CS/34099-4, DE-AC04-77CS34099. King of Prussia, PA: Thermacore.

Evans, G. H., and O. A. Plumb. 1982. Laminar mixed convection from vertical heated surface in crossflow. *J. Heat Transfer* 104: 554–558.

Evans, G. H., and O. A. Plumb. 1983. Turbulent mixed convection from vertical heated surface in a crossflow. *Proc. ASME-JSME Thermal Engineering Joint Conference*, vol. 3. Honolulu, HI, March 20–24, pp. 47–53.

Francia, G. 1962. A new collector of solar energy—theory and experimental verification—calculations of efficiencies. Paper no 594B, SAE, October (see also U.N. Conference of New Sources of Energy—Rome 196—paper S/71).

Galanter, S. A. 1980. *Transpired Solar Air Heaters*. Final Report, vol. 1: Investigation of transpired porous bed solar air heaters. UCLA-ENG-8076. Los Angeles: UCLA School of Engineering and Applied Science.

Gordon, N. R. 1979. *Plastic Honeycomb for Solar Coverplate*. PNL-2953, Battelle Pacific NW Lab., EY-76-C-03-1830.

Gribik, and Osterle. 1986. Transactions of ASME *J. Solar Energy* 108: 78.

Gupta, G. L., and H. P. Harg. 1967. Performance studies of solar air heaters. *Solar Energy* 11(1): 25–31.

Hedstrom, J. C. 1982. Performance of a solar air conditioning system utilizing boiling collectors. Presented at DOE Active Program Review, Washington, DC, October 27–28. LA-UR-82-3045.

Hewitt, H. C. 1980. Wind effect of collectors. *Proc. Annual DOE Active Solar Heating and Cooling Contractors' Review Meeting*. Incline Village, NV, March 26–28. Washington, DC: U.S. Department of Energy, pp. 9–58, 59.

Hewitt, H. C., Jr., and E. I. Griggs. 1979. *Wind Effect of Collectors*. Final report, DOE/CS/35364-T1. Tennessee Technological University. DE-AC04-78CS35364.

Hewitt, H. C., Jr., and N. Onur. 1980. Study of wind effect on collector performance. ASME paper no. 80-C2/Sol-4.

Hollands, K. G. T. 1965. Honeycomb devices in flat-plate solar collectors. Presented at the Solar Energy Society Conference. Phoenix, AZ, March 15–17. *Solar Energy* 9(3): 15–165.

Hollands, K. G. T., G. D. Raithby, B. B. Russel, and R. G. Wilkinson. 1979. *Methods for Reducing Heat Losses from Flat Plate Solar Collectors, Phase III*. Final report. U.S. Department of Energy.

Hollands, K. G. T., G. D. Raithby, and T. E. Unny. 1978. *Methods for Reducing Heat Losses from Flat Plate Solar Collectors, Phase II*. Final report, COO/2597-4. U.S. Department of Energy.

Hollands, K. G. T., T. E. Unny, G. D. Raithby, and L. Konicek. 1975. Free convective heat transfer across inclined air layers. ASME paper no. 75-HT-55.

Hollands, K. G. T., and E. C. Shewen. 1981. Optimization of flow passage geometry for air-heating, plate-type solar collectors. *J. Solar Energy Eng.* 103: 32–330.

Hollands, K. G. T., and J. L. Wright. 1983. Heat loss coefficient and effective product for flat-plate collectors with diathermanous covers. *Solar Energy* 30(3): 211–216.

Hottel, H. C. 1954. The performance of flat-plate solar energy collectors. *Proc. Course Symposium*, MIT, August 1950. Cambridge: MIT Press, pp. 58–79.

Howell, J. R., and R. B. Bannerot. 1977. Optimum solar collector maximizing cycle work output. *Solar Energy* 19: 149.

Hull, J. R., and W. W. Schertz. 1986. *Analysis of Heat Pipe Absorbers in Evacuated-Tube Solar Collectors*. ANL-86-16. Argonne, IL: Argonne National Laboratory.

Hunt, J. A. 1981. *Development of a Direct Absorption High Temperature Gas Receiver*. LBL-13234. Berkeley, CA: Lawrence Berkeley Laboratory.

Jones, G. F., and N. Lior. 1978. Isothermal flow distribution in solar collectors and collector manifolds. *AS/ISES, Proc. Annual Meeting*, vol. 2.1. August 28–31, Denver, CO, pp. 362–372.

Jürges, W. 1924. Die Warmeubergang an einer ebenen Wand. *Beihefts Sum Gesundheits-Ingenieur*, vol. 1, suppl. 19.

Kays, W. M. 1966. *Convective Heat and Mass Transfer*. New York: McGraw-Hill.

Keenan, J. H. 1941. *Thermodynamics*. New York: Wiley.

Keenan, J. H., G. N. Hatsopoulos, and E. P. Gyftopoulos. 1980. Principles of thermodynamics. *Encyclopedia Britannica*.

Kennard, E. H. 1938. *Kinetic Theory of Gases*. New York: McGraw-Hill.

Kestin, J. 1980. Availability: The concept and associated terminology. *Energy* 5: 679–692.

Kline, S. J., and F. A. McClintock. 1953. Describing uncertainties in single sample experiments. *Mech. Eng.* 75: 3–8.

Koenig, A. A., and M. Marvin. 1981. Convection heat loss sensitivity in open cavity solar receivers. Final report, DOE contract no. EG77-C-04-3985.

Kraabel, J. S. 1983a. An experimental investigation of natural convection from a side-facing cubical cavity. *Proc. ASME-JSME Thermal Engineering Joint Conference*. Honolulu, HI, March 20–24, pp. 299–306.

Kreith, F., and J. F. Kreider. 1978. *Principles of Solar Energy*. Washington, DC: Hemisphere.

Kuehn, T. H., and R. J. Goldstein. 1976. An experimental and theoretical study on natural convection in the annulus between horizontal concentric cylinders. *J. Fluid Mech.* 74: 695–719.

Kuehn, T. H., and R. J. Goldstein. 1976. Correlating equations for natural convection heat transfer between horizontal circular cylinders. *Int. J. Heat Mass Transfer* 19: 1127–1135.

Kuehn, T. H., and R. J. Goldstein. 1978. An experimental study of natural convection heat transfer in concentric and eccentric horizontal cylinders. *J. Heat Transfer* 100: 635–640.

Kuehn, T. H., and R. J. Goldstein. 1980. A parametric study of Prandtl number and diameter ratio effects on natural convection heat transfer in horizontal cylindrical annuli. *J. Heat Transfer* 102: 768–770.

Kutscher, C. F., R. L. Davenport, D. A. Dougherty, R. G. Gee, P. M. Masterson, and E. K. May. 1982. *Design Approaches for Solar Industrial Process Heat Systems: Nontracking and Line-Focus Collector Technologies.* Golden, CO: SERI.

Kuzay, T. M., et al. 1971. *Solar Collectors of Solar One.* Newark, DE: Institute of Energy Conversion, University of Delaware.

Lalude, O. A., and H. Buchberg. 1971. Design of honeycomb porous bed solar air heaters. *Solar Energy* 13: 223–242.

LeBru, A., P. Michel, and R. Torrenti. 1984. Performance of large flat plate solar collector arrays. *Proc. International Energy Agency Workshop of Design and Performance of Large Solar Thermal Collector Arrays.* San Diego, CA.

Lee, T. S., N. E. Wijeysundera, and K. S. Yeo. 1984. Free convection fluid motion and heat transfer in horizontal concentric and eccentric cylindrical collector systems. *Solar Eng.,* D. Y. Lowani, ed. New York: ASME, pp. 194–204.

Leffingwell, J. W., and J. L. Schwendeman. 1979. *Medium Temperature Air Heaters Based on Durable Transparent Films.* Final report, ALO/4147, Monsanto, EG-77-C-04-4147.

Lindsey, J. L. 1976. *Force and Pressure Tests of Solar Collector Models in the Vought Corporation Systems Division Low Speed Wind Tunnel.* SAND76-7007. Albuquerque, NM: Sandia Laboratories.

Lior, N., J. O'Leary, and D. Edelman. 1977. Optimized spacing between rows of solar collectors. *AS/ISES, Proc. Annual Meeting,* June 6–10, Orlando, FL.

Lior, N., and H. Yeh. 1980. Collector array studies. *Proc. Annual DOE Active Solar Heating and Cooling Contractors' Review Meeting.* Incline Village, NV, March 26–28. Washington, DC: U.S. Department of Energy, pp. 9–40, 41.

Lior, N., et al. N.d. *Optimal configuration and flow-rate of a flat-plate solar collector array.* Draft final report. Unpublished.

Löf, G. O. G. 1946. *Solar Energy Utilization for House Heating.* WPB Report P-25357. Office of the Publication Board, Department of Commerce.

Löf, G. O. G. 1950. Solar energy collectors of overlapped-glass-plate type. *Proc. Space Heating with Solar Energy.* Cambridge: MIT Press, pp. 72–86.

Löf, G. O. G. 1954. Performance of solar energy collectors of overlapped-glass-plate type. MIT Solar Heating Symposium, *Open Space Heating with Solar Energy,* MIT.

Löf, G. O. G., and T. D. Nevens. 1953. Heating of air by solar energy. *Ohio J. Sci.* 53: 272.

Loth, J. L. 1979. *Grooved Foamglas Air Heater.* Final report, DE-FG04-77CS34087. West Virginia University.

Lumsdaine, E. 1969. Solar heating of a fluid through a semi-transparent plate: Theory and experiment. *Solar Energy* 12: 457–467.

Marshall, K. N., S. A. Greenberg, K. G. T. Hollands, and R. K. Wedel. 1977. *Optimization of Thin-Film Transparent Plastic Honeycomb-Covered Flat Plate Solar Collectors.* Final report, Lockheed, EY-76-C-03-1256.

Mattox. D. L. 1979. *Evaluation of Heat Transfer Enhancement in Air-Heating Collectors.* Final report. Washington, DC: U.S. Department of Energy.

McAdams. W. H. 1954. *Heat Transmission,* 3rd ed. New York: McGraw-Hill.

McCargo, M. 1979. Optimization of thin film transparent plastic honeycomb covered flat plate solar collector. *Solar Heating and Cooling Research and Development Project Summaries.* DOE/CS-0010. Washington, DC: U.S. Department of Energy, pp. 1–40.

McEligot, D. M., and C. A. Bankston. 1980. Forced convection in solar collectors. *Proc. International Solar Energy Society Meeting.* Atlanta, GA.

McMordie, R. K. 1981. Convection losses from a cavity receiver. *AIChE 20th National Heat Transfer Conference,* August 2–5.

Menuchin, Y., S. Bassler, G. F. Jones, and N. Lior. 1981. Optimal flow configuration in solar collector arrays. Reprint of paper presented at American Section of International Solar Energy Society.

Meyer, B. A., J. W. Mitchell, and M. M. El-Wakil. 1980. Convective heat transfer in trough and CP collectors. *Proc. Annual Meeting, American Section of ISES,* vol. 3.1. Phoenix, 437–440.

Miller, K. W. 1954. *Solar Heat Trap.* U.S. Patent 2,680,437, June 3.

Mirenayat, H. 1981. Etude experimentale du transfert de chaleur par convection naturelle dans une cavite isotherme ouverte. D. Engr. thesis. University of Poitiers.

Moan, K. L., and D. C. Beekley. 1980. *Evaluation of Sunpak Tube Vacuum Quality.* Final report, A00/4286-21, Owens-Illinois, DE-AC04-78CS30131.

Moan, K. L., and D. C. Beekley. 1982. *Evaluation of Sunpak Tube Vacuum Quality.* Final report, DE-AC04-78CS30131.

Morton, R. E. 1981. Design of collector system piping layouts. *Proc. Line Focus Thermal Energy Technology Development Seminar for Industry.* SAN 80-1666. Albuquerque, NM: Sandia National Laboratory.

Morton, R. E. 1983. Field layout studies. *Proc. Distributed Solar Collector Summary Conference—Technology and Applications,* Robert L. Alvis, ed. Contract DE-AC04-76DP00789. Albuquerque, NM: Sandia National Laboratories, pp. 24–31.

Muller, T., and J. Sullivan. 1981. *Further Development of a Low-Cost Solar Panel.* Final report for period 9/28/79-5/31/80, ALO/2032-2, Acurex, DE-AC04-79AL12032.

Neeper, D. A., 1980. Analysis of matrix air heaters. *Proc. International Solar Energy Society Meeting.* Atlanta, GA.

Nelson, E. V., and T. K. Muller. 1980. Further development of a low-cost solar panel. *Proc. Annual DOE Active Solar Heating and Cooling Contractors' Review Meeting.* Incline Village, NV, March 26–28. Washington, DC: U.S. Department of Energy, pp. 9–7, 8.

Oosthuizen, P. H., and R. K. Leung. 1978. *Combined Convective Heat Transfer from Vertical Cylinders in a Horizontal Flow.* ASME paper no. 78-WA/HT-45.

Ortabasi, U. N.d. *An Evacuated Tubular Collector Utilizing a Heat Pipe.* Final report, COO-2608-3. Corning Glass Company, EY-76-C-02-2608.

Ortabasi, U. 1978. Research on evacuated tubular solar collectors utilizing a heat pipe. *Solar Heating and Cooling Research and Development Project Summaries.* Washington, DC: U.S. Department of Energy, DOE/CS-0010, Corning Glass Company, EY-76-C-02-2608, pp. 1–16.

Ozoe, H., K. Yamamoto, and S. W. Churchill. 1979. Three dimensional numerical analysis of natural convection in an inclined channel with a square cross section. *AIChE J.* 25: 709–716.

Ozoe, H., K. Fujii, S. W. Churchill, and N. Lior. 1982a. A theoretically based correlation for natural convection in horizontal rectangular enclosures heated from below with arbitrary aspect ratios. *Proc. Seventh International Heat Transfer Conference,* vol. 2. Munich, pp. 25–262.

Ozoe, H., P. K. B. Chao, S. W. Churchill, and N. Lior. 1982b. *Laminar Natural Convection in an Inclined Rectangular Box with the Lower Surface Half Heated and Half Insulated.* 82-HT-72. New York: American Society of Mechanical Engineers.

Ozoe, H., K. Fujii, N. Lior, and S. W. Churchill. 1983. Long rolls generated by natural convection in an inclined, rectangular enclosure. *Int. J. Heat Mass Transfer* 26(10): 1424–1438.

Ozoe, H., A. Mouri, M. Ohmuro, S. W. Churchill, and N. Lior. 1985. Numerical calculations of laminar and turbulent natural convection in water in rectangular channels heated and cooled isothermally on the opposing vertical walls. *Int. J. Heat Mass Transfer* 28(1): 125–138.

Payne, P. R., and J. P. Brown. 1979. *Experiments with Structural Concrete Blocks Which Double as Solar Air Heaters*. Final report, Payne DE-FG04-77CS34138.

Pearson, J. T. 1979. *Forced and Natural Convection Studies of Solar Collectors for Heating and Cooling Applications*. Semiannual report for the period 3/12/79–9/10/79, AL0/5366-2. Albuquerque, NM: U.S. Department of Energy, Albuquerque Operations Office.

Pearson, J. T. 1980. *Forced and Natural Convection Studies of Solar Collectors for Heating and Cooling Applications*. Semiannual report for the period 9/11/79–3/10/80, ALO/5366-3. Albuquerque, NM: U.S. Department of Energy, Albuquerque Operations Office.

Pearson, J. T. 1983. *Forced and Natural Convection Studied on Solar Collectors for Heating and Cooling Applications*. Final report, ALO/5366-4. Albuquerque, NM: U.S. Department of Energy, Albuquerque Operations Office.

Perrot, M., et al. 1965. Contribution a l'étude des caracteristiques techniques des structures cellulaires et leur application à la conversion thermique a basse temperature. *Bull. du Comples* 8: 143–151.

Peterka, J. A., J. M. Sinou, and J. E. Cermak. 1980. *Mean Wind Forces on Parabolic Trough Solar Collectors*. SAND80-7023, contract 13-2412. Albuquerque, NM: Sandia National Laboratories.

Raghuraman, P., and D. Kon. 1981. *On Reducing Convective Losses from Cover Glazings of Solar Thermal Collectors*. ASME reprint 81-WA/Sol-3. New York: American Society of Mechanical Engineers.

Randall, D. E., D. D. McBride, and R. E. Tate. 1980. Steady-state wind loading on parabolic trough solar collectors. ASME publication 80-C2/Sol-20. New York: American Society of Mechanical Engineers.

Randall, K. R., J. W. Mitchell, and M. M. El-Wakil. 1979. Natural convection heat transfer characteristics of flat plate enclosures. *J. Heat Transfer* 101: 12–125.

Rask, D. R. 1979. Low-cost solar air heater. *Proc. 3rd Annual Solar Heating and Cooling Research and Development Branch Contractors' Meeting*. Washington, DC, September 24–27. Washington, DC: U.S. Department of Energy, pp. 71–72.

Rask, D. R., and L. J. Mueller. 1977. *Low-Cost Solar Air Heater*. Final report, COO-2929-13, Honeywell EY-76-C-02-2929.

Reynolds, W. C., W. M. Kays, and S. J. Kline. 1960. A summary of experiments on turbulent heat transfer from a non-isothermal flat plate. *ASME J. Heat Transfer* 82: 341–348.

Rhee, S. J., and D. K. Edwards. 1983. Comparison of test results for flat-plate, transpired flat-plate, corrugated, and transpired corrugated solar air heaters. *J. Solar Energy Eng.* 105: 231–236.

Rhee, S. J. 1980. *Transpired Solar Air Heaters, Effect of Wall Suction on Laminar Entrance Flow with Application to Solar Air Heaters*, vol. 3. Final report, UCLA-ENG-8078. Los Angeles: UCLA, School of Engineering and Applied Science.

Schlichting, H. 1960. *Boundary Layer Theory*, 4th ed. New York: McGraw-Hill.

Schinkel, W. M. M., and C. J. Hoogendoorn. 1983. Natural convection in collector cavities with an isoflux absorber plate. *J. Solar Energy Eng.* 105: 19–22.

Scholten, W. B. 1983. Comparison of exergy delivery capabilities of solar collectors. Prepared for presentation at the Panel Meeting on Second Law and Irreversibility Considerations in Solar Cooling. Washington, DC.

Scholten, W. B. 1984. *Investigation of the Exergy Delivery Capabilities of Solar Collectors.* DOE contract DE-AC01-83CE-30784.

Scholten, W. B. 1985. Exergy performance of solar collectors as determined from standard test data. Presented at ASME Solar Energy Division Meeting. Knoxville, TN, March 25–28.

Scott, J. R. 1980. *Transpired Solar Air Heaters, Experimental and Analytical Investigation of a Glass-Tube Porous Bed Solar Air Heater,* vol. 2. Final report, UCLA-ENG-8077. Los Angeles: UCLA School of Engineering and Applied Science.

Segall, R. N., N. Lior, and H. Yeh. 1983. *Forced Convection on Solar Collector Arrays.* Topical report, ALO-5319-5. Albuquerque, NM: DOE, Albuquerque Operations Office.

Selcuk, K. 1971. Thermal and economic analysis of the overlapped glass plate solar-air heater. *Solar Energy* 13: 165–191.

Sharp, J. K., and C. J. Chiang. 1983. Siting tradeoffs for parabolic trough fields. *Proc. Distributed Solar Collector Summary Conference—Technology and Applications,* Robert L. Alvis, ed. Contract DE-AC04-76DP00789. Albuquerque, NM: Sandia National Laboratories, pp. 32–42.

Sharp, J. 1981. Design of collector subsystem piping layout II. *Proc. Line Focus Thermal Energy Technology Development Seminar for Industry.* SAND80-1666 Albuquerque, NM: Sandia National Laboratory.

Shockley, K. A., J. T. Pearson, and D. P. DeWitt. 1983. Heat transfer characteristics of a back-corrugated absorber surface for solar air collectors. *J. Solar Energy Eng.* 105: 86–91.

Shockley, K. A. 1981. Experimental study of heat transfer characteristics for flat plate and back-corrugated solar air collector absorbers. M. S. thesis. School of Mechanical Engineering, Purdue University.

Siebers, D. L. 1983. Experimental mixed convection heat transfer from a large, vertical surface in a horizontal flow. Ph.D. dissertation, Department of Mechanical Engineering, Stanford University.

Siebers, D. L., R. G. Schwind, and R. J. Moffat. 1982. Experimental mixed convection heat transfer from a large, vertical plate in a horizontal flow. *Proc. 7th International Heat Transfer Conference 3.* Munich. September 6–10, pp. 477–482.

Siebers, D.L., and J. S. Kraabel. 1984. Estimating convective energy losses from solar central receivers. *Proc. International Energy Agency Workshop of the Design and Performance of Large Solar Thermal Collector Arrays.* San Diego, CA.

Simpson, D. R. 1977. *Low-Cost Solar Collector of a Packed Bed Design.* Final report, AOO-2972-2, Lehigh University EY-76-S-02-2972.

Simpson, D. R. 1977. Low-cost solar collector of a packed bed design. *Solar Heating and Cooling Research and Development Project Summaries.* DOE/CS-0010, Washington, DC: U.S. Department of Energy, pp. 1–36.

Skartvedt, G., D. Pedregra, R. McMordie, J. Kidd, J. Anderson, and R. Jones. 1980. *Evaluation of Solar Collector for Heat Pump Applications.* Final report, DOE/CS/35351-T1. American Heliothermal Corporation, U.S. Department of Energy.

Sparrow, E. M., J. W. Ramsey, and E. A. Mass. 1979. Effect of finite width of heat transfer and fluid flow about an inclined rectangular plate. *ASME J. Heat Transfer* 101: 199–204.

Sparrow, E. M., and K. K. Tien. 1977. Forced convection heat transfer at an inclined and yawed square plate—Application to solar collectors. *ASME J. Heat Transfer* 99: 507–512.

Sparrow, E. M., and K. K. Tien. 1979. Local heat transfer and fluid flow characteristics for airflow oblique or normal to a square plate. *Int. J. Heat Mass Transfer* 22: 349–360.

Stoddard, M. C., and G. Evans. 1984. Solar One convection loss experiments. *Proc. International Energy Agency Workshop on the Design and Performance of Large Solar Thermal Collector Arrays.* San Diego, CA, June 1984.

Tabor, H. 1958. Radiation, convection, and conduction coefficient in solar collectors. *Bull. Res. Counc. of Israel* 6C: 155–176.

Tabor, H., and Z. Weinberger. 1980. Nonconvecting solar ponds. *Solar Energy Handbook,* J. F. Kreider and F. Kreith, eds. New York: McGraw-Hill, ch. 10.

Thermacore, Inc. 1981. *Development of a Technological Base for Solar Steam Generation.* Final report, prepared for Southern California Gas Co., Los Angeles, CA.

Thomas, J. R., Jr. 1979. *Heat Conduction through Partial Vacuum.* U.S. Department of Energy Report ALO/5367-1.

Thomas, J. R., Jr. 1979. *Helium Penetration in Evacuated Solar Collectors: Theory and Effect of Their Performance.* ASME paper 79-WA/SOL-17.

Thomas, J. R., Jr. 1980. *Heat Conduction in Partial Vacuum.* Final technical progress report, DOE/CS/15367-T2, AC04-78CS15367.

Tieleman, H. W., R. E. Akins, and P. R. Sparks. 1980. *An Investigation of Wind Load on Solar Collectors.* Appendix I: *Data Listing for Top and Bottom of Collector. Investigation of Wind Loads on Solar Collectors.* Appendix II: *Net Pressure Coefficients.* NBS VPI-E-80-1.

Vliet, G. C. 1969. Natural convection local heat transfer on constant-heat-flux inclined surfaces. *ASME J. Heat Transfer* 91: 511–516.

Weiner, M. D., and D. K. Edwards. 1963. Simultaneous conduction, convection, and radiation in a porous bed. *Proc. Heat Transfer and Fluid Mechanics Institute.* Stanford, CA: Stanford University Press, pp. 236–250.

Whillier, A. 1964. Performance of black-painted solar air heater of conventional design. *Solar Energy* 8(1): 31–37.

Whillier, A. 1967. Design factor influencing solar collector performance. *Low Temperature Engineering Application of Solar Energy. ASHRAE Guide to Fundamentals.* New York, p. 27.

Wideman, D. C., and J. R. Thomas, Jr. 1980. Heat conduction in rarefied gases between concentric cylinders: Arbitrary accommodation of both surfaces. *Proc. 1980 Heat Transfer and Fluid Mechanics Institute,* Gerstein and Choudhury, eds. Stanford, CA: Stanford University Press, p. 186.

Wilhelm, W. G. 1981. *Low Cost Solar Energy Collection for Cooling Applications.* BN 31360. Washington, D.C. U.S. Department of Energy, Department of Energy and Environment.

Wilhelm, W. G. 1982. *The Use of Polymer Film and Laminated Technology for Low Cost Solar Energy Collectors.* BN 31360. Prepared for presentation at the Symposium of Polymers for Solar Energy Utilization, National Meeting of the American Chemical Society, Las Vegas, NV.

Winston, R. 1970. Light collection within the framework of geometrical optics. *J. Opt. Soc. Am.* 60: 245.

Woo, K. N. 1983. Performance of the compound cylindrical concentrator. *Proc. Annual Meeting of ASES,* vol. 6. Minneapolis, MN, pp. 465–470.

Yao, L. S., and F. M. Chen. 1981. A horizontal flow past a partially heated infinite vertical cylinder. *Journal of Heat Transfer* 103: 546–541.

Young, M. F., and T. R. Ulrich. 1983. Mixed convection heat transfer from a vertical heated cylinder in a crossflow. *J. Heat Transfer.*

Yung, S. C., and R. B. Oetting. 1969. Free convection heat transfer from an inclined heated flat plate in air. *ASME J. Heat Transfer* 91: 192

Zarea, M., and E. R. Mayer. 1984. Modeling and simulation of a solar thermal plant using parabolic trough collectors. Presented at the IEA Workshop on the Design and Performance of Large Solar Thermal Collector Arrays. San Diego, CA, June 11–13.

9 Collector Engineering Research and Development

Charles F. Kutscher

Beginning in 1977, the Energy Research and Development Administration (the predecessor of DOE) initiated a National Program for Research and Development in Solar Heating and Cooling. Collector research, development, and commercialization constituted the largest single element of this program. This chapter describes that research as well as subsequent DOE-sponsored collector projects that relate to general engineering research and development rather than to specifically optical, thermal, or material research. Included is design development as well as engineering issues such as wind loads and corrosion. The emphasis here is on collectors designed for temperatures below 300°C—namely, flat plate, evacuated tube, and line-focus types—although work on parabolic dishes and heliostats done as part of the DOE Solar Thermal Technologies Program will also be briefly discussed.

A considerable amount of the collector research resulted from a series of 11 Requests for Proposals (RFPs) and Program Research and Development Announcements (PRDAs) issued in 1977 by ERDA, the forerunner of DOE. Out of 1,200 proposals received, approximately 140 were selected for funding. Many of the research projects cited in this chapter were part of that program.

9.1 Nonconcentrating Collectors

Flat plate collectors are the most common collector type used for heating domestic hot water and supplying building space heat. They can be designed to heat either liquid or air. Evacuated tube collectors have most often been employed for solar cooling applications where the higher temperatures needed to operate absorption and Rankine chillers would result in very low flat plate collector efficiencies. Although they use reflectors, they typically have very low concentration ratios and are thus included in this section.

9.1.1 Design Development

A number of government research contracts were awarded to determine ways for lowering the costs and improving the performance of flat plate collectors. Much of this research was directed to materials improvement such as in the areas of absorber coatings and glazings and is covered elsewhere in this book. However, there were also a number of studies of innovative designs, and key concepts will be reviewed here.

Los Alamos National Laboratory investigated the use of mild steel as a lower cost alternative to copper (Moore 1974). They built an absorber by seam

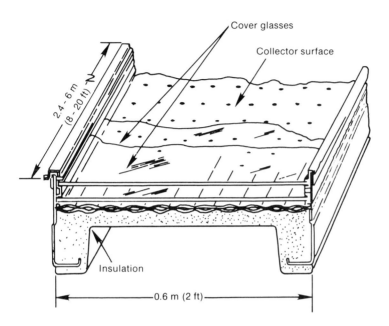

Figure 9.1
LANL structurally integrated solar collector unit.

welding two mild steel sheets and pressure expanding them to form flow passages. Further cost reductions were realized by structurally integrating the absorber with the roof (see figure 9.1). This type of collector array was built and has operated successfully at the National Security and Resources Study Center in Los Alamos, New Mexico. For a time it was marketed in California by Colt Industries.

A black liquid collector was studied by Battelle Memorial Institute (Landstrom 1978). This design limits the stagnation temperature (since the absorbing fluid drains out when the pump is not operating), thereby allowing the use of low cost polymers. A considerable portion of their effort was devoted to developing an acceptable black fluid. They built and tested two collectors utilizing extruded acrylic absorbers and found that they were significantly less expensive than a conventional single glazed flat plate collector of comparable performance. Figure 9.2 shows a cross section of the collector design. A significant question with this collector type is the long-term stability of the acrylic/black fluid combination. A somewhat similar concept employing polycarbonate in an integrated absorber/glazing design was manufactured by Ramada Energy Systems but was eventually abandoned due to manufacturing difficulties. A well-behaved black fluid was ultimately developed by Sperry-

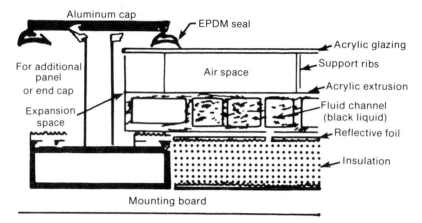

Figure 9.2
Battelle acrylic black liquid solar collector.

Univac (Anderson et al. 1980). Their formulation consisted of a colloidal carbon dispersion with a xanthan gum derivative suspending agent.

The Solar Energy Research Institute investigated the use of fiberglass-reinforced concrete (GRC) as an inexpensive absorber (Kutscher et al. 1984a). It was shown analytically that a high heat removal factor could be obtained in spite of the low thermal conductivity by using closely spaced water passages (molded into the absorber) in a 2-cm-thick plate. Though costs proved to be very low, problems with proper sealing of the passages and header attachment were encountered.

There has been considerable interest in the use of plastic absorbers as a means of cost reduction. An early contract was awarded to CALMAC Manufacturing Corporation for the design and construction of a collector utilizing an EPDM absorber (MacCracken 1978). Costs were lower than for conventional collectors, but performance was also less.

Acurex Corporation investigated the use of thin plastic films for an absorber in hopes that such a design could be manufactured in large quantities by high speed printing and laminating (Muller 1978). They experimented with various polymers. One design consisted of a combination of different layers of polyester films as shown in figure 9.3. Though costs were promising, problems with leakage (due to inadequate bonding of the materials) were never resolved, and it was decided that better polymer materials would be needed.

J. M. Bradley (1977) developed a freeze-tolerant plastic absorber consisting of a spiral of carbon-black-impregnated cross-linked polyethylene tubing. The polyethylene not only withstood 100 freeze/thaw cycles but also held up to

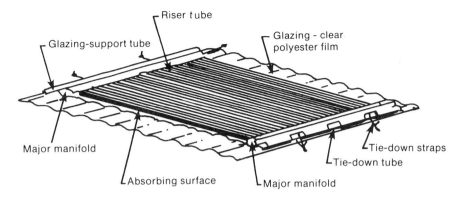

Figure 9.3
Acurex thin-film plastic absorber.

five years of dry stagnation under a single glazing. Researchers at Lawrence Livermore Laboratory successfully demonstrated the use of hypalon bags in the form of shallow solar ponds to supply batch hot water (Casamajor and Parsons 1979).

Perhaps the most extensive government attempt at the construction of a thin film plastic collector was the work performed at Brookhaven National Laboratory (Wilhelm 1984). Their absorber design consists of two layers of a teflon-aluminum foil laminate attached by beads of RTV sealant with the teflon layers facing each other to form the water barrier (see figure 9.4). Water flows down through the collectors at atmospheric pressure. The collector box is constructed by using rigid foam insulation behind the absorber, a sheet of steel around the perimeter, and a sheet of Tedlar glazing wrapped around the entire box.

Prototypes of this collector proved to be extremely lightweight (16 lbs for an 8×3 ft panel) and exhibited very high performance ($F_R \tau \alpha = 0.95$, $F_R U_L = 5.3$ W/m^{2}K) in tests at SERI (Kutscher et al. 1984a). In a large-scale field test, however, delamination of the absorbers resulted in leaking. BNL also experimented with thin stainless steel sheets in place of the teflon-aluminum foil combination but believed that the sealing problems with the plastic model could ultimately be resolved.

The collectors described above all use liquid heat transfer fluids; however, a considerable amount of research has also been done on air collectors. Payne, Inc. built and tested integral wall air heaters utilizing concrete blocks (Payne 1979). Other work on building-integrated air collectors was done by Total Environmental Action, Inc. (Kohler 1978), Alcoa Laboratories (Aluminum

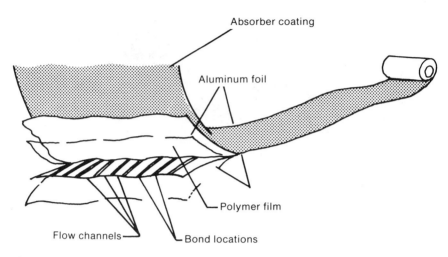

Figure 9.4
Brookhaven laminated thin-film absorber.

Co. of America 1979), and by Mississippi State University (Forbes 1978). West Virginia University built a lightweight air collector containing blackened grooved foamed glass as the absorber (Loth 1978).

The use of a transpired absorber to increase heat transfer area was explored at UCLA and is shown in figure 9.5 (Buchberg and Edwards 1981). Design of this type can improve efficiency and also reduce fan power. Collier (1979) showed that crushed glass could be used as a transpired absorber. Collier and Arnold (1980) have summarized the work of various researchers in the area of mesh-absorber air heaters.

Sunmaster Corporation developed and marketed a drainable version of the Owens-Illinois evacuated tube collector. In 1980 they received a DOE contract to improve their absorber and improve manufacturing processes (Frissora 1981). They attempted to develop a thin low-cost nickel-based foil surfaced on one side with a highly absorptive coating and on the reverse with a low emissivity coating. Problems were encountered in attempting to bond the foil to a glass tube with continuous contact. However, researchers at Los Alamos National Laboratory successfully developed electrochemical coatings and bonded foil absorbers to glass tubes (Grimmer and Avery 1982). Sunmaster also lowered costs by changing from a 4-ft to a 6-ft tube.

While most evacuated tube collectors utilize water as the working fluid, a number of research projects were aimed at developing an air-cooled evacuated collector. Use of air as the heat transfer fluid eliminates problems such as freezing, leakage, and boiling. Under contract to NASA, Owens-lllinois de-

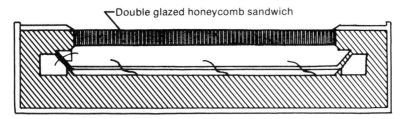

Figure 9.5
Air collector utilizing porous absorber.

signed, built and tested an air-cooled panel utilizing a selective surface which exhibited a very low loss coefficient (Moan 1978). General Electric also developed an air-cooled version of their evacuated tube collector (Hanson 1978). Their design, shown in figure 9.6, used GE TC-100 tubes in conjunction with foam insulation and light gauge galvanized steel.

Solaron Corporation took a different approach to using evacuated tubes for air heating. They designed and built several air heating collectors employing a conventional flat plate absorber, but having a row of evacuated fluorescent-type tubes in place of a double glazing (Jones 1980). They found that although the evacuated tube array effectively suppressed convection, thermal radiation was still significant at higher temperatures. Use of an infrared-reflective coating marginally reduced radiant heat losses but at the cost of significantly reduced transmittance. They made no attempt to commercialize the concept due to the lack of a market for high temperature air heating.

The long thin geometry of evacuated tubes, which makes plumbing difficult, lends itself well to the use of heat pipes for transferring the collected energy to the working fluid. Thermacore, Inc. worked with General Electric to design a collector consisting of 5/16-in. carbon steel heat pipes containing trimethylborate and located within GE evacuated tubes (Ernst 1978). Their design allowed loop fluid to flow directly over the heat pipe condenser with a resultant temperature drop between heat pipe and bulk fluid of only 7°C. Tests at Los Alamos National Laboratory indicated that this collector performed somewhat better than the standard GE TC-100 collector at higher temperatures (Neeper 1981).

9.1.2   Engineering Issues

A considerable amount of work has focused on the use of inexpensive planar reflectors to increase the irradiance on a field of flat plate or evacuated tube collectors. Grimmer et al. (1977) presented results for specular, diffuse, and combination specular/diffuse reflectors used in conjunction with a vertical

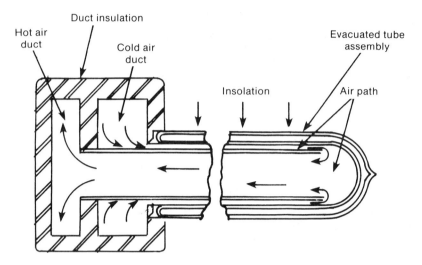

Figure 9.6
General Electric evacuated tube air collector.

wall-integrated collector. Another study by Kaehn et al. (1978) indicated that augmentation of an array of tilted flat plate collectors with planar reflectors can increase annual energy collected during a heating season by 40% to 50%, approximately the same improvement over a simple titled collector array as for the Grimmer configuration. When using planar reflectors, system designers must keep in mind that the stagnation temperature of the collectors will be increased and that the collector materials must be capable of withstanding this elevated temperature.

A considerable amount of work has been done to improve the structural integrity of nonconcentrating collectors. Texas A&M University performed a study (Chevalier 1979) of wind loads on flat plate solar collector arrays. Solar panels were placed in a wind tunnel, and it was found that the force coefficient is a function of aspect ratio, wind angle, clearance ratio, thickness-to-height ratio, and inclination angle. Roof support connections were found to be a critical design area. Based on both wind tunnel and field tests, a theoretical analysis for predicting wind loads was developed.

The effect of hail on flat plate collector glazings has long been a concern. Altas Corporation studied hail effects to determine the extent of this potential hazard (Cox and Armstrong 1980). They developed a general hail risk model based on data from the National Weather Service, including statistical distributions of hailstone size, number of hailstones per unit area per storm, and number of hail days per year in a given location.

These data were used to calculate the maximum probable hail stone size in four locations in the "hail belt." The results were used to determine the proper parameters for impact testing, and this testing revealed that in severe hail areas 3/16-in. tempered glass provides more cost-effective hail protection than cover screens. Altas also took a look at the problem of glass breakage due to vandalism and concluded that its incidence is negligible (De Winter 1978).

Most work on the corrosion of flat plate solar collectors has focused on internal corrosion involving the interaction between the heat transfer fluid and the absorber material. Giner, Inc. performed a study to determine the proper conditions for using glycol heat transfer fluids with aluminum absorbers (Swette et al. 1981). They found that the major cause of corrosion is the presence of heavy ion impurities (chloride, copper, and iron ions) in the glycol and that corrosion inhibitors similar to the ones found in commercial automotive coolants (nitrates, silicates, borates, phosphates, and sodium MBT) are effective in stopping aluminum corrosion at temperatures below 100°C. At temperatures up to 160°C, a 2-wt % zinc powder added to the fluid acts as a heavy metal scavenger and as a sacrificial anode and can stop pitting corrosion even if chloride ions are present.

Los Alamos National Laboratory investigated the corrosion of copper in glycol solutions (Avery 1983). They found that even if the glycol solution were severely degraded, corrosion of the copper might still be acceptable over a 20-year life.

The effects of atmospheric corrosion on flat plate collector enclosures were studied by ECA, Inc. as part of the Solar Reliability and Materials Program at Argonne National Laboratory (Cheng 1983). Candidate materials were exposed on outdoor test racks at nine National Solar Data Network (NSDN) sites representing mild marine, mild industrial, and rural environments. The materials evaluated included galvanized steel, aluminized steel, aluminum, and white polyester painted steel. The analyzed data indicated that the first three materials would last more than 20 years in all nine sites. The painted steel would probably require repainting within 5 years in a mild marine environment and within 5 to 10 years in a mild industrial or rural environment.

Collector array configurations were investigated at the University of Pennsylvania (Lior 1980). Optimum configuration was studied as a function of total area, interrow spacing, row orientation, collector inclination, collector height, wind effects, location, and cost. Graphs were developed to allow the designer to choose optimum parameters. It was shown that performance can be improved by spacing the rows closer together than was typical practice, thereby allowing some row-to-row shading.

9.2   Line Focus Concentrators

To obtain reasonable collector efficiencies at temperatures up to 300°C, it is necessary to reduce the absorber heat loss area relative to the collector aperture area. This is most easily accomplished by focusing the solar radiation onto a fluid-carrying pipe with a parabolic reflector, a compound parabolic reflector, or a Fresnel lens.

9.2.1   Design Development

Very little work on parabolic trough collectors was performed as part of the DOE Heating and Cooling Program. In one contract, however, Williams and Skaggs (1979) developed a means for forming parabolic reflectors without the use of molds or expensive forming equipment. Their technique involved the application of forces and moments to the edges of a flat sheet followed by oven baking. In another contract Honeywell (1979) developed a tensioned membrane reflector for line-focus concentrator applications.

The most commonly used line-focus collector is the parabolic trough. Most of the government-funded design development work in troughs has been directed by Sandia National Laboratories, Albuquerque. The Sandia program involved in-house testing and systems studies as well as funding manufacturers for trough development, either directly or as part of industrial process heat (IPH) field test construction projects.

The major manufacturers involved in this program were Acurex, Solar Kinetics, Foster Wheeler, BDM, and Suntec. Much of the design work by these companies involved improving the optical and thermal performance of their collectors. These efforts, together with work by Custom Engineering, Inc., culminated in the Modular Industrial Solar Retrofit (MISR) program in which six modular systems each containing about 2,500 m^2 of the latest collector designs were built and subjected to qualification tests. Performance data were collected to allow prediction of annual energy, and modifications were made to improve reliability and tracking capabilities. In particular, recommendations were made to improve reflective surface durability and provide better wind resistance (Cameron and Dudley 1984).

Much of the trough design work falling outside of the Sandia program was aimed at achieving large cost reductions. Researchers at SERI found that by lowering the design operating temperature of troughs from 315° to 150°C, they could lower design costs and still capture most of the industrial process heat market available at the higher temperature. This relaxation of design requirements allows use of less rigid, lighter weight reflectors, larger receiver pipes, and multiple-row drive systems employing wire rope (Gee 1983). Fig-

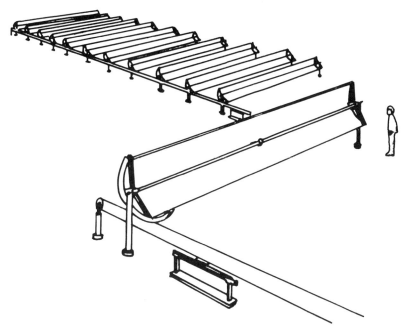

Figure 9.7
Parabolic through array employing multiple row drive system.

ure 9.7 shows a multiple row drive system. This allows for a number of collector rows to be operated by a single drive motor. An analysis performed at SERI also showed that for the supply of industrial steam, significant improvements in delivered energy cost could be obtained by boiling water directly in the collectors (Murphy and May 1982).

Other researchers have sought to lower trough costs by using thin sheets and foils for the reflector. In one study an adjustable trough was developed that allowed for "tuning" by varying the tension on a thin reflective sheet (Dame 1983). For low concentration ratios (on the order of 3), an inexpensive inflated cylindrical reflector can be used in place of the more difficult parabolic shape, and this concept was studied at Lawrence Livermore Laboratory (Gerich 1977) and at Monsanto Research Corp. (Ball et al. 1981) (see figure 9.8). An aluminum foil/polyester laminate was selected as the best reflector material.

The use of a Fresnel lens concentrator for solar cooling applications was employed by Northrup Corp. The original Northrup design had problems with tracking (both mechanical and electronic problems) and with swivel joint leakage, and DOE supplied funding to develop an improved model (Waller

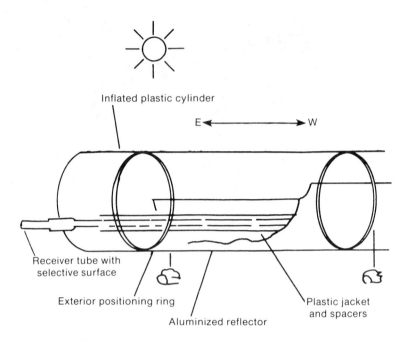

Inflated plastic cylinder

E ←————————————→ W

Receiver tube with
selective surface

Exterior positioning ring

Aluminized reflector

Plastic jacket
and spacers

Figure 9.8
Inflated cylindrical concentrator.

1978). The combination of a wide-angle lens with a subreflector was studied
by Solar Technology, Inc. (Sletten 1980). A study of a low concentration
Fresnel lens collector was also undertaken by Swedlow, Inc. (Holdridge 1978).

For solar cooling applications, the compound parabolic concentrator
(CPC) offers several potential advantages over the parabolic trough or Fresnel
lens concepts, including ability to collect diffuse radiation, insensitivity to
reflector specularity, and feasibility of a nontracking design for concentration ratios less than 2. Unlike other concentrating devices, the CPC is a nonimaging device and does not produce a true focal line, which is why it can
concentrate some diffuse radiation as well as direct. A great deal of development of this concept has taken place at Argonne National Laboratory and
the University of Chicago since 1977. The latest design combines evacuated
tubes with the CPC reflectors and boasts excellent performance, but costs of
this design are likely to be higher than for troughs (McGarity 1982).

Whereas the Argonne work has involved designs with a concentration ratio
of 1.5, other researchers have studied designs with higher concentration ratios.
Direct Energy Corp. developed a unit with a concentration ratio of 2.0 that
utilizes an asymmetric reflector shape (Sterne 1979). Costs were kept down in

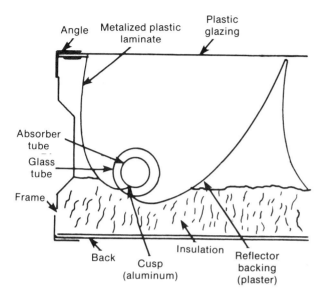

Figure 9.9
Direct Energy Corp.'s asymmetric CPC collector

this design by using a plastic outer glazing, a mild steel absorber, and a nonevacuated glass tube around the absorber (see figure 9.9). Chamberlain Manufacturing Corporation developed a CPC design utilizing GE evacuated tubes and having a concentration ratio of 2.6 (Ballheim 1980).

Early work on line-focus collectors included concepts utilizing a number of flat mirrors concentrating radiation onto an absorber pipe with either the pipe or the mirrors moving to maintain focus. AAI Corp. built collectors of this type utilizing both tracking concepts and tested them in field demonstration projects (Wilkening 1981). A fixed mirror concept was also studied at Georgia Institute of Technology (Williams 1974). These concepts have generally been abandoned due to their higher complexity and lower efficiency (resulting from cosine losses) compared with troughs.

9.2.2 Engineering Issues

A major concern with line-focus collectors is the wind loading that results from their large surface area. These collectors thus require rigid support structures. Sandia National Laboratories studied the foundation requirements for ground-mounted parabolic troughs and recommended economical designs (Auld 1980).

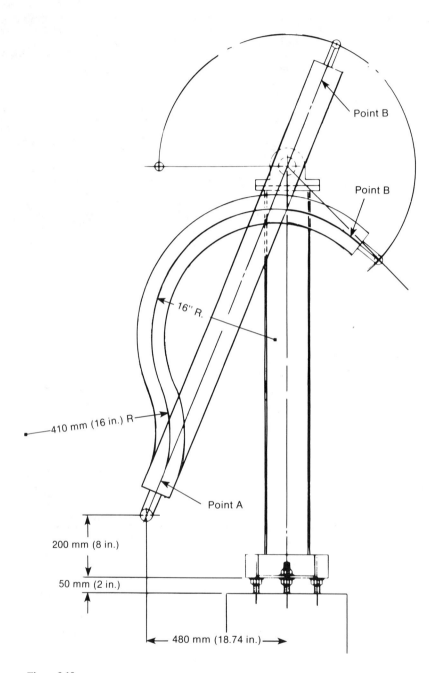

Figure 9.10
Flexible hose installation for tracking collector. Source: Boyd (1980).

Sandia also conducted wind tunnel tests to determine the wind loads on parabolic trough arrays (Randall, McBride, and Tate 1980). Their tests showed that interior collector modules experience a 50% to 60% reduction in peak wind loads compared with those on the edge of the field. A similar reduction on outside rows can be achieved by using a fence or berm with a height equal to three-fourths of the maximum collector height. The effects of wind loading on parabolic troughs were also studied by Colorado State University (Peterka et al. 1980). An assessment of existing wind loading studies discusses these effects, particularly in regard to wind fences and berms (Murphy 1981).

Because tracking collectors move, they require a flexible hose for connecting the moving absorber pipe and the stationary header pipe. These flex hoses have often leaked in the field and have thus been the subject of research at Sandia. It was found that proper installation is the key to success. It is important that these hoses not be flexed beyond their minimum bending radius and that they not be subjected to any torsional stresses during installation or operation. This can be accomplished with the proper use of pipe elbows such that the connecting pipes at each end of the flex hose both lie in a plane perpendicular to the moving receiver tube as shown in figure 9.10 (Boyd 1980).

Because parabolic trough reflectors must have high specular reflectivity for proper concentration, dirt accumulation is a significant concern with these collectors. A study funded by Sandia to determine the effect of dirt accumulation on trough collectors at operating industrial process heat projects found considerable degradation (Morris 1982). In a separate study Bergeron (1982) found that a spray wash followed by a deionized rinse is an effective cleaning technique. Also simply turning the collectors face up during rain showers has been shown to do a reasonable job of cleaning parabolic trough collectors.

Proper layout of a field of parabolic trough collectors is more complicated than for stationary collectors due to row-to-row shading effects, end losses, and the freedom of orientation. A study at Sandia devoted to parabolic trough arrays resulted in a design handbook (Harrigan 1983). A broader handbook produced by SERI covers the layout of parabolic troughs as well as other collector types (Kutscher et al. 1982).

9.3   Point-Focus Collectors

To supply higher temperatures than those that can be delivered by line-focus devices, it is necessary to concentrate the solar radiation even further. So-called "point-focus" collectors typically use a parabolic-shaped dish to focus the solar radiation onto a small receiver. They must always aim directly at the sun and therefore track in two axes.

9.3.1 Design Development

Work on point-focus parabolic dishes was managed for DOE by the Jet Propulsion Laboratory in Pasadena. Development was aimed at supplying both thermal and electric energy. Prototype concentrators were developed in 1982 by Acurex Corp. and by GE. (The GE-designed concentrator was constructed by Ford Aerospace Communications Corp.) Solar Kinetics Inc. built 114 parabolic dishes for use in the Solar Total Energy Project (STEP) in Shenandoah, Georgia. These collectors produce steam, which is used to run a turbine (supplying 400 kW of electricity) as well as supply thermal energy for cooling and for an industrial process (JPL 1982).

JPL studied a number of alternatives for reducing dish costs. These concepts included stamped sheet metal panels, sheet-molding compound, sagged glass, and stressed skin honeycomb structures. They concluded that the stamped sheet metal approach (used at Shenandoah) is best from a structural standpoint but requires further work for high concentration applications (JPL 1983). More recently a hydrostatically loaded membrane was considered as a lower cost concentrator alternative (SERI 1984).

In all of the above concepts, reflectors are arranged on the concave surface of a parabolic dish looking much like a television satellite antenna (see figure 9.11). A different concept was developed by Rensselaer Polytechnic Institute and produced by Power Kinetics, Inc. and was used to supply process steam at the Capitol Concrete Products plant in Topeka, Kansas. This design consists of long mirror slats arranged in parallel fashion on a frame that is curved in one direction only (see figure 9.12). The whole assembly rotates to track solar azimuth, and the individual mirrors move to account for solar altitude. Interestingly this type of collector was first advanced by rocket pioneer Robert Goddard in 1934 (Hagemann 1962).

In the Shenandoah project the heat transfer fluid is heated in the cavity-type receiver of each dish before flowing to a steam generator. An alternative for electricity generation is to locate a heat engine in each individual receiver so that each dish supplies direct electric output. The bulk of the receiver research at JPL has been aimed at the latter application. The receivers developed used three different heat engine cycles: Rankine (manufactured by Ford and AiResearch), Brayton (AiResearch and Sanders), and Stirling (Fairchild). A summary of the receiver characteristics is given in table 9.1 (JPL 1982). The

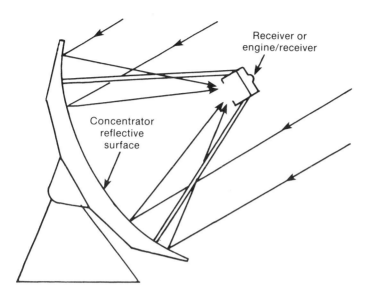

Figure 9.11
Typical parabolic dish.

Figure 9.12
PKI point-focus collector at Capitol Concrete Products plant.

Table 9.1
Point-focus collector receiver characteristics

	Rankine		Brayton		Stirling
Engine cycle	Ford	AiResearch	AiResearch	Sanders Assoc.	Fairchild
Working fluid	Toluene	Steam	Air	Air	Helium
Fluid outlet Temperature [°C (°F)]	400 (750)	705[a] (1,300)	815 (1,500)	1,370 (2,500)	815 (1,500)
Aperture diameter [cm (in.)]	38 (15)	22.8 (9)	25.4 (10)	19.7 (7.75)	27.9 (11)
Integral hybrid design	No	No	No	No	Yes
Efficiency (%)[b]	70 to 90	80 to 92	70 to 80	up to 90	85 (est.)
Maximum pressure, [Pa (psi)]	5.5 (790)	14 (2,000)	0.25 (38)	0.7 (100)	14 (2,000)
Material	Metal	Metal	Metal	Ceramic	Metal
Buffer storage	Yes[c]	No	No	Yes	No

a. This is the capability for the receiver; a typical engine would use a lower flux.
b. Temperature dependent.
c. 135 kg_m (300 lb_m) of copper acts as integral buffer storage.

latest design of a module incorporating a Stirling engine is shown in figure 9.13 (SERI 1984).

9.3.2 Engineering Issues

Like other solar collectors having a large cross-sectional area, wind loads are a serious concern with parabolic dishes. A study by JPL found that existing wind load information was sufficient for determining the effects on a single parabolic dish; however, it was concluded that wind tunnel tests on model heliostat arrays cannot be extrapolated to a field of dishes (Roschke 1984).

The concept of using a clear plastic bubble around a dish to protect it from the wind and allow a less expensive structure was studied at SERI. It was concluded that in order to protect the bubble from the rising hot air plume above the receiver as well as the possibility of beam "walk-off" (movement of the concentrated light beam off of the receiver due to tracking failure), the bubble would have to be uneconomically large (Murphy 1984).

The problem of beam walk-off was also studied at JPL from the standpoint of potential damage to the receiver enclosure. It was found that the only materials that did not slump or shatter when exposed to the concentrated beam resulting from tracking failure were two grades of medium-grain extruded graphite (Jaffe 1984).

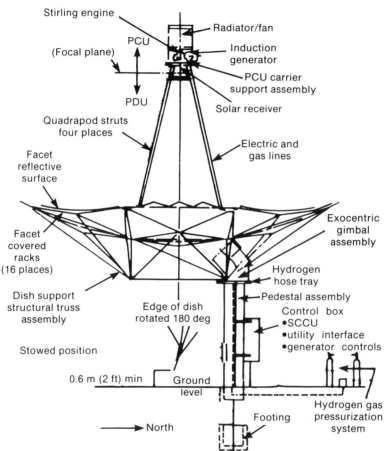

Overall module features
Diameter: 11 m (36 ft 1 in.)
Height: 12 m (39 ft 4 in.)
Weight: 7980 kg (17,560 lb)

Stirling engine

Radiator/fan

PCU

(Focal plane)

Induction
generator

PCU carrier
support assembly

PDU

Solar receiver

Quadrapod struts
four places

Electric and
gas lines

Facet
reflective
surface

Exocentric
gimbal
assembly

Facet
covered
racks
(16 places)

Hydrogen
hose tray

Dish support
structural truss
assembly

Edge of dish
rotated 180 deg

Pedestal assembly

Control box
•SCCU
•utility interface
•generator controls

Stowed position

0.6 m (2 ft) min

Ground
level

North

Footing

Hydrogen gas
pressurization
system

Figure 9.13
Parabolic dish collector incorporating a Stirling engine.

9.4   Heliostats

Solar central receivers utilize an array of ground-mounted mirrors called heliostats to concentrate solar radiation onto a receiver at the top of a centrally located tower. Research on receivers has basically been aimed at improving optics and minimizing heat losses and is covered elsewhere in this book. We will address the design and other engineering issues associated with heliostats.

9.4.1   Design Development

Most of the government-funded research in heliostats was done in support of the 10 MW_e central receiver project near Barstow, California. The heliostats used for this pilot plant were manufactured by Martin Marietta. They consisted of 12-mirror panels, each 1×3 m, of 3-mm low-iron float glass forming a sandwich with aluminum honeycomb and steel with a total reflective area of 39.9 m^2 and weighing 1,875 kg (Battleson 1981) (see figure 9.14).

A number of advanced heliostat concepts have been pursued in order to achieve a better cost/performance ratio than for the pilot plant. So-called second generation heliostats were manufactured by Arco Power Systems, Boeing, McDonnell Douglas, and Martin Marietta. These heliostats were tested at Sandia's Central Receiver Test Facility (CRTF) in 1983. They demonstrated improved tracking accuracy, higher reflectivity, and improved life expectancies (SERI 1984).

One major approach to heliostat cost reduction has been to use lighter weight materials such as plastic reflectors in place of glass. Since lighter weight structures are more susceptible to wind loads, the concept of enclosing lightweight heliostats in clear inflated plastic bubbles was pursued. Boeing built a prototype plastic-enclosed heliostat shown in figure 9.15 (Battleson 1981). SERI studied polymer enclosures both analytically and through bench models. They found that previously considered materials for enclosures had poor tear resistance and identified Tefzel and Halar as alternative materials (Murphy 1984). In general, these plastic enclosures have the disadvantage of relatively short life, and for heliostat applications, the transmissivity loss occurs twice. It was concluded that larger versions of the second-generation heliostats and stretched membrane heliostats (see below) would be more cost-effective than plastic-enclosed designs.

A considerable research effort at SERI has been devoted to the development of lightweight heliostats using a stretched membrane for the reflector. By laminating two layers together while one of them is under tension, it is possible

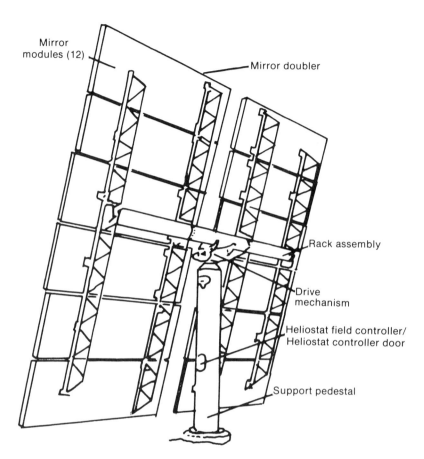

Figure 9.14
Martin Marietta's heliostat.

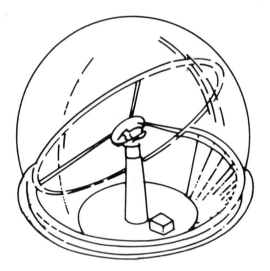

Figure 9.15
Boeing's plastic-enclosed heliostat.

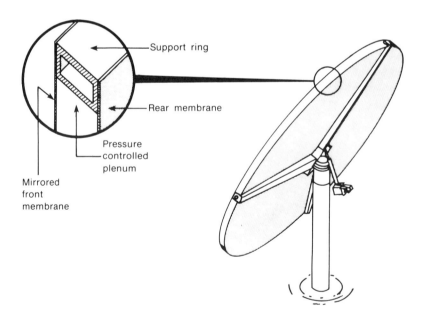

Figure 9.16
Stretched membrane heliostat.

to produce a heliostat with concentration (see figure 9.16). Several proto-types were built, including two 1-m modules with variable focus and tension capabilities (Murphy 1985).

9.4.2 Engineering Studies

The effect of wind on heliostat arrays has been an important concern. Follow-up studies to the Second-Generation Heliostat Development Program by Martin Marietta (Sandia 1982) and McDonnell Douglas (Dietrich et al. 1982) indicated that it is more important to design heliostats for survival strength than for stiffness. Optimum heliostat size in both studies was about 95 m². They also concluded that delivered energy costs could be reduced by lowering the operational wind requirements. Wind tunnel tests at Colorado State University indicated that wind loads are considerably reduced for heliostats located away from the edge of a field and that a fence can provide considerable protection (Cermak 1978).

Specular reflectivity of the heliostats is critical for plant performance, and cleaning is thus a major concern. Heliostats at the Barstow pilot plant are cleaned by rain or with high pressure spray equipment. Foster Miller Associates developed a truck-mounted automated cleaning system that combines a water spray and brushing device. Testing of this system at Sandia's CRTF indicated that it produces better results than the techniques being used at Barstow (SERI 1984).

References

Aluminum Company of America. 1979. *Non-concentrating Collector for Solar-heating Low-Temperature Air Heater, Development of Solar Wall Assembly*. Alcoa Center, PA: Alcoa Labs. March.

Alvis, R. L. 1983. Custom Engineering, Inc. MISR QTS, description and test results. *Proc. Distributed Solar Collector Summary Conference—Technology and Applicatiors*, SAND83-0137C. Albuquerque, NM: Sandia National Laboratories. March.

Anderson, J. H., S. O. Jensen, and J. E. Koviac, 1980. Solar collector studies for solar heating and cooling applications. ALO-5355-72. Sunsource Systems Company, Burnsville, Minnesota and Sperry-Univac, St. Paul, MN. January.

Auld, H. E., and J. B. West. 1980. *Solar Collector Foundation Designs*. SAND 80-7070. Albuquerque, NM: Sandia National Laboratories. September.

Avery, J. G., and J. J. Krall. 1982. Corrosion prevention and fluid maintenance in active solar systems: The state of the art. *Solar Engineering and Contracting Magazine*, 3 part series (February–April).

Ball, G. L., III, J. W. Leffingwell, C. E. McClung, and J. L. Schwendeman. 1981. *Low-Cost Mirror Concentrator Based on Inflated, Double-Walled, Metallized, Tubular Films*, Report no. MRC-DA-944; DE81027813. Dayton, OH: Monsanto Research Corp. DOE contract no. DE-AC04-78AL04227. July.

Ballheim, R. W. 1980. 3x *Compound Parabolic Concentrating (CPC) Solar Energy Collector. Final Technical Report.* Report no. DOE/CS/04239-Tl. Waterloo, IA: Chamberlain Manufacturing Corp. DOE Contract No. DE-AC04-78CS04239. April.

Battleson, K. W. 1981, *Solar Power Tower Design Guide: Solar Thermal Central Receiver Power Systems. A Source of Electricity and/or Process Heat.* SAND81-8005. Albuquerque, NM: Sandia National Laboratories. April.

Bergeron, K. D., and J. M. Freeze. 1981. *Cleaning Strategies for Parabolic Trough Collector Fields: Guidelines for Decisions.* SAND-81-0385. Albuquerque, NM: Sandia National Laboratories.

Boyd, W. E. 1980. Insulated metal hose for tracking receiver application. Proc. Line-Focus Solar Thermal Energy Technology Development Seminar. SAND-80-1666. Albuquerque, NM: Sandia National Laboratories. September.

Bradley, J. M. 1977. Development of a freeze-tolerant solar water heater using cross-linked polyethylene as material of construction. Final report, October 15. COO/2956-77/8.

Buchberg, H., and D. K. Edwards. 1981. *Transpired Solar Air Heaters, 1977–October 1980,* vol. 5. UCLA-ENG-8080. Los Angeles, CA: University of California School of Engineering and Applied Science.

Cameron, C. P. and V. E. Dudley. 1984. Testing of modular industrial solar retrofit industrial process steam systems: Design and performance of large solar thermal collector arrays. Proc. International Energy Agency Workshop. San Diego, CA, June.

Casamajor, A. B., and R. E. Parsons. 1979. Design guide for shallow solar ponds. Lawrence Livermore National Laboratory. UCRL-52385. Rev. 1.

Cermak, J. E., J. A. Peterka, and A. Kareem. 1978. *Heliostat Field-Array Wind-Tunnel Test,* CER78-79 JEC-JAP-AK2. Fort Collins, CO: Colorado State University.

Cheng, C. F. 1983. *Atmospheric Corrosion of Batten and Enclosure Materials for Flat-Plate Solar Collectors, Final Report.* ANL-83-30, DE83012304. Lisle, IL: ECA, Inc.

Chevalier, H. L., and D. J. Norton. 1979. *Wind Loads on Solar Collector Panels and Support Structure.* DOE/CS/35130, DE82014983. College Station, TX: Texas A&M University.

Collier, R. K., and F. H. Arnold. 1980. Comparison of transpired beds for solar collector applications. Proc. 1980 Annual Meeting of the American Section of the International Solar Energy Society. Phoenix, AZ. June.

Collier, R. K. 1979. The characterization of crushed glass as a transpired air heating solar collector material. Presented at the 1979 Annual Meeting of ISES, Atlanta, GA. May.

Cox, M., and P. R. Armstrong. 1980. *Need for and Evaluation of Hail Protection Devices for Solar Flat Plate Collectors, Final Report.* ALO-4291-2. Santa Cruz, CA: Altas Corp.

Dame, R. E. 1983. *Line-Focusing Adjustable Parabolic Solar-Concentrating Collector. Final Report.* DOE/CS/15072-TS; DE84001590. Silver Spring, MD: (available from NTIS).

Dietrich, J. J., et al. 1982. *Optimization of the Second Generation Heliostat and Specification.* SAND-82-8181, DE82018317. Albuquerque, NM: Sandia National Laboratory.

Ernst, D. M. 198l. *Cost Effective Solar Collectors Using Heat Pipes.* DOE/CS/34099-4, DE82011239. Lancaster, PA: Thermacore, Inc.

Forbes, R. E., and R. W. McClendon. 1979. *Addition of Inexpensive Solar Air-Heaters to a Pre-Engineered Metal Buildtng.* ALO-4086-1. Mississippi State, MS: Mississippi State University.

Frissora, J. R. 1981. Design of improved coating and construction of an evacuated solar receiver. *Proc. 1981 Solar Heating and Cooling Contractors' Review Meeting.* Washington, DC: U.S. Department of Energy.

Gee, R. C. 1983. *Low Temperature Parabolic Troughs: Design Variations and Cost Reduction Potential.* SERI/TR-253-1662. Golden, CO: Solar Energy Research Institute.

Gerich, J. W. 1977. *An Inflated Cylindrical Solar Concentrator for Producing Industrial Process Heat.* UCID-17612, Livermore, CA: Lawrence Livermore Laboratory.

Grimmer, D. P., and J. S. Avery. 1982. Bonding Solar-Selective Absorber Foils to Glass Receiver Tubes for Use in Evacuated Tubular Collectors: Preliminary Studies. *Solar Energy* 29, 2.

Grimmer, D. P., K. S. Zinn, I. C. Herr, and B. E. Wood. 1977. Augmented solar energy collection using different types of planar reflective surfaces: Theoretical calculations and experimental results. Proc. 1977 Annual Meeting of the American Section of the International Solar Energy Society. Orlando, FL. June.

Hagemann, E. R. 1962. R. H. Goddard and solar power. *Solar Energy* 6, 2.

Hanson, K. L. 1978. Non-concentrating collectors for solar heating and cooling medium temperature air heater. *Proc. 3rd Annual Solar Heating and Cooling Research and Development Branch Contractors' Meeting.* CONF-780983. Washington, DC: U.S. Department of Energy.

Harrigan, R. W. 1981. *Handbook for the Conceptual Design of Parabolic Trough Solar Energy Systems—Process Heat Applications.* SAND81-0763. Albuquerque, NM: Sandia National Laboratories.

Holdridge, D. 1978. Development of a 10 × lens concentrator. *Proc. 3rd Annual Solar Heating and Cooling Research and Development Branch Contractors' Meeting.* CONF-780983. Washington, DC: U.S. Department of Energy.

Jaffe, L. D. 1983. *Solar Tests of Aperture Plate Materials for Solar Thermal Dish Collectors.* N-84-22010, JPL-5105-121. Pasadena, CA: Jet Propulsion Laboratory.

Jet Propulsion Laboratory. 1983. *Solar Thermal Technology, Annual Evaluation Report. Fiscal Year 1982,* vol. 2. Technical, report no. DOE/JPL-1060-81. Pasadena, CA: Jet Propulsion Laboratory.

Jet Propulsion Laboratory, June 1982, *Solar Thermal Technology, Annual Technical Progress Report, FY 1981,* vol. 2. *Technical,* JPL 82-60, DOE/JPL-1060-53. Pasadena, CA: Jet Propulsion Laboratory.

Jones, D. E., L. E. Shaw, and G. O. G. Löf. 1981. *Development of a High Performance Air Heater through Use of an Evacuated Tube Cover Design.* Final report, DOE/CS/34193-T1. Englewood, CO: Solaron Corp.

Kaehn, H. D., M. Geyer, D. Fong, F. Vignola, D. K. McDaniels, K. W. Boeer, and G. E. Franta (eds.). 1978. Experimental evaluation of the reflector-collector system. Proc. 1978 Annual Meeting of the American Section of the International Solar Energy Society, vol. 2.1. Denver, CO. August.

Kohler, J. T., P. L. Temple, J. J. Coleman, and P. W. Sullivan. 1978. *Development of a Site-Fabricated, Building Integrated Air Collector.* Interim report. *September 30, 1977–March 31, 1978.* ALO-4123-T3. Harrisville, NH: Total Environmental Action (TEA).

Kutscher, C. F., R. Davenport, R. Farrington, G. Jorgensen, A. Lewandowski, and C. Vineyard. 1984. *Low-Cost Collectors/Systems Development Progress Report.* SERI/RR-253-1750. Golden, CO: Solar Energy Research Institute.

Kutscher, C. F., R. L. Davenport, D. A. Dougherty, R. C. Gee, P. Michael Masterson, and E. Kenneth. May 1982. *Design Approaches for Solar Industrial Process Heat Systems.* SERI/TR-253-1356. Golden, CO: Solar Energy Research Institute. August.

Kutscher, C. F., and R. Gee. 1984. A design procedure for solar industrial process heat systems. *Workshop on the Design and Performance of Large Solar Thermal Collector Arrays.* SERI/SP-271-2664. Washington, DC: International Energy Agency.

Landstrom, D. K., S. G. Talbert, G. H. Stickford, Jr. R. D. Fischer, and R. E. Hess. 1978. *Development of a Low-Temperature, Low cost, Black-Liquid Solar Collector. Final Report.* ALO 4097-1. Columbus, OH: Battelle Memorial Institute. October.

Lior, N., and H. Yeh. 1980. Collector array studies. *Proc. Annual DOE Active Solar Heating and Cooling Contractors' Review Meeting.* Washington, DC: U.S. Department of Energy. March.

Loth, J. L. 1978. Grooved foamglass solar air heater. *Proc. 3rd Annual Solar Heating and Cooling Research and Development Branch Contractors' Meeting.* CONF-780983. Washington, DC: U.S. Department of Energy. September.

MacCracken, C. 1978. *Design and Installation Package for the Sunmat Flat Plate Solar Collector.* CONF-780983. Englewood, NJ: Calmac Mfg. Corp. March.

McGarity, A. E. 1982. Simulation modeling of solar heating systems with CPC collectors. *Progress in Solar Energy: The Renewable Challenge,* DE82905269, Boulder, CO: American Solar Energy Society. (Work performed at Argonne National Laboratory). June.

Moan, K. L. 1978. Development and delivery of 334 sq. ft. of prototype air cooled collector. *Proc. 3rd Annual Solar Heating and Cooling Research and Development Branch Contractors' Meeting.* CONF-780983. Washington, DC: U.S. Department of Energy.

Moore, S. W. 1974. Structurally integrated steel solar collector development. *Proc. Workshop on Solar Collectors for Heating and Cooling Development of Buildings.* CONF-740811-3. November.

Morris, V. L. 1982. *Final Report on Solar Collector Materials Exposure at IPH Site Environments,* vol. 2. SAND81-7028/II, DE82011688. Albuquerque, NM: Sandia National Laboratories.

Muller, T., and J. Sullivan. 1982. *Further Development of a Low-Cost Solar Panel.* DOE/AL/12032-3, DE83002875. July.

Murphy, L. M. 1985. *Structural Design Considerations for Stretched Membrane Heliostat Reflector Modules with Stability and Initial Imperfection Considerations.* Golden, CO: Solar Energy Research Institute.

Murphy, L. M. 1983. *Polymer Enclosed Thermal Power Dishes: An Initial Feasibility, Engineering and Cost Performance Assessment.* SERI/SP-253-2197. Golden, CO: Solar Energy Research Institute. December.

Murphy, L. M. 1981. *An Assessment of Existing Studies of Wind Loading on Solar Collectors.* SERI/TR-683-812. Golden, CO: Solar Energy Research Institute.

Murphy, L. M., and E. K. May 1982. *Steam Generation in Line Focus Solar Collectors: A Comparative Assessment of Thermal Performance, Operating Stability and Cost Issues.* SERI/TR-632-1311. Golden, CO: Solar Energy Research Institute, pp. 15–78. April.

Neeper, D. A. 1980. Active solar collector and materials research. *Proc. DOE Active Solar Heating and Cooling Contractors' Review Meeting.* LA-UR-80-852, pp. 9–28. March.

Payne, P. R., and J. P. Brown. 1979. *Experiments with Structural Concrete Blocks Which Double as Solar Air Heaters, Final Report.* DOE/CS/34138-T1. Annapolis, MD: Payne, Inc.

Peterka, J. A., J. M. Sinou, and J. E. Cermak. 1980. *Mean Wind Forces on Parabolic Trough Solar Collectors.* SAND80-7023. Fort Collins, CO: Department of Civil Engineering, Colorado State University. May.

Randall, D. E., D. D. McBride, and R. E. Tale. 1980. *Steady-State Wind Loading on Parabolic Trough Solar Collectors.* SAND-79-2134, New York: American Society of Mechanical Engineers. March.

Rask, D. R., et al. 1977. *Low-Cost Solar Air Heater.* COO-2929-13. Minneapolis, MN: Honeywell Inc. August.

Roschke, E. J. 1984. *Wind Loading on Solar Concentrators: Some General Considerations.* DOE/JPL-1060-66, DE85000337. May.

Sandia National Laboratory. 1982. *Second-Generation Heliostat Optimization Studies. Final Report.* SAND-82-8175; DE82017058. Albuquerque, NM: Sandia National Laboratories. May.

Sletten, C. J., et al. 1980, *Image Collapsing Concentrators. Final Report.* DOE/CS/34163-T1 Bedford, MA: Solar Energy Technology, Inc. September.

Smith, Gary A. 1978. Dual curvative acoustically damped concentrating collector. DOE contract no. EM-78-C-04-4196. Proc 3rd Annual Solar Heating and Cooling Research and Development Branch Contractors' Meeting. Washington, DC. September.

Solar Energy Research Institute. 1984. *Solar Thermal Technology. Annual Evaluation Report. Fiscal Year 1983.* SERI/PR-253-2188. Golden, CO: Solar Energy Research Institute. August.

Sterne, K. E. 1979. *Optimized Concentrating/Passive Tracking Solar Collector. Final Report.* DOE/AL/04181-TI. Irvine, CA: Mithra Company. January.

Swette, L. L., and F. H. Cocks. 1981. *Study of Aluminum Corrosion in Aluminum Solar Heat Collectors Using Aqueous Glycol Solution for Heat Transfer. Final Report.* DOE/CS/31072-2; DE81027582. July.

Waller, R. 1978. Development of a second generation concentrating collector. *Proc. 3rd Annual Solar Heating and Cooling Research and Development Branch Contractors' Meeting.* CONF-780983. Washington, DC: U.S. Department of Energy. September.

Wilhelm, W. G. 1984. Thin polymer film collectors as a contribution to the solar industry. *Proc. Annual Meeting of the American Solar Energy Society.* Boulder, CO: American Section of the International Solar Energy Society. (Paper available from Brookhaven National Laboratory BNL-34937.) June.

Wilkening, H. A. 1981. AAI Corporation receiver design experience in concentrating solar collectors. *Joint Pressure Vessels and Piping Materials Nuclear Engineering and Solar Conference.* CONF-810625. New York: American Society of Mechanical Engineers. June.

Williams, J. R., and S. F. Hutchins. 1974. Development of a Solar Heat Supply System with Fixed Mirror Concentrators. *Proc. Solar Collectors for Heating and Cooling of Buildings.* NSF-RA-N-75-019. Washington, DC: National Bureau of Standards.

Williams, O. G., and R. Skaggs. 1978. Solar parabolic trough forming process. DOE contract no. EG-77-6-04-4158. Pro. 3rd Annual Solar Heating and Cooling Research and Development Branch Contractors' Meeting. September. Washington, DC.

Williamson, R. T. 1980 Versatile transparent polymer collector. *Proc. Annual Meeting of the American Section of the International Solar Energy Society (AS/ISES),* vol. 3.1., pp. 461–464. CONF-800604-P2. Boulder, CO: American Section/International Solar Energy Society. June.

de Winter, F. 1978. Development of cost effective techniques and concepts for the protection of glazing against breakage caused by hail, vandals, and thermal stresses. *Proc. Annual DOE Active Solar Heating and Cooling Contractors' Review Meeting.* CONF-780983. Washington, DC: U.S. Department of Energy.

10 Solar Pond Research and Development

John R. Hull and Carl E. Nielsen

10.1 Introduction

10.1.1 Overview

Solar ponds combine low-cost solar energy collection with long-term storage for a variety of low temperature thermal applications. The term *solar pond* describes several different concepts, all of which use water as both a direct absorber of solar radiation and a thermal storage medium for collected energy. Most commonly, convection is suppressed at or near the top of the pond by a means that is partially transparent to solar radiation. Among the various convection suppressing methods, the salt gradient has received the most attention.

As shown in figure 10.1, a salt gradient solar pond is a body of water that typically has three regions (from top to bottom): surface zone, gradient zone, and lower zone. The lower zone is a homogeneous, concentrated salt solution. It is covered by the gradient zone, a thermal insulating layer that contains a salt gradient such that water closer to the surface is always less salty than the water below it. The surface zone is a homogeneous layer of low salinity brine or freshwater. If the salt gradient is large enough, there will be no convection in the gradient zone when heat is absorbed on the bottom, because the hotter, saltier water at the bottom of the gradient will be denser than the colder, less salty water above it. Because water is transparent to visible light but opaque to infrared radiation, energy in the form of sunlight reaches the darkened bottom, is absorbed there, and can escape only via conduction. The thermal conductivity of water is moderately low, and if the gradient zone is relatively thick, heat escapes upward from the lower zone very slowly. This makes the solar pond both a thermal collector and a long-term storage device.

Solar ponds, specifically salt-gradient ponds, enjoy a number of advantages compared with other solar thermal technologies. They are relatively simple in concept, consisting mainly of an excavation in the ground, water, and some salt. In many locations excavation is inexpensive, and salt and water are readily available, resulting in low total cost and the use of indigenous resources. In some instances the cost may be low enough to justify converting the low temperature heat from the pond to electricity, even though the total efficiency is only between 1% and 2%. Also operating costs are expected to be low. While fluid pumping is usually necessary to extract heat from the solar pond, collection and storage are completely passive.

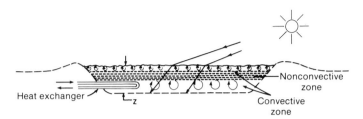

Figure 10.1
Schematic representation of a salt-gradient solar pond.

Naturally occurring salt-gradient solar ponds and lakes have been found in many wind-sheltered parts of the earth. The high temperatures [up to 70°C (158°F)] observed in some of these natural solar ponds, coupled with the abundance of salt lakes and salt deposits throughout the world, strongly suggest that in many locations artificial solar ponds are technically feasible with minimal environmental impact. However, as is typical of many simple solar technologies, there are subtle effects, such as the hydrodynamics of double-diffusive convection, that play a major role in the behavior of solar ponds. These effects must be carefully considered in any successful pond design and better understood before use of solar ponds becomes widespread. Both small and large prototype solar ponds have been made to work, but these efforts have not yet demonstrated the economic competitiveness of the technology. Although many aspects of solar pond design and operation are well understood, there are still many basic questions that must be answered before performance is maximized and cost minimized.

10.1.2 Scope

Reviews of solar pond technology have been given by Nielsen (1980a), Tabor and Weinberger (1980), Tabor (1981), Fynn and Short (1983), and French et al. (1984). This chapter updates and fills some gaps in these previous reviews. Although interesting solar pond activities have been conducted in many parts of the world, this discussion emphasizes activities in the United States and focuses on research conducted to provide a basic understanding of solar pond technology. The discussion is limited to nonconvecting solar ponds, specifically salt-gradient solar ponds (except in section 10.9). Shallow solar ponds, waterbags, ponds with seasonal storage and external collectors, and evaporation suppression on freshwater ponds are not discussed. Even with these restrictions, the space limitation of this chapter prohibits a detailed treatment of all solar pond research. The references and discussion in this chapter have been selected to illustrate the state of the art and are not meant to be exhaustive. For a more complete listing of solar pond literature, the reader is directed to the bibliography compiled by the Solar Energy Research Institute (SERI 1984).

10.1.3   General Solar Pond Practice

Fortunately most of the construction and general operation of solar ponds employs well-established technologies, including constructing earth berms, installing pond liners, pumping fluids, and delivering heat to a load. Swimming pool techniques are often used to control biological organisms and maintain pond transparency. Details of construction and operation procedures are found in the review articles mentioned in subsection 10.1.2.

A unique requirement of solar ponds is that a salt gradient must be established and maintained near the top of the pond. A simple and versatile method to establish or modify the gradient, adopted almost universally, is the redistribution method of scanning injection described by Zangrando (1980). Redistribution involves pumping fluid from one level of the pond (or external tank) to another level of the pond in such a way that the injected fluid mixes with the native fluid, thus changing its density.

As an example of the injection technique, consider the establishment of a gradient zone that will have constant gradient and occupy the upper half of the pond. Initially the pond is filled three-quarters full with a concentrated salt solution. Freshwater is injected at what will be the bottom of the gradient zone. Because freshwater is less dense than the brine, it will rise to the surface, mixing with the brine. To a first approximation, all the water above the injection point will be of uniform concentration. As the pond fills, the point of injection should rise twice as fast as the pond surface rises so that a uniform gradient is established below the injection level. In practice, some of the injected water will reach the surface before it is completely mixed, forming a weak gradient there. The velocity of the injected water should be high enough to mix as completely as possible with the water in its path before the buoyancy force begins to make it rise. However, the velocity should be low enough to minimize entrainment of fluid from below the injection level. The gap separating the injection diffuser plates is generally about 2–3 mm (0.08–0.12 in.), and water velocities as high as 2.6 m/s (8.5 ft/s) have been successfully used during gradient establishment. The rate at which the injection level is raised can be varied to create any desired salinity profile.

A modification of this technique is layered injection in which the diffuser is raised in discrete intervals (e.g., every few centimeter). The injection velocity should be high enough that some of the fluid from below is mixed with the injected fluid to smooth out the discontinuities between injection levels. If the injection velocity is too low, or if the interval spacing is too large, discrete layers may form. If the discrete layers are too thick, diffusion between layers will not form a strong enough gradient by the time the pond becomes hot. Once a large temperature difference exists across two gradient regions separated by a thin, nearly homogeneous layer, a salt gradient will not form in the homogeneous layer, and it will become convective. Unless artificially eliminated, such internal convective layers usually persist until the pond cools.

The injection technique can also be used to eliminate internal convective layers within the gradient zone or to modify the gradient profile (Nielsen 1976, 1979, 1983a; Zangrando 1980). For these purposes it may be desirable to deposit the injected brine or water relatively close to the injection level with a minimum of general stirring. This can be accomplished by using a diffuser with a very thin gap, typically 0.1–1 mm (0.004–0.04 in.) (Nielsen 1979, 1983a). Even when the injected fluid has a density very different from that of the pond at the injection level, equilibration of the thin injected stream with the background occurs before it is displaced more than a few centimeters up or down by the buoyancy forces. Quantitatively predictable salinity profile changes can be produced in this way.

10.2   Thermal Performance Estimation

Proper engineering of a solar pond requires that the size of the pond be carefully chosen to match the intended load. Several of the different design methods available to predict thermal performance are discussed in this section.

10.2.1   Steady-State Model

In addition to providing a good first approximation of the thermal performance of a solar pond, a steady-state analysis provides physical insight into the effects of parameter variation for a given design. The predicted heat storage zone temperature is the same as the mean value predicted by a model driven by arbitrary periodic inputs (see subsection 10.2.2).

Calculations of steady-state or mean-value solar pond performance were made by Weinberger (1964), Rabl and Nielsen (1975), and Nielsen (1980a). The two earlier papers omitted the effect of the convective surface zone and therefore overestimated the pond temperature. A recent analysis by Kooi (1979) shows that pond performance can be described by an equation of the same form as the Hottel-Whillier-Bliss equation for flat plate solar collectors.

Energy collected per unit area is given by

$$q = (z_2 - z_1)^{-1} \left[\int_{z_1}^{z_2} h(z) \, dz \right] H - (z_2 - z_1)^{-1} K \Delta T, \tag{1}$$

where

q = the energy input rate per unit area into the volume below z_2,

z = depth below the surface,

H = insolation at the pond surface,

$h(z)$ = the fraction of insolation penetrating to depth z,

K = thermal conductivity of water,

ΔT = the total temperature difference between the top and bottom of the gradient zone,

z_1 = the depth of the top of the gradient zone,

z_2 = the depth of the bottom of the gradient zone.

This equation describes pond energy balance, but in a rather unfamiliar way because the first term on the right includes the effect of radiation absorbed in the gradient zone and the second term is *not* the heat conducted upward through the z_2 plane. The factor multiplying H on the right side of equation (1) corresponds to the absorptance-transmittance product in the Hottel-Whillier-Bliss equation:

$$\alpha\tau = (z_2 - z_1)^{-1} \int_{z_1}^{z_2} h(z) \, dz.$$

Kooi compares the performance characteristics of solar ponds with those of flat plate collectors. He discusses the effect of varying surface zone thickness and shows the importance of using the optimum gradient boundary depth z_2 to obtain maximum output with a given value of $\Delta T/H$. Equation (1) shows that energy collected depends directly on $h(z)$, and it is therefore essential to use in the calculation the $h(z)$ characteristic of possible real pond operation rather than an ideal $h(z)$ evaluated for distilled water. It may be emphasized that since $h(z)$ appears only in an integral between z_1 and z_2, the form of $h(z)$ for $z < z_1$ (i.e., close to the surface) is irrelevant. Thus, if $h(z)$ is approximated for calculation by an analytic form, the quality of the fit is important only between z_1 and z_2. Kooi (1981) and Hull (1983) have discussed the modifications necessary to include the effects of a partially reflecting floor.

An equivalent mean-value performance calculation given by Nielsen (1980a) emphasizes the temperature profiles in the gradient zone. Results are summarized in a performance chart from which the effect of varying surface zone depth z_1 and the dependence on gradient zone lower boundary z_2 can be read directly.

The above analyses are made for zero heat loss to the earth. To incorporate heat loss to the earth into the calculation for efficiency e, the definition of q in equation (1) is modified to include both heat removed for use and heat lost to the earth. With the heat loss term carried to the right side,

$$e = (z_2 - z_1)^{-1} \left[\int_{z_1}^{z_2} h(z)\,dz \right] - (z_2 - z_1)^{-1} \frac{K\Delta T}{H} - \frac{g\Delta T}{H}, \qquad (2)$$

in which g is the ground heat loss coefficient.

Hull et al. (1984) calculated g assuming a constant temperature heat sink at distance D_g below the bottom of the pond:

$$g = \left(\frac{1}{D_g} + \frac{bP}{A} \right) K_g, \qquad (3)$$

where b is a constant ($0.5 < b < 1.4$) that depends on sidewall geometry and insulation strategy, K_g is the effective ground thermal conductivity, P is the pond perimeter, and A is the pond surface area.

For a given pond operating temperature, the steady-state model readily yields the gradient zone thickness necessary to provide maximum efficiency. Results from the model indicate that increasing the thickness of the upper convecting zone decreases efficiency and that thermal performance is very sensitive to changes in the water transparency and to ground losses.

10.2.2 Fourier and Computer Models

One-dimensional analytical models also exist that predict solar pond thermal performance with insolation, ambient temperature, and load conditions that vary sinusoidally through the year. These models assume that the pond is large enough that perimeter heat loss is negligible. Rabl and Nielsen (1975) presented results for the solar pond with either perfect insulation from the ground or with an infinitely deep ground. Crevier and Moshref (1981) extended these results to include a water-table heat sink at some fixed depth below the pond. The results of these models are valid after initial transients have decayed. The analysis can be used on smaller ponds by incorporating the perimeter heat loss as a reduction of the usable heat from the load.

For these models the sinusoidal input parameters are

$$H(t) = \bar{H} + \tilde{H} \cos \omega t \qquad \text{(insolation)},$$

$T_a(t) = \bar{T}_a + \tilde{T}_a \cos(\omega t - \delta_a)$ (ambient temperature),

and

$U(t) = \bar{U} + \tilde{U} \cos(\omega t - \delta_u)$ (heat extraction rate),

where $\omega = 2\pi/\text{yr}$, δ is a phase angle, and $t = 0$ normally occurs near the summer solstice. The output of the models is the temperature of the heat storage zone:

$T(t) = \bar{T} + \tilde{T} \cos(\omega t - \delta)$,

where $\bar{T}$, $\tilde{T}$, and δ are functions of the input parameters. The output pond temperature $T(t)$ and the load profile $U(t)$ can be used as input parameters to a load model (e.g., an electricity-generating turbine).

Where the load profile or insolation cannot easily be put in the form of a simple sine function, as with a peaking plant, thermal-network type computer models of solar ponds can be used. Different models of this type, such as SOLPOND (Jayadev and Henderson 1979; Henderson and Leboeuf 1980), have been developed by numerous researchers and used to study solar pond behavior in a wide variety of circumstances. In many cases the representation becomes a simple one-dimensional set of finite-difference equations that can be readily solved with a microcomputer. Results of these studies show that for a given location, the performance of the pond is very sensitive to water transparency, heat loss to the ground, and load distribution over the year. Predictions from some of these models have been compared with data obtained from operating solar ponds in the United States, and within the accuracy of the data, agreement is satisfactory.

The output of solar pond models can be used in estimating the economics of different solar pond designs. Lin (1982) conducted a cost study for many different applications, taking into account regional climates and geographical differences throughout the United States, and he concluded that solar ponds could supply several quads ($> 10^{18}$ J) per year. Drost and Johnson (1983) coupled a power plant optimization model to SOLPOND to study a conceptual solar pond design for the Great Salt Lake. Other design feasibility studies of solar pond electricity production include SERI (1981), May, Leboeuf, and Waddington (1982, 1983), and Carmichael et al. (1985).

10.2.3 Ground Heat Loss

A problem that has occurred in several experimental solar ponds in the United States is unexpectedly high ground heat loss in moist but unsaturated, clay soils. Very careful measurements by Hull, Kamal, and Nielsen (1983) and Hull et al. (1984) have shown that the soil effective thermal conductivity can range from 2.0 to 2.4 W/m °C (14–17 Btu in./h ft^2 °F) and does not decrease with time. Because solar pond efficiency is very sensitive to ground heat loss, it is desirable to predict the expected soil conductivity for any proposed site. (The problem of measuring thermal conductivity of the soil is discussed is subsection 10.6.5.) Unfortunately, general analysis of combined heat and moisture transport in a porous medium is difficult, and at solar pond temperatures, the highly temperature-dependent vapor transport is expected to be very important. Leboeuf and Johnson (1984) coupled SOLPOND with a model that incorporates the effects of moisture transport into the soil thermal conductivity. Although the results of this study are preliminary, there is qualitative agreement with observed data.

A separate problem associated with ground heat loss is the effect of a moving water table on pond efficiency. The first term in equation (3) assumes a constant temperature sink at the depth of the water table, essentially assuming an infinitely large subterranean flow. Hull (1985) has shown that for horizontal sheet flows in either steady-state or transient conditions, this problem is equivalent to that of a thin-walled heat exchanger. In steady state,

$$\frac{q}{q_{max}} = N[1 - \exp(-N)], \tag{4}$$

where q is the actual heat loss from the pond bottom, q_{max} is that predicted by the first term in equation (3), and

$$N = \frac{\dot{m}cD_g}{K_gL}, \tag{5}$$

where L is the length of the pond, c is the specific heat of water, and $\dot{m}$ is the mass flow rate per width of the water table. Duyar and Bober (1984) have examined this problem numerically.

10.3   Gradient Zone Stability

Although many different effects are important to the operation of the solar pond, stability of the gradient zone is essential. The gradient zone is an example of a double-diffusive system, with the slowly diffusing salt stabilizing against the faster diffusing heat. The complex dynamical system is in non-equilibrium, and under some circumstances the potential energy stored in the temperature gradient can be released, resulting in mixing.

10.3.1   Theory of Uniform Gradients

Oceanographers (Turner 1973) originated analyses of double-diffusive systems, and their results have often been applied to understand solar pond hydrodynamic behavior. A useful understanding of the physical mechanisms involved with the gradient zone can be achieved by linear stability analysis (e.g., Baines and Gill 1969). Uniform initial gradients of salinity and temperature are assumed in an infinite horizontal fluid slab of depth d. Constant salinity and temperature reservoirs with stress-free boundaries are above and below the slab. The gradient zone is characterized by

$$R_S = \frac{g\beta\Delta S d^3}{\nu\kappa_T}, \tag{6}$$

$$R_T = \frac{g\alpha\Delta T d^3}{\nu\kappa_T}, \tag{7}$$

where R_S is the salinity Rayleigh number, R_T is the thermal Rayleigh number, g is the acceleration of gravity, β is the relative density expansion coefficient with respect to salinity, α is the absolute value of the relative density expansion coefficient with respect to temperature, ΔS is the difference in salinity across the zone, ΔT is the difference in temperature across the zone, ν is the dynamic viscosity, and κ_T is the thermal diffusivity. For static stability $R_T < R_S$. Dynamic stability in terms of R_S and R_T is indicated in figure 10.2. The boundary between stability and instability is given by

$$R_T = \frac{(\text{Pr} + \text{Le}^{-1})}{(\text{Pr} + 1)}R_S + \frac{27}{4}\pi^4\frac{(1 + \text{Le}^{-1})(\text{Pr} + \text{Le}^{-1})}{\text{Pr}}, \tag{8}$$

where $\text{Pr} = \nu/\kappa_T$ is the Prandtl number, $\text{Le} = \kappa_T/\kappa_S$, and κ_S is the salinity diffusivity. This boundary is indicated by the solid line in figure 10.3. Because $\text{Pr} = 3\text{--}10$ and $\text{Le} = 30\text{--}140$ for solar ponds, instability can occur even when the density distribution is statically stable. Linear stability analysis predicts that the instability takes the form of overstable oscillations. Oscillations attributed to solar radiation, which may be related to this instability, have been observed in the gradient zone of a solar pond by Almanza and Bryant (1983). For large Rayleigh numbers ($>10^{10}$) characteristic of solar ponds, equation (8) can be written as

$$R_T = \frac{(\text{Pr} + \text{Le}^{-1})}{(\text{Pr} + 1)}R_S, \tag{9}$$

which was originally suggested by Weinberger (1964) as the stability criterion

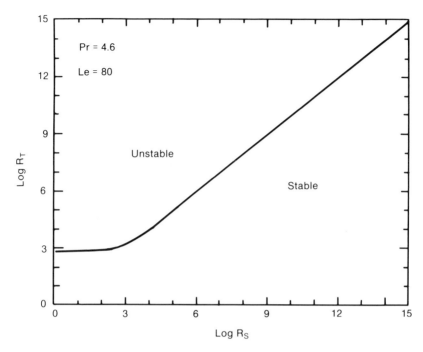

Figure 10.2
Stability diagram of a salt-stratified fluid slab heated from below. R_S is the salinity Rayleigh number, R_T is the thermal Rayleigh number, Pr is the Prandtl number, and Le is the ratio of thermal diffusivity to salt diffusivity.

for solar ponds. Thus the R_S required for dynamic stability is only moderately greater than R_T. In solar pond practice the required salinity gradient is likely to be much higher than that needed for dynamic stability, in order to maintain the gradient zone boundaries at a constant position, as discussed in subsection 10.4.2. Wall effects also may require a larger value of R_S for stability.

In some double-diffusive systems both the Soret and Dufour effects can play a significant role in the system stability. Rothmeyer (1980) has shown that while the Soret effect increases salt transport in sodium chloride ponds, the effects on gradient zone stability are negligible.

10.3.2 Solar Pond Experience

Because of radiation absorption within the gradient zone, diurnal and seasonal cycling, and physical properties that are not constant with temperature and salinity, neither the temperature gradient nor the salt gradient is likely to be uniform in a solar pond. Zangrando (1979) studied the stability of the

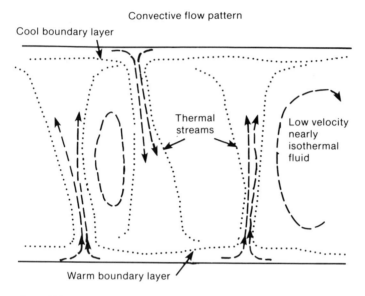

Figure 10.3
Schematic representation of fluid flow in the surface zone of a solar pond. The cool boundary layer occurs at the free surface, and the warm boundary layer occurs at the top of the gradient zone. Similar circulation occurs in the lower zone in the absence of heat extraction.

gradient zone under variable gradient conditions in an experimental solar pond. In these experiments instability resulted in the formation of one or more internal convective zones located about the weakest points in the salinity gradient. These internal sublayers were from 3 to 10 cm (1 to 4 in.) thick.

Walton (1982) analyzed the condition of a uniform temperature gradient with a variable salt gradient for large thermal and saline Rayleigh numbers. The analysis showed that when convection occurs it takes the form of an overstable mode and is confined to a region that is vertically $R_S^{-1/14}$ times the depth of the gradient zone, where R_S is calculated over the total gradient zone. The analysis also showed that equation (9) could be used to estimate instability anywhere in the gradient zone, as long as the Rayleigh numbers are calculated from the local values of the gradients. Walton's analysis agrees with Zangrando's experimental results within the accuracy of the measurements. Zangrando and Bertram (1985) studied the same problem in a series of numerical calculations. The numerical results approach Walton's asymptotic solutions at large R_S but differ significantly at smaller R_S ($< 10^8$).

Laboratory experiments by Tsinober, Yahalone, and Shlien (1983) have shown that when a salt gradient is heated at a point, a hot plume is formed with a series of layers around and above it. Such heating sources may arise in a solar pond from radiation absorption on leaves or other debris suspended in the gradient zone. Plumes from local heating in a pond would be likely to result in increased salt transport or formation of internal convecting zones. This phenomenon has, however, not yet been observed in experimental solar ponds.

10.3.3 Side Wall Effects

Side wall effects can influence solar ponds in several ways. Local thermal convection where the gradient zone meets a sloping side wall heated by the sun has been observed in a 250-m^2 (2700-ft^2) pond (Nielsen 1976). This convection is expected to increase the salt transport rate, and there is some evidence of transport in excess of the diffusion rate in this pond. Much more serious side wall effects were reported by Collins (1983) in the 2,000-m^2 (22,000-ft^2) sloping wall pond at Alice Springs. On one occasion following rain, the storage zone temperature dropped within four days by more than 20°C (36°F), which Collins attributed at the time to side wall heat losses. It now seems more likely that what occurred was a gradient breakdown and formation of internal convective zones. The average salt transport rate over several years was four to five times that expected based on molecular diffusion through a continuous gradient. If the actual gradient thickness were much less than the total thickness because of internal convective zones, then much larger salt transport would be expected. These problems were considered so serious that the pond has been replaced by one with insulated vertical walls. The side wall effects in the Alice Springs pond are not fully understood, and more study of side wall processes is needed.

Laboratory experiments (e.g., Suzukawa and Narusawa 1982) have likewise shown that heating of a side wall results in convective cells forming at the wall and propagating into a fluid stablilized by a salinity gradient. The influence of such wall processes on stability is not known quantitatively. It is evident that they have a destabilizing influence, and it is the operating practice in Israel to make the ratio R_S/R_T much greater than equation (9) would require for stability, to reduce the likelihood of internal convective zones being created by side wall effects or by any other disturbances.

A less dangerous effect at sloping side walls is the departure from one-dimensional diffusion caused by the slope. Surfaces of uniform salinity must be perpendicular to the walls. Thus there is a downward curvature of iso-density surfaces near the wall, and a buoyancy excess that results in a slow upward boundary flow at a rate controlled by the salt diffusion rate.

10.4   Gradient Zone Erosion

10.4.1   General Considerations

One of the more important problems in the operation and maintenance of a salt-gradient solar pond is the stability of the horizontal boundaries that define the gradient zone. In most experimental salt-gradient ponds the gradient zone erodes from above and below. Unless preventive or corrective measures are taken, the result of this erosion is a thin gradient zone. Such a thin gradient zone will reduce the pond thermal efficiency below optimum and will increase the molecular diffusive flux of salt across the gradient zone. In addition the erosion process itself increases salt transport upward. The choice of reduced efficiency or higher maintenance will increase the cost of the energy obtained from the pond.

A rich variety of hydrodynamical phenomena is associated with the zone boundaries (e.g., Kamal and Nielsen 1982 and references therein). Differences between species and thermal diffusivities, wind-induced or externally induced fluid flows, and convective plumes seem to have a role in determining the behavior. The effects of heat removal, as discussed in section 10.5, also have an important influence. The hydrodynamics associated with zone boundaries are not yet fully understood; however, significant progress has been made in recent years.

10.4.2   Equilibrium Relationship

Two methods of examining the stability of a gradient zone boundary are often employed. The first is Nielsen's empirical relationship for equilibrium (Nielsen 1979):

$$\frac{dT}{dz} = A\left(\frac{dS}{dz}\right)^{1.6} \tag{10}$$

When dT/dz is in °C m^{-1} and dS/dz is in kg m^{-4}, the constant A is 5×10^{-3} °C m^{-1} (m^{-4}kg)$^{-1.6}$ [1.6°F ft^{-1} (ft^{-4}4lb)$^{-1.6}$]. This value was determined from observations on laboratory tanks (Nielsen 1979, 1983a) and solar ponds (Nielsen 1979, 1983a; Hull et al. 1982; Wittenberg 1984) using NaCl solutions. It is somewhat remarkable that such a simple relation fits the data as well as it does because the processes at zone boundaries are certainly influenced by many other properties of the system, such as viscosity and thermal expansion coefficient, that vary with temperature (Nielsen 1978). The significance of dT/dz for boundary behavior is as a measure of the strength of thermal convection in the convective zone. The relation of equation (10) therefore does not apply to those zone boundaries, such as the surface zone boundary, where there are other causes of fluid motion, such as wind shear, in addition to thermal convection.

If the left side of equation (10) is made larger than the right side, corresponding to an increase in convective motion, gradient zone erosion occurs; if the left side is made smaller, the gradient zone grows. Rates of erosion and growth are limited by salt diffusion within the gradient zone.

A second method of examining zone boundary stability uses the density stability ratio

$$R_\rho = \frac{\beta dS/dz}{\alpha dT/dz},\tag{11}$$

where α is the absolute value of the thermal expansion coefficient and β is the salinity expansion coefficient. This parameter has been popular with several researchers because it allows a direct comparison between solar pond phenomena and the large body of results from the double-diffusive layer experiments conducted by oceanographers. Based on laboratory studies of relatively thin gradient zones conducted in a 3-m (10-ft)-deep tank, Newell and Von Driska (1986) have suggested that gradients with $R_\rho > 10$ are stable to erosion, gradients with $R_\rho < 6$ are unstable, and those with $6 < R_\rho < 10$ are in a transition zone.

A difficulty in applying either of the above two criteria directly to solar ponds is that even in the absence of wind, the salinity gradients and temperature gradients vary significantly at the boundaries, both during the day and from day to day.

10.4.3 Experimental Observations

The effort to understand the physical mechanisms behind gradient zone erosion has prompted a number of laboratory tank studies. Meyer, Grimmer, and Jones (1982) observed the flow patterns immediately below the zone boundary. Plumes of cold fluid were repeatedly visible descending from the interface. With heat fluxes in the range of 50 to 90 W/m^2 (16–29 Btu/h ft^2), plume velocities ranged from 0.1 to 0.2 cm/s (0.003–0.006 ft/s), and the average plume velocity appeared to increase with increasing heat flux. Plume spacing ranged from 3 to 6 cm (1–2 in.). The experiment was not designed to observe corresponding plumes rising from below.

In a series of experiments on laboratory thermohaline systems heated from below, Poplawsky, Incropera, and Viskanta (1981) observed that the boundary layers had a thickness that varied from 1 to 3 cm (0.4–2 in.). The boundary layers were characterized by a nearly uniform temperature distribution, but a nonlinear vertical salt distribution. With a heating rate of up to 230 W/m^2 (73 Btu/h ft^2), there was a continual erosion of the gradient zone; however, in general, the growth rate of the convecting zone decreased with increasing R_ρ for the range $3 < R_\rho < 12$.

Newell and Von Driska (1985) reported the results of a series of experiments that measured heat and salt flux across relatively thin gradient zones at the large R_ρ values (as high as 32) that are applicable to solar ponds. Previously oceanographers have been concerned mostly with $R_\rho < 7$. Results of these measurements indicate that extrapolation of the empirical correlations developed by the oceanographers to large R_ρ is not justified. Moreover the experiments were in agreement with early pond observations by Nielsen and Rabl (1975) that for initially thin gradient zones, significant growth of the gradient zone can occur under reduced temperature gradient conditions, with the growth rate controlled by the diffusion of salt. Further, whereas at small R_ρ the erosion behavior always followed a unique pattern, at large R_ρ the growth and erosion may follow a hysteresis pattern.

Results corresponding to many of these laboratory observations have been observed both in laboratory tanks and in outside solar ponds in a series of careful measurements made by Kamal and Nielsen (1982). Interpreting the measurements and observations results in the physical picture of the fluid behavior shown in figure 10.3. The convective zone was uniform in temperature to within approximately 0.01°C (0.02°F) except for localized thermal streams that were approximately 0.1°C (0.2°F) warmer or cooler than the background. The position of the temperature gradient boundary fluctuated vertically by approximately 1 cm (0.4 in.) about an equilibrium value, and the temperature gradient extended on the average about 1 cm (0.4 in.) into the region of uniform salinity. This is equivalent to saying that the gradient zone as determined by the temperature profile is thicker than the gradient zone as determined by the salinity profile. The salinity gradient boundary also fluctuated in position but by a smaller amount than the temperature gradient boundary. These fluctuations have been interpreted as deformations of the boundary caused by the impact of the thermal streams. A heat transfer analysis (Kamal and Nielsen 1982) of the convective streams at a heat flux of 40 W/m^2 (13 Btu/h ft^2) indicates that the fluid flow of the thermal streams is randomly distributed but laminar, with a maximum fluid velocity less than 1 cm/s (0.03 ft/s) and a stream distribution of approximately 600/m^2 (56/ft^2).

A number of experimenters have tried to accurately measure the salt flux across a zone boundary. The most accurate method keeps track of the salt inventory in the adjacent convecting zone with a gravimetric measurement. The largest source of error is then the determination of the interface position.

10.4.4 Theoretical Models

Several theoretical models have been developed to help understand gradient zone erosion and predict solar pond hydrodynamics. Meyer (1983) proposed a one-dimensional model based on an empirical correlation between salt and heat flux developed earlier by oceanographers. The reasoning of the model involves some of the dynamics occurring at the zone boundary. The proposed model assumes the simultaneous growth of a thermal boundary layer and salinity boundary layer from the diffusive core of the gradient zone into the convective zone. Because of the high thermal diffusivity, the thermal boundary layer outdistances the layer caused by salinity. Because the temperature distribution in the thermal layer is unstable, at some thickness the layer will break down, and a buoyant element will be released as a thermal plume. The thickness of the boundary layer at breakdown is calculated from stability considerations. The model assumes that both boundary layers are fully mixed into the convecting zone at breakdown. This model has had reasonable qualitative agreement with many laboratory experiments and some solar pond data. Its major weaknesses are its inability to account for gradient zone growth and its neglect of the influence of hydrodynamic conditions in the convecting zone. In addition the correlation between heat flux and salt flux was developed for a thin diffusive boundary layer that separates two mixed layers and may not be applicable to a boundary layer separating a mixed layer and a large diffusive layer.

A different one-dimensional model was developed by Bergman, Incropera, and Viskanta (1982) to help understand a series of laboratory tank experiments. In this model the boundary layer thickness is assumed to be arbitrarily thin, with negligible heat or salt transport across the boundary. The key feature of the model is the representation of gradient zone erosion by an entrainment velocity u_e given by the correlation

$$u_e = 0.2u_* \text{Ri}^{-1}, \tag{12}$$

where

$$u_* = \left(\frac{g\alpha q H}{\rho c}\right)^{1/3},$$

$$\text{Ri} = \frac{g(\partial \rho/\partial m)\Delta m H}{\rho u_*^2},$$

g = acceleration of gravity,

α = thermal expansion coefficient,

q = heat flux,

H = height of the lower convecting layer,

ρ = fluid density,

c = specific heat,

m = salt mass fraction,

Δm = difference in m across the interface.

The parameter u_* is the friction velocity associated with turbulence in the convecting zone, and Ri is the bulk Richardson number. The density jump across the interface is calculated from the initial salinity profile as affected by the conservation of salt and the effects of entrainment. The model gave reasonable qualitative agreement with the tank experiments at high heat flux [83–125 W/m^2 (26–40 Btu/h ft^2)]. Major weaknesses of the model are the absence of a mechanism for gradient zone growth and the neglect of diffusion across the boundary layer. The exact form of the correlation in equation (12) has been further examined by Bergman et al. (1983). Depending on whether the system is temperature or salt stratified, the exponent of Ri can be -1 or $-3/2$ (Turner 1968). As discussed in subsection 10.5.2, there may also be a R_ρ dependence on the value of the exponent.

10.4.5 Zone Boundary Maintenance

Several methods have been developed to maintain artificially the position of gradient zone boundaries. The most obvious method is to reestablish periodically the eroded part of the gradient, using the scanning injection method described by Zangrando (1980). Nielsen (1983a) has controlled erosion at the upper zone boundary by pumping salt from the bottom of the pond into the upper part of the gradient zone, steepening the salt gradient there. The resulting increase in salt flux upward because of the pumping, and the locally larger salinity gradient is less than that obtained if erosion is allowed to occur with periodic repair.

Controlling erosion at the lower boundary is generally easier than at the upper boundary because the temperature gradient at the lower boundary is smaller. Nielsen (1983a) has prevented erosion at the lower boundary by extracting heat from the heat storage zone, thereby reducing or even reversing the temperature gradient in the gradient zone immediately above the boundary. Hull (1984) demonstrated that the position of the lower zone boundary can be stabilized totally by passive means. The lower boundary of a 1000-m^2 (10,800-ft^2) solar pond was continuously maintained for more than three years by a salt pile located at the bottom of the pond, with the top of the pile determining the position of the boundary. In this case the temperature gradient at the boundary was small and the salinity gradient large, avoiding any tendency of the boundary to move upward.

10.4.6   Surface Effects

In general, the hydrodynamics of the upper boundary is more complicated than that of the lower boundary because of the strong influence of the pond surface. Daily, seasonal, and intermittent effects combine to influence the boundary behavior. Influencing parameters include temperature gradient, salinity gradient, ambient temperature fluctuations, insolation, evaporation rate, surface winds, rainfall, and winter ice formation. Rainfall is generally beneficial because it decreases the salinity of the surface zone, thereby strengthening the salt gradient at the top of the gradient zone. Ice formation on the surface typically decreases transmission of sunlight into the pond because of scattering from air bubbles trapped in the ice, reflection from snow, or absorption by dirt accumulated on the surface. At the same time it is beneficial because it prevents surface wind mixing and transports salt downward in the pond (Nielsen 1982). When salt water freezes, a number of effects combine to produce ice consisting of relatively freshwater above a brine of increased salinity. In addition to direct stabilizing or destabilizing effects, surface influences can cause the upper convecting layer to develop into several discrete sublayers, complicating the hydrodynamics.

Several investigators have applied results from oceanography and limnology to model the surface effects in solar ponds. Cha, Sha, and Schertz (1982) modeled surface effects by incorporating augmented thermal and solutal diffusivities into the conservation equations for heat and salt. In addition to the molecular diffusivities, there were contributions from wind mixing and double-diffusive convection. The wind contribution was based on a turbulent wave orbital model that depends on the friction velocity. The double-diffusive contribution was based on an empirical correlation developed by oceanographers. In the numerical solution of the equations, mixing occurred whenever the gradient did not meet the Weinberger criterion of equation (9). Results of the study showed that strong surface cooling could cause gradient zone erosion comparable to that of wind mixing.

Atkinson and Harleman (1983) formulated a similar solar pond model based on a wind-mixing algorithm that had been developed earlier for limnology studies. In this model surface wind shear produces turbulent kinetic energy, which is transported downward and is partially entrained into the gradient zone. The entrainment velocity depends on the friction velocity according to a function that decreases as the bulk Richardson number increases. Here Ri is proportional to the height of the upper mixed layer and the density jump across the upper zone boundary. Results of this model compared well with field data from the small experimental solar pond at Wooster (Shah, Short, and Fynn 1981).

Schladow (1983) also calculated pond behavior using a computer model developed to simulate the dynamics of small- to medium-sized lakes and reservoirs. He found that the relative importance of the two most important processes, wind stirring and convective overturn resulting from heat conduction and evaporation, was very site dependent. At Alice Springs, where there is relatively little wind but a high evaporation rate, he found that convective overturn was the dominant influence determining the thickness of the surface zone. Calculations for another site with more wind and higher humidity and rainfall indicated that wind would be more important.

There have as yet been only very limited comparisons of the various calculational results with actual pond data, mainly because hardly any surface zone data are available with all the important quantities measured accurately enough for definitive comparison.

10.5   Heat Extraction

10.5.1   General Consideration

Two methods have been used to extract energy from solar ponds. The first circulates freshwater or glycol through a submerged heat exchanger in the pond. This method is limited in transferring heat by natural convection within the lower zone (LZ), and the surface area of the submerged heat exchanger is determined accordingly. The second method circulates brine from the LZ through a heat exchanger that is external to the pond. This brine withdrawal method requires less heat exchange surface, but the rate of energy extraction is limited by a requirement for hydrodynamic stability of the gradient zone (GZ).

Because conventional heat exchanger design is fairly well developed, most of the research on solar pond heat extraction has been devoted to understanding the effects of brine withdrawal on GZ erosion. There is also a need to know the horizontal limits of the effects of a single withdrawal or injection port so that heat exchange systems for larger ponds can be designed. If diffuser ports must be located at opposite ends of the pond, then piping may become expensive. The withdrawal and return ports are located on the same side of the large 210,000-m^2 (2,300,000-ft^2) pond in Israel, which suggests that horizontal limits on diffuser ports will probably not be a restricting factor for solar ponds with smaller surface areas. A further problem that has received little attention is the effect of corrosion on solar pond heat extraction systems. This is particularly important in electricity-generating ponds, where the heat exchanger is a major component cost.

The brine withdrawal method is illustrated in figure 10.4. Hot brine is withdrawn from the pond near the top of the LZ, passed through an external heat exchanger, and cooled brine is returned (injected) to the pond near the bottom of the LZ. The main advantage of this method, compared with the submerged heat exchanger method, is that the heat exchange surface area is smaller. A major requirement of the brine withdrawal method is that the withdrawal and reinjection does not erode the GZ. A further requirement is that in pumping hot brine, corrosion-resistant construction in the pump and external heat exchanger must be used. Also, if the external system is not totally

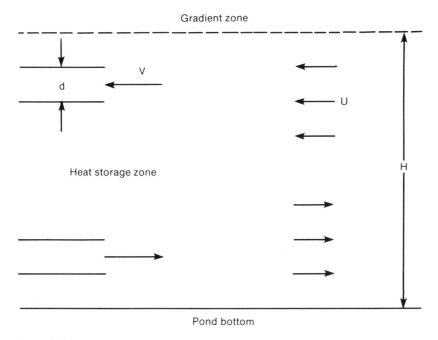

Figure 10.4
Representation of bulk flow in the lower (heat storage) zone during heat removal via brine extraction and reinjection.

closed to the environment, air bubbles could be introduced into the LZ, changing the corrosion potential and possibly damaging the GZ. In both the brine withdrawal and submerged heat exchange methods, heat removal may produce a temperature stratification in the LZ.

The effect on boundary stability of brine extraction and reinjection in the LZ is not completely understood. Zangrando (1982) gives a survey of related research in this area. The basic phenomenon appears to be that motion of the extracted fluid may cause turbulence in the LZ. The turbulent disturbance causes mixing of the LZ and the GZ near the GZ lower boundary and results in erosion of the GZ. Reinjection of brine is much more likely to cause turbulence than extraction. However, if the reinjection brine is cold enough to produce good stratification, then the turbulence produced is confined to the bottom of the LZ and does not affect the GZ.

The stability of a stratified flow is governed mainly by the Richardson number. Physically, it is the local conditions at the interface that should determine Ri, but these are very difficult to measure, and the variety of length and velocity scales used by different researchers makes it difficult to relate results obtained under different experimental conditions. Instead, the bulk Richardson number is sometimes used, defined by

$$\text{Ri} = g\left(\frac{1}{\rho}\right)\left(\frac{\Delta\rho}{\Delta z}\right)\left(\frac{H}{U}\right)^2, \tag{13}$$

where g is the acceleration of gravity, ρ is the density, U is the horizontal velocity, and H is the depth of the mixed layer.

A further consideration for brine extraction is that the flow in the pond adjacent to an injection or extraction diffuser should be laminar rather than turbulent. The parameter characterizing this aspect of the flow is the Reynolds number:

$$\text{Re} = \frac{Vd}{v}, \tag{14}$$

where V is the velocity of the fluid at the entrance or exit of the diffuser plates, v is the fluid kinematic viscosity, and d is the separation of the diffuser plates at the diffuser outlet. Another parameter used is R_ρ (discussed in subsection 10.4.2), calculated at the GZ lower boundary.

10.5.2   Laboratory Experiments

Several of the one-dimensional hydrodynamic models of the solar pond (see subsection 10.4.4) use an entrainment velocity that is correlated with Ri to account for erosion from wind mixing or convection. It is natural to extend this concept to entrainment at the GZ lower boundary based on Ri derived from the heat extraction flow. A careful series of experiments performed at SERI to determine the entrainment velocity u_e as a function of Ri for solar pond conditions have been reported by Zangrando, Green, and Fisher (1984). The test facility consisted of a 2-m (80-in.) deep by 5-m (200-in.) long by 1-m (40-in.)-wide tank, open to the ambient at the top, heavily insulated on sides and bottom, with bottom heating of up to 100 W/m^2 (34 Btu/h ft^2). Line diffusers produced two-dimensional flows in the tank.

As discussed in subsection 10.4.4, most correlations for u_e are of the form

$$\frac{u_e}{U} = A\,\mathrm{Ri}^{-n}, \tag{15}$$

where A is a constant on the order of unity, and n is between 1 and 1.5. For the SERI experiments a bulk Richardson number was defined as in equation (13), but using a bulk velocity,

$$U = \frac{q}{H}, \tag{16}$$

where q is the volumetric flow rate per unit width in the diffusers. The results of the experiments show correlations between u_e and Ri as given in equation (15) in the range $10^5 < \mathrm{Ri} < 10^9$, but both n and A depend on R_ρ. For $R_\rho > 25$, $n = 1.5$; for $20 < R_\rho < 25$, $n = 1$. The magnitude of n decreases further with decreasing R_ρ.

Several other observations from these experiments illustrate some of the physical phenomena behind the correlations. First, interface movement was not a continuous process, but rather the entrainment appeared intermittent. This suggests that the interface cycles first develop by diffusion to a critical condition, then collapse to form a sharpened interface. Second, the bulk recirculation flow appeared capable of greatly reducing the entrainment effect of the convective flow caused by bottom heating, even though the bulk velocity was much less than the convective velocity.

10.5.3 Solar Pond Experience

So far results of heat extraction in experimental outdoor solar ponds have not been as carefully analyzed as the laboratory experiments, but the observed phenomena correlate well with the laboratory results. In brine withdrawal systems laminar flow at the diffuser entrance appears necessary. Also a sufficiently high Ri is required for stability between the GZ lower boundary and the brine return level. This requirement is usually met by a sufficiently large temperature difference between withdrawal and injection ports to establish good thermal stratification in this layer.

The present heat extraction system in the 2,000-m^2 (22,000-ft^2) Miamisburg solar pond (Wittenberg and Etter 1982) uses the brine withdrawal method. The initial diffusers in this system were horizontal plastic pipes with equally sized holes drilled in the pipe sides for fluid flow. The flow distribution through the holes was uneven, causing a high velocity flow through the holes nearest the pump and erosion of the GZ lower boundary. The pipes were replaced by circular horizontal plate diffusers. Heat extraction using these diffusers also resulted in GZ erosion, even though Ri was estimated as 5.6 (Wittenberg and Etter 1982). Unfortunately, this estimate was based on a $\Delta\rho/\Delta z$ term calculated from the salt and temperature gradients in the GZ. Later analysis in terms of conditions in the lower zone, where there was only a few degrees temperature difference over the entire lower zone depth, yielded a much smaller Ri that was compatible with the gradient zone erosion observed.

The 400-m^2 (4,300-ft^2) research solar pond at The Ohio State University has used a brine withdrawal system to transfer heat from the LZ through a counterflow heat exchanger to the pond surface. The diffusers are circular, separated by 20 cm (8 in.) horizontally and only 9 cm (3.5 in.) vertically. With a temperature difference of 5° to 15°C (9°–27°F) between withdrawal and injection diffusers, the heat extraction flow moved the GZ lower boundary down to an equilibrium position within 5 cm (2 in.) of the extraction level (Nielsen 1983a). However, when the fluid flow rate was doubled, from 30 to 60 L/min (8–16 gal/min), the GZ eroded. This erosion can be attributed to an increase in Re at the diffuser entrance, with the local turbulence causing the erosion. Later experiments at 47 L/min (12 gal/min) in which the temperature difference was $<0.6°C$ (1.1°F) produced rapid erosion of the GZ (Nielsen 1983b). This erosion can be attributed to a decrease in Ri between the GZ lower boundary and the injection diffuser. With $\Delta T > 4°C(7°F)$, the boundary reached an equilibrium 10 cm (4 in.) above the extraction diffuser. In general, this distance is expected to depend on temperature and salinity gradients in the gradient zone at the boundary.

The 156-m^2 (1,680-ft^2) Wooster pond has used a brine withdrawal system, connected by a shell-and-tube heat exchanger to a heat pump, to heat a greenhouse (Fynn, Short, and Shah 1980). The withdrawal and injection diffusers are horizontal plastic pipes with holes drilled in the sides. The total area of the holes is equal to the flow area of the pipe, but holes farther from the pump are made progressively larger to provide a more uniform flow. The withdrawal tube is 20 cm (8 in.) below the GZ lower boundary, and the injection tube is 10 cm (4 in.) above the pond bottom. As much as 112 W/m^2 (38 Btu/h ft^2) of heat has been extracted from this pond without noticeable erosion of the GZ.

Probably the greatest heat withdrawal rate per unit area employed to date was during the intermittent operation in Israel in 1983 of a 2-MW power unit using heat from the 40,000-m^2 (430,000-ft^2) pond. The heat requirement was about 40 MW thermal, corresponding to a heat extraction rate of about 1,000 W/m^2 (340 Btu/h ft^2) of pond surface area. This pond was built as an intermediate step between the 7,000-m^2 (75,000-ft^2) Ein Bokek pond and the 210,000-m^2 (2,260,000-ft^2) pond now used, together with the 40,000-m^2 pond, to run the 5-MW peaking power station. One of the major development studies carried out with the 40,000 m^2 pond was the design of a heat extraction system. Brine is withdrawn from the side of the pond nearest to the power unit, using an array of very large and carefully designed circular flow distributors, and after supplying heat to the power station, it is returned to the same end of the pond. Withdrawal and return from points close together had previously been carried out in the 400-m^2 (4,300-ft^2) pond at The Ohio State University, as noted above, but not everyone had anticipated that this would be possible on ponds as large as 210,000 m^2.

Because the amount of dissolved oxygen in water decreases with both temperature and salinity, and because the stagnant waters of the GZ prevent the introduction of large amounts of oxygen to the LZ, corrosion of heat exchanger components in contact with LZ brine should be minimal. However, the different temperature and salinity regimes of the solar pond create several unusual opportunities for corrosion. Corrosion problems with heat exchangers are exemplified by the early experience at the Miamisburg solar pond with a submerged, copper/tube heat exchanger (Harris, Etter, and Wittenberg 1981). The original solder joints, which consisted of 95% tin and 5% antimony, dissolved after 10 months. Replacement brazed joints, which consisted of 56% silver, showed no signs of corrosion. A further corrosion problem with this heat exchanger occurred when the top of the exchanger was exposed to oxygen-rich, low salinity water, while the copper-supporting structure rested in oxygen-free brine. A differential aeration type of corrosion cell was initiated using the copper of the heat exchanger as the electrical conductor, leading to total dissolution of the supporting structure. While the corrosion occurred during an abnormal circumstance in pond operation (no corrosion problems have been observed at the Farm Science Review solar pond, which uses a submerged copper heat exchanger), the Miamisburg experience emphasizes the need to avoid creating galvanic couples in the heat exchange loop.

A method to avoid completely both corrosion and gradient zone erosion problems is the use of a submerged heat exchanger with plastic components. Hull, Scranton, and Kasza (1985) have reported the successful operation of a relatively low-cost, submerged, plastic-tube heat exchanger at the 1,000-m^2 (10,800-ft^2) solar pond at Argonne National Laboratory. At the maximum withdrawal rate of 65 W/m^2 (22 Btu/h ft^2), no deterioration of the gradient zone was detected. Polypropylene was used as the tube material, but the best grades of the more thermally conductive high density polyethylene should also be suitable for many solar pond applications.

10.5.4   Advanced Heat Exchange Methods

For electricity-generating applications, low temperature thermal-conversion devices such as the solar pond suffer from low efficiency because of the inherent limitations of the Carnot cycle and from temperature drops across heat exchanger walls. To date, solar ponds have used the organic Rankine cycle to produce electricity. Heat from brine withdrawn from the LZ is used to evaporate the working fluid in a conventional shell-and-tube boiler. Because of the low Rankine cycle efficiency, a large surface area is required for both the boiler and condenser, and these two components account for roughly 50% of the conversion-system cost.

Using direct-contact heat exchange has been suggested to achieve significant cost reductions. Direct-contact heat exchange had been suggested earlier for use in electricity production from geothermal energy (Jacobs and Boehm 1980). The major advantages are a lack of fouling on the heat exchange surface, a smaller boiler size, and increased efficiency caused by the absence of a solid heat exchange wall. Inherent problems are loss of working fluid by its partial solubility in brine, increased parasitic losses if brine pressure needs to be increased to match that of the working fluid, the possible presence of non-condensibles (coming from the brine) in the condenser, and a possible increase in turbine corrosion if any brine is carried over into the working fluid. Wright (1982, 1984) analyzed a number of working fluids in solar pond systems and concluded that pentane was the most promising fluid and that direct-contact heat exchange had the potential to reduce solar pond plant costs by 25%.

10.6   Special Instrumentation

Solar pond instrumentation can be broadly categorized into normal monitoring and research needs. Normal monitoring in a pond usually involves measuring the temperature distribution in the pond, the salinity distribution, the radiation input, the heat extracted, water pH, and perhaps the heat lost to the earth. Research studies require similar information, sometimes with greater accuracy or better spatial resolution. Research may also require study of particular processes, such as the nature of the flow pattern in convective regions. Care must be taken when selecting instruments for solar pond use because they may be subject to large temperature variations as well as to the corrosive effects of hot salt water. In some instances, such as temperature measurement, instruments are commercially available. In other instances, special instruments have had to be developed.

10.6.1 Temperature

Temperature profiles provide some of the most essential information about the state of a pond. They give the thickness of the various zones, show the response of the surface zone to rain or wind and the effects of heat extraction on the storage zone, and provide the temperature record that, together with radiation input and heat extraction data, is necessary for calculation of pond performance.

In practice, temperature profiles have been measured either from an array of fixed sensors or from a single sensor scanned between the surface and the bottom of the pond. At first glance the array of sensors may appear more attractive. Depth corresponding to every reading is known and is constant, and temperatures are readily recorded with a multichannel data-logger. However, unless extreme care is taken with both initial calibration and accuracy of reading, this system gives a rather poor representation of the profile. Absolute errors tend to be about 0.5°C (0.9°F), corresponding to an uncertainty of 10°C/m (5.5°F/ft) in the local gradient if a 5-cm (2-in.) spacing is used. Absolute temperature is usually not needed to better than 0.5°C (0.9°F), but a gradient error of this magnitude is undesirable. An even more serious problem has arisen in some ponds in which, to limit the number of sensors and readings required, the sensors were spaced by 10 cm (4 in.) or more. The result was that internal convective zones could exist undetected within the gradient zone. In one case when a 10-cm (4-in.) spacing was used, there were five undetected internal zones present, and the actual gradient thickness was probably not over one-third of the apparent thickness as indicated by the distance between the top and bottom of the supposed gradient zone. If fixed sensors are used, the spacing must be not greater than 5 cm (2 in.), at least over the gradient region. The 5-cm (2-in.) figure arises from the observation that internal convective zones thinner than about 5 cm (2 in.) are transient: They either disappear or they become thicker.

The record of a scan with a single sensor gives a continuous profile in which positions of zone boundaries and thin internal convective zones (if any exist) are shown in detail. A simple and reliable way to obtain such a record is to suspend a thermocouple from a cable wrapped around a drum. As the drum turns, the thermocouple output is recorded on a strip chart recorder, with the chart movement synchronized with the thermocouple movement. Alternatively, the data from a moving thermocouple can be recorded at suitable intervals using a digital system.

10.6.2 Salinity

Salinity profiles give more reproducible zone boundary locations than temperature profiles because sometimes a level observed for a temperature boundary fluctuates several centimeters. Thus changes in pond configuration, such as growth of the surface zone from wind, are best evaluated from salinity profiles. Determination of the salinity profile is necessary to monitor stability conditions and to keep track of salt inventory. Because salt diffusivity is low, accurate salinity knowledge is necessary to calculate the salt flux in the pond. Because density is a known function of temperature and salinity and because accurate temperature measurements are readily made, a measurement of density and temperature is often substituted for a direct measurement of salinity.

The most accurate density measurement is made gravimetrically. A sample fluid is extracted from a fixed position in the pond, and a known volume of the sample is accurately weighed at a known temperature in the laboratory. Nielsen and Kamal (1981) have successfully used this technique to calculate salt flux in a solar pond. This type of measurement is sensitive enough to detect relatively small leaks in the pond liner. The method requires fixed sampling locations to achieve a spatial reproducibility commensurate with the accuracy of the density determination.

A number of density-measuring techniques are available for scanning devices that traverse the entire depth of the pond. Zangrando (1984) has evaluated many of these techniques and concludes that the hydrometer and vibrating U-tube are the most appropriate of the commercially available devices, although methods based on magnetic float and speed of sound hold the most promise for future development. If a pond area can be sheltered from the wind, a small weight suspended from a sensitive balance can give accurate density measurements in a scanning system.

A popular method of determining salinity is to measure electrical conductivity, which is a function of both salinity and temperature. Conductivity measurements give reasonable accuracy at low salinity, but the salinity dependence is not as strong at high concentrations. The response of a conductivity probe depends on the surface condition of the electrode, which changes with time on exposure to saltwater, and periodic recalibration is necessary for meaningful results. Several solar ponds use electrodeless conductivity probes, which have better long-term stability because there is no exposed electrode. These probes measure the mutual inductance between two coils whose windings, but not centers, are encased in plastic. Unfortunately, the spatial resolution of these electrodeless probes is a few centimeters.

To better understand the gradient zone boundary, low noise conductivity probes with spatial resolutions of about 1 mm (0.04 in.) were investigated. Grimmer et al. (1983) reported satisfactory laboratory results using a point-electrode probe. In this dual-electrode device, electric current passes from a central point electrode to a secondary electrode coiled around the probe tip to shield the point electrode against noise from stray current effects. Stray currents arise if there are other metals (internal battery effect) or other conductivity probes (cross-talk effect) in the system. The current density at the probe tip is much greater than that at the secondary electrode because of the large difference in surface areas. The resulting electrical conductivity measurement is thus more heavily weighted in the region of the tip than elsewhere. The platinum-tipped probe is calibrated in situ by measuring the specific gravity of fluid samples withdrawn from the vicinity of the probe when it is in a convecting region.

A tetrapolar conductivity probe has also been successfully used in laboratory experiments (Bergman et al. 1983). With this device two outer electrodes generate current and two inner electrodes measure the resulting electric field. With the two inner electrodes 2 mm (0.08 in.) apart, a spatial resolution of 5 mm (0.2 in.) was obtained. This tetrapolar probe shares with the two-electrode probe the problem that response drifts with time, so that it needs frequent recalibration.

10.6.3   Transmittance

The thermal performance of a solar pond depends strongly on the transparency of the pond brine, and prediction of performance is possible only when the transmission function is known. In practical operation, monitoring of transmission is needed to know when corrective measures may be required to maintain good water clarity.

Two different kinds of observation have been used for determining the optical quality of pond brine: laboratory measurements of the transmission through absorption cells containing samples from various depths in the pond, and field measurements of the fraction of the solar energy penetrating to various depths in the pond. Both kinds of observation are useful; both have limitations and problems.

In most cases scattering by suspended particulate material, inorganic or organic, is the main cause for pond transparency poorer than that of clean water or brine. (The effect of dissolved NaCl or other colorless salts is relatively small.) When attenuation of transmitted radiation is the result of scattering, laboratory absorption cell measurements give no quantitative information about the transmission function in the pond because the geometry is different. Multiple small-angle scattering in an absorption cell may cause a significant fraction of the incident light to exit the cell through the side wall and not intercept the sensor. In the solar pond, however, a large fraction of this scattered light will propagate downward into the heat storage zone. For example, transmission through 20-cm (8-in.)-thick pond samples may be only 70% of that through 20-cm (8-in.)-thick distilled water samples, and yet the transmission to 1.0-m (39-in.) depth in the pond [relative to 1.0 m (39 in.) of distilled water] may be very much more than the $(0.7)^5$ expected if Beer's law applied. However, in those special situations in which absorption in the pond water results primarily from colored dissolved substances, spectrophotometric measurements on carefully filtered samples can be related to transmission in the pond (Giulianelli and Naghmush 1982).

Solar energy penetration into the pond is measured by comparing the reading of an instrument submerged in the pond with the reading of a similar reference instrument normally located above the pond. The reference instrument is essential for good data because even on a clear day there can be appreciable short-term variations in solar flux. The spectral composition of the radiation changes with path length in the water, and therefore instruments with flat spectral response, such as thermal pyranometers, must be used if absolute transmission is to be obtained. Photovoltaic instruments also have been used, and useful data can be obtained with them if the readings are corrected by a calibration function that depends on depth in the pond and angle of the radiation. Special care must be taken in the calibration curve to properly account for the change in spectral composition of the sunlight when the diffuse sky component changes.

Aside from the obvious problems of leaks and corrosion, thermal and optical problems are also associated with using submerged instruments. Because temperature varies with depth in the pond, it is necessary either to have a response that is independent of temperature or to know the temperature dependence. Commercial pyranometers are temperature compensated, commonly by an auxiliary semiconductor circuit, but the time response of the compensating circuit is usually much slower than the instrument response to radiation. It follows that readings at a fixed depth are readily obtainable, but it requires a carefully designed procedure to get good data from a scan in depth covering a range in temperature. Photovoltaic cells are also temperature sensitive, some less than others. The change in optical environment from air to water must also be considered. There is a decrease in reflectivity from the cover because of the smaller difference in index of refraction between water and glass than between air and glass. If a hemispherical cover is used, a diverging lens effect decreases the intensity at the sensor when the instrument is immersed in water.

The preceding considerations indicate why both laboratory and field measurements are valuable. Field measurements of radiation penetrating to various depths provide the only generally satisfactory way of determining the pond transmission function. Very simple laboratory measurements of white light beam transmission through pond samples in a fairly long absorption cell provide by far the most sensitive indication of the amount of dirt present at various levels, and of the trend, whether toward deterioration or improvement.

10.6.4   Heat Extracted

Complete systems for measuring the quantity of heat extracted are available commercially, but it may be simpler and much less costly to assemble the necessary components as elements of a computer-controlled pond monitoring system. All that is required is to measure water volume and temperature difference. Ordinary water meters sold in quantity for recording use of water in public water systems satisfactorily measure the volume of pond brine circulated in a pump system. Such meters are available with both a recording mechanical dial and an electrical signal output that can be used with the signal from a thermocouple pair giving temperature difference to provide a measurement of extracted heat.

10.6.5   Soil Properties

As noted earlier in section 10.2, heat loss to the ground has an important influence on pond performance. This heat loss involves the effective heat transport coefficient of the soil, which depends on moisture content and vapor transport; heat loss depends also on the distance to the heat sink and on possible groundwater movement. It would be desirable either to measure directly the heat loss to the ground or to determine the parameters on which it depends so that it could be calculated.

In situ measurement of the moisture content would thus be relevant to ground heat loss. Electrical conductivity probes have been installed under several ponds primarily for leak detection. Unfortunately, the conductivity (impedance is actually measured) is a strong function of both moisture and salinity, and no meaningful results have yet been reported. Moisture-sensitive resistance blocks have been placed in the soil under the TVA pond (Chinery and Siegel 1983). These sensors measure the relative humidity and must be calibrated for each soil type.

A recent development that might be applicable to solar ponds is using time-domain reflectivity to measure simultaneously water content and electrical conductivity (and therefore salinity) independent of soil type (Dalton et al. 1984). The method measures the attenuation and return time of a pulse sent to a waveguide consisting of two parallel metallic rods embedded in the soil.

When temperature and soil moisture are such that the water vapor transport component is important, the direct measurement of heat flux is extremely difficult because conventional heat flux sensors form a solid barrier to the vapor flow, significantly distorting the results. An effective thermal conductivity can be estimated from temperature measurements if the heat capacity is known (Hull, Kamal, and Nielsen 1983); however, this technique is most easily used if the heat flow is one dimensional and the thermal conductivity isotropic. Recently, Blaney and Nielsen (1985) measured the heat transport of soil in situ using buried heat sources and temperature sensors. The results were easy to analyze and internally very consistent. When future ponds are constructed, it may be worth considering installing one or more heat sources adjacent to temperature sensors underneath the pond to measure the heat transport coefficient as the pond operates.

It is suspected that the temperature and moisture gradients in the soil under the pond may give rise to significant horizontal vapor transport, making interpretation of heat fluxes derived from temperature measurements alone difficult. An array of heat flux sensors placed on the floor of the pond, rather than underneath, might give useful information about heat loss from the pond and also, when combined with temperature data from underneath the pond, yield a value for the soil heat transport coefficient. Such measurements have not yet been made.

10.6.6 Flow Visualization

A number of different laboratory experiments have employed various flow visualization techniques to gain a general understanding of the hydrodynamic phenomena associated with solar pond systems. Meyer, Grimmer, and Jones

(1982) used thymol blue as a tracer in the fluid to closely observe thermal plumes forming at and impinging on the gradient zone boundary. The method involves placing thymol blue in the lower convective zone. Because thymol blue is a pH indicator, the color of the solution can be altered locally by the creation of ions at the surface of a grid electrode. The movement of this dyed fluid then reveals the details of the flow.

Shadowgraph visualization has been used with temperature and salt concentration measurements to investigate gradient zone erosion and instabilities (Poplawsky, Incropera, and Viskanta 1981). This technique can be used to detect the location and thickness of zone boundaries and to observe some of the flows occurring in the convecting zones.

A Mach-Zehnder interferometer has also been used to visualize erosion of the gradient zone in laboratory experiments (Lewis, Incropera, and Viskanta 1982a, b). Although this type of system provides horizontally averaged information, the boundary between the convecting zone and the gradient zone is sharply defined, and details such as the evolution of slight indentations of the boundary, which are caused by impinging thermal plumes, are readily observed. This type of interferometer can also be used to infer salt concentration and mass density distributions in the gradient zone, according to the number of interference fringes between a given location and a reference point.

10.7   Pond Liners

The construction of a solar pond is similar to that of other containment systems and water reservoirs. Except in locations where salt contamination is not a problem, the large quantities of salt needed for a solar pond require that the embankment be sealed to prevent contamination of the surrounding ground and water table and to conserve the salt inventory. An additional reason for good containment of the brine solution is to prevent water from entering the soil under the pond. Under most circumstances an increase in soil moisture content will increase the effective thermal conductivity of the soil and decrease pond thermal efficiency.

10.7.1   Exposed Membrane Liners

All solar ponds constructed in the United States have synthetic membrane liners. The advantage of a membrane liner is that it has low permeability $[<10^{-11}$ m/s $(<3.3 \times 10^{-11}$ ft/s)$]$ and can be relatively easily installed. Different possible liner materials include chlorinated polyethylene, Hypalon®, XR-5®, high density polyethylene, and ethylene propylene diene monomer.

Some of the solar ponds built in the United States have operated for many years without construction-related problems (e.g. Zangrando 1979; Nielsen 1980b; Hull et al. 1982), while others have experienced leakage through the pond liner. The original liner in the Wooster pond delaminated in places, with the membrane material extruding through the scrim (Fynn and Short 1983). The Miamisburg pond leaked when the insufficiently compacted containment berm settled, resulting in large stresses on the liner (Harris, Etter, and Wittenberg 1981). The TVA pond leaked because of failure of factory seams in the primary liner (Chinery, Siegel, and Irwin 1983). The LANL pond was disrupted by gas formation under the pond, which lifted the liner (Jones et al. 1983).

10.7.2 Soil Liners

Because a membrane liner is a large fraction of the total construction cost, soil liners have been suggested to reduce the cost of large solar ponds. Soil liners are made by mixing soil with other materials to create a soil with high compaction and low permeability. A crystallized salt bed is frequently used in industrial evaporation ponds for mineral recovery. Several feasibility studies have proposed treated clays or dikes around dried lake beds. Because all natural solar ponds have soil liners, this option should be feasible, at least in some locations. However, most well-known soil liners have permeabilities of 10^{-8} m/s (3.3×10^{-8} ft/s) (Fynn and Short 1983), and the resulting loss of 30 cm (12 in.) of concentrated brine per year is probably too high for economic pond use, except near a salt lake. [The effective permeability of the gradient zone because of molecular diffusion is about 10^{-9} m/s (3.3×10^{-9} ft/s).] These types of pond liners, though promising, have received little research attention. Major problems of soil liners are the degradation of liner strength with salinity and increasing temperature (Marsh 1983), the possibility of gas generation under the liner, and the possible chemical interaction of the soil with the pond brine, resulting in a possible decrease of pond transparency and the need for a major water treatment program.

10.7.3 Buried Membranes

A buried membrane technique was employed in the 40,000-m^2 (430,000-ft^2) and 210,000-m^2 (2,260,000-ft^2) ponds in Israel. A thin membrane is laid down, covered with a layer of clay, and a second membrane and another layer of clay are placed on top of this. The membranes are in strips that are overlapped at the edges but not sealed. Detailed information has not been published, but cost is said to be about half that of the usual exposed membrane liners, and permeability is low enough to be satisfactory at the solar pond site near the Dead Sea, where brine replacement cost is not of concern. A major accomplishment in their use of this lining technique is the development of an appropriate installation procedure. Buried thin membrane liners had been used earlier in India (Jain 1973), but the use of multiple membrane layers with solar ponds is new.

10.8   Alternative Salts and Designs

Because of its wide availability and well-known properties, sodium chloride (NaCl) has been the most frequently used salt in U.S. solar ponds. Cost of NaCl depends on production and transportation and varies widely from location to location. In some locations the use of low-cost alternative salts could significantly improve solar pond economics. Sodium sulfate (Na_2SO_4) has the potential for widespread availability as a waste product from flue gas desulfurization at coal-fired power plants, from agricultural waste water treatment, and from NaCl production at certain lakes. Additional alternative salts include magnesium chloride ($MgCl_2$), potassium chloride (KCl), calcium chloride ($CaCl_2$), ammonium nitrate (NH_4NO_3), potassium nitrate (KNO_3), sodium carbonate (Na_2CO_3), sodium bicarbonate ($NaHCO_3$), disodium phosphate (Na_2HPO_4), and borax ($Na_2B_4O_7$).

Ochs et al. (1981), Webb (1981), and Schell and Leboeuf (1983) have studied physical properties for some of the alternative salts, but most of the properties needed to determine dynamic stability are still unknown, especially for salt combinations. Experience with alternative-salt solar ponds is also limited, although the largest solar pond [36,000 m^2 (390,000 ft^2)] operated in the United States used a magnesium brine (Ochs, Johnson, and Sadan 1981). This pond used a thin gradient zone to keep the winter temperature of a deep brine pond above 13°C (55°F), preventing precipitation of epsomite salt ($MgSO_4$) and thus increasing yearly magnesium production. Mangussi, Saravia, and Lesino (1980) and Herrmann, Merriam, and Rahmani (1984) have operated Na_2SO ponds. An 850-m^2 (9,100-ft^2) pond using mixed $MgCl_2$ and NaCl was operated briefly at the Great Salt Lake (Newell, Pande, and Boehm 1980).

Several of the alternative salts have solubilities that increase strongly with temperature. These salts have been proposed for saturated solar ponds, in which the salinity is maintained at saturation at all levels in the pond, and a stable density gradient is achieved once a temperature gradient is established. Ochs, Stojanoff, and Day (1980) have reported experience with a $Na_2B_4O_7$ saturated pond at the Desert Research Institute, and a KNO_3 pond was studied at the University of New Mexico (Salamah 1982). In theory a saturated pond requires little maintenance because the diffusive salt flux upward is balanced by a downward flux caused by precipitation. In practice, operation is more complicated. During a rapid cooling period followed by crystallization, the $Na_2B_4O_7$ pond experienced a dramatic increase in pond bottom reflectivity, resulting in a deleterious positive feedback cycle of further cooling and crystallization. This effect is expected to be important for salts with a high degree of supersaturation. In addition many alternative salts have several hydrate forms, and pond chemistry is likely to be complicated if low-cost impure salts are used.

For electricity-generating applications most solar pond development has centered on using the thermal energy of the pond to drive an organic Rankine cycle turbine. A possible alternative is to use a dialytic battery (reverse electrodialysis) operating between the low concentration salt solution from the top of a saturated solar pond and the high concentration solution from the bottom. The resulting product of medium concentration solution is returned to the pond, where the concentration differences are regenerated. Mehta (1982) investigated this concept and concluded that the electrical resistance of present-day membranes, which are produced for electrodialysis and not for reverse electrodialysis, is too high and solution compartments are too thick, resulting in uneconomically high capital costs for the system.

10.9 Saltless Ponds

In many locations there would be a ready market for smoothly working, low-cost solar ponds if the potential environmental problem of salt pollution could be eliminated. Several varieties of saltless solar ponds have been proposed as alternatives to expensive brine containment and recovery systems. A saltless pond uses a transparent convection-suppressing means located above a storage layer of freshwater (or perhaps locally available brackish water). Hull (1980) examined using closely spaced membranes within the water of the pond. Borrowing from work on convection suppression in flat plate collectors, Lin (1983) experimented with air-filled honeycombs floating on top of freshwater. Dyksterhuis and Ortabasi (1983) experimented with honeycombs partially filled with oil on top of freshwater. Using transparent polymer gel as an insulating layer was first studied by Shaffer (1978), and subsequently Wilkins, Yang, and Kim (1981) experimented with transparent polymer gels floating over a pond fluid.

The major advantage of saltless ponds is the absence of large quantities of salt, eliminating a major cost component (salt and possibly liner), as well as the environmental threat. Saltless ponds also minimize surface evaporation and requirements for makeup water, which is a problem in arid climates. The pond fluid is no longer open to the air, and it should be easier to maintain good transparency. Maintenance should be less than for salt-gradient solar ponds, because fluid instability and wind-mixing problems do not occur.

Each of the saltless pond concepts holds promise, but they also have several difficult problems that do not occur with salt-gradient ponds. Foremost is the necessity of low cost in what could be a relatively complicated engineered structure. The convection-suppressing mechanisms must be able to withstand extremes in weather conditions over a large area device. Keeping the pond surface clean is also a difficult problem. Dust and debris accumulating on surface membranes or gels drastically reduce transparency and may be extremely difficult to remove. Effective mechanical cleaning systems may be costly.

10.10   Practical Applications

Solar ponds collect and store solar energy for use at temperatures below the boiling point of the storage zone brine. They cost less than conventional flat plate collectors in almost all locations, and much less in favorable locations. As Kooi (1979) shows, the solar pond is more efficient than the best flat plate collector for large $\Delta T/H$; however, the advantage of the pond in collecting energy on days of low insolation when $\Delta T/H$ is too large for the flat plate collector to operate must be averaged with the disadvantage of pond losses at night, when the flat plate collector is turned off. Typically the average energy gain per unit area of the pond is comparable to that of a flat plate collector working with the same average $\Delta T/H$, and the pond advantages are found in the lower cost and the large intrinsic heat storage capacity. Pond limitations are that they cannot be mounted on roofs or tilted for high latitude use and that heat losses to the earth make uninsulated small ponds inefficient. Because of these losses, ponds only a few hundred meters square for individual house or water heating are not economically competitive.

10.10.1   Thermal Applications

In locations where salt must be transported and where impermeable plastic liners must be used, estimated pond construction and operating costs indicate that the cost of pond heat will be approximately $2\textcent/kWh_t$. This is much less than the usual cost of electric resistance heat, and substantially less than the cost of heat from propane or fuel oil. Thus, for example, for large-scale water heating or grain drying in the midwestern United States, where electricity or propane is now the energy source, the pond is a promising lower-cost alternative. In this and other process heat applications the pond has higher efficiency when used at lower temperature as a preheater and may be the most valuable when used in this mode, combined with conventional energy sources to reach the final temperature needed. Similar efficiency considerations indicate the suitability of a pond as the heat source for a heat pump to heat a greenhouse (Fynn and Short 1983).

One solar pond that has been constructed and used for a thermal application is the 2,000-m² (22,000-ft²) pond at Miamisburg, Ohio, constructed by the city for spring and autumn heating of the outdoor municipal swimming pool. The Miamisburg pond has been able, after initial difficulties from liner leaks, to supply the heat required, even though the average solar radiation in this part of Ohio is only 160 W/m² (54 Btu/h ft²). Swimming pool heating is a very favorable solar application because the heat is delivered at a low temperature.

The solar pond concept can be readily incorporated into selected mining operations at saline lakes or salt beds. Ochs, Johnson, and Sadan (1981) reported the use of a thin gradient zone to increase the winter yield of magnesium at the Great Salt Lake. The elevated temperature due to the salt gradient prevented crystallization of epsomite in a large holding pond. Lesino et al. (1982) reported the commercial use of a 400-m² (4,300-ft²) salt-gradient solar pond using sodium sulfate. A mixture of sodium sulfate and sodium chloride from a mine was put into the bottom of the solar pond where it dissolved as the pond heated to a temperature greater than 40°C (104°F). Sufficient salt was added to saturate the pond solution at the bottom with respect to sodium sulfate. The heated brine was then pumped from the bottom of the solar pond to a cooling pond, where a large fraction of the sodium sulfate crystallized. The relatively pure sodium sulfate decahydrate was then used in an industrial process.

10.10.2 Power Generation

Low temperature solar heat usually compares most favorably with heat from other energy sources when used directly as heat. The low efficiency of thermodynamic conversion to produce shaft work or electricity excludes solar pond thermal power except in very favorable sites where solar pond heat is unusually low cost or other energy sources are unusually high cost.

The new 5-MW solar pond peaking power plant at the Dead Sea in Israel is in a very favorable site, and it is estimated that electricity can be produced there at 10¢–15¢/kWh. Source and sink temperature difference is around 50°C

(122°F), leading to a Carnot efficiency of 14% and an actual practical efficiency of around 5%. (Heat exchanger ΔT is 10°–15°C (18°–27°F), and parasitic power is 20% of mechanical output.) This means that when power station costs are accounted for, heat must be supplied by the pond at less than 0.5¢/kWh to yield electricity at 10¢. However in this site, as in other similar ones where the cost of pond heat is very low, there is no demand for low temperature heat as heat, and the only use for the pond is for power, providing that it can be produced at low enough cost.

Central Australia is an example of a region where relatively small solar pond power stations in the 20–200 kW range may be useful because of the very high cost of electricity from other sources, which in some cases may exceed $1.00/kWh. Because of the favorable potential, solar pond power station work aimed toward demonstration of practical systems for rural Australia was started at Alice Springs and has been under way there for several years (Collins 1983).

Similar situations exist in other places in the world, for example, in parts of the arid and sparsely populated Qinghai and Xinjiang provinces of western China.

10.11   Conclusions

During the past decade several significant advances have been made in understanding the basic phenomena associated with solar ponds. Salt-gradient solar ponds have been demonstrated to work in many diverse climates, and they appear to have economic potential in a wide variety of applications that depend on the availability of land, salt, water, and sunshine. Thermal behavior, including design optimization and some aspects of heat loss to the ground, is now relatively well understood. A good understanding of gradient zone stability and the fundamentals of zone boundary erosion is available for existing pond designs. Instrumentation has been developed to study most aspects of solar pond behavior, both in the laboratory and in the field. Several alternative solar pond designs have been shown to have some potential.

Despite the success of past research, there are still important problems that remain to be solved before solar ponds can successfully enter the commercial marketplace. Verification of theories and procedures that have been developed, based on results obtained in laboratory tanks and small-scale ponds, requires operating and monitoring a large pond to ensure that scale dependence does not affect the results and to develop operating techniques appropriate for large ponds. Remaining research areas can be divided into questions that require testing on a large pond and those that can be resolved in the laboratory prior to field testing.

Scale-dependent hydrodynamic issues require testing in a solar pond that is large enough to exhibit three-dimensional effects. This increase in pond size from that of existing solar ponds in the United States is critical to the understanding of wind-driven mixing of the surface layer, erosion of the gradient zone lower boundary due to energy extraction, and overall maintenance of the gradient zone clarity and stability. Specific research areas are (1) evaluating surface wind mixing effects and correlating them with climatic conditions; (2) evaluating surface structures to suppress wind mixing; (3) improving gradient zone boundary maintenance techniques; (4) determining how the distributor dimension, fluid speed, and fluid buoyancy influence the effect of fluid injection into the gradient zone; (5) determining the performance of multiple injection ports for gradient establishment and maintenance; and (6) evaluating interface erosion due to a recirculation flow in the heat storage zone caused by different energy extraction geometries.

Scale-independent issues that can be investigated in the laboratory include long-term material performance, operational procedures, instrumentation, and chemistry of brines and soils. Specific research areas are (1) determining the properties of synthetic liners under pond conditions; (2) developing suitable leak detection and repair techniques; (3) evaluating biochemical processes affecting water clarity; (4) evaluating the detailed mechanisms of pond-to-ground heat transfer; (5) developing ground heat loss monitoring instrumentation; (6) determining soil permeability under pond conditions; and (7) developing soil impermeabilization techniques.

References

Almanza, R., and H. C. Bryant. 1983. Oscillatory motions in the nonconvective layer of a solar pond. *ASME J. Solar Energy Eng.* 105: 375–378.

Atkinson, J. F., and D. R. F. Harleman. 1983. A wind-mixed layer model for solar ponds. *Solar Energy* 31: 243–259.

Baines, P. G., and A. E. Gill. 1969. On thermohaline convection with linear gradients. *J. Fluid Mech* 37: 289–306.

Bergman, T. L., F. P. Incropera, and R. Viskanta. 1982. A multi-layer model for mixing layer development in a double-diffusive thermohaline system heated from below. *Int. J. Heat Mass Transfer* 25: 1411–1418.

Bergman, T. L., D. R. Munoz, F. P. Incropera, and R. Viskanta. 1983. Correlation for entrainment of salt-stratified fluid by a thermally driven mixed layer. ASME Paper 83-WA/HT-76.

Blaney, D. L., and C. E. Nielsen. 1985. Measurement of heat transfer through earth in situ. In *Proc. Int. Solar Energy Soc.* Montreal, Canada. pp. 1227–1231.

Carmichael, A. D., M. J. Markow, P. L. Dintrans, A. M. Salhotra, E. E. Adams, D. H. Marks, S. Mukherji, and D. L. Thurston. 1985. *A State-of-the-Art Study of Nonconvective Solar Ponds for Power Production.* EPRI/AP-3842. Palo Alto, CA.

Cha, Y. S., W. T. Sha, and W. W. Schertz. 1982. Modeling of the surface convective layer of salt-gradient solar ponds. *ASME J. Solar Energy Eng.* 104: 293–298.

Chinery, G.T., and G. R. Siegel. 1983. Design, construction and cost of TVA's 4000 m² (1-acre) nonconvecting salt gradient solar pond. ASME-SED. Orlando.

Chinery, G. T., G. R. Siegel, and W. C. Irwin. 1983. Gradient zone establishment and maintenance at TVA's 4000-m² (1-acre) nonconvecting salt gradient solar pond. In *Proc. Am. Solar Energy Soc.* Minneapolis, MN, pp. 399–404.

Collins, R. B. 1983. Alice Springs solar pond project. In *Proc. Int. Solar Energy Soc.* Perth, Australia, pp. 775–779.

Crevier, D., and A. Moshref. 1981. The floating solar pond. In *Proc. Am. Sec. Int. Solar Energy Soc.* Philadelphia, PA, pp. 796–800.

Dalton, F. N., W. N. Herkelrath, D. S. Rawlins, and J. D. Rhoades. 1984. Time-domain reflectometry: Simultaneous measurement of soil water content and electrical conductivity with a single probe. *Science* 224: 989–990.

Drost, M. K., and B. M. Johnson. 1983. Optimization of a solar pond power plant at Great Salt Lake. In *Proc. Am. Solar Energy Soc.* Minneapolis, MN, pp. 423–428.

Duyar, A., and W. Bober. 1984. The bottom heat loss of a solar pond in the presence of moving ground water. *ASME J. Solar Energy Eng.* 106: 335–340.

Dyksterhuis, F. H., and U. Ortabasi. 1983. Honeycomb stabilized saltless solar pond. In *Proc. Int. Solar Energy Soc.* Perth, Australia, pp. 1030–1034.

French, R. L., D. H. Johnson, G. F. Jones, and F. Zangrando. 1984. *Salt-Gradient Solar Ponds: Summary of U.S. Department of Energy Sponsored Research.* JPL-84-74. Pasadena, CA: Jet Propulsion Laboratory.

Fynn, R. P., T. H. Short, and S. A. Shah. 1980. The practical operation and maintenance of a solar pond for greenhouse heating. In *Proc. Am. Soc. Ag. Eng.* Kansas City, MO, pp. 531–535.

Fynn, R. P., and T. H. Short. 1983. *Solar Ponds: A Basic Manual.* Special Circular 106, Ohio Agricultural Research and Development Center.

Giulianelli, J. L., and A. M. Naghmush. 1982. The spectrophotometric method of determining solar energy penetration profiles for solar ponds. *International Solar Pond Letters* 1: 7–9.

Grimmer, D. P., G. F. Jones, J. Tafoya, and T. J. Fitzgerald. 1983. Development of a point-electrode conductivity salinometer with high spatial resolution for use in very saline solutions. *Rev. Sci. Inst.* 54: 1744–1748.

Harris, M. J., D. E. Etter, and L. J. Wittenberg. 1981. Observations regarding materials and site preparation for salt gradient solar ponds. In *Proc. Am. Sec. Int. Solar Energy Soc.* Philadelphia, PA, pp. 787–790.

Henderson, J., and C. M. Leboeuf. 1980. *SOLPOND—A Simulation Program for Salinity Gradient Solar Ponds.* SERI/TP-351-599. Golden, CO: Solar Energy Research Institute.

Herrmann, C. C., M. F. Merriam, and R. Rahmani. 1984. Solar pond using sodium sulfate. In *Proc. Am. Solar Energy Soc.* Anaheim, CA, pp. 255–257.

Hull, J. R. 1980. Membrane stratified solar ponds. *Solar Energy* 25: 317–325.

Hull, J. R., Y. S. Cha, W. T. Sha, and W. W. Schertz. 1982. Construction and first year's operational results of the ANL research salt gradient solar pond. In *Proc. Am. Solar Energy Soc.* Houston, TX, pp. 197–202.

Hull, J. R. 1983. Calculation of solar pond thermal efficiency with a diffusely reflecting bottom. *Solar Energy* 29: 385–389.

Hull, J. R., J. Kamal, and C. E. Nielsen. 1983. Time dependence of ground heat loss from solar ponds. In *Proc. Am. Solar Energy Soc.* Minneapolis, MN, pp. 381–385.

Hull, J. R., K. V. Liu, W. T. Sha, J. Kamal, and C. E. Nielsen. 1984. Dependence of ground heat loss upon solar pond size and perimeter insulation: Calculated and experimental results. *Solar Energy* 33: 25–33.

Hull, J. R. 1984. Passive stabilization of gradient zone boundaries in solar ponds. In *Proc. Am. Solar Energy Soc.* Anaheim, CA, pp. 259–264.

Hull, J. R. 1985. Solar pond ground heat loss to a moving water table. *Solar Energy* 35: 211–217.

Hull, J. R., A. B. Scranton, and K. E. Kasza. 1985. Solar pond heat removal using a submerged heat exchanger. In *Proc. Int. Solar Energy Soc.* Montreal, Canada, pp. 1505–1509.

Jacobs, H. R., and R. F. Boehm. 1980. Direct contact binary cycles. *Source-Book on the Production of Electricity from Geothermal Energy*, J. Kestin, ed. DOE/RA/28320-2. Washington, DC: U.S. Department of Energy, pp. 413–471.

Jain, G. C. 1973. Heating of solar pond. In *Proc. The Paris Congress on Solar Energy*. Paper EH-61.

Jayadev, T. S., and J. Henderson. 1979. Salt concentration gradient solar ponds—Modeling and optimization. In *Proc. Int. Solar Enery Soc.* Atlanta. GA, pp. 1015–1019.

Jones, G. F., K. A. Meyer, J. C. Hedstrom, J. S. Dreicer, and D. P. Grimmer. 1983. Design, construction, and initial operation of the Los Alamos National Laboratory salt-gradient solar pond. In *Proc. ASME-SED* Orlando, FL, pp. 157–164.

Kamal, J., and C. E. Nielsen. 1982. Convective zone structure and zone boundaries in solar ponds. In *Proc. Am. Solar Energy Soc.* Houston, TX, pp. 191–196.

Kooi, C. F. 1979. The steady state salt gradient solar pond. *Solar Energy* 23: 37–45.

Kooi, C. F. 1981. Salt gradient solar pond with reflective bottom: Application to "saturated" pond. *Solar Energy* 26: 113–120.

Leboeuf, C., and D. H. Johnson. 1984. *Effects of Soil Conditions on Solar Pond Performance.* SERI/TP-253-2157. Golden, CO: Solar Energy Research Institute.

Lesino, G., L. Saravia, J. Mangussi, and R. Caso. 1982. Operation of a 400 m² sodium sulphate solar pond in Salta, Argentina. *International Solar Pond Letters* 1: 12–13.

Lewis, W. T., F. P. Incropera, and R. Viskanta. 1982a. Interferometric study of mixing layer development in a laboratory simulation of solar pond conditions. *Solar Energy* 28: 389–401.

Lewis, W. T., F. P. Incropera, and R. Viskanta. 1982b. Interferometric study of stable salinity gradients heated from below or cooled from above. *J. Fluid Mech.* 116: 411–430.

Lin, E. I. H. 1982. *Regional Applicability and Potential of Salt Gradient Solar Ponds in the United States.* DOE/JPL-1060-50. Pasadena, CA: Jet Propulsion Laboratory.

Lin, E. I. H. 1983. Outdoor performance of a honeycomb-covered solar pond model. In *Proc. Am. Solar Energy Soc.* Minneapolis, MN, pp. 411–416.

Mangussi, J., L. Saravia, and G. Lesino. 1980. The use of sodium sulfate in solar ponds. *Solar Energy* 25: 475–477.

Marsh, H. E. 1983. Investigation of indigenous water, salt and soil for solar ponds. In *Proc. Intersoc. Energy Conv. Eng. Conf.*, pp. 1976–1981.

May, E. K., C. M. Leboeuf, and D. Waddington. 1982. *Conceptual Design of the Truscott Brine Lake Solar Pond System.* SERI/TR-253-1833. Golden, CO: Solar Energy Research Institute.

May, E. K., C M. Leboeuf, and D. Waddington. 1983. *Conceptual Design of a 20-Acre Salt-Gradient Solar Pond/System for Electric Power Production at Truscott, Texas.* SERI/TR-253-1868. Golden, CO: Solar Energy Research Institute.

Mehta, G. D. 1982. Performance of present-day ion-exchange membranes for power generation using a saturated solar pond. *J. Membrane Science* 11: 107–120.

Meyer, K. A., D. P. Grimmer, and G. F. Jones. 1982. An experimental and theoretical study of salt-gradient pond interface behavior. In *Proc. Am. Solar Energy Soc.* Houston, TX, pp. 185–190.

Meyer, K.A. 1983. A numerical model to describe the layer behavior in salt-gradient solar ponds. *ASME J. Solar Energy Eng.* 105: 341–347.

Newell, T., J. Pande, and R. Boehm. 1980. Development of performance information for large scale solar pond applications. In *Proc. Am. Sec. Int. Solar Energy Soc.* Phoenix, AZ, pp. 376–380.

Newell, T. A., and P. M. Von Driska. 1986. Double diffusive effects on solar pond gradient zones. *ASME J. Solar Energy Eng.* 108: 3–5.

Nielsen, C. E., and A. Rabl. 1975. Operation of a small salt gradient solar pond. *ISES Meeting, Extended Abstracts.* Los Angeles, CA, pp. 271–272.

Nielsen, C. E. 1976. Experience with a prototype solar pond for space heating. In *Proc. Sharing the Sun: Solar Technology in the Seventies*, vol. 5. Winnipeg, Canada, pp. 169–182.

Nielsen, C. E. 1978. Equilibrium thickness of the stable gradient zone in solar ponds. In *Proc. Am. Sec. Int. Solar Energy Soc.* Denver, CO, pp. 932–935.

Nielsen, C. E. 1979. Control of gradient zone boundaries. In *Proc. Int. Solar Energy Soc.* Atlanta, GA, pp. 1010–1014.

Nielsen, C. E. 1980a. Nonconvective salt-gradient solar ponds. In *Solar Energy Technology Handbook*, W. C. Dickinson and P. N. Cheremisinoff, eds. New York: Marcel Dekker, ch. 11.

Nielsen, C. E. 1980b. Design and initial operation of a 400 m^2 solar pond. In *Proc. Am. Sec. Int. Solar Energy Soc.* Phoenix, AZ, pp. 381–385.

Nielsen, C. E., and J. Kamal. 1981. The 400 m^2 solar pond: One year of operation. In *Proc. Am. Sec. Int. Solar Energy Soc.* Philadelphia, PA, pp. 758–762.

Nielsen, C. E. 1982. Salt transport and gradient maintenance in solar ponds. In *Proc. Am. Solar Energy Soc.* Houston, TX, pp. 179–184.

Nielsen, C. E. 1983a. Experience with heat extraction and zone boundary motion. In *Proc. Am. Solar Energy Soc.* Minneapolis, MN, pp. 405–4l0.

Nielsen, C. E. 1983b. Practical zone boundary control in solar ponds. In *Proc. Int. Solar Energy Soc.* Perth, Australia, pp. 780–784.

Ochs, T. L., C. G. Stojanoff, and D. L. Day. 1980. One year's experience with an operating saturated solar pond. In *Proc. Am. Sec. Int. Solar Energy Soc.* Phoenix, AZ, pp. 391–394.

Ochs, T. L., S. C. Johnson, and A. Sadan. 1981. Application of a salt gradient solar pond to a chemical process industry. In *Proc. Am. Sec. Int. Solar Energy Soc.* Philadelphia, PA, pp. 809–811.

Ochs, T. C., G. Stojanoff, D. L. Day, E. Eckert, J. Langeliers, and D. Wruck. 1981. *Analysis of Saturated Solar Pond Characteristics.* DOE/CS/30174-T2, Washington, DC: U.S. Department of Energy.

Poplawsky, C. J., F. P. Incropera, and R. Viskanta. 1981. Mixed layer development in a double-diffusive, thermohaline system. *ASME J. Solar Energy Eng.* 103: 351–359.

Rabl, A., and C. E. Nielsen. 1975. Solar ponds for space heating. *Solar Energy* 17: 1–12.

Rothmeyer, M. 1980. The Soret effect and salt-gradient solar ponds. *Solar Energy* 25: 567–568.

Salamah, A. I. 1982. An experimental study of a KNO_3 salt gradient solar pond. Ph.D. dissertation. University of New Mexico.

Schell, D., and C. M. Leboeuf. 1983. *The Behavior of Nine Solar Pond Candidate Salts.* SERI/TR-253–1512. Golden, CO: Solar Energy Research Institute.

Schladow, S. G. 1983. Stability of salt gradient solar ponds. In *Proc. Int. Solar Energy Soc.* Perth, Australia, pp. 1127–1131.

SERI. 1981. *Conceptual Design of the Truscott Brine Lake Solar Pond System: Volume I—Utility Independent Scenario.* SERI/TR-731-1202. Golden, CO: Solar Energy Research Institute.

SERI. 1984. *Solar Ponds: A Selected Bibliography.* SERI/SP-271-2470. Golden, CO: Solar Energy Research Institute.

Shaffer, L. H. 1978. Viscosity stabilized solar ponds. In *Proc. Int. Solar Energy Soc.* New Delhi, India, pp. 1171–1175.

Shah, S. A., T. H. Short, and R. P. Fynn. 1981. Modeling and testing a salt gradient solar pond in northeast Ohio. *Solar Energy* 27: 393–401.

Suzukawa, J., and U. Narusawa. 1982. Structure of growing double-diffusive convection cells. *ASME J. Heat Trans.* 104: 248–254.

Tabor, H. 1981. Solar ponds. *Solar Energy* 27: 181–194.

Tabor, H., and Z. Weinberger. 1980. Nonconvecting solar ponds. In *Solar Energy Handbook*, J. F. Kreider and F. Kreith, eds. New York: McGraw-Hill, ch. 10.

Tsinober, A. B., Y. Yahalone, and D. J. Shlien. 1983. A point source of heat in a stable salinity gradient. *J. Fluid Mech.* 135: 199–217.

Turner, J. S. 1968. The influence of molecular diffusivity on turbulent entrainment across a density interface. *J. Fluid Mech.* 33: 639–656.

Turner, J. S. 1973. *Buoyancy Effects in Fluids.* Cambridge: Cambridge University Press, ch. 8.

Walton, I. C. 1982. Double-diffusive convection with large variable gradients. *J. Fluid Mech.* 125: 123–135.

Webb, J. 1981. *Optical Transparency of Inexpensive Salt Solutions for Construction of Density-Gradient Solar Ponds.* SERI/RR-641-615. Golden, CO: Solar Energy Research Institute.

Weinberger, H. 1964. The physics of the solar pond. *Solar Energy* 8: 45–56.

Wilkins, E., E. Yang, and C. Kim. 1981. The gel pond. In *Proc. 16th IECEC Conf.* Atlanta, GA. pp. 1726–1731.

Wittenberg, L. J., and D. E. Etter. 1982. Heat extraction from a large solar pond. ASME Paper No. 82-WA/Sol-31.

Wittenberg, L. J. 1984. Interface movement in salt-gradient solar ponds. In *Proc. ASME-SED.* Las Vegas, NV.

Wright, J. D. 1982. Selection of a working fluid for an organic Rankine cycle coupled to a salt-gradient solar pond by direct-contact heat exchange. *ASME J. Solar Energy Eng.* 104: 286–292.

Wright, J. D. 1984. Direct-contact preheater/boilers for solar pond power plants. In *Proc. ASME-SED.* Las Vegas, NV.

Zangrando, F. 1979. Observation and analysis of a full scale experimental salt-gradient solar pond. Ph.D. dissertation, University of New Mexico.

Zangrando, F. 1980. A simple method to establish salt gradient solar ponds. *Solar Energy* 25: 467–470.

Zangrando, F. 1982. *Heat and Mass Extraction from Solar Ponds: Analysis and Development of a Laboratory Facility.* SERI/TR-252-1569. Golden, CO: Solar Energy Research Institute.

Zangrando, F. 1984. *Survey of Density Measurement Techniques for Application in Stratified Fluids.* SERI draft report. Golden CO: Solar Energy Research Institute.

Zangrando, F., and L. A. Bertram. 1984. The effect of variable stratification on linear doubly diffusive stability. *J. Fluid Mech.* 151: 55–79.

Zangrando, F., J. Green, and E. Fisher. 1984. Heat and mass exchange in a partially stratified fluid. In *Proc. ASME/AIChE National Heat Transfer Conf.* Niagara Falls, NY, pp. 452–457.

11 Cost Issues and Opportunities

John A. Clark

11.1 Overview

During the early 1970s the federal government accelerated its support of national efforts in solar energy largely as a result of a foreseeable decline in the availability of conventional energy resources. The pace of federal involvement quickened following the oil embargo in the fall of 1973. Within eight years total federal support for solar energy programs increased from less than $50 million to more than $800 million annually. A lead research laboratory for solar energy, the Solar Energy Research Institute (SERI), and four regional solar energy centers were in place by 1980 (the latter were discontinued by 1982). The principal federal agencies that contributed to this effort were the National Aeronautics and Space Administration (NASA), the National Science Foundation (NSF), Research Applied to National Needs (RANN), the Federal Energy Administration (FEA) (1970–80), the Energy Research and Development Administration (ERDA) (1974–77), the U.S. Department of Energy (DOE) (1978–present), and the U.S. Department of Housing and Urban Development (HUD) (1975–80). In addition virtually all the national laboratories established some kind of visible activity in solar energy. Private industry, research institutions, and universities also became actively involved in solar energy, usually as a contractor to the federal government. It is interesting to note in this connection that several of the smaller private industrial organizations that survived the commercial competition of the last five years did not participate directly as federal contractors in solar energy, while many of those who did participate have either abandoned their solar efforts or have gone out of business entirely.

A significant federal effort during this period was a program to accelerate the commercialization of solar energy and a companion activity, the Solar Demonstration Program (U.S. House of Representatives 1973; U.S. Senate 1976; U.S. House 1979; ERDA 1975, 1976; ERDA 1975; ERDA 1977; HUD 1977). A partial summary of results and a description of these activities are given in Freeborne, Mara, and Lent (1979), HUD (1976–78), Christensen (1981), and DOE (1978). The Solar Demonstration Program, in which solar collectors played a major technical role, was very controversial in terms of its direct effectiveness in promoting the commercialization of solar energy. Today there appears to be little visible evidence of its positive effectiveness, although certain indirect benefits can be identified. It is probably more realistic to view the demonstration program as a national experiment, and many important lessons were learned—especially regarding the operation of complete solar energy systems. Some of these lessons have been summarized (Noreen et al. 1980). A critical assessment of this program is also available (Spielvogel 1980).

The principal outcome of this decade of federal solar involvement may be the recognition that by 1985 solar energy systems (including solar collectors) had largely become consumer products. As such, these systems must respond to market forces regarding cost and customer acceptability. The market penetration of solar systems can be advanced only when the customer views the product as economically preferable to conventional energy supply systems and sources. Because a solar collector represents the single largest cost component in a solar energy system, this market condition applies most to this component.

During this ten year period other important events occurred that are directly related to (and also the result of) the evolution of collector technologies and costs. Realistic solar applications have been identified and simplified, and reliable, cost-effective solar products and systems have been developed for residential and commercial markets. It is instructive to compare the perceptions of solar applications that prevailed at both the beginning and at the end of the decade. In 1975 solar applications were considered to be applicable to space heating and cooling on new residential construction, new commercial construction, or retrofit installations on commercial and institutional buildings. Most systems were very complex compared with 1985 installations, and none were, or expected to be, cost effective. The exception was solar swimming pool heating, a system that was largely ignored by solar programs and left to private industry to develop, which it did. Heating domestic water by solar energy (DHW) was usually considered in 1975 to be an adjunct to space heating and cooling. By 1985, however, the situation had completely changed. The principal application for solar energy in the United States had become medium temperature, flat plate collectors [110°–180°F (48°–82°C)] for heating domestic hot water, mostly in residential, retrofit installations. The second most widely used solar application is low temperature, flat plate collectors [less than 110°F (43°C)] for swimming pool heating. Solar space heating in 1983 placed a distant third, and solar space cooling was essentially nonexistent. This situation is borne out by the statistical summary prepared by DOE (1984) that lists the results for a total annual shipment in 1983 of 16,828,000 ft² (1,563,940 m²) of collectors (see tables 11.1–11.3).

This change in emphasis on the types of solar applications that prevailed may also be illustrated by comparing systems proposed in 1975 with systems actually operating in 1985. Figure 11.1 shows a typically conceived, 1974 liquid solar system for heating and cooling (McGarity 1975), and figures 11.2

Table 11.1
Solar end-use sectors, 1983

End use	Percentage of total shipment
Hot water	55.4 (50.20% liquid; 0.78% air)
Pool heating	28.8 (25.77% nonmetallic)
Space heating	12.4 (8.19% liquid; 3.14% air)
Space cooling	0.1
Other	3.3
	100.0

Source: DOE (1984).

Table 11.2
Solar market sectors, 1983

Market sector	Percentage of total shipment
Residential	70.0
Commercial	18.1
Industrial	9.9
Other	2.0
	100.0

Source: DOE (1984).

Table 11.3
Distribution by type of collector, 1983

Type of collector	Percentage of total shipment
Medium temperature flat plate	67.5
Low temperature flat plate	28.8
Other	3.7
	100.0

Source: DOE (1984).

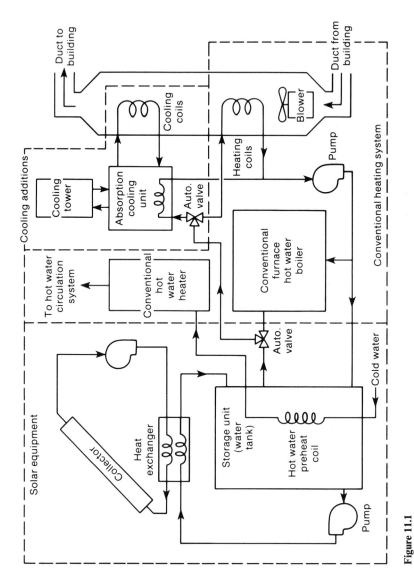

Figure 11.1
Typical solar heating and cooling system using water as the collector circulation fluid and a water tank for heat storage. The conventional furnace operates in a parallel mode with the collector and storage unit. Source: McGarity (1975).

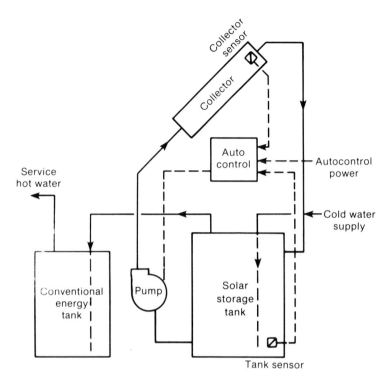

Figure 11.2
Preheater system schematic (1984). Source: Honikman and Bendt (1980).

and 11.3 show the kind of system widely used for domestic hot water heating in 1985 (Honikman and Bendt 1980). The drain-back solar DHW system is another simpler system in current use. The figures show the shift in emphasis that has occurred in both applications and in the relative simplicity of the systems.

Although air-space-heating systems made up only 3% of the total 1983 shipment of collectors, many of these installations were much simpler than the ones proposed and built during the early part of the decade. Rock-bed storage and its associated controls and air-handling unit were generally included in solar air-space-heating systems in the mid-1970s. Most of the solar collectors in these systems ranged from 400 to 600 ft² (37–56 m²), providing annual solar fractions above 50%. At present, however, and largely as a result of a need to reduce the total initial investment and to improve system reliability, many air-space-heating systems do not include built-in storage and have simple, demand-type, on/off controls with a small air blower to force

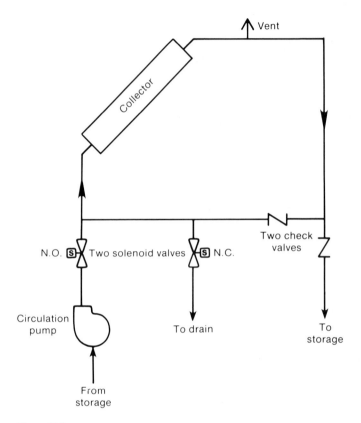

Figure 11.3
Standard direct drain-down method (1984). Source: Honikman and Bendt (1980).

airflow. Collectors are also much smaller, ranging from 100 to 300 ft² (9–28 m²) for 1,000 to 2,000 ft² (93–186 m²) residences (Energy Research 1984), and the corresponding annual solar fraction is reduced 10% to 30%. In many installations having built-in storage during most of the heating season, the storage system was ineffective, especially those linked with reasonably sized, cost-effective collector arrays. Accordingly, eliminating the built-in storage and its associated equipment and controls was an important step in simplifying these systems and in improving their cost-effectiveness and reliability. Since storage is inherent in the thermal mass of buildings, basement areas, and furnishings, the effect of thermal storage is present in retrofit installations at no additional cost.

The boiling collector containing a refrigerant has recently penetrated the market for domestic hot water and some space heating (Solar Research 1984).

These highly efficient, simple, but elegant units have inherent freeze protection and are cost effective.

Manufacturers of solar collectors and solar systems have also recognized that solar conversion applications must be engineered to complement other energy sources in an installation (Starpak Solar Systems 1984). This is especially true for domestic hot water systems in the northern half of the United States. There the solar contribution is mostly seasonal (April to October), but thermal recovery systems can be retrofitted to conventional gas furnaces to provide an additional energy contribution during the cold season (October to April). This complementary matching of solar energy with thermal recovery (and conservation) provides for 12-month energy gains and contributes significantly to the cost effectiveness of the combined systems.

By 1985 federal and state tax credits were an important factor in the total cost effectiveness of solar collectors and solar conversion systems. Many in the industry, however, were ambivalent about this issue. The tax credits clearly helped a great deal in marketing and in closing sales, but the credits also tended to encourage the manufacture and sale of systems of marginal quality that were sold because of exaggerated performance claims. In the long run, marginal systems retard the market penetration of solar conversion systems and harm the industry. Some form of third-party evaluation, certification, or accreditation may become necessary to protect both the public and reliable producers when government-supplied tax credits are involved in solar purchases. Meanwhile, anticipating a reduction or removal of tax credits, manufacturers in 1985 were striving to simplify their collectors and systems as much as possible to reduce costs. These cost reductions ideally apply to all phases of the business from purchased materials and components, production, shipping, installation, and servicing (Energy Research 1984; Solar Research 1984; Starpak Solar Systems 1984) to marketing, sales, and other overhead functions.

From 1975 to 1985 the solar industry both began and matured. The decade began in an atmosphere of crisis; the major participant in national energy matters was designated to be the federal government. Different ideologies of energy policy were also evident in such debates as those between the promoters of "soft energy paths" and others (U.S. Senate, Joint Hearings, December 9, 1976, app. III, especially pp. 463–531). By the end of the decade, ideological considerations had largely been replaced by practical issues. The role of the federal government was becoming limited to policymaking on taxation and economic issues of various kinds, supporting basic research and high risk ventures, disseminating public information, conducting statistical data surveys, and providing assistance in developing codes and standards. Solar energy systems became consumer products that had to be competitive in the marketplace with other (and sometimes complementary) energy sources. A private, productive industry has developed to supply these products. The loss of the 40% federal tax credit at the end of 1985 caused a serious dislocation in the solar industry.

11.2 Types of Collectors and Scope of the Study

The types of collectors currently being manufactured include flat plates (liquid and air), evacuated tubes (liquid), compound parabolic concentrators (liquid), parabolic troughs (liquid), parabolic dishes (liquid), and heliostat mirrors with a central receiver (liquid). In 1983 most collector manufacturing activity was for low and medium temperature flat plate collectors (DOE 1984). Of the total 1983 production of collectors that has been noted [16,828,000 ft^2 (1,563,940 m^2)], 96.3% was for these types. Special concentrating collectors amounted to only 2.58% of the total production, and evacuated tubes made up only 0.46%. The remaining production, 0.66%, was for parabolic troughs and parabolic dishes. Heliostat production for the central tower receiver in Barstow, California, represented a special case, one not involved in the EIA Summary (DOE 1984); it was not strictly a commercial market product.

These data show that commercial activity in manufacturing and shipping solar collectors in 1985 was essentially limited to flat plate collectors. Therefore this cost evaluation study focuses on flat plate collectors; some cost data on evacuated tube collectors are included when possible. Most cost data for 1975–85 are available only for flat plate collectors, however.

11.3 Factors Influencing Collector Cost Definition

The costs of a solar collector, which may be described in a number of ways, must be clearly defined in order to make meaningful comparisons and evaluate trends. To the consumer the significant cost is the net,[1] installed cost of complete system, including the collector, storage (if any), piping, valves, controls, and other associated equipment. The actual (net) cost of the collectors in this system will be about one-half of the total (net) system cost, a significant portion that must be carefully considered in evaluating collector versus system costs. The installed cost of the collectors is the sum of the costs of a chain of production and commercial activity, beginning with the basic materials and flowing through manufacture, shipping, dealer arrangements to the actual installation. This cost/price chain is illustrated in table 11.4.

Table 11.4
Cost/price chain for a solar collector incorporated into a total system (approximate)

	Single collector ($N = 1.0$)
Materials	$0.60X$
Production tooling	$0.01X$
Labor	$0.29X$
Overhead	$0.10X$
Subtotal, manufacturing	$1.0X$
Markup (profit, fee)	$2.0X$ to $3.0X$
Total, FOB	$3.0X$ to $4.0X$

1. *Manufacturer's (FOB) cost* $(1.0X)$/price $(3.0X$ to $4.0X)N$

A	B
2. *Installation + equipment* $(2.0X$ to $3.0X)N$	6. *Shipping cost* $(0.1X)N$
↓	↓
3. *Total customer cost* $(5.0X$ to $7.0X)N$	7. *Commercial distributors, dealers* $(1.4X$ to $1.90X)N$
	↓
C	8. *Markup* $(1.50X$ to $2.50X)N$
	↓
	9. *Installation + equipment* $(2.0X$ to $3.0X)N$
	↓
	10. *Total customer cost* $(5.0X$ to $7.5X)N$

4. *Retail (kit) price* $(3.0X$ to $4.0X)N$
↓
5. *Installation + equipment* $(1.5X$ to $2.50X)N$
(Total: $4.5X$ to $6.50X)N$

Note: Numbers refer to cost/price levels; magnitude of cost/price levels ($1.0X$, etc.) of basic manufacturer's FOB cost/price level of X/($3.0X$ to $4.0X$) are best estimates and will vary according to the manufacturer and the region. Discounts from distributors and dealers can be expected, depending on business conditions. N is the number of collector panels in an installation. Results are for ranges of conditions and are drawn from experience with DHW systems.

Path A: manufacturer installs system $(1, 2, 3)$.
Path B: dealer installs system $(1, 6, 7, 8, 9, 10)$.
Path C: owner installs system from kit $(1, 4, 5)$.

As the table indicates, at least ten cost/price levels can be identified in the chain of events from collector manufacturing to the installation of a system. The progression of costs shown in table 11.4 is intended to indicate typical circumstances; they will vary according to business conditions and regions. It is important, however, to recognize that the chain of costs that exists in this kind of commercial enterprise results in system installed costs that are $5N$ to $7N$ times the manufacturing costs of a single solar collector, where N is the number of collectors in the system. The final cost to the consumer of course also includes the cost for storage, pumps, valves, controls, and installation, which are added to the cost of the collectors.

To summarize, for a commercial installation the cost sequence is approximately as follows for a two-collector DHW system:

Manufacturing cost (single collector)	$1.0X$
Manufacturer's markup (single collector)	$2.0X - 3.0X$
Manufacturer's FOB price (two collectors)	$6.0X - 8.0X$
Installation and equipment (including two collectors)	$4.0X - 6.0X$
Total installed system cost (including two collectors)	$10.0X - 14.0X$

Table 11.4 also indicates the three general distribution paths for collectors: manufacturer to customer by means of kits or customer installations (path C), manufacturer to customer through professional installations (path A), and manufacturer to distributor and/or dealer with professional installations (path B). Not all manufacturers follow all three; the most common path is from manufacturers to distributors and dealers. The most successful companies appear to be those that manufacture original equipment and provide a complete line of services, including engineering design, R&D, manufacture, installation, service, parts supply, marketing and sales, and assistance in financial arrangements for customers (Clark 1981).

Because of the complexity of the cost/price paths and the many possible business arrangements involved in purchasing and installing a solar energy system, it is necessary to establish a single, identifiable, reasonable basis for analyzing and evaluating collector costs. The basis chosen for this study is the FOB price a manufacturer assigns to the collector. That price reflects all the manufacturer's technical and managerial operations and represents a system of common business activities and accounting procedures common to most manufacturers. As such, it is a fair and reasonable basis for comparison and evaluation.

However, two additional considerations are included. The first is the need to formulate meaningful cost/price comparisons over time on the basis of "real" dollars to allow for the effects of inflation. This can be done by using the Producer Price Intex (PPI) found in the *Statistical Abstract of the United States*. Specific PPI information used here is derived from data given in Thuesen and Fabrycky (1984). Manufacturers' published price data (mostly for the mid-1970s) are converted to 1975$ or 1985$. The first set of data allows prices to be compared among the manufacturers that existed (or would have existed) at the beginning of the decade. The projections of prices in 1985$ provides a basis for comparison with current collector prices as well as a basis for evaluating the effects of a decade of experience in the manufacture of solar collectors.

The second consideration in cost/price analyses is the need to present both the economic and thermal performance characteristics of a collector together in some meaningful way. For example, high-cost collectors might also be high in performance and thus require less total area. These collectors could be the most cost effective. The method to use in determining this relationship is to calculate the ratio of useful output per square foot to FOB price (in 1985$) per square foot, in a manner similar to that proposed by Deffenbaugh and Carper (1977) and Sims (1976). This ratio is called the performance–cost ratio (PCR) in this study, and it will be evaluated where collector efficiency data are given for summer $[I = 250$ Btu/h ft^2 (788 W/m^2), $\Delta T/I = 0.2]$ and winter $[I = 200$ Btu/h ft^2 (630 W/m^2), $\Delta T/I = 0.30]$ condition.[2] Other bases for this kind of comparison can be used, such as all-day performance of side-by-side collectors having identical loads, but those comparisons are not within the scope of this study. In any case the PCR results given here can provide a basis for comparative evaluation.

11.4   FOB Collector Costs: Mid-1970s and Projections to 1985

The FOB cost of a collector is determined by the five principal items shown in table 11.4: materials, tooling, labor, overhead, and markup. About 50% to 60% of the production costs are for materials, and about 25% to 30% for labor. Thus the cost of collectors is usually influenced primarily by these two principal cost items. Both of these costs will reflect changes that were caused by general inflationary trends in the economy. The materials used in solar collectors are usually already in high volume production and hence are also usually at their lowest real cost. These include copper (tube and sheet), steel (tube and sheet), aluminum (tube, channels, extrusions, and sheet), glass, plastic, and thermal insulation. The real future cost changes in these commodities can be expected to be minimal. Labor costs will show similar trends. Naturally future technological changes and market expansion will influence each of these costs.

Some representative costs for five collectors in 1977$ are given in table 11.5. The 1977 costs were published in *Solar Engineering* (May 1977); they are projected to 1985$ on the basis of the PPI.

These results illustrate the effects of material costs, labor costs, markup, and market outlet on the FOB price/costs. The mean FOB price is about $12.00/ft^2 ($129/m^2) (in 1977$), and it varies both higher and lower depending on the absorber material and the number of glazings. The corresponding mean price in projected 1985$ is about $21.00/ft^2 ($226/m^2), which is reasonably close to the average manufacturer's published costs for 1985 (see tables 11.9 and 11.10).

Deffenbaugh and Carper (1977) report the results of a 1977 cost study on 16 different collector designs in which various kinds and numbers (one or two) of covers and absorber coatings were used. Each of the 16 collectors had the same kind of absorber, an aluminum tube-in-sheet construction. The collectors were manufactured by the Southwest Research Institute, San Antonio, Texas. The results for eight of these designs, which correspond to 1985 designs,

Table 11.5
Representative collector prices (1977$ and projected 1985$)

Manufacturer	Type	FOB unit selling price ($/ft^2)[a] 1977$ (published)	1985$ (projected)
Sunworks	Liquid, copper, single glazed,		
	(1) commercial sale	8.50–10.00	15.00–17.50
	(2) retail sale	12.00–14.00	21.00–24.50
Northrop	Liquid, aluminum, commercial sale	8.00–10.00	14.00–17.50
Revere	liquid, copper, double glazed, retail sale	11.50	20.00
Solar Energy Products	Liquid, aluminum single glazed	10.00–12.00	17.50–21.00
Solaron (Enersol)	Air, steel, single glazed, dealer sales	12.05	21.00

Source: *Solar Enginneering*, May 1977 (for 1977$); 1985 costs are projected on the basis of the PPI (Thuesen and Fabrycky 1984) and rounded to the nearest $0.50.
a. To convert to $/m^2, multiply table entry by 10.76.

are summarized in table 11.6. Note that the cost data are for the collector materials only (in 1977$/ft^2) and are projected to 1985$ per square foot.[3]

For a given collector these data show that the 1977 cost difference for a selective coating (black chrome) over a nonselective coating (black paint) was $0.80/ft^2 ($8.61/m^2), or a projected 1985 cost difference of $1.40/ft^2 ($15.06/m^2). The increased cost in 1977$/ft^2 (projected to 1985$/ft^2) for an additional cover (Deffenbaugh and Carper 1977) is (1) plastic, 0.54 (0.94); (2) float glass, 0.74 (1.30); and (3) crystal (water white) glass, 1.02 (1.78). The actual 1985 costs for the cover materials depend on the quantity purchased, the sizing, and discounts but are approximately these: (1) plastic (Tedlar®), $0.46–0.58/ft^2 ($4.95–$6.24/m^2); float glass (Clark, 1984a and b) $0.57–$0.64/ft^2 ($6.13–$6.89/m^2);[4] crystal (water white) glass, $0.90–$1.10/ft^2 ($9.68–$11.84 m^2). Hence the costs of these materials have not followed trends but rather have been reduced in real terms since 1977. The same is true for absorber coatings. Note that the costs in table 11.6 are somewhat higher than the supplier's FOB price because of the cost of shipping. The Phelps Dodge Company (1982) quotes the 1982 FOB prices per square foot for black paint (polyester base) at $0.35/ft^2 ($3.77/m^2) and for black chrome at $1.35/ft^2 ($14.53/m^2) for a difference of $1.00/ft^2 ($10.76/m^2) in the unit price. Projected to 1985$, this difference becomes $1.13/ft^2 ($12.15/m^2) which is significantly below the $1.40/ft^2 ($15.06/m^2) price projected from the data in Deffenbaugh and Carper (1977). This reduction in "real" price for selective coatings is probably a result of improvements achieved in the coating process since 1977 as well as higher production volumes. The uncertainty inherent in projecting

Table 11.6
Collector material costs per square foot (1977$ and 1985$)

Design Number[a]	Number of covers	Cover material	Absorber coating	Material cost/ft^{2b}	
				1977$	1985$[c]
1	1	Plastic	Black paint	5.97	10.44
2	1	Float glass	Black paint	6.17	10.76
3	1	Crystal glass	Black paint	6.45	11.27
6	1	Plastic	Black chrome	6.77	11.83
8	1	Float glass	Black chrome	6.97	12.16
10	1	Crystal glass	Black chrome	7.25	12.67
12	2	Crystal glass	Black paint	7.47	13.06
16	2	Crystal glass	Black chrome	8.27	14.66

a. Source: Deffenbaugh and Carper (1977).
b. To convert to $/m^2, multiply table entry by 10.76.
c. Cost to manufacturer at plant site: supplier FOB plus shipping.

cost/prices for a single commodity on the basis of the PPI is another factor contributing to these differences.

The material cost data (1985 \$/ft^2) in table 11.6 are considerably higher (about twice) than comparable costs for 1985 volume production. The reasons for this are unclear, although the costs listed in table 11.6 are for manufacturing a single collector and so are probably higher than costs for those produced in volume.

Sims (1976) gives a detailed breakdown of 1976 collector costs, including 28 separate parts and assembly operations that were involved in the construction of the single-cover, 24-ft^2 (2.23-m^2) NASA/MSFC liquid collector. These are summarized in table 11.7.

These results, projected to 1985\$ are considerably higher than those for a similar type of collector from manufacturer V, table 11.9 (in 1985\$) whose overhead was about 10%. Significant variations in price are introduced in the manufacturer's fee or markup, and this distorts price comparisons considerably. When the cost data in table 11.7; are adjusted to include an "overhead/fee" of 10% (rather than 21.71%) and the costs projected to 1985\$, the FOB unit cost becomes \$16.95/ft^2 (\$182.38/m^2). Hence in this case the in-plant

Table 11.7
Cost breakdown of the NASA/MSFC collector

Part or operation	Weight (lb)[a]	Cost (1976\$)	Labor (h)	Percent of FOB
Absorber	71.6	86.618	0.0333	34.73
Frame	26.0	4.982	0.6341	2.00
Glass	40.2	9.740	0.8889	3.90
Pan, barrier	20.17	4.072	0.0844	1.63
Insulation, foam	16.00	16.000	0.1667	6.41
Miscellaneous, scrap, etc.	0.24	13.598	0.4377	5.45
Subtotal, materials	174.21	135.010	—	54.12
Subtotal, direct labor	—	37.898	3.7898[c]	15.19
Subtotal, materials, direct labor	—	172.968	3.7898	69.31
Subtotal, indirect labor	—	18.934	1.8934	7.59
Overhead/fee	—	54.158	—	21.71
Subtotal, without tooling		246.00(\$10.25/ft^2)		98.61
Tooling[b]	—	3.45	—	1.39
Total, including tooling	FOB	249.45(10.39/ft^2)		100.00

Source: Sims (1976).
a. To convert to kilograms, multiply table entry by 0.454
b. Tooling costs of \$6,896 (1976 \$) assumed to be amortized over a production volume of 2,000 collectors (author). Labor at \$10/h. Weight/ft^2 is 7.26 lb/ft^2 (347.61 Pa).
c. Including 1.5447 h miscellaneous assembly.

manufacturing cost of a 1985 comparable collector is significantly less than that listed in table 11.7. An explanation of this difference might be found in comparing the weight per unit area of each collector. Manufacturer V produces a collector weighing 2.34 lb/ft^2 (112.04 Pa), whereas the one described in table 11.7 weighs 7.26 lb/ft^2 (347.60 Pa). This is a weight ratio of 3.10. The ratio of the unit manufacturing costs (in 1985\$) is 2.45, which is close enough to suggest that the apparently large, "real" cost reduction occurring from 1976 to 1985 was caused by reducing the amount of material in the collector. This is another indication that current manufacturers are seeking increased cost effectiveness through simplicity and weight reduction. The PCR in 1985\$ of manufacturer V's collector (table 11.9) is virtually identical to that of the NASA/MSFC collector in table 11.7. Also the percentage cost for in-plant manufacturing for manufacturer V, shown in table 11.8, is about the same as that shown in the last column in table 11.7 if the overhead/fee for the NASA/MSFC collector is set at 10%, the same as that for manufacturer V. The large difference in cost between FOB (1985\$) prices appears to be the result of markup, which, particularly in the 1985 solar industry, varies widely and is influenced by local market conditions. Thus it appears that savings in the manufacturing process are offset by increased markups to produce a 1985 FOB price not too different from 1975 FOB prices.

Some measure of the trend toward simpler collector designs, without sacrificing performance (as measured by the PCR and thermal efficiency) and reliability, may be seen in comparing 1976 and 1985 designs and thermal performance data for collectors of the same general type. These collectors are shown in figures 11.4, 11.5, 11.6, and 11.7. Figures 11.4 and 11.5 show the NASA/MSFC collector (Sims 1976) discussed in connection with the data in tables 11.7 and 11.8. Figure 11.6 shows a cross section of the collector of manufacturer V (table 11.9). The relative complexity of the 1976 design, compared with that of 1985, is evident. Glass covers have been replaced by a

Table 11.8
In-plant manufacturing costs of manufacturer V compared with those of NASA/MSFC for a 10% overhead operation (%)

	Manufacturer V (1985)	NASA/MSFC (1976)
Materials	65.00	62.21
Tooling	2.00	1.60
Direct labor	23.00	26.19
Overhead	10.00	10.00
	100.00	100.00

Table 11.9
Summary of manufacturers' FOB prices and performance-cost ratios for 1975–85

Collector Number	Manufacturer (date)	Covers Type	Covers Number	Area gross (ft²)	Absorber Material	Coating	Total weight (lb_m)	lb_m/ft²	FOB price ($/ft²) 1975$[a]	FOB price Published year	FOB price 1985$[a]	Price ($/lb_m) 1975$[a]	Price Published year	Price 1985$[a]	PCR Btu/h/1985$[a] Summer	Winter
1	A (1975)	FP-L	GL	1	21.46	Al/Cu	FB	105	4.89	11.86	11.86	22.71	2.43	2.43	4.64	—	—
2	B (1975)	FP-L	GL	2	21.46	Al/Cu	FB	135	6.29	13.98	13.98	26.71	2.22	2.22	4.25	—	—
3	C (1975)[b]	FP-L	G	1	11.67	Al	FB	46.68	4.00	6.94	6.94	13.26	1.74	1.74	3.22	9.37	11.31
4	D (1975)[b]	FP-L	G	2	11.67	Al	FB	64.19	5.50	8.48	8.48	16.21	1.54	1.54	2.95	5.55	6.61
5	E (1975)	FP-L	GL	1	21.00	Cu	S	110	5.24	10.57	10.57	20.58	2.02	2.02	3.93	7.90	5.83
6	F (1975)	FP-A	GL	1	21.65	Cu	S	110	5.07	10.25	10.25	19.96	2.02	2.02	3.93	—	4.00
7	G (1975)	FP-L	G	1	30.54	Al	FB	99	3.24	8.68	8.68	16.56	2.68	2.68	5.12	—	—
8	H (1975)	FP-L	G	2	30.54	Al	FB	142	4.65	9.66	9.66	18.46	2.08	2.08	3.97	—	—
9	I (1975)	FP-L	G	2	25.27	Al	S	94	3.92	10.42	10.42	19.91	2.66	2.66	5.08	7.79	5.83
10	J (1976)	FP-L	GL	1	24.00	Steel	S	174.21	7.26	9.81	10.25	19.21	1.35	1.41	2.65	7.83	5.29
11	K (1976)	FP-L	P	1	26.30	Cu	FB	45	1.71	12.95	13.52	25.34	7.56	7.90	14.81	6.26	3.92
12	L (1976)	FP-L	P	2	26.30	Cu	FB	45	1.71	13.86	14.50	27.18	8.12	8.48	15.86	5.52	3.83
13	M (1976)	FP-L	GL	1	18.07	Al	FB	106	5.88	11.34	11.84	22.16	1.93	2.02	3.77	6.54	4.51
14	N (1976)	FP-L	GL	2	18.07	Al	FB	130(est)	7.25	14.16	14.83	27.76	1.97	2.06	3.86	5.22	3.60
15	O (1976)	FP-L	GL	2	21.00	Cu	S	150	7.14	—	—	—	—	—	—	—	—
16	P (1977)	FB-L	G	1	18.00	Cu	FB	—	—	9.83	10.94	19.13	—	—	—	—	—
17	Q (1977)	FP-L	G	1	18.00	Cu	S	—	—	12.46	13.87	24.28	—	—	—	6.69	4.78
18	R (1978)	FP-L	P	1	31.27	Al	FB	87	2.78	9.53	11.43	18.60	2.69	3.23	5.26	—	—
19	S (1978)	FP-L	P	1	31.27	Cu	FB	122.5	3.92	10.92	13.10	21.30	2.19	2.63	4.27	—	—
20	T (1982)	FP-A	GL	1	32.00	Al	FB	105.6	3.30	8.77	15.00	16.46	2.66	4.55	5.00	6.04	3.67
21	U (1985)	FP-A	GL	1	32.00	Al	S	98	3.08	11.97	22.91	22.91	3.88	7.43	7.43	4.47	2.53
22	V (1985)	FP-L	P	1	32.00	Al/Cu	S	75	2.34	12.86	24.44	24.44	5.49	10.43	10.43	7.32	5.34
23	W (1985)	FP-L	GL	1	32.00	Cu	S	114	3.59	12.81	24.48	24.48	3.57	6.82	6.82	6.39	4.57
24	X (1985)	FP-A	GL	1	35.60	Al	S	98	2.75	11.03	21.07	21.07	4.02	7.68	7.68	4.45	2.77
25	AA (1976)	ET-L	GL	1	28.00	Steel	FB	100(est)	3.57	19.15	20.00	37.47	5.36	5.60	10.45	3.22	2.50
26	BB (1976)	ET-L	GL	1	17.40	Cu	S	60	3.45	14.36	15.00	28.12	4.17	4.35	8.15	5.04	3.78
27	CC (1977)	LC-L	P	1	40.00	Steel	S	180	4.50	16.85	18.75	32.77	3.75	4.17	7.29	3.59	2.56

Note: Key: G = glass (ordinary); GL = low-iron glass; P = plastic; S = selective; FB = flat black; FP-L = flat plate, liquid; FP-A = flat plate, air; ET-L = evacuated tube, liquid; LC-L = light concentrator, liquid; PCR = performance-cost ratio.

Conversion factors: ft²/10.76 = m²; lb_m/2.205 = kg; (lb_m/ft²) 4.88 = kg/m²; ($/ft²)10.16 = ($/m²).

a. Cost projected by PPI.

b. Collectors mounted as part of roof or wall structure.

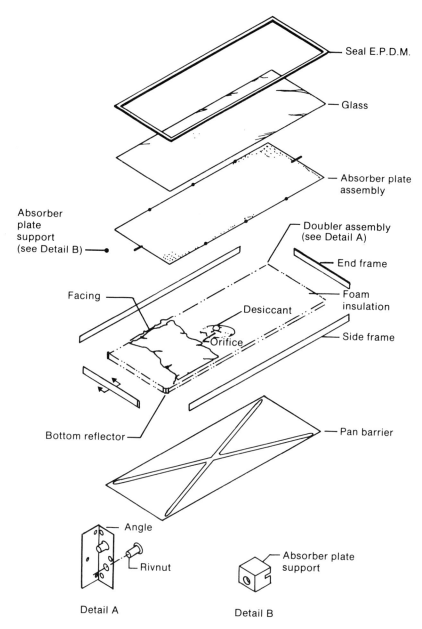

Figure 11.4
Exploded view of NASA/MSFC's solar collector (1976).

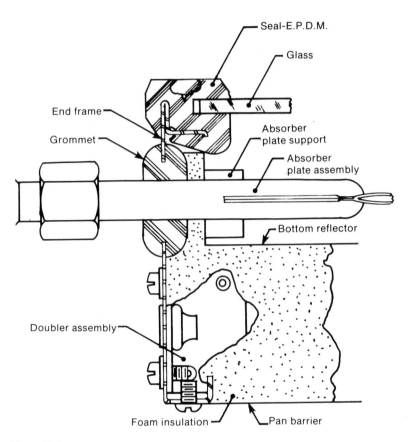

Figure 11.5
Cross section of NASA/MSFC's solar collector (1976).

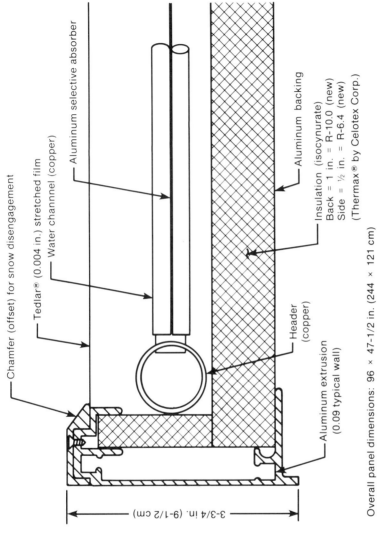

Chamfer (offset) for snow disengagement

Tedlar® (0.004 in.) stretched film

Water channel (copper)

Aluminum selective absorber

Aluminum backing

Insulation (isocynurate)
Back = 1 in. = R-10.0 (new)
Side = ½ in. = R-6.4 (new)
(Thermax® by Celotex Corp.)

Header (copper)

Aluminum extrusion (0.09 typical wall)

3-3/4 in. (9-1/2 cm)

Overall panel dimensions: 96 × 47-1/2 in. (244 × 121 cm)

Figure 11.6
Cross section of solar collector of manufacturer V (1985).

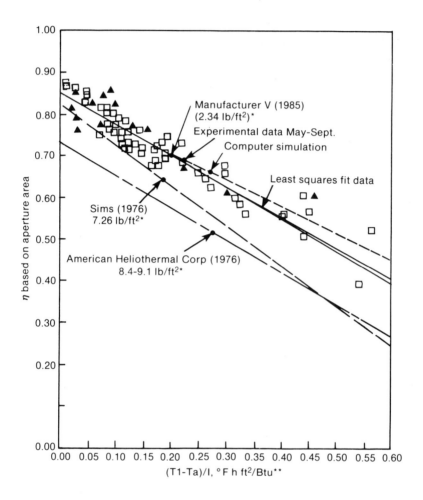

* To convert to kg/m², multiply by 4.88
** To convert to °C m²/W, multiply by 0.176

Figure 11.7
Thermal efficiencies of comparable solar collectors (1-cover, selective) (1976, 1985).

single, thin, stretched plastic film [0.004 in. (0.01 cm) thick], and the general enclosure is greatly simplified. The result is a dramatic reduction in weight, from 7.26 lb/ft² (347.61 Pa) (1976) to 2.34 lb/ft² (112.04 Pa) (1985) and a corresponding reduction in costs. The thermal performance of these two collectors, based on aperture area, is given in figure 11.7. Note the improvement in performance in the 1985 designs. Figure 11.7 also shows the thermal efficiency of a single glass cover, selective-coated absorber (liquid) collector (American Heliothermal Corp. 1976) designed in 1976. Its weight per unit area ranges from 8.40 to 9.10 lb/ft² (402.19–435.71 Pa).

Reducing a collector's weight results in additional cost savings in both shipping and installation. Lighter collectors may be installed by hand by a small crew, without cranes or other large machinery.

11.5   Summary of Manufacturers' FOB Prices and Performance— Cost Ratios for 1975–85

A survey was made of the prices and general characteristics of 27 collectors with one- and two-cover designs and selective and nonselective coatings on the absorber. Both liquid and air collectors were included in the survey, although the majority were liquid collectors. Three evacuated-tube or light-concentrating collectors were also included. The survey covered 1975–85, although most of the data applied to the 1975–78 period. These summary data are tabulated in table 11.9. Most of the information listed was taken from manufacturers' bulletins and price sheets and from Solar Vision Inc. (1982). The date of the information is given in the second column of the table. For some of the collectors, certain data could not be obtained or were not available. Manufacturers' FOB published price data for different dates were projected to equivalents for both 1975 and 1985 with the PPI. This provides a common basis for comparing the data. The price data in the table are FOB prices per square foot of collector (gross) area and FOB prices per pound of collector panel. To compare both collector FOB price and thermal performance, the performance–cost ratio (PCR) was developed, which is defined as

$$PCR = \frac{\text{useful energy per square foot}}{\text{cost per square foot (1985\$)}} \quad \left(\frac{\text{Btu/h}}{\$}\right). \tag{1}$$

This definition is arbitrary, but it does provide a measure for performance/cost comparisons. Note, however, that a realistic comparison of collectors on the basis of PCR can be made only when the collectors have equal reliability, durability, and maintenance costs. This equality is difficult to determine, especially when novel collector designs are included in the comparison. There is legitimate disagreement among manufacturers on this point, particularly regarding cover plate materials, absorber coatings, and fabrication methods. Currently, some disagreement exists regarding glass versus plastic for cover plates. Glass, used as a fabrication material for over 6,000 years, possesses excellent optical properties, and when it is tempered, it is quite durable. Its main disadvantage in solar collector design is its cost and weight. Over the past ten years plastic materials [polyvinyl fluoride (TEDLAR ®), and polycarbonate (LEXAN ®), etc.] have been used as well. These materials have good optical properties, although they have higher infrared transmission properties than glass. However, they are much lighter than glass and can be made in very thin sheets, which are easily fabricated in collector manufacture. Some manufacturers question the durability (optical and mechanical) of these materials, however. While it is true that less field testing has been done on plastic covers than on glass, these materials have shown excellent durability after about ten years of exposure to normal sunlight and environmental conditions. Furthermore maintenance problems appear to be no different than those for glass, including potential problems with high winds, hail, and heavy, wet snow loading. Manufacturers choosing these materials could confidently provide performance warranties to their customers on parts and maintenance.

The useful energy per square foot is determined from collector efficiency data for typical summer and winter conditions. The cost data are the FOB prices per square foot provided by the manufacturer and/or projected by the PPI. The values given in table 11.9 for the PCR are in 1985$ but correspond reasonably well to similar values given by Sims (1976) when they are each placed on a common price-year basis.

Manufacturer X reported a 1985 in-plant manufacturing cost of $3.00/ft^2 ($32.28/m^2) distributed among materials, direct labor, tooling, and overhead essentially as those cited in table 11.8. The $3.00/ft^2 ($32.28/m^2) manufacturing cost is the lowest found in the survey, but it is an indication of the economy that can be obtained by careful design and proper selection of manufacturing processes and materials. The collector of manufacturer X is rated by the Solar Rating and Certification Corporation (SRCC), Washington, DC, and performs well in the field. The corresponding 1975 manufacturing cost would be approximately $1.60/ft^2 ($17.22/m^2).

The manufacturers are not listed by name because the purpose of the survey was to examine collector costs as a class. Price projections over several years is probably not an adequate way to represent the collector price of a current manufacturer. Also some of the manufacturers are no longer in business, and

Table 11.10
Average collector FOB prices and performance parameters (1975$ and 1985$)

Collector type	$/lb[a,b] 1985$	FOB ($/ft^2)[c] 1975$	1985$	PCR (1985$) Summer	Winter
Flat plate, liquid					
Selective, 1-cover	5.67	11.43	22.13	7.43	5.31
Selective, 2-cover	5.08	10.42[e]	19.91[b]	7.79	5.83
Flat black, 1-cover	4.05[d]	10.36	20.01	6.15	3.80
Flat black, 2-cover	3.82[d]	13.12	25.70	5.76	3.98
Flat plate, air					
Selective, 1-cover	7.56	11.46	21.95	4.46	2.65
Flat black, 1-cover	5.00	8.72	16.46	6.04	3.67
Evacuated tube, liquid					
Selective, 1-cover	7.02	16.77	32.77	3.95	2.95

a. For comparison, the average U.S. supermarket price for T-bone steak in 1985 was $3.95/lb.
b. To convert to $/kg, multiply table entry by 2.205.
c. To convert to $/m^2, multiply table entry by 10.76.
d. Manufacturer L not included.
e. Single sample with plastic covers, which probably is the reason for price lower than 1-cover collector.

those who are have probably made significant design changes in their collectors since 1975.

The data in table 11.9 can probably best be evaluated by examining average values of the various quantities for similar types of collectors. In some cases, such as for air collectors and evacuated-tube collectors, the data are not sufficient for meaningful averaging; nevertheless, the results are presented because they are useful for comparisons with the other types. The average data are given in table 11.10.

11.6   Current Assessment of Prices/Costs and Business Conditions

The assessment of collector costs/prices from 1975 to 1985 must take into account the changes in market conditions that occurred in this period. In the early to mid-1970s, a real commercial market existed only for low temperature, largely nonmetallic, flat plate collectors, where were used mostly for swimming pool heating in California, the Southwest, Texas, and Florida. These collectors (and systems) were inexpensive, relatively simple, and were not included in the federal and state tax credit subsidies. The market for these products peaked in 1980 (DOE, 1984) at 12.233 million ft^2 (1.137 million m^2) per year, and has declined steadily since then. In 1983 the annual shipments for these collectors was 4.853 million ft^2 (0.451 million m^2); this sharp decline indicated a saturating market and could have been an effect of the depressed national economy during that period. A steady, perhaps increasing, market for these collectors will doubtless be reached at or slightly below current production levels, although such projections will be influenced greatly by economic conditions. These low temperature collectors basically have always been consumer products.

On the other hand, medium temperature collectors were not consumer products at the beginning of the decade. These collectors experienced growth in production and technical development in the mid-1970s largely because of federal procurements (Clark 1981). The principal producers were medium to large corporations and suppliers motivated by the opportunity to develop new products, expand markets of current products (glass, plastics, metals, insulation, etc.), and respond to a clear national need to realign energy resources. Of course a large number of very small commercial enterprises also responded to these new business opportunities. At the end of 1985 most of the larger corporations and suppliers and many of the smaller businesses dropped out of the solar collector market, but for different reasons. The larger corporations did not obtain a sufficiently high rate of return on invested capital and turned to other business activity, and many of the smaller companies simply failed by being unable to maintain a profit. Companies that survived the competition by 1985 did so by carefully selecting their products and markets, by incorporating their solar operations with other energy products and other commercial products, and by being willing to accept a lower rate of return on invested capital. Corporate indebtedness is common, and the value for stock shares (if any) is weak, though in some cases increasing. The 1985 solar market can be described as primarily one of medium temperature collectors (and systems) for retrofit hot water applications; retrofit space heating applications were a small but growing part. The market for solar products was essentially consumer oriented in 1985 and very market sensitive. This was probably the single most important characteristic of the solar industry and a significant contrast to the business conditions for medium temperature collectors (and systems) that existed in 1975.

It is reasonable to assume that the next important economic culling of the solar industry will occur when the federal and state tax credits are reduced and/or eliminated. Many of 1985's solar product companies may not be able to survive the resulting economic pressures. The survivors will be those that have tightened their corporate structure to operate with minimum overhead costs, that have reduced their indebtedness significantly during the "tax credit" era, that have simplified their products and systems and improved system reliability as much as possible, that have incorporated solar products into consumers' general energy requirements, and that are chiefly original equipment manufacturers capable of many functions from product design to marketing and service, with a well-defined dealer organization supplemented by in-house sales. As things stood in 1985, the solar industry was lean but healthy with reasonably good prospects for enterprises that manage and plan well.

There are many uncertainties in these projections, however. The influence of the scheduled deregulation of natural gas prices is one of them. Any increase in these energy prices will assist the solar industry and, in the context of this chapter, solar collector manufacturers. Price reductions would have the opposite effect. Solar energy has been cost competitive under tax credits with natural gas in most parts of the country. It has been generally cost competitive nationally with electricity and fuel oil for several years. This situation can be expected to prevail in the absence of tax credits if solar system costs can be maintained or reduced and expected escalations in the costs of oil and electricity occur. A life-cycle view of the economics of solar conversion systems also gives a better perspective on this issue.

An analysis of collector costs shows that their "real" FOB prices have not generally changed much in ten years. However, on the basis of available information, what has changed significantly are "real" in-plant manufacturing costs for collectors. The relatively constant "real" FOB prices suggest that there have been proportional increases in manufacturers' markups on collectors since 1975. This appears to be a result of the shift from the earlier, federal procurement type of market to 1985's consumer product market, as well as a recognition of where the real costs are. Accordingly some downward adjustments in manufacturers' markups may be a key to future corporate survival, if these adjustments can be made without damaging corporate profits.

Notes

1. Net cost is the actual dollar cost to the consumer, including the effects of tax credits, discounts, etc.

2. I is the solar irradiance in the plane of the collector in English and SI units, and ΔT is the difference in temperature between the inlet collector fluid and the ambient, with units (°F h ft^2/Btu).

3. The 1985$ costs per square foot are current costs and do not represent real or discounted costs.

4. Pattern glass, low iron, $\frac{1}{8}$-in. thick, minimum 91% transmissivity, FOB.

References

American Heliothermal Corp., Denver, CO. 1976. Specification AHC 101-76A. Miromit Flat Plate Solar Collectors.

Christensen, D. L. 1981. Solar commercial demonstration program assessment (outline and checklist of positive factors). Personal communication to John A. Clark. February.

Clark, J. A. 1981. The solar industry in the United States: Its status and prospects, 1981. In *Solar Technology Assessment Project*, vol. 11. Cape Canaveral, FL: Florida Solar Energy Center. April.

Clark, J. A. 1984a. Personal communication with AFG Industries. Kingsport, TN.

Clark, J. A. 1984b. Personal communication with Libby-Owens-Ford Glass Company, Toledo, OH. December.

Deffenbaugh, D. M. and H. J. Carper. 1977. *Cost-Effective Selection of Flat-Plate Solar Collectors*. Internal report. San Antonio, TX: Southwest Research Institute.

Energy Research, Inc. 1984. Novi, Michigan, personal communication to John A. Clark.

Energy Research and Development Administration (ERDA). 1975. *National Program for Solar Heating and Cooling*. ERDA-23, (March) and ERDA-23A (October).

Energy Research and Development Administration (ERDA). 1975, 1976. *A National Plan for Energy Research, Development and Demonstration: Creating Energy Choices for the Future.* ERDA-48, June 28, 1975, and ERDA-76-1, April 15, 1976.

Energy Research and Development Administration (ERDA). 1977. *Program Research and Development Announcement (PRDA): Non-Concentrating-Collectors for Solar Heating and Cooling Applications.* PRDA-EG-77-D-29-0001 (January 27).

Freeborne, W. E., G. Mara, and T. Lent. 1979. The performance of solar energy systems in the residential solar demonstration program. Presented at the *Second Annual Solar Heating and Cooling Operational Results Conf.*, November 27–30, 1979 Colorado Springs, CO.

Honikman, T. C., and P. Bendt. 1980. Domestic solar water heating. In *Economics of Solar Energy and Conservation Systems*, vol. 2, F. Kreith and R. E. West, eds. Boca Raton, FL: CRC Press.

McGarity, A. E. 1975. *Solar Heating and Cooling: An Economic Assessment*. Washington, DC: National Science Foundation.

Noreen, D., R. Lechevalier, M. Choi, and T. Morehouse. 1980. *Comparison of Conventional and Solar Water Heating Products and Industries*. Golden, CO: Science Applications, Inc. July 11.

Phelps Dodge Solar Enterprises Brochure. 1982. 1590 South Sinclair St., Anaheim, CA.

Sims, W. H. 1976. Considerations in the development of a high performance per unit cost solar collector. In *Proc. Solar Cooling and Heating: A National Forum.* Clean Energy Research Institute, University of Miami, Coral Gables, FL, December 13–15, 1976.

Solar Engineering Magazine. May 1977: 22.

Solar Research, Inc. 1984. Brighton, Michigan, personal communication to John A. Clark.

Solar Vision, Inc. 1982. Solar products specification guide. *Solar Age.*

Spielvogel, L. G. 1980. The solar bottom line. *ASHRAE Journal* (November): 38–44.

Starpak Solar Systems, Inc. 1984. Novi, Michigan, personal communication to John A. Clark.

Thuesen, G. J., and W. J. Fabrycky. 1984. *Engineering Economy*. 6th ed. Englewood Cliffs, NJ: Prentice Hall.

U.S. Department of Energy (DOE). May 1978. *Solar Heating and Cooling Demonstration: Project Summaries.* DOE/CS-0009.

U.S. Department of Housing and Urban Development. 1977. *Solar Energy Program Cycle 3, H-8000.* January 24.

U.S. Department of Housing and Urban Development, Office of Policy Development and Research. 1976, 1977, 1978. *Solar Heating and Cooling Demonstration Program, A Descriptive Summary of HUD Cycle (I through IVA) Solar Residential Projects.*

U.S. Department of Energy (DOE). 1984. *Solar Collector Manufacturing Activity, 1983*. Energy Information Administration. DOE/EIA-0174 (83). June.

U.S. House of Representatives, Committee on Science and Astronautics. June 7–12, 1973. *Solar Energy for Heating and Cooling*. Washington, DC: GPO.

U.S. House of Representatives, Subcommittee on Energy and Power. January 10–11, 1979. *Solar Commercialization*. Washington, DC: GPO.

U.S. Senate, Joint Hearings, Committee on Small Business and Interior and Insolar Affairs. December 9, 1976. *Alternative Long-Range Energy Strategies*. Washington, DC: GPO.

12 Reliability and Durability of Solar Collectors

William Freeborne

12.1 Introduction

The solar collector has to survive an environment that can impose moisture, differential thermal expansion, thermal shock, stagnation freezing, projectile impact, and ultraviolet (UV) exposure. The result of this exposure can be leaks, outgassing, degradation, distortion, disintegration, and breakage.

This section will deal primarily on experiences in the Department of Housing and Urban Development (HUD) solar demonstration program but will draw on other references where appropriate. Most of the text relates to unsuccessful or problematic situations in an effort to promote more positive solutions in the future. Liquid collectors and air collectors are treated separately in the following sections.

12.2 Liquid Collectors

12.2.1 Collector Description

Active space heating systems in the HUD program used a variety of collector designs in a wide range of collector areas per system. Thirty-three separate manufacturers provided over 68,000 ft² of collector area, with four of the manufacturers supplying 59% of the total square footage (38,000) in 10 to 26 projects each. The remaining 29 manufacturers supplied 112 to 2,618 ft² of collector panels each in one to five projects (see figure 12.1). Most collector panels were either factory-built modular (made of glass or plastic glazing, metallic absorber plates and tubing, wooden or metallic frames, and insulation) or factory fabricated site-assembled (with components consisting of wooden framing, glass glazing, and a copper absorber plate with plywood sheathing and steel or aluminum backing). Very few systems (only about 6% of the total area) were site built. Two systems (4% of the total area) used tubular collectors (Christopher and Houser 1981).

A sample of grants shows that 79% used single glazed collectors and 21%, double glazed; most collectors used glass. Fifty-nine percent used collectors with flat black absorber coating, whereas 41% employed some form of selective surface. Most of the grants using selective surface collectors employed single glazing, with only two grants combining double glazing and a selective surface.

Analyses of materials used for absorbers in liquid collectors indicates that most of those installed in the program used copper for both absorber plates

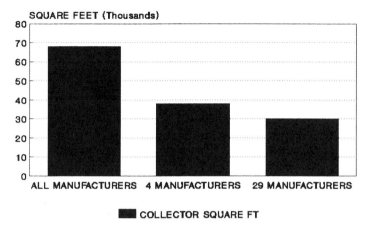

Figure 12.1
Area of liquid-heating collectors supplied by largest manufacturers.

(66%) and absorber tubing (79%). Few collectors used steel or aluminum for plates and tubing (20% steel plates, 16% steel tubing; 14% aluminum plates, 5% aluminum tubing).

Insulation materials used include glass fiber (30%), glass fiber and iso-cyanurate (33%), polyurethane and/or isocynanurate (21%), mineral wool (10%) and other (6%) (see figure 12.2).

12.2.2 General Experiences

The most severe collector problem in the HUD program to date has involved the durability of certain collector models or materials (Moore 1981; Greenberg et al. 1982; Freeborne and Mara 1982; BE&C Engineers 1983; Dubin and Roman 1983). In particular, there is a serious question about the ability of wood collector components to withstand temperatures over 200°F. Wood deterioration can create potential fire and structural hazards. Less hazardous, but nevertheless damaging, durability problems have involved the deterioration of other collector components, including steel absorber plate coating, aluminum absorber tubes, plastic glazing, and insulation under stagnation temperatures.

Two of the four major manufacturers mentioned above have had major durability problems indicative of a programwide trend where some collectors have had significant problems and others minimum deterioration. Also, 11 of the 29 minor manufacturers have had reported collector durability problems.

Besides experiencing hardware durability problems, a large number of projects have had difficulties with connections between collectors and sup-

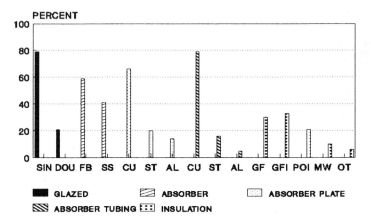

Figure 12.2
Liquid-heating collectors characteristics and materials.

ply/return piping and, in the case of collectors mounted integral with the roof surface, with the interface between collectors and roof structures. However, builders did not appear to experience great difficulty adapting collectors to the overall design of the house. Each of these findings are discussed in more detail in the following subsections.

12.2.3 Durability

A number of collectors installed in HUD demonstration projects had problems withstanding high temperatures, both during operation and particularly under stagnation (no flow) conditions.

Two major collector manufacturers either designed heat-dumping features into their systems or required dumping procedures in their operating guidelines to protect collectors from overheating. Others selected materials that could survive stagnation temperatures. Problems related to exposure to high temperatures varied from relatively minor absorber plate discoloration and small outgassing deposits on glazing interiors to total equipment failures requiring replacement.

Collectors produced by two of the four major manufacturers had fairly significant durability problems. Similar problems with collector integrity were experienced by other manufacturers. The principal HUD problems are described below supplemented by information from other programs.

12.2.4 Wood Deterioration

The most serious temperature-related problem involves the use of wood as a component in solar energy collectors. Problems with wood deterioration were

Figure 12.3
Construction using site-assembled collector.

visible in one manufacturer's factory-assembled collector and another's site-assembled collector (see figure 12.3).

When exposed to temperatures in excess of 200°F for an extended period, plywood can become a fire hazard due to reduced ignition temperatures. In one project, a flat plate collector with wood framing and plywood backing ignited a fire in the roof of an unoccupied home. Analysis after the fact identified a number of contributing factors that caused the plywood to catch fire (see figure 12.4). The collector was double-glazed with a selective surface absorber and foamed-in-place polyurethane insulation on the back of the absorber plate. It was flush mounted on the roof sheathing and flashed in as part of the roof's weather surface. The roof served as a cathedral ceiling of an interior room with about 9 in. of fiberglass insulation between the roof sheathing and the interior surface. Operating conditions were such that the collectors were in stagnation for at least part of each day during the summer. Absorber plate temperatures apparently became too hot for the foam insulation which charred and melted, losing all of its insulating capacity. When the insulation failed, the plywood backing was exposed to, but did not touch, the hot absorber plate; the radiating heat was sufficient to cause the plywood to

Figure 12.4
Fire-damaged collectors.

gradually char through. The same process continued with the roof sheathing; finally, when the charring plywood sheathing was exposed to the air in the ceiling cavity, a fire started.

In many projects using one site-assembled collector design, plywood cores between absorber plates and metal backing plates deteriorated significantly. In some cases the damage has been so extensive that the plywood turned black, at times resembling laminated charcoal (see figures 12.5 and 12.6), and losing most of its structural strength.

The prevalence of this problem in many applications with flat plate collectors using wood caused concern about the safety of other kinds of systems with wood components. Although most references indicate that the ignition temperature for wood is in the range of 392°F (200°C), this critical temperature drops when wood is exposed to high temperatures over a long period of time. The Forest Products Laboratory points out that ignition temperatures can drop to as low as 212°F (100°C) when the wood is exposed to high temperatures for periods of months. In addition to reduced ignition levels, wood exposed to temperature in the 150° to 200°F (65.5°–93.3°C) range also can lose a large portion of its structural capacity. The same concern exists for site-built

Figure 12.5
Fire-damaged collectors.

air systems using wooden plenums in the roof structure for heat transfer. Program experience with these systems is discussed in section 12.3.

12.2.5 Cover Plates

Glazing must withstand, a wide range of conditions in addition to transmitted solar energy and its function as an insulator. An outer covering must withstand physical loads, weather (including hail), ultraviolet light, and abrasion. An inner glazing must resist temperatures as high as 300°F (148.8°C) as well as thermal cycling. Table 12.1 (HUD Intermediate Minimum Property Standards Supplement, 1977) lists properties of typical cover plate materials.

One study (Herzenberg, Silberglitt, and LaPorta 1981) indicated that glass breakage due to stress, heat, thermal gradients, and abrasion or hail was very rare.

In the HUD program some plastic glazing or heat trap materials sagged or melted when exposed to stagnation temperatures for a prolonged period (see figures 12.7 and 12.8). Fiberglass-reinforced polyester (FRP) has survived in greenhouses for years (Ward and Oberoi 1980). Poly (vinyl fluoride and

Figure 12.6
Fire-damaged collectors.

ethylene-terephtahalate) have withstood weather. Polyvinyl fluoride has dis-
integrated under UV exposure. Ethylene terephthalate has been accepted for
use as a inner glazing. Poly(methyl methacrylate) withstands UV but does not
survive extended temperatures. Polycarbonate covers have suffered stress
cracking, distortion, and deformation (LaPorta 1981; Herzenberg, Silberglitt,
and LaPorta 1981).

12.2.6   Absorber Plates and Tubes

Absorber plates and tubes generally have been one of the materials listed in
table 12.2 (HUD Intermediate Minimum Property Standards Supplement
1977). In the HUD program there were problems with steel absorber plate
coatings and aluminum absorber tubes to the extent that removal of the
complete collectors was required (see figures 12.9 and 12.10). Steel absorber
plates and tubes are subject to corrosion especially if tubes are used in
drain-down or drain-back systems where oxygen is continually injected. Aluminum tubes are also subject to internal corrosion if dissimilar materials and
improperly treated fluid is used.

Table 12.1
Properties of typical cover plate materials

Property	Material								
	Poly(vinyl fluoride)	Poly(ethylene terephthalate)	Polycarbonate	Fiberglass reinforced plastics	Poly(methyl methacrylate)	Fluorinated ethylene-propylene	Clear lime glass (float)	Sheet lime glass	Water White glass
% Solar transmittance (for thickness listed below)	92–94	85	82–89	77–90	89	97	83–85	84–87	85–91
Maximum operating temperature (°F)	227°	220°	230°–270°	200°	180°–190°	248°	400°	400°	400°
Tensile strength (psi) (ASTM D-638)	13,000	24,000	9,500	15,000–17,000	10,500	2,700–3,100	4,000 annealed 10,000 tempered	4,000 annealed 10,000 tempered	4,000 annealed 10,000 tempered
Thermal expansion coefficient (in./in./°F × 10^{-6})	24	15	37.5	18–22	41.0	8.3–10.5	4.8	5.0	4.7–8.6
Elastic modulus (psi × 10^9) (D-638)	0.26	0.55	0.345	1.1	0.45	0.5	10.5	10.5	10.5
Thickness (in.)	0.004	0.001	0.125	0.040	0.125	0.002	0.125	0.125	0.125
Weight (lb/ft²) for above thickness	0.028	0.007	0.77	0.30	0.75	0.002	1.63	1.63	1.65
Refractive index	1.45	1.64	1.59	—	1.49	1.34	1.52	1.52	1.52

Sources: Grimmer, D. P., Moore, S. W., Practical aspects of solar heating: A review of materials used in solar heating applications (LA-UR-75-1952, paper presented at SAMPE Meeting, October 14–16, 1973, Hilton Inn). Kobayashi, T., and Sargent, L., A survey of breakage-resistent materials for flat-plate solar collector covers (paper presented at U.S. Section-ISES Meeting, Ft. Collins, Colorado, August 20–23, 1974). Scoville, A. E., An alternate cover material for solar collectors (paper presented at ISES Congress and Exposition, Los Angeles, California, July, 1975). Clarkson, C. W., and Herbert, J. S., Transparent glazing media for solar energy collectors (paper presented at U.S. Section-ISES Meeting, Ft. Collins, Colorado, August 21–23, 1974). *Modern Plastics Encyclopedia, 1975–1976* (McGraw-Hill). Toenjes, R. B., Integrated solar energy collector final summary report. (LA-6143-MS, Los Alamos Scientific Laboratory, Los Alamos, New Mexico, November, 1975).

Figure 12.7
Plastic glazing or heat trap materials sagged when exposed to stagnation temperatures for a prolonged period.

Figure 12.8
Plastic glazing or heat trap materials sagged when exposed to stagnation temperatures for a prolonged period.

Table 12.2
Characteristics of absorptive coatings

Material	Property				
	Absorptance (α)[a]	Emittance (ε)	α/ε	Breakdown temperature [°F (°C)]	Comments
Black chrome	0.87–0.93	0.1	~0.9		
Alkyd enamel	0.9	0.9	1		Durability limited at high temperatures
Black acrylic paint	0.92–0.97	0.84–0.90	~1		
Black inorganic paint	0.89–0.96	0.86–0.93	~1		
Black silicone paint	0.86–0.94	0.83–0.89	~1		Silicone binder
PbS/silicone paint	0.94	0.4	2.5	662 (350)	Has a high emittance for thicknesses, $> 10\,\mu m$
Flat black paint	0.95–0.98	0.89–0.97	~1		
Ceramic enamel	0.9	0.5	1.8		Stable at high temperatures
Black zinc	0.9	0.1	9		
Copper oxide over aluminum	0.93	0.11	8.5	392 (200)	
Black copper over copper	0.85–0.90	0.08–0.12	7–11	842 (450)	Patinates with moisture
Black chrome over nickel	0.92–0.94	0.07–0.12	8–13	842 (450)	Stable at high temperatures
Black nickel over nickel	0.93	0.06	15	842 (450)	May be influenced by moisture at elevated temperatures
Ni–Zn–S over nickel	0.96	0.97	14	536 (280)	
Black iron over steel	0.90	0.10	9		

Sources: G. E. McDonald, Survey of coatings for solar collectors (NASA TMX-71730, paper presented at Workshop on Solar Collectors for Heating and Cooling of Buildings, November 21–23, 1974, New York City) G. E. McDonald, Variation of solar-selective properties of black chrome with plating time (NASA TMX-71731, May 1975). S. W. Moore, J. D. Balcomb, and J. C. Hedstrom, Design and testing of a structurally integrated steel solar collector unit based on expanded flat metal plates (LA-UR-74-1093, paper presented at U.S. Section-ISES Meeting, Ft. Collins, Colorado, August 19–23, 1974). D. P. Grimmer, and S. W. Moore, Practical aspects of solar heating: A review of materials use in solar heating applications (paper presented at SAMPE Meeting, October 14–16, 1975, Hilton Inn). R. B. Toenies, Integrated solar energy collector final summary report (LA-6143-MS, Los Alamos Scientific Laboratory, Los Alamos, New Mexico, November 1975). G. L. Merrill, Solar heating proof-of-concept experiment for a public school building (Honeywell Inc., Minneapolis, Minnesota National Science Foundation Contract No. C-870). D. L. Kirkpatrick, Solar collector design and performance experience (for the Grover Cleveland School, Boston, Massachusetts, paper presented at Workshop on Solar Collectors for Heating and Cooling of Buildings, November 21–23, 1974, New York City).

a. Dependent on thickness and vehicle to binder ratio.

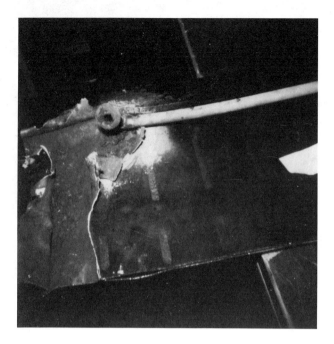

Figure 12.9
Corrosion of steel absorber plate coatings and aluminum absorber tubes.

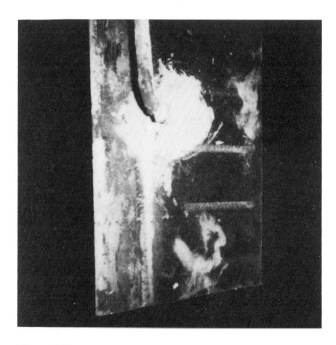

Figure 12.10
Corrosion of steel absorber plate coatings and aluminum absorber tubes.

12.2.7   Absorber Coatings

Characteristics of absorptive coatings are listed in t⌐
mediate Minimum Property Standards Supplement 1
able to resist stagnation temperatures, thermal sho⌐⌐, ⌐⌐⌐⌐⌐⌐ ⌐
moisture, and ultraviolet light.

There has been reports of outgassing, selective coating color fading, and
black paint peeling and cracking (Herzenberg, Silberglitt, and LaPorta 1981).
Outgassing was reported on the glazings of virtually all of the solar collectors
in the NBS durability program, but changes could not be discerned by thermal
performance measurements (Waksman, Thomas, and Streed 1984).

In the HUD program many instances of absorber plate peeling or discoloration were also encountered (see figure 12.11). Absorber plate deterioration
problems were due primarily to poor product design and quality control. For
example, some absorber plate problems were traced to the uneven coating of
the surface during manufacture, others to the use of coating compounds that
deteriorate or discolor at high temperatures.

12.2.8   Heat Transfer Fluids

The most typical heat transfer liquids are cited in table 12.4 (HUD Intermediate Minimum Property Standards Supplement 1977). Most systems used
water (with or without additives), water mixed with glycol (ethylene or propylene), or silicone (Herzenberg, Silberglitt, and LaPorta 1981). Water can cause
galvanic corrosion in the presence of dissimilar metals, especially when used

Table 12.3
Properties of typical absorber substrate materials

Property	Material			
	Aluminum	Copper	Mild carbon steel	Stainless steel
Elastic modulus, tension (psi $\times 10^6$)	10	19	29	28
Density (lb/cu in.)	0.098	0.323	0.283	0.280
Expansion coefficient [(68°–212°F) in./in.°F $\times 10^{-6}$]	13.1	9.83	8.4	5.5
Thermal conductivity [(77°–212°F) Btu/h·ft^2·°F·ft]	128	218	27	12
Specific heat [(212 °F) Btu/lb·°F]	0.22	0.09	0.11	0.11

Note: Standard specifications or manufacturer's literature should be consulted for specific types
or alloys.

Figure 12.11
Absorber plate peeling or discoloration.

in drain-down or drain-back systems where oxygen may be present. Water/glycol solutions if exposed to air through an air vent or vacuum breaker at high temperatures can turn acidic over time. Glycol solutions can leak through joints where water would not. Silicone liquids will leak through pump seals and joints where water would be retained. Silicone requires larger pumps due to high viscosity and low specific heats (Ward and Oberoi 1980).

All liquid systems are subject to freezing. In one survey freezing problems have been reported in $\frac{1}{3}$ of the water systems and $\frac{1}{5}$ of the glycol systems. Most of the failures have been in the collector, but a number of failures have occurred in the heat exchanger where water side exposure to cold glycol has happened (Herzenberg, Silberglitt, and LaPorta 1981).

12.2.9   Insulation

See table 12.5a through 12.5g (Herzenberg, Silberglitt, and LaPorta 1981) for properties of various insulation materials. Insulation characteristics needed are (Herzenberg, Silberglitt, and LaPorta 1981)

1. low thermal conductivity,

Table 12.4
Examples of typical heat transfer liquids

	Water	50% ethylene glycol/water	50% propylene glycol/water	Silicone fluid	Aromatics	Paraffinic oil
Freezing point [°F (°C)]	32 (0)	−33 (−36)	−28 (−33)	−58 (−50)	−100 to −25 (−73 to −32)	—
Boiling point [°F (°C) (at atm. pressure)]	212 (100)	230 (110)	—	None	300–400 (149–204)	700 (371)
Fluid stability	Requires pH or inhibitor monitoring	Requires pH or inhibitor monitoring	Requires pH or inhibitor monitoring	Good	Good	Good
Flash point [°F (°C)]^a	None	None	600 (315)	600 (315)	145–300 (63–149)	455 (235)
Specific heat [(73°F), Btu/(lb·°F)]	1.0	0.80	0.85	0.34–0.48	0.36–0.42	0.46
Viscosity (cstk at 77°F)	0.9	21	5	50–50,000	1–100	—
Toxicity	Depends on inhibitor used	Depends on inhibitor used	Depends on inhibitor used	Low	Moderate	—

Note: These data are extracted from manufacturers literature to illustrate the properties of a few types of liquid that have been used as transfer fluids.
a. It is important to identify the conditions of tests for measuring flash point. Since the manufacturers literature does not always specify the test, these values may not be directly comparable.

Table 12.5a
Physical and structural properties of various insulation materials

Type of insulation	Compressive strength (psi)	Density (lb/ft^3)
Cellular plastic		
Extruded polystyrene	40	1.7
		2.4
Molded polystyrene	12	1.0
		1.8
Polyurethane	14	1.85
	34	2.7
Polyisocyanurate	17	1.8
		2.3
Ureaformaldehyde		0.6
		0.9
Glass insulation		
Glass fiber batts		0.6
		1.0
Glass fiber board	130	4.5
		9.0
Foam glass	100	8.5
Loose fill		
Vermiculite		4.0
		10.0
Perlite		2.0
		11.0
Cellulose		2.2
		3.0
Miscellaneous		
Rock wool		1.5
		2.5
Calcium silicate	250	13.0
Insulating concrete	125	12.0
		88.0
	225	
Powdered limestone	11,110	60.0
Foam rubber		5.5

Source: Versar Inc., and Burt Hill Kosar Rittlemann Associates, 1980. *Final Report: Survey and Evaluation of Available Thermal Insulation Materials for Use in Solar Heating and Cooling Systems,* work performed under DOE Contract No. DE-AC04-78CS3563. (Available from NTIS, Springfield, VA, 22161.)

Table 12.5b
Thermal properties of various insulation materials

Type of insulation	Resistance (R/in.) at 75°F	Lower temperature limit (°F)	Upper temperature limit (°F)
Cellular plastic			
Extruded polystyrene	3.7		165
	5.0		
Molded polystyrene	4.0		165
Polyurethane	5.5	−320	250
	7.14		
Polyisocyanurate	6.67	−423	250
Ureaformaldehyde	4.2	−300	415
	a		b
Glass insulation			
Glass fiber batts	3.16		1,000
			c
Glass fiber board	4.0		d
Foam glass	2.63	−459	900
Loose fill			
Vermiculite	2.4		1,000
	3.0		
Perlite	2.5		1,200
	3.7		
Cellulose	3.2		180
	3.7		
Miscellaneous			
Rock wool	3.2		1,800
	3.7		
Calcium silicate	2.6		1,500
Insulating concrete	1.2		1,000
Powdered limestone	1.3		480
Foam rubber			

Source: Versar Inc., and Burt Hill Kosar Rittlemann Associates, 1980, *Final Report: Survey and Evaluation of Available Thermal Insulation Materials for Use in Solar Heating and Cooling Systems*, work performed under DOE Contract No. DE-AC04-78CS3563. (Available from NTIS, Springfield, VA, 22161.)
a. HUD specifications recommend derating the R value by 28% because of shrinkage upon curing.
b. Decomposes at this temperature.
c. At 400°F, organic binders may break down and binder may deposit on cover plate. Facing may be flamable above 180°F.
d. Discolors and chars from 300–400°F after exposure of 96 hours. Laboratory results were conducted at National Bureau of Standards, "Solar Energy Systems: Test Methods for Collector Insulations," NBSIR 79-1908, October, 1979.

Table 12.5c
Effects of temperature on various insulation materials

Type of insulation	Coefficient of expansion/°F	Dimensional stability	Thermal cycling	Outgassing	Toxic gases emitted	Corrosive gas emitted	Odor
Cellular plastic							
Extruded polystrene	3.5×10^{-5}	Good		Yes	Yes (CO)	[a]	None
Molded polystrene	3.5×10^{-5}	Good			Yes (CO)	[a]	None
Polyurethane	3.5×10^{-5}	Poor		[b]	Yes (CO)	[c]	None
Polyisocyanurate	3.5×10^{-5}	Poor		[b]	Yes (CO)	[c]	None
Ureaformaldehyde		Poor	[d]				Yes [e]
Glass insulation							
Glass fiber batts		Good					None
Glass fiber board		Good				[c]	None
Foam glass	4.6×10^{-6}	Good	[g]		Yes (CO, H_2S)	None	
Loose fill							
Vermiculite					No	None	
Perlite					No	No	
Cellulose				[f]	Yes (CO)	Yes [h]	
Miscellaneous							
Rock wool		Good		[i]		[c]	None
Calcium silicate		Good				[j]	
Insulating concrete		Good			No		
Powdered limestone							
Foam rubber		Poor				[k]	

Source: Versar Inc., and Burt Hill Kosar Rittlemann Associates, 1980, *Final Report: Survey and Evaluation of Available Thermal Insulation Materials for Use in Solar Heating and Cooling Systems*, work performed under DOE Contract No. DE-AC04-78CS3563. (Available from NTIS, Springfield, VA, 22161.)

Note: The notes identified with an asterisk represent laboratory results conducted at National Bureau of Standards, "Solar Energy Systems: Test Methods for Collector Insulations," NBSIR 79-1908, October, 1979.

*a. Mild pitting with copper and brass in an oven at 212°F for 30 days. None for stainless steel and aluminum.

*b. In a heated chamber (570°F), polyurethane lost nearly 24% of its original weight, resulting in an effective decrease of cover plate transmittance of 8% in 3 hours.

*c. Mild pitting with copper, brass and stainless steel in an over at 212°F for 30 days. None for aluminum.

*d. Cracked, then crumbled in 30-cycle test.

e. Formaldahyde may linger on until cured.

f. At 400°F organic binders may break down and binder may deposit on cover plate. Facing may be flamable above 180°F.

*g. Cracks in 45-cycle test.

*h. Severe pitting with stainless steel in an oven at 212°F for 30 days. Mild pitting with brass, copper, and aluminum.

i. Between 200°–450°F, binder may break down.

*j. Mild pitting with copper in an oven at 212°F for 30 days. None for stainless steel, brass, and aluminum.

*k. Mild pitting with brass and copper in an oven at 212°F for 30 days. None for stainless steel and aluminum.

Table 12.5d
Fire resistance of various insulation materials

Type of insulation	Flame spread[a]	Fuel contributed[a]	Smoke developed[a]
Cellular plastic			
Extruded polystyrene	25		
Molded polystyrene	<25	5	10
		80	400
Polyurethane	<75	14	116
		25	233
Polyisocyanurate	25	0	55
		5	200
Ureaformaldehyde	0	0	0
	25	30	10
Glass insulation			
Glass fiber batts	15	5	0
	20	15	20
Glass fiber board			
Foam glass	5		0
Loose fill			
Vermiculite	0	0	0
Perlite	0	0	0
Cellulose	15	0	0
	40	40	45
Miscellaneous			
Rock wool	15	0	0
Calcium silicate	0		0
Insulating concrete	0	0	0
Powdered limestone			
Foam rubber			

Source: Versar Inc., and Burt Hill Kosar Rittlemann Associates, 1980, *Final Report: Survey and Evaluation of Available Thermal Insulation Materials for Use in Solar Heating and Cooling Systems,* work performed under DOE Contract No. DE-AC04-78CS3563. (Available from NTIS, Springfield, VA, 22161.)
a. According to ASTM E-84. Laboratory results were conducted at National Bureau of Standards, "Solar Energy Systems: Test Methods for Collector Insulations," NBSIR 79-1908, October, 1979.

Table 12.5e
Effects of moisture on various insulation materials

Type of insulation	Moisture absorption (vol %)	Permeability (Perm-in.)	Corrosive solutions created
Cellular plastic			
Extruded polystyrene	1	0.4	a
		0.9	
Molded polystyrene	<2	1.2	a
		3.0	
Polyurethane	Low	1.2	b
		3.0	
Polyisocyanurate	<1.5	2.0	b
		3.0	
Ureaformaldehyde	High	4.5	
	c	100.0	
Glass insulation			
Glass fiber batts	0.1	100.0	
	0.2		
Glass fiber board	2		b
Foam glass	0.2	0	None
Loose fill			
Vermiculite	0	High	None
Perlite	Low	High	None
Cellulose	5	High	d
	20		
Miscellaneous			
Rock wool	2	>100	b
Calcium silicate	High	High	e
	d		
Insulating concrete		Varies	
Powdered limstone			
Foam rubber	Low	Low	f

Source: Versar Inc., and Burt Hill Kosar Rittlemann Associates, 1980, *Final Report: Survey and Evaluation of Available Thermal Insulation Materials for Use in Solar Heating and Cooling Systems*, work performed under DOE Contract No. DE-AC04-78CS3563. (Available from NTIS, Springfield, VA, 22161.)

Note: The notes identified with an asterisk represent laboratory results conducted at National Bureau of Standards, "Solar Energy Systems: Test Methods for Collector Insulations," NBSIR 79-1908, October 1979.

*a. Mild pitting with copper and brass in an oven at 212°F for 30 days. None for stainless steel and aluminum.

*b. Mild pitting with copper, brass and stainless steel in an over at 212°F for 30 days. None for aluminum.

*c. Moisture absorbed 8.8% by mass at 90% RH and 23°C for 30 days.

*d. Severe pitting with stainless steel in an oven at 212°F for 30 days. Mild pitting with brass, copper and aluminum.

*e. Mild pitting with copper in an oven at 212°F for 30 days. None for stainless steel, brass, and aluminum.

*f. Mild pitting with brass and copper in oven at 212°F for 30 days. None for stainless steel and aluminum.

Table 12.5f
Environmental concerns for various insulation materials

Type of insulation	Ultraviolet degradation	Resistance to bacteria and fungus
Cellular plastic		
Extruded polystyrene	Yes	Good
Molded polystyrene	Yes	Good
Polyurethane	Yes	Good
Polyisocyanurate	Yes	Good
Ureaformaldehyde	Yes	Good
Glass insulation		
Glass fiber batts		Good
Glass fiber board		Good
Foam glass		Good
Loose fill		
Vermiculite		Good
Perlite		Good
Cellulose		Poor a
Miscellaneous		
Rock wool	No	Good
Calcium silicate		Poor a
Insulating concrete		Good
Powdered limestone		
Foam rubber		Good

Source: Versar Inc., and Burt Hill Kosar Rittlemann Associates, 1980, *Final Report: Survey and Evaluation of Available Thermal Insulation Materials for Use in Solar Heating and Cooling Systems*, work performed under DOE Contract No. DE-AC04-78CS3563. (Available from NTIS, Springfield, VA, 22161.)
a. ASTM D 3273. Laboratory results were conducted at National Bureau of Standards, "Solar Energy Systems: Test Methods for Collector Insulations," NBSIR 79-1908, October, 1979.

Table 12.5g
Miscellaneous properties of various insulation materials

Type of insulation	Chemical reaction problems	Resistance to boiling water	Recovery drying
Cellular plastic			
Extruded polystyrene	Yes		Good
Molded polystyrene	Yes		
Polyurethane	Yes	Poor	Fair
Polyisocyanurate			
Ureaformaldehyde			
Glass insulation			
Glass fiber batts		Poor	Poor
Glass fiber board		Poor	Poor
Foam glass		Fair	Good
Loose fill			
Vermiculite			
Perlite			
Cellulose			
Miscellaneous			
Rock wool		Poor	Poor
Calcium silicate		Good	Good
Insulating concrete		Good	Good
Powdered limestone			
Foam rubber			

Source: Versar Inc., and Burt Hill Kosar Rittlemann Associates, 1980, *Final Report: Survey and Evaluation of Available Thermal Insulation Materials for Use in Solar Heating and Cooling Systems*, work performed under DOE Contract No. DE-AC04-78CS3563. (Available from NTIS, Springfield, VA, 22161.)

2. resistance to high temperature degradation (i.e., dimensional stability, minimal outgassing, and high ignition point),

3. nontoxicity from outgassing products,

4. low moisture and water absorption,

5. Nonpromotion of absorber corrosion.

In the HUD program, collectors using incompletely cured polyurethane and polyisocyanurate insulation have expanded and bulged due to expansive growth of the insulation (see figure 12.12). In one design the frame pulled away from the glazing, causing leaks into some of the collectors.

12.2.10   Sealants

Table 12.6 (HUD Intermediate Minimum Property Standards Supplement 1977) gives the characteristics of typical sealants. Sealants must be resilient, able to withstand operating and stagnation temperatures, moisture, external

Figure 12.12
Bulging caused by swelling of incompletely cured insulation.

weathering, UV light, and the stress and strain cycling that comes with the expansion and contraction of adjacent components.

Sealant leakage has caused moisture to enter the collector and promote cover plate condensation plus absorber plate corrosion. Outgassing from sealants on to cover plates is the other prevalent problem (Herzenberg, Silberglitt, and LaPorta 1981).

12.2.11 Housing

Housings must be structurally sound, weathertight, fire resistant, and capable of being exposed to microclimatic effects (e.g., atmospheric pollutants, high humidity, UV radiation, and diurnal temperature fluctuations). Fabrication materials include galvanized or painted steel, aluminum, plastics, and wood products (Argonne National Laboratory 1981). Caution must be taken when using wood housings, as was discussed earlier.

12.2.12 Collector/Piping Connections

Poor integration of collector arrays with transport piping systems resulted in reduced or uneven flow rates and occasional leakage of transfer fluid in a number of projects.

Table 12.6
Comparative characteristics and properties of sealing compounds

	Oil base	Butyls		Acrylics		Polysulfides		Polyurethanes		Silicones
		Skinning type	Nonskinning type	Solvent-release type	Water-release type	One-component	Two-component	One-component	Two-component	One- and two-component
Chief ingredients	Selected oils, fillers, plasticizers, binders, pigment	Butyl polymers, inert reinforcing pigments, nonvolatile plasticizers and polymerizable dryers	Butyl polymers, inert reinforcing pigments, nonvolatizing and nondrying plasticizers	Acrylic polymers with limited amounts of fillers and plasticizers	Acrylic polymers with fillers and plasticizers	Polysulfide polymers, activators, pigments, inert fillers, curing agents, and nonvolatilizing plasticizers	Base polysulfide polymers, activators, pigments, plasticizers, fillers. Activator: accelerators, extenders, activators	Polyurethane prepolymer, filler pigments and plasticizers	Base: polyurethane prepolymer, filler, pigment, plasticizers. Activator: accelerators, extenders, activators	Silicone polymer, pigment and fillers
Primer required	In certain applications	None	None	None	None	Usually	Usually	Usually	Always	Usually
Curing process	Solvent release, oxidation	Solvent release, oxidation	No curing; remains permanently tacky	Solvent release	Water evaporation	Chemical reaction with moisture in air and oxidation	Chemical reaction with curing agent	Chemical reaction with moisture in the air	Chemical reaction with curing agent	Chemical reaction and/or moisture
Tack-free time (h)	6	24	Remains indefinitely tacky	36	36	24	36–48	36	24	0.3–1
Cure time days[a]	Continuing	Continuing	N/A	14	5	14–21	7	14	3–5	1–7
Maximum cured elongation (%)	15	40	N/A	60	Not available	300	600	300	400	100–450
Recommended max. joint width movement (%)	23% decreasing with age	+7.5	N/A	+10	+5	±25	±15	±15	±25	±25
Maximum joint width (in.)	1	3/4	N/A	3/4	5/8	1	1	3/4	1	5/8
Resiliency	Low	Low	Low	Low	Low	High	High	High	High	High
Resistance to compression	Very low	Moderate	Low	Very low	Low	Moderate	Moderate	High	High	High
Resistance to extension[b]	Very low	Low	Low	Very low	Low	Moderate	High	Moderate	High	High

Table 12.6 (continued)

	Oil base	Butyls		Acrylics		Polysulfides		Polyurethanes		Silicones
		Skinning type	Nonskinning type	Solvent-release type	Water-release type	One-component	Two-component	One-component	Two-component	One- and two-component
Service temp. range (°F)	−20°–150°	−20°–180°	−20°–180°	−20°–180°	−20°–180°	−40°–200°	−60°–200°	−25°–250°	−40°–250°	−85°–500°
Normal application temperature range (surface) (°F)	+40°–+120°	+40°–120°	+40°–120°	+40°–120°	+40°–120°	+40°–120°	+40°–120°	+40°–120°	+40°–120°	0°–+160°
Weather resistance	Poor	Fair	Good	Very good	Not available	Very good	Very good	Very good	Very good	Excellent
Ultraviolet resistance, direct	Poor	Good	Good	Very good	Not available	Good	Good	Poor to good	Poor to good	Excellent
Cut, tear, abrasion resistance	N/A	N/A	N/A	N/A	N/A	Good	Good	Excellent	Excellent	Fair
Life expectancy (years)c	5–10	10+	10+	20+	Not available	20+	20+	20+	20+	20+
Hardness shore A	20–80	20–40	N/A	20–40	30–35	25–35	24–45	25–45	25–45	20–60
Applicable specifications	TT-C-598C	TT-S-001657	NAAMM SS-1a-68	TT-S-00230C	none	TT-S-00230C	TT-S-00227e	TT-S-00230C	TT-S-00227e	TT-S-00230C TT-S-001543 TT-S-00227E

a. Cure time as well as pot life are greatly affected by temperature and humidity. Low temperatures and low humidity create longer pot life and longer cure time; conversely, high temperatures and high humidity create shorter pot life and shorter cure time.

b. Resistance to extension is better known in technical terms as modulus. Modulus is defined as the unit stress required to produce a given strain. It is not constant but changes in values as the amount of elongation changes.

c. Life expectancy is directly related to joint design, workmanship and conditions imposed on any sealant. The length of time illustrated is based on joint design within the limitations outlined by the manufacturer, and good workmanship based on accepted field practices and average job conditions. A violation of any one of the above would shorten the life expectancy to a degree. A total disregard for all would render any sealant useless within a very short period of time.

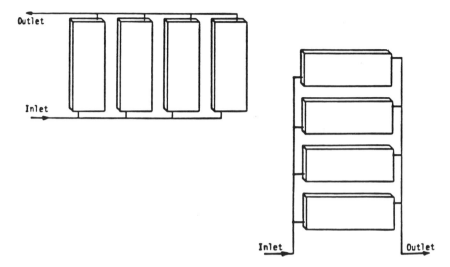

Figure 12.13
Direct-return collectors manifold.

Some collector manufacturers recommend that the panels be piped in a direct-return configuration (see figure 12.13). However, collectors piped for direct-return frequently could not achieve balanced flows. In some of these installations, repairs were made by replumbing to a reverse-return configuration (see figure 12.14). However, relying only on a reverse-return piping layout for a balanced flow may be inadequate, especially in larger arrays. A number of manufacturers did not specify flow configuration balancing techniques.

In a few installations collector panels and piping were poorly aligned, resulting in low flow rates and potential freezing problems. In others connections between the collector array and supply/return piping or between individual collector panels were inadequate. Typical installation errors included the use of dissimilar metal, which can lead to galvanic corrosion (see figure 12.15), the use of solder that could not withstand high temperatures (see figure 12.16), and failure to adequately compensate for expansion.

Some collectors have inlet and outlet connections perpendicular to the plane of the collector, pointing at the roof surface (see figure 12.15). Collectors with this type of connection have experienced air blockage and drainage problems because the connections between collectors form loops that create air traps and cannot be drained.

Most collector panels have internal manifolds, connected to each other by hoses and hose clamps, pipe unions, or permanent hard-pipe connections (see figure 12.17). Selecting the method of connection between collectors was

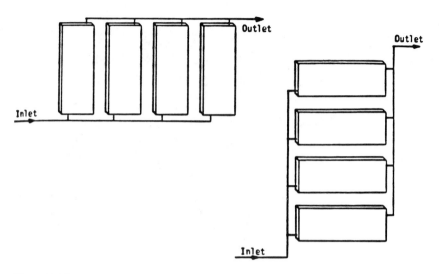

Figure 12.14
Reverse-return collector manifold.

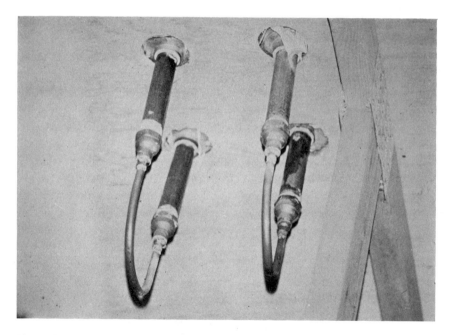

Figure 12.15
Galvanic corrosion due to dissimilar metals.

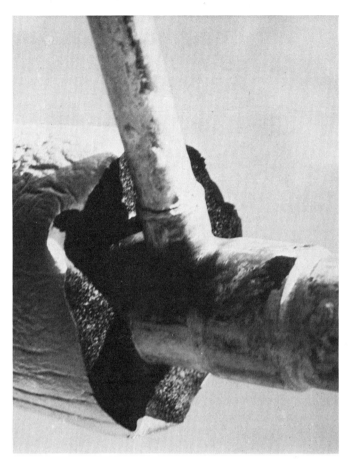

Figure 12.16
Improperly chosen solder.

Figure 12.17
Typical intercollector connection.

difficult at times due to their close proximity to each other (see figure 12.12).
Hose connections split or leaked on a number of projects. Collector hookups
to external manifolds occasionally resulted in failure to adequately compensate for expansion (see figure 12.18).

Solder joints between some collector fluid passageways and the absorber
plate created problems due to improper cleaning and applications of fluxes.
Flux corrosion has been found in this area on several collectors. Improper
soldering of fluid passageways and the absorber plate has also resulted in loss
of heat transfer capacity where bonding has not been complete.

Plumbing with 50/50 lead-tin solder is not adequate for high temperature
service. When high temperatures are experienced (e.g., during stagnation), this
solder becomes plastic; the thermal expansion of the copper tubing will often
be sufficient to pull the joints apart, causing a leak. Joints in copper tubing in
or near the collector arrays should be soldered with 95/5 tin-antimony solder
or silver solder.

12.2.13   Collector/Roof Structure Integration

HUD projects using site-assembled systems have also experienced problems
with the uneven mounting of collector panels from rafter to rafter. In at least

Figure 12.18
Inadequate allowance for thermal expansion of manifold.

Figure 12.19
Improperly sealed collector assembly.

one project uneven mounting of site-assembled collectors seriously impeded the flow rate of the transfer fluid. The only available solution, apart from complete system removal and reinstallation, was to add additional flow balancing devices.

Some installers have used sealing or flashing material (e.g., ordinary roof cement) not capable of withstanding exposure to high temperatures (see figure 12.19). Others did not seal panels or flash arrays tightly enough, leaving room for rainwater penetration.

Site-assembled collectors have often been shipped without adequate guidelines for installing battens supporting collector glazings or stops to hold glass panels in place. In some cases, the design allowed the stacking of site-assembled collectors two or three rows high without providing adequate glass stops between rows or at the foot of the array. In more than one of these projects the builder or installer had to improvise glass supports or glass stops (see figures 12.20 and 12.21). In one of the projects the collector supplier provided the builder with copper battens adequate to support a double-glazed cover plate, but the installation employed single glazing. The builder improvised by placing plywood spacers on top of the copper battens to help support the single cover plate. In time the wood spacers deteriorated due to

Figure 12.20
Improvised glass supports or glass stops.

high temperatures, causing the cover plates to bow and permitting rainwater
to penetrate.

12.2.14   Collector/House Integration

In general, builders participating in the HUD program seemed to have little
trouble integrating the solar energy collectors into the design of the house.
The variety of collector designs and models available allowed grantees to
make a number of trade-offs to accommodate building design, siting, or cost
requirements. In one grant the grantee believed that tilting the collector at the
recommended design angle would make the house look too unusual and slow
the sale. So he chose a less-than-optimal tilt angle, substituting collectors with
a selective surface absorber coating for those with flat black surfaces that were
originally specified.

Most houses were designed to accommodate the collectors at the prescribed
tilt. The orientations and tilt angles used in most projects were in the general
range of recommended rules of thumb, which state that collectors in a space-
heating system should be tilted at an angle equaling latitude plus 10 to 15 deg
and oriented within plus or minus 30 deg of true South. Some orientations of
35 deg from true South provided reasonable performance because the size of

Figure 12.21
Improvised glass supports or glass stops.

the collector array was increased. However, experience has shown that orientation outside of the general recommended range should be adopted only in special situations.

Shading of collectors has also not proved to be a big problem in HUD demonstration grant homes. Grantee builders who were developing part or all of a subdivision had substantial leeway to prevent potential shading problems. A few relatively infrequent occurrences do suggest, however, that builders need to be as conscious of preventing on-site shading as they are wary of off-site shading. In one or two townhouse developments, the structure of one unit actually shaded collectors on another unit. In another single-family structure, large oak trees located to the southeast and southwest created quite a bit of shading.

12.3   Air Collectors

12.3.1   Collector Description

Air-space-heating systems used two basic types of collector designs. Of a sample of 51 grants using air systems, 92% used factory-built collector panels,

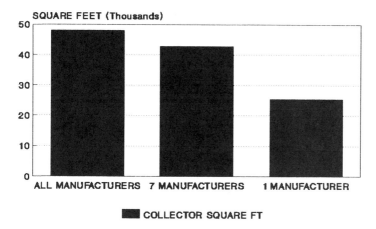

Figure 12.22
Area of air-heating collectors supplied by largest manufacturers.

while 8% employed some form of site-built model in which collectors were integral with the roof structure, with glazing mounted directly over the rafters (Christopher and Houser 1981). Nineteen collector manufacturers provided factory-built models, with seven supplying 89% of the collector square footage used. One manufacturer alone accounted for 53% of the total factory-built collector square footage (see figure 12.22).

In a sample of 51 grants, 51% used double-glazed collectors and 47% single-glazed (one grant, accounting for 2% of the sample, used triple-glazed panels). Nearly all of the collectors (96%) used flat black absorber coating, and only 4% employed some form of selective surface. Sixty-two percent of the factory-built systems used absorber plates made of steel or galvanized steel, 30% used aluminum and only 4% copper. Insulation was fiberglass, polyisocyanurate, or insulating wool (see figure 12.23).

12.3.2 General Experiences

Air collectors have had some heat deterioration problems. Wood and other materials have durability problems in site-built collectors, for instance. Air collectors have had serious problems, however, with reductions in airflow rates and with heat losses from collector leaks in collector arrays. Both factory-built and, particularly, site-built panels have also had rainwater penetrate inside collector glazing or the roof itself.

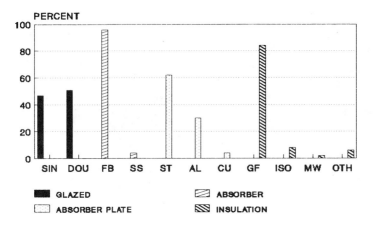

Figure 12.23
Air-heating collector characteristics and materials.

12.3.3   Durability of Collectors

In general, factory-built air collectors have not experienced operating temperatures as high as found in liquid collectors. The use of air as a transfer fluid
and the high heat losses from some of the collector arrays have reduced the
stagnation temperatures reached in air systems, so in general the arrays are
less vulnerable to temperature induced deterioration.

The lack of wood in most factory-built air collectors further reduces the
potential material deterioration found in some liquid collectors with plywood
sheathing. Air collectors, however, have experienced deterioration of some
types of plastic cover plates under exposure to high temperatures, UV radiation, and severe weather conditions (e.g., hail). Metal absorber surface coatings
have also discolored in some applications.

Many air collectors in the program experienced problems with outgassing
similar to those encountered in liquid installations. Deterioration of sealants,
gaskets, absorber coatings, and insulation have to led to outgassing in a
number of air systems, although, just as with liquid systems, the short-term
and long-term effects of outgassing are not precisely known. There has been
a polysulphate odor in one of the systems under negative pressure, however.

Some site-built attic configurations use roof rafters as part of the collector
structure, creating an attic plenum for the collection and transport of solar
heat (see figure 12.24). In several systems the rafters in the attic were painted
flat black to help absorb solar heat more efficiently. Due in part to experiences
with plywood deterioration in some site-assembled liquid systems, questions
were asked about the effects of long-term exposure to high temperatures in

Figure 12.24
Use of roof rafters as part of collector.

wooden roof rafters and braces. Inspections of these systems disclosed some wood drying and potential structural deterioration resulting from temperatures above 150°F (65.5°C). These systems usually use plywood and gypsum board as heat absorbers in the attic space. The plywood surfaces were generally cracked, but the cores exhibited no delamination or deterioration. Some gypsum board cores were crumbly and powdery.

12.3.4   Reductions in Airflow

A number of projects using air collectors reported problems with low flow rates through the array. Often reductions below design airflow rates were traced to a number of design or installation defects.

In some projects the collector panels themselves were at fault. Poorly sealed factory-built or site-built panels suffered air leakage and heat loss in a number of sites. In other projects, the configuration of the collector array led to reduced airflow. Collectors with ductwork configured in a reverse-return layout generally exhibited more balanced airflows than those arrayed in direct-return.

Figure 12.25
Improperly secured cover plates.

Poorly sealed connections between collector and transport ductwork led to reduced airflow in a number of systems, often due to the use of the wrong sealing materials. For example, some fiberboard materials have been unable to withstand high temperatures and have had to be replaced with metallic duct connections. Duct assembly was also sometimes at fault. The use of unsupported flexible ductwork led to high pressure drops across collector arrays in a few projects; in other projects wooden plenums could not be made leak-tight. In some systems repair work was hampered by the inaccessibility of collector-to-ductwork connections made under the roof surface. In at least one project part of the ceiling had to be removed to gain access to these connections.

12.3.5   Water Leakage

In both factory-built and site-built systems, poor sealing and flashing design and installation practices led not only to air leakage but also to rainwater penetration. Rain penetration in turn led to condensation which adversely affected collection efficiency.

A number of factory-built and site-built systems have experienced rainwater penetration inside the collector panel, causing rust or other damage on the

Figure 12.26
Improper allowance for roof drainage.

absorber plate and condensation on the absorber and glazing surfaces. For example, cover plates were not always secured properly with rivets, and caulking failed causing leaks (see figure 12.25).

In addition to leaks in poorly sealed collector panels, water leakage also resulted from inadequate flashing where collector-to-transport duct connections penetrated the roof. As was the case with many site assembled liquid systems, some designs for site-built air systems did not provide adequate instructions for sealing, flashing, and drainage. In one system the grantee followed design specifications, but the bottom edge configuration of the completed collector installation did not allow water to run (see figures 12.26 and 12.27). The only way to solve this problem was to reconfigure the bottom edge. Where guidelines were absent, more than one grantee had to field fabricate flashing or collector seals.

Where site-assembled arrays with factory-built collectors were directly mounted over living areas, rain penetration due to poor collector design or assembly caused damage to interior ceilings requiring roof and ceiling repairs as well as collector reflashing and resealing.

Figure 12.27
Improper allowance for roof drainage.

12.4 Conclusions and Recommendations

In the HUD program 31% of the problems with liquid systems involved collectors and 23% of the problems with air systems were with collectors. Many of these occurrences were such that thermal performance or operation were affected only by a minimal amount. Other occurrences did cause shutdown of the system or did require repair or replacement of the collectors.

When selecting or designing a collector, it is most important to be aware of stagnation conditions and freezing conditions. The materials and combinations of materials should be selected for survival in these conditions as well as for other environmental exposure. References particularly helpful in evaluating collector parameters are Argonne National Laboratory (1981), Ward and Oberoi (1980), Metz and Holton (1982), and Herzenberg, Silberglitt, and LaPorta (1981).

References

Argonne National Laboratory. 1981. *Final Reliability and Materials Design Guidelines for Solar Domestic Hot Water Systems.* ANL/SDP-11, Argonne, IL: Argonne National Laboratory.

Baum, B., and M. Bennette. 1983. *Solar Collectors.* DOE/CS/35359-Tl. Washington, DC: Department of Energy.

BE&C Engineers, Inc. 1983. *Final Report of the Management Support Contractor for the Residential Solar Heating Demonstration,* vols 1–5. Seattle, WA: BE&C Engineers, Inc.

Christopher, P. M., and A. D. House. 1981. *Residential Solar Data Center: Data Resources and Reports.* NBSIR 81-2369. Gaithersburg, MD: National Bureau of Standards.

Dubin, F. S., and K. Roman. 1983. Solar systems maintenance. *ASHRAE Journal:* 33–37.

Freeborne, W., and G. Mara. 1982. *The Use of Active Solar Heating and DHW Systems in Single Family Homes: Technical Findings and Lessons Learned from the HUD Solar Demonstration Program.* Draft. Washington, DC: Department of Housing and Urban Development.

Greenberg, J., S. Berry, P. Chen, P. Cooke, and C. Yancey. 1981. *Analysis of Reliability and Maintainability of Residential and Commercial Solar Systems Included in the National Solar Heating and Coding Demonstration Program.* National Bureau of Standards letter report to the Department of Energy. This report also appears as Cole, M. H., et al. 1982. NBS/HUD projects data analysis, appendix C. *An Assessment of the Field Status of Active Solar Systems.* DOE/SF/11485-1.

Herzenberg, S., R. Silberglitt, and C. LaPorta. 1981. *Materials Research Needs to Improve the Performance and Durability of Solar Heating and Cooling Systems.* Washington, DC: DHR, Inc.

HUD. 1977. *HUD Intermediate Minimum Property Standards Supplement, Heating and Domestic Hot Water Systems.* Washington, DC: Department of Housing and Urban Development.

Jorgensen, G. 1984. *A Summary and Assessment of Historical Reliability and Maintainability Data for Solar Hot Water and Space Conditioning Systems.* SERI/TR-253-2120. Golden, CO: Solar Energy Research Institute.

King, T., and E. Kline. 1979. *Active Solar Energy System, Design Practice Manual.* New York: The Ehrenkrantz Group/Baltimore, MD: Mueller Associates, Inc.

Kutscher, C. F. 1984. *An Investigation of Means for Lowering Costs of Active Solar Hot Water/ Space Heating Systems.* SERI/TP-253-2309. Golden, CO: Solar Energy Research Institute.

LaPorta, C. 1981. *Handbook of Experience.* SERI contract for Solar Energy Industries Assoc. Washington, DC: SEIA.

Meeker, J., and L. Boyd. 1981. The great, the good and the unacceptable. *Solar Age* 6(10): 29–36.

Metz, E., and J. Holton. 1982. *Performance Criteria for Solar Heating and Cooling Systems in Residential Buildings.* Gaithersburg, MD: National Bureau of Standards.

Moore, D. C. 1981. Lessons learned from the HUD solar demonstration program. *Proc. 1981 AS/ISES Annual Meeting*, vol. 4.1, B. H. Glenn and G. E. Franta, eds. Newark, DE: American Section of the International Solar Energy Society, Inc.

Versar, Inc. 1980. *Survey and Evaluation of Available Thermal Insulation Materials for Use on Solar Heating and Cooling Systems.* DOE/CS/3563-TI. Springfield, VA: Burt, Hill, Kosar, Rittleman Associates.

Waksman, D., W. Thomas, and E. Streed. 1984. *NBS Solar Collector Durability/Reliability Test Program: Final Report.* Gaithersburg, MD: National Bureau of Standards.

Ward, D. S., and H. S. Oberoi. 1980. *Handbook of Experiences in the Design and Installation of Solar Heating and Cooling Systems.* SPSP-10. Fort Collins, CO: Colorado State University.

Wolosewicz, R. M., and J. Vresk. 1982. *Reliability and Design Guidelines for Combined Solar Space Heating and Domestic Hot Water Systems.* ANL/SDP-12. Argonne, IL: Argonne National Laboratory.

13 Environmental Degradation of Low-Cost Solar Collectors: Research Issues and Opportunities

Fred Loxsom and Eugene Clark

13.1 Introduction

If solar energy is to become a commercially important energy source, solar energy systems must become cheaper. Solar energy systems that reduce the cost of the collector, a major cost of the total system, can meet this requirement. Low-cost collectors made with thin-film polymers provide the opportunity for achieving lower-cost systems; however, environmental degradation of thin-film polymers raises an issue of concern.

Assuming that displaced heat is worth $12/10^6$ Btu ($11.60/GJ) and that the solar system must produce a cost to annual savings ratio of less than ten, Andrews (1983) estimates that a complete solar thermal system, which collects 1×10^5 Btu/ft^2 per year (1.1 GJ/m^2 per year), can cost no more than $12/ft^2 ($130/m^2). Conventional active solar thermal systems are now overpriced by at least a factor of four (Andrews 1983). Because of high materials, production, assembly, and installation costs, conventional metal/glass solar thermal systems cannot meet this installed cost target. Collector materials cost alone is about $5/ft^2 ($55/m^2), and the added costs of assembly, marketing, and installation increase the installed cost of the conventional collector to about $20/ft^2 ($220/m^2) (Andrews, 1983).

Thin-film polymeric (plastic) materials have been shown to provide high energy collection efficiency at low cost. Collector designs based on thin films offer two major advantages. They use a small amount of material, and they have low production and assembly costs.

The Brookhaven National Laboratory built a water heating, thin-film-laminated flat plate collector with a measured efficiency near the best of conventional flat plate efficiencies. This collector is designed to heat water at atmospheric pressure. The total material requirement of the structurally complete, insulated collector is only 0.8 lb/ft^2 (3.5 kg/m^2). If this collector were produced using commercial materials and high speed, laminated film–production techniques, the installed collector cost, including all commercial markups, is estimated to be only $5.70/ft^2 ($63/m^2) (Wilhelm 1983). If the Brookhaven design were to use advanced printing techniques to create the selective surface and state-of-the-art "oriented" polymer films, the estimated installed cost of the collector would drop to $3.60/ft^2 ($50/m^2) (Wilhelm 1983). A thin-film air heating collector that uses a roof or wall for structural support and for back insulation could be even cheaper than the Brookhaven design.

Reducing the installed collector cost to less than $4/ft^2 ($43/m^2) is dramatic progress; however, even this low cost leaves only $8/ft^2 ($86/m^2) for the balance of the solar system. To reach the $12/ft^2 ($130/m^2) cost goal, the remainder of the solar energy system must be simplified and redesigned to take advantage of the opportunities offered by thin-film materials. The complete $12/ft^2 ($130/m^2) system based on separate flat plate collectors has not yet been demonstrated.

One related thin-film system that has been shown to be technically and economically competitive is the salt-gradient solar pond (SGSP). By combining the collector with storage and using the earth as additional storage and structural support for a thin-film pond liner, the SGSP becomes a complete solar energy system that requires only a tiny mass of manufactured liner material per pond area.

A series of SGSP projects in Israel has demonstrated that pond liner, heat extraction, and maintenance costs per area can be very low for large ponds (Raviv 1984; JPL 1982). Based on experience with a 50-acre (20-ha) SGSP at the north end of the Dead Sea, Ormat Corporation estimates that a 20-acre (8-ha) SGSP system built on a site with free salt and free land will cost less than $3/ft^2 ($32/m^2). Under ideal conditions a very large SGSP system may cost as little as $1/ft^2 ($11/m^2) (JPL 1982). Such a pond can provide annual storage with annual collection efficiency near 20%. In a warm, sunny climate the SGSP can produce an annual thermal output of about 1×10^5 Btu/ft^2 yr (1.1 GJ/m^2 yr) at temperatures near 170°F (77°C).

Low-cost brine is available in many locations in the south and west, so salt ponds are not restricted to dry salt beds (Manning, Wisneski, and Clark 1983). Even land costs as high as $10,000/acre ($4,049/ha) will only add about 10% to the total solar energy cost. A large SGSP at a sunny southern site is an example of a large-scale system with first cost per annual benefit ratio of less than four (Manning, Wisneski, and Clark 1983).

These examples illustrate that solar energy systems designed to take advantage of specialized thin-film polymers offer important cost reduction opportunities. Thin films offer the solar energy industry the same advantages that integrated circuits offered to the electronics industry—namely, very low mass of manufactured material and very high performance, yet very low production and assembly costs. However, uncertainty about the lifetimes of exposed thin films is a significant barrier to commercialization. Wilhelm (1983) extrapolated thin-film flat plate collector lifetimes as approaching ten years by using accelerated exposure tests; however, these tests do not reproduce the combination of cycling stresses that occur during system operation and therefore cannot predict operating lifetimes with certainty (Carroll and Schissel 1980; McKellar and Allen 1979).

This chapter accepts thin films as a cost reduction opportunity and surveys the recent literature seeking answers to the following questions:

1. What is the nature of the polymer degradation process?

2. What stresses originating in the environment contribute to the degradation of performance and shorten the lifetimes of solar collectors?

3. What are the effects of these environmental stresses on performance and durability of collectors?

4. Can currently available plastic collector materials provide durable service?

5. What additional information about the environment would assist in identifying and developing solar collector materials that are durable, efficient and low in cost?

Mechanisms and Agents of Environmental Degradation

In order to provide acceptable life-cycle benefits, solar collectors must function efficiently long enough to return more than the initial investment. In this section we consider environmental factors that degrade the performance of the collector and degradation mechanisms. The primary degradation mechanism is photooxidation that degrades optical efficiency as well as mechanical and electrical properties; the near ultraviolet (UV) radiation is the primary environmental agent that initiates this degradation. Subsection 13.2.2 describes current information on spatial and temporal variations in UV irradiance.

The optical and mechanical properties of collector materials (especially the plastic materials), however, can degrade even in the absence of UV radiation. Fundamental oxidation processes can often proceed in the dark; high temperatures and UV only accelerate the damaging processes. Also contributing to the damage process are moisture, wind, dust, and atmospheric pollutants. The "weathering" effects of these environmental factors are not simply additive. Their simultaneous effects can be far greater than the sum of their individual effects.

The effects of environmental factors also depend on the operation and design of the collector, particularly the temperature and moisture cycles and associated mechanical stresses in collector components.

13.2.1 Mechanisms of Polymer Degradation

Polymer degradation or "weathering" caused by the combined effects of sunlight, oxygen, and other environmental stresses begins when the polymer absorbs a harmful photon. The photon produces an excited electronic molecular state in the polymer with energy that must ultimately be dissipated in photochemical processes that change the chemical, optical, mechanical, and electrical properties of the polymer.

The pathways by which the photon can be absorbed and its energy dissipated are multiple and complex. The photochemistry of synthetic polymers has been an active field of research for decades but has become far more active during the past five years. UV photon absorption is now considered an industrial tool for creating polymers with new and desirable properties. The application of light as a manufacturing tool has accelerated understanding of degradation pathways. Two annual review series for specialists (Applied Science Ltd. 1979; The Royal Society of Chemistry 1967) have been established to document the voluminous new research results in photochemistry of polymers. Recent texts in photochemistry of polymers are available for the graduate student and researcher (Davis and Sims 1983; McKellar and Allen 1979).

This section describes only enough of the photodegradation mechanisms to identify those environmental stresses and operational conditions that, acting together, contribute to the degradation of polymers in solar collectors.

Technical discussions of polymer degradation caused by light distinguish three types of processes: photodegradation (which can proceed in the absence of air), photooxidation (which requires air), and photothermal degradation (in which the photon energy is added to significant thermal energy because of the elevated polymer temperature). All three of these processes usually occur simultaneously, and each can accelerate the rates of the other reactions.

Polymers are divided into two classes to describe their photodegradation (McKellar and Allen 1979). Pure type A polymers, which include polyethylene, polypropylene, polyvinylchloride, nylon, and polystyrene, contain only C–C, C–H, C–O, C–N, and C–Cl bonds and so cannot absorb light of wavelength longer than 0.19 μm. Type A polymers do not absorb sunlight with $\lambda \leq 0.29$ μm because it is almost completely screened by the atmosphere (discussed in following section); however, all *commercial* samples of these polymers do absorb the near UV (0.29 μm $\leq \lambda \leq$ 0.40 μm), which penetrates the atmosphere. These polymers contain light-absorbing centers (chromophores), which are impurities introduced during the high temperature manufacturing and processing of the polymer. The primary UV absorbing groups are carbonyl and hydroperoxide groups (McKellar and Allen 1979). Unsaturation, oxygen–polymer–charge transfer complexes, metallic impurities, and catalyst residues also contribute (McKellar and Allen 1979).

Type B polymers (polyesters, polystyrene) have a UV-absorbing site that forms part of the polymer backbone and so can be damaged even when impurities are not introduced.

Photodegradation processes for specific polymers are an active area of research and the relative importance of alternative damage pathways is often controversial. Some of the features are generally accepted. The mechanisms often involve free radicals (quasimolecular systems with an odd number of electrons). These damage processes are similar to those that occur in oxidation or thermal degradation. Often the only difference lies in the photon initiation step.

In type A polymers the carbonyl groups are easily excited to singlet and triplet states by the absorbed photon, and the excited state initiates the damaging chemical reactions. These reactions of excited carbonyl groups are considered central to the damage pathways and are described in photochemistry texts as Norrish reactions (McKellar and Allen 1979). For current purposes the important aspect of the Norrish reactions is that they result in scission of the bonds in the main polymer chain. Accompanying the chain scission is an increase in the crosslinking of these chains, resulting in embrittlement.

The Norrish reactions and other reaction sequences of excited states allow decay through chemical reactions with O_2 and H_2O and air pollutants as well as with the polymer. These reaction products can also absorb near UV photons and so can accelerate the weathering process. Many of the processes occur near the polymer surface where O_2 and H_2O are available by diffusion. The damage process results in surface cracking, which increases the depth and rate of availability of the O_2 and H_2O (Blaga and Yamasaki 1976).

Because the products of photooxidation are often chromophores or oxidants, the rate of photooxidation of some polymers is an exponentially increasing function of UV exposure (Benachour and Rogers 1983). In this sense photooxidation is often "auto catalytic." All types of deformation (uniaxial, biaxial, and cyclic stretching) increase the rate of photooxidation of low density polyethylene films (Benachour and Rogers 1983). This effect is attributed primarily to the propagation of polymer chain defects into microcracks and secondarily to stress (stored molecular energy enhancing reaction rates).

The net effect of the photochemical processes is embrittlement, loss of transparency, and an increase in electrical conductivity of the polymer. The photodegradation can cause warpage of rigidly shaped polymers (e.g., mirror support surfaces). These processes also cause the evolution of volatile gases, which subsequently form an opaque condensate on the inside of glazings (Ranby and Rabek 1975). These gases can also delaminate multiple film surfaces, cause bubbles in reflective coatings, and react chemically with the reflective coating (Brauman, MacBlane, and Mayo 1983).

An additional effect of the cross-linking, chain scission, and loss of volatile gases is the creation of void spaces in the polymer. These spaces not only admit oxygen and atmospheric pollutants, they also increase the permeability of the polymer to water. Often the purpose of the polymer in the collector is to provide an impermeable barrier to water, oxidants, and corrosive gases (e.g., caulks, mirror edge seals, front surface films for second surface mirrors; see Brauman, MacBlane, and Mayo 1983; Berry and Dursch 1983; Wischmann 1983; Rogers 1983).

Polymers can be effectively stabilized against photodegradation by adding UV absorbers (carbon black) or reflective screens (titanium and zinc dioxide). These options are not available for transparent glazings, but in this case stability can be achieved by adding quenchers, free-radical scavengers, and antioxidants (McKellar and Allen 1979; Ranby and Rabek 1975). Long-term benefits of these stabilizers have been limited because they are often lost as a result of outgassing and leaching. Methods of attaching the stabilizer to the polymer structure are now available (Applied Science Ltd. 1979).

A recent strategy for suppressing photooxidation is "orientation" of the polymer. Oriented polymers are those in which the polymer chains are aligned, an effect that can be achieved by uniaxial stretching. The benefit of orientation is that it suppresses the entry of atmospheric oxygen into the polymer by increasing the density of the polymer (Benachour and Rogers 1983). As noted earlier, stress and deformation effects compete with this benefit, but these can be relieved during manufacture of the oriented film.

13.2.2 Ultraviolet Irradiance

By definition, ultraviolet light has a wavelength shorter than the visible limit at 0.40 m. Ten percent of the extraterrestrial solar irradiance is ultraviolet. Because of absorption and scattering in the atmosphere, a variable but smaller fraction of the incident irradiance at the earth's surface is ultraviolet. This subsection surveys current information on ultraviolet irradiance at sites in the United States.

Because of absorption by oxygen and nitrogen species, essentially no irradiance with wavelength shorter than 0.295 μm reaches the surface. Absorption by ozone in the Hartley band (0.220–0.295 μm) is so strong that although the total ozone content of the atmosphere is equivalent to a vertical column of about 3 mm at standard temperature and pressure, only about one in 10^{40} photons reaches the surface (Davis and Sims 1983).

Near UV (0.295 μm–0.40 μm) is also absorbed by ozone and strongly scattered and absorbed by the other atmospheric constituents, by clouds, and by natural and synthetic aerosols. However, a significant and variable fraction of the near UV reaches the surface. Near UV, usually acting in concert with oxygen, water, and elevated temperatures, is the primary materials degradation agent.

Figures 13.1 and 13.2 illustrate the sensitivity of direct-normal near UV solar irradiance to changes in air mass (m), site elevation, and atmospheric turbidity. These curves result from previously unpublished Trinity University simulations using a spectral transmission model developed at the Solar Energy Research Institute (SERI) (Bird 1982). (Because of the structure of the SERI model, the near UV is defined in these figures as irradiance in the wavelength range 0.297–0.390 μm.) The simulations assume that the sky is cloud free and that the vertical column ozone is 0.25 cm (0.1 in.).

The low turbidity case is defined by the angstrom wavelength coefficient $\alpha = 1.0274$ and particulate load parameter $\beta = 0.0490$ so that the turbidity optical depth at 0.5 μm ($T_{a,0.5}$) is 0.10. This would be typical of very clear

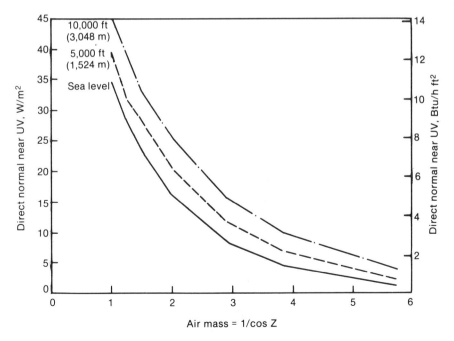

Figure 13.1
Sensitivity of direct-normal near UV to air mass and elevation (cloud free, low turbidity).

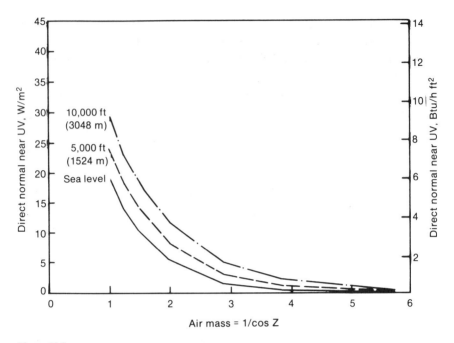

Figure 13.2
Sensitivity of direct-normal near UV to air mass and elevation (cloud free, high turbidity).

days, which usually occur only in the winter (Kacoroski 1984). The high turbidity case represents a very hazy day of limited visibility, which would usually occur only in summer or during a temperature inversion episode in an urban area (Kacoroski 1984). The angstrom turbidity parameters are defined as $\alpha = 1.0274$ and $\beta = 0.0490$ so that $T_{a,0.5} = 0.51$.

Figures 13.3 and 13.4 result from using the same SERI model to compute the total (beam plus diffuse) near UV irradiance on the horizontal. In these simulations the surface albedo is assumed to be 0.27.

For cloud-free conditions figures 13.1 through 13.4 lead to the following conclusions:

1. The direct (beam) near UV is quite sensitive to air mass m and turbidity. Station elevation also has a significant effect.

2. At air masses near one, the diffuse UV is negligible. In the sea level, high turbidity case, at $m = 1$ the diffuse is less than 5% of the direct normal UV.

3. At air mass = 3 the near UV is much reduced and is dominated by diffuse.

4. Because of canceling effects resulting from increased diffuse, site elevation has little effect on total horizontal UV at air mass = 1.5.

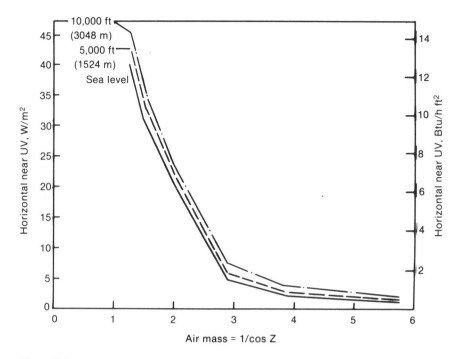

Figure 13.3
Sensitivity of total horizontal near UV to air mass and elevation (cloud free, low turbidity).

Atmospheric water vapor absorption bands are all in the infrared; therefore, except that increased water vapor increases turbidity (Shettle and Fenn 1979), water vapor has no direct effect on UV.

Total atmospheric ozone is variable. Seasonal variations are well understood; maximum ozone concentration occurs in the spring and the minimum occurs in late fall. These seasonal variations are small at latitudes of less than 35 deg, but significant above 45 deg (Robinson 1966). The seasonal ozone variations cause the spectrum of UV to shift seasonally; the UV irradiance for wavelengths shorter than 0.33 μm increases significantly in the spring. This short UV region is particularly damaging to some polymers (McKeller and Allen 1979; Koller 1965).

Human activity can deplete ozone levels. The effects of two trace gases, NO and Cl, have aroused particular concern. Soil bacteria release NO, and its rate of release is increased by the use of fertilizer. NO is also released by jet aircraft. Atmospheric chlorine has been added to the atmosphere by the freons previously used in spray-can propellants and still used in refrigerants. These gases act catalytically to remove ozone and so may continue to deplete ozone

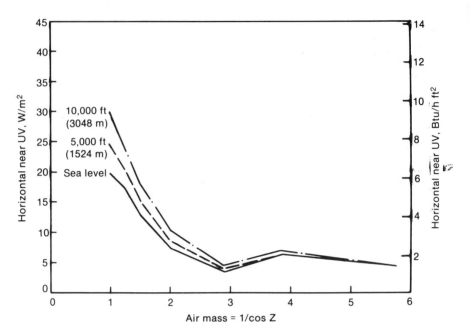

Figure 13.4
Sensitivity of total horizontal near UV to air mass and elevation (cloud free, high turbidity).

for many years after their introduction into the atmosphere. Some atmospheric transport models predict very long residence times for Cl and NO, and so predict significant systematic ozone reductions. However, experimental validation of these models is very difficult because natural variability of ozone caused by frontal activity can change ozone concentrations by 20% in a few days, and rapid change can occur across regions of greater than 1,000 miles (1,610 km) (Fleagle and Businger 1980).

Rao and Takashima (1985) have described two simple models for calculating the attenuating effects of ozone (and other cloud-free atmosphere processes) on UV irradiance. They compared the predictions of their model with measurements taken at the South Pole (Baker-Blocker et al. 1984). These measurements are of special interest for model validation because the South Pole atmosphere is almost dust free and during summers is almost cloud free.

UV irradiance data, including the effects of clouds, are necessary in order to begin the process of defining maps of UV irradiance. Such maps would be an input to the development of a photodegradation index. Until recently measurements of UV suitable for this purpose were almost nonexistent in the United States or other western countries. In contrast, the UV climatologies of the Soviet Union and Australia have been observed and analyzed for more than two decades. Monthly maps of the spatial and spectral distribution of UV across the Soviet Union are available (Coulson 1975).

Before 1979 UV measurements were made in the United States only at a few research sites (e.g., Mauna Loa Observatory, Washington, DC) or at materials degradation test sites in Arizona and Florida. Instrumentation, calibration, and UV measurement procedures varied from site to site.

Although these intermittent or spatially isolated measurements were not sufficient to construct a broad national understanding of UV climatology, they did establish that the near UV constitutes about 5% of the global insolation during sunny periods at these sites (Davis and Sims 1983; McKellar and Allen 1979).

In 1978 the U.S. Department of Energy (DOE) established an eight-station network of geographically well-distributed Solar Meteorological Research and Training (SMRT) sites. In 1979 these eight university sites began measuring and recording a nearly continuous, standardized set of quality controlled, minute values of irradiance parameters including UV as well as direct, diffuse, and global insolation (Hulstrom 1981). The UV measurements were made with the Eppley ultraviolet pyranometer, which uses a Weston selenium barrier photovoltaic sensor and includes a filter that limits response to the spectral range of 0.295–0.385 μm. Although funding for the SMRT program was terminated in 1981, a few of the sites have continued to operate.

Table 13.1 lists seasonal and annual summaries of mean daily horizontal UV measured at each of the sites. Although Honolulu receives the greatest UV irradiance, all sites, including Fairbanks, Alaska, receive significant summer UV. Figures 13.5, 13.6, and 13.7 illustrate the daily global insolation, the daily UV, and the daily ratio of UV to global at the Davis, California, SMRT site during 1980 (Hatfield et al. 1981). The pattern of temporal variability of the UV is similar to that of the global insolation. At the Davis site the daily UV/global ratio is approximately 0.05, with a minimum of about 0.04 occurring during the summer and a winter maximum of about 0.08.

Table 13.2 lists the ratio of daily UV/global for selected spring and summer days of 1980 at a wide range of latitudes and elevations (Hatfield et al. 1981). These data were taken using one instrument set in a mobile van and illustrate that the daily UV ratio is not strongly dependent on latitude or elevation.

Figure 13.7 illustrates that the 5% UV-to-global ratio is an estimate that is not always accurate. If this ratio could be accurately defined as an empirical function of a widely available climate parameter, available national maps of solar irradiance could be transformed into maps of UV irradiance. One irradiance parameter has repeatedly proved to be useful in similar correlation

Table 13.1
Mean daily values of UV radiation for summer, winter, and the year for eight sites in the
United States (kJ/m² day)

	Year	April–September	October–March	Annual value
Fairbanks, AK	1980	619	120	369
	1981	626	113	370
Corvallis, OR	1981	974	350	662
Davis, CA	1981	926	424	675
Honolulu, HI	1981	1117	775	946
Ann Arbor, MI	1980			
	1981	788	368	578
	1982			
San Antonio, TX	1981	994	589	792
Atlanta, GA	1981	988	572	780
Albany, NY	1981	816	358	587

Source: Compiled by Gerd Wendler of the Geophysical Institute University of Alaska, from data
provided by SMRT Program principal investigators.

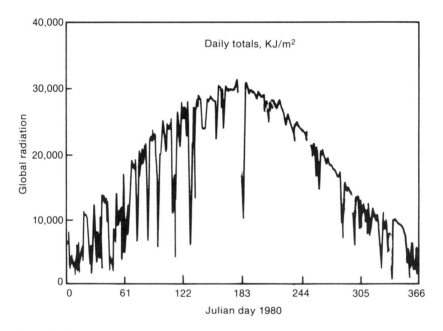

Figure 13.5
Daily global radiation measurements for 1980, Davis, California. Source: Hatfield et al. (1981).

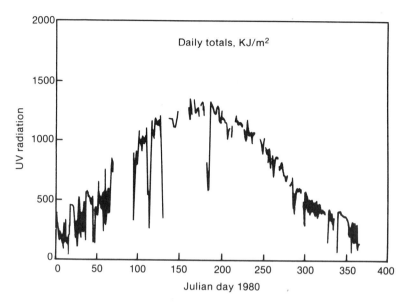

Figure 13.6
Daily UV measurements for 1980, Davis, California. Source: Hatfield et al. (1981).

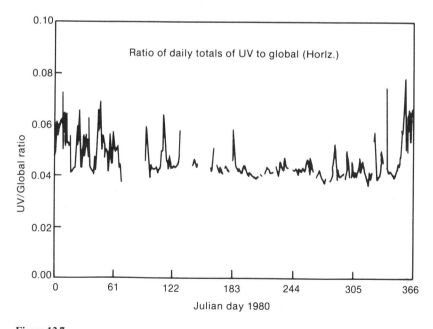

Figure 13.7
Daily ratio of UV-to-global radiation measurements for 1980, Davis, California. Source: Hatfield et al. (1981).

Table 13.2
Ultraviolet-to-global ratios for selected locations in the United States from data collected with
the mobile van during 1980

Location	Julian date	UV/global ratio
Brawley, CA	154	0.04
Weslaco, TX	160	0.05
Temple, TX	166	0.05
Manhattan, KS	174	0.05
Lincoln, NE	174	0.04
St. Paul, MN	184	0.05
Fargo, ND	192	0.04
Sidney, MT	196	0.04
Beartooth Pass, WY	203	0.06
Kimberley, ID	208	0.05

Source: Hatfield et al. (1981).

problems. This parameter is the ratio of ground level global irradiance to
extraterrestrial irradiance on the same horizontal plane. The monthly mean
of this atmospheric transmittance or clearness index (K_T) is known to provide
a nearly complete description of the nonspectral solar climatology at any site
(Duffie and Beckman 1980). Associated with a particular value of K_T at a given
latitude is a unique distribution of hourly beam and hourly diffuse irradiance.
A similar approach appears promising for UV irradiance.

Figure 13.8 illustrates the relationship between the daily UV-to-global
insolation ratio and daily K_T. The curve in figure 13.8 is the best fit polynomial
through all data with zenith angle < 70 deg taken at the Trinity University
SMRT site from January 1980 to March 1982. There is very little scatter in
the data about this curve; the root-mean-square (rms) difference between the
correlation curve and the daily ratio of UV to global is only 0.003 in this data
set.

Figure 13.9 is the same as figure 13.8 except that the hourly UV-to-global
ratio is correlated with the hourly clearness index k_t. The rms difference
between the hourly UV-to-global ratio data, and the correlation curve is only
0.004.

Figures 13.8 and 13.9 establish the relationship between UV and global
irradiance for a wide range of San Antonio conditions including clear and
heavily overcast periods. The UV-to-global ratio can exceed 0.07 under heavy
clouds. [Note that because the highest zenith angle included in the data used
to produce figures 13.8 and 13.9 is 70 deg $(m = 2.9)$, any extremely low daily

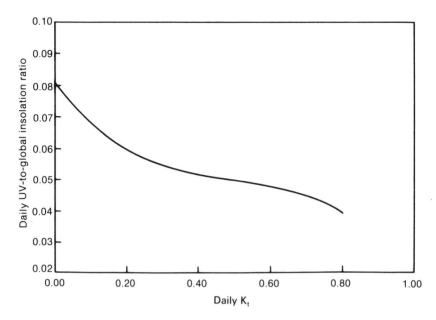

Figure 13.8
Ratio of daily UV to global versus K_T.

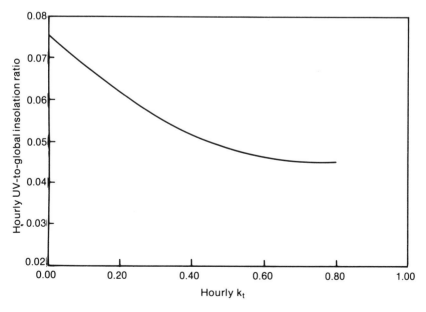

Figure 13.9
Ratio of hourly UV to global versus k_t.

and hourly clearness index values shown must be associated with heavy cloud cover.]

Figure 13.8 and 13.9 have not been established as universal; however, they do represent the type of analysis that can be used to estimate horizontal UV irradiance from the extensive data base of horizontal solar irradiance.

Applying this approach to estimating the UV irradiance incident on track-ing surfaces or tilted surfaces requires short-term data on the beam and diffuse components of UV and data on the distribution of UV diffuse radiance across the sky. Although data of this type have been taken at materials degradation sites, almost no data of this type have been published.

In addition to developing algorithms that convert insolation on the horizontal into UV on tilted and tracking surfaces, another important, but not sufficiently well-defined, characteristic of the UV is spectral content. All photons within the near-UV range are not equally damaging to all polymers. For example, photons with $\lambda = 0.33$ μm are three times more damaging to unstabilized polypropylene than photons with wavelengths between 0.37 μm and 0.40 μm (McKellar and Allen 1979). It is known that the scattering and absorption processes described above cause the UV spectrum to vary with air mass, surface orientation, and season (McKellar and Allen 1979; Koller 1965).

Seasonal ozone variations have a small effect on the total near UV but cause significant changes at the short wavelength end of the near UV. Based on eight years of measurements of horizontal UV at wavelengths between 0.290 and 0.315 μm, Coblentz has shown that the maximum percentage of this short UV wavelength occurs in the summer (McKellar and Allen 1979; Koller 1965). This is in contradiction to the trend of figure 13.7 and can only be explained as an ozone effect. Detailed spectral data in the UV have been taken by SERI (Bird and Hulstrom 1982) and at DSET laboratories near Phoenix, Arizona; however, these data have not been analyzed for the purpose of developing algorithms for spectral UV on tracking and tilted surfaces.

A preliminary application of figures 13.8 and 13.9 suggests that the long-term average annual horizontal near UV irradiance in the northeastern United States is about 70% of the near UV received in the desert Southwest. On the basis of annual horizontal near UV dose alone, polymers would last about 40% longer in the Northeast than in the desert Southwest. As noted in subsection 13.2.1, estimates of lifetimes based only on UV dose will underestimate the lifetime difference.

13.2.3 Other Environmental Factors

Elevated temperatures contribute to materials aging through photothermal degradation, materials deformation associated with increased plasticity at high temperature, outgassing from materials, and stresses associated with expansion and contraction of collector components. These effects are accelerated if the collector is allowed to stagnate. Because stagnation temperature is a function of insolation level, collector properties, and ambient temperature, it is difficult to assess on a regional basis; however, it is mostly a function of the peak insolation and ambient temperature at the time of stagnation. Improper installation or design can lead to stresses or sagging of collector components; these problems can be aggravated by differential thermal expansion of these components.

Moisture works together with photodegradation and thermal stress to weather surfaces and to damage materials exposed by weathering. Rain can directly erode material surfaces.

Dirt accumulation is a problem for glazings and reflectors and depends on many factors, including amount of airborne dust and other airborne materials, amount of rainfall, and surface properties of glazings or reflectors. Plastic glazings and reflector surfaces are sensitive to chemical pollutants, so proximity to industrial pollution should be considered in collector design and selection. Plastic surfaces are also more likely to accumulate dust.

Wind loading is usually the major structural design consideration for the collector and collector supports. Murphy (1981) has assessed existing studies of wind loading on solar collectors. He concludes that load reductions of three or more can result from well-designed wind shield fences and from self-shielding in large collector arrays.

Hail damage is no longer considered to be a serious problem for solar collectors (Löf and French 1979; de Winter 1980); 3/16-in. (4.8-mm) thick tempered glass is considered to be sufficient to prevent hail damage in all U.S. climates (Landstrom et al. 1978).

Freezing presents a design problem which, although regional in nature, is potentially a problem in nearly every U.S. location. Cycles of freezing and thawing can also contribute to weathering of exposed surfaces, which then can be penetrated at the microscopic level by moisture.

13.2.4 Tests and Indices for Predicting Environmental Degradation

Although understanding of photodegradation processes is advancing rapidly, the development of long-life polymers for solar collectors remains largely an empirical art. Because the great majority of candidate materials will fail almost immediately, new materials are first tested by inexpensive screening tests in controlled laboratory conditions. Laboratory weathering devices have become quite sophisticated and can produce programmed cycles of temperature, spectral UV irradiance, and moisture (McKellar and Allen 1979; Ranby and

Rabek 1975). Samples that pass the screening tests are then tested outdoors, using standardized accelerated test devices and procedures that concentrate the light by a factor of eight and can provide water spray cycles (McKellar and Allen 1979).

The previous discussion illustrates that the lifetime of the polymer in service in a collector will depend on the polymer history, mechanical installation conditions (e.g., tension), operating conditions, rainfall, the site dry-bulb temperature, dew point temperature, aperture solar irradiance, and aperture spectral UV. This complex of conditions is very difficult to duplicate or accelerate. It is not surprising that most polymer materials that appear promising in screening tests and in accelerated outdoor weathering tests actually fail prematurely when installed in an operating collector. The lengthy, expensive, yet not definitive tests required to identify durable new materials are a major barrier to the development of durable thin-film collectors.

Promising new analytic approaches to these problems have been identified. Performance prediction modeling is now being applied to photovoltaic encapsulation (Carroll and Schissel 1980). Computer-controlled dynamic degradation tests can be applied to polymer degradation in the same way that dynamic mechanical analysis has been long applied to mechanical degradation testing (Carroll and Schissel 1980). Limited success in computer simulation of polymer degradation has already been demonstrated (Guillet, Somersall, and Gordon 1983).

These approaches will require the development of a complete cross-correlated statistical distribution description of all the environmental stress factors acting on the installed polymer. Erbs (1984) has developed such a statistical description of all the temperature, moisture, and insolation parameters relevant to the performance of thermal solar systems. His methods require only easily available monthly average data as inputs. Extension of Erbs methods to the polymer degradation problem requires only the development of a relationship between aperture broad spectrum irradiance and aperture UV.

13.3 Environmental Degradation of Conventional Collector Components

Studies conducted for DOE and the American Society of Heating, Refrigeration, and Air Conditioning Engineers (Ward, Oberoi, and Weinstein 1982) revealed material deficiencies of solar collectors as perceived by solar product users.

For flat plate collectors, the deficiences identified were the following: faulty seals around cover/glazing, leading to moisture buildup inside the collector;

absorber coatings that peel away from their substrates and/or have discolored and otherwise degraded; plastic glazings with poor optical and structural quality; outgassing from sealants and insulation; and collector stagnation, leading to deterioration of absorber surface coatings.

For evacuated-tube collectors, the deficiencies identified were collector stagnation, thermal shock caused by rapid approach to stagnation temperatures, and dirt accumulation on reflecting surfaces.

For concentrating collectors, the deficiencies identified were poor durability of reflector surfaces and attaching glues and dirt accumulation on reflecting surfaces.

Almost all these problems result from environmental effects on materials. Solar collector materials most strongly affected by environmental factors are plastics and thin-film coatings; however, glass and other collector materials are also affected.

Soda-lime-silica glass is conventionally used as a solar collector glazing. Its main disadvantages are its high cost, heavy weight, low impact strength, and high reflectivity. Measurements of the transmittance of soda-lime glass show no difference in the total transmittance of exposed and sheltered samples after 40 years of exposure in a semiarid climate (Lind and Hartman 1980). Although glass is not damaged by UV radiation, dirt accumulation can seriously reduce its effective transmissivity (Garg 1974). These dirt effects become more serious as the collector becomes more nearly horizontal (Garg 1974).

Plastics are used as solar collector glazings, reflectors, lenses, honeycombs for convection suppression, absorber plates, housing, pipings, and structural support members. Plastics are competitive with more traditional materials such as glass, wood, and metals. In extended outdoor use, plastics may gradually discolor, chalk, crack, become brittle, and lose their structural integrity. Blaga (1978) described the advantages and disadvantages of plastics. Some of their advantages are their high strength-to-weight ratios, high resistance to erosion, easy formation into intricate shapes, and low cost. Some of their disadvantages are that they soften at elevated temperatures, have low mar resistance, are sensitive to UV radiation, and have low fire resistance. The plastics that are commonly used in solar applications (Blaga 1978) are polymethyl methylacrylate (PMMA) (acrylic), polycarbonate (PC) (e.g., Lexan), glass-fiber-reinforced polyester (GRP) (e.g., Kalwall or Filon), polyvinyl fluoride (PVF) (e.g., Tedlar or Kynar), fluorinated ethylene propylene (FEP) (e.g., Teflon), polyethylene terephthalate (PET) (e.g., Mylar), cellulose acetate buyrate (CAB), polyvinyl chloride (PVC), and plastic foams (polystyrene, phenolformaldehyde, ureaformaldehyde, polyurethane, and polyisocyanurate).

Surface coatings or surface treatments are used on absorber plates to increase absorption and decrease emission, and on glazings to decrease reflectance of insolation or to increase reflectance of IR radiation. Surface treatments may be classified as follows (Zerlaut 1980):

1. Nonselective black absorber coatings, including paint (e.g., 3M's Nextel Black Velvet and PPG's Duracon Super 600), fused porcelains, and metal conversion coatings (e.g., anodized aluminum).

2. Selective solar absorber coatings, including absorber-reflector coatings (e.g., black nickel, black chrome), multilayer interference stacks (e.g., Honeywell AMA coating), powdered semiconductor/reflector combinations, wavefront discriminators, and resonant scattering systems.

3. Surface etching of glass.

4. Antireflection coating of glass.

5. Heat mirror coating of glass or plastic.

Although porous silica antireflection coatings are not currently used on commercial glazings, they can reduce two-surface reflectance on glass to 2% or less, are cheap to apply, and, with periodic water washing, maintain their low reflectance outdoors (Cathro, Constable, and Solaga 1981). These films have been successfully applied to polycarbonate and acrylic by simple dipping (Cathro, Constable, and Solaga 1981).

The environmental degradation of specific collector components is addressed below; polymers and polymer surface treatments are emphasized.

13.3.1 Collector Glazing

A study carried out by DHR, Inc., for DOE (Herzenberg and Silberglitt 1981) identifies 12 characteristics of an ideal glazing: high transmissivity, high IR reflectivity or, as a second choice, high IR absorptivity, high tensile strength, high impact strength, thermal expansion coefficient matched with frame, abrasion resistance, moderate Young's modulus (if too low, then the glazing sags; if too high, then there is a high thermal stress), durability, low dirt adherence, low cost, and light weight.

Another survey (Baum and Gage 1979) included the following additional factors: weather resistance to UV, SO_2, and moisture; retention of mechanical properties for temperatures from $-40°$ to $300°F$ ($-40°-149°C$); resistance to hydrolysis; resistance to stress fatigue from thermal cycling; and ease of maintenance.

Glass has many of these desirable characteristics but has low impact resistance, is expensive, is too heavy, and does not have high IR reflectance. The

DHR study (Herzenberg and Silberglitt 1981) suggested that the high weight of glass is its major deficiency because a glass glazing requires an expensive supporting structure.

Plastics are light, cheap, and have high tensile and impact strength; however, plastic glazings present problems of durability and low mar and abrasion resistance. Of these problems, durability is considered to be dominant. Durability problems caused by thermal and mechanical stresses (resulting from unequal expansion of the plastic and its supporting frame) contribute to short lifetimes for plastic glazings, but durability problems caused by UV degradation appear to be more significant. Baum and Gage (1979) distinguish between outer and inner glazings. The outer glazing must withstand greater abrasion, weathering, and UV radiation. The inner glazing must withstand high temperatures. Their cost and lifetime survey of all commercially available transparent polymers, including those with the best available UV protection, led to the following interesting conclusions:

1. Plastic glazings in the price range of $0.01 to $0.05/ft^2 ($0.11–$0.54/m^2) could not survive in outdoor use for 20 years.

2. Polycarbonate, fluorocarbons, and silicone products that may last 20 years cost between $0.44 and $2.00/ft^2 ($4.74–$21.53/m^2).

3. Acrylics may be cheap enough and sturdy enough to be viable candidates for low-cost glazings; they have unprotected lifetimes of up to 10 years and could be upgraded to 20 years.

Other research projects reached the following conclusions:

1. Inexpensive plastics (e.g., polyethylene) have short lifetimes (less than five years), but with internal UV absorbers and an external UV absorber coat, lifetimes could be extended to ten years (Blaga 1978; Baum and Gage 1979).

2. PMMA (acrylic) has particularly good glazing properties (Blaga 1978). It has transmission properties as good as glass (Sherr 1972); it is weather resistant and is not very sensitive to UV damage (Rainhart and Schimmel 1975; Sherr 1972; Brydson 1975; Mark, Gaylord, and Bikales 1964). The chief disadvantages of acrylics are that they have low softening points and may distort if used as an inner glazing (Saunders 1972).

3. PC has a very high impact strength (Sherr 1972) and a higher heat resistance than acrylic, but has low mar resistance and does not weather well. A Hughes Aircraft Co. project (Landis and Mosher 1979; Bilow et al. 1981) included attempts to develop PC sheets with improved UV and thermal stability.

4. GRP has good glazing properties, but does not weather well outdoors

(Parkyn 1970; Blaga 1975). Covering the surface of a GRP sheet with PVF improves its resistance to outdoor exposure (Blaga and Yamaski 1977; Dickman 1976). A UV-stabilized form of GRP has been developed and evaluated (Martino 1976; White 1977). Results indicate improved performance over other grades of GRP glazings.

FEP and PVF films have high thermal stability, high chemical stability, durability, and high transmissivity (Brydson 1975; Martino 1976; Mark, Gaylord, and Bikales 1970; Edlin 1958). The durability has been tested by outdoor exposure tests in Florida (Mark, Gaylord, and Bikales 1970). A Brookhaven National Laboratories (BNL) project for developing a "polymer film solar collector" (Wilhelm and Andrews 1982) used a PVF glazing (DuPont's Tedlar™). The main problem encountered was the nonsymmetrical shrinking of the plastic film during a thermal activation stage of collector manufacture and resultant stress loading of the supporting frame.

5. Abrasion-resistant coatings (e.g., DuPont's Abcite AC, acrylic sheet) have been developed for use with plastic glazings.

One of the problems associated with using plastic films as collector glazings is that these films must be supported; if they are supported by stretching them over a frame, they can sag under high temperature conditions (e.g. stagnation). A Battelle study (Gordon 1979) investigated plastic honeycomb material as support for a plastic film glazing. Under some conditions a honeycomb core coverplate made by sandwiching the plastic honeycomb between two plastic films is as efficient as a collector glazed with two sheets of glass.

13.3.2 Absorber Plates

An ideal absorber plate should have high absorptivity, low emissivity, a low thermal expansion coefficient, a low thermal softening point, high durability, and low cost. The absorption and emission properties are associated with the surface of the absorber and usually involve a selective surface coating or thin film. Problems of mechanical rigidity at high temperatures and structural failures under freezing conditions are associated with the plate itself. Collectors made of traditional materials have a metal absorber plate with a selective surface coating. Low-cost collectors substitute plastics for the metal of the plate. The low-cost plastic absorber plate is subject to mechanical and thermal stress. Because the absorber plate can use carbon black absorber, it is not as subject to UV degradation as transparent plastic glazlng .

Selective coatings for absorber plates have reasonably good absorption and very good emission values; however, they are usually expensive and not enough is known about their durability in use. Long-term exposure to UV radiation, moisture, and high temperature can lead to bleaching, oxidation, and flaking or crazing. A study of outgassing by a nonselective absorber coating (Tortorello and Wolf 1979) demonstrated that some coatings begin to deteriorate at temperatures as low as 177°F (81°C). Some coatings outgas and deposit material on the glazing.

Black chrome is widely used as a selective surface coating. It is a tandem coating. It can be produced by electroplating an initial layer of bright nickel (the IR reflector) and an overlay of chromium oxide (the insolation absorber). It can also be produced by a physical vapor deposition (PVD) process. The PVD process uses aluminum as the reflector and sputtered chrome oxide as the absorber.

Durability tests on black chrome, black nickel, and black iron have been carried out by Honeywell, Inc. (Lin and Zimmer 1977). Black chrome is determined to be the most durable of the commercially available selective coatings; a lifetime of 30 years was suggested for this coating. According to field experience (Sparkes and Raman 1978), black chrome is the most durable and efficient selective surface and can be applied with uniform quality. A single-cover system with a black chrome surface can be used in place of a double-glazed system with a nonselective surface. The absorber coating is sufficiently more durable than that of the inner glazing, so this trade-off is practical. Durability tests carried out by Berry Solar Products (Rughunatheen 1981) indicate that black chrome on nickel-plated copper, multilayer coatings on aluminum, black nickel on nickel foil, and conversion coatings on stainless steel are all very durable, but that other surfaces show some change of absorptivity and emissivity under prolonged exposure to environmental conditions.

Although black chrome selective surfaces are cost effective, they do add to the cost of the system; therefore less expensive selective surface coatings are being investigated. Honeywell has developed a selective paint (Lin and Zimmer 1977). The paint has a solar absorptance equal to 0.92 and an emittance of 0.10. The estimated cost of this paint on a square foot basis is an order of magnitude less than plated coatings.

13.3.3 Reflectors

Reflective films (e.g., aluminized mylar), surfaces of polished aluminum, and surfaces of polished glass are used as solar reflectors (Ward, Oberoi, and Weinstein 1982). Each of these reflector materials has advantages and disadvantages. Polished glass and aluminum have higher spectral quality; mylar and acrylic reflector films have lower optical quality but are lighter and less expensive.

Polished glass does not suffer from UV degradation and is scratch resistant; however, it is relatively fragile. Aluminum with an anodized finish does not suffer from UV degradation but decreases in specularity over time and may be easily scratched. Specular films can be coated with polymer layers to protect them from UV degradation and oxidation. One problem with such films is that they are somewhat permeable; therefore moisture, oxygen, and pollutants may be able to reach the metal surface beneath.

Attachment to a supporting surface can be a problem for polymer film reflector materials because of differential thermal expansion. Loss of attachment or delamination of the reflective layers and the supporting layers may be caused by moisture permeability of the outside protective layer, UV radiation, thermal cycling, or a combination of causes.

UV degradation has been an especially severe problem for metallized polymer films because the light passes through the film twice and because the film must maintain very strict dimensional stability to maintain its image-forming character. Chemical interactions between the polymer degradation products and metal film lead to failure of the reflector at the polymer-metal interface (Bethea et al. 1981). Accelerated outdoor tests have identified aluminized Teflon as promising. After nearly three years in an eight times concentrator the specular reflectance had not decreased (Hampton and Lind 1978; Lind and Ault 1978). Extrapolation of this result to normal service lifetimes is problematic. SERI has developed a new metallized acrylic (PMMA) reflector film that appears to be resistant to UV degradation (Neidlinger and Schissel 1985). SERI is also developing a new method of testing reflectors and their degradation (Masterson and Lind 1983).

Dust is a serious problem for mirrors because the dust changes the direction of reflection, and so the specular reflectance is drastically decreased. Glass mirrors in Albuquerque, New Mexico, were found to have lost as much as 15 reflectance units because of dust accumulation (Roth and Pettit 1980).

Electrostatic charging effects increase the dust adhesion on metallized polymers (Bethea et al. 1981). Morris exposed samples of glass and FEK reflectors on exposure racks at six industrial sites across the United States. (FEK refers to a 3M product, FEK-244. FEK is a trade designation and not an abreviation. This product has been replaced by ECP-305, which has a silver surface and a 96% reflectance.) He found the decrease in the specular reflectance of FEK and direct normal transmittance of the glass to be well correlated at each site. For a one-month exposure the maximum rate of decrease of reflectance and transmittance was about 1% per day at each site. The average rate of decrease for a total of one month was highly site dependent and varied from 0.17% to 0.4% per day (Morris 1982). Deffenbaugh (1984) has correlated measured thermal performance of a parabolic trough collector (with FEK reflector) with measured effects of dust on specular reflectance and concluded that in San Antonio, Texas (one of the least dusty cities in the sample), a parabolic trough collector that is carefully pressure-washed once a month will still collect about 30% less energy per year because of dust and dirt effects on specular reflectance and receiver transmittance.

13.3.4 Collection Housings

A comparison of plastic, metal, paper, and ceramic materials for collector housings (Baum and Gage 1979) led to indeterminate results. Ceramic substrates (e.g., glass-reinforced gypsum) were ruled out because of their high density and other problems, but metals may continue to be competitive with plastics. Plastics evidently require fillers, which increase their weight and cost.

13.4 Lower-Cost Collector Design Concepts

Solar collector design is the subject of another chapter; however, it is clear from previous sections that the rate of environmental degradation of a solar collector is affected by collector design, production, installation, and operation as well as collector materials and environmental stress characteristics of the site. For completeness, selected polymer solar collector design projects are briefly described here.

1. Lawrence Livermore National Lab completed a study of an inflated, plastic, cylindrical concentrator system (Gerich 1978). Materials problems were encountered; plastic materials degraded because of UV radiation and wind exposure. The results indicated that Kynar (PVF) is a promising material for solar applications.

2. Battelle began the development of a low-cost, high efficiency, low temperature black liquid collector (Landstrom et al. 1978). The collector designs use plastics suitable for black liquid use.

3. Acurex Corporation carried out a project (Nelson, Adams, and McLeod 1979) with the intention of developing a low-cost solar collector made of several polymer films.

4. Fafco also developed a low-cost design (Fafco, Inc. 1980) that uses low-cost materials. The design uses cylindrical channels in a parallel array; an extruded concentric cylindrical glazing is associated with each channel.

5. Brookhaven National Lab developed a low-cost collector (Wilhelm and Andrews 1982; Wilhelm 1983), with the goal of an installed cost less than $5/ft^2

($53.80/m^2). The collector design uses an absorber formed from a laminate of polymer film and aluminum foil, a polymer film glazing, and a nonpressurized liquid absorber.

6. The shallow solar pond, the salt-gradient solar pond, and the integral or "batch" domestic solar water heater can combine the cost advantages of the polymer film collector with the additional benefits of integrating collection and storage. The abundant literature on these collector/storage concepts is described in other chapters.

13.5   Research Priorities

Several studies attempted to define the priorities for research in this area. A recent BNL study (Fthenakis and Leigh 1984) concluded that the most valuable collector improvement would be a decrease in glazing reflectivity. Following this characteristic in order of decreasing importance are increased absorber absorptance, decreased absorber emittance, and decreased glazing emittance.

A recent report on solar materials (Herzenberg and Silberglitt 1981) included the following recommendations:

1. Improving photodurability of polymer glazings should be "the first priority of the flat plate materials program."

2. Investigating the impact of polymer soiling on the transmittance of promising glazing polymers also deserves attention.

3. Studying polymers to be used as absorber or structural components should also be a part of this research effort.

4. Developing low-cost, durable, selective absorbers is a second critical materials research area. "If cheap, durable, selective absorbers could be applied to polymer substrate, high performance, light weight, low-cost collectors could be achieved."

Based on discussion in previous sections, we add the following research recommendations:

1. Easy methods of evaluating the spectral UV irradiance at the collector aperture should be developed. The development of these methods appears to be primarily a problem of analysis of existing data.

2. Fast test methods that are good predictors of durability of operating collectors should be developed. Promising new approaches to durability testing of collector materials are described in subsection 13.2.4. These new analytic test methods will require the development of a more complete descrip-

tion of the total environment of the operating component of the collector. A promising statistical method of describing the total operating environment is described in subsection 13.2.4.

References

Andrews, J. W. 1983. Economics of solar heating systems. In *Polymers in Solar Energy Utilization*, C. G. Gebelein, D. J. Williams, and R. D. Deanin, eds. ACS Symposium Series 220. Washington, DC: American Chemical Society, pp. 19–26.

Applied Science, Ltd. 1979. *Developments in Polymer Photochemistry*. (Series begins in 1979.) London.

Baker-Blocker, A., J. J. De Luisi, and E. Dutton. 1984. Received ultraviolet radiation at the South Pole. *Solar Energy* 32: 659.

Baum, B., and M. Gage. 1979. *Solar Collectors*. AL0-5359-T1. Washington, DC: U.S. Department of Energy.

Benachour, D., and C. E. Rogers. 1983. Effects of deformation on the photodegradation of low-density polyethylene films. In *Polymers in Solar Energy Utilization*, C. G. Gebelein, D. J. Williams, and R. D. Deanin, eds. ACS Symposium Series 220. Washington, DC: American Chemical Society, pp. 307–330.

Berry M. J., and H. W. Dursch. 1983. Optical, mechanical, and environmental testing of solar collector plastic films. In *Polymers in Solar Energy Utilization*, C. G. Gebelein, D. J. Williams, and R. D. Deanin, eds. ACS Symposium Series 220. Washington, DC: American Chemical Society, pp. 99–114.

Bethea, R. M., M. T. Barriger, P. F. Williams, and S. Chin. 1981. Environmental effects on solar concentrator mirrors. *Solar Energy* 27(6): 497–511.

Bilow, N., R. I. Akawie, D. I. Basilius, and A. L. Landis. 1981. *Development of Non-glass Glazings and Surface Coatings*. Summary report, DOE/CS/35301-T2. Hughes Aircraft Co., USDOE.

Bird, R. E. 1982. *A Beer's Law Based, Simple Spectral Model for Direct Normal and Diffuse Horizontal Irradiance*. SERI-TR-215-1781. Golden, CO: Solar Energy Research Institute.

Bird, R., and R. Hulstrom. 1982. *Terrestrial Solar Spectral Data Sets*. SERI/TR-642-1149. Golden, CO: Solar Energy Research Institute.

Blaga, A. 1975. Durability of GRP composites. Batiment International/*Build. Res. Prac.* 1: 10.

Blaga, A., and R. S. Yamasaki. 1976. Surface microcracking induced by weathering of polycarbonate sheet. *J. Material Science* 11: 1513.

Blaga, A., and R. S. Yamasaki. 1977. Outdoor durability of a common type (tetachlorophtate acid based) fire retardent glass fiber-reinforced polyester (GRP) sheet. *RILEM Mat. Struct.* 10(58): 197.

Blaga, A. 1978. Use of plastics in solar energy applications. *Solar Energy* 21: 331.

Brauman, S. K., D. B. MacBlane, and F. R. Mayo. 1983. Reactivity of polymers with mirror materials; In *Polymers in Solar Energy Utilizationr*, C. G. Gebelein, D. J. Williams, and R. D. Deanin, eds. ACS Symposium Series 220. Washington, DC: American Chemical Society, pp. 125–142.

Brydson, J. 1975. *Plastic Materials*. London: Butterworth.

Carroll, W. F., and P. Schissel. 1980. *Polymers in Solar Technologies: An R&D Strategy*. SERI/TR-334-601. Golden, CO: Solar Energy Research Institute.

Cathro, K. J., D. C. Constable, and T. Solaga. 1981. Durability of porous silica antireflection coatings for solar collector cover plates. *Solar Energy* 27(6): 491–496.

Coulson, K. L. 1975. *Solar and Terrestrial Radiation*. New York: Academic Press.

Davis, A., and D. Sims. 1983. *Weathering of Polymers*. New York: Applied Science Publishers.

Deffenbaugh, D. M. 1984. The performance of parabolic trough solar collectors. M.S. thesis. Trinity University.

de Winter, F. 1980. Evaluation of hail protection needs and devices for solar flat plate collectors. In *Proc. Ann. DOE Active Solar Heating and Cooling Contractors Review Meeting*. Incline Village, NV: USDOE.

Dickman, W. C. 1976. *Performance of Sohio Solar Water Heating Systems Using Large Area Plastic Collectors*. SAND 1038-76. Livermore, CA: Lawrence Livermore Laboratory.

Duffie, J. A., and W. A. Beckman. 1980. *Solar Engineering of Thermal Processes*. New York: Wiley.

Edlin, F. E. 1958. Plastic glazing for solar energy absorption collectors. *Solar Energy* 2: 3.

Erbs, D. G. 1984. Models and applications of weather statistics related to building heating and cooling loads. Ph.D. thesis. Madison, WI: University of Wisconsin-Madison.

Fafco, Inc. 1980. *A Coaxial Extrusion Conversion Concept for Polymeric Flat Plate Solar Collectors*. Final Report. USDOE.

Fleagle, R. G., and J. A. Businger. 1980. *An Introduction to Atmospheric Physics*. New York: Academic Press.

Fthenakis, V. M., and R. W. Leigh. 1984. The value of improvements in the absorbing and glazing surfaces of solar devices. *Solar Energy* 32: 367.

Garg, H. P. 1974. Effect of dirt on transparent covers in flat plate collectors. *Solar Energy* 15: 299.

Gerich, J. W. 1978. An inflated cylindrical solar concentrator. In *Proc. AS/ISES 1978 Annual Meeting* 2.1: 889

Gordon, N. R. 1979. *Plastic Honeycomb for Solar Coverplates*. PNL-2953. Richland, WA: Pacific Northwest Laboratories.

Guillet, J. E., A. C. Somersall, and J. W. Gordon. 1983. New approach to the prediction of photooxidation of plastics in solar applications. In *Polymers in Solar Energy Utilization*, C.G. Gebelein, D. J. Williams, and R. D. Deanin, eds. ACS Symposium Series 220. Washington, DC: American Chemical Society, pp. 217–230.

Hampton, H. L., and M. A. Lind. 1978. *Weathering Characteristics of Potential Solar Reflector Materials: A Survey of the Literature*. PNL-2824/UC-62. Richland, WA: Pacific Northwest Laboratories.

Hatfield, J. L., P. J. Smietana, Jr., J. J. Carroll, R. G. Flocchini, and R. H. Hamilton. 1981. *Solar Energy, Implementation of a Research and Training Site at Davis, California*. Annual report, DOE grant DE-FG03-79ET20187.

Herzenberg, S., and R. Silberglitt. 1981. *Materials Research Recommendations to Improve the Performance and Durability of Solar Heating and Cooling Systems*. USDOE.

Hulstrom, R. L. 1981. *Solar Energy Meteorological Research and Training Site Program*. SERI/SP-642-947. Golden, CO: Solar Energy Research Institute.

JPL. 1982. *Regional Applicability and Potential of Salt-Gradient Solar Ponds in the United States*, vol. 2. DOE/JPL-1060-50. Pasadena, CA: Jet Propulsion Laboratory.

Kacoroski, C. C. 1984. Atmospheric turbidity over San Antonio, Texas: July 1981–December 1983. M.S. thesis. Trinity University.

Koller, L. R. 1965. *Ultraviolet Radiation*. New York: Wiley.

Landis, A. L., and A. J. R. Mosher. 1979. *Development of Non-glass Glazings and Surface Coatings*. Technical report. Hughes Aircraft Co., USDOE.

Landstrom, D. K., S. G. Tablert, G. H. Strickford, R. D. Fischer, and R. E. Hess. 1978. *Development of a Low-Temperature, Low-Cost, Black Liquid Collector*. ALO/4097-1.

Lin, R. J. H., and P. B. Zimmer. 1977. *Optimization of Coatings for Flat Plate Solar Collectors.* C000-2930-12.

Lind, M. A., and L. E. Ault. 1978. *Summary Report of the Solar Reflective Materials Technology Workshop.* PNL-2763/UC-62. Richland, WA: Pacific Northwest Laboratories.

Lind, M. A., and J. S. Hartman. 1980. Natural aging of soda-lime-silicate glass in a semi-arid environment. *Solar Energy Materials* 3: 81–95.

Löf, G., and R. R. French. 1979. Hail resistance of solar collectors with tempered glass covers. In *Pre-Conf. Proc. of the Sec. Solar Heating and Cooling Systems Operational Results Conference.*

Luck, R. M., and M. A. Mendelsohn. 1983. The reduction of solar light transmittance in thermal solar collectors as a function of polymer outgassing. In *Polymers in Solar Energy Utilization,* C. G. Gebelein, D. J. Williams, and R. D. Deanin, eds. ACS Symposium Series 220. Washington, DC: American Chemical Society, pp. 81–99.

Manning, R. A., T. P. Wisneski, and E. E. Clark. 1983. *Conceptual Study of a Salt-Gradient Solar Pond in the Crude Oil Production Industry.* Final report. Prepared for Texas Energy and Natural Resources Advisory Council. TENRAC/EDF-086.

Mark, H. F., N. G. Gaylord, and N. M. Bikales. 1964. *Encyclopedia of Polymer Science and Technology,* vol. 1. New York: Wiley Interscience, p. 304.

Mark, H. F., N. G. Gaylord, and N. M. Bikales. 1970, 1971. *Encyclopedia of Polymer Science and Technology* New York, NY: Wiley Interscience, vol. 13, p. 654; vol. 14, p. 522.

Martino, R. 1976. Run for the sun: Solar heating opens a vast new construction market. *Mod. Plast.* 53: 52.

Masterson, K. D., and M. A. Lind. 1983. *Matrix Approach for Testing Mirrors—Parts 1 and 2.* SERI/TR-255-1504. Golden, CO: Solar Energy Research Institute.

McKellar, J. F., and N. S. Allen. 1979. *Photochemistry of Man-Made Polymers.* London: Applied Science, Ltd.

Morris, V. L. 1982. *Final Report Solar Collector Materials Exposure to the IPH Site Environment.* SAND 81-7028. Also Deffenbaugh, D. F. 1984. The performance of parabolic trough solar collectors. M.S. thesis, Trinity University.

Murphy, L. M. 1981. *An Assessment of Existing Studies of Wind Loading on Solar Collectors.* SERI/TR-632-812. Golden, CO: Solar Energy Research Institute.

Nelson, E. V., C. S. Adams, and A. H. McLeod. 1979. *Development of a Low-Cost Solar Panel Using Laminated Polymer Films.* ALO/4121-2.

Neidlinger, H. H., and P. Schissel. 1985. *Polymer Synthesis and Modification Research During FY 1985.* SERI/TR-255-2590. Golden, CO: Solar Energy Research Institute.

Parkyn, B., ed. 1970. *Glass Reinforced Plastics.* London: Iliffe Books, ch. 6.

Rainhart, L. G., and W. P. Schimmel, Jr. 1975. Effect of outdoor aging on acrylic sheet. *Solar Energy* 17: 259.

Ranby, B., and J. F. Rabek. 1975. *Photodegradation, Photo-Oxidation and Photostabilization of Polymers.* London: Wiley.

Rao, C. R. N., and T. Takashima. 1985. Measured and computed values of clear-sky ultraviolet irradiances at the South Pole. *Solar Energy* 34(4/5): 435–437.

Raviv, A. June 1984. Personal communication. Yavne, Israel: Ormat Turbines, Ltd.

Robinson, N. 1966. *Solar Radiation.* Amsterdam: Elsevier.

Rogers, C. E. 1983. Effects of photodegradation on the sorption and transport of water in polymers. In *Polymers in Solar Energy Utilization,* C. G. Gebelein, D. J. Williams, and R. D. Deanin, eds. ACS Symposium Series 220. Washington, DC: American Chemical Society, pp. 231–242.

Roth, E. P., and R. B. Pettit. 1980. The effect of soiling on solar mirrors and techniques used to maintain high reflectivity. In *Solar Materials Science*, L. E. Murr, ed. New York: Academic Press, ch. 6.

The Royal Society of Chemistry. 1967. *Photochemistry*. (Series published annually since 1967.) London.

Rughunatheen, K. 1981. Thermal cycling testing of selective absorber surfaces. Presented at SEIA Annual Conference, Las Vegas, NV.

Saunders, A. P. 1972. *Plexiglass DR Thermal Stability Test*. National Center for Energy Management and Power, NTIS.

Sherr, A. E. 1972. A bright future for glazing plastics. *SPE J*. 28: 24.

Shettle, E. P., and R. W. Fenn. 1979. *Models for the Aerosols of the Lower Atmosphere and the Effects of Humidity Variations on Their Optical Properties*. AFGL-TR-79-0214. Paper no. 676.

Sparkes, H. R., and K. Raman. 1978. Lessons learned on solar system design problems from the HUD Solar Residential Demonstration Program. In *Proc. of the Solar Heating and Cooling Systems Operational Results Conference*. Colorado Springs, CO. SERI/TP-49063. Golden, CO: Solar Energy Research Institute.

Tortorello, A. J., and R. E. Wolf. 1979. Outgassing studies of some solar absorber coatings. In *Proc. Sec. Ann. Conf. on Absorber Surfaces for Solar Receivers*. SERI/TP-49-182. Golden, CO: Solar Energy Research Institute.

Ward, D. S., H. S. Oberoi, and S. D. Weinstein. 1982. *How to Solve Material and Design Problems in Solar Heating and Cooling*. DE/NBM-3006116.

White, J. S. 1977. Weatherability of fiberglass solar collector covers. *Polymer News* 3: 239.

Wilhelm, W. G., and J. W. Andrews. 1982. *The Development of Polymer Film Collectors: A Status Report*. BNL-51582. Brookhaven, NY: Brookhaven National Laboratory.

Wilhelm, W. G. 1983. Polymer film and laminate technology for low-cost solar energy collectors. In *Polymers in Solar Energy Utilization*, C. G. Gebelein, D. J. Williams, and R. D. Deanin, eds. ACS Symposium Series 220. Washington, DC: American Chemical Society, pp. 27–38.

Wischmann, K. B. 1983. Protective coatings and sealants for solar applications. In *Polymers in Solar Energy Utilization*, C. G. Gebelein, D. J. Williams, and R. D. Deanin, eds. ACS Symposium Series 220. Washington, DC: American Chemical Society, pp. 115–124.

Zerlaut, G. A. 1980. Fundamental material considerations for solar collectors. In *Solar Energy Technology Handbook, Part A. Engineering Fundamentals*, W. C. Dickinson and P. N. Cheremisinoff, eds. New York: Marcel Decker, ch. 13.

II ENERGY STORAGE FOR SOLAR SYSTEMS

14 Overview

C. J. Swet

14.1 Scope

A large number of existing and emerging energy storage technologies are potentially suitable for solar energy conversion systems. Among these storage technologies are

electrochemical storage in batteries;

chemical storage as hydrogen fuel;

mechanical storage in flywheels, compressed air, and pumped hydro;

superconducting magnetic energy storage;

thermal energy storage (TES) in sensible heat, latent heat, and reversible chemical reactions.

Any of them might plausibly be used in at least one application of solar thermal energy (photothermal) conversion processes. For example, dedicated storage for a solar thermal electric power plant might be in the form of TES positioned between the solar collector and the heat engine, or of a flywheel between the engine and the electrical generator, or possibly of batteries between the generator and the load. Another solar thermal power plant, without dedicated storage might feed directly to a regional electrical grid that contains compressed air, pumped hydro, or superconducting magnetic energy storage serving the entire grid. Thus it may be legitimate, although perhaps digressive and certainly cumbersome, to examine in this book all technologically credible storage options for solar thermal systems.

More realistic considerations militate for a narrower technological scope. Some of these conceivably applicable energy storage technologies are intended mainly for nonsolar applications such as industrial processes, utility load management, or vehicle propulsion. Although their use in certain kinds of solar thermal systems may be technically feasible, it is seldom found to be economically attractive. Others are economical only in capacities that greatly exceed the storage requirements of presently contemplated solar thermal systems. Typically they serve electrical grids, are located at a distance from the grid-feeding solar thermal power plant, and are charged primarily by nonsolar energy sources. All of these minimally solar associated storage technologies are documented elsewhere (PSE&G 1976; LANL 1981; DOE 1977–1983). Of the technologies that realistically can be considered for dedicated storage in solar thermal systems, only chemical energy storage (as thermally or thermochemically produced hydrogen fuel) and TES (for which the principal inputs and outputs are thermal energy) have been closely linked developmentally with such systems and are commonly viewed as solar heat technologies. Hydrogen production and storage, however, are examined elsewhere in this series. Therefore the technical scope of this part of the book can reasonably be limited to TES in its three generic forms: sensible heat, latent heat, and reversible thermochemical reactions.

Narrowing the focus to TES limits the scope but does not define it. The various applications and purposes of solar TES call for many different ranges of temperature, capacity (energy), charging and discharging rates (power), and duration. They also call for diverse modeling techniques, test and evaluation procedures, cost goals, and methods of integration into the overall system, as well as a broad selection of design concepts and materials within each of the three technological categories of TES. In this overview chapter TES is examined in an overall systems context to provide perspective for the presentations in subsequent chapters, as well as by technological categories. In it all of the principal applications and purposes, and their corresponding ranges of temperature, capacity, rates, and durations, are identified. Most of the other chapters, in keeping with the components and materials theme of this volume, treat storage as a distinct and separately evaluated subsystem except where it is necessary to explain the rationale of storage design concepts or operating strategies. In the special case of storage for passive solar buildings, where the distinctions between storage and other system elements tend to be blurred, the major treatment is in volume 7. Attention is paid here only to a few of the more novel materials and configurations for passive solar storage.

The balance between breadth of coverage and depth of treatment varies from chapter to chapter, depending largely on the professional experience and interests of each author and his access to recent work. Chapter 15, on storage concepts and designs, examines all three technological categories of TES over their entire temperature ranges, with an extensive coverage of the various storage media, containment and heat transfer methods, and subsystem design concepts. Each of the many items, though, is only briefly touched upon. Chapter 16, on analytical and numerical modeling, contains a comprehensive review of modeling techniques for low temperature storage in liquids, rock beds, and phase change units, but has little on the considerable body of analytical work on thermochemical storage and none on the fairly recent analyses of stratification in very high temperature molten salts with radiant heat transfer across the thermocline. Chapter 17, on test and evaluation, focuses primarily on ASHRAE and NBS methods for water, rock bed, and phase change storage units in residential applications. Chapter 18, on research and development, deals exclusively with R&D sponsored by DOE on low temperature sensible heat, latent heat, and thermochemical storage for building applications. Chapter 19, on issues and opportunities, is intended to identify important things that remain to be learned and applied about solar TES. It is organized more by application than by technology, and the examined topics span the entire range of TES temperatures and technologies, heavily emphasizing the overall systems aspects of storage.

Costs are examined quantitatively in only three chapters. Chapter 15 presents representative prices of a number of TES materials and components. Chapter 18 reports the cost results of two R&D projects out of approximately eighty reported. Chapter 19, in order to illustrate the importance of its selected topics, heavily emphasizes both subsystem costs and the effect of TES cost and performance on overall system economics.

14.2 Purposes and Strategies of Solar TES

It is well known that the basic need for solar TES stems from the intermittent, variable, and often poorly predictable availability of solar radiation. Less widely understood are the specific purposes and strategies of TES that are linked with certain applications, overall system characteristics, and operating philosophies. Principal among these are buffering, delivery period displacement, delivery period extension, and yearly averaging, as illustrated in figure 14.1.

Buffering (figure 14.1a) uses a very small thermal store, usually intended only to smooth transients in the solar thermal input, bridging brief periods of reduced insolation due to cloud passage. It finds use in solar power and process heat installations that serve primarily as "fuel savers", operating in parallel with conventionally fueled heat sources and contributing to the thermal input only during daylight hours. By masking the solar interruptions, it reduces the frequency of large and rapid, hence inefficient, fluctuations in firing rates. Buffer stores also can be attached to small dish-mounted engine generators that have no backup heat source, to reduce the number of start–stop cycles and consequently to minimize the engine replacement frequency. In commercial space-heating applications, where only daytime heating is required, the thermal mass of conventional building structures often suffices as a buffer store to prevent excessive swings in indoor temperature. In some applications buffer storage is used to accommodate fluctuations in the load rather than in the solar source. An example is the use of "coolness storage" in solar thermal space-cooling systems, in order to absorb variations in the cooling load and consequently reduce the chiller-cycling frequency. Because

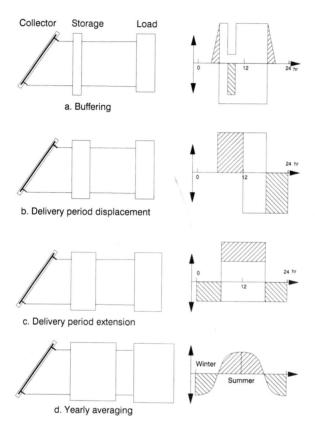

Figure 14.1
Purposes of storage.

of its small capacity (typically less than one-hour duration of delivery at full load), buffer storage has little effect on either the solar fraction or the required solar collection area. (Note that "storage capacity" is expressed here in terms of discharge duration capability at full load rather than in units of energy, in order to avoid irrelevant considerations of the overall system size. This should not be confused with the use of "storage duration" to indicate the length of time during which unused thermal energy is "bottled up" in storage.)

Delivery period displacement (figure 14.1b) involves the use of a larger store that shifts some or all of the energy delivery from the hours of available sunlight to a later period of greater energy demand or value. In a space-cooling system it may store heat collected in the morning for afternoon or evening use by an absorption chiller, thus augmenting or extending the chiller output during the period of highest cooling demand. In a solar thermal power plant it may store much of the heat produced early in the day in order to delay electrical power generation until it is needed to meet peak loads later in the day. Storage for this purpose does not necessarily increase either the solar fraction or the required collection area beyond no-storage values, and the design charge/discharge rate ratio may be more or less than unity depending on the specific scheduling requirements. Typical capacities are three to six hours at full load.

Delivery period extension (figure 14.1c) uses a still larger store that enables the solar source to meet system needs both during and beyond the normal hours of sunlight, sometimes including one or more subsequent days of low insolation. It always increases the solar fraction and the required solar collection area beyond no-storage values, and the design charge/discharge rate ratio always exceeds unity. This usually is called *diurnal storage*, even when the capacity exceeds overnight requirements, and is commonly encountered in space heating applications. Typical capacities are 8 to 24 hours.

Yearly averaging storage (figure 14.1d) is intended to maximize the solar collector utilization, thereby minimizing its size and cost, by storing its excess heat production in seasons of low load and/or high insolation. It also is known as annual cycle storage, seasonal storage, trans-seasonal storage, and inter-seasonal storage. In most cases the store is just large enough to accept all of the annually produced excess collector heat output without dumping. Often it has the additional purpose of achieving extremely high solar fractions, which in large installations frequently can be achieved more economically than with diurnal storage (see treatment of relative costs in chapter 19). Sometimes it has the goals of unity solar fraction and elimination of the need for a backup heat source. Although large scale space heating (Chuard, 1983) is its usually contemplated application, it has been investigated for crop drying and solar thermal power and has been suggested for irrigation pumping (see chapter 19). The design charge/discharge rate ratio always is less than unity.

14.3 Technical Approaches and Terminology

This cursory presentation of solar TES technologies is intended mainly to introduce nonspecialists to the basic concepts and techniques that later will be examined more fully and to familiarize them with some specialized terminology.

The characterizing feature of TES, which distinguishes it from other kinds of energy storage, is that its principal inputs and outputs are thermal energy, either as "hotness" or as "coolness" i.e., as a heat source or sink. The three basic technical approaches to TES, which may be applied singly or in combinations, are sensible heat storage, latent heat storage, and storage as the potential heat of recombination in reversible thermochemical reactions.

14.3.1   Sensible-Heat Storage

Thermal energy can be stored in the sensible heat (measurable temperature change) of substances that experience a change in internal energy, the stored quantity being the product of mass, temperature change, and mean specific heat capacity over that change, minus losses. The stored energy may be as "hotness" due to a temperature increase above some baseline value (Baylin 1981), or as "coolness" due to temperature reduction below ambient (General Electric 1977). It may be stored in a liquid medium (the material in which thermal energy is stored is commonly called the storage *medium*), a solid medium, or in some combination of liquid and solid media (a *dual medium*).

Liquid media, listed in ascending order of temperature suitability, are water, petroleum and synthetic oils, molten salts, liquid metals, and molten glassy slags. Water, oils, and the lower temperature (molten nitrates) salts can maintain *natural stratification* in favorably configured vessels, separating the upper hotter liquid from the lower cooler liquid by their differential densities. The horizontal hot/cold interface zone, called a *thermocline*, moves downward as hot charging liquid (heated by the collector) enters at the top and cooler liquid leaves the bottom; during discharge the hotter liquid leaves the top, cooler liquid enters the bottom, and the thermocline moves upward. This permits the use of a single full tank rather than separate "hot" and "cold" tanks that are alternately full and empty. Natural stratification is not possible in liquid metals because of their high thermal conductivity, or in very high temperature molten salts and glassy slags because of radiant heat transfer across the thermocline. Storage in stratified water and oil has been extensively modeled, and ASHRAE test and evaluation procedures are available for residential hot water storage units.

Solid media, listed in ascending order of temperature suitability, are concrete masonry, rocks, metals (principally cast iron), sand, and refractories. Typically they are immobile and exchange heat with moving air or other gases, an exception being irradiated collector/storage walls that discharge primarily by natural convection. In packed bed configurations natural stratification with moving thermoclines can be maintained in a manner roughly analogous to that in liquid tanks, with the added potential capability of horizontal stratification. In monolithic media the thermocline maintenance and motion are controlled by appropriate combinations of thermal path length and conductivity. Rock beds have been extensively modeled, and ASHRAE and NBS test and evaluation procedures for them are available.

Commonly used dual media include the combination of groundwater with sand, pebbles, and porous rock in *natural aquifers*, and with pebbles in excavated pits (*artificial aquifers*). For higher temperatures rock/oil combinations in constructed tanks are used, with most of the storage mass being in the relatively inexpensive solid medium. The presence of the solid medium enhances natural stratification by suppressing conduction and unwanted motion of the liquid, which usually is also the heat transport and transfer fluid. The thermal performance of aquifers and oil/rock units has been extensively modeled and evaluated by testing.

14.3.2   Latent-Heat Storage

Thermal energy can be stored nearly isothermally in some substances as the latent heat of phase transition: as heat of *fusion* (solid–liquid transition), heat of *vaporization* (e.g., liquid–vapor), or heat of *solid–solid crystalline phase transformation*. All such substances are *phase change materials* (PCMs), but unless otherwise stated the term commonly denotes heat of fusion materials. Often the charging temperature is considerably higher than the transition temperature, which is higher than the delivered temperature, which introduces a substantial sensible-heat storage component.

Generic types of PCMs, listed in ascending order of temperature suitability (with considerable overlap) are organic materials such as clathrates, paraffins and polymers; salt hydrates; inorganic compounds and eutectic mixtures; and metals and alloys. Although salt hydrates are commonly viewed as heat of fusion PCMs, their thermal energy actually is stored mainly as heat of solution (rehydration of the thermally dehydrated salt), which is properly described as a reversible thermochemical reaction. For some of them special measures must be taken to induce *nucleation* (prevent *supercooling*) and to avoid *phase segregation* due to *incongruent melting*.

PCMs can be stored in bulk, with indirect heat exchange (HX) by means of an embedded exchanger or direct heat exchange by means of an immiscible heat transfer–transport fluid (Michaels 1982). They can be *encapsulated* in metal or plastic containers, with many such being immersed in a liquid bath or exposed to a gas stream for heat transfer. The capsules may also be in the form of irradiated tubes or panels in passive solar applications. Unencapsulated pellets of form-stable cross-linked polymers may constitute a packed bed with direct heat transfer to a flowing gas or immiscible liquid, storing heat of fusion without melting or agglomeration (Baylin 1981). Granules of organic or solid–solid PCMs can be dispersed (*microencapsulated*) in building material such as concrete or wallboard, and individually encapsulated (*macroencapsulated*) PCMs may be inserted or embedded in such materials in order to add a latent heat component to the otherwise sensible heat solid storage medium (Swet 1980). Many techniques have been devised for modeling the moving phase boundary in PCMs, and ASHRAE test and evaluation procedures are available for residential latent-heat storage units.

14.3.3 Thermochemical Storage

Heat added to certain substances can alter them chemically, remain stored in them as nonthermal heat of potential chemical recombination and then be released as thermal energy by their reconstitution. The alteration can be *thermal decomposition* of a compound into two or more recombinable species, as in the high temperature breaking down of calcium carbonate into calcium oxide and carbon dioxide gas (Smith 1979). At lower temperatures it may be the dehydration of a solid salt hydrate into water vapor and a lower hydrate or the anhydrous salt (Bakken 1981), or it may be the concentration of a weak solution by boiling off the solvent (Rocket Research 1982). Typically the driving (charging) temperature is substantially higher than the discharge temperature; when an ambient heat source is added, such systems can become *thermochemical heat pump storage systems* (Swet 1981).

The heats of reaction typically are about an order of magnitude greater than the sensible heat in the heated stored reactants; therefore it is possible to store thermal energy thermochemically for indefinitely long periods of time with little loss. However, there seldom is economic justification for this capability, and the primary advantages of thermochemical storage usually lie in its high storage density and opportunities for heat pumping in both heating and cooling applications. Several varieties of thermochemical heat pump storage systems have been modeled and tested; higher temperature systems have been analyzed but tested only on the component level. There are not standard test and evaluation procedures.

14.4 TES Applications

The various applications of photothermal conversion, and the variety of system characteristics that have evolved for each application, provide a useful framework in which to examine solar TES technologies (Anon 1979). Organization of the material by overall system characteristics improves understanding of why the many different TES concepts, configurations, and materials were developed, allowing the more detailed technical treatments in subsequent chapters to concentrate more on what was done than on the systems considerations that motivated the doing.

Apart from solar fuels and chemicals production, which is not considered here, there are four broad categories of solar thermal applications:

Buildings (space-heating and cooling and domestic hot water)

Process heat (industrial and agricultural)

Cooking

Power generation (electrical and mechanical, including water pumping)

Tables 14.1 through 14.3 present the relationships between these applications and solar TES technologies. They divide each application into subcategories that define more precisely the diverse system characteristics with which the TES must be compatible, and for each subcategory they present appropriate TES concepts, designs, and materials. To minimize clutter, the listed TES technologies are defined only to the level needed for comparison with the material in section 14.3 and with their fuller exposition in later chapters.

14.5   Retrospective

This overview is the most recently written of the six chapters comprising part II. They span a writing period of approximately three years, with several midstream changes in authorship and corresponding shifts in points of view. On the whole, all of the completed chapters are seen to complement each other well, without excessive duplication or too many topics fallen through cracks. However, since some of these chapters received independent final approval years apart, there now become visible deficiencies that no amount of in-process effort by the editor could have prevented. Two of the more noteworthy blemishes are noted here.

In chapter 15, subsection 15.2.1, it was suggested that natural stratification in chilled water tanks may be unfeasible because of the relatively small amount of buoyant effect. After that chapter had been finally approved, this author learned of recent analytical and experimental work (Wildin 1985) on diffuser design that seems to prove conclusively that natural stratification is indeed feasible for chilled water in certain tank geometries. This revelation reduces somewhat the economic incentive to develop improved phase change coolness storage systems.

It is regrettable that the R&D reported in chapter 18 was limited to DOE-sponsored work on TES for buildings applications. A more penetrating reexamination of the reported activity shows a substantial fraction of it to be

Table 14.1
TES options for building applications

Space heating (only)
Active solar
Residential scale   Air-heating collectors
(diurnal storage)   60°C   Rock beds (sensible)
                           30°–40°C   Encapsulated PCM (latent)
                                         Bulk solid/solid PCM (latent)
                           Liquid-heating collectors
                           50°C   Stratified water in thermocline tank (sensible)
                                         Moistened replaced earth in lined pit, indirect HX
                                         (sensible)
                           30°–40°C   Bulk PCM with indirect HX (latent)
                                         Bulk PCM with direct HX and immiscible fluid (latent)
                                         Encapsulated PCM
                           Large-scale collectors (yearly averaging storage)
                           50°–90°C   Water in tanks, excavated pits, mined caverns (sensible)
                                         Natural and artificial aquifers (sensible)
                                         Undisturbed earth, clay, rock (sensible)
Passiver solar   30°–45°C   Water walls (sensible)
                                         Masonry Trombe walls (sensible)
                                         Encapsulated PCM in irradiated tubes or panels
                                         (latent)
                                         Encapsulated PCM in architectural components
                                         (latent)
                                         Dispersed PCM in architectural components (latent)
                                         Dispersed solid/solid PCM in architectural com-
                                         ponents (latent)

Space cooling (only)
Closed cycle systems   Hot-side storage for Rankine and liquid absorption chillers
                           90°–120°C   Water in vented or pressurized tanks (sensible)
                                         Bulk PCM with indirect HX (latent)
                                         Bulk form-stable cross-linked polymer PCM, direct
                                         HX (latent)
                           Cold-side storage for Rankine, liquid absorption, and zeolite chillers
                           5°–10°C   Water in naturally stratified tank (sensible)
                                         Water in multiple, partitioned, or flexibly
                                         Divided tanks (sensible) encapsulated PCM (latent)
                                         Bulk PCM with indirect HX (latent)
                                         Bulk PCM with direct HX and immiscible fluid (latent)
                           Open cycle systems (desiccant chillers)
                           90°C   Rock beds for solid adsorption systems (sensible)
                           60°C   Strong salt solution (thermochemical)

Space heating and cooling (thermochemical heat pumps with reactant storage)
                           100°–130°C   Driving temperature
                                         Intermittent solid/vapor systems
                                         Continuous or intermittent liquid/vapor systems

Domestic hot water (DHW)
Naturally stratified   Supplementary PCM
deliverable water in   55°C   Encapsulated
thermocline tank   Bulk, between double walls of water tank (latent)
(sensible)

Table 14.2
TES options for process heat and cooking

Industrial and agricultural process heat (Kriz 1983)		
Process hot water	60°–90°C	Deliverable water in thermocline tank or two tanks (sensible)
Low pressure steam	100°–130°C	Pressurized water with indirect HX (sensible)
		Petroleum oil in thermocline tank or two tanks (sensible)
		Pressurized water delivered as steam (latent)
		Continuous liquid–vapor thermochemical heat pump (T/C)
"100 pound" steam	170°C	Petroleum oil in thermocline tank or two tanks (sensible)
		Pressurized water delivered as steam (latent)
High pressure saturated steam (Dubberly 1983)	300°C	Petroleum oil in thermocline tank or two tanks (sensible)
		Petroleum oil/rocks (dual medium) in thermocline tank (sensible)
		Encapsulated PCM with evaporative HX (latent)
		Bulk PCM with indirect HX (latent)
		Pressurized water above ground or underground (latent)
Crop drying (air heating)	50°–70°C	Rock beds (sensible)
		Liquid desiccant, yearly averaging (thermochemical)
Solar cooking (Alward 1982)		
	100°–200°C (hotboxes and ovens)	Heated stone and cooking vessel placed in "haybox" (sensible)
		PCM in oven
	150°–250°C (line-focusing collectors)	PCM under hot plate (latent)
	200°–300°C (point-focusing collector)	Solid–gas–solid container removed to cooking area (thermochemical)

Table 14.3
TES options for solar thermal power

Small power plants and water pumps		
Organic Rankine	100°C	Water in thermocline tank or two tanks (sensible)
	300°C	Petroleum oil in thermocline tank (sensible)
Steam Rankine with organic fluid receiver (Dubberly 1983)	375°C	Synthetic oil with trickle charge (sensible)
Dish mounted engine generators (buffer storage only) (Manvi 1981)		
Organic Rankine	400°C	Bulk PCM with indirect HX (latent)
Stirling and air Brayton	800°C	Bulk PCM with indirect HX (latent)
Advanced air Brayton	1,370°C	Graphite (sensible)
		Encapsulated PCM (latent)
Larger power plants (typically 3–8 h of storage)		
Steam Rankine with organic fluid receiver (Kearney 1985)	300°C (evaporation only)	Petroleum oil in thermocline tank or two tanks (sensible)
		Petroleum oil/rocks (dual medium) in thermocline tank (sensible)
Steam Rankine with water-steam receiver (Dubberly 1983)	300°C (evaporation stage)	Petroleum oil thermocline tank or two tanks (sensible)
		Petroleum oil/rocks (dual medium) in thermocline tank (sensible)
		Encapsulated PCM with evaporative HX (latent)
		Bulk PCM with indirect HX (latent)
		Bulk PCM with direct HX (latent)
		Pressurized water above ground or underground (latent)
	540°C (superheat stage)	Molten draw salt in thermocline tank or two tanks (sensible)
		Air/rocks (sensible)
		Bulk PCM with direct HX (also for evaporation stage) (latent)
		Solid or liquid decomposition (also for evaporation stage) (thermochemical)
Steam Rankine with molten draw salt receiver	540°C (evaporation and superheat stages)	Molten draw salt in thermocline tank or two tanks (sensible)
Steam Rankine with liquid metal receiver	540°C (evaporation and superheat stages)	Liquid sodium in one tank, mixed, buffer only (sensible)
		Liquid sodium in two tanks (sensible)
		Air/rocks (sensible)
Brayton with gas-cooled receiver	800°C	Refractory or cast-iron chequerworks in pressure vessel (sensible)
		Bulk PCM with indirect HX (latent)
		Solid or liquid decomposition (thermochemical)
Brayton with liquid-cooled receiver	800°C	VHT molten salt in two tanks (sensible)
		VHT molten salt/refractory (dual medium) in thermocline tank (sensible)
	1,100°C	Bulk glassy slag, liquid and solid bead storage, direct HX (sensible and latent)

of doubtful value and certainly of less than enduring interest; more selective reporting would have permitted the inclusion of comparably important higher temperature R&D, making for better balance with the wider scope of other chapters. Fortunately much of the high temperature storage R&D is covered elsewhere.

Annotated References

Alward, Ron. 1982. *Solar Cooker Manual.* Technical report T138. Brace Research Institute, Macdonald College of McGill University.

Anon. 1979. *Proc. Solar Energy Storage Options.* DOE CONF-790328.

Baken, K. 1981. System TEDIPUS, high capacity thermochemical storage/heat pump. *Papers Presented at the International Conference on Energy Storage Held at Bedford Hotel, Brighton, U.K. April 29–May 1, 1981.*

Baylin, F., and Merino, F. 1981. *A Survey of Sensible and Latent Heat Thermal Energy Storage Projects.* SERI/RR-355-456.

Chuard, P., and Hadorn, J. C. 1983. *Central Solar Heating Plants with Seasonal Storage. Heat Storage Systems: Concepts, Engineering Data and Compilation of Projects.* IEA A24389/3.

Dubberly, L. J., J. Gormley, W. Lang, D. Liffengren, A. McKenzie, and R. Porter. 1983. *Comparative Ranking of Thermal Storage Systems. Volume II Cost and Performance of Thermal Storage Concepts in Solar Thermal Systems,* Phase I. SERI/TR-631-1238.

General Electric CD&D. 1977. *Cool Storage Assessment Study.* EPRI EM-468.

Kearney, D., et al. 1985. Design and preliminary performance of solar electric generating station I at Daggett, California. In *Intersol 85 Proceedings of the Ninth Biennial Congress of the International Solar Energy Society,* pp, 1424–1428. E. Bilgen, and K. G. T. Hollands, eds. Pergamon Press. Contains a two-tank petroleum oil storage system for a steam Rankine power plant with distributed trough collectors containing the storage fluid. Fuller description of the storage system by W. Joseph and J. Hogsett in *Extended Abstracts* of the congress.

Kriz, T., C. Christensen, H. Gaul, J. Leech, A. Rabl, S. Sillman, C. J. Swet, and J. Ullman. 1983. *Thermal Energy Storage for Process Heat and Building Applications.* SERI/TR-231-1780.

Los Alamos National Laboratory. 1981. *Proc. Mechanical, Magnetic and Underground Energy Storage 1981 Annual Contractors' Review.* DOE CONF-810833. See also 1983 updates in CONF-830974 (listed earlier) on flywheels, compressed air, and magnetic storage.

Manvi, R. 1981. Dish-mounted latent heat buffer storage. In CONF-810940 (listed earlier).

Michaels, A. 1982. A review of latent heat storage technology. In *Solar Storage, Proc. 3rd SOLERAS Workshop,* J. Williamson and B. Khoshaim, eds. CONF-820312.

Public Service Electric and Gas. 1976. *An Assessment of Energy Storage Systems for Use by Electric Utilities.* EPRI EM-264.

Rocket Research Co. 1982. *Sulfuric Acid/Water Chemical Heat Pump/Chemical Energy Storage,* vols. 1–2. RRC-82-R-813. Brookhaven, NY: Brookhaven Laboratories.

Smith, R. D. 1979. *Chemical Energy Storage for Solar Thermal Conversion.* Prepared by Rocket Rsearch Co. for Sandia Labs. SAND79-8198.

Swet, C. J. 1980. Phase change storage in passive solar architecture. *Proc. 5th National Passive Solar Conf.,* AS/ISES, pp. 282–6.

Swet, C. J. 1981. Solar applications of thermochemical heat pumps—progress and prospects. *Proc. Solar World Forum,* vol. 1. New York: Pergamon.

U.S. Department of Energy. 1977–1983. *Proc. Thermal Energy Storage Contractors' Information Exchange Meeting.* (Titles vary sightly from year to year.) 1977: CONF-770955; 1978: 781142 and 781231; 1979: 791232; 1980: 801055; 1981: 810940; 1982; 1983: 830974. Annual reports on projects sponsored by the Storage Division.

Wildin, M., and Truman, C. 1985. A summary of experience with stratified chilled water tanks. In *ASHRAE Transactions 1985* vol. 91, pt 1.

15 Storage Concepts and Design

C. J. Swet

15.1 Introduction

Chapter 14 identified storage needs, functions, and applications, and volume 4 of this series examines various aspects of building energy storage in an overall systems context. This chapter identifies technical concepts and designs of storage subsystems that have evolved in response to the identified needs, functions, and applications, with minor attention to overall system implications. The emphasis is on concepts and designs that have evolved within the past decade, although many of these are variants of older storage techniques, such as water tanks and packed beds, that are widely used in nonsolar applications. The scope is limited to technical approaches that have been explored beyond the conjectural stage, although not necessarily to the point of ascertaining technical and economic feasibility. Some initially promising approaches that later were found to lack merit are included for cautionary purposes. Space constraints prevent anything approaching full characterization of each generic concept, or even bare identification of many specific designs that embody those concepts; omissions and relative extent of exposition are based on the author's sense of relative importance (present and potential), general familiarity, and comprehensibility. Readers must rely heavily on the many listed references.

15.2 Sensible-Heat Storage

Thermal energy can be stored as "hotness" or "coolness" in the sensible heat of various substances, the stored quantity being the product of mass, temperature change, and mean heat capacity $\bar{c}_p$, minus losses. Commonly cited generic advantages include proven reliability and ease of monitoring state of charge. Commonly used figures of merit are $\$/\bar{c}_p$ (for relative storage media cost), $\rho_{min}\bar{c}_p$ (volumetric heat capacity, where ρ_{min} is the minimum mass density, for relative container size), and thermal diffusivity (for relative size of heat exchange components).

15.2.1 Liquids

Advantages of sensible-heat storage in liquids include pumpability and high volume utilization (no voids other than ullage space). Their pumpability suits them for heat transport as well as storage, thereby facilitating heat exchange, reducing constraints on system geometry, and making possible positive methods of temperature separation such as diaphragms. Table 15.1 broadly characterizes representative media used in all-liquid sensible-heat storage subsystems or as the liquid component in mixed media subsystems (see subsection 15.2.3).

Water Water usually is the preferred liquid for temperatures between 0° and 100°C (32°–212°F). It is plentiful, essentially free, nontoxic, nonflammable, easily pumped, has excellent thermal properties, and is only mildly corrosive in the absence of oxygen. It is used mainly in building applications where storage subsystem concepts and designs divide conveniently into those most suitable for passive space heating, domestic water heating, diurnal storage in active space heating and cooling systems, and seasonal storage. Concepts for process heat storage in water tend to be similar to those for large domestic hot water systems and large diurnal space-heating installations.

SUBSYSTEMS FOR PASSIVE HEATING   Typically these are collector-storage units of the type known generically as *water walls*, installed in south walls behind glazing, as reviewed in volume 7. They also can be temperature-stabilizing vessels in either shaded or direct gain interior spaces, or they can serve as roof ponds. Table 15.2 characterizes some water wall storage container concepts and designs, of which one of the most promising is depicted in figure 15.1 (Jones 1984). Interior spaces can contain drums or cylinders such as those that comprise water walls, or cans, bottles, or other miscellaneous water containers (Bainbridge 1979). Roof ponds are reviewed in volume 7.

SUBSYSTEMS FOR DOMESTIC WATER HEATING   These subsystems are distinctive in that the storage medium also is the delivered product. The simplest containers are black plastic "pillows" that are filled with water in the morning, allowed to heat in the sun all day, then drained for evening use of the heated contents. They are commercially available in 5-gal (19-L) and 15-gal (57-L) capacities, costing (1982) about $20 and $33, respectively (Solar Components 1982). Vented steel tanks are used in gravity hot water systems or as combination draindown/preheat vessels, but in most buildings in the United States all of the storage elements are under mains pressure and conform generally to standards for conventional nonsolar hot water tanks (Skartvedt 1981). Systems aspects, stratification strategies, and representative tank and heat exchanger configurations are reviewed in volume 7.

DIURNAL SUBSYSTEMS FOR ACTIVE HEATING AND COOLING   These are constructed tanks with enough capacity to meet specified thermal loads for periods from overnight to several days. For residences they typically are single tanks of 2 to 20 m³ (70 to 707 ft³); for larger buildings individual tank volumes

Table 15.1
Liquid sensible-heat thermal storage media

Medium	Temperature range [°C (°F)]	Specific gravity	Specific heat	FOB cost [$/kg ($/lb)]	Remarks
Water	0/100 (0/212)	1.0	1.0	nil	2 atm at 120°C (248°F), 3 atm at 134°C (273°F)[b]
Organic fluids					
Caloria HT43 (petroleum)	<0/316 (<0/600)	0.69	0.66	1.06 (0.48)	Exxon product (representative)[b,c]
Therminol T-66 (modified terphenyl)	<0/343 (<0/649)	0.75	0.66	6.23 (2.83)	Monsanto product (representative) used only in R&D facilities[b,c]
Syltherm 800 (silicone)	<0/399 (<0/750)	0.66	0.50	13.55 (6.16)	Dow-Corning product (representative) used only with solid media[b,c]
Molten salts					
Hitec or HTS (NaNO$_2$, NaNO$_3$, KNO$_3$)	m.p. 142/540 (288/1,004)	1.68	0.37	1.34 (0.61)	Hitec (DuPont TM)[c,d]
Draw salt (NaNO$_3$, KNO$_3$)	m.p. 220/550 (428/1,022)	1.73	0.37	0.97 (0.44)	[c,d,f]
Mixed chlorides (Mg, Na, K)	m.p. 385/900+ (725/1,652+)	1.63	0.25	0.73 (0.33)	[b,c,e,f]
Mixed carbonates (Li, Na, K)	m.p. 400–700/900+ (752–1,292/1,652+)	~1.8	~0.4	0.6/2.2 (0.3/1)	Properties and cost depend on % Li[b,c,e,f]
Liquid metal					
Sodium	m.p. 78/760 (208/1,400)	0.96	0.30	2.42 (1.10)	Special alloys required[c]

a. All data approximated from Dubberly et al. (1980); 1980 prices.
b. Carbon steel okay.
c. Inert atmosphere blanket required.
d. Carbon steel okay up to 316°C (601°F).
e. Containment techniques and materials uncertain above 900°C (1,652°F).
f. Also used for heat of fusion storage (see table 15.6).

Table 15.2
Water wall storage components

Description	Approximate unit size	Approximate FOB cost [$/gal ($/m³)]	Note	Remarks
Horizontal steel drums, stacked	30 or 50 gal (0.1 or 0.2 m³)	0.1–0.5 (1979) (30–150)	a	Volume/projected area ratio much larger than needed
Flat steel tanks	Various	0.3–1.5 (1979) (90–450)	a	Usually custom built
Vertical FRP cylinders	12- and 18-in. (30- and 46-cm) diameter up to 10 ft (3 m) high	1 to 2 (1982) (300–600)	b	Transparent, optional dye in water for improved solar absorption
Vertical steel culverts	12-in. (30-cm) diameter and up	0.3 to 0.5 (1979) (90–150)	a	Corrugations aid convective heat transfer
Vertical PVC pipe	6- and 12-in. (15- and 30-cm) diameter	0.1 to 0.2 (1979) (30–60)	a	Poor heat transfer through walls
Molded PE module ("Crystalite")	72-in. (183-cm) thick, 4-ft (1-m) wide, 4- and 8-ft (1- and 2-m) high		c	Translucent, integral glazing
Acrylic "aquarium" ("Transwall")	7-in. (18-cm) thick, 52-in. (132-cm) wide, 23-in. (58-cm) high [30 gal (0.1 m³)]		d	Two 3/8-in. (0.9-cm) acrylic plates with immersed 3/8-in. (0.9-cm) gray acrylic absorber plate; approximately 15% light transmission
Molded PE module with tongue, stacked	See figure 15.1	R&D status	e	Functions as convective liquid diode to reduce night losses (see figure 15.1 for operating principle; development by One Design, Inc, and evaluation by LANL

a. Bainbridge (1979).
b. Solar Components (1984).
c. One Design, Inc. (1984).
d. McClellard et al. (1981).
e. Jones (1984).

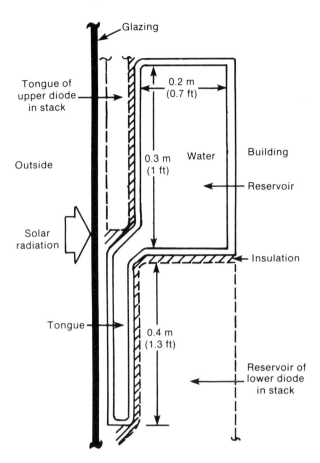

Figure 15.1
Stacked liquid convective diode for water wall. During the day convection in the tongue heats water in the reservoir; at night reverse convection is suppressed and losses are minimal. Source: Adapted from Jones (1984).

commonly exceed 100 m^3 ($3,500 \text{ ft}^3$) (roughly the upper limit for prefabrication and delivery in one piece), and multiple tanks are not uncommon. Table 15.3 broadly characterizes them according to function; systems aspects are reviewed in volume 6. Numerous configurations have been used or investigated for indoor, outdoor, and buried locations, for new and retrofit installations, and for factory and on-site assembly (Cole et al. 1980; Baylin 1979; Baylin and Merino 1981; Lee et al. 1979). Many are adaptations of concepts previously developed for nonsolar applications: buried concrete utility vaults, septic tanks, storage vessels of various sizes and geometries for water, oil, and chemicals, and the use of earth augering techniques [holes up to 15 ft (4.6 m) in diameter and 100 ft (30.5 m) deep] (Pickering 1976). A preinsulated fiberglass tank was developed for shipment in two telescoping sections to minimize volume (Hudson et al. 1981), and lightweight concrete tanks were developed for on-site casting in two or more bolted sections (Buckman et al. 1980). However, neither concept proved to be economically attractive. The least expensive options for residential uses at moderate temperatures are site-assembled membrane-lined structures such as that shown in figure 15.2 (Bourne 1981; Forbes 1981). Heat exchange with a nonfreezing collector fluid usually occurs by an immersed or external indirect heat exchanger; direct contact exchange with an immiscible fluid has also been demonstrated (Ward et al. 1978), but it is not economically competitive.

Natural stratification (theory reviewed in chapter 16 and systems aspects in volume 6) is best in tall, slim cylinders, and can be enhanced by diffusers. Nonconducting vertical concentric baffles have been shown analytically to maintain nearly plug flow at throughputs as high as one volume change per hour (Lin and Sha 1979). However, tall tanks cannot always fit within available space, and the benefits of stratification in space-heating systems are greatest, hence more easily maintained, at much lower collector flow rates. A functionally proven design based on low flow rates, low fluid velocities, and optimally located heat addition and removal is shown in figure 15.3 (van Koppen and Thomas 1978). In chilled water storage tanks with temperature swings between $0°$ and $10°C$ ($32°$ and $50°F$), there is much less buoyant effect (hence poorer natural stratification) than in space-heating storage tanks with swings between perhaps $40°$ and $70°C$ ($104°$ and $158°F$). Thus cool storage usually must rely on a single large tank with mixed temperature, two or more smaller tanks with mixed but different temperatures, or design approaches such as that illustrated in figure 15.4 (Tamblyn 1980). In small cool storage subsystems the complexities of mix prevention can seldom be economically justified.

Table 15.3
Diurnal storage water tanks for active heating and cooling

Function	Maximum temperature [°F (°C)]	Suitable materials	Remarks
Cold-side storage for absorption or vapor compression chiller	50 (10)	C, FRP, LF, LW, S, W	Temperature separation essential but natural stratification difficult; two tanks or positive separation may be favored; minimal insulation required
Evaporator-side storage for solar-assisted heat pumping	104 (40)	C, FRP, LF, LW, S, W	Large size may favor outdoor or buried location; minimal insulation required and can be uninsulated and ground coupled if buried
Storage for space heating	158 (70)	C, FRP, LW, S	Sometimes used also for hot- or cold-side storage in summer; moderate insulation required
Hot-side storage for absorption chiller	203 (95)	C, S	Dual tanks (one small for high temperature; one large for lower temperatures) may be justified in large installations; heavily insulated
Hot-side storage for vapor compression chiller	212+ (100+)	S	Pressurized, heavily insulated

Note: C = concrete with plastic linear—dense or foamed; block or poured; precast or cast on site. FRP = fiberglass reinforced polymer. LF = foamed polystyrene with plastic membrane liner—site assembled (Forbes 1981). LW = wood frame with plastic membrane liner—site assembled (Bourne 1981). S = steel, epoxy lined. W = wood stave or rectangular wood; insulated only on top; unlined unless periodically emptied.

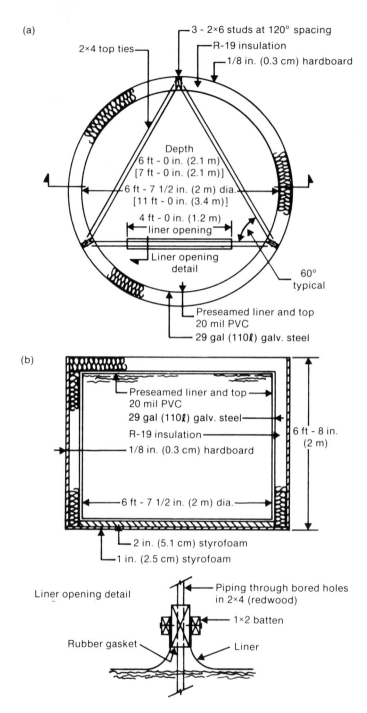

(a)

3 - 2×6 studs at 120° spacing

R-19 insulation

1/8 in. (0.3 cm) hardboard

2×4 top ties

Depth
6 ft - 0 in. (2.1 m)
[7 ft - 0 in. (2.1 m)]

6 ft - 7 1/2 in. (2 m) dia.
[11 ft - 0 in. (3.4 m)]

4 ft - 0 in. (1.2 m)
liner opening

Liner opening
detail

60°
typical

Preseamed liner and top
20 mil PVC

29 gal (110ℓ) galv. steel

(b)

Preseamed liner and top
20 mil PVC

29 gal (110ℓ) galv. steel

R-19 insulation

1/8 in. (0.3 cm) hardboard

6 ft - 8 in.
(2 m)

6 ft - 7 1/2 in. (2 m) dia.

2 in. (5.1 cm) styrofoam

1 in. (2.5 cm) styrofoam

Liner opening detail

Piping through bored holes
in 2×4 (redwood)

1×2 batten

Rubber gasket

Liner

Figure 15.2
Membrane-lined water tank: (a) plan; (b) section. Dimensions are for 1,500 (and 5,000) gal (6 and 19 m³). Source: Bourne (1981).

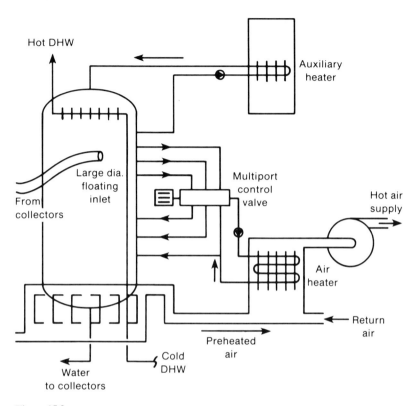

Figure 15.3
Exploitation of natural stratification in a water tank. Source: Adapted from van Koppen and Thomas (1978).

Representative designs and specifications for steel, FRP, and concrete tanks are in Cole et al. (1980). Cost estimation guides for conventional steel tanks and their lining, insulation, jacketing, supports, foundations, and heat exchangers are in Lawrence (1980).

SUBSYSTEMS FOR SEASONAL STORAGE   These subsystems have sufficient capacity to permit year-round use of the solar collectors by storing heat from summer to winter. (Seasonal storage may also be used to make winter-chilled water available for summer cooling, but this application will not be considered here since it is not strictly solar. Solar ponds represent yet another kind of seasonal storage in water—saturated salt solutions—but are viewed as collectors; see chapter 10 of this volume.) As discussed in chapter 19, seasonal storage usually can be justified economically only for large installations that require storage volumes of at least 1,000 m³ (3.5×10^4 ft³). A number of

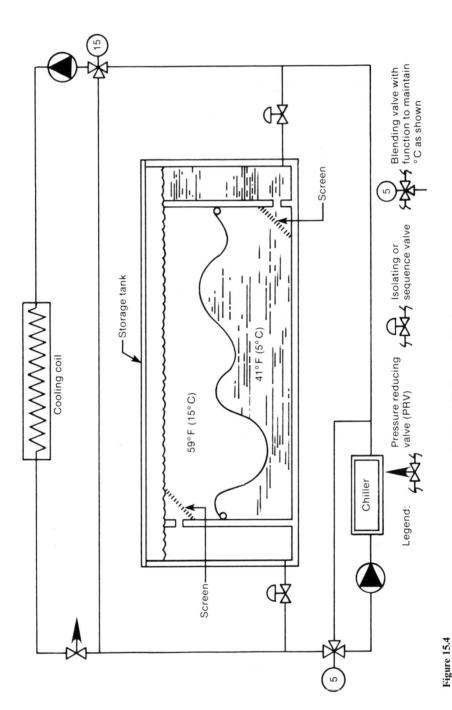

Figure 15.4
Floating membrane for positive temperature separation in a chilled water tank. The membrane
is of coated fabric. Source: Tamblyn (1980).

containment concepts have been used or explored (PNL 1981; Chuard and Hadorn 1983; Margen 1980); most of those concepts are also suitable for nonsolar heat sources. Those currently in use include steel and concrete tanks, excavated pits, and excavated rock caverns (figure 15.5). Among concepts still being explored are natural caverns, abandoned mines, and partitioned lakes (figure 15.6). Steel tanks currently are limited to about 10^5 m^3 (3.5×10^6 ft^3), above which circumferential stress makes the lower wall too thick for relieving on-site weld stress. A concept that would remove size constraints is being explored, based on a dish-shaped configuration in which the side walls experience only relatively small membrane stresses along their contour (Chuard and Hadorn 1983). The side walls have approximately the contour of a vertical quarter ellipse, being joined tangentially to the planar bottom plate and suspended from supports at the top. Estimated costs and suitable size ranges for these "water only" seasonal storage concepts are plotted in figure 19.3 in chapter 19. Aquifer storage, which is in mixed liquid/solid media, is discussed in subsection 15.2.3.

Organic Fluids Organic fluids are used as liquid-only storage media in Rankine systems for power generation (Hogsett n.d.; Leonard and Hunke 1981; Bacconnet et al. 1981; Koehne et al. 1981) and irrigation pumping (Matteo and Rafinejad 1980; Krivokapiche et al. 1983), and these fluids have been examined for process heat applications (Dubberly et al. 1981). Ordinarily only the relatively inexpensive petroleum-based oils can be justified; synthetic oils have been used only in research facilities (Leonard 1978) and for buffer storage in higher temperature collector fluid circuits (Leonard and Hunke 1981). They permit much higher storage temperatures at atmospheric pressure than does water, but their heat transfer properties are poorer, they are flammable, and they must be protected by inert blankets from exposure to air. Fluid replacement requirements due to thermal degradation can be substantial near the maximum operating temperatures listed in table 15.1, as shown in table 15.4. Table 15.4 also shows that these rates are lower for all-liquid (bulk) storage than when the fluid is in direct contact with solids (see subsection 15.2.3).

In all known installations the storage vessels are heavily insulated, vertical, cylindrical, steel tanks, either above grade or in excavated depressions, with temperature separation either by natural thermocline (stratification) or by uniform but different temperatures in two or more tanks. In principle the total tank volume of two- and three-tank configurations would be 2 and 1.5 times, respectively, that of a single thermocline tank of equal capacity, but the actual differences are narrowed by unusable volume in the thermocline tank due to

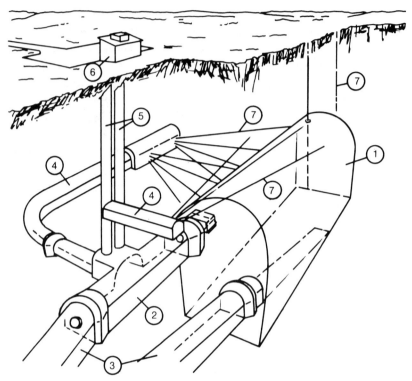

Three-dimensional sketch of the demonstration cavern

1 = Storage cavern
2 = Machine room
3 = Transport tunnel
4 = Research tunnel
5 = Access shaft
6 = Entrance building
7 = Research bore-holes

Figure 15.5
Seasonal hot water storage in an excavated rock cavern at Avesta, Sweden. Source: Martna (1981).

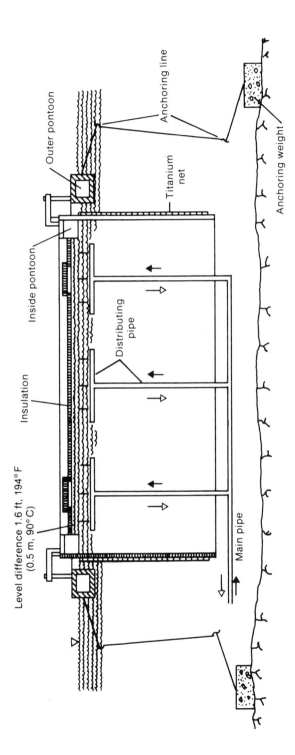

Figure 15.6
Seasonal hot water storage in a partitioned natural lake. Source: Margen (1980).

Table 15.4
Organic fluid loss rates due to thermal degradation (% by weight in 2,500 h)

	Caloria T43		Therminol T-66		Syltherm 800	
	Temperature [°F (°C)]	%	Temperature [°F (°C)]	%	Temperature [°F (°C)]	%
In bulk storage	601 (316)	7	606 (319)	0.65	750 (399)	38
	475 (246)	0.02	574 (301)	0.18	531 (277)	0.03
			538 (281)	0.042		
	Granite and sand		Granite and sand		Taconite	
In solid/liquid mixed media						
	601 (316)	27	601 (316)	5.7	750 (399)	56
	556 (291)	4.2	556 (291)	1.0	500 (260)	0.5
	475 (246)	0.06	475 (246)	0.02		

Source: Dubberly et al. (1983).

thickness of the thermocline and placement of diffusers. The natural thermocline method tends to be favored in smaller installations where containment costs dominate and soil loading permits high height/diameter (H/D) ratios. In operating systems H/D ranges from 1.75 for a 54-m^3 (1,908-ft^3) tank (Bacconnet et al. 1981) to 3.5 for a 189-m^3 (6,678-ft^3) tank (Hogsett n.d.). Multiple tanks may be preferable in larger systems where the storage medium cost is dominant, since all of the fluid is usable and lower H/D due to soil loading limits would make thermocline thickness a larger fraction of the tank volume. A two-tank subsystem has been installed at Daggett, California, containing 4,000 m^3 (140,000 ft^3) of HT43 with a temperature swing between 304° and 240°C (579° and 464°F) (Hogsett n.d.). However, there is a unit in Getafe, Spain, with only 23 m^3 (813 ft^3) of oil in three 12.5-m^3 (442-ft^3) tanks (Koehne et al. 1981). It has been suggested that sunken concrete pits, insulated by the earth, may be an economical two-tank configuration in some process heat applications (Kriz et al. 1983).

Molten Salts   Molten draw salt is widely viewed as a superior receiver/ storage fluid for high temperature Rankine power systems. Cost studies for 100-MW$_e$ systems show that it also is superior as the superheat storage medium with water/steam receivers (Dubberly et al. 1983) and for both evaporation and superheat storage with liquid sodium receivers (Dubberly et al. 1981). All known molten salt installations are of the two-tank type, with heavily insulated vertical cylindrical tanks (Munoz Torralbo 1981; Tanaka 1980), this configuration being preferred largely because the heavy salt causes soil loading limits to be reached with low H/D ratios (Scott 1979). Although the hot tank temperature typically exceeds 316°C (601°F), internal insulation (see figure 15.7a) can avoid the need for stainless steel. Hitec has been used as the receiver/storage fluid where freezing might occur on the cold side (Munoz Torralbo 1981). Experiments and analyses indicate that direct-contact salt–air heat exchange may be only about one-half as costly as with finned tubes (Bohn 1983).

The higher temperature chloride and carbonate salts appear to have potential merit as receiver/storage fluids in Brayton cycle power systems and for solar fuels production, but their containment vessels must remain full in order to prevent excessive thermal cycling. At temperatures above 900°C (1,652°F) sharp natural thermoclines are hard to maintain because of radiative heat transfer between zones. Efforts to solve the high temperature containment and thermocline maintenance problems have led to the tentative configuration shown in figure 15.7b, with an insulated buoyant "raft" and a double wall containing insulation packed with magnesia pellets to suppress convection (Copeland 1983). With direct-contact salt-air heat exchange at these temperatures the cost of heat transfer is expected to be less than one-fifth that with finned tubes (Bohn 1983).

Liquid Metal   Liquid sodium has been used as the receiver/storage fluid in a 0.5-MW_e IEA Rankine power station in Spain (Kalt 1979) and has been proposed for 100-MW_e plants (Johnson et al. 1979). It is suitable only for two-tank configurations or in a very small pressurized buffer tank; natural thermocline concepts are unfeasible because of liquid sodium's high thermal conductivity. Cost studies for 100-MW_e plants with liquid metal receivers (Dubberly et al. 1983) show that it is competitive as the primary storage medium only for less than 1 h of storage.

15.2.2   Solids

Typical advantages of solid sensible heat storage media include low-cost, wide operating temperature range, inexpensive or nonexistent containment structure, and the virtual absence of spill and leakage hazards. Table 15.5 characterizes representative media used in all-solid sensible heat storage subsystems or as the solid component in mixed media subsystems (see subsection 15.2.3).

Undisturbed Earth and Rock   Where subsurface conditions permit, these media are attractive alternatives to water for seasonal storage (see subsection 15.2.1) in active solar space-heating systems. They are essentially free except for possibly assigned land costs, and ordinarily need no containment structure and little excavation or insulation, although the cost of heat exchanger emplacement may be high. Figure 15.8 schematically depicts a 23,000-m^3

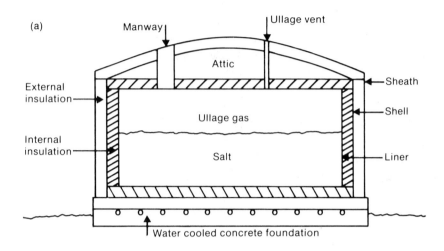

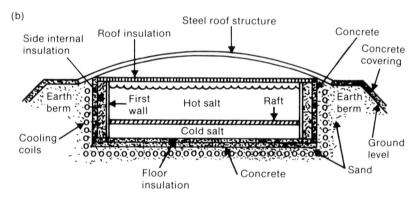

Figure 15.7
Sensible-heat storage in molten salts: (a) tank with internal insulation for molten salt at 540°C (1,004°F); (b) tank with internal insulation and raft thermocline maintenance for molten salt at 950°C (1,742°F). Source: Wells and Nassopoulos (1982).

Table 15.5
Solid sensible-heat thermal storage media

Medium	Temperature range [°F (°C)]	Specific gravity	Specific heat	FOB cost [$/kg ($/lb)]	Remarks
Water (reference)	32/212 (0/100)	1.0	1.0	nil	
Earth or clay	32/212 (0/100)	1.2/1.8	0.2/0.6	nil	
Concrete		2.24	0.27	0.04 (0.02)	
Sand		2.60	0.24	0.02 (0.01)	$0.03/kg ($0.07/lb) pulverized
Rock		2.64	0.21	0.02 (0.01)	
Sodium sulfate	36/1623 (0/884) m.p.	2.68	0.26		a
Glass		2.72	0.20	0.04 (0.02)	
Magnesia		2.87	0.292	0.24 (0.11)	
Alumina		3.04	0.286	0.37 (0.17)	
Taconite		3.62	0.22	0.09 (0.04)	
Cast iron, gray		7.19	0.16	0.48 (0.22)	b
Slag					

Note: Costs rounded from Dubberly et al. (1983), 1980 prices. Other properties approximated from various sources.
a. Small solid–solid latent-heat contributions at 340°F (171°C) and 466°F (241°C).
b. Meehanite cast iron suitable for higher temperatures; FOB cost $2.90/kg ($1.32/lb).

(30,100-yd^3) earth store in the Netherlands for a group of 96 solar-heated houses (Chuard and Hadorn 1983c). The medium is saturated sand with clay and peat layers, with an annual temperature swing between 25° and 60°C (77° and 140°F) with heat pump boosting. The relatively small, short-term water storage tank makes this a two-tank configuration, with the advantages discussed in chapter 19. Each pair of plastic heat exchange tubes is driven into the earth by a vibrating lance that, when removed, allows the soil to close around the tubes. In locations with groundwater movement or more porous soil, vertical screens of bentonite can be injected around the perimeter of the nominal storage volume to reduce thermal losses.

The 85,000-m^3 (111,200-yd^3) "Sunclay Project" in Sweden (Chuard and Hadorn 1983, pp. 149–150) contains 612 pairs of 16-mm (0.63-in.) diameter vertical HDPE pipes 35-m (115-ft) deep, with 170-mm (6.7-in.) pipe spacing

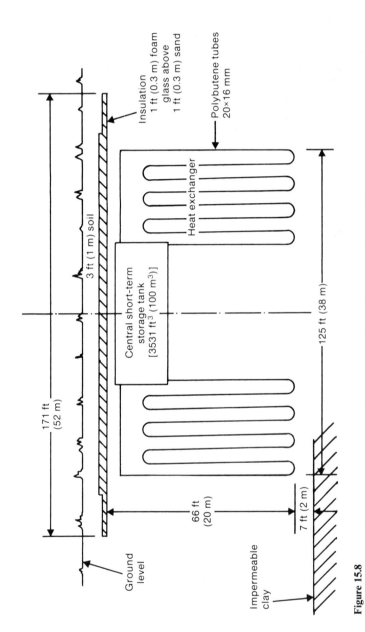

Figure 15.8
Large-scale seasonal earth storage subsystem. Source: Adapted from Chuard and Hadorn (1983, pp. 141–144).

and 2-m (6.6-ft) pair spacing. Each pair is inserted into a driven steel pipe which then is removed, and the clay closes in. The only insulation is a 0.3-m (0.98-ft) layer of expanded burnt clay on top. Another Swedish installation (Chuard and Hadorn 1983, pp. 153–154) uses "heat piles" in clay and silt to provide both foundations and heat storage for one house, with 1,500 m^3 (53,000 ft^3) of storage volume, a temperature swing between 7° and 30°C (45°–86°F) with heat pump boosting, and 5 m^3 (177 ft^3) of mineral wool top insulation .

Figure 15.9 illustrates the "Soil Therm" store for a single-family dwelling in France which uses plate heat exchangers in trenches filled with wet sand in polyethylene bags to ensure good thermal contact and prevent drying of the surrounding soil (Vachaud and Person 1981). The indicated "humidificateur" is an underground irrigation pipe to control the sand wetness.

In Luleå, Sweden, multiple boreholes in granite bedrock provide 10^5 m^3 (1.3 × 10^5 yd^3) of storage with a temperature swing between 10° and 70°C (50° and 158°F) and heat pump boosting, for solar heating of a school building (Chuard and Hadorn 1983, pp. 167–170). (Figure 19.3 in chapter 19 presents unit costs for the larger installations. It should be noted that all of these stores use end-of-season heat pump boosting.)

Processed Earth and Rock Several concepts for seasonal storage in excavated and replaced moist earth have been explored (Yuan and Bloom 1977; Walker et al. 1981; Chuard and Hadorn 1983, pp. 145–148). All involve lining an excavated pit with a sheet plastic moisture barrier and then replacing the moistened earth to surround a buried network of plastic heat exchange tubes. The designs vary mainly in the amount and location of manufactured thermal insulation (none are insulated on the bottom) and the tubing layout, and storage volumes are between 1,000 and 4,000 m^3 (3,280 and 13,700 ft^3) for small buildings.

Crushed granite or river rocks are used in the rock bed diurnal stores that are used in nearly all air-based active space-heating systems. Rock beds usually are approximately cubical, with temperature swings between about 30° and 60°C (86° and 140°F) and volumes of 3.5 to 10 m^3 (115 to 330 ft^3) in residential installations. Their containers may be wood, as illustrated in figure 15.10, or masonry or concrete. Wood is least expensive but unsuited for damp or buried locations. For best thermocline maintenance airflow typically is downward while charging and upward while discharging. Where low headroom dictates squat configurations, it is hard to prevent channeling; for such situations various horizontal flow concepts have been explored, but they present problems of unwanted stratification and of air bypassing due to rock

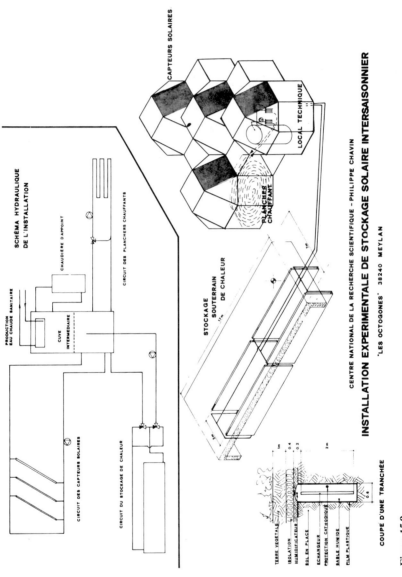

Figure 15.9
Small-scale seasonal earth storage subsystem. Source: Vachaud and Person (1981).

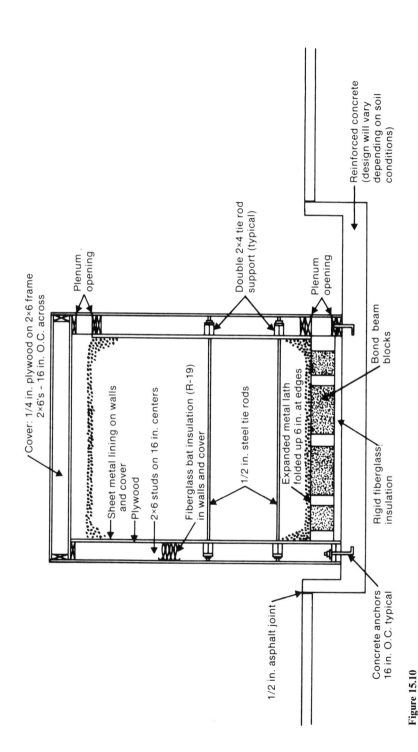

Figure 15.10
Wooden rock bin. Source: Cole et al. (1980).

settling. Construction details and representative specifications for a wooden rock bin are in Cole et al. (1980), and guides for cost estimation are in Lawrence (1980). Rock bed stores of various configurations are also used in passive and hybrid heating systems, with both natural circulation and forced airflow, and for storage of night coolness.

Large high temperature [500°C (932°F)] rock beds for seasonal storage have been studied in which crushed rock with air distribution manifolds fill a row of parallel trenches in the ground (Riaz 1977). Surrounding and intervening native earth serves as insulation and supplemental storage medium, and the top has an impermeable covering. A dewatering pump maintains underground dryness if needed. Many questions remain unanswered concerning operating costs and integrating the store with a high temperature solar heat source and with process heat or thermal power loads.

Cost studies have been made of a high temperature rock bed store for a 100-MW$_e$ liquid sodium receiver Rankine power system (Dubberly et al. 1981). The store would contain 64 air–rock/air–sodium heat exchanger modules, as illustrated in figure 15.11, with temperature swings between 385° and 563°C (725° and 1,045°F). It would be supported on the sides by an earth berm, and the total depth below the roof, including 1.2 m (3.9 ft) of sand and 0.15 m (0.49 ft) of insulating concrete, would be 9.5 m (31.2 ft). This was found to be the most economical concept for storage durations longer than about four hours. A similar configuration was examined for a closed cycle air Brayton system, with air–rock and air–air heat exchange and a temperature swing between 474° and 788°C (885° and 1,450°F) (Dubberly et al. 1983). It was found to be not competitive.

Concrete and Other Building Materials   The use of these materials for thermal storage in passive solar applications is discussed at length in volume 7 and needs no elaboration here.

Sand   Conceptual design studies have been made of sand as a high temperature [to 500°C (932°F)] thermal storage medium for solar Rankine power systems, examining fixed and fluidized beds and separate hot and cool stores with mechanical and pneumatic sand transport to a heat exchanger (Turner and Awaya 1978). More detailed design and cost studies have been made of a moving bed pulverized sand store for a 100-MW$_e$ water–steam (Dubberly et al. 1983) and liquid sodium (Dubberly et al. 1981) receiver Rankine power system. The sand flows by gravity over heat exchange tubes and is lifted by an Archimedes-type screw lift to minimize abrasion (Wright et al. 1981). The concept was found to be not economically competitive in either application.

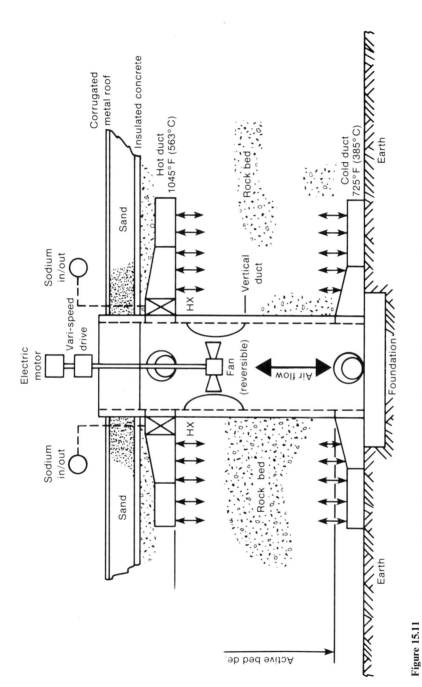

Figure 15.11
Fan and heat exchanger module for a high temperature rock bed store. Source: Dubberly et al. (1981).

Refractories   Magnesia bricks have been examined as sensible heat storage media for closed cycle helium (Boeing 1977) and air (Davidson et al. n.d.; Dubberly et al. 1983) Brayton power systems. This concept stems from the well-known use of firebrick "checquerworks" in blast furnace air preheaters. Magnesia bricks were judged superior for the helium system at 500 psi (3.45 MPa) and 816°/553°C (1,501°/1,027°F); alumina was selected for air at 816°/467°C (1,051°/873°F). In the most extensively studied configuration the bricks are stacked in horizontal internally insulated steel pressure vessels as shown in figure 15.12, with spheres of the same material in the hemispherical ends. In an alternate configuration they are stacked in vertical prestressed cast iron vessels (Westinghouse 1979) (see figure 15.13), with estimated savings of about 10% in tank materials cost and substantially lower field erection costs (Dubberly et al. 1983). In all versions the bricks are of the hexagonal Freyn type, which, when in place, impart circular gas flow in the axial direction.

Metals and Slags   Cast iron bricks have been considered as an alternative to refractories in closed cycle Brayton power systems (Boeing 1977; Dubberly et al. 1983) contained in steel or prestressed cast iron vessels. Gray cast iron was judged satisfactory for the helium working fluid (Boeing 1977), but for the air system it wag judged unsuitable at 816°C (1,501°F) primarily due to high growth and scaling (Dubberly et al. 1983) and the more expensive "Meehanite" iron was selected for system costing. Even with Meehanite the estimated reduction in overall system cost due to substitution of the denser iron for the refractory was several percent for the horizontal steel vessel configuration. Cost comparisons of iron and refractories in prestressed cast iron vessels were not made.

Heat storage in hollow steel ingots and in welded slabs was investigated for steam Rankine power systems (Turner 1977) but was found to be economically infeasible. Storage of high temperature solar process heat in slag from the same metallurgical process, in which the factured slag is piled over air channels in earth-covered mounds, has also been studied (Curto and Gillespie 1981).

15.2.3   Combined Liquids and Solids

In most of the concepts involving combinations of liquid and solid sensible heat storage media, the liquid permeates a porous, fractured, or granular solid substance that occupies most of the storage volume. The liquid typically is also the heat transfer and transport fluid, and permits natural thermocline maintenance as it flows through the mixed bed. Advantages and shortcomings of this generic approach can best be identified in specific applications.

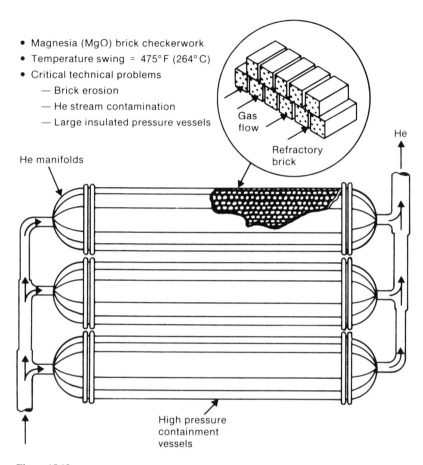

- Magnesia (MgO) brick checkerwork
- Temperature swing = 475°F (264°C)
- Critical technical problems
    - Brick erosion
    - He stream contamination
    - Large insulated pressure vessels

Gas flow

Refractory brick

He

He manifolds

High pressure containment vessels

Figure 15.12
High temperature sensible-heat storage in refractory bricks for a closed cycle air Brayton system.
Source: Boeing Engineering and Construction (1977).

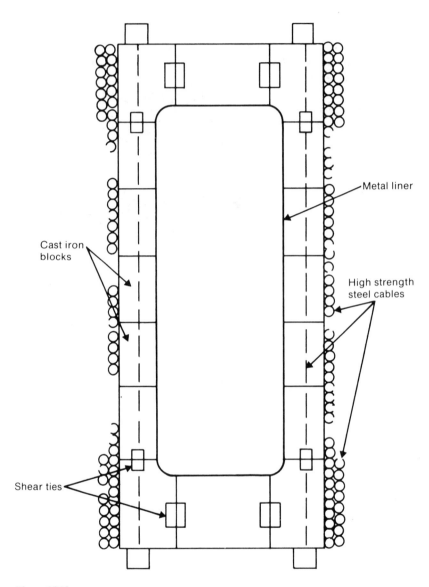

Figure 15.13
Basic features of a prestressed cast iron vessel. Source: Westinghouse Advanced Energy Systems Division (1979).

Natural Aquifers Natural aquifers are underground strata of saturated porous media such as sand and gravel or sandstone. Depending on many site-dependent factors—including permeability, type and degree of confinement, depth below surface, chemical composition, homogeneity, regional flow velocity, and ease of well drilling—they may provide economical large-scale seasonal storage of heat or coolness (PNL 1981; Chuard and Hadorn 1983; Meyer and Hausz 1980). Figure 15.14 illustrates investigated concepts for storage of heated water in unconfined or "free" aquifers, namely, those in which there is no impermeable layer between the water level and the surface. The same options are available for confined aquifers, which tend to be deeper and can contain hotter water because of the hydrostatic head imposed by the upper confining layer. An experimental single well installation (concept b in figure 15.14 is operational in Switzerland; see Chuard and Hadorn 1983, pp. 125–126). It is unconfined, in saturated silt and sand, with 6 m (19.7 ft) of earth above the water level and a nominal 60,000-m^3 (78,500-yd^3) volume defined by a 19-m (62-ft) vertical distance between 45-m (148-ft) diameter radial drains. The temperature swing is between 25° and 80°C (77° and 176°F). The well diameter is sufficiently large to permit insertion of the radial drains without excavation. In an experimental doublet installation (concept e-1 in figure 15.14) near Mobile, Alabama, the two wells are 244 m (800 ft) apart and penetrate a 21-m (69-ft)-thick confined aquifer of sand (85%), silt, and clay (Melville et al. 1981). Up to 58,000 m^3 (75,850 yd^3) of water at temperatures up to 82°C (180°F) have been injected in half-year cycles, with energy recovery factors of 40% to 58%. The use of two externally interconnected doublets (four wells) has been proposed for summer injection of solar heated water and winter injection of environmentally chilled water (Davison and Harris 1975).

Artificial Aquifers These are excavated pits containing water, uniformly sized pebbles, and internal piping, functioning like the free aquifer system illustrated in figure 15.14b but with more predictable performance. An experimental installation in France has a 250-m^3 (8,834-ft^3)-lined vertical cylindrical pit, configured as such to facilitate validation of a mathematical model (Torrenti 1979). The conceptual design for a much larger unit, shown in figure 15.15, is from a German study (Chuard and Hadorn 1983, pp. 123–124). Although the volumetric energy density of artificial aquifers is only about 70% that of excavated water pits with floating covers, their overburden can support heavy loadings without reinforcement and permit their use where land area must be used for other purposes.

Block-Filled Rock Caverns As an alternative to excavated rock caverns completely filled with water, the cavern can be mined in such a way that the

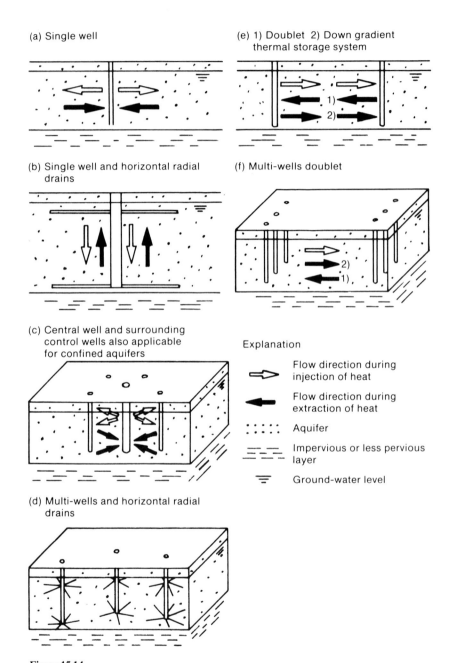

Figure 15.14
Thermal energy storage in free (unconfined) aquifers Configurational options for injection and withdrawal wells. Source: Adapted from Chuard and Hadorn (1983).

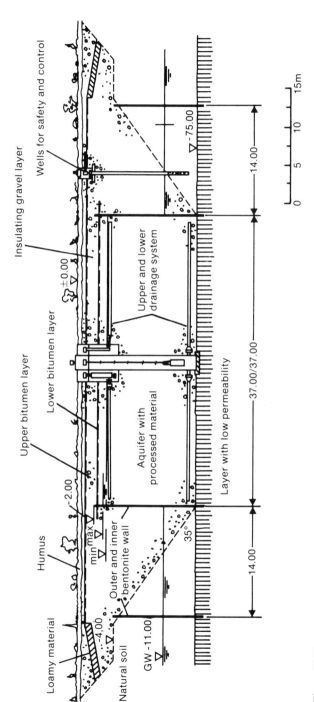

Figure 15.15
Artificial aquifer for seasonal storage. Source: Chuard and Hadorn (1983, pp. 123–124).

cavity is largely filled with and structurally supported by demucked blasted rock (Lindblom 1981). Although the cavern must be larger for equal storage capacity, excavation costs are lower because the blasted rock need not be removed.

Other Water/Rock Concepts   Thomason "Solaris" solar-heated houses typically have an uninsulated horizontal steel water tank buried in a rock bed (Thomason and Thomason 1973). The tank contains water heated by the collectors and transfers its heat to the surrounding rocks, in which it is stored, then transferred by forced air to the house. The advantages of this concept are unclear, but sponsored analyses of several variants of it showed the Thomason version to be the most cost-effective (Rockwell International 1979).

A solar-heated greenhouse at Rutgers University has a porous concrete floor poured over and supported by a 23-cm (9-in.)-deep pebble bed separated from the ground by a 2-cm (0.79-in.) styrofoam pad and a polyethylene water barrier (Mears et al. 1980). The pebble bed is saturated with slowly moving heated water from the collectors, thus storing the heat and transferring it to the floor.

Oil/Solids Thermocline Concepts   Oil-saturated packed beds are functionally identical to artificial aquifers, but the solids are mainly for purposes of cost reduction rather than for structural support. The liquid medium typically is a petroleum oil such as Caloria HT43, and the solid medium may be crushed rock, sand, glass, sodium sulfate pellets (Lee et al. 1979) taconite, or slag (Dubberly et al. 1983). More expensive synthetic organic fluids seldom can be justified, even though they may occupy only a small space.

Containment typically is in an insulated vertical cylindrical steel tank, usually with a smaller H/D ratio than for all-liquid natural thermocline stores because of the higher mass density solid media. The Barstow 10-MW$_e$ steam power station uses HT43 and crushed granite with sand at $218°/296°C$ $(424°/565°F)$ in a 3,058-m^3 (100,000-ft^3) tank with 0.88 H/D (Hallett and Gervais 1979). Excavated concrete pits lined with insulating concrete have been examined for process heat application (Dubberly et al. 1983). Media cost per unit capacity is lower than that of all-liquid organic fluid stores because of the substantially cheaper solids, but the larger volume tends to offset this advantage. Storage cost studies for water/steam receiver systems show that oil/rock is the most cost-effective sensible heat storage concept for the evaporation stage in a 100-MW$_e$ Rankine system (Dubberly et al. 1983). Potential problem areas are fluid degradation (see table 15.4), thermocline instability due to channeling, heat exchanger fouling, and excessive tank wall stresses caused by "ratcheting" (i.e., progressive solids settling with repeated thermal cycling, due to unequal thermal expansion). Rock fracture with repeated thermal cycling also may be a problem.

Oil-rock storage in solution mined salt caverns has also been explored (Dubberly et al. 1983), but was found not to be competitive with other concepts for 100-MW$_e$ water steam receiver systems.

Oil/Solids Trickle Charge   This approach is intended to permit the economical combined use of expensive high temperature synthetic organic receiver fluids with cheap solid storage media. It was developed for (but not used in) the Shenandoah solar total energy project (Hunke et al. 1979), using Syltherm 800 and taconite at 260°/399°C (500°/750°F) in multiple insulated steel tanks. Liquid does not remain in the tanks; it is introduced at the top of each tank by a distribution manifold and allowed to trickle down through the taconite pebble bed, thereby heating the solid bed for charging and extracting heat from it for discharging. Strictly viewed, it is a solid sensible heat storage concept with liquid heat transfer. The concept was validated by tests, which showed good thermocline stability in the nearly dry solid medium, but degradation rates of the hot fluid in contact with the bed were found to be excessive (as may be inferred from table 15.4). Subsequently another fluid has become available that has much lower degradation rates under like conditions (Dubberly et al. 1983).

15.3   Latent-Heat Storage

Thermal energy can be stored nearly isothermally in the latent heat of transition of various substances: as heat of fusion (solid–liquid phase change), heat of vaporization, or heat of solid–solid crystalline transformation. Commonly cited advantages of latent-heat over sensible-heat storage are smaller volume, lower required collector temperature, and lower heat loss due to lower storage temperature, although their importance or even their existence depends on overall systems considerations that are touched upon in chapter 19 and reviewed in more detail elsewhere in these volumes. Disadvantages can include high cost, corrosiveness, and poor thermal conductivity of the media, although these often can be minimized or circumvented. Commonly used figures of merit are $\$/\Delta H_{lat}$ and $\rho_{min}\Delta H_{lat}$ where ΔH_{lat} is the latent heat of transition per unit mass. These become meaningful, though, only when the usually accompanying sensible-heat contributions due to temperature excursions above and below the transition point are included. Most of the experience has been with heat of fusion media and subsystems, which will be discussed in greatest detail.

15.3.1 Heat of Fusion

Table 15.6 identifies and broadly characterizes some representative heat of fusion materials (commonly called *phase change materials,* or PCMs) that have been used or extensively investigated for solar applications. The groupings by generic chemical composition also classify these materials, with some overlap, according to ranges of transition temperature and suitable applications. The media are reviewed first, and then storage subsystem concepts and designs based on their use.

Ice   Ice is an excellent medium for seasonal storage of coolness from the winter ambient environment, and many ways of improving on traditional methods of forming and storing naturally produced ice are being explored. It is the obvious choice for solar-driven refrigeration systems in which it is the intended product. However, it seldom is used for cold side storage in solar-driven air-conditioning systems because its low freezing point penalizes chiller performance.

Organic PCMS   Advantages of organic materials include congruent melting (they melt and freeze repeatedly without phase segregation and consequent degradation of their heat of fusion), self-nucleation (they crystallize with little or no supercooling), and (usually) noncorrosiveness. Typical disadvantages compared to inorganic materials are their lower mass density (hence larger container volume), poorer thermal conductivity (hence larger and more expensive heat exchangers), flammability, and typically higher price. Three classes of organic PCMs have received continuing interest: clathrate hydrates, paraffin waxes, and polymers.

Clathrate hydrates are continuous solid water structures containing closed cavities within which are guest molecules which do not interact strongly with water. These guest molecules stabilize the "ice" structure and change its transition temperature. Clathrates such as tetrahydrofuran are of particular interest because they melt at temperatures appropriate for cold side storage in solar space cooling systems with attractively high heats of fusion, and their waters of hydration (tetrahydrofuran has 17 waters) minimize fire hazard (Kauffman and Grundfest 1973; General Electric 1977). Their high cost in currently produced quantitites has thus far discouraged commercial use.

Paraffin waxes are available in a wide range of transition temperatures covering most building heating and cooling applications (Mancini 1980), but commercial waxes at plausibly competitive prices melt over a trajectory of perhaps 10°C (18°F) in the 40° to 60°C (104° to 140°F) range. Fire hazard can be reduced by encapsulation or incorporation into building materials (see

Table 15.6
Heat of fusion storage media

Medium	Melting temperature [°C (°F)]	Heat of fusion [kJ/kg (Btu/lb)]	Specific heat (solid–liquid)	Specific gravity (solid–liquid)	Thermal condition [W/mK (Btu in./h ft² °F)]	Cost [$/kg ($/lb)]	Remarks
Ice	0	334 (765)	1.26 s / 1.01	0.92 s / 1.01	0.62 s (43) / 2.26 l (157)	nil	
Organic materials							
Tetrahydrofuran clathrate hydrate	4.5 (40)	225 (515)					
Paraffin wax P116E	47 (117)	209 (479)	0.69 s / 0.601	0.82 s / 0.771	0.14 s (10)	1.45 (0.66)	Sunoco product
Cross-linked HDPE	132 (270)	230 (527)	0.6 s / 0.61	0.96 s / 0.901	0.36 s (25) / 0.36 l (25)		
Hydrated salts							
Na₂SO₄/NaCl/NH₄Cl H₂O (Glaubers salt eutectic)	13 (55)	146 (334)	0.34 s / 0.641	1.47 s			
CaCl₂·6H₂O (calcium chloride hexahydrate)	27 (81)	171 (392)	0.34 s / 0.531	1.71 s / 1.531	1.09 s (76) / 0.54 l (37)	0.44 (0.20)	Dow product in nucleated and stabilized form
Na₂SO₄·10H₂O (Glaubers salt)	31 (88)	225 (515)	0.42 s / 0.791	1.46 s / 1.331	2.25 s (156)		
Na₂HPO₄·12H₂O (disodium phosphate dodecahydrate)	35 (95)	200 (458)	0.44 s / 0.721	1.43 s / 1.371	0.51 s (35) / 0.48 l (33)		

Table 15.6 (continued)

Medium	Melting temperature [°C (°F)]	Heat of fusion [kJ/kg (Btu/lb)]	Specific heat (solid–liquid)	Specific gravity (solid–liquid)	Thermal condition [W/mK (Btu in./h ft² °F)]	Cost [$/kg ($/lb)]	Remarks
$Na_2S_2O_3 \cdot 5H_2O$ (sodium thiosulfate pentahydrate— "hypo")	48 (118)	209 (479)	0.35 s	1.73 s	0.57 s f(40) (photograde)	1.17 (0.53)	$0.33/kg ($0.73/lb) 98% pure
$Mg(NO_3)_2/MgCl_2 \cdot 6H_2O$ (magnesium nitrate/magnesium chloride eutectic)	57 (135)	121 (277)	0.41 s 0.59 l	1.60 s 1.52 l	0.68 s (47) 0.54 l (37)	1.21 (0.55)	Dow product in nucleated form; melts congruently
$CH_3CO_2Na \cdot 3H_2O$ (sodium acetate trihydrate)	58 (136)	264 (605)	0.68 l				
$Mg(NO_3)_2 \cdot 6H_2O$ (magnesium nitrate hexahydrate)	89 (192)	146 (334)	0.44 s 0.38 l	1.56 s 1.44 l	0.87 s (60) 0.50 l (35)	1.21 (0.55)	Dow product in nucleated form; melts congruently
$NH_4Al(SO_4)_2 \cdot 12H_2O$ (ammonium alum)	94 (201)	234 (536)					Used with Rankine chiller (Nishiyama et al. 1979)
$MgCl_2 \cdot 6H_2O$ (magnesium chloride hexahydrate)	120 (248)	169 (387)	0.38 s 0.68 l	1.56 s 1.44 l		0.53 (0.24)	Dow product nucleated
Inorganic compounds and eutectics							
$NaNO_2/NaNO_3$ (draw salt)	220 (428)		0.37 l	1.73 l	0.57 l (40)		Carbon steel OK
$NaNO_3$ (sodium nitrate)	307 (585)	182 (417)	0.45 s 0.44 l	2.26 s 1.90 l	0.57 s (40) 0.61 l (42)		Carbon steel OK
$NaOH$ (sodium hydroxide)	318 (604)	316 (724)	0.48 s 0.50 l	2.03 s 1.76 l	0.92 s (64) 0.92 l (64)		Carbon steel OK if 8% $NaNO_3$ and 0.2% MnO_2 added (Thermkeep™)

						Comments
KCl/NaCl/MgCl (mixed chlorides)	385 (725)	292 (669)	0.23 s / 0.25 l	2.25 s / 1.63 l	1.6 s (111) / 1.01 (69)	Carbon steel OK in inert atmosphere
LiCO$_3$/KCO$_3$ (mixed carbonates)	505 (941)	328 (751)				Also higher m.p. eutectic with less Li (Dubberly et al. 1983) 304 S.S. OK
CaF$_2$/KF/NaF (mixed fluorides)	682 (1,260)	559 (1,280)				
Metals and alloys						
Aluminum/silicon	579 (1,074)	515 (1,179)	0.40 s / 0.34 l			SiC or graphite containment may be needed (Birchenall 1980)
Aluminum	660 (1,220)	400 (916)	0.31 s	2.57 s / 2.39 l	211 s (14643) / 911 (6315)	Incompatible with stainlesses and Inconels above m.p. (Webb and Pohlman 1979)

Note: s = solid; l = liquid.
a. Values rounded, from various sources.
b. Values for salt hydrates are for pure substances, except where listed as commercial products. Required nucleators and stabilizers will reduce heat of fusion and change other properties.

the discussion toward the end of this subsection 15.3.1), and poor conductivity in bulk storage can be enhanced by fins or metal matrices (Bailey et al. 1978; deJong and Hoogendoorn 1980).

Of the various polymers with potential for phase change storage, only high density polyethylene (HDPE) has been extensively explored. When partially cross-linked by chemical means or by electron beam radiation, it remains essentially form-stable through the "melting" process, retaining a rubbery resilient consistency. In laboratory tests a packed bed containing 110 kg (243 lb) of 3-mm (0.12-in.) diameter pellets in a water-glycol mixture retained porosity over hundreds of melt-freeze cycles (Salyer and Davison 1980). However, approximately 5,000 kg (11,103 lb) of pellets in water coalesced into an impermeable mass after only a few cycles (Cole 1983). The merit of this material as a form-stable pellet for packed beds remains in doubt.

Salt Hydrates   These PCMs have the same general range of transition temperatures and applications as organic materials, and often are preferred because of their higher mass density and thermal conductivity and their lower cost and fire hazard. Although their latent heat is partly as heat of solution, they generally are viewed only as heat of fusion materials. Drawbacks common to many salt hydrates are incongruent or semicongruent melting, supercooling, and corrosiveness, all of which can be largely corrected or circumvented, but at extra cost and with permanent performance penalty (Lane et al. 1976; Kando and Telkes 1978; Baylin 1979; Abhat 1983).

Many methods have been devised for preventing the phase separation that occurs with repeated cycling of incongruently and semicongruently melting salt hydrates (Kando and Telkes 1978; Telkes 1974), with varying degrees of success. Among them are

1. additives for crystal habit modification, to prevent the growth of large crystals, together with thickening agents (Marks 1983),

2. addition of polymer matrices to retain all solids in uniform suspension (Page et al. 1981; van Galen 1980; MacCracken 1982),

3. addition of a thickening agent and hydraulic cement (Michaels 1982),

4. extra water to remelt anhydrous salt or lower hydrates (Biswas 1977; Furbo 1980),

5. agitation by stirring (MacCracken et al. 1980; Greene and Watson 1979) or by introducing air or an immiscible liquid (Barlow 1983; Furbo 1978; Helshoj 1981; Arrhenius et al. 1983),

6. tumbling in a rotating drum (Herrick and Zarnoch 1979).

Supercooling usually can be greatly reduced by adding nucleators, such as borax to Glaubers salt and strontium chloride to calcium chloride hexahydrate. If satisfactory nucleating agents cannot be found, supercooling can be avoided by agitation, "cold fingers," or by limiting the degree of temperature rise above the transition point. Methods of preventing phase separation can also be effective means of maintaining uniform distribution of added nucleators.

Not shown in table 15.6 are reciprocal salt pairs that undergo double decomposition and phase change at the transition temperature. An example is the system $NaOH \cdot H_2O-KBr$, with a measured transition temperature of 58°C (136°F) and a theoretical enthalpy change (heat of fusion plus heat of reaction) of 289 kJ/kg (125 Btu/lb) (Sharma et al. 1982). Another investigated system is $Ba(OH)_2 \cdot 8H_2O-2KNO_3$, with a transition temperature of 65°C (149°F) and an enthalpy change of 207 kJ/kg (89 Btu/lb) (Reiter and Rota 1984). This concept is still in the research phase.

Inorganic Compounds and Eutectics   These higher melting point PCMs have been investigated mainly for thermal power and process heat applications. Those listed do not supercool appreciably and their heats of fusion do not degrade with cycling.

Metals and Alloys   The principal advantage of metals and alloys is their high thermal conductivity, which in some cases may offset their high cost and difficulty of containment. They are potentially superior in buffer storage configurations with required high discharge rates. Thermal properties of many alloys with transition temperatures between 343° and 956°C (649° and 1,753°F) have been measured, but fully satisfactory containment methods and materials have not yet been found (Birchenall 1980; Webb and Pohlman 1979).

Bulk Storage with Indirect Heat Exchange   In most versions of this generic concept the storage unit is basically a shell and tube heat exchanger in which the shell encloses a fixed amount of PCM. Typically the configurations are "tube intensive" to compensate for poor heat transfer from the solidifying salt during discharge. Figure 15.16 shows a commercially available all-plastic unit for corrosive salt hydrates, with a stirring pump to prevent phase separation. Units containing organic or higher temperature noncorrosive PCMs typically are of carbon steel. Figure 15.17 is a conceptual design for fluoride salt eutectic with a maximum temperature of 816°C (1,501°F)in a buried concrete tank.

To reduce the heat exchanger "tube intensiveness," and therefore its cost, "passive" methods of improving heat transfer to the tubes have been investigated. Besides those for paraffin (see the discussion at the beginning of this

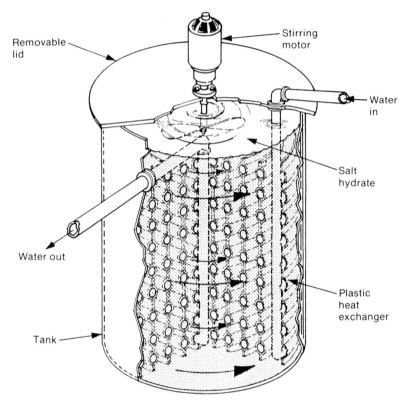

Figure 15.16
Calmac "Heat Bank" latent-heat storage unit for sodium thiosulfate pentahydrate. Source: Baylin
(1979).

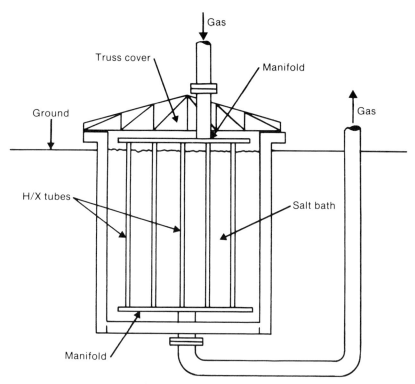

- Fusible fluoride salt bath
- Temperature swing 432° F (222° C)
- Critical technical problems material compatibility

Figure 15.17
High temperature latent-heat storage in fluoride salts for a closed cycle air Brayton system.
Source: Boeing Engineering and Construction (1977).

subsection 15.3.1), finned tubes and embedded metal wools have been investigated for high temperature carbonate mixtures, producing 20% to 30% increases in flux density (Maru et al. 1978). Cost studies for a 100-MW$_e$ water steam receiver Rankine power system showed Thermkeep in a finned tube unit to be more economical for evaporation storage than sensible heat (Dubberly et al. 1983).

Active methods of heat transfer improvement have also been explored. Tumbling Glaubers salt in a rolling drum, with heat transfer through the drum wall to and from air, also improved phase stability (see subsection 15.3.3). Prevention of solid PCM build-up on tubes by "slipperiness," vibration, and ultrasonics was attempted without success for higher temperature units (LeFrois et al. 1978). Experiments with tube scraping for removal of the slushy frozen state of an off-eutectic mixture of sodium nitrate and sodium hydroxide were promising, and a subscale prototype unit was built but never tested (LeFrois and Venkatasetty 1976).

Bulk Storage with Direct Heat Exchange   All of these concepts involve direct contact of the PCM with an immiscible heat transfer fluid. In space-heating applications the PCM typically is an incongruently melting salt hydrate, the immiscible fluid being either a light oil or a vaporizing liquid. In one version oil floats above sodium acetate trihydrate in a sealed vessel; when stirred, it transfers heat between the PCM and an immersed tubular heat exchanger as illustrated in figure 15.18 (Greene and Watson 1979). The agitation also prevents phase separation. In other versions a circulating pump forces the oil through a piping network so that oil droplets upwell through the salt (Barlow 1983; Furbo 1978; Helshoj 1981). Figure 15.19 shows one way of controlling the oil circulation, in this design through sodium thiosulfate pentahydrate with extra water. An additive to the oil prevents salt carryover and ultimate blockage of the circulation system. The PCM-oil approach has not been extensively explored for coolness storage, but the natural melting during shutdowns and the *heating* of any salt-bearing oil as it cools the building should offer opportunities for simplification.

One investigator was led to the PCM-oil concept by failure to develop a satisfactory method of pumping saturated salt solutions through an external heat exchanger, using the extra water (mentioned earlier in this subsection) in a hydrate (Kauffman et al. 1977). A later investigator also was unable to prevent solidification and heat exchange blockage, and also was led to the same concept (Bell 1981). Conversely, a DOE-sponsored experimental program concluded that PCM-oil concepts were technially and economically

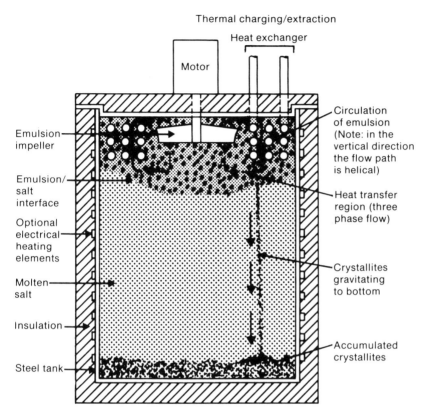

Figure 15.18
Kohler Corp.'s direct heat exchange latent-heat storage unit with stirred immiscible fluid. Source:
Greene and Watson (1979).

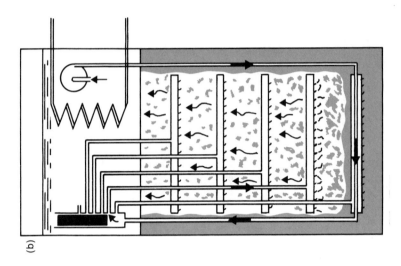

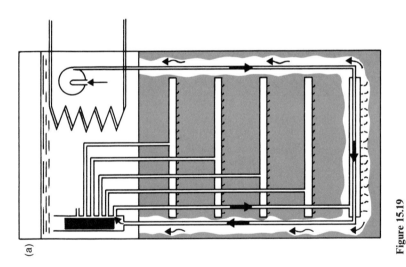

Figure 15.19
Effex Innovation's A/S direct heat exchange latent-heat storage unit with pumped immiscible fluid: (a) charging in progress; (b) discharging in progress. Source: Helshoj (1981).

questionable (Wright and Bohn 1982) after such systems had proved to be commercially successful (Barlow 1983).

The concept illustrated in figure 15.20 is based on the use of a vaporizing immiscible fluid for heat transfer (Arrhenius et al. 1983). It would completely preclude salt carryover, avoid the need for stirring or mechanical circulation, and reduce the required heat exchanger surface through condensing heat transfer.

In a proposed design for large thermal power stations, molten lead falls through molten fluoride salts in a 30,000-m^3 (3.9×10^4-yd^3) tank to extract the heat of fusion and transfer it to load via a conventional heat exchanger (see 1974 patent by F. P. Bundy on p. 107 in Bramlette et al. 1976). In another high temperature concept, droplets of molten salt fall as in a shot tower through a less dense heat transfer fluid in a vertical counterflow heat exchanger, thereby releasing their heat of fusion and forming solid pellets that can be mechanically transported to a separate storage vessel, and then conveyed as needed to a second shot tower for remelting and storage as a liquid (Nichols and Green 1977). An engineering prototype was constructed and partially tested of a unit containing a molten eutectic salt (mixed chlorides or draw salt) and a molten lead-bismuth eutectic (Alario and Haslett 1982). Droplets of molten salt rise through the heavier downflowing molten metal in an open-topped heat exchanger, and the overflowing solid salt pellets return by gravity to the molten salt container. Less successful was the engineering test prototype of a sodium nitrate-water reflux condenser heat exchanger in which condensate was to bubble up through molten salt, flash into steam, condense onto conventional heat exchanger tubes, then be pumped back for reinjection into the salt (LeFrois 1979). The salt hydrolyzed.

Storage in molten glassy slag at 1,400°C (2,552°F) was studied for a central receiver closed cycle argon Brayton power system (Bruckner and Hertzberg 1982). As shown in figure 15.21, droplets of stored molten slag solidify as they fall through a countercurrent flow of gas and then are mechanically conveyed either to storage or up to the receiver for remelting. As originally conceived, the slag would release both its sensible and latent heat to the gas, but it was found that although the molten slag would solidify in the exchanger, the gas would receive virtually no heat of fusion because of delayed crystallization. Cost of the storage subsystem was not found to be competitive, but possible overall economic benefits of the extremely high temperature were not quantified in the study.

Macroencapsulation Encasing the PCM in multiple containers small enough to limit the thermal path through the PCM to a few centimeters is an

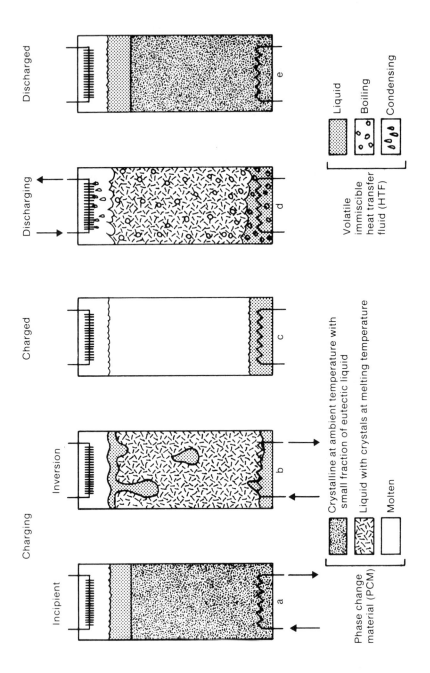

In the charging mode (a,b) heat is introduced at the bottom of the storage tank by a heat exchanger or electric resistance heater (or by injection of HTF vapor from a heat source); this leads to gradual melting of the crystalline PCM. Since the molten PCM in the case used for illustration is less dense than the HTF an instability develops, the HTF breaks through the column of PCM and sinks to the bottom of the tank (b). The rise of the liquid PCM and the HTF inversion help distribute the heat input and accelerate the charging. After complete charging (c) the liquid PCM is stratified on top of the HTF.

In the discharge mode the cold medium to be heated (e.g., the water or air) is allowed to pass through the heat exchanger-condenser shown at the top of the tank (the condenser may also be external and remote). This lowers the vapor pressure of HTF in the tank, causing the liquid HTF to boil. As heat is withdrawn by condensation of HTF vapor, the PCM begins to crystallize, continuously liberating latent heat of melting. The condensed HTF rains back into the boiling liquid, and the vaporization-condensation cycle is continued until all heat above ambient is withdrawn (e).

Figure 15.20
Direct heat exchange latent-heat storage unit with vaporizing immiscible fluid. Source: Arrhenius et al. (1983).

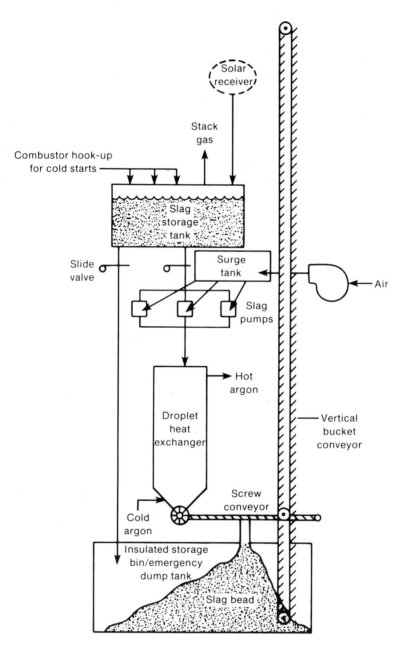

Figure 15.21
Molten slag storage for very high temperature closed-cycle argon Brayton cycle. Source: Bruckner and Hertzberg (1982).

alternative to tube-intensive heat exchange concepts. It also can facilitate complete melting and freezing in latent heat equivalents of waterwalls (table 15.2), and permit incorporation of the PCM in building components (Swet 1980). Table 15.7 identifies some of the many commercially available packagings for passive heating and for active heating and cooling. Figure 15.22 shows one of the more thoroughly engineered packages for south wall mounting.

Figure 15.23 illustrates a higher temperature macroencapsulation concept for large-scale solar process heat (Dubberly et al. 1983). The stacked steel cans contain Thermkeep, and the heat transfer fluid is biphenyl. The cans accept heat by condensation of fluid evaporated by the charging heat exchanger and release it by evaporation of sprayed liquid, which then condenses on the discharge heat exchanger tubes.

Microencapsulation   This generic approach involves PCM permeation of a structural matrix, thereby avoiding the need for individually encapsulating and inserting each PCM element. Its potential attractiveness in passive heating applications has stimulated much exploration (Swet 1980), but none of the investigated concepts has yet reached commercial status. One investigator successfully infused liquid calcium chloride hexahydrate into sealed porous concrete blocks, but chemical incompatibilities (perhaps avoidable) caused cracking (Chahroudi 1977). Another investigator dispersed the same PCM in polyester wallboard by creating a stable emulsion in the liquid plastic resin, but attempts to prevent water gain were unsuccessful (Lane and Rossow 1976). Granular solid Gulfwax was dispersed in cement concrete mixes in quantities up to 40% by volume without excessive loss of strength, but problems of sealing and of finding sufficiently inexpensive waxes with the desired transition temperature were not solved (Mumma and Liu 1980). Infusion of salt hydrates into porous glass beads, for addition to Portland cement and polymer concrete mixes, poisoned the mixes and did not prevent weight change (Sansone et al. 1978). Currently under development are methods of dispersing polyglycols and paraffin waxes in a rubber matrix which can be used as wall or floor tile or ground up and mixed into cementitious building materials (Salyer et al. 1984).

15.3.2   Heat of Vaporization

The latent heat of vaporization or condensation per unit mass of a fluid typically is nearly an order of magnitude greater than its heat of fusion, but its isothermal storage potential is seldom exploited in closed systems because of the extremely large changes in pressure or volume. It has frequently been used, however, in "steam accumulators" that store thermal energy by adding

Table 15.7
Encapsulated PCM products

Configuration	Containment material	Approximate dimensions [in. (cm)]	PCM [°C (°F)]	Manufacturer
Tube	Polypropylene	2.1 (5.3) diameter × 32 (81) long 2.1 (5.3) diameter × 44 (112) long	Glaubers salt 31 (88)	Calortherm
	High density polyethylene steel	3.5 (8.9) diameter × 72 (183) long 4.5 (11.4) diameter × 72 (183) long 4.75 (12.1) diameter × 24 (61) long	Calcium chloride 27 (81) Glaubers salt 7, 18–21, 31 (45, 64–69, 88)	Thermol 81 Boardman
Can	Lined steel	4.25 (10.8) diameter × 7 (18) high	Calcium chloride 27 (81)	Texxor
Chub	Laminated plastic	2 (5) diameter × 20 (51) long	Glaubers salt 13 (55)	U. Del. IEC
Pouch	Flexible plastic	12 (30) × 24 (61) rectangle 3/4 (2) thick	Glaubers salt 23, 30 (73, 86)	CMI HeatPac
Tray	Plastic		Glaubers salt	Valmont
Wall pod	Transparent fiberglass reinforced plastic	48 (122) wide × 16 (41) high 2 (5) thick	Calcium chloride 27 (81)	Solar Comp. Corp.
Wall panel	Molded plastic	14 (36) wide × 22 (56) high 2.2 (5.6) thick	Calcium chloride 27 (81)	Dow Chem Co.
Pellet (tablet)	Latex	1/2 (1.3) diameter convex	Glaubers salt 23 (73) Calcium chloride 27 (81) P116 wax 47 (117)	Pennwalt

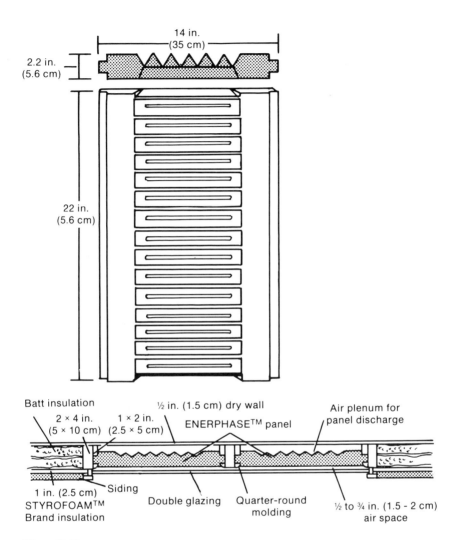

Figure 15.22
Dow "Enerphase" latent-heat storage panel for stud wall cavity installation. Source: Dow Chemical Co. (Brochure Form No 179-7484-83R).

674

C. J. Swet

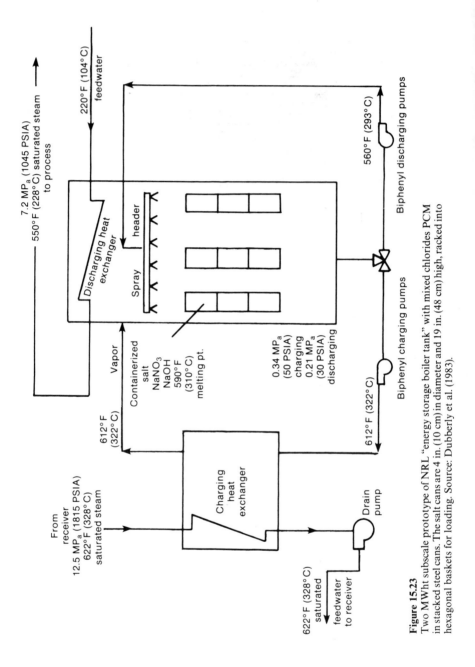

Figure 15.23

Two MWht subscale prototype of NRL "energy storage boiler tank" with mixed chlorides PCM in stacked steel cans. The salt cans are 4 in. (10 cm) in diameter and 19 in. (48 cm) high, racked into hexagonal baskets for loading. Source: Dubberly et al. (1983).

heat to water and steam in a pressure vessel and then releasing the added enthalpy as lower pressure steam. This concept has been used for buffer storage, with indirect heat addition, in a 30-kW solar Rankine system (Koehne et al. 1981), and with injected saturated steam in 1-MW$_e$ systems (Tanaka 1980). Containment typically is in steel pressure vessels, although prestressed cast iron vessels appear to be more economical in larger sizes (see subsection 15.2.2 and figure 15.13). Cost studies show that pressurized hot water in lined excavated rock caverns is the most economical storage concept, given favorable geological conditions, for large solar process steam delivery systems (see figure 15.24), and for evaporation heat storage in 100-MW$_e$ water-steam receiver Rankine systems (Dubberly et al. 1980).

For air-based solar space heating and cooling systems, storage methods have been investigated that utilize the heat of adsorption (condensation) of water vapor in humid ambient air passing through beds of desiccant material (Close and Dunkle 1977; Gopal et al. 1979). Heats of condensation and evaporation also are exploited in closed cycle solar sorption heat pump/ storage systems, which are reviewed in subsection 15.4.1 as thermochemical heat pump/storage systems.

15.3.3   Heat of Solid–Solid Transformation

Latent heat storage in a maintained solid state is a compelling notion because it would eliminate the need for encapsulation in packed beds and would facilitate dispersion in building materials. This was the incentive for developing form stable HDPE (subsection 15.3.1), which really is a heat of fusion material. Searches for substances with usefully large latent heats of solid–solid crystalline transformation at desired transition temperatures and plausibly competitive cost have been unsuccessful in most cases (Leffler 1977; Busico et al. 1980; Baylin and Merino 1981). However, several polyhydric alcohols and mixtures thereof show promise over a wide range of transition temperatures, although their rather high costs may be justifiable only in special circumstances (Benson 1983). Some of these materials are characterized in table 15.8. The listed values for thermal conductivity can be substantially increased by adding small amounts of graphite or aluminum (Benson 1984). Remaining problems include flammability, water solubility, and possible long-term toxicity, all of which would be alleviated or possibly eliminated by incorporation into building materials.

Anhydrous sodium sulfate has two solid–solid transitions (table 15.5) that contribute slightly to its effective heat capacity as a solid sensible heat storage material. Anhydrous sodium hydroxide has a more substantial solid–solid transition (see table 15.5) only a few degrees below its solid–liquid transition

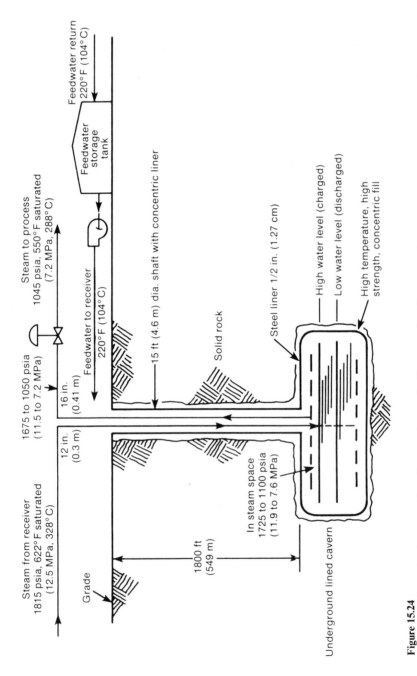

Figure 15.24
Latent heat of vaporization storage in underground pressurized hot water for solar process steam.
Source: Dubberly et al. (1983).

Table 15.8
Polyalcohol solid state phase change materials

Material	Transition temperature [°C (°F)]	Transition enthalpy [kJ/kg (Btu/1b)]	Specific gravity	Thermal condition [W/mK (Btu/in./h ft² °F)]
Pentaerithritol (PE)	188 (370)	616 (269)	1.34 [170°C (338°F)] 1.22 [205°C (401°F)]	1.88 [170°C (338°F)] 0.86 [205°C (401°F)]
Pentaglycerine (PG)	89 (192)	318 (139)	1.19 [65°C (149°F)] 1.12 [95°C (203°F)]	0.83 [65°C (149°F)] 0.71 [95°C (203°F)]
Neopentyl glycol (NPG)	48 (118)	272 (119)	1.05 [27°C (81°F)] 0.98 [60°C (140°F)]	0.53 [27°C (81°F)] 0.47 [60°C (140°F)]
67NPG 33TMP[a]	24 (75)			
15PE 15TME[b] 70NPG	17 (63)			

a. TMP = trimethylol propane.
b. TME = trimethylol ethane.

temperature, and temperature swings typically are large enough to include both.

15.4   Storage in Reversible Thermochemical Reactions

Heat added to certain substances can alter them chemically, remain stored in them as nonthermal heat of potential chemical reconstitution, and then be released as thermal energy by their reconstitution. Commonly cited advantages of thermal energy storage in reversible thermochemical reactions are high storage energy density (much higher on both mass and volume bases than that achievable with sensible or latent heat storage), indefinitely long storage duration at near-ambient temperature, and heat-pumping capability. Drawbacks may include complexity, high first cost, toxicity, flammability, and poor second law efficiency due to unrecoverable work input. As with latent heat storage, however, the importance or existence of these considerations depends strongly on the specific application and system concept. The various concepts and designs are grouped here according to whether the process is sorption/desorption, decomposition/recomposition of a compound, or liquid phase.

15.4.1   Thermochemical Heat Pump/Storage Systems

Many kinds of sorption heat pump concepts for space-heating, space-cooling, refrigeration, and process heat include energy storage within the thermodynamic cycle. Generically they often are called *chemical heat pumps* or *thermochemical heat pump/storage systems*, whether the process is adsorption (involving only heats of condensation and evaporation) or absorption (also involving heats of reaction). Their basic principles of operation, advantages, limitations, and potential performance are briefly discussed in section 19.6 of chapter 19 and depicted in figures 19.7, 19.8, 19.9, and 19.11 of that chapter. More extensive treatments are in Wettermark (1980), Oelert et al. (1982), Raldow (1981), Offenhartz (1981), and Nonnenmacher and Groll (1981). Table 15.9 identifies representative reactions.

It should be noted that the two-tank intermittent system shown in the simplified schematic of figure 19.7 in chapter 19 provides space heating during the day and night (storage) with an ideal COP of 2 (heat pumping). As shown, however, cooling can be provided only at night; for day and night cooling another "A" tank must be added, with additional complexities. Continuous systems such as that shown in figure 19.8 operate much like conventional water/lithium bromide chillers or gas-fired absorption heat pumps, but with stored solution and working fluid. Solution in varying strengths and amounts

Table 15.9
Thermochemical heat pump/storage systems

Working fluid	Sorbent	Cycle	Reactions	References and remarks
Water	Sulfuric acid	Continuous or intermittent	$H_2SO_4 \cdot H_2O(l) \leftrightarrow H_2SO_4(l) + H_2O(g)$ $H_2O(g) \leftrightarrow H_2O(l)$	Subscale prototype (Clark 1978; Clark 1979; Rocket Research 1982a, b)
	Sodium sulfide	Intermittent	$Na_2S \cdot 5H_2O(s) \leftrightarrow Na_2S(s) + 5H_2O(g)$ $H_2O(g) \leftrightarrow H_2O(l)$	Commercial (Bakken 1981)
	Zeolite	Intermittent	$Zeolite \cdot H_2O(s) \leftrightarrow Zeolite(s) + H_2O(g)$ $H_2O(g) \leftrightarrow H_2O(l)$	Commercial ice maker (Tchernev 1978) engineering development heat pump (Alefeld 1981)
Methanol	Calcium chloride	Intermittent	$CaCl_2 \cdot 2CH_3OH(s) \leftrightarrow CaCl_2(s) + 2CH_3OH(g)$ $CH_3OH(g) \leftrightarrow CH_3OH(l)$	Developmental test prototype space heating and cooling (Offenhartz 1980)
Ammonia	Ammonium nitrate	Intermittent	$2NH_4NO_3 \cdot 3NH_3(l) \leftrightarrow 2NH_4NO_3(s) + 3NH_3(g)$ $NH_3(g) \leftrightarrow NH_3(l)$	Developmental test prototype space heating and cooling (Martin Marietta 1981)
	Sodium thiocyanate	Continuous or intermittent	$NaSCN \cdot NH_3(l) \leftrightarrow NaSCN(l) + NH_3(g)$ $NH_3(g) \leftrightarrow NH_3(l)$	Performance studies (McLinden and Klein 1983)
	Calcium chloride	Intermittent	$CaCl_2 \cdot 8NH_3(s) \leftrightarrow CaCl_2 \cdot 4NH_3(s) + 4NH_3(g)$ $NH_3(g) \leftrightarrow NH_3(s)$	Operational prototype ice maker (Worsoe-Schmidt 1979)

Note: s = solid; l = liquid; g = gas.

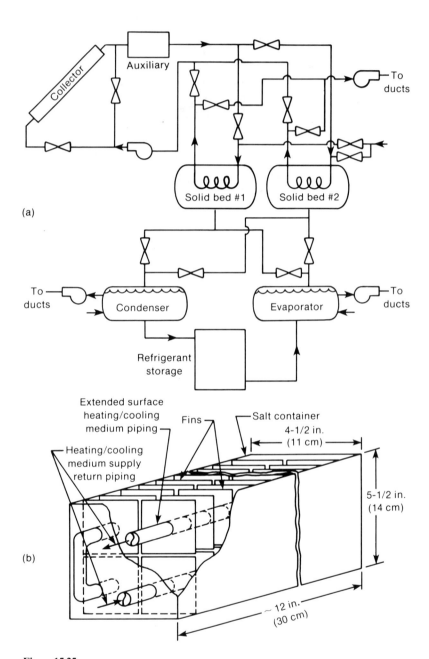

Figure 15.25
Methanol/calcium chloride thermochemical heat pump/storage system: (a) simplified schematic;
(b) salt bed design. Source: Offenhartz (1980, 1981a).

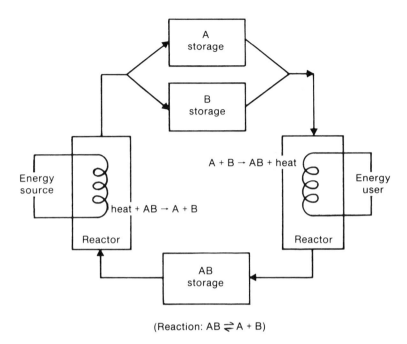

(Reaction: AB $\rightleftharpoons$ A + B)

Figure 15.26
Hypothetical thermochemical storage reaction. Source: Swet (1981).

may be stored in one tank as shown, or in separate weak and strong solution tanks (or possibly in a single segregated tank) with no working fluid storage.

Figure 15.25 shows the schematic design of a methanol/calcium chloride system with two salt beds to permit space heating or cooling day and night. Also shown is the salt bed configuration used in a developmental test prototype of that system with a storage capacity of approximately 10^5 kJ (95,238 Btu). The temperature boost is from about 5° to 45°C (40° to 113°F) with a collector temperature of 130°C (266°F). Heating and cooling COPs are 1.5 to 1.6 and 0.5 to 0.6, respectively.

15.4.2   Thermal Decomposition of Compounds

The thermal decomposition type of thermochemical storage is illustrated by the hypothetical reversible thermochemical reaction $AB \leftrightarrow A + B$ shown in figure 15.26, with a "turning temperature" $T^* = H/S$ above which the equilibrium shifts to the right (endothermic decomposition) and below which the shift is to the left (exothermic recombination). Catalysis may or may not be required. Such reactions are most suitable for high temperature processes,

Table 15.10
Reversible decomposition reactions for thermochemical storage

	Reaction	Equilibrium temperature, t^* [°C (°F)]	Theoretical [kJ/kg (Btu/lb)]	H_3 [MJ/m³ (Btu/ft³)]	Reference
	$\rightarrow$ endothermic $\leftarrow$ exothermic				
$CaCO_3(s)$	$\leftrightarrow CaO(s) + CO_2(g)$	835 (1,535)	4,094 (1,788)	2242 (66,587)	(Smith 1979; Rocket Research 1977)
$SO_3(g)$	$\leftrightarrow SO_2(g) + \frac{1}{2}O_2(g)$	782 (1,440)	2,830 (1,236)	795 (23,611)	(Smith 1979; Rocket Research 1977)
$NH_4HSO_4(s)$	$\leftrightarrow NH_3(g) + H_2O(g) + SO_3(g)$	467 (873)	4,550 (2,042)	3100 (92,070)	(Smith 1979; Rocket Research 1977; Prengle and Sun 1976)
$Ca(OH)_2(s)$	$\leftrightarrow CaO(s) + H_2O(g)$	447 (837)	3,375 (1,474)	3126 (92,842)	(Smith 1979; Rocket Research 1977; Rockwell 1980)
$CS_2(g)$	$\leftrightarrow C(s) + 2S(s)$	427 (837)	3,522 (1,538)	1939 (57,588)	(Smith 1979; Rocket Research 1977)

Note: s = solid; l = liquid; g = gas.

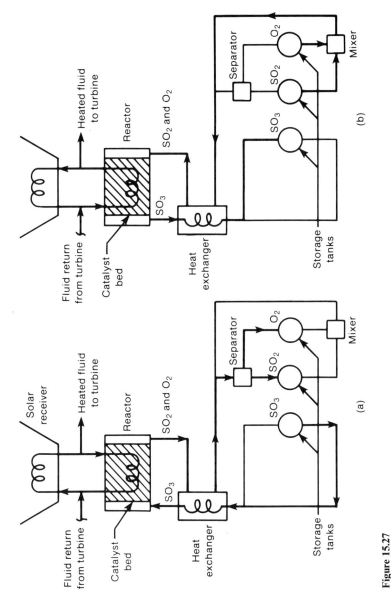

Figure 15.27
Simplified schematic of SO_2/O_2 thermochemical storage system for a central receiver power plant: (a) storage charging——endothermic decomposition; (b) storage discharging——exothermic recombination. Source: Swet (1981).

since ΔS must be fairly low to facilitate reactant separation and the desired high value of ΔH is achievable only with a correspondingly high value of T^*. Table 15.10 identifies some representative reactions, among which the sulfur dioxide/oxygen system (see figure 15.27) has been most thoroughly examined for solar Rankine power (Smith 1979). Despite their typically high energy storage densities and inexpensive reactants, the high power-related costs (primarily reactors, heat exchangers, and compressors) make them uncompetitive for short-term storage; their primary potential role appears to be in applications where both long duration near-ambient storage and long distance thermal energy transport are required. In some configurations heat pumping also can be incorporated to advantage.

15.4.3 Liquid Phase Reactions

The absence of solid or gaseous phases would greatly simplify thermochemical energy storage, but few prospective reactions appear to be in sight. Thermally driven catalyzed Diels-Alder reactions (diene + dienophile Diels-Alder adduct) are potentially attractive because they can be entirely liquid phase at modest temperatures between 150° and 250°C (302° and 482°F) (Lenz and Hegedus 1979). Although some possibly suitable Diels-Alder reactions have been identified, this approach has not yet been extensively explored.

References

Abhat, A. 1983. Low temperature latent heat thermal energy storage: Heat storage materials. *Solar Energy* 30: 313

Alario, J., and R. Haslett. 1982. High temperature active heat exchanger research for latent heat storage. *Proc. DOE Physical and Chemical Storage Annual Contractors' Review Meeting.*

Alefeld, G. 1981. A zeolite heat pump, heat transformer and heat accumulator. *Papers presented at the International Conference on Energy Storage, Brighton, U.K. April 29–May 1, 1981.* Paper D1. Cranford, Bedfordshire, U.K.: BHRA Fluid Engineering.

Arrhenius, G., J. Hitchin, E. Jensen, and A. Tsai. 1983. Latent heat exchange by direct contact vaporization: a new concept in energy storage and retrieval. *Opportunities in Thermal Storage R&D*, W. Hausz and B. Berkowitz, eds. EPRI EM-3159-SR. Paper 24.

Bacconnet, E., M. Dancette, and J. Malherbe. 1981. 100 kW solar power plant in Corsica-Sofretes/Bertin. *Proc. Solar World Forum, ISES,* pp, 2941–2949.

Bailey, J. A., J. Mulligan, and C. Liao. 1978. *Reserach on Solar Energy Storage Subsystems Utilizing the Latent Heat of Phase Change of Certain Organic Materials.* Final report to ERDA.

Bainbridge, D. A. 1979. Water wall passive systems—For new and retrofit construction. *Proc. 3rd National Passive Solar Conf. AS/ISES.* pp. 473–478.

Bakken, K. 1981. System tepidus, high capacity thermochemical storage/heat pump. *Papers presented at the International Conference on Energy Storage, Brighton, U.K. April 29–May 1, 1981.* Paper C1. Cranford, Bedfordshire, U.K.: BHRA Fluid Engineering.

Barlow, W. 1983. Thermal storage utilizing off-peak electro-thermal charging of phase change materials. *Opportunities in Thermal Storage R&D*, W. Hausz and B. Berkowitz, eds. EPRI EM-3159-SR. Paper 20.

Baylin, F. 1979. *Low Temperature Thermal Energy Storage: A State-of-the-Art Survey*. SERI/RR-54-164. Golden, CO: Solar Energy Research Instiute.

Baylin, F., and F. Merino. 1981. *A Survey of Sensible and Latent Heat Thermal Energy Storage Projects*. SERI/RR-355-456. Golden CO: Solar Energy Research Institute.

Bell, M. A. 1981. Low grade heat storage using sodium acetate solution. *Papers presented at the International Conference on Energy Storage, Brighton, U.K. April 29–May 1, 1981*. Cranford, Bedfordshire, U.K.: BHRA Fluid Engineering. Paper J2.

Benson, D. K. 1983. Organic polyols: Solid state phase change materials for thermal energy storage. *Opportunities in Thermal Storage R&D*, W. Hausz and B. Berkowitz, eds. EPRI EM-3159-SR. Paper P19.

Benson, D. K. 1984. Solid state phase change materials for thermal energy storage in passive solar heated buildings. *Proc. Passive and Hybrid Solar Energy Update*, pp. 74–79.

Birchenall, C. E. 1980. Heat storage in alloy transformations. *Proc. DOE Thermal and Chemical Storage Annual Contractors' Review Meeting*, McLean, VA, October 14–16, 1980. CONF-801055.

Biswas, D. R. 1977. Thermal energy storage using sodium sulfate decahydrate and water. *Solar Energy* 19: 99.

Boeing Engineering and Construction. 1977. *Advanced Thermal Energy Storage Concept Defintion Study for Solar Brayton Power Plants*, vol. 1, SAN/1300-1.

Bohn, M. S. 1983. Experiment and economic analysis of an air/molten salt direct-contact heat exchanger. *Proc. DOE Physcial and Chemical Energy Storage Annual Contractors' Review Meeting*. Arlington, VA, September 12–14, 1983. CONF-830974, pp. 59–66.

Bourne, R. C. 1981. *Membrane-Lined Foundations for Liquid Thermal Storage*. DOE/ET/20111-1.

Bramlette, T. T., R. Green, J. Bartel, C. Ottesen, C. Schafer, and T. Brumleve. 1976. *Survey of High-Temperature Thermal Energy Storage*. SAND 75-8063.

Bruckner, A. P., and A. Hertzberg. 1982. *High Temperature Integrated Thermal Energy Storage System for Solar Thermal Applications*. SERI/STR-231-1812. Golden, CO: Solar Energy Research Institute.

Buckman, R. W. Jr., 1980. *Lightweight Concrete Materials and Structural Systems for Water Tanks for Thermal Storage*. COO-4703-26.

Busico, V., C. Carfagna, V. Salerno, M. Vacatello, and F. Fittipaldi. 1980. The layer perovskites as thermal energy storage systems. *Solar Energy* 24: 575.

Chahroudi, D. 1977. Thermocrete and thermotile building components with isothermal heat storage. *Proc. Second Annual Thermal Energy Storage Contractors' Information Exchange Meeting*. Gatlinburg, TN, September 29–30, 1977. CONF-770955.

Chen, J., R. Nelson, and F. Polinski. 1981. *Pelletization and Roll Encapsulation of Thermal Energy Storage Materials*. ORNL/TM-8543.

Chuard, P., and J. C. Hadorn. 1983. *Central Solar Heating Plants with Seasonal Storage—Heat Storage Systems: Concepts, Engineering Data and Compilation of Projects*. IEA A24389/3.

Clark, E. C. 1978. *Final Report—Phase II Sulfuric Acid-Water Chemical Heat Pump and Storage System*. Rocket Research Co. Report RRC-78-R-595 for DOE contract no. EY-76-C-03-1183.

Clark, E. C. 1979. *Final Report—Phase II-A Sulfuric Acid and Water Chemical Heat Pump/ Chemical Energy Storage Program*. Rocket Research Co. Report RRC-79-R-627 for Sandia contract no. 18-4958.

Close, D. J., and R. Dunkle. 1977. Use of adsorbent beds for energy storage in drying of heating systems. *Solar Energy* 19: 233.

Cole, R. L., K. Nield, R. Rohde, and R. Wolosewicz. 1980. *Design and Installation Manual for Thermal Energy Storage.* 2d. ed. ANL-79-15.

Cole, R. L. 1983. Personal communication.

Copeland, R. J. 1983. Advanced high-temperature molten-salt storage research. *Proc. DOE Physical and Chemical Energy Storage Annual Contractors' Review Meeting,* Arlington, VA, September 12–14, 1983, CONF-830974, pp. 54–58.

Curto, P., and A. Gillespie. 1981. Solar cogeneration for copper smelting. *Proc. 1981 Annual Meeting, AS/ISES.* pp. 732–736.

Davidson, W. S. N.d. *Closed Brayton Cycle Advanced Central Receiver Solar-Electric Power System,* vol. 2. Final report. SAN/1726-1.

Davision, R. R., and W. B. Harris. 1975. Storing sunlight undergound—The Solaterre System. *Chemtech.* (December): 736–741.

deJong, A. G., and C. J. Hoogendoorn. 1980. Improvement of heat transport in paraffins for latent heat storage systems. *Thermal Storage of Solar Energy—Proc. International. TNO-Symposium,* Amsterdam, The Netherlands November 5, 1980. C. den Ouden, ed. The Hague: Martinus Nijhoff.

Dubberly, L. J., J. Gormley, W. Lang, D. Liffengren, A. McKenzie and R. Porter. 1981. *Cost and Performance of Thermal Storage Concepts in Solar Thermal Systems Phase 2: Liquid Metal Receiver.* SERI/TR-XP-0-9001-1-B. Golden, CO: Solar Energy Research Institute.

Dubberly, L. J., J. Gormley, W. Lang, D. Liffengren, A. McKenzie, and R. Porter. 1983. *Comparative Ranking of Thermal Storage Systems, Volume II Cost and Performance of Thermal Storage Concepts in Solar Thermal Systems, Phase I,* vol. 2. SERI/TR-631-1283. Golden, CO: Solar Energy Research Institute.

Forbes, R. E. 1981. An inexpensive site-assembled thermal storage tank, *Proc. AS/ISES 1981 Annual Meeting,* pp. 555–556.

Furbo, S. 1980. Heat storage with an incongruently melting salt hydrate as storage medium based on the extra water prinicple. *Thermal Storage of Solar Energy. Proc. International. TNO-Symposium,* Amsterdam, The Netherlands, November 5, 1980. The Hague: Martinus Nijhoff.

Furbo, S. 1978. *Investigation of Heat Storages with Salt Hydrate as Storage Medium Based on the Extra Water Principle.* Thermal Insulation Laboratory, Technical University of Denmark (Meddelelse no. 80).

General Electric CR&D. 1977. *Cool Storage Assessment Study.* EPRI EM-468.

Greene, N. D., and W. Watson. 1979. *Proc. Solar Energy Storage Options,* San Antonio, TX, March 19–20, 1979. M. B. McCarthy, ed. CONF-790328-P1,2,3, pp. 465–472.

Hallet, R. W., Jr., and R. Gervais. 1979. The solar ten megawatt pilot plant. *Proc. ISES Solar Jubilee Congress.* pp. 1137–1140.

Helshoj, E. 1981. A high-capacity, high-speed latent heat storage unit. *Proc. Solar World Forum, ISES,* pp. 703–709.

Herrick, C. S., and K. Zarnoch. 1979. Heat storage capability of a rolling cylinder using Glauber's salt. *Thermal Energy Storage Fourth Annual Review Meeting.* Tyson's Corner, VA, December 3–4, 1979. CONF-791232, pp. 239–260.

Hogsett, J. 1990. Design and preliminary performance of a two-tank oil storage system for an electric generating plant. *Proc. Intersol 85, ISES,* forthcoming.

Hudson, W. T. 1981. *Final Report* (draft). Independent Living, Inc. DOE contract no. EM-78-C-02-4699.

Hunke, R. W., W. Mertz, and A. Poche. 1979. Definitive design of the solar total energy large scale experiment at Shenandoah, Georgia. *Proc. ISES Silver Jubilee Congress.* pp. 1117–1120.

Johnson, T. L., L. Glasgow, W. Thomson, and A. Frangos. 1979. Sodium cooled solar central receiver power station. *Proc. ISES Silver Jubilee.* pp. 1146–1150.

Jones, G. F. 1984. Liquid convective diodes. *Proc. Passive and Hybrid Solar Energy Update.* CONF-8409118. p. 88.

Kalt, A. 1979. The International Energy Agency Small Solar Power Plant Project. *Proc. ISES Silver Jubilee.* p. 1077.

Kando, P. F., and M. Telkes. 1978. *Characterization of Sodium Sulfate Dekahydrate (Glauber's Salt) as a Thermal Energy Storage Material.* C00-4042-16.

Kauffman, K., and I. Grundfest. 1973. *Congruently Melting Materials for Thermal Energy Storage.* University of Pennsylvania report for NSF RANN NCEMP-20.

Kauffman, K. W., H. Lorsch, and D. Kyllonen. 1977. *Thermal Energy Storage by Means of Saturated Aqueous Solutions.* TID-28330.

Koehne, R., M. Kraft, and A. Pérez Vidal. 1981. Two years' operation of the 30/50 kW$_E$ solar farm at Getafe, Spain. *Proc. Solar World Forum, ISES.* pp. 2999–3005.

Krivokapich, G., D. Fenton, G. Abernathy, and J. Otts. 1983. Operational characteristics of the solar thermal power system near Willard, New Mexico. *ASME Journal of Solar Energy Engineering* 105: 268.

Kriz, T., C. Christensen, H. Gaul, J. Leech, A. Rabl, S. Sillman, C. J. Swet, and J. Ullman. 1983. *Thermal Energy Storage for Process Heat and Building Applications.* SERI/TR-231-1780. Golden, CO: Solar Energy Research Institute.

Lane, G. A., J. Best, E. Clarke, D. Glew, G. Karris, S. Quigley, and H. Rossow. 1976. *Isothermal Solar Heat Storage Materials.* NSF/RANN/SE/C906/FR/76/1.

Lane, G. A., and H. Rossow. 1976. Encapsulation of heat-of-fusion materials. *Proc. 2nd Southeastern Conference on Application of Solar Energy.* CONF-760423, pp. 442–450.

Lawrence, W. 1980. *Capital Cost Estimates of Selected Advanced Thermal Energy Storage Technologies.* ANL/SPG-11.

LeFrois, R. 1979. Active heat exchange system development for latent heat thermal energy storage. *Thermal Energy Storage Fourth Annual Review Meeting,* Tyson's Corner, VA, December 3–4, 1979. CONF-791232.

LeFrois, R. T., and H. Venkatasetty. 1976. Inorganic phase change materials for energy storage in solar thermal program. *Proc. Sharing the Sun,* AS/ISES and Solar Energy Soc. of Canada, vol. 8, pp. 107–132.

LeFrois, R. T. 1978. *Active Heat Exchanger System Development for Latent Heat Thermal Energy Storage System.* Honeywell Topical Report 78336 for NASA-Lewis contract DEN 3-38.

Lee, C., L. Taylor, J. DeVries, and S. Heibein. 1979. *Solar Applications of Thermal Energy Storage.* Final report H-C0199-79-753F. Hittman Associates, Inc.

Leffler, A. J. 1977. The use of solid state phase transitions for thermal energy storage. *Proc. of Second Annual Thermal Energy Storage Contractors' Information Exchange Meeting,* Gatlinburg, TN, September 29–30, 1977. CONF-770955, pp. 170–171.

Lenz, T. G., and L. Hegedus. 1979. Moderate temperature thermochemical systems—An application of Diels-Alder chemistry. *Proc. Silver Jubilee Congress, ISES,* pp. 154–158.

Leonard, J. 1978. Operating experience at the DOE/Sandia Midtemperature Solar Systems Test Facility. *Proc. 13th IECEC,* pp. 1662–1667.

Leonard, J., and R. Hunke. 1981. The Shenandoah Total Energy Project. *Proc. 1981 Annual Meeting, AS/ISES,* pp. 334–338.

Lin, E. I. H., and W. Sha. 1979. Effects of baffles on thermal stratification in thermocline storage tanks. *Proc. ISES Silver Jubilee Congress,* pp. 586–590.

Lindblom, U. 1981. The block-filled cavern and pond—New systems for low-cost seasonal thermal energy storage. *Proc. International Conference on Seasonal Thermal Energy Storage and Compressed Air Energy Storage,* Seattle, WA, October 19–21, 1981. CONF-811066, vol. 1.

MacCracken, C. D. 1982. Effect of additivies on performance of hydrated salt TES systems. *Proc. DOE Physical and Chemical Storage Annual Contractors' Review Meeting.*

MacCracken, C. D. 1980. *Bulk Storage of PCM.* Final report by Calmac Mfg. Corp. for DOE Office of Solar Applications.

Mancini, N. A. 1980. Use of paraffins for thermal storage. *Thermal Storage of Solar Energy. Proc. International TNO-Symposium,* Amsterdam, The Netherlands, November 5, 1980. C. den Ouden, ed. The Hague: Martinus Nijhoff.

Margen, P. 1980. Swedish storage practice and developments relating to district heating systems. Presented at Thermal Storage Seminar, Novemeber 13–14, Toronto.

Marks, S. B. 1983. The effect of crystal size on the thermal energy storage capacity of thickened Glaubers salt. *Solar Energy* 30: 45.

Martin Marietta Denver. 1981. *Thermal Storage for Solar Cooling Using Paired Ammoniated Salt Reaction.* DOE-CS-34700-2.

Martna, J. 1981. The Avesta test plant for storage of hot water in an unlined rock cavern. *Proc. International Conference on Seasonal Thermal Energy Storage and Compressed Air Energy Storage,* Gatlinburg, TN, September 29–30, 1977. CONF-770955, vol. 1, p. 186.

Maru, H., J. Dullea, A. Kardas, and L. Paul. 1978. *Molten Salt Thermal Energy Storage Systems.* C00-2888-3.

Matteo, M., and D. Rafinejad. 1980. Design, construction, and operation of a solar-powered 150 kW irrigation facility. *Proc. 1980 Annual Meeting, AS/ISES,* pp. 519–523.

McClelland, J. F., R. Mercer, L. Hodges, R. Szydlowski, P. Sidles, R. Struss, J. Hull, and D. Block. 1981. TRANSWALL—A modular visually transparent thermal storage wall. Status report. *Proc. ISES Solar World Forum,* p. 1881.

McLinden, M., and S. Klein. 1983. Simulation of an absorption heat pump solar heating and cooling system. *Solar Energy* 31: 473–482.

Mears, D. R., W. Roberts, J. Simpkins, P. Keudall, J. Cipolletti, and H. Janes. 1980. The Rutgers system for solar heating of commerical greenhouses. *Proc. 1980 Annual Meeting, AS/ISES,* pp. 59–63.

Melville, J. G. 1981. Aquifer storage using the doublet well configuration. *Proc. International Conference on Seasonal Thermal Energy Storage and Compressed Air Energy Storage,* Seattle, WA, October 19–21, 1981. CONF-811066, vol. 1, p. 103.

Meyer, C. F., and W. Hausz. 1980. *Guidelines for Conceptual Design and Evaluation of Aquifer Thermal Energy Storage.* Battelle PNL-3581.

Michaels, A. I. 1982. A review of latent heat storage technology. *Solar Storage. Proc. Third SOLERAS Workshop,* J. S. Williamson and B. H. Khoshaim, eds. MRI/SOL-1101. CONF-820312.

Mumma, S. A., and T. C. Liu. 1980. PCM-concrete energy storage and building element. *Proc. 1980 Annual Meeting, AS/ISES,* pp. 243–246.

Munoz Torralbo, A. 1981. A Spanish "power tower" solar system. The project CESA-1. *Proc. Solar World Forum, ISES,* pp. 2931–2935.

Nichols, M. C., and R. Green. 1977. *Direct Contact Heat Exchange for Latent Heat-of-Fusion Energy Storage Systems.* SAND77-8665.

Nishiyama, E., M. Sugihara, J. Kai, K. Kashiwamura, and M. Ohtsubo. 1979. Test result of the solar powered Rankine cycle refrigerator installed in the experimental house. *Proc. Silver Jubilee Congress, ISES,* pp. 691–695.

Nix, R. G. 1983. Thermochemical energy systems research. *Proc. DOE Physical and Chemical Energy Storage Annual Contractors' Review Meeting.* Arlington, VA, September 12–14, 1983. CONF-830974, pp. 67–71.

Nix, R. G. 1982. Reversible chemical reactions for energy storage in a large-scale heat utility. *Proc. 17th IECEC.*

Nonnenmacher, A., and M. Groll. 1981. Chemical heat storage and heat transformation using reversible solid-gas reactions. *Papers Presented at the International Conference on Energy Storage*, Brighton, U.K. April 29–May 1, 1981. Cranford, Bedfordshire, U.K.: BHRA Fluid Engineering. Paper C3.

Oelert, G. 1982. *Thermochemical Heat Storage State-of-the-Art Report*. Document D2: 1982. Stockholm: Swedish Council for Building Research.

Offenhartz, P. 1980. A heat pump and thermal storage system for solar heating and cooling based on the reaction of calcium chlroide and methanol vapor. *ASME J. of Solar Energy Eng.* 102(1): 59–65.

Offenhartz, P. 1981a. A chemical heat pump based on the reaction of calcium chloride and methanol for solar heating, cooling, and storage. *Proc. DOE Thermal and Chemical Storage Annual Contractors' Review Meeting*. McLean, VA, October 14–16, 1980. CONF-801055.

Offenhartz, P. 1981b. *Thermal Storage Studies for Solar Heating and Cooling: Applications Using Chemical Heat Pumps*. DOE/CS/30248-F.

One Design, Inc. 1984. *The Enlightenment of the Wall* (brochure). Winchester, VA.

Page, J. K. R., R. Swayne, I. Mead, and C. Hayman. 1981. Thermal storage materials and components for solar heating. *Proc. Solar World Forum, ISES*, pp. 723–730.

Pickering, E. E. 1976. *Residential Hot Water Energy Storage Subsystems*. NSF RA-N-75-095.

PNL. 1981. *Proc. International Conference on Seasonal Thermal Energy Storage and Compressed Air Energy Storage*, Seattle, WA, October 19–21, 1981. CONF-811066, vol. 1.

Prengle, H. W., and C. H. Sun. 1976. Operational chemical storage cycles for utilization of solar energy to produce heat or power. *Solar Energy* 18: 561–567.

Raldow, W. 1981. Thermal efficiencies of chemical heat pump configurations. *Solar Energy* 27: 307.

Reiter, F., and R. Rota. 1984. Low temperature latent heat storage by reciprocal salt pairs. *Solar Energy* 32: 499.

Riaz, M. 1977. *Rock Bed Heat Accumulators*. C00/4009-1.

Rocket Research Co. 1982a. *Sulfuric Acid/Water Chemical Heat Pump/Chemical Energy Storage —Final Report*, vol. 1. Report no. RRC-82-R-813 for BNL contract no. 494588-S.

Rocket Research Co. 1982b. *Sulfuric Acid/Water Chemical Heat Pump/Chemical Energy Storage—Final Report*, vol. 2. Report no. RRC-82-R-813 for BNL contract no, 494588-S.

Rockwell International. 1979. *Hybrid Thermal Storage with Water*. ESG-DOE-13259.

Rockwell International. 1980. *Solar Energy Storage by Reversible Chemical Processes Final Reprot*. SAND79-8199.

Salyer, I. O., and J. Davison. 1980. Development of an optimum process for electron beam crosslinking of high density polyethylene pellets. *Proc. DOE Thermal and Chemical Storage Annual Contractors' Review Meeting*, McLean, VA, October 14–16, 1980. CONF-801055.

Salyer, I. O., D. Duvall, C. Griffen, S. Molnar, R. Chartoff, and D. Miller. 1984. Advanced phase-change materisls for passive solar storage applications. *Proc. Passive and Hybrid Solar Update*. DOE/CONF-8409118, pp. 61–71.

Sansone, M. J. 1978. *Encapsulation of Phase Change Material in Concrete Masonry Construction. Progress Report No. 2*. Brookhaven report 50896.

Scott, O. L. 1979. Internally insulated thermal storage system development program. *Thermal Energy Storage Fourth Annual Review Meeting*. CONF-791332, held in Tyson's Corner, VA, Dec. 3–4, 1979, pp. 141–156.

Sharma, S. K., C. Jotshi, and A. Singh. 1982. Evaluation of PCM and double decomposition reactions for thermal energy storage. *Solar Storage, Proc. of the Third SOLERAS Workshop*,

J. S. Williamson and B. H. Khoshaim, eds. MRI/SOL-1101. CONF-820312 (appears in preliminary issue only).

Shigeishi, R., C. Langford, and B. Hollebone. 1979. Solar energy storage using chemical potential changes associated with drying of zeolites. *Solar Energy* 23: 489.

Skartvedt, G. 1981. Solar water heating. In *Solar Energy Handbook*, J. Kreider and F. Kreith, eds. New York: McGraw Hill, ch. 11.

Smith, R. D. 1979. *Chemical Energy Storage for Solar Thermal Converison*. SAND79-8198.

Solar Components Corp. 1982. *Solar Catalog.*

Swet, C. J. 1980. Phase change storage in passive solar architecture. *Proc. 5th National Passive Solar Conference, AS/ISES*, pp, 282–286.

Swet, C. J. 1981. Energy storage for solar applications. In *Solar Energy Handbook*, J. Kreider and F. Kreith, eds. New York: McGraw-Hill, ch. 6.

Tamblyn, R. T. 1980. Thermal storage: Resisting temperature blending. *ASHRAE Journal* (January): 69.

Tanaka, T. 1980. Solar thermal electric power systems in Japan. *Solar Energy* 25: 97.

Tchernev, D. I. 1978. Solar energy applications of zeolites. In *Natural Zeolite Occurrence, Properties, Use*, L. B. Sand and Mumpton, eds. Pergamon Press. pp. 479–485.

Telkes, M. 1974. Storage of solar heating cooling. *ASHRAE Trans.*, vol. 80, pt. 2.

Thomason, H. E., and H. J. L. Thomason, Jr. 1973. Solar houses/heating and cooling progress report. *Solar Energy* 15: 27.

Torrenti, R. 1979. Seasonal storage in solar-heating systems. *Proc. ISES Silver Jubilee*, p. 624.

Turner, R. H. 1977. Thermal energy storage in solids. *Proc. Second Annual Thermal Energy Storage Contractors' Information Exchange Meeting*, Gatlinburg, TN, September 29–30, 1977. CONF-770955.

Turner, R. H., and H. Awaya. 1978. High temperature thermal energy storage in moving sand. *Proc. 13th IECEC*, pp. 923–927.

Vachaud, G., and J. P. Person. 1981. Field experimentation of the "soil therm" interseasonal storage system of solar energy in the subsoil. *Proc. International Conference on Seasonal Thermal Energy Storage and Compressed Air Energy Storage*. Seattle, WA, October 19–21, 1981, CONF-811066, vol, 1. p. 203.

van Galen, E. 1980. Experimental results of a latent heat storage system based on sodium acetate trihydrate in a stabilizing colloidal polymer matrix, tested as a component of a solar heating system. *Thermal Storage of Solar Energy. Proc. International TNO-Symposium*, Amsterdam, The Netherlands, November 5, 1980. C. den Ouden, ed. The Hague: Martinus Nijhoff, pp. 147–156.

van Koppen, C. W. J., and J. P. S. Thomas. 1978. *Preliminary Performance of the Heating System in the Solar House of the Eindhoven Univeristy of Technology*. Eindhoven University of Technology report no. WPS3-78.11.R291. Eindhoven, The Netherlands.

Walker, W. R. 1981. *Studies of Heat Transfer and Water Migration in Soils*. Final report, DOE/CS/30139.

Ward, J. C., W. Loss, and G. Löf. 1978. Direct contact liquid-liquid heat exchanger: Pilot plant results. *Proc. 1978 Annual Meeting, AS/ISES*, pp. 413–419.

Webb, J. D., and S. Pohlman. 1979. *Reliability and Durability Study of a Thermal Receiver Utilizing ASI Type 316 Stainless Steel in Contact with Molten Aluminum*. SERI/TR-31-338. Golden, CO: Solar Energy Research Institute.

Wells, P., and G. Nassopoulos. 1982. Molten salt thermal energy storage subsystem for solar thermal central receiver plants. *Proc. Sixth Annual Thermal and Chemical Storage Contactors' Review Meeting*, Washington, DC, September 14–16, 1981. CONF-810940.

Westinghouse Advanced Energy Systems Division. 1979. *Evaluation of Prestressed Cast Iron Pressure Vessels for Coal Gasifier.* DOE FE-3013-1.

Wettermark, G., ed. 1980. *Proceedings from the International Seminar on Theromchemical Energy Storage.* Document D25:1980. Stockholm: Swedish Council for Building Research.

Worsoe-Schmidt, P. 1979. A solar-powered solid-absorption refrigeration system. *Int'l. J. of Refrigeration* 2(2): 75–84.

Wright, R. L., T. Suchocki, and D. Schluderberg. 1982. Conceptual design of the moving bed thermal energy storage system for a commerical-scale (100 MWe) solar central receiver power plant. *Proc. Sixth Annual Thermal and Chemical Storage Annual Contractors' Review Meeting.* Washington, DC, September 14–16, 1981. CONF-810940.

Wright, J. D., and M. Bohn. 1982. Direct-contact thermal storage research. *Proc. of the DOE Physical and Chemical Storage Annual Contractors' Review Meeting.*

Yuan, S. W., and A. Bloom. 1977. *Long Duration Earth Storage of Solar Energy.* Final report to ERDA (under NSF grant no. AER 75-18608-A02).

16 Analytical and Numerical Modeling of Thermal Energy Storage

John R. Hull

16.1 Introduction

16.1.1 Why Model?

A mathematical model of a physical system consists of a set of coupled equations relating the system's physical parameters. The equations used are based on the known laws of nature plus a set of assumptions that emphasize the most important aspects of the system. Choosing the proper assumptions is the most important task of a modeler. Analytic solutions are preferable for elucidating the effects of parameter variations, but except for the most simple models, it is unusual to find a set of equations that has a closed form solution. Numerical solutions are sufficiently accurate for a large number of complicated systems, however.

Modeling of engineered systems and components is generally considered desirable in the development and commercialization of most technologies. An accurate model allows the behavior of the system to be examined for many sets of parameters that cannot be reproduced experimentally. Modeling reduces developmental costs when a device is in the conceptual stages, when experiments are expensive, or when investigating the parameter values of interest is likely to lead to destruction of the device. Modeling energy technologies is particularly valuable, because levelized energy costs must be calculated to compare various competing technologies. To calculate these costs, system performance must be predicted for many diverse climates and for many different load conditions. See Duff and Wynn (1981) for some other advantages of modeling solar energy systems, as well as for some general modeling approaches.

16.1.2 Scope

In this chapter we analyze the present state of the art in modeling thermal energy storage. We focus on the different modeling methods available and their applicability to the various important problems associated with thermal storage. Results generated by the different models do not receive much attention here, except to illustrate a particular applicability. The focus is also restricted to active system thermal performance. The important topics of energy storage modeling in passive solar systems and economics are discussed in other chapters. Technologies that are in the early developmental stage, where extensive modeling is either inappropriate or has not been pursued, are not mentioned.

In practical applications, energy storage devices are valuable only when they are combined with an energy input and energy load in a complete system. As expected, energy storage components are an integral part of many energy system models, but usually in a simplified form. The focus here is mainly on the storage component and not on the system. Because heat transfer between the storage material and the working fluid that comes from the energy source or goes to the load is one of the most critical parameters of an energy storage system, modeling of basic heat transfer processes must also enter the discussion. This is especially important in phase change material systems.

A great number of computer programs now model some aspect of energy storage. Many of these programs are available to the public, and some are well documented and relatively easy to use. Space limitations prohibit our cataloging these programs here. Kriz and Swet (1982) have briefly summarized a few of the computer models. The references and discussion in this chapter have been selected to illustrate the current state of the art, and they are not meant to be comprehensive.

16.1.3 General Storage Model

All of the elements for a general model of a thermal storage system are shown schematically in figure 16.1 The storage unit is divided into numbered cells, which are indicated by the dashed boxes. Each cell has its own temperature and heat capacity. The thermal performance of the system is then determined by specifying the thermal conductance U between each of the cells and between the cells and the sources and sinks in the system. The value of the conductance may depend on temperature, possibly on other parameters, and may range from zero to infinity.

The particular configuration shown in figure 16.1 might represent a domestic hot water heater, heated by a submerged heat exchanger, which is connected to an external energy source such as a solar collector. Cells 3 and 4 represent the water in the tank, with water from the mains entering the tank in cell 3 at d and leaving the tank to the load from cell 4 at f. Typically cell 3 would represent the bottom half of the tank, and cell 4 the top half. Path e indicates a fluid connection between the two cells. Cells 1 and 2 represent the solid part of the heat exchanger. Heat enters the system via $U_{1,b}$ and $U_{2,c}$, which connects the heat transfer fluid at points b and c to cells 1 and 2, respectively. Heat enters the tank liquid via $U_{1,3}$ and $U_{2,4}$. The system may leak heat to the environment, which is at ambient temperature T_a. For example, $U_{3,a}$ determines the leakage through the wall insulation from the bottom of the tank. If the heat exchanger is totally submerged within the tank, $U_{1,a}$ and $U_{2,a}$ will be zero.

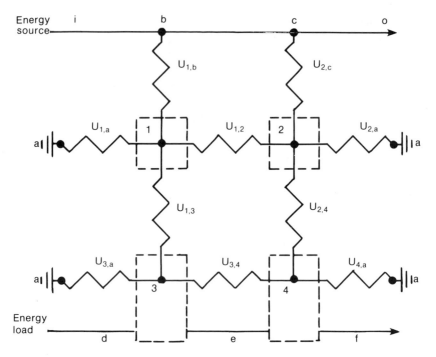

Figure 16.1
Schematic diagram of an example of a thermal network storage model.

To translate the schematic model into a mathematical model, the energy balance for each cell must be written. For example, for cell 1 the energy balance is

$$\frac{dE_1}{dt} = U_{1,2}(T_2 - T_1) + U_{1,3}(T_3 - T_1) + U_{1,a}(T_a - T_1) + U_{1,b}(T_b - T_1), \quad (1)$$

where E is the energy content, t is time, T is the temperature, and the subscripts of E and T designate the cell number. The equation of state relates E and T for a single cell; for example,

$$E_1 = r_1 c_1 V_1 T_1, \quad (2)$$

where r is the density of material in the cell, c is the specific heat, and V is the volume of the material. If more than one type of material is present in the cell, then the product of r, c, and V must be summed for each component. If a phase change occurs within the cell, then E will also depend on the amount of material in each phase. The specific heat may be a function of temperature as well as of phase.

An energy balance must also be written for the fluid in the heat exchanger; for example,

$$\frac{dE_b}{dt} = U_{1,b}(T_1 - T_b) + m_S r_i c_i T_i - m_S r_b c_b T_b, \tag{3}$$

where m_S is the volumetric flow rate in the heat exchanger connected to the energy source. Because fluid from outside the system is allowed to flow into the tank, the energy balance for cells 3 and 4 is similar to equation (3); for example,

$$\frac{dE_3}{dt} = U_{1,3}(T_1 - T_3) + U_{3,4}(T_4 - T_3) + U_{3,a}(T_a - T_3)$$

$$+ m_L r_d c_d T_d - m_L r_3 c_3 T_3, \tag{4}$$

where m_L is the volumetric flow rate to the load. In both equations (3) and (4), the effects of thermal expansion have been ignored.

All that is needed to complete the model is specification of the initial temperature conditions and the time-dependent input conditions m_S, T_i, m_L, T_d, and T_a. A mixing algorithm is needed for liquids with a temperature-dependent density. A control algorithm is also often needed to specify behavior for times when the storage cannot meet the demand.

Most models for energy storage are fashioned around the general format listed above. Major differences between these models arise from how the cells are linked together and what solution scheme is used. The storage device can be divided into as many cells as desired, and the model's ability to solve the resulting equations acts as the main constraint on complexity. The number of cells is usually limited by the size and speed of the computer used, so many models will employ only one computational cell. The advantage here is that solutions are relatively easy to obtain and that varying a relatively few parameters usually reveals specific areas where the device can best be optimized, at least in the early stages of design. When only one computational cell (or a few) is used, details about the internal behavior of the device are lost because many different effects must be lumped into one parameter. For example, the heat losses that occur in the energy transfer from collector to storage and from storage to load may be accounted for by increasing the U value of the heat loss predicted from storage to ambient.

Most solar energy systems and many energy storage devices experience a yearly cycle of operating conditions, and the important factor for economic calculations is usually the yearly efficiency of the system. When the technology is fairly well developed and the results of a detailed model have been validated against actual performance data for a number of different situations, a detailed model may be used to generate a more simple design model. The detailed model can predict the performance of a given system, at 15-minute or 1-hour time increments, using actual or typical weather data for a given location. The model can be run for many sets of parameters to find the optimal configuration for maximum yearly efficiency. Once the system is optimized, a simpler model presents the performance results in terms of fewer parameters. This simple model makes performance predictions on a monthly or yearly basis. Because variations in weather are similar in many different parts of the world, the results of the simple model are often valid for many other locations. The simple model usually does not make reliable daily predictions, but it is usually accurate to within a few percent on long-term predictions. The simple model is a great boon to a designer or system buyer because it can expedite economic calculations.

A classic example of this procedure is the use of TRNSYS by Klein et al. (1976) to produce f-Chart. TRNSYS (University of Wisconsin-Madison 1979) is a general computer simulation program for solar energy systems. The code is based on a modular approach, in which the system simulation is performed by serial solution to the governing equations of independently modeled components. Extensive experience in simulating solar heating systems helped identify and eliminate details in the component models that had only a small effect upon long-term system performance. The relationships between system performance and the remaining parameters were then presented in graphical form, resulting in f-Chart (Beckman et al. 1977). Based on monthly averaged weather data, the f-Chart method can quickly determine the optimal parameters of a standard solar heating system.

16.1.4 Model Validation

One of the most important, and unfortunately most often neglected, tasks in the development of a model is its validation. Validation occurs when the predicted results of the system output parameters and internal temperatures agree with experimental results, given the same input conditions. The validation process can help to refine a particular model in a number of ways, but its main function is to show that the model is self-consistent and free of errors. If a given design is still in the conceptual stage and experimental results cannot be obtained, then the model should be tested against either another well-validated model or under conditions in which an exact mathematical solution is available.

Often some or all of the model parameters (e.g., the U values) will not be known a priori. An adaptation of the validation process can help to determine these values if experimental data are available and if the solution time of the model is not too great. The technique is to vary the unknown parameters of the model until a best fit, usually in the least-squares sense, is obtained with the data. There are a number of well-known algorithms to optimize this procedure (e.g., see Bevington 1969). If, after the parameters have been optimized, the predicted results still do not represent the data well, then the model might have to be revised. Unless the system is linear in terms of the fitted parameters, this fitting procedure is useful only when the number of parameters is manageable (e.g., less than 10).

16.1.5   Thermal Network Models and Transient Response Functions

The similarity of the symbol for thermal conductance in figure 16.1 to an electrical resistor is no coincidence. The physical variables of most thermal problems correspond directly to the variables of electric circuits. Thermal resistance (the inverse of conductance) corresponds to electrical resistance, heat capacity corresponds to electrical capacitance, and so on. Systems like the one indicated in figure 16.1 are often referred to as *thermal network models* and sometimes as *lumped-parameter models*. The result of this correspondence is that all of the advanced techniques devised for solving electric network problems can be used for thermal problems. Furthermore a heat transfer problem can be solved with a resistor and capacitor model of the system, or equivalently with an analog computer. Most modeling of thermal storage systems has employed a digital computer. However, Beard et al. (1978) achieved very accurate results using an electrical resistance-capacitance model to simulate heat transfer from a hemispherical pool of water to the surrounding soil, and Sheridan et al. (1967) used an analog computer to study stratification in hot water storage tanks.

Another modeling alternative to finite-difference approximations is to use transient-response functions, or convolution models. A convolution model is based on the principle of linear superposition, in which the inlet temperatures and flows are described as a series sum of particular basic functions. The transient response of the system is solved analytically for the basic function input, and the solution to an actual input is then the series sum of response functions that corresponds to the basic functions describing the input. Riaz (1978) used impulse, step, ramp, and sinusoidal basic functions to describe the behavior of a rock bed. The advantages of convolution models include a freedom from discretization error and an absence of *numerical* stability problems. The accuracy of the results are limited only by how well the sum of basic functions approximates the actual inlet conditions.

16.2   Sensible-Heat Storage in Liquids

16.2.1   Hot Water Storage

Hot water tanks for diurnal storage applications are undoubtedly the most common energy storage device, and as expected, modeling of hot water storage has received more attention than any other kind of energy storage modeling. The performance of a water tank system depends on the amount of water present, the shape and material composition of the tank, the amount of external insulation, the sizes and locations of the inlets and outlets, and the water flow rate. Thermodynamic and hydraulic processes within the tank can be complex. Hydraulic disturbances caused by flow into the tank can be very difficult to model. Modeling the internal natural flows induced in storage tanks by the vertical side walls can also be quite complicated, even with no external flow (Hess and Miller 1982), but simplified models can usually represent this behavior accurately enough (Jaluria and Gupta 1982). A number of computer programs now model hot water tanks with various degrees of success and complexity. The main details and validation for some of these models have been reviewed by Kuhn et al. (1980).

An accurate simulation of the complex hydraulic behavior in water storage tanks can only be given by a properly designed, transient, three-dimensional mathematical model such as COMMIX-SA (Sha et al. 1980). COMMIX-SA implements a first-order, finite-difference approximation of the full Navier-Stokes equations. The conservation equations for mass, momentum, and energy are solved using a modified ICE (implicit continuous-fluid Eulerian) technique in a staggered mesh of either rectangular or cylindrical coordinates. Any porous or solid structures within a cell are incorporated into the model. As demonstrated by Hull et al. (1981), COMMIX-SA is capable of modeling very subtle aspects of fluid behavior.

Sha and Lin (1978) and Lin and Sha (1979) have used COMMIX-SA to examine the thermal behavior of a wide variety of storage tank designs. As shown in figure 16.2, the output of this type of code can provide insight into the hydraulics of a given system. Three-dimensional codes such as COMMIX-SA are very valuable for modeling the detailed performance of components and for understanding the physical processes that occur in different designs and for different operating conditions. However, such programs are too complex and time-consuming to use for routine applications in system design.

A consideration that makes one-dimensional models for system design attractive is that storage tanks can be created in which stratification approaches the theoretical maximum (Cole and Bellinger 1982). Except for

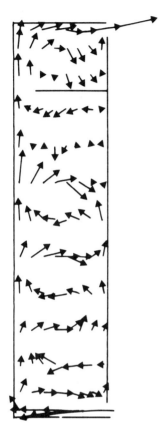

Figure 16.2
Flow pattern during discharge of a water storage tank with horizontal baffle, as modeled by COMMIX-SA. Arrows represent the velocity vector of the fluid, where the center of the fluid cell is at the tail end of the arrow and the velocity is proportional to the length of the arrow.

hydraulic effects near the inlets, flow in the tank is essentially one-dimensional. As Hull et al. (1982) pointed out, good stratification is so easy to achieve in practice that there is no reason not to incorporate it into system design from the very beginning.

A number of one-dimensional computer models, based on the general storage model concepts discussed in subsection 16.1.3, have been formulated to model water tank storage. A very widely used model for water tanks is the Duffie and Beckman (1974) model employed in TRNSYS. The governing equations of this model are based on the energy balance for an n-section tank. The equation for section i is

$$dE_i = (mrc)_S \left[FS_i(T_{S,0} - T_i) + (T_{i-1} - T_i) \sum_{j=1}^{i-1} FS_j \right]$$

$$+ (mrc)_L \left[FL_i(T_{L,r} - T_i) + (T_{i+1} - T_i) \sum_{j=i+1}^{i-1} FL_j \right]$$

$$+ U_{i,a}(T_a - T_i). \tag{5}$$

The variables have the same meaning as in previous equations; subscript o refers to outlet (e.g., solar collector outlet) and subscript r refers to return. Here, the flow from both the source S and the load L passes through the tank, and the heat conduction between tank segments is ignored. The source and load control functions, FS and FL, are defined such that water entering at a tank inlet will enter the tank segment that has the same temperature as the entering fluid. This model would represent a system with a variable position inlet such as the vertical porous distribution manifold developed by Loehrke et al. (1979). For fixed inlet positions the model ignores any mixing that may occur as the fluid finds its proper level in the tank. This would not be a concern if only a few segments were used in the model, but it would be important if an accurate simulation were desired.

A model that avoids this limitation was developed by Sharp and Loehrke (1979). This model is similar to the TRNSYS model, but thermal conduction between segments is included, and the water enters the tank at the correct spatial locations. Two adjacent segments are allowed to mix completely whenever the temperature of the upper segment is less than the temperature of the lower segment.

Both of these models, however, ignore the details of hydraulic disturbances at the inlets. Han and Wu (1978a) successfully incorporated the effects of such disturbances into a one-dimensional model. This viscous-entrainment model was used to predict the performance of a simple tank system (Han and Wu

1978b). The work is an interesting example of the all-too-common phenomenon in which considerable effort is expended to model behavior arising from poor design precepts. In the early days of storage tank design research, little attention was paid to the details of the inlet. Water would often be allowed to gush vertically into the experimental tanks at a high velocity, causing a large amount of mixing. Now it is recognized that in a good tank design, the inlets can be constructed easily and inexpensively so that the disturbance is confined to the area immediately around the inlet, which is equivalent to one segment or less in a tank model. If a viscous-entrainment model is needed to model a storage tank system, the tank should be redesigned.

Phillips and Dave (1982) used the computer model of Sharp and Loehrke (1979) to validate a very simplified analytical storage tank model that incorporated the effects of stratification. They found that the behavior of a water storage tank connected to a solar collector could be well characterized by two parameters, the stratification coefficient K_S and the effectiveness of the collector–heat exchanger E. The coefficient K_S is defined as the ratio of the actual useful energy gain to the energy gain that would be achieved if there were no stratification. For most operating conditions, K_S is found to be a simple function of E and the Fourier number M, which is the ratio of conduction to forced convection in the storage tank. The usefulness of this model is that the system performance can be modeled by a single equation of the Hottel-Whillier-Bliss type involving only one unknown parameter, the mean storage temperature. However, the validation results of Phillips and Dave (1982) are somewhat puzzling in that they claim that the stratification coefficient is independent of the circulation number C_N, which is the number of times the entire storage mass is circulated through the system. For $C_N > 1$, one expects the stratification coefficient to decrease dramatically as C_N increases (e.g., see Rademaker 1981). A very large C_N would, of course, be equivalent to a fully mixed tank. The performance of the collector-tank system would also be expected to depend on the load dynamics, and it is not clear whether this model is adaptable to varying loads.

16.2.2 Stratification

The density of most fluids is a function of temperature, and density usually decreases as the temperature increases. As a result fluid in storage tanks will tend to stratify with the hotter fluid on top of the colder. One can take advantage of this thermocline formation to improve the thermal performance of a water storage tank connected to a solar collector. Pumping cold fluid from the bottom of the tank to the solar collector causes the heat loss from the collector to be lower. Pumping hot water from the top of the tank to the load means that less auxiliary heating will be required.

The ramifications of stratification in storage tank systems are still not fully appreciated. Most storage tank models underestimate the effects of stratification, and most system designers using the models examine operating conditions that do not fully utilize the stratification effect. Operating under conditions that are optimized for fully mixed tanks, most models predict an increase in performance of about 5% (e.g., Han and Wu 1978b) when stratification is used. When the system is reoptimized, incorporating the effects of stratification from the beginning, the predicted efficiency is higher. For example, Sharp and Loehrke (1979) optimized for the mass flow rate through the system and found an increase in efficiency over that of a fully mixed tank of 6% to 16% depending on application. Veltkamp (1981) claims that an increase in efficiency of 20% is possible with an optimized control strategy employing conventional flat plate collectors with glass covers.

All the models discussed in subsection 16.2.1 will always underestimate the performance of a stratified tank. The reason is that hydrodynamic models based on finite difference equations suffer from numerical diffusion, an effect that arises from the finite spatial size of the computational cells and the assignment of a single, homogeneous temperature to each cell. The result of numerical diffusion is the prediction of a temperature profile that is equivalent to increased mixing in the tank. This effect is illustrated by the two computational cells shown in figure 16.3. The lower cell occupies the spatial location $0 < x < 1$, and the upper cell occupies $1 < x < 2$. Initially both cells are at a hot temperature T_0. Cold water T_L enters at the bottom of the lower cell at $x = 0$. For purposes of this example, real thermal diffusion has been ignored. In figure 16.3 the actual temperature profile is indicated by the solid line, and the predicted temperature profile, calculated according to equation (5), by the dashed line. The temperature profile at the end of the first computational time step is shown in figure 16.3a. Note that the predicted profile, because of the difference in average temperatures, now exhibits a temperature gradient between the two cells. A change in energy will be predicted for the upper cell during the next time step, as shown in figure 16.3b. In actuality, as the solid line shows, there is no temperature gradient at the cell's boundary, and the actual temperature of the upper cell will remain at T_0. As shown in figures 16.3c and 16.3d, the process continues in subsequent time steps. In this example the effect of numerical diffusion is the prediction of considerable mixing in what should be a perfectly stratified tank.

The effects of numerical diffusion can be somewhat alleviated if the storage tank is divided into many segments. Sharp and Loehrke (1979) report the

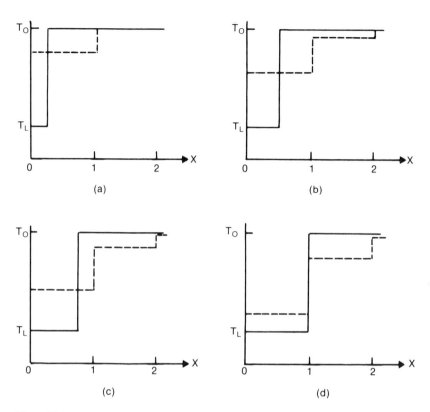

Figure 16.3
Two-cell example of the effects of numerical diffusion in predicting fictitious mixing in a stratified system. Solid lines are actual temperature profiles. Dashed lines are profiles predicted by the numerical model.

predicted efficiency of a stratified tank increasing with a 40-segment model. Kuhn et al. (1980) found that at least 100 segments were necessary t⟩ accurately predict the temperature profile in a well-stratified tank. The problem with increasing the number of segments is that the computational time then also increases dramatically. The time interval of each time step must be decreased accordingly for numerical stability. The result is that the computational time is proportional to the square of the number of segments. Modifications of the usual first-order, finite-difference equations can minimize numerical diffusion without excessively increasing computational time. So far these methods have not been used to model water storage tanks.

A model that overpredicts the effects of stratification is the extended SOLSYS model developed by Kuhn et al. (1980). In this algebraic model, the temperature profile in the tank is represented by a series of step functions. Thermal diffusion within the tank is ignored. Full mixing between step-function segments is assumed whenever higher density fluid is above lower density fluid. An example of the behavior predicted by this model is shown in figure 16.4. The model should work well if the temperature of the incoming fluid from the collector and the load remain relatively constant over time. Otherwise, the relatively large number of step-function segments within the tank will greatly increase the computation time required.

The importance of a proper control strategy to take advantage of stratification has been stressed by Rademaker (1981), Veltkamp (1981), and Cole and Bellinger (1982). A proper strategy, however, runs counter to traditional thinking, which holds that collectors capture more heat when more of the cooling medium is pumped through them, all other factors being equal. This desire for a very high flow rate is checked only by the cost of pumping, and the typical result is that the storage mass is circulated three or four times daily through the collector and the storage tank system. Because high circulation results in poor stratification, the effects of stratification become relatively insignificant in such a conventional system. An optimal control strategy in a perfectly stratified tank, on the other hand, would always deliver water to the collector at the coldest temperature and water to the load at the hottest temperature. In other words, the "heat front," the boundary between the hot and cold water, should always remain in the tank. The initially cold water in the tank at the beginning of the day plus the water delivered to the load should be circulated through the collector exactly once each day. Veltkamp (1981) has shown that the actual optimum occurs at a slightly higher rate. Of course this optimal control strategy requires that the heat load and solar input for the day be known ahead of time.

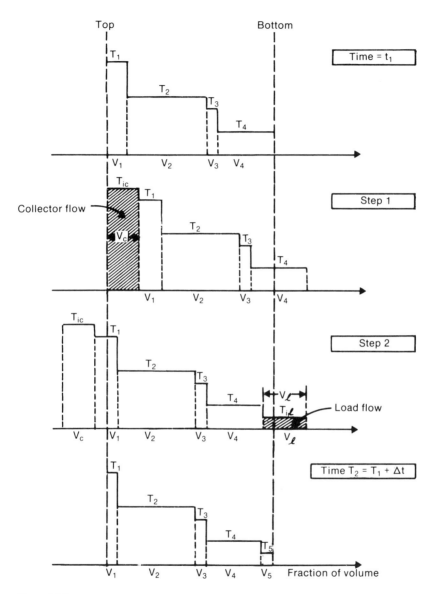

Figure 16.4
Example of updating the temperature profile in Extended SOLSYS. Source: Kuhn et al. (1980).

A common criticism of stratification in storage tanks is that the reduced flow in the solar collector reduces the effectiveness of the heat transfer enough to offset any gains made by stratification. Lunde (1980) presents examples in which no benefit is gained by stratification and concludes that well-mixed storage tanks are preferred. This is, however, another example of using system parameters optimized for well-mixed tanks to predict the behavior of stratified tanks. It should not be too difficult to find solar collectors and heat exchangers that have highly effective heat transfer at reduced-flow conditions. Cole and Bellinger (1982) suggest that in many cases a simple change in plumbing in the existing collector array, from a parallel to a series configuration, would be beneficial.

Cole and Bellinger (1982) have also developed an analytical expression that accurately represents the temperature profile of a stratified storage tank during charge and discharge. The theory concerns a small mixed zone that forms at the inlet. After the "front" of convecting water moves past the boundary of this mixed zone, the temperature profile around this front develops according to molecular diffusion. The equations involve the dimensionless parameters

$$T = \frac{(T' - T_{in})}{(T_0 - T_{in})},$$

$$x = \frac{x'}{x_0},$$

$$t = \frac{t'}{t_F},$$

where T' is the actual temperature, T_{in} is the inlet temperature, T_0 is the initial tank temperature, x' is the actual vertical coordinate, x_0 is the tank height, t' is the actual time, and t_F is the time it would take for the inlet fluid to completely fill the tank. The equation describing the temperature is

$$T = T_1 + T_2, \tag{6}$$

where T_1 is the temperature development of the water proper. T_1 is given by

$$T_1 = \frac{1}{2}\left(1 + \mathrm{erf}\frac{x - t}{A}\right), \tag{7}$$

where

$$A = 2F^{1/2}(t + c)^{1/2},$$

where F is the Fourier number and c is an empirical coefficient describing the size of the mixed zone. T_2 describes the effect of the wall heat capacity and is given by

$$T_2 = \frac{H(a-1)}{2} \frac{1}{a} \exp\left(\frac{B^2 + 2xB}{A^2} - Dt\right)(x-B)\left[\operatorname{erf}\left(\frac{x+B}{A}\right)\right.$$
$$\left. - \operatorname{erf}\left(\frac{x-t+B}{A}\right)\right]$$
$$+ \frac{A}{\pi^{1/2}}\left\{\exp\left(\frac{-x^2}{A^2} + Dt\right) - \exp\left[-\left(\frac{x-t^2}{A}\right)\right]\right\}, \tag{8}$$

where the normalized film coefficient is

$$H = \frac{hLt_M}{rcA},$$

where h is the film coefficient, L is the inside circumference of the tank, r is the density of the tank wall, c is the wall specific heat, A is the cross-sectional area, and

$$t_M = t_F \sum \frac{r_i c_i A_i}{r_w c_w A_w},$$

where the subscript w refers to the water in the tank. Other coefficients in equation (8) are

$$a = \frac{t_M}{t_F},$$

$$B = \frac{HA^2}{(2a)},$$

$$D = \frac{Ht_F}{t_M}.$$

As shown in figure 16.5, equation (6) predicts the temperature profiles during a tank discharge fairly well.

16.2.3 Seasonal Storage

In a seasonal storage system the storage volume is large enough to supply heat to the load for about a month or more. The possibility of storing heat during the summer months when insolation is high and later using the heat during the winter months when insolation is low is especially attractive in

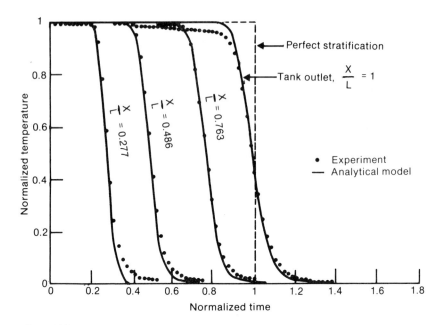

Figure 16.5
Comparison of analytical model with stratified-storage discharge experiment. Source: Cole and Bellinger (1982).

regions with little winter insolation. If the seasonal storage system is large enough, 100% of heating needs could be supplied, and an auxiliary heating system would not be needed. The size of a solar collector array required for seasonal storage can be 30% to 70% of that required for diurnal storage for the same amount of energy delivered. In most seasonal storage systems to date, water has been the storage medium. The economic question is whether the reduction in the cost of the solar collectors is offset by the increase in the cost of the storage tank.

Hooper (1979) has stressed that accurate sizing is much more important for a seasonal storage system than for a diurnal system, especially if no backup heater is installed. Moreover overestimating the system's performance by as little as 10% would result in underheating the load for as much as two to three weeks.

The size of seasonal storage also determines its cost. Unlike solar collectors, where the cost is proportional to the energy collected (collector area), the cost of a volume of water storage decreases as the size of the tank increases (costs are proportional to tank surface area). Heat losses per unit of energy stored also decrease as storage volume increases. Thus seasonal storage is best suited for large loads such as district heating. Because storage is essentially charged and discharged only once each year, the cost of each unit of stored energy is greater than that of a diurnal system, averaged over the year. It is important that the size of storage be correctly estimated so that no collectable heat is rejected and no excess capacity is purchased. Both McGarity (1979) and Sillman (1981) indicate that an economic optimum exists for a storage system in which the size of the tank is such that all the heat collected in the summer is stored. The economics of this system are thus particularly sensitive to accurate modeling.

Seasonal storage systems require more accurate modeling than diurnal systems and are comparatively more difficult to model. A detailed simulation requires weather data for an entire year. Computational times are large, and many more computational errors accumulate than in daily or weekly simulations. Many small effects, which are not important for small cycle times, become important for yearly cycle times. Heat loss to the ground, discussed in subsection 16.5.2, becomes especially important.

It is desirable to generate seasonal storage models that can use daily or even monthly time steps and still faithfully represent the system's behavior averaged over a year. Details of system performance must disappear from the model, and it is important that the modeler be careful to validate that the averaged predictions correspond to the actual performance data. There are many pitfalls to avoid in generating a simplified model. For example, Braun et al. (1981) have devised a design method, analogous to f-Chart, for solar-space-heating systems with seasonal storage based on a detailed TRNSYS simulation. Throughout the parameter variation process, the TRNSYS simulation showed at most a 2% increase in the system's efficiency over that of a completely mixed tank when stratified storage was used. As indicated by the discussion of numerical diffusion in subsection 16.2.2, models like TRNSYS always underestimate the efficiency of stratification, even for a daily simulation. Moreover, errors associated with numerical diffusion will compound over a year. Although molecular diffusion will certainly lessen the degree of stratification over many months, the effect of stratification in well-insulated seasonal storage should not be negligible. Optimization of seasonal storage systems that incorporate stratification is an important task for future development.

Several computer codes have been specifically designed for seasonal storage systems. MINSUN (Chant and Biggs 1983), developed as part of the International Energy Agency program, has several efficient subroutines for modeling seasonal storage: SST for stratified water storage, DST for direct earth or rock heating, and AST for aquifers. MINSUN uses daily time intervals, produces system performance and economic data, and is capable of optimization. Breger (1982) and Breger and Michaels (1983) have used MINSUN to study a seasonal heat storage system coupled with different types of solar collectors. A somewhat simpler model, which also uses daily time intervals, is the SASS code developed at SERI (Baylin and Sillman 1980). This code was used by Michaels et al. (1983) to examine seasonal heat storage in an aquifer using a variety of solar collector options.

16.2.4   Cold Water and Other Liquids

The models developed for hot water storage can usually be applied with minor changes to the models of sensible heat storage systems that use other liquids. Often only the thermal physical parameters of the fluid need to be changed. Gross and Harrigan (1981) found the stratification behavior of high temperature storage tanks to be similar to that found by Cole and Bellinger (1982) for hot water. Cold water storage tanks for air-conditioning applications can be modeled by most hot water storage tank models. However, the problem of stratification is tricky because the thermal expansion coefficient is small for cold water and actually becomes negative for temperatures between 32° and 39°F (0° and 4°C).

16.3   Sensible-Heat Storage in Solids

Solid sensible-heat systems can usually be classified as one of the following types of beds: packed, fluidized, or solid. A packed bed consists of small, modular units of the storage material stacked side by side in a container. The shape of the units is such that much storage volume is void, allowing the flow of air, or any other fluid, to progress through the bed from one end of the container to the other. Heat is transferred by direct contact between the working fluid and the storage material, eliminating the expense of heat exchange equipment. A good example of a packed bed is the rock bed (see subsection 16.3.1). In a fluidized bed, small, solid particles, such as sand, are suspended in a working fluid, such as air. Heat transfer in and out of storage can be very fast, and the fluidized solids can be transported almost like a liquid. In a solid bed, storage consists of a container of solid material interspersed with pipes. The pipes carry the heat transfer fluid, and heat is transferred in and out of storage by conduction between the pipes and the rest of the solid.

16.3.1   Rock Beds

Rock beds have been used extensively for thermal storage in space-heating systems that use solar air collectors. The rock bed is a logical choice for forced air heating systems, because it functions as a direct-contact heat exchanger as well as an energy storage medium. Types of rock beds include vertical, horizontal, and U-shaped; the vertical type is usually preferred because of its superior performance. Only the vertical rock beds are discussed here.

The performance of a rock bed is characterized by volume, face velocity, the pressure drop across the rock bed, the size of the rocks, and the degree of stratification. If the pressure drop is too large, then the parasitic power needed to run the fans becomes too large. These parasitic losses are generally much higher than those of water tank storage systems. If smaller rocks are used to improve the heat transfer between the rocks and the air, the pressure drop increases proportionally. On the other hand, too low a pressure drop can result in channeling, in which the airflow travels predominantly through only a small fraction of the bed, usually at the edges. Channeling may also occur if the rocks are not packed homogeneously, although it can be minimized by breaking up the flow and inducing turbulence in the inlet plenum (Solaron 1982). Because there is no hydraulic motion in the rocks, temperature stratification is relatively easy to achieve, provided that there is sufficient heat transfer between the rocks and the air. During charging, hot air from the collector enters from the top and returns to the collector from the bottom of the rock bed. In the discharge mode cold air is blown up through the bottom of the rock bed and exits to the space to be heated from the top of the rock bed. In a well-designed rock bed the width of the thermocline may be as small as several rock diameters, and heated air going to the load exits the rock bed at nearly the same temperature as the top-most layer of rocks.

An important rock bed modeling problem is the degradation of stratification that occurs during quiescent periods. The problem is particularly important because there is usually a minimum usable temperature in forced-air systems. Once the top of the rock bed falls below this temperature, auxiliary heat must be used. The degradation is caused by internal conduction between the rocks and natural convection within the bed. Natural convection is a three-dimensional problem caused by horizontal temperature gradients that result from channeling and heat loss from the sides of the rock bed. The effect is most noticeable near the upper plenum. Despite careful precautions, measurements show that uniform flow is rarely achieved within 12 in. (30 cm) of the inlet (Kuhn et al. 1980). The three-dimensional aspects of this problem suggest that a porous media computer code such as COMMIX-SA would greatly aid our understanding of rock beds. A three-dimensional model would also be useful in helping to improve the performance of rock beds with nonvertical flows. To date, no published results of such models are available.

Almost all of the rock bed modeling has been devoted to one-dimensional models. Fortunately, the same constraints that are necessary for the one-dimensional models to be valid, such as uniform plug flow and negligible side heat loss, also result in better rock bed performance. Most of the models are a particular form of the general model discussed in subsection 16.1.3 and are based around the Schumann (1929) model for a packed bed. The heat capacity of air is typically 0.1% that of the rocks, and it is usually omitted in the models. The equations for the rock bed then reduce to the two coupled equations (Hughes et al. 1976),

$$\frac{dT_a}{d(x/L)} = \text{NTU}(T_b - T_a), \tag{9}$$

$$\frac{dT_b}{d(t/t_s)} = \text{NTU}(T_a - T_b), \tag{10}$$

where x is the position along the bed in the flow direction, L is the length of the bed, T is the temperature, t is the time, subscript a refers to air, subscript b refers to bed, and t_s is a time constant, given by

$$t_s = r_b c_b (1 - e) \frac{AL}{mc_a},$$

where r is the density, c is the heat capacity, e is the void fraction of the packed bed, A is the cross-sectional area of the bed, and m is the mass flow rate of the air. NTU is the number of heat transfer units, given by

$$\text{NTU} = \frac{hAL}{mc_a},$$

where h is the volumetric heat transfer coefficient between the rocks and the air. For cases in which the temperature within the individual rocks is not uniform, a modified NTU based on the Peclet and Biot numbers, can be used.

A term is usually added for heat loss from the sides of the bed, and the equations are then solved by a finite-difference approximation. The model requires a large number of segments and short time steps for accuracy, which results in excessive solution times for long-term simulation. Von Fuchs (1979) used a polynomial interpolation algorithm to represent the temperatures in this model and greatly decreased the computational time. Hughes et al. (1976) found that for NTU > 10, the performance of a rock bed system was essentially the same as that for an NTU of infinity. For an infinite NTU the temperature of the air and rock at a given spatial location must be the same. The model can be simplified to one equation,

$$\frac{dT_a}{d(t/t_s)} = -L\frac{dT_a}{dx} + \frac{UPL}{mc_a}(T_{env} - T_a),\tag{11}$$

where U is the heat loss coefficient from the bed to the surroundings, P is the perimeter of the bed, and T_{env} is the environmental temperature. Because NTU is greater than 10 in most practical system designs, this infinite NTU model is much more efficient and is often used in TRNSYS. The finite difference approximation to the model reintroduces a smearing of the temperature front. Sowell and Curry (1980) point out that the width of the thermocline predicted by this model may be very different from the actual value in some circumstances. The only basic improvement in this model has been to add a term describing heat conduction between the rocks (Riaz 1978; Coutier and Farber 1982).

Most differences between models arise from the particular correlation used for the heat transfer coefficient and for calculating the pressure drop across the rock bed. Several correlations exist for the packed bed heat transfer correlation h. A commonly used correlation is that of Löf and Hawley (1948),

$$h = 652\left(\frac{G}{D_e}\right)^{0.7},\tag{12}$$

where h is in $Wm^{-3}\,{}^{\circ}C^{-1}$, G is the superficial mass velocity in $kg\,m^{-2}\,s^{-1}$, and D_e is the equivalent spherical diameter of the rocks in m, given by

$$D_e^3 = \frac{6 \times \text{net volume of rocks}}{\pi \times \text{number of rocks}}.$$

Kuhn et al. (1980) surveyed possible correlations for h that apply to rock bed thermal storage. The correlations are functions of the Prandtl number, the Reynolds number, and (sometimes) the void fraction. There is considerable variation among the values for Reynolds numbers between 20 and 400, the usual range of applicability for rock beds. Most of these correlations were *not* developed for the type of rocks preferred for rock bed designs. Chandra and Willits (1981) presented a correlation that matched data collected from a small rock bed. As long as the resulting NTU value is sufficiently large, the results of most modeling applications will not depend strongly on h. Moreover, unless very accurate results are needed, determining a precise value for h is not critical.

A more critical correlation relates the pressure drop dP across the rock bed to the other design parameters. Pressure drop is an important parameter in the economic calculation for parasitic power. Numerous correlations exist for

dP, and many of the more likely candidates have been discussed by Kuhn et al. (1980), who suggest the correlation

$$dP = \frac{LG^2}{rgD_e}\left[3.02\frac{(1-e)}{e^3}a + 283\frac{(1-e)^2a^2}{D_ee^3}\right],\tag{13}$$

where g is the acceleration of gravity, and a is the shape factor of the rock bed. Chandra and Willits (1981) suggest the form

$$dP = \frac{m^2}{rD_e^3e^{2.5}}\left[185\left(\frac{GD_e}{m}\right) + 1.7\left(\frac{GD_e}{m}\right)^2\right],\tag{14}$$

where m is the viscosity of air.

Because the bulk of the thermal mass of the rock bed is immobile, there is no mixing of different levels caused by differences in density. This lack of mixing makes the rock bed equations particularly appropriate for solution by the transient response function method discussed in subsection 16.1.5. Riaz (1978) solved the rock bed equations with this method for a large number of input functions. Saez and McCoy (1982) used a moment method to solve the equations for step function inputs. Both of these models included a term for conduction between rocks. Sowell and Curry (1980) presented a bidirectional convolution model of equations (9) and (10) based on triangular pulses. This solution is available in the form of a TRNSYS subroutine.

16.3.2 High Temperature Storage

High temperature, sensible heat storage is useful for power-generation solar technologies, such as power towers and focusing collectors, as well as for a host of industrial applications. In general, the same storage models that apply to space heating can also be used for high temperature storage. However, conduction terms often become the most important factors. For example, the rock bed models discussed in subsection 16.3.1 are appropriate for modeling packed beds of iron spheres, but as Pomeroy (1979) pointed out, the large thermal conductivity of the iron and the liquid metal heat transfer fluid makes the axial conduction term one of the dominating factors. The thermoclines of oil and rock storage systems (Radosevich and Wyman 1983) often degrade because of conduction along the side walls of the container. In both the rock and iron systems, the heat capacity of the heat transfer fluid cannot be neglected. The basic design of high temperature storage devices may be somewhat different from rock beds. Szego and Schmidt (1978) used an NTU model to analyze a solid sensible heat storage system composed of solid slabs with two different heat transfer fluids flowing countercurrently.

16.4   Latent-Heat Storage

16.4.1   Phase Change Heat Storage

Phase change heat storage systems gain or lose heat by changing phase (e.g., a solid melting to a liquid) without changing temperature. The important feature of phase change materials (PCM) is that the amount of energy absorbed or released during the phase change is large; in a sensible heat storage medium, the same amount of energy would cause the temperature to rise many tens of degrees. Phase change storage systems are desirable when small storage volumes or a limited range of storage temperatures are required. Various types of phase transitions, such as melting–freezing, solid–solid, and boiling–condensing, and many phase change materials have been investigated. However, all commercial applications to date have been the melting–freezing type, and this discussion is limited to this one.

PCMs exhibit several features that make both modeling and the use of models in designing efficient subsystems difficult. Many PCMs are corrosive and must be encapsulated or kept in a container. Materials such as waxes are not corrosive but do need to be contained in the liquid state. Often the cost of the container and the heat exchanger in PCM systems is much higher than the cost of the PCMs themselves. The melting behavior of PCMs can be complicated, which contributes to problems in design as well as modeling. Incongruent melting will occur for some salt hydrates in which the salt is not completely soluble in its water of hydration. In this case the undissolved salt will precipitate and settle out of the system because of density differences. Suspension agents such as thickeners or polymeric matrices are often used to prevent this segregation of anhydrous salts. Semicongruent melting occurs when a PCM has two or more hydrate forms, each with a different solid composition and melting temperature. If the salt with the higher melting point forms, it will settle out of the system in the same way as the salt in the incongruent melting case. In both cases the storage capacity of the PCM system decreases over time. A behavior problem associated with freezing is supercooling, where the temperature of the liquid PCM can drop below the normal melting point without a change in phase. Nucleating agents can prevent this, but the degradation of these agents is also a common problem.

Typically the solid and liquid phases of PCMs have different physical properties. These differences not only change the basic model parameters but also the type of thermal behavior in the system. During the charge cycle, natural convecting liquid PCM is adjacent to the heat exchanger (or container wall), and heat transfer is good. During heat extraction, the solid phase can form around the heat exchanger, and since the thermal conductivity of the solid is often low, the overall heat transfer coefficient decreases as extraction proceeds. If there is a difference in thermal expansion, the solid PCM may break away from the wall of the heat exchanger, leaving an air gap and decreasing the heat transfer coefficient further. To extract heat at a reasonable rate often requires a large heat exchange area, and the thermal mass of the heat exchanger cannot be ignored in modeling PCM systems.

Fouda et al. (1980) have suggested that direct-contact heat exchange may eliminate many PCM design problems because the propagation of bubbles through the PCM results in a more homogeneous mixture and a high rate of heat transfer. There are still design problems because one must use a heat transfer fluid that has low solubility in the PCM and does not form stable emulsions. Modeling direct-contact heat exchange has its own unique problems. The velocity of the drops through the PCM must be calculated, and the heat transfer within the drop also affects performance. Droplet models include rigid, internal circulation, and well mixed.

PCM systems are difficult to model in detail because of the nonlinearities associated with a moving liquid–solid interface. Many of the difficulties involve basic problems in heat transfer. During freezing, if the solid stays attached to the heat exchanger, a model can be readily developed by accounting for conductive heat transfer through the solid and developing a convective-to-solid heat transfer coefficient at the interface. Hale and Viskanta (1980) have indicated that for freezing from above, natural convection at the solid–liquid interface may decrease the rate of solidification with time if the Rayleigh number is sufficiently high. Sparrow et al. (1980) and Ramsey and Sparrow (1980) have demonstrated that natural convection plays a very strong part in the melting cycle of a PCM system. A cylindrical funnel forms around a vertical heat exchanger during melting, and heat transfer to the melt varies along its length. Carlsson and Wettermark (1980) have indicated that the limiting step in melting occurs at the boundary between the solid and the liquid, so it is important to calculate the size of this interface variable correctly. For horizontal heat exchangers Hale and Viskanta (1980) have shown that the behavior of the interface above the exchanger is considerably different from that below the exchanger. An array of heat exchange surfaces would present a formidable modeling task.

Numerical solutions of melting–freezing systems require extensive calculations and are especially difficult when convection is present. Despite the demonstration that natural convection in the liquid must be considered for good agreement with data, most models ignore its effects completely. Even if convection is neglected, however, detailed models of PCM systems can still be complicated. Several complex models of PCM systems exist, but none are streamlined enough to be incorporated into TRNSYS-type system codes. Except in a relatively few restrictive designs, streamlined models have not exhibited good agreement with experimental results.

Morrison and Abdel-Khalik (1978) modified the rock bed model in TRNSYS to study PCM storage with air and liquid solar collectors. The change that was needed is a modification of equation (10) to

$$\frac{du}{dt} = \frac{UP}{rA}(T_f - T_{\text{PCM}}), \tag{16}$$

where T_f is the temperature of the heat transfer fluid, T_{PCM} is the temperature of the PCM, and u is the enthalpy of the PCM, given by

$$u = c(T_{\text{PCM}} - T_{\text{ref}}) + dHf, \tag{17}$$

where c is the specific heat, assumed constant for both phases; dH is the heat of fusion; T_{ref} is the melting point temperature; and f is the fraction of PCM melted. This modification, as well as others like it, is sometimes referred to as the *enthalpy model*. Green and Vliet (1981) used a similar approach, with the NTU value calculated from the heat transfer coefficient for several simple geometries. They neglected the sensible heat storage and conduction terms, which permitted a single resistance term to be used to describe the heat transfer. This one-dimensional model agreed only qualitatively with the data. Offenhartz et al. (1979) used a TRNSYS-compatible NTU model but included a thermal hysteresis loop to simulate supercooling by making the melting point upon heating greater than the freezing point upon cooling.

Bailey et al. (1977) developed a two-dimensional finite-difference model, based on the enthalpy method, to examine PCMs in an aluminum honey-comb. Henze and Humphrey (1981) used a quasi-linear transient, thin-fin equation to study a similar system and found good agreement with experimental values for the fraction of melt and the shape of the solid–liquid interface.

Much work has been devoted to solving the pure conduction problem with a phase change for different geometries and conditions. A good example is the work of Solomon (1979), who derived an analytical expression for the melt time from several simple geometries, assuming radial melting and a uniform melt thickness. Saxena et al. (1982) presented similar results but allowed for variations in temperature along the axial direction. Although analytical solutions permit quick computations to be made for system optimization, the neglect of convection effects in the melt limits these models to very restrictive designs. Even if the melt were to maintain a uniform thickness, the heat transfer between the liquid and the solid should depend on the thickness of the melt.

Marshall and Dietsche (1982) developed an NTU model for a system composed of encapsulated paraffin dispersed in a water tank. The temperature dependence of the specific heat and conductivity was fitted by a parabolic function to the actual data. The heat transfer coefficient also varied, depending on the film thickness and whether melting or freezing was taking place. The researchers found that a model that used a single node for the bulk of the wax could well represent the data in this design. This decreased the computational time so that the model could be used easily for hour-by-hour, long-term simulations.

16.4.2   Desiccant Cooling

Desiccant cooling systems use a drying agent to remove moisture from humid air. The dry air can then be cooled in evaporative coolers and sensible heat exchangers to meet both sensible and latent air-conditioning needs. Solar energy or other energy sources regenerate the desiccant. A number of different designs have been investigated and have been the subjects of various kinds of modeling (Lavan et al. 1982). Nelson et al. (1978) have used an NTU model in TRNSYS to investigate a rotary dehumidifier and regenerator. Worek and Lavan (1982) have investigated the performance of a cross-cooled dehumidifier.

In addition to the usual energy balance equations, a model of a desiccant system must consider the mass balance of the moisture in the air and the desiccant. Typical mass relationships between the airstream and the desiccant matrix in the wall (after Worek and Lavan 1982) are

$$\frac{dY}{dx} = \frac{KW}{m}(Y_w - Y), \tag{18}$$

and

$$\frac{dw}{dt} = \frac{KWL}{M}(Y - Y_w), \tag{19}$$

where Y is the humidity ratio of air, x is the spatial coordinate in the direction of the airflow, K is the mass transfer coefficient, L is the length of the airstream, W is the width of the airstream, m is the mass flow rate of dry air, w is the mass ratio of water to desiccant in the desiccant wall, t is time, M is the mass of desiccant in the wall, and Y_w is the humidity ratio of air in equilibrium with the wall. An additional modification is the inclusion of the heat of adsorption in the energy balance equations for the airstream and the wall.

Typical simplifying assumptions in desiccant models include constant air-flow, negligible heat and moisture capacity of the air, neglect of axial diffusion terms, and description of heat and moisture film resistances by lumped transfer coefficients. Capillary-porous materials frequently exhibit hysteresis behavior in the adsorption/desorption cycle. Because the equilibrium relations of the desiccant materials are usually nonlinear, the system of governing equations is nonlinear, and solutions are usually obtained by numerical techniques. If the equations are approximated by a linearized form, they can often be decoupled and solved with the NTU method. Mathiprakasam and Lavan (1980) have shown that Laplace transform solutions of linearized equations for periodic steady-state operation of both fixed-bed and rotating adiabatic dehumidifiers agree well with the nonlinear solutions over a broad range of system parameters and inlet conditions.

The relationship between Y_w and the temperature is one of the key desiccant properties. The decrease in Y_w at high temperatures allows the desiccant to be regenerated, but the details of this relationship determine which solar collector system is appropriate for a particular desiccant, as well as the maximum theoretical efficiency of the system. Jurinak and Mitchell (1984) have shown in parametric studies of desiccant properties that changes in the shape of the adsorption isotherm can vary the performance of a counterflow rotary humidifier as much as fourfold.

16.4.3   Ice Storage

Freezing water can be an effective method of cold storage on either a diurnal or a seasonal basis. The ice can be manufactured via a heat pump or through various systems that take advantage of the winter cold (Gorski et al. 1982). If the ice is formed around a heat exchanger, then the situation is equivalent to freezing a PCM, and many of the modeling considerations discussed in subsection 16.4.1 would apply. Because there has been significant commercial interest independent of thermal storage in ice, many engineering correlations and models have been developed that are particular to it. Lunardi (1981) has summarized many of the analytical and finite-difference modeling methods that apply to melting and freezing ice.

Several frequently used types of ice formation involve exposing water to cold, ambient air. Ice blocks can be built in layers from thin slabs of water that freeze. Or water droplets can be sprayed into the air to form small crystals of ice that accumulate in slush piles. Artificial snowmaking for ski slopes is an example of the latter method. Radiation to a clear sky, as well as evaporative cooling, allows ice to form at ambient temperatures well above the freezing point. Because the amount of ice that forms depends on system design, and because the critical temperature range occurs during a significant portion of the winter in many locations, accurate modeling is desirable. Hobbs (1974) has discussed many of the models that describe ice crystal formation from the liquid or vapor phase.

16.5 Soils

Soils are important in a number of different thermal storage applications. Earth beds can act as thermal storage reservoirs either alone or coupled with heat pumps. Earth can enhance the thermal capacity of storage tanks and other devices using seasonal storage, such as solar ponds. The heat losses from storage tanks placed into or on top of the ground are greatly influenced by the thermal properties of the surrounding soil.

Heat transfer in soils is often strongly related to the transfer of moisture. The thermal behavior of soils can vary considerably, depending on type, moisture content, temperature, initial conditions, and geometry. Although reasonable estimates can be made for the thermal behavior of most soils, proper modeling of soils' heat transfer properties involves several unsolved basic problems in heat and mass transfer through porous media.

Philip and de Vries (1957) developed a detailed model of heat and moisture transfer in soils to study conditions suitable for plant growth. This model has been validated for a large number of cases at relatively low temperatures [$T < 140°F$ ($40°C$)] and is frequently used to model heat transfer in soils. In this mechanistic model both liquid and vapor transport are considered. The heat transport equation has terms for conduction, sensible transport from liquid convection, and latent transport from vapor convection. An average thermal conductivity is calculated from the shape and type distribution of the soil particles and the fraction of water and air present. The transport of water vapor involves two terms: The first depends on temperature gradient, and the second on the moisture gradient. The transport equation for liquid water depends on hydraulic conductivity as well as on moisture gradient and temperature gradient. Adapting these equations, Walker et al. (1981) developed a two-dimensional thermal storage computer model. The model uses a predictor-corrector method that solves the nonlinear parabolic partial differential equations and is second-order correct in time.

16.5.1 Earth Beds

Earth bed thermal storage consists of a volume of earth containing heat exchange pipes. The advantage of this system is that, except for the pipe, the only cost associated with storage is for earth moving. With some soils, pipes may be vibrated into the ground, making this a very low-cost system. The storage temperatures are relatively low, less than 104°F (40°C), so a heat pump is usually required to deliver heat to the load. The most effective use of earth beds is as a long-term or annual storage device. This application minimizes the size of the associated solar collector, resulting in a low-cost system. The earth bed can also act as ground coupling to a solar heat pump system. In this case the earth bed is heated during the summer from the surrounding ground. When the regular storage temperature is below the ground temperature, the ground acts as the source for the heat pump. Metz (1982) has described the use of ground-coupled storage tanks with series solar-assisted heat pump systems.

Most models of ground heat storage systems employ only the pure conduction equations. The effects of moisture are usually ignored, and the thermal conductivity of the ground is assumed constant in time. Analytical models can be used for general analysis in highly idealistic conditions (see, e.g., Shelton 1975), whereas numerical models are more useful for modeling realistic conditions, such as the effects of water tables and time-varying, far-field soil temperatures. Metz (1983) has presented a simple computer program of this type for modeling three-dimensional, underground heat flow. One version of this program is compatible with TRNSYS. More ambitious models that incorporate both heat and mass transfer (e.g., Benet and Jouanna 1981) are also being developed, but these models require considerably more computational time than the pure conduction models.

16.5.2   Heat Loss from Tanks

Heat loss to the ground is important to consider for seasonal heat storage devices. Many of the special considerations applying to seasonal storage in the ground also apply here. An important economic calculation is the optimum amount of insulation needed for these devices. Williams et al. (1980) calculated the optimal insulation distribution around cylindrical tanks using the method of Lagrange multipliers with a simple, steady-state heat transfer model. This model included variable insulation thicknesses and nonuniform soils. Using a similar thermal model, Hull et al. (1984) showed that better insulation strategies could be developed by extending insulation below the tank at the perimeter rather than by insulating the tank bottom.

Much development work remains in modeling soils for thermal storage applications. Most two- and three-dimensional pure conduction models for soil thermal engineering leave much to be desired. Because the soil around the storage device is likely to be hot, there will be horizontal and vertical temperature gradients at the edges of the tank. If there is any moisture in the soil, convection currents will develop and greatly increase the heat loss from the tank. Conduction models often assume a constant temperature sink at the level of the water table. A moving water table would have a pronounced effect on thermal performance and could easily be incorporated into the simple conduction models. Even if a multidimensional Philip and de Vries model were used, there would still be modeling uncertainties. At high temperatures, vapor heat transport is much more significant, and no soil model has been validated for these high temperatures.

16.6 Aquifers

Aquifers are extensive strata of permeable rock, sand, or gravel that can yield large quantities of water over a long time. Naturally occurring aquifers make very inexpensive thermal storage devices, because the only capital expense involved is to establish injection and supply wells and because pumping costs are relatively low. During charging of a heat storage aquifer, water is pumped from the supply well to the heater and then down the injection well. When heat is needed during discharge, hot water is pumped from the injection well and used for heating; the cooled water is returned to the aquifer via the supply well. Aquifers can be used for either hot or cold seasonal storage but must be applied to large loads, such as district heating, to be economical. Michaels et al. (1983) studied the performance of several different types of solar collectors with aquifer storage. Because of the high pressures that occur deep in the ground, water storage temperatures as high as 392°F (200°C) are sometimes feasible.

The porous aquifer acts as a rock bed heat exchanger with very small rocks. Ideally the temperature wave moves through the aquifer with a very sharp thermocline. In practice, fingering and natural convection are often observed (Molz et al. 1983). Thermal front fingering is a result of the heterogeneity of the aquifer, as water tends to flow by forced convection into high permeability layers. The fingering is smeared out by conduction, resulting in a diffuse hot region with a low recovery factor. Modeling this effect is difficult because of the lack of a priori knowledge of the aquifer's permeability. Natural convection of the injected hot water is caused by both a density difference and a viscosity difference. Convection is driven by the density difference, where hot, less dense fluid below a cold, more dense fluid results in an unstable buoyancy force. The convective flow is enhanced by the viscosity difference, as the hot, less viscous water flows faster than the cold, more viscous water.

Models of aquifer thermal energy storage have relied on the previous extensive work done to formulate hydrology models to predict the behavior of groundwater systems. The three-dimensional equation of groundwater flow for a homogeneous aquifer (Papadopulos and Larson 1978) is given by

$$\vec{\nabla} \cdot \left[r \frac{k}{\mu} (\vec{\nabla} - r\vec{g}) \right] - q = \frac{d}{dt}(nr), \tag{20}$$

where $\vec{\nabla}$ is the gradient operator, p is pressure, k is intrinsic permeability, r is density, μ is viscosity, $\vec{g}$ is the gravitational acceleration vector, n is porosity, and q is the mass flow rate per unit volume from sources and sinks. The corresponding energy equation is

$$\vec{\nabla} \cdot \left[\frac{rkH}{\mu} (\vec{\nabla}p - r\vec{g}) \right] + \vec{\nabla} \cdot \vec{K} \cdot \vec{\nabla}T - q_L - qH = \frac{d}{dt}[nrU + (1-n)(rc)T] \tag{21}$$

where H is enthalpy, $\vec{K}$ is hydrodynamic thermal dispersivity (thermal conductivity plus hydrodynamic dispersivity tensor), T is temperature, q_L is rate of heat loss across boundaries, U is internal energy, and c is specific heat. The water density is a function of both temperature and pressure. Viscosity is a function of temperature. The equations are usually solved by a finite-difference or a finite-element scheme.

Tsang et al. (1981) have used a computer code to accurately simulate the results of several aquifer thermal energy storage field tests. This model has permeabilities that may be both direction and temperature dependent. Vertical deformation of the rock is calculated using a one-dimensional consolidation model. The two- or three-dimensional versions of the code can accurately model natural convection effects (Buscheck et al. 1983). Some simplified linear models developed by Tsang have been incorporated into MINSUN.

With high temperature aquifer storage, corrosion and fouling of the heat exchangers can become problems. The solubility of most minerals is higher at high temperatures, and the injected hot water will bring the more soluble minerals to the surface. This effect also changes the permeability of the aquifer. Accurate geochemical models are necessary to help solve these problems.

16.7 Thermochemical Storage

Thermochemical energy storage uses reversible chemical reactions to store and later release thermal energy. This form of storage is marked by both high energy density and loss-free, long-term storage at ambient temperatures.

Individual chemical reactions are appropriate for a particular temperature range. However, because there are so many different reversible chemical reactions, they span a wide range of temperatures (Mar and Bramlette 1980). Many of the products and reactants are gases and liquids that can be transported in pipelines. Thermochemical storage is still in the development stage, and only a few systems have been studied in detail.

In addition to thermal storage and thermal energy transport, thermochemical reactions can be used in chemical heat pump applications. A chemical heat pump is similar to a mechanical heat pump, but it eliminates the need to turn thermal energy into mechanical energy. A chemical heat pump is simply an absorption cycle heat pump with built-in energy storage via absorber-refrigerant separation. Any absorption cycle is potentially a chemical heat pump, but in most cycles, large-scale separation is impractical because of thermodynamic or cost constraints.

For the well-known absorption cycle reactions, detailed models are available. For other thermochemical reactions, general thermodynamic analysis is available (e.g., Raldow 1981), but modeling of these systems is still in a relatively primitive stage. The basic chemical reaction is governed by the change in free energy

$$dG = dH - T\,dS, \tag{22}$$

where G is free energy, H is enthalpy, T is temperature, and S is entropy. At the generation temperature (i.e., the temperature of the solar collector), dG must be negative, but at the end-use temperature, dG must be positive. In many cases of proposed thermochemical storage systems, the enthalpy as a function of temperature and concentration and the boiling temperature as a function of concentration and vapor pressure are not known for all possible modes of operation. These basic thermodynamic data must be generated before detailed modeling is possible.

Simple TRNSYS-compatible models have been developed for a few thermochemical systems. Offenhartz et al. (1979) simulated H_2SO_4/H_2O and $MgCl_2/H_2O$ systems using a simple batch process model. Here the entire tank was assumed to be heated to the boiling point before any water could be recovered. Offenhartz (1981) used a continuous process model to simulate H_2SO_4/H_2O, NH_4NO_3/NH_3, and $CaCl_2/CH_3OH$ systems.

16.8 Conclusions

Over the last ten years, modeling of thermal energy storage components has advanced significantly in many areas. Detailed, accurate models have been developed for water storage tanks, rock beds, and aquifers. Advances have been made in modeling phase change systems, heat transfer in soils, and thermochemical systems. Many simplified storage subroutines have been incorporated into system computer codes like TRNSYS and MINSUN. However, much work still remains to be done, both in developing and using more powerful models for developing technologies and in understanding the limitations of existing models. More attention also must be paid to proper validation. Too often, validation of models against experimental data has stopped with very simple systems (e.g., a mixed water tank), whereas the models are used to predict much more complicated systems (e.g., a stratified tank). Validation must be extensive enough to determine the limitations of a model. The work of Kuhn et al. (1980) is an example of a good validation procedure.

An undesired, but common, phenomenon associated with modeling engineered systems is blind faith in a particular model. An important example of this phenomenon in energy storage modeling has appeared in studies of the effect of stratification on water storage tank performance. As discussed in subsection 16.2.2, many model users have claimed that the effect of stratification was insignificant when, in actuality, either the model used artificially reduced stratification or the system parameters were not optimized for stratified behavior. Models that correctly optimize for stratification show significant performance increases for the system. Appropriate mathematical or computer models still do not relieve the system designer from responsibility for understanding the details of the particular system to be designed or optimized. Model users must understand the limitations of their models when comparing different technologies.

Improvements are needed in several areas. The effects of stratification are still not properly accounted for in one-dimensional computer models of storage tanks. Optimization studies based on stratification are still lacking. The effects of convection in the liquid phase of PCM models need to be modeled in detail, and models of the summarized behavior need to be incorporated into system codes. The results of fundamental research of heat transfer in soils need to be incorporated into energy storage models. Modeling energy storage in soils with both heat and moisture transport has just begun. The details of the effects of fluid mechanics on chemical reactions have yet to be considered in thermochemical models.

As new thermal storage technologies emerge and as currently developing technologies mature, mathematical models must develop with them to enhance their optimization. Subroutines that are compatible with existing system codes must be developed that accurately describe the thermal performance of the storage device. As computational resources become faster and less expensive, more detailed models can be used in component designs, and performance maps for system models can be established more quickly. Three-dimensional thermohydrodynamic computer codes such as COMMIX-SA will also be more feasible. So far the power of the results that such codes are capable of producing has been largely unappreciated.

Acknowledgments

The author is indebted to A. I. Michaels, R. L. Cole, and D. Morrison for reviewing the manuscript. A. I. Michaels provided guidance throughout the writing, and A. J. Gorski provided useful discussions about ice storage.

References

Bailey, J. A., J. C. Mulligan, and C.-K. Liao. 1977. *Research on Solar Energy Storage Subsystems Utilizing the Latent Heat of Phase Change of Certain Organic Materials.* Final report, EY-76-S-05-5101.

Baylin, F., and S. Sillman. 1980. *System Analysis Techniques for Annual Cycle Thermal Energy Storage Solar Systems.* SERI/RR-721-676. Golden, CO: Solar Energy Research Institute.

Beard, J. T., F. A. Iachetta, and L. V. Lilleleht. 1978. *Annual Collection and Storage of Solar Energy for the Heating of Buildings.* Final report, ORO/5136-78/2, Charlottesville, VA: University of Virginia.

Beckman, W. A., S. A. Klein, and J. A. Duffie. 1977. *Solar Heating Design by the f-Chart Method.* New York: Wiley.

Benet, J. C., and P. Jouanna. 1981. Theoretical model of heat and mass tranfer in non-saturated soils with phase change. In *Thermal Storage of Solar Energy*, C. den Ouden, ed. The Hague, The Netherlands: Martinus Nijhoff, pp. 233–248.

Bevington, P. R. 1969. *Data Reduction and Error Analysis for the Physical Sciences.* New York: McGraw-Hill, pp. 134–254.

Braun, J. E., S. A. Klein, and J. W. Mitchell. 1981. Seasonal storage of energy in solar heating. *Solar Energy* 26: 403–411.

Breger, D. 1982. *A Solar District Heating System Using Seasonal Storage for the Charleston, Boston Navy Yard Redevelopment Project.* ANL-82-90. Argonne, IL: Argonne National Laboratory.

Breger, D. S., and A. I. Michaels. 1983. *A Seasonal Storage Solar Energy Heating System for the Charlestown, Boston Navy Yard National Historic Park, Phase II: Analysis with a Heat Pump.* ANL-83-58. Argonne, IL: Argonne National Laboratory.

Busheck, T. A., C. Doughty, and C. F. Tsang. 1983. Prediction and analysis of a field experiment on a multilayered aquifer thermal energy storage system with strong buoyancy flow. *Water Resour. Res.* 19: 1307–1315.

Carlsson, B., and G. Wettermark. 1980. Heat-transfer properties of a heat-of-fusion store based on $CaCl_2 6H_2O$. *Solar Energy* 24: 239–247.

Chandra, P., and D. H. Willits. 1981. Pressure drop and heat transfer characteristics of air-rockbed thermal storage systems. *Solar Energy* 27: 547–553.

Chant, V. G., and R. C. Biggs. 1983. *Tools for Analysing Central Solar Heating Plants with Seasonal Storage.* CENSOL1. National Research Council of Canada.

Cole, R. L., and F. O. Bellinger. 1982. *Natural Thermal Stratification in Tanks*. Phase 1 final report, ANL-82-5. Argonne, IL: Argonne National Laboratory.

Coutier, J. P., and E. A. Farber. 1982. Two applications of a numerical approach of heat transfer process within rock beds. *Solar Energy* 29: 451–462.

Duff, W., and B. Wynn. 1981. Modeling of solar-thermal systems. In *Solar Energy Handbook*, J. F. Kreider and F. Kreith, eds. New York: McGraw-Hill, ch. 17.

Duffie, J. A., and W. A. Beckman. 1974. *Solar Energy Thermal Processes*. New York: Wiley, pp. 215–239.

Fouda, A. E., G. J. G. Despault, J. B. Taylor, and C. E. Capes. 1980. Solar storage systems using salt hydrate latent heat and direct contact heat exchange. *Solar Energy* 25: 437–444.

Gorski, A. J., W. W. Schertz, A. S. Wantroba, A. E. McGarity, and E. H. Buyco. 1982. *Ice Production and Storage for Seasonal Applications Utilizing Heat Pipe Technology*. ANL-82-87. Argonne, IL: Argonne National Laboratory.

Green, T. F., and G. C. Vliet. 1981. Transient response of a latent heat storage unit: An analytical and experimental investigation. *ASME J. Solar Energy Eng.* 103: 275–280.

Gross, R. J., and R. W. Harrigan. 1981. Status report on thermocline thermal energy storage studies at Sandia National Laboratories. In *Proc. DOE Thermal and Chemical Storage Annual Contractor's Review Meeting*. McLean, VA. CONF-801055, pp. 104-108.

Hale, N. W., Jr., and R. Viskanta. 1980. Solid-liquid phase-change heat transfer and interface motion in materials cooled or heated from above or below. *Int. J. Heat Mass Transfer* 23: 283–292.

Han, S. M., and S. T. Wu. 1978a. Computer simulation of a solar energy system with a viscous-entrainment liquid storage tank model. In *Application of Solar Energy*, S. T. Wu et al., eds. Huntsville, AL: UAH Press, University of Alabama, pp. 165–183.

Han, S. M., and S. T. Wu. 1978b. Experimental and numerical study of liquid thermal storage tank models. In *Proc. Am. Sec. Int. Solar Energy Soc.*, Denver, CO. K. W. Böer and G. F. Franta, eds., pp. 265–270.

Henze, R. H., and J. A. C. Humphrey. 1981. Enhanced heat conduction in phase-change thermal energy storage devices. *Int. J. Heat Mass Transfer* 24: 459–474.

Hess, C. F., and C. W. Miller. 1982. An experimental and numerical study on the effect of the wall in a thermocline-type enclosure. *Solar Energy* 28: 153–161.

Hobbs, P. V. 1974. *Ice Physics*. London: Oxford University Press, pp. 461–629.

Hooper, F. C. 1979. Annual cycle storage for building heating. In *Proc. Solar Energy Storage Options*, San Antonio, TX. M. B. McCarthy, ed., pp. 11–21.

Hughes, P. J., S. A. Klein, and D. J. Close. 1976. Packed bed thermal storage models for solar air heating and cooling systems. *ASME J. Heat Transfer* 98: 336–338.

Hull, J. R., K. V. Liu, Y. S. Cha, H. M. Domanus, and W. T. Sha. 1981. Solar pond salt gradient instability prediction by means of a thermohydraulic computer code. In *Proc. Am. Sec. Int. Solar Energy Soc.*, Philadelphia, PA. B. H. Glenn and G. F. Franta, eds., pp. 812–816.

Hull, J. R., R. L. Cole, and A. B. Hull. 1982. *Energy Storage Criteria Handbook*. CR082.034. Port Hueneme, CA: Naval Civil Engineering Laboratory.

Hull, J. R., K. V. Liu, W. T. Sha, J. Kamal, and C. E. Nielsen. 1984. Dependence of ground heat loss upon solar pond size and perimeter insulation. *Solar Energy* 33: 25–33.

Jaluria, Y., and S. K. Gupta. 1982. Decay of thermal stratification in a water body for solar energy storage. *Solar Energy* 28: 137–143.

Jurinak, J. J., and J. W. Mitchell. 1984. Effect of matrix properties on the performance of a counterflow rotary dehumidifier. *ASME J. Solar Energy Eng.* 106: 638–645.

Klein, S. A., W. A. Beckman, and J. A. Duffie. 1976. A design procedure for solar heating systems. *Solar Energy* 18: 113–127.

Kriz, T. A., and C. J. Swet. 1982. *Solar Storage for Process Heat and Building Application: A Review of Prior Assessments.* SERI/TR-231-1595. Golden, CO: Solar Energy Research Institute.

Kuhn, J. K., G. F. von Fuchs, and A. P. Zob. 1980. *Developing and Upgrading of Solar System Thermal Energy Storage Simulation Models.* Final report, BCS-40319. Seattle, WA: Boeing Computer Services Co.

Lavan, Z., J.-B. Monnier, and W. M. Worek. 1982. Second law analysis of desiccant cooling systems. *ASME J. Solar Energy Eng.* 104: 229–236.

Lin, E. I. H., and W. T. Sha. 1979. Effects of baffles on thermal stratification in thermocline storage tanks. In *Proc. Int. Solar Energy Soc.*, Atlanta, GA. K. W. Böer and B. H. Glenn, eds., pp. 586–590.

Loehrke, R. I., J. C. Holzer, H. N. Gari, and M. K. Sharp. 1979. Stratification enhancement in liquid thermal storage tanks. *J. Energy* 3: 129–130.

Löf, G. O. G., and R. W. Hawley. 1948. Unsteady-state heat transfer between air and loose solids. *Indust. Engng. Chem.* 40: 1061–1070.

Lunardi, V. J. 1981. *Heat Transfer in Cold Climates.* New York: Van Nostrand Reinhold, pp. 353–533.

Lunde, P. J. 1980. *Solar Thermal Engineering.* New York: Wiley, pp. 280–310.

Mar, R. W., and T. T. Bramlette. 1980. Thermochemical storage systems. In *Solar Energy Technology Handbook—Part A*, W. C. Dickinson and P. N. Cheremisinoff, eds. New York: Marcel Dekker, pp 811–839.

Marshall, R., and C. Dietsche. 1982. Comparisons of paraffin wax storage subsystem models using liquid heat transfer media. *Solar Energy* 29: 503–511.

Mathiprakasam, B., and Z. Lavan. 1980. Performance predictions for adiabatic desiccant dehumidifiers using linear solutions. *ASME J. Solar Energy Eng.* 102: 73–79.

McGarity, A. E. 1979. Optimum collector-storage combinations involving annual cycle storage. In *Proc. Solar Energy Storage Options*, San Antonio, TX. M. B. McCarthy, ed., pp. 165–173.

Metz, P. D. 1982. The use of ground-coupled tanks in solar-assisted heat-pump systems. *ASME J. Solar Energy Eng.* 104: 366–372.

Metz, P. D. 1983. A simple computer program to model three-dimensional underground heat flow with realistic boundary conditions. *ASME J. Solar Energy Eng.* 105: 42–49.

Michaels, A. I., S. Sillman, F. Baylin, and C. A. Bankston. 1983. *Simulation and Optimization Study of a Solar Seasonal Storage District Heating System: The Fox River Valley Case Study.* ANL-83-47. Argonne, IL: Argonne National Laboratory.

Molz, F. J., J. G. Melville, O. Guven, and A. D. Parr. 1983. Aquifer thermal energy storage: An attempt to counter free thermal convection. *Water Resour. Res.* 19: 922–930.

Morrison, D. J., and S. I. Abdel-Khalik. 1978. Effects of phase-change energy storage on the performance of air-based and liquid-based solar heating systems. *Solar Energy* 20: 57–67.

Nelson, J. S., W. A. Beckman, J. W. Mitchell, and D. J. Close. 1978. Simulations of the performance of open cycle desiccant systems using solar energy. *Solar Energy* 21: 273–278.

Offenhartz, P. O'D. 1981. *Thermal Storage Studies for Solar Heating and Cooling: Applications Using Chemical Heat Pumps.* Final report, DOE/CS/30248-F. Newton, MA: EIC Labs, Inc.

Offenhartz, P. O'D., J. M. Marston, and J. I. Watts. 1979. *Analysis of Advanced Thermal Storage Subsystems for Solar Heating and Cooling.* Final report, COO-4483-F. Newton, MA: EIC Corp.

Papadopulos, S. S., and S. P. Larson. 1978. Aquifer storage of heated water. Part II: Numerical simulation of field results. *Ground Water* 16: 242–248.

Philip, J. R., and D. A. de Vries. 1957. Moisture movement in porous materials under temperature gradients. *Trans. Am. Geophysical Union* 38: 222–232.

Phillips, W. F., and R. N. Dave. 1982. Effects of stratification on the performance of liquid-based solar heating systems. *Solar Energy* 29: 111–120.

Pomeroy, B. D. 1979. Thermal energy storage in a packed bed of iron spheres with liquid sodium coolant. *Solar Energy* 23: 513–515.

Rademaker, O. 1981. On the dynamics and control of (thermal solar) systems using stratified storage. In *Thermal Storage of Solar Energy*, C. den Ouden, ed. The Hague, The Netherlands: Martinus Nijhoff, pp. 61–72.

Radosevich, L. G., and C. E. Wyman. 1983. Thermal energy storage development for solar electrical power and process heat applications. *ASME J. Solar Energy Eng.* 105: 111–118.

Raldow, W. 1981. Thermal efficiencies of chemical heat pump configurations. *Solar Energy* 27: 307–311.

Ramsey, J. W., and E. M. Sparrow. 1978. Melting and natural convection due to a vertical embedded heater. *ASME J. Heat Transfer* 100: 368–370.

Riaz, M. 1978. Transient analysis of packed-bed thermal storage systems. *Solar Energy* 21: 123–128.

Saez, A. E., and B. J. McCoy. 1982. Dynamic response of a packed bed thermal storage system—A model for solar air heating. *Solar Energy* 29: 201–206.

Saxena, S., S. Subrahmaniyam, and M. K. Sarkar. 1982. A preliminary model for phase change thermal energy storage in a shell and tube heat exchanger. *Solar Energy* 29: 257–263.

Schumann, T. E. W. 1929. Heat transfer: a liquid flowing through a porous prism. *J. Franklin Institute* 208: 405–416.

Sha, W. T., and E. I. H. Lin. 1978. Three dimensional mathematical model of flow stratification in thermocline storage tanks. In *Applications of Solar Energy*, S. T. Wu et al., ed. Huntsville, AL: UAH Press, University of Alabama, pp. 185–202.

Sha, W. T., E. I. H. Lin, R. C. Schmitt, K. V. Liu, J. R. Hull, J. J. Oras, Jr., and H. M. Domanus. 1980. *COMMIX-SA-1: A Three-Dimensional Thermohydrodynamic Computer Program for Solar Applications*. ANL-80-8. Argonne, IL: Argonne National Laboratory.

Sharp, M. K., and R. I. Loehrke. 1979. Stratified thermal storage in residential solar energy applications. *J. Energy* 3: 106–113.

Shelton, J. 1975. Underground storage of heat in solar heating systems. *Solar Energy* 17: 137–143.

Sheridan, N. R., K. M. Bullock, and J. A. Duffie. 1967. Study of solar processes by analog computer. *Solar Energy* 11: 69–77.

Sillman, S. 1981. Performance and economics of annual storage solar heating systems. *Solar Energy* 27: 513–528.

Solaron Corp. 1982. *Cost Effective Improvements of Rockbeds Used in Solar Air Heating System*. Final report, 31-109-38-5692. Denver, CO: Solaron Corp.

Solomon, A. D. 1979. Melt time and heat flux for a simple PCM body. *Solar Energy* 22: 251–257.

Sowell, E. F., and R. L. Curry. 1980. A convolution model of rock bed thermal storage units. *Solar Energy* 24: 441–449.

Sparrow, E. M., R. R. Schmidt, and J. W. Ramsey. 1978. Experiments on the role of natural convection in the melting of solids. *ASME J. Heat Transfer* 100: 11–16.

Szego, J., and F. W. Schmidt. 1978. Transient behavior of a solid sensible heat thermal storage exchanger. *ASME J. Heat Transfer* 100: 148–154.

Tsang, C. F., T. Buscheck, and C. Doughty. 1981. Aquifer thermal energy storage: A numerical simulation of Auburn University field experiments. *Water Resour. Res.* 17: 647–658.

University of Wisconsin-Madison. 1979. *TRNSYS Simulation Manual*. Madison, WI: University of Wisconsin, Solar Energy Laboratory.

Veltkamp, W. B. 1981. Thermal stratification in heat storage. In *Thermal Storage of Solar Energy*, C. den Ouden, ed. The Hague, The Netherlands: Martinus Nijhoff, pp. 47–59.

von Fuchs, G. 1979. Rock bed computer model. In *Proc. Solar Energy Storage Options*, San Antonio, TX. M.B. McCarthy, ed., pp. 547–559.

Walker, W. R., J. D. Sabey, and D. R. Hampton. 1981. *Studies of Heat Transfer and Water Migration in Soils*. Final report, DE-AC02-79CS30139. Fort Collins, CO: Colorado State University.

Williams, G. T., C. R. Attwater, and F. C. Hooper. 1980. A design method to determine the optimal distribution and amount of insulation for in-ground heat storage tanks. *Solar Energy* 24: 471–475.

Worek, W. M., and Z. Lavan. 1982. Performance of a cross-cooled desiccant dehumidifier prototype. *ASME J. Solar Energy Eng.* 104: 187–196.

Williams, G. T., C. R. Attwater, and F. C. Hooper. 1980. A design method to determine the optimal distribution and amount of insulation for in-ground heat storage tanks. *Solar Energy* 24: 471–475.

Worek, W. M., and Z. Lavan. 1982. Performance of a cross-cooled desiccant dehumidifier prototype. *ASME J. Solar Energy Eng.* 104: 187–196.

17 Testing and Evaluation of Thermal Energy Storage Systems

Robert D. Dikkers

17.1 Development of Thermal Performance Test Methods

The basis for American Society of Heating, Refrigeration, and Air Conditioning Engineers (ASHRAE) Standard 94-77 (1977) was provided by the studies reported by the National Bureau of Standards (NBS) (Kelly and Hill 1975; Hill et al. 1976; Hill et al. 1977; Yang and Lee 1974).The theoretical analysis of storage devices subject to step and sinusoidal variations of inlet temperature was presented for latent heat storage devices. Two alternative methods of testing storage devices were identified; they were constant input and output heating rate and variable input and output heating rate. The step change of temperature method was selected because it appeared to be the most fundamental approach. Also the methods would indicate the same relative performance and the step change procedure allows measurement of thermal storage capacity that can be used as a basis of comparison for different storage devices.

The ASHRAE Standard differs from the NBS procedure. The major difference is that the NBS procedure requires four charge and four discharge tests, whereas the ASHRAE Standard requires only two of each. The NBS procedure required the tests to be performed at each of two step changes of temperature, whereas only one step change of temperature is required for the ASHRAE procedure. The NBS procedure also specifies a means of checking the validity of the test data that the ASHRAE Standard does not require. The ASHRAE Standard has relaxed some instrument accuracy and test conditions to more realistically attainable values (ΔP across storage device of ± 25 Pa or 0.1 in. of water) instead of ± 2.5 Pa (0.01 in. of water), ambient air temperature constant within $\pm 2°C$ (3.6°F) instead of $\pm 1°C$ (1.8°F), and dew point temperature for air systems constant within $\pm 1°C$ (1.8°F) instead of $\pm 0.5°C$ (0.9°F). The NBS procedure requires reporting an estimate of the heat transfer fluid pumping power which is not required by the ASHRAE Standard.

The draft standard was submitted to ASHRAE Standard Committee in January 1977 and was subsequently approved as 94–77 at the Chicago meeting of ASHRAE. In validating the Standard 94–77 test procedure for use with phase change materials (PCMs), NBS concluded there were problems and recommended changes in 1979. Calmac Manufacturing Corporation [under a Department of Energy (DOE)/Argonne National Laboratory contract] concluded the standard was only good for sensible heat storage devices, and that it was off by 2 : 1 in calculating total storage or efficiency for latent heat storage devices. Also standby losses were too small for practical measurement. A new committee, SPC 94.1R, formed in 1981 undertook work to develop a revised test procedure for phase change materials using constant heat input and withdrawal, a recirculating system, and permitting specification of time, flow rate, and temperature differences. A revised method of test for latent heat storage devices was completed and approved in January 1985 as ASHRAE Standard 94.1-1985. A separate standard (94.3) for sensible heat storage devices was approved by ASHRAE in 1986.

The ASHRAE 94–77 procedures provide the testing methods for thermal performance evaluation of thermal storage devices. The storage unit is initialized by flowing a constant-temperature heat transfer fluid through its ports. Then the unit is subjected to a step change of temperature, and the rate of charging or discharging is measured. The rate of charging or discharging is the temperature difference between the inlet and the outlet multiplied by the mass flow rate and the heat capacity of the heat transfer fluid. The charge and discharge capacities are the time integrals of the charge and discharge rates, respectively. A correction to the charge capacity for the thermal losses, but none for the discharge capacity, is specified. The charge and discharge capacities, C_c and C_d, respectively, are determined from

$$C_c = w_c C_{pf} \int_0^{t_c} (T_{in} - T_{out}) dt - Lt_c \left[T_i + \frac{T_{in} - T_{out}}{2} - T_a \right], \tag{1}$$

$$C_d = w_d C_{pf} \int_0^{t_d} (T_{out} - T_{in}) dt, \tag{2}$$

where

C_c = charge capacity of the thermal storage device [J (or Btu)],

C_d = discharge capacity of the thermal storage device [J (or Btu)].

The mass flow rates for the test are determined from the theoretical storage capacity (TSC) and the time required for a particular test. The TSC is the sum of all sensible and latent heats theoretically contained in the storage unit from the melting point minus $\Delta T/2$ to the melting point plus $\Delta T/2$. The step change of temperature T is specified as 15°C (27°F) for units having a liquid heat transfer fluid and 35°C (63°F) for units having air as the heat transfer fluid. Test durations of two and four hours are required. The mass flow rates are specified for charge and discharge by

$$w_c = \frac{TSC}{t_c \Delta T C_{pf}}, \tag{3}$$

$$w_d = \frac{TSC}{t_d \Delta T C_{pf}}, \tag{4}$$

where the subscripts c and d denote the charge and discharge tests, respectively.

Equations (3)and (4) imply a sensible heat transfer fluid; heat transfer fluids that change phase are not allowed.

Thermal losses are measured by a steady-state test. The storage unit is subjected to a constant inlet temperature 25°C (45°F) above ambient temperature, and after steady state is attained, the temperature difference between the inlet and the outlet is measured as an average value for one hour.

The heat loss rate is

$$L = \frac{w_L C_{pf}}{(3,600 \text{ s})(25°C)} \int_0^{3,600\,s} (T_{in} - T_{out})dt, \tag{5}$$

where

L = heat loss rate [J/s °C (Btu/h °F)],

W_L = mass flow rate of the transfer fluid for the heat loss test [kg/s (lb/h)],

C_{pf} = specific heat of the transfer fluid [J/kg °C (Btu/lb °F)],

T_{in} = temperature of the transfer fluid entering the storage device [°C (°F)],

T_{out} = temperature of the transfer fluid leaving the storage device [°C (°F)].

The required mass flow rate of the transfer fluid for the heat loss test is

$$w_L = \frac{TSC_L}{C_{pf}(14,400s)(25°C)}, \tag{6}$$

where TSC_L = theoretical storage capacity of the thermal storage device for the heat loss test [J (Btu)].

TSC_L is the amount of energy that could be stored in the device if all the device components would undergo an increase in temperature of 25°C (45°F) from an initial temperature equal to t_a, the ambient air temperature.

A PCM storage unit is required to undergo a specified number of cycles before testing (30 cycles—ASHRAE 94-77; 5 cycles—ASHRAE 94.1-1985). Such cycling is intended to establish a condition of the storage unit representative of the "as installed" condition. Since the standards specifically state that it does not include factors relating to cost, life, reliability, or considerations of interfacing with other heating, ventilating, and air conditioning (HVAC) units, the cycling should not be construed as a test of material stability Indeed, there is no test before the cycling to establish how much degradation occurs after cycling. The ASTM material tests would not be sufficient because PCMs lose capacity from loss of moisture, which is in turn dependent upon the packaging.

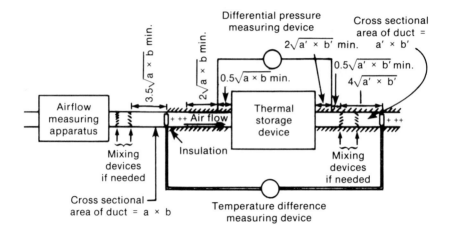

Figure 17.1
Representative test configuration for a thermal storage device using air as the transfer fluid.

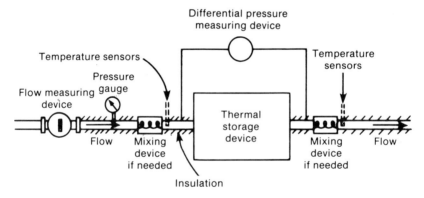

Figure 17.2
Representative test configuration for a thermal storage device using a liquid as the transfer fluid.

For testing air systems, the representative equipment is shown in figure 17.1. The instrumentation required, specific dimensions and dimensional relationships are provided in the standard. Flow mixers may be required upstream of the temperature measurements to ensure accuracy. If the flow is in an open loop, air-reconditioning devices may be necessary to maintain the required storage device inlet conditions. Flow in a closed loop requires an air-reconditioning device.

For testing liquid systems the schematic of the equipment layout is shown in figure 17.2. As with air systems, flow mixers upstream of the temperature measuring devices may be necessary to ensure accurate temperature measurements. The instrumentation requirements and dimensional relationships are provided in the standard. If the flow is in an open loop, a transfer fluid reconditioning apparatus may be necessary to maintain the required storage device inlet temperature. Flow in a closed loop requires a transfer fluid reconditioning device.

17.2 Evaluation of Thermal Storage Devices

17.2.1 Sensible Storage

Water The applicability of ASHRAE Standard 94-77 to water tanks as sensible-heat thermal storage devices was reported in Hunt et al. (1978, 1979). The heat loss characteristics and energy charge or discharge were quantitatively obtained through use of the measurement methods in the standard. Dimensionless parameters were recommended (temperature difference ratio and normalized charge or discharge thermal capacity times) for comparison of thermal performance of different storage devices. Also a stagnant heat loss test evaluated in the study was suggested as an alternate test technique. The tests in the standard consist of the following:

1. one test to determine a heat loss factor for the thermal storage device (see figure 17.3),

2. two tests to determine the response characteristics (charge capacity) of the device to a step increase in the entering transfer fluid temperature,

3. two tests to determine the response characteristics (discharge capacity) of the device to a step decrease in the entering transfer fluid temperature.

The heat loss test requires passing the transfer fluid through the storage device with an inlet temperature of 25°C (45°F) above the ambient air temperature. After steady-state conditions are obtained, measurements are made of the average temperature difference between the inlet and outlet transfer

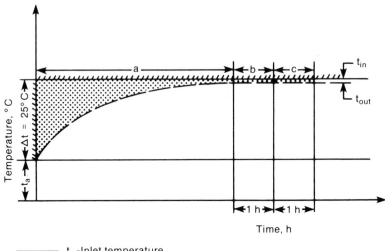

Figure 17.3
Time-temperature variation of the transfer fluid during the heat loss test.

fluid and the average ambient air temperature over a one-hour period. Steady-state conditions require the inlet and outlet transfer fluid temperatures to vary by less than $\pm 0.5°C$ ($\pm 0.9°F$) during a one-hour period.

The concept of test fill time is defined as

$$t_f = \frac{TSC}{w_{tf}c_{pf}\Delta t},$$

(7)

where

t_f = test fill time (s),

TSC = theoretical storage capacity of the thermal storage device for the temperature increase Δt [J (or Btu)],

w_{tf} = mass flow rate of the transfer fluid [kg/s (or lb/h)],

Δt = temperature step of the inlet transfer fluid from an initial temperature t_i to a final temperature $t_i \pm \Delta t$ [°C (°F)].

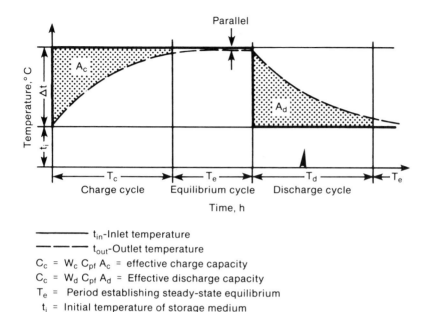

Figure 17.4
Time-temperature variation of transfer fluid during a charge and discharge test cycle.

A water tank with no heat loss to the ambient and with perfect stratification of the storage medium would be completely charged or discharged in the time defined by equation (7). Since such an ideal storage device does not exist, the fill time defined above is found to be less than time required to completely charge or discharge actual storage devices.

Figure 17.4 represents the charge and discharge cycles of a thermal storage device undergoing the transient response tests in accordance with the Standard.

The measurements required for the charge and discharge test computations are the temperature difference between the entering and leaving transfer fluid over the test fill time, the transfer fluid flow rate, and the ambient air temperature. It should be noted that the charge and discharge capacities are a function of the specific heat of the transfer fluid and hence a function of the transfer fluid used. It should be noted in reference to figure 17.4, a single charge and discharge test are shown to be performed in series with each other. For charge tests, the inlet transfer fluid is introduced representing the mode in which the device would normally function within any solar system.

For discharge tests the transfer fluid is introduced in the typical operating mode of the thermal storage tank being tested. Discharge tests performed by

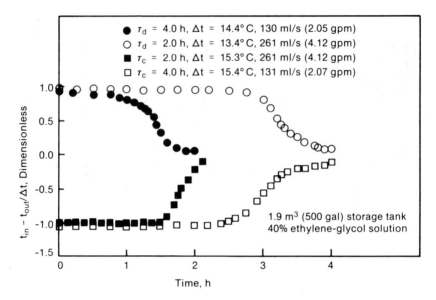

Figure 17.5
ASHRAE recommended plot of time variation in transfer fluid temperature.

introducing the colder transfer fluid opposite to the normal mode of operation modifies the stratification in the tank, and test results drastically different from other discharge tests can occur.

ASHRAE Standard 94-77 suggests a plot be provided showing the time variation in transfer fluid temperatures for each test. Such a graph for those tests involving the ASHRAE-recommended test fill times and temperature step changes is shown in figure 17.5. The abscissa represents time from $\tau = 0$ to $\tau_{c,d}$ and the ordinate is a dimensionless quantity

$$\frac{t_{out} - t_{in}}{\Delta t}. \tag{8}$$

In equation (8) the initial step change in transfer fluid inlet temperature, Δt, is a constant. However, the difference between the transfer fluid outlet and inlet temperatures decreases with time, yielding the asymptotic approach at the end of the test time (data from Hunt et al. 1978, 1979). For a thermal storage tank with perfect stratification and no heat loss to the ambient, the temperature of the inlet transfer fluid t_{in} and the outlet transfer fluid t_{out} would remain constant for the entire test period (time for one entire tank change), at which time they would become equal. A curve of test results for such a device would be represented by a rectangle.

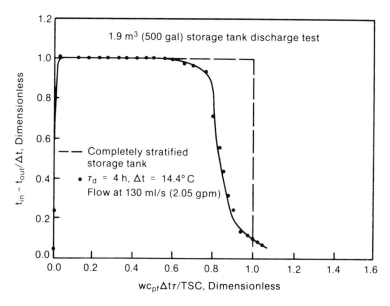

Figure 17.6
Plot of dimensionless temperature versus dimensionless time for a 4-h transient discharge test.

The concept of a dimensionless plot showing the thermal storage capabilities of storage devices relative to those of an ideal device was proposed in Hill et al. (1977). Such a dimensionless plot is shown in figure 17.6 for the discharge test where $\tau_d = 4$ h (Hunt et al. 1978). The ordinate represents the difference between the inlet and outlet transfer fluid temperature normalized by the step change in temperature Δt (identical concept utilized in figure 17.5). The abscissa represents a dimensionless time defined by

$$\frac{WC_{pf}\Delta T\tau}{\text{TSC}}.$$

(9)

A value of dimensionless time equal to 1.0 corresponds precisely to the test fill time or fluid dwell time of a perfectly insulated, completely stratified water tank. It is noted that the area under the curve on such a plot represents the amount of energy capable of being stored relative to the ideal device. Therefore the area corresponding to the energy storage capability of the ideal device is 1.0 and all real devices have an area representatively smaller than 1.0. Table 17.1 shows the listing of transient tests performed on the thermal storage tank and their respective energy capabilities (or performance coefficients) based on a dimensionless analysis (Hunt et al. 1978).

Table 17.1
Rating coefficients for the 1.9-m³ (500-gal) thermal storage tank

Test description	Performance coefficient (%)
Charge-2	88
Discharge-2	84
Discharge-2[a]	55
Charge-4	88
Discharge-4	84
Charge-1	85
Discharge-1	79
Charge-$\frac{1}{2}$	86
Discharge-$\frac{1}{2}$	85

a. Cool inlet transfer fluid introduced into top of the thermal storage tank.

The utility of such a dimensionless analysis becomes apparent when one refers to table 17.2. This table reveals a variety of "effective" charge and discharge capacities. For example, the two-hour charge test performed resulted in a C_c equal to 9.29×10^7 J (88,000 Btu) where the two-hour discharge test resulted in a C_d equal to 7.63×10^7 J (72,317 Btu) [there existed a 2°C (4°F) difference between the temperature step change achieved in the two tests]. Numerical identification following test description in table 17.2 designates the test period involved, τ_c or τ_d, and where C_c designates the "effective" charge capacity with heat loss accounted for, C_d designates the "effective" discharge capacity, and Q designates the quantity of heat loss over the charge period.

By utilizing the "effective" charge and discharge capacities obtained via the ASHRAE Standard in a dimensionless analysis, a meaningful overall rating coefficient can be obtained, even though the step change in entering fluid temperature called for in the Standard is not met exactly. In addition, since the transfer fluid specific heat appears in the abscissa of the dimensionless plot, the performance coefficient of the device should be independent of the transfer fluid (that was not verified).

Because of the nature of the transient tests performed, the larger the degree of stratification, the larger will be the charge and discharge capacities and hence the better the rating. In order to demonstrate the relationship between test results obtained in accordance with the Standard and the predicted performance of a water tank, some simple simulations using the computer program TRNSYS (Klein et al. 1975; University of Wisconsin 1977) were reported in Hunt et al. (1979).

Table 17.2
Results of the transient tests performed on the 1.9-m^3 (500-gal) thermal storage tank

Test description	C_c	Q	C_d
Charge-2	9.29	0.16	—
Discharge-2	—	—	7.36
Discharge-2[a]	—	—	5.72
Charge-4	9.24	0.21	—
Discharge-4	—	—	8.26
Charge-1	8.91	0.07	—
Discharge-1	—	—	9.49
Charge-$\frac{1}{2}$	8.03	0.06	—
Discharge-$\frac{1}{2}$	—	—	7.11
All units, 10^7 J			

a. Cool inlet transfer fluid introduced into top of the thermal storage tank.

TRNSYS is a digital computer program written in FORTRAN and developed at the University of Wisconsin. The modular way in which solar energy systems are constructed has been carried over into the structuring of the program itself. To use the program, the user specifies which components he desires to include in a system and how they are to be connected. The inputs and outputs of component subroutines correspond to the inputs and outputs of the modeled hardware (e.g., temperatures, flow rates) and are simply connected, via the executive program, to the outputs and inputs of other component subroutines.

In modeling the tank using TRNSYS, the primary variable to specify is the number of nodes or segments into which to divide the tank. This governs the degree of stratification that will result. If only one node or segment is chosen, the tank will be modeled as uniformly mixed at any instant of time. By making a number of repeated computations, it was found that the experimental results could be reproduced almost exactly by using a 20-segment tank (Hunt et al. 1979). Once it was verified that TRNSYS could accurately simulate the behavior of this tank, it was used to model a simple solar space-heating system incorporating the tank.

With the tank modeled as a one-segment unit, approximately 75% of the space-heating load was met by solar energy. The performance coefficient for a one-node tank in accordance with tests from ASHRAE Standard 94-77 and the definition of performance coefficient based on Hunt et al. (1979) was only 64%. When the tank was modeled as a 20-node unit, the percentage of the load supplied by solar energy was only increased two percentage points to 74%, even though the performance coefficient for the tank itself was increased

24 points to 88%. The strategy for TRNSYS calculations was discussed in Cole et al. (1978) where a constant flow rate with about three to five fill times were shown to not be optimum for a stratified tank (and also applies to charge periods).

Additional simulations were completed with various assumptions. A drastic improvement in the stratification in the storage tank, corresponding to a major improvement in the performance coefficient for the tank, does not show improvement in system performance. The above simulations need to be repeated for a variety of different storage devices and systems in order to fully understand the implications of the ASHRAE Standard 94-77 test results for system design and performance. However, it appears, based on the initial modeling done, that a large variation in rating of water tanks in accordance with the Standard does not necessarily mean a large variation in performance of a system in which they are used.

Rock   The applicability of ASHRAE Standard 94-77 to a pebble bed and a phase change unit utilizing sodium sulfate decahydrate using air as the transfer fluid was reported in Jones and Hill (1979). The tests in the standard have been noted in the previous section on water. The set of equations (1) through (6) and figures 17.3 and 17.4 also discussed in the previous section on water as the thermal storage medium apply to the rock materials (sensible heat storage devices). Modifications for applications to thermal phase change materials are discussed in the next section.

The heat loss tests reported in Jones and Hill (1979) were carried out in accordance with Standard 94-77, except for using a temperature difference of 35°C (63°F) rather than 25°C (45°F) as specified. The reason for this was that it was more convenient to do the heat loss test at the end of a charge test, which was done at 35°C and would not compromise the test results. All test cycles consisted of a charge test followed by a heat loss test and finally by a discharge test.

For the heat loss test the inlet fluid temperature was set at 35°C (63°F) above ambient temperature, and the outlet fluid temperature was allowed to reach a steady-state condition. The difference between inlet and outlet fluid temperatures was averaged over a several-hour period. The heat loss coefficient L was then calculated. A major problem with the heat loss test was found to be the stability and measurement of the low temperature difference across the unit. Typical ranges of inlet-outlet temperature difference were 1.5° to 3°C (3° to 6°F). The fluid inlet temperature controller was required to maintain a steady temperature only within ± 1°C (± 2°F), which made it very difficult to accurately measure the 1.5°–3°C (2.7°–5.4°F) temperature difference. In order to overcome these problems, the heat loss rate was averaged over several hours rather than the one-hour test period, as specified.

Transient tests were performed in accordance with the Standard except that the recommended flow rates were not used. Both thermal energy storage units were designed for specific flow rates, and these specified flow rates were used in the tests. In general, the flow rates specified in the Standard are much higher than experienced in most solar systems using air as the transfer fluid. The Standard specifies fill times of two and four hours, whereas in a typical system the flow rate is such that the fill time is approximately six hours. The flow direction was top to bottom with test flow rates approximately 0.42 and 0.21 m^3/s (900 and 450 ft^3/min).

Transient charge tests were performed by first circulating room temperature air through the device until the inlet and exit temperatures were within 0.5°C (1.0°F). The flow rate was then adjusted to the desired value and the air heaters were then turned on starting the test. Test fluid inlet temperature was generally attained within five minutes, meeting the requirements of the Standard that 90% of the step change be achieved within 2% of the test fill time. The charge test continued until the exit fluid temperature reached steady-state conditions at which time the heat loss test was started.

The transient discharge tests were performed following the heat loss tests. The discharge test was started by simply cutting off the heaters and introducing unconditioned room temperature air.

Charge–heat loss–discharge test cycles were reported, four at the flow rate recommended by the manufacturer and three at one-half the recommended flow rate. A heat loss test was included in all test cycles and resulted in an overall heat loss coefficient of 95 kJ/(h °C) (50 Btu/h °F) for both flow rates. An approximate overall heat loss coefficient was estimated to be 36 kJ/(h °C) (19 Btu/h °F). Heat loss during the transient tests generally was on the order of 3% to 8% of the energy transferred during a test fill time period.

Results for the charge and discharge tests of the pebble-bed are shown in table 17.3. The charge capacity determined ranged from 258 to 320 MJ (2.44 × 10^5 to 3.03 × 10^5 Btu) and the discharge capacity from 247 to 293 MJ (2.34 × 10^5 to 2.78 × 10^5 Btu). Charge and discharge capacities are highly dependent on test temperatures. The performance factors for the charge tests ranged from 0.79 to 0.87 and from 0.80 to 0.86 for the discharge tests. In general, the performance factor for the charge tests was higher at higher flow rates due to less heat loss over the shorter fill time. Moisture exchange with the airstream was not taken into account in the calculation of charge or discharge capacity.

744 Robert D. Dikkers

Table 17.3
Pebble-bed test results

Cycle	Fill time (h)	Flow rate (kg/h)	Initial temperature (°C)	Input temperature (°C)	Heat loss (MJ)	Theoretical storage capacity (MJ)	Charging capacity (MJ)	Total storage capacity (MJ)	Performance factor
1 C	5.54	1742	22.1	61.3	14.3	381.0	320.0	—	0.84
1 D	5.64	1712	62.2	26.9	—	342.0	293.0	—	0.86
2 C	5.42	1782	24.2	60.1	13.9	349.0	300.0	—	0.86
2 D	5.34	1808	58.9	24.5	—	334.0	268.0	—	0.80
3 C	Initial conditions not recorded								
3 D	5.40	1788	58.8	24.9	—	328.0	—	—	—
4 C	Malfunction in data system								
4 D	Malfunction in data system								
5 C	10.78	896	23.6	59.8	29.3	351.0	284.0	—	0.81
5 D	11.14	866	57.2	25.3	—	309.0	262.0	—	0.85
6 C	10.7	898	23.0	56.5	26.0	325.0	258.0	—	0.79
6 D	11.0	879	55.0	24.6	—	296.0	247.0	—	0.84
7 C	5.3	1810	23.0	56.9	13.3	329.0	287.0	—	0.87
7 D	5.4	1781	57.9	25.4	—	315.0	254.0	—	0.81

Note: The dash denotes erroneous or insufficient data.

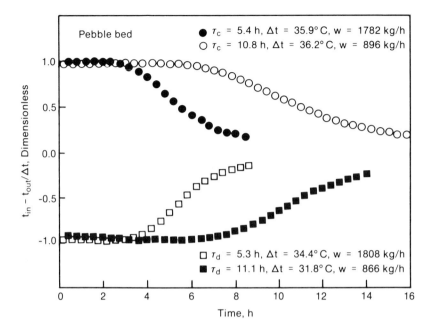

Figure 17.7
ASHRAE-recommended plot of time variation in transfer fluid temperature for the pebble-bed thermal energy storage test unit.

A plot of charge and discharge test results at both flow rates as required by the Standard is shown in figure 17.7. The dimensionless plot of the same two charge tests is shown in figure 17.8. Note that the results for the two different flow rates fall nearly on the same curve, indicating that the thermal performance of the pebble-bed is not greatly affected by flow rate.

Pebble-beds are known to exchange moisture with the air stream, which has a significant effect on the performance of the bed (Close and Pryor 1976; Close and Dunkle 1977). In general, moisture is released from the pebble-bed during charging and absorbed during discharging. The mechanism of moisture exchange is a sorption process in which moisture is exchanged between the air and the fine pores of an absorbent. During the process there is an energy exchange equal to the latent heat of vaporization of the water. The quantity of water finally held by a given sorbent when equilibrium has been reached is dependent only on the relative humidity of the contacting air-water vapor mixture. This ratio is largely independent of the temperature of operation and of the vapor pressure at which the water exists in the air mixture.

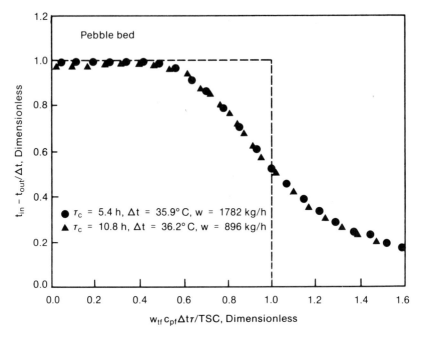

Figure 17.8
Plot of dimensionless temperature versus dimensionless time for charge tests on the pebble-bed thermal energy storage test unit at two flow rates.

The moisture exchange in a pebble-bed should generally behave as follows:

1. Initially the bed will be at thermal and moisture equilibrium with the cool precharging airstream (assuming entering temperature and humidity are held constant).

2. Charging, which is accomplished with air of high temperature and low relative humidity, will raise the temperature of the rocks and extract moisture from them. Energy is stored in both the sensible temperature increase of the rocks and in the "dryness" of the bed.

3. Discharging, which is accomplished with air of low temperature and high relative humidity, will lower the temperature of the rocks and deposit moisture in them. Energy removed from the bed is contributed by the sensible heat of the rocks and the heat of vaporization as the water changes from vapor to liquid.

Section 8.2 of the Standard suggests that an energy balance approach be used for rating a thermal energy storage device when there is a net moisture exchange between the airstream and the device. In contrast, the suggestion for rating based on sensible-heat exchange only when the thermal energy storage device is used for heating purposes was presented in Close and Pryor (1976). That recommendation seems to be the most feasible for pebble-beds since the moisture exchange process is one that will occur in a random fashion and primarily based on the system design and geographical location of the system.

From analysis of the phenomenon of moisture exchange with porous beds, it appears that the present Standard should include requirements to cover this situation. Inlet humidity conditions for tests must be established so that the various thermal energy storage devices can be rated equitably. In order for the moisture exchange process to be controlled properly in the test, Standard 94-77 requires modification. Section 8.2 states that the inlet and exit dewpoint temperatures should be equal and remain constant, which is impossible for some devices. Three alternate approaches for establishing the inlet humidity conditions are

1. Use a closed test loop and a drier to remove all significant quantities of moisture from the air and the bed prior to starting the test.

2. Use a closed test loop, but initially have the bed at equilibrium at a specified condition prior to a transient test. This approach would simulate a closed loop with a solar collector such as exits in a solar heating system.

3. Use an open test loop with the inlet humidity conditions controlled to a specified condition.

The major change contained in the proposed new standard (ASHRAE 94.3) for sensible thermal energy storage devices is to provide wider freedom to the supplier of the test device in the choice of cycling time, capacity, maximum flow rate, and temperature range. Another important revision is in making the standby heat loss test more practical and accurate.

17.2.2 Latent Heat

Phase change materials and systems presently are in a prestandards state. For this maturing technology a document intended to be used as a guide in applying the existing codes and standards to phase change systems was prepared at NBS (Greenberg and Reeder 1984). By analyzing existing regulatory documents, issues relevant to the application of phase change thermal energy storage systems in building construction were determined. Through the standards mechanism, requirements are clearly stated for the designer and installer, and the building code official is given some reasonable assurance that the health, safety, and general welfare of the public is protected. The establishment of standards, however, especially for an evolving technology is a developmental process that normally occurs over a relatively long time.

A great number of materials are candidates for phase change thermal storage applications. Phase change materials potentially appropriate for residential heat storage application can be broadly classified into two groups: salt hydrates and organic materials. A listing of approximately 6,000 salts relevant to energy storage along with data compilations outlining their physical properties was published by NBS (Janz et al. 1978). However, only a very few salts have been studied in depth. Salt hydrates have great potential for applications in solar thermal storage. The crystalline structures of the salt hydrates change phase or melt in three ways: congruent melting, incongruent melting, and semicongruent melting (see Shelpuk et al. 1976). Salt hydrates are also subject to supercooling, which can degrade the full thermal potential of the material, although proper container design or nucleating additives can reduce or eliminate this problem. A list of promising salt hydrate materials, a discussion of selection criteria, and characteristics for use as thermal energy storage materials are contained in Shelpuk et al. (1976), University of Maryland (1973), Hale et al. (1971), and Lane et al. (1976).

Organic materials have also been identified for potential use as phase change materials. If the criteria for selection of the salt hydrates are applied, potential candidates are quickly limited to a very few. Other candidate materials are generally in the paraffin classification (Shelpuk et al. 1976). Those materials generally melt congruently and therefore have no cycling problems, as do many of the salt hydrates. The paraffins have reasonable levels of latent-heat capacities and are commercially available. However, they shrink considerably upon solidifying, have low thermal conductivity, and in many cases are flammable.

Phase change materials must undergo a large number of cycles of melting and solidification. Tests on the reliability/durability and changes in properties are required for standards; this function is under the American Society for Testing and Materials (ASTM). Test methods for possible modes of degradation that may affect service lives must be considered. In the NBS report (Ings and Brown 1982a) the effects of crystal growth, crystal segregation, supercooling, corrosion and thermal decomposition are discussed. The generic basis for the development of performance tests for inorganic phase change materials was also described.

An NBS feasibility study of calcium aluminate hydrates and calcium aluminate hydrates containing other ions was conducted to determine their utilization as PCMs as energy storage media (Ings and Brown 1982b). The energy liberated on hydration of each compound was measured using conduction calorimetry and the dehydration temperature was measured using differential scanning calorimetry. Of the compounds investigated, $3CaO\ Al_2O_3\ 3CaSO_4$ $32H_2O$ liberated the largest amount of energy upon rehydration. Initially, this value was about 100 cal/g. However, after 18 cycles of hydration and dehydration the value dropped to about 70 cal/g.

In the NBS studies cited earlier (Jones and Hill 1979) on evaluation of ASHRAE Standard 94-77, phase change device tests were conducted. The phase change device consisted of 726 plastic trays containing a Glauber saltwater mixture (sodium sulfate decahydrate). The phase change temperature of the material is 32°C (89°F). The trays were arranged in an array 26 trays high, 4 trays wide, and 7 trays in the flow direction. Air was the heat transfer fluid used with this device.

The nominal thermal energy storage capacity of the unit was 264 MJ (250,000 Btu). A theoretical energy storage capacity of 414 MJ (392,000 Btu) was calculated using the manufacturer's specifications and a temperature swing of 35°C (63°F) chosen in accordance with Standard 94-77 and used during the test. The thermal energy storage capacity is a combination of the latent-heat and sensible-heat capacity of the storage material. Table 17.4 summarizes the thermal energy storage contribution of the various components. The theoretical latent contribution was approximately 65% of the total capacity.

An important exception to the ASHRAE test requirements (Standard 94-77) for this test program was that the phase change device was not cycled through

Table 17.4
Theoretical storage capacity of the phase change thermal energy storage device

Component	Mass per tray (kg/tray)	Temperature range High (°C)	Temperature range Low (°C)	Specific heat (kJ/kg°C)	Energy storage capability (kJ/tray)	Percent of total energy storage capability
$Na_2SO_4-10H_2O$	1.49					
Fusion		32	32	251.8	373.9	65.3
Sensible heat, solid		32	21	1.93	31.6	5.5
Sensible heat, liquid		56	32	3.52	125.9	22.0
Plastic	0.25	56	21	3.46	30.3	5.3
Wood	0.13	56	21	2.52	11.5	2.0
Insulation	0.09	—	21	2.52	—	—
Total					573.2	100.0

Note: kJ/kg at 32°C.

the phase change temperature 30 times prior to testing. Tests performed at NBS were primarily to evaluate the testing techniques and not to evaluate the particular unit. Five complete charge–heat loss–discharge test cycles were completed, three at the flow rate recommended by the manufacturer and two at one-half the recommended flow rate. Airstream pressure drop across the device was 0.21 kPa (0.85 in. H_2O) at the recommended flow rate.

A heat loss test was included in all five test cycles and resulted in an overall heat loss coefficient of 85.5 kJ/(h °C) (45 Btu/h °F) for both test flow rates. An estimation of the heat loss factor utilizing basic heat transfer theory, the material properties, and dimensions of the phase change container (and assuming no air leakage) yielded an overall heat loss coefficient of 17 kJ/(h °C) (9 Btu/h °F). Heat loss during the transient tests generally was on the order of 5% to 10% of the energy transferred during a test fill time period.

The results from the charge and discharge tests for five test cycles of the phase change unit are shown in table 17.5. The charge capacity was found to be approximately 220 MJ (2.1×10^5 Btu) using the fill time as defined in Standard 94-77. A more meaningful measure of performance proposed by the authors is to divide the charge capacity by the theoretical storage capacity, which yields an average value of 0.56. (This compares to values of between 0.80 and 0.90 for a similarly sized water tank tested at NBS.) Transient test results for a phase change device are in general highly dependent on the initial temperature, the phase change temperature, and the temperature step function.

The following are advantages of defining the performance factor (PF) for the device on the ratio of the charge or discharge capacity to the theoretical storage capacity rather than the charge capacity (C_c) and discharge capacity (C_d) used in ASHRAE Standard 94-77:

1. The performance factor is a dimensionless number, which is independent of the physical size of the unit.

2. The performance factor tends to normalize the test results in more consistent values. Test variations of temperature and flow rate do not affect the value of the performance factors nearly as much as they do the charge and discharge capacities presently used.

3. The performance factor relates the performance of the unit to the theoretical maximum performance of any completely sensible-heat storage device.

ASHRAE Standard 94-77 suggests a plot be provided showing the time variation in transfer fluid temperature for each test. Figure 17.9 shows such a

Table 17.5
Test results from phase change thermal energy storage device

94-77 fill time (h)	Modified fill time (h)	Flow rate (kg/h)	Initial temperature (°C)	Input temperature (°C)	Heat loss (MJ)	Theoretical storage capacity (MJ)	94-77 charge/discharge capacity (MJ)	Modified charge/discharge capacity (MJ)	Total storage capacity (MJ)	94-77 performance factor	Modified performance factor
—	—	1,728	20.4	54.1	11.0	411.2	—	—	366.4	—	—
Power failure											
7.13	8.3	1,744	26.1	58.3	14.6	413.1	227.2	237.1	287.5	0.55	0.57
6.75	23.6	1,842	57.1	24.9	—	411.6	155.3	278.9	—	0.38	0.68
—	—	1,729	25.1	58.8	13.1	418.7	—	—	—	—	—
7.00	31.1	1,820	58.0	26.6	—	410.6	133.4	—	—	0.33	—
16.1	20.0	911	25.6	51.2	22.0	381.0	220.0	236.6	238.4	0.58	0.62
18.6	82.2	914	48.6	28.1	—	362.7	99.3	—	—	0.27	—
17.3	21.2	888	26.4	50.5	23.2	375.3	207.1	218.9	219.0	0.55	0.58
17.2	53.0	889	49.0	25.5	—	371.6	127.1	—	—	0.34	—

Note: The dash denotes erroneous or insufficient data.

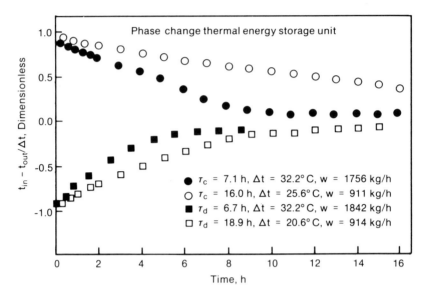

Figure 17.9
ASHRAE-recommended plot of time variation in transfer fluid temperature for the phase change
thermal energy storage test unit.

plot for those tests involving the ASHRAE-recommended test fill times and
temperature step changes.

An analysis was derived for the performance of sensible-heat, latent-heat
devices for thermal energy storage. Any latent-heat thermal energy storage
device will also store some energy in the form of sensible heat. The curve for
an ideal device with both latent- and sensible-heat contributions is difficult to
visualize; however, the fill time for such an ideal device can be calculated.

The theoretical storage capacity (TSC) of the combined latent-sensible
device is

$$\text{TSC} = Mc_{pf}\Delta t + MH_L, \tag{8}$$

where M = mass of storage material (kg), c_{pf} = sensible specific heat of the
material, [J/kg °C (Btu/lb °F)], and H_L = latent heat of the material at the
phase change temperature, t_m [J/kg (Btu/lb)].

The fill time components can be calculated from the following expression:

$$t_{F,x} = \frac{\text{TSC}}{w_{tf}c_{pf}\Delta t_x}. \tag{9}$$

For the sensible-heat portion, Δt_x is the difference between the inlet fluid temperature t_{in} and initial storage temperature t_i. For the latent-heat portion, Δt_x is the difference between the inlet fluid temperature t_{in} and the phase change temperature t_m. Consequently, for the combined latent-sensible device,

$$t_F = \frac{Mc_p(t_{in} - t_i)}{w_{tf}c_{pf}(t_{in} - t_i)} + \frac{MH_L}{w_{tf}c_{pf}(t_{in} - t_m)}, \tag{10}$$

or

$$t_F = \frac{Mc_p(t_{in} - t_i) + MH_L[(t_{in} - t_i)/(t_{in} - t_m)]}{w_{tf}c_{pf}(t_{in} - t_i)}. \tag{11}$$

Note that if the phase change temperature were exactly halfway between the initial temperature t_i and the input temperature t_{in}, equation (11) would reduce to

$$t_F = \frac{Mc_p + 2MH_L}{w_{tf}c_{pf}}. \tag{12}$$

Also note that if the sensible-heat contribution were small and could be assumed to be zero,

$$t_F = \frac{2MH_L}{w_{tf}c_{pf}}. \tag{13}$$

In other words, the fill time would be exactly twice that called for in ASHRAE Standard 94-77.

It is recommended that the fill time for latent-heat storage devices be calculated using equation (11) or (12) (denoted hereafter as modified fill time) instead of equation (7) as is presently required in Standard 94-77. In this way latent-heat devices will be evaluated more equitably with sensible-heat devices.

The data from the transient tests on the phase change device were reanalyzed using the modified fill time and the results are also shown in table 17.5. The modified charge capacity (the charge capacity calculated using the modified fill time) was calculated to average 230 MJ (2.2×10^5 Btu) compared to 220 MJ (2.1×10^5 Btu) using the fill time of equation (7). The performance factor for the charge tests averaged 0.59 (compared to 0.56). Because the input temperature was so close to the phase change temperature during the discharge tests, the modified fill time for most of the tests averaged several days. The tests were not run that long, and as a result the discharge capacities and corresponding performance factors could not be calculated using the modified fill times.

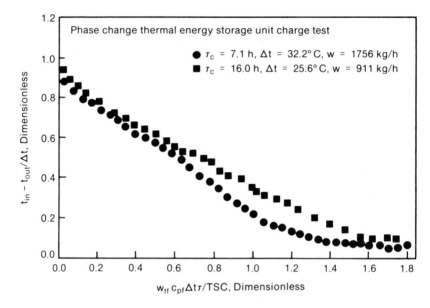

Figure 17.10
Plot of dimensionless temperature versus dimensionless time for charge tests on the phase change thermal energy storage test unit at two flow rates, using the ASHRAE Standard 94-77 fill time.

The test results for the transient tests were plotted in accordance with the completely dimensionless plot described above. Figure 17.10 is such a plot for charge tests at two flow rates using the Standard 94-77 fill time. A second plot using the modified fill time is shown in figure 17.11.

Argonne National Laboratory (ANL) contracted with the Calmac Manufacturing Corporation for development and tests with ASHRAE 94-77 Standard. Bulk PCM storage devices of 100 kWh (350,000 Btu) for 7°, 46°, and 115°C (45°, 115°, and 240°F) temperature levels were considered. The devices were all-plastic tank/heat exchangers with phase change materials surrounding the heat exchanger tubes. The heat exchanger tubes are evenly spaced throughout the tank. NBS representatives inspected and approved the test setup.

ASHRAE 94-77 test procedures were found to be impractical because total charge and discharge could not be achieved before exceeding the TSC limitation. Also the standby loss temperature differences between the inlet and outlet were too small to measure accurately.

Tests run using the NBS revision of ASHRAE 94-77 were somewhat improved but still did not achieve full charge or discharge. Because facilities to provide constant temperature water were difficult to provide with sufficient

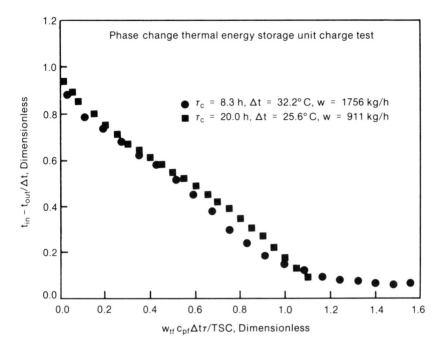

Figure 17.11
Plot of dimensionless temperature versus dimensionless time for charge tests on the phase change thermal energy storage test unit at two flow rates, using the modified fill time.

accuracy, large tanks of water maintained at specific temperatures throughout were required.

With a constant input electric heater, the recirculating charging mode was achieved. The discharge mode required continual regulation of ambient air-flow volume through a cooling coil to achieve constant heat discharge; that was accomplished with required accuracy by regulation of a manual damper. These tests were performed according to the proposed ASHRAE 94.1R Standard. The results indicated maximum efficiencies of 95% to 97% with 24-hour total cycle times.

An ANL report (Cole, 1983) discusses a latent-heat storage device test project that uses cross-linked, high density polyethylene (HDPE) as the phase change material. The material has solid–solid phase changes at 130°C (266°F), which is a reasonable source temperature for solar-powered air-conditioning. The design of the storage device is unique because it uses a boiling and condensing heat transfer fluid (water). The advantage of latent-heat transfer is that it imposes smaller losses due to the second law of thermodynamics than does sensible-heat transfer. The self-encapsulated high density polyethylene pellets were developed by the University of Dayton and Monsanto Research Corporation for the Department of Energy (Salyer et al. 1978; Botham et al. 1978; Salyer 1983). The polyethylene stores about 184 J/g (44 cal/g) and has a nominal melting point of 130°C (266°F). Melting behavior of the polyethylene has been modified by cross-linking of the molecules so that the material does not flow at temperatures as high as 150°C (302°F).

A test loop was constructed and prepared for testing. The test procedure used was similar to the proposed replacement for ASHRAE Standard 94-77. That is, heat input to and removal from the storage device was at a constant rate and variable temperature instead of the variable rate and constant temperature specified by 94-77. Completion of the project was expected to provide results for (1) development of a new type of thermal storage device, (2) proof of the feasibility of cross-linked HDPE as a phase change material, and (3) proof of the feasibility of the boiling-and-condensing fluid heat transfer mechanism. The test results were not encouraging since the HDPE pellets became fused into a single mass.

ANL has constructed a 75 m^3 (20,000 gal) underground water/ice tank insulated with 23 m (9 in.) of polystyrene foam (Gorski and Schertz 1982). Cooling coils or plates filled with R-22 refrigerant and submerged in the water are connected to condensing plates on the roof of a small building. During the winter of 1981–82, approximately 16,400 kg (36,000 lb) of ice were formed without requiring any external power source. Estimating the amount of ice formed was challenging because of the difficulty of measuring irregularly shaped objects in a tank of ice water. Considerable difficulty in releasing the ice from the cold surfaces was experienced, and the natural buoyancy of ice caused mechanical problems. The freezing rate was found to be a function of the cumulative freezing degree hours and the temperature difference between the tank and the outdoor air (McGarity 1982).

Phase change material slurries are under development at ANL (Kasza and Chen 1982, 1983; Kasza 1983). The theoretical advantages are (1) increased fluid-to-solid surface heat transfer, (2) increased heat capacity of the heat transfer fluid resulting in lower pumping power, (3) end-use temperature closer to the heat-source temperature, and (4) use of an integral storage concept.

The ANL report (Cole et al. 1983) describes several alternatives to Standard 94-77 and discusses the advantages and disadvantages of each alternative. Two proposed replacement standards, and experimental results for each were considered for preparing recommendations. The proposed ASHRAE Standard (Draft) 94.1-77R and the revised ASHRAE Standard 94.1-1985 drew extensively upon the suggestions of the ANL results as a basis for development.

From Cole et al. (1983) "ASHRAE Standard 94-77," "Method of Testing Thermal Storage Devices Based on Thermal Performance," was developed to allow manufacturers of thermal energy storage devices to compare their products. For thermal energy storage products based on latent-heat storage, the Standard was found to be inadequate for comparison of these devices.

Standard 94-77 had several obvious inadequacies when applied to latent-heat devices. The temperature range of the test is not adequately specified. In some conditions, this could presumably lead to a temperature range that does not include the melting point of the PCM. Also the method of calculating the theoretical storage capacity (TSC) is not clear. This problem is not so important in a sensible-heat storage device, as the calculation is much more straightforward than for a device containing a PCM. In addition there is no specification as to how the unit should be cycled prior to testing. Finally there is no test for degradation of the storage device. This can be especially important for PCMs that experience incongruent or semicongruent melting.

The difficulty with Standard 94-77 results from improperly measuring the heat loss coefficient, as well as the use of this coefficient in subsequent calculations (Marshall 1981). In addition, the calculation of the TSC makes several simplifying assumptions, such as not including the heat capacity of the heat exchanger and assuming ideal behavior for the PCM, which leads to confusing and erroneous results. A further problem with Standard 94-77 is that it is not clear how the computed performance coefficients relate to real system performance.

The second proposed (Cole et al. 1983) standard is that recommended to Project Committee SPC 94.1. This proposed standard, which was approved as Standard 94.1-1985, differs considerably from Standard 94-77 in that the theoretical storage capacity is not part of the analysis and that constant energy input and output are respectively used in the charge and discharge tests.

The most significant change from the old ASHRAE Standard 94-77 is that the new standard specifies constant heat input and removal rates with variable input temperatures instead of the constant inlet temperatures and variable heat input and removal rates specified by ASHRAE 94-77. Separate standards are proposed for latent-heat storage devices and sensible-heat storage devices because the sensible-heat storage devices are less amenable to testing with constant heat input and removal rates. Unlike the old standard, Standard 94.1-1985 allows the storage unit supplier to specify the design charge and discharge rates, temperatures at which the storage device is considered fully charged and discharged, and flow rates for the charge and discharge tests.

The new standard specifies tests at half, one, two, and four times the design charge and discharge rates. Test cycles include standby periods during which heat is lost to the ambient air. Three types of loss coefficient are identified: standby loss in the fully charged condition, standby loss in the discharged condition, and overall loss. The equipment required and instrumentation accuracy are similar to the requirements of ASHRAE 94-77.

Some of the problems are corrected by Standard 94.1-1985. Instead of basing the total charge upon the product of the inlet–outlet temperature difference times time with a given flow, it is based upon a constant heat input at a recirculating flow rate and an operating temperature range selected by the supplier. This approach imposes no limitation on achieving a full charge or discharge, whereas 94-77 did limit the charge and discharge because of the limiting TSC equation and the lack of full operating temperature range throughout. The recirculating system and the constant heat input and removal are more realistic.

The results of testing all latent storage devices can be graphed on a single sheet, similar to the Hottel-Whillier solar collector chart. A curve of efficiency versus logarithmic charging rate with points at one-half, one, two, and four times the design heat input and removal rates displays where various storage devices are most efficient and at what efficiency level (see figure 17.12). The heat loss in standby is incorporated in the efficiency figures since a single cycle incorporates a typical charge, standby, and discharge period. No separate standby loss determination is required. The new standard does not mention TSC. Since the flow rates are specified by the storage device supplier, TSC is not necessary, and the flow rates will be more representative of the flow rates the manufacturer intends for field use. The charge and discharge capacities and the storage efficiency are measured and reported as functions of the charging rate. These parameters are more useful to a system designer than is the TSC.

The ASHRAE 94-77 procedure required each part of the test to begin with a steady-state condition. Such a condition is difficult to determine accurately because the outlet temperature approaches the steady state asymptotically. The new standard avoids asymptotic approaches to a steady state because it uses constant charge and discharge rates instead of constant inlet temperatures. Parts of a test end when the outlet temperature (which changes relatively rapidly) reaches specified temperature.

Three of the four problems associated with the thermal loss measurements are avoided by the new procedure. The problem of determining when a steady state has been reached is avoided because the proposed procedure does not rely on steady states. The problem of the inlet (test) controller being unable to maintain the inlet temperature with sufficient accuracy is also avoided by eliminating the reliance on steady states. The incorrect calculation of the loss

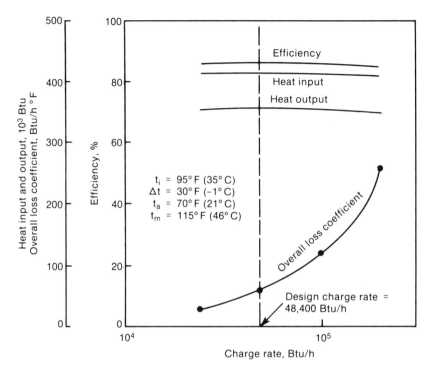

Figure 17.12
Heat input, heat output, storage efficiency, and overall loss coefficient versus charge rate (ASHRAE Standard 94.1-1985).

coefficient reported in Marshall (1981a, b) is corrected. The fourth problem, that of the difficulty of measuring small temperature differences, is only partly solved. The measurement of small temperature differences is replaced by the difficulty of accurately determining the difference between the heat input to the storage device and the heat removed from the storage device. Since this inaccuracy is comparable in size to the inaccuracy of measuring small temperature differences, there is only a small improvement in the accuracy of the thermal loss measurement.

Criticism of the ASHRAE test procedure is only partly valid because it lacks a mathematical model, but no adequate mathematical model exists (Marshall 1981a, b). Were an adequate mathematical model available, a computer code could be used to adjust its parameters until the response of the model to inputs would match the test results. Until such a model becomes available, the proposed test procedure provides information, such as thermal capacity, storage efficiency, and thermal loss coefficients, that is immediately and directly usable to system designers. Measure of thermal degradation is not included under ASHRAE policy; such data come from other sources (e.g., ASTM).

References

American Society of Heating, Refrigeration, and Air Conditioning Engineers, Inc. 1977. *Standard 94-77: Methods of Testing Thermal Storage Devices Based on Thermal Performance.* New York: ASHRAE.

ASHRAE. 1983. *Standard 94.1-77R (Draft), Method of Testing Active Latent Heat Storage Devices Based on Thermal Performance.* Atlanta, GA: ASHRAE.

ASHRAE. 1985. *Standard 94.1-1985: Method of Testing Active Latent Heat Storage Devices Based on Thermal Performance.* Atlanta, GA: ASHRAE.

ASHRAE. n.d. *ASHRAE 94.3-77R, Method of Testing Active Sensible Thermal Energy Storage Devices Based on Thermal Performance.* Atlanta, GA: ASHRAE.

Botham, R. A., G. H. Jenkins, G. L. Ball, III, and I. O. Salyer. 1978. *Form-Stable Crystalline Polymer Pellets for Thermal Energy Storage—High Density Polyethylene Intermediate Products: Final Report.* ORNL/Sub-7398/4. Oak Ridge, TN: Oak Ridge National Laboratory.

Close, D. J., and R. V. Dunkle. 1977. Use of absorbent beds for energy storage in drying of heating systems. *Solar Energy* 19: 233–238.

Close, D. J., and T. L. Pryor. 1976. The behavior of absorbent energy storage beds. *Solar Energy* 18: 287–292.

Cole, R. L. 1983. *Thermal Storage Device Based on High-Density Polyethylene: Interim Report.* ANL-83-52. Argonne, IL: Argonne National Laboratory.

Cole, R. L., and F. O. Bellinger. 1982. Storing solar energy in thermally stratified tanks. In *Proc. 17th IECEC.* Paper no. 829344, pp. 2074–2079.

Cole, R. L., J. R. Hull, Y. Lwin, and Y. S. Cha. 1983. *Comparison of Testing Methods for Latent Heat Storage Devices.* ANL-83-89. Argonne. IL: Argonne National Laboratory.

Gorski, A. J., and W. W. Schertz. 1982. Winter results at Argonne ice storage and test facility. In *Proc. Second Annual Workshop on Ice Storage for Cooling Applications.* Argonne, IL: Argonne National Laboratory.

Greenberg, J., and B. C. Reeder. 1984. *Phase Change Thermal Energy Storage and the Model Building Codes.* NBSIR 84-2909. Gaithersburg, MD: National Bureau of Standards.

Hale, D. W., et al. 1971. *Phase Change Materials Handbook.* NASA CR-61363. Huntsville, AL: Lockheed Missiles and Space Company for NASA—George C. Marshall Space Flight Center.

Hill, J. E., G. E. Kelly, and B. A. Peavy. 1977. A method of testing for rating thermal storage devices based on thermal performance. *Solar Energy* 19: 721–732.

Hill, J. E., E. R. Streed, G. E. Kelly, J. C. Geist, and T. Kusuda. 1976. *Development of Proposed Standards for Testing Solar Collectors and Thermal Storage Devices.* National Bureau of Standards Technical Note 899. Gaithersburg, MD: National Bureau of Standards.

Hunt, B. J., T. E. Richtmyer, and J. E. Hill. 1978. *An Evaluation of ASHRAE Standard 94-77 for Testing Water Tanks for Thermal Storage.* NBSIR 78-1548. Gaithersburg, MD: National Bureau of Standards.

Hunt, B. J., T. E. Richtmyer, J. E. Hill, and E. A. Franklin. 1979. *Testing of water tanks for thermal storage according to ASHRAE Standard 94-77. ASHRAE Transactions* 85, pt. 1: 481–501.

Ings, J. B., and P. W. Brown. 1982. *Factors Affecting the Service Lives of Phase Change Storage Systems.* NBSIR 81-2422. Gaithersburg, MD: National Bureau of Standards.

Ings, J. B., and P. W. Brown. 1982. *An Evaluation of Hydrated Calcium Aluminate Compounds as Energy Storage Media.* NBSIR 82-2531. Gaithersburg, MD: National Bureau of Standards.

Janz, G. J., et al. 1978. *Physical Properties Data Compilations Relevant to Energy Storage 1. Mortar Salts: Eutectic Data.* NSRDS-NBS 61, pt. 1. Gaithersburg, MD: National Bureau of Standards.

Jones, D. E., and J. E. Hill. 1979. *Testing of Pebble-Bed and Phase-Change Thermal Energy Storage Devices According to ASHRAE Standard 94-77.* NBSIR 79-1737. Gaithersburg, MD: National Bureau of Standards.

Jones, D. E., and J. E. Hill. 1979. *An Evaulation of ASHRAE Standard 94-77 for Testing Pebble-Bed and Phase Change Thermal Energy Storage Devices.* ASHRAE DE 79-4, No. 3.

Kasza, K. E. 1983. Development of enhanced heat transfer/transport/storage slurries for thermal system improvement. In *Proc. DOE Physical and Chemical Energy Storage Annual Contractors' Review Meeting,* CONF-830974. Springfield, VA: National Technical Information Service.

Kasza, K. E., and M. M. Chen. 1982. *Development of Enhanced Heat Transfer/Transport/Storage Slurries for Thermal System Improvements.* DOE/ANL-82-50. Argonne, IL: Argonne National Laboratory.

Kasza, K. E., and M. M. Chen. 1983. Development of enhanced heat transfer/transport/storage slurries for thermal system improvement. In *Opportunities in Thermal Storage R&D,* W. Hausz and B. Berkowitz, eds. EM-3159-SR. Palo Alto, CA: Electric Power Research Institute.

Kelly, G. E., and J. E. Hill. 1975. *Method of Testing for Rating Thermal Storage Devices Based on Thermal Performance.* NBSIR 74-634. Gaithersburg, MD: National Bureau of Standards.

Klein, S. A., P. I. Cooper, T. L. Freeman, D. M. Beekman, W. A. Beckman, and J. A. Duffie. 1975. A method of simulation of solar processes and its application. *Solar Energy* 17: 29–37.

Lane, G. A., et al. 1976. *Isothermal Solar Heat Storage Materials—Final Project Report.* NSF-C906. Midland. MI: Dow Chemical Company.

Marshall, R. H. 1981a. Experimental experience with the ASHRAE/NBS procedures for testing a phase change thermal storage device. In *Proc. Internation Conference on Energy Storage,* Brighton, UK, April 19–May 1, 1981, pp. 129–143.

Marshall, R. H. 1981b. A theoretical study of ASHRAE 94-77 for testing thermal storage devices. In *Thermal Storage of Solar Energy,* C. den Ouden, ed. The Hague, the Netherlands: Martinus Nijhoff.

McGarity, A. E. 1982. Heat pipe models for computer simulation of a passive seasonal ice cooling system. In *Proc. Second Annual Workshop on Ice Storage for Cooling Applications.* Argonne, IL: Argonne National Laboratory.

Salyer, I. O. 1983. Thermal energy storage in cross-linked pellets of high-density polyethylene. In *Opportunities in Thermal Storage R&D,* W. Hausz and B. J. Berkowitz, eds. EM-3159-SR. Palo Alto, CA: Electric Power Research Institute.

Salyer, I. O., R. A. Botham, G. H. Jenkins, and G. L. Ball. 1978. Form-stable crystalline polymer pellets for thermal energy storage. In *Proc. 13th Intersociety Energy Conversion Engineering Conference.* Warrendale, PA: Society of Automotive Engineers, Inc.

Shelpuk, B., et al. 1976. *Technical and Economic Feasibility of Thermal Storage—Final Report.* C00/2591-76/1. RCA Advanced Technology Laboratories for Energy Research and Development Administration, Division of Solar Energy.

University of Maryland. 1973. *Conservation and Better Utilization of Electric Power by Means of Thermal Energy Storage and Solar Heating—Final Summary Report.* NSF/RANN/SE/FI 27976/PR75/5.

University of Wisconsin. 1977. *TRNSYS, A Transient Simulation Program.* Report 38, Madison, WI: Solar Energy Laboratory, Engineering Experiment Station, University of Wisconsin.

Yang, W. J., and C. P. Lee. 1974. Dynamic response of solar heat storage systems. In *Proc. ASME Winter Annual Meeting.* Paper 74-WA/HT-22. New York.

Bibliography

Lane, G., ed. 1990. *Solar Heat Storage: Latent Heat Materials, Vol. II-Technology and Applications*. Boca Raton, FL: CRC Press Inc.

MacCracken, C. D. 1980. Salt hydrate thermal energy storage system for space heating and air conditioning. In *Proc. Annual DOE Active Solar Heating and Cooling Contractors' Review Meeting*. CONF-800340. Springfield. VA: National Technical Information Service

MacCracken, C. D. 1981. Effect of additives on performance of hydrated salt TES systems. In *Proc. Thermal and Chemical Storge Annual Contractors' Review Meeting*. CONF-810940. Springfield, VA: National Technical Information Service.

MacCracken, C. D., J. M. Armstrong, M. M. MacCracken, and B. M. Silvetti. 1980. *Bulk Storage of PCM, Final Report December 1977 to June 1980*. DOE/CS/34698-T1. Springfield, VA: National Technical Information Service.

Masoero, M. 1981. *Annual Storage of Naturally-Frozen Ice for Building Air-Conditioning*. PU/CEES 127. Princeton, NJ: Princeton University.

Michaels, A. I. 1982. A review of latent heat storage technology. In *Proc. Solar Storage Workshop*. Kansas City, MO: Midwest Research Institute.

Wright, J. D. 1981. The design and economics of direct-contact salt hydrate storage systems. In *Proc. Second World Congress of Chemical Engineering*, Montreal, October 4–9, 1981. CONF-811007-3.

Wright, J. D., M. S. Bohn, and R. S. Barlow. 1983. *Oil/Salt Hydrate Direct-Contact Heat Exchange Experiments*. SERI/TR-252-1614. Golden, CO: Solar Energy Research Institute.

18 Storage Research and Development

Allan I. Michaels and C. J. Swet

18.1 Introduction

Storage technologies suitable for solar thermal applications have progressed substantially during the past decade under the auspices of many different research and development (R&D) programs throughout the world. Some of the programs have been governmental, such as those sponsored by the U.S. Department of Energy (DOE); its predecessor, the Energy Research and Development Administration (ERDA); and the National Research Council of Canada. Others have been intergovernmental, such as those funded by the European Economic Community and those coordinated by the International Energy Agency (IEA). Still others have been supported by the military and by private-sector organizations such as the Electric Power Research Institute, while a large number of individual companies have developed specific marketable products. There has also been a diversity of sponsorship along application lines within primary funding agencies, with jurisdictional divisions according to whether the storage is for building applications or for higher temperature uses such as solar thermal power. Additionally there are thermal energy storage (TES) programs that, while not designated as solar activities, develop concepts and materials appropriate for solar requirements.

In this chapter we cannot adequately characterize all storage programs worldwide, much less provide substantive reviews of their individual R&D projects. We cannot even do justice to all of the DOE-sponsored solar TES activities. Rather, we must deal principally with the TES program of the Solar Heating and Cooling of Buildings (SHACOB) activity in what is now the DOE Office of Solar Heat Technologies, with which A. I. Michaels was closely associated. To a lesser extent we also deal with some of the complementary R&D sponsored by the TES program in what is now the DOE Office of Utility Technologies, which was managed for four years by C. J. Swet.

Because solar thermal energy storage R&D at DOE has been fragmented, it might be helpful to consider some of the history of these programs before proceeding to the technical content of this chapter. There have been three distinct storage programs and a number of scattered projects, first at ERDA and then at DOE. All began in 1975 when ERDA inherited solar and TES projects from the Atomic Energy Commission (AEC), the National Science Foundation (NSF), and the National Aeronautics and Space Administration Lewis Research Center (NASA-LeRC). Those inolving basic research went to the Office of Energy Research, while the high temperature solar projects became the nucleus of the Solar Thermal Power Branch of the Solar Energy

Division. None of these activities will be dealt with here. The low temperature solar TES projects inherited from NSF, along with some responses to unsolicited proposals, formed the initial phase of the SHACOB TES program. Other TES projects that were not specifically solar oriented went to the TES program in the Energy Storage Division (STOR) under the Office of Conservation. Subsequently the STOR TES program acquired management responsibility from NSF for several newly awarded studies on solar TES, and in an agreement with the solar division, that program undertook the sponsorship of exploratory R&D on advanced storage concepts and materials with potential solar applications. With some exceptions the SHACOB program concentrated on relatively mature storage technologies. In 1977 a massive, directed competitive procurement by the SHACOB branch resulted in a large number of projects that constituted the bulk of SHACOB TES activities for several years. Additional SHACOB TES projects were conducted or subcontracted mainly by Argonne National Laboratories (ANL), the lead laboratory, in areas not covered by the competitive procurements. After ERDA dissolved into DOE in 1977, both the SHACOB and STOR TES programs underwent a series of changes in name and organizational affiliation, but their identities and complementary functions have remained essentially unchanged. At present there is very little DOE funding for TES related to SHACOB.

The TES programs reviewed in this chapter included some modeling and evaluation and test procedures, which are only briefly noted here since they properly belong in chapters 16 and 17, respectively. Additional insights on selected SHACOB and other solar TES projects are offered in chapters 15 and 19, and in chapter 25 of part III.

18.2 The SHACOB Storage Program

From 1975 to 1982 DOE's SHACOB storage program sponsored over 40 projects with more than $7.5 million in funds (see table 18.1). It accounted for virtually all of the storage R&D for space heating and cooling supported by the DOE Office of Solar Heat Technologies and its predecessors at DOE and ERDA. The projects fall naturally into three distinct groupings according to the phase of the SHACOB activities in which they were procured: initial projects, the large solar solicitation, and additional projects.

18.2.1 Initial Projects

Some of these were continuations of projects inherited from NSF, while others were in response to unsolicited proposals. Most of the contracts were awarded before the overall state-of-the-art assessment was completed and an integrated

Table 18.1
Budget authorizations for active solar heating and cooling projects monitored by Argonne National Laboratory

	FY82	FY81	FY80	FY79	FY78	Before FY78	Total
1. Alabama, Univ. of (models)					40	25	65
2. Altas (HEX)						60	60
3. Altas (gas heater)			35	95	66		196
4. ANL 49500 (PCM support)	30	130	180	200	100	105	745
5. 49506 (PCM test)	60	90	100				250
6. 49507 (HDPE)	200	105	150				455
7. 49565 (manual)				100	110	192	402
8. 49572-00 (tanks)			150	100	70		320
9. 49572-10 (rock bed)[a]			200				200
10. 49572-20 (pond)			80				80
11. 49589 (strat.)		[c]	150				150
12. 49590 (IEA)	95[d]	70	100				265
13. Boeing (models)			76	60	206	58	400
14. Colorado State Univ. (solar-assisted heat pump)					95		95
15. Calmac (PCM)			55	200	200		455
16. Colorado State Univ. (DCLLHX)				93	86	107	286
17. Colorado State Univ. (SOIL)				60			60
18. Delaware, Univ. of (PCM)					84		84
19. Dynatech (rock bed, cooling)					67	32	99
20. EIC (sys. anal.-I)					48	29	77
21. EIC (sys. anal.-II)				26			26
22. Franklin (self-pump)					64	39	103
23. Hittman Assoc. (TES surv.)						55	55
24. Hittman Assoc. (cooling TES)			35				35
25. GE Research (ROT. drum)					150		150
26. GE Space (micro-PCM)						160	106
27. Independent Living (fib. tank)					130		130
28. Martin-Marietta (CHP)			66	210	152		428
29. NBS (stds.)				100			100
30. North Carolina State (PCM)						94	94
31. Nebraska (membrane I)				44	68		112
32. Nebraska (membrane II)			236				236
33. Nebraska (STRAT.)				24			24
34. Pennwalt (micro-PCM)[b]		[b]					[b]
35. Rockwell (rock bed, tank)					67	42	109
36. Science Appl., Inc. (sys. anal.)			112	90			202
37. Toronto (seasonal TES)				76	70	67	213
38. Virginia, Univ. of (seasonal TES)					47	118	165
39. Westinghouse (concrete tank)			20	150	120		290
40. ZIA (rock bed. H.P.)					107		107
Total	385	395	1,745	1,628	2,147	1,129	7,429

a. Includes $120K subcontract to Solaron.
b. Funded by ORNL P.O. to ANL, $100K in FY81.
c. Funded by ORNL P.O. to ANL, $100K in FY81.
d. Includes $35K subcontract to C. A. Bankston.

SHACOB plan was formulated. They are grouped here for convenience into four categories: heat transport and exchange, annual storage with water as the storage medium, phase change TES with paraffin as the storage medium, and miscellaneous contracts. Subsections under these main categories begin with the title of the contracted work, followed by the contractor's name and the name of the principal investigator (in parentheses).

Heat Transport and Exchange

INVESTIGATION OF METHODS TO TRANSFER HEAT FROM SOLAR LIQUID HEATING COLLECTORS TO HEAT STORAGE TANKS (ALTAS CORPORATION; F. DE WINTER, PRINCIPAL INVESTIGATOR)   This was a comprehensive survey and analysis of the most cost-effective methods of transferring heat from the solar collector to a water storage system. Emphasis was on heat transport and exchange that would either minimize or eliminate problems of freezing and corrosion. Initially data were compiled on commercially available heat transfer fluids and heat exchangers. The remainder of the project consisted of detailed analyses and evaluation of specific concepts, including evaporating collector-condensing exchange, double-loop exchange, liquid–liquid exchange, and traced (heat exchanger jacketed) tank systems. With the double-loop and trace-tank systems, computer modeling was used in operational simulation and design optimization. One finding was that double-loop systems become more cost effective than single-loop systems when the collector area exceeds 110 to 180 ft^2 (10–16 m^2). Another was that when double-walled heat exchangers are dictated by the use of toxic collector fluids in domestic hot water systems, it may be easier to avoid excessive heat exchanger performance penalties by using traced tanks (in which there are in effect two walls between the fluids) than by using coils immersed in the tanks (in which concentric vented tubes might be needed). Equations were developed for the straightforward calculation of heat exchanger penalties in a single-loop system (Horel and de Winter 1978).

DIRECT-CONTACT LIQUID–LIQUID HEAT EXCHANGERS FOR SOLAR HEATED AND COOLED BUILDINGS (COLORADO STATE UNIVERSITY SOLAR ENERGY APPLICATIONS LABORATORY; S. KARAKI, PRINCIPAL INVESTIGATOR)   The objective was to establish the technical feasibility and economic practicability of using direct-contact, liquid–liquid heat exchangers (DCLLHE) in solar heating and cooling systems. In this concept, heat from solar collectors is transferred to water storage by the direct contact of droplets of collector fluid with water in the storage tank. A dense liquid that is heavier than and immiscible in water is circulated through the collector array and dispersed downward through the storage vessel in the form of droplets. Heat transfer is rapid and complete because the droplets are small, with a high surface-to-volume ratio. The droplets collect at the bottom of the tank (which is conical to minimize the fluid volume in the collector) and flow back continuously to the collector as a liquid. After a fluid-testing (including for toxicity) process, diethyl phthalate was selected, followed by glass pilot-plant tests, materials compatibility studies, and design, construction, and tests of a full-scale system in the Solar Houses I and III facilities. Finally, simulations were run on a TRNSYS model validated with experimental data for five U.S. locations. Improvements in annual collector efficiencies were found to be 1 to 2 percentage points, which were partially offset by the higher (almost doubled) parasitic pumping power needed for the denser, more viscous fluid. Experimenters concluded that this system's performance advantage over a conventional system is not large enough to overcome the slightly higher capital and operating costs associated with it (Buchan, Majestic, and Billau 1976; Ward et al. 1977; Karaki and Brothers 1980). For work on direct-contact heat exchange in other programs, see subsection 15.3.1 in chapter 15 and the following references: Greene and Watson (1979), Barlow (1983), Furbo (1978), Helshoj (1981b), Arrhenius et al. (1983), Kauffman et al. (1977), Bell (1981), and Wright and Bohn (1982).

Annual Storage in Water

SOLAR SPACE-HEATING SYSTEMS USING ANNUAL HEAT STORAGE (UNIVERSITY OF TORONTO; F. HOOPER, PRINCIPAL INVESTIGATOR)   The objectives were to investigate solar space-heating systems utilizing annual (seasonal) storage of solar-heated water to establish the relationship between the primary weather variables and the optimum system configuration and to develop a simple algorithm that would allow a designer to calculate the optimum system configuration. An hourly annual storage solar energy program (ASTEP) was developed to predict performance, incorporating a simplified storage heat loss model derived from and validated by more detailed calculations. From this, and given certain cost assumptions, optimum configurations were determined. The ASTEP performance predictions generally agreed with measured data from two operating annual storage systems, Provident House and Aylmer House, both of which had been funded by Canadian agencies. The general usefulness of this program led to the development of simple methods for initially approximating optimum system configurations for any location or building type, using preprocessed weather data for the typical meteorological year in ten U.S. cities. The analysis, which considered buildings of more than 10,760 ft^2 (1,000 m^2) of floor area and insolation typical of the northern United

States, indicated that annual storage systems providing 100% of the space-heating load from the solar source in an average year appear to be more cost-effective than conventional, short-term storage systems. This study considered only storage in well-insulated buried tanks, with separate (not considered) provisions for domestic water heating (Hooper et al. 1980).

ANNUAL COLLECTION AND STORAGE OF SOLAR ENERGY FOR THE HEATING OF BUILDINGS (UNIVERSITY OF VIRGINIA; J. T. BEARD, PRINCIPAL INVESTIGATOR) The objective of this contract was to demonstrate the feasibility of using a collector-covered excavated pond (actually a pit) for the annual collection and storage of solar heat for space heating. A 27,400-gal (103,572-L) square pit was dug with sloping lower sides and vertical walls 4 ft (1.2 m) above grade, insulated, braced, and bermed. The bottom and sides were fitted with a one-piece, factory-made, 20-mm polyvinylchloride (PVC) liner, and the top was covered by a 1-ft (0.3-m)-thick floating layer of styrofoam pellets for insulation. Supported by the styrofoam and upper walls was a nearly horizontal, open water channel ("trickle") solar collector with a 576-ft^2 (53-m^2) sheet metal absorber having north–south corrugations. Water was pumped from the pit to manifolds at the north and south ends of the absorber, flowing down the valleys of the corrugations to the center, where it exited hot and flowed through the beads back to the reservoir. Double glazings of 6-mil greenhouse film were kept separated by air pressure from a fan, with the lower film supported by the corrugated absorber plate. A 12-ft (3.6-m)-high vertical reflector on the north end augmented the interception of solar radiation. A thermal model, validated by initial test data, simulated performance for a full year. Early in the first heating season it became evident that losses were greatly exceeding the collected energy, so heat extraction was reduced to simulate the use of a heat pump. Researchers concluded, however, that the system as tested was not adequate for its intended use (Beard et al. 1980). A larger system, similar in concept but with more efficient collectors, was successfully demonstrated in Sweden.

Phase Change Storage in Paraffins

RESEARCH ON SOLAR ENERGY STORAGE SUBSYSTEMS UTILIZING THE LATENT HEAT OF PHASE CHANGE OF CERTAIN ORGANIC MATERIALS, PHASE II (NORTH CAROLINA STATE UNIVERSITY, RALEIGH; J. A. BAILEY, PRINCIPAL INVESTIGATOR) An earlier phase I study with NSF funding developed, tested, and modeled performance of TES devices incorporating pure paraffin in aluminum honeycomb structures. This project was a one-year continuation to test the performance of low-cost, commercial paraffin waxes in low-cost containers. From a survey of petroleum product manufacturers, one candidate material was chosen: Sunoco wax number 3420. An analytical procedure known as the "enthalpy method" was used in a computer program to predict the performance of devices containing phase change materials that melt over a temperature trajectory rather than isothermally. The container intrastructures considered included honeycombs, a single-finned tube, and a multiple-tube corrugated plate. To validate the models and gain experience in constructing and testing such devices, three experimental units were made and tested. Researchers concluded that unrefined hydrocarbons (waxes) represent an "ideal" class of PCM for TES, that they can be readily encapsulated in various configurations that enhance heat transfer from the working fluid to the wax, that analytical models of these configurations have been validated experimentally, and that the analytical model and experimental results can be represented in a self-consistent, nondimensionalized form to show the relationship between device performance and important design (geometric) parameters (Bailey et al. 1977). See deJong and Hoogendoorn (1980) and Mancini (1980) for reports of R&D elsewhere on paraffins.

TWO-COMPONENT THERMAL STORAGE MATERIAL STUDY, PHASE II (GENERAL ELECTRIC COMPANY, SPACE DIVISION; A. TWEEDIE, PRINCIPAL INVESTIGATOR)   An earlier phase I study funded by NSF established techniques for microencapsulation of commercial-grade paraffins in plastic films, but it was unsuccessful in attempts to pump the microparticles as water slurries. The objective of this project was to study the performance and durability of larger ($>500\ \mu$m) encapsulated pellets in a packed bed with water. Small samples were thermally cycled successfully more than 3,000 times, and flow tests at a constant water temperature in a subscale packed bed were successful, though the temperature cycling of the bed inlet resulted in clogging of the test assembly. Experimenters concluded that the walls could withstand internal volume changes of the paraffin with thermal cycling through the melting point but that more rigid walls are needed for a packed bed. The walls could possibly be made more rigid if they were thicker or if materials other than nylon were used. Researchers also found that the bed must be initially free of unencapsulated wax and of subsize capsules, which become trapped in the end screens of the bed. More significantly, economic studies showed that large-scale production costs of the capsules must be reduced considerably for this concept to become viable in solar applications. To compete with a water-only system, the cost of the encapsulated paraffin must be between $0.25 and $0.40/lb ($0.55 and $0.88/kg); cost projections for large-scale production, however, were between $0.90 and $1.40/lb ($1.98 and $3.09/kg) (Mehalik and Tweediet 1979).

Miscellaneous Contracts

A HANDBOOK AND GUIDE FOR THERMAL ENERGY STORAGE APPLICATIONS IN SOLAR HEATING AND COOLING SYSTEMS (ARGONNE NATIONAL LABORATORY; W. W. SCHERTZ AND R. L. COLE, PRINCIPAL INVESTIGATORS)   This contract covered the preparation of an initial version of a design and installation manual for state-of-the-art, commercially available heat exchanger, water tank, and rock bed subsystems directed to small residential or commercial builders and HVAC contractors as well as to the more sophisticated "do-it-yourselfers." Not itself an R&D project, this work is mentioned because it was a part of the SHACOB storage program (Cole et al. 1979). A second edition, including phase change storage, was prepared with follow-on funding (Cole et al. 1980).

DEVELOPMENT OF METHODS FOR EVALUATION AND TEST PROCEDURES FOR SOLAR COLLECTORS AND THERMAL STORAGE DEVICES (NATIONAL BUREAU OF STANDARDS; J. HILL, PRINCIPAL INVESTIGATOR)   This effort was initially funded by NSF and continued under ERDA and DOE sponsorship (Jones and Hill 1979). The work is reviewed in chapter 17 and is mentioned here only because it was a part of the SHACOB storage (and collector) programs.

18.2.2   The Large Solar Solicitation

Eighteen new R&D projects in six technical areas of TES for SHACOB were initiated by awards made in 1977 and 1978 as the result of a massive, directed ERDA solicitation. This solicitation evolved from a state-of-the-art assessment completed in 1976 followed by an integrated SHACOB program plan. The TES projects reviewed here are grouped according to the categories under which they were procured in the solicitation: heat transfer and exchange, short-term storage in water tanks, rock bed storage, phase change for heating, TES for solar-driven air conditioning, and general TES modeling and guidelines.

Heat Transfer and Exchange

SELF-CONTROLLING, SELF-PUMPING HEAT CIRCULATION SYSTEM STUDY (THE FRANKLIN INSTITUTE RESEARCH LABORATORIES; G. P. WACHTELL, PRINCIPAL INVESTIGATOR)   The objective of this work was to investigate and evaluate self-controlling, self-pumping heat circulation systems that transport heat downward from the solar collector to storage or directly to the load, without relying on utility power. The transport problem is trivial if the TES or user is above the collector, since natural convection (thermosiphoning) will suffice. Flat plate collectors with outlet temperatures of 120° to 160°F (49°–71°C) were assumed. The maximum allowable difference between the mean collector and mean TES temperatures, because of self-pumping, was assumed to be $10°F$ ($-12°C$). The schemes evaluated were these (circulants are given in parentheses):

a. external power cycles (air, liquid),

b. internal power cycles (vapor/liquid),

c. vapor bubble lift (liquid),

d. vapor expansion (liquid),

e. entrainment of condensate by vapor (vapor/liquid),

f. injector (liquid),

g. pulsejet (liquid),

h. fluidyne (liquid),

i. osmotic pumping (vapor/liquid),

j. capillary pumping (vapor/liquid),

k. negative thermal expansion coefficient for downward thermosiphoning (gas and condensed phase),

l. magnetic forces on ferrofluids (liquid),

m. electrostatic forces on dielectric fluids (liquids).

Other schemes that are not strictly self-pumping were considered: photovoltaic power (air, liquid, vapor/liquid), methods employing small amounts of utility power to return condensate in latent-heat (vapor/liquid) systems, and optical ducting of solar radiation directly to TES. Researchers found that external power cycles, which use indirect heat exchangers to draw heat from the collector outlet for the heat engine and then to transfer rejected heat to the collector inlet, must have efficiencies at least 20% to 30% of Carnot. Rankine and Stirling cycles were found feasible for air, liquid, and vapor/liquid circulants, but Brayton was practical only for liquids. Internal power cycles, in which the circulant directly powers a heat engine, were found suitable only for latent heat transport, in which vapor generated in the collector releases its heat on condensing in the TES. Of the studied schemes that do not employ a power cycle apparatus, three were recommended for development and demonstration: c, e, and f (Wachtell 1978). Currently self-pumping vapor systems for hybrid space heating, somewhat similar to scheme e, are being developed at the Los Alamos National Laboratory (Neeper and Hedstrom 1985), while an all-liquid concept has been developed in Japan (Sawada et al. 1985) and a bubble pump has been tested in Canada (Hoffman and Chan 1985).

Short-Term Storage in Water Tanks

COMPUTER MODELING OF THERMAL STRATIFICATION IN WATER TANKS (UNIVER-
SITY OF ALABAMA, HUNTSVILLE; S. T. WU, PRINCIPAL INVESTIGATOR)   This was
a modeling project, involving solar domestic hot water systems and tanks,
with three objectives: quantitatively determine the effects of various degrees
of temperature stratification on overall system performance, using TRNSYS
(a 7% gain was indicated); develop an improved TRNSYS-compatible stratification code after an evaluation of existing codes against available experimental data; and provide experimental data for the validation of more elegant
two- and three-dimensional codes being developed at Argonne National
Laboratory (Wu and Han 1980). This and other modeling activities are
reviewed more fully in chapter 16.

HYBRID THERMAL STORAGE WITH WATER (ROCKWELL INTERNATIONAL, ATOMICS
INTERNATIONAL DIVISION; M. J. MORIARTY AND S. J. NALBANDIAN, PRINCIPAL
INVESTIGATORS)   For the purposes of this study, a hybrid thermal storage
system consists of a water tank surrounded by a bed of inexpensive materials,
such as rock, intended to augment the storage capacity while insulating the
tank. This concept has been used in the Thomason SOLAR is houses, where
the rock bed acts as a heat exchanger with room air. The main objectives were
to develop parametric engineering data to aid in the evaluation and selection
of hybrid storage systems, to establish storage system size and configuration,
to evaluate the use of hybrid materials such as rocks and earth, and to select
tank types and configurations. Four distinct hybrid storage concepts were
identified: water tank in rock bin, buried tank (earth storage), above ground,
and below grade with passages for air heat exchange. Computer codes were
developed and used for comparative performance analyses, as reviewed in
chapter 16. Of the four concepts only the water tank and rock bin combination
showed a distinct performance/cost advantage; the highest performance was
obtained with an annular water tank. None of the geometries examined
showed prospects of economic superiority over the basic Thomason concept,
which used a conventional cylindrical steel tank (Nalbandian and Sudar 1979).

DEVELOPMENT OF A GAS BACKUP WATER HEATER PROPERLY INTEGRATED WITH
SOLAR-HEATED, DOMESTIC HOT WATER STORAGE TANKS (ALTAS CORPORATION;
D. J. MORRISON, PRINCIPAL INVESTIGATOR)   The objective of this work was to
develop a compact, gas-fired backup heater mounted on top of the solar tank,
with gas heat supplied to the backup tank by a heat pipe. An engineering
prototype unit was successfully developed, in the course of which a compact
combustion chamber was developed, ignited by an electronic pilot, with an

83% (typical) firing efficiency at 50,000 Btu/h (14,650 W), operating with natural draft. The flame tips are not farther than 0.5 in. (1.3 cm) from the surfaces of the heat exchangers, which are a single bank of finned tubes containing the heat pipe fluid (water, chosen because it is nontoxic). Frost protection is provided by only partly filling the finned tubes. Because water was used, it was necessary to use a copper heat pipe configuration; the engineering prototype was all copper.

It was also necessary to separate the upflowing vapor and the downflowing condensate in the heat pipe for stable operation. To avoid "burnout," a slow ignition sequence was used. A novel thermal diode was developed to transfer excess solar heat to the top tank with minimal temperature difference between the tanks, thereby fully integrating them.

A detailed computer model was made of the entire system to determine overall performance and optimize individual component designs. A TRNSYS subroutine for stratified tanks evolved from this process (Morrison et al. 1980).

DEVELOPMENT OF AN IMPROVED WATER TUNK FOR THERMAL STORAGE (INDE-
PENDENT LIVING, INCORPORATED; W. T. HUDSON, PRINCIPAL INVESTIGATOR)
The objective of this contract was to develop an insulated fiberglass tank in two nested mating sections for easy shipping and handling. A 1,000-gal (3,780-L) prototype with an estimated life of 20 to 30 years using 200°F (93°C) water was successfully developed. Each section consisted of an inner and an outer fiberglass shell with 3 to 4 in. (8–10 cm) of polyurethane foam insulation between them; the two sections could be joined by a resealable clamping device on mating fiberglass flanges. The inner shells were formed by hand lay-up and the outer shells were sprayed on. Much of the effort consisted of selecting the materials, composition percentages, and thicknesses of the inner shells, which were composed of four layers of resin and glass, each with the same resin but with different resin-to-glass ratios. The resin selected was Derakane 470 and the total thickness was approximately 0.150 in. (0.38 cm). Although the concept proved to be technically feasible, the estimated cost for large-volume production did not appear to be competitive (Hudson et al. 1981).

MEMBRANE-LINED FOUNDATION SYSTEMS FOR LIQUID THERMAL ENERGY STORAGE
(UNIVERSITY OF NEBRASKA; R. C. BOURNE, PRINCIPAL INVESTIGATOR)   This title applies to the first phase of the project, which was procured through the large solar solicitation in 1978. Follow-on work, procured in 1980, was entitled "Low-Cost Membrane-Lined Water Storage Tanks," and R. Youngberg was the principal investigator. However, they are reviewed here as one project.

The continuing objective of the project was to evaluate the potential of membrane-lined foundations (water-containing structures) for solar heating

TES. Specific tasks were to survey existing membrane-lined storage projects, evaluate alternative membrane liner materials, develop and evaluate improved designs for heat transfer to and from membrane-lined storage containers, and evaluate concepts for enhancing stratification, including baffles, multiple tanks, and variations in tank configurations. Results of phase 1 indicated that membrane-lined containers generally may be constructed and installed for less than a third of the cost of steel tanks, with fewer corrosion problems, more convenient construction scheduling, and improved heat transfer configurations. Phase II continued the development of stratification-enhancing features, including design development, full-scale field testing, and cost-effectiveness studies of two preferred alternatives. It also continued the development of concepts for auxiliary heating of storage, including sizing studies, design development, and full-scale testing. Of the two preferred alternatives, the below-slab foam wall concept was found to be inferior, and emphasis was shifted to the woodframe basement corner (WBC) concept. A 1,500-gal (5,670-L) tank was constructed and remained full, but not continuously hot, for two years, at the end of which time no significant degradation was apparent (Bourne 1981; Youngberg and Bourne 1982; Bourne 1982).

LIGHTWEIGHT CONCRETE MATERIALS AND STRUCTURAL SYSTEMS FOR WATER TANKS FOR THERMAL STORAGE (WESTINGHOUSE ADVANCED ENERGY SYSTEMS DIVISION; R. W. BUCKMAN, PRINCIPAL INVESTIGATOR)   The objective of this work was to develop tanks made of precast or cast-on-site foamed concrete elements assembled on site. Structural strength was provided by an inner shell of lightweight [approximately 100 lb/ft^3 (1,600 kg/m^3)] steel-reinforced cellular concrete with glass fibers added to improve mechanical properties. Insulation was provided by an outer shell of 25 lb/ft^3 (400 kg/m^3) cellular concrete, self-bonded to the inner shell, and monomer moisture barrier was applied to the inner and outer surfaces. Two 500-gal (1,890-L) prototype designs were developed. The first consisted of two roughly hemispherical sections joined by a bolted, reinforced concrete flange. It had about 4 in. (10 cm) of insulating concrete, which was found by thermal performance testing to be insufficient. The second had two similar sections consisting only of the inner shell; the outer insulating shell was poured in one piece about the joined inner sections, with a minimum thickness of about 9 in. (23 cm). Performance objectives were met, but it was concluded that an internal membrane liner would be required to ensure a long life for the tank. The ASHRAE 94-77 test for measuring heat loss from storage was found to be unfeasible, so a stored fluid temperature decay test was developed and used successfully. Economic analysis showed that factory casting of the elements would be excessively expensive because of high shipping costs and that estimated installed costs for large-scale production would be \$1.84/gal (\$0.49/L) for 500-gal (1,890-L) tanks and \$0.59/gal (\$0.16/L) for 2,000-gal (7,560-L) tanks (Buckman 1982).

Rock Bed Storage

ROCK BED STORAGE FOR COOLING (DYNATECH R&D COMPANY; E. C. GUYER, PRINCIPAL INVESTIGATOR)   The objective of the contract was to investigate and quantify, on a regional level, the cost and performance of using rock beds chilled by night air or air-cooled by an evaporative chiller as a means of cooling a building. The concept was investigated for residential and commercial buildings in six climatic regions of the United States, combined with solar air heating as a supplement to or replacement for conventional air conditioning, and combined with conventional heating. Life-cycle costs of these night-effect systems were compared with those of conventional vapor compression air conditioners, and consideration was given to the systems ability to produce satisfactory thermal comfort, compatibility with solar air-heating systems, energy efficiency, and compatibility with conventional air conditioning systems. Key results were these: (1) Night-effect rock bed cooling can provide satisfactory thermal comfort in residences in large areas of the United States; (2) commercial buildings and other structures with high latent/sensible cooling are generally not suitable for this concept; (3) the economics would be greatly enhanced by the development of more efficient, less expensive air-handling units and lower-cost bins for rock beds; (4) where applicable, the night effect currently can be economically justified only as an addition to a solar air-heating system; and (5) a seasonal coefficient of performance (COP) of about 5 should be achievable for a well-designed system, given high blower and flow control efficiency and effective system control (Guyer et al. 1978).

SERIES-PARALLEL SOLAR AUGMENTED ROCK BED HEAT PUMP (CALIFORNIA STATE UNIVERSITY AT FULLERTON; E. F. SOWELL, PRINCIPAL INVESTIGATOR)   The goal of this work was to determine analytically the technical and economic feasibility of a solar-assisted, air source heat pump in which all the air that passes through the rock bed also passes through the evaporator and condenser. Heating and cooling performance was simulated with TRNSYS for Sacramento, Albuquerque, and New York, and economic comparisons with a stand-alone heat pump were made for Albuquerque. In Albuquerque the system investigated was found to use approximately 28% less energy annually than a conventional, high performance heat pump alone. Seasonal COPs were 3.02 for heating and 3.01 for cooling. However, the investment cost was nearly six times higher, and the life-cycle cost (30 years) was 2.1 times higher. Parasitical fan power had a pronounced effect on COP. In Albuquerque, eliminating fan power would have raised the seasonal heating COP from 3 to 4, whereas the cooling COP would have increased from 3 to 8. Investigators concluded that the high cost of the solar components would make the concept economically unattractive (Sowell and Othmer 1979; Sowell et al. 1979). The modeling is reviewed more fully in chapter 16.

ROCK BED STORAGE WITH HEAT PUMP (ZIA ASSOCIATES, INC.; H. E. REMMERS, PRINCIPAL INVESTIGATOR)   The objective of this contract was to investigate the technical and economic feasibility of coupling a heat pump to the rock bed storage unit of a conventional solar air-space-heating system in a single-family residence. Operating characteristics of available components were used to define representative systems in Albuquerque and Madison. Performance was simulated with TRNSYS, and economic comparisons were made with the LCCA (life-cycle cost analysis) program. The ASHRAE bin method was used to predict performance. Conclusions drawn for Albuquerque and Madison are considered valid for most parts of the United States except those with exceptionally high electric utility rates. Principal among the conclusions of the study were these: (1) The most economical system is a stand-alone heat pump sized for the maximum cooling load; (2) load-side storage is the preferred configuration; (3) load-side storage improves the performance of an oversized heat pump by reducing cycling frequency and augmenting dehumidification; and (4) the ASHRAE bin model predicts higher heating SPFs than TRNSYS with a detailed load model (Remmers and Mills 1979).

Phase Change Storage for Solar Heating

ENGINEERING DESIGN FOR THERMOCRETE CENTRAL STORAGE UNITS FOR LOW TEMPERATURE SOLAR APPLICATIONS (SUNTEK RESEARCH ASSOCIATES; C. TILFORD, PRINCIPAL INVESTIGATOR)   This was to have been an applications and systems design study for the Thermocrete materials that concurrently were being developed under contract to the STOR TES program (Chahroudi 1979). Before any substantive work could be done, however, problems and delays in the materials development project cast doubt on the wisdom of proceeding with this type of study before the material could be fully characterized, and Suntek shifted the SHACOB funds to nonstorage work. The Thermocrete materials development project is briefly mentioned in section 18.3.

PROTOTYPE DESIGN, CONSTRUCTION, AND TESTING OF THE ROLLING CYLINDER THERMAL STORAGE SYSTEM (GENERAL ELECTRIC COMPANY, CORPORATE RESEARCH AND DEVELOPMENT; C. S. HERRICK, PRINCIPAL INVESTIGATOR)   This project, although funded by the SHACOB TES program, was technically managed and monitored by the STOR TES program. The concept involves containment of an incongruently melting salt hydrate phase change material such as Glauber's salt in a slowly rotating horizontal drum, with heat transfer through the drum wall to and from air in an enclosing shroud. Tumbling of the salt prevents phase separation during melting and prevents encrustation on the drum wall as it freezes. The original objective was to design and construct a prototype unit with a capacity of about 200,000 Btu (190 MJ), suitable for air-based residential solar space heating, and then to test the prototype under a follow-on contract. The contract was terminated, however, after a year of design studies and preliminary assessments resulted in a recommendation to build and test a subscale unit before constructing the prototype. The preferred single-drum configuration was not attractive. For it to deliver 22,500 Btu/h (6592 W) (fairly typical for a residence), the unit must be 2 ft (0.6 m) in diameter, 36 ft (11 m) long, and have a capacity of about 10^6 Btu (948 MJ), requiring 0.25 kW of blower power and 0.5 kW for the rotation drive (Thornton and Herrick 1979).

SALT HYDRATE THERMAL ENERGY STORAGE SYSTEM FOR SPACE HEATING AND AIR-CONDITIONING (CALMAC MANUFACTURING CORPORATION; C. D. MACCRACKEN, PRINCIPAL INVESTIGATOR)   This project consolidated two awards to Calmac: on phase change storage for solar heating and on thermal energy storage for solar-driven air conditioners. The objective was to develop, test, and prepare for commercial distribution three different phase change material (PCM) salt hydrate/heat exchanger modules with on-site filling techniques to cover coolness storage for solar-driven chillers [40°–50°F (4°–10°C)], storage for space heating [100°–120°F (38°–49°C)], and hot-side storage for solar-driven chillers [200°–250°F (93°–121°C)]. The concept, which had been developed for ice storage, involves bulk storage of the PCM in a vented plastic container, with heat transfer to and from the PCM by means of a liquid in immersed coiled polyethylene tubes. For the coolness storage unit, a 45°F (7°C) m.p. eutectic of sodium sulfate, ammonium chloride, potassium chloride, and water was used. This PCM, previously developed by Maria Telkes, also contained a clay thickener to prevent phase segregation, borax for nucleation, and a small amount of detergent. For space heating the PCM used photograde sodium thiosulfate pentahydrate ("hypo"), with a melting point of 115°F (46°C), although a less pure "storage grade" was judged satisfactory. The high-temperature PCM was magnesium chloride hexahydrate [240°F (115°C) m.p.]. Cycling and other tests were performed in 1-, 30-, and 90-gal (4-, 113-, and 340-L) test units. The thickened 45°F (7°C) PCM lost nearly 30% of its initial heat of fusion after repeated melt–freeze cycles, through phase separation. Because of its high water vapor pressure, a covering layer of oil was required to prevent evaporation. The 115°F (46°C) PCM required intermittent stirring to prevent solids settling, but without stirring the settled solids redissolve by reheating to 150°F (65°C). The 240°F (115°C) PCM needed no stirring but required an oil cover, and the plastic container was found to require reinforcement because it softened at the elevated temperature. The ASHRAE 94-77 standard for testing thermal energy storage devices was found to be unsatisfactory, and additional tests were made using a modified procedure suggested by NBS (see chapter 17). Improvements resulting from this work have been incorporated into commercial units (MacCracken et al. 1980).

Thermal Energy Storage for Solar-Driven Air Conditioners

THERMAL STORAGE FOR SOLAR COOLING USING PAIRED AMMONIATED SALT REACTORS (MARTIN MARIETTA CORPORATION, AEROSPACE DIVISION; F. A. JAEGER, PRINCIPAL INVESTIGATOR)   This project, although funded by SHACOB, was technically managed and monitored by the STOR TES program and coordinated with another Martin Marietta project on more fundamental investigations of ammoniated salt reactions funded by STOR. As originally proposed, the concept was an intermittent thermochemical heat pump (see chapters 15 and 19) with ammonia as the working fluid, the novel aspect being that the lower temperature reaction would be resorption-desorption rather than condensation-evaporation. The initial objectives were to select a preferred solid salt for the high temperature reaction (magnesium and manganese chloride were candidates) and a preferred solid or liquid ammoniate for the low temperature reaction (calcium chloride and ammonium chloride were candidates) to demonstrate concept feasibility with subscale system testing and to demonstrate system performance with full-scale system testing. In the salts selection phase, materials compatibility and heat transfer were experimentally investigated while reaction kinetics were being studied in the coordinated STOR-funded project. All of the solid salts had unacceptably poor heat transfer properties, so the original concept was revised to one involving a liquid high temperature ammoniate (ammonium nitrate with one to three ammonias); condensation and evaporation of ammonia occurred in the lower temperature vessel. Using the liquid ammoniate, however, meant that the required solar input temperature and the temperature of the delivered heat of reammoniation varied with the ammonia concentration; with a solid ammoniate, the processes would have been isothermal (see thermochemical heat pumps in section 18.3). A full-scale prototype was built and then tested with separate IR&D funds available to Martin Marietta. Heating and cooling COPs calculated from the test data were 1.3 and 0.77, respectively. Considerable corrosion of carbon steel elements was observed (Jaeger and Fox 1981). Solar-driven refrigerators using solid calcium chloride and ammonia have been developed elsewhere with some moderate success; these used more advanced salt bed designs with better heat and mass transfer properties.

DEVELOPMENT OF INTERMEDIATE TEMPERATURE THERMAL ENERGY STORAGE SYSTEMS (UNIVERSITY OF DELAWARE; J. R. MOSZYNSKI, PRINCIPAL INVESTIGATOR) This project was also funded by SHACOB but technically managed and monitored by the STOR TES program. As initially awarded, the objective was to develop an optimized system providing thermal storage between 194° and 356°F (90°–180°C), but early in the effort the upper limit was extended to 572°F (300°C) to allow for possible high temperature Rankine chillers. The project began by identifying suitable processes snd materials and selecting storage/containment combinations for properties and compatibility determination. Design, construction, and testing of a model system was then to have begun. Primary attention was paid to solid/liquid phase change materials; heat of vaporization was considered briefly and discarded as a possible candidate. Two initially promising salts, $FeSO_4 \cdot 7H_2O$ and $Fe_2(SO_4)_3 \cdot 9H_2O$, were discarded because of extreme supercooling and poor rehydration kinetics. PCMs found to be potentially suitable were (1) a eutectic of $FeCl_3$ and NaCl, with 306°F (152°C) m.p. and 172 kJ/kg (75 Btu/lb) heat of fusion; (2) zinc chloride, with 594°F (312°C) m.p. and 355 kJ/kg (155 Btu/lb) heat of fusion; and (3) a eutectic of NaOH and Na_2CO_3, with 541°F (283°C) m.p. and ~412 kJ/kg (180 Btu/lb) heat of fusion. Polyethylene and magnesium chloride hexahydrate also were identified but found to be under investigation by others. The listed materials were studied calorimetrically to determine melting point, heat of fusion, specific heats of the liquid and the solid, and any subcooling. All were subjected to repeated heating and cooling cycles. For the $FeCl_3$–NaCl eutectic, thermal diffusity was determined and compatibility with prospective containment materials was investigated (Moszynski et al. 1979). The project was discontinued because none of the PCMs identified had sufficient promise.

General TES Modeling and Guidelines

DEVELOPING AND UPGRADING OF SOLAR SYSTEM THERMAL ENERGY STORAGE SIMULATION MODELS (BOEING COMPUTER SERVICES, INC.; J. KUHN, PRINCIPAL INVESTIGATOR) The objectives of this work were to collect, standardize, and link existing TES models from the literature and other contractors; correlate TES models with available TES component data; develop streamlined versions of validated TES component models; and provide DOE and industry with a competent solar TES simulation tool. Only rock bed and water tank models were investigated (Kuhn et al. 1980). For a fuller review of this work, see chapter 16.

ANALYSIS OF ADVANCED THERMAL ENERGY STORAGE SUBSYSTEMS FOR SOLAR HEATING AND COOLING (CONTRACT NUMBER EG-77-C-02-4483), AND THERMAL STORAGE STUDIES FOR SOLAR HEATING AND COOLING; APPLICATIONS USING CHEMICAL HEAT PUMPS (CONTRACT NUMBER DE-ACO2-79CS30248) (EIC CORPORATION; P. O'D. OFFENHARTZ, PRINCIPAL INVESTIGATOR)   The work done under the second contract was a continuation and elaboration of work done under the first; therefore, both are reviewed here as a single effort. The modeling effort is reviewed more fully in chapter 16, here the emphasis is on results obtained from TRNSYS system simulations using the developed TES subsystem models and the conclusions drawn from them. Simulations using various storage concepts were carried out for four applications: space and domestic hot water (DHW) heating, solar-assisted heat pumping, absorption air conditioning, and thermally activated (thermochemical) heat pumping. Annual cycle (seasonal) storage for residential space heating in Madison, using a water tank and using the water/sulfuric acid thermochemical system (storage only: COP = 1), required about the same solar collection area for comparable performance; however, the water tank volume was about four times that of the water/acid system. For solar-assisted heat pumping, only water TES systems were modeled; system performance was found to be fairly insensitive to the storage capacity. Two types of absorption air-conditioning systems were studied: water/lithium bromide and water/calcium chloride (which was built-in chemical storage and was being investigated by EIC under the SHACOB cooling program). Cold-side storage appeared preferable to hot-side, providing it is of the phase change type. Intermittent-type thermochemical heat pumps (with built-in storage) may compete with water-lithium bromide-plus-storage systems if they take advantage of daytime heat rejection without a cooling tower and heat-pumping capability in the winter. The need for better thermochemical heat pump storage models was apparent, which led to the follow-on, second effort. An improved model was made of the water/acid heat pump/storage system in the continuous mode, and new models were made of the ammonia/ammonium nitrate (as discribed earlier in this subsection) and methanol/calcium chloride (being developed by EIC under the STOR TES program; see section 18.3) systems. The performance of both the water/acid and ammonia/ammonium nitrate systems was nearly identical (a finding contradicted by simulations at the University of Wisconsin). Based on week-long summer and winter simulations for Washington, DC, and Madison, annually the acid system outperformed the methanol system, which in turn outperformed the lithium bromide system (with external storage and direct solar heat in winter) only in Washington, where heating load dominates (Offenhartz et al. 1979; Offenhartz et al. 1981).

18.2.3   Additional SHACOB TES Projects

A number of additional projects, based on unsolicited proposals, were initiated by the Argonne National Laboratory (ANL) under the SHACOB program to supplement the projects acquired by the large solar solicitation.

Stratification Modeling (ANL; W. T. Sha and E. I. H. Lin, principal investigators)   The COMMIX 3-d numerical model was adapted to water tank, rock bed, and salt-gradient pond applications in several separate projects, as reviewed in chapter 16 with documentation.

Rock Bed Development

COST-EFFECTIVE IMPROVEMENTS OF ROCK BEDS USED IN SOLAR-AIR-HEATING SYSTEMS (ANL SUBCONTRACT TO SOLARON CORPORATION; D. JONES, PRINCIPAL INVESTIGATOR)   The objectives of this subcontract were to obtain reliable thermophysical data for COMMIX rock bed model validation, to test alternative rock bed configurations and baffle structures, and to design and test uniform-flow horizontal and vertical rock beds. Although the work was not completed because of an early contract termination, a cam-lock foam panel design proved to be successful, and it was found that flow distribution in a vertical rock bed might be greatly, and inexpensively, improved by using a concrete block to obstruct the upper plenum velocity jet. In horizontal rock beds, baffles were found superior to, although slightly more expensive than, "sand traps" as a means of preventing air channeling along the lid and walls. The importance of deflection devices in the inlet plenum was demonstrated; the double-frame, alternating baffle design is an inexpensive way to prevent flow maldistributions. The performance of horizontal units was found generally to be 5% higher than that of vertical units because of the longer flow path and the improved flow distribution due to baffles. Pressure drop data were compared to correlations reported in the literature, and critical parameters were developed and presented as a dimensionless energy graph to simplify performance comparisons of different rock bed configurations (Jones 1982).

782 Allan I. Michaels and C. J. Swet

Seasonal Storage

TRANSSEASONAL STORAGE OF ENERGY IN MOIST SOILS (COLORADO STATE UNI-
VERSITY; W. D. KEMPER, PRINCIPAL INVESTIGATOR) This effort was partly
funded by DOE. It consisted partially of theoretical analyses and laboratory
measurements of simultaneous heat transfer and water migration through
partially saturated soils, including the development and validation of a model
of these processes. The model's ability to predict temperatures over extended
periods suggested that long-term soil drying was accurately accounted for.
This work also included the construction, testing, and feasibility assessments
of a heat storage reservoir near Solar House III at CSU. The reservoir was
an 8-ft (2-m)-high by 19-ft (60-m)-diameter vertical cylindrical volume of moist
earth containing heating coils and enclosed in a polyethylene film moisture
barrier. It was buried with its top 7 ft (2 m) below grade in a larger volume of
dried, replaced soil; the conductivity of the soil, after a 70-day heating period,
was about twice that expected. Economic analysis suggested that a solar
residence with that type of ground-coupled storage could be cost-effective in
conditions similar to that experienced in Colorado. Researchers concluded
that excavating and predrying the surrounding soil was unnecessary (Kemper
and Walker 1978).

PARTICIPATION IN INTERNATIONAL ENERGY AGENCY (IEA) PROGRAMME ON CEN-
TRAL SOLAR HEATING PLANTS WITH SEASONAL STORAGE (ANL; A. I. MICHAELS,
PRINCIPAL INVESTIGATOR) The objectives are to coordinate and lead U.S.
participation in the IEA program and to evaluate and provide technical input
to investigations of large-scale, long-term TES systems. Specific technical
activities of the U.S. participants in TES aspects of the program have included
performance simulations and economic studies of annual storage in water
tanks by SERI (Sillman 1981); a feasibility study on seasonal storage in water
tanks at the Boston Navy Yard by ANL and MIT (Breger and Michaels 1983);
a feasibility study on seasonal storage in an aquifer for a hypothetical new
community in Madison by ANL and SERI (Michaels et al. 1983); and adaptation of the IEA storage model to U.S. system models.

Latent-Heat TES

TESTING LATENT-HEAT STORAGE UNITS FOR SOLAR APPLICATIONS (ANL; R. L. COLE,
PRINCIPAL INVESTIGATOR) The objective of this work was to develop an
improved testing procedure to supersede ASHRAE standard 94-77, which
several investigators had found deficient for phase-change storage units
(see subsection 18.2.2) (Cole et al. 1983). This work is reviewed more fully in
chapter 17.

DEVELOPMENT OF THERMAL ENERGY STORAGE DEVICES USING CROSS-LINKED HIGH DENSITY POLYETHYLENE (ANL; R. L. COLE, PRINCIPAL INVESTIGATOR) The objective of the contract was to develop latent heat storage devices for solar-driven chillers operating at 266° to 302°F (130°–150°C), using pebble-bed configurations. A prototype unit was designed and built for use with pellets of form-stable, cross-linked high density polyethylene (HDPE) in 50–50 water/ethylene glycol heat transfer fluid. The form-stable pellets had been developed under the STOR/TES program (see also chapter 15). A test loop was also constructed for use in a procedure similar to the intended replacement for ASHRAE 94-77. A later modification was to use a boiling and condensing heat transfer fluid (water). In initial testing the device functioned well, with better-than-predicted charging and discharging rates, but on repeated cycling, the bed compacted and blocked flow (Cole 1983). Funding was discontinued before diagnostic and corrective steps could be taken.

PELLETIZATION AND ROLL ENCAPSULATION OF PHASE CHANGE MATERIALS (PENNWALT CORPORATION; J. CHEN, PRINCIPAL INVESTIGATOR) This project was selected and funded by the STOR TES program and monitored by the SHACOB program. Its objective was to develop coated capsules of phase change materials (PCMs) for use in packed beds (the same intention as that for uncoated, form-stable HDPE pellets) or for incorporation into building materials. Seven PCMs covering the 73° to 243°F (23°–117°C) m.p. range were pelletized into half-inch tablets with a laboratory-sized compacting machine and encapsulated with organic coating formulations in a roll coater. The PCMs were Glauber's salt eutectic [73°F (23°C)], calcium chloride hexahydrate [84°F (29°C)], disodium hydrogen phosphate dodecahydrate [97°F (36°C)], Sunoco P-116 wax [116°F (47°C)], sodium thiosulfate pentahydrate [120°F (49°C)], magnesium nitrate hexahydrate [185°F (85°C)], and magnesium chloride hexahydrate [243°F (117°C)]. Candidate PCMs for coolness storage [45°–50°F (7°–10°C)] were unsatisfactory for pelletization. After small-scale and packed-bed thermal cycling tests in aqueous solutions of calcium chloride and ethylene glycol, the 185° and 243°C (85° and 117°C) PCMs were discarded because of wall degradation at those temperatures, and the 97°F (36°C) PCM was also eliminated because of its excessive supercooling and high cost. The remaining four were judged good candidates for scale-up tests and possible commercialization, with estimated encapsulated costs in large-scale production of less than 40¢/lb (88¢/kg). Capsules of the Glauber's salt eutectic also were exposed to air for 95 days with only 1.4% weight gain, suggesting good possibilities for use in building materials (Chen et al. 1982). The 73°, 84°, and 117°F (23°, 29°, and 47°C) capsules are now commercially available.

SOLID–SOLID PHASE CHANGE MATERIALS (SERI; D. K. BENSON, PRINCIPAL INVES-
TIGATOR)   This work, ongoing at present, is sponsored by the Passive and
Hybrid Division of the Office of Solar Heat Technologies. Its objectives are
to evaluate the technical and economic feasibility of using solid-state phase
change materials for thermal energy storage in passive solar architectural
applications and to develop a better understanding of the molecular processes
involved in solid-state transformations. Most closely examined for passive
heating is a polyhydric alcohol, neopentyl glycol (NPG), in various mixtures
with pentaglycerine for adjusting the transition temperature and suppressing
subcooling and with graphite to improve thermal conductivity. This material
is now being marketed for TES by Eastman Chemical Company. Transition
temperatures are between 79° and 118°F (26°–48°C), with latent heats of
solid–solid transition ranging from 76 to 199 kJ/kg. Performance simulations
of Trombe walls with solid-state PCMs indicate that a 81°F (27°C) transition
temperature is optimal, but preliminary cost estimates indicate that although
such a wall could be one-fourth as thick and one-ninth as heavy as a comparably performing concrete wall, it would cost nearly twice as much. Work
continues to improve the performance of these materials and to evaluate other
potential passive solar applications for them (Benson et al. 1985).

ADVANCED PHASE CHANGE MATERIALS FOR PASSIVE SOLAR STORAGE APPLICA-
TIONS (UNIVERSITY OF DAYTON RESEARCH INSTITUTE; I. O. SALYER, PRINCIPAL
INVESTIGATOR)   This is ongoing work at present and is sponsored by the
Passive and Hybrid Division of the Office of Solar Heat Technologies. Its
purpose is to identify suitable PCMs and develop methods of incorporating
them into cementitious and other building materials. The search has narrowed
to two chemically different systems: the polyethylene glycols (PEGs) and
crystalline alkyl hydrocarbons of about 18 to 20 carbon atoms (paraffin
waxes). Two basic methods of incorporating them into building materials have
been investigated: mechanical dispersion (mixing) directly into concrete, cement, plaster, etc., and dispersion into rubber. The rubber containing the
PCM may be used directly as floor or wall tile, made into pellets for packed-
bed stores, or ground up for mixing into cementitious materials. Future plans
are to complete investigations of humidity effects, start aging tests in selected
construction materials, and prepare test panels for thermal conductivity and
other critical parameters (Salyer et al. 1984).

ENERGY STORAGE IN INORGANIC COMPOUNDS (NATIONAL BUREAU OF STANDARDS;
L. J. STRUBLE, PRINCIPAL INVESTIGATOR)   This is ongoing work sponsored by
the Passive and Hybrid Division of the Office of Solar Heat Technologies to
characterize the phase change properties of ettringite and related compounds.

The progress and significance of this work are reviewed in chapter 25 of part III.

Stratified Water Tank Testing and Analysis

STRATIFIED STORAGE MEASUREMENT AND ANALYSIS (ANL; R. L. COLE, PRINCIPAL INVESTIGATOR) This project was initially funded by the SHACOB program and then continued with STOR TES program funding. Objectives were to experimentally validate the COMMIX-SA computer code, examining the effects of baffle and diffuser configurations and tank shapes suggested by COMMIX simulations, and prepare design guidelines based on the experimental results. All experiments were on vertical cylindrical tanks; the experiments were to have extended to horizontal tanks, but funds terminated before that could be done. Although the testing was not complete enough to fully validate COMMIX, a stratification index was developed that allows different tanks to be compared; a simple one-dimensional thermal diffusion theory was developed that yields good results in the central region of the tank. The best stratification was obtained with dual radial-flow diffusers that are merely circular plates, with no baffles, followed closely by a single horizontal baffle and by the side inlet and outlet configuration. There is a critical value of Richardson number (about 0.31–0.48) below which rapid mixing occurs, but this permits a wide range of flow rates. Although thermal diffusion broadens the thermocline, the rate of broadening is sufficiently slow for most tanks to remain stratified in diurnal cycling; deeper tanks decrease the detrimental effect. Vertical baffles are detrimental to stratification. Particularly stressed was the importance of a solar collector operating strategy that allows more than one tank volume per day to pass through the collector or collector exchanger, with the flow rate varied to maintain the water entering the tank at a constant temperature, only a few degrees above the load's requirement (Cole and Bellinger 1982).

THE MONITORING OF THERMAL STRATIFICATION OF THE STORAGE TANK OF THE MABEL LEE HALL SOLAR HOT WATER SYSTEM (UNIVERSITY OF NEBRASKA; E. E. ANDERSON, PRINCIPAL INVESTIGATOR) The objective of this work was to obtain an experimental data base, for a variety of operating modes, to validate numerical models of natural stratification in rectangular tanks. Water storage is in a buried rectangular tank 16 by 32 by 9 ft (5 by 10 by 3 m) deep [32,000 gal (120,960 L)], sealed with a PVC thin-film preseamed liner and insulated with styrofoam. Internal heat exchangers transfer heat from liquid collectors to storage and from storage to preheat DHW. With air-heating collectors (previously installed for comparison with liquid-heating collectors for DHW preheating) an external air-to-liquid heat exchanger is used. Fourteen different operating modes may be examined, with high-, low-, and no-flow conditions. The testing was not completed (Anderson 1981).

Active Solar-Cooling Studies

EVALUATION OF THERMAL STORAGE CONCEPTS FOR SOLAR-COOLING APPLICA- TIONS [SCIENCE APPLICATIONS, INC. (SAI); P. J. HUGHES, PRINCIPAL INVES- TIGATOR] The objective of this work was to compare the thermal and economic performance of solar heating and cooling systems with conventional alternatives, where various thermal storage concepts have been incorporated into the solar systems. The evaluations were made for both residential-size [3-ton (2,722-kg)] and small commercial-size [25-ton (22,680-kg)] systems. The residential analysis considered energy requirements for space heating, space cooling, and water heating; the commercial building analysis considered only space cooling. For the commercial building, ten different TES/solar systems were considered: five each for water-LiBr absorption and Rankine chillers. The residential analysis considered four systems, all with an absorp- tion chiller. All 14 systems were simulated using TRNSYS in Miami, Phoenix, Fort Worth, and Washington, DC. Trade-offs included cold-side versus hot- side storage, single- versus multiple-stage storage, and phase change versus sensible-heat storage in water. The essential findings were these: cold-side, latent-heat storage generally appears most promising; multiple-stage units cannot be justified; for residential systems, the variations caused by differences in storage options are up to 15% in solar fraction and 20% in life-cycle cost, and comparable variations for commercial systems with absorption chillers; in commercial systems with Rankine chillers, the variations are much smaller; and none of the solar systems studied, regardless of storage options, is attrac- tive on a life-cycle basis with the assumptions used (Hughes et al. 1981). Some of the more critical assumptions are so constraining, however [i.e., cooling- only for the commercial building, 205°F (96°C) max storage temperature for Rankine chillers, unoptimized storage configurations], that the quantitative findings are suspect.

SOLAR-COOLING THERMAL ENERGY STORAGE STUDY (HITTMAN ASSOCIATES, INC.; H. M. CURRAN, PRINCIPAL INVESTIGATOR) The objective of the contract was to identify the thermal storage requirements for solar-cooling systems using absorption, Rankine, and desiccant technologies, and to identify research and development needs. Included in the work was a definition of the operating characteristics of DOE-funded solar-cooling equipment and solar collectors suitable for space cooling. Data on existing TES systems came from the final report of an earlier Hittman study entitled "Solar applications of thermal energy storage" (Hittman number H-C0199-79-753F, January 1979) and from elsewhere in the literature. The basic design requirements identified for hot-side and cold-side storage included storage temperature range, energy storage capacity, length of storage time, charge time per cycle, discharge time per cycle, thermal energy losses, and acceptable cost level. Since SAI had been assigned the task of developing a general simulation program for the use of TES in solar cooling systems (as mentioned earlier in this subsection) the Hittman effort was limited to an analysis aimed at defining the basic required storage characteristics. These storage requirements were examined in light of then-available technology, and these comparisons helped researchers determine storage R&D needs (Curran and DeVries 1981).

18.3 Complementary R&D at DOE/STOR on TES for Solar Heating and Cooling

The TES program in what is now the DOE Office of Utility Technologies (the STOR TES program) was directed at many solar and nonsolar applications. The solar applications included solar thermal power and solar fuels as well as solar heating and cooling of buildings. A substantial fraction of its sponsored R&D, however, was either aimed at or suitable for solar space heating and cooling, much of it being coordinated with the DOE SHACOB program as well as with other organizations supporting comparable efforts. The SHACOB-related effort falls into three main categories: phase change storage, seasonal storage, and thermochemical heat pump storage. Some of the more significant work in each of these categories is mentioned briefly in the subsections that follow.

18.3.1 Phase Change Storage

Two projects on Glauber's salt were supported at the University of Delaware Institute for Energy Conversion. One was intended to demonstrate conclusively and document thoroughly the efficacy of the Telkes clay stabilizing technique in preventing phase segregation and heat of fusion reduction after many cycles; it succeeded in doing neither (Kando 1978). The other was to demonstrate the durability of laminated plastic "chubs" containing Glauber's salt formulations; good results were obtained (Frysinger 1979). Calmac Manufacturing Corporation successfully adapted a polymer gel stabilizing technique to sodium thiosulfate pentahydrate, which allowed designers to eliminate the mixing pump (MacCracken 1982). A plastic tube containing this formulation is now being marketed for passive applications. The Dow Chemical Company investigated various methods of encapsulating calcium chloride hexahydrate and other low temperature PCMs (Lane et al. 1978), which resulted in the commercial use of that material in HDPE tubes for passive applications. Support for the development of the Pennwalt coated pellets (see subsection 18.2.3) resulted in three commercial products. In 1976 an effort at Villanova University to identify attractive solid–solid transition materials was unsuccessful (Leffler 1977), as was a search at ORNL for better low temperature PCMs (Johnson 1982). Work by Suntek Research Associates in 1976 and 1977 on thermocrete (see subsection 18.2.2) was the first federally sponsored attempt to develop methods of incorporating PCMs into building materials (Chahroudi 1977); whether its developmental difficulties could have been overcome with continued support remains conjectural. Lockheed Huntsville successfully tested the performance of corrugated plastic wall panels filled with the Dow calcium chloride hexahydrate formulation, but long-term water gain from the atmosphere was excessive (Fletcher 1983). Work at SERI (Wright 1981) and at Clemson University (Edie, Melsheimer, and Mullins 1979) on direct heat transfer using immiscible fluids failed to prevent salt carryover, and the same problem was encountered at The Franklin Institute Research Laboratories when they tried that approach after inability to pump solidifying saturated salt solutions (Kauffman, Lorsch, and Kyllonen 1977). As noted in subsection 18.2.3, The Monsanto Research Company (with further work by the University of Dayton Research Institute) developed a form-stable HDPE phase change pellet (Botham et al. 1977).

18.3.2   Seasonal Storage

George Washington University studied the technical and economic feasibility of storage in the earth for a single-family residence, using a different design approach from that used by Colorado State University (CSU) (see subsection 18.2.3). This project was inherited from NSF, and a small follow-on contract was added to it. A multiproject program on aquifer storage of heated and chilled water included large-scale experiments (Melville, Molz, and Güven 1983), modeling (Tsang and Doughty 1983), case studies (Walton and McSwiggen 1983), and economic evaluations (Reilly, Brown, and Huber 1981); nonaquifer seasonal storage concepts were identified by PNL (Blahnik 1981). This work was coordinated with the IEA activities both on seasonal storage for SHACOB (see subsection 18.2.3) and on nonsolar seasonal storage. Rocket Research Company examined the feasibility of seasonal storage using a water/ sulfuric acid thermochemical heat pump (Clark 1979).

18.3.3   Thermochemical Heat Pumps

Alternative concepts to those investigated by Martin Marietta (see subsection 18.2.2) were explored by Chemical Energy Specialists (water/magnesium chloride, intermittent) (Greiner 1977), EIC Corporation (methanol/calcium chloride, intermittent) (Offenhartz et al. 1980), and Rocket Research Company (water/sulfuric acid, intermittent and continuous) (Clark 1978, 1982). Sandia Livermore Laboratories conducted many supporting physical/chemical experiments (Carling 1979; Mar and Carling 1979; Nissen 1979; Carling, Wondolowski, and Macmillan 1980), and TRW conducted cost-effectiveness studies that suggested competitiveness with most nonsolar options (Gorman et al. 1981). For discussions of the potential of these concepts, see chapter 19.

18.4   Appraisal of DOE Storage Programs for Solar Space Heating and Cooling

This cannot be a final summing up since DOE support of storage R&D for solar space heating and cooling continues, albeit at a level currently much reduced. Neither can it be a final judgment of all the work done to date since the impact of some projects has not yet been fully assessed. Rather, it is a brief review of the major accomplishments and shortcomings that have by now become apparent, if we view the SHACOB and STOR programs as a whole.

Some of the accomplishments are clearly evident in the commercialization of concepts developed with DOE support. Notable are marketed products developed by, or resulting directly from, sponsored R&D on phase change storage by Dow, Pennwalt, SERI, and Calmac (see subsections 18.2.2 and 18.3.1). Potentially important are recent negotiations with the Rocket Research Company by a Japanese firm for manufacturing rights to the water/ sulfuric acid thermochemical heat pump (see subsection 18.3.3) for industrial process heat. Less directly apparent were the results of sponsored development, testing, and cost studies of site-assembled, membrane-lined water storage vessels (see subsection 18.2.2). Although vessels of this type were already on the market before the DOE-sponsored work began, that work has influenced subsequent designs and has greatly increased recognition of the merit of this concept. The investigations of seasonal storage in aquifers (see subsection 18.3.2) is known to have been of value in the development of currently operational systems in China, and the IEA studies on very large water-containment vessels (see subsection 18.2.3), in combination with DOE participation in nonsolar seasonal storage R&D appears to be leading to operational use in district heating systems in Europe. Seasonal storage is economically more attractive than diurnal storage for large solar-heating systems in many situations, but its lower overall system costs are not sufficient to overcome the current cost disadvantage (attributed mainly to high collector costs) relative to nonsolar systems. In the case of thermochemical heat pump storage systems for solar space heating and cooling, their potential economic advantage over most nonsolar concepts (see subsection 18.3.3) seems to make them leading contenders among active solar concepts, but the evidence is not sufficiently robust. Unfortunately, support was discontinued before harder evidence could be accumulated.

Some negative results can also be viewed as accomplishments, in that they put to rest notions that were found to be either technically or economically unfeasible. New concepts for diurnal water tank construction and geometry (see subsection 18.2.2) were found not to be economically competitive with membrane-lined vessels. Pumping saturated salt solutions for latent heat storage (see subsection 18.3.1) was found to be technically unfeasible after the notion had been widely postulated. Other negative results early in the program, such as the high cost of wax encapsulation (see subsection 18.2.1), were lessons later researchers learned and profited from such as the successful development of pelletized and roll-encapsulated wax (see subsection 18.2.3).

Another kind of accomplishment is our enlarged understanding of the nature and control of stratification in water storage vessels (see subsections 18.2.2 and 18.2.3). This is having an immediate impact in nonsolar storage applications. It is also instigating widespread changes in the design philosophy for liquid collector systems. Increased understanding of phase change processes, including solid–solid transitions (see subsections 18.2.3 and 18.3.1), seems likely to have greatest impact in higher temperature applications.

Still another accomplishment has been the discernment through this work of previously unrecognized problems and underestimated effects. One problem not orginally considered significant, but which proved often to be a major contributor to the poor performance of the overall field system, is the control of heat losses from storage tanks. Insulated tanks frequently were found to lose more than twice as much energy as predicted, which proved especially troublesome in heating systems with outdoor tanks or with indoor hot storage tanks in cooling systems. Work is still needed to identify and eliminate the causes of high heat losses, some of which involve water absorption in insulation (particularly on underground tanks) and losses through piping connections. Another discernment was of the importance of the interactive or coupling effects of the TES subsystem on the overall performance of the solar system and consequent cost effectiveness. System efficiency is clearly extremely sensitive, in many cases, to the correct selection of storage temperature, capacity, heat transfer medium, and charge-discharge rates.

Significant remaining issues and opportunities for further R&D are discussed in chapter 19.

References

Anderson, E. E. 1981. *The Monitoring of Thermal Stratification of the Storage Tank of the Mabel Lee Hall Facility.*

Arrhenius, G., J. Hitchin, E. Jensen, and A. Tsai. 1983. Latent heat exchange by direct contact vaporization: a new concept in energy storage and retrieval. *Opportunities in Thermal Storage R&D*, W. Hausz and B. Berkowitz, eds. EPRI EM-3159-SR. Paper 24.

Bailey, J. A., J. Mulligan, and C. Liao. 1977. *Research on Solar Energy Storage Subsystems Utilizing the Latent Heat of Phase Change of Certain Organic Materials.* COO-5101.

Barlow, W. 1983. Thermal storage utilizing off-peak electro-thermal charging of phase change materials. *Opportunities in Thermal Storage R&D*, W. Hausz and B. Berkowitz, eds. EPRI EM-3159-SR. Paper 20.

Beard, J. T. and F. Iachetta. 1980. *Annual Collection and Storage of Solar Energy for the Heating of Buildings.* ORO/5136-78/2.

Bell, M. A. 1981. Low grade heat storage using sodium acetate solution. *Papers Presented at the International Conference on Energy Storage, Brighton. U.K. April 29–May 1, 1981.* Cranford, Beds, U.K.: BHRA Fluid Engineering. Paper J2.

Benson, D. K., J. Webb, R. Burrows, J. McFadden, and C. Christensen. 1985. *Materials Research for Passive Solar Systems: Solid-State Phase-Change Materials.* SERI/TR-255-1828. Golden, CO: Solar Energy Research Institute.

Blahnik, D. E. 1981. Preliminary assessment of promising nonaquifer STES. In *Proc. of International Conference on Seasonal Thermal Energy Storage and Compressed Air Energy Storage.* CONF-811066, 1: 71-79.

Botham, R. A., G. Jenkins, G. Ball III, and I. Salyer. 1977. *Form-Stable Crystalline Polymer Pellets for Thermal Energy Storage: Phase I.* ORO/5159-10.

Bourne, R. C. 1981. *Membrane-Lined Foundations for Liquid Thermal Storage.* DOE/ET/20111-1.

Bourne, R. C. 1982. *Installation Manual: Membrane-Lined Storage for Solar Heating Systems.*

Breger, D. S., and A. I. Michaels. 1983. *A Seasonal Storage Solar Energy District Heating System for the Charlestown.* Boston Navy Yard National Park: Phase II. ANL-83-47. Argonne, IL: Argonne National Laboratory.

Buchan, R. M., L. R. Majestic, and R. Billau. 1976. *Toxicological Evaluation of Liquids Proposed for Use in Direct Contact Liquid–Liquid Heat Exchangers for Solar Heated and Cooled Buildings.* COO-2867-1.

Buckman, R. W. 1982. *Lightweight Concrete Materials and Structural Systems for Water Tanks for Thermal Storage.* COO-5900.

Carling, R. W. 1979. *Dissociation Pressure Measurements on Salts Proposed for Thermochemical Energy Storage.* SAND79-8033. Albuquerque, NM: Sandia National Laboratory.

Carling, R. W., A. T. Wondolowski, and D. C. Macmillan. 1980. *Heat of Formation of $CaCl_{12} \cdot 2CH_3OH$ and $CaCl_2 \cdot 2C_2H_5OH$ by Solution Calorimetry.* SAND80-8689. Albuquerque, NM: Sandia National Laboratory.

Chahroudi, D. 1977. Thermocrete and thermotile building components with isothermal heat storage. In *Proc. Second Annual Thermal Energy Storage Contrators' Information Exchange Meeting.* CONF-770955, pp. 147–151.

Chahroudi, D. 1979. Engineering design for thermocrete central storage units for low temperature solar applications. In *Proc. 3rd Annual Solar Heating and Cooling Research and Development Branch Contractors' Meeting.* CONF-780983, pp. 214–215.

Chen, J., R. Nelson, an F. Polinski. 1982. *Pelletization and Roll Encapsulation of Thermal Energy Storage Materials.* COO-5900.

Clark, E. C. 1978. *Sulfuric Acid and Water Chemical Heat Pump/Storage Program Phase II Final Report.* Rocket Research Company report RRC-78-R-595 prepared for Sandia Corp. under DOE contract EY-76-C-03-1185.

Clark, E. C. 1979. *Sulfuric Acid and Water Chemical Heat pump/Storage Program Phase II-A Final Report.* Rocket Research Company report RRC-79-R-627 prepared under Sandia contract 18-4958.

Clark, E. C. 1982. *Sulfuric Acid/Water Chemical Heat Pump/Chemical Energy Storage,* vols. 1 and 2. Rocket Research Company report RRC-82-R-813 prepared under BNL contract 494588-S.

Cole, R. L. 1983. *Thermal Storage Device Based on High Density Polyethylene, Interim Progress Report.* ANL-83-52. Argonne, IL: Argonne National Laboratory.

Cole, R. L., and F. O. Bellinger. 1982. *Natural Thermal Stratification in Tanks. Phase I Final Report.* ANL-82-5. Argonne, IL: Argonne National Laboratory.

Cole, R. L., K. Nield, R. Rohde, and R. Wolosewicz. 1979. *Design and Installation Manual for Thermal Energy Storage.* ANL-79-15. Argonne, IL: Argonne National Laboratory.

Cole, R. L., K. Nield, R. Rohde, and R. Wolosewicz. 1980. *Design and Installation Manual for Thermal Energy Storage,* 2d ed. ANL-79-15. Argonne, IL: Argonne National Laboratory.

Cole, R. L. 1983. *Comparison of Testing Methods for Latent Heat Storage Devices.* ANL-82-89. Argonne, IL: Argonne National Laboratory.

Curran, H. M., and J. DeVries. 1981. *Options for Thermal Energy Storage in Solar Cooling Systems.* Hittman report H-C1007/060-80-976 for DOE SHACOB program.

deJong, A. G., and C. J. Hoogendoorn. 1980. Improvement of heat transport in paraffins for latent heat storage systems. *Thermal Storage of Solar Energy. Proc. International TNO-Symposium,* Amsterdam, The Netherlands, November 5, 1980. C. den Ouden, ed. The Hague: Martinus Nijhoff.

Edie, D. D., S. S. Melsheimer, and J. C. Mullins. 1979. Immiscible fluid-heat of fusion heat storage system. In *Thermal Energy Storage Fourth Annual Review Meeting.* CONF-791232, pp. 391–399.

Fletcher, J. W. 1983. Heat storage building materials for passive solar applications. In *Proc. DOE Physical and Chemical Energy Storage Annual Contractors' Review Meeting.* CONF-830974, pp. 15–21.

Frysinger, G. R. 1979. Life and stability testing of packaged low-cost energy storage materials. In *Thermal Energy Storage Fourth Annual Review Meeting.* CONF-791232, pp. 277–281.

Furbo, S. 1980. Heat storage with an incongruently melting salt hydrate as storage medium based on the extra water principle. *Thermal Storage of Solar Energy. Proc. International TNO-Symposium,* Amsterdam, The Netherlands, November 5, 1980. The Hague: Martinus Nijhoff.

Gorman, R., P. Moritz, T. O'Gorman, and W. Standley. 1981. *Chemical Heat Pump Cost Effectiveness Evaluation.* BNL 51484.

Greene, N. D., and W. Watson. 1979. *Proc. of Solar Energy Storage Options,* M. B. McCarthy, ed. CONF-790328-P1,2,3, pp. 465–472.

Greiner, L. 1977. The chemical heat pump (hydrated salt heat pump). In *Proc. Second Annual Thermal Energy Storage Contractors' Information Exchange Meeting.* CONF-770955, pp. 289–295.

Guyer, E. C. 1978. *Rock Bed Storage for Cooling*. COO-4481-1.

Helshoj, E. 1981. A high-capacity, high-speed latent heat storage unit. *Proc. Solar World Forum, ISES*, pp. 703–709.

Hoffman, E., and K. Chan. 1985. The bubble action pump solar heating system. In *Intersol 85 Extended Abstracts*, p. 125.

Hooper, F. C. 1980. *Solar Space Heating Systems Using Annual Heat Storage, Final Report*, vol. 1. DOE-CS-32939-12.

Horel, J. D., and F. de Winter. 1978. *Investigation of Methods to Transfer Heat from Solar Liquid-Heating Collectors to Heat Storage Tanks*. DOE/CS/31238.

Hudson, E. T. 1981. *Development of an Improved Water Tank for Thermal Energy Storage*. Final report on DOE contract DE-AC0278CS34699.

Hughes, P. J., J. Morehouse, M. Choi, N. White, and W. Scholten. 1981. *Evaluation of Thermal Storage Concepts for Solar Cooling Applications*.

Ives, C. J. M., W. Wight, and A. Andreos. 1985. Development of heat pipe solar DHW systems. In *Intersol 85 Extended Abstracts*, p. 126.

Jaeger, F. A., and E. C. Fox. 1981. *Thermal Storage for Solar Cooling Using Paired Ammoniated Salts Reactors*. DOE-CS-34700-2, MCR-81-581.

Johnson, J. S., Jr. 1982. New physical-chemical reactions useful for TES. In *Proc. DOE Physical and Chemical Storage Annual Contractors' Review Meeting*.

Jones, D. 1982. *Cost Effective Improvements of Rockbeds Used in Solar Air Heated Systems*. COO-5692.

Jones, D. E., and J. E. Hill. 1979. *Testing of Pebble Bed and Phase-Change Thermal Energy Storage Devices According to ASHRAE Standard 94-77*. NBSIR-79-1737. Gaithersburg, MD: National Bureau of Standards.

Kando, P. F. 1978. *Characterization of Sodium Sulfate Dekahydrate (Glauber's Salt) as a Thermal Energy Storage Material*. COO-4042-16.

Karaki, S., and P. Brothers. 1980. *Direct Contact Liquid–Liquid Heat Exchanger for Solar Heated and Cooled Buildings*. COO-2867-8.

Kauffman, K. W., H. G. Lorsch, and D. M. Kyllonen. 1977. *Thermal Energy Storage by Means of Saturated Aqueous Solutions*. TID-28330.

Kemper, W. D., and W. R. Walker. 1978. *Trans-seasonal Storage of Solar Energy*. COO/4546-3.

Kuhn, J. K. 1980. *Developing and Upgrading of Solar System Thermal Energy Storage Simulation Models*. Prepared under contract DE-AC02-77CS34482.

Lane, G. A., A. Kott, G. Warner, P. Hartwick, and H. Rossow. 1978. *Macro-Encapsulation of Heat Storage Phase-Change Materials for Use in Residential Buildings*. ORO/5217-8.

Leffler, A. 1977. The use of solid state phase transitions for thermal energy storage. In *Proc. Second Annual Thermal Energy Storage Contractors' Information Exchange Meeting*. CONF-770955, pp. 170–171.

MacCracken, C. D. 1982. Effects of additives on performance of hydrated salt TES systems. In *Proc. the DOE Physical and Chemical Storage Annual Contractors' Review Meeting*.

MacCracken, C. D. 1980. *Salt Hydrate Thermal Energy Storage System for Space Heating and Air Conditioning*. Final report on DOE contract DE-AC02-78CS34698.

Mancini, N. A. 1980. Use of paraffins for thermal storage. *Thermal Storage of Solar Energy. Proc. International TNO-Symposium*, Amsterdam, The Netherlands, November 5, 1980. C. den Ouden, ed. The Hague: Martinus Nijhoff.

Mar, R. W., and R. W. Carling. 1979. *Thermochemistry of Hydrated Salts*. SAND79-8503. Albuquerque, NM: Sandia National Laboratory.

Mehalik, E., and A. Tweedie. 1979. *Two Component Thermal Storage Material Study: Phase II, Final Report.* COO-2845-78/2.

Melville, J. G., F. J. Molz, and O. Güven. 1983. Thermal energy storage in confined aquifers using the doublet well configuration. In *Proc. DOE Physical and Chemical Energy Storage Annual Contractors' Review Meeting.* CONF-830974, pp. 229–233.

Michaels, A. I., S. Sillman, F. Baylin, and C. A. Bankston. 1983. *Simulation and Optimization Study of a Solar Seasonal Storage District Heating System: The Fox River Valley Case Study.* ANL-83-47. Argonne, IL: Argonne National Laboratory.

Morrison, D. J. 1980. *Development of a Gas Backup Heater for Solar Domestic Hot Water Systems, Final Report.* DOE/CS/34696-1.

Moszynski, J. R. 1979. *Development of Intermediate Temperature Thermal Storage Systems, Final Report.* COO-5760.

Nalbandian, S. J., and S. Sudar. 1979. *Hybrid Thermal Storage with Water, Final Report.* ESG-DOE-13259.

Neeper, D. A., and J. C. Hedstrom. 1985. A self-pumping vapor transport system for hybrid space heating. In *Intersol 85 Extended Abstracts*, p. 102.

Nissen, D. A., 1979. *The Physical Properties of $NH_4Cl \cdot 3NH_3$.* SAND79-8049. Albuquerque, NM: Sandia National Laboratory.

Offenhartz, P. O'D. 1981. *Thermal Storage Studies for Solar Heating and Cooling: Applications Using Chemical Heat Pumps. Final Report.* DOE/CS/30248-F.

Offenhartz, P. O'D., F. Brown, R. Mar, and R. Carling. 1980. A heat pump and thermal storage system for solar heating and cooling based on the reaction of calcium chloride and methanol vapor. *J. of Solar Energy Eng.* 102(1): 59–65.

Offenhartz, P. O'D., J. Marston, and J. Watts. 1979. *Analysis of Advanced Thermal Storage Subsystems for Solar Heating and Cooling, Final Report.* COO-4483-F.

Reilly, R. W., D. R. Brown, and H. D. Huber. 1981. The cost of heat storage from a seasonal source. In *Proc. International Conference on Seasonal Thermal Energy Storage and Compressed Air Storage*, vol. 1. CONF-811066, pp. 429–439.

Remmers, H. E., and G. L. Mills. 1979. *Rock Bed Storage with Heat Pump, Final Report.* COO-4704-3.

Salyer, I. O., D. Duvall, C. Griffen, S. Molnar, R. Chartoff, and D. Miller. 1984. Advanced phase change materials for passive solar storage applications. In *Proc. Passive and Hybrid Solar Energy Update.* DOE/CONF-8409118, pp. 61–71.

Sawada, T., J. Jackudo, and K. Iwamura. 1985. In *Intersol 85 Extended Abstracts*, p. 129.

Sillman, S. 1981. The value of storage in annual cycle solar heating systems. In *Proc. International Conference on Seasonal Thermal Energy Storage and Compressed Air Storage*, vol. 1. CONF-811066, pp. 417–424.

Sowell, E. F., and P. W. Othmer. 1979. *Series-Parallel Solar Augmented Rock Bed Heat Pump, Final Report.* COO-4697-2.

Sowell, E. F. 1979. *Series-Parallel Solar Augmented Rock Bed Heat Pump: Convolution Model of Rock Bed.* COO-4697-3.

Thornton, R. F., and C. S. Herrick. 1979. *Thermal Energy Storage Subsystem for Solar Heating and Cooling Applications: Rolling Cylinder Thermal Storage, Interim/Final Report.* ORO-5759-11.

Tsang, C. F., and C. Doughty. 1983. Numerical analysis of the Mobile, Alabama, aquifer test facility. In *Proc. DOE Physical and Chemical Energy Storage Annual Contractors' Review Meeting.* CONF-830974, pp. 234–239.

Wachtell, G. P. 1978. *Self Controlling Self Pumping Heat Circulation System Study, Final Report.* COO-4484-07.

Walton, M., and P. L. McSwiggen. 1983. Investigation of thermal energy storage and heat exchange capacity of water-filled mines—Ely, Minnesota. In *Proc. DOE Physical and Chemical Enerey Storage Annual Contractors' Review Meeting.* CONF-830974, pp. 264–270.

Ward, J. C., W. M. Loss, and G. O. Löf. 1977. *Direct Contact Liquid–Liquid Heat Exchangers for Solar Heated and Cooled Buildings: Pilot Plant Results.* COO-2867-2.

Wright, J. D. 1981. *The Design and Economics of Direct Contact Salt Hydrate Storage Systems.* SERI/TP-631-1163. Golden, CO: Solar Energy Research Institute.

Wright, J. D., and M. Bohn. 1982. Direct-contact thermal storage research. *Proc. DOE Physical and Chemical Storage Annual Contractors' Review Meeting.*

Wu, S. T., and S. M. Han. 1980. *The Numerical and Experimental Studies of Liquid Storage Tank Thermal Stratification for a Solar Energy System.* DOE/CS/34479.

Youngberg, R. J., and R. C. Bourne. 1982. *Membrane-Lined Foundation Systems for Liquid Thermal Storage.* Final report on contract DE-AC03-80CS30227.

19 Issues and Opportunities

C. J. Swet

19.1 Background and Introduction

In chapters 14 through 18 energy storage has been treated primarily as a separate entity, in consonance with the components and materials theme of this volume. These chapters dealt with collectors as more or less autonomous components within the larger solar system. They cover progress in the areas of concepts, designs, theory, modeling, testing, evaluation, research, and development of energy storage subsystems. Generally, these chapters have been neither judgmental nor speculative, and attention to economics has been limited mainly to component and subsystem costs.

Broadly stated, this chapter identifies what remains to be learned and applied about energy storage, viewing storage in an overall system context. More specifically, it identifies storage-related issues and opportunities that could have significant impact on overall system life-cycle cost (microeconomics), the solar energy industry (macroeconomics), or other important figures of merit.

Some of the issues identified here involve choices between storage subsystem technical approaches, such as latent versus sensible heat; others involve choices between ways of using or deploying storage within the system, such as hot versus cold side storage for chillers.

Some of the opportunities involve new ways of effectively exploiting storage in existing applications or system concepts, such as annual storage for active solar heating; others involve ways in which storage might enhance the attractiveness of less thoroughly explored applications, such as solar cooking.

Ideally, this chapter would cover all applications and technologies for solar thermal energy conversion in which significant storage-related issues and opportunities can be identified. Practical considerations limit it to items that, in the author's judgment, are most important or least recognized.

Most of the identified issues and opportunities are organized by application —namely buildings, agriculture and industry, electrical power generation, and cooking. Within the buildings category more specific breakdown is necessary since the appropriateness or importance of an item may depend on climate, site-peculiar conditions, and the building use, size, and design. A few items that are not readily identified by application appear under a "miscellaneous" heading.

An attempt has been made to avoid judging the relative importance or merits of competing overall solar system concepts, but occasionally storage considerations favor one concept over others. In some cases a particular storage-related approach may be said to improve greatly the competitiveness of a well-known overall system concept, but it should not be assumed that the overall concept can therefore be made cost-effective. Commonly perceived issues are mentioned briefly in a few cases for purposes of reinforcement, or are reexamined to place them in perspective.

19.2   Building Applications

19.2.1   Annual Storage for Active Space and Water Heating

Annual storage is treated at greater length than other storage issues in this chapter because of its importance and many subtleties. Annual storage is an overall system design strategy for achieving very high solar fractions. It enables year-round use of the solar collectors by storing heat from summer to winter, often being called *seasonal* or *interseasonal* storage for that reason. Sometimes it is substantially more economical than diurnal (day-to-day) storage, and in certain situations may make active solar heating competitive with conventional heat sources. However, few of the many interrelated factors that collectively determine its competitiveness are commonly recognized.

Presented here are some insights, examples, and rules of thumb intended as a guide for deciding whether to dismiss the notion of annual storage or to pursue it further. Emphasis is on competition with diurnal storage, presupposing that active solar heating can plausibly be justified on the basis of high fuel prices. Much of the illustrative material is from a SERI study (Sillman 1981a, b) that was limited to systems with water storage in constructed tanks, but the indicated trends apply generally.

Unconstrained Operation   Usually systems with annual storage are most economical when they operate in an unconstrained fashion, namely if their thermal stores are just large enough to accept all of the surplus collected heat throughout the year. The representative curves in Figure 19.1 show why the optimum is likely to be either diurnal storage or unconstrained annual storage. In the diurnal region (region *A*) the solar fraction for a collector of given area is seen to increase rapidly with increasing storage capacity (expressed in this example as water tank volume), indicating a correspondingly rapid increase in the value added by storage. In the intermediate region (region *B*) value continues to be added at a lower and nearly constant rate until unconstrained operation marks the beginning of the annual storage region (region *C*). Beyond that point a larger store provides no additional storage

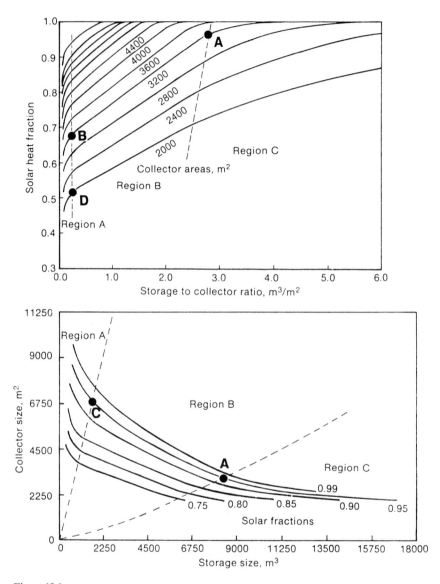

Figure 19.1
System performance and collector storage trade-off curves for a typical annual storage system
Source: Stillman (1981).

capability, although some value continues to be added at a decreasing rate due to system temperature reduction and consequently improved solar collection efficiency. Because of the nearly linear intermediate region, optimal conditions tend to be found only at its boundaries. This general pattern appears to be common to all active solar heating systems with sensible heat storage. (Note that the optimal storage/collector ratio for diurnal storage is shown as 0.25 m^3/m^2, which is typical.)

Figures of Merit   For annual storage to compete economically with diurnal storage, two criteria must be satisfied:

1. The value added to the system by annual storage must exceed the additional storage cost.

2. The selected solar fraction must be provided at lower cost by increased storage capacity rather than by increased solar collection area.

For annual storage also to be cost effective, it must cause the system's life-cycle cost to be lower than that of any available (presumably nonsolar) option, but that criterion is not examined here.

    Value added to the system is determined by the performance curve slopes in the nearly linear intermediate region, as indicated earlier. This can be illustrated by comparing an annual system (point A in figure 19.1a) with a diurnal system (point B in figure 19.1a) having the same collector area. For the total heating load (3.89 TJ annually in this example), the net added energy is

$$\frac{(3.891 \times 10^{12})J \times (0.95 - 0.66)}{(8,740 - 800)m^3} = 140\frac{MJ}{m^3}.$$

If purchased fuel costs 6¢/kWh, the allowable storage cost is

$$\frac{140\,MJ/m^3(\$0.06)}{3.6\,MJ/kWh} = \frac{\$23.40}{m^3}.$$

Satisfaction of the second criterion is determined by the collector/storage trade-off curve slopes as illustrated in figure 19.1b, which is a rearrangement of the data presented in figure 19.1a, and no cost assumptions need be made. The trade-off between design points A and C is

$$\frac{(7,000 - 3,200)\,m^2}{(8,740 - 1,750)\,m^3} = 0.56\frac{m^2}{m^3}.$$

Thus the allowable storage cost per cubic meter is 0.56 times the collector cost per square meter.

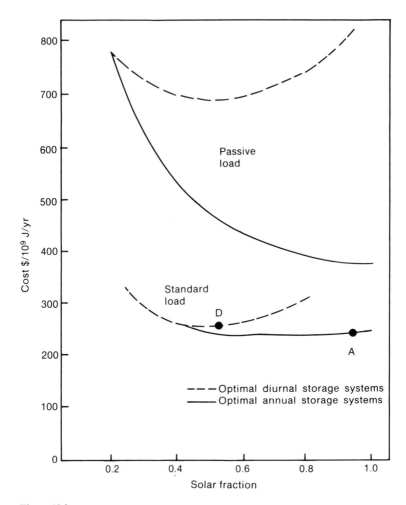

Figure 19.2
System cost per unit heat delivered versus solar fraction for space-heating systems.

Near-Unity Solar Fraction   Usually systems with annual storage are most competitive if they also provide a solar fraction of near unity (with unconstrained operation). The lower curves of figure 19.2 illustrate this for the example in figure 19.1, using estimated costs (see the figure caption) of collectors and thermal stores at the "optimal" intermediate region boundaries. Points *A* and *D* in figure 19.2 correspond to like-lettered points in figure 19.1a. The nearly flat cost curve for annual storage is typical, in shape, for active solar-heating systems. It signifies more than the indicated nearly constant return with increasing solar fraction: When attendant reductions in amounts of purchased fuel are also considered, it becomes clear that the system's life-cycle cost must be minimal at essentially unity solar fraction.

In this example annual systems happen to cost less per unit delivered energy than diurnal systems over a wide range of solar fractions, which should not be considered typical. For other examples the shape of the annual system cost curve may be somewhat different, as will be discussed later, but whenever fuel prices are high enough to favor active solar heating, then the most cost-effective system with annual storage will operate at or near 100% solar. (It should be kept in mind that "unity solar fraction" and "100% solar" are fictitious unless the stated system performance is based on a "worst" rather than a "typical" meteorological year.)

Storage Subsystem Technology   The choice of a storage subsystem design concept can profoundly affect the competitive position of a system with annual storage since differences in cost per unit capacity can be great. The choice of a basis for storage cost estimation is equally important, as will be shown. All of the commonly considered concepts (water in steel or concrete tanks, pits, caverns, or natural aquifers; earth, clay, or rock) have been characterized elsewhere in this book and need not be reviewed.

Annual storage in reversible chemical reactions (Lawrence 1980) and with phase change materials (PCM) (Kriz 1983) have been looked at cursorily but will not be considered here because of the many uncertainties. Figure 19.3 presents approximate unit costs and appropriate size ranges from an IEA study (Hadorn and Chuard 1983) to compare with the methods of cost estimation used to develop the curves in figure 19.2.

In figure 19.1 the unit cost of storage is $30.50/m^3 at point *A*, using the estimating method of figure 19.2. This is substantially higher than the allowable cost of $23.40/m^3 for the assumed unit cost of purchased fuel. In figure 19.3, a steel tank of the required size costs approximately $100/m^3, although 38% of that cost is accounted for by tubing, valves, pumps, instruments, and

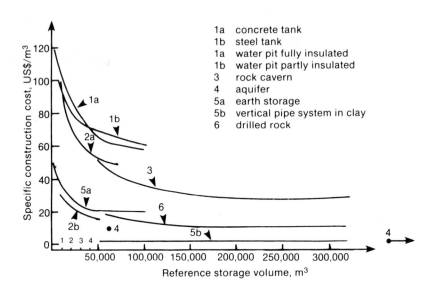

Figure 19.3
General comparison of specific costs based on a reference volume cost level and currency exchange rate as of July 1980.

controls, which presumably were included elsewhere in the SERI estimate of overall system cost.

However, figure 19.3 also shows that a partly insulated water pit (slightly larger to allow for higher losses) could be substituted for the tank at about $30/m³, or perhaps $18.60/m³ without accessories. A diurnal storage tank would, of course, be too small for such a substitution. It is important to bear in mind that concept selection for large annual stores tends to be influenced at least as much by factors such as required volume, available space, temperature swings, and local hydrogeological conditions as by subsystem cost alone.

Climate Annual storage is most competitive with diurnal storage in locations with sunny summers and cloudy winters, where the seasonal variation in insolation is greatest and the winter insolation is so low that the inherently poor winter collection efficiency of annual systems is of little consequence. Cold winters have little effect, and sunny winters are disadvantageous to annual storage because of less seasonal variation in insolation and because the poor collection efficiency has greater effect.

Climate has a pronounced effect on the collector/storage trade-off ratio, which has a breakeven (lowest allowable) value of 0.1 to 0.3 m²/m³ for the configurations and cost assumptions of Sillman (1981a, b). For the system example in figure 19.1, the ratios in four different climates were found to be 0.55 in Boston, Massachusetts; 0.77 in Bismark, North Dakota; 0.70 in Medford, Oregon; and 0.17 in Albuquerque, New Mexico By that criterion seasonal storage is favored strongly everywhere except Albuquerque.

Nature of the Thermal Load   Annual storage tends to be most competitive with diurnal storage when space heating is the only system thermal load, namely, when the water-heating load (if any) is met by a separate collector/ storage system or by nonsolar means. That is because the collectors in diurnal systems are used least effectively if there is no summer water-heating load. Annual systems become less competitive when the space- and water-heating systems are combined, although both strategies are penalized by inherently lower performance than that attainable with separately optimized systems.

The penalties depend on the relative loads, which in turn depend in a highly interactive way on climate and building characteristics. If the water-heating fraction of the total annual load is that normally encountered in a residence of standard construction in a northern climate, the penalty suffered by annual systems is not much greater than for diurnal systems, and it can be decreased by using a "dual-tank" store, as will be shown later. Even with fractions up to one-half, the use of two tanks generally can keep the decrease in net added energy (lowered allowable storage cost) to a few percent.

If water heating is the only active solar load, there is little incentive to consider annual storage. Seasonal load variations then tend to be small, and the benefit from interseasonal transfer of surplus heat collected during periods of relatively high insolation is largely offset by thermal losses from the store, which are larger than for space heating because of the higher delivery temperature. Actually there is little incentive in this situation to strive for high annual solar fractions by any means; typically the most economical water heating systems meet only about half the load in winter but nearly all of it in summer.

Building Type and Size   Single-family dwellings can be the best prospects for annual storage, depending on whether they are heated individually or communally in a district. If individually solar heated, their thermal stores are too small to benefit greatly from cost economies of scale, and much too small to exploit any of the lower unit cost storage subsystem design concepts indicated in figure 19.3. The main drawback, though, is excessive heat loss due to high surface/volume ratios; their allowable unit cost of storage is typically 30% to 50% lower than that of centrally heated districts of many detached houses.

Apartment buildings tend to be good prospects because of their larger sizes, even though there is some penalty due to their typically high water-heating fraction. Commercial buildings seldom benefit from annual storage because daytime-only occupancy minimizes the need for storage and high internal gains often make space heating a secondary economic consideration.

Building Design   In buildings with substantial passive solar contributions or extremely energy-conserving designs ("superinsulated"), the cost-effectiveness of any kind of active solar heating is less than in buildings with more conventional designs. For space heating only (see the upper curves in figure 19.2) the disadvantage tends to be especially severe for diurnal systems because on sunny winter days there is little need for active solar heat and the collectors are poorly utilized; for annual systems the disadvantage becomes much less at high solar fractions since the larger store damps out daily load fluctuations.

With combined loads (see figure 19.4) there is some active load every day, so both the absolute and relative disadvantages of annual storage are less. The spread in allowable unit cost of annual storage attributable to differences in building design seldom exceeds 20% for large systems.

Collector Type   Using evacuated tube collectors or other high efficiency fixed collectors instead of flat plate collectors improves the performance of annual systems more than it does that of diurnal systems, especially in locations with sunny winters. For the example of figure 19.1, this would increase the allowable unit cost of annual storage from $23.40 to $30 while reducing collector and storage sizes by about 16% and 4%, respectively. However, it can seldom be economically justified because of the much higher collector unit cost.

Dual Tank Systems   Both the cost-effectiveness of annual systems and their competitiveness with diurnal systems can nearly always be improved substantially by dividing the total storage capacity into two thermal stores: one large, and the other only 3% to 5% of the total. This technique has been examined most thoroughly for water tank storage (Sillman 1981a; Cha et al. 1979) but, in principle, it could apply to nearly any containment method or sensible-heat storage medium.

The smaller tank functions more or less as a diurnal store, preferentially accepting higher temperature collector outputs (especially in the winter) and storing them to meet relatively immediate space- or water-heating needs. The benefits are greatest where there are cloudy winters (which tend to favor annual storage anyway) and for combined loads (to offset the penalty thereof). For the example of figure 19.1, the benefits would be approximately as tabulated in table 19.1

Stratification   In diurnal stores that use water or other liquids as the storage medium, the benefits of temperature stratification (thermoclines) are clear, as discussed toward the end of this chapter. In the case of annual stores, however,

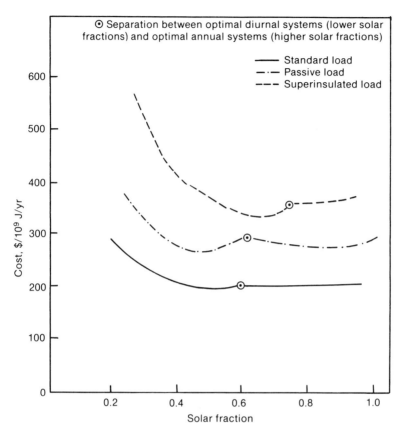

Figure 19.4
System cost per unit heat delivered versus solar fraction for combined space heat and hot water systems.

Table 19.1
Benefits of two-tank annual storage systems

		Collector (m^2)	Storage (m^3)	Collector/storage trade-off	Energy added by storage (MJ/m^3)
Space heat only	1 tank	3,120	8,740	0.55	140
	2 tanks	2,800	8,000	0.53	175
Combined loads	1 tank	3,830	8,880	0.47	124
(HW = 0.184 total)	2 tanks	3,550	7,840	0.55	170

there may not always be a net benefit (Hadorn and Chuard 1983). Although it is always advantageous to remove cooler bottom water to the collectors and to deliver hotter top water to the load, in some situations large thermal losses from the upper portion may outweigh that benefit. This applies especially to water pits with relatively large and poorly insulated tops (Hansen 1981). A fairly modest study should resolve this question.

Design Methods A simplified model of annual storage systems has been developed that uses bimonthly steps and a utilizability formula for collection efficiency (Baylin and Sillman 1980). It yields accurate results for single tank systems with all building types, and for dual tank systems with all but the most energy-conserving buildings. Another simple method for sizing systems near the point of unconstrained operation is also based on a utilizability method for calculating collector efficiency, with storage temperature assumed to follow a sinusoidal pattern over the year (Drew and Selvage 1980). Yet another method based on utilizability uses monthly steps and allows for a finite load heat exchanger but does not apply to two-tank systems (Braun et al. 1981). It has been incorporated into f-Chart 4.1 and is readily available. With these "tools and cost data," the feasibility of annual storage in a given situation can readily be assessed.

19.2.2 Latent-Heat Storage for Residential Active Space Heating

Despite lingering contention the question of latent- versus sensible-heat storage is virtually a nonissue in terms of likely impact on this market. It is mainly a question of how well phase change storage can compete with sensible-heat storage in water since the market for air-based systems in which it might compete with rock beds is small. Situations in which phase change storage has any real prospect of being more economical than low-cost water tanks seem to be too few to justify much further study or development.

System performance is not significantly improved by the nearly isothermal storage characteristics of phase change materials. Simulations at the University of Wisconsin (Morrison and Abdel-Khalik 1978) showed that on winter days the collectors typically deliver too little heat to storage to melt much of the material; in spring and fall they tend to keep storage so fully charged that the material remains melted and only its sensible heat is used. The only real advantage was found to be a storage volume about one-half that of water for equivalent performance.

Consistent results were obtained for Madison, Wisconsin, and Albuquerque, New Mexico, under assumed conditions most favorable to latent-heat storage: no supercooling, no void volume, no additives, lowest plausible transition temperature, and unity effectiveness heat exchange. Similar results were obtained at SERI (Kriz 1983) for Madison and Ft. Dodge, Kansas, and by Brown-Boveri (Ziegenbein 1979) for Heidelberg, Federal Republic of Germany. The Brown-Boveri simulation was validated experimentally.

In the sizes of interest, the toughest competition probably will be site-assembled, membrane-lined water tanks, which have estimated installed costs (see figure 19.5) of only $75/m^3 (1979$) for a volume of 6 m^3 (Bourne 1981). It is unlikely that the unit cost of a latent-heat storage vessel would be lower, so optimistically, assuming no voids, the breakeven cost for the contents of a 3-m^3 latent-heat vessel would be, at most, 7.5¢/L. This is an implausibly low unit cost for any known or contemplated phase change system (i.e., the material, plus its additives and encapsulation or heat exchanger).

By far the cheapest phase change system presently known is Glaubers salt in plastic "chubs," at about 31¢/L (1978$) (Lawrence 1980) which melts at too low a temperature for conventional heat delivery systems. The cheapest such system with an acceptably high melting point would cost about 60¢/L. With Glaubers salt in chubs, a 3-m^3 latent-heat storage unit would cost $700 more than a 6-m^3 water tank, making the break-even cost of occupied floor space at least $470/m^2. Even if the phase change material were free, the cost of its stabilization and encapsulation or heat exchanger would make the break-even floor space value greater than that of most basements.

19.2.3 Latent-Heat Storage in Passive Solar Architecture

Many opportunities in passive solar architecture have been identified (Swet 1980) and a number of products are available commercially. There are PCM-filled plastic panels designed for placement in wood-frame south walls to absorb solar radiation. There are also plastic tubes and metal cans intended either for stacking in south walls or for storing convectively transferred room heat. The potential benefits of incorporating phase change materials in structural components such as concrete blocks are widely recognized, in a general way, and need no emphasis here, but relatively little has been accomplished in this specific area, either in product development or in detailed applications studies.

19.2.4 Storage for Residential Solar-Assisted Heat Pumps

This subsection revives an issue that had widely been judged moribund in the late 1970s, after detailed performance and cost studies indicated that solar-assisted electric heat pumps are unlikely to ever compete economically with stand-alone air-to-air heat pump systems. A critical examination of the modeled configurations and economic assumptions suggests that a more imagina-

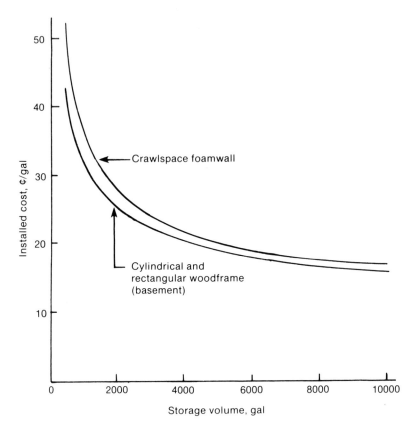

Figure 19.5
Enclosure cost versus volume for membrane-lined water tanks.

tive reevaluation, especially of storage-related aspects, may greatly reduce the perceived gap in cost-effectiveness.

Simulations at the University of Wisconsin in 1976 (Freeman et al. 1979) compared the performance of series systems ("true" solar-assisted heat pumps, in which stored solar heat feeds the heat pump), parallel systems (active solar heating with electric heat pump backup), solar-only systems, and stand-alone air-to-air heat pumps. For heating they showed that parallel systems used the least amount of purchased energy, followed by series, solar-only, and stand-alone in that order, regardless of storage capacity, climate, or heat pump coefficient of performance (COP). Series performed better than parallel for cooling, and on an annual basis, the two were found to perform comparably. It was tentatively concluded that the extra complexity of series ("true") solar-assisted heat pumps could not be justified.

A study by SAI in 1979 (Hughes and Morehouse 1979) indicated that the year-round performance of series and parallel systems would be comparable and substantially better than that of stand-alone systems, based on anticipated improvements in heat pump technology. It also indicated that life-cycle costs of series and parallel systems would be comparable, but about two and one-half times that of a stand-alone system, regardless of climate and parametric variations. With caveats regarding unexplored opportunities, the study recommended that neither parallel nor series systems should continue to be developed unless justified by noneconomic factors.

It was assumed in the SAI study that thermal storage would be in buried concrete water tanks of $10-14$ m^3, costing \$311/m^2, installed. It also was assumed that storage would be isolated by a heat exchanger from an anti-Freeze collector fluid. By assuming a membrane-lined water tank in the basement instead, costing only \$55 to \$61/m^3, and a drainback collector system, the capital cost could be reduced by about 30% without considering reductions in collector size made possible by removal of the heat exchanger.

Yearly maintenance and repair costs were assumed to be 5% of the total system capital cost, which is realistic for electric heat pumps but probably too high for the "static" and much more costly solar and storage elements. Adjustment of this assumption might further reduce the cost-effectiveness gap by as much as 72%. Using latent- rather than sensible-heat storage probably would increase the storage subsystem cost but would have a more pronounced effect on system performance than in solar-only space-heating applications, with a probable net reduction in system cost.

In addition to such straightforward modifications, there are clearly productive opportunities for exploiting storage in heat pump systems with refrigerant-filled collectors, such as those developed by Charters et al. in

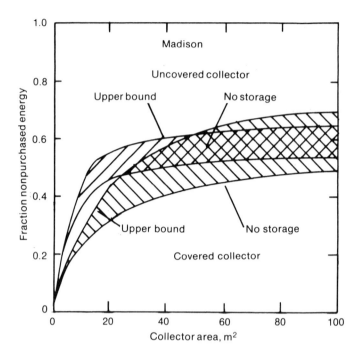

Figure 19.6
Effects of storage on performance of solar-assisted heat pump with refrigerant-filled collector.

Australia (1981). Figure 19.6 plots the potential performance benefits of thermal storage on the hot (condenser) side of such a system (O'Dell et al. 1984), some or all of which could be in the form of building thermal capacitance. With latent-heat storage the indicated upper bounds for covered and uncovered collectors could be approached. There also may be intriguing opportunities for providing storage on the cold (evaporator/collector) side, such as by integrating the evaporator coils into a water wall or a masonry storage wall.

19.2.5 Storage for LiBr Absorption Chillers with Active Solar Heating

In situations where space cooling is the dominant load, the combination of a conventional water/LiBr (lithium bromide) absorption chiller with active solar heating has the best chance of being competitive. It involves two inter-related storage issues: hot-side-only storage versus combined hot-side and cold-side, and latent-heat versus sensible-heat storage.

These issues have often been addressed, but usually in fragmented fashion and with inconsistent results (Hedden and Cassel 1979; Duff et al. 1978; Ward et al. 1979). Contrasting conclusions have even been drawn from the same operating experience with the same system (Duff et al. 1978; Ward et al. 1979). They were examined methodically and fairly thoroughly in a comparative cost and performance study by SAI (Hughes et al. 1981) which is the focus of this discussion.

The SAI study modeled conventional solar absorption cooling systems in single-family dwellings and commercial buildings in Fort Worth, Texas, Miami, Florida, Phoenix, Arizona, and Washington, D.C. It indicated that the choice of storage option can cause variations in annual system performance (heating plus cooling) of 15% and up to 20% in life-cycle cost, but that the life-cycle cost was generally more than double that of nonsolar alternatives. Cold-side latent-heat storage appeared most promising for commercial buildings in all four locations, and for single-family dwellings in the more strongly cooling-dominated locations, although there were some unexplained anomalies.

These results alone do not provide a strong incentive for further development of latent-heat coolness storage concepts, but the study had some serious shortcomings which, if corrected, would strengthen the argument for further development and might substantially reduce the apparent gap in cost-effectiveness. Some of these shortcomings and their possible implications are discussed here.

The modeled commercial building was for daytime-only occupancy, and its heating load was ignored because of high internal gains. If an apartment building had been modeled instead, there would have been a greater need for storage, a more significant heating load, and more use of the solar equipment.

Latent-heat storage for the commercial building was modeled by multiple (up to six) standard Calmac units, thereby unrealistically increasing cost, complexity, and preempted floor space (to which no value was assigned). Other configurational options, not yet commercially available, may be substantially more economical, especially on the cold side. An example would be the use of a single, large membrane-lined tank containing chubs in a water bath, as suggested for somewhat higher temperature latent-heat storage earlier in this chapter. Another would be the use of chubs stacked in an air duct enlargement, as has been investigated experimentally for coolness storage with electric air conditioners (Rizzuto 1980). Examination of such alternatives may suggest that emphasis should be placed on the development of latent-heat coolness storage concepts that are adaptable to a variety of sizes and geometries.

An antifreeze fluid collector loop isolated from storage by a heat exchanger is modeled in all cases, even though freezing is unlikely to be a severe problem in some of the examined locations and can be avoided by drainback. As discussed previously, that imposes a cost and performance penalty; in this case its elimination may also permit the use of flat plate collectors instead of the more expensive evacuated tube collectors that were modeled.

19.2.6 Storage for Rankine Chillers with Active Solar Heating

This subsection applies to situations in which space cooling is the dominant load and the building is large enough to justify the use of turbine-driven machinery. As discussed previously, it involves issues of hot- and cold-side storage and of latent- and sensible-heat storage, deriving from another aspect of the same SAI study (Hughes et al. 1981) that assessed storage options for LiBr systems.

The study concluded that variations in performance and life-cycle cost of the modeled Rankine system are insignificantly small but that cold-side storage of some type may be needed because of inherent inabilities of Rankine systems to modulate capacity and to cycle efficiently. Only a commercial building was examined. As with absorption systems the life-cycle cost was found to be much greater than that of nonsolar alternatives.

The study shortcomings identified in the previous section apply here also, with the same implications. However, three additional considerations that apply only to the Rankine aspects of the study should be highlighted.

Backup power for cooling (heating was ignored) was properly assumed to be an electric motor, but no consideration was given to the possibility of using cold-side storage for electrical demand reduction or to exploit time of day pricing. Had heating not been ignored, consideration could also have been given to use of either the hot- or cold-side store for the same purpose in the heating mode. Storage could also be used for these purposes with nonsolar electric air conditioners or heat pumps, of course, but it seems clear that having both sources of power could more frequently avoid on-peak electric power usage.

More significant, perhaps, is the fact that the performance model did not allow the boiler firing water (storage outlet) temperature to exceed 210°F (99°C). Somewhat higher temperatures probably would increase the Rankine cycle efficiency more than they would degrade collector efficiency, with a net gain in overall performance at increased cost. Greatly higher temperatures might justify the use of more efficient, but perhaps only slightly more expensive, concentrating collectors, which can reasonably be considered because of the relatively large building. Evaluation of such trade-offs may be especially rewarding if the implications of shifting all storage to the cold side are examined as well.

If it is feasible to delete hot-side storage, it may be worthwhile to consider refrigerant evaporation in the collector itself, thereby eliminating the refrigerant "boiler" and the temperature drop in its heat exchanger. This may not be as drastic or uncertain a departure from usual practice as it may seem at first, since collectors of this type are commonly used in solar water-heating systems and are being developed for solar-assisted heat pumps.

19.2.7   Storage for Zeopower-Type Sorption Chillers with Active Solar Heating

Latent-heat coolness storage with solid/gas sorption chillers such as the roof-mounted water/zeolite system developed by Zeopower (Hughes et al. 1981) is most suitable for situations in which cooling is the dominant load. However, the thoughts expressed in this subsection can be extended to apply to any intermittent solid/gas sorption heat pump that has a cooling function.

The Zeopower system operates on a diurnal cycle as illustrated in figure 19.7 In the cooling season it heats water during the day and removes heat from the building at night, requiring coolness storage for daytime air-conditioning. In the heating season it provides space and water heating during the day and requires thermal storage for space heating at night. In current designs a single water tank is used for storing either heat or coolness. In its present configuration the system performance is better than that of a LiBr chiller, and the projected costs in volume production are claimed to be lower.

A cursory study at SERI (Kriz et al. 1983) suggests that the heat and coolness might be stored more advantageously in a phase change storage unit than in the water tank, with an increase in system performance of 10% and reductions of 90% in storage volume, 65% in storage cost, and 12% in the annual cost of delivered energy. Since heating loads are relatively small, the sensible heat in the phase change material and its liquid bath (if chubs or other capsules are used with water) will suffice ordinarily.

The reported reduction in storage cost is probably unrealistic, since high unit costs for tanks were assumed. However, the extremely large volume reduction of about 7 m^3 would convert to a savings of perhaps 3 m^2 in preempted floor space, the cost of which was not estimated but may be quite substantial if storage is indoors since houses in cooling-dominated regions tend not to have basements with cheap floor space. If the generic system concept is as promising as it seems to be, this is another strong incentive to pursue vigorously the development of latent-heat coolness storage concepts.

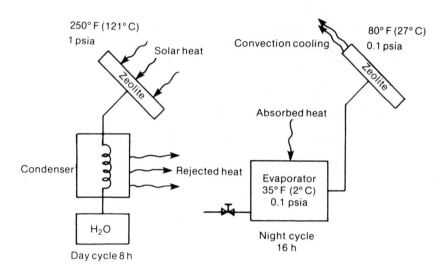

(a) Day and night cycles for cooling

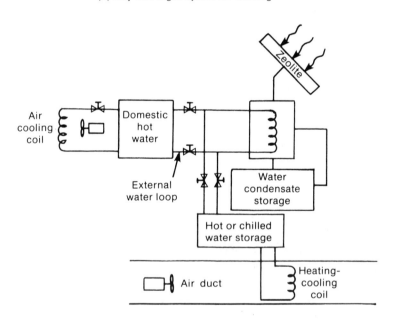

(b) Combined heating/cooling system

Figure 19.7
Water/zeolite system.

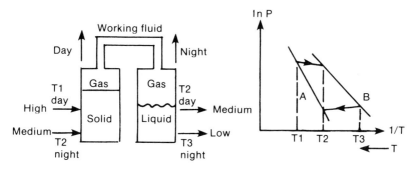

Figure 19.8
Intermittent chemical heat pump with solid absorbent.

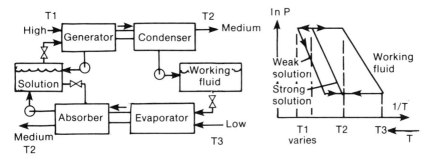

Figure 19.9
Continuous chemical heat pump with internal storage.

19.2.8 Sorption Heat Pumps with Integral Storage for Active Heating and Cooling

In situations where space heating is the dominant load, closed-cycle sorption heat pumps with integral storage (sometimes called chemical heat pumps or thermochemical heat pumps) are most likely to compete with other solar options. The technical feasibility of this generic concept has been established experimentally (Offenhartz n.d.; Clark 1979), and performance simulations (McLinden and Klein 1981) indicate substantial performance advantages over conventional water/LiBr chillers with separate thermal storage and active solar heating. A comparative cost-effectiveness evaluation (Gorman et al. 1981) suggests that such systems may be competitive with most nonsolar options in residential applications, yet DOE-sponsored development in this partially explored area effectively ceased in 1980.

Figures 19.8 and 19.9 depict, rather simplistically, the two basic technical approaches: (1) intermittent systems, which may have either solid or liquid

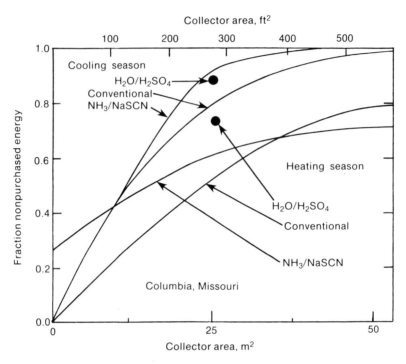

Figure 19.10
Seasonal performance comparisons of $NH_3/NaSCH$ and H_2O/H_2SO_4 sorption heat pumps with conventional $H_2O/LiBr$ chiller.

*ab*sorbents or solid *ad*sorbents (e.g., zeolite), and (2) continuous systems, which always have liquid absorbents. Figure 19.10 compares the simulated heating and cooling performance of a representative continuous system using two different working fluid/absorbent pairs with that of a water/LiBr chiller with active solar heating and separate storage (McLinden and Klein 1981). These simulations were done using ambient air as the low temperature source/sink; with groundwater, the superiority of sorption heat pumps in the heating mode would be more pronounced still.

Other choices of fluid pairs, chemical inventories, heat exchanger effectiveness, and control strategies may further affect relative performance (up or down) by a number of percentage points. A lot remains to be learned about the benefits of configurational options such as multistaging (for larger systems) or the substitution of latent heat coolness storage for multiple solid absorbent beds in intermittent systems for cooling. Only through more exploration, both analytical and experimental, can the true potential of this promising approach be ascertained.

19.2.9   Seasonal Storage of Environmental Coolness

Seasonal storage of environmental coolness involves chilling or freezing a fluid, usually water, by direct or indirect exposure to the winter ambient environment, and its interseasonal storage for space cooling in the summer. It is not in itself a solar area, and the notion is being actively pursued. It is mentioned briefly here to highlight the fact that there are opportunities for coupling it in the same system with seasonal storage of solar heat, especially in natural aquifers (Davidson et al. 1975).

19.3   Agricultural and Industrial Applications

19.3.1   Annual Storage for Crop Drying

It is widely recognized that solar drying of field crops can seldom be cost-effective using only the solar radiation available during the typically short drying season. Annual storage of heat in salt-gradient solar ponds is a well-known approach to the solution of this problem and need not be emphasized here. Often, though, there are situations that call for other approaches: The soil may be too moist for good heat retention, or land area may be more valuable for other purposes. In such cases it is worth considering annual storage schemes involving the use of desiccants that can be more compact and do not require storage at elevated temperatures.

An experimentally verified scheme for desiccant annual storage (Fletcher 1979) is shown schematically in figure 19.11. Its required solar collection area is about 10% of that which would be needed without annual storage. Waste heat is available during nondrying periods for space heating or other uses. With maximum $CaCl_2$ concentrations below 50% and typical bulk salt prices, the cost-effectiveness of this specific scheme may be marginal, but many potentially rewarding variations of this generic approach remain virtually unexplored.

There may be inexpensive ways of inhibiting crystallization of the calcium chloride, thereby permitting higher concentrations and reducing the amount required (salt is the dominant cost item). It may be discovered that a substantially cheaper desiccant, such as sulfuric acid, which also has a higher energy storage potential, may be preferable despite its greater hazard and corrosivity. In large vessels the exposed containment surface per unit volume is relatively small, and the need for pumping can be avoided. Ways may also be discovered to eliminate the separate collector without reintroducing excessive moisture during regeneration.

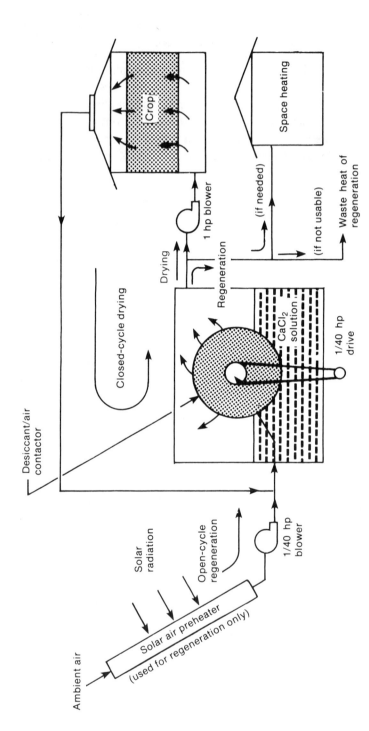

Figure 19.11
Simplified schematic of desiccant crop-drying system.

19.3.2   Annual Storage for Irrigation Pumping

Solar thermal irrigation pumping systems for the sunbelt in the United States or for landholdings of more than a few hectares in the Third World are commonly visualized as consisting of many medium temperature trough collectors, a thermal store of several hours of capacity, an organic Rankine cycle heat engine, possibly an electric generator and motor, and a pump. It is generally recognized that such systems can seldom be cost-effective since the pump is idle much of the year and the availability of profitable on-site uses of the collector output during nonpumping seasons is risky.

Annual storage in solar ponds, with low temperature organic Rankine power generation, is a fairly thoroughly explored alternative that need not be emphasized here. It is not always appropriate, though, as pointed out in the preceding section. A potentially attractive alternative also involves the collection and storage of solar thermal energy throughout the year for use during a short irrigation season, but its feasibility does not depend on low latitudes, favorable subsurface characteristics, or the absence of strongly competing land uses.

In this alternative approach a much smaller number of high temperature point-focusing collectors (perhaps only one) replaces the many medium temperature troughs. They (or it) charge a thermochemical storage device that stores the thermal energy at ambient temperature as potential heat of chemical recombination, then releases it at the required temperature when needed for pumping. The water/CaO reaction is one of several that appear suitable for this application and have been thoroughly studied (Rockwell International 1980; Wyman 1979). Its material cost per unit of stored energy is less than 14¢/kWh(t), even with allowance for unrecovered sensible heat and possibly lost heat of condensation, but the poor thermal conductivity and high charging temperature (about 500°C) require large and expensive heat exchangers.

The objective is to make the additional cost of annual storage less than the savings in collector cost, as illustrated in the simplified comparison of table 19.2.

Costs have been estimated for a diurnal water/CaO storage system of hundreds of megawatt-hours of capacity (Gilbert Associates, Inc. 1979). Energy-related costs (media and containment) were \$1/kWh and power-related costs (heat exchanger, blower, etc.) were \$100–\$150/kW. Based on these estimates, the total cost of a 720-kWh annual storage system would not exceed \$1.21/kWh, or about one-third of the allowable cost.

Table 19.2
Comparison of solar irrigation pumping systems with diurnal and annual storage

	One-dish (annual)	Many troughs (diurnal)
Available solar energy (kWh/m²-yr)	2,000	2,000
Fraction intercepted	1 (2 axis)	0.9 (1 axis)
Collector efficiency	0.8 (dish at 550°C)	0.6 (trough at 300°C)
Thermal transport efficiency	0.9 (short)	0.8 (long)
Storage efficiency	0.5	0.9
Pumping season, months	2	2
Pump utilization during season	1 (continuous)	0.5 (12 h/day)
Collector utilization	1	0.167 (2/12)
Collector area (from above, relative)	1	11.1
Energy to engine (kWh/m²-yr)	720	65
Collector array cost ($/m²)	200	200 (with diurnal storage)
($/kWh)	0.28	2.80
Allowable annual storage cost ($/kWh)	2.52	

19.3.3 Sorption Heat Pumps with Integral Storage for Solar Process Heat

Sorption heat pumps with integral storage can be used for solar process heat supply in industrial applications much as they can for solar space heating. Although it is clearly a better match of concept with application (Swet 1981), this notion does not appear to have been explored systematically.

Industrial processes commonly shut down on weekends and holidays and operate for two or more shifts per day, which permits full exploitation of storage. The choice of chemical combinations is less constrained than in residential applications, and the typically greater loads better justify complexities such as multistaging for higher performance. If ambient air is the low grade heat source, then the annual COP tends to be higher since many plants have twelve-month operating years.

In much of the sunbelt in the United States, water can be the working fluid (and one of the storage media) without freeze protection. The higher delivery temperatures typically call for high performance collectors with flatter temperature-efficiency curves than those commonly used for space heating, thereby minimizing the overall thermal performance penalty imposed by the

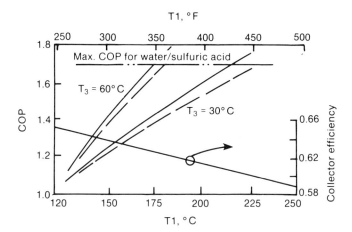

T1 = solar collector outlet temperature
T2 = process heat delivery temperature = 120°C
T3 = waste stream temperature, °C
Solid lines represent heat pump COP.
Broken lines represent overall system COP adjusted for collector efficiency degradation with increasing temperature.
No allowance made for heat exchanger temperature drops, parasitic power, or losses from storage.

Figure 19.12
Theoretical performance of a continuous chemical heat pump for solar process heat supply.

need for collector temperatures higher than those required by the industrial process. Perhaps most important, waste heat streams are often available, either to serve as the low grade heat source or to increase its temperature.

These last two points are highlighted in figure 19.12 which sets bounds on the potentially achievable COP of a single-stage system for low pressure steam delivery. A single-stage cycle such as that shown in figure 19.8 may be suitable in such an application, perhaps using water and sulfuric acid, although other configurations or chemical combinations may be superior. A prototype water/sulfuric acid heat pump has been developed for industrial waste heat upgrading (nonsolar, without storage), which could provide a reliable basis for cost and performance estimation in solar process heat applications (Rocket Research Co. 1982) using cost/value criteria such as those developed at SERI for more conventional types of solar process heat storage concepts (Kriz et al. 1983)

19.4   Electric Power Generation Applications

19.4.1   Very High Temperature Storage in Molten Salts

For solar Brayton power systems with direct absorption receivers to become feasible, means must be developed to store molten salts at a temperature of about 1,100°C and to transfer their sensible heat by direct contact to the Brayton working fluid. Efforts are underway in both of these areas (Copeland and Coyle 1983; Bohn 1983) as well as on receiver concepts, and projected cost/value figures on the storage aspects appear encouraging.

These capabilities may also be valuable for solar hydrogen production and other high temperature processes (Copeland and Coyle 1983). The present work on direct contact heat exchange is basic and has many applications; its continuation should not hinge on justification of the high temperature molten salt endeavor. At present, detailed justification data for the direct absorption receiver and high temperature molten salt efforts do not appear to be readily available.

19.5   Cooking Applications

19.5.1   Cooking with Stored Solar Heat

Solar-cooking devices of many kinds are available commercially in developing countries at a low price. In most regions, however, they have not gained wide acceptance because the cooking cannot be done indoors, in the evening or early morning (when meals are commonly prepared), or on cloudy days.

Indoor cooking requires transport of heat by some means from an outdoor source but is not by itself a sufficient inducement. Cooking in the evening or early morning requires thermal storage of short duration (less than one day), along with some kind of transport. Cooking on cloudy days with solar heat from previous days is a less absolute requirement since backup fuel can usually be found and cannot readily be met by sensible-heat or latent-heat storage. Concepts have been proposed that address these problems, but they require development and new approaches.

Descriptions and assessments of many currently used and proposed systems for solar cooking have been compiled (Alward 1982). Some of the described systems involve transport of heat from an outdoor collector to an indoor cooking device while the sun is shining, but few have progressed beyond the conceptual design stage. Some involve solar preheating of the food along with a sensible-heat storage mass such as rock, then hand-carrying both to an indoor insulated "haybox" where cooking continues without additional external heat.

One partially developed concept uses vapor transport at about 300°C in a heat pipe from a trough collector to an indoor cooking unit with a phase change storage mass (Swet 1973). Another (see figure 19.13) provides indefinitely long thermochemical storage and offers the option of using a single collector for many families (Hall et al. 1978).

19.6 Miscellaneous

19.6.1 Thermal Energy Transport

Although the transport of heat to and from storage may not be strictly a storage topic, the two are closely coupled and failure to consider them jointly may prevent an optimal choice of storage concept. In large fields of numerous collectors, the collector-storage piping can account for a large fraction of the storage volume, parasitic power, and thermal losses. Where the thermal load is distant (perhaps kilometers in a large industrial complex) from the collector-storage system, the need for a common approach is equally important. In some cases such considerations may justify investigation of thermochemical storage/transport concepts (Nix 1983).

19.6.2 State of Charge Monitoring

A need exists for inexpensive means of determining the state of charge of latent-heat storage units, where temperature may not be an accurate or reliable indication. It appears to be most challenging when the PCM is encapsulated, making it difficult to sense the condition of individual capsules; it is perhaps most acute if the backup charging source is electrical resistance heat with time-of-day pricing, where storage is also used for load management. Regardless of its value to marketed systems, there is a need for such a capability in developmental and evaluation testing.

19.6.3 Media Disposal, Replacement, and Hazard Reduction

Although these items demand attention for economic, legal, and moral reasons, frequently only lip service is granted them until uncomfortably late stages of design. They are widely recognized, and adequate technical and economic approaches exist; this brief mention is to emphasize the need for addressing them thoroughly and realistically early in the design process.

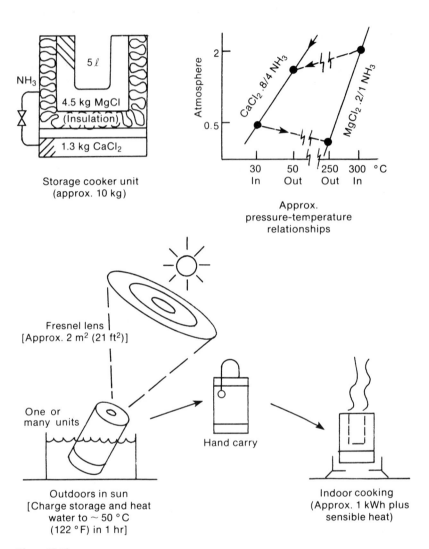

Figure 19.13
Representative chemical heat pump for cooking with stored solar heat.

19.6.4 Stratification

This subject divides conveniently into three aspects:

1. description and understanding of stratification in quiescent tanks,
2. assessment of the benefits of stratification,
3. stratification enhancement by operational techniques.

All of these aspects have been or are being intensively investigated, as reported in earlier chapters. However, the basic hydrodynamics remain incompletely understood, largely because federal sponsorship of appropriate computer codes was terminated before they could be fully validated. The modest additional effort required to close this loop would enable designers to predict the effects of parametric variations quickly and with much greater confidence. Perhaps equally critical is the problem of technology transfer regarding the third aspect. For example, it has been known at least since the mid-1970s that, for residential heating systems, the flow rate between storage and the collector (or heat exchanger) should be only one volume change per day, rather than the customary five or more. This reduces pumping energy requirements significantly as well as enhancing stratification. But the inferior practice continues, presumably because these available valuable guides have not come to the attention of enough system designers and developers.

19.6.5 Solar Ponds

Research and development on solar ponds has been covered in part I of this volume, under the heading Solar Collectors. Some applications of solar ponds were mentioned briefly earlier in this chapter. Solar ponds were not considered in the section on annual storage for space and water heating mainly because their selection is governed by factors other than those that apply to the more commonly considered storage concepts. This brief mention is to emphasize that solar ponds, whether thought of as collectors or stores, are viable candidates in a wide variety of low temperature applications (Leboeuf 1981).

References

Alward, R. 1982. *Solar Cooker Manual.* Technical Report T138. Brace Research Inst.

Baylin, F., and S. Sillman. 1980. *Systems Analysis Techniques for Annual Cycle Thermal Energy Storage Systems.* SERI/RR-721-676.

Bohn, M. 1983. Experiment and economic analysis of an air/molten salt direct/contact heat exchanger. *Proc. DOE Thermal and Chemical Energy Storage Annual Contractors' Review Meeting.* CONF-830974.

Bourne, R. 1981. *Membrane-Lined Foundations for Liquid Thermal Storage.* DOE/ET/20111-1.

Braun, J. E., et al. 1981. Seasonal storage of energy in solar heating. *Solar Energy* 26, 5: 403–411.

Cha, B., D. Conner, and R. Mueller. 1979. A two-tank seasonal storage concept for solar space heating of buildings. *Proc. Second Miami International Conference on Alternative Energy Sources,* Miami, FL, December 1979.

Charters, W., C. Dixon, and L. Taylor. 1981. The University of Melbourne Solar Boosted Heat Pump Programme. *Proc. Solar World Forum,* vol. 1. Pergamon Press.

Clark, E. C. 1979. *Sulfuric Acid and Water Chemical Heat Pump/Chemical Energy Storage Progam, Phase II-1 Final Report.* RRC-79-R-627. Rocket Research Co. Prepared for Sandia Livermore.

Copeland, R., and R. Coyle. 1983. Advanced high-temperature molten-salt research. *Proc. DOE Physical and Chemcial Energy Storage Annual Contractors' Review Meeting.* CONF-830974.

Davison, R., W. Harris, and J. Martin. 1975. Storing sunlight underground. *Chemtech* 5 (December): 736–741.

Drew, M., and R. Selvage. 1980. Sizing procedure and economic optimization methodology for seasonal storage solar systems. *Solar Energy* 25: 80–91.

Duff, W., and J. Leflar. 1978. Simulation and design of evacuated tubular solr residential air conditioning systems and comparison with actual performance. *Proc. International Solar Energy Conference,* New Delhi.

Fletcher, J. 1979. Annual cycle solar regeneration of desiccants for crop drying. Presented at USDA *Conference on Use of Solar Energy for Crop Drying.* Georgia Institute of Technology.

Freeman, T., et al. 1979. Performance of combined solar-heat pump systems. *Solar Energy* 22, 2: 125–135.

Gilbert Associates, Inc. 1979. *An Assessment of the Use of Chemical Reaction Systems in Electric Utility Applications.* EPRI EM-1094.

Gorman, R., P. Moritz, T. O'Gorman, and W. Standley. 1981. *Chemical Heat Pump Cost Effectiveness Evaluation.* Final report, BNL 51484. TRW.

Hadorn, J.-C., and P. Chuard. 1983. *Central Solar Heating Plants with Seasonal Storage—Cost Data and Cost Equations for Heat Storage Concepts.* IEA Solar Heating and Cooling Programme Task VII.

Hall, C., C. J. Swet, and L. Temanson. 1978. Cooking with stored solar heat. *Proc. International Solar Energy Conference,* New Delhi.

Hansen, P. 1981. Analytical description of the heat losses from underground thermal seasonal stores. *Proc. International Conference on Seasonal Thermal Energy Storage and Compressed Air Energy Storage,* vol. 1. CONF-811066.

Hedden, R., and D. Cassel. 1979. Economics of hot versus cold water storage for solar cooling. *Proc. Solar Energy Storage Options,* vol. 1, pt. 1. CONF-790328-P1.

Hughes, P., and J. Morehouse. 1979. *Comparison of Solar Heat Pump Systems to Conventional Methods for Residential Heating, Cooling, and Water Heating.* SAI report no. 80-906-WA.

Hughes, P. J., et al. 1981. *Evaluation of Thermal Storage Concepts for Solar Cooling Applications.* SERI/TR-09083-1. SAI.

Kriz, T., C. Christensen, H. Gaul, J. Leech, A. Rabl, S. Sillman, C. J. Swet, and J. Ullman. 1983. *Thermal Energy Storage for Process Heat and Building Applications.* SERI/TR-231-1780.

Lawrence, W. 1980. *Capital Cost Estimates of Selected Advanced Thermal Energy Storage Technologies, Final Report.* ANL/SPG-11. A. D. Little, Inc.

Leboeuf, C. 1981. *Application of Solar Ponds to District Heating and Cooling.* SERI/TR-731-1036.

McLinden, M., and S. Klein. 1983. Simulation of an absorption heat pump solar heating and cooling system. *Solar Energy* 31, 5: 473–482.

Morrison, D., and S. Abdel-Khalik. 1978. Effects of phase-change energy storage on the performance of air-based and liquid based solar heating systems. *Solar Energy* 21, 5: 377–383.

Nix, R. 1983. Thermochemical energy systems research. *Proc. DOE Thermal and Chemical Energy Storage Annual Contractors' Review Meeting.* CONF-830974.

O'Dell, M., J. Mitchell, and W. Beckman. 1984. Design method and performance of heat pumps with refrigerant-filled solar collectors. *Journal of Solar Energy Engineering* 106, 2. ASME.

Offenhartz, P. 1981. A chemical heat pump based on the reaction of calcium chloride and methanol for solar heating, cooling and storage. *Proc. DOE Thermal and Chemical Storage Annual Contractors' Review Meeting.* CONF-801055.

Rizzuto, J. 1980. Design and demonstration of a storage-assisted air conditioning system. *Proc. DOE Thermal and Chemical Storage Annual Contractor's Review Meeting*, CONF-801055.

Rocket Research Co. 1982. *Sulfuric Acid/Water Chemical Heat Pump/Chemical Energy Storage*, vols. 1 and 2. RRC-82-R-813. Brookhaven Labs.

Rockwell International. 1980. *Solar Energy Storage by Reversible Chemical Processes.* Final report, SAND79-8199.

Sillman, S. 1981a. *The Trade-off between collector Area, Storage Volume, and Building Conservation in Annual Storage Solar Heating Systems.* SERI/TR-721-907. Golden, CO: Solar Energy Research Institute.

Sillman, S. 1981b. The value of storage in annual cycle solar heating systems. *Proc. International Conference on Seasonal Thermal Energy Storage and Compressed Air Energy Storage*, vol. 1. CONF-811066. October.

Swet, C. J. 1973. A prototype solar kitchen. Paper 73-WA/Sol-4. Presented at the *ASME Winter Annual Meeting*, Detroit, 1973.

Swet, C. J. 1980. Phase change storage in passive solar architecture. *Proceedings of the Fifth National Passive Solar Conference.* AS/ISES.

Swet, C. J. 1981. Solar applications of thermochemical heat pumps—Progress and prospects. *Proc. Solar World Forum*, vol. 1. Pergamon Press.

Ward, D., W. Duff, J. Ward, and G. Löf. 1979. Integration of evacuated tubular solar collectors with lithium bromide absorption cooling systems. *Solar Energy* 22: 335.

Wyman, C. 1979. *Thermal Energy Storage for Solar Applications: An Overview.* SERI/TR-34-089.

Ziegenbein, B. 1979. Investigation of latent heat of fusion storage for solar heating systems. *Proc. International Solar Energy Society Silver Jubilee Congress*, vol. 1. Pergamon Press.

III MATERIALS FOR SOLAR TECHNOLOGIES

20 Overview

Stanley W. Moore

20.1 Introduction and Background

Before fiscal year 1974 little federal funding for terrestrial solar materials research was available through the National Science Foundation (NSF). Other agencies that funded some terrestrial solar energy research were the National Aeronautics and Space Administration (NASA) and the Department of Housing and Urban Development (HUD). Additional research efforts, and perhaps many of the earlier efforts, were university supported. There was also some support in foreign countries by national laboratories and here in the United States by the American Society of Heating, Refrigeration, and Air Conditioning Engineers (ASHRAE).

Some of the early materials research included the work carried out by Tabor (1967) and the study of radiative characteristics of materials for solar heat collection or rejection by the University of California (Edwards et al. (1960). Many of these early researchers' efforts provided the groundwork for concepts and research areas that have only recently matured and have been commercialized.

20.2 Materials' Status in the 1970s

20.2.1 Absorbers

Many early efforts to develop selective coatings were based on Tabor's chemical treatments (Tabor 1955), oxidized Cu and Co (Hottel and Unger 1959; Kokoropoulos, Salam, and Daniels 1959), and metal blacks (Vershinin and Kozyrev 1959). Investigations were carried out on electroplated and vapor-deposited metals, anodized ceramics, and paints for space radiators.

Several commercial selective surfaces were in use, including copper black on copper (Australia), nickel black on galvanized steel (Israel), and aluminum oxide on aluminum. Performance and durability improvements were also investigated as were new cost-effective coatings.

20.2.2 Glazings and Reflectors

A variety of glazing materials were used in early solar collectors. Except for glass, most were later found to be unacceptable or uneconomical for solar applications. Of the common polymers available, many like polyvinylchloride, polyethylene, polycarbonate, and polyester were unsatisfactory. Moderate success was achieved with polymethyl methocrylate and the fluorocarbons.

Many of these materials later became marketable solar products with formulation changes, additions of ultraviolet (UV) inhibitors, or screening agents. Antireflective coatings on glass were tried on one early solar home, that of George Löf, as an experiment. The early reflectors of glass or coated polymers also lacked adequate durability for outdoor applications.

20.2.3 Insulations

Like many other solar components the early insulations used in solar applications were the easily obtained, commercially available products. Some, when protected from moisture, performed adequately. Others, such as the polystyrene foams and polyurethane foams, had problems with swelling and shrinking, temperature tolerance, UV degradation, outgassing of noxious or toxic chemicals, and smoking and smoldering.

20.2.4 Heat Transfer Fluids and Corrosion

In the early 1970s few operational liquid systems existed, so few problems were encountered. Many early domestic hot water systems were installed in non-freezing environments and in many cases the water supply contained few corrosive impurities. Glycol breakdown, high chloride impurities, flux inclusions in joints, and other problems that were not severe in low temperature domestic water systems became disasters in high temperature systems.

20.2.5 Sealants and Gaskets

Many early collector designs used low temperature elastomers for sealants. Problems with these sealants usually surfaced rapidly. Other relatively high temperature gasket and sealant materials were available in the early 1970s. These included ethylene-propylene-diene monomers (EPDM), silicones, butyls, and neoprenes, none of which were designed or formulated specifically for long-life, high temperature solar collectors. Many creep, compression set, and outgassing problems resulted, which, when solved, eventually led to improved solar products.

20.2.6 Materials Performance

In 1977 NBS assessed and studied the performance of materials in operational solar energy systems (Skoda and Masters 1977). Their assessment included materials that were available before the escalation and development of solar materials beginning about 1972. Most of the materials used before 1977 were conventional commercial materials for solar applications. Many of these materials were inadequate for the solar environment.

20.3 Initiatives and Plans

20.3.1 Early Incentive Activities

Fiscal year 1971 marked the first year of increased federal funding for terrestrial solar energy research. NSF through its Research Applied to National Needs (NSF-RANN) program was the principle agency of the federal government for developing solar applications. NSF was assigned the task of developing a five- to ten-year plan to develop terrestrial applications of solar energy. This plan, which was implemented in fiscal year 1974, initiated a major upward trend in solar energy use. Fiscal year 1975 represented the first year of the plan and had a solar budget of $50M, of which $17 million was allocated to development of heating and cooling systems for buildings and or agricultural purposes. A concentrated effort to advance the state of the art of materials for solar energy systems began with the formation of the Energy Research and Development Administration (ERDA) and budgets for solar materials R&D began to grow.

The National Energy Act (DOE 1978), which was passed by Congress on October 15, 1978, after nearly a year and a half of deliberation, provided further stimulation to the advancement of solar materials.

20.3.2 Assessment of Needs

An Assessment of Solar Heating and Cooling Technology (Balcomb and Perry 1977) was conducted in 1975 and 1976 to determine the status of current solar technology and establish those areas for potential research. This assessment was conducted in five 2-day meetings that covered the following areas: solar collectors, thermal energy storage, solar air-conditioning and heat pumps, systems and controls, and nonengineering aspects. These meetings, held from December 4, 1975, to April 6, 1976, brought together experts from large and small industries, universities, and government laboratories and organizations. The resulting assessment established the state of the art of solar technology, identified problem areas, and established objectives.

20.3.3 Materials Program's Plans

The results of the foregoing assessment were factored into and established the Interim Report National Program Plan for Research and Development in Solar Heating and Cooling, ERDA 76-144, which later became the National Plan, DOE CS-0008.

This plan was implemented through a number of program research and development announcements (PRDA) (PRDA 1977) and requests for proposals (RFPs). These solicitations resulted in the selection and funding of approximately 100 materials research projects.

More recently DHR, Inc. (Herzenberg, Hien, and Silberglitt 1983), was funded by DOE to evaluate the state of the art of solar energy systems and identify the needs of further materials research. The study identified those needs and assigned priorities to areas appropriate for federal support.

20.4   Materials Progress since the Onset of Federal Support

Chapter 21 describes in detail the materials required for glazings, heat mirrors, and selective absorbers. Although the chapter focuses on only three basic components or concepts, it points out the vast amount of research conducted since the early 1970s. Chapter 25 extends the discussion of materials investigations carried out on solar materials. Many of the concepts and materials investigations were extremely successful and have led to commercialization of products. Many others, while not successful enough to reach commercialization, have contributed to the understanding of performance durability and cost of solar materials. Many research projects were in the high risk category and resulted in failure.

This section presents some of the major federally funded projects and their contriutions that have advanced the state of the art of solar materlals— namelly, absorbers, glazings, insulations, heat transfer fluids and corrosion, sealants and gaskets, and durability testing. A summary report (Reisfeld et al. 1983) describes many of these projects.

20.4.1   Absorbers

Absorber surfaces have received the primary emphasis and attention during the early apportioning of federal support. Most of the theory and modeling that has been applied to absorber materials is discussed in chapter 22. The analytical studies carried out were largely based on theory developed many years ago. Among these were developments of interference film theory, Drude free electron theory, effective medium theory, and antenna theory, all of which occurred prior to the large solar research efforts. These theories have been used extensively in the evaluations of various coatings, particularly selective absorber coatings. Some of the DOE-supported absorber programs are presented here.

Electroplated Selective Surfaces

HONEYWELL, INC.   Honeywell (Mar et al. 1975) studied many different types of selective coatings, beginning in 1974 under Contract NSF-C-957 and continuing through 1982. They completed their study of electroplated coatings in 1979 (Lin 1979).

Three of the most promising selective-absorbing platings studied and their typical optical properties were black nickel (nickel/zinc sulfide), $\alpha_s = 0.95$, $\varepsilon = 0.07$; black chrome (chrome oxide), $\alpha_s = 0.95$, $\varepsilon = 0.10$; and black iron (iron oxide), $\alpha_s = 0.85$, $\varepsilon = 0.10$. Black nickel had superior optical characteristics; however, performance must be measured by the best combination of optical properties, cost, and durability. Black iron was the lowest-cost plated coating and showed good durability but had relatively low solar absorptance. The use of polymer overcoats to increase both the solar absorptance and the durability of plated coatings was investigated and showed promise. For black chrome, a polymer overcoat could eliminate the need for a nickel-plated substrate without degrading durability. Eliminating the nickel plating would result in significant cost reduction. For black iron, a polymer overcoat could improve durability and significantly increase solar absorptance. The primary disadvantage of a polymer overcoat is an increase in ε caused by infrared (IR) absorption in the polymer.

Honeywell completed their study of black chrome in August 1979 with the completion of their report on plating bath formulation as it relates to thermal stability and durability.

Sandia National Laboratoris (Pettit and Sowell 1979) and the Harshaw Chemical Company (Benning 1980) also investigated plating parameters that might increase the temperature capability of black chrome. Sandia carried their efforts to the point of establishing the accepted process control method for high temperature black chrome.

NASA-LEWIS RESEARCH CENTER   In the early 1970s the NASA-Lewis Research Center (McDonald 1975) began investigating four high performance, moderate-cost, widely available coatings. Two of these—black copper and black nickel—were known to have selective properties. The selective properties of the other two—black chrome and black zinc—were discovered at the National Aeronautics and Space Administration (NASA) Lewis Research Center, Cleveland, Ohio. The broad selective characteristics of all four coatings are very similar.

NASA-Lewis investigated black chrome in depth to determine the preparatory methods that would produce its optimum optical solar selective properties. Black chrome appeared to have the potential and later proved to be the most stable of the early coatings.

Vapor-Deposited Coatings

PENNSYLVANIA STATE UNIVERSITY   Objectives of Pennsylvania State University's project included developing a basic understanding of sputtered film preparation and characterization of the solar-absorbing properties of black germanium films.

Semiconductor films with an appropriate band gap (~ 0.5 to 1.24 eV) have a high absorption coefficient in the solar spectrum and high transmission in the IR thermal-emission region and thus have a potential for high selectivity. Such high selectivity would result if the absorbing layer were thin enough, so that the IR emissivity would be controlled by an appropriate substrate, yet still have high total absorptance. However, because of their high refractive index, semiconductor films have high reflectance above their absorption edge that limits their total absorption. A coating that would retain the advantages of the semiconductor's selectivity and also reduce the reflectance from the present values of 30%–50% to a level of less than 2% would be a significant advancement.

This study described the preparation and characterization of structurally anisotropic black germanium films by rf-sputtering and postdeposition etching. By controlling the sputtering preparatory conditions of noncrystalline germanium films, the surface microstructure could be drastically altered by simply etching in 30% H_2O_2. The total reflectance of the resulting surface was reduced from about 45% to less than 1%. Furthermore the films appear black over a wide range of angles of incidence (up to 45 deg from the normal). This flat-black appearance was caused by a dense array of aligned, needlelike protrusions that have an extremely high aspect ratio and both a cross-sectional area and a separation between the needles within an order of magnitude of the wavelength of solar radiation. The final coating surface was quite fragile and subject to handling damage.

Some preliminary work on obtaining black silicon films was also described. Black germanium films achieved a solar absorptance α_s of greater than 0.99 in the best cases. Variation of α_s and ε_{IR} (IR remittance) with the thickness of the film and other preparation conditions was described. The lowest ε_{IR} obtained was 0.44 associated with an α_s of 0.98. Because the material was an indirect band-gap substance, it was questionable if going to a thinner film could reduce ε_{IR} without seriously affecting α_s. The microstructures were classified and understood on the basis of a structure zone model.

TELIC CORPORATION   The Telic Corporation's goal was to develop the materials and technology for producing inexpensive and durable selective coatings in a continuous or semicontinuous mode on plates or metal strips by planar magnetron reactive sputtering. An important adjunct to this goal was the preparation of a continuously coated strip to be provided to interested manufacturers for the fabrication of collectors and testing in the field by firms working in the solar industry.

The investigation was based on promising results that were previously achieved with $Al_2O_3/Mo/Al_2O_3$ coatings deposited by cylindrical magnetrons. In that work, coatings with reactive sputtered Al_2O_3 layers were found to be stable to about 350°C (662°F) in air and about 450°C (842°F) in vacuum. Coatings with Al_2O_3 layers deposited by direct rf-sputtering from alumina targets in argon were stable to about 550°C (1,022°F) in air and 700°C (1,292°F) in vacuum. The present program involves examining M-layer materials, other than molybdenum, that might provide improved thermal stability in air when used with active sputtered Al_2O_3 layers. The studies of continuous deposition onto a strip were described in their final report (Thorton, Lamb, and Tenfold 1980).

$Al_2O_3/M/Al_2O_3$ coatings with M-layers of chromium, molybdenum, nickel, and tantalum were deposited and tested, along with coatings having Pt/Al_2O_3 cermet M-layers, which were deposited during another project. The M-layer metal was also used for the low emittance base layer. The substrates were glass and stainless steel plates. In all cases hemispherical absorptances from 0.90 to 0.95 with room temperature emittances of about 0.10 were achieved on glass substrates by selecting the individual layer thicknesses. The coatings were subjected to thermal testing in both air and vacuum and to humidity and moisture testing. Extensive Auger depth profile analysis was used to determine failure mechanisms.

The results of the thermal tests are consistent with the observations on the previous cylindrical-magnetron-deposited coatings and confirm the importance of the deposition procedures for the Al_2O_3 layers. In particular, the reactively sputtered Al_2O_3 layers deposited thus far are poor diffusion barriers. Thus coatings with M-layers of chromium, molybdenum, nickel, tantalum, or Pt/Al_2O_3 cermet and reactively sputtered Al_2O_3 typically undergo absorptance changes of greater than 1%, in air or vacuum, between 300° and 450°C (572° and 842°F). The failure mechanism is apparently diffusion of the metal layers into the Al_2O_3.

By contrast, the Al_2O_3 layers deposited by direct rf-sputtering of alumina provide good barriers to metal diffusion. Thus coatings with M-layers of chromium, molybdenum, nickel, or Pt/Al_2O_3 cermet, and rf-sputtered Al_2O_3, are stable in air up to 500°–600°C (932°–1,112°F) and in vacuum to at least 650°–700°C (1,202°–1,292°F). Failure apparently occurs by oxygen that diffuses through the top Al_2O_3 layer and attacks the M-layer. Thus no failure was observed over the temperature range examined for the platinum coatings in air [up to 600°C (1,112°F)] and for all coatings with the rf-sputtered Al_2O_3 in vacuum [up to 650°–700°C (1,202°–1,292°F)].

All coatings (except the molybdenum-based ones) that were tested at the Lockheed Palo Alto Research Laboratory were stable in 12-day humidity tests at 95°C (203°F) and 95% relative humidity. The molybdenum failure appeared to be related to the tendency of this metal to form a hydroxide. All the coatings failed a perhaps unreasonably severe moisture test that was established at Telic. The problem in this case appeared to be water take-up by the amorphous sputtered Al_2O_3 layers.

The results achieved thus far confirm those originally achieved for the $Al_2O_3/Mo/Al_2O_3$ coatings; that is, stability in air at 300°C (572°F) for reactive sputtered Al_2O_3 and at 500°C (932°F) for rf-sputtered Al_2O_3. Coatings with reactive sputtered Al_2O_3 and M-layers of molybdenum and nickel have shown no degradation during 1,000-hour tests in air at 350°C (662°F).

Chemical Conversion Coatings

LOS ALAMOS NATIONAL LABORATORY   A need exists within the solar industry for low-cost, low temperature selective coatings for collectors or passive applications; chemical conversion coatings may fill this need. The Los Alamos Solar Group conducted a joint program with the Los Alamos Materials Technology Group to investigate these coatings. The initial phases of this program focused on studying the variables associated with producing the coatings. Optical properties resulting both from these investigations and for some of the substrate materials studied are presented in Moore, Clements, and Doty (1983).

The program focused on using commercial blackening agents and optimizing the coating parameters to obtain the best properties for substrate materials. The resulting coating might be a chromate, phosphate, sulfate, oxide, or a combination of these. Optimizing coating parameters has produced the properties shown below for the various substrates studied.

Determination of durability and weatherability limits or capabilities has followed optical optimization.

Selective Paints   Selective paint investigations were aimed at producing an inexpensive, durable paint that would have moderately good selectivity combined with extremely low cost.

HONEYWELL, INC.   Honeywell (McChesney, Zimmer, and Lin 1982) undertook the development of selective paints in 1974 and continued their studies until 1982. Their study eventually led to the investigation of two distinct types of paint: thickness sensitive (one requiring its application to a low emittance substrate), and thickness insensitive (one containing a metallic flake and not requiring application to a low emittance substrate).

Honeywell investigated dip, spray, roll, lamination, and gravure processes and used them to produce final samples. They selected high speed gravure coating as the most promising process for solar foil fabrication. The foil was produced as a 10,000-ft^2 (930-m^2) sample with optical properties of $\alpha_s = 0.90$ and $\varepsilon_{TH} = 0.15$ for the 28 PVC F-6331/6247 silicone/epoxy coating, and $\alpha_s = 0.90$ and $\varepsilon_{TH} = 0.07$ for the 28 PVC F-6331/SR-125 silicone coating. Total material and fabrication cost was \$0.24/ft^2 (\$2.58/m^2), with a \$0.15/ft^2 (\$1.61/m^2) projected cost for 1,000,000 ft^2 (93,000 m^2). These costs include the protective film over the paint coating, paint-coating material, aluminum foil, adhesive, release liner, and fabrication. Honeywell demonstrated that the processes described in table 20.1 can produce the optical quality shown.

Development and optimization of thickness-insensitive solar selective paints (TISP) were not completely successful. A more or Less conventional approach previously pursued used aluminum flakes as the low emitting substrate in the solar paint. The major problem in the development of TISP coatings has been the inability to replicate the optical properties of the original coatings. The feasibility of this concept was demonstrated when test coatings showed a capture ratio ($\alpha_s/\varepsilon_{TH}$) of 3 : 1 with $\alpha_s = 0.89$ and $\varepsilon_{TH} = 0.31$ when applied to a low emittance substrate. The reasons for this nonreproducibility were largely unknown. All coating ingredients were suspect, but the primary concern focused on the aluminum flake pigment because its behavior in test coatings was unpredictable.

A new method of achieving better control of coating components was conceived and preliminary development initiated. The new concept was described as an engineered pigment approach. The engineered pigment approach uses metal foil particles coated with thickness-sensitive solar selective paint (TSSP) instead of uncoated aluminum flakes in a liquid TSSP coating.

Table 20.1
Thickness-sensitive selective paint results

Process	Size production typical	Typical range		Best obtained	
		α_s	ε	α_s	ε
Dip	Small–medium	0.88–0.90	0.06–0.14	0.90	0.06
Spray	Small–large	0.91–0.93	0.13–0.32	0.92	0.13
Roll	Small–medium	0.86–0.89	0.09–0.14	0.88	0.14
Gravure	Medium–large	0.90–0.92	0.07–0.20	0.90	0.07

The approach offers many advantages over the use of uncoated aluminum flakes: control of particle flatness, size, and thickness; control of the optical selectivity of each particle; and control of the liquid TSSP coating surrounding the coated particles. Unfortunately, this development came late in the program and could not be optimized.

LOS ALAMOS NATIONAL LABORATORY   Los Alamos continued some of Honeywell's efforts in TISPs to develop better properties. The two approaches they studied were chemically converted metallic flake and a continuation of the engineered pigment approach started by Honeywell.

Engineered-pigment TISP. Both particle orientation and size affect the solar absorptance and thermal emittance of a surface. Material samples were prepared by measuring the properties of foil, then chopping it into small square particles.

Uncoated aluminum foil and selective coated nickel foil were cut in square sizes from 0.008 to 0.062 in. (0.052–0.403 cm) to study size and surface roughness effects. An α_s of 0.94 and an ε of 0.24 can be obtained by this process, which is quite encouraging for a TISP system. Two factors control the optical properties—particle orientation and the ratio of surface area of the cut edges to that of the flat, selective coated surface. Further optimization of this concept should result in even better properties; however, this approach will probably never be as cost-effective as the leafing flake approaches.

Chemically converted flake/powder materials. Chemical conversion of sheet materials was pursued in an earlier program and reported by Moore, Clements, and Doty (1983). Some of the same blackening agents were used on flake and powder materials in an attempt to develop a TISP. Table 20.2 describes cleaned and chemically converted sheet material properties that were achieved previously as compared with the powder/flake results. As can be seen in the table, the cleaned base-line emittance properties for sheet are always lower than those for the flake material. This first attempt to blacken the flakes did not produce the selectivity that was obtained from the sheet. While these results are encouraging, the processes that hold the most promise will not be known until further optimization studies are completed. Variations in bath immersion times, temperatures, solution concentration, flake cleaning, sensitization, and processing all have a bearing on the final properties.

Evacuated Tube Absorbers

LOS ALAMOS NATIONAL LABORATORY   Currently selective absorbers are deposited by physical vapor deposition (PVD) directly onto glass receiver tubes

Table 20.2
Cleaned and chemically converted sheet and flake substrates

	Cleaned sheet property, α_s/ε	Chemically converted sheet property,[a] α_s/ε	Powder or substrate flake property, α_s/ε	Best chemically converted flake property in FY83,[b] α_s/ε
Aluminum	0.096/0.029	0.902/0.157	0.120/0.169	0.741/0.518
Copper	0.179/0.023	0.893/0.081	0.274/0.152	0.851/0.411 (0.927/0.617)
Brass	0.181/0.030	0.923/0.081	0.198/0.152	0.841/0.300 (0.921/0.641)
Nickel	0.344/0.039	0.937/0.076	0.384/0.252	0.849/0.488 (0.889/0.688)
Stainless	0.393/0.097	0.806/0.119	0.426/0.251	0.796/0.291 (0.830/0.326)
Zinc	0.245/0.038	0.954/0.079	0.441/0.437	Not tried
Mild steel	0.496/0.064	0.925/0.075	0.793/0.470	Not tried
Tin	—	—	0.572/0.370	Not tried

a. See Moore et al. (1983).
b. Numbers in parentheses are the highest α_s and resulting ε. Other properties are the best selectivity obtained to date.

of evacuated collectors. Two alternative methods for depositing a selective absorber coating on glass receiver tubes were studied. One method involved electrodeposition of a selective film on the tubes (Grimmer 1978). Another involved bonding a foil coated with a selective absorber to the glass tube (Grimmer and Avery 1981).

The electrodeposited films showed the best α and ε, after vacuum baking at elevated temperatures, of any film known on glass. However, problems were found with the diffusion of copper into the electrodeposited black chrome in the chromium oxide/copper film. Electrostatic bonding of aluminum foils coated with an excellent selective coating was successful. Less successful were bonding techniques using glass frit.

OWENS-ILLINOIS, INC.   The Owens-Illinois (Spanoudis 1980) project sought to increase the cost-effectiveness of evacuated tube solar collectors by using antireflection (ar) coatings.

Both surfaces of the cover tube were successfully ar-coated using a method found in the patent literature. ar coating increased the transmittance from 0.92 to 0.98. Several thin-film materials were effective as ar overcoats for the CrO_x/Al selective coating. Prototype collector tubes with a MgF_2 ar coating had absorptivities of 0.89 and emissivities of 0.058 with excellent thermal stability.

Thermal degradation in a CrO_x/Al coating was concluded to be a reaction between the aluminum and the CrO_x, which accelerates with increasing temperature. The reaction, which became more evident as the aluminum thickness decreased, was impeded by barrier coatings between the aluminum and CrO_x and by isolating the CrO_x/Al coating through the use of undercoats and topcoats.

The most thermally stable coating was obtained by substituting copper for aluminum. Some work remains in preventing premature degradation of the copper during glass sealing and annealing. An excellent selective coating proprietary to the vendor was applied by Denton Vacuum of Cherry Hill, New Jersey. A new selective coating was prepared in which a metal and a dielectric were coevaporated in reverse concentration gradients onto metallized coupons. Coatings with high absorptivity were obtained with Al + MgF_2, Cu + MgF_2, and Cr + MgF_2. Additional development work was required to apply these coatings onto tubes.

This was a comprehensive and valuable research effort (Parts and Conine 1981). A caveat ought be mentioned: It was determined that the most thermally stable coating obtained was CrO_x applied over copper instead of aluminum. These CrO_x/Cu films were applied by PVD. However, CrO_x/Cu films applied by electrodeposition will not show this same stability; even in vacuum the copper will migrate at high temperature into the presumably more open structure of the electroplated copper.

WESTINGHOUSE ELECTRIC CORPORATION   Although the primary objective of Westinghouse's study was the design and manufacturing analysis of an evacuated tube solar collector designed to be both efficient and cost-effective, the final report (Beecher et al. 1979) on the study contains a valuable section on vacuum deposition of selective coatings.

A multilayered graded film of selective black chromium oxide with an absorptivity of about 0.9 and an emissivity of 0.03 was vapor deposited. When the film was heated to a temperature of 400°C (752°F) in a gettered vacuum for as little as 24 hours, however, irreversible changes occurred both between and within the coating layers, which resulted in a decrease of α to about 0.73 and an increase of ε to 0.14. This work on the formation of PVD black chrome films represents the most detailed presentation of these types of films in the literature.

Miscellaneous Absorbers

BATTELLE MEMORIAL INSTITUTE   The primary objective of this program (Landstrom, Talbert, and McGinnis 1981) was to develop a low temperature, black-liquid-type solar collector that was cost-effective, highly efficient, and nonconcentrating. The program was to demonstrate the feasibility of the concept, to determine if such a collector would have specific advantages when used as a thermal energy source for a heat pump or other low temperature application, and to provide necessary technical and economic data to evaluate and compare a black liquid collector with conventional flat plate collectors. Some specific materials goals were to evaluate several candidate plastic materials and solar collector designs suitable for black liquid use; to recommend suitable plastics, coatings, designs, and applications that have the potential for long life and good performance; to investigate overall system performance for variations in design, materials, and specific uses; and to monitor and incorporate improved black liquids.

One important accomplishment of this program has been the collection of data about problems associated with the long-term use of plastic materials in solar applications. Certain polymeric materials exhibit long life under solar exposure. However, because test conditions do not always duplicate service conditions, the materials must be investigated under actual collector operating conditions to be realistic. This is a vast area of research, and this program considered only those candidate polymeric materials that were most suitable for direct application to solar collectors. Additional research in polymeric protection and enhancement of collector lifetime is still necessary.

CALMAC CORPORATION   CALMAC (Hottel and Unger 1959) developed, tested, and delivered a flat plate collector [300 ft^2 (28 m^2)] that used a flexible, nonmetallic absorber plate. The collector consisted of a flexible grid of 30 closely spaced elastometer twin tubes (material: EPDM) cemented in an insulation board base. This grid served the same purpose as the metal absorber plate used in the more conventional flat plate collectors. Marshall Space Flight Center monitored the equipment performance.

POLYSET, INC.   The Polyset program investigated (Bradley 1977) the feasibility of using carbon-black-impregnated, cross-linked polyethylene as a material for solar collector absorbers; such a material would be able to withstand repeated freezing of water within the absorber.

The program investigated various cross-linked polyethylene formulations and geometries. Variations were made in density, melt index, carbon loading, cross-linking agent and loading, antioxidant loading, tube-wall thickness, and tube diameter. Ten-foot tube specimens made from various cross-linked polyethylene formulations were filled with water at various pressures, placed in a deep freeze, then thawed and frozen again for 100 freeze/thaw cycles, or until the tube specimen failed. Tube diameters were measured before and after each freezing to determine how much distention the freezing caused and how much permanent distention was caused by the strains of repeated freezings. Five tube specimens containing water as high as 80 psi survived 100 freeze/thaw cycles.

Also a flat plate collector was fabricated using a single tube of carbon-black-reinforced, cross-linked polyethylene as an absorber surface, which was in the form of a flat spiral coil; this collector was tested for performance at the Los Alamos National Laboratory. The collector was constructed of 403 ft (123 m) of spirally wound 0.336-in. (0.853-cm)-o.d. tubing with a 0.047-in. (0.119-cm) thick wall made of 0.034 lb/in^3 (0.933 g/cm^3) resin. The tubing had a melt index of 2.33 and was carbon-black loaded to 40 parts by weight per 100 parts resin (phr), with 0.5 phr loading of Age Rite Resin D$^®$ and 1.5 phr loading of cross-linking agent Varox$^®$. The resulting collector was 42 in.^{2} (273 cm^2).

The program proved the feasibility of using carbon-impregnated, cross-linked polythylene for a freeze-tolerant collector.

SPERRY-UNIVAC, INC. The Sperry-Univac program (Jensen 1980) studied solar collectors that use heat transfer fluid to collect solar energy and evaluated their success potential in the medium temperature range. A significant result of this contract was the formulation of a superior black liquid for heat transfer. This fluid has since been extensively tested with good results.

The stable black liquid, composed of colloidal carbon, dispersants, and water, or water and propylene glycol, was developed. Under laboratory conditions exceeding the limits of anticipated normal use, there was no noticeable deterioration of the stability of the solution. Tests included cycling with 3,600 hours of heating at 160°F (70°C), 1,100 hours cooling at -15°F (-26°C), and 72 hours at 300°F (149°C) at 1.5 atm.

The absorptance of such a solution using colloidal carbon and water, or water and propylene glycol, is 100% by conventional measurement. Collector designs incorporating low emittance films and high insulating values will be important to retain the benefits of this high absorptance. The liquid developed was within the common limits of corrosivity, toxicity, and flammability. Collector designs using commonly available materials have been made and used efficiently and appear to be cost-effective.

20.4.2 Glazings

Chapters 21 and 25 describe the various glazing materials and coatings developed worldwide. One area that has received considerable attention is that of heat mirrors. Heat mirror materials have a range of application, from energy-conserving window glazings to a heat-trapping option for solar collectors similar to that of selective absorbers. Studies of ITO coatings, tin oxides, cadmium orthostannites, thin metal films, and micro grids have advanced the state of the art in recent years. The early heat mirrors had rather low integrated solar transmittances (τ_s), with a τ_s maximum of approximately 0.77. There are now more forms of data reporting available because an increasing number of investigators who enter the field have varied backgrounds and interests. An example of this might be the reporting of solar transmittance of 0.95 over a wavelength of 0.4 to 1.4 μm. To be meaningful, solar transmittance must be integrated over the entire solar spectrum, that is, 0.3 to 2.5 μm per ASTM E891-82 or a similar standardized spectrum. Some investigators report transmittances over even narrower bands, or even the peaks obtained, and refer to them as solar transmittances. This comparison is not intended to be critical of anyone's research but only to serve as a caution to consider the optical properties of the various materials and coatings. Caution is also advised for the measurement of any optical component, be it a glazing coating, absorber, or reflector. Occasionally misinterpretations have been made when the data are reported only for a coating, while the normally included representative substrate is disregarded.

The form of data used must be relevant to the specific application of the glazing material. An example is the use of both normal–normal $(n-n)$ or normal–hemispherical $(n-h)$ measurements. For passive or flat plate applications, and where the incident energy is the only consideration, the $(n-h)$ transmittance is of primary importance. Where specular effects must be considered, as in the case of solar concentrators, then the $(n-n)$ properties are required. These property descriptions and measuring equipment are covered in more detail in chapters 23 and 24. Some of the DOE-funded glazing projects are presented here.

Glass

HONEYWELL, INC. The Honeywell program (Lin, Lee, and Zimmer 1980) concerned development of large-scale production processes for manufacture of low-cost organic and etched AR coatings on glass. Scale-up of the etched glass process was extended to include the latest low-iron, soda-lime glass being used in the solar market.

The etched ar coating process had previously been scaled up to etch full-sized glass panes, but the chemical control of the etching solution remained a problem. Analytical methods including high pressure liquid chromatography (HPLC), flame-excited atomic absorption, IR spectroscopy, Coulter counter techniques, and liquid index of refraction measurements were examined for application to this problem. A phenomenological model for the key reaction species in the etching process was presented, along with a monitor control scheme using refractive index measurements, to set the initial reactant concentrations and HPLC for tracking in-process concentrations.

The environmental durability of the organic ar coating of fused FEP-120 Teflon dispersion was also improved by pretreating with a hydrofluoric acid etch/silane coupling agent. Improved coating/processing techniques were developed, culminating in the demonstration of scale-up feasibility of this process for glass panels up to 1 × 1 ft (0.3 × 0.3 m) with good batch sample-to-sample repeatability.

OWENS-ILLINOIS, INC.   This Owens-Illinois project (Spanoudis 1980) sought to increase the cost-effectiveness of evacuated tube solar collectors by increasing the high temperature resistance and the absorptivity of the selective coating by use of ar coatings. Methods were investigated for reaching these goals. Both surfaces of the cover tube were successfully ar coated using a method found in the patent literature. ar coating increased the transmittance from 0.92 to 0.98.

Plastics

ACUREX CORPORATION   This multiphased project (Muller and Sullivan 1982) was aimed at producing a low-cost solar panel (LCSP) constructed of several polymer films bonded together by high speed manufacturing to make a flexible absorber structure. Although the project fell short of developing the commercial film collector, extensive material research was accomplished. This work could be the starting point for future work on thin-film solar collectors.

Material research included a rigorous material search, survivability testing of possible solar panel materials, bonding and adhesive testing of the most likely candidate materials, and the fabrication of a full-scale prototype solar collector. The final phase of the LCSP development project concentrated on the material requirements and worked toward commercialization of the LCSP concept.

FAFCO CORPORATION   High initial costs of solar collectors are caused by the collectors' complexity of components, excessive weight, installation problems, corrosion protection, and glazing breakage. This project (FAFCO 1980) addressed these problems by using polymeric substances integrated into a single structure resulting in a simple, mass-producible, low-cost collector design.

The project was divided into two parts, with associated tasks as follows:

Part 1. (a) Environment definition—computer simulation and thermal experiments were used to define the operating environment for candidate materials; (b) materials search—candidate materials were chosen from the literature; (c) materials evaluation—materials tests were performed to determine suitability; (d) processability tests—lab tests were performed to confirm processability of materials combinations; (e) process experiments—existing machinery was used to test processing methods and variables; and (f) process evaluation—output rate, process reliability, and energy consumption were evaluated.

Part 2. (a) Prototype fabrication, (b) prototype evaluation, (c) application study, and (d) systems integration study.

During FAFCO's previous research efforts, a design concept was conceived that was thought to have potential for improving on cost, installation, durability, and efficiency.

The concept called for a radical departure from conventional box-type, flat plate collectors. The glazing, absorber, and insulation were integrated into a single structure, thus eliminating the box frame and its functions of support and enclosure. The result was a simple, low weight, low-cost, flat plate collector that could be mass produced.

HUGHES AIRCRAFT COMPANY The purpose of the Hughes Aircraft program (Bilow et al. 1980) was to develop polycarbonate polymers and copolymers with glass transition temperatures higher than 150°C (302°F), good optical, physical, and mechanical properties, and markedly improved resistance to UV radiation. A second objective was to develop a surface treatment that would minimize reflectance if a sufficiently good polycarbonate could be developed during the course of the program.

Eight types of polycarbonates were synthesized during this investigation to develop at least one that had significantly improved solar radiation stability relative to bisphenol A polycarbonate, without a serious loss of other desirable properties. Of these new materials, one was derived from a cycloaliphatic diol, two were derived from bicycloaliphatic diphenol, one was derived from a perfluoroalkylene-substituted diphenol, and three were derived from diphenols that had all of their positions ortho to the phenolic hydroxyl groups substituted by either methyl groups or chlorine atoms.

Each polymer was synthesized several times to provide products with various mean molecular weights; this was necessary to compare their relative ease of fabrication. Many of the polymers could not be processed effectively enough to yield specimens suitable for evaluation, because of either their poor melt flow characteristics or discoloration caused by degradation at their high melting points. In some cases polymer films were effectively cast from solution, even though compression molding was not satisfactory.

Thermal properties of the new polymers were studied by thermal differential analysis, thermomechanical analysis, and thermogravimetric analysis; relative solar radiation stability was studied by exposing samples to simulated solar radiation in an Atlas Weather-Ometer, which also provided nine minutes of rain per hour.

Of the various polycarbonates studied, the one derived from 4, 4'-hexafluoro-isopropylidenediphenol (bisphenol AF) showed the greatest promise, since at least one sample remained unchanged through the course of the Weather-Ometer test.

LAWRENCE LIVERMORE NATIONAL LABORATORY   The Lawrence Livermore project (Gerich 1979) investigated the technical feasibility of using an inflated, plastic, cylindrical concentrator to produce industrial process heat. Optical and heat transfer computer codes were developed to model the concentrator's performance, a prototype was built and tested, and several plastic film materials were tested for suitability in the project.

The plastic materials degraded from UV light and wind exposure. Those plastics that worked best were also the most expensive (e.g., Teflon). A protective coating was necessary for the aluminized plastic reflector. Results indicated that Kynar® (polyvinylidiene fluoride) is a promising plastic material.

MONSANTO RESEARCH CORPORATION (PROJECT 1)   A potentially low-cost solar mirror concentrator of inflatable plastic was designed and built based on segments of a cylinder joined along the length of the collector on a plane passing through the axis of the absorber tube. This design resulted in a savings of approximately 40% in window and mirror material and in a savings of about 20% of the land occupied by a single collector when compared with a fully cylindrical collector. This project's (Schwendeman et al. 1980) type of construction permitted the attaching of the mirror/window envelope to the collector without disturbing the collector frame, absorber tube, or the associated plumbing.

Aluminum foil/plastic laminates were used as an alternative to aluminized polyester films because of their potential low cost and durability. Specially UV-stabilized and polyester-scrim-reinforced flexible polyvinyl chloride (PVC) films were developed for use as the outer cover material; inflation maintained the collector's shape.

The window material was fabricated from clear 4-mil PVC film laminated to 5-mil Mylar® film with a UV stabilizer added. Accelerated indoor testing, performed to determine the behavior of the plastic, was followed by outdoor testing of the full unit.

This work complements that of Lawrence Livermore National Laboratory. Monsanto developed and evaluated a potentially low-cost, durable, UV-stable, reinforced, flexible PVC film, in conjunction with work done on the mediumt temperature solar air heater program.

MONSANTO RESEARCH CORPORATION (PROJECT 2)   The Monsanto project's (Leffingwell and Schwendeman, 1979) objective was to make available a low-cost, durable, medium temperature solar air heater by optimizing specially stabilized and reinforced PVC film.

The special UV-stabilized and scrim-reinforced flexible PVC film was developed, characterized, and tested on a solar air heater according to the ASH-RAE test 93. Solar energy transmission of early precontract 20-mil-thick film was determined to be about 83% after two months of outdoor Florida exposure, dropping 78% after two years. Outdoor exposure and artificial UV-aging tests indicate that a ten-year lifetime is possible.

RESEARCH TRIANGLE INSTITUTE   The Research Triangle Institute's project (Pitt 1980) evaluated several types of organosilicones for use as collector glazing materials. Organosilicones were selected on the basis of their commercial availability and their superior weathering characteristics.

Materials included silicone rubbers and reinforced silicone rubber films. The resistance of the materials to heat and UV radiation was determined using accelerated exposure conditions. Polydimethylsiloxane fluids were also exposed to UV radiation to determine the amount of cross-linking that resulted. Five silicone materials were exposed to accelerated solar irradiation for a total dose of 2×10^6 L, corresponding to 11 years' exposure to sunlight in Arizona and a longer duration in localities of greater cloud cover and higher altitudes. The effect of such accelerated solar exposure on mechanical properties, chemical properties, and solar transmission was determined. The abrasion resistance and dirt retention of silicone materials were compared with those of glass and poly(methylmethacrylate). Finally, an economic analysis was made of silicones as compared with glass and other polymeric glazings. The tendency of silicones to collect dirt impedes their application as a glazing. Another impediment is that there are no clear cost advantages over glass.

SPRINGBORN LABORATORIES   Springborn (Baum and Binette 1983) evaluated 20 weather-resistant, low-cost, collector glazing materials and 12 housing materials that would have lifetimes of up to 20 years under varying conditions of stress and high temperature [300°F (149°C)]. Methods of bonding UV-screening polymer films were developed and tested; ar coatings and etching processes were also investigated.

A broad information search was carried out in five areas: glazings, housing materials, acrylic coating, etching processes, and ar coatings. An extensive list of all (known) U.S. transparent polymers was developed in addition to tables of plastic, ceramic, and metallic materials that could function as housing. Also, a compilation was made of commercially available solvent and water-based acrylic coatings for use as a UV protective coating for the glazing.

Twenty transparent polymers were chosen as possible glazings and exposed in the Atlas Weather-Ometer®. These glazing materials were also exposed outdoors at Hazardville, Connecticut, in the EMMAQUA in Arizona, and under the "wet" RS-4® sunlamp. (EMMAQUA refers to the test device, equatorial mount with mirrors for acceleration, used with a periodic spray of distilled water.) Solar optical transmission and tensile properties were measured periodically. Several acrylic coatings containing UV absorbers were investigated as protective coatings for glazings and the coated glazings were exposed in the EMMAQUA. Tedlar 20® Halar 500®, with strong absorption in the UV, and two commercial films containing UV absorbers, Tedlar UT® and Korad 201-R®, were laminated by several different processes to four promising glazing materials and exposed in the Weather-Ometer. ar coatings and surface etching processes were explored as a means of increasing transmission by reducing reflection.

Fluorocarbon polymers; polymethylmethacrylate; UV-stabilized, scrim-reinforced, cross-linked ethylene/vinyl acetate copolymer; and acrylic/polyester laminate are promising as glazings for solar collectors. Interpretation of the weathering data is based on the need for high original transmission, as well as percentage of transmission and tensile properties retained.

Several acrylic-coating systems containing UV absorbers are promising as a means of protecting polymeric glazings against weathering degradation. Longer-term aging is needed to find the best coating systems.

Methods of bonding UV-screening polymer films to several polymer glazings have been developed. Tedlar 400XRB160SE® (polyvinyl fluoride) film containing a UV absorber and Halar® (ethylene/chlorotrifluoroethylene) film are promising for protecting a less stable polymeric glazing substrate. The laminates prepared using these two films as top layers over peroxide cross-linked ethylene vinyl acetate copolymer show little degradation after one year in the Weather-Ometer®.

Magnesium fluoride ar coating increased the light transmission characteristics of several plastic sheets but was not weather stable. Etching of the glazing sheets did not improve transmission.

UNIVERSITY OF WISCONSIN   Development and evaluation of certain coatings on polymeric substrates were the objectives of this investigation (Klein, Yu, and Yen 1981). The particular coatings were intended to downconvert the UV component of solar radiation into the visible region and simultaneously inhibit photodegradation of plastic glazings.

Polycarbonate films (Lexan®) were irradiated with a high pressure mercury vapor lamp. The radiation corresponding to the solar spectrum was simulated by filtering the wavelengths less than 3,000 Å using Pyrex® glass. The polycarbonate film degraded when exposed to such radiation with a decrease in the transmittance in the UV region. The coating of cerium/polyvinylalcohol on polycarbonate film not only protected the film from photodegradation but also downconverted the UV component of solar energy to the visible region. The variation of complex composition that afforded the best protection to polycarbonate was studied as a function of time; also investigated was the complex formation between cerium and polyvinylalcohol present in polyvinylbutyral in ethanol solution, a formation that leaves a transparent film on Lexan® upon coating. This coating protects Lexan® from sun damage and also provides resistance to acid rain. Cerous chloride is a better salt than cerous nitrate for the complex formation because the nitrate ion strongly absorbs UV radiation and turns yellow, thereby decreasing transmittance in the visible region.

The three initial coating schemes of surface carboxylation by n. ric acid followed by cerous salt formation; carboxylated latex coating followed by cerous salt formation; and αw-dicarboxylic acid cerous salt coated by solvent casting were all tried and failed. Future efforts should study shifts toward 400 nanometers and durability properties of cerium or other downconverter binders or coatings.

Honeycombs

BATTELLE PACIFIC NW LABORATORY   Battelle Pacific NW Laboratory's program (Gordon 1979) evaluated commercially available plastic honeycomb materials to determine materials and sizes to provide structural support for thin plastic films used for collector coverplates. Studies were conducted on thermal and physical stresses on the bond between cover and films; prototype honeycomb/film covers were fabricated and compared in efficiency with double glass coverplates.

This study demonstrated that a solar coverplate can be fabricated with a plastic honeycomb core and plastic skins. An acceptable adhesive bond line between the components can be maintained by controlled application of a thixotropic adhesive.

The ability of a honeycomb, constrained between two sheets, to reduce thermal losses was demonstrated by solar efficiency studies. Consequently the daily average efficiency of a solar collector with a honeycomb core coverplate can be equivalent to one with a dual-pane, low-iron glass coverplate during adverse weather such as high winds [above 15–20 mph (7–9 m/s)] or overcast skies. The glass coverplate may be up to 12% more efficient when the sky is clear and there is no wind.

The honeycomb core coverplates used in this study were not necessarily the optimum design; their efficiency could very likely be improved by using a more transparent honeycomb with a better aspect ratio and by using different skin materials or combinations of materials.

The study indicated that honeycomb core coverplates can be competitive with dual-pane glass coverplates in locations where weather conditions are not ideal or at high collector operating temperatures when heat losses through the coverplate become significant.

These coverplates are equally effective for active and passive solar systems. They may even provide a particular advantage in a greenhouse because they are better insulators than glass. Also, by using a honeycomb that can provide shading from the summer sun, it may be easier to control the greenhouse temperature to avoid overheating. Additional research is needed to optimize the materials and geometry of a honeycomb core coverplate. Additional testing is also needed to measure efficiencies at higher operating temperatures.

UNIVERSITY OF CALIFORNIA AT LOS ANGELES   The University of California at Los Angeles (Buchberg 1979) developed glass honeycombs for suppressing natural convection within flat plate collectors. The four parts of the study were

1. studies of corrugated and plain glass sheet arrays, study of fabrication methods, and measurement of transmittance and emittance of collectors with such covers;

2. studies of prototype collector performance using honeycombs made of thin-walled glass cylinders, studies of the effect of cell-wall thickness, and studies of the effect of glass optical properties on honeycomb transmittance;

3. studies of possible methods for large-scale manufacture of honeycombs;

4. studies of design applications.

A properly shaped glass honeycomb placed between a nonselective absorber and the coverglass of a flat plate solar collector gives collection efficiencies significantly higher than those of conventional flat plate units—even those with selective absorbers—collecting solar energy at temperatures required for heating and cooling buildings. Three basic glass honeycomb shapes were analyzed and tested: thin-walled cylindrical glass tube honeycombs in square or hexagonal arrays; corrugated thin glass sheets stacked peak-to-trough to form double-sinusoid-shaped cells; and flat, thin glass sheets stacked to form long paralleled slots.

A continuous, hot rolling mill was used to corrugate commercial Micro-Sheet ® glass, thus demonstrating a key step needed for commercialization of glass honeycomb fabrication. Experimental-scale [61 × 61 cm (24 × 24 in.)] collectors and scaled-up collectors were fabricated and tested outdoors to verify the analytical/numerical, performance-prediction algorithms developed during the program.

LOCKHEED MISSILES AND SPACE CO. Lockheed's program (Marshall et al. 1977) derived designs for efficient, high temperature, flat plate collectors using convection-and-irradiation-inhibiting cellular structures between the absorber face and transparent cover. The goals of the program were to establish the effectiveness of plastic cellular structures in the reduction of thermal losses, to complete a systematic design and test leading to selection of an optimized flat plate collector system, and to develop manufacturing techniques for cost-effective, large-scale production of the plastic cellular structures.

Several plastic materials including Mylar ®, Tedlar ®, Lexan ®, Kapton ®, and FEP Teflon ® were selected for evaluation. Honeycomb sections of these materials were fabricated using various film thicknesses and aspect ratios (L/D) between 1 and 10. Optical properties of both the plastic films and honeycomb sections were determined, and performance characteristics of collectors using those materials were established on the basis of analytical models. Results from optical property measurements showed that reradiation losses from flat-black absorbers can be reduced significantly when the absorbers are covered with properly designed thin-film plastic honeycomb. Results also showed that all the plastic honeycomb materials except Kapton ® have excellent transmission of solar energy at high angles of solar incidence.

Initial analytical studies indicated that plastic, honeycomb-covered collectors could provide significantly better performance at high temperatures than comparable nonhoneycomb systems. To confirm these predictions, full-scale collectors were constructed using Mylar ®, Lexan ®, Tedlar ®, and Kapton ® honeycombs. Honeycombs over both flat-black and selective-black absorbers were included in the experimental studies. Only limited testing was conducted using FEP Teflon ® because this material did not lend itself to economical honeycomb fabrication using the expansion method.

Lockheed researchers tested both honeycomb and nonhoneycomb collectors under ambient weather conditions over a three-month period in the

Lockheed Solar Collector Test Facility. They evaluated collector performance over the temperature range of 104°F (40°C) to 250°F (121°C). Results from their tests confirmed that properly designed honeycomb collectors can provide a significant improvement in performance over nonhoneycomb systems. Instantaneous efficiencies above 50% were obtained for operating temperatures up to 230°F (110°C). Comparison of diurnal operation between honeycomb and nonhoneycomb systems showed even larger increases in performance by the honeycomb collectors.

Of the materials used, the overall performance of Lexan was best, followed closely by Mylar®. Tedlar® honeycomb exhibited mechanical instability problems, and although Kapton® provided better high temperature stability, it had a lower diurnal performance than the other materials because of its solar absorption characteristics at high incident angles. Kapton®'s performance, however, compared favorably with the nonhoneycomb systems.

Of the aspect ratios tested, an L/D of 5 appeared to be the optimum in terms of collector performance and honeycomb cost. However, additional cost and performance trade-off studies are required to determine the optimum L/D ratio for specific collector applications.

The program clearly showed that plastic cellular structures placed between the absorber surface and transparent cover can lead to significant improvement in thermal performance for the full operational range associated with heating and cooling of buildings. Whereas improved performance was demonstrated, some of the plastic honeycombs lacked structural stability at the high temperatures often encountered during "no-flow" conditions.

Miscellaneous

BROOKHAVEN NATIONAL LABORATORY   Brookhaven (Wilhelm, 1981) conducted an R&D program for a low-cost solar collectors that could stimulate a solar industry and promote a solar-heating market.

The initial part of the work, performed in the early 1980s, established the need for low-cost solar collectors as a critical factor in stimulating a solar industry and in promoting a solar-space-heating market in the interest of displacing fossil fuel. This early analysis placed a fairly severe constraint on collector and system cost based on considerations of payback and cash flow as well as life-cycle costing. Manufacturing costs of $1/ft² ($11/m²) and installation costs of $5/ft² ($55/m²) were indicated as necessary goals. Because Brookhaven was unable to find commercial products with these characteristics, researchers were led to an in-house effort to identify solar collector designs consistent with these goals.

The very low-cost goals for these collectors were to be achieved from a design that permitted very thin material to be incorporated into the construction of the collector. This was accomplished through the use of laminate technology, high performance polymer film, and a nonpressurized liquid absorber. The mass production of this kind of collector could yield a factory cost of $1.40/ft^2 ($13/m^2) and possibly less.

The most significant component of the design is the absorber heat exchanger. This unique component differs from conventional absorbers by the dramatic reduction in material it requires and its method of liquid transport.

Conventional absorbers are constructed of heavy metal and require some pressurization to function. The laminate absorber is made with thin polymer film and aluminum foil for a strong watertight envelope that requires no pressurization. The absorber functions in a flowback mode; in operation, water is pumped to the upper portion of the absorber and is permitted to flow back through the absorber envelope and piping to storage by gravity forces alone. Initial field tests of this concept pointed out a number of problem areas.

CENTER FOR THE ENVIRONMENT AND MAN   The Center for the Environment and Man (Shaffer 1978) investigated various thickeners to increase the viscosity of water so that it would not convect when heated from the bottom of the collector. This type of thickener would have made possible an alternate approach to the salt-gradient solar pond.

The first phase of the study was an investigation of various natural, semi-synthetic, and synthetic polymeric materials that could be used for increasing viscosity.

Although this project was seriously impaired by the untimely death of the original principal investigator, L. H. Shaffer, a few thickening agents for increasing viscosity were identified.

20.4.3 Insulations

Although little is said in the following chapters about insulations, these materials have presented some serious problems in collectors. Some have caused fires, smoked or smoldered, outgassed noxious vapors, not maintained physical shape, degraded in thermal resistance, absorbed moisture, suffered UV degradation, or outgassed products that condense on glazings. Chapter 23 does describe the important physical properties required of insulating materials.

Only a few insulation research projects have received DOE funding. There is nevertheless great potential for new concepts. Multiple layers of low emittance foils used independently or in combination with other insulating materials appears to be an area worthy of study. The DOE-funded projects are as follows:

DOW CORNING CORPORATION   Dow Corning (Rabe 1980) demonstrated a cost-effective silicone foam insulating system that offered good insulating properties, resistance to outgassing and moisture retention, thermal stability, and flame resistance.

Three types of silicone foam were developed. In general, they were equal to commercial insulation materials in thermal conductivity, better in thermal stability (outgassing), but poorer in mechanical strength and higher in cost.

SOLAR TURBINES INTERNATIONAL   Solar Turbines International (Wilcoxon, Sorathia, and Gagliani 1979) designed a project to characterize and optimize flame-resistant, no-smoke-emitting, open-cell polyimide foams with improved insulating capability, resistance to outgassing and moisture retention, thermal and dimensional stability, and cost-effectiveness for solar applications.

This project resulted in the optimization of two distinct types of advanced insulating materials, a flexible and a rigid foam, and the introduction of a third material, a high density foam. These materials offer unique safety features as well as excellent thermal and physical properties.

The specific properties that characterize these polyimide materials as candidates for use in solar energy systems are their ability to be produced as a lightweight flexible foam, a load-bearing rigid foam, or a density-molded configuration; their excellent weather resistance; their absence of wicking in the flexible or rigid insulating foams; and their maximum continuous-use temperature tolerance to 316°C (600°F) with short-term use to 482°C (900°F).

VERSAR, INC.   Versar's project (Versar 1980) objectives were to survey and document the properties of existing insulating materials, to tabulate and evaluate the properties of the materials, and to evaluate multifunction uses of various thermal insulation materials.

Desirable insulation properties for collector applications were reviewed, and all commercially available forms of insulation were evaluated for collector applications. Desirable insulation properties for both above- and below-ground transport and storage were also discussed, but commercially available forms of insulation were not evaluated.

20.4.4   Heat Transfer Fluids and Corrosion

Chapter 23 describes the currently used heat transfer fluids in detail, and chapter 24, describes the corrosion problems being encountered. A large number of the problems can be immediate and catastrophic, but they can be avoided through good engineering practices. It appears that an acceptable system can result through good materials, fluids and inhibitor (if needed) selection combined with good construction practices and compatible system materials. Periodic fluid monitoring is necessary to ensure a long life for the system.

Most of the DOE-funded projects have been completed, with the only active corrosion projects being those sponsored by the U.S. Army Corps of Engineers Construction Engineering Research Lab. The projects that were DOE funded are presented here.

ARGONNE NATIONAL LABORATORY   The objectives of this project (Wolosewicz 1980) were to design and fabricate a solar fluid corrosion test loop and to perform corrosion studies on aluminum, steel, and copper when these materials are exposed to heat transfer fluids being subjected to simulated stagnation conditions.

Solar collector materials have exhibited shorter than expected lifetimes. The collector environment includes temperatures of $80°-160°F$ ($27°-71°C$) under operating conditions and extremes of $30°-300°F$ ($1°-149°C$) during no-flow conditions (stagnation). Temperature extremes were evaluated as a possible cause of material failures. Tests for corrosion in operating systems were compared with tests conducted at stagnation temperatures.

Experiments were conducted in specially designed loops to evaluate metal corrosion and degradation of solar fluids and inhibitors under simulated operating and stagnation conditions in solar collector systems. Some aspects of the metal corrosion of solar components in circulating aqueous solar fluids below boiling were investigated in other DOE-sponsored research. This program complemented the other corrosion studies.

BATTELLE MEMORIAL INSTITUTE   Battelle created a data base to help users in developing cost-effective solar collectors through control of corrosion. The data base identified types of corrosion problems in collectors and set down the factors affecting corrosion. The overall program included a review of collector corrosion processes, a study of corrosion in multimetallic systems, and a determination of interaction between different waters and chemical antifreeze additions.

The Battelle report (Diegle et al. 1979) served as a reference to the state of the art at the time of publication and helped identify research needs that were pursued by DOE in the ensuing years.

DOW CORNING CORPORATION   Dow Corning (Marinik 1980) determined the cost-effectiveness of three silicone fluids as solar collector heat transfer fluids.

Physical data and performance data from actual collectors were also gathered and compared with those collectors using heat transfer fluids currently available.

The study concluded that silicones are essentially nontoxic, long lived, and noncorrosive compared with water/glycols or hydrocarbon-based oils. The negative aspects of silicones (i.e., cost, leakage, and lowered system efficiency) were not totally addressed, and thus a complete assessment of the relative merits of silicones versus other fluids did not evolve.

EIC CORPORATION   The EIC Corporation's project (Koch, Schnaper, and Brummer 1980) was designed to select electrochemically deposited polymeric coatings for solar collector heat exchangers, to define conditions for their application, to test their effectiveness for inhibition of corrosion, and to characterize them in terms of other properties (thermal stability, abrasion resistance, expected lifetime, and ease of reconditioning).

Polyphenylene oxide films did not stand up to ethylene glycol solutions. Cinnamaldehyde was resistant to the environment but was not rated successful in resisting the pinholes that result in pitting.

The results of this work indicated that polymeric coatings were not a technically feasible method of corrosion prevention in active solar systems.

UNIVERSITY OF FLORIDA   The University of Florida (Ingley 1980) investigated the use of vegetable oils as collector heat transfer fluids. The themophysical properties, cost, and performance of several natural oils were determined. Significant problems in terms of flammability and degradation occurred during the work. This appears to be an unpromising class of heat transfer fluids for active solar applications.

GINER, INC.   Giner's (Swette and Cocks 1981) objective was to determine a means whereby light-gauge aluminum solar collector panels, operating with aqueous glycol heat transfer fluids, could survive over long periods without suffering significant corrosion problems. Characteristic changes of glycol solutions caused by aging at elevated temperatures were studied. In addition the corrosive effects of those aged glycol solutions on aluminum and the pitting potential of the aluminum were investigated. A corrosion control manual was prepared that defines the conditions under which aluminum can be used successfully with inhibited glycols as a heat transfer fluid.

HOUSTON CHEMICAL CORPORATION   The Houston Chemical Corporation (Wisnewski 1979) tested material compatibility between aqueous base heat-transport fluids (deionized water, hard water, propylene glycol, etc.) and collector materials (copper, steel, aluminum, etc.).

LOS ALAMOS NATIONAL LABORATORY   Los Alamos (Reisfeld and Neeper 1982) has long been interested in the operational characteristics of fluids and in long-term corrosion in active solar systems. The special interest in corrosion in mild steel systems dates from the generation of the National Security and Resources Study Center (NSRSC) project that was proposed in 1973 and completed in 1977. Thus began a long-standing program of evaluating the potential long-term corrosion effects of leading collector fluids in resistance-welded, pressure-expanded, mild steel collectors such as the Los Alamos collector used on the NSRSC.

In 1974 two corrosion test loops were constructed to study the long-term effects of using glycol/water/inhibitor fluids in mild steel collectors. Two different designs of mild steel, resistance-welded, and pressure-expanded collectors were used. One had a serpentine up/down pattern that might trap air; the other was a diagonal flow arrangement. The systems were constructed with storage tanks and a controller and operated much as a home system would be operated. The fluid used in both loops was 85% ethylene glycol, 15% water, and 4 oz/gal NALCO® 39L inhibitor. Only the necessary makeup fluid was added, with no complete fluid change. When any system controller or pump failures occurred, the collectors were allowed to stagnate; the collectors also underwent stagnation for one entire summer. In-line corrosion test pots were installed in the system containing mild steel resistance-welded samples simulating the collector fabrication techniques.

The latest evaluation of the system and fluid was performed after seven years of operation. The in-line coupons were essentially corrosion free, pH had not appreciably decreased, and the reserve alkalinity of one of the two loops had decreased only slightly.

In addition to the ongoing testing, a project was undertaken to validate the considerable laboratory data on corrosion from DOE-funded research. A field data collection system was set up using the most commercially significant designs and metal/fluid combinations solar systems. Each site had removable corrosion coupons and automated data collection systems. The same metal/fluid combinations are being tested in an accelerated laboratory procedure by Olin Corporation under a DOE contract monitored by Los Alamos.

This project also compiled and reviewed all the literature addressing corrosion, fluids, fluid degradation, and active solar reliability. In addition unpublished data from a variety of private sector sources were included. This information was used to generate a state-of-the-art paper that described the factors affecting active system reliability and lifetime, including design and installation, fluid/metal compatibility, and fluid lifetime. The report appeared as a three-part series in *Solar Engineering and Contracting*, (Avery and Krall 1982) in February, March, and April 1982.

As of October 1982 the field sites were operational and Olin Corporation's laboratory tests were in progress. Methodology to compare lab and field data was also developed.

MONSANTO RESEARCH CORPORATION   Monsanto conducted a survey of heat transfer fluid properties to evaluate commercially available products and developmental materials for solar applications.

Information on fluids was collected with the new data, gaps in information were identified, and tests were performed to measure flammability and toxicity of certain fluids. A report (Parts and Conine 1981) was produced containing all available information in a readily usable format for designers of solar energy systems.

THE OLIN CORPORATION   Olin's project (Smith 1982) provided manufacturers, designers, engineers, and installers of solar equipment with the relative corrosion potential of various metal/fluid combinations by conducting both accelerated and longer-term tests on the corrosive potential in various metal/fluid interactions.

Researchers performed accelerated lab tests on most commercially significant metal/fluid combinations in aqueous, nonaqueous, and glycol environments. Lab tests were also performed on galvanic couples, solder fluxes, degraded glycols, reinhibition formulations, and metal/fluid combinations present in the Los Alamos National Laboratory field site program.

In combination with quantitative field data, Olin's work should result in a laboratory test that is semiquantitative in predicting corrosion rate and lifetime for nearly all potentially useful metals and fluids.

THERMACORE, INC.   Thermacore's objective was to develop high temperature, high performance solar receivers using heat-pipe absorbers in evacuated tubes. Fluid/material compatibility studies identified the best candidates. The first prototype used a steel heat-pipe absorber with a trimethylborate working fluid in standard General Electric TC-100® tubes. The second prototype used alloy steel pipes and water as the working fluid with better performance.

This project (Ernst 1978) is of interest to the solar materials field because of the studies of fluid/metal compatibility in heat-pipe applications. The end product will be directly applicable to efficient use with absorption and Rankine cycle chillers. Evacuated tubular solar collectors were selected as the only economical nonconcentrating approach capable of efficient operation of chillers.

Three heat-pipe fluld/vessel combinations were identified and tested for durability at design and stagnation conditions for periods exceeding 33,000 hours. Testing was carried out at the lower end of the temperature range by freeze/thaw testing several types of water heat pipes. This work resulted in the issuance of U.S. Patent No. 4,248,295.

Two heat-pipe collectors were tested using trimethylborate/1010 steel and copper/water heat pipes. Both collectors showed improved performance compared with the standard General Electric TC-100.

A cost analysis showed that in volume production the heat pipes could be made for $1.50 each (1978$) and would be cost-effective for the performance achieved in collector testing.

20.4.5 Sealants and Gaskets

Although sealants and gaskets are not specifically covered in chapter 24, many of their properties are discussed in the sections on absorber substrates and adhesives. In addition to the DOE-funded project presented here, there has been considerable cooperation between the sealant and gasket manufacturers in developing acceptable products for collectors. User requirements of long life, higher temperature tolerance, good UV stability, low creep, low compression set, and low outgassing are encouraging manufacturers to try to formulate new or modify existing products to meet these requirements. The DOE-sponsored projects are as follows:

MEGA ENGINEERING, INC. Although the main object of this project (Mega Engineering 1979) was the design and construction of a low-cost, high temperature, liquid-heating, flat plate collector, the project also produced considerable information on adhesive materials. Candidate cements were tested to verify their conductivities, and additives were evaluated. Bonding techniques were also tested. The use of high temperature, low outgassing, high conductivity adhesives can produce collectors that can be fabricated using aluminum, steel, or other economically competitive plate materials bonded to copper tubing.

Filled adhesives and cements using "low absorption" aluminum oxide powder or beryllia combined with various epoxy or inorganic adhesive binders have produced high conductivity, high service temperature materials. These adhesives have also exhibited low electrical conductivity combined with high structural strength. These adhesives permit bonding of different metals such as copper to either aluminum or galvanized sheet steel without danger of galvanic action.

Many of these high conductivity adhesives require no elevated cure cycles and therefore may be used by semiskilled labor in conventional fabrication techniques. Typical room-temperature materials such as Hysol's EA 934 epoxy and Castall's RTV material HTC-1100 were evaluated in prototype collectors at the Smithsonian Solar Physics Observatory, and elevated cure epoxy collectors (3M 3476[®] adhesives) were tested at the Florida Solar Research Center. Performance data, in terms of collector efficiency (based on useful gain over the solar incident energy), were compared with other existing collector systems and found to be equal to or better than similar brazed or Roll-Bond[®] collectors (i.e., efficiencies ~ 0.7).

Prototype collector absorber plates were formed with copper circulating tubes bonded to a flat sheet of aluminum alloy. Three collectors were tested, each with a different high service temperature, low outgassing, high conductivity adhesive (one with a room-temperature cure epoxy, Hysol EA 934[®]; one with an RTV[®] adhesive, Castall HTC 1100[®]; and one with a high temperature cure epoxy, 3M-3476[®] material). These adhesives have been used previously for high temperature applications in heat exchangers and engines. Other similar adhesives have been developed for high vacuum aerospace applications.

Finally, a cost/performance summary was conducted (using the University of Wisconsin's f-Chart program) to determine the economic viability of this concept. Note that the intent of this study was not to select specific adhesives; rather, its intent was to prove the feasibility of using general types of adhesives for collectors.

PRODUCTS RESEARCH & CHEMICAL CORPORATION   The Products Research & Chemical Corporation (Morris and Schubert 1980) evaluated high-temperature-resistant sealants (for solar collector use) based on silicones, fluorocarbons, acrylic elastomers, chlorinated polyethylene, ethylene propylene terpolymers, chlorinated polyethers, and polythioethers.

New formulations, organic modifiers, stabilizers, and pigmentation were studied and evaluated for adhesion quality and ease of application. Several sealant compositions (an EPDM rubber, a Viton[®] fluoroelastomer, and silicone polymers) performed satisfactorily in the initial screening consisting of high temperature stability and adhesion retention tests. Further tests of these candidate materials included determining adhesion retention under UV/water/heat conditions, fogging temperature, low temperature flexibility, and physical properties. Four silicone-based materials appeared to be suitable candidates for collector sealants: Dow Corning 90-006-02[®] and 3120[®], GE 1200[®], and PRC-PR-1939[®].

WESTINGHOUSE ELECTRIC CORPORATION Westinghouse (Mendelsohn et al. 1980) surveyed and evaluated characteristics of collector sealants and studied the control and effects of breathing in collectors. This work covered investigation and laboratory testing of commercially available sealants on modes of degradation, long-term durability, and the limitations of elastomers.

The second phase of the project concentrated on developing improved sealant materials for high temperature solar use. This second round of testing commercial sealants showed that many improvements have been made in the available sealants.

Another major area of the project involved a study of the effects of design and materials on the durability of solar collectors. Factors such as design, fabrication, materials of construction, seals and sealing techniques, and absorber plate coatings were observed on actual field units removed from service. Such phenomena as leakage, corrosion, and formation of deposits on glazing and absorber plate were noted.

Properties of several desiccants were evaluated to mitigate the deleterious effects of water on collector life. Absorbents for organic degradation products of sealants were also investigated to protect the glazing and absorber plate from deposited coatings.

20.4.6 Durability Testing

Evaluation of materials for terrestrial solar use began in 1975 with the funding of the IITRI materials exposure program. The commercial sector, as well as spacecraft materials investigations, provided extensive background information and results upon which to base the program. This and subsequent programs described next represent only a small portion of the efforts that have gone into durability testing because they do not represent commercial interests or PV-applicable program results.

Chapter 23 describes the types of durability testing being conducted, as well as the test methods, measurements, and equipment, and reviews some of the results. The major investigators and principle publications are referenced exhaustively.

Many programs investigating materials durability have been carried out by Sandia National Laboratories, McDonnell Douglas Astronautics, Co., Jet Propulsion Laboratory, and Battelle PNL. Of the many durability and exposure programs that were conducted under DOE funding, the only program remaining for low temperature application is the LANL exposure testing program. DSET Laboratories were involved in part of this program of exposing commercial materials. The DOE-funded projects were as follows:

BERRY SOLAR PRODUCTS   The Berry Solar Products project (Righunathan 1981) concentrated on providing continuous electrodeposits of improved black chrome on copper, aluminum, and stainless steel; the project also included accelerated durability testing.

This project focused primarily on durability testing of absorbers. The testing procedure consisted of simultaneous exposure of materials to solar irradiance and thermal cycling. About 100 commercially available solar absorptive coatings, both selective and nonselective, were tested for durability by simultaneously exposing them in a collector box with a coverplate to 78 temperature cycles between 450 ± 35°F (232 ± 20°C) and 200 ± 25°F (93 ± 14°C) and to solar radiation of about 22,000 L. Changes in the absorptance and emittance values were measured. Some of the coatings were also exposed to additional solar radiation of 156,000 L at Desert Sunshine Exposure Tests (DSET), Phoenix, Arizona, in EMMA under 8 : 1 concentration. (EMMA refers to outdoor accelerated testing devices using mirrors to concentrate solar energy.)

Results of testing selective coatings have shown that black chrome on nickel-plated copper, multilayer coatings on aluminum, black nickel on nickel foil, and conversion coatings on stainless steel are the most durable coatings under the afore mentioned test procedure. Black chrome on zincated aluminum can also be a viable coating if the zincating is done properly. Nickel coatings on aluminum and copper do not withstand humid environments; copper oxide coatings are also not durable.

DSET LABORATORIES, INC.   In the early 1980s DSET Laboratories, Inc. (Zerlaut and Anderson 1984), under contract to SERI and more recently to Los Alamos, began developing information for a solar materials design handbook. This program subjects materials to both real-time and-accelerated exposure tests, measures optical and physical properties at periodic intervals, and reports the retention (or loss) of properties as a function of exposure. Its primary objectives were to develop new materials information for a current data base and to validate the existing data base.

Several features distinguish this program from previous materials-testing programs: All materials are commercially available and were purchased from distributors or manufacturers, who did not have knowledge of their purpose; exposure intervals are based on the total UV radiation received; and most of the test methods employed simulate actual end-use conditions of solar device materials by using special fixtures designed for that purpose.

All of the materials in the program—selective and nonselective absorbers, reflectors, polymeric glazings, glass glazings, FRPs, and reflectors—were commercial state-of-the-art materials in 1981.

IIT RESEARCH INSTITUTE   The IITRI program (Gilligan 1980; Gilligan et al. 1980) was the first broad-scale attempt to obtain exposure data on many solar materials under identical exposure conditions and test methods.

IITRI exposed many samples of collector materials to direct and concentrated sunlight to determine their resistance to sun and weather. Optical and mechanical tests were performed to determine deterioration and mechanisms of degradation.

This program exposed more than 100 different types of materials, consisting of

1. transparent sheet materials—acrylics, polycarbonates, cellulose acetate butyrates, and inorganic glasses;

2. fiber glass-reinforced sheet materials—fiber glass-reinforced polyesters, and fiber glass-reinforced acrylics;

3. transparent thin-film materials—polyesters, fluorocarbons, and laminated or coextruded films;

4. absorber plate coatings and materials—black paints, selective absorbers, and organic absorber materials;

5. reflector materials—rigid second-surface plastic mirrors, inorganic glass second-surface mirrors, metallized plastic films, and aluminum reflector surfaces;

6. sealing tapes and strapping materials.

Three exposure sites were selected for comparison: Chicago, Phoenix, and Miami. All accelerated testing was conducted at Phoenix. Exposure times extended to 34.5 months for the longest exposure.

Although not all materials were tested, tests included mechanical tests (tensile, elongation, impact, abrasion, and peel tests) and optical tests [normal–normal or normal–hemispherical (N–N or N–H) transmittance, reflectance (N–N, N–H or bidirectional) absorptance, and emittance tests.]

Program results ranged from complete material failure in a few months to essentially no degradation over the maximum 34.5-month test period.

LOCKHEED MISSILES AND SPACE CO.   Lockheed's study (Greenburg 1980; Osiecki 1982) was divided into two phases. Phase I was the sampling and limited environmental exposure of a large number of current and developmental selective absorber surface finishes. Phase II was the extended environmental exposure of surface finishes chosen from those of phase I and the subsequent analysis of the degraded (environmentally exposed) and nondegraded finishes to determine the degradation mechanisms in the coatings.

The surfaces evaluated in phase I were black chrome on copper, aluminum, stainless steel, and mild steel substrates; black chrome on nickel-plated copper; nickel/chromium oxide on nickel foil; chromate conversion on stainless steel; evaporated/sputtered coatings; paints—thickness-sensitive and thickness-insensitive; and copper oxide on copper on mild steel.

Phase I testing included the following: 95% humidity at 30°C (86°F) for 30 days; -40°C to $+120$°C (-40°F to $+248$°F), 60 cycles; 12,000 equivalent sun hours [at 5 suns, UV illumination operates at 130°C (266°F)]; and 250°C (482°F) for 200 hours.

Because the phase I tests showed significant degradation to only the copper oxide surface, the more severe phase II tests were undertaken.

Samples of black chrome, copper oxide, nickel oxide, chromate conversion coating, and a thickness-sensitive paint were exposed to 90°C (194°F)/95% relative humidity (RH), 90°C (194°F)/low RH for up to 5,000 hours for the phase II program. The optical, chemical, and surface morphological characteristics of the samples were monitored at various times during the exposures.

The copper oxide surface was the least environmentally stable finish. During exposure to the 90°C (194°F)/95% RH environment, the spectral reflectance of the finish progressively increased below the near infrared and decreased at longer wavelengths, the copper oxidized from the 0 and $+1$ valence states to $+2$, and the finish became more porous. Similar but smaller changes were noted in the samples exposed to the natural environment in a collector box for 9.5 months. The properties of the samples exposed to 90°C (194°F)/low RH did not change significantly.

The black chrome surface changed slightly during the first 1,250 hours of exposure to the 90°C (194°F)/low RH environment. The spectral reflectance increased slightly at wavelengths greater than about 1.0 μm, the chromium was either oxidized from the 0 valence state or reduced from the $+6$ valence state to a $+3$ valence state, and the surface morphology became grainier during environmental exposure. Similar changes were observed for the samples exposed to the natural environment. The sample exposed to the 90°C (194°F)/95% RH environment did not show the change in spectral reflectance or the chromium reduction reaction.

The properties of the nickel oxide, chromate conversion coating, and thickness-sensitive paint did not change significantly during environmental exposure.

LOS ALAMOS NATIONAL LABORATORY   The objectives of the optical materials investigation (Moore 1983) included operation of a high altitude, materials exposure facility; exposure of chemical conversion coatings for passive or low temperature selective surface applications; determination of collector materials durability and reliability; and monitoring DOE contracts for research of optical materials.

In the exposure test facility outdoor test boxes were fabricated per ASTM E781-81 to satisfy stagnation conditions for simulation of single-glazed selective, double-glazed flat black, single-glazed flat-black, film-glazed selective uninsulated, and open exposure conditions.

Site measurements included total global insolation at horizontal, 36 and 45 deg vertical orientations, altazimuth solar direct beam, 36 deg total UV, horizontal IR, relative humidity, ambient temperature, and collector absorber or glazing temperatures.

Exposure testing was conducted on developmental and commercial materials. Exposure of developmental materials includes DOE contractor materials and near-commercial industrial materials. Commercial materials are a part of the DSET/SERI materials test program in which Los Alamos National Laboratory served as the high altitude exposure facility. The materials exposed included selective and nonselective absorbers, glazing materials including ar- and IR-treated materials, and reflector materials. Optically promising materials from the chemical conversion and selective paint projects at Los Alamos were included in the outdoor exposure efforts.

NATIONAL BUREAU OF STANDARDS   The National Bureau of Standards (NBS) (Clark et al. 1980; Masters et al. 1981) was involved in many aspects of the DOE Active Heating and Cooling Program. Their work in establishing standards and serving as a coordinator of research projects resulted in the generation of much basic solar materials research.

A few of the studies undertaken at NBS are described next.

Optical transmittance measurements. This study compared methods of calculating the transmittance of coverplate materials and the reflectance of absorber materials. The study was limited to evaluation of factors that influence the results with Method A of the American Society for Testing and Materials (ASTM) E 424.

Coverplate material. Laboratory studies were performed to obtain data needed for the development of standards to evaluate the performance and durability of coverplates for flat plate solar collectors used in solar heating and cooling systems. Coverplate materials were evaluated to assess their durability after exposure to heat aging, natural weathering, and accelerated weathering.

Survey of materials performance. This study was performed to obtain data on the performance of materials in operational solar energy systems, to identify and assess available standards for evaluating materials, to provide recommendations for the development of test standards for materials, and to provide guidelines to aid in the selection of materials for use in solar energy systems. Field inspections of 25 operational solar energy systems were performed and a questionnaire was sent to 459 manufacturers and installation contractors to obtain materials performance data.

Absorber materials. This study was performed to aid in the development of accelerated test methods needed for the evaluation of absorber materials and to incorporate the methods into draft standards for consideration as consensus standards by ASTM. Performance requirements for absorber materials were identified, and laboratory and field studies were performed to measure performance according to the requirements. The data obtained, using 12 absorber materials, were used as the technical basis for two draft standards. The report of this study presents the results of the research and includes the proposed draft standards.

Plastic containment materials. This project was centered around establishing a date base for materials performance and developing test methods and standards for plastic containment materials.

20.5 Conclusion

Chapter 26 addresses the economic issues of solar materials in terms of the overall solar system. The chapter covers low conductance glazings, durable polymer films, new solid absorbants, and high temperature receiver/storage fluids. Some of the more classic glazing systems are discussed, namely, high resistance, switchable, and low-cost glazings. The study points out that many of the systems have highly questionable enconomic justification. It also points out the need for a consistent form of economic evaluation of many research materials and concepts. For instance, many factors besides energy—such as glare reduction, direction of view, comfort, or aesthetics—enter into the desirability of certain products. A realistic economic value should be placed on as many factors as possible. The chapter finds that many studies and research projects have highly questionable economic justification. For instance, having received large federal support, the information, products, and research developments have overshadowed the work needed in other areas. Results of the electroplated, vapor deposited, chemical conversion, paints, black liquids, and polymeric absorbers are well documented. However, a selective coating that can be applied in situ, that is durable and costs virtually nothing, and that has high selectivity and temperature capability has not yet evolved. The closest to attaining these characterisitcs might be the selective paints.

In the area of polymeric absorbers the cross-linked polyethylene study has shown the material to be surprisingly durable both under glass and in open exposure. This could be an area where further study might be warranted in developing low-cost collectors.

There have been many promising results from the research conducted on films, polymers, glazings, and coatings. Some of the primary projects that appear to show promise in the long run are etched ar coating on glass, fluorinated polycarbonate, UV-stabilized and scrim-reinforced PVC film, and scrim-reinforced silicone. All of these must prove cost effective for the use intended, including initial cost, installation cost, maintainenance cost, and have durability and high prformance capability. Aesthetic values and other side issues, be they benefits or deteriments, must also be considered. Many of the advanced heat mirror coatings, switchable coatings, and high-R glazings presently undergoing study might also become cost effective.

Two basic new insulating materials have evolved from the DOE-sponsored projects. These were a series of silicone foams and rigid and flexible polyimide foams. Both have higher temperature capabilities than conventional foams, better resistance to outgassing or moisture retention, and better thermal stability. Other systems making use of foil low emittance properties should be studied further.

Various corrosion test programs have produced a number of reports that describe corrosion problems related to solar applications. From these have evolved field surveys, test loop results, state-of-the-art studies, protective coatings, and fluids studies of glycols, hydrocarbons, silicones, vegetable oils, water, inhibitors, and so on. It appears that with the information presently available, and with routine monitoring, well-engineered systems can operate without serious corrosion problems.

The progress that has been made since 1972 has been tremendous. An entire solar industry has grown out of the support and initiatives created by DOE and its predecessors. The solar industry has grown from a state where few solar-specific materials were in the marketplace and conventional materials were used for many unsuitable applications to one where categories of materials for each specific solar use are now available.

References

Avery, J. G., and J. J. Krall. 1982. Corrosion prevention and fluid maintenance in active solar systems: The state of the art. *Solar Engineering and Contracting* 1(2–4): 22–26, 31–36, 22–26.

Balcomb, J. D., and J. E. Perry. 1977. *Assessment of Solar Heating and Cooling Technology.* LA-6379-MS. Los Alamos, NM: Los Alamos National Laboratory.

Baum, B., and M. Binette. 1983. *Solar Collectors.* Final report. DOE/CS/35359. Washington, DC: U.S. Department of Energy.

Beecher, D. T., H. E. Ferree, M. C. Chuang, D. W. Feldman, and C. M. Rively. 1979. *Low-Cost Evacuated Tube Solar Collector.* Interim report.

Benning, A. C. 1980. Harshaw Plating Process HT SOL CRX 1079. The Harshaw Chemical Company, Cleveland, OH.

Bilow, N., R. I. Akawie, D. I. Basiulis, and A. L. Landis. 1980. *Development of Non-glass Glazings and Surface Coatings.* Final report. DOE/CS/35301. Washington, DC: U.S. Department of Energy.

Bradley, J. M. 1977. *Development of a Freeze-Tolerant Solar Water Heater Using Cross-Linked Polyethylene as a Material for Construction.* Final report, C00/2956-77/8. Washington, DC: U.S. Department of Energy.

Buchberg, H. 1979. Transparent glass honeycomb structures for energy loss control. *Solar Heating and Cooling Research and Development Project Summaries.* DOE/CS-0010. Washington, DC: U.S. Department of Energy, p. 1-12.

Clark, E. J., W. E. Roberts, J. W. Grimes, and E. J. Embree. 1980. *Solar Energy Systems— Standards for Coverplates for Flat Plate Solar Collectors.* NBS technical note 1132. Gaithersburg, MD: National Bureau of Standards, p. 157.

Diegle, R. B., J. A. Beavers, and J. E. Clifford. 1979. *Corrosion Problems with Aqueous Coolants.* Final report. Washington, DC: U.S. Department of Energy.

Edwards, D. K., K. E. Nelson, R. D. Roddick, and J. T. Gier. 1960. *Basic Studies on the Use and Control of Solar Energy.* NSF G 9505. University of California, Department of Engineering. Los Angeles, CA.

Ernst, D. M. 1978. *Cost-Effective Solar Collectors Using Heat Pipes.* Interim report. Leola, PA: Thermacore, Inc.

FAFCO, Inc. 1980. *A Coaxial Extrusion Conversion Concept for Polymeric Flat Plate Solar Collectors.* Final technical report. Washington, DC: U.S. Department of Energy.

Gerich, J. W. 1979. An inflated cylindrical solar concentrator. In *Proc. 3rd Annual Solar Heating and Cooling Research and Development Branch Contractors' Meeting.* Washington, DC, September 24–27, 1978. Washington, DC: U.S. Department of Energy, pp. 86–87.

Gilligan, J. E. 1980. *Exposure Testing and Evaluation of Solar Collector Materials.* DOE/CH-90034-TI (DE81027311). Washington, DC: U.S. Department of Energy.

Gilligan, J. E., J. Brzuskiewicz, J. E. Bruzskiewicz, and R. Mell. 1980. *Handbook of Materials for Solar Energy Utilization—Low-Temperature Applications.* DOE/CF/90034-TI (DE81027310). Washington, DC: U.S. Department of Energy.

Gordon, N. R. 1979. *Plastic Honeycomb for Solar Coverplate.* PNL-2953. Richland, WA: Battelle Pacific Northwest Laboratory.

Greenburg, S. A. 1980. *Evaluation of Selective Solar Absorber Surfaces.* LMSC-D-766220, CDE 82004140. Washington, DC: U.S. Department of Energy.

Grimmer, D. P. 1978. Solar-selective absorber coatings on glass substrates. In *Proc. AS/ISES Annual Meeting.* Denver, CO, August 28–31, 1978. Newark, DE: ASES, vol. 2, p. 280 (LA-UR-78-2478).

Grimmer, D. P., and J. G. Avery. 1981. Bonding solar selective absorber foils to glass receiver tubes for use in evacuated tubular collectors: Preliminary studies. Presented at the Solar Rising 1981 Annual ISES Meeting, Philadelphia, PA, May 26–30, 1981 (LA-UR-81-1023).

Herzenberg, S. A., L. K. Hien, and R. Silberglitt. 1983. *Active Solar Energy System Materials Research Priorities: A Subcontract Report*. SERI/STR-255-1782. Golden, CO: Solar Energy Research Institute.

Hottel, H. C., and T. A. Unger. 1959. The properties of a copper oxide-aluminum selective black surface absorber of solar energy. *Solar Energy* 3(3): 10.

Ingley, H. A. 1980. Vegetable oils: A superior liquid coolant. *Active Solar Heating and Cooling System Development Projects*. Washington, DC: U.S. Department of Energy, p. 13-7.

Jensen, S. O. 1980. Evaluation of absorbent liquid collectors. *Active Solar Heating and Cooling System Development Projects*. Washington, DC: U.S. Department of Energy, p. 9-34.

Klein, A., H. Yu, and W. Yen. 1981. Inhibition of photo-initiated degradation of polycarbonate by cerium (III) overcoating. *J. Appl. Polymer Science*, 26: 2381–2389.

Koch, V. R., G. H. Schnaper, and S. B. Burmmer. 1980. *Corrosion Protection of Solar Collector Heat Exchangers with Electrochemically Deposited Films*. Final report, C00/2496-3. Washington, DC: U.S. Department of Energy.

Kokoropoulos, P., E. Salam, and F. Daniels. 1959. Selective radiation coatings: Preparation and high temperature stability. *Solar Energy* 3(4): 19.

Landstrom, D. K., S. G. Talbert, and V. D. McGinnis. 1981. *Development of a Low-Cost Black Liquid Solar Collector, Phase II*. Final report. Washington, DC: U.S. Department of Energy.

Leffingwell, J. W., and J. L. Schwendeman. 1979. *Medium-Temperature Air Heaters Based on Durable Transparent Films*. Final report, ALO/4147. Washington, DC: U.S. Department of Energy.

Lin, R. J. H., J. C. Lee, and P. B. Zimmer. 1980. *Low-Cost Solar Antireflection Coatings*. Final report. Washington, DC: U.S. Department of Energy.

Lin, R. J. H. 1979. *Black Chrome Coatings Development*. Honeywell, Inc., DOE Contract No. DE-AC04-78CS 14287. Washington, DC: U.S. Department of Energy.

Mar, H. Y. B., J. H. Lin, P. B. Zimmer, R. E. Peterson, and J. S. Gross. 1975. *Optical Coatings for Flat Plate Solar Collectors*. Final report, September 16, 1974–September 16, 1975, C00/2625-75/1. Roseville, MN: Honeywell, Inc.

Marinik, J. A. 1980. Development of superior liquid coolants. In *Proc. Annual DOE Active Solar Heating and Cooling Contractors' Review Meeting*. Incline Village, NV, March 26–28, 1980. Washington, DC: U.S. Department of Energy, pp. 13-2–13-3.

Marshall, K. N., S. A. Greenburg, K. G. T. Hollands, and R. K. Wedel. 1977. *Optimiziation of Thin Film Transparent Plastic Honeycomb Covered Flat Plate Solar Collector*. Final report. Washington, DC: U.S. Department of Energy.

Masters, L., J. Seiler, E. Embree, and W. Roberts. 1981. *Solar Energy Systems—Standards for Absorber Materials*. NBSIR 81-188278. Gaithersburg, MD: National Bureau of Standards.

McChesney, M. A., P. B. Zimmer, and R. J. H. Lin. 1982. *Optimization of Solar-Selective Paint Coatings: Final Report for Period September 15, 1980, to June 15, 1982*. Roseville, MN: Honeywell, Inc.

McDonald, G. E. 1975. Survey of coatings for solar collectors. In *Proc. Workshop on Solar Collectors for Heating and Cooling of Buildings*. New York, November 21–23, 1974. NSF-RA-N-75-019, pp. 407–410.

Mega Engineering. 1979. *Final Report Study and Anaylsis of a Low-Cost Cement-Bonded Flat Plate Solar Collector*. ALO-41221T1. Washington, DC: U.S. Department of Energy.

Mendelsohn, M. A., R. M. Luck, F. A. Yeoman, and F. W. Navish, Jr. 1980. *Collector Sealants and Breathing*. ALO-15362-1. Pittsburgh, PA: Westinghouse Electric Corp.

Moore, S. W., J. S. Clements, and W. R. Doty. 1983. *Chemical Conversion Surfaces for Solar Energy Applications*. LA-9842-MS. Los Alamos, NM: Los Alamos National Laboratory.

Moore, S. W. 1983. *Los Alamos Optical Materials Reliability, Maintainability, and Exposure Testing Program*. LA-9735-MS. Los Alamos, NM: Los Alamos National Laboratory.

Morris, L., and R. J. Schubert. 1980. *Development of 400°F Sealants for Flat Plate Solar Collector Construction and Installation*. Final report. Livermore, CA: Lawrence Livermore National Laboratory.

Muller, T. K., and J. Sullivan. 1982. *Further Development of a Low-Cost Solar Panel*. Final report. DE-AC04-79AL12032. Mountain View, CA: Acurex Corp.

Osiecki, R. A. 1982 *Evaluation of Selective Solar Absorber Surfaces, Phase II*. LMSC-D876763. Palo Alto, CA: Lockheed Palo Alto Research Laboratories.

Parts, L., and D. L. Conine. 1981. *Superior Heat Tranfer Fluids for Solar Heating and Cooling Applications—Results of Acute Oral Toxicity Determinations*. Final report for September 15, 1980–April 30, 1981, MRC-DA-1096 (vol. 1) (DE82002758). Washington, DC: U.S. Department of Energy.

Pettit, R. B., and R. R. Sowell. 1979. Recent developments regarding electroplated black chrome solar coatings. In *Proc. Second Annual Conference on Absorber Surfaces for Solar Receivers*, January 24–25, 1979, Boulder, CO. SERI/TP-49-182, pp. 33–40.

Pitt, C. G. 1980. *Non-glass Glazings*. Final report, RTI/1708/00-1F, DOE Contract No. DE-AC04-78CS 30132. Washington, DC: U.S. Department of Energy.

PRDA. 1977. *Solar Collector Materials and Fluids for Solar Heating and Cooling Applications*. Program Research and Development Announcement EG-77-D-29-0003. Washington, DC: U.S. Department of Energy.

Rabe, J. A. 1980. Development of improved insulation materials. In *Proc. Annual DOE Active Solar Heating and Cooling Contractors' Review Meeting*. Incline Village, NV, March 26–28, 1980. Washington, DC: U.S. Department of Energy, pp. 13-4–13-5.

Raghunathan, K. Beatty, C. C., and Cotsworth J. L. 1982. *Accelerated Durability Testing and Evaluation of Potential Improvements of Coatings for Solar Absorbers*. Final report, DOE contract no. DE-AC04-78CS34293. Berry Solar Products, Edison, N.J.

Reisfeld, S. K., and D. A. Neeper. 1982. *Solar Energy Research at Los Alamos: October 1, 1981—March 31, 1982*. LA-9474-PR. Los Alamos, NM: Los Alamos National Laboratory.

Reisfeld, S. K., J. G. Avery, D. P. Grimmer, J. J. Krall, and S. W. Moore. 1983. *A Solar Materials Annotated Bibliography*. LA-9581-MS. Los Alamos, NM: Los Alamos National Laboratory.

Schwendaman, J. L., G. L. Ball, III, L. W. Leffingwell, and C. E. McClung. 1980. *Low-Cost Mirror Concentrators Based on Double-Walled, Metallized, Tubular Films*. ALO-4227-6. Washington, DC: U.S. Department of Energy.

Shaffer, L. H. 1978. Viscosity-stabilized solar ponds. *Solar Heating and Cooling Research and Development Project Summaries*. DOE/CS-0010. Washington, DC: U.S. Department of Energy, p. 1-14.

Skoda, L. F., and L. W. Masters. 1977. *Solar Energy Systems—Survey of Materials Performance*. NBSIR 77-1314. Gaithersburg, MD: National Bureau of Standards.

Smith, E. F., III. 1982. *The Corrosion Resistance of Metallic Solar Absorber Materials in a Range of Heat Transfer Fluids*. First annual report. The Olin Corp., Metals Research Lab., New Haven, CT.

Spanoudis, L. 1980. *Evaluation of Solar Selective Coating Stability*. A00/4285-20. Washington, DC: U.S. Department of Energy.

Swette, L. L., and F. H. Cocks. 1981. *Aluminum Corrosion in Uninhibited Ethylene Glycol-Water Solutions*. Final report for July 30, 1979–March 31, 1981, July 1981, DOE/CS/31072-2. Washington, DC: U.S. Department of Energy.

Tabor, H. 1967. Selective surfaces for solar collectors. *Low Temperature Engineering Application of Solar Energy*. ASHRAE, New York, ch. 4, pp. 41–52.

Tabor, H. 1955. Selective radiation I. wavelength discrimination. *Trans. Conference on the Use of Solar Energy* 2(1)sec A: 24.

Thorton, J. A., J. L. Lamb, and A. S. Tenfold. 1980. *Development of Selective Surfaces*. Telic 80-1. Santa Monica, CA: Telic Company.

U.S. Department of Energy. 1978. *The National Energy Act*. DOE/OPA-0003. Office of Public Affairs. Washington, DC: U.S. Department of Energy.

Versar. 1980. *Survey and Evaluation of Available Thermal Insulation Materials for Use on Solar Heating and Cooling Systems*. Final report. DOE/CS/35363-T1. Washington, DC: U.S. Department of Energy.

Vershihin, O. E., and B. P. Kozyrev. 1959. Determination of spectral coefficients of diffuse reflection of infrared radiation from blackened surfaces. *Optics and Spectroscopy* 6: 345.

Wilcoxon, A. L., V. A. Sorathia, and J. Gagliani. 1979. *Development of Polyimide Materials for Use in Solar Energy Systems*. Final report, August 1, 1978–January 1979, DOE/CS/35305-T1. Washington, DC: U.S. Department of Energy.

Wilhelm, W. G. 1981. Flat plate solar collector utilizing polymeric film for high performance and very low cost. Presented at ISES International Congress, Brighton, England, August 23–28, 1981.

Wisnewski, J. P. 1979. Testing of aqueous fluids for use in collectors of solar energy. *Solar Heating anc Cooling Research and Development Project Summaries*. Washington, DC: U.S. Department of Energy, p. 1-31.

Wolosewicz, R. M. 1980. Metal corrosion and scale deposition associated with solar fluid degradation in solar energy systems. *Active Solar Heating and Cooling System Development Projects*. Washington, DC: U.S. Department of Energy, p. 13-1.

Zerlaut, G. A., and T. E. Anderson. 1984. Testing and evaluation materials from solar applications. In *Proc. SPIE—The International Society for Optical Engineering*, August 21–23. *Optical Materials Technology for Energy Efficiency and Solar Energy Conversion*, vol. 502-14, pp. 152–160.

21 Materials for Solar Collector Concepts and Designs

Richard Silberglitt and Hien K. Le

21.1 Introduction

The two major goals of solar collector design are to maximize collector efficiency and to increase collector lifetime. Materials research can contribute significantly to the realization of these goals. One way to accomplish this is to identify and quantify properties one wishes to have for a given collector component, conduct a material selection and testing campaign, and finally, manufacture the component using the best material. Many of the materials available off-the-shelf do not quite fit the desired material property profile, so further material research and development is needed.

This chapter describes how, by using common sense, sophisticated understanding of materials science, and appropriate application of "design tricks" and "innovative concepts," materials research and development can be productively brought into play to develop materials that meet or exceed the property requirements for a given solar collector component.

Rather than broaden the scope of this chapter by looking at every solar collector material, we will consider a few solar collector components that are most amenable to innovative materials research (i.e., glazings, heat mirrors, and absorbers). The following text uses each of these selective components as a case study to illustrate the materials concepts and design approaches that are most useful for solar applications.

21.2 Case Study 1: Glazings

Glazings are the top cover of a solar collector. A glazing has three major functions: to minimize convective and radiant heat loss from the absorber, to transmit the incident solar radiation to the absorber plate with a minimum of loss, and to protect the absorber plate and tubing systems from the outside environment. A typical glazing design consists of one or two plates of a transparent material such as glass or polymer. An ideal solar glazing would have good optical properties [high transmissivity and high infrared (IR) reflectivity or absorptivity], good mechanical properties (high tensile strength, impact strength, softening point, melting point, and a thermal expansion coefficient matched to the collector frame), durability (retaining its original

Information for the case studies has been derived from the report *Active Solar Energy System Materials Research Priorities*, SERI/STR-255-1782 (January 1983), by Stephen A. Herzenberg, Le Khac Hien, and Richard Silberglitt, with some updates by the authors.

optical and mechanical properties, resisting abrasion, and collecting little dirt upon outdoor exposure), low cost, and light weight. In addition rigid glazings must be self-supporting, and stretched films must be able to withstand wind, snow loading, and high temperatures without sagging to the absorber. Most glazings are made out of glass. However, glass is expensive, heavy, and easily breakable, tempering notwithstanding. The relatively high weight of glass and the care necessary to install a glass-glazed collector can also dramatically increase collector cost. Polymeric materials (e.g., plastic sheets and films) are lightweight, hard to break, and often very cheap. They are especially inexpensive when used as a stretched thin-film glazing. Polymers would make excellent glazings, except for the fact that they are not as durable as glass when exposed to the collector environment (see table 21.1).

Much research can be done to improve the durability of polymer glazings. One can investigate the basic degradation mechanisms; concentrate on designing new polymers, laminates, blends, or additives to overcome degradation; look into new processing methods designed to optimize properties and lifetimes; and conduct extensive laboratory and field testing of available and promising materials. The high pay-off, more innovative avenues of glazing R&D are photodegradation mechanisms, UV stabilization, and polymer processing.

21.2.1 Impact of Research on Photodegradation Mechanisms

Photodegradation begins when UV photons, with sufficient energy to break many polymeric bonds, strike absorbing groups (or chromophores) in the repeating unit of the polymer. Degradation may also occur when impurities are introduced during processing or outdoor exposure. The excited state formed in the polymer either dissipates the photon energy harmlessly (by fluorescence, phosphorescence, or more commonly, by releasing energy as heat), or undergoes chain scission, oxidation, or some other degradation process. The vulnerability of a polymer to outdoor light depends largely on its UV absorptivity, the percentage of absorbed photons which lead to degradation, the average number of chain scissions induced by each destructive photon, and the nature of the degradation products.

In general, UV absorption and subsequent degradation occurs rapidly in polymers with chromophores in their repeat unit. These are unsuitable for application as the outermost layer of a polymer glazing. For such polymers UV protection can be achieved simply by making them the innermost layer of a laminate. Strength, UV resistance, and lower costs of raw materials are achievable in a glazing since with this concept a laminate is also stronger than a single-film or polymer sheet. Many such laminated films are on the market:

Table 21.1
Properties of solar collector glazing materials

Material	Solar transmittance[a] (%)	Tensile properties			Remarks
		Tensile strength (1b/in.^{2})	Elongation at break[b] (%)	Impact strength (ft-1b/in.)	
Glass[c]					
Libby-Owens-Ford 1/8-in. regular float					Estimated life depends primarily on avoiding induced mechanical failure; exposure increases transmittance
Initial	82.6[d]	—	—	—	
Exposed	84.3/32[e]	—	—	—	
Estimated life: indefinite					
Pittsburgh Plate Glass 1/8-in. tempered					Estimated life depends primarily on avoiding induced mechanical failure
Initial	84.0	—	—	—	
Exposed	83.0/24	—	—	—	
Estimated life: indefinite					
Libby-Owens-Ford 1/8-in. regular soda-lime					Estimated life depends primarily on avoiding induced mechanical failure
Initial	82.5	—	—	—	
Exposed	81.9/24	—	—	—	
Estimated life: indefinite					
Self-supporting plastic sheet					
Rohm & Haas Plexiglas V-100 (Type 045) 1/8-in. molded acrylic					
Initial	87.4	12,000	9.0	0.72	
Exposed	87.3/34	9,800/12	8.0/12	0.63/12	
Estimated life: 5–10 yr					

Rohm & Haas Plexiglas "G"				
1/4-in. cast acrylic				
Initial	85.8	8,600	5.0	0.6
Exposed	83.5	10,200/12	5.7/12	0.6/12
Estimated life: 5–10 yr				
U.S. Noramont S-300				
3/16-in. cast acrylic				
Initial	85.2	11,600	5.9	0.64
Exposed	85.4[a]/24	8,000/24	5.0/12	0.61/24
Estimated life: 5–10 yr				
U.S. Noramont S-GPA				
3/16-in. cast acrylic				
Initial	84.6	10,600	6.0	0.61
Exposed	84.3/24	9,000/12	5.0/12	1.0/12
Estimated life: 5–10 yr				
Rohm & Haas Tuffak				
1/4-in. extruded polycarbonate				
Initial	77.5(83.9)[f]			
Exposed	69.3(77.7)/12			
Estimated life: 1 yr				
Mobay Merlon				
1/8-in. polycarbonate				
Initial	81.1	10,800	112	17.7
Exposed	76.0/12	8,900/12	78/12	18.6/12
Estimated life: <4 yr				
Mobay Merlon with protective coating				
1/8-in. polycarbonate				
Initial	82.8	10,200	106	—
Exposed	82.9/12	8,200/12	82/12	—
Estimated life: <8 yr				

Table 21.1 (continued)

Material	Solar transmittance[a] (%)	Tensile properties			Remarks
		Tensile strength (lb/in.²)	Elongation at break[b] (%)	Impact strength (ft-lb/in.)	
Eastman Chemical UVEX 3/16-in. cellulose acetate butyrate					Limited data; strong temperature dependence
Initial	80.3	5,900	84	2.8	
Exposed	80.8/6	—	—	0	
Estimated life: 4–8 yr					
Plastic films					
Kalwall Sunlite Regular acrylic-fortified polyester, fiberglass-reinforced, 0.04 in.					
Initial	79.9	10,400	2.0	—	
Exposed	84.5/12	8,100/12	1.7/12	—	
Estimated life: 4–6 yr					
Kalwall Sunlite Premium modified polyester, fiberglass reinforced 0.04 in.					Degrades rapidly in sunny, moist climates
Initial	87.5	14,100	3.0	—	
Exposed	83.1/12	10,600/12	3.1/12	—	
Estimated life: 2–7 yr					
Filon Type S48 acrylic-fortified polyester, Tedlar[g] coated, fiberglass-reinforced, 0.03 in.					
Initial	88.2	16,600	2.0	9.1	
Exposed		15,500/12	2.8/12	9.3/12	
Estimated life: 2–4 yr					

		MD[h]	TD[i]	MD	TD	High temperatures decrease life
American Acrylic Corp. antique glass acrylic-alkyd, fiberglass-reinforced, 0.06 in.						
Initial	89.4	10,200		2.0		8.6
Exposed	86.4/12	8,300/12		1.6/12		7.2/12
Estimated life: 3–10 yr						
American Acrylic Corp. crystal acrylic fiberglass-reinforced. 0.06 in.						
Initial	89.9	9,500		3.0		3.6
Exposed	85.4/12	5,600/12		1.9/12		3.0/12
Estimated life: 1.5–3 yr						
Thin film plastic DuPont Tedlar SE polyvinylfluoride 0.004 in.						
Initial	90.7	13,000		137		—
Exposed	85.5/24	11,200/24		156/12		—
Estimated life: 2–5 yr						
3-M Company Scotchgard #10, polyester						
Initial	86.9	24,000	14,000	59	55	—
Exposed	86.1	13,000/6	6,900/6	25/6	19.5/6	—
Estimated life: 1.5–2 yr						
Penwalt Coporation Kynar 450, fluoro-carbon, 0.0038 in.						
Initial	82.5	5,400	4,600	126	288	—
Exposed	82.1/24	6,250/124	4,600/12	56/12	23/12	—
Estimated life: 1–5 yr						

Table 21.1 (continued)

Source: Data extracted from an unpublished Illinois Institute of Technology report on solar transmittance and tensile and impact strengths for glass, self-supporting plastic sheets, and plastic films. The data include initial and past exposure values and estimated service lives. Service life is determined by estimating the exposure required to decrease solar transmittance to 70%, regardless of initial value, or to produce mechanical failure.

a. Unless otherwise noted, value is solar normal transmittance and is a conservative estimate of solar hemispherical transmittance.

b. Elongation at break is extension in length induced by tensile stress at rupture or break, which is generally expressed as a percent of original length.

c. Glass is an excellent glazing material, so only a few types were tested. AFG Corporation and the Glass Division of Combustion Engineering Corporation are among suppliers of low irons glass with transmittance values of 89% to 92%.

d. Initial refers to unexposed control samples.

e. Exposed refers to exposed samples. Applicable exposure period is in months: 84.3/32 means solar transmittance was 84.3% after 32-month exposure.

f. Notations 77.5(83.9) is % solar normal and solar hemispherical transmittance.

g. Trademark of E.I. DuPont de Nemours & Company.

h. Machined direction.

i. Transfer direction.

e.g., Tedlar over FRP, which is made by Filon Corporation (Jorgenson 1982); and Tedlar over polycarbonate, which is part of the Ramada Energy Systems predominantly polymeric collectors.

Alternatively, one can also consider the polymer structure and relate it to the polymer resistance to photodegradation, so as to identify the more promising materials. PMMA and copolymers have no oxygen in the polymer backbone chain. This means that they should be less susceptible to UV photons, and hence more stable. Similarly the fluorocarbons [e.g., polyvinyl fluoride (Tedlar) and polyvinylidene fluoride (Kynar) which have very strong F–C bonds] should also be less susceptible to UV degradation. A high degree of cross-linkage would also confer resistance to UV degradation in materials such as methyl-silicone glass resins. Indeed, all of these materials have been shown to be very promising, and many have been used in the polymer laminate glazings.

21.2.2 Pay-off from UV Stabilization Improvement

UV stabilization refers to the reduction of UV radiation degradation in a polymer. A variety of stabilizing approaches have been developed over the years (Lappin 1971; Wiles and Carlsson 1980–81), including UV absorbers, which absorb the potentially damaging radiation and then dissipate it without transferring energy to the base polymer (e.g., hydroxybensophenones; see Kloppfer 1976); excited state quenchers, which remove energy from chromophores in their excited state before the onset of irreversible degradation; and radical scavengers, which combine with alkyl or peroxide radicals and interrupt the oxidation chain reaction. It is suspected that the hindered amines stabilize in part through scavenging by a nitroxide radical. The nitroxide radical is then regenerated by a second scavenging reaction and other anti-oxidants, which actually break down oxidation products, in particular, the hydroperoxides (Usilton 1978; Carlsson, Chan, and Wiles 1980; Felder, Schumacher, and Sitek 1980).

To stabilize a polymer, it is common practice to simply blend the ultraviolet stabilizers with the base polymer to be protected. However, stabilizers are often incompatible with the base polymer, and slowly diffuse onto the surface of a film. They then evaporate, are hydrolyzed, or form nontransparent precipitates with atmospheric gases or dirt (Lappin 1971; Gupta 1978). This very serious problem—stabilizer bleaching—has been dealt with in a variety of ways. One simple approach is to use only high molecular weight stabilizers. Another is to modify the stabilizer structure to reduce its mobility. For example, modifications of the polymer structure of benzatriazoles has enabled Ciba-Geigy to extend the time by more than a factor of 10 until 50% of the stabilizer is lost at 280°C (436°F) (Dexter and Winter 1981). However, it is questionable whether the above two approaches can provide the compatibility needed in solar glazings exposed for 10–20 years.

Approaches in the last 10–20 years include stabilizer-base polymer copolymerization, stabilizer polymerization, and mixing the stabilizer into the monomer mixture prior to polymerization.

The first alternative is to copolymerize the stabilizer to the base polymer. This approach could theoretically completely prevent the stabilizer from leaching to the surface of a thin film. The approach has been researched by both the PV encapsulant program and private industry (Bailey and Vogl 1976; Tirrel 1980; Karrer 1981; Gupta 1978). JPL successfully copolymerized benzophenon and abenzatriazole UV absorbers with methyl methacrylate (MMA) (Gupta 1982). Because completely copolymerized stabilizers might be extremely expensive and might have insufficient mobility (particularly a problem for excited-state quenchers and radical scavengers), JPL combined the copolymerized acrylic UV absorbers with commercially available acrylics by blending and grafting. The blended copolymers did not separate even after continuous exposure to liquid water for seven days at 30°C (86°F). Outdoor exposure of grafted films indicated that the chemical attachment is permanent, and resistance to degradation is excellent. Promising avenues for further research include durability of these films, new degradation mechanisms that may be introduced by the stabilizer in the backbone or side chain, and additional stabilizer/polymer combinations. Similarly the optimization of copolymerized and blended stabilizer concentrations, to minimize cost and allow sufficient stabilizer migration for protection of the outer surface of the polymer, should also be very interesting.

The second alternative is to polymerize only the stabilizer, and then blend the base polymer and the stabilizer chains (Lappin 1971). This approach is essentially an extension of the idea of incorporating high molecular weight stabilizers. How much it will extend the time needed to complete separation of the stabilizer and base polymer has not been established, however.

The third approach is to retard stabilizer leaching, as has been attempted in West Germany (Hosch n.d.). Methyl methacrylate has been polymerized in the presence of sterically hindered amines. Without slowing polymerization, the monomer mixture additive improved PMMA UV-stability and retarded leaching rates substantially when compared with PMMA blended with the same stabilizer during extrusion. Copolymers of 83% MMA and 12% acrylate proved particularly stable, retaining 80% transmission (at 300 nm) for 1,000 hours under a UV lamp. With no stabilizer, the copolymer transmission dropped from 76% to 49%. This technique would be more likely to prevent surface degradation of sheet material than polymerization or copolymerization of the stabilizer. This approach could be applicable to solar panels because of the stability of the PMMA and copolymers produced, the high UV transmission of the material, and the slow separation of polymer and stabilizer.

Another approach is to use stabilizers in combination. Research by Dexter and Winter (1981) and Gupta (1982) has demonstrated that the incorporation of many stabilizers into a material can dramatically lengthen its lifetime. Since the conditions in which glazings must perform are highly variable (temperature, strain, humidity, UV light, etc.), combinations of stabilizers may be particularly desirable. Combinations of UV absorbers, the hindered amines, and other antioxidants might be especially effective. It has also been observed that stabilizer migration can be strongly affected by the orientation of the polymer. Thus the use of biaxially oriented films might also help retard stabilizer leaching.

21.2.3   Effect of Processing Variables on Degradation

Processing affects the materials properties, and thence the durability of polymers. Research has shown that processing at high shear rates (low temperature, rapid draw) causes mechanochemical rupture of polymer bonds and the formation of macroalkyl radicals which begin oxidation reactions (Mellor, Moir, and Scott 1973). This results in increased chromophore concentration, which leads to higher initial weathering rates and mechanical property deterioration. Since high temperature processing produces too much thermal oxidation, the optimal draw temperature is generally an intermediate one. Although there is clear evidence of processing-induced chromophore incorporation, there is no clear evidence that the amount of chromophore incorporation during processing significantly affects durability. Most experimental evidence suggests that the concentration of peroxides and ketones builds up so rapidly during the onset of degradation that initial chromophore concentration is irrelevant to polymer lifetime. Furthermore, while peroxide and carbonyl buildup is more rapid in polymers drawn quickly at low temperatures than in those drawn at lower shear rates, the separation between times to embrittlement is much smaller (Carlsson, Garton, and Wiles 1978). Even in well-processed materials there are regions near the surface where initial chromophore concentration is large and where embrittlement occurs relatively quickly. These arguments suggest that for a material in which initial chromophore concentration is important, careful control of the processing rate can yield much better resistance to long-term degradation. This is a very valuable insight for materials research.

21.3   Case Study 2: Heat Mirrors

Heat mirrors perhaps best exemplify the usefulness of innovative concepts in solar materials research. The concept of a heat mirror is simple. For solar collector application a heat mirror is a wavelength-selective material that exhibits good transmittance over the visible spectrum and low emittance (or high reflectance) throughout the infrared spectrum. This implies that a heat mirror, when placed on the inner side of a collector glazing, will reflect most IR radiation from a nonselective absorber coating back toward the absorber repeatedly. Thus it effectively helps the absorber coating to eventually capture most, if not all, of the solar energy incident upon it, just as a selective absorber does. Although in principle one could achieve equivalent benefits from a selective absorber or a heat mirror, the latter may be preferable in certain applications.

Heat mirrors are far more effective in collectors with evacuated glazings. With selective absorbers convection currents carry heat to the inner glazing where it is absorbed and then re-radiated by a high emissivity material. Heat mirrors, on the other hand, are placed outside the air-filled space between absorber and inner glazing. Top losses in a vacuum-glazed heat-mirror coated collector are only about a third of those in a vacuum-glazed selective absorber collector (Benson 1984).

In addition heat mirrors place no constraints on the absorber design or manufacturing process and are themselves easily adaptable to high speed, continuous production. On the other hand, selective absorber coatings can slightly constrain both manufacturing process and collector design. For example, Acurex Corporation eliminated coated foils as an option in its low-cost collector work because thin metal layers would have constrained the polymeric absorber tubes in the Acurex collector as they expanded and contracted due to changes in fluid temperature and pressure (Muller 1982).

The desired properties of a heat mirror for solar collector applications are high solar transmittance and low reflectance in the UV, visible and near-infrared spectrum (e.g., at wavelength $\lambda < 2\ \mu$m); low solar transmittance and high reflectance in the infrared spectrum ($\lambda > 2\ \mu$m); high durability; low cost; and ease of deposition on a variety of substrates. These desired properties are generally found in three classes of materials: doped semiconductors (or doped metal oxides), thin metal films, and metal microgrids. The more innovative research concepts that have been instrumental in developing these types of materials are discussed below.

21.3.1   Innovation in Doped Semiconductor Heat-Mirror Research

Doped semiconductor heat mirrors are essentially materials that have a band gap—for example, wide enough to exhibit high solar transmittance—and a large enough number of free charge carriers to move the plasma wavelength λ to the near infrared, so as to exhibit high IR reflectivity in the appropriate wavelengths of the solar spectrum.

The physics of doped semiconductor heat mirrors is complicated. Haacke (1982) has defined a quality factor Q_r that includes all the semiconductor parameters affecting the steepness of the reflectivity. Ideally

$$Q_r = \frac{N\mu^2 m^*}{\varepsilon_0 \varepsilon} - 1$$

should be as large as possible; here N is the number of charge carriers, μ the electron mobility, m^* the effective mass, ε_0 the vacuum dielectric constant, and ε the dielectric constant of the semiconductor. The implications of the Haacke model for heat-mirror design are numerous. A high carrier concentration of 10^{20} to 10^{21} cm^{-3}, which can be obtained by doping, will yield an impurity-conduction band spacing about 0.3–0.4 eV, close to the photon energy at the reflectivity step wavelength ($\sim 2\ \mu$m). A high electron mobility, which can be achieved by careful control of the film deposition parameters, will improve Q by increasing infrared reflectivity. Although these general approaches to improving particular coatings are well understood, theory is not yet advanced sufflciently to guide the selection of new conducting oxides for laboratory research. The interdependence of electronic parameters (e.g., the inverse relation between m^* and μ) makes this selection complex. This is an extremely important area because of the large number of alternative heat-mirror materials that merit careful examination (Lampert 1981).

The three most carefully researched doped semiconductors, or transparent conducting oxides, are indium tin oxide ($In_2O_3:SnO_2$, or more commonly, ITO), doped tin oxide (SnO_2), and cadmium orthostannate ($CdSnO_4$). Several ternary semiconductors have also been examined.

The major limitation of transparent conducting oxides for collector applications has been their low solar transmittance. However, table 21.2 shows that several ITO, tin oxide, and $CdSnO_4$ coatings have solar transmissivities of about 0.9 and above, and IR reflectivities of 0.8 to 0.9. In general, the ITO coatings exhibit slightly superior selectivity due to their higher electron densities and mobilities (which result in greater IR reflectivity). The ways by which properties are improved for each one of these major doped semiconductors are highlighted below.

Table 21.2
Properties of doped semiconductor conducting-oxide heat mirrors

Material	Deposition technique	Sheet resistance (Ω/sq)	Thickness (microns)	Bulk resistivity (Ω/cm)	Mobility (cm²/V-s)	Carrier density (cm³)	Band gap (eV)	Average τ_{vis}	τ_{solar}	R_{IR} or ε_{IR}	Reference
In₂O₃:Sn	Plasma reactive vacuum evaporation, 350°C	25	0.4	1×10^{-3}	—	—	—	—	0.96 0.4–1.6 μm	—	Nath and Bunshah (1980)
In₂O₃:Sn	Plasma reactive vacuum evaporation (~300°C)	~25	—	$\sim 4 \times 10^{-3}$	—	$\sim 10^{21}$	—	—	0.9– 0.94 0.4– 1.2 μm	—	Nath et al. (1980)
In₂O₃:Sn	Reactive evaporation	~21	0.5	—	—	—	—	>0.85[a]	—	—	Machet et al. (1981)
In₂O₃:Sn	rf-sputtering in Ar 600°C	—	0.35	—	—	—	—	—	0.90[b] (AM2)	(0.081, 121°C)	Fan and Bacher (1979)
In₂O₃:Sn	rf-sputtering in Ar 550°C	2.6	<0.3	-1×10^{-4}	38	7×10^{20}	—	>0.8[a]	~0.8	0.92, 10 μm	Fan and Hachner (1976)
In₂O₃:Sn	rf-sputtering in Ar 600°C	3	0.35	1×10^{-4}	—	—	—	—	0.90 (AM2)	0.83, 10 μm	Fan (1979)
In₂O₃:Sn	Reactive sputtering, Ann, 300°–500°C	~30–120	0.13	—	—	—	—	~0.86 (0.4–0.7 μm)	—	—	Smith et al. (1980)
In₂O₃	Thermal evaporation	<10	2.7	—	74	4.7×10^{20}	3.56	0.41 (0.3–0.6 μm)	—	—	Pan and Ma (1980)
In₂O₃:Sn	CVD organometallic 500°–555°C	50	0.165–0.5	4.3×10^{-4} 7.1×10^{-4}	—	—	—	0.83–0.89	—	—	Kane et al. (1975)
In₂O₃:Sn	sp pyro 600°–800°C	20[c]	—	—	—	5×10^{20}	—	0.91, 0.59 μm	0.79	0.90, 10 μm	Kane et al. (1975)
In₂O₃:Sn	sp hydro	15–20	—	—	—	1.5×10^{21}	—	~0.9	~0.8	0.85	Kostlin (1974)
In₂O₃:F	Reactive low plating, Ann, 450°C	—	1.5	—	13	7×10^{2}	—	~0.8	~0.8 (0.4–1.2 μm)	>0.8	Avaritsiotis and Howson (1981)
In₂O₃:Sn (AR w/MgF₂)	e-beam evaporation	—	0.4	3×10^{-4}	—	—	—	—	—	≥0.9	Hamberg et al. (1982)
SnO₂:F[d]	CVD organometallic (250°C)	10	—	—	41.2	2.25×10^{20}	—	—	0.8	—	DeWaal and Simonis (1981)
SnO₂:F[d]	sp hydro 500°– 570°C	4	1.0	4×10^{-4}	37	4.4×10^{20}	—	0.75	0.75	(0.15)	Van der Leij (1979)
SnO₂:F[d]	sp pyro >400°C	—	0.3	—	20	6×10^{20}	—	0.85	0.8	>0.8, 10 μm	Frank et al. (1982)
Cd₂SnO₄	rf-sputtering	—	<0.3	—	—	—	—	—	0.89	(0.20, 77°C)	Haacke (1977)
Cd₂SnO₄	rf-sputtering, Ann, 420°C	26–43	<0.3	—	—	—	—	—	0.86	(0.12, 77°C)	Haacke (1977)
Cd₂SnO₄	sp pyro 450°C	—	0.22	—	—	$\sim 2.5 \times 10^{19}$	—	>0.85	—	—	Ortiz (1982)

a. Includes substrate.
b. $T_{glass} = 0.906$.
c. A range of 10 to 10^{15} has been reported.
d. All compounds designated SnO₂ may or may not be doped.

Indium Tin Oxide   The ITO coating with the highest reported solar transmissivity at the time of this writing has $\tau_s > 0.95$ until the Sn fraction increases to more than 10% by weight, and it remains above 90% until the Sn fraction reaches 20% (Nath et al. 1980). Coatings have been deposited by plasma-assisted reactive evaporation with various tin fractions and substrate temperatures. Transmission is relatively insensitive to substrate temperature between room temperature and 250°C (482°F), while the electrical resistivity decreases steadily, especially above 300°C (572°F). TEM micrographs and the above coating properties suggest that low substrate temperatures produce stoichiometric oxide films, whereas higher deposition temperatures induce oxygen deficiency, high carrier concentration ($N = 10^{21}/cm^3$), and increases in conductivity. The IR reflectivity of these films is proprietary. However, data suggest that for this process, the optimum deposition parameters (in collector applications) would be a substrate temperature between 200° and 300°C (392°–572°F), and a Sn fraction between 10% and 20% (Nath et al. 1980). It has also been shown that films with lower, but reasonable, visible transmissivity and low resistivity could be achieved as well with simple reactive evaporation as with plasma-assisted activation (Habermeier 1981).

Highly transparent $In_2O_3 : SnO_2$ films have been reactively ion-plated at 300°–350°C (572°–662°F) (Machet et al. 1981). Other highly solar or visible transparent ITO films have been reactively sputtered and then annealed at 300°–500°C (572°–932°F) (Smith et al. 1980); thermally evaporated from an In, In_2O_3 source (to produce an undoped, metal-rich film) (Pan and Ma 1980); deposited by CVD using organometallic reagents (Kane, Schweizer, and Kern 1975); laid down by CVD spray pyrolysis (Van Boort and Groth 1968) and spray hydrolysis (Kostlin 1974); and reactively ion-plated in the presence of fluorine (Avaritsiotis and Howson 1981).

Antireflection or texturing of ITO might make its optical properties equivalent to the best selective absorbers by eliminating local, visible region reflectance maxima that degrade its solar transmittance. This approach has been explored in Sweden and Germany (Hamberg, Hjorstberg, and Granqvist 1982; Frank and Kostlin 1982; Frank et al. 1982).

German researchers, reporting on the advantages of activated reactive evaporation, have proposed the production of multiple-layer oxide heat mirrors. A four-layer transparent oxide stack ($MgF_2/In_2O_3/SiO_2/MgF_2/glass$) had a reflectivity of below 1% between 450 and 750 μm (Ebert 1982). The quoted deposition rates (5 Å/s) and the quality of the coatings produced (due to the low bombarding energy of plating ions) suggest this process should be examined more closely.

A final possible ITO (or other) heat-mirror design might disperse large conducting flakes in a solar transparent binder. In contrast to the case for selective absorbing paints, the long-wave IR (>2 μm) emissivity typical of organic binders should not damage the selectivity of a heat-mirror paint. As long as the flake is large enough that its dimensions neither inhibit electron mobility nor introduce scattering effects, a heat-mirror paint might well have excellent properties. In addition to a large flake, some way of inducing even distribution, and alignment parallel to the substrate, would probably be required.

Tin Oxides   Of the tin oxide films described in the literature, those doped with fluorine have the best properties (Van der Leij 1979; De Waal and Simonis 1981; Frank and Kostlin 1982; Frank et al. 1982). Fluorine doping results in twice the electron mobility, and twice the electron mean free path, of antimony and phosphorus doping.

No deposition of a homogeneous antireflection layer of SnO_2 has been reported in the literature. However, it has been proposed to grade SnO_2 content from 0% to 100% in a dielectric layer between the glass substrate and heat-mirror coating. Visible reflectivity of 15%, IR reflectivity of 90% (at 10 μm), and electrical resistivity of 5 Ω/in.^{2} per 0.5 μm, has been reported in a fluorine-doped, CVD-deposited, SiO_2–SnO_2/SnO_2 coating (Fan 1979).

Cadmium Orthostannate   The optical properties of the best cadmium ortho-stannate coatings approach those of the best ITO ($\tau_s = 0.86$, $\varepsilon_{IR} = 0.12$) (Haacke 1977). Optimal properties to date are obtained in films prepared by re-sputtering in oxygen and then annealed (Haacke 1982). Although these films can be deposited onto plastic substrates (e.g., polycarbonate) with good adhesion by reactive sputtering in an Ar/O_2 atmosphere (with no subsequent anneal), transmissivity and reflectivity suffer on the order of 10% each (Haacke 1982). $CdSnO_4$ can also be deposited by ion plating at high deposition rates (500 Å/min) at room temperature (Lampert 1981) and by CVD spray pyrolysis at 450°C (842°F) (Ortiz 1982). The former process has been used to coat polyester, and the latter to produce coatings with visible transmissivity of greater than 0.85.

A related ternary semiconductor, $CdIn_2O_4$, has shown promising IR reflectivity (0.93) and solar transmissivity that oscillates between 0.78 and 0.92 (Haacke 1982). An antireflection layer might bring τ_s up to 0.9.

A variety of other transparent oxide films have been suggested (e.g., LaB_6 and BiO_2). Researchers in the heat-mirror field are fond of pointing out that at least some of these alternatives must be as good as ITO, SnO_2, and

Cd_2SnO_4. However, it is still unclear if these three unusual materials, which have somewhat accidentally become the focus of heat-mirror research, are indeed the best available choices.

21.3.2 Design Concepts in the Development of Thin-Metal Films as Heat Mirrors

Thin-metal films can be made into transparent heat-mirror fllms because the free electrons of the metal produce the desired infrared reflectivity, while transmittance occurs at solar spectral wavelengths that are below the plasma wavelength. In selecting metal films for heat-mirror applications, it is also desirable to have a low lattice damping constant so as to get a rapid transition from low to high reflectance (Lampert 1981).

The simplest type of thin-metal film heat mirror is a single-layer metal film. Since refractory heavy metals have a high lattice damping content, most metal films examined as possible heat mirrors have been noble metals (especially gold, silver, and copper). The properties of these single-layer metal films are seen in table 21.3. None of these individual thin-metal films has acceptable solar transmissivity. Even if the bulk optical constants of metals were valid for very thin metal films, the best obtainable τ_s, R_{IR} product would be about 0.7 (Karlsson et al. 1981). Since island formation, which can substantially degrade conductivity and IR reflectivity, has made bulk constants impossible to achieve in thin films, an antireflecting layer is required to produce high selectivity (Karlsson et al. 1981). These materials characteristics have led to the realization that single-layer metal films are severely restricted as heat-mirror materials and shifted attention to multilayered thin-metal films. Deposition of multilayer films was pioneered at MIT, where heat mirrors have been applied to solar collectors (Fan and Hachner 1976; Fan 1979). In addition Japanese researchers at Teijin, Ltd., have produced terabutyl titanate, (TBT)/Ag/TBT, coatings on polyethylene terephthalate using a low-cost hydrolysis process (Fuschillo 1975). Optical properties can be manipulated by changing the percentage of TBT-n-hexane in the coating solution and by changing layer thicknesses. The deposition process is compatible with roll coating. However, the reflectivity step of the Teijin films climbs over a broad range, substantially degrading the optical properties. Another possibility is to use the metal nitrides, which are very stable and can be antireflected. Calculations suggest that the solar transmissivity of refractory metal nitride films might increase to 0.75, and the IR reflectivity to a similar value, through the deposition of ZnS dielectric layers on either side of the metal.

Table 21.3
Properties of metal film and multilayer heat mirrors

Material	Deposition technique	Sheet resistance (Ω/sq)	Thickness (Å)	Average R_{vis}	Average τ_{vis}	τ_{solar}	R_{IR} or ε_{IR}	Reference
Ag	CVD	—	100	—	—	0.41	0.95, 10 μm	Groth and Kauer (1965)
Cr	rf sputtering	—	30	0.16	0.60	0.58	0.14, 0.8–2 μm	Fuschillo (1975)
Ag	Vacuum evaporation	—	—	—	—	0.42	~0.95	Ortiz (1982)
Cu	Vacuum evaporation	—	—	—	—	0.6	0.8	Ortiz (1982)
Au	Vacuum evaporation	—	—	—	—	0.35	0.85	Ortiz (1982)
Al	Vacuum evaporation	—	—	—	—	0.4	0.7	Ortiz (1982)
Ni	Vacuum evaporation	—	—	—	—	0.55	0.3	Ortiz (1982)
TBT/Ag/TBT	sp hydro	—	400/140/400	—	0.75 (0.4–0.7 μm)	—	0.97	Chiba et al. (1982)
In_2O_3: Sn	rf sputtering, etched microgrid	$6^{[a]}$ $2 \times 10^{-4\,[c]}$	$0.35^{[b]}$	—	—	0.9	0.83	Fan (1979)
Zn/Ag/ZnS	Vacuum evaporation	10	520/100/770	—	—	0.68	(0.06)	Koltun and Faiziev (1974)

a. Before etching.
b. Thickness in microns.
c. Bulk resistivity in Ω/cm.

21.3.3   Development of Metal Microgrids

The concept of metal microgrid heat mirrors is extremely interesting. A metal microgrid can be designed to exhibit heat-mirror properties through careful selection of the grid size and wire diameter. The grid will reflect wavelengths much greater than the mesh spacing (the infrared wavelengths), while appearing transparent for wavelengths close to or shorter than the spacing, acting essentially as an electromagnetic filter or antenna.

The metal microgrid examined by Fan (1979) had excellent optical properties ($\tau_s = 0.9$, $R_{IR} = 0.83$). However, since it was an ITO coating, the extent to which the microgrid design enhanced its selectivity is unclear. In general, despite initial theoretical interest in this approach, microgrids have proved difficult to produce in either suspended wire form or photoetched thin films (Haacke 1982), and much work remains to be done to fully explore their possibilities.

21.4   Case Study 3: Selective Absorbers

Selective absorbers are materials that are used as coatings, paints, or adhesive-backed foils on the absorber plate to help it maximize the conversion of solar to thermal energy. The term *selective* is used to indicate that the absorber has a very high absorptance in the solar spectrum and a very low emissivity in the blackbody spectrum.

The ideal selective absorber properties are high selectivity (e.g., absorbtivity greater than 95%, emissivity less than 0.1%), high durability, ease of application, and high compatibility with the various types of absorber substrates available (e.g., metals and polymers).

A large number of materials have been used as selective absorbers (see table 21.4 which lists the most important selective absorbers available). From a materials point of view, these selective absorbers can be classified in five categories: intrinsic absorbers, metal-dielectric composites, multilayer absorbers, semiconductor metal tandems, and textured surfaces. The development of each one of these selective absorber types embodies many innovative designs and concepts, which are described below.

21.4.1   Intrinsic Absorbers

Intrinsic absorbers are conceived as materials that inherently have the desired spectral response of ideal absorbers. Transition metals doped with C, N, O, and B are thought to be good candidates for intrinsic absorbers, because they localize electrons in the incompletely filled d-shell of the metallic ion, and thus

Table 21.4
Properties of candidate-selective absorbers

Absorber[a]	Type	Deposition technique	Maturity[b]	α	$\varepsilon\,(T, ^\circ C)$	T stability[c] ($^\circ$C)	1982 estimated commercial cost ($/m²)
ZrB_2/Si_3N_4	Intrinsic	CVD	1	0.93	0.08–0.09 (102)	500 (air)	
Cr–Cr_2O_3 (black chrome)	Graded composite	Electroplating	5	0.92–0.97	0.04–0.06 (100)	400 (air)	Coating: 8.50 Foil: 15.00
Cr–Mo–Cr_2O_3	Graded composite	Coelectroplating	1	0.96–0.97		400 (air)	>Black chrome
Ni–Al_2O_3/Al_2O_3	Graded composite	Anodic oxidation	5	0.92–0.97	0.1–0.26 (65)	300 (air)	
Zn–ZnO	Graded composite	Anodic oxidation	1	0.98	0.18 (100)	<300 (air)	
Cu_2O–CuO–Cu	Composite/tandem	Anodic oxidation	1	0.95	0.34 (100)	130 (air)	2.50
NiO–Ni–$Cr/NiCrO_x$	Graded composite	Chemical conversion	5	0.97–0.99	0.07–0.1 (100)	250 (air)	Foil: 8.00
SS–C	Graded composite	Reactive magnetron sputtering	3	0.94	0.03–0.1 (100)	400 (vacuum) 200 (air)	5.00
SS–C (on rough sputtered Cu)	Textured composite	Reactive magnetron sputtering	3	0.9	0.04 (67)	450 (vacuum)	5.00
SS–SSO/SSO	Antireflective composite	Reactive magnetron sputtering	1	0.89–0.93	0.08 (20)	150 (air)	
SS–SSO/Al_2O_3	Antireflective composite	Reactive magnetron sputtering	1	0.9	0.07 (20)	150 (air)	

Cr–Al$_2$O$_3$	Graded composite	Dual source magnetron sputtering	1	0.92	0.09 (20)		
Mo–MoO$_2$/Si$_3$N$_4$	Composite/intrinsic	CVD	1	0.91	0.11 (500)	500 (vacuum) 300 (air)	
Fe,Mn,Cu oxides in Paint binder	Pigment in binder (Thickness sensitive)		3	0.9	0.24 (100)	200 (air)	1.00
Oxides in paint binder	Pigment in binder (thickness insensitive)		1	0.9	0.31 (100)	200 (air)	1.00
Al$_2$O$_3$/Pt–Al$_2$O$_3$/Al$_2$O$_3$	3-layer composite	rf Magnetron sputtering	3	0.91–0.93	0.08–0.1 (20)	>600 (air)	10.00 (on Cr)
Soot in paint binder	Pigment in binder		1	0.9	0.3 (100)		
AMA (M = Cr)	3-layer	Reactive magnetron sputtering	3	0.95	0.12 (20)	300 (air)	2.00–5.00
AMA (M = Ni)	3-layer	Reactive magnetron sputtering	3	0.91	0.08 (20)	350 (air)	2.00–5.00
AMA (M = Ta)	3-layer	Reactive magnetron sputtering	3	0.89	0.12 (20)	300 (air)	2.00–5.00
Proprietary	3-layer	Electron Beam evaporation	3	0.92–0.96	0.05–0.08	250 (air)	Foil: 7.00–8.00
Te	Textured tandem	Angled-vapor deposition	1	0.92	0.03		
a–Si/Si$_3$N$_4$	AR tandem	CVD	1	0.75	0.8 (500)	500 (air)	
Cu$_2$S	Tandem	Chemical spray deposition	1	0.89	0.25 (100)	139 (air)	2.50
CoO	Textured graded tandem	Electroplated Co: heat oxidation	1	0.98	0.2 (100)	425 (air)	
CoO–FeO$_3$–Co$_3$O$_4$	Textured tandem	Electoplated Co: heat oxidation	1	0.9	0.07 (100)	300 (air)	

Table 21.4 (continued)

Absorber[a]	Type	Deposition technique	Maturity[b]	α	ε $(T, °C)$	T stability[c] (°C)	1982 estimated commercial cost ($/m²)
Ge–CaF$_2$	Semiconductor in a dielectric	Sputtering	1	0.65–0.72	0.01–0.1 (100)		
Ge	Textured tandem	rf sputtering: H$_2$O$_2$ etching	1	0.98–0.99	0.58		
Copper	Textured	Sputter etch	1	0.9–0.95	0.08–0.11	300 (air)	
SS	Textured	Sputter etch	1	0.9–0.96	0.22–0.26	350 (air)	
Ni	Textured	Sputter etch	1	0.9–0.95	0.08–0.11	250 (air)	

Note: Conversion factors: $1°F = (9/5)°C + 32$; $1\,m^2 = 10.76\,ft^2$.
a. Absorber layers are separated by a slash. Constituents of composite layers are separated by a dash.
b. Maturity of absorber coatings: 5 = commercial, 3 = development, 1 = research.
c. Temperature stability and commercial cost for most absorbers are not well-known.

shift the plasma edge to a longer wavelength (Ehrenreich and Seraphin 1976; Call 1979; Lampert 1979; Herzenberg and Silberglitt 1982).

Materials which approximate intrinsic absorber behavior are ZrB_2 and NiN_x. ZrB_2 was developed by Randich, Allred, and Pettit. In the non-optimized sample the absorptivity of ZrB_2 is 0.67–0.77, while emissivity is 0.08–0.09. The absorptivity can, however, be increased to 0.93 by the use of a S_3N_4 antireflective coating (Randich and Allred 1981). NiN_x, which in strict terms is not an intrinsic absorber because it must be coated onto a metallic substrate to show selectivity, was developed by Sikkins (1981, 1982). The absorptivity and emissivity of NiN_x are 0.84 and 0.039, respectively.

21.4.2 Semiconductor-Metal Tandems

The concept involved here is to place a semiconductor with a band gap near the long wavelength end of the solar spectrum (0.58 eV) and a high absorption coefficient onto a metal substrate which provides the requisite low emissivity. Further enhancement of the tandems can be achieved by an antireflective layer, textured surface, or dielectric host matrix to reduce the front surface reflection that results from the high refractive index of the semiconductors.

Both silicon and tellurium have been explored as possible tandem selective absorber materials. The silicon work (Booth, Allred, and Seraphin 1979) has focused on amorphous material because of its high absorption coefficient and the fact that CVD-deposited amorphous silicon (a-Si) has broadened band tails that allow significant absorption, down to about 0.9 eV. Absorption at longer wavelengths also increases at higher temperatures. Carbon-doped a-Si on Mo with an Si_3N_4 antireflecting layer has achieved $\alpha = 0.75$, $\varepsilon = 0.08$ with excellent durability to at least 500°C (932°F) in air. The tellurium tandems have been made by both angled-vapor deposition and gas evaporation (Cocks and Peterson 1980). Tellurium (Te) is characterized by a very small band gap (0.35 eV), which ensures absorptivity throughout the solar spectrum, but a large index of refraction ($n = 3$), which leads to large front surface reflection. Surface texturing has successfully eliminated the latter problem, yielding Te tandem selective absorbers with $\alpha = 0.95$, $\varepsilon = 0.06$ (gas evaporation), and $\alpha = 0.92$, $\varepsilon = 0.03$ (angled-vapor deposition). Research to establish the durability of these coatings is necessary before their commercial potential can be evaluated.

Other semiconductor-metal tandems that have been characterized include Cu_2S, chemical spray deposited on aluminum in air (Gadgil et al. 1981), and germanium either incorporated in a CaF_2 dielectric or with a textured surface (Gittleman, Sichel, and Arie 1979; Swab, Krishnaswamy, and Messier 1980). Neither has shown acceptable selectivity, although textured surface Ge has high absorptivity and the Cu_2S coating has an estimated (1982) cost of only $2.70/m^2 ($.25/ft^2).

A unique tandem selective absorber consisting of a heat mirror deposited on a black substrate has also been developed in Holland. It consists of fluorine-doped tin oxide over black enameled steel (Van der Leij 1979; De Waal and Simonis 1981). It is highly resistant to mechanical scratching and to chemical attack, is stable up to 350°C, and has been used in several European Solar Pilot Test facilities.

21.4.3   Metal-Dielectric Composites

Metal-dielectric composites are conceived from effective medium theories—such as the Maxwell-Garnet theory—that attempt to derive the optical properties from the microstructure of composites consisting of metal particles suspended in a dielectric medium. By varying the type, size, shape, and distribution of the metal in the dielectric medium (which can be an oxide), it is believed that ideal selective absorber properties can be achieved.

A very large number of metal-dielectric composite selective absorbers have been deposited in a large variety of ways (Granqvist 1981). The most prominent of these is black chrome, which consists of a Cr_2O_3 layer over a small Cr particle/Cr_2O_3 host composite. Electroplated black chrome has achieved $\alpha = 0.95, \varepsilon = 0.05$, which remained unchanged after 5,600 hours in air at 350°C (662°F). Its microstructure and optical properties have been extensively studied. The principal research issue for black chrome in low temperature applications is cost reduction. The major parameters affecting the (1982) cost of about $9/m^2 are thickness of the 0.1–0.2 mil Ni underlayer, the Cr/Cr_2O_3 coating deposition time, and the plating current (and associated power supply capital costs). Several alternative metallic composite coatings with properties competitive with those of black chrome have also been investigated.

In Sweden anodized coatings, consisting of metallic nickel grains embedded at the root of a 0.7 μm porous Al_2O_3, have been developed with absorptivity in the 0.92–0.97 range and emissivity in the 0.1–0.26 range. The Swedish coatings also display considerable temperature and humidity durability (Andersson, Hunderi, and Granqvist 1980).

Ergenics produces a chemically converted black nickel foil, about 0.3 μm thick with about 200 Å of $NiCrO_x$ near the surface, that grades to NiO (Lampert 1980). It may also contain some elemental Ni or Cr. It is deposited on 0.5 mils of electroformed nickel in a proprietary acidic oxidizing continuous production line. Optical properties and durability appear to be excellent (Mason and Brendel 1982).

Sputtered cermets (graded metal carbides) include an Australian prototype consisting of three layers with different homogeneous concentrations of elemental metal (Harding and Window 1979). Reactive sputtering in Ar and CH_4 on Cu substrates has yielded coatings with $\alpha = 0.93$, $\varepsilon = 0.04$, and a deposition time of a few minutes at most. These coatings degrade above 200°C (392°F) in air, in part because they become homogeneous. But they are extremely stable in a vacuum, which makes them strong candidates for evacuated tube applications.

Even better candidates for high temperature application might be the graded cermet coatings of Pt/Al_2O_3 and Mo/Al_2O_3 that have been produced by coevaporation at Cornell University (Nyberg, Craighead, and Buhrman 1982), and the Pt/Al_2O_3 coating sputtered at Telic Corporation (Thornton and Lamb 1981) These coatings have shown even better optical and durability properties than the metal carbides. Pt/Al_2O_3 has achieved $\alpha = 0.93$, $\varepsilon = 0.08$, and it is stable after 700 hours in air at 600°C. Mo/Al_2O_3 has achieved $\alpha = 0.992$, $\varepsilon(200°C) = 0.08$, and it is stable in air at 400°C and in a vacuum at 750°C.

Spectrally selective Ni–C films have been produced by reactive sputtering in an Ar–CH_4 mixture (Sikkins 1982). These films consist of small Ni particles (25–50 Å) embedded in an electrically insulating amorphous carbon matrix. Absorptivity is moderate (0.08–0.90), and emissivity extremely low (0.028–0.045 at 150°C). The films are stable at greater than 400°C in vacuum.

Beyond these commercial or near-commercial competitors to black chrome, a number of other approaches have been explored in the laboratory. These include coelectroplated black chrome-molybdenum (Smith and Ignatiev 1981), chemical vapor deposited black molybdenum (Gesheva, Seshan, and Seraphin 1981; Chain, Seshan, and Seraphin 1981; Seshan et al. 1981), electroplated cobalt (Smith, Ignatiev, and Zajac 1980), and anodically oxidized ZnO (Hohaul et al. 1981) and copper black (Potdar et al. 1981).

A significant effort has also been devoted to making selective paints that absorb much like metal-dielectric composite absorbers. In particular, Honeywell has developed thickness-sensitive and thickness-insensitive solar-selective paints composed of an iron, manganese, and copper oxide pigment in a silicone binder (McKelvey, Zimmer, and Lin 1979; McChesney 1981). These paints have reasonable absorptivity and might be very cheap at approximately $1.00/m^2$ ($0.10/ft^2$) in 1982 dollars.

The thickness-insensitive paint incorporates an aluminum flake in the binder to suppress infrared emissivity. The best laboratory properties achieved with this design to date are a $\alpha = 0.92$, $\varepsilon = 0.31$. An alternative design might incorporate a precoated metal flake in a binder, thereby requiring only the optimization of one particle size, shape, and distribution (Moore 1982a).

21.4.4   Textured Surface Absorbers

Textured surface absorbers provide selectivity through a combination of multiple reflections and a graded index of refraction resulting from surface voids. However, texturing must be extremely fine and deep to achieve selectivity. The depth of pores or peaks must be approximately ten times the spacing between peaks (Moore 1982). Not surprisingly, low-cost texturing techniques (e.g., chemical etching) tend to produce coatings with poor selectivity. The high technology equipment usually necessary to produce closely spaced, tall needles or cones (e.g., sputtering or CVD equipment) is relatively expensive. Dendritic growth is probably the most promising approach to the production of cheap, highly selective coatings. We note, however, that many selective absorbers employ texturing to enhance selectivity (Smith, Ignatiev, and Zajac 1980).

21.4.5   Multilayer Absorbers

Multilayer absorbers achieve selectivity via destructive interference of the light reflected from the front surface of each layer. Telic Corporation has magnetron-sputtered a large variety of such coatings, sandwiching a metal or metal oxide layer between two layers of rf- or reactively deposited Al_2O_3 (Thornton, Penfold, and Lamb 1980). Excellent selectivity and durability have been achieved for a variety of coatings (see table 21.4).

The Optical Coating Laboratory (OCLI) in Santa Rosa, California, has also developed three-layer interference coatings using physical vapor deposition. Applied by electron beam evaporation to a 60-in.-wide foil, OCLI's best absorber endured extended exposure at 500°C (932°F) in air, and 90°C (194°F) at 95% relative humidity (Williams n.d.). OCLI estimated that the coating would be sold for about \$7.50–\$8.50/m^2 (\$0.70–\$0.80/ft^2) in 1982 dollars. Another commercial multilayer coating (containing aluminum) with high selectivity ($\alpha = 0.96, \varepsilon = 0.07$) has been reported (Raghunathan 1981). The coating survived 161 thermal cycles between 93.3° and 232°C (200° and 450°F) in an outdoor test collector.

21.4.6   Deposition Technologies

Selective absorbers can be deposited on the absorber substrate by a variety of methods. These include electroplating, rf-magnetron sputtering, chemical vapor deposition, pigment dispersion, chemical conversion, anodic oxidation, electron beam evaporation and similar processes.

Worth mentioning is the dispersion of a solar-absorbing pigment in a paint binder, which is a process considered likely to produce the cheapest low temperature absorbers, presuming adequate durability can be achieved. Achieving durability with the most selective thickness-sensitive paints is not simple, however, when the coating must be so thin the IR-absorbing binder does not increase emissivity and must also contain enough pigment to attain high absorptivity. Essentially a small amount of binder must provide good adhesion and retain its properties at high temperatures for at least a few years. Of course one advantage of paints is the facility with which they may be repaired or replaced. However, because substrate preparation is critical to paint lifetime, on-site painting may result in much shorter lifetimes than factory application by the manufacturer. Therefore the lifetime may be considerably shorter than with other absorbers.

Similarly chemical conversion coatings have been touted as possible cheap, adherent absorbers for flat plate collectors or for passive applications. Although most such coatings do not have good optical properties, Ergenic's black nickel does have high selectivity and appears to be quite durable. A number of alternatives examined by Moore also have moderate selectivity (Moore 1982b).

21.5   Conclusions

Materials research can contribute significantly to collector development, as seen in the description of selected innovative research concepts and designs for three solar collector components: glazings, heat mirrors, and selective absorbers.

The principles of materials development are straightforward, and they remain the same throughout the examples shown: First, design the desired properties; then identify the interplay of materials parameters and desired properties; and finally, apply this understanding to materials research.

References

Andersson, A., O. Hunderi, and C. G. Granqvist. 1980. Nickel pigmented anodic aluminum oxide for selective absorption of solar energy. *Journal of Applied Physics* 51(1): 754.

Avaritsiotis, J. N., and R. P. Howson. 1981. Fluorine doping of In_2O_3 films employing ion-plating techniques. *Thin Solid Films* 80: 63.

Bailey, D., and O. Vogl. 1976. Polymeric ultraviolet absorbers. *Journal of Macromolecular Science: Reviews in Macromolecular Chemsitry* C14(2): 267.

Benson, D. 1984. Evacuated glazings for flat plate collectors. Progress report, SERI/PR-255-1894. July.

Booth, D. C., D. D. Allred, and B. O. Seraphin. 1979. Stabilized CVD amorphous silicon for high temperature photothermal solar energy conversion. *Solar Energy Materials* 2: 107–214.

Call, P. 1979. *National Program Plan for Absorber Surfaces R&D*. SERI/TP-31-103. Solar Energy Research Institute, Golden, CO, Jan. 1979.

Carlsson, D, J., K. H. Chan, and D. M. Wiles. 1980. Polypropylene photostabilization by tetramethylpiperidine species. *ACS Organic Coatings and Plastics Chemistry Proceedings*, vol. 42. American Chemical Society 179th National Meeting, Houston, TX, March 23–28, 1980, p. 555.

Carlsson, D. J., A. Garton, and D. M. Wiles. 1978. Some Effects of production conditions on the photo-sensitivity of polypropylene fibers. *ACS Advances in Chemistry Series* 169: 56.

Chain, E. E., D. Seshan, and B. O. Seraphin. 1981. Composition and microstructure of black molybdenum photothermal converter layers deposited by the pyrolysis of $Mo(CO)_6$. *Journal of Applied Physics* 52: 1356.

Chiba, K., S. Sobajima, U. Tonemura, and N. Suzuki. 1982. Optical coatings for energy efficiency and solar applications. *Proc. SPIE—The International Society for Optical Engineering*, January 28–29, 1982, C. M. Lampert, ed. Los Angeles, CA, vol. 324, p. 23.

Cocks, F. H., and M. J. Peterson. 1979. *Journal Vacuum Science Technology* 16(4).

Cocks, F. H., M. J. Peterson, and P. L. Jones. 1980. *Thin Solid Films* 70: 297.

De Waal, H., and F. Simonis. 1981. Tin oxide coatings: Physical properties and applications. *Thin Solid Films* 77: 253.

Dexter, M., and R. A. E. Winter. 1981. Ciba-Geigy Corporation. U.S. Patent No. 4,278,589, July 14, 1981.

Ebert, J. 1982. Optical thin films. *Proc. SPIE—The International Society for Optical Engineering*, January 26–27, 1982, R. I. Seddon, ed. vol. 325, p. 29.

Ehrenreich, H., and B. O. Seraphin. 1976. *The Fundamental Optical Properties of Solids Relevant to Solar Energy Conversion*, November 20–23, 1975. NSF Symposium report under Grant #DMR 75-18134, Washington, DC. The National Science Foundation, Tucson, AZ, 1976.

Fan, J. C. C. 1979. Optics in solar energy utilization II. *Proc. SPIE—The International Society for Optical Engineering*, vol. 85, p. 39.

Fan, J. C. C., and F. H. Bacher. 1979. *Journal of the Electrochemical Society* 122: 1719.

Fan, J. C. C., and F. J. Hachner. 1976. Transparent heat mirrors for solar energy applications. *Applied Optics* 15(4): 1012.

Felder, B., R. Schumacher, and F. Sitek. 1980. Hindered amine light stabilizers: A mechanistic study. *Proc. ACS Organic Coatings and Plastics Chemistry*, vol. 42. American Chemical Society 179th National Meeting, Houston, TX, March 23–28, 1980, p. 561.

Frank, G., and H. Kostlin. 1982. Electrical properties and defect model of tin-doped indium oxide layers. *Applied Physics* A27: 197.

Frank, G., E. Kauer, F. J. Schmitte, and H. Kostlin. 1982. Optical coatings for energy efficiency and solar applications. *Proc. SPIE—The International Society for Optical Engineering*, January 28–29, 1982, C. M. Lampert, ed. Los Angeles, CA, vol. 329, p. 58.

Fuschillo, N. 1975. Semi-transparent solar collector window systems. *Solar Energy* 17: 159.

Gadgil, S, B., R. Thangaraj, J. V. Iyer, B. K. Gupta, and O. P. Agnihotri. 1981. Spectrally selective copper sulphide coatings. *Solar Energy Materials* 5(2): 129.

Gesheva, K. A., D. Seshan, and B. O. Seraphin. 1981. Composition and microstructure of black molybdenum photothermal converter layers deposited by the pyrolytic hydrogen reduction of MoO_2Cl_2. *Thin Solid Films* 79: 39.

Gittleman, J. I., E. K. Sichel, and Y. Arie. 1979. Characterization of black Ge selective absorbers. *Solar Energy Materials* 1: 93.

Granqvist, C. G. 1981. *Journal de Physica* C1, 42(1).

Groth, R., and E. Kauer. 1965. Thermal insulation of sodium lamps. *Philips Technical Review* 26(4/5/6): 105.

Gupta, A. 1978. Photodegradation of polymeric encapsulants of solar cell modules. *JPL Low-Cost Solar Array Project 5101-77*, August 10, 1978.

Gupta, A. 1982. Jet Propulsion Laboratory, private communication.

Haacke, G. 1977. Evaluation of cadmium stannate films for solar heat collectors. *Applied Physics Letters* 30(8): 380 April 15, 1977.

Haacke, G. 1982. Optical coatings for energy efficiency and solar applications. *Proceedings of SPIE—The International Society for Optical Engineering*, January 28–29, 1982, C. M. Lampert, ed. Los Angeles, CA, vol. 324, p. 10.

Habermeier, H. U. 1981. Properties of Indium tin oxide thin films prepared by reactive evaporation. *Thin Solid Films* 80: 157.

Hamberg, I., A. Hjorstberg, and C. G. Granqvist. 1982. Optical coatings for energy efficiency and solar applications. *Proc. SPIE—The International Society for Optical Engineering*, January 28–29, 1982, C. M. Lampert, ed. Los Angeles, CA, vol. 324, p. 31.

Harding, G. L., and B. Window. 1979. Cylindrical magnetron sputtering system for coating solar selective surfaces to batches of tubes. *Journal of Vacuum Science Technology* 16(6): 2105.

Herzenberg, S. A., and R. Silberglitt. 1982. Optical coatings for energy efficiency and solar applications. *Proc. SPIE—The International Society for Optical Engineering*, January 28–29, 1982, C. M. Lampert, ed. Los Angeles, CA, vol. 324, p. 92.

Hohaul, C., O. T. Inal, L. E. Murr, A. E. Torma, and I. Gundiler. 1981. *Solar Energy Materials* 4(3).

Hosch, L. n.d. Rohm GmbH, German Patent No. 2,913,853.

Inal, O. T., J. C. Mabon, and C. V. Robino. 1981. *Thin Solid Films* 83.

Jorgensen, G. J. 1982. Glazing survey. SERI Interoffice Memorandum, June 2, 1982.

Kane, J., H. P. Schweizer, and W. Kern. 1975. Chemical vapor deposition of transparent electrically conducting layers of indium oxide doped with tin. *Thin Solid Films* 29: 155.

Karlsson, B., E. Valkonen, T. Karlsson, and C. G. Ribbing. 1981. Materials for solar transmitting heat-reflecting coatings. *Thin Solid Films* 86(1): 91.

Karrer, F. 1981. Ciba-Geigy Corporation, U.S. Patent No. 4,210,612, June 30, 1981.

Kloppfer, W. 1976. *Journal of Polymer Science*, Symposium 57: 205.

Koltun, M. M., and S. A. Faiziev. 1974. *Geliotekhnika* 10(3): 58. See also *Applied Solar Energy* 10(3/4): 42 (English translation).

Kostlin, H. 1974. Double-glazed windows with very good thermal insulation. *Philips Technical Review* 34(9): 242.

Lampert, C. N. 1979. Coatings for enhanced photothermal energy collection, I. Selective absorbers. *Solar Energy Materials* 1(5/6): 319, and 2(1): 1.

Lampert, C. M. 1980. *Plating and Surface Finishing* (November): 52.

Lampert, C. M. 1981. Heat mirror coatings for energy-conserving windows. *Solar Energy Materials* 6(1): 1.

Lappin, G. R. 1971. Ultraviolet radiation absorbers. *Encyclopedia of Polymer Science and Technology*, vol. 14. New York: Wiley-Interscience, pp. 125–148.

Machet, J., J. Guille, P. Saulnier, and S. Robert. 1981. Deposition of conducting and transparent thin films of indium tin oxide by reactive ion plating. *Thin Solid Films* 80: 149.

Mason, J. J., and T. A. Brendel. 1982. Optical coatings for energy efficiency and solar applications. *Proc. SPIE—The International Society for Optical Engineering*, January 28–29, 1982. C. M. Lampert, ed. Los Angeles, CA, vol. 324, p. 139.

McChesnay, M. A. 1981. Selective paint development progress reports. April 1, May 2, and May 29, 1981.

McKelvey, W. D., P. B. Zimmer, and R. J. H. Lin. 1979. *Solar Selective Paint Development*. Final report, June 1978–December 1979. DOE/CS/34287-T1. December 1979.

Mellor, D. C., A. B. Moir, and G. Scott. 1973. Effect of processing conditions on the UV stability of polyolefins. *European Polymers* 9(3): 219.

Moore, S. W. 1982a. Optical coatings for energy efficiency and solar applications. *Proceedings of SPIE—The International Society for Optical Engineering*, January 28–29, 1982, C. M. Lampert, ed. Los Angeles, CA, vol. 324, p. 148.

Moore, S. W. 1982b. Los Alamos, private communication.

Muller, T. 1982. Acurex, private communication.

Nath, P., and R. F. Bunshah. 1980. *Thin Solid Films* 69: 63.

Nath, P., R. F. Bunshah, B. M. Basol, and O. M. Staffsud. 1980. Electrical and optical properties of tin-doped indium oxide films prepared by activated reactive evaporation. *Thin Solid Films* 72(3): 463.

Nyberg, G., H. G. Craighead, and R. A. Buhrman. 1982. Optical coatings for energy efficiency and solar applications. *Proc. SPIE—The International Society for Optical Engineering*, January 28–29, 1982, C. M. Lampert, ed. Los Angeles, CA, vol. 324, p. 117.

Ortiz, R. A. 1982. Spray deposition and characterization of cadmium stannate films for solar cells. *Journal of Vacuum Science Technology* 20(1): 7.

Pan, C. A., and T. P. Ma. 1980. High-quality transparent conductive indium oxide films prepared by thermal evaporation. *Applied Physics Letters* 37(2): 163.

Pettit, C. N., R. R. Sowell, and I. J. Hall. 1982. Black chrome solar selective coatings optimized for high temperature applications. *Solar Energy Materials* 7(2): 153.

Potdar, H. S., N. Pavaskar, A. Mitra, and A. P. B. Sinha. 1981. Solar selective copper-black layers by an anodic oxidation process. *Solar Energy Materials* 4(3): 291.

Raghunathan, K. 1981. Thermal cycling testing of solar absorptive surfaces. Presented to the *Solar Energy Industries Association Annual Conference* November 1–4, 1981.

Randich, E., and D. D. Allred. 1981. CVD ZrB_2 as a selective absorber. *Thin Solid Films* 83: 393.

Seshan, K., P. D. Hillman, K. A. Gesheva, E. E. Chain, and B. O. Seraphin. 1981. On the mechanism of growth and the hydrogen reduction of CVD black molybdenum thin films. *Journal of Materials Research Bulletin* 16: 1345.

Sikkins, M. 1981. Physical background of spectral selectivity. *Solar Energy Materials* 5(1).

Sikkins, M. 1982a. Optical coatings for energy efficiency and solar applications. *Proceedings of SPIE—The International Society for Optical Engineering*, January 28–29, 1982, C. M. Lampert, ed. Los Angeles, CA, vol. 324, p. 131.

Sikkins, M. 1982b. Properties of spectrally selective no-C films produced by reactive sputtering, parts 1 and 2. *Solar Energy Materials* 6(4).

Smith, G. B., A. Ignatiev, and G. Zajac. 1980. Solar selective black cobalt: Preparation, structure, and thermal stability. *Journal of Applied Physics* 51(8): 4186.

Smith, G. B., and A. Ignatiev. 1981. Black chromium-molybdenum: A new stable solar absorber. *Solar Energy Materials* 4(2): 119.

Smith, J. F., A. J. Aronson, D. Chen, and W. H. Class. 1980. Reactive magnetron deposition of transparent conductive films. *Thin Solid Films* 72: 469.

Stephens, R. B., and G. D. Cody. 1979. Inhomogeneous surfaces as selective solar absorbers. *Solar Energy Materials* 1: 397.

Swab, P., S. V. Krishnaswamy, and R. Messier. 1980. *Journal of Vacuum Science Technology* 17(1): 362.

Thornton, J. A., and J. L. Lamb. 1981. Sputter deposited Pt–Al$_1$O$_3$ selective absorber coatings. *Thin Solid Films* 83: 377.

Thornton, J. A., and J. L. Lamb. 1982. *Evaluation of Vacuum Deposition-Sputter Deposited Cermet Selective Absorber Coatings.* Interim technical progress report on contract No. XP-9-8260-1. Solar Energy Research Institute. July 30, 1982.

Thornton, J. A., A. S. Penfold, and J. L. Lamb. 1980. *Development of Selective Surfaces.* Final technical progress report, September 11, 1978–March 31, 1980. TELIC-80:1, contract no. DE-AC04-78CS35306. August 15, 1980.

Tirrel, D. A. 1980. Preparation of polymeric ultraviolet stabilizers from salicylate esters, hydroxybenzophenones, a-cyano-B-phenyleinnamates and hydroxyphenylbenzotriazoles. *ACS Organic Coatings and Plastics Chemistry Proceedings*, vol. 42. American Society 179th National Meeting, Houston, TX, March 23–28, 1980, pp. 3–18.

Usilton, J. J. 1978. *ACS Advances in Chemistry Series* 69: 116.

Van Boort, H. J. J., and R. Groth. 1968. Low pressure sodium lamps with indium oxide filter. *Philips Technical Review*, 29(1), p. 17.

Van der Leij. 1979. Ph.D. dissertation. Delft University Press, Delft, The Netherlands.

Wiles, D. M., and D. J. Carlsson. 1980–81. *Polymer Degradation and Stabilization* 3: 67.

Williams, R. n.d. OCLI, private communication.

22 Theory and Modeling of Solar Materials

Carl M. Lampert

22.1 Introduction

This chapter covers the theory and modeling of solar materials used for passive and active solar heating and other applications at high temperatures. Theory and modeling are restricted to microstructure or intrinsic optical properties of materials; systems are excluded. Most of the research on solar materials has been analytical or developmental, with very little new theoretical work available. Most optical theories and microstructure models are based on work done by physicists quite some time ago, even in the last century. Solar materials under current research are not covered elsewhere (Lampert 1989). Such materials excluded are fluorescent concentrators, optical switching materials, transparent insulation, spectral splitting coatings, cold mirrors, angle-selective films, radiative cooling surfaces, holographic films, and chromophoric absorber fluids. Desiccant and heat storage materials are covered elsewhere. This study covers the primary solar materials including selective absorbers, transparent low emittance coatings, and antireflection and reflector films.

Solar absorbers are modeled by several theories. Classical interference-film theory is used to model semiconductor/metal tandems. Effective-medium theories are used to model graded composites and particulate absorbers. Optical-trapping surfaces are modeled by scattering theory. Multilayer transparent low emittance and antireflection films rely on interference-film theory. Semiconductor low emittance films and reflectors are characterized by Drude free-electron theory. Finally, mesh low emittance films are covered by antenna theory. Many topics are still not covered by a good theoretical model, and many more materials are left unmodeled. One very promising area of study is intrinsic absorbers. Modeling of weathering and degradation of materials is also an important area of necessary theoretical work (Carlsson 1988).

22.2 Absorber Surfaces

Solar absorber surfaces have been researched the most extensively over the last decade compared to the other collector materials. Much of the theory and modeling of materials has also been confined to this area. There are two categories of absorbers: selective and nonselective. The selective absorber has optical properties that vary distinctly from the solar to the thermal-infrared spectral regions. The optical properties of a selective absorber are responsible

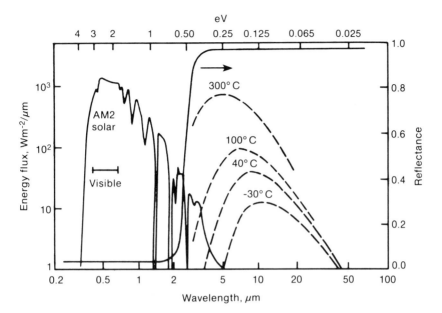

Figure 22.1
Spectral relationship between an ideal selective surface, solar energy distribution (AirMass 2), and blackbody spectra [$-22°-572°F$ ($-30°-300°C$)].

for its high solar absorptance α_s and low, thermal-infrared emittance ε_t. An idealized spectral response or a selective absorber is depicted in figure 22.1.

In contrast to the selective absorber, the nonselective absorber has optical properties that vary slightly over the solar and thermal-infrared spectra. There are numerous examples of nonselective absorbers. Most of them belong to the categories of paint, conversion coatings, and carbon-filled plastics. The application of nonselective absorbers has been confined to low temperature processes. The theory and modeling of these materials have not been explored.

Solar-selective absorbers can be classified as semiconductor/metal tandems, interference multilayers, and optical-trapping surfaces. Other less practical but theoretically interesting selective absorbers are particulate, intrinsic, and quantum-size-effect absorbers. Detailed reviews have been written on all classes of solar absorbers, giving insight beyond the scope of this work (Lampert 1979b; Call 1979; Herzenberg and Silberglitt 1982; Seraphin 1979a; Koltun 1980; Meinel and Meinel 1976; Agnihotri and Gupta 1981; Bogaerts and Lampert 1983; Murr 1980). Niklasson and Granqvist (1983a) provide an excellent annotated bibliography on solar absorbers that covers to the end of 1981.

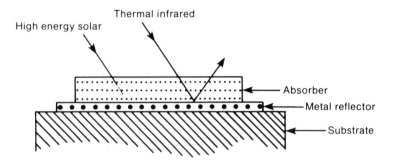

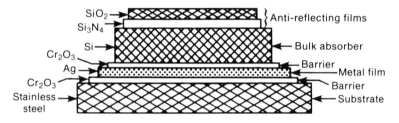

Figure 22.2
Semiconductor metal tandems: (a) schematic cross section of a tandem showing wavelength selectivity; (b) high temperature silicon tandem with greater than 932°F (500°C) stability. Sources: Lampert (1979) and Seraphin (1979a).

22.2.1 Semiconductor/Metal Tandems

One of the most widely used selective absorber types is the semiconductor/metal tandem. The semiconductor may be an effective medium, consisting of a composite of oxide and metal. Also selective paint absorbers are basically tandems. The semiconductor part of the tandem provides high absorptance in the solar region, and it is transparent in the infrared because the energy is less than the semiconductor's absorption edge. See figure 22.2 for examples of tandems. In the infrared the underlying metal layer provides the tandem with a low thermal emittance (high reflectance). A simple tandem can be an oxidized metal sheet. Usually the thickness and optical properties of the oxide are not optimum or stable enough to be a good selective absorber.

Many metals have been oxidized by chemical conversion to form absorber surfaces. Coatings have been formed on stainless steel, copper, titanium, and aluminum. Tandems can be designed for high temperature operation, such as Si/Ag which is stable to 932°F (500°C) (Seraphin 1979a). Tandem selective absorbers can be made by a variety of methods, including chemical, electro-

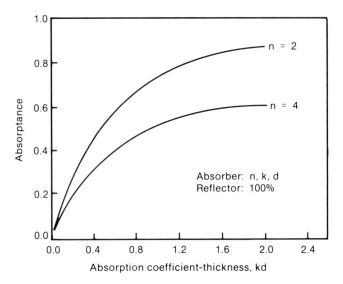

Figure 22.3
Absorptance of an absorber-reflector tandem solar absorber stated as a function of absorption coefficient thickness product kd for different refractive indices of absorber layer. Source: Seraphin (1979a).

chemical, chemical vapor deposition (CVD), and physical-vapor deposition (PVD) processes. The theoretical treatment of tandem absorbers is straightforward and is a direct result of interference-filter theory (Heavens 1965). This type of absorber makes use of an intrinsic semiconductor film that is transparent at longer wavelengths but absorbs strongly at wavelengths less than $W_t = 1.24/E_g$, where W_t is the transition wavelength and E_g is the band gap of the semiconductor in electron volts.

The basic requirements for a tandem absorber have been modeled (Seraphin 1976). Assume an absorber of thickness d and optical constants N_a, K_a is placed on a metal reflector with reflectance R. If we assume that the interface formed obeys the Fresnel equations and that no interference takes place from the beams reflected from the air/semiconductor and semiconductor/metal interfaces, this assembly can be modeled easily. An example of this calculation is shown in figure 22.3. In this figure the absorption coefficient $K = 4\pi k/w$ is used as a product of thickness d of the semiconductor. As a result of this model an efficient tandem should have properties in the solar region of $Kd > 1$ and $n < 2$ for the absorber, and $0.6 < R < 1.0$ for the reflector. In the thermal infrared the absorber should have $Kd \sim 0$ for any n, and the reflector should have $R = 1.0$ for any n. For the absorber many semiconductors have the proper Kd and E_g, but a low n is a strong restriction since most semiconductors have high n values in the solar region. Because of this restriction in many cases it is important to antireflect the semiconductor with another low index, quarter-wavelength coating. The advantage of using semiconductors is that given different energy gaps, different W_t values can be obtained for a range of collector-surface operating temperatures, making this design very versatile.

Theoretical work on tandem absorbers has covered both the absorber and reflector portions of the tandem. Many absorbers listed in table 22.1 have had their optical constants derived by Kramers-Krönig analysis (Abeles 1972) or have had their reflectance spectra compared to calculated spectra using Kramers-Krönig analysis. Absorbers that exhibit intrinsic selectivity (do not require a low emittance metal layer) have been modeled using Drude (Sikkens 1981) and Drude-Lorentz (Roux et al. 1982) free-electron models. The semiconductor portion of the tandem absorber has been theoretically optimized, including graded materials (Ritchie and Window 1977; Trotter, Craighead, and Sievers 1979; Trotter and Sievers 1980; Window, McKenzie, and Harding 1980). Fully graded absorbers have been modeled, including small-index mismatches to free space. The effect of the emittance of real metals used in the tandem has been modeled and bounded by lower limits (Sievers 1980; Trotter and Sievers 1979). Calculations have been presented for real metals and ideal absorbers on real metals (Sievers et al. 1979; Sievers 1980). Examples are shown in figure 22.4. A Fermi model has been used to characterize the infrared emittance function of several tandems, including Si/Ag, black chrome, and black nickel (Soule and Smith 1977). Also of interest to tandem-absorber theory is quantum-size effects (QSE) in thin films. Anomalous absorption is noted in metal films less than 10–20 Å thick and from 100–500 Å thick for degenerate semiconductors. QSE relates the influence of the geometrical dimensions of the sample to that of the distribution of electron states. This distribution can be optimized by size effects to interact strongly with incident electromagnetic radiation. QSE has been experimentally verified for InSb/Al and InSb/Ag (Burrafato et al. 1977).

22.2.2 Graded-Composite/Metal Tandems

In graded-composite/metal tandems, graded composites replace the semiconductor absorber in the semiconductor/metal tandem. The metal portion of the tandem is essentially unchanged when a graded composite is used as an absorber. Composites include $Cr-Cr_2O_3$, $Ni-Al_2O_3$, $Zn-ZnO$, $Pt-Al_2O_3$, $Mo-Al_2O_3$, and metal carbides. In composites the components are phase separated and interdispersed within the film microstructure. In some cases grading from the top to the bottom surface of the film can also exist (Lampert

Table 22.1
Modeling of intrinsic and tandem absorbers

Material	Model/theory	Deposition method	References
TiC_xN_y	Optical constants	Reactive rf and sputtering of Ti	Karlsson et al. (1982)
TiN_x	Rough-surface	Reactive sputtering of Ti	Lafait et al. (1981)
TiN_x ($0.5 \leq x \leq 1.0$)	Optical constants	Reactive sputtering	Rivory et al. (1981)
TiN_x, TiN_xC_y	Band structure	Reactive sputtering/anneal in Ar, C_2H_2	Sigrist et al. (1981)
TiN_x, ZrN, HfN	Optical constants	CVD	Karlsson et al. (1983)
TiN_x, TiN_xCu	Drude-Lorentz	Reactive sputtering	Roux et al. (1982)
NiC_x	Optical constants	Reactive sputtering of Ni in N_2, CH_4	Sikkens and Francken (1979)
NiN_x, NiC_x	Drude	Reactive sputtering in N_2, CH_4	Sikkens (1981)
NiC_x	Drude/Raleigh	Reactive sputtering of Ni in Ar CH_4	Sikkens (1982)
FeC_x	Optical constants	dc reactive sputtering of Ni in Ar + CH_4	Ritchie and Harding (1979)
Stainless steel-carbon	Reflectance, multilayer, and graded	dc reactive sputtering	McKenzie, McPhedron, and Briggs (1982)
Stainless steel-carbon	Reflectance	dc reactive sputtering in C_2H_2	Craig and Harding (1983)
α-C:H	Microstructure	Reactive sputtering Ar, C_2H_2	McKenzie and Briggs (1981)
α-Si:H	Optical constants	Reactive sputtering	Karlsson et al. (1982)
Stainless steel	Oxidation-degradation	Oxidized	Karlsson et al. (1982)
CuO/Ag	Optical constants	Thermal oxidation	Blattner et al. (1977)
CuO, Cu_2O	Optical constants	Thermal oxidation and chemical oxidation	Roos and Karlsson (1983)
Cu_2O, CuO, Cr_2O_3, Fe_2O_3	Optical constants	Thermal oxidation of metals	Karlsson et al. (1982); Karlsson and Ribbing (1982)
ZnO	Structural	Electrodeposition	Murr et al. (1980)
Al_2O_3–Ni pigmented	EMT: MG, BR	Anodized Al	Andersson et al. (1980); Granqvist et al. (1979); Granqvist (1980)
FeO_x	Radiative transfer	Paint	Moore et al. (1976)
PbS	Optical constants	Paint	Williams et al. (1963)

Abbreviations: EMT = effective-medium theory; MG = Maxwell-Garnett theory; BR = Bruggerman theory.

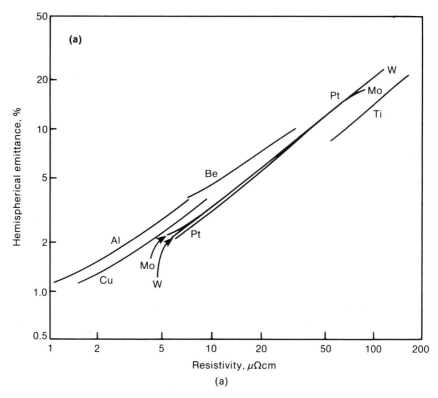

Figure 22.4
Low hemispherical emittance limits for metals and absorbers: (a) emittance lower bounds determined by Drude modeling, plotted as a function of resistivity; (b) calculated lower bounds for high temperature absorbers ($a = 0.94$). The hatched region signifies emittance values that are unobtainable. The three data points in the hatched region used room-temperature optical constants to calculate 932°F (500°C) values. Source: Sievers (1980).

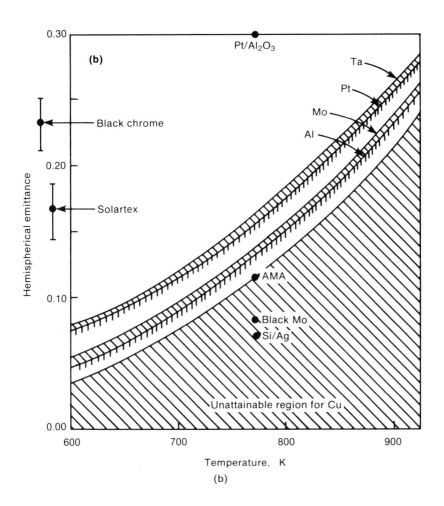

(b)

1980). Resonant scattering in composites depends on particle size, shape, and the effective index of refraction of the components. The relationship of different effective-medium theories has been derived for homogeneous absorbers (Niklasson, Granqvist, and Hunderi 1981). The bounds of various effective-medium theories have also been investigated and determined (Bohren and Gilra 1979). For metallic-particle absorbers the scope of effective-medium theory has been determined (Smith 1979). Many effective medium theories have been used to describe composites, the most notable being the Maxwell-Garnett (Garnett 1904, 1906; Van de Hulst 1980) and Bruggerman (1935) theories. Most modeling has been performed on $Cr–Cr_2O_3$ and electrodeposits of black chrome.

The Maxwell-Garnett (MG) theory is used to predict the absorption properties of extremely fine particles in a solid matrix. The MG theory was developed from Mie-scattering theory, which describes the scattering properties of spherical particles larger than those handled by Rayleigh theory (Kerker 1969). The major difference between the MG and Bruggerman theories is self-consistency. In MG theory one component (a metal) is treated as a scatterer embedded in the matrix of another component. The theory takes into account the modification of the applied electric field in the matrix by the fields of the metal particles. But the metal particles must not interact or the theory breaks down; hence the metal-packing fraction f must be low. In Bruggerman's theory the two components are treated equally as part of an effective medium with dielectric properties equivalent to the composite material.

Both theories can be developed from scattering theory (Niklasson, Granqvist, and Hunderi 1981). A random-unit cell is embedded in an effective medium. The extinction of the random-unit cell should be the same as if it were replaced with an effective dielectric permeability. An optical theorem for absorbing media (Bohren and Gilra 1979) relates the extinction of the cell C compared to that of the surrounding media, with the amplitude of scattering in the direction of the incident beam $S(o)$:

$$C = 4\pi Re\left[\frac{S(o)}{k^2}\right],\tag{1}$$

where

Re = Reynold's number,

k = wave vector of the effective medium ($k = 2\pi e^{1/2}/w$),

w = wavelength in vacuum.

For the effective medium, if $C = 0$, then $S(o) = 0$.

To derive the MG equations, one must use explicit results for $S(o)$ from classical scattering theory. For a separated grain structure as in MG theory, the random-unit cell is represented as a sphere with dielectric permeability e_A coated with a dielectric with permeability e_B and where, in general, the complex dielectric constant e is related by $e = (n - ik)^2 = e_i + e_2$. The details of the scattering of the coated sphere are developed as

$$S^{mg}(o) = \frac{1}{2} \sum_{n=1}^{\infty} (2n + 1)(\alpha_n + \beta_n), \tag{2}$$

where α_n and β_n contain Bessel functions and their derivatives. Expanding $S^{mg}(o)$ in terms of (kb) (Rivory et al. 1981),

$$S^{mg}(o) = i(kb)^3 \frac{(e_B - \bar{e})(e_A + 2e_B) + f(2e_B + \bar{e})(e_A - e_B)}{(e_B + 2\bar{e})(e_A + 2e_B) + f(2e_B - 2\bar{e})(e_A - e_B)} + \varphi[(kb)^5], \tag{3}$$

where the filling factor is given by $f = (a/b)^3$, where a is the radius of the inner sphere and b is the radius of the outer sphere.

In the small-sphere limit the leading terms in the above equation equal zero. Then one can obtain the standard MG result:

$$\left[\frac{\bar{e} - e_B}{\bar{e} + 2e_B} \right] = f \left[\frac{e_A - e_B}{e_A + 2e_B} \right]. \tag{4}$$

In this case $\bar{e}$ is the MG effective dielectric permeability. Solving for $\bar{e}$,

$$\bar{e} = e_B \frac{e_A + 2e_B + 2f(e_A - e_B)}{e_A + 2e_B - f(e_A - e_B)}. \tag{5}$$

If ellipsoids are considered instead of spheres, another expression can be derived. The ellipsoids must be identical in shape and oriented with one of their principal axes parallel to the external field. The effective dielectric permeability becomes

$$\frac{\bar{e} - e_B}{L_B \bar{e} + (1 - L_B)e_B} = f \left[\frac{e_A - e_B}{L_A e_A + (1 - L_A)e_B} \right], \tag{6}$$

where L_A and L_B are the depolarization factors for the ellipsoids.

In the Bruggeman theory an inhomogeneous two-phase material is treated as a system of spherical particles. These particles are composed separately of pure phases A and B. The theory solves the local electric field around a typical two-phase element, a sphere of either phase in an effective medium. The probability that the sphere is A is given as (f), and the probability that the sphere is B is given as $(1 - f)$. The Bruggeman theory's result can be derived from the same series expansion as for MG theory according to

$$S^B(o) = i(kb)^3 \frac{e - \bar{e}}{e + 2\bar{e}} + \phi[(kb)^5].$$ (7)

This is the result for Mie-scattering theory, and using the small-sphere limit as before, one can obtain

$$f\left[\frac{e_A - \bar{e}}{e_A + 2\bar{e}}\right] = (1 - f)\left[\frac{\bar{e}_B - e}{e_B + 2\bar{e}}\right].$$ (8)

The application of various effective-medium theories is shown in table 22.2, along with various structural models. In figure 22.5 are examples of MG and Bruggerman models used to fit the reflectance data for electrodeposited black chrome. For comparison, experimental data are shown in figure 22.6.

22.2.3 Particulate Absorbers

Related to composite absorbers are particulate absorbers, which contain small absorbing metallic particles. Although very few are practical, they serve as simple examples for applying various effective-medium theories. A list of particulate-absorber research is detailed in table 22.3.

22.2.4 Optical-Trapping Surfaces

A textured top surface can be created by oriented growth structures or surface roughening. If a surface is roughened properly with the proper period, it will appear rough and absorbing in the solar region but smooth in the infrared. Oriented growth structures, by virtue of their geometry, can efficiently absorb incoming radiation by multiple absorption and reflection. By this method metals with low intrinsic absorption can be made to have high absorption. Surface texturing can also be used to antireflect surfaces to increase absorption, as was done for the $Pt-Al_2O_3$ absorber (Craighead et al. 1981). Modeling has been performed on general surfaces with periodic and random textures (Botten and Ritchie 1977; Cody and Stephens 1978; Stephens and Cody 1979; Wirgin 1981a, b). Cavity (Hollands 1963 and corrugated absorbers have also been modeled (Petit 1981; Stephens and Cody 1977; Demichelis and Russo 1979). The Cody-Stephens theory has been used to model etched Al–Si alloys (Niklasson and Craighead 1982).

22.2.5 Multilayer Absorbers

Both two- and three-layer interference absorbers have been modeled or designed by interference-filter theory. Computer modeling of multilayer optical films has been available for many years (Liddell 1981). Several excellent

Table 22.2
Modeling of composite absorbers

Material	Model/theory	Deposition method	References
$Cr-Cr_2O_3$	EMT: MG, BR	Electrodeposition	Berthier and Lafait (1979a, b)
$Cr-Cr_2O_3$	EMT: MG, Drude	Electrodeposition	Hogg and Smith (1977)
$Cr-Cr_2O_3$	EMT: MG Spheroids	Electrodeposition	Ignatiev et al. (1979a); Ignatiev (1980); Ignatiev et al. (1979b)
$Cr-Cr_2O_3$	Structural	Electrodeposition	Lampert (1978, 1980); Lampert and Washburn (1979)
$Cr-Cr_2O_3$	Structural	Electrodeposition	Inal et al. (1981)
$Cr-Cr_2O_3$	EMT: MG, BR	Electrodeposition	Lafait et al. (1979)
$Cr-Cr_2O_3$	Structural	Electrodeposition	Ritchie et al. (1980)
$Cr-Cr_2O_3$	EMT	Electrodeposition	Valignat et al. (1980)
$Cr-Cr_2O_3$	Optical constants	Electrodeposition	Window et al. (1978,1979)
$Cr-Cr_2O_3$	Structural	Electrodeposition	Zajac et al. (1980)
$Cr-Cr_2O_3$	EMT: MG, BR	Electrodeposition	Zajac and Ignatiev (1982)
$Cr-Cr_2O_3$, air–MgO	Optical constants	rf-sputtered composite/cosputtered	Fan (1978)
Cr–dielectric	EMT: MG, spheroids	Theory	Granqvist and Hunderi (1979)
$Ni-Al_2O_3$, $Pt-Al_2O_3$	EMT: MG	e-beam evaporated, two sources	Craighead et al. (1979)
$Ni-Al_2O_3$, $Ag-Al_2O_3$	Two sources	e-beam evaporated	Buhrman and Craighead (1980)
$Ni-Al_2O_3$, $Au-MgO$			
$Ni-Al_2O_3$	EMT: MG, MMG	e-beam evaporated, two sources	Ashcroft et al. (1979)
$Ni-Al_2O_3$	EMT: MG	e-beam evaporated, two sources	Craighead and Buhrman (1977)
$Ni-Al_2O_3$, $V-Al_2O_3$	EMT: MG	e-beam, two sources	Craighead and Buhrman (1978)
$V-MgO$, $Fe-Al_2O_3$, $Fe-MgO$			
$Co-Al_2O_3$	EMT: MG	e-beam evaporated	Niklasson and Granqvist (1981, 1983)

Table 22.2 (continued)

Material	Model/theory	Deposition method	References
Ge–Al$_2$O$_3$	Reflectance	Cosputtered e-beam evaporated	Gittleman (1976)
Pt–Al$_2$O$_3$ Textured	EMT: MG	Two sources, ion Etched CF$_4$	Craighead et al. (1981)
Al–Al$_2$O$_3$	Optical constants	e-beam evaporated	Eriksson et al. (1982)
Au–Al$_2$O$_3$, Au–MgO	EMT: BR	Theory	Granqvist (1979)
W–MgO, Au–CaF$_2$ Si–MgO, Au–MgO	Optical constants	From composites	Gittleman et al. (1977)
Na–MgO, Au–MgO	EMT: MG	Ion implanted	Abouschacra et al. (1981a, b)
Ni–NiS	EMT: MG	Electrodeposition	Rajogopalan et al. (1965)
Pt–Al$_2$O$_3$	EMT: MG, BRH, BR, L	rf-sputtered/cosputtered	Penfold (1983)

Abbreviations: EMT = effective-medium theory; MG = Maxwell-Garnett theory; BR = Bruggerman theory; MMG = multiple Maxwell-Garnett theory; BRH = Bruggerman-Hanoi model; L = laminar cermet model.

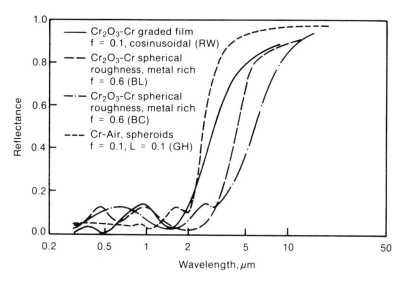

Figure 22.5
Maxwell-Garnett and Bruggerman modeling of black chrome showing various modifications. Shown in this plot are cosinusoidal grading using MG theory (RW), uniform spheroids in air using MG theory (GH), modified MG model (BL), and Bruggerman model (BC) for metal-rich cermets. Sources: Ritchie and Window (1977), Granqvist and Hunderi (1979) and Berthier and Lafait (1979b).

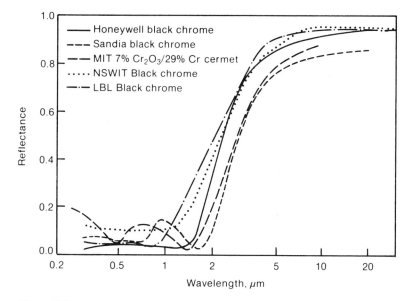

Figure 22.6
Spectral reflectance for different types of black chrome surfaces. Source: Lampert (1979)

Table 22.3
Modeling of particulate absorbers

Material	Model/theory	Deposition method	References
Au	Continuum Capacitor (IR)	Gas evaporated in He	Doland et al. (1977)
Au	EMT: BR, MG, Hu	Gas evaporated in air	Granqvist and Hunderi (1977)
Au	Mie	Gas evaporated	McKenzie (1976)
Au	Continuum/spheroid model/capacitor	Gas evaporated in He	O'Neill et al. (1977); O'Neill and Ignatiev (1978); O'Neill et al. (1978a)
Au	Radiative-transfer	Gas evaporated Ar + O_2	Lee (1977)
Au	EMT	—	O'Neill et al. (1978b)
Al, C	Scattering	—	Hottel et al. (1967)
Cr	EMT: MG	Gas evaporated in Ar + air	Granqvist and Niklasson (1978); Niklasson and Granqvist (1978)
Cr	EMT: shape effects	—	Granqvist (1977); Granqvist and Hunderi (1978); Hunderi and Granqvist (1979)
Ni	EMT: MG	Evaporated in Ar Oxidated in air	Niklasson and Granqvist (1979)
Mo	EMT	Gas evaporated in Ar + O_2	Fantini (1981)

Abbreviations: EMT = effective-medium theory; MG = Maxwell-Garnett theory; BR = Bruggerman theory; MMG = multiple Maxwell-Garnett theory.

works provide further information on theory and modeling of multilayer films (Heavens 1973; MacLeod 1986; Hass, Francombe, and Hoffman 1977; Knittle 1976; Demichelis et al. 1982; Eckertova 1986). Graphical and transmission-line analogies have been developed for solar-absorber designs (Tabor, Weinberger, and Harris 1964; Thornton and Tran 1978). In the following section on multilayer low emittance, interference-filter theory is outlined. An example of the effect of various layers on selective absorption is depicted schematically in figure 22.7.

22.3   Transparent Low Emittance Coatings

Transparent low-emittance films are wavelength-selective coatings that exhibit spectral selectivity between the solar and infrared or visible and near-infrared spectra. Depending on design, most low-emittance films are highly transparent in the visible or solar energy regions. Characteristically low-emittance films are highly reflecting in the near-infrared or infrared, providing a very low thermal-emitting surface. There are three categories of transparent low emittance coatings: those based on highly doped transparent semicoductors, three-layer interference films, and conductive grids. Complete descriptions of these types are given in several reviews (Lampert 1981; Haacke 1977a; Vossen 1977; McPhedron and Maystre 1977). Modeling of various interference and doped-semiconductor-type low emittance coatings is shown in table 22.4. There are well-established theories and design equations for interference films. The theory is detailed in a number of works (Koltun 1980; Heavens 1973; MacLeod 1969; Hass, Francombe, and Hoffman 1977; Knittle 1976; Demichelis et al. 1982; Liddell 1981). It will be outlined briefly next.

22.3.1   Interference Films

Consider the reflectance and transmittance of a thin-film filter (nonmagnetic) with thickness d and refractive index $\bar{n} = n - ik$ at wavelength w. We will follow the effect of an incident wave as it propagates from the incident medium into the thin film and then into the substrate. Using well-developed matrix formalism, the theory of interference films can be developed (Heavens 1973; MacLeod 1969; Hass, Francombe, and Hoffman 1977; Knittle 1976; Liddell 1981). The characteristic matrix $\mathbf{M}$ is developed from elastic-network theory (Heavens 1973). Matrix $\mathbf{M}$ transforms amplitude and phase from one side of a four-pole network to the other. The electric- and magnetic-field strengths of the incident media i and secondary media s (which can be a film or substrate) are related by

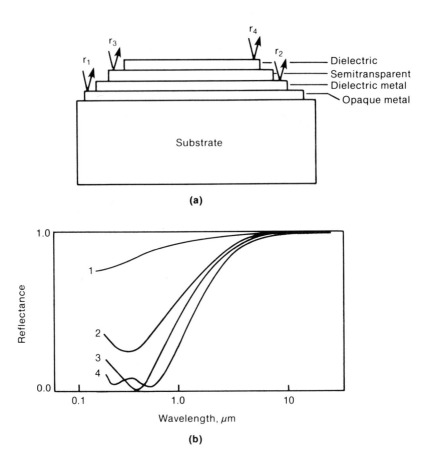

(a)

(b)

Figure 22.7
(a) Schematic cross section of a four-layer interference absorber consisting of two quarter-wavelength dielectric layers separated by metal reflective layers; (b) variation in reflectance with the addition of each successive layer starting with the opaque metal layer. Source: Seraphin (1976).

Table 22.4
Modeling of transparent low emittance coatings

Material	Model/theory	Deposition method	References
$In_2O_3:Sn$	Drude	rf-sputtering in $O_2 - Ar$	Ohhata et al. (1977)
$In_2O_3:Sn$	Drude	rf-sputtering	Yoshida (1978)
$In_2O_3:Sn$	Drude	Reactive e-beam evaporated	Hamberg and Granqvist (1983)
$In_2O_3:Sn$ Cd_2SnO_4	Drude	Reactive sputtering in $O_2 + Ar$	Howson et al. (1981)
$In_2O_3:Sn,$ $SnO_2:F$	Drude and ionized impurity scattering	CVD	Frank et al. (1983)
$In_2O_3:Sn,$ $In_2O_3:Sn$	Drude	Sputtering	Avaritsiotis and Hawson (1980)
In_2O_3	Bandgap widening	Reactive e-beam evaporated	Granqvist (1984)
$In_2O_3:Sn$	Optical constants	Reactive rf-sputtering in $O_2 + Ar$	Ohhata et al. (1979)
$In_2O_3:Sn$	Drude and ionized impurity scattering	Reactive e-beam evaporated	Hamberg and Granqvist (1984)
$In_2O_3:Sn$	Burstein effect	PVD	Mizuhashi (1980)
$SnO_2:F$	Drude	CVD	Grosse (1979); Grosse et al. (1982)
$SnO_2:Sb$	Drude	Reactive sputtering $O_2 + Ar$	Noguchi et al. (1978)
$SnO_2:F$	Drude	CVD	Lampert (1982)
$SnO_2:F,$ $SnO_2:Sb$	Drude	CVD	Simonis et al. (1979)
$SnO_2:Sb$	Drude	CVD	Arai (1960)
$TiO_2/Ag/TiO_2$	Optical constants	rf-sputtering	Granqvist (1981); Fan and Bachner (1976)
$TiO_2/Ag/TiO_2$	Optical constants	Solgel/PVD	Chiba et al. (1983)
$In_2O_3:Sn$	Drude	rf-sputtering	Fan and Bachner (1975)

Abbreviations: PVD = physical-vapor deposition; CVD = chemical-vapor deposition; e-beam = electron beam.

$$\begin{bmatrix} 1 \\ Y \end{bmatrix} = \begin{bmatrix} E \\ H \end{bmatrix}_i = \mathbf{M} \begin{bmatrix} E \\ H \end{bmatrix}_s, \qquad \mathbf{M} = \begin{bmatrix} \cos b & i/\bar{n} \sin b \\ i\bar{n} \sin b & \cos b \end{bmatrix}, \tag{9}$$

where Y is the admittance and b is the phase thickness ($b = 2\pi\bar{n}d/w$). For off-normal incidence at refracted angle c, the parameters transform from $\bar{n} \to \bar{n} \cos c$ for s polarizations (TE), $\bar{n} \to \bar{n}/\cos c$ for p polarizations (TM), $b \to 2\pi\bar{n}d/w \cos c$, where cps $c = [1 - (\sin c_0/\bar{n})^2]^{1/2}$, where c_0 is the incidence angle.

For a nonabsorbing dielectric (with $\bar{n} = n_d$ and $k = 0$) ($k = 0$), the admittance is given by

$$Y_D = \frac{\bar{n}_s + i n_d \tan b_0}{1 + i(\bar{n}_s/n_d)\tan b_0}. \tag{10}$$

For metallic films the admittance is

$$Y_m = \frac{\bar{n}_s - i k_m \tanh b_m}{1 + i(\bar{n}_s/k_m \tanh b_m)}. \tag{11}$$

If multiple-film layers are considered, then $\mathbf{M}$ becomes a product of representative matrices for each film layer l:

$$\mathbf{M} = \prod_{l=1}^{M} M_1. \tag{12}$$

Considering the single-layer case again, the reflection coefficient r between the incident medium and the film interface is simply

$$r = \frac{Y_f - Y}{Y_f + Y}, \tag{13}$$

where $Y_f = n_f$. The reflectance R is

$$R = rr^*,$$

where r^* is the complex conjugate of r. The reflectance can differ in value depending on the propagation direction through the film, either beginning from the top surface or through the substrate.

The transmittance of the interference system is defined as the ratio of radiation entering the substrate to that incident on the surface:

$$T = (1 - R)\left[\frac{\text{Re}(E \cdot H^*)_s}{\text{Re}(E \cdot H_i^*)}\right], \tag{14}$$

where H^* is the complex conjugate of the magnetic field H.

Reducing transmittance is also possible in the substrate material, so the total transmission (T_t) is

$$T_t = Te^{-a_s d}. \tag{15}$$

The transmission of the substrate caused by absorption is

$$T_0 = e^{-a_s d}. \tag{16}$$

The absorption coefficient and refractive index for transparent media are, respectively,

$$a_s = \frac{4\pi l_s}{w \cos c} \quad \text{and} \quad \bar{n}_s = n_s - ik_s,$$

where c is the refracted angle as the beam traverses the film-substrate interface.

If multiple reflections are taken into account, the total reflectance is

$$R = R_i + \frac{R_0 T_t^2}{(1 - R_0 R_s T_0)^2}, \tag{17}$$

where R_0 is the reflectance of the bare substrate, R_i is the incident reflectance of the film, and R_s is the secondary reflectance from the film-substrate interface.

A good discussion of multilayer low emittance coatings and their dependence on material selection is available (Pracchia and Simon 1981). Modeling of the optical constants of the popular $TiO_2/Ag/TiO_2$ system compared to experiment has also been performed (Granqvist 1981; Fan and Bachner 1976; Chiba, Sobajima, and Yatabe 1983).

22.3.2 Doped Semiconductors

Highly doped semiconductors, such as $SnO_2:F$, $In_2O_3:Sn$, and Cd_2SnO_4, can transmit the visible portion of the solar spectrum but appear as good reflectors in the thermal infrared. The abrupt change in reflectance has been modeled using classical Drude theory (Drude 1900). This theory was originally developed to describe the electronic behavior of metals usings a free-electron gas model. The electronic behavior of highly doped semiconductors is similar to that of metals. Drude theory uses a well-defined plasma edge to characterize the material. The plasma edge is caused by the excitation of free carriers by electromagnetic radiation. All charge carriers are moving in a potential field in the media. To properly characterize this, an effective electron mass m^* must be defined. The effective mass $m^* = m_r m_e$, where m_e is the rest mass and m_r is the relative mass of the charge carriers in the field. For a Drude reflector the plasma frequency is expressed as

$$W_p = \left(\frac{Ne^2}{e_b e_0 m^*} \right)^{1/2} - Y^2, \tag{18}$$

where

Y = relaxation frequency ($Y = e/um_e$),

u = carrier mobility,

e = electron charge,

e_b = dielectric constant associated with bound carriers at very high frequency,

e_0 = dielectric constant of air,

N = carrier density.

For a Drude reflector the dielectric constant is expressed as

$$e_1 = e_b \frac{1 - (1 + Z^2)}{L^2 + Z^2}, \tag{19}$$

$$e_2 = e_b - \frac{Z(1 + Z^2)}{L(L^2 + Z^2)}, \tag{20}$$

where

$Z = Y/W_p$,

$L = W_p/W_p$,

W = frequency.

The ratio of the relaxation frequency to plasma frequency, Y/W_p, serves as a measure of wavelength selectivity.

The complex dielectric constant is related to the optical constant by

$$e = e_1 - e_2 = (n - ik)^2, \tag{21}$$

where $e_1 = (n^2 - k^2)$ and $e_2 - 2nk$. Knowing this, n and k can be obtained in terms of frequency W. Furthermore, by applying the Fresnel equations, the reflectance can be derived.

One should note that in the Drude equation, as $W \gg W_p > Y$, n becomes constant and k approaches 0, implying that the material is transparent at short wavelengths, and for $W \gg Y > W_p$, n and k are about equal and large in magnitude, implying a high reflectance. An example of Drude modeling is shown in figure 22.8. Parameters used in the model are $e_b = 4$, $N = 5 \times 10^{20}$ cm^{-3}, effective mass ratio $= 0.3$, and $u = 17$–35 v/cm^2 s. Figure 22.9 shows experimental results for the best research-grade films (Kostlin, Jost, and Lems 1975; van der Leij 1979; Haacke 1977b). The SnO$_2$:F experimental

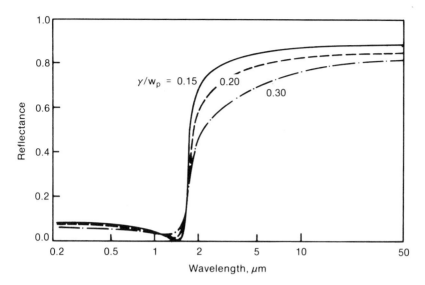

Figure 22.8
Drude modeling for doped SnO_2. Experimental parameters used in the model are $e_b = 4$, $N = 5 \times 10^{20}$ cm^{-2}, effective mass ratio $= 0.3$, $u = 17\text{-}35$ v/cm^2 s. Source: Abeles (1972).

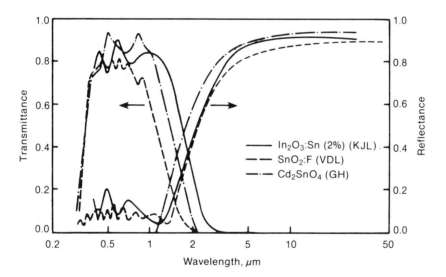

Figure 22.9
Spectral normal transmittance and reflectance of heat-mirror coatings based on In_2O_3 : Sn, SnO_2 : F, and Cd_2SnO_4. Sources: Trotter and Sievers (1979, 1980) and Sievers et al. (1979).

films have properties of solar transmittance $T_s = 0.75$, emittance (100°C) = 0.15, electrical resistivity = 0.67 × 10^{-3} ohm-cm, N = 5.6 × 10^{20} cm^{-3}, u = 37 V cm^{-2} s, and Y/W_p = 0.14. The modeled response agrees well with that obtained by experiment.

The effect of bandgap widening in In$_2$O$_3$:Sn and In$_2$O$_3$ has been explained by modeling of the Burnstein shift (Burnstein 1954), including electron–electron and electron–impurity scattering at high doping levels (Granqvist 1984; Hamberg et al. 1984).

22.3.3 Mesh Low Emittance Coatings

The theory of mesh low emittance coatings has been developed, and several experimental films have been made (McPhedron and Maystre 1977; Sievers 1980; Maystre and Nevieve 1979; Pramanik, Sievers, and Silsbee 1979). One design involves chemical etching of the surface of an In$_2$O$_3$:Sn film to form a regular grid pattern (Fan, Bachner, and Murphy 1976). The etched grid consists of square openings of d = 2.5 μm wide a line width of r = 0.6 μm.

For this grid, T_s increased from 0.8 to 0.9, while reflectance decreased from 0.91 to 0.83. The theoretical infrared reflectivity for grids has been determined from antenna theory (Kontorovich et al. 1962). If the perfectly conducting (nonmagnetic) parallel wires of a crossed grid are joined at their intersections and if $a \gg r$ and $W \gg a$, where $a = \ln[d/2\pi r]$, the reflection coefficient is given by

$$R = |R|e^{ip} = \left[1 + \left(\frac{am_1}{\cos c} \right) \left(1 - \sin^2 \left(\frac{c}{2} \right) \right) \right]^{-1}, \qquad (22)$$

where

c = angle of incidence,

p = phase angle,

$m_1 = 2\pi/W$,

W = frequency,

R = reflectance at W.

For an opaque material the solar transmission is given by

$$T_s = \frac{d^2}{d^2 + r^2}. \qquad (23)$$

The best modeled results for metallic grids were obtained for aluminum and magnesium, where the line width would have to be about 0.1 μm to give

$T_s = 0.90$ and $R_{IR} = 0.90$ (Pramanik, Sievers, and Silsbee 1979). So metallic grids, because of their intrinsic absorption and reflection in the solar region, appear less promising than doped semiconductor grids.

22.4 Antireflection Films

The theoretical development of antireflection films is quite old and follows the matrix method outlined in the interference low-emittance film section. Work on antireflection films used in solar designs consists mainly of developmental research for absorbers and photovoltaics. Specialized antireflection coatings for solar absorbers (Seraphin 1979b; Donnadieu and Seraphin 1978; Seraphin 1976) and transparent low emittance coatings (Hamberg and Granqvist 1983) have been devised. Chemistry modeling has been performed on sol-gel-based TiO_2–SiO_2 for silicon (Brinker and Harrington 1981). The considerations of single- versus double-layer antireflection films have been detailed (Donnadieu and Seraphin 1978). Silicon exynitride films for ZrB_2, TiN, HfC, Te, Si, and Ge absorbers have been fabricated with varying indices of silicon nitride ($n = 1.46$ to 3.4). Interference-film theory has been used to model plasma-assisted silicon nitride and silicon-oxynitride films (Wilson 1984; Sexton 1984). Systems studied include SiO_2–SiN_x, Al_2O_3–SiN_x, SiN_xTiO_2, and Si_3N_4 (Sexton 1984). Gradient-index coatings have been modeled by inhomogeneous film theory and applied to chemical-etched, phase-separated glass (Minot 1976), including angular reflectance effects (Minot 1977). Further work on gradient-index antireflection coatings includes modelings of $AlO(OH)_2$ on plastics and etched glass (Lee and Debe 1980). Two techniques were used to model these surfaces, the McCrackin numerical technique and an analytical method (McCrackin and Colson 1964; Jacobson 1966).

22.5 Reflectors

Most of the reflector research for solar-energy use is developmental. Theoretical concerns were fairly isolated. The theory of metals is well-known and modeled by Drude free-electron theory (Drude 1900). An excellent review including properties of many transition metals is available (Ehrenrich 1966). Island growth in sputtered silver reflector films has been modeled and correlated to specularity properties. Theoretical analysis of island growth predicts island spacing on the order of 140–160 nm (Dake and Hartman 1983). Alloys of aluminum and rare earth metals have been researched to reduce the intrinsic absorption in aluminum by reduction of interband transition strengths. Drude modeling is presented for alloys of Al–Gd and Al–Sm (Trotter 1980).

References

Abeles, F., ed. 1972. *Optical Properties of Solids*. Amsterdam: North-Holland.

Abouchacra, G., G. Chassague, and A. Delmas. 1981a. *J. Physique* 42: C-1-327.

Abouchacra, G., G. Chassague, and A. Delmas. 1981b. In *Proc. ISES*. Brighton, England.

Agnihotri, O. P., and B. K. Gupta. 1981. *Solar Selective Surfaces*. New York: Wiley.

Andersson, A., O. Hunderi, and C. G. Granqvist. 1980. *J. Appl. Phys.* 51: 754.

Arai, T. 1960. *J. Phys. Soc. Japan* 15: 916.

Ashcroft, N. W., R. A. Buhrman, H. G. Craighead, W. Lamb, A. J. Sievers, R. Smalley, D. M. Trotter, Jr., J. W. Wilkins, L. Wojcik, and D. M. Wood. 1979. *Proc. Second Annual Conference on Absorber Surfaces for Solar Receivers*, Boulder, CO, P. J. Call, ed. SERI/TP-49-182, Conf. 790120. Golden, CO: Solar Energy Research Insititute, p. 77.

Avaritsiotis, J. N., and R. P. Howson. 1980. *Thin Solid Films* 65: 101.

Berthier, S., and J. Lafait. 1979a. *J. Physique* 40: 1093.

Berthier, S., and J. Lafait. 1979b. *J. Physique* 40: C1-285.

Blattner, R. J., C. A. Evans, Jr., and A. J. Braundmeier, Jr. 1977. *J. Vac. Sci. Technol.* 14: 1132.

Bogaerts, W., and C. M. Lampert. 1983. Materials for photothermal solar energy conversion. *J. Mat. Sci.* 18: 2847.

Bohren, C. F., and D. P. Gilra. 1979. *J. Colloid Interf. Sci.* 72: 215.

Botten, L. C., and I. T. Ritchie. 1977. *Opt. Commun.* 23: 421.

Brinker, C. J., and M. S. Harrington. 1981. *Solar Energy Mat.* 5: 159.

Bruggerman, D. A. G. 1935. Berechnung verschiedener physikalischer, konstanten van heterogenen substanzen. *Ann. der Phys.* 24: 636.

Buhrman, R. A., and H. G. Craighead. 1980. Composite film selective absorbers. *Solar Materials Science*, L. E. Murr, ed. New York: Academic Press, p. 277.

Burnstein, E. 1954. *Phys. Rev.* 93: 632.

Burrafato, G., et al. 1977. Thin film solar acceptors. *Heliotechnique and Development*, M. Ketlani and J. Sossors, eds. New York: Develop Analysis Assoc., Inc.

Call, P. 1979. *National Program Plan for Absorber Surfaces R&D*. SERI/TP-31-103. Golden, CO: Solar Energy Research Institute.

Carlsson, B. 1989. *Solar Materials Research and Development*. International Energy Agency Task 10 report, D16: 1989. Stockholm: Swedish Council for Building Research.

Chiba, K., S. Sobajima, and T. Yatabe. 1983. *Solar Energy Mat.* 8: 371.

Cody, G. D., and R. B. Stephens. 1978. *A.I.P. Conf. Proc.* 40: 225.

Craig, S., and G. L. Harding. 1983. *Thin Solid Films* 101: 97.

Craighead, H. G., R. Bartynski, R. A. Buhrman, L. Wojcik, and A. J. Sievers. 1979. *Solar Energy Mat.* 1: 15.

Craighead, H. G., and R. A. Buhrman. 1977. *Appl. Phys. Lett.* 31: 423.

Craighead, H. G., and R. A. Buhrman. 1978. *J. Vac. Sci. Technol.* 15: 269.

Craighead, H. G., R. E. Howard, J. E. Sweeney, and R. A. Buhrman. 1981. *Appl. Phys. Lett.* 39: 29.

Dake, L. S., and J. S. Hartman. 1983. *Solar Energy Mat.* 7: 43.

Demichelis, F., G. Ferrari, E. Minetti, Mezzatti, and W. Petrotlo. 1982. *Appl. Opt.* 21: 1854.

Demichelis, F., and G. Russo. 1979. *Appl. Phys.* 18: 307.

Doland, C., P. O'Neill, and A. Ignatiev. 1977. *J. Vac. Sci. Technol.* 14: 259.

Donnadieu, A., and B. O. Seraphin. 1978. *J. Opt. Soc. Am.* 68: 292.

Drude, P. 1900. *Phys. Z.* 1: 61.

Eckertova, L. 1986. *Physics of Thin Films.* 2d ed. New York: Plenum.

Ehrenrich, H. 1966. *Optical Properties and Electronic Structure of Metals and Alloys,* F. Abeles, ed. Amsterdam: North-Holland.

Eriksson, T. S., A. Hjortsberg, and C. G. Granqvist. 1982. *Solar Energy Mat.* 6: 191.

Fan, J. C. C. 1978. *Thin Solid Films* 54: 139.

Fan, J. C., and F. J. Bachner. 1975. *J. Electrochem. Chem. Soc.* 122: 1719.

Fan, J. C. C., and F. J. Bachner. 1976. *Appl. Opt.* 15: 1012.

Fan, J. C., F. J. Bachner, and R. A. Murphy. 1976. *Appl. Phys. Lett.* 28: 440.

Fantini, A. M. C., J. R. Moro, and A. Abramvich. 1981. *J. Physique* 42: C1-317.

Frank, G., E. Kauer, H. Kostlin, and F. J. Schmitte. 1983. *Solar Energy Mat.* 8: 387.

Garnett, J. C. M. 1904. Colours in metallic glasses, in metallic films. *Phil. Trans. Roy. Soc.* 203A: 385.

Garnett, J. C. M. 1906. Colours in metal glasses, in metallic films and in metallic solution. *Phil. Trans. Roy. Soc. London* 205A: 237.

Gittlemann, J. I. 1976. *Appl. Phys. Lett.* 28: 370.

Gittlemann, J. I., B. Abeles, P. Zanzucchi, and Y. Arie. 1977. *Thin Solid Films* 45: 9.

Granqvist, C. G. 1977. *Phys. Scripta* 16: 163.

Granqvist, C. G. 1979. *J. Appl. Phys.* 50: 2916.

Granqvist, C. G. 1980. *J. Appl. Phys.* 51: 3359.

Granqvist, C. G. 1981 *Appl. Optics* 20: 2606.

Granqvist, C. G. 1984. *Proc. SPIE* 502: 2.

Granqvist, C. G., A. Andersson, and O. Hunderi. 1979. *Appl. Phys. Lett.* 35: 268.

Granqvist, C. G., and O. Hunderi. 1977. *Phys. Rev. B.* 16: 3513.

Granqvist, C. G., and O. Hunderi. 1978. *Appl. Phys. Lett.* 32: 793.

Granqvist, C. G., and O. Hunderi. 1979. *J. Appl. Phys.* 50: 1058.

Granqvist, C. G., and G. A. Niklasson. 1978. *J. Appl. Phys.* 49: 3512.

Grosse, P., F. J. Schmitte, G. Frank, and H. Kostlin. 1982. *Thin Solid Films* 90: 309.

Grosse, P. 1979. *Freie Elektronen in Festkorpern.* Berlin: Springer.

Haacke, G. 1977a. *Ann. Rev. Mat. Sci.* 7: 73.

Haacke, G. 1977b. *Appl. Phys. Lett.* 30: 380.

Hamberg, I., and C. G. Granqvist. 1983. *Proc. SPIE* 428: 2.

Hamberg, I., and C. G. Granqvist. 1984. *Appl. Phys. Lett.* 44: 721.

Hamberg, I., C. G. Granqvist, K. F. Berggren, B. F. Sernelius, and L. Engstrom. 1984. *Phys. Rev. B.* 30: 3240.

Hass, G., M. H. Francombe, and R. W. Hoffman. 1977. *Physics of Thin Films* 9: 73.

Heavens, O. S. 1965. *Optical Properties of Thin Solid Films.* New York: Dover.

Heavens, O. S. 1973. *Thin Film Physics.* London: Methuen.

Herzenberg, S. A., and R. Silberglitt. 1982. Low temperature selective absorber research. *Proc. SPIE* 324: 92.

Hogg, S. W., and G. B. Smith. 1977. *J. Phys. D: Appl. Phys.* 10: 1863.

Hollands, K. G. T. 1963. *Solar Energy* 7: 108.

Hottell, H. C., A. T. Sarofin, and E. J. Fahiman. 1967. *Solar Energy* 11: 2.

Howson, R. P., M. I. Ridge, and C. A. Bishop. 1981. *Thin Solid Films* 80: 137.

Hunderi, O., and C. G. Granqvist. 1979. *Thin Solid Films* 57: 303.

Ignatiev, A. 1980. In *Solar Materials Science*, L. E. Murr, ed. New York: Academic Press, p. 151.

Ignatiev, A., P. O'Neill, and G. Zajac. 1979a. *Solar Energy Mat.* 1: 69.

Ignatiev, A., P. O'Neill, C. Doland, and G. Zajac. 1979b. *Appl. Phys. Lett.* 34: 42.

Inal, O. T., M. Volayapetre, L. E. Murr, and A. E. Torma. 1981. *Solar Energy Mat.* 4: 333.

Jacobson, R. 1966. In *Progress in Optics V*, E. Wolf, ed. Amsterdam: North-Holland, p. 249.

Karlsson, B., T. Karlsson, and C. G. Ribbing. 1982. *Proc. SPIE* 324: 156.

Karlsson, B., C. G. Ribbing, A. Roos, E. Valkonen, and T. Karlsson. 1982. *Physica Scripta* 25: B26.

Karlsson, B., R. P. Shinshock, B. O. Seraphin, and J. C. Haygarth. 1983. *Solar Energy Mat.* 7: 401.

Karlsson, B. J. E. Sundgrem, and B. O. Johansson. 1982. *Thin Solid Films* 87: 181.

Kerker, M. 1969. *The Scattering of Light and Other Electromagnetic Radiation*. New York: Academic Press.

Knittle, K. 1976. *Optics of Thin Films*. New York: Wiley.

Koltun, M. M. 1980. *Selective Optical Surfaces for Solar Energy Converters*. New York: Allerton Press. (Translated from Russian from Izdatel'Stvo *Nauka*.)

Kontorovich, M. I., V. Y. Petrumkin, N. A. Yesephkino, and M. I. Astrakhan. 1962. *Radio Eng. Electro Phys* 7: 223.

Kostlin, H., R. Jost, and W. Lems. 1975. *Phys. Stat. Sol.* A29: 87.

Lafait, J., S. Berthier, and J. M. Behagel. 1979. In *Sun II, Proc. ISES*, vol. 3, Atlanta, GA. New York: Pergamon, p. 1922.

Lafait, J., J. M. Behaghel, S. Berthier, and J. Rivory. 1981. *J. Physique* 42: L1-133.

Lampert, C. 1978. *Proc. SPIE* 161: 84.

Lampert, C. M. 1979. Coatings for enhanced photothermal energy collection. *Solar Energy Mat.* 1: 319, 2: 1.

Lampert, C. 1980. *Thin Solid Films* 72: 73.

Lampert, C. M. 1981. *Solar Energy Mat.* 6: 1.

Lampert, C. M. 1982. *Ind. Eng. Chem. Prod. Res. Dev.* 21: 612.

Lampert, C. M. 1989. Advances in solar optical materials. In *Advances in Solar Energy*, vol. 5, K. W. Böer, ed. New York: Plenum.

Lampert, C., and J. Washburn, 1979. *Solar Energy Mat.* 1: 81.

Lee, C. W. 1977. *Proc. ISES*, American Section—1977 Meeting, Cape Canaveral, FL, p. 4.11.

Lee, P. K., and M. K. Debe. 1980. *Photo. Sci. Eng.* 24: 211.

Liddell, H. M. 1981. *Computer-Aided Techniques for the Design of Multilayer Filters*. Bristol, United Kingdom: Adam Hilger.

MacLeod, H. A. 1986. *Thin Film Optical Filters*. 2d ed. New York: Macmillian.

Maystre, D., and M. Nevieve. 1979. *J. Opt.* 9: 301.

McCrackin, F. L., and J. P. Colson. 1964. *NBS Tech. Note 242*.

McKenzie, D. R. 1976. *Opt. Soc. Am.* 56: 249.

McKenzie, D. R., and L. M. Briggs. 1981. *Solar Energy Mat.* 6: 97.

McKenzie, D. R., R. C. McPhedron, and L. M. Briggs. 1982. *Solar Energy Mat.* 7: 75.

McPhedron, R. C., and D. Maystre. 1977. *Appl. Phys.* 14: 1.

Meinel, A. B., and M. P. Meinel. 1976. *Applied Solar Energy—An Introduction.* Reading, MA: Addison-Wesley.

Minot, M. J. 1976. *J. Opt. Soc. Am.* 66: 519.

Minot, M. J. 1977. *J. Opt. Soc. Am.* 67: 1050.

Mizuhashi, M. 1980. *Thin Solid Film* 70: 91.

Moore, C. S., T. S. Ashley II, and H. A. Blum. 1976. In *Proc. ISES Conf. Sharing the Sun,* vol. 6, Winnipeg, Canada, p. 187.

Murr, L. E., ed. 1980. *Solar Materials Science.* New York: Academic Press.

Murr, L. E., O. T. Inal, and M. Valayapetre. 1980. *Thin Solid Films* 72: 111.

Niklasson, G. A., and H. G. Craighead. 1982. *J. Appl. Phys.* 54: 5488.

Niklasson, G. A., and C. G. Granqvist. 1978. In *Proc. ISES,* New Delhi, India. New York: Pergamon, p. 870.

Niklasson, G. A., and C. G. Granqvist. 1979. *J. Appl. Phys.* 50: 5500.

Niklasson, G. A., and C. G. Granqvist. 1981. In *ISES, Proc. Solar World Forum.* Brighton, England, p. 221.

Niklasson, G. A., and C. G. Granqvist. 1983a. *Solar Energy Mat.* 7: 501.

Niklasson, G. A., and C. G. Granqvist. 1983b. *J. Mat. Sci.* 18: 3475.

Niklasson, G. A., C. G. Granqvist, and D. Hunderi. 1981. Effective medium models for the optical properties of inhomogeneous materials. *Appl. Opt.* 20: 26.

Noguchi, S., M. Mizuhashi, and H. Sakata. 1978. *Ashi Garasu Kenkyu Hokoku* 28: 25.

O'Neill, A. P., and A. Ignatiev. 1978. *Phys. Rev. B.* 18: 6540.

O'Neill, A. P., C. Doland, and A. Ignative. 1977. *Appl. Opt.* 16: 2822.

O'Neill, A. P., A. Ignatiev, and C. Doland. 1978a. *Solar Energy* 21: 465.

O'Neill, A. P., A Ignatiev, and C. Doland. 1978b. *A.I.P. Conf. Proc.* 40: 288.

Ohhata, Y., and S. Yoshida. 1977. *Oyo Butsuri* 46: 43.

Ohhata, Y., F. Shinoki, and S. Hoshida. 1979. *Thin Solid Films* 59: 255.

Pajasova, L., A. Abrahams, I. Gregora, and M. Zavetova. 1980. *Solar Energy Mat.* 4: 1.

Penfold, A. S. 1983. *Proc. SPIE* 428: 151.

Petit, R. 1981. *J. Physique* 42: C1.

Pracchia, J. A., and J. M. Simon. 1981. *Appl. Opt.* 20: 251.

Pramanik, D., A. J. Sievers, and R. H. Silsbee. 1979. *Solar Energy Mat.* 2: 81.

Rajagopalan, S. R., K. S. Indira, and K. S. G. Doss. 1965. *J. Electroanal. Chem.* 10: 465.

Ritchie, I. T., and G. L. Harding. 1979. *Thin Solid Films* 57: 315.

Ritchie, I. T., and B. Window. 1977. *Appl. Opt.* 16: 1438.

Ritchie, I. T., S. K. Sharma, J. Valignat, and J. Spitz. 1980. *Solar Energy Mat.* 2: 167.

Rivory, J., J. M. Behaghel, S. Berthier, and J. Lafait. 1981. *Thin Solid Films* 78: 161.

Roos, A., and B. Karlsson. 1983. *Solar Energy Mat.* 7: 467.

Roux, L., J. Hanns, J. C. Francois, and M. Sigrist. 1982. *Solar Energy Mat.* 7: 299.)

Seraphin, B. O., ed. 1976. *Optical Properties of Solids: New Developments.* Amsterdam: North-Holland.

Seraphin, B. O. 1979a. *J. Vac. Sci. Technol.* 16: 193.

Seraphin, B. O., ed. 1979b. Solar energy conversion—Solid state aspects. *Topics in Applied Physics*, vol 31. Berlin: Springer Verlag.

Sexton, F. W. 1984. *Solar Energy Mat.* 7: 1.

Sievers, A. J. 1980. *Solar Materials Science*, L. S. Murr, ed. New York: Academic Press.

Sievers, A. J., D. M. Trotter, Jr., R. A. Buhrman, and H. G. Craighead. 1979. *Sun II. Proc. ISES*, vol. 3, Atlanta, GA. New York: Pergamon, p. 1878.

Sigrist, M., G. Chassaing, J. C. Francois, P. Gravier, L. Armand, R. Pierrisnard, L. Roux, D. Pailhavey, P. Kayoun, and J. Chevallier. 1981. *J. Physique* 42: C1–453.

Sikkens, M., and J. C. Francken. 1979. *Sun II, Proc. ISES*, vol. 3, Atlanta, GA. New York: Pergamon, p. 1907.

Sikkens, M. 1981. *J. Physique* 42: 1–465.

Sikkens, M. 1982. *Solar Energy Mat.* 6: 415.

Simonis, F., M. van der Leij, and C. J. Hoogendoorn. 1979. *Solar Energy Mat.* 1: 221.

Smith, G. B. 1979. *Appl. Phys. Lett.* 35: 668.

Soule, D. E., and D. N. Smith. 1977. *Appl. Opt.* 16: 2818.

Stephens, R. B., and G. D. Cody. 1977. *Thin Solid Films* 45: 19.

Stephens, R. B., and G. D. Cody. 1979. *Solar Energy Mat.* 1: 397.

Tabor, H., H. Weinberger, and J. Harris. 1964. In *Symp. on Thermal Radiation of Solids*. NASA SP55, p. 525.

Thornton, B. S., and Q. M. Tran. 1978. *Solar Energy* 20: 371.

Trotter, D. M., Jr. 1980. *Solar Energy Mat.* 3: 317.

Trotter, D. M., Jr., H. G. Craighead, and A. J. Sievers. 1979. *Solar Energy Mat.* 1: 63.

Trotter, D. M., Jr., and A. J. Sievers. 1979. *Appl. Phys. Lett.* 35: 374.

Trotter, D. M., Jr., and A. J. Sievers. 1980. *Appl. Opt.* 18: 711.

Trotter, D. M., Jr., and A. J. Sievers. 1982. *Solar Energy Mat.* 7: 281.

Valignat, J., J. Spitz, and I. T. Ritchie. 1980. *Rev. Phys. Appl.* 15: 397.

Van der Leij, M. 1979. Ph.D. dissertation. Delft, The Netherlands: Delft University Press.

Van de Hulst, H. C. 1980. *Light Scattering by Small Particles*. New York: Dover.

Vossen, J. L. 1977. *Thin Films* 9: 1.

Williams, D. A., T. A. Lappin, and J. A. Duffie. 1963. *Trans. ASME* 85: 213.

Wilson, A. D. 1984. *Solar Energy Mat.* 10: 9.

Window, B., D. McKenzie, and G. Harding. 1980. *Solar Energy Mat.* 2: 395.

Window, B., I. T. Ritchie, and K. Cathro. 1978. *Appl. Opt.* 17: 2639.

Window, B., I. T. Ritchie, and K. Cathro. 1979. *Thin Solid Films* 57: 309.

Wirgin, A. 1981a. *Compte Rendues Acad. Sci.* 292: 945.

Wirgin, A. 1981b. *J. Physique* 42: C1.

Yoshida, S. 1978. *Appl. Opt.* 17: 145.

Zajac, G., G. B. Smith, and A. Ignatiev. 1980. *J. Appl. Phys.* 51: 5544.

Zajac, G., and A. Ignatiev. 1982. *J. Appl. Phys.* 41: 435.

23 Testing and Evaluation of Solar Materials

Hien K. Le and Richard Silberglitt

A typical solar collector will use metals, polymers, and ceramics for its construction and some sort of a heat transfer fluid for its operation. This section reviews the properties of these solar collector materials and the techniques which are used to characterize them.

23.1 Overview of Solar Collector Materials

23.1.1 Metals

The major metallic components of a solar collector are the collector enclosure or box, the absorber substrate, the flow pipes for the heat transfer fluids, and the external structural support. The minor metallic components are the various types of screws, nuts, bolts, connectors, and similar pieces that hold the collector together. Although these minor parts might play a disproportionate role in reducing collector life considering their size, as in the galvanic corrosion caused by dissimilar metals, we will not be concerned with them here. The common flat plate collector metals are copper, steel, and aluminum. Typical alloy grades are CDA122 and CDA110 for copper; 1100, 3001, 3003, and 6061 for aluminum; and mild steel and 440 carbon steel for steel. Copper is the most expensive material at about $2.50/kg, followed by stainless steel at $2/kg, aluminum at $1.80/kg, and mild steel at $.40/kg (1983 prices). The materials properties that must be characterized for metallic solar materials, especially absorber substrates and flow pipes, are corrosion resistance, thermal conductivity, and strength to weight ratio. Values for thermal conductivity, modulus of elasticity, modulus of rigidity, and density of selected absorber substrate sheet metals are shown in table 23.1.

Thermal conductivity controls the ability of an absorber substrate to efficiently transfer the solar energy captured by the absorber coating to the heat transfer fluid. However, this does not automatically mean that a material such as copper, which has a thermal conductivity eight times greater than steel, is a better absorber substrate material (Kruger 1977). The critical factor that must be considered here is the design of the absorber (the positioning of the flow tubes, the thickness of the absorber, the dimensions of the fins, etc.) in relation to the overall efficiency of the collector. If design is taken into consideration, the inherent advantage of a higher thermal conductivity might just be marginal, so the material selection process must appeal to other criteria such as contribution of the candidate material to the overall collector cost-effectiveness. Some calculations shown in table 23.2 illustrate the point.

       Hien K. Le and Richard Silberglitt

Table 23.1
Properties of selected absorber substrate materials

Materials	Thermal conductivity (W/mK)	Modulus of rigidity (GPa)	Modulus of elasticity (GPa)	Density (kg/m^3)
Copper	377.0	119	45.0	8,900
Aluminum	206.4	71	26.2	2,710
Steel	45.0	207	79.3	8,000
Stainless steel	16.1	190	73.1	7,750
Brass	104.4	106	40.0	8,550

Source: Jorgensen, J., 1982, Internal memorandum, Solar Energy Research Institute, Golden, CO, March.

Table 23.2
Effect of absorber substrate design on collector thermal efficiency factor (Fr)

Material	Tube diameter (mm)	Tube spacing (mm)	Absorber thickness (mm)	Flow rate (kg/hr)	Thermal efficiency factor
Copper	2.5	5.1	1.0	4.5	0.9321
Aluminum	2.5	5.1	1.0	4.5	0.9297
Steel	2.5	5.1	0.06	4.5	0.9186
Steel	0.5	5.1	1.5	4.5	0.9268
Steel	2.5	2.0	1.5	4.5	0.9186
Steel	2.5	5.1	1.5	50.0	0.9464

Source: Kruger, P., 1977, "Solar Absorber Plate Materials," *Solar Age* 2(5), p. 16.

Good corrosion resistance is required for metallic solar collector components if collector lifetime is to be maximized. The relevant issue here is the resistance of solar materials to the corrosive environment existing inside and about a solar collector installation. Such an environment is complex. It is characterized by the collector operating conditions (e.g., temperature extremes, stagnation temperatures, heat transfer rates, fluid flow rates, flow regime), the chemical composition of the heat transfer fluid (which usually contains water, antifreeze and assorted corrosion inhibitors), the presence of factors that will influence the corrosive process (galvanic coupling between dissimilar metals in the collector), and the installation of corrosion monitoring or prevention devices. The properties of metals that will influence their corrosion behavior within the collector environment are chemical composition, thermal history, the state of stresses resulting from fabrication and thermal

Table 23.3
Average corrosion rates for copper, stainless steel, and plain carbon steel obtained in the recirculating system

Corrodent	Alloy	Average corrosion rate[a]	
		mg/cm^2/yr	μm/yr
EG/distilled water	Copper 122	152	171
	T444 SS	<1	<1
EG/20% ASTM water	Copper 122	90	101
	T444 SS	<1	<1
	1018 steel	3,930	5,040
EG/ASTM water	Copper 122	160	180
	T444 SS	<1	<1
	1018 steel	890	1,140
EG/ASTM water	Copper 122	212	239
+50 ppm Cl	T444 SS	<1	<1
PG/distilled water	Copper 122	71	79
	T444 SS	<1	<1
PG/20% ASTM water	Copper 122	70	79
	T444 SS	<1	<1
	1018 steel	2,930	3,760

Source: Diegle, R. B., J. A. Beavers, and J. E. Clifford, 1980, *Final Report on Corrosion Problems with Aqueous Coolants*, DOE/CS/10510-T11, Battelle Columbus Laboratories, Columbus, OH.
a. Rates are averages from triplicate tube specimen. They were extrapolated to annual rates in a linear manner.

cycling, and the surface finish and treatment. As can be readily seen, predicting the corrosion behavior of metallic solar collector components from a knowledge of environmental conditions and metal properties can be an enormous task. DOE has sponsored a great deal of research aimed at such characterization since 1976, and the literature generated by this effort is voluminous. The more general results presented below are drawn from a DOE-sponsored study by Battelle Columbus Laboratories (Diegle, Beavers, and Clifford 1980).

In a recirculating system using various heat transfer fluid compositions, stainless steel exhibits the lowest corrosion rate, followed by copper and carbon steel (table 23.3). Aluminum, under the same conditions, is subject to serious pitting corrosion, with pit density in the 1.8×10^5 pit/cm^2 range (table 23.4). Crevice corrosion experiments also show that aluminum suffers from minor to severe attacks (table 23.5), whereas the copper and steel are unattacked. Manifestations of other types of corrosive attack (e.g., general corrosion, stress corrosion cracking, intergranular corrosion, erosion corrosion, and cavitation corrosion) are minimal to none for all three types of materials.

Table 23.4
Maximum pit depths and approximate pit densities for aluminum alloys exposed in the recirculating system

Corrodent	Alloy	Exposure time,[a] h	Maximum pit depth,[b] μm	Maximum pit density, cm^{-2}
EG/distilled	1100	260	5	$> 1.8 \times 10^5$
water	3003		5	$< 9 \times 10^3$
	6061		19	$> 1.8 \times 10^5$
EG/20% ASTM	1100	440	6	$> 1.8 \times 10^5$
water	3003		8	$> 1.8 \times 10^5$
	6061		20	$> 1.8 \times 10^5$
EG/ASTM	1100	285	72	$> 1.8 \times 10^5$
water	3003		127	$> 1.8 \times 10^5$
	6061		150	$> 1.8 \times 10^5$
PG/distilled	1100	234	3	$< 9 \times 10^3$
water	3003		4	$< 9 \times 10^3$
	6061		13	$> 1.8 \times 10^5$
PG/20% ASTM	1100	404	3	$> 1.8 \times 10^5$
water	3003		5	$< 9 \times 10^3$
	6061		14	$> 1.8 \times 10^5$

Source: Diegle, R. B., J. A. Beavers, and J. E. Clifford, 1980, *Final Report on Corrosion Problems with Aqueous Coolants*, DOE/CS/10510-T11, Battelle Columbus Laboratories, Columbus, OH.
a. Time at operating temperature, 100°C.
b. Determined by means of a microscope with a calibrated focus adjustment. Pit depths and pit densities were obtained by examining about 125 cm^2 of each type of specimen.

Table 23.5
Summary of results of crevice corrosion experiments

Solution	Alloy	Severity of crevice attack[a] Metal–metal interface	Metal–PTFE interface
EG/ASTM	1100	Moderate	Minor
water	3003	Minor	None
	6061	Minor	Severe
	Cu 122	None	None
	T444	None	None
PG/ASTM	1100	None	Moderate
water	3003	None	Minor
	6061	Minor	Severe
	Cu 122	None	None
	T444	None	None

Source: Diegle, R. B., J. A. Beavers, and J. E. Clifford, 1980, *Final Report on Corrosion Problems with Aqueous Coolants*, DOE/CS/10510-T11, Battelle Columbus Laboratories, Columbus, OH.
Note: Duplicate specimens were exposed in nonflowing aerated solutions at 100°C for 50 days. Chloride concentration was 50 ppm.
a. Terms are relative.

Other properties such as moduli of elasticity and rigidity, density, and the like are very well characterized for most if not all commonly used metals and alloys. With the proper design the commercially available copper, steel, and aluminum alloys can meet or exceed the structural requirements for use in a solar collector.

23.1.2 Polymers

The major components of a solar collector in which polymers may be used are the cover plate glazings and insulating materials. The minor components are the gaskets, sealants, adhesives, connecting flexible hoses, and similar bits and pieces. All polymer solar collectors, such as Brookhaven National Laboratory's Fluoroplastic Film Solar Flat-Plate Collector Panel (Wilhelm 1981), Acurex Solar Corporation's Experimental Low-Cost Solar Panel (Nelson and Muller 1980), and those commercially available from manufacturers like FAFCO, Ramada Energy Systems Ltd., and Bio-Energy Systems, Inc., are a class of their own and place even more materials properties requirements on polymers than those components listed previously.

The polymer properties that must be characterized are dependent on application. Solar transmittance, abrasion resistance, impact strength, resistance to warping and embrittlement, resistance to thermal aging, and resistance to weathering, are the major ones for cover plate glazings. For insulating materials, the relevant properties are thermal conductivity, thermal resistance, heat transfer, specific heat, resistance to moisture, vapor permeability, and resistance to weathering. For polymeric absorber substrates, thermal stability and weathering resistance are the critical properties. For sealants, the properties that must be characterized are adhesion, compression set, hardness, tensile strength, and ultimate elongation. For adhesives, the important properties are strength (flexural, shear, and tensile), thermal stability, resistance to aging, and moisture resistance. For flexible coupling hoses, one needs to know compression set, creep, extension recovery, strength, hardness, tear resistance, compatibility with heat transfer fluids, thermal stability, and resistance to weathering.

Due to the very large number of polymers that can be and are used in solar collectors, only representative polymer properties will be listed here. (The reader is advised to refer to the sections on glazings, absorber substrates, and adhesives in chapter 25 for a listing of the major polymers in use in solar collectors.) Table 23.6 lists some commercially available polymers for glazing applications. Tables 23.7a through 23.7g list the properties of various insulating materials.

Table 23.6
Commerically available polymers for solar applications

Materials' trade name and manufacturer	Relevant Properties			
	Transmittance (%)	U-value	Other	Cost ($/ft²)
Polycarbonate sheets				
Danpalon™ (Poligal Industries)	75	0.52–0.21		4.0
Exolite™ (CY/RO)	74	0.58		6.5
Lexan® (GE)	84	—	Stable up to 200°F	3.15–4.5
Thomas Insu-Wall (Inter American Marketing Systems)	81	0.36		3.0
TEM Glaze (Ramada Energy Systems Ltd.)	80	0.58		3.0
Fiberglass-reinforced polyester sheets				
Sunlite HP® (Solar Component Corporation)	84–88	—		0.99
Crystallite-T™ (Lasco Industries)	86	—		0.97
Filon Solar E™	85	—	Stable up to 290°F	1.36
Polyethylene films				
602 (Monsanto) Visqueen (Ethyl Corporation)	87–90	—	UV inhibitors up to 130°F	0.06
Fluorocarbon films				
Teflon® (DuPont)	96	—	Stable up to 400°F	0.37
Tedlar® PVF (DuPont)	90	—	10 year + lifetime	0.45
Acrylic polyester films				
Flexigard® (3M)	87	—	Stable up to 240°F	0.74
Polyester films				
Sungain™ (3M)	96	—	Stable up to 220°F	0.70

Source: Kohler, J., and V. Reno, 1983, "Engineer Guide to Plastic Glazings," *Solar Age* 8(8), p. 40.

Table 23.7a
Pysical and structural properties of various insulation materials

Type of insulation	Compressive strength (psi)	Density (lb/ft^3)
Cellular plastic		
Extruded polystyrene	40	1.7
		2.4
Molded polystyrene	12	1.0
		1.8
Polyurethane	14	1.85
	34	2.7
Polyisocyanurate	17	1.8
		2.3
Ureaformaldehyde		0.6
		0.9
Glass insulation		
Glass fiber batts		0.6
		1.0
Glass fiber board	130	4.5
		9.0
Foam glass	100	8.5
Loose fill		
Vermiculite		4.0
		10.0
Perlite		2.0
		11.0
Cellulose		2.2
		3.0
Miscellaneous		
Rock wool		1.5
		2.5
Calcium silicate	250	13.0
Insulating concrete	125	12.0
		88.0
	225	
Powdered limestone	11,110	60.0
Foam rubber		5.5

Source: Versar Inc., and Burt Hill Kosar Rittlemann Associates, 1980, *Final Report: Survey and Evaluation of Available Thermal Insulation Materials for Use in Solar Heating and Cooling Systems,* work performed under DOE Contract No. DE-AC04-78CS3563. (Available from NTIS, Springfield, VA, 22161.)

Table 23.7b
Thermal properties of various insulation materials

Type of insulation	Resistance (R/in.) at 75°F	Lower temperature limit (°F)	Upper temperature limit (°F)
Cellular plastic			
Extruded polystyrene	3.7		165
	5.0		
Molded polystyrene	4.0		165
Polyurethane	5.5	−320	250
	7.14		
Polyisocyanurate	6.67	−423	250
Ureaformaldehyde	4.2	−300	415
			a
Glass insulation			
Glass fiber batts	3.16		1,000
			b
Glass fiber board	4.0		c
Foam glass	2.63	−459	900
Loose fill			
Vermiculite	2.4		1,000
	3.0		
Perlite	2.5		1,200
	3.7		
Cellulose	3.2		180
	3.7		
Miscellaneous			
Rock wool	3.2		1,800
	3.7		
Calcium silicate	2.6		1,500
Insulating concrete	1.2		1,000
Powdered limestone	1.1		480
Foam rubber			

Source: Versar Inc., and Burt Hill Kosar Rittlemann Associates, 1980, *Final Report: Survey and Evaluation of Available Thermal Insulation Materials for Use in Solar Heating and Cooling Systems,* work performed under DOE Contract No. DE-AC04-78CS3563. (Available from NTIS, Springfield, VA, 22161.)
Note: Laboratory results were conducted at National Bureau of Standards, "Solar Energy Systems: Test Methods for Collector Insulations," NBSIR 79-1908, October 1979.
a. Decomposes at this temperature.
b. At 400°F organic binders may break down and binder may deposit on cover plate. Facing may be flamable above 180°F.
c. Discolors and chars from 300–400°F after exposure of 96 hours.

Table 23.7c
Effects of temperature on various insulation materials

Type of insulation	Coefficient of expansion/°F	Dimensional stability	Thermal cycling	Outgassing	Toxic gases emitted	Corrosive gas emitted	Odor
Cellular plastic							
Extruded polystyrene	3.5×10^{-5}	Good		Yes	Yes (CO)	[a]	None
Molded polystyrene	3.5×10^{-5}	Good			Yes (CO)	[a]	None
Polyurethane	3.5×10^{-5}	Poor		[b]	Yes (CO)	[c]	None
Polyisocyanurate	3.5×10^{-5}	Poor		[b]	Yes (CO)	[c]	None
Ureaformaldehyde		Poor	[d]				Yes [e]
Glass insulation							
Glass fiber batts		Good					None
Glass fiber board		Good				[c]	None
Foam glass	4.6×10^{-6}	Good	[g]	[f]	Yes (CO, H_2S)	None	
Loose fill							
Vermiculite					No	None	
Perlite					No	No	
Cellulose					Yes (CO)	Yes [h]	
Miscellaneous							
Rock wool		Good				[c]	None
Calcium silicate		Good				[j]	
Insulating concrete		Good		[i]	No		
Powdered limestone							
Foam rubber		Poor				[k]	

Table 23.7c (continued)

Source: Versar Inc., and Burt Hill Kosar Rittlemann Associates, 1980, *Final Report: Survey and Evaluation of Available Thermal Insulation Materials for Use in Solar Heating and Cooling Systems*, work performed under DOE Contract No. DE-AC04-78CS3563. (Available from NTIS, Springfield, VA, 22161.)

Note: The notes identified with an asterisk represent laboratory results conducted at National Bureau of Standards, "Solar Energy Systems: Test Methods for Collector Insulations," NBSIR 79-1908, October 1979.

*a. Mild pitting with copper and brass in an oven at 212°F for 30 days. None for stainless steel and aluminum.

*b. In a heated chamber (570°F), polyurethane lost nearly 24% of its original weight, resulting in an effective decrease of cover plate transmittance of 8% in 3 hours.

*c. Mild pitting with copper, brass and stainless steel in an oven at 212°F for 30 days. None for aluminum.

*d. Cracked, then crumbled in 30-cycle test.

e. Formaldahyde may linger on until cured.

f. At 400°F organic binders may break down and binder may deposit on cover plate. Facing may be flammable above 180°F.

*g. Cracks in 45-cycle test.

*h. Severe pitting with stainless steel in an oven at 212°F for 30 days. Mild pitting with brass, copper and aluminium.

i. Between 200°–450°F, binder may break down.

*j. Mild pitting with copper in an oven at 212°F for 30 days. None for stainless steel, brass, and aluminum.

*k. Mild pitting with brass and copper in an oven at 212°F for 30 days. None for stainless steel and aluminium.

Table 23.7d
Fire resistance of various insulation materials

Type of insulation	Flame spread[a]	Fuel contributed[a]	Smoke developed[a]
Cellular plastic			
Extruded polystyrene	25		
Molded polystyrene	<25	5	10
		80	400
Polyurethane	<75	14	116
		25	233
Polyisocyanurate	25	0	55
		5	200
Ureaformaldehyde	0	0	0
	25	30	10
Glass insulation			
Glass fiber batts	15	5	0
	20	15	20
Glass fiber board			
Foam glass	5		0
Loose fill			
Vermiculite	0	0	0
Perlite	0	0	0
Cellulose	15	0	0
	40	40	45
Miscellaneous			
Rock wool	15	0	0
Calcium silicate	0		0
Insulating concrete	0	0	0
Powdered limestone			
Foam rubber			

Source: Versar Inc., and Burt Hill Kosar Rittlemann Associates, 1980, *Final Report: Survey and Evaluation of Available Thermal Insulation Materials for Use in Solar Heating and Cooling Systems*, work performed under DOE Contract No. DE-AC04-78CS3563. (Available from NTIS, Springfield, VA, 22161.)
a. According to ASTM E-84. Laboratory results were conducted at National Bureau of Standards, "Solar Energy Systems: Test Methods for Collector Insulations," NBSIR 79-1908, October 1979.

Table 23.7e
Effects of moisture on various insulation materials

Type of insulation	Moisture absorption (vol %)	Permeability (in.)	Corrosive solutions created
Cellular plastic			
Extruded polystyrene	1	0.4	a
		0.9	
Molded polystyrene	<2	1.2	a
		3.0	
Polyurethane	Low	1.2	b
		30	
Polyisocyanurate	<1.5	2.0	b
		3.0	
Ureaformaldehyde	High	4.5	
	c	100.0	
Glass insulation			
Glass fiber batts	0.1	100.0	
	0.2		
Glass fiber board	2		b
Foam glass	0.2	0	None
Loose fill			
Vermiculite	0	High	None
Perlite	Low	High	None
Cellulose	5	High	d
	20		
Miscellaneous			
Rock wool	2	>100	b
Calcium silicate	High	High	e
Insulating concrete		Varies	
Powdered limestone			
Foam rubber	Low	Low	f

Source: Versar Inc., and Burt Hill Kosar Rittlemann Associates, 1980, *Final Report: Survey and Evaluation of Available Thermal Insulation Materials for Use in Solar Heating and Cooling Systems*, work performed under DOE Contract No. DE-AC04-78CS3563. (Available from NTIS, Springfield, VA, 22161.)
Note: The notes identified with an asterisk represent laboratory results conducted at National Bureau of Standards, "Solar Energy Systems: Test Methods for Collector Insulations," NBSIR 79-1908, October 1979.
*a. Mild pitting with copper and brass in an oven at 212°F for 30 days. None for stainless steel and aluminum.
*b. Mild pitting with copper, brass and stainless steel in an oven at 212°F for 30 days. None for aluminum.
*c. Moisture absorbed 8.8% by mass at 90% RH and 23°C for 30 days.
*d. Severe pitting with stainless steel in an oven at 212°F for 30 days. Mild pitting with brass, copper and aluminum.
*e. Mild pitting with copper in an oven at 212°F for 30 days. None for stainless steel, brass, and aluminum.
*f. Mild pitting with brass and copper in an oven at 212°F for 30 days. None for stainless steel and aluminum.

Table 23.7f
Environmental concerns for various insulation materials

Type of insulation	Ultraviolet degradation	Resistance to bacteria and fungus
Cellular plastic		
Extruded polystyrene	Yes	Good
Molded polystyrene	Yes	Good
Polyurethane	Yes	Good
Polyisocyanurate	Yes	Good
Ureaformaldehyde	Yes	Good
Glass insulation		
Glass fiber batts		Good
Glass fiber board		Good
Foam glass		Good
Loose fill		
Vermiculite		Good
Perlite		Good
Cellulose		Poor [a]
Miscellaneous		
Rock wool	No	Good
Calcium silicate		Poor [a]
Insulating concrete		Good
Powdered limestone		
Foam rubber		Good

Source: Versar Inc., and Burt Hill Kosar Rittlemann Associates, 1980, *Final Report: Survey and Evaluation of Available Thermal Insulation Materials for Use in Solar Heating and Cooling Systems*, work performed under DOE Contract No. DE-AC04-78CS3563. (Available from NTIS, Springfield, VA, 22161.)
a. ASTM D 3273. Laboratory results were conducted at National Bureau of Standards, "Solar Energy Systems: Test Methods for Collector Insulations," NBSIR 79-1908 October 1979.

Table 23.7g
Miscellaneous properties of various insulation materials

Type of insulation	Chemical reaction problems	Resistance to boiling water	Recovery drying
Cellular plastic			
Extruded polystyrene	Yes		Good
Molded polystyrene	Yes		
Polyurethane	Yes	Poor	Fair
Polyisocyanurate			
Ureaformaldehyde			
Glass insulation			
Glass fiber batts		Poor	Poor
Glass fiber board		Poor	Poor
Foam glass		Fair	Good
Loose fill			
Vermiculite			
Perlite			
Cellulose			
Miscellaneous			
Rock wool		Poor	Poor
Calcium silicate		Good	Good
Insulating concrete		Good	Good
Powdered limestone			
Foam rubber			

Source: Versar Inc., and Burt Hill Kosar Rittlemann Associates, 1980, *Final Report: Survey and Evaluation of Available Thermal Insulation Materials for Use in Solar Heating and Cooling Systems*, work performed under DOE Contract No. DE-AC04-78CS3563. (Available from NTIS, Springfield, VA, 22161.)

23.1.3 Ceramics

It might seem surprising, but ceramics are also solar collector materials. The selective absorber coatings on the absorber substrate—be they intrinsic absorbers such as ZrB_2 or NiN_x, semiconductor metal tandems such as Cu_2S on aluminum, metal dielectric composites such as Cr_2O_3 on Cr or $NiCrO_x$, or multilayer absorbers such as the $Al_2O_3/Metal/Al_2O_3$ systems—are ceramic materials in the strict sense of the term. Similarly some of the heat mirror materials, such as indium tin oxide ($In_2O_3 : SnO_2$) and cadmium orthostannate (Cd_2SnO_4), are also ceramics. Many insulation materials are ceramic, such as glass fiber batts, glass fiberboard, foam glass, and calcium silicate. Desiccants are also ceramics, as exemplified by silica gel and the salt hydrates. Although a ceramic flat plate collector has yet to be built, ceramic materials have been considered for absorber substrate applications. These materials are concrete reinforced with glass (GRC), or polymer (Polysil®). Insulating mate-

Table 23.8
Currently used heat transfer fluids, based on chemical type

Fluid type	Flat plate collector use (%)
Petroleum-based aliphatic hydrocarbons	2.3
Synthetic aliphatic hydrocarbons	2.3
Aromatic hydrocarbons	1.6
Ethylene glycol/water	18.6
Propylene glycol/water	17.0
Water	29.5
Ethers	0.8
Esters	0.8
Silicones	8.5
Fluorocarbons	0.8
Other (air)	17.8
	100.0

Source: Diegle, R. B., J. A. Beavers, and J. E. Clifford, 1980, *Final Report on Corrosion Problems with Aqueous Coolants*, DOE/CS/10510-T11, Battelle Columbus Laboratories, Columbus, OH.

rials can also be ceramics such as glass fiber batts and boards, foam glass, calcium silicate, or even powdered limestone (see table 23.7). Most thermal storage materials are salt hydrates of Ca, Mg, and Na and can also be classified as ceramics.

The critical material properties that must be characterized for selective absorbers are absorptivity, emissivity, and optical durability. For heat mirrors they are transmissivity, infrared reflectivity, and durability. Desiccant properties of interest are heat of adsorption, isotherm behavior, thermal cycling hysterisis, water capacity, moisture diffusivity, and thermal capacitance. The critical absorber substrate properties have been enumerated in the previous section.

The various types of selective absorbers, heat mirrors, thermal storage materials, and desiccants are extensively listed in chapter 25. Table 23.7 shows some of the properties of the various insulating materials.

23.1.4   Other Solar Collector Materials

Heat transfer fluids and corrosion inhibitors are also collector materials. The most widely used heat transfer fluids are water, glycol/water solutions, and air. Table 23.8 lists heat transfer fluids by chemical type. Their critical properties for solar collector applications are density, specific heat, viscosity, thermal conductivity, boiling point, resistance to thermal cycling, and temperature extremes.

Corrosion inhibitors are materials that interfere with the oxidation or reduction half-reactions of the corrosive process. Typical inhibitors are salt hydrates such as $NaNo_2 \cdot Na \cdot B_4O_7 \cdot 10H_2O$, $Na_2SiO_3 \cdot 6H_2O$ and $Na_3PO_4 \cdot 12H_2O$. In actual use, an inhibitor formulation consisting of a few inhibitors, each in a few grains/liter concentration, is added to the heat transfer fluid.

23.1.5 Summary

Table 23.9 lists the materials properties that should be characterized for each type of material application. The list of properties is derived from a 1977 survey of solar materials by Skoda and Masters of the National Bureau of Standards (now NIST).

Although the matrix "materials-properties" to be characterized is large, most physical properties are known if the material is commercially available, or they are under intensive evaluation if the material is under development. Solar material characterization and testing consists primarily of durability and reliability testing under actual or simulated solar collector environments for known materials, and property characterization for new, developmental ones.

23.2   Review of Material Characterization Research

23.2.1   Characterization of Optical Properties

The reflectance, absorptance, transmittance, and emissivity of glazings, heat mirrors, and selective absorbers are important solar optical properties that must be characterized. Reflectance ρ is the fraction of the incident radiation that is reflected from the sample surface. Absorptance α is the fraction absorbed. And transmittance τ is the fraction transmitted through (the cover plate or glazing). Emissivity is the ratio of the thermal radiation emitted by a surface to that emitted by a blackbody at the same temperature. The optical property is said to be spectral if a single wavelength of incident light is used to make the measurement. It is said to be total if all wavelengths are included. If the total solar spectrum is used, the measured value is referred to as solar, as in solar transmittance. Furthermore, depending on the geometry of measurement, the incident and emerging beams should be specified as normal (e.g., $90°$ to the surface), directional (e.g., nonnormal), hemispherical (e.g., over a 2π angle over the surface), or bidirectional (e.g., measurement at specific, arbitrary

Table 23.9
Materials' properties of importance to solar collectors

Materials	Optical properties			Physical properties																Resistance					Compatibility					Safety										
	Absorptance	Reflectance	Transmittance or Emissivity	Abrasion resistance	Adhesion	Boiling point	Burst strength	Compression set	Densities	Dimensional stability	Electrical conductivity	Embrittlement	Expansion coefficient	Flexual strength	Freezing point	Hardness	Impact strength	Permeability	Specific heat	Stress cracking	Swelling/shrinking	Tensile strength	Thermal conductivity	Ultimate elongation	Water absorption	Elevated temperature	Depressed temperature	Thermal cycling	Solar radiation	Moisture	Pollutants	Ozone	Exposed stresses	Outgassing	Adjoining materials	Heat transfer fluid	Soil constituent	Flammability	Toxicity	Structural adequacy
Metals																																								
Absorber substrate	×			×					×	×			×	×			×		×	×			×			×	×	×	×	×			×		×	×				×
Enclosure				×						×			×						×	×		×	×		×	×	×	×		×	×	×	×	×	×					×
Pipings				×			×					×	×	×				×					×				×	×							×	×				
Flexible couplings							×					×	×					×								×	×	×												
Support structure												×	×	×			×					×					×	×									×			×
Polymers																																								
Absorber substrate										×			×				×		×	×		×	×		×	×	×	×	×	×	×		×	×	×	×				×
Adhesive					×								×				×						×											×	×			×		
Glazing	×	×	×													×			×	×					×	×			×	×		×		×						×
Insulation				×					×		×	×						×		×	×	×	×			×	×	×		×	×		×		×					
Pipings							×	×										×		×	×	×	×			×	×	×			×	×	×		×	×	×	×		
Flexible couplings							×	×										×		×	×	×				×	×	×			×	×	×			×	×	×		
Ceramics																																								
Absorber substrate										×							×			×					×	×	×	×	×	×					×					
Desiccant										×															×										×					
Heat mirror		×	×																							×		×	×	×	×				×					
Insulation											×												×			×	×	×	×	×										
Selective absorbers	×	×		×	×						×							×				×	×		×	×	×	×	×	×					×					
Heat transfer fluids						×									×				×																					

Source: Skoda, L. F., and L. W. Masters, 1977, *Solar Energy Systems Survey of Materials Performance*, NBSIR77-1314, National Bureau of Standards, Washington, DC.

Table 23.10
Materials optical requirements: appropriate measurements and criteria

	Minimum acceptable values	
Use/property	Tracking[a]	Fixed
Flat plate collectors		
Cover/transmittance (τ_s)[b]	>0.85	>0.85, $\Delta\theta_A > \pm 45°$
Absorber/absorptance (α_s)	>0.92	>0.92, $\Delta\theta_A > \pm 60°$
Reflector (augmentor)/reflectance (ρ_s)	>0.80	>0.80, $\Delta\theta_A > \pm 60°$
Infrared suppression/emissivity (ε_H)[c]	<0.15	<0.15
Focusing collectors		
Cover/transmittance (τ_N, s)[b]	>0.85	>0.80, $\Delta\theta_A > \pm 50°$
Reflector/specular reflectance (ρ_θ, s)	>0.80	>0.80
Receiver enclosure/transmittance (τ_N, s)	>0.85	>0.85
Receiver surface/absorptance (α_s)	>0.90	>0.90
Infrared suppression/emissivity (ε_H)[c]	<0.10	<0.10[c]
Central receivers		
Heliostat/specular reflectance (τ_θ, s)	>0.80	$\Delta\Psi < 10$ mr[d]
Receivier surface/absorptance (α_s)	>0.95	

Source: Versar Inc., and Burt Hill Kosar Rittlemann Associates, 1980, *Final Report: Survey and Evaluation of Available Thermal Insulation Materials for Use in Solar Heating and Cooling Systems*, work performed under DOE Contract No. DE-AC04-78CS3563. (Available from NTIS, Springfield, VA, 22161.)
a. Unless otherwise given, no requirement for minimum acceptance angle, $\pm \Delta\theta_A$.
b. Subscript notation: s = solar, N = normal (H = hemispherical is implied otherwise), θ = angle of incidence.
c. Value measured at use temperature is required.
d. Induced divergence.

incidence and emergence angles). Table 23.10 gives the desired optical properties for various solar applications. These optical properties can be measured in the laboratory using either calorimetric or spectrophotometric methods. They may also be measured in the field with the help of portable instruments. Table 23.11 lists available portable instruments for the measurement of solar absorptance and hemispherical emissivity. Commercially available spectrophotometers are the Beekman's DK-1, DK-2, model 5270; Perkin-Elmer's Hitachi SF-34; and Varian's Cary Model 17 and 19. With attachments such as absolute integrating spheres, heated cavities, and parabolic or elipsoid mirrors, spectral normal hemispherical reflectance ρ can be measured in the spectral range of 2 to 30 μm. The total emissivity can be readily derived from the relation $\varepsilon = 1 - \rho$ which can be simply derived from Kirchoff's law, and from the expression

$$\varepsilon_H = \frac{\int_0^\infty \varepsilon_\lambda E_b(\lambda, T)\, d\lambda}{\sigma T^4},$$

Table 23.11
Characteristics of available instrumentation for solar absorptance and emissivity measurements

Manufacturer	Cost ($)	Measurement	Minimum sample size (cm)	Radiation source	Detector	Reproducibility (%)	Comments
Absorptance measurement							
Devices and Services Company Model SSR-ER	7,245	s	2 diameter	Tungsten-Halogen	Four-detector/Filter Combination	0.3	0.02 absorptance units all coatings
Gier-Dunkle Inst. (Dynatech) Model MS-251	15,000	s	1.3 diameter	Xenon lamp	Thermopile	0.5	0.03 absorptance units for black chrome; no reference required
International Technology Corp. (Willey Corp.) Model 2150 Alpha Meter	1,825	s	1.2 diameter	Projector lamp	Silicon	0.5	Errors as large as 0.11 absorptance units for black chrome; calibration reference required
Emissivity measurement							
Devices and Services Company Model AE	1,000	N	6 × 6 plate	Heated cavity	Differential thermocouple	1.0	80°C emissivity; 0.03 emissivity units for black chrome
Gier-Dunkle Inst. (Dynatech) Model DB-100	20,000	H or N	2 diameter	Heated	Thermocouple	0.2	100°C emissivity; 0.02 emissivity units for black chrome
International Technology Corp. (Willey Corp.) Model 2138 Ambient Emissionmeter	5,000	H	3 diameter	Heated cavity	Thermocouple	1.0	Linear relationship for black chrome

Sources: Gilligan, J. E., and H. Betz, 1979, "Appropriate Optical Measurements and Criteria for Solar Materials," *Proc. American Electroplaters' Society, Second Coatings for Solar Collectors Symposium*, St. Louis, MO, October; Pettit, R. B., and A. R. Mahoney, 1980, "Portable Instrumentation for Solar Absorptance and Emittance Measurements," *Proc. Line Focus Solar Thermal Energy Technology Development Conference*, Albuquerque, NM, September.

where

$\varepsilon_\lambda = (1 - \rho_\lambda)$ (normal or hemispherical),

$E_b(\lambda, T) = $ blackbody intensity distribution $= c_1 \lambda^{-5}/[\exp(c_2/\lambda T) - 1]$,

$\lambda = $ wavelength,

$\sigma = $ Planck's constant,

$c_1 = $ first radiation constant,

$c_2 = $ second radiation constant.

In the calorimetric method a sample is heated electrically in an enclosure that is maintained at a constant temperature. At equilibrium the energy radiated by the sample, $A_s \varepsilon_H \sigma T_s^4$, is equal to the electrical energy EI input to the sample minus any other conductive or radiative heat losses Q_L in the enclosure:

$$A_s \varepsilon_H \sigma T_s^4 = EI - Q_L,$$

so that

$$\varepsilon_H = \frac{EI - Q_L}{A_s \sigma T_s^4},$$

where

$A_s = $ surface area of the sample,

$E = $ voltage supplied to the heater,

$I = $ current supplied to the heater,

$Q_L = $ heat losses,

$T_s = $ sample surface temperature.

Alternatively, dynamic or nonequilibrium measurement is also possible. The procedure used in this case is to heat up the sample and then to cut off the electric power and let the sample cool off, while monitoring the sample temperature. The emissivity is found from the dynamic energy balance:

$$MCp\frac{dT_s}{dt} = A_s \varepsilon_H \sigma T_s^4 + Q_L(T_s, t),$$

where

$M = $ mass of the sample,

$Cp = $ heat capacity,

$t = $ time.

Although the calorimetric method yields accurate values for ε_H as a function of temperature, it is time-consuming and experimentally difficult. Spectrophotometric methods are, in contrast, simpler to use, yielding an accuracy of 1% or better. Compared to the portable instruments, both laboratory techniques are much better in terms of providing absolute optical properties and a range of precisely controllable variables. The cost of a laboratory optical measurement setup can run to $100,000 or more, however, and highly trained personnel are also required. It is thus not surprising that many published data on the optical properties of solar materials are taken with portable instruments.

23.2.2 Characterization of Mechanical Properties

The mechanical properties of structural components of solar collectors should ideally be well characterized. This is the case for the metallic components such as external support, frames, and absorber substrates. The mechanical properties of polymers, however, are not well characterized for solar collector applications and are the main concern of research in this area.

The mechanical properties of interest for polymers have been listed in table 23.9. Of these, tensile strength, elongation at break, impact strength, and abrasion resistance are considered most important for cover plate materials. Tensile strength at break is the ratio of the tensile force at breakage to the original surface area. Elongation at break is the ratio of the change in gage length to the original length, measured in percentages. The impact strength is defined as the ratio of the energy needed to break a notched specimen to its notch thickness. Abrasion resistance is measured by a wear index, which is the ratio of the weight loss after a thousand abrasive cycles to the total number of test cycles carried out.

For polymeric adhesive films and tapes, peel strength must also be characterized. Peel strength is the resistance to peeling off, which is measured by the force required to peel a tape of known surface area from its substrate. Standard ASTM test methods are available to measure these properties (e.g., ASTM Methods D638–58T or D1709–59T for tensile strength, ASTM Test Method D256 for impact strength, and ASTM Test Method D1044 for abrasion resistance). DOE sponsored an extensive test and characterization effort using such methods at the IIT Research Institute. The materials tested were commercially available polymer sheets and transparent thin films, glasses and fiberglass-reinforced polymer sheets intended for use as cover plates, various types of polyimide, fluorocarbon, glass cloth, and aluminized sealing tapes and adhesives (Gilligan et al. 1980).

23.2.3   Other Characterizations

The characterization of the surface morphology of selective absorber coatings, heat mirrors, and glazings is essential to the understanding of the critical properties of these components. Standard surface characterization techniques (SEM, TEM, EDS, SIMS, ESCA, etc.) are used for these characterizations.

References

Diegle, R. B., J. A. Beavers, and J. E. Clifford. 1980. *Final Report on Corrosion Problems with Aqueous Coolants*. DOE/CS/10510-T11. Battelle Columbus Laboratories, Columbus, OH.

Gilligan, J. E., J. Brzuskiewicz, J. E. Brzuskiewicz, S. J. Gaumer, and H. T. Betz. 1980. *Handbook of Materials for Solar Energy Utilization Low Temperature Applications*, DOE/CH/90034-TI, IIT Research Institute, Chicage, IL.

Kruger, P. 1977. Solar absorber plate materials. *Solar Age* 2(5): 16.

Nelson, E. V., and T. K. Muller. 1980. Further development of a low-cost solar panel *Proc. Annual DOE Active Solar Heating and Cooling Contractors' Review Meeting*, March 26–28, pp. 9-7–9-8.

Skoda, L. F., and L. W. Masters. 1977. *Solar Energy Systems Survey of Materials Performance*. NBSIR77-1314. Washington, DC: National Bureau of Standards.

Wilhelm, W. G. 1981. Low cost solar flat-plate collector development. *Proc. Annual DOE Active Solar Heating and Cooling Contractors' Review Meeting*, March 26–28, pp. 4-4–4-6.

24 Exposure Testing and Evaluation of Performance Degradation

Thomas E. Anderson

24.1 Performance and Degradation

24.1.1 Introduction and Background

The performance of a material or product is dependent on its ability to resist degradation processes induced by a particular environment. For example, the efficiency (performance) of a solar collector can deteriorate from its initial design criteria as a result of changes in component material properties (degradation). If the transmission properties of the cover plate material change as a result of exposure to ultraviolet radiation, the performance of the solar collector will also change. For those materials used in outdoor applications, exposure testing (weathering) is the rather broad field of engineering endeavor that seeks to determine the type and degree of response to the natural climatic environments of interest.

The primary elements that contribute to degradation are solar radiation (particularly the ultraviolet wavelengths), moisture (dew, rain, humidity), temperature (primarily the time-averaged temperature of the exposed surface), and both natural and man-made pollutants (particularly aerosols, acid rain, etc.). The synergistic effect of these elements will invariably result in aesthetic, optical, and/or physical changes in the properties of materials.

Shortly after 1900 the first formal outdoor weathering facility was established near Miami, Florida. The area was selected by the paint industry to induce practical weathering resistance in their products because the subtropical climate of southern Florida was considered to produce the most severe degradation in paint coatings.

The first exposures were conducted at 90 deg south, facing the equator, using linseed oil-based coatings on wood panels to realistically simulate their end use on domestic and commercial wood structures. Even during these early days in the development of outdoor weathering tests, the time differential to predict the service life of coatings was considered to be "forever" and all concerned felt the need to develop an accelerated method. Within a few years a comparative evaluation was made using identical coatings on wood panels exposed at both 45 deg and 90 deg south. Although some of the differences could be attributed to "time of wetness" of the panels mounted at 45 deg south, the rate of acceleration in the generation of typical failure modes was greater by a factor of two for the panels exposed at 45 deg. Hence the first "accelerated" test method was an actual real-time weathering exposure in southern Florida at 45 deg south, facing the equator. Since these exposures were conducted miles from the ocean, the condition of their exposure was thereafter referred to as *direct weathering inland*; the term is still used today to describe the condition of outdoor exposure testing.

The plastics industry as well as others followed the paint industry's precedent of sending samples to Florida. However, the development of new and varied polymers soon proved that one climatic condition was inadequate as a universal testing site. A hot, arid climate, such as that of Arizona with its daily temperature fluctuation averaging 38°F (20°C), had a different effect on many plastics from that of the warm, humid climate of southern Florida, and in many cases materials failed more rapidly. Hence Arizona became a de facto international reference site for hot, arid climates during the early 1950s.

Today, after nearly 80 years of exposure testing, there are numerous outdoor testing sites throughout the world. On an international basis the warm, moist, subtropical environment of southern Florida and the hot, dry desert environment of Arizona are still considered to be among the most severe climates for materials testing. However, in recent years concern has been expressed for materials behavior in specific microenvironments, in that their response to stimuli may be different than that experienced in more pristine environments. Consequently a comprehensive testing program should also include exposure to industrial pollutants, acid rain, or other site-specific factors to which the material may be subjected in its end-use application.

24.1.2   Solar Materials Test Method Development

The outdoor durability of most materials is generally determined by subjecting them to various real-time exposure tests in the Florida and Arizona environments, or to a variety of laboratory and outdoor accelerated test methods. Most of these exposure tests are applicable for a wide range of materials, and in many cases the results will accurately predict the performance of the materials in an outdoor environment.

However, materials that are subjected to specific end-use conditions, such as interior automotive fabrics or solar absorbers and glazings, were generally not exposed to realistic conditions when subjected to standard test methods (ASTM, SAE, etc.). Consequently a need developed for more severe exposure tests to more closely simulate the actual end-use conditions of the material in order to develop meaningful data.

NBS   In 1977 a program was initiated by the National Bureau of Standards (NBS) to help provide an experimental basis for the development of consensus standards for assessing the reliability and durability of solar collectors and their materials (Waksman et al. 1984). In this program eight different types of flat plate solar collectors and small-scale cover and absorber materials specimens, representative of those in use at that time, were exposed outdoors at four sites located in different climatic regions. Periodic measurements were made of their performance as a function of exposure time. Laboratory aging tests were conducted concurrently on specimens of the same materials to provide a basis for comparison with the outdoor exposure tests.

For solar applications, temperature is a key consideration. To achieve realistic exposure conditions for absorbers and glazings, several different test fixtures were designed and constructed. NBS employed a variety of ersatz collectors for evaluating the outdoor durability of solar materials. Short of exposing a full-size collector, ersatz collectors afford an economical and realistic method for such testing. Ersatz collectors are defined as test boxes that have an optical (glazings and receiver) and thermal (receiver and insulation) environment that provides no cooling for the receiver but is otherwise identical to a flat plate collector.

These data showed that the outdoor exposure of glazings and absorbers at temperatures representative of stagnation conditions is needed for assessing performance, particularly if the materials are polymeric in nature .

This effort has resulted in the publication of several procedures that can be used to assess the thermal performance and durability of solar energy systems and the components and materials used in their construction. The American Society for Testing and Materials (ASTM) has developed several standards concerned with the reliability and durability of cover plate and absorptive materials based to a considerable extent on technical data generated by NBS.

ASTM   In 1978 ASTM Committee E-44 on Solar Energy Conservation was formed for the purpose of developing standards and test methods applicable to converting solar energy to directly usable energy forms.

Subcommittee E44.04 on Materials Performance has developed standards related to the reliability and durability of materials used in solar energy applications. Task Group .01 on Absorptive Materials has written standard practices for evaluating absorber materials for thermal applications (E744-80) and for evaluating solar receiver materials exposed to simulated stagnation conditions (E781-81). Task Group .02 on Cover Plates (glazings) has written standard practices for evaluating (E765-80) and cleaning (E962-83) cover plate materials, as well as for exposure testing under simulated operational (E782-81) and stagnation modes (E881-82). Task Group .04 on Reflective Materials and Optical Properties has written a test method for determining solar optical properties, E903-82, which is an extensively revised and updated version of E424-71, method A.

A complete listing of the current standards and test methods relating to solar energy can be found in the *1987 Annual Book of ASTM Standards* (American Society for Testing and Materials 1987, vol. 12.02).

ISO   Headquartered in Geneva, the International Standards Organization (ISO) issues voluntary International Standards (IS) on nearly every technical field associated with world commerce. New ISO standards are derived, whenever possible, from existing national standards used by its member countries.

The technical Committee on Solar Energy (Thermal Applications), ISO/ TC180, was organized at an international meeting held in Sydney, Australia, in May 1981. Its secretariat is held by the Standards Association of Australia (SAA), which is the ISO member body in Australia. The committee's breakdown and the secretariats responsible for the individual programs of work are nomenclature—Canada (SCC); climate—Germany (DIN); materials—France (AFNOR); component-thermal performance/reliability and durability (colors) —Israel (SII); and systems-thermal performance/reliability and durability —the United States (ANSI).

Under the auspices of the American National Standards Institute (ANSI), the ISO member body for the United States, a U.S. Technical Advisory Group (TAG) was organized in the spring of 1981 to represent the technical positions of U.S. industry and the U.S. solar community of interest in ISO/TC180 activities. The U.S TAG is made up of technical experts selected from and appointed by the various standards organizations with pertinent interests in the fields represented by the TC180 program of work. The U.S. TAG currently has 15 members that represent the American Society of Heating, Refrigeration and Air Conditioning Engineers (ASHRAE), the American Society for Testing and Materials (ASTM Committee E-44), the Solar Energy Industries Association (SEIA), and the American Solar Energy Society (ASES). This representation ensures that the U.S. delegations to technical committee meetings and meetings of its subcommittees and working groups are properly briefed on U.S. positions as represented by their constituent organizations.

24.1.3   Evaluation Methods

Visual   The evaluation of materials by an exposure facility or testing organization should begin with the condition of the samples as received. All materials should be checked for damage (tears, scratches, cracks, bends, etc.), and materials showing any damage should be immediately reported and noted.

Damaged samples are almost always either the result of inadequate protection of the materials in packaging for shipment or improper handling of the package during shipment.

To properly perform visual and instrumental evaluations of different materials, it is important to have an additional set of unexposed control samples. This set of samples should be carefully maintained in a temperature-controlled environment. They should be utilized only during the periodic visual inspections. Many companies request that inspections be performed using the masked areas of their materials. Since masks sometimes fail to prevent the intrusion of moisture and heat buildup, these areas sometimes exhibit slight changes in appearance when compared to unexposed controls. Therefore, the use of unexposed controls provides a more accurate means to assess the visual changes in a material's appearance.

Inspections are performed according to ASTM, AATCC, or SAE Standard Practices or Test Methods. Visual evaluation criteria is in accordance with photographic reference standards such as the Federation of Societies of Coating Technology's *Pictorial Standards of Coating Defects*, ASTM Reference Standards, DIN Standards, ISO Standards, or others as requested.

Generally, the weathered samples are compared to the pictorial standards in the relevant document, and a subjective numerical rating is assigned to the samples. A rating of 10 means that the sample is in the same condition as when it was received (no change); decreasing numbers indicate increasing severity of the property being evaluated.

The most commonly used rating scale for visually inspecting a material sample for general appearance, color change, and other physical or optical attributes is shown in table 24.1.

Subjective visual evaluations can be of value when assessing the durability of materials. They can generally be performed at minimal cost in a short period of time. Although most such inspections can be performed in the field, it is

Table 24.1
Rating scale for inspection

Rating number	Attribute evaluation	Degree of failure (change)
10	As received	None
9	Excellent	Very slight
8	Good	Slight
7	Good to Fair	Slight to considerable
6	Fair	Considerable
5	Fair to Poor	Considerable to severe
4	Poor	Severe
3	Poor to very poor	Severe to very severe
2	Very poor	Almost complete
1	Extremely poor	Complete

recommended that whenever possible the samples should be evaluated indoors under controlled lighting in order to minimize rating variations caused by the variety of viewing conditions that can exist outdoors.

Materials intended for solar applications can be visually evaluated for a number of different attributes. For example, absorbers can be rated for delamination, cracking, pitting, or blistering; glazings for fiber show, haze, discoloration, or dirt retention; mirrors for bloom, water spotting, dirt retention, or specularity. This type of rating system can be very effective in establishing a relative ranking of material performance.

Optical Knowledge of the optical properties of potential solar device materials is essential in making the proper choice of materials for a particular application, in being able to predict the optically dependent performance of the solar device, and in assessing environmentally induced changes in such materials. Relevant optical data would include transmittance values for glazings, absorptance and emittance values for absorber materials, and reflectance values for mirror materials.

Solar transmittance, absorptance, and reflectance should be measured in accordance with ASTM E903. This spectrophotometric method details the use of integrating spheres for making optical property measurements and provides for uniform reporting of results. However, if a spectrophotometer/integrating sphere facility is not available, other methods such as ASTM E424 method B may be employed.

The integrating sphere geometrics most commonly used are shown schematically in figure 24.1. Since the resultant spectra are considered to be absolute (Zerlaut and Anderson 1981), no adjustment is necessary to account for a reference standard.

The Edwards-type sphere (A) uses a center-mounted sample, and it is used primarily for obtaining off-angle optical data on reflector and absorber materials. The detector is mounted so that the entrance port, specimen, and detector form a right triangle, with the detector looking only at the edge of the specimen. The specimen holder is a spherical segment graduated in 5-deg increments from -40 to $+40$ deg with respect to the entrance beam. The specular component of a reflector material can be determined by subtracting the diffuse-only reflectance value (measured at a 0-deg angle of incidence) from the total reflectance value measured at, for example, a 7-deg angle of incidence. Although the Edwards-type sphere provides very accurate and precise data, the 1-in. diameter specimen size requirement can result in limited applications of its use.

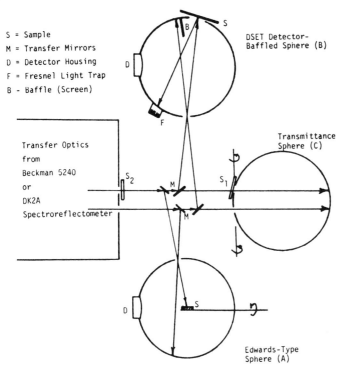

S = Sample
M = Transfer Mirrors
D = Detector Housing
F = Fresnel Light Trap
B - Baffle (Screen)

DSET Detector-
Baffled Sphere (B)

Transfer Optics
from
Beckman 5240
or
DK2A
Spectroreflectometer

Transmittance
Sphere (C)

Edwards-Type
Sphere (A)

Figure 24.1
Optical diagram of MIS spectroreflectometer/spectrotransmissometer.

The detector-baffled sphere (B) uses a wall-mounted specimen configuration, and it therefore has virtually no sample size limitations. However, the hemispherical reflectance data thus obtained are restricted to a fixed angle of incidence, usually 20 deg. This sphere geometry is very useful for measuring the hemispherical reflectance of absorbers, and subtracting the value thus obtained from unity to determine absorptance. The use of an interchangeable sphere-completing plug and a Fresnel light trap permits the measurement of total hemispherical reflectance and diffuse-only reflectance for determining the specular component of reflector specimens. This sphere can also be used for measuring the hemispherical reflectance of glazing specimens by allowing the transmitted component to pass completely through the sphere.

Although total reflectance properties of a reflector material may not change significantly as a result of exposure testing, the specular reflectance properties may change dramatically. Therefore evaluation of reflector materials should include a specularity measurement. It has been suggested that mirrors used in heliostat applications be measured at a cone angle of 1.5 to 4.0 mrad, with those for trough applications measured at about 15 mrad.

The transmittance sphere can be used for both hemispherical and near-normal transmittance measurements, depending on the entrance geometry employed. With the glazing specimen mounted directly against the entrance port of the sphere (S_1), total hemispherical transmittance values are obtained. These data are generally quite adequate for most nonconcentrator applications. However, with the specimen mounted at a known distance away from the entrance port (S_2), a near-normal transmittance value is obtained. These data are much more sensitive in detecting changes in the glazing due to weathering-induced degradation such as crazing or haze development. This type of measurement would be applicable in certain concentrator applications where the specular transmittance properties of the glazing are important considerations, or in assessing the degree of physical changes within the glazing. Both the hemispherical and near-normal transmittance measurements are usually made at a fixed (7-deg) angle of incidence.

Emittance is defined as the ratio of the radiance of a specimen to that of a blackbody radiator at the same temperature. Various thermal performance calculations require the measurement of this important variable.

Emittance measurements of absorbers, solar control films, and other surfaces are often performed at room temperature using inspection-meter techniques, as described in ASTM E408. These methods are generally suitable for production control and are primarily intended for rapid nondestructive measurements on large surfaces.

The two basic types of commercially available instruments used for determining emittance are those that measure radiant energy reflected from the specimen (method A) and those that measure radiant energy emitted from the specimen (method B). An example of a method A instrument would be the Gier-Dunkle Infrared Reflectometer Model DB100. A method B instrument would include the Devices and Services Model AE emissometer.

Although an instrument such as the DB100 actually measures total normal emittance, these values can be corrected to total hemispherical emittance. Most of the method B instruments actually give an intermediate value between total normal and total hemispherical emittance. However, when method A data are corrected to hemispherical values, the correlation with method B data is generally quite good.

Although emittance measurements at room temperature have considerable utility in following changes in absorber materials as a function of exposure level, the ability to measure emittance at or near absorber plate operating temperatures would be more desirable. Spectral emittance data would be useful in predicting failure modes in selective coatings and can be normalized against various energy distributions of Planck's law to give total emittance values for any surface temperature. A heated cavity, temperature-controlled specimen holder, and a recording infrared spectrophotometer would generally be required for determining spectral emittance values, which makes the cost for such measurements considerably higher than that incurred by using the inspection-meter techniques.

Physical When evaluating such physical properties as the impact resistance of a coating, the tear strength of a fabric, or the tensile properties of a plastic, great consideration must be given to the relevancy of the test to establish the suitability of a material for a particular product application. The physical tests that are performed during the evaluation stages of the program must support this objective if the program is to prove successful.

Physical testing involves measuring the output response of a material under given boundary conditions and in a given surrounding environment when it is subjected to external input, physical excitations. These excitations may be such things as force, heat, or voltage, and may be in simple or complex forms. Boundary conditions may be considered as geometric constraints placed on the specimen. The time form of the excitation functions, the boundary conditions, and the surrounding environment have a strong influence on the nature of the material response. The major task therefore is to express this complex behavior in terms of a few identifiable, measurable, and useful parameters.

Most of the ASTM test methods covering physical testing of materials are somewhat arbitrary by nature. These tests have been developed to evaluate materials, particularly plastics, in terms of fundamental properties. These methods may be scientifically sound, but they can be misleading in that although the input excitation and boundary conditions of the test are controlled so that "materials properties" can be compared or used as a basis for acceptance, it does not necessarily mean that these methods are relevant to the actual end-use application. Failure to subject the materials to extended time, temperature, and environmental conditions can result in misapplication.

In evaluating the physical properties of a material for a known application, the following considerations should be made: (1) Analyze the application to determine the physical disturbances to which the material will be subjected;

(2) estimate the shape and magnitude of the expected input functions, the boundary conditions, and the environment to which the material will be exposed, as well as the expected service life of the material; (3) select fundamental or practical standard tests or develop special tests to match the requirements; (4) evaluate the physical properties of the material using exaggerated and suppressed inputs and environments to establish limits for acceptable and unacceptable performance.

In performing general evaluations of the physical properties of materials, the technologist should (1) develop a working hypothesis of the various physical attributes believed to be inherent in the material; (2) select test procedures and specimens that can be employed over a wide range of input excitations and boundary conditions, avoiding tests that merely provide a single data point or that are poorly defined; (3) evaluate physical properties over a range of times and temperatures; (4) measure physical changes during and after environmental or thermal aging; (5) present data in a form useful to the designer; and (6) establish criteria for acceptable and unacceptable performance (Evans 1981).

24.1.4 Solar Materials Durability Testing

IITRI Program The first comprehensive solar materials exposure program was initiated in 1976 by the U.S. Energy Research and Development Administration (ERDA) through the IIT Research Institute (IITRI) (DOE contract EY-76-C-02-2672). The scope of the program was to evaluate potential solar utilization materials before and after exposure to outdoor weathering tests in different geographical locations.

The basic intent of this effort was to provide users of solar devices with practical design data at the materials level which related to their cost-effectiveness and performance. Absorbers, glazings, and mirrors were included in this program, as well as various sealing tapes and strapping materials. Optical and physical property data were obtained initially and after several exposure intervals.

Samples were exposed to standard conventional and accelerated exposure tests. Conventional weathering was conducted at 45 deg from the horizontal, facing south, both direct to the weather and under glass (ASTM D1435 and G27) for up to five years. Accelerated testing was conducted using the EMMA test method (ASTM E838) for up to one year. The exposure sites employed for the conventional weathering tests were located in Chicago, Illinois; Phoenix, Arizona; and Miami, Florida. All accelerated testing was performed in Phoenix, Arizona.

Optical property data included transmittance, reflectance, and emittance measurements. Physical testing included tensile, elongation, impact, abrasion, and peel tests. Degradation mechanisms were also studied as part of this program.

Interim results of the IITRI program were published in April 1980. Data from up to about three years of exposure are contained in the *Handbook of Materials for Solar Energy Utilization* (DOE/CH/90034-T1 Handbook). This handbook lists the materials, relevant properties, weathering effects, source, and cost information. A companion volume, entitled *Exposure Testing and Evaluation of Solar Utilization Materials* (DOE/CH/90034-T1) contains general discussions and R&D data developed in connection with the preparation of the handbook. The materials remaining on exposure at the conclusion of the IITRI program eventually became part of the SERI handbook program. It is anticipated that the SERI data will be used in updating the IITRI handbook.

NBS Efforts in the development of reliability/durability tests for solar collectors and their materials have been hampered by the lack of real-time and accelerated degradation data that can be correlated to actual end-use conditions. In 1977 the U.S. Department of Energy initiated the Solar Collector Reliability/Durability Test Program at the National Bureau of Standards (NBS) to help generate the data required to develop methods for predicting the long-term durability and reliability of flat plate solar collectors and their materials.

Many important findings have resulted from the work conducted by the NBS, and the NBS, as previously mentioned, has made significant contributions to the development of ASTM and other standards pertinent to testing materials for solar applications (Waksman 1985; Thomas et al. 1990). For example, NBS data have shown that, for polymeric cover materials, outdoor exposure at sites having a combination of high humidity and high solar radiation generally produce more severe changes than other test sites, and that an outdoor exposure of two or more calendar years in a simulated collector test fixture is required to induce degradation that is detectable without using sophisticated analysis techniques. Much of the research conducted at the NBS is summarized in two reports, NBSIR 84-2845 and NBSIR 83-2782 (see Trechsel and Collins 1984; Yancey 1983).

Most of the ATSM standards that pertain to solar materials evaluation are intended to be used as screening procedures, so they do not provide information about the service of candidate materials.

Reliability theory and life-testing analysis appears to offer considerable promise as a means for making quantitative service life predictions for materials used in solar heating and cooling applications. Demonstration of the applicability of these techniques for solar materials is the primary emphasis of durability related research currently in progress at the NBS. The mathematical models being developed should be capable of predicting the service life of materials simultaneously exposed to several different environmental stresses (see Martin et al. 1984).

SERI's IPH Program   In early 1980 the Solar Energy Research Institute (SERI), acting for the Department of Energy (DOE), selected five industrial sites for possible DOE cost participation toward the installation of solar collectors to provide Industrial Process Heat (IPH). The five sites were located in Perry, Iowa; Shelbyville, Tennessee; Santa Isabel, Puerto Rico; Oxnard, California; and Santa Cruz, California. To help them decide on which of these sites might receive the funding, SERI decided to conduct a short-term conventional weathering program to study the site-specific environmental effects on possible solar collector materials. To conduct this study, SERI selected DSET Laboratories, Inc. of Phoenix, Arizona (SERI subcontract XH-O-9087-1). The project's plan included selection of materials applicable to the type of collector proposed for each of the IPH sites. Flat plate, enhanced flat plate, evacuated tube, and parabolic trough collectors were potential candidates at one or more sites. Therefore the materials selected included polymeric and glass glazings, selective and nonselective absorbers (including polymeric types), and first-and second-surface mirrors.

All material candidates were fully characterized initially and after each exposure interval. This included visual inspections as well as determination of appropriate solar optical properties. Glazing and absorber samples were mounted into small food service pans, approximately three inches by five inches in size. The pans were selected as a means of elevating sample temperatures to simulate exposure conditions that would occur in solar collectors. All pans had small holes drilled in the bottom to allow the local moisture and air pollutants at the exposure site to enter the exposure pan.

Absorber test pans were constructed by first placing a thin layer of insulation (Owens-Corning SI-100) and then the test absorber material into each test pan. Insulation and absorber were held in place using General Electric's RTV-101 high temperature silicone sealant. After curing of the absorber sealant, a Solatex glass cover was sealed to the top of the pan. Dial gage stem thermometers were installed in selected test pans for monitoring of absorber temperatures by plant personnel.

Glazing test pans were constructed in a similar manner. However, the absorber installed in all test pans was a selectively coated material from the Berry Solar Corporation.

Once sample preparation was completed, the reflector samples and test pans were mounted to surfaced particle board. Sample exposure racks were constructed of steel and shipped with the samples to the five exposure sites.

Exposure racks and samples were installed at each site by DSET personnel. The racks were installed facing due south, at tilt angles of 10 deg less than latitude for the Iowa, Tennessee, and California sites. The Puerto Rico samples were exposed at latitude angle.

All samples at each site started exposure at the same time. One set of samples was returned after approximately two, four, six and ten months of exposure. All samples were visually inspected on site at each return interval, and those samples remaining on exposure were washed with deionized water.

All data from this program were combined and analyzed to determine which sites, if any, exhibited any significant effect on the materials after exposure. For selected materials, plots were prepared to graphically show the changes in material performance as a function of exposure duration.

After a ten-month exposure period, only the reflector materials at the Stauffer Chemical Plant at Oxnard, California, showed any significant permanent change from the initial unexposed condition. At most other sites, some minor spotting and pitting was evident only on reflector materials; however, no significant change in any spectral measurements was observed.

At the Oxnard and Santa Cruz test sites, a high degree of dirt retention was observed on all reflector and outer glazing samples. This dirt retention did produce a decrease in optical performance when measured prior to washing. However, after the samples were washed at DSET, little, if any, difference could be found in optical performance between washed and unwashed sample halves.

SERI's Handbook Program   Since early 1980, DSET Laboratories, Inc. has been involved in developing information for a solar materials design handbook (SERI subcontract XP-9-8215-1). The scope of this program is to subject materials to both real-time and accelerated exposure tests, measure optical and physical properties at periodic intervals, and report the retention (or loss) of properties as a function of exposure. The primary objectives of this program are to develop new materials information for a current data base and to validate the existing data base.

There are several features that distinguish this program from previous materials testing programs: (1) All materials are commercially available and were purchased from distributors or manufacturers without their knowledge of the purpose; (2) exposure intervals are based on the total ultraviolet radiation received; (3) most of the test methods employed simulate actual end-use conditions of solar device materials by utilizing special fixtures designed for that purpose.

Solar absorber, glazing, and mirror candidates are being exposed at latitude in New River, Arizona (desert); Homestead, Florida (subtropical); Los Alamos, New Mexico (high altitude); and Compton, California (industrial pollution). Accelerated weathering tests have included the EMMA and EMMAQUA test methods (ASTM E838), fluorescent/condensation (ASTM G53), and Xenon Weather-o-meter (ASTM G26).

Diagnostic measurements have included solar transmittance and reflectance, total emittance, and on selected polymeric candidates, infrared transmittance and tensile/elongation properties. All solar optical data have been normalized to air mass 1.5.

In addition to recording environmental conditions at each exposure site, a considerable effort has been expended on monitoring the temperature of absorber and glazing materials in order to validate the choice of test methods employed in the program.

A modified ersatz collector is being employed in this program for evaluating the durability of solar absorber materials to real-time exposure conditions (figure 24.2). The collectors were constructed of 24-gauge galvanized steel with outside dimensions of approximately 673 by 673 by 83 mm deep. A 50-mm layer of Owens Corning SI 100 insulation was placed in the bottom of the collector, above which was mounted either a selective or nonselective absorber plate. In our tests Olympic Plating BCO-91 on copper was used as the selective plate, and Rustoleum Bar-B-Que Black on copper was used as the nonselective plate. The entire exterior frame of the ersatz collector was painted with Rustoleum Bar-B-Que Black. The removable cover plate is ASG Sunadex tempered glass.

The same type of ersatz collector can be used to evaluate glazing materials. However, in the interest of minimizing costs and in providing specimens suitable in size for spectrophotometric measurements, smaller ersatz collectors are employed. Commercially available institutional cafeteria roast pans of 0.25 mm aluminum are used for the real-time exposure of glazing materials. These collectors are 200 by 305 by 65 mm and use Rustoleum Bar-B-Que Black on copper as a common absorber over 40 mm of Owens Corning SI 100 insulation. These collectors are mounted in a standard automotive black box test fixture for exposure testing (see figure 24.3).

Figure 24.2
E-44 ersatz collectors.

Figure 24.3
Roast-pan collectors mounted in an automotive black box.

Figure 24.4
Mini-loaf collectors mounted to an EMMAQUA target board.

Another type of ersatz collector, referred to as a mini-loaf pan, is used for accelerated exposure testing of glazing materials. The mini-loaf pan is 65 by 118 by 25 mm and is constructed of 0.5-mm aluminum. This collector is equipped with a selective absorber plate or Berry Solar-strip, but no insulation is employed. The size conveniently fits the sample holders of the EMMAQUA Test Method machines (ASTM Standard E 838-817), Xenon Weather-o-meter (ASTM Standard G 26–77) and QUV (ASTM Standard G 53–77) accelerated weathering devices (see figure 24.4).

The optical property data indicate that absorber materials are much less susceptible to local climatological differences when exposed in the relatively pristine environment of a collector. The ersatz collectors used in the program were not sealed and were "breathable" fixtures. Although plate temperatures varied significantly from site to site, the absorptance a measurements listed in table 24.2 show little if any effect of site variations.

The data listed in table 24.3 show the effect on the temperature of glazing specimens exposed using the roasting-pan ersatz collector test method versus standard exposure practice (ASTM D 1435) for the same glazings. Typical clear-day values for June and December are shown, and the data clearly indicate the effectiveness of the ersatz collector method in elevating the tem-

Table 24.2
Comparative solar absorptance (a) values of absorber specimens exposed two years at latitude in ersatz collectors

Absorber[a]	Initial value	Arizona	California	Florida	New Mexico
#9	(0.940)	0.898	0.895	0.897	0.895
#10	(0.935)	0.914	0.907	0.909	0.907
#11	(0.946)	0.921	0.921	0.917	0.915
#12	(0.952)	0.941	0.942	0.936	0.940
#14	(0.936)	0.915	0.907	0.911	0.906
#15	(0.940)	0.925	0.931	0.910	0.930

a. Absorbers #9 through #12 are nonselective paints; #14 and #15 are selective foils.

Table 24.3
Comparative glazing temperatures for at latitude exposure in New River, Arizona (°C)

	Standard exposure		Roast-pan collector	
	June	December	June	December
Glazing				
Acrylic	45°	28°	92°	78°
Polycarbonate	44°	28°	89°	78°
FRP-2[a]	45°	28°	81°	68°
Glass-1	43°	26°	76°	62°
Ambient	42°	25°	42°	25°

a. Fiber-reinforced plastic.

perature of the glazing specimens to a more realistic level. All temperatures were recorded using shaded copper constantan thermocouples mounted to the back side of the glazing specimens.

The mini-loaf collectors used for the accelerated evaluation of glazing specimens gave maximum temperatures of 180°F (82°C) and 195°F (91°C), respectively, for the FRP-2 and Glass-1 specimens, using the EMMAQUA test methods. Although the Xenon Weather-o-meter temperature data were similar to those from EMMAQUA, the fluorescent weathering cabinet data showed maximum specimen temperatures of only 113° to 122°F (45°–50°C). This was undoubtedly due to the fact that no radiant energy in the infrared region was available for elevating the temperature of the ersatz collector absorber plates.

The data obtained thus far show that the ersatz collector test method will induce more severe changes than conventional test methods in the optical

Table 24.4
Comparative solar transmittance (N/N) of glazing materials exposed for two years at latitude in New Mexico

	Initial value	Standard exposure	Roast-pan collector
FRP-1	(0.793)	0.775	0.714
FRP-2	(0.746)	0.679	0.315
Glass-1	(0.902)	0.907	0.862
Glass-2	(0.895)	0.886	0.872
Polyvinyl Fluoride	(0.883)	0.881	0.816

properties of glazing materials. Table 24.4 lists near-normal/near-normal spectral transmittance results for glazing materials exposed for two years at Los Alamos, New Mexico.

Optical property data from accelerated weathering methods showed the same general trend as the real-time results. The data show that the use of mini-loaf ersatz collectors generally produces more severe results on EMMAQUA than the use of standard exposure techniques; however, the fluorescent weathering cabinet data show very little difference between the use of mini-loaf collectors and standard test procedures.

All optical property data from this program was obtained in a timely manner upon removal of the test specimens from the field in order to minimize recovery effects from dark-time storage. An interim status report has been submitted to LANL, with a final report to be published in late 1987.

Summary The Solar Heating and Cooling Act of 1974 called for programs to develop, demonstrate, and promote the use of solar heating and cooling systems. The intent of this act was to stimulate use of solar energy equipment and to assist industry with technology development and systems demonstration. The objective of the research and development programs was to identify potentially cost-effective systems for the heating and cooling of buildings, as well as for agricultural and industrial process heating.

In order to produce solar systems that would meet the above objective, the proper materials, components, and methodologies were needed by the solar energy industry. This chapter highlights only those programs in which the author was intimately involved. However, numerous other research and development programs have been conducted in support of the national effort to develop solar energy systems.

For example, the Los Alamos National Laboratories (LANL) has conducted or directed research efforts in such areas as solar absorber materials, chemical conversion coatings, adhesives and sealants, heat transfer fluids, and corrosion control. LANL has published a bibliography of various solar related programs that includes a description of the project, objectives of the work, and, in some cases, methodology and results. Also included is a listing of the important publications that resulted from each project activity (see Reisfeld 1983).

References

American Society for Testing and Materials. 1987. Nuclear (11), Solar and Geothermal Energy. *1987 Annual Book of ASTM Standards*, vol. 12.02. Philadelphia, PA.

Evans, R. E. 1981. A rationale for testing to establish suitability for an end-use application. In *Physical Testing of Plastics—Correlation with End-Use Performance*, R. E. Evans, ed. ASTM STP 736. American Society for Testing and Materials, pp. 3–14.

Martin, J. W., Waksman, D., Bentz, D. P., Lechner, J. A., and Dickens, B. 1984. A preliminary stochastic model for service life predictions of a photolytically and thermally begraded polymeric cover plate material. *Proc. Third International Conference on the Durability of Building Materials and Components*. VTT symposium 50, vol. 3. Technical Research Centre of Finland, Espoo, Finland. August.

Reisfeld, S. K., Avery, J. G., Grimmer, D. P., Krall J. J., and Moore, S. W. 1983. *A Solar Materials Annotated Bibliography*. LA-9581-MS. Los Alamos National Laboratory, Los Alamos, NM. January.

SERI Subcontract No. XH-0-9087-1.

SERI Subcontract No. XP-9-8215-1/LANL Subcontract 9-L34-Q8088-1.

Thomas, W. C., Dawson III, A. G., and Waksman, D. 1990. Use of integrated day-long stagnation temperatures for measuring changes in solar collector materials properties. NBI: WCT-0153. *ASME Solar Energy Eng.* forthcoming.

Trechsel, H. R., and Collins, B. L. 1984. *Test Methods and Standards Development for Active Solar Heating and Cooling Systems*. NBSIR 84-2845. National Bureau of Standards, Gaithersburg, MD. May.

U.S. Department of Energy. n.d. *Exposure Testing and Evaluation of Solar Utilization Materials*. DOE contract no. EY-76-C-02-0578-034.

Waksman, D., W. C. Thomas, and E. R. Streed. 1984. NBS Technical Note 1196, NBS Solar Collector Durability/Reliability Test program: Final Report, U.S. Department of Commerce/National Bureau of Standards. September.

Waksman, D. 1985. *Solar Collector Materials Durability and Testing Procedures*. Presented at the U.S. Department of Energy Conference "Solar Buildings: Realities for Today-Trends for Tomorrow." Washington, D.C. March 18–20, 1985.

Yancey, C. W. C. 1983. *Materials Research Activities at the National Bureau of Standards (1975–1982) Pertaining to Active Solar Heating and Cooling Systems*. NBSIR 83-2872. National Bureau of Standards, Gaithersburg, MD. November.

Zerlaut, G. A., and Anderson, T. E. 1981. Multiple-integrating sphere spectrophotometer for measuring absolute spectral reflectance and transmittance. *J. Applied Optics* 20: 3797–3804x.

25 Solar Materials Research and Development

Richard Silberglitt and Hien K. Le

25.1 Introduction

Since the early 1970s solar heat technology has been improving significantly, with improved materials playing a crucial role in this development. Today's solar hot water, space-heating, and cooling and process heat systems are far superior to those of 1974. Innovations coming from materials research have included optically selective glazing, absorber films and coatings, lightweight polymeric glazings and absorbers, more effective and compact insulation, anticorrosive heat transfer fluids and additives, and phase change thermal storage materials. The federal solar R&D program has played a key role in supporting these innovations. Despite these advances, however, the solar systems have further cost, efficiency, and durability targets to meet before they will be broadly competitive with conventional heating alternatives. Materials research will continue to be important in this context because it can lead to lower unit area costs, higher collection, transport, and storage efficiencies, and longer system lives. This chapter will review the materials R&D that led to improvements in solar technology, as well as indicate future prospects for this research. The focus will be on solar collector materials, specifically materials for glazings, heat mirrors, absorber surfaces, absorber adhesives, absorber substrates, heat transfer fluids, thermal storage, and desiccants. Each of these materials categories will be treated separately in the following text.

25.2 Polymer Glazings

Research to improve polymer durability includes the following areas: basic study of degradation mechanisms; design of new polymers, laminates, blends or additives to overcome degradation; investigation of processing methods designed to optimize properties and lifetimes; and laboratory and field testing of promising materials. Figure 25.1 shows the research alternatives at each of these levels that are relevant to improving the durability of polymers for solar glazing applications.

The logic of figure 25.1 is as follows: Levels 1 through 3 represent efforts to understand and overcome the weaknesses of existing polymers, to surmount the fundamental technical barriers to improved polymer glazing performance.

This chapter is based on the report *Active Solar Energy System Materials Research Priorities*, SERI/STR-255-1782 (January 1983), by Stephen A. Herzenberg, Le Khac Hien, and Richard Silberglitt, and some updating of its contents.

Level 4 applies the knowledge in levels 1 through 3 to the engineering of improved polymers. Levels 5 and 6 describe processing industry efforts to optimize and improve polymer-manufacturing techniques. Levels 7 and 8 check how well the desired glazing characteristics have been achieved. In addition testing feeds back into the first three levels, suggesting fundamental mechanistic work that needs to be done to overcome remaining barriers.

25.2.1 Polymer Degradation Mechanisms

Current understanding of polymer degradation mechanisms, as applied to solar collector glazings, draws heavily from polymer research and development in other applications such as piping and structural components. Consequently the following review discusses polymer degradation generally.

Photodegradation Degradation begins when UV photons, with sufficient energy to break many polymeric bonds, strike absorbing groups (chromophores) in the repeating unit of the polymer. Photodegradation may also occur when impurities are introduced during processing or outdoor exposure. The excited state formed in the polymer either dissipates the photon energy harmlessly (by fluorescence, phosphorescence, or, more commonly, by releasing energy as heat) or undergoes chain scission, oxidation, or some other degradation process. The vulnerability of a polymer to outdoor light depends largely on its UV absorptivity, the percentage of absorbed photons that leads to degradation, the average number of chain scissions induced by each destructive photon, and the nature of the degradation products.

It is generally known that UV absorption and subsequent degradation occurs rapidly in polymers with chromophores in their repeat unit. These are unsuitable for application as the outermost layer of a polymer glazing. Such polymers may, nonetheless, be prime candidates for application as the innermost layer of a laminate. Polycarbonate, for example, has excellent strength and temperature properties and will probably last close to 20 years in an active collector if screened sufficiently from UV light. In polymers without UV-absorbing groups in the monomer, the two most significant chromophores are generally carbonyl and peroxide groups. These groups both initiate photodegradation and are produced during this process. UV photodegradation is therefore an autoaccelerating process unless it is limited by some external factor, such as radical mobility or oxygen diffusion rate.

Although the generalities of photodegradation are well understood, mechanisms for specific polymers are rarely completely known (Wiles 1978). For example, despite extensive study of polymethyl methacrylate (PMMA) by photovoltaic (PV) encapsulant researchers and others, the nature of the chro-

R&D level	Definition				
VIII	Field testing	Development of standard testing procedures	Building up of a field-testing data base	Collection of data from privately owned systems	
VII	Accelerated testing	Building up screening-test data base	Development of service-life predictions	Quantification of environments in which glazings must perform	Development of standard testing procedures
VI	General processing improvements	Reducing processing chromophore incorporation	Reducing impurity content of processed polymers	Improvement of processing to reduce surface flaws or other surface problems	
V	Process parameter —property correlation	Process chromophore incorporation and UV degradation rates	Impurity content (catalyst, monomer residue, etc.) and polymer properties and durability	Effect of extrusion on polymer surface properties	Cost reduction of thermoplastic production processes

					Surface abrasion and bulk strength properties
IV	Laboratory production of polymers with the desired combination of properties	Development of improved polymer blends	Development of improved composites	Development of polymer laminates	Development of improved copolymers
III	Combating primary environmental sources of attack	Improvement of UV stabilizers	Improvement of base polymers	Combating polymer soiling	Development of abrasion-resistant polymers
II	Correlation of microscopic degradation with macroscopic property degradation	Degradation and strength property reduction	Degradation and temperature-dependent property changes (Tg and melting point)	Degradation and transmittance loss	Volatile evolution and dirt pickup
I	Study of degradation mechanisms	Study of photodegradation under UV light in air	Study of degradation under conditions of combined stress	Study of dust adhesion mechanisms	Study of polymer surface properties and abrasion

Figure 25.1
Polymer glazing research alternatives.

mophore that initiates degradation is not yet established (Gupta 1978). The degradation mechanisms of glazing candidates are understood in varying degrees. PMMA has been studied in considerable detail, but its temperature resistance may restrict its use in collectors, except as the UV screen in a laminate. Degradation in some of the leading flurocarbons, such as polyvinylidene fluoride (Mavis 1980), polyvinyl fluoride-Tedlar (Wilson 1982), and silicones has also been examined extensively.

One of the principal barriers to increased understanding of photodegradation mechanisms for specific polymers is the inadequacy of analytical techniques for exploring the kinetics of oxidation, radical propagation, and radical termination in the solid phase. As pointed out by Garton, Carlsson, and Wiles (1980), new instrumentation is needed to study such issues as radical site migration, the molecular weight distribution of reaction products, and the effect of temperature and morphology on reaction rates. These factors profoundly influence UV degradation and decisions regarding stabilization strategies.

Degradation under Combined Stress   Little has been published on polymer degradation at the microscopic level under active solar glazing conditions: exposure to UV radiation, high and cycling temperatures, moisture, ozone and other gases, and atmospheric pollutants. The microscopic and macroscopic evidence that has been reported suggests that the presence of a second type of stress is often necessary to induce rapid degradation; for example, dimensional (temperature or moisture) cycling enormously accelerates the surface microcracking of polycarbonate exposed to UV radiation (Blaga and Yamasaki 1976).

Dust Adhesion and Soiling   Mechanisms of dust adhesion to polymers have been the subject of research at Sandia National Laboratories–Albuquerque and at the Jet Propulsion Laboratories (JPL) as part of the photovoltaic encapsulant, reflector, and heliostat research program (White 1981). Work at JPL suggests that some of the harder polymers, including the acrylics, have adequate soil resistance (Maag 1982). In general, because UV stabilization research will reduce the number of low molecular weight photodegradation products and radicals that diffuses to the surface of a polymer glazing, it will lead naturally to the development of polymers that attract and hold little dirt. Abrasion resistance of most polymer glazings appears adequate for nonconcentrating applications except in extreme environments (e.g., the desert).

Correlation of Microscopic Degradation with Macroscopic Property Degradation   Data for correlation between microscopic characteristics (average molecular weight, polymer free-volume, etc.) and macroscopic property degradation are not available for most polymers. Such correlations are well developed for elastomers as a result of extensive research supported by the space program(Carroll and Schissel 1980). Analogous correlations for thermoplastic materials are being developed at JPL (Hong 1982). The ultimate goal of this work is to develop regression equations for acrylics and fluorocarbons that correlate microscopic polymer characteristics with mechanical and optical properties of interest, and correlate changes in microscopic characteristics with exposure times in different environments. Polyvinylidene fluoride (Kynar) and PMMA have been studied using various spectroscopic techniques. Kynar appears to undergo a crystal reorientation process when heated at 80°C (185°F) leading to an increase in scattering and a gradual reduction in transmissivity. PMMA (with a copolymerized UV absorber) has remained unchanged in the conditions under which it has been exposed.

25.2.2 Approaches to the Stabilization of Glazing Polymers Exposed to Combined Stress

Development of UV Stabilizers   A variety of UV stabilizers have been developed over the years, including UV absorbers which absorb the potentially damaging radiation and then dissipate it without transferring energy to a base polymer, such as a hydroxybenzophenone (Kloppfer 1976); excited-state quenchers which remove energy from chromophores in their excited state before the onset of irreversible degradation (Lappin 1971); radical scavengers which combine with alkyl or peroxide radicals and interrupt the oxidation chain reaction; and other antioxidants which actually break down oxidation products, in particular, the hydroperoxides.

In common practice, ultraviolet stabilizers are simply blended with the base polymer to be protected. However, stabilizers are often incompatible with the base polymer and slowly diffuse onto the surface of a film, then evaporate, are hydrolyzed, or form nontransparent precipitates with atmospheric gases or dirt (Lappin 1971; Gupta 1978; Winslow 1982). The most common approach to dealing with this problem is through the incorporation of high molecular weight stabilizers. Another approach has been to modify stabilizer structure. For example, modification of benzatriazoles has enabled Ciba-Geigy to extend the time until 50% of the stabilizer is lost at 280°C (436°F) by more than a factor of 10 (Dexter and Winter 1981). However, it is questionable whether or not the above two approaches can provide the compatibility needed in solar glazings that might be exposed for 10 to 20 years.

A variety of new approaches to preventing stabilizer leaching have been explored in the last 10 to 20 years. These include stabilizer-base polymer copolymerization, stabilizer polymerization, and stabilizer mixed into the monomer mixture prior to polymerization. Both of the first two approaches rely on polymerizable UV stabilizers. A large number of these stabilizers have been synthesized by university and industry researchers, including polymerizable benzophenones, benzatriazoles, and hindered amines (Bailey and Vogl 1976; Tirrel 1980; Karrer 1981).

Copolymerization of the stabilizer to the base polymer could theoretically completely prevent the stabilizer from leaching to the surface of a thin film. The approach has been researched by both the federal PV encapsulant program and private industry (Bailey and Vogl 1976; Tirrel 1980; Karrer 1981; Gupta 1982). JPL successfully copolymerized benzophenone and benzatriazole UV absorbers with methyl methacrylate. Because completely copolymerized stabilizers can be extremely expensive and can have insufficient mobility (particularly for excited-state quenchers and radical scavengers), JPL combined the copolymerized acrylic-UV absorbers with commercially available acrylics by blending and grafting. The blended copolymers did not separate even after continuous exposure to liquid water for seven days at 30°C (86°F). Outdoor exposure of grafted films indicates that the chemical attachment is permanent and resistance to degradation is excellent.

A possible cheaper alternative to stabilizer-base polymer copolymerization is polymerization of only the stabilizer, and subsequent blending of base polymer and stabilizer chains (Lappin 1971). This approach is essentially an extension of the idea of incorporating high molecular weight stabilizers. How much it will extend the time needed to complete separation of the stabilizer and base polymer has not been established.

A third approach to retarding stabilizer leaching has been reported in West Germany (Hosch n.d.). Methyl methacrylate (MMA) has been polymerized in the presence of sterically hindered amines. Without slowing polymerization the monomer mixture additive improved PMMA UV stability and retarded leaching rates substantially, when compared with PMMA blended with the same stabilizer during extrusion. Copolymers of 83% MMA and 12% acrylate proved particularly stable, retaining 80% transmission (at 300 nm) for 1,000 hours under a UV lamp. With no stabilizer the copolymer transmission dropped from 76% to 49%. The technique would be more likely to prevent surface degradation of sheet material than polymerization or copolymerization of the stabilizer. The approach could be applicable to solar panels because of the stability of the PMMA and copolymers produced, the high UV transmission of the material, and the slow separation of polymer and stabilizer.

Beyond stabilizer-polymer compatibility the area of stabilization research most relevant to solar applications is the study of stabilizers used in combination. Research has demonstrated that the incorporation of many stabilizers into a material can dramatically lengthen its lifetime (Dexter and Winter 1981; Gupta 1982; Winter 1982). Since the conditions in which glazings must perform are highly variable (e.g., temperature, strain, humidity, or UV light), combinations of stabilizers may be particularly desirable. Combinations of UV absorbers, hindered amines, and other antioxidants might be especially effective.

Development of Improved Polymers for Glazings   These materials must be highly transparent, moderately cheap, and inherently somewhat durable. Candidates identified to date include (1) PMMA and copolymers, which have no hydrogens on the polymer backbone chain; (2) the fluorocarbons (e.g., polyvinyl fluoride or Tedlar, and polyvinylidene fluoride or Kynar), which can be cheap enough in thin-film form and have better temperature stability than the acrylics; and (3) cross-linked methyl silicone glass resin, which is used as a coating on aircraft windows (Winslow 1982). Silicon has low weight and excellent UV and temperature resistance but is somewhat expensive. Less UV-stable polymers such as polycarbonate or polyester can be used if stabilization approaches become an order of magnitude more effective, or they can be used in a laminate with a UV screen. At least one of the commercial stabilized polyesters uses one of the advanced UV stabilization techniques described later. A number of laminated films have also been commercialized: Tedlar over FRP made by Filon Corporation (Jorgenson 1982), Tedlar over a vinyl polymer made by Environmental Structures (Jorgenson 1982), and Tedlar over polycarbonate, which is part of the Ramada Energy Systems predominantly polymer collector.

25.2.3   Applied Research on the Effects of Processing on Glazing Polymer Stability

The greatest body of research conducted in this area deals with the effect of processing variables on UV degradation, through its influence on initial chromophore density. In particular, processing at high shear rates (low temperature, rapid draw) has been shown to cause mechanochemical rupture of polymer bonds and the formation of macroalkyl radicals that begin oxidation reactions (Mellor, Moir, and Scott 1973). This results in significantly increased chromophore concentration, which leads to higher initial weathering rates and mechanical property deterioration. Since high temperature processing produces too much thermal oxidation, the optimal draw temperature is generally an intermediate one. Although there is evidence of processing-induced chromophore incorporation, it is not clear that the amount of chromophore incorporation during processing significantly affects durability. Most experimental evidence suggests that the concentration of peroxides and ketones builds up so rapidly during the onset of degradation that initial chromophore concentration is irrelevant to polymer lifetime. Furthermore, as Carlsson, Garton, and Wiles (1978) have shown, although peroxide and carbonyl buildup is more rapid in polymers drawn quickly at low temperatures than in those drawn at lower shear rates, the separation between times to embrittlement is much smaller. Even in well-processed materials there are regions near the surface where initial chromophore concentration is large, and where embrittlement occurs relatively quickly.

Process adaptation has not been pursued to any significant extent, both because there is little evidence of substantial payoff and because of the limited resources of the small companies that dominate the polymer-processing industry. With respect to cost reduction, the normal course of technological development will, in general, lead to modernization of equipment and techniques for polymer processing independent of the solar application. Except for the exploration of cheap processes tailored to the solar industry (e.g., the DOE-sponsored effort at Acurex Corporation to develop a lightweight collector with a polymeric glazing and absorber, and a thin metal frame), there is little government action in this area.

25.2.4 Other Glazing Research Areas

Accelerated testing has several purposes, including (1) screening of large numbers of polymers to identify promising materials, (2) in-depth comparisons between competing materials, and (3) the development of service-life predictions. Screening tests have become increasingly sophisticated in recent years, and a great deal of test data exists on polymer glazing materials. Screening tests aimed at selecting flat plate collector glazings have been carried out by Springborn Laboratories (Baum 1982).

Field testing is the final step in research on polymer glazing materials. Development of standard testing procedures is rather straightforward, referring merely to the testing of alternative materials in similar systems under identical conditions. The most critical area of effort is to build up a field testing data base. At present very little field data exist. The data available do not deal with polymers as design elements in solar collectors or otherwise exposed at higher than ambient temperatures (Gilligan 1980; Gilligan et al. 1980).

25.2.5 Prospects for the Future

Much work remains to develop an ideal solar glazing. Such a glazing would have the following characteristics: good optical properties [high transmissivity and high infrared (IR) reflectivity, or failing the latter, high IR absorptivity], good mechanical properties (high tensile strength, impact strength, softening point, melting point, and thermal expansion coefficient matched to the collector frame), durability (retaining its original optical and mechanical properties, resisting abrasion, and collecting little dirt upon outdoor exposure); and low cost and light weight. In addition rigid glazings must be self-supporting, and the stretched films must be able to withstand wind, snow loading, and high temperatures without sagging to the absorber.

Research that appears to be most promising to the realization of these goals is concerned with cheap, transparent, and moderately durable polymers such as polymethyl methacrylate (PMMA), polycarbonate, thin-film fluorocarbons, and methyl silicone glass resin; the degradation of promising laminated combinations of UV-stable and temperature-resistant materials in these classes (e.g., PMMA over polycarbonate); and the development of stabilizers.

25.3   Heat Mirrors

Research to improve heat mirrors for active collector applications progresses from a basic study of the optical properties of semiconductors, metal films, and dielectrics through exploration of candidate heat mirrors and their optical and durability properties. Research also includes optimization of process parameters, examination of competing deposition processes, and laboratory and field testing. The range of possible heat-mirror research activities is illustrated in figure 25.2. Moving vertically through figure 25.2, one progresses from basic studies to the development and testing of innovative heat mirrors. The horizontal direction represents the research options at each stage of this progression.

Level 1 provides the knowledge of fundamental physics necessary for exploration of new heat-mirror concepts. Levels 2 and 3 have become increasingly active areas of research since 1975 (Lampert 1981). At first only a few large band-gap semiconductor heat mirrors and thin-metal films were examined, but then a variety of new dielectric/metal/dielectric multilayer films, ternary compounds, and nitride refractory metal coatings were investigated. These can be grouped into three classes: semiconductors, conducting microgrids, and metal films (usually antireflected or incorporated in multilayer films). Level 4 has been examined as researchers seek to make the critical step from laboratory materials to heat mirrors that can be produced in large-volume commercial processes and still exhibit excellent stability and selectivity. Closely related research has attempted to adapt various manufacturing processes to cheap, large-scale production (level 5). A final area of heat-mirror

R&D level	Definition						
7	Field testing			Development of field-testing data base	Data collection from private systems		Development of service life prediction models
6	Accelerated testing	Screening tests	Aging under high temperature	Aging under photon flux	Aging under combined stresses	Humidity testing in air	
5	Study of alternative deposition techniques	CVD: hydrolysis	CVD: pyrolysis	Vacuum evaporation	Sputtering	Ion plating	Chemical dip processes
4	Optimization of coating deposition parameters		Transparent conductors	Metal microgrids	Metal films		
3b	Study of coating degradation mechanisms		Transparent conductors	Metal microgrids	Metal films		
3a	Study of coating optical properties		Transparent conductors	Metal microgrids	Metal films		

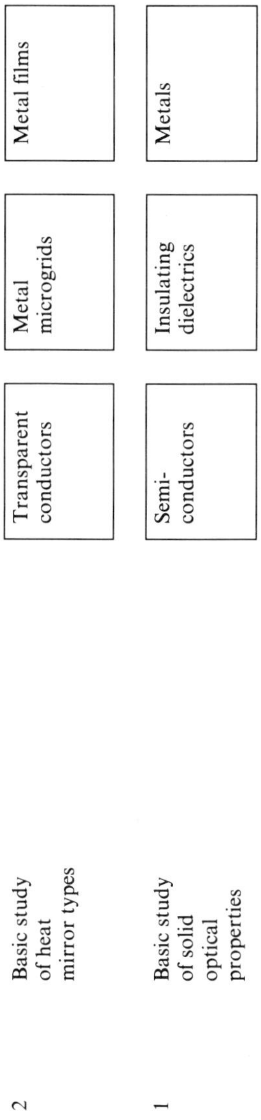

Figure 25.2
Heat-mirror research alternatives.

research is testing of optimized coatings under simulated or actual service conditions (levels 6 and 7). This work determines whether absorbers behave as predicted in the laboratory when exposed in actual components and field conditions. Very little testing of candidate heat mirrors has been reported.

25.3.1 Development of Doped Semiconductor Heat Mirrors

The most widely investigated heat mirrors are doped, wide-band-gap (E_g) semiconductors, also known as conducting oxide heat mirrors (Haacke 1982) The three most carefully researched doped semiconductors are indium tin oxide ($In_2O_3 : SnO_2$, or ITO), doped tin oxide (SnO_2), and cadmium orthostannate ($CdSnO_4$). Several ternary semiconductors have also been examined.

Indium Tin Oxide   The ITO coating with the highest reported solar transmissivity at the time of this draft has $\tau_s > 0.95$ until the Sn fraction increases to more than 10% by weight, and remains above 90% until the Sn fraction reaches 20% (Nath et al. 1980). Coatings have been deposited by plasma-assisted reactive evaporation with various tin fractions and substrate temperatures. Transmission is relatively insensitive to substrate temperature between room temperature and 250°C (482°F), while the electrical resistivity decreases steadily, especially above 300°C (572°F). TEM micrographs and the above coating properties both suggest that low substrate temperatures produce stoichiometric oxide films, while higher deposition temperatures induce oxygen deficiency, high carrier concentration ($N = 10^{21}/cm^3$), and increases in conductivity. The IR reflectivity of these films is proprietary. However, data suggest that for this process, the optimum deposition parameters (for collector applications) would be a substrate temperature of between 200° and 300°C (392° and 572°F), and an Sn fraction of between 10% and 20% (Nath et al. 1980). In related research it has been found that films with reasonable visible transmissivity and low resistivity can be achieved as well with simple reactive evaporation as with plasma-assisted activation (Habermeier 1981). However, these films have considerably lower visible transmissivity (~ 0.8).

Highly transparent $In_2O_3 : SnO_2$ films have been reactively ion plated at 300° to 350°C (572°–662°F) (Machet et al 1981). These coatings had an electrical resistivity of 10^{-5} Ω/cm (which remained constant with Sn content of 10%–50%) and visible transmissivity greater than 85% (including the glass substrates), and they could be deposited at 4 to 6 $\mu m/h$ (a 6-min deposition time for a 0.5-μm coating). This is a higher deposition rate than that achieved by sputtering but also a higher resistivity. Examination of the IR reflectivity of the coating and optimization of process parameters, coupled with microstructural investigation, need to be pursued.

Other highly solar or visible transparent ITO films have been reactively sputtered and then annealed at 300° to 500°C (572°–932°F) (Smith et al. 1980); thermally evaporated from an In, In_2O_3 source to produce an undoped, metal-rich film (Pan and Ma 1980) deposited by CVD using organometallic reagents (Kane, Schweizer, and Kern 1975); laid down by CVD spray pyrolysis (Van Boort and Groth 1968) and spray hydrolysis (Kostlin 1974); and reactively ion plated in the presence of fluorine (Avaritsiotics and Howson 1981).

Antireflection or texturing of ITO might make its optical properties equivalent to the best selective absorbers, by eliminating local visible region reflectance maxima that degrade its solar transmittance. This approach has been explored in Sweden, where ITO/MgF_2 coatings with a visible transmission 1% greater than that of uncoated glass have been produced (Hamberg, Hjorstberg, and Granqvist 1982). The transmittance below $0.42\ \mu m$ and above $0.67\ \mu m$ are below that of bare glass, so that the integrated solar transmittance could be less. MgF_2 treatment to bare glass would produce solar transmittances about 5% greater than the glass $+$ ITO $+$ MgF_2. The optical properties of a different antireflected ITO film have also been reported (Frank et al. 1982; Frank and Kostlin 1982). A module composed of 18 evacuated tubes with this ITO coating, without an additional glass coverplate, showed an $\alpha\tau = 0.77$ and a collector loss coefficient of about half that of conventional flat plate collectors with two glass plates and a selective absorber. The investigators estimate that an antireflection coating would raise this value to 0.82 or 0.83.

German researchers, reporting on the advantages of activated reactive evaporation, have proposed the production of multiple-layer oxide heat mirrors. A four-layer transparent oxide stack ($MgF_2/In_2O_3/SiO_2/MgF_2/glass$) had reflectivity of below 1% in the range 450 to 750 μm (Ebert 1982). The quoted deposition rate (5 Å/s) and the quality of the coatings produced (due to the low bombarding energy of plating ions) suggest this process should be examined more closely.

Tin Oxide Of the tin oxide films described in the literature, those doped with fluorine have the best properties (e.g., solar transparency of 0.8) (De Waal and Simonis 1981). Thermal degradation at 250°C (480°F) in vacuum reduces electron mobility and presumably degrades optical properties, although these were not measured after aging. The authors suggest the degradation of electronic properties may be due to reduction of SnO_2. Another contributing factor might be the evolution of fluorine gas and the reintroduction of electron-deficient Sn-scattering centers.

Spray-coated Sn_2O_3:F films have been reported with $\tau_s = 0.75$ and $\varepsilon_{IR} = 0.15$ (Van der Leij 1979). Related films have even higher electron mobility and IR reflectivity approaching 0.9 (Haacke 1982). The solar transmissivity of the latter coatings was not reported. SnO_2 films with reasonably high conductivity and electron mobility, IR reflectivity of about 0.8, and solar transmissivity of slightly greater than 0.8 have also been reported (Frank et al. 1982; Frank and Kostlin 1982).

No deposition of a homogeneous antireflection layer of SnO_2 has been reported in the literature. However, visible reflectivity of 15% and IR reflectivity of 90% (at 10 μm) has been obtained in a 0.5 μm, fluorine-doped, CVD-deposited, SiO_2–SnO_2/SnO_2 coating that utilizes graded SnO_2 content from 0% to 100% in a dielectric layer between the glass substrate and heat-mirror coating (Gordon 1982).

Cadmium Orthostannate   The optical properties of the best cadmium orthostannate coatings approach those of the best ITO ($\tau_s = 0.86$, $\varepsilon_{IR} = 0.12$) (Haacke 1977). Optimal properties are obtained in films prepared by rf-sputtering in oxygen and then annealed. Although these films can be deposited onto plastic substrates (e.g., polycarbonate) with good adhesion by reactive sputtering in an Ar/O_2 atmosphere (with no subsequent anneal), transmissivity and reflectivity suffer on the order of 10% each (Haacke 1982). $CdSnO_4$ can also be deposited by ion plating at high deposition rates (500 Å/min) at room temperature (Lampert 1981), and by CVD spray pyrolysis at 450°C (842°F) (Ortiz 1982). The former process has been used to coat polyester, and the latter to produce coatings with visible transmissivity of greater than 0.85.

A related ternary semiconductor, $CdIn_2O_4$, has shown promising IR reflectivity (0.93) and solar transmissivity that oscillates between 0.78 and 0.92 (Maag 1982). An antireflection layer might bring τ_s up to 0.9.

25.3.2   Thin-Metal-Film Heat Mirrors

Thin-metal films of Ag, Cr, Cu, Au, Al and Ni have been examined and do not have acceptable solar transmissivity for heat-mirror applications (Groth and Kauer 1965; Fuschillo 1975). It has been estimated that refractory metal nitrides can have τ_s as high as 0.5 to 0.6 depending on film thickness but might have corresponding IR reflectivity of only about 0.6, due to their low electron mobility (Karlsson and Ribbing 1982). Therefore attention in this area shifted to multilayer metal-dielectric films. Most of the coatings sold commercially (by Southwall, OCLI, and Teijin) and metal-dielectric multilayer films.

Deposition of multilayer films was pioneered at MIT, where heat mirrors have been applied to solar collectors (Fan and Bachner 1976; Fan 1979). Solar transmissivity of greater than 0.72 has been matched with IR reflectivity of 0.95 (using low temperature sputtering, which is compatible with polymeric substrates). No degradation has occurred in these films after 48-hour exposure at 200°C (392°F). Small changes have been observed at 300°C. Japanese researchers at Teijin, Ltd., have produced terabutyl titanate (TBT)/Ag/TBT coatings on polyethylene terephalate using a low-cost hydrolysis process (Chiba et al. 1982). Optical properties can be manipulated by changing the percentage of TBT-*n*-hexane in the coating solution, and by changing layer thicknesses. The deposition process is compatible with roll coating. However, the reflectivity step of the Teijin films climbs over a broad range, substantially degrading the optical properties.

The metal nitrides are very stable and can be antireflected. Calculations suggest that the solar transmissivity of refractory metal nitride films can increase to 0.75, and the IR reflectivity to a similar value, through the deposition of ZnS dielectric layers on either side of the metal.

25.3.3   Metal Microgrid Heat Mirrors

The metal microgrid examined by Fan (1979) had excellent optical properties ($\tau_s = 0.9$, $R_{IR} = 0.83$). However, since it was an ITO coating, the extent to which the microgrid design enhanced its selectivity is unclear. In general, despite initial theoretical interest in this approach, microgrids have proved difficult to produce in either suspended wire form or photoetched thin films (Haacke 1982).

25.3.4   Other Research Areas

Deposition Technologies   Rf- and reactive sputtering have been employed most frequently to deposit heat mirrors. In general, only reactive sputtering can provide the desired deposition rates, although it is more difficult to control. In particular, cathode poisoning can result in the formation of oxide coatings on the metallic target. Furthermore, during deposition of tin-doped In_2O_3, metallic inclusions occur, and oxygen annealing is necessary to increase solar transmission (Lampert 1981). During reactive and rf-sputtering (and evaporation), alloy diffusion in In–Sn and Sn–Sb targets occurs due to their low melting points. Special measures must be taken to ensure alloy composition regularity. The advent of magnetron sputtering has permitted greater bombarding densities, has increased deposition rates, and has enabled more straightforward low temperature deposition. Of particular interest to active solar applications is the adaptation of magnetron sputtering to coating the inside of evacuated tubes (Walker and McKenzie 1981). Although the authors tested the process with a sputtered carbide selective absorber, the efficiency gains associated with heat mirrors applied to evacuated tubes suggest a logical step might be the deposition of ITO. Magnetron sputtering has also been adapted to deposition on a range of plastic substrates (polyester, polycarbonate, Teflon, Kapton) and has been integrated with continous-roll-casting equipment, as have a variety of other sputtering and vacuum evaporation processes (Howsen and Ridge 1982; Kitler and Richie 1982).

Other physical vapor deposition technologies applied to heat-mirror coating include evaporation and ion beam sputtering. Deposition rates are difficult to control with evaporation, and expensive source material is usually wasted (Lampert 1981). However, it is less capital intensive than sputtering. During ion beam sputtering an ion beam of inert or reactive gas at high energy is directed at target material which is sputtered onto a nearby substrate. Better control over deposition angle and substrate temperature is achieved than with sputtering. Less material is wasted, and the level of electron bombardment of the substrate is low. Polyester substrates, ion beam sputtered with In_2O_3:Sn, achieved a maximum temperature of only 80°C (176°F) (Lampert 1981).

Ion plating is a hybrid process in which material (metal) is evaporated or sputtered onto the substrate. The substrate then acts as the cathode of the discharge and is bombarded with ions (sputter etched). Reactive ion plating has successfully deposited various oxide heat mirrors on polyester substrates at near room temperature. Film properties approach those of high temperature sputtered films. Ion plating combined with a roll coater apparatus can continuously deposit conductive oxides on polyester film substrates (Lampert 1981).

Chemical vapor deposition has excellent properties for glass coating, but hydrolysis requires temperatures of 400° to 800°C (752°–1472°F) and is therefore unacceptable for coating onto a polymeric film.

Plasma-assisted pyrolytic CVD, however, has been used to deposit ITO at room temperature. A related chemical process has been used to deposit SnO_2 from tetramethyl tin at 160 Å/min (Lampert 1981). This and other chemical processes offer the potential of very low cost coatings. A number of organometallic dip coating solutions and their possible optical applications, including heat mirrors have been reviewed (Dislich and Hussman 1981).

25.3.5 Prospects for the Future

In the early 1980s heat-mirror research reached the stage attained by selective absorber research a few years earlier. A large number of materials with promising optical properties await further examination. Fine-tuning of heat-mirror optical properties, together with improvements in coating durability, could make heat mirrors ready for routine incorporation into solar collectors.

In this context the goal of heat-mirror research is to develop materials with optical properties equivalent to those of the best selective absorbers. These properties would include solar transmissivity and IR reflectivity of 0.90 to 0.95. The heat mirrors most likely to achieve these objectives are the transparent oxides, in particular, an antireflected or textured indium tin oxide. As noted previously, an antireflected ITO had visible transmission 1% greater than that of uncoated glass, suggesting that this approach might produce excellent collector heat mirrors. However, reasonably selective $CdSnO_4$ and $SnO_2:F$ coatings have also been reported and a number of other transparent oxides have been investigated.

25.4   Selective Absorbers

Selective absorber research spans a wide spectrum from basic study of candidate material optical properties to field testing of prototype absorbers. The relevant research activities can be thought of as comprising a hierarchy from basic optical studies, through conceptualization and development of the various classes of absorbers, through detailed characterization of the optical and durability characteristics of specific candidate absorber materials, through optimization of properties and evaluation of manufacturing options, to laboratory and field testing of actual absorbers.

This range of possible absorber research activities is illustrated in figure 25.3. Moving vertically through figure 25.3, one progresses from basic studies to the development and testing of innovative absorbers. The horizontal direction represents typical research options at each stage of this progression.

Basic optical properties research, shown in level 1 of figure 25.3, provides the knowledge of fundamental physics necessary for exploration of new selective absorber concepts.

Selective absorber conceptualization (level 2) and initial exploration has been a fertile area, with literally hundreds of different absorber combinations being suggested and/or studied (Call 1979; Lampert 1979; Herzenberg and Silberglitt 1982; Moore 1982). These fall into five principal classes: materials with intrinsic solar selectivity; tandems made by depositing a semiconductor absorber over an underlying metallic reflector; composites that consist of small metallic particles embedded in a dielectric medium; materials with a textured surface; and multilayer materials that rely on interference effects to achieve selectivity.

Characterization research (levels 3 and 4) involves the detailed study of the optical properties and degradation characteristics of specific selective absorber systems. Such research examines the dependence of optical properties and

R&D level	Definition								
7	Field testing				Field-testing data base	Data collection from private systems			
6	Accelerated testing		Screening testing	Aging testing in air	Humidity testing in air	Aging under photon flux	Service-life prediction models		
5	Study of alternative deposition techniques	Electro-plating	Sputtering	Chemical vapor deposition	Dispersion in a paint binder	Chemical conversion	Anodic oxidation	Other physical vapor deposition	Textured surface processing
4	Optimization of coating deposition parameters		Intrinsic absorbers	Semi-conductor metal tandems	Metal-dielectric composites	Textured surface absorbers	Multilayer absorbers		
3b	Study of coating degradation mechanisms		Intrinsic absorbers	Semi-conductor metal tandems	Metal-dielectric composites	Textured surface absorbers	Multilayer absorbers		

		Intrinsic absorbers	Semi-conductor metal tandems	Metal-dielectric composites	Textured surface absorbers	Multilayer absorbers
3a	Study of coating optical properties	Intrinsic absorbers	Semi-conductor metal tandems	Metal-dielectric composites	Textured surface absorbers	Multilayer absorbers
2	Basic study of selective absorber types	Intrinsic absorbers	Semi-conductor metal tandems	Metal-dielectric composites	Textured surface absorbers	Multilayer absorbers
1	Basic study of optical properties of solids		Pure metals	Carbides, nitrides, and oxides	Semi-conductors	Insulating dielectrics

Figure 25.3
Selective absorber research alternatives.

durability on both coating microstructure and processing history. For example, an extensive effort (Pettit, Sowell, and Hall 1982) has established that the optical properties and durability of electrodeposited black chrome are correlated with deposition variables, such as Cr^{3+} and acetic acid content. Parallel microstructural investigations have established that highly selective black chrome consists of a layer of Cr_2O_3 over a matrix of small chromium particles embedded in Cr_2O_3. The most durable coatings have a graded chromium content, a high proportion of metal in the composite matrix, large Cr particles, and a small void ratio (Ignatiev et al. 1979; Holloway et al. 1980; Zajac, Smith, and Ignatiev 1980; Inal et al. 1981).

There are often several options for manufacturing selective absorber coatings using a given material (level 5). Research to evaluate the relative efficacy, cost, flexibility, and other important characteristics of methods such as sputtering, chemical vapor deposition, and anodic oxidation is essential for overall cost/performance optimization.

A final area of selective absorber research is testing of optimized coatings under simulated or actual field conditions (levels 6 and 7). This work establishes a data base to determine whether absorbers behave as predicted when exposed in actual components and field conditions.

In the following sections the major achievements of selective absorber research are highlighted.

25.4.1 Intrinsic Absorber Research

Since early theoretical consideration of the solar selective absorber field (Ehrenreich and Seraphin 1976), researchers have hoped that a single, intrinsically selective material might be found that would supersede more inherently complex composites, tandems, or multilayer absorbers. Attention has focused on transition metals doped with elements (C, N, O, B) that might localize electrons in the incompletely filled d-shell of the metallic ion and shift the plasma edge to longer wavelengths. In addition to approaching the desired intrinsic reflectivity profile, many of these alloys also oxidize at relatively high temperatures.

ZrB_2 has been demonstrated to be an extremely selective intrinsic absorber (Randich and Allred 1981). It oxidizes very slowly below 527°C (981°F) and forms a passivating amorphous film, even when it does oxidize below this temperature. ZrB_2 coatings, 15 to 30 μm thick, chemical vapor deposited at 30 μm/h and at 950°C (1,742°F) by Sandia National Laboratory, have non-optimized absorptivity of 0.67 to 0.77 and emissivity of 0.08 to 0.09. Antireflecting with Si_3N_4 increases absorptivity to 0.93. The coating has aged 1,000 hours in air at 500°C (932°F), with no resulting change in optical properties.

The characteristics of a cheaper alternative intrinsic absorber, NiN_x, have been reported (Sikkins 1981, 1982). Sputtering 0.067-μm thick NiN_x onto polished Cu, at a nitrogen pressure of 0.13 Pa, produced a coating with $\alpha = 0.84$ and ε (150°C) = 0.039. At this low nitrogen pressure the coating is believed to have nitrogen atoms interstitially dissolved in the Ni f.c.c. lattice. For reasons that are not completely understood, the conductivity grades with increasing distance from the substrate. (In strict terms NiN_x is not an intrinsic absorber because it must be coated onto a metallic substrate to show selectivity.)

25.4.2 Semiconductor-Metal Tandem Selective Absorber Research

A considerable number of absorbers of this type have been investigated in recent years. All the typical metal substrates provide the requisite low emissivity. The challenge is to find a semiconductor with a band gap near the long wavelength end of the solar spectrum (0.58 eV) and a high absorption coefficient, and an antireflecting layer, textured surface, or dielectric host matrix that would sufficiently reduce the front surface reflection resulting from the high refractive index of semiconductors.

Both silicon and tellurium have been explored as possible tandem selective absorber materials. Silicon work (Booth, Allred, and Seraphin 1979) has focused on amorphous silicon (a-Si) material because of its high absorption coefficient and the fact that CVD-deposited a-Si has broadened band tails that allow significant absorption down to about 0.9 eV. Absorption at longer wavelengths also increases at higher temperatures. Carbon-doped a-Si on Mo with an Si_3N_4 antireflecting layer has achieved $\alpha = 0.75$, $\varepsilon = 0.08$, with excellent durability to at least 500°C (932°F) in air. The tellurium tandems have been made by both angled-vapor deposition and gas evaporation (Cocks and Peterson 1979; Cocks, Peterson, and Jones 1980). Tellurium is characterized by a very small band gap (0.35 eV), which ensures absorptivity throughout the solar spectrum. It also has a large index of refraction ($n = 3$), which leads to large front surface reflection. Surface texturing has successfully eliminated the latter problem, yielding Te tandem selective absorbers with $\alpha = 0.95$, $\varepsilon = 0.06$ (gas evaporation), and $\alpha = 0.92$, $\varepsilon = 0.03$ (angled-vapor deposition). Research to establish the durability of these coatings is necessary before their commercial potential can be evaluated.

Other semiconductor-metal tandems that have been characterized include Cu_2S, chemical spray deposited on aluminum in air (Gadgil et al. 1981), and germanium, either incorporated in a CaF_2 dielectric or with a textured surface (Gittleman, Sichel, and Arie 1979; Swab, Krishnaswamy, and Messier 1980). Neither has shown acceptable selectivity, although textured surface Ge has high absorptivity and the Cu_2S coating has an estimated (1982) cost of only $2.70/m^2 ($0.25/ft^2).

A heat-mirror/black substrate selective absorber of fluorine-doped tin oxide over black enameled steel has been developed in Holland (Van der Leij 1979; De Waal and Simonis 1981). It is highly resistant to mechanical scratching and chemical attack, is stable up to 350°C, and has been used in several European Solar Pilot Test Facilities.

25.4.3 Metal-Dielectric Composite Selective Absorbers

A very large number of metal-dielectric composite selective absorbers have been deposited in a large variety of ways (Granqvist 1981). The most prominent of these is black chrome electroplated from the Harshaw Chromonyx bath. This coating, which consists of a Cr_2O_3 layer over a small Cr particle/ Cr_2O_3 host composite, has achieved $\alpha = 0.95$, $\varepsilon = 0.05$, unchanged after 5,600 hours in air at 350°C (662°F) (Ignatiev et al. 1979; Holloway et al. 1980; Zajac, Smith, and Ignatiev 1980; Inal et al. 1981; Pettit, Sowell, and Hall 1982). Black chromes electroplated by other techniques (e.g., by Berry Solar) have also been studied and appear similar in microstructure and of acceptable durability for low temperature applications (Raghunathan 1981). Thus electroplated black chrome represents a benchmark for composite absorber research. The principal research need for black chrome, for low temperature applications, is cost reduction. The major parameters affecting the (1982) cost of about $9/m^2 are a Ni underlayer with a thickness of 0.1 to 0.2 mil; the Cr/Cr_2O_3 coating deposition time; and the plating current and associated power supply capital cost.

The most promising of several alternative metallic composite coatings appear to be an anodically oxidized Ni/Al_2O_3 developed in Sweden, a chemically converted black nickel foil made by Ergenics in this country and Europe, and magnetron-sputtered cermets (graded metal carbides) developed at Sydney University in Australia and Telic Corporation in California. The first two are sold commercially; a license for the production of sputtered carbide-evacuated tubes has been sold to a Japanese company, Nitto Kohki.

The Swedish anodized coatings consist of metallic nickel grains embedded at the root of a 0.7-μm porous Al_2O_3. They have absorptivity in the 0.92 to 0.97 range and emissivity in the 0.1 to 0.26 range and have shown considerable temperature and humidity durability (Andersson, Hunderi, and Granqvist 1980). Further research is needed, however, to optimize coating properties, establish long-term durability, and adapt the coating to deposition on a foil.

Ergenics chemically converted, black nickel foil is about 0.3 μm thick with about 200 Å of $NiCrO_x$ near the surface, which grades to NiO (Lampert 1980).

It may also contain some elemental Ni or Cr. It is deposited on 0.5 mils of electroformed nickel in a proprietary acidic oxidizing continuous production line. Optical properties and durability appear to be excellent (Mason and Brendel 1982). Future research should focus on characterizing microstructure, understanding degradation mechanisms, and minimizing coating thickness to reduce cost.

Sputtered cermets include an Australian graded metal carbide, consisting of three layers with different homogeneous concentrations of elemental metal (Harding and Window 1977). Reactive sputtering in Ar and CH_4, on Cu substrates, has yielded coatings with $\alpha = 0.93$, $\varepsilon = 0.04$, and a deposition time of, at most, a few minutes. These coatings degrade above 200°C (392°F) in air, in part because they become homogeneous. But they are extremely stable in vacuum, making them strong candidates for evacuated-tube applications.

Even better candidates for high temperature application might be the graded cermet coatings of Pt/Al_2O_3 and Mo/Al_2O_3 that have been produced by coevaporation at Cornell University (Nyberg, Craighead, and Buhrman 1982) and (for the former) sputtered at Telic Corporation (Thornton and Lamb 1981, 1982). These coatings have shown even better optical and durability properties than the metal carbides. Pt/Al_2O_3 has achieved $\alpha = 0.93$, $\varepsilon = 0.08$, and is stable after 700 hours in air at 600°C. Mo/Al_2O_3 has achieved $\alpha = 0.992$, $\varepsilon(200°C) = 0.08$, and is stable in air at 400°C and stable in a vacuum at 750°C.

Another development was the production of spectrally selective Ni–C films by reactive sputtering in an Ar–CH_4 mixture (Sikkins 1982). These films consist of small Ni particles (25–50 Å) embedded in an electrically insulating amorphous carbon matrix. Absorptivity is moderate (0.88–0.90), and emissivity extremely low (0.028–0.045 at 150°C). The films are stable up to more than 400°C in vacuum.

Beyond these commercial or near-commercial competitors to black chrome, a number of other approaches have been explored in the laboratory. These include coelectroplated black chrome molybdenum (Smith and Ignatiev 1981), chemical vapor deposited black molybdenum (Gesheva, Seshan, and Seraphin 1981; Chain, Seshan, and Seraphin 1981; Seshan et al. 1981), electroplated black cobalt (Smith, Ignatiev, and Zajac 1980), and anodically oxidized ZnO (Homhual et al. 1981; Inal, Mahon, and Robino, 1981) and copper black (Potdar et al. 1981). Of these, both black cobalt and black molybdenum appear promising for medium concentration applications (troughs, dishes).

A significant effort has also been devoted to making selective paints, which absorb much like metal-dielectric composite absorbers. In particular, Honeywell has developed thickness-sensitive and thickness-insensitive solar selective paints composed of an iron, manganese, and copper oxide pigment in a silicone binder (McKelvey, Zimmer, and Lin 1979; McChesney 1981; Moore 1982a). These paints have reasonable absorptivity and might be very cheap at approximately $1/m^2 ($0.10/ft^2) in 1982 dollars.

The thickness-insensitive paint incorporates an aluminum flake in the binder to suppress infrared emissivity. The best laboratory properties achieved with this design to date are $\alpha = 0.92$, $\varepsilon = 0.31$ (Moore 1982a). An alternative design might incorporate a precoated metal flake in a binder, thereby requiring only the optimization of one particle size, shape, and distribution (Moore 1982a). Further research is needed to improve understanding of the optical properties of these thickness-insensitive paints, to empirically optimize metal flake size and density, and to explore the dependence of durability on paint constituents, substrate preparation, and deposition conditions.

25.4.4 Multilayer Absorbers

Multilayer absorbers achieve selectivity via destructive interference of light reflected from the front surface of each layer. Telic Corporation has magnetron sputtered a large variety of such coatings, sandwiching a metal or metal oxide layer between two layers of rf- or reactively deposited Al_2O_3 (Thornton, Penfold, and Lamb 1980). Excellent selectivity and durability have been achieved for a variety of these coatings. Reactively sputtered Ni and Cr coatings have the greatest promise for active applications. Although their durability does not match that of rf-sputtered coatings, their potentially higher deposition rates may permit continuous in-line Al_2O_3/Metal/Al_2O_3 (AMA) sputtering, which reduces equipment capital cost per square foot of absorber produced. Al_2O_3 reactive deposition rates have reached 1,000 Å/min. The cost of AMA coatings sputtered in high volume line processes has been projected to be as low as $2.37/m^2 ($0.22/ft^2) in 1978 dollars.

The Optical Coating Laboratory (OCLI) in Santa Rosa, California, has also developed three-layer interference coatings using physical vapor deposition. Applied by electron beam evaporation to a 60-in.-wide foil, OCLI's best absorber has endured extended exposure at 500°C (932°F) in air and 90°C (194°F) at 95% relative humidity (Williams 1982). OCLI estimated that the coating would be sold for $7.50 to $8.50/m^2 ($0.70–0.80/ft^2) in 1982 dollars. One other commercial multilayer coating (containing aluminum) with high selectivity ($\alpha = 0.96$, $\varepsilon = 0.07$) has been reported (Raghunathan 1981). The coating has survived 161 thermal cycles in the range 93.3° to 232°C (200°–450°F) in an outdoor test collector.

25.4.5 Textured Surface Absorbers

Textured surface absorbers provide selectivity through a combination of multiple reflections and graded index of refraction, resulting from surface voids. However, texturing must be extremely fine and deep to achieve selectivity. The depth of pores or peaks must be approximately ten times the space between peaks (Stephens and Cody 1979). Not surprisingly, low-cost texturing techniques (e.g., chemical etching) tend to produce coatings with poor selectivity. The sputtering or CVD equipment usually necessary to produce closely spaced, tall needles or cones is relatively expensive. Dendritic growth is probably the most promising approach to the production of clean, highly selective coatings (Lampert 1982). We note, however, that many existing selective absorbers employ texturing to enhance selectivity (Smith, Ignatiev, and Zajac 1980).

25.4.6 Other Research Areas

Deposition Techniques A variety of deposition techniques can be used to produce selective absorber coatings, including electroplating, rf magnetron sputtering, chemical vapor deposition, pigment dispersion, chemical conversion coating, anodic oxidation, and electron beam evaporation. As black chrome illustrates, electroplating can produce coatings with durability up to 300° to 400°C (572°–752°F) in air (Ignatiev et al. 1979; Holloway et al. 1980; Zajac, Smith, and Ignatiev 1980; Inal et al. 1981; Pettit, Sowell, and Hall 1982). However, extending electroplated absorber durability is complicated because each new coating demands a separate effort to understand complex bath chemistry and to develop precise quality control. As noted previously, rf magnetron sputtering also produces coatings with excellent durability.

Although gas injection rates and partial pressures must be carefully controlled and unwanted reactions may introduce impurities, reactive sputtering produces coatings with sufficient durability for both flat plate and evacuated-tube applications (Thornton, Penfold, and Lamb 1980). Since reactive sputtering has higher deposition rates and can easily be adapted to the coating of graded absorbers, it has an excellent combination of properties for low temperature, low-cost applications.

Chemical vapor deposition allows very precise control of deposition conditions and constituents and produces highly stable coatings for medium and high temperature vacuum applications. The coatings are highly stable because the substrate is held at high temperature during the deposition process (Booth, Allred, and Seraphin 1979; Nyberg, Craighead, and Buhrman 1982). However, no detailed estimates of CVD coating's commercial cost have been published, making assessment of the promise of the technology for solar applications difficult.

Dispersion of a solar-absorbing pigment in a paint binder has the potential to produce extremely low-cost, low temperature absorbers, presuming adequate durability can be achieved. Achieving durability with the most selective thickness-sensitive paints is not simple, however. The coating must be thin so that the IR absorbing binder does not increase emissivity, but it must also contain enough pigment to attain high absorptivity (McKelvey, Zimmer, and Lin 1979; McChesney 1981; Moore 1982a).

Chemical conversion coatings have been touted as possible cheap, adherent absorbers for flat plate collectors or for passive applications. Although most such coatings do not have good optical properties, Ergenics' black nickel does have high selectivity and appears to be quite durable (Mason and Brendel 1982). A number of alternatives examined by Moore also have moderate selectivity (Moore 1982b).

Anodic oxidation has many of the same characteristics as electroplating, although it is less energy intensive and its bath chemistry may be less complex. However, the leading anodized coatings have not been subjected to durability testing, comparable to that of electroplated black chrome.

Electron beam evaporation, gas evaporation, angled-vapor deposition, and other physical vapor deposition processes have many of the same advantages as sputtering. However, these processes have only been applied to the deposition of one or two developmental absorbers.

Testing   Although a number of selective coatings have been thermally cycled (Raghunathan 1981) and NBS has tested two selective and many nonselective absorbers (Masters et al. 1981), very little durability data exist. Testing— including extensive aging in air either 225°C (437°F) or above, thermal aging under photon flux, exposure to high humidity at a moderate temperature of 100°C (212°F), or temperature cycling—has been reported for only a few coatings. Additional data on specific commercial coatings were collected by DSET under contract to SERI. More durability testing of developmental selective absorbers is needed.

25.4.7   Prospects for the Future

It is clear from the preceding text that there are already several selective absorbers with adequate to outstanding optical properties and promising durability for active solar heating and cooling applications. Metal-dielectric composites such as black chrome and black nickel consistently show the greatest selectivity, but individual coatings of other types (e.g., the sputtered

AMA metal carbides) approach the selectivity of the best composites. The greatest high temperature durability, both in air and vacuum, has been exhibited by multilayer absorbers and cermets sandwiched between two dielectric diffusion barriers. However, optimized laboratory black chrome has successfully weathered extended exposure at 350°C (662°F) without significant loss of selectivity. Highly selective and durable solar collector paints are still quite far from becoming commercial products, requiring theoretical and empirical studies of their optical properties, degradation mechanisms, and property dependence on processing and substrate preparation. Similarly much research remains to be done on infrared-emitting binders for selective paints. The future should see the further development, characterization, and testing of many of these new selective absorber concepts, using black chrome as the principal benchmark for comparison.

25.5   Absorber Substrates

Although not the focus of research, the absorber substrate may well be the most important component of a flat plate collector. It must perform all or some of these functions: (1) quickly and efficiently transfer the energy collected by the absorber to the heat transfer system (tubes and heat transfer fluids), (2) provide mechanical support to the heat transfer system, (3) act as direct conduit for the heat transfer fluid in the event tubings are not used, and (4) contribute to the overall structural strength and integrity of the collector over its life cycle.

Metals such as copper, aluminum, and steel are used as absorber substrate materials because of their proven low-cost, high durability, and good thermo-mechanical properties. Metals, however, are being increasingly challenged by polymers as absorber substrate materials in advanced collector designs. These designs use polymers and high speed extrusion or lamination manufacturing processes to drastically reduce collector materials and manufacturing costs while projecting performance comparable to metal flat plate collectors of conventional design. Examples of such designs are Brookhaven National Laboratory's Fluoroplastic Film Solar Flat-Plate Collector Panel (Wilhelm 1982), Acurex Solar Corporation's Experimental Low-Cost Solar Panel (Nelson and Muller 1980), and those commercially available from manufacturers such as FAFCO, Ramada Energy Systems, Ltd., and Bio-Energy Systems, Inc.

Besides metals and polymers, ceramics also appear to be useful as absorber substrate materials. Ceramics are promising because of their low cost and the possibility of manufacturing them using the high speed, high volume injection molding process (McDermott 1981; Whalen and Johnson 1981).

Table 25.1
Candidate absorber substrate materials

Candidate material	Material cost (1982 dollars)			Thickness (10^{-2} m)	Density (kg/m³)	Temperature limits (°C)			Thermal conductivity (W/mK)	Coefficient of thermal expansion (m/mK)	Strength[a]
	($/kg)	($/m²)	(kg/m²)			Min	Max	Continuous			
Metal sheet											
Copper (Centaur Metals)	2.45	21.81	8.90	0.10	8,900				377.0		$E = 119$ GPa, $G = 45$ GPa
Aluminum (Reynolds Al)	1.89 (2.62)	5.12	2.71	0.10	2,710				206.0		$E = 71$, $G = 26.2$ GPa
Steel (Reynolds A1)	0.34 (0.845)	2.71	8.00	0.10	~8,000				~45.0		$E = 207$, $G = 79.3$ GPa
Stainless steel (18–8) (Reynolds A1)	1.89 (2.62)	14.65	7.75	0.10	7,750				~16.1		$E = 190$, $G = 73.1$ GPa
Nickel (200)	18.30	164.24	8.98	0.10	8,975				58.8		
Metal foils											
Aluminum foil				0.012 (5 mil)	2,710				225.3	2.4×10^{-5}	$E = 71$ GPa, $G = 26$ GPa
Copper foil (Connie)	8.95	9.56	1.07	0.012 (5 mil)	8,900						$E = 199$ GPa, $G = 44.7$ GPa
Steel foil (Ni or C)				0.012 (5 mil)	~8,000						$E = 207$, $G = 79.3$ GPa
Stainless steel foil				0.012 (5 mil)	7,750						$E = 190$, $G = 73.1$ GPa
Foamed plastics											
Polyurethane	7.04	5.01	0.71	2.54 (1 inch)	28			121	0.03	4.9	413.8
Polyurethane	1.63	13.75	8.43	2.54	332			149	0.07	4.9	8,965.5
Rigid plastics											
Polycarbonate	4.06	14.93	3.68	0.3175	1,158			121	0.19	6.8×10^{-5}	93.1/ 2,207
Glass-reinforced polyester (Fiberglas) (premixed chopped glass)	0.94	5.78	6.15	0.3175	1,937			163	0.54	2.9×10^{-5}	179.3/11,034
Polypropylene, black: 30% mineral filled	0.66	2.32	3.51	0.3175	1,107			129	~0.13	2.9×10^{-5}	58.6/ 2,759
Polypropylene, black: 30% glass-coupled	1.26	4.60	3.65	0.3175	1,151			138	~0.13	3.6×10^{-5}	55.2/ 5,862
Polysulfone	8.82	34.72	3.94	0.3175	1,240			150	0.12	5.4×10^{-5}	106.2/ 2,690
Asbestos-filled phenolic					1,700			150	0.35	3.3×10^{-5}	44.8/

Material											
Glass-reinforced vinyl ester	~3.10 (for resin)	10.33	3.33	0.3175	~1,050			>200	1.73	3.1 × 10⁻⁵	202 MPa flexural 8,750 MPa flexural modulus
Thin-film plastics											
Fluorinated ethylene-propylene (Teflon)	13.23	2.84	0.21	0.01 (4 mil)	1,380	−72	107	107	0.25	16.8 × 10⁻⁵	/ 690
Polyvinylidene fluoride (Kynar)	11.03	1.94	0.18	0.01 (4 mil)	1,760	−80	150	135	0.21	11.6 × 10⁻⁵	/ 1,393
Polyester (Mylar)	2.34	0.32	0.14	0.01 (4 mil)	1,386	−70	150		0.15	1.7 × 10⁻⁵	
Polyaryl sulfone (Astrel)					1,360			230			
Polyxylylene (Parylene)					1,120			180			
Polycholorotrifluoroethylene (Aclar)					2,200			200			
Polysulfone (Udel)					1,700			170			
Polycarbonate (Lexan)					1,345			180			
Ethylene-chlorotrifluoro-ethylene (Halar 500)	18.74	2.52	0.13	0.01 (4 mil)				180			
Ethylene-tetrafluoroethylene copolymer (Tefzel 280)	19.85	3.37	0.17	0.01 (4 mil)	1,700			199		8.6 × 10⁻⁵	
Kel-F	44.10	8.82	0.20	0.01 (4 mil)	2,000					3.5 × 10⁻⁵	
Sylgard 184 (Silicone)	19.89	2.09	0.11	0.01 (4 mil)	1,050	−269				4.5 × 10⁻⁵	
Polyimide (Kapton)	99.23	7.04	0.07	0.005 (2 mil)	1,420		400	252	0.16	2.0 × 10⁻⁵	
Rubbers											
Neoprene	1.33	0.16	0.12	0.01 (4 mil)	1,230	−40	120	80–95			
Hydrocarbon rubber (Nordel)	1.08	0.09	0.09	0.01 (4 mil)	860	−50	170	145			
Synthetic rubber (Hypalon) (chlorosulfonated polyethylene elastomer)	1.61	0.19	0.12	0.01 (4 mil)	1,170	−20	150	120–135			
Ethylene/acrylic elastomer (Vemac)	2.46	0.27	0.11	0.01 (4 mil)	1,100	−45		165			
Fluoroelastomer (Viton) (fluorocarbon elastomer)	19.98	3.70	0.19	0.01 (4 mil)	1,850	−20	260	204			
Silicone rubber	5.99	0.90	0.15	0.01 (4 mil)	1,500	−57		315			

Table 25.1 (continued)

Candidate material	Material cost (1982 dollars)			Thickness (10⁻² m)	Density (kg/m³)	Temperature limits (°C)			Thermal conductivity (W/mK)	Coefficient of thermal expansion (m/mK)	Strength[a]
	($/kg)	($/m²)	(kg/m²)			Min	Max	Continuous			
Fabrics/textiles											
Kevlar weave (Kevlar 29)	22.05–77.18	7.23–25.30	0.33	0.075	437 (fabric) 1,440 (raw material)	−46	204	135–150		−2.0 × 10⁻⁵	
Nomex: cloth (style X-630)	26.46–45.20	6.28–10.73	0.24	0.069	346		600		0.035		
Nomex: paper (type 410)	21.09	2.28	0.11	0.0125	866		220	204			
Silicone-coated glass cloth	11.04	5.38	0.49	0.03 (0.012 in.)	1,625	−45	250				
Ceramics											
Concrete	0.02	0.95	47.50	1.905 (0.75 in.)	2,491			540	0.13	1.1 × 10⁻⁵	4.10/
Glass-reinforced concrete	0.15	6.70	44.67	1.905 (0.75 in.)	2,343			400		0.9 × 10⁻⁵	16.20/13,800
Polymer concrete (Polysil[H])	0.11	4.57	41.55	1.905 (0.75 in.)	2,180			93	1.67	2.4 × 10⁻⁵	65.90/
Borosilicate cellular glass (Foamsil-28)	5.58	27.50	4.93	2.54 (1 in.)	194		425		0.07	0.3 × 10⁻⁵	0.76/
Soda-lime silicate cellular glass (Foamglas)	0.95	3.33	3.51	2.54 (1 in.)	138		425		0.07	0.8 × 10⁻⁵	0.55/
Soda-lime gass	0.06	0.47	7.83	0.3175	2,463					0.9 × 10⁻⁵	89.70/
Fusion glass	0.73	5.64	7.73	0.3175	2,497		274	246		0.9 × 10⁻⁵	
Float glass	0.45	3.57	7.93	0.3175	2,497			38			
Porcelain	0.86–8.60	58.16–581.60	67.63	1.905 (0.75 in.)	2,300–3,550			1,000	1.05	0.4 × 10⁻⁵	78.5/ 117.2/
Polymer-cement concrete	0.38	16.23	42.71	1.905 (0.75 in.)	2,242			245		2.4 × 10⁻⁵	8.6/
Polymer concrete (Lone Star)	0.39	17.43	44.69	1.905 (0.75 in.)	2,346			260	12.1–20.3		

a. For metal foils: E = modulus of elasticity; G = modulus of rigidity. For plastics: strength = flexural strength, MOR (MPa)/MOE (MPa).

Table 25.1 lists the materials which might be used as absorber substrates and their properties: metal sheets, metal foils, foamed and rigid plastics, thin-film plastics, rubbers, fabrics/textiles, and ceramics. The table was compiled by G. Jorgensen of SERI and augmented with data from a survey of organic film materials that were considered for solar sail application by the Jet Propulsion Laboratory (1978).

The range and scope of research activities aimed at developing an absorber substrate with optimized properties are shown in figure 25.4. Level 1 research investigates the relations between microstructure and thermomechanical properties for metals, polymers, and ceramics. Level 2 research characterizes the degradation mechanism and kinetics of promising materials identified at level 1 under the effect of heat (thermal degradation), light (photodegradation), or corrosion (chemical degradation). Level 3 research draws upon the results of level 2 to devise approaches to combat degradation. These approaches include varying the chemical composition and microstructure, as well as applying protective coatings or paints. Level 4 research attempts to develop substrates that are optimized in chemical composition, microstructure, and ease of manufacturing. Level 5 research subjects the optimized substrate to an optimization of its manufacturing processes, with particular emphasis on ways to control composition and microstructure, and enhance yield. Level 6 research subjects the substrates to a series of compatibility tests involving the adhesives, the absorber, and the heat transfer fluid. Research at levels 7 and 8—accelerated testing and field testing—is aimed at providing the data base for absorber substrate comparison and selection. Actual absorber substrate research is, however, very limited. This is because most solar collectors use metals as absorber substrates, and metal properties and capabilities are well known. A few of the absorber substrate research areas that have been active are highlighted below.

25.5.1 Development of Polymeric Absorber Substrates

Only a few polymeric collectors are available for medium temperature [120°C (248°F)] applications (e.g., Tem Tech Energy Systems TES6000 Series). None are available at higher temperatures. Polymers, however, can sustain high temperatures. The fluorocarbons are stable up to 280°C (536°F), the polyesters to 176°C (348.8°F); and the silicon-based polymers to 280°C (536°F). Through copolymerization and alloying, it is possible to tailor a polymer for a given temperature application in the 130° to 300°C (266°–572°F) range (Hong 1982). An all-polymer solar collector was designed at Brookhaven National Laboratory (Wilhelm 1982). In this collector the absorber substrate was a laminate of two outer aluminum foils and two inner polymer films. The foils and films

R&D level	Definition				
8	Field testing		Development of standard testing procedure	Design and acquisition of field-testing data base	Collection of data from third-party systems
7	Accelerated testing	Design and acquisition of screening-test data base	Development of service-life prediction model	Quantitative characterization of service environment	Development of standard accelerated testing procedures
6	Compatibility testing		With candidate adhesives specifically on • Thermal expansion • Peel strength • Combined conductivity	With candidate absorbers in • Direct deposition • With adhesives	With candidate heat transfer fluids (corrosion compatibility testing)
5	Manufacturing optimization		Control of compositional fluctuation during manufacturing processes	Control of microstructural deviation during manufacturing	Design of measures to enhance yield

4 Laboratory development of optimized substrate	Development of optimum chemical composition	Development of optimum physical microstructure	Development of optimum processing methodology: polymerization steps, processing steps, etc.
3 Investigation of approaches to combat degradation	Protection through change in chemical composition	Protection through change in physical microstructure	Protection by other means: coatings, paints, total system design
2 Characterization study of substrate degradation (mechanism and kinetics study)	Chemical degradation (corrosion)	Photodegradation	Thermal degradation
1 Basic study to correlate chemicophysical microstructure to thermomechanical properties	Metals	Polymers	Ceramics

Figure 25.4
Absorber substrate research alternatives.

were laminated together at intervals along their length so as to form flow channels for the heat transfer fluid in a high speed, low-cost continuous-roll manufacturing process. The other components of the collector were a light metal frame, a glazing film, and a rigid back that also acts as insulation. Both the absorber substrate and glazing were stretched over the frame to give additional structural strength to the collector.

25.5.2   Study of Absorber Substrate–Selective Absorber Interactions

Work at the University of Houston showed that the substrate can play a key role in absorber surface degradation, through thermal activation of oxidation and diffusion processes (Bacon 1982; Ignatiev, Zajac, and Smith 1982; Smith, Zajac, and Ignatiev 1982). Five substrate materials for black chrome were studied. It was shown that 304 SS enhances the optical durability of the black chrome absorber surface, whereas Cu, Au, Ni, and Cr contribute to its degradation.

25.5.3   Ceramic Absorber Substrates

Glass-reinforced concrete (GRC) and polymer bonded silica (Polysil) have been considered as absorber substrate materials. Evaluation of a GRC being considered for solar collector trough application showed that, when un-stiffened, flat areas of GRC are prone to warping and curvature, arising from differential shrinkage during drying (Slemmons et al. 1981). Although no undesirable effects were produced by thermal cycling between $-17°C$ ($1.4°F$) and $+9°C$ ($33.8°F$), thermal cracking at wider temperature ranges should not be excluded because the thermal expansion coefficient usually increases with temperature (Kingery, Bowen, and Uhlmann 1976). The thermal stability of Polysil at high temperature has also not been determined, though its heat capacity and coefficient of thermal expansion are known (Gunasekaran and Grelika 1977).

25.5.4   Prospects for the Future

It is anticipated that the solar collector industry will continue in the near future to rely predominantly on metal substrates. However, some polymeric substrate materials have already appeared on the market, and the research on polymeric and ceramic absorber substrates promises to provide a broader range of future options.

25.6   Adhesives

Adhesives can be used for three different purposes in a solar collector: to bond an absorber foil to the metal substrate, headers, or tubes; to bond the layers of laminated glazings; and to bond laminated polymer substrates. Acurex Company's all polymer Low Cost Solar Panel (LCSP) is an adhesive-bonded laminate that consists of an outer layer of UV-stabilized polyester (PET), a middle layer of HYTREL (a du Pont polyester elastomer), and an inner layer of PET. The selling price of this LCSP was estimated to be below $21.50/m^2 ($2/ft^2) (Nelson and Muller 1980). Another experimental design that also depends on adhesive for absorber structural integrity is the Brookhaven National Laboratory's thin-film, flat plate collector, which has an even lower projected manufacturing cost: $12/m^2 ($1.20/ft^2) (Wilhelm 1982)

The critical requirement for adhesives is temperature stability: Excluding swimming pool collectors, flat plate absorbers operate at 50° to 120°C (122°–248°F) and stagnate at 150° to 200°C (300°–392°F). Glazings operate and stagnate at slightly below half these temperatures. Evacuated tube absorbers operate at up to 250°C (483°F), stagnate at up to 350°C (662°F), and are extremely unlikely to use absorber foils. If polymeric collectors, selective absorber foils, and laminated glazings are to have low life-cycle costs, the bond between polymer layers, or foil and polymer layers, must be permanent.

Figure 25.5 shows the range and scope of research activities aimed at developing an adhesive with optimized properties for solar applications. At level 1 the bonding mechanisms of interest are those between polymers and polymers, polymers and metals, and metals and metals. The results of level 1 research are a deeper understanding of the mechanisms of adhesive bonding and an identification of adhesives, or classes of adhesives, that appear promising for solar collector application. Level 2 research investigates the degradation mechanisms operating in these promising adhesives as they are exposed to conditions similar to those encountered in operating solar collectors. Such conditions include operating and stagnation temperatures up to 350°C (662°F); thermal cycling and shock; exposure to H_2O, O_2, and O_3; mechanical stresses from substrate thermal expansion; exposure to sunlight; and any combination of these factors. Using the understanding of adhesive degradation mechanisms gained in level 2 research, level 3 research attempts to design adhesives that are optimized as regards durability, compatibility, and ease of application. Accelerated testing and field testing of these optimized adhesives (research at levels 4 and 5) then follow. The more important adhesive research activities are highlighted below.

25.6.1 Degradation Mechanisms of Adhesives under Stress

Research in this area aims at characterizing the behavior of the adhesives when subjected to sunlight and thermal, chemical, and mechanical stresses. Concerning thermal stresses, the Jet Propulsion Laboratory (1978), in connection

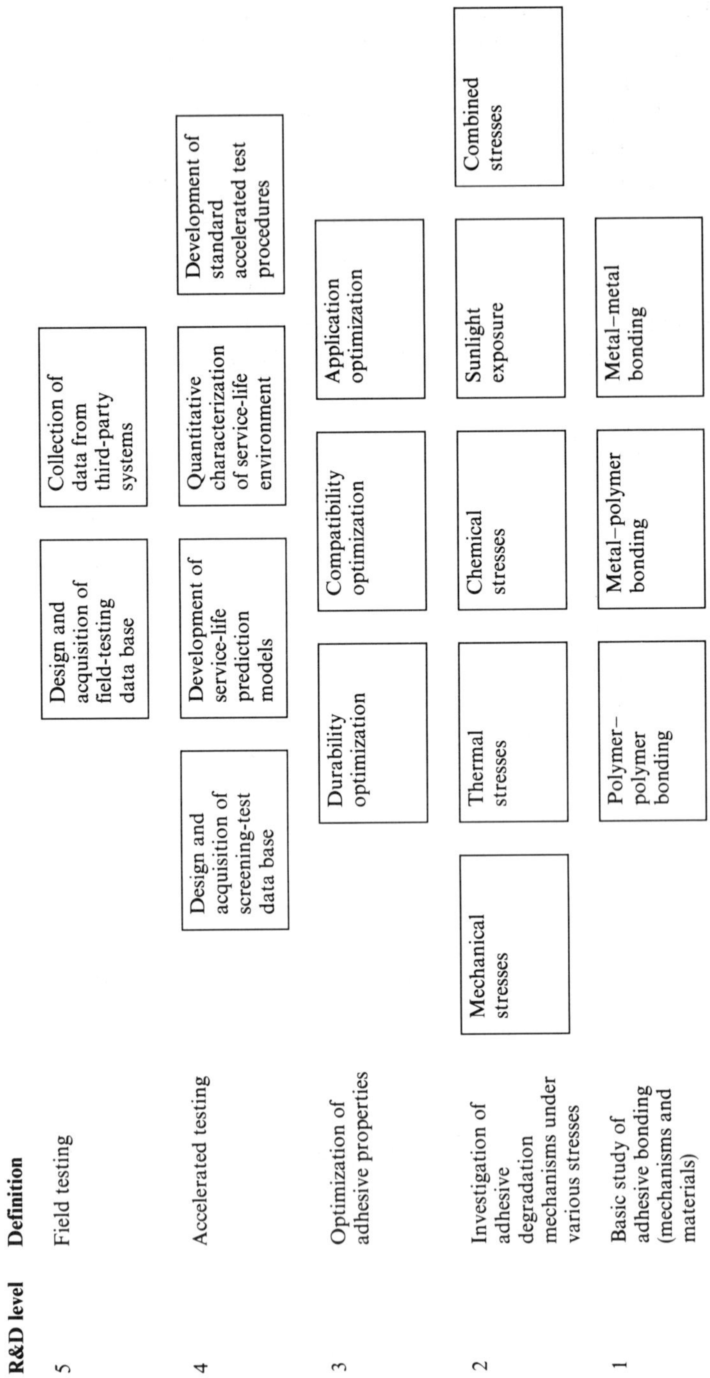

R&D level   **Definition**

5   Field testing

     Design and acquisition of field-testing data base

     Collection of data from third-party systems

4   Accelerated testing

     Design and acquisition of screening-test data base

     Development of service-life prediction models

     Quantitative characterization of service-life environment

     Development of standard accelerated test procedures

3   Optimization of adhesive properties

     Durability optimization   Compatibility optimization   Application optimization

2   Investigation of adhesive degradation mechanisms under various stresses

     Mechanical stresses   Thermal stresses   Chemical stresses   Sunlight exposure   Combined stresses

1   Basic study of adhesive bonding (mechanisms and materials)

     Polymer–polymer bonding   Metal–polymer bonding   Metal–metal bonding

Figure 25.5
Adhesive research alternatives.

Table 25.2
High temperature adhesives evaluated by Jet Propulsion Laboratory

Adhesive designation	Type	Manufacturer	Comments
NR 150A2G	Polyimide	du Pont	Good for use at 250°C $T_s = 280°-300°C$
NR 150B2G	Polyimide	du Pont	Good for use at 310°C $T_g = 350°-375°C$
NR 150 B252X and NR-055X	Polymide	du Pont	Experimental resins, no advantage over NR 150B2G
2080B	Polyimide	Upjohn	Poor handling properties, precipitation of resin on exposure to air
TR 150-25	Polyimide-amide	Thermo-Resist	Availability uncertain
TR 800-25	Polyimide-amide	Thermo-Resist	Availability uncertain
PPQ 401	Polyphenylquinoxaline	Acurex	Unsatisfactory bond strength
XYLOK 210	Aralkylether + phenol	Ciba-Geigy	Inadequate thermal stability embrittlement on aging at 310°C
Sheldahl 3P	Polyimide-amide + polyester	Sheldahl	Inadequate thermal stability bond failure on aging 24 h at 310°C

Source: Jet Propulsion Laboratory, 1978, "Solar Film Materials and Supporting Structures for a Solar Sail—A Preliminary Design," vol. 4, JPL Document 720-9, October.

with its Solar Sail Project, surveyed polyimides and polyphenylquinoxaline and phenolic aralkyl ethers as potential high temperature adhesives. Major requirements on the adhesives were no failure in bond after curing and intact bond after aging for 24 hours at 315°C (599°F). The results of JPL's evaluations are shown in table 25.2. In terms of overall properties such as lap shear strength, ease of bonding, and thermal stability between 130°C (266°F) and more than 350°C (662°F), du Pont's adhesive, NR150B2G, was found to be best for solar application. Du Pont discontinued BR150 adhesives production; however, an equivalent polyimide is available from the Ciba-Geigy Corporation This adhesive, XU218, is thermally stable up to 350°C (662°F) due to its predominantly aromatic structure and the absence of benzylic hydrogens in the cycloaliphatic portion of the molecule, and adjacent carbon-bearing hydrogens that might otherwise provide a low energy oxidative degradation pathway (Bateman, Gersey, and Neiditch 1975). It is also soluble in many common organic solvents and has excellent chemical resistance.

In order to achieve adhesive bonding of a thin glass mirror to a parabolic steel trough, the Acurex Solar Corporation, in an approach similar to JPL's, screened eight epoxies, one acrylic, one anaerobic, and three urethanes, for

Table 25.3
Shear strengths and processing data for sample adhesives

Adhesive	Shear strength (psi)		Pot life (min)	Viscosity (cps)	Cure time[a] (h)	Handling time[b] (h)	Remarks[b]
	At room temperature	at 180°F					
Epoxies							
Armstrong A-271	2,800	2,500	60	14,000	400	20	Cures in 30 min at 200°F
Epibond 1210/9861	2,800	2,600	35	thick	48	6	Bonds well to unprimed surfaces
Epon 828/Versamid 140	2,350	1,500	90	25,000	24	4	Requires very clean adherend surfaces
Key 93-88	1,000	500	20	paste	24	4	Appears to contain polysulfide
Mor-Ad L-142	3,000	2,000	10	9,000	0.5 at 180°F	—	Exotherms
Newport NB-74	2,300	500	90	20,000	24	6	Cures in 40 min at 200°F
Cyanamid FM 73	5,360	3,940	—	—	1.0 at 250°F and 40 psi	—	Film adhesive on polyester carrier; store below 40°F
McCann MA 229	5,000	3,500	—	—	1.0 at 200°F and 20 psi	—	Film adhesive on scrim or mat carrier; store at 0°F; autoclave and/or vacuum bag
Urethanes							
Fiber Res in FR-8503	1,100	460	20	25,000	50	6	Postcure at 140°C tends to foam in moist air
Pliogrip (modified 6000)	1,400	500	10	2,500	24	0.5	Needs priming
Uralane 5757	2,000	700	20	2,500	200	6	Postcure at 200°F
Anaerobic							
Loctite 312	4,000	—	—	1,000	24	0.5	Will not cure until components are assembled
Acrylic							
MacTac IB-2100	<1,000	—	—	—	—	—	Pressure sensitive adhesive on polyester film carrier; difficult to adjust assembly once mirror is placed on adhesive

Source: Jet Propulsion Laboratory, 1979, "Solar Film Materials and Supporting Structures for a Solar Sail—A Preliminary Design," vol. 4, JPL Document 720-9, October; Hornberger, L. E., and J. L. Hull, 1982, "Adhesive Bonding of Thin Glass Mirror to Parabolic Steel Trough," *Optical Coating for Energy Efficiency and Solar Applications*, SPIE, vol. 324, pp. 86–90.
Note: Conversion factor: °C = (°F − 32)/1.8.
a. Room temperature cure time, unless noted.
b. Samples were cured in a vacuum bag to apply pressure and estimate voids, unless noted.

thermal durability at up to 800°C (180°F) (Hornberger and Hull 1982). The results were inconclusive, mainly because the company wanted to look at every property and not just thermal stability. The preliminary results are shown in table 25.3. The liquid epoxies and liquid urethane adhesives show good promise for solar application; the first because of their superior properties and the second because of rapid processing. Acurex also identified a urethane-based adhesive with a cross-linking cure agent as best suited for its Low Cost Solar Panel (Nelson and Muller 1980). This adhesive has high peel strength and high thermal durability and is compatible with HYTREL, as well as high speed lamination equipment.

It has been reported that a silicon-based pressure-sensitive adhesive, DC282 (produced by Dow Corning), has exhibited excellent thermal stability as evidenced by the absence of edge delamination after 200 days at 200°C (392°F) on aluminum and mild steel substrates. It also has a peel strength above 200 g/cm for the test duration (Mason and Brendel 1982). Even better results were reported for a silicone with heat stabilization, SR6574 (produced by General Electric Silicones), and an acrylic-based silicone, 9460 ISOTAC Tape (produced by 3M United Kingdom Ltd.). It is estimated that a peel strength of 200 g/cm and the ability to withstand 200°C (392°F) for 200 days would be equivalent to a real-lifetime of more than 20 years in a flat plate collector environment.

25.6.2 Optimization of Adhesive Properties for Solar Applications

Application optimization aims to produce an effective adhesive having maximum ease of application. This can best be done through use of structure-property relationships (Hoffman, Hammond, and Althouse 1981). For example, the prepolymer and curing agents control pot life and working time (Richter and Macosko 1973; Lipshitz and Macosko 1976), while the concentration and chemical structure of the hard and soft segments control the modulus and elastic properties of the adhesive (Cooper and Estes 1979; Hoffman 1980).

Compatibility optimization aims at making the adhesive usable with a wide range of materials. The best compatibility is possible when the adhesive and the laminates to be bonded together are made from the same base materials, as was the case with the JPL solar sail where both the sail film, KAPTON, and the adhesive were polyimide (Jet Propulsion Laboratory 1978).

Durability optimization aims to maximize adhesive life while preserving good adhesion. Work done under DOE's Low-Cost Array Project/Chemical Bonding Technology for a Terrestrial Solar Cell Module has identified chemical polar function groups that, when added to polymeric adhesives such as modified styrene polymers, silane, or ethylene vinyl acetate, substantially increase adhesion durability (Plueddemann 1979). These groups include trimethoxysilylpropylamine and trimethoxysilane.

25.6.3   Prospects for the Future

Although adhesive R&D is probably not as crucial to the development of solar collectors as that on glazings or selective absorbers, such research is needed to determine the suitability of the promising adhesives for collector applications. The principal issue is investigation of the degradation processes (thermal, mechanical, chemical, and photo) occurring in these adhesives when they are exposed to the collector environment. In addition the future should see new or improved adhesives for solar collector use based on such developments as chemical and physical modifications and adhesive tailoring for polymeric laminates. Basic research on the nature of adhesive bonding is not critical to this effort because of the vast amount of knowledge that already exists with respect to adhesives.

25.7   Heat Transfer Fluids and Corrosion

Field evidence indicates that corrosion is a serious problem in a large fraction of commercial solar heating and cooling systems (USDOE 1981; Herzenberg, Silberglitt, and La Porta 1981). Most of the reported problems have resulted from poor materials selection, flawed design, or improper installation, rather than fundamental materials limitations. Design research and information transfer activities are therefore the most direct approaches to eliminating corrosion of metal absorbers and transport piping. Nonetheless, there are a variety of materials research alternatives that could contribute to making conventional solar system designs more resistant to corrosion and more immune to the mistakes of inexperienced designers and installers.

The corrosiveness of an active solar system is affected by both the heat transfer fluid and piping material properties. Inert, nonelectrolytic fluids, such as silicone and hydrocarbon oils, generally ensure that both uniform and localized corrosion rates are slow. However, if a transfer oil system is water-pressure-tested and not dried properly, localized corrosion may occur at places where residual water collects (Avery and Krall 1981). Polar fluids, such as water and the two glycols, promote corrosion and must be protected by inhibitors, scavengers, or buffers, unless they are extremely pure and are used in combination with corrosion-resistant metals. The common transport metals vary substantially in their vulnerability to galvanic, pitting, and crevice corrosion. Copper and stainless steel (SS) are relatively resistant, and aluminum and carbon steel progressively less so.

Fluid research alternatives include basic study of different types of corrosion, applied exploration of corrosion in particular solar systems, examination of corrosion prevention strategies (including the search for a noncorrosive fluid with heat transfer properties similar to those of water), exploration of corrosion detection strategies, and field testing to confirm the expected durability of fluid-transport-piping material combinations. The stages of research, and the alternative research activities at each stage, are represented in figure 25.6.

Basic study of corrosion (level 1) provides the knowledge of basic chemistry necessary to understand the basic mechanisms of different types of corrosion and the conditions that will accelerate each type.

Study of corrosion mechanisms (level 2) in solar systems with different transport loop materials, fluids, and designs isolates the specific types of corrosion to which different systems are vulnerable. Ideally this work would establish corrosion mechanisms and reaction rates as a function of impurity ion levels, pH, dissolved oxygen, and flow rate. However, since localized corrosion rates are a sensitive function of so many parameters, including piping geometry and other system-specific factors, only broad estimates of absolute corrosion rates can generally be derived. More definitive conclusions can usually be drawn about relative corrosion rates in different systems.

Corrosion prevention research (level 3) isolates the most promising materials approaches to preventing or delaying corrosion. These approaches include the incorporation of additives into a fluid, the development of protective coatings for transport piping, improvement of the corrosion resistance of piping materials, and synthesis of a noncorrosive fluid with good heat transfer properties.

Corrosion detection research (level 4) identifies fluid properties (e.g., pH, and conductivity), or other phenomena (e.g., the corroding of a thin wire filament) that correlate with corrosion rates or the onset of serious degradation, and develops devices that track these characteristics and warn operators of the need to begin preventive maintenance.

Field testing (level 5) exposes optimized piping-fluid-additive-design combinations in operating collectors, verifies their durability, and suggests areas for additional research at more basic levels.

25.7.1 Metallic Corrosion Research of Relevance to Solar Collectors

All corrosion reactions involve the oxidation of a metal, that is, an increase in its valence charge, and the consumption of electrons by a reducing agent.

R&D level	Definition				
V	Field testing	Monitoring of simulated field systems	Collection of data from privately owned systems	Collection of data from manufacturers	
IV	Corrosion detection research	Corrosion early warning devices	Fluid conductivity measurement		
III	Combating corrosion	Protective coatings	Inhibitors	Buffers	Scavengers
II	Study of corrosion with different metal, fluid, and design combinations	Aluminum systems	Stainless steel systems	Copper systems	Carbon steel systems

Development of a noncorrosive fluid with the heat transfer properties of water

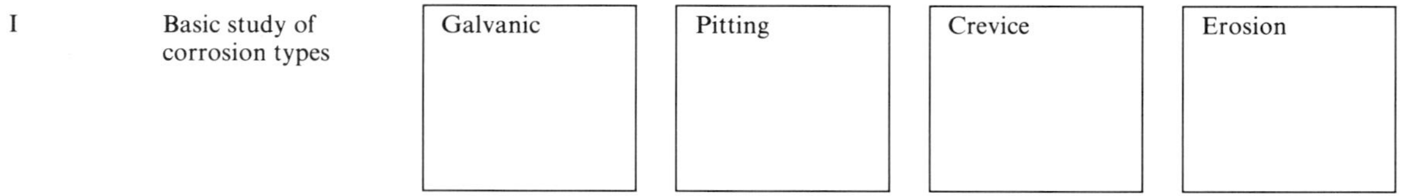

| I | Basic study of corrosion types | Galvanic | Pitting | Crevice | Erosion |

Figure 25.6
Heat transfer fluid and corrosion research alternatives. For the sake of completeness, level II should be represented as a 4 × 3 × 3 × 4 matrix (i.e., there are four metals, three types of fluids, three kinds of design—drain-down, drain-back, and closed-loop—and four relevant types of corrosion—uniform, pitting, crevice, and erosion). With the restriction that glycols and non-aqueous fluids are not used in drain-down and drain-back designs, research could reasonably be performed on any permutation of these four variables. But it is easier to illustrate just one dimension of this matrix than to show all possible choices.

In most cases the oxidized metal ion will then form a precipitate oxide, such as $Al(OH)_2$.

Three different types of localized corrosion predominate in solar collectors. Uniform corrosion is not generally of serious concern since uniform penetration rates will generally be less than 1 mpy (mil per year). Typical solar collector piping wall thicknesses vary from 20 to 36 mils.

In localized corrosion, attack is limited to a particular area, and thus the possibility of premature piping penetration is greatly increased. Galvanic corrosion occurs near the junction between dissimilar and electrically connected (either directly or through an electrolyte) metals. Pitting corrosion is a localized "pin-hole" attack that occurs at stagnant sites in a fluid transport system or at flaws or defects in passivating oxide films. Crevice corrosion occurs in any restricted area where small stagnant pools and an occluded cell can form. Erosion corrosion results when high velocity fluid strips the protective oxide film from a metal, exposing the underlying layer to oxidation. (Erosion corrosion should not be a major problem at the low fluid velocities typical of solar systems, although copper and SS depend on passivating oxide films for protection and are susceptible to it.)

The general results of the solar corrosion research conducted for various materials-fluid combinations are summarized in table 25.4. Stainless steel, especially type 444, has excellent corrosion resistance and could probably survive 20 years without inhibitors in draw-down, drain-back, and glycol

Table 25.4
Corrosion resistance of alternate fluid-metal design combinations

Metal	Fluid			
	Water: open-loop design	Water: closed-loop design	Glycol: closed-loop design	Heat transfer oil: closed-loop design[a]
Uniform corrosion				
Stainless steel	5	5	5	5
Copper	4	5	4	5
Aluminum	3[b]	3[b]	3[b]	4
Carbon steel	1	2	2	4
Localized corrosion				
Stainless steel	5	5	5	5
Copper	5	5	4	5
Aluminum	2[b]	2[b]	2[b]	4
Carbon steel	—	—	—	—

Note: Corrosion resistance is weighted on a five-point scale, with five indicating the greatest corrosion resistance.
a. More research needs to be done with heat transfer oil to corroborate these evaluations.
b. Considerable research has been done with aluminum to corroborate these evaluations.

systems. Copper is also relatively corrosion resistant, especially to localized attack, although it is unlikely to last 20 years without the addition of carefully selected inhibitors in glycol systems, or open-loop water designs. Aluminum can probably last at least 10 years, even in open designs, if the entire solar loop is aluminum, the concentration of heavy metal and chloride ions is low, and carefully selected additives are included in the water or antifreeze. Aluminum system contamination with copper or SS components (heat exchangers, valves, filters, etc.) or failure to monitor additive concentration periodically will cause catastrophic corrosion. Mild steel has corroded uniformly and rapidly in the laboratory tests in which it has been exposed, although it has survived a seven-year field test in a glycol solution.

Stainless steel has the best track record of the metal alloys exposed to simulated solar collector conditions. Simulated drain-back tests at Olin (Smith and Davis 1981) and tests at two other laboratories (Tabony 1977; Diegle, Beavers, and Clifford 1980) have induced no localized SS corrosion and uniform corrosion rates of less than 0.05 mpy. Even vaporized high temperature glycol exposure and thermal cycling have produced very low corrosion rates. However, since it is known that SS can be susceptible to pitting attack (Fontana and Greene 1978), more extended testing is needed to confirm its resistance to localized attack in the presence of heavy ion concentrations. Such research is especially relevant to the development of a thin SS absorber with absorber tubes designed to withstand repeated freezing.

DOE and the Industrial Corrosion Research Association (INCRA) have supported extensive corrosion testing of copper (Smith and Castillo 1981). After six months of testing in simulated low temperatures up to 34°C (93°F) and drain-back environments, copper 122 had suffered no localized attack and had corroded uniformly at a rate asymptotically approaching less than 0.1 mpy in water. Corrosion in simulated low temperature glycol systems reached rates of just over 1 mpy after six months. Thirty-day tests in vaporized high temperature ethylene glycol environments resulted in corrosion rates of approximately 13 mpy at 163°C (325°F) and about 20 mpy at 204°C (399°F). The highest temperature corrosion rate, however, appeared to be logarithmic, suggesting that the 20 mpy figure may be an overestimate. Longer-term tests are needed to settle this issue and to determine whether or not the protective oxide film that forms on copper can provide extended protection against glycols at high temperature.

Pitting corrosion is the largest threat to aluminum alloys, except in the most severe conditions (e.g., with Fe^{3+}, Cu^{2+}, and Cl^- ions present), where uniform attack prevails (Wong and Cocks 1980; Smith and Castillo 1981). Aluminum appears resistant to pitting in pure glycols. The maximum estimated penetration in 20 years in a 35 ethylene glycol is 20 mils (Wong, Swette, and Cocks 1979). However, the presence of any one of the three ions listed above increases glycol pitting corrosion to unacceptable levels.

Carbon steel has proved vulnerable to high uniform attack rates in water and glycol solutions that average corrosion rates in the range 40 to 200 mpy in water, propylene, and ethylene glycol (Tabony 1977; Diegle, Beavers, and Clifford 1980). High temperature tests induced carbon steel corrosion rates of 2.9 to 25.5 mpy (Smith and Davis 1981).

Although hydrocarbon and silicone oils do not themselves promote corrosion, residual water can cause localized attack in nonaqueous systems. The precautions that must be taken to prevent corrosion in active transport loops filled with heat transfer oils were reviewed (Avery and Krall 1981).

25.7.2 Corrosion Prevention Research of Relevance to Solar Collectors

There are a number of approaches to preventing corrosion. Corrosion inhibitors interfere with the oxidation or reduction half-reaction. Buffers neutralize acid decomposition products to maintain the neutrality or alkalinity of a solution. Scavengers both cathodically protect the metal and pick up damaging heavy metal and chloride ions. Protective coatings of a nonconducting film reduce the overall rate of both oxidation and reduction reactions. One might also search for an inert fluid with better heat transfer properties than silicone or hydrocarbon oils.

Since SS appears sufficiently resistant to corrosion with additives, scavengers, or coatings, we will discuss only copper and aluminum corrosion prevention research. As noted earlier, copper, without additives, has sufficient corrosion resistance for well-designed systems. Medium temperature tests have indicated that a range of commercial inhibitors reduce corrosion rates in water and the two glycols to well below 1 mpy (Smith and Castillo 1981). Higher temperature tests are needed to confirm the 20-year durability of these inhibitors.

Research at Giner, Inc., in Massachusetts, has examined a number of corrosion inhibitors. They have also developed an optimized formulation containing phosphates, borates, nitrates, silicates, and sodium mercaptobenzothiazole. However, subsequent testing showed even this latter inhibitor to be effective only up to 130°C (266°F); at 140°C (284°F) it protects aluminum for about 4,000 to 5,000 hours, after which uniform corrosion rates top 1 mpy, and pitting becomes evident (Wong and Cocks 1980; Smith and Castillo 1981). At 180°C (350°F) to 190°C (284°F) the inhibitor loses its effectiveness in about 3,000 hours. The addition of zinc powder substantially increases the corrosion resistance of aluminum (Wong and Cocks 1980). The zinc cathodically protects the aluminum and picks up heavy metals in solution. The maximum expected 20-year pit depth with a Zn/impurity ion ratio of 100 is 11 mils and 4 mils, respectively, for Al-1100 and A1-3003 at 130°C (266°F). Maintenance of the inhibitor and zinc powder concentration would require considerable diligence, but the approach appears quite promising.

There has been considerable interest in the development of low-cost corrosion monitoring or detection devices that might warn an operator of the need for preventive maintenance. A number of possibilities have been suggested by experts in the field: immersing a thin filament of the most vulnerable transport piping material in the transport fluid so that it will turn on a warning light upon corroding and breaking; immersing a thin wire in the fluid and measuring its change in resistance as it corrodes and its cross-sectional area decreases; or building a corrosion rate measurement device employing linear polarization techniques. Rockwell International (Mansfield 1982) has evaluated a number of corrosion monitoring devices for application to flat plate solar collectors and has recommended use of a thin-film pitting fuse.

25.7.3 Prospects for the Future

Corrosion and other fluid-related research indicates that some significant gaps in knowledge remain. However, most of what needs to be known to protect the common solar system metals and designs against corrosion is already available to the diligent designer. Certainly, more precise specification of the ideal fluid, inhibitor, and scavenger for different metals and designs would be desirable, since it would reduce monitoring and maintenance requirements. However, information transfer seems the most immediate need. In particular, the importance of designing solar systems to exclude oxygen from the environment has not been fully appreciated by solar designers. The exclusion of oxygen, plus the application of existing technology for the prevention and monitoring of corrosion (e.g., from automotive and industrial applications) to solar systems, should be sufficient to eliminate corrosion as a serious problem.

25.8 Thermal Storage Materials

Due to the intermittent nature of solar energy, solar systems must include a thermal storage subsystem if they are to supply energy at night or during cloudy periods. At present, most designs store collected heat in either water storage tanks or, in the case of air systems, in rock beds. However, a range of alternative thermal storage materials has been investigated that would require storage columns less than a tenth the size of water or rock storage systems with equal capacities. These systems could reduce installation and capital costs of the storage subsystem. If they could be made cheap enough, they might also make a larger storage capacity economical, thereby making increases possible in energy collected per unit area of collector. Storage systems with the ability to operate in specific temperature ranges could also improve system efficiency by allowing collectors to operate in highly efficient temperature ranges. A workshop sponsored by the Electric Power Research Institute (EPRI) reviewed approaches to thermal energy storage (Hausz and Berkowitz 1983).

All advanced thermal storage systems rely on a phase or chemical transition to store energy. The energy stored in the transition leads to the superior storage capacity (per unit volume and weight) of these approaches as compared to sensible-heat storage, which is hoped to lead ultimately to lower cost. Commonly the transition is a solid–liquid phase change, as in the case of salt hydrates (Telkes 1980a, b). However, solid–solid phase transitions (Benson, Burrows, and Webb 1986), thermochemical reactions (Mar 1980a, b; McBride 1981; Nonnemacher and Groll 1981), and photochemical reactions (Hautala, King, and Kutal 1979) have also been considered for solar collector applications. The aim of research on advanced storage materials is to prove their ability to cycle energy into and out of storage, efficiently and repeatedly, at a reasonable cost and with a long lifetime.

Research to improve solar energy storage materials includes the following areas: basic study of the phase transitions or reactions in alternative types of materials, thermodynamic studies of transitions and heat transfer in promising materials, study of storage material degradation, exploration of approaches to overcoming degradation, consideration of manufacturing and packing issues, and accelerated testing and field testing. Figure 25.7 indicates specific research alternatives at each of these levels.

25.8.1 Research on Heat of Fusion Materials

Widely varying amounts of research have been conducted on alternative solar energy storage materials. Heat of fusion materials, particularly the salt hydrates, have been explored most completely. Upon melting, hydrates of the form $AB \cdot nH_2O$ change to another hydrate, $AB \cdot mH_2O$, and some or all of the latter molecule dissolves in $(n - m)$ moles of water.

A review of the empirical literature on salt hydrate materials (and selected hydrates of bromides and iodides) and the efforts made to correlate heat of fusion with fundamental thermodynamic parameters (Telkes 1980a, b) proposes a method of predicting the entropy and heat of fusion (S_f) of salt compounds from the entropies of fusion of their component elements. Calculated values agree with observed values by $\pm 30\%$, and often $\pm 10\%$. Given the vast number of alternative salt compounds, these theoretical estimates of heats of transition are a useful tool for identifying promising materials without expensive laboratory testing. Of course cost considerations are also crucial and may rule out many of these materials for practical applications.

25.8.2 R&D on Solid–Solid Phase Change Materials (PCM)

A number of solid–solid (S–S) phase change materials have been investigated. Order–order transformations account for the S–S transformation in simple plastic materials. However, these transformations do not account for the total specific heat in compounds with large enthalpies of transition. It has been demonstrated that hydrogen bonds between nearest neighbors in the crystal phase are reversibly broken at the transition temperature in certain hydrocarbon molecular crystals (Benson, Burrows, and Webb 1986). One extremely interesting property of these S–S PCM's is that phase change materials of the pentaerythritol series can be alloyed together to produce materials with lower transition temperatures. The potential advantages of these materials (besides their high heats of fusion and the ability to modify transition temperature to match the application) include their possible resistance to phase separation and their compatibility with containment materials. In addition the lower transition temperatures may make them useful for both passive and active solar applications (Benson, Burrows, and Shinton 1986; Benson, Burrows, and Webb 1986; Benson and Burrows 1987; Christensen and Benson 1986).

25.8.3 R&D on Photo-Excitable Isomers

Reversible thermochemical reactions (Mar 1980a, b; McBride 1981; Nonnemacher and Groll 1981) and molecules that can be photochemically excited to high energy isomers (Hautala, King, and Kutal 1979) (and the energy extracted during the exothermic reverse reaction) have also been considered for active solar energy storage. Photochemical isomers can only be applied in complex, integrated photochemical storage (and collection) evacuated-tube collectors because they absorb only in the short wavelength end of the solar spectrum (Hautala, King, and Kutal 1979).

25.8.4 Other Research Areas

A large number of screening studies have been conducted on candidate solar energy storage materials, as described in a review of university and commercial activity in this area (Michaels 1982). Table 25.5 contains the data in this review

R&D level	Definition					
7	Field testing		Development of a field test data base	Collection of data from privately owned systems		
6	Accelerated testing	Cycling tests	Prototype system efficiency measurements	Container durability testing	Development of testing procedures	
5	Manufacturing and packaging	Packaging and efficiency	Packaging and phase separation	Packaging and nucleation	Container-storage material compatibility	
4	Approaches to extending life of candidate materials	Nucleation phenomena	Thickness and phase separation	Mechanical mixing	PCM pelletization	Formation of cross-linked hydrogels

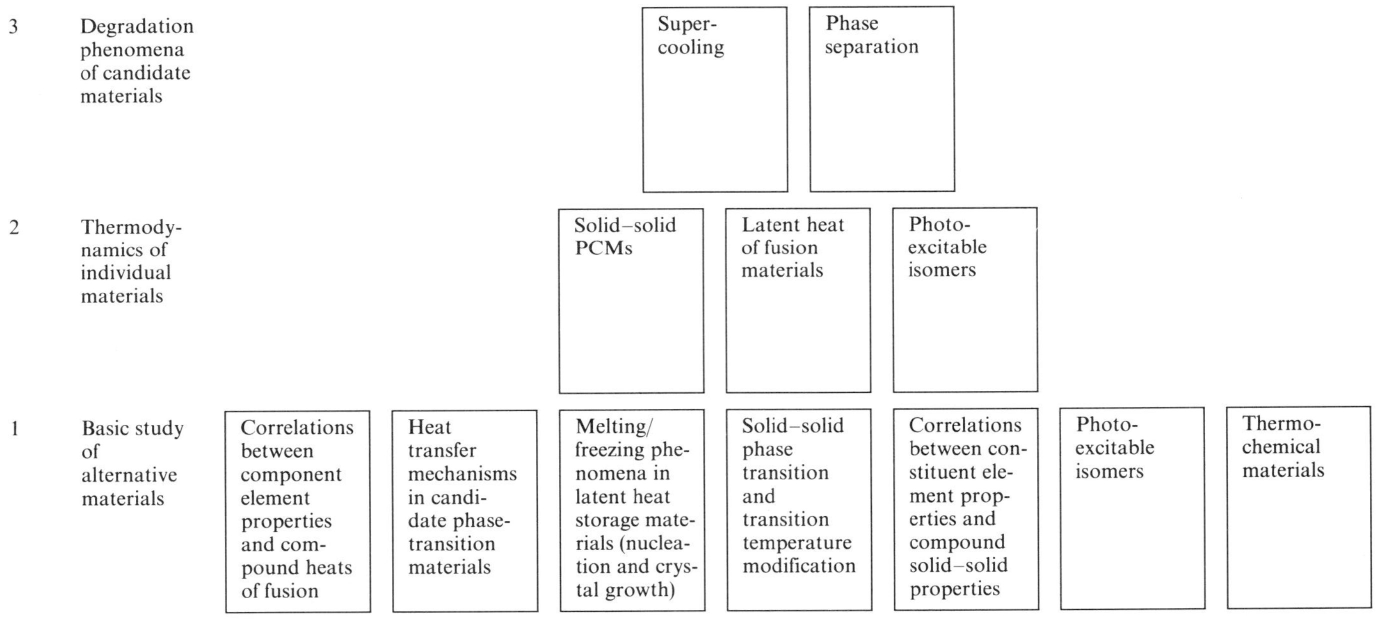

Figure 25.7
Thermal storage research alternatives.

Table 25.5
Properties of representative active solar thermal storage materials

Liquid–solid phase change material	Trade name	Manufacturer (or developer)	Transition temperature [°C (°F)]	Heat of transition (10^3 J/kg)	Storage material or system cost (if available)
Na_2SO_4–$10H_2O$ with Borax plus a thickening agent	(Not yet commercial)	University of Delaware	13 (55)	93	$2.45[a] (estimate)
Cross-linked HDPE pellets	(Not yet commercial)	University of Dayton Monsanto	130 (766)	170–190	$7.27[a] $13.00 (estimate)
$Na_2S_2O_3$–$5H_2O$ (excess water)	(Not yet commercial)	Effex Innovation A/5, Denmark	48 (115)	101	NA
"Ammonium Alum"	(Not yet commercial)	J. Kai et al.	94 (201)	250	$1.04[b]
Modified $CaCl_2$–$6H_2O$	TESC-81	DOW Chemical	27 (81)	191	NA
Eutectic Mixture of $Mg(NO_3)6H_2O/Mg$ Cl_2–$6H_2O$	TESC-135	DOW Chemical	57 (135)	135	$4.07[b]
$Mg(NO_3)_2$–$6H_2O$	TESC-190	DOW Chemical	89 (190)	163	$3.37[b]
$MgCl_2$–$6H_2O$	TESC-240 (not yet available)	DOW Chemical	117 (240)	167	$3.30[b]
$Na_2S_2O_3$–$5H_2O$ (excess water)	Heat Bank 115	CALMAC	45 (113)	169	$2.60[b]
Modified Na_2SO_4–$10H_2O$	Boardman tube	Boardman Energy Systems	31 (89)	222 (weight may not include tube weight)	NA
Modified Na_2SO_4–$10H_2O$	Boardman tube	Boardman Energy Systems	18–21 (64–70)	200	$14.45[a] (cost of filled tubes plus stacking ring)

Source: Benson, D. 1981, "Solid-Solid Phase Change Materials for Thermal Energy Storage," Solar Energy Research Institute memorandum, October.
a. Storage system cost per unit of stored energy ($/MJ).
b. Storage material cost per unit of stored energy ($/MJ).

and also includes data from other sources. Many of the salt hydrates in table 25.5 are already commercial and appear to have dealt somewhat effectively with the problems of phase separation and supercooling. For example, mechanical mixing and novel thickening agents have been employed to stabilize sodium thiosulfate and other leading salts. PCM encapsulation to increase storage-transport medium surface area contact and prevent phase segregation has also been considered. Seven phase change materials have been encapsulated by Pennwalt Corporation through physical compaction and roll-spray coating (Michaels 1982). Sodium thiosulfate pentahydrate, disodium hydrogen phosphate dodecahydrate, wax, and calcium chloride hexahydrate all survived more than 2,600 cycles, although the pentrahydrate suffered severe supercooling.

Another approach to preventing salt hydrate phase segregation employs cross-linked water-soluble synthetic polymers which form a stable gel network (Page and Swayne 1981). A Glauber's salt, containing 4% Borax nucleator and approximately 8% polymeric stabilizer, retained its original heat capacity for 2,000 cycles.

An alternative liquid–solid phase transition material to the salt hydrates has been explored at Monsanto and the University of Dayton (Michaels 1982). Cross-linked, high density polyethylene pellets retain their shape and over 85% of their thermal storage capacity after 500 cycles.

The leading S–S phase transition materials examined at the Solar Energy Research Institute (SERI) are members of the pentaerythritol homolog series. The properties of a number of alloys of this series have been described (Benson, Burrows, and Webb 1986).

Manufacturing and packaging considerations are being actively explored by the private sector. Tubes, trays, stacks, and barrels of storage materials have been developed, and the number of alternatives, available containment shapes, and sealing techniques continues to grow. Studies in Europe of the compatibility of metals, plastics, and leading storage materials have found stainless steel compatible with a range of salt hydrates, copper compatible with all but $Na_2S_2O_3 \cdot 5H_2O$, and mild steel compatible with all but $Zn(NO_3) \cdot 6H_2O$ (Heine 1981). Aluminum corroded in the presence of all but sodium thiosulfate pentahydrate and lauric acid.

Accelerated testing is essential to confirming the laboratory properties of candidate storage materials. ASHRAE 94-77 has proved unreliable for latent-heat storage devices. The test results in estimates of storage capacity that are higher than theoretically possible (Marshall 1981). A modified procedure has been developed at Argonne National Laboratory (ANL).

25.8.5   Prospects for the Future

The literature on thermal storage materials suggests that, by and large, available materials are adequate for existing system needs. The leading storage materials are the S–S phase transition materials because of their potential for adjusting transition temperatures through alloying, their resistance to phase separation, their broad corrosion compatibility, and their high heats of transition. Efforts to develop theoretical or empirical procedures for predicting heat of transition and study of possible degradation mechanisms are the research areas where advances are likely to be made.

25.9   Desiccants

Desiccants are materials that can remove water or other vapors from a process stream. They are extensively used in a variety of commercial separation and drying applications (Davison Chemical 1982a). In solar-cooling systems desiccants can be used as dehumidifiers, permitting air-conditioning with a simple combination of an evaporative cooler and a heat exchanger, eliminating the use of compressors or a working fluid other than water. The desiccant is typically used in a cycling mode in which it removes water from the air at ambient temperature. It is then heated to drive the water out in order to "regenerate" its dehumidification property.

This type of cooling system is particularly suited to the use of solar energy because air can be used as the regeneration medium and there is no need for hermetic sealing (e.g., an open system that allows the desiccant to flow over the solar-collecting surface in contact with ambient air is possible). However, solar desiccant cooling systems must compete economically with vapor compression and other commercial space-cooling systems, requiring a much larger coefficient of performance (COP) than is common in existing desiccant industrial air dryers. Obtaining a high COP requires high efficiency components (i.e., heat exchangers and evaporative coolers) (Jurinak and Beckman 1980) and a desiccant material that provides a high degree of dehumidification with a minimum of heat input for regeneration.

Desiccant research can be divided into several categories, as indicated in figure 25.8. At the most fundamental level, investigation of adsorption and absorption phenomena can provide the theoretical understanding necessary to accurately model the properties of various classes of desiccants. More applied research involves the empirical study of adsorption mechanisms in real systems and correlation of performance with changes in material properties that affect these mechanisms, eventually leading to optimization of

Figure 25.8
Desiccant research alternatives.

R&D level	Definition				
5	Application to solar cooling systems		Regeneration methods	Cooling, storage, other process needs	Staging of desiccants
4	Optimization of desiccant materials		Porous materials (e.g., Si gel, clay)	Molecular cavities (e.g., molecular sieves, zeolite)	Hydrated halide salts (e.g., LiBr, LiCl, CaCl)
3	Correlation of material properties with adsorption (absorption) effectiveness		Porous materials (e.g., Si gel, clay)	Molecular cavities (e.g., molecular sieves, zeolite)	Hydrated halide salts (e.g., LiBr, LiCl, CaCl)
2	Investigation of adsorption (absorption) mechanisms in real systems		Porous materials (e.g., Si gel, clay)	Molecular cavities (e.g., molecular sieves, zeolite)	Hydrated halide salts (e.g., LiBr, LiCl, CaCl)
1	Basic study of adsorption/absorption phenomena	Van der Waals adsorption	Capillary attraction	Heat and vapor transport	Liquid/vapor equations of state

desiccant material properties. Finally, materials research may be needed to properly integrate the desiccant into a solar-cooling system. Research in these areas is described below.

25.9.1   Porous Desiccant R&D

Highly porous materials can provide a large surface area for absorption of a monomolecular layer, followed by a capillary attraction due to the condensed liquid in the capillary forming a meniscus concave to the vapor phase. Examples are silica gel (a granular, amorphous form of silica synthetically manufactured via chemical reaction between sulfuric acid and sodium silicate) and a number of naturally occurring clays.

Silica gel possesses an extremely good combination of low regeneration temperature and high maximum water capacity. It comes in a variety of grades with different pore sizes, the adsorption properties of which are well known (Davison Chemical 1982b). The major problem in using silica gel in desiccant cooling systems is the resultant rise in temperature of the desiccant bed and a corresponding loss in adsorption effectiveness. This creates a need for cooling during adsorption, which introduces parasitic power requirements. Research on optimization of the properties of silica gel would have to focus on reducing these cooling needs.

25.9.2   Molecular Cavity Desiccant R&D

These are crystalline materials with a network of interconnecting cavities of uniform size, separated by narrower openings, or pores, of equal uniformity that adsorb via Van der Waals attraction. Examples are the molecular sieves (crystalline-metal aluminosilicates possessing a zeolite structure in which various combinations of sodium, potassium, or calcium can be incorporated and in which combinations of silica and alumina tetrahedra can be stacked in either simple cubic or diamond structure) and natural zeolites themselves.

The molecular sieves, as well as their naturally occurring counterpart—zeolite—while having higher heats of adsorption than silica gel, higher regeneration temperatures, and lower maximum capacity, possess a uniquely steep adsorption isotherm at low vapor pressure. This makes them ideal in applications where the low relative humidity range of the adsorption spectrum is crucial. One might try using a molecular sieve desiccant bed in tandem with a silica gel in a desiccant cooling system. There are a wide variety of molecular sieves available with almost optimum properties (Davison Chemical 1982a), so that, except for engineering for specific applications, research on these materials has not been an active area.

25.9.3   Salt Solution Desiccant R&D

Hydrated salt or other solvent solutions absorb and desorb water in response to changes in temperature; examples are $LiBr/H_2O$, $LiCl/H_2O$, $CaCl/H_2O$, and triethylene glycol/H_2O. These solutions are typically used in an open-cycle absorption chiller with water as the refrigerant and a tilted solar collector over which the strong solution flows as the regeneration source.

These differ from the previous two classes in both operation and application. They absorb rather than adsorb water, have minimal cooling requirements, and operate at extremely low regeneration temperatures when compared to solid desiccants, in the vicinity of $60°C$ ($140°F$). Most important, these materials (LiCl and CaCl are the most popular) are particularly well suited to open-cycle systems using the simplest possible form of solar regeneration, a flat plate nonselective collector open to the ambient air (Robison and Griffiths 1981, 1982; Siebe et al. 1982). The key materials variable is the vapor pressure at absorption temperatures—the lower the vapor pressure, the larger is the capacity to absorb water. The key research problem is to characterize the vapor pressure of solutions of combinations of LiCl and CaCl. LiCl by itself has a much lower vapor pressure, but there is some evidence that combinations of LiCl and CaCl provide vapor pressures even lower than LiCl alone (Robison 1983). Since CaCl is an order of magnitude cheaper and desiccant cost can be a substantial fraction of system cost, this is an important problem. Two groups worked on optimizing the properties of LiCl/CaCl/H_2O solutions for open-cycle solar desiccant cooling systems (Wood 1983).

25.9.4   Other Research Areas

The discovery of a new class of solid desiccants, manganese oxides (Turner and Buseck 1981), introduced the potential for materials engineering of a different sort. These MnO_2 systems have a characteristic tunnel structure that is uniform and long-range ordered but leads to a much lower heat of adsorption than the three-dimensional cage like structure of the zeolites and synthetic molecular sieves. Moreover the tunnels provide for capillary attraction like the silica gel pores, but within a uniform crystal structure. The tailoring of the surface energetics of water adsorption in these materials, by doping with small quantities (about 1%) of foreign cations, was explored at ANL (USDOE 1982).

It has been demonstrated via a model calculation (Barlow and Collier 1981) that a staging of desiccants (use of several desiccants), each within a limited temperature range, could improve system cooling capacity by 10% and thermal COP by 6.8% over that which would be obtained with a silica gel bed. (This is for the design point; seasonal values are likely to be much lower.) In fact the research on tailoring of MnO_2 tunnel structures was motivated by their potential use in such a staged desiccant bed.

25.9.5 Prospects for the Future

There are a number of existing desiccant materials that are adequate for solar applications. However, some optimization research can improve the efficiency or cost-effectiveness of solar desiccant cooling systems, for example, that on the properties (primarily vapor pressure) of $LiCl/CaCl/H_2O$ mixtures. The work on tailoring the properties of MnO_2 tunnel structures for staged desiccant beds is also important because of the demonstrated potential of a staged desiccant for moderate improvement in cooling system efficiency. Moreover research on desiccant-matrix properties could be useful in the development of new and more effective desiccant-substrate combinations, such as silica gel bonded to plastic films for parallel passage dehumidifiers.

References

Andersson, A., O. Hunderi, and C. G. Granqvist. 1980. Nickel pigmented anodic aluminum oxide for selective absorption of solar energy. *Journal of Applied Physics* 51(1): 754.

Avaritsiotics, J. N., and R. P. Howson. 1981. Fluorine doping of In_2O_3 films employing ion-plating techniques. *Thin Solid Films* 80: 63.

Avery, J. G., and J. J. Krall. 1981. Corrosion prevention and fluid maintenance in active solar systems: The state-of-the-art Los Alamos Scientific Laboratory. LA-UR-81-3339.

Bacon, D. A. 1982. The influence of Substrate material on thermal stability of selective solar absorbing black chrome. M.S. thesis. the University of Houston, Physics Department. August.

Bailey, D., and O. Vogl. 1976. Polymeric ultraviolet absorbers, *Journal of Macromolecular Science: Reviews in Macromolecular Chemistry* C14(2): 267.

Barlow, R., and K. Collier. 1981. Optimizing the performance of desiccant beds for solar regenerated cooling. *Proc. 1981 AS/ISES Conference*, vol. 4.1. Philadelphia, PA, pp. 496–500.

Bateman, J. H., W. Gersey, Jr., and D. S. Neiditch. 1975. Soluble polyimides derived from phenylindane diamine: A new approach to heat resistant protective coatings. *Coatings and Plastics* (preprints of papers presented at the Chicago ACS Meeting) 35(2): 77.

Baum, B. 1982. Springborn Laboratories, private communication.

Benson, D. K., R. W. Burrows, and Y. D. Shinton. 1986a. Composite materials for thermal energy storage. U.S. Patent No. 4,572,864, February 25, 1986.

Benson, D. K., R. W. Burrows, and J. D. Webb. 1986b. Solid state phase transitions in pentaerythritol and related polyhydric alcohols. *Solar Energy Materials* 13: 133–152.

Benson, D. K., and R. W. Burrows. 1987. Phase change thermal energy storage material. U.S. Patent No. 4,702,853, October 27, 1987.

Blaga, A., and R. S. Yamasaki. 1976. Surface microcracking induced by weathering of polycarbonate sheet. *Journal of Materials Science* 11(8): 1513.

Booth, D. C., D. D. Allred, and B. O. Seraphin. 1979. Stabilized CVD amorphous silicon for high temperature photothermal solar energy conversion. *Solar Energy Materials* 2: 107–214.

Call, P. 1979. *National Program Plan for Absorber Surfaces R&D*. SERI/TP-31-103. Golden, CO: Solar Energy Research Institute. January.

Carlsson, D. J., A. Garton, and D. M. Wiles. 1978. Some effects of production conditions on the photo-sensitivity of polypropylene fibers. *ACS Advances in Chemistry Series* 169: 56.

Carroll, W. F., and P. Schissel. 1980. *Polymers in Solar Technologies: An R&D Strategy*, SERI/TR-334-601. Golden, CO: Solar Energy Research Institute, July.

Chain, E. E., D. Seshan, and B. O. Seraphin. 1981. "Composition and Microstructure of Black Molybdenum Photothermal Converter Layers Deposited by the Pryolysis of $Mo(CO)_6$," *Journal of Applied Physics* 52: 1356.

Chiba, K., S. Sobajima, U. Tonemura, and N. Suzuki. 1982. Optical coatings for energy efficiency and solar applications. *Proc. SPIE—The International Society for Optical Engineering*, vol. 324, C. M. Lampert, ed. Los Angeles, CA, p. 23. January 28–29, 1982.

Christensen, C. B., and D. K. Benson. 1986. The energy performance of buildings with distributed thermal storage. Proceedings of the Tenth CIB Congress, Advancing Building Technology, International Council for Building Research, Studies and Documentation, Washington, DC. September.

Cocks, F. H., and M. J. Peterson. 1979. *Journal of Vacuum Science Technology* 16(4).

Cocks, F. H., M. J. Peterson, and P. L. Jones. 1980. *Thin Solid Films* 70: 297.

Cooper, S. L., and G. M. Estes, eds. 1979. Multiphase polymers—Segmented polymers *Advances in Chemistry Series* 176: 3–180.

Davison Chemical (Division of W. R. Grace and Co.). 1982a. *Davison Molecular Sieves*. Baltimore, MD, pp. 10–11.

Davison Chemical (Division of W. R. Grace and Co.). 1982b. *Davison Silica Gels*. Baltimore, MD.

De Waal, H., and F. Simonis. 1981. Tin oxide coatings: Physical properties and applications. *Thin Solid Films* 77: 253.

Dexter, M., and R. A. E. Winter. 1981. Ciba-Geigy Corporation. U.S. Patent No. 4,278,589. July 14, 1981.

Diegle, R. B., J. A. Beavers, and J. E. Clifford. 1980. *Corrosion Problems with Aqueous Coolants*. DOE/CS/10510-TM. Columbus, OH: Battelle Columbus Laboratories. April 11, 1980.

Dislich, H., and E. Hussman. 1981. Amorphous and crystalline dip coatings obtained from organometallic solutions: Procedures, chemical processes and products. *Thin Solid Films* 77: 129.

Ebert, J. 1982. Optical thin films. *Proc. SPIE—The International Society for Optical Engineering*, vol. 325, R. I. Seddon, ed. p. 29, 26–27 Jan. 1982.

Ehrenreich, H., and B. O. Seraphin. 1976. *The Fundamental Optical Properties of Solids Relevant to Solar Energy Conversion*. November 20–23, 1975. NSF symposium report under grant no. DMR 75-18134. Washington, DC: The National Science Foundation.

Fan, J. C. C. 1979. Optics in solar energy utilization II. *Proc. SPIE—The International Society for Optical Engineering*, vol. 85, p. 39.

Fan, J. C. C., and F. J. Bachner. 1976. Transparent heat mirrors for solar energy applications. *Applied Optics* 15(4): 1012.

Fontana, M. G., and N. D. Greene. 1978. *Corrosion Engineering*. New York: McGraw-Hill.

Frank, G., E. Kauer, F. J. Schmitte, and H. Kostlin. 1982. Optical coatings for energy efficiency and solar applications. *Proc. SPIE—The International Society for Optical Engineering*, vol. 324, C. M. Lampert, ed. Los Angeles, CA. January 28–29, 1982, p. 58.

Frank, G., and H. Kostlin. 1982. Electrical properties and defect model of tin-doped indium oxide layers. *Applied Physics* A27: 197.

Fuschillo, N. 1975. Semi-transparent solar collector window systems. *Solar Energy* 17: 159.

Gadgil, S. B., R. Thangaraj, J. V. Iyer, B. K. Gupta, and O. P. Agnihotri. 1981. Spectrally selective copper sulphide coatings. *Solar Energy Materials* 5(2): 129.

Garton, A., D. J. Carlsson, and D. M. Wiles. 1980. *Developments in Polymer Photo-Chemistry*, vol. 1, N. S. Allen, ed. Applied Science Publishers.

Gesheva, K. A., D. Seshan, and B. O. Seraphin. 1981. Composition and microstructure of black molybdenum photothermal converter layers deposited by the pyrolytic hydrogen reduction of MoO_2Cl_2. *Thin Solid Films* 79: 39.

Gilligan, J. E. 1980. *Exposure Testing and Evaluation of Solar Utilization Materials*. IIT Research Institute. DOE report CH/900 34-T1 (DE81027311). April.

Gilligan, J. E., J. Brzuskiewicz, J. E. Brzuskiewicz, S. J. Gaumer, and H. T. Betz. 1980. *Handbook of Materials for Solar Energy Utilization—Low Temperature Applications*. IIT Research Institute. DOE report CH/90034-T1 (DE81027310). April.

Gittleman, J. I., E. K. Sichel, and Y. Arie. 1979. Characterization of black Ge selective absorbers. *Solar Energy Materials* 1: 93.

Gordon, R. G. 1982. Deposition methods for coating gas and the like. U.S. Patent No. 4,206,252. June 3, 1982.

Granqvist, C. G. 1981. *Journal de Physique* C1, 42(1).

Groth, R., and E. Kauer. 1965. Thermal insulation of sodium lamps. *Philips Technical Review* 26(4/5/6): 105.

Gunasekaran, M., and R. B. Grelika. 1977. *Development of Polymer-Bonded Silica (Polysil) for Electrical Applications*. Final report. EPRI document EL-488. Palo Alto, CA: Electric Power Research Institute. May.

Gupta, A. 1978. Photodegradation of Polymeric Encapsulants of Solar Cell Modules. *JPL Low-Cost Solar Array Project 5101-77*. August 10, 1978.

Gupta, A. 1982. Jet Propulsion Laboratory, private communication.

Haacke, G. 1977. Evaluation of cadmium stannate films for solar heat collectors. *Applied Physics Letters* 30(8): 380.

Haacke, G. 1982. Optical coatings for energy efficiency and solar applications. *Proc. SPIE—The International Society for Optical Engineering*, vol. 324, C. M. Lampert, ed., Los Angeles, CA, p. 10 January 28–29, 1982.

Habermeier, H. U. 1981. Properties of indium tin oxide thin films prepared by reactive evaporation. *Thin Solid Films* 80: 157.

Hamberg, I., A. Hjorstberg, and C. G. Granqvist. 1982. Optical coatings for energy efficiency and solar applications. *Proc. SPIE—The International Society for Optical Engineering*, vol. 324, C. M. Lampert, ed. Los Angeles, CA, p. 31. January 28–29, 1982.

Harding, G. L., and B. Window. 1979. Cylindrical magnetron sputtering system for coating solar selective surfaces to batches of tubes. *Journal of Vaccum Science Technology* 16(6): 2105.

Hausz, W., and B. J. Berkowitz, ed. 1983. Opportunities in thermal storage R&D. EPRI report EM-3159-SR. July.

Hautala, R. R., B. R. King, and C. Kutal. 1979. The norbornadien-quadricyclene energy storage systems. *Solar Energy Chemical Conversion and Storage*, R. R. Hautala, et al., ed. Clifton, NJ: Humana Press, pp. 333–369.

Heine, D. 1981. Chemical compatibility of construction materials with latent heat storage materials. *International Conference on Energy Storage*. BRHA Fluid Engineering, Brighton, U.K., April 29,–May 1, 1981, p. 185.

Herzenberg, S. A., and R. Silberglitt. 1982. Optical coatings for energy efficiency and solar applications. *Proc. SPIE—The International Society for Optical Engineering*, vol. 324, C. M. Lampert, ed. Los Angeles, CA, p. 92. January 28–29, 1982.

Herzenberg, S. A., R. Silberglitt, and C. La Porta. 1981. *Materials Research Needs to Improve the Performance and Durability of Solar Heating and Cooling Systems.* DOE contract no. DE-AC01-80ER30008. Washington, DC: DHR, Inc. September.

Hoffman, D. M. 1980. *Org. Coat. Plast. Prep.* 43: 894.

Hoffman, D. M., H. G. Hammond, and L. P. Althouse. 1981. Application of sturcture property relationship to develop seven segmented polyurethane adhesives. *Society of Plastic Engineers, Inc., Technical Papers* 27: 314–315.

Homhual, C., O. T. Inal, L. E. Murr, A. E. Torma, and I. Gündiler. 1981. Microstructural and mechanical property evaluation of zinc oxide coated solar collectors. *Solar Energy Materials,* 4(3): 309.

Holloway, P. H., K. Shanker, R. B. Pettit, and R. R. Sowell. 1980. *Thin Solid Films* 72: 121.

Hong, D. 1982. Jet Propulsion Laboratory, private communication.

Hornberger, L. E., and J. L. Hull. 1982. Adhesive bonding of thin glass mirror to parabolic steel trough. *Optical Coating for Energy Efficiency and Solar Applications.* SPIE, vol. 324, pp. 86–90.

Hosch, L. n.d. Rohm GmbH, German patent no. 2,913,853.

Howsen, B. P., and M. I. Ridge. 1982. Optical coatings for energy efficiency and solar applications. *Proc. SPIE—The International Society for Optical Engineering,* vol. 324, C. M. Lampert, ed. Los Angeles, CA, p. 16. January 28–29, 1982.

Ignatiev, A. P. O'Neill, D. Doland, and G. Zajac. 1979. *Applied Physics Letters* 34(42).

Ignatiev, A., G. Zajac, and G. B. Smith. 1982. Optical coatings for energy efficiency and solar-applications. *Proc. SPIE—The International Society for Optical Engineering,* vol. 324, C. M. Lampert, ed. Los Angeles, CA, p. 170. January 28–29, 1982.

Inal, O. T., J. C. Mahon, and C. V. Robino. 1981. *Thin Solid Films* 83.

Inal, O. T., M. Valayapetre, L. E. Murr, and A. E. Torma. 1981. Microstructural and mechanical evaluation of black chrome coated solar collectors—II. *Solar Energy Materials* 4(3): 333.

Jet Propulsion Laboratory. 1978. Solar film materials and supporting structures for a solar sail—A preliminary design. JPL Document 720-9, vol. 4. October.

Jorgensen, G. J. 1982. Glazing survey. Solar Energy Research Institute memorandum. June 2, 1982.

Jurinak, J. J., and W. A. Beckman. 1980. A comparison of the performance of open cycle air conditioners utilizing rotary desiccant dehumidifiers. *Proc. 1980 AS/ISES Conference* vol. 3.1. Phoenix, AZ, pp. 215–219.

Kane, J., H. P. Schweizer, and W. Kern. 1975. Chemical vapor deposition of transparent electrically conducting layers of indium oxide doped with tin. *Thin Solid Films* 29: 155.

Karlsson, B., and C. G. Ribbing. 1982. Optical coatings for energy efficiency and solar applications. *Proc. SPIE—The International Society for Optical Engineering,* vol. 324, C. M. Lampert, ed. Los Angeles, CA, p. 52. January 28–29, 1982.

Karrer, F. 1981. Ciba-Geigy Corporation, U.S. patent no. 4,210,612. June 30, 1981.

Kingery, W. D., H. K. Bowen, and D. R. Uhlmann. 1976. *Introduction to Ceramics.* 2d ed. New York: Wiley, p. 593.

Kitler, W. C., Jr., and I. T. Richie. 1982. Optical thin films. *Proc. SPIE—The International Society for Optical Engineering,* vol. 325, R. I. Sheldon, ed. p. 61. January 26–27, 1982.

Kloppfer, W. 1976. *Journal of Polymer Science,* Symposium, 57: 205.

Kostlin, H. 1974. Double-glazed windows with very good thermal insulation. *Philips Technical Review* 34(9): 242.

Lampert, C. M. 1979. Coatings for enhanced photothermal energy collection. I: Selective absorbers. *Solar Energy Materials* 1(5/6): 319, and 2(1): 1.

Lampert, C. M. 1980. *Plating and Surface Finishing* (November): 52.

Lampert, C. M. 1981. Heat mirror coatings for energy-conserving windows. *Solar Energy Materials* 6(1): 1.

Lampert, C. 1982. Lawrence Berkeley Laboratory, private communication.

Lappin, G. R. 1971. Ultraviolet radiation absorbers. *Encyclopedia of Polymer Science and Technology*, vol. 14. New York: Wiley-Interscience, pp. 125–148.

Lipshitz, S. D., and C. W. Macosko. 1976. Rheological changes during a urethane network polymerization. *Polymer Engineering Science* 16(2): 803.

Maag, C. 1982. Jet Propulsion Laboratory, private communication.

Machet, J., J. Guille, P. Saulnier, and S. Robert. 1981. Deposition of conducting and transparent thin films of indium tin oxide by reactive ion plating. *Thin Solid Films* 80: 149.

Mansfied, F. B. 1982. *Evaluation of Techniques for Corrosion Monitoring in Flat-Plate Collectors*. Final report on SERI subcontract No. XD-9-8204-1. March.

Mar, R. 1980a. Materials science issues encountered during the development of thermochemical concepts. *Solar Materials Science*, L. E. Murr, ed. New York: Academic Press, p. 459.

Mar, R. 1980b. Application of reversible chemical reactions to solar thermal energy systems. *Solar Materials Science*, L. E. Murr, ed. New York: Academic Press, p. 439.

Marshall, R. 1981. Experimental experience with the ASHRAE/NBS procedures for testing a phase-change thermal storage device. *International Conference on Energy Storage*, BRHA Fluid Engineering, Brighton, U.K., p. 129. April 29–May 1, 1981.

Mason, J. J., and T. A. Brendel. 1982a. Maxorb—A new selective surface of nickel. *Optical Coatings for Energy Efficiency and Solar Application*, SPIE, vol. 324, pp. 143–145.

Mason, J. J., and T. A. Brendel. 1982b. Optical coatings for energy efficiency and solar applications. *Proc. SPIE—The International Society for Optical Engineering*, vol. 324, C. M. Lampert, ed. Los Angeles, CA, p. 139. January 28–29, 1982.

Masters, L. W., J. F. Seiler, E. J. Embree, and W. E. Roberts. 1981. *Solar Energy Systems—Standards for Absorber Materials*. NMSIR 81-2232. January.

Mavis, C. L. 1980. *Status and Recommeded Future of Plastic-Enclosed Heliostat Development*. SAND 80-8032. Albuquerque, NM: Sandia National Laboratories. October.

McBride, J. R. 1981. "Chemical Heat Pump Cycles for Energy Storage and Conversion," *International Conference on Energy Storage*, BRHA Fluid Engineering, Brighton, U.K., p. 29. April 29–May 1, 1981.

McChesney, M. A. 1981. Selective paint development progress reports. April 1, May 2, and May 29, 1981.

McDermott, J. 1981. Ceramics: The future beyond plastics. *Technology* 1: 18–29.

McKelvey, W. D., P. B. Zimmer, and R. J. H. Lin. 1979. *Solar Selective Paint Development*. Final report, June 1978–December 1979. DOE/CS/34287-T1, December.

Mellor, D. C., A. B. Moir, and G. Scott. 1973. Effect of processing conditions on the UV stability of polyolefins. *European Polymers* 9(3): 219.

Michaels, A. I. 1982. *Proc. Solar Storage Workshop*, Jeddah, Saudi Arabia, March 21–24, 1982. Report no. CP-270-1521. Golden, CO: Solar Energy Research Institute.

Moore, S. W. 1982a. Optical coatings for energy efficiency and solar applications. *Proc. SPIE—The International Society for Optical Engineering*, vol. 324, C. M. Lampert, ed. Los Angeles, CA, p. 148. January 28–29, 1982.

Moore, S. W. 1982b. Los Alamos, private communication.

Nath, P., R. F. Bunshah, B. M. Basol, and O. M. Staffsud. 1980. Electrical and optical properties of tin-doped indium oxide films prepared by activated reactive evaporation. *Thin Solid Films* 72(3): 463.

Nelson, E. V., and T. K. Muller. 1980. Further development of low-cost solar panel. *Proc. Annual DOE Active Solar Heating and Cooling Contractors' Review Meeting*, Lake Tahoe, NV. Report #CONF800340, pp. 9-7–9-8.

Nonnemacher, A., and M. Groll. 1981. Chemical heat storage and heat transformation using reversible solid gas reactions. *International Conference on Energy Storage*, BRHA Fluid Engineering, Brighton, U.K., p. 47. April 29–May 1, 1981.

Nyberg, G., H. G. Craighead, and R. A. Buhrman. 1982. Optical coatings for energy efficiency and solar applications. *Proc. SPIE—The International Society for Optical Engineering*, vol. 324, C. M. Lampert, ed. Los Angeles, CA, p. 117, January 28–29, 1982.

Ortiz, R. A. 1982. Spray deposition and characterization of cadmium stannate films for solar cells. *Journal of Vacuum Science Technology* 20(1): 7.

Page, J. K. R., and R. E. H. Swayne. 1981. Phase-change thermal storage for solar applications. *International Conference on Energy Storage*, BRHA Fluid Engineering, Brighton, U.K., p. 105. April 29–May 1, 1981.

Pan, C. A., and T. P. Ma. 1980. High-quality transparent conductive indium oxide films prepared by thermal evaporation. *Applied Physics Letters* 37(2): 163.

Pettit, R. B., R. R. Sowell, and I. J. Hall. 1982. Black chrome solar selective coatings optimized for high temperature applications. *Solar Energy Materials* 7(2): 153.

Plueddemann, E. P. 1979. Chemical bonding technology for terrestrial solar cell modules. JPL document 5101–132. Pasadena, CA: Jet Propulsion Laboratory. September 1, 1979.

Potdar, H. S., N. Pavaskar, A. Mitra, and A. P. B. Sinha. 1981. Solar selective copper-black layers by an anodic oxidation process. *Solar Energy Materials* 4(3): 291.

Raghunathan, K. 1981. Thermal cycling testing of solar absorptive surfaces. Paper presented to the *Solar Energy Industries Association Annual Conference*, November 1–4, 1981.

Randich, E., and D. D. Allred. 1981. CVD ZrB$_2$ as a selective absorber. *Thin Solid Films* 83: 393.

Richter, E. B., C. W. Macosko. 1973. Kinetics of fast (RIM) urethane polymerization. *Polymer Engineering and Science* 13: 1012.

Robison, H. 1983. Coastal Carolina College, University of South Carolina, private communication.

Robison, H., and W. Griffiths. 1981. Low cost open cycle chemical heat pump powered by solar energy. *Proc. International Passive and Hybrid Cooling Conference*, Miami Beach, FL. November 6–10, 1981.

Robison, H., and W. Griffiths. 1982. *Open Cycle Chemical Heat Pump and Energy Storage System*. Final report on SERI subcontract ZE-0-9185-1, p. 306. January.

Seshan, K., P. D. Hillman, K. A. Gesheva, E. E. Chain, and B. O. Seraphin. 1981. On the mechanism of growth and the hydrogen reduction of CVD black molybdenum thin films. *Journal of Materials Research Bulletin* 16: 1345.

Siebe, D. A., G. A. Buck, M. L. Breslauer, and B. D. Wood. 1982. *Performance Characteristics of an Open Flow Liquid Desiccant Solar Collector/Regenerator in a Hot Arid Climate*. Report to SERI under subcontract XE-0-9179-2. February. (Arizona State University College of Engineering and Applied Sciences Report CR-R-82007.)

Sikkins, M. 1981. Physical background of spectral selectivity. *Solar Energy Materials* 5(1): 55.

Sikkins, M. 1982a. Optical coatings for energy efficiency and solar applications. *Proc. SPIE—The International Society for Optical Engineering*, vol. 324, C. M. Lampert, ed. Los Angeles, CA, p. 131. January 28–29, 1982.

Sikkins, M. 1982b. Properties of spectrally selective no-C films produced by reactive sputtering, parts 1 and 2. *Solar Energy Materials* 6(4): 403.

Slemmons, A. J., et al. 1981. *Conceptual Design of a Glass-Reinforced Concrete Solar Collector*. SAND81-7011. Albuquerque, NM: Sandia National Laboratories. July.

Smith, E. F., and A. P. Castillo. 1981. Corrosion of copper under simulated solar service conditions. *Corrosion/81: International Corrosion Forum.* NACE. April 6–10, 1981.

Smith, III, E. F., and E. Davis. 1981. The effect of elevated temperatures on the compatibility of glycol solutions with metals used in solar absorbers. *Corrosion/81: International Corrosion Forum.* NACE. April 6–10, 1981.

Smith, G. B., and A. Ignatiev. 1981. Black chromium-molybdenum: A new stable solar absorber. *Solar Energy Materials* 4(2): 119.

Smith, G. B., A. Ignatiev, and G. Zajac. 1980. Solar selective black cobalt: Preparation, structure, and thermal stability. *Journal of Applied Physics* 51(8): 4186.

Smith, G. B., Zajac, and A. Ignatiev. 1982. High flux photochemical changes in black chrome solar absorbing coatings. *Solar Energy* 29: 279–289.

Smith, J. F., A. J. Aronson, D. Chen, and W. H. Class. 1980. Reactive magnetron deposition of transparent conductive films. *Thin Solid Films* 72: 469.

Stephens, R. B., and G. D. Cody. 1979. Inhomogeneous surfaces as selective solar absorbers. *Solar Energy Materials* 1: 397.

Swab, P., S. V. Krishnaswamy, and R. Messier. 1980. *Journal of Vacuum Science Technolgoy* 17(1): 362.

Tabony, J. H. 1977. *Inhibitor Analysis for a Solar Heating and Cooling System.* Sourthern University, LA, DOE/NASA CR-150513. June 1, 1977.

Telkes, M. 1980a. Thermal storage in salt-hydrates. *Solar Materials Science,* L. E. Murr, ed. New York: Academic Press, p. 377.

Telkes, M. 1980b. Thermodynamic basis for selecting heat storage materials. *Solar Materials Science,* L. E. Murr, ed. New York: Academic Press, p. 405.

Thornton, J. A., and J. L. Lamb. 1981. Sputter deposited $Pt-Al_2O_3$ selective absorber coatings. *Thin Solid Films* 83: 377.

Thornton, J. A., and J. L. Lamb. 1982. *Evaluation of Vacuum Deposition-Sputter Deposited Cermet Selective Absorber Coatings.* Interim technical progress report on contract no. XP-9-8260-1 with the Solar Energy Research Institute, Golden, CO, July 30, 1982.

Thornton, J. A., A. S. Penfold, and J. L. Lamb. 1980. *Development of Selective Surfaces.* Final technical progress report, September 11, 1978–Mach 31, 1980. TELIC-80-1, contract no. DE-AC04-78CS35306, August 15, 1980.

Tirrel, D. A. 1980. Preparation of polymeric ultraviolet stabilizers from salicylate esters, hydrozybenzophenones, a-cyano-B-phenyleinnamates and hydroxyphenylbenzotriazoles. *ACS Organic Coatings and Plastics Chemistry Proc.,* vol. 42. Presented at the American Chemical Society 179th National Meeting, Houston, TX, pp. 3–18. March 23–28, 1980.

Turner, S., and P. B. Buseck. 1981. Todorokites: A new family of naturally occurring manganese oxides. *Science* 212: 1024.

USDOE. 1981. *An Assessment of the Field Status of Active Solar Systems.* Washington, DC: U.S. Department of Energy. pp. 19–21 and app. A.

USDOE. 1982. Field Task Agreement 49532 between the U.S. Department of Energy and Argonne National Laboratory, February 1982–September 1982. (Principal Investigators, W. R. McIntire and A. Fraioli.)

Van Boort, H. J. J., and R. Groth. 1968. Low pressure sodium lamps with indium oxide filter. *Philips Technical Review* 29(1): 17.

Van der Leij. 1979. Ph.D. dissertation. Delft University Press, Delft, The Netherlands.

Walker, S. J., and D. R. McKenzie. 1981. Magnetron sputtering of solar coatings inside tubes. *Journal of Vacuum Science Technology* 19(3): 700.

Whalen, T. J., and C. F. Johnson. 1981. Injection molding of ceramics. *American Ceramics Society Bulletin* 60(2): 216–220.

White, S. 1981. Soiling and cleaning meeting Summary. Albuquerque, NM: Sandia National Laboratories. August.

Wiles, D. M. 1978. Photostabilization of macromolecules by excited state quenching. *Pure and Applied Chemistry* 50: 291.

Wilhelm, W. G. 1982. Low-cost solar flat-plate collector development. *Proc. Annual DOE Active Solar Heating and Cooling Contractors' Review Meeting*, September 1981. Washington, DC: U.S. Department of Energy, pp. 4-4–4-6. February.

Williams, R. 1982. OCLI, private communication.

Wilson, J. D. C. 1982. DuPont, private communication.

Winslow, F. 1982. Bell Laboratories, private communication.

Winter, R. 1982. Ciba-Geigy Corporation, private communication.

Wong, D., and F. H. Cocks. 1980a. The use of zinc to control aluminum corrosion in aqueous glycol solutions. *Corrosion* 36(11): 587.

Wong, D., and F. H. Cocks. 1980b. *Study of Aluminum Corrosion in Aluminum Solar Heating and Cooling Systems Using Aqueous Glycol Solutions for Heat Transfer*. Annual technical progress report, DE-AC04-79-31072. August.

Wong, D., L. Swette, and F. H. Cocks. 1979. Aluminum corrosion in uninhibited ethylene glycol water solutions. *Journal of the Electrochemical Society* 126(1).

Wood, B. 1983. Arizona State University, private communication.

Zajac, G., G. B. Smith, and A. Ignatiev. 1980. *Journal of Applied Physics* 51: 5544.

26 Solar Materials Issues and Opportunities

C. J. Swet

26.1 Introduction

In chapters 21 through 25 solar materials are treated primarily as separate entities, in consonance with the components and materials theme of part III. Major portions of parts I and II deal similarly with collectors and storage, respectively, as more or less autonomous elements within the larger solar thermal system. Chapters 2 through 5 deal with concepts and designs, theory and modeling, test and evaluation, and research and development of solar materials. These chapters are mainly reports of progress and the present state of knowledge in these areas. Generally they have been neither judgmental nor speculative, and attention to economics has been limited mainly to the cost of raw materials and their processing.

Broadly stated, this chapter identifies what still needs to be learned and accomplished in the field of solar materials, viewing the materials in an overall system context. More specifically, its purpose is to identify materials-related issues and opportunities that are potentially significant in terms of impact on overall system life-cycle cost (microeconomics), on the solar energy industry (macroeconomics), or on other important figures of merit. Some of the issues identified here involve choices between technical approaches, such as variable absorptance versus variable reflectance for switchable glazings; others involve the justification of a developmental goal, such as 20-year life for polymeric glazings on flat plate collectors. Some of the opportunities involve new or not widely recognized ways of exploiting certain materials, such as the use of ettringite-like solid desiccants for thermal storage in open-cycle, air-based active solar-heating systems.

Obviously only a few of the many significant issues and opportunities can be presented here, and still fewer can be treated in any depth. Selected for fullest discussion are those that appear to be most important and least recognized. Briefly mentioned are some related items that provide additional perspective, and some others that are adequately dealt with elsewhere but appear to need reinforcement. It should be borne in mind that most of the topics were selected because of their controversial nature and that the treatment is deliberately speculative in order to stimulate either resolution or further exploration.

26.2   Glazings with Very Low Thermal Conductance

The materials considered here are aerogel and evacuated glazings, both of
which were reviewed in earlier chapters. They are being developed under
Department of Energy (DOE) sponsorship as advanced transparent insulators
for building envelopes. Their purpose is to substantially improve passive
solar-heating efficiency, with consequently fuller exploitation, on a national
scale, of the solar resource. The focus here is on their potential value. Would
the economic value of their superior performance be sufficiently compelling,
in enough situations, to significantly enlarge the national residential market
penetration by passive solar-heating components and systems? Would their
value significantly exceed that of other kinds of transparent insulators, such
as multifilm glazings, that already are commercially available and presumably
will be undergoing improvement in the same time frame with little or no
federal support? No hard answers will emerge from the discussion that follows,
but it will be shown that the potential national benefit of federally supported
research and development on aerogel and evacuated glazing for passive applications may be considerably smaller than apparently had been envisioned
when those programs were initiated.

Justification of the aerogel and evacuated glazing development efforts appears to have been based largely on the results of a 1982 DOE-sponsored
study (Neeper and McFarland 1982) that investigated (among other things)
the potential performance benefits of postulated glazing improvements in
passive solar-heating applications. In the study a representative one-story
house was modeled in Albuquerque, New Mexico, Columbia, Missouri, and
Caribou, Maine. The model was varied to have one Trombe wall or one direct
gain system facing south, east, or north. The study also modeled the thermal
performance of variously glazed windows with the same facings for a nonsolar
house in the same locations. Scrutinized with hindsight, the simulation results
offer less than fully persuasive arguments for the federally sponsored development of either aerogel or evacuated glazings for building envelopes.

Table 26.1 presents results for a 300-ft^2 (27.9-m^2) concrete Trombe wall with
a flat black absorbing surface. In the first column the designation "standard"
refers to standard double glass, "multifilm" refers to advanced (in 1982)
commercially available multifilm glazing, and "evacuated" to a postulated
evacuated double glazing with infrared reflective coating. Although aerogel
was not specifically considered in the study, its targeted thermal conductance
and solar transmittance are about the same as for evacuated glazing (Selkowitz 1983). It is seen that glazing *that was then available* could produce a net
gain at any wall orientation in any of the three climates, and (in principle, with

Table 26.1
Effects of glazings on performance of a Trombe wall

Glazing	Orientation	Albuquerque		Columbia		Caribou	
		Annual SSF (%)	Annual savings (kBtu/ft²)	Annual SSF (%)	Annual savings (kBtu/ft²)	Annual SSF (%)	Annual savings (kBtu/ft²)
Standard	South	63	72	29	37	13	30
R = 2.2	East	29	33	9	11	3	7
τ = 0.7	North	0	0	−3	−4	−7	−16
Multifilm	South	75	86	41	52	25	57
R = 4.0	East	39	45	18	23	11	25
τ = 0.58	North	9	10	4	5	1	2
Evacuated	South	89	102	58	73	41	94
R = 12.0	East	55	63	32	40	25	57
τ = 0.49	North	22	25	17	21	13	30

Source: Neeper and McFarland (1982).

Trombe walls on more than one side of the house) 100% solar savings fraction in Albuquerque and 81% in Columbia. The additional heating energy saved by evacuated over multifilm glazing would at most be 18 kBtu/ft^2 yr (187 MJ/m^2 yr) in Albuquerque, 21 kBtu/ft^2 yr (218 MJ/m^2 yr) in Columbia, and 37 kBtu/ft^2 yr (384 MJ/m^2 yr) in Caribou. At 7¢/kWh the corresponding energy cost savings would be 37¢, 43¢, and 76¢/ft^2 ($3.98, $4.64, and $8.19/m^2), and expressed as present worth (20-year system life, 10% discount rate, 5% general inflation rate), the allowable marginal costs of evacuated over multi-film glazing would be $2.30, $2.70, and $4.75/ft^2 ($24.8, $28.8, and $50.6/m^2). Considering that glazed Trombe walls may cost perhaps $20/ft^2 ($216/m^2) more than ordinary walls and that the choice is more likely to be solar versus nonsolar than evacuated versus multifilm glazing, this is minimal economic leverage.

Other factors, not considered in the 1982 study, conspire to diminish the significance of its results concerning Trombe wall glazing. All of the numbers in table 26.1 would have been larger had the Trombe wall been modeled with a low emissivity coating, which is fairly common practice, but the improvement would have been least for the glazings with lowest thermal conductance. A similar pattern of improvement would have resulted had the glazings been modeled with an antireflective (in the visible spectrum) coating on the outer pane. [Another study by Leigh and Fthenakis, 1981, who examined water walls rather than Trombe walls, showed that a reduction in glazing reflectivity from 0.07 to 0.04 would result in savings of 6.7 kBtu/ft^2 yr (69 MJ/m^2 yr) in Albuquerque, compared with standard double glazing.] Had the house in Caribou been modeled with two stories and/or superinsulation, which would have been more appropriate for that severe climate, the lower heating load would have reduced the incentive for better passive solar heating. Most important, better multifilm glazings have become commercially available since 1982, and still better ones presumably will appear on the market by the time evacuated or aerogel glazings are fully developed. A 1984 report on evacuated glazing (Benson 1984) referred to a commercial glazing with an R-value of 7.27 and a solar transmittance of about 0.36. According to Lawrence Berkeley Laboratory studies (Selkowitz 1983, 1984), "New developments, such as using low conductance gases and two low-E plastic inserts, should make it possible to build R-5 to R-15 windows having a solar transmittance of 50% to 60%."

Figure 26.1 plots some of the tabulated Trombe wall results for Caribou, spotting in also the only reported data points for a direct gain system in that location. Connecting the direct gain points are plausible curves I sketched in

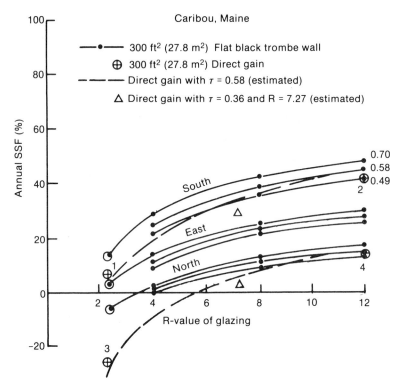

Figure 26.1
Annual solar savings fraction in Caribou, Maine, as a function of the R-value of the glazing, with optical transmittance as a parameter. Selected points for the performance of a direct gain system are shown as (1) south, $\tau = 0.7$; (2) south, $\tau = 0.49$; (3) north, $\tau = 0.7$; (4) north, $\tau = 0.49$. Source: Adapted from Neeper and McFarland (1982).

the absence of intermediate data. Also added are points representing the reported commercial glazing with an R-value of 7.27 and a transmittance of 0.36. Concerning the simulation results for Trombe walls and for direct gain systems the 1982 study authors remarked, "The most striking fact is that an R-12 glazing provides positive energy benefits even for north-facing direct gain in Caribou! This means that, with a high-R glazing, *any part of the building envelope could be used for net heating benefits, even in the most severe winter climate.*" More striking to me is the indication that such may already be so and that in most situations the additional energy savings due to further improvements are of modest benefit. Were a strong point to be made about the doubtful "north-facing in Caribou" situation, one might counter by asking how heavy an impact that situation might have on the national energy situation.

Evacuated or aerogel glazing for windows in nonsolar houses may, in certain circumstances, have greater microeconomic impact than when used with Trombe walls or direct gain systems. Since windows cost less per unit aperture area, the same degree of performance improvement can more strongly influence decisions concerning their number, size, and placement. Typically, though, the heating energy savings of windows with R-12 glazing are little better than those of windows with currently available multifilm glazing. Figure 26.2 was prepared from tabulated results of the 1982 study, with plausibly shaped curves sketched in between published data points. About their tabulated data the authors wrote, "We can see that the R-12 window material always has a better net performance than an R-30 wall, and that south-facing windows with the R-12 material are significantly better than standard double glazing only in the severe climate." Figure 26.2, however, strongly suggests that *existing* windows may perform better than R-30 walls in all but one of the examined cases. In that singular case the energy price that must be paid for a northern view (or for not having R-12 glazing) would not exceed perhaps 15 kBtu/ft^2 yr (155 MJ/m^2 yr), and the allowable extra cost for R-12 glazing would be perhaps \$1.90/ft^2 (\$20.5/m^2).Thus to individual buyers the economic justification for R-12 glazing would be quite weak even in Caribou, and the resulting impact on national solar resource utilization would be minimal.

A more recent study (Neeper et al. 1984) specifically examined the potential performance benefits of aerogel windows, with results that seem to offer even less justification for federally sponsored research and development. In table 26.2 (which was included in the study report only "for programmatic purposes," with many caveats and qualifications) the designation "20 mm aerogel low-iron" refers to a window with the currently targeted R-value of 12 and solar transmittance of 0.50. The designation "20A FUTURE?" refers to a hypothetical aerogel window in which the aerogel optical transmission (not counting losses attributable to the encasing glass panes) has somehow been increased from 0.59 (the present value) to 0.88. About it the authors remark, "The researchers may be able to determine if this is a reasonable goal." The table indicates that, except for "20A FUTURE?" in Albuquerque, it is not even potentially possible for *any* north-facing window to be a net gainer. It also indicates that in south-facing windows multifilm glazing always out performs "20 mm aerogel" and would never be penalized more than 17.6 kBtu/ft^2 yr (182 MJ/m^2 yr) relative to "20A FUTURE?" For north-facing windows the worst penalty would be 10.7 kBtu/ft^2 yr (111 MJ/m^2 yr) relative to "20 mm aerogel" and 18.4 kBtu/ft^2 yr (191 MJ/m^2 yr) relative to "20A FUTURE?"

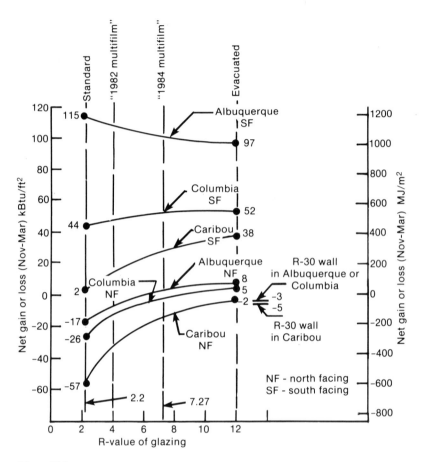

Figure 26.2
Performance of improved glazing as a window. Source: Adapted from Neeper and McFarland (1982).

Table 26.2
Net energy transmitted by vertical windows from November to March (kBtu/ft^2)

	South				North			
	Albuquerque, NM	Madison, WI	Caribou, ME	Buffalo, NY	Albuquerque, NM	Madison, WI	Caribou, ME	Buffalo, NY
$\Delta T/H$ (°F ft^2 h/Btu)[a]	0.39	1.09	1.34	1.79	3.51	5.72	6.80	4.76
Single float glass	116.4	−35.9	−66.3	−73.6	−75.6	−132.6	−154.5	−113.8
Double float glass	139.6	26.7	7.8	−12.8	−29.2	−58.3	−69.7	−48.3
low-iron	153.1	34.2	14.7	−8.7	−27.7	−56.9	−68.4	−46.7
5 mm aerogel float	116.8	24.2	8.9	−8.7	−22.8	−46.1	−55.3	−38.0
low-iron	127.3	30.0	14.2	−5.5	−21.6	−45.0	−54.2	−36.8
20 mm aerogel float	72.0	25.4	18.6	6.0	−4.8	−13.3	−16.7	−10.2
low-iron	77.6	28.5	21.5	7.7	−4.2	−12.7	−16.1	−9.6
+ Glass-heat mirror-glass[b] (g-ph-g)	121.2	43.5	32.3	10.9	−7.4	−21.2	−26.8	−16.1
Glass-poly-poly-glass[b] (g-a-a-g)	150.3	56.5	43.1	16.2	−7.1	−22.8	−29.2	−16.9
20A FUTURE?	154.6	71.1	60.7	30.7	4.2	−4.6	−8.4	−0.9
R18 Opaque[c] white wall	−6.6	−8.7	−9.9	−7.7	−5.6	−8.7	−9.9	−7.7

Source: Neeper et al. (1984).
Note: Assumed room temperature is 68°F (20°C). Conversion factor: 1 kBtu/ft^2 = 10.36 MJ/m^2.
a. Anisotropic sky model. Ground reflectance is 0.3.
b. Net energy obtained using shading coefficients and U-values for half-inch gap widths from Rubin and Selkowitz (1981). Glass is float glass, heat mirror is single-coated 4 mil polyester, poly is 4 mil polyester with both sides antireflective coated.
c. Shown for comparison with windows.

Possibly other solar thermal uses of aerogel or evacuated glazing could better justify their federally sponsored development. An obvious candidate is their use as cover glass for flat plate collectors in applications that call for extreme durability (see section 26.4 for a discussion of glazing durability requirements in relatively benign environments). For this use the research and development goal presumably would be to achieve optical transmission comparable to that of conventional two-glass covers. Comparisons of the thermal conductance and solar transmittance of aerogel windows and standard double-glass windows suggest that an aerogel thickness of about 6 mm would provide equal transmittance with about 35% lower conductance. Deletion of the inner sandwich pane, together with increased aerogel thickness, could further improve the aerogel glazing performance advantage. This may be feasible due to the protected inner surface environment but would preclude the possibility of perhaps greater improvements by evacuation or the introduction of low molecular weight gasses. In the case of evacuated glazing (without aerogel) comparable improvements in transmittance might be achieved by deleting the infrared reflective coating, which also would simplify its manufacture. The consequently increased front end losses would tend to be offset by the selective absorber. Investigations along these lines may be at least as rewarding as those underway for building window applications, *provided* applications analyses confirm the existence of a sufficiently large market.

26.3   Switchable Glazings for Building Cooling Load Reduction

The materials considered here are electrochromic coatings for glass, which were reviewed in earlier chapters of this book. Their function is to reduce unwanted transmission of solar radiation, on command, without excessively impairing solar transmission when gain is desired for passive heating. Building cooling load reduction is the primary economic justification, although glare reduction and daylighting aspects may also be beneficial. There are two conceptually different technical approaches: variable absorptance and variable reflectance. At present the technology of variable absorptance coatings is approaching commercial status, having been developed entirely within the private sector. Variable reflectance coatings represent a relatively immature technology that offers potentially greater performance benefits: Research and development of this approach has been sponsored by DOE since 1983. In some situations both approaches appear to be potentially cost-effective; that is, the economic value of their cooling load reduction seems likely to exceed their additional cost. The main issue considered here is whether the potential value of variable reflectance coatings on glass is sufficiently higher than that of variable absorptance coatings to justify federally supported development. A related question, only touched upon here, centers on the marginal benefit of switchable glazing over combinations of heat-absorbing glass and fixed overhangs.

Two DOE-sponsored studies (Neeper and McFarland 1982; Parker and Shadis 1984) investigated the potential benefits of switchable glazings in building envelope applications. The results of both studies showed substantial annual energy savings in certain situations, but relatively small additional savings for variable reflectance coatings over variable absorptance coatings. However, there are large quantitative disparities that cannot readily be reconciled because of the quite different scenarios.

Neeper and McFarland (1982) examined the possible effects of switchable glazings on the performance of Trombe walls and direct gain systems in a representative house, considering both the impairment of passive solar-heating and the beneficial reduction of cooling loads imposed by the passive systems. They also examined the effects of switchable glazing when used as a window during the summer. Figure 26.3 and table 26.3 present simulation results for a south-facing Trombe wall. Figure 26.3 shows that for the assumed switched and unswitched conditions, the imposed sensible cooling load is

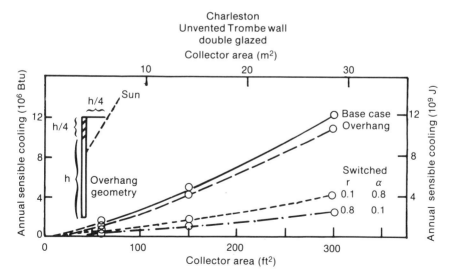

Figure 26.3
Annual added sensible cooling load versus area of a Trombe wall, for Charleston, South Carolina. The effect of the switched glazing is shown for two pairs of values of reflectance and absorptance. Source: Neeper and McFarland (1982).

Table 26.3
Effects of a switchable glazing on a south-facing Trombe wall

City	Glazing	Annual SSF (%)	Annual solar savings (kBtu/ft^2)	Annual solar cooling[a] (kBtu/ft^2)	June–September solar cooling[a] (kBtu/ft^2)
Albuquerque,	Standard	63	72.7	34.2	25.5
NM	Switchable	62	71.0	7.6	5.0
Charleston,	Standard	62	35.9	41.2	28.4
SC	Switchable	59	34.1	8.4	6.6
Columbia,	Standard	30	37.5	29.7	24.6
MO	Switchable	29	36.0	6.1	5.0

Source: Neeper and McFarland (1982).
Note: Conversion factor: 1 kBtu/ft^2 = 10.36 MJ/m^2. 300-ft^2 (27.8-m^2) flat-black, double-glazed Trombe wall. Standard glazing: $\tau = 0.7$. Switched glazing: outer pane has $\tau = 0.1$, $r = 0.8$ when room temperature is $> 73°F (23°C)$ but returns to normal when room temperature $< 70°F (21°C)$.
a. Cooling requirments due to the solar system.

reduced 80% by the variable reflectance coating and about 67% by the variable absorptance coating. (It is unclear whether these reductions include the 12% reduction indicated for the fixed overhang or are for a totally unshaded wall.) Nowhere else in Neeper and McFarland (1982) are there data on the performance of variable absorptance coatings. In table 26.3 it is seen that the maximum annual cooling load reduction (variable reflectance, in Charleston) of 32.8 kBtu/ft^2 (340 MJ/m^2) is accompanied by a passive heating penalty of 1.8 kBtu/ft^2 (18.6 MJ/m^2). Assuming an electric heat pump with a seasonal cooling COP of 3 and a seasonal heating COP of 2, the maximum annual savings due to switchable glazing would be 2.9 kWh/ft^2 (22.6 kWh/m^2). Assuming a 20-year system life, a 5% general escalation rate, a 10% discount rate, and electricity at 7¢/kWh, the present worth (allowable extra cost) compared to clear glazing would be \$1.3/ft^2 (\$14.0/m^2) and the net worth of variable reflective compared to variable absorptive coatings would be only about 21¢/ft^2 (\$2.24/m^2). Considering also that switchable glazing for Trombe walls offers no possible additional advantages of glare reduction or daylighting, this application appears to offer little incentive for the development of either kind of coating.

Table 26.4 shows that for direct gain systems the net annual benefit of switchable glazings can be much greater than for Trombe walls. In Albuquerque, using the same assumptions and arithmetic as above, the present worth of variable reflectance glazing would be about \$3.9/ft^2 (\$42.0/m^2), or approximately 63¢/ft^2 (\$6.8/m^2) more than that of variable absorptance glazing if the same ratio of cooling load reduction (67/80) applies. Although a

Table 26.4
Effects of switchable glazing on south-facing direct gain

City	Glazing	Annual SSF (%)	Annual solar savings (kBtu/ft²)	Cooling due to solar system (kBtu/ft²)	
				Annual	June–September
Albuquerque,	Standard	57	65.9	150	66.2
NM	Switchable	35	39.8	20.2	13.1
Charleston,	Standard	53	30.9	131	72.2
SC	Switchable	31	17.6	23.4	16.6
Columbia,	Standard	23	28.5	103	61.5
MO	Switchable	11	14.1	16.8	12.9

Source: Neeper and McFarland (1982).
Note: Conversion factor: 1 kBtu/ft² = 10.36 MJ/m². 300-ft² (27.8 m²) double-glazed aperture. The optical properties are the same as those of table 26.3.

marginal cost goal of \$3.90 may be achievable for switchable glazing, the differential of 63¢ seems to offer slight justification for developing a second switchable glazing technology.

Table 26.5, which presents the effects of switchable glazing when used as a window, considers only the summertime reduction of imposed cooling load. The net annual energy savings may be greater if the remaining months are cooling-dominated or less if they require much daytime heating. The maximum summer cooling load reduction (a west-facing window in Albuquerque) is seen to be 67.7 kBtu/ft² (700 MJ/m²), corresponding to a present worth of perhaps \$4.1/ft² (\$43.1/m²) and a differential of possibly 67¢/ft² (\$7.20/m²) if indeed the reduction ratio 67/80 applies. Here again the *economic* justification for developing the second technology seems doubtful.

Parker and Shadis (1984) considered only window applications, with annual simulations of modeled buildings. They modeled a double-glazed house in Madison, Wisconsin, the same house with single glazing in El Paso, Texas, and a 40,000 ft² (3,710 m²) office building with single glazing in El Paso.

Tables 26.6 through 26.9 summarize the results in terms of present worth, assuming a general inflation rate of 5% and discount rates of 10% for the residence and 20% for the commercial building. Note that in El Paso all of the tabulated net worths are plausibly achievable, even considering that the cost of manufacture must be lower than the values shown, which are what the buyer would be willing to pay. This also is the case in Madison for a 20-year system life and electricity at 7¢ to 10¢/kWh. For the residence in El Paso the differential present worth for variable reflectance may be achievable in some cases, but in all other scenarios it appears to be hopelessly low.

Table 26.5
Effects of switchable glazing when used as a window

		South		West	
City	Glazing	Solar incident	Energy transmitted	Solar incident	Energy transmitted
Albuquerque,	Standard	140	74.5	148	90.8
NM	Switchable	140	24.6	148	23.1
Charleston,	Standard	102	56.6	103	62.7
SC	Switchable	102	12.1	103	12.7
Columbia,	Standard	124	68.6	133	82.0
MO	Switchable	124	22.2	133	20.8

Source: Neeper and McFarland (1982).
Note: Units: $kBtu/ft^2$. Conversion factor: 1 $kBtu/ft^2$ = 10.36 MJ/m^2. The optical properties are the same as those of table 26.3. The incident solar radiation and the net energy transmitted by the window (including conduction) are given for the months of June through September.

Table 26.6
Present worth of double-glazed electrochromic windows in Madison, Wisconsin, 1,400-ft^2 residence

Electricity cost (¢/kWh)	Absorbing windows		Reflecting windows	
	10 years	20 years	10 years	20 years
5	1.70	2.74	1.85	2.99
7	2.39	3.86	2.55	4.11
10	3.40	5.48	3.62	5.86

Source: Parker and Shadis (1984).
Note: Units: $/ft^2$. Conversion factor: 1$/ft^2$ = 10.8 $/m^2$.

Table 26.7
Present worth of single-glazed electrochromic windows in El Paso, Texas, residence compared to ordinary clear glazings

Electricity cost (¢/kWh)	Absorbing windows		Reflecting windows	
	10 years	20 years	10 years	20 years
5	3.55	5.73	4.48	7.23
7	5.02	8.10	6.33	10.22
10	7.18	11.59	9.03	14.58

Source: Parker and Shadis (1984).
Note: Units: $/ft^2$. Conversion factor: 1$/ft^2$ = 10.8 $/m^2$.

Table 26.8
Present worth of electrochromic windows in speculative office building, E1 Paso, Texas, compared to clear glass

Window	Low energy prices		Medium energy prices		High energy prices	
	10 years	20 years	10 years	20 years	10 years	20 years
Electrochromic heat-absorbing glass	5.58	6.96	7.60	9.48	10.74	13.39
Electrochromic reflecting glass	5.75	7.23	7.90	9.84	11.17	13.92

Source: Parker and Shadis (1984).
Note: Units: $/ft^2. Conversion factor: 1 $/ft^2 = 10.8 $/m^2.

Table 26.9
Marginal present worth of reflecting electrochromic glazing over absorbing electrochromic glazings in a speculative office building, E1 Paso, Texas

Glazing lifetime	Low energy prices	Medium energy prices	High energy prices
10 years	0.211	0.296	0.423
20 years	0.2664	0.372	0.529

Source: Parker and Shadis (1984).
Note: Units: $/ft^2. Conversion factor: 1 $/ft^2 = 10.8 $/m^2.

Comparing the results of these two studies is tricky at best. Neeper and McFarland (1982) assume double glazing throughout, with "coloring" of the outer glazing whenever the room temperature exceeds 73°F (23°C) and return to normal when it falls below 70°F (21°C). Parker and Shadis (1984) assume double glazing only in Madison, with constant coloring year-round in all cases. Neeper and McFarland (1982) look only at effects on loads, whereas Parker and Shadis (1984) assume COPs (unstated) and compute purchased energy demand. The building models are different and the climates are dissimilar. For windows Neeper and McFarland (1982) consider a single window in isolation; Parker and Shadis (1984) model them as part of the building. With all this in mind one may select the two most similar climates (Columbia and Madison), use consistent economic assumptions, and compare present worths of variable reflective window glazings relative to standard double glazing. For Neeper and McFarland (1982) (Columbia) the present worth is estimated as $2.0 to $2.6/ft^2 ($21.6 to $28.0/m^2); for Parker and Shadis (1984) (Madison) it is listed as $4.11/ft^2 ($44.3/m^2). For the latter the ratio of present worths (absorptive/reflective) is 0.94; for the former it is estimated as 0.84.

Table 26.10
Present worth of single glazed electrochromic windows in El Paso, Texas, compared to clear glass and standard heat-absorbing glass windows

Window	Residence	Office building
Clear glass	0	0
Standard heat absorbing	3.75	3.03
Variable absorptance electrochromic	8.10	9.48
Variable reflectance electrochromic	10.22	9.84

Source: Adapted from Parker and Shadis (1984).
Note: Units: $/ft^2. Conversion factor: 1 $/ft^2 = 10.8 $/m^2. Assumptions: 20-year system life, 5% general inflation rate, 10% discount rate for residence, 20% discount rate for office building, 7 ¢/kWh electricity cost.

In Parker and Shadis (1984) the present worths of switchable glazings were determined relative to standard heat absorbing glazing as well as to ordinary clear glass. The results for a 20-year system life and electricity at 7¢/kWh are presented in table 26.10, showing that the allowable costs above those of heat-absorbing glass·are 30% to 46% less than above clear glass. These results presumably did not include the effects of any sizable fixed overhangs, which would further reduce the marginal present worth of switchable glazings. Studies of cooling load reduction by means of fixed south overhangs (Lau 1982; Siminovitch et al. 1982; Jones 1981) show that in mid-latitudes reduction of the aperture-imposed load can be as high as 50%. Conceivably, then, combinations of standard heat-absorbing glass and fixed overhangs could in some situations reduce the marginal present worth of switchable glazings to the point of noncompetitiveness.

As a final point the most recent available (March 1985) update on the development of variable reflectance electrochromic glazing (Rauh 1985) casts further doubt on the economic justification for that approach. It reports that based on accumulated results, a solar throughput of > 80% in the transmissive state and < 10% in the blocking state is feasible. It goes on to say, however, that with present materials a window with those capabilities can be designed in which about 50% of the light blockage is reflective. No indication is given of hopes or expectations to exceed that percentage.

26.4 Durability of Polymeric Film Glazing for Flat Plate Collectors

Recommendations to DOE on materials research priorities for active systems (Herzenberg 1983) emphasize the need for polymeric glazings with improved

Table 26.11
Polymer glazing properties and AHAC system cost-effectiveness

Property(ies) changed	Glazing 1[a]	Glazing 2[a]	Change in system efficiency[f] (%)	Economic value ($/m²)[b] 1 $/ft² = 10.8 $/m²
Solar transmissivity	$\tau_i = \tau_{av} = 0.91$	$\tau_i = \tau_{av} = 0.81$	10–20	42–86[c] 22–43[d] (4–8) (2–4)
Optical property durability	$\tau_i = \tau_{av} = 0.91$	$\tau_i = 0.91\ \tau_{av} = 0.81$[e]	10–20	26–52[c] 13–26[d] (2.4–4.8) (1.2–2.4)

Source: Herzenberg (1983).
Note: Conversion factor: 1 $/ft² = 10.8 $/m². $/ft² are in parentheses.
a. Properties not listed are assumed to be $\tau_i = \tau_{av} = 0.91$; $\alpha_{IR_i} = \alpha_{IR_{av}} = 0.95$.
b. Calculated assuming a 10% discount rate and that competing systems are designed to collect the same amount of energy in each year of their 20-year lifetime.
c. Installed cost of collector plus transport piping = $430/m² ($40/ft²).
d. Installed cost of collector plus transport piping = $215/m² ($20/ft²).
e. Optical properties are assumed to degrade linearly throughout the 20-year period.
f. Estimated from Hottel-Whillier efficiency plots for existing low–medium temperature systems (S. Moore, private communication).

durability for both medium temperature and high temperature flat-plate collectors. They outline a research program with the goal of a 20-year glazing life with minimal degradation, to match the typically assumed economic lifetime of solar collector systems. This is an exceptionally demanding technical goal, since it applies principally to stretched thin-film glazings which currently have estimated service lives of only two to five years 3 (see Herzenberg 1983, ch. 3.2, table 3.2.1). Although it may be valid for the long payback times associated with conventional collector systems that cost $20 to $40/ft² (1.8–2.7 m²), some rethinking may be in order if the much lower costs and consequently faster paybacks projects by the Brookhaven National Laboratory for ultralightweight integrated polymeric collectors (Wilhelm 1985) are indeed realistic.

Table 26.11 presents the economic basis for assigning a high priority to glazing durability improvement, indicating the costs attributable to sacrifices in optical properties and their durability. For the assumed 20-year economic lifetime and a collector system cost of $40/ft² ($430/m²), the total elimination of optical property degradation is seen to justify additional glazing costs of $2.4 to $4.8/ft² ($25.8 to $51.6/m²). Currently available 4-mil Tedlar 400se film has essentially acceptable initial properties and costs only about $0.4/ft² ($4.3/m²), so it might be said that a "20-year Tedlar" film would have an allowable cost up to twelve times that of "two-to-five-year Tedlar." That may indeed be an achievable goal. However, the projected installed cost of advanced integrated polymeric collectors is less than $4/ft^2 ($43/m^2) (Andrews et al. 1983); for such collectors the allowable increase in glazing cost might be as low as 50%. Since some degradation presumably will persist regardless of cost, the actual allowable increase would be still smaller. Considering that the competitive prospects of active solar systems hinge largely on the availability of ultra-low-cost collectors, the recommended goal seems unrealistic.

Less demanding goals for the improvement of polymeric glazing durability can more readily be justified. Corresponding to the extremely low system cost projections by Brookhaven are payback times of seven years or less, which are expected to stimulate a residential market penetration of about 20%. For such a quick payback the assigned economic lifetime could legitimately be reduced to perhaps ten years, so "lifelong" glazing could cost less and be more readily developed. Since the balance-of-system (piping, storage, etc.) would have an inherently longer actual lifetime and would account for a large fraction of the total initial installed cost, it may be found economical to replace a complete individual collector unit whenever its polymeric film glazing (which in the Brookhaven designs is an essential structural element) becomes excessively degraded. Similar reasoning may be applied regarding durability improvement of low-cost polymer film/metal foil absorbers.

Additional perspective can be gained by reviewing the polymer reflector research plan for the National Solar Thermal Technology Program (DOE 1984). Currently available silvered polymeric films for low-cost concentrators have a high (94%–95%) initial reflectivity but useful lifetimes of less than a year. The near-term (by 1990) research and development goal is 93% reflectivity averaged over a five-year life, although the overall system life is much longer.

26.5 New Solid Absorbents

From 1982 until 1984 the National Bureau of Standards had been evaluating ettringite (Candlot's salt) and related Friedel salts as solar thermal energy storage materials (Ings and Brown 1982; Struble and Brown 1984). Ettringite is a naturally occurring highly hydrated calcium alumino-sulfate

$$[Ca_6Al_2(OH)_{12}](SO_4)_3 \cdot 26H_2O$$

that dehydrates partially and apparently reversibly at fairly low temperatures with interestingly high heats of reaction. It is readily synthesized at low cost, and various chemical substitutions can be made to tailor the dehydration temperatures without changing the ettringitelike structure.

The sponsoring DOE passive solar components and materials program views these solid desiccants primarily as potentially superior phase change storage materials that partially "melt" by releasing liquid water of hydration as their temperatures rise through a narrow temperature band, then "freeze" when decreasing temperature causes them to rehydrate as a single solid phase. Their enthalpy change over the range of temperatures typically encountered in passive solar applications is expected to be at least twice that of glaubers salt. This would represent a significant advance in phase change thermal energy storage technology, and continued evaluation clearly is justified. However, investigation of other applications and modifications of these materials may also prove to be rewarding.

One potentially promising additional use of ettringitelike materials is in open-cycle air-based, active solar-heating systems such as that shown schematically in figure 26.4. The use of solid adsorbent beds in this fashion has been investigated by Close and Dunkle (1977) and Verdonschot (1980) using silica

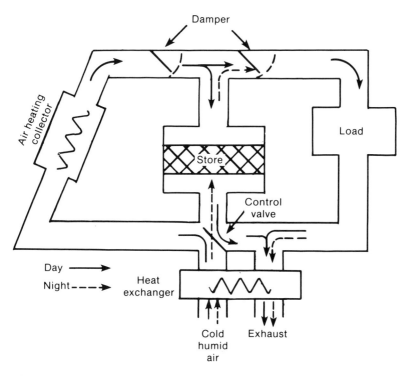

Figure 26.4
Open-cycle air-based active solar-heating system with absorbent bed storage. Source: Adapted from Close and Dunkle (1977).

gel and by Shigeishi et al. (1979) using zeolites. Close concluded that with silica gel the scheme could not compete economically with more conventional concepts using rock bed stores; cost savings due to reduced bed volume would be outweighed by the much higher cost per unit mass of the storage medium. A comparable situation appears to limit the potential cost-effectiveness of zeolites, which have the additional drawback of requiring higher drying temperatures. With ettringitelike materials, however, there is cause for optimism even though they have not yet been fully characterized. They cost far less per unit mass than either silica gel or zeolites, and their mass density is much higher. Their maximum weight percentage of water transfer is substantially greater, and most of the dehydration appears to occur at lower temperatures. Since they are *ab*sorbents rather than *ad*sorbents their energy storage capacity per mole of water is greater due to the higher heat of reaction (apart from the heat of condensation of the water), and their equilibrium water vapor pressure does not depend on the amount of absorbed water. For an adequate assessment of their potential in this application, dynamic as well as equilibrium sorption/desorption data must be obtained.

These materials may also be found superior to desiccants that are commonly used in open-cycle solar-cooling systems, but the case for them is less clear than in heating applications. Their apparent ability to dehydrate at lower temperatures would tend to be offset by their higher (than adsorbents) heat of reaction (Collier 1980). The net benefit, if any, could be ascertained by cost and performance analyses after fuller characterization.

Perhaps the most exciting opportunities for exploiting ettringite-like materials lie in the area of closed-cycle, thermochemical heat pumping and storage (see part II, chapters 15 and 19). Each of the extensively explored solid-gas systems (e.g., sodium sulfide/water, calcium chloride/methanol, calcium chloride/ammonia) has a sharply limited temperature boosting or transforming range that is dictated by its specific pressure temperature characteristics. These ranges seldom are optimal for commonly encountered requirements, so there are often unavoidable severe cost and performance penalties. With ettringitelike materials it should be possible to optimize more fully, especially in building applications, by chemical substitutions such as iron for the aluminum and carbonate for the sulfate. It may also be possible to extend their operational suitability even further by substituting for water other polar compounds such as methanol or ammonia, which additionally could avoid problems of freezing and of high volumetric flow rates associated with high vacuum. Although these projections are quite speculative, the payoffs could be large and the cost of evaluating them would be modest.

26.6   High Temperature Receiver/Storage Fluids

Applications have been identified that require solar-heated liquids at temperatures up to 2,102°F (1,150°C) (Luft 1985). These applications include combined Brayton/Rankine cycles for power generation, heated air for industrial processes, and cogeneration. Still higher temperature uses of solar-heated liquids have also been explored: 2,552°F (1,400°C) for a proposed closed-cycle argon Brayton cycle for power generation (Bruckner and Hertzberg 1982). Molten nitrate salts are suitable for temperatures up to perhaps 1,040°F (560°C), above which there is excessive decomposition. Molten carbonates, and methods of containing them, are being actively investigated for temperatures up to about 1,742°F (950°C). Molten carbonates, chlorides, and hydroxides are stable at temperatures as high as 2,012°F (1,100°C) with acceptably low vapor pressures—namely, pressurized tanks are not required, but their containment problems in that temperature regime have not yet received programmatic attention. No suitable liquids have been identified specifically for the identified 2,102°F (1,150°C) applications, but molten slag at 2,552°F (1,400°C) was the suggested receiver/storage/heat transfer liquid for the proposed argon Brayton system, and its containment problems were fairly thoroughly investigated.

- In addition to pushing the temperature limit of molten salts upward, it seems prudent to search concurrently for molten slags with suitably low viscosity at temperatures well below 2,552°F (1,400°C). Direct contact heat exchange of molten slag droplets with air can deliberately produce "glass beads" that may be stored for indefinitely long periods or immediately mechanically conveyed to the receiver for remelting. If storage capacity of the melt is not more than a few hours, the liquid can be stored below the receiver and above the heat exchanger, permitting gravity flow to and from storage. These potential advantages appear to be worth an intensive search for suitable slags.

References

Andrews, J. W., P. LeDoux, P. Metz, and W. Wilhelm. 1983. *Development of Low-Cost Polymer Film Solar Collectors: Annual Report.* Brookhaven report no. BNL 51693.

Benson, D. K. 1984. *Laser Sealed Evacuated Window Glazings.* SERI/TP-225–2454.

Bruckner, A. P., and A. Hertzberg. 1982. *High Temperature Integrated Thermal Energy Storage System for Solar Thermal Applications.* Final subcontract report, SERI/STR-231–1812. Golden, CO: Solar Energy Research Institute.

Close, D. J., and R. V. Dunkle. 1977. Use of adsorbent beds for energy storage in drying of heating systems. *Solar Energy* 19: 233.

Collier, R. K. 1980. Desiccant and other cooling systems. *Solar Cooling*, M. Nazer et al., eds. MRI/SOL-0601 (DOE CONF-8004244-), pp. 89–106.

Herzenberg, S. A. 1983. *Active Solar Energy System Materials Research Priorities, A Subcontract Report*. SERI/STR-255–1782. Golden, CO: Solar Energy Research Institute.

Ings. J. B., and P. W. Brown. 1982. *An Evaluation of Hydrated Calcium Aluminate Compounds as Energy Storage Media*. National Bureau of Standards report no. NBSIR 82–2531, 15 pp.

Jones, R. W. 1981. Summer heat gain control in passive solar heated buildings: Fixed horizontal overhangs. Presented at International Passive and Hybrid Cooling Conference, Miami Beach, FL. November 11–13, 1981.

Lau, A. 1982. Design and effectiveness of fixed overhangs: Development of a novel design tool. *Proc. 7th National Passive Solar Conference*, J. Hayes and C. B. Winn, ed. ASES, pp. 393–398.

Leigh, R. W., and V. Fthenakis. 1981. *The Value of Technical Improvements in Passive Solar Collection Storage Devices*. Brookhaven report no. BNL 51486.

Luft, W. 1985. *SERI Solar Energy Storage Program: FY1984 Annual Report*, SERI/TR-231-2593. Golden, CO: Solar Energy Research Institute.

Neeper, D. A., and R. D. McFarland. 1982. *Some Potential Benefits of Fundamental Research for the Passive Solar Heating and Cooling of Buildings*. Report no. LA-9425-MS. Los Alamos National Laboratory.

Neeper, D. A., R. McFarland, J. Hedstrom, and G. Lazaruz. 1984. Potential performance benefits of advanced components and materials research. *Proc. Passive and Hybrid Solar Energy Update*, Washington, DC, September 5–7, 1984, DOE CONF-8409118, pp. 21–26.

Parker, A. J., Jr., and W. J. Shadis. 1984. Analysis of the cost/performance of advanced passive materials, *Proc. Passive and Hybrid Solar Energy Update*. DOE CONF-8909118, pp. 5–9.

Rauh, R. D. 1985. Variable transmittance electrochromic windows. *Preliminary Proceedings of the Solar Buildings Conference*. DOE/CONF-850388, pp. 142–148.

Rubin, M., and S. Selkowitz. 1981. Thermal performance of windows having high solar transmittance. *Proc. Sixth National Passive Solar Conference*, Portland, OR. September 8–12, p. 141.

Selkowitz, S. E., A. Hunt, C. Lampert, and M. Rubin. 1984. Advanced optical and thermal technologies for aperture control. *Proc. Passive and Hybrid Solar Energy Update*. DOE CONF-8409118, pp. 10–19.

Selkowitz, S. E., C. Lampert, and M. Rubin. 1983. Advanced optical and thermal technologies for aperture control. *Proc. Passive and Hybrid Solar Energy Update*. DOE CONF-830980, pp. 208–216.

Shigeishi, R. A., C. Langford, and B. Hollebone. 1979. Solar energy storage using chemical potential changes associated with drying of zeolites. *Solar Energy* 23: 489.

Siminovitch, M. J., D. Bergeson, and M. McCulley. 1982. Thermal protection of the solar aperture in the cooling season—A quantitative performance based study. *Proc. 7th National Passive Solar Conference*, J. Hayes and C. B Winn, eds. ASES, pp. 883–888.

Struble, L. J., and P. W. Brown. 1984. *An Evaluation of Ettringite and Related Compounds for Use in Solar Energy Storage*. National Bureau of Standards report no. NBSIR 84–2942.

U.S. Department of Energy. 1984. *National Solar Thermal Technology Program Five Year Research and Development Plan* (draft).

Verdonschot, J. K. M. 1980. Thermal storage system based on the heat of adsorption in air-based solar heating systems. *Thermal Storage of Solar Energy*, C. den Ouden, ed. The Hague: Martinus Nijhoff Publishers. pp. 283–294.

Wilhelm, W. G. 1985. Polymer film solar collector development. *Preliminary Proc. Solar Buildings Conference*. Washington, DC, March 18–20, 1985. DOE/CONF-850388, pp. 50–57.

Contributors

Thomas E. Anderson

Thomas E. Anderson has been involved in the exposure testing and evaluation of materials for solar application at Desert Solar Exposure Testing Laboratories (DSET) since 1972. He has extensive experience in outdoor and accelerated materials testing and instrumentation and has published a number of papers. He also holds several patents for the development of various lacquers and coatings.

Charles A. Bankston

Charles A. Bankston, the editor-in-chief of this series, has been active in solar energy research and development for more than 15 years. Bankston was professionally associated with the Los Alamos National Laboratory, where he conducted directed research in heat transfer, nuclear propulsion, advanced drilling, geothermal energy, and solar energy, from 1958 to 1982. In 1982 Bankston founded, and is the president of CBY Associates, an energy consulting firm. While at Los Alamos, Bankston was responsible for technical direction of the U.S. DOE program for solar collector and materials research—a program that involved over 100 research contracts. His personal research in solar energy applications includes studies of large collector arrays, concentrating collectors, seasonal thermal energy storage, and systems analysis and optimization. Bankston has published numerous paper and articles, and is an associate editor of *Solar Energy Engineering*. He received his DSc degree in mechanical engineering from the University of New Mexico.

Eugene Clark

As a Fulbright Scholar at Kings College, University of London, Eugene Clark discovered the critical importance of energy to the United States. As president of Applied Solar Engineering for 10 years, he participated in the cooling design of many major commercial and institutional buildings. As professor of physics, he directed Trinity University's graduate program in solar energy applications. Clark has been principal investigator in the evaluation of passive cooling resources and assessment of the performance of select residential passive cooling systems (including roof ponds) in hot climates of the United States. He has degrees from the University of Florida in physics with graduate work in astrophysics.

John A. Clark

John A. Clark has been involved with active solar thermal energy research since 1972. He is professionally affiliated with the University of Michigan, Ann Arbor, and is president of Central Solar Energy Research Corp. and director of research at Starpak Energy Systems. His major accomplishments in solar energy research include studies in the performance of boiling collectors, parabolic trough concentrators, thermal storage systems, economic analysis of the performance of solar thermal and energy avoidance systems, manufacture methodology and cost analysis of wind energy, photovoltaic and thermal concentrator systems. His contributions to the solar energy literature has been in economic analysis and thermal storage.

Francis de Winter

Francis de Winter has been involved with spacecraft power and solar thermal energy research since 1966. From 1966 to 1974 he was professionally affiliated with the Spacecraft Power Section of the Jet Propulsion Laboratory in Pasadena, California. From 1974 to 1989 he was president of the Altas Corporation in Santa Cruz. De Winter originated the "heat exchanger factor and penalty" concepts, which describe the effect of a heat exchanger interposed between a solar collector system and a water storage tank. These have become an integral part of the Hottel-Whillier collector model. Among his other achievements are the PV power design for the *Venus–Mercury Flyby* spacecraft, and he worked on the (nonsolar) RTG power system for the *Voyager* spacecraft. He has developed new system designs for the solar heating of swimming pools, and for the heating of domestic water for the single-family home using an integrated gas backup heater. He was formerly chairman (1978, 1982, 1984, 1985), vice chairman (1980), and treasurer (1981, 1989) of ASES, chairman of the Energy Advisory Committee of city of Santa Cruz, and is currently finance chairman of the ISES/ASES Solar Congress, Denver. He received his BS and MS from the the Massachusetts Institute of Technology in 1958 and 1960, respectively, as well as a mechanical engineer degree in 1961.

Robert D. Dikkers

Robert D. Dikkers is assistant chief of the Structures Division of the Center for Building Technology, National Institute of Standards and Technology (formerly the National Bureau of Standards). From 1974 to 1986 he was responsible for the management and coordination of solar heating and cooling research activities. He served as chairman of the ANSI Steering Committee on Solar Energy Standards Development from 1976 to 1978 and received the Department of Commerce Silver Medal Award for his significant contributions to the development of national performance criteria and standards for solar energy systems in 1979. Dikkers received his BS and MS degrees in civil engineering from Northwestern University and the California Institute of Technology, respectively.

William Freeborne

William Freeborne has been involved with both active and passive solar energy research since 1975. He served as the technical specialist for the solar residential demonstration program which was predominant at the Department of Housing and Urban Development from 1975 to 1980. He currently is the chairman of the American Society of Testing and Materials E-44 subcommittee on active and passive systems. Freeborne has a BSME from Union College, Schenectady, New York, and a masters in engineering administration from George Washington University.

John R. Hull

John R. Hull is staff scientist at Argonne National Laboratory. Since 1980 he has worked on diverse energy-related research projects, including thermal energy storage, evacuated-tube solar collectors, and solar ponds. Hull received a BS and PhD in physics from Iowa State University.

David W. Kearney

David W. Kearney has been involved with solar thermal energy research and development since 1978. Kearney is vice president of advanced technology with Luz Development and Finance Corporation in Los Angeles, part of the Luz group which is the major world developer of large 30–80 MWe solar thermal electric power plants. Kearney managed the Industrial Process Heat Program as well as the Solar Thermal, Ocean and Wind Division at the Solar Energy Research Institute, leaving in 1981 to consult in the energy field before joining Luz in 1984. He holds degrees in mechanical engineering, receiving a PhD from Stanford University in 1970.

Charles F. Kutscher

Charles F. Kutscher is a senior engineer at the Solar Energy Research Institute which he joined in 1978; he currently manages the Solar Heating Program. Kutscher designed and built SERI's Desiccant Cooling Test Laboratory, managed the production of a design handbook for solar industrial process heat systems, and directed an effort to analyze and develop stretched-membrane parabolic concentrators. He has participated as a speaker in numerous solar design workshops and has written chapters in several books dealing with solar energy design. He received a BS in physics from the State University of New York at Albany and an MS in nuclear engineering from the University of Illinois at Champaign-Urbana.

Carl M. Lampert

Carl M. Lampert has been actively engaged in solar energy research. Since 1974 he has worked at the Lawrence Berkeley Laboratory where he is staff scientist responsible for the research and development of new optical materials and coatings for glazing and solar energy applications. Lampert is scientific advisor and technical program planner for the U.S. Department of Energy-Solar Buildings Division. He is the U.S. representative to the International Energy Agency (IEA) program on solar materials. Lampert has lectured in the U.S., Europe, Venezuela, the West Indies, the Middle East, China, Japan, Thailand, Indonesia, and India. He has published over 60 papers and has chaired and organized 9 conferences, all on the materials science and optics of energy-related materials. He is also general editor of the *Journal of Solar Energy Materials*. He is also a consultant for the United Nations Development Program in China, India, and Yugoslavia and for private firms in the fields of commercial coatings, electronics, and aerospace materials. He received his PhD in materials science from the University of California at Berkeley.

Hien K. Le

Hien K. Le has specialized in solar materials and energy conversion and conservation issues. From 1982 to 1987, as a senior engineer at DHR Inc., then a subsidiary of QuesTech Inc., Le assessed and prioritized for the U.S. Department of Energy research on active solar energy system materials (polymer glazings, heat mirrors, selective absorbers, absorber subtrates, adhesives, fluid and corrosion, thermal storage materials, and desiccants) and photovoltaics cells (silicon, III-V abd II-VI materials). Le holds BS and MS degrees in metallurgical engineering from the Ohio State University. He currently works as a computer consultant in the Washington. DC area.

Noam Lior

Noam Lior is professor and graduate group chairman of mechanical engineering and applied mechanics at the University of Pennsylvania. He has been intensively involved in active, passive, and solar thermal energy research since 1973. Some of the specific projects include heat transfer and fluid mechanics of solar collectors and salt-gradient solar ponds, heat transfer in phase change thermal storage, development of a novel solar-powered/fuel-assisted hybrid Rankine cycle, open-cycle OTEC, solar heating and cooling, solar retrofit, and applications of solar energy in the urban environment. He designed and brought into operation the first solar-heated home in Philadelphia (a retrofit), and served as director of the American Solar Energy Society, chairman of its Engineering Division, and associate editor of the ASME *Journal of Solar Energy Engineering*. In the solar energy field he was consultant to ERDA, DOE, Argonne National Laboratory, and the Solar Energy Research Institute. He has published over 80 papers and is the editor of *Measurements and Control in Water Desalination* (Elsevier, 1986).

Fred Loxsom

Fred Loxsom, a biophysicist by training, has worked actively and productively in the field of solar energy since joining the staff of Trinity University in 1977. Loxsom has directed research and published widely in passive solar systems for buildings as well as in physics and biophysics. He has been responsible for much of acquisition and analysis of data from Trinity's passive solar test facility and its solar meteorological site. Loxsom received his PhD in physics from Dartmouth College in 1969.

Allan I. Michaels

Allan I. Michaels is assistant manager of the DOE Tribology Project and the Tribology Section at Argonne National Laboratory (ANL). He directed the Thermal Energy Storage (TES) Program of the DOE Office of Solar Applications and the ANL Solar TES Program from 1975 to 1984. In 1979 he organized and chaired a large and pivotal workshop, "Solar Energy Storage Options," which explored the TES technical status and development needs for six solar energy applications. From 1979 to 1984 he was leader of the U.S. delegation to the IEA Solar Heating and Cooling Program, Task VII on Central Solar Heating Plants with Seasonal Storage. He holds a PhD in materials science from Stanford University.

Stanley W. Moore

Stanley W. Moore has been involved with both active and passive solar energy research since 1973. He is a staff member of the Los Alamos National Laboratory Solar R&D program and has served as associate group leader. At Los Alamos he has established the passive solar test facility and the materials exposure test facility and has served as project manager for the active and passive mobile/modular solar home projects. Moore has authored numerous Los Alamos and technical society reports. He received a BS in mechanical engineering from Kansas State University.

Carl E. Nielsen

Carl E. Nielsen is professor emeritus of physics at the Ohio State University, where he continues to be active in the solar pond research he began there in 1973. He received his PhD from the University of California at Berkeley in 1941 and has worked in cosmic rays, high energy accelerators, and plasma physics and controlled fusion. Nielsen has consulted with solar pond workers throughout the world; he is the author of papers on solar ponds and pond processes and of numerous review chapters on solar ponds. He is coauthor, with John R. Hull and Peter Golding, of *Salinity-Gradient Solar Ponds* (CRC Press, 1989).

Ari Rabl

Ari Rabl is research scientist with the Centre d'Energétique of the Ecole des Mines in Paris. He has also worked at Argonne National Laboratory, the Solar Energy Research Institute, and Princeton University. His solar research has been concerned with seasonal storage, solar collectors, insolation models, and design methods. He holds a PhD in physics from the University of California at Berkeley.

Richard Silberglitt

Richard Silberglitt is senior scientist at Technology Assessment and Transfer, Inc., a high technology company that specializes in advanced materials, tribology, and manufacturing technology. He became interested in solar energy in 1976 as senior staff officer of the National Academy of Sciences' CONAES study. He has consulted for DOE's Office of Conservation and Renewable Energy since 1977, chaired comprehensive independent technical evaluations of materials research needs for solar thermal collectors and photovoltaic cells for DOE from 1979 to 1983, and has served on technical review panels for DOE, the Solar Energy Research Institute, and the National Bureau of Standards (now NIST). He holds a BS in physics from Stevens Institute of Technology, and an MS and PhD in theoretical solid state physics from the University of Pennsylvania.

Charles J. Swet

Charles J. Swet has been involved with all aspects of solar energy research for more than 20 years. He has been professionally affiliated with the U.S. Department of Energy, the Johns Hopkins University Applied Physics Laboratory, and various companies in the aerospace, textile, chemical, steam power generation, and shipbuilding industries. He formerly managed DOE's thermal and thermochemical energy storage R&D. Swet has authored more than 50 papers and has lectured widely on solar conversion, storage, and applications. He received a BS in naval architecture and marine engineering from the Massachusetts Institute of Technology in 1946.

Roland Winston

Roland Winston is professor and chairman designate in the department of physics at the University of Chicago. In connection with designing Cerenkov radiation detectors for a high energy experiment in 1965, Winston discovered the ideal nonimaging concentrator now called the compound parabolic concentrator. He extended the principles of nonimaging concentration to the design of solar collectors in 1973 and has led research groups in solar energy at Argonne National Laboratory and at the University of Chicago. In addition to nonimaging optics and solar energy collection, Winston is also active in experimental high energy physics conducted at the Fermilab National Accelerator Laboratory. He is a member of the board of directors of the American Solar Energy Society. He was an Alfred P. Sloan Fellow from 1967 to 1969, a Guggenheim Fellow from 1977 to 1978, and received the Charles Greeley Abbot Award of the American Solar Energy Society in 1987. Winston has authored over 100 publications and has coauthored (with W. T. Welford) two books, *The Optics of Nonimaging Concentrators* (Academic Press, 1978) and *High Collection Nonimaging Optics* (Academic Press, 1989). He holds 14 U.S. letter patents on nonimaging concentration of radiant energy. Winston received the PhD in physics from the University of Chicago in 1963.

Index

Abbot, Charles Greeley, 7
Absorbents, solid, 1056–1058
Absorbers
  absorption by, 65
  coatings for, 549, 597, 838
  in collector design development, 442–446
  convection from, 411–412
  designs for, 30–32
  dimpled, 413–414
  durability of, 544–548, 600–601
  heat exchangers in, 855
  heat transfer in, 131–147
  loss in, 75–76
  materials for, 242–243, 538–539, 831, 834–844
  modeling of, 904–919
  for parabolic reflectors, 86
  plate emittance of, 232
  radiation from, 399–400
  selective, 891–899, 906–908, 991–1001
  stagnation temperature testing of, 236–240
  substrates for, 1001–1008
Absorption, 62–65
Accelerated testing, 1027
Acceptance angle, 65–68
Adhesives, R&D for, 1008–1014
Aerogel glazing, 1045
Agricultural storage applications, 817–821
Air collectors, 445–446
  convection from, 410–414
  reliability and durability of, 570–576
  testing, 213–216, 270, 273–276
Air cooling, storage for, 778–779, 787–788
Airflow reductions in air collectors, 573–574
All-day collector performance, 216–221
Alloys for storage, 661
Angular multiplexing, 336
Annual storage
  for building applications, 797–806
  for crop drying, 817
  for irrigation pumping, 819–820
  in water, 767–768
Antireflection films, 927
Aplanatism, 321
Aquifers, 617, 651, 722–723
Architecture for latent heat storage, 807
Arrays, collector, 161–163
  performance of, 215–216
  R&D for, 449
  testing, 297–303
Arrhenius equation, 21
Artificial aquifers, 617, 651
ASHRAE standards
  for collectors, 16, 188–199, 206–207, 216–219, 288, 290, 292
  for thermal storage systems, 731–735, 738, 741–742, 749–760

ASTEP program, 767
ASTM testing, 294, 957–958
Atkinson, J. F., 485–486
Atmospheric corrosion, 449

Bankston, Charles, 17
Beam walk-off, 458
Beckman, W. A., 65
BEST calculations, 220
Black chrome surface, 601
Black liquid collectors, 145, 443–444
Block-filled rock caverns for storage, 651, 654
Boiling collectors, 142–145, 516–517
Boiling disease, 380
Boltzmann equation, 403
Bouguer's law, 62
Boundary conditions, 424
  and convection, 125
  maintenance of, 484–485
Brightness distributions, 81–83
Brine withdrawal solar pond method, 487–491
Broadband focusing, 335–336
Buffering for storage, 613, 675
Buildings, storage applications for, 797–817
Bulk storage, 661–667
Business conditions, 533–535

Cabot, Godfrey L., 7
Cadmium orthostannate, 888–889, 988
Carnot efficiency, 364
Cavities, convection from, 391–398
Cavity receivers, 103, 107, 157–158, 428
Central receivers, 89–93, 374
Ceramics, 946–947, 1001, 1008
Channel flow, 375–380
Charge monitoring, 823
Chemical conversion absorber coatings, 838
Chemical heat pumps, 678
Cherenkov couplers, 338
Circumsolar region, 81
Clark, Eugene S., 21
Clark, John, 19
Climate and storage, 802–803
Coatings, 523, 549, 597, 836–838, 919–927
Collapse collector tests, 270, 273–276
Collectors, 3, 445–446
  arrays of, 161–163
  concentration ratio and acceptance angle
    for, 65–68
  concentrator receiver design for, 156–161
  concepts and designs for, 12
  costs of, 5, 9, 18–19, 33, 518–535
  durability and reliability of (see Durability
    of collectors)
  efficiency of, 25–28, 116–117, 122, 133–134,
    184, 186, 196, 294, 361, 532

Collectors (cont.)
environmental degradation of (see
  Degradation)
evacuated (see Evacuated collectors)
fire tests for, 276–281
flat plate, 9, 13, 29–33, 68–69, 100, 380–381
heat transfer in (see Heat transfer)
line-focus, 347–348, 450–455
literature on, 4–6
materials for, 228–236, 240–243, 524–525,
  933–937, 946–948
modeling of, 12–14, 99
nontracking (see Nontacking collectors)
nontracking concentrators, 38–41, 67
optical efficiency of, 56–65, 67, 362
performance of, 8–10, 163–164, 215–216
point-focus, 455–459
R&D for, 4–8, 16–17, 442–459
rain effects on, 246–256, 574–576
for space heating, 516–517
standards for, 16, 188–199, 206–207, 216–
  219, 288, 290, 292
structural tests for, 260–276
technology for, 10–11
testing, 14–16, 183–221, 245–246, 297–303
theory and modeling for, 12–14
thermal cycling for, 256–260
thermal energy balance for, 100–117
thermal research for (see Thermal research)
thermal theory of, 99
tracking (see Tracking collectors)
transient effects in, 154–156
Coma aberrations, 321
Combined stresses on polymers, 978–981
COMMIX-SA model, 698, 785
Compound parabolic concentrators, 39–41,
  69–78
and convection, 128
for cooling applications, 452
and edge ray principle, 327
evacuated, 344–345
integrated concentrators, 345–347
nonevacuated, 342–344
Compounds, thermal decomposition of, 681–
  684
Computers
for analysis, 119
for arrays, 162
for central receivers, 91
for convection computation, 123
for design, 9–10
for flow distribution, 384
for liquid storage models, 696, 700–701,
  709
for point-focus systems, 351
for ray tracing, 120
for solar pond models, 473–474

for storage systems, 740–742, 767, 772–773,
  785
for thermal/flow behavior in collectors, 138
Concave mirrors in image–forming systems,
  324–326
Concentrating collectors, 103–109
design of, 156–161
flow over, 376
testing, 203–207
Concentration ratio, 65–68, 87
Concentrators. See also Compound parabolic
  concentrators
designing, 332–333
image-forming, 318–324
integrated, 345–347
line-focus, 450–455
nonimaging, 328–335
nontracking, 38–41
tracking, 41–50
Conduction, 400–407
Cone optics, 87–88
Connections, durability of, 560–566
Conservation of energy principles, 360
Convection
and collector heat transfer, 121–131
in cylinders, 127–128, 157–158, 391–398,
  425–426
external, 388–398
in heated fluids, 136–138
internal, 407–429
in solar ponds, 468
from wind, 152–153
Converging lenses, 308
Convex lenses, 310
Convolution models, 697
Cooking applications, 822–823
Cooling
applications for, 512
CPC for, 452
desiccant, 718–719
storage for, 613, 817
Copper for absorbers, 31
Corrosion, 449
of connections, 563
by heat transfer fluids, 549–550, 832, 856–
  861, 947, 1014–1021
of metal collector components, 934–935
with solar ponds, 487, 491–492
Cost, 511–517
of collectors, 5, 9, 18–19, 33, 518–535
vs. performance, 413
Coupling photomultipliers, 338
Covers
absorption by, 62–65
durability of, 543–544
loss through, 117–121
material for, 240–242

Crop-drying applications, 817
Cylinders, convection from, 127–128, 157–158, 391–398, 425–426
Cylindrical collectors, 100, 103
Cylindrical glass envelopes, 119–120

Dave, R. N., 701
Decomposition, thermochemical, 618
Degradation, 15, 579–580, 596–597
  of absorber plates, 544–548, 600–601
  of adhesives, 1009–1013
  evaluation of, 958–964
  of glazing, 598–600, 975–981
  from helium, 402–407
  of housings, 603
  of lower-cost collectors, 20–21, 603–604
  and performance, 955–973
  from photodegradation, 581–584, 875–881, 975–981
  of reflectors, 601–603
  research for, 604–605
  of stratification, 711
  testing for, 595–596, 956–958, 964–972
  from ultraviolet irradiance, 584–594, 602, 881–883, 975–981
Delivery period displacement, 614–615
DELSOL program, 91, 351
Density-measuring for solar ponds, 494
Density stability ratio for gradient zones, 481
Department of Energy, testing by, 245–246
Deposition technologies
  for absorbers, 898–899, 999–1000
  for heat mirrors, 989–991
Desiccants, 718–719, 1028–1032
Design development
  for annual storage, 806
  for heat mirrors, 889–890
  for heliostats, 460–463
  for line-focus concentrators, 450–453
  for nonconcentrating collectors, 442–447
  for point-focus collectors, 456–458
Detector-baffled sphere, 961
Dewar flask absorber design, 36
Diffuse radiation, 111, 146, 216–221
Diffusion in solar ponds, 479
Dimpled absorbers, 413–414
Direct absorption receivers, 161
Direct heat exchange, 492, 664–667, 716
Diurnal storage, 615, 626, 630
Doped semiconductors, 885–889, 923–926, 986–988
Double-exposure collectors, 149–150
Drude theory, 923–924
Dual storage systems, 616, 804
Duffie, J. A., 65
Durability of collectors, 19–20, 221–230, 538–539, 570–572. See also Degradation

absorber coatings, 549
absorber plates and tubes, 544–548, 600–601
airflow reductions, 573–574
conclusions about, 240–245
cover plates, 543–544
heat transfer fluids, 549–550
house integration, 569–570
housing, 560
insulation, 550–559
materials for, 832, 863–868
piping connections, 560–566
roof structure integration, 566–569
rupture and collapse tests, 270, 273–276
sealants, 559–560
test results for, 231–240
water leakage, 574–576
wood deterioration, 540–543

Earth beds, 639–646, 720–721
Edge ray principle, 324, 327–328
Edwards-type sphere, 960
Efficiency
  of collectors, 25–28, 116–117, 122, 133–134, 184, 186, 196, 294, 361, 532
  optical, 56–65, 67, 362
  of solar ponds, 468, 473
Electrical circuit analogies, 113–117
Electrical conductivity measurements for solar ponds, 494–495
Electroplated absorber surfaces, 834–835
Electrostatic charging, 602–603
Emissivity, 948–950
Emittance, measuring, 962–963
Engineering issues
  for collectors, 447–449
  for heliostats, 463
  with line-focus concentrators, 453–455
  with point-focus collectors, 458
Enthalpy model, 717
Environment
  degradation by (see Degradation)
  and saltless ponds, 502–503
Equilibrium relationship in gradient zones, 480–481
Ericcson, John, 7
Erosion in solar ponds, 480–486
Etendue, 309
  calculating, 331–332
  generalized, 314–318
Evacuated collectors, 32, 35–38
  absorbers for, 840–842
  drainable version of, 446–447
  internal conduction in, 402–403
  reflectors for, 78–79
Evacuated CPCs, 344–345

Evacuated enclosures, heat transfer in, 150–151
Evacuated glazing, 1045
Evaluation
  of collector materials, 933–937, 946–948
  of collectors, 14–16
  of degradation, 958–964
  of thermal storage systems, 735–760, 789–791
Exergy, 364–367
External flow, 369–370
External heat transfer, 386–398
External receivers, 107, 159–161

Fermat's principle, 313–314
Fiberglass reinforced concrete absorbers, 444
Field efficiency of central receivers, 92
Figures of merit for storage systems, 799
Finned absorbers, 141
Fin-tube problem, 135
Fire hazards, 541–542
Fire tests, 276–281
First law of thermodynamics, 360–363
Flashing, leakage from, 576
Flat absorbers, loss in, 75–76
Flat plate collectors, 9, 13, 29–33, 68–69, 100, 380–381
Flex hoses for tracking collectors, 455
Fluid mechanics, 369–384
  convection in, 136–138
  in parallel channels, 380–384
  for solar ponds, 498–499
Foam insulation systems, 856
Focal length, 308
Forced convection heat transfer coefficient, 407–414
Fourier models for solar ponds, 473–474
Freeborne, William, 20
Freezing
  plastic absorbers for, 444–445
  protection from, 550, 595
  in solar ponds, 485
Fresnel equations, 59–60
Fresnel lenses, 47, 93–95, 353, 451–452
Fresnel mirror troughs, 338
Fresnel reflectors, 89–93
Friesem, A. A., 95
Full-irradiance conditions, collector performance under, 188–207
Fusion storage, 617, 656–671, 1022–1023

Gasket materials, 832, 861–863
Gaussian distribution, 85
Gaussian optics, 310
Geometric optics, 305–326
Glass
  materials for, 844–846

vs. plastic, 532, 598–599, 875
Glaubers salt, 807
Glazings
  environmental degradation of, 598–600, 975–981
  with low thermal conductance, 1041–1048
  materials for, 831–832, 844–855, 874
  photodegradation of, 875–881
  polymer, 974–983, 1054–1056
  and processing variables, 883
  switchable, 1048–1054
  ultraviolet stabilization of, 881–883
Government programs, 833–834
  and collector research, 8
  for convection, 417
  and costs, 511, 517–518
  for research, 358–359
Graded-composite/metal tandem absorbers, 908–914
Gradient zones. See Salt gradients
Grooved cavities for gap loss, 76
Ground heat loss with solar ponds, 474–475, 497–498

Hail
  collector tests for, 264, 270
  protection from, 448–449, 595
Harleman, D. R. F., 485–486
Heat deterioration, 571
Heated fluids, convection in, 136–138
Heat exchangers, 855
Heat extraction from solar ponds, 486–492, 497
Heat field collector tests, 299–303
Heat loss coefficient U, 414–429
Heat mirrors, 884–891, 983–991
Heat pump storage, 678–681
  in building applications, 807–810
  thermochemical, 789
Heat transfer, 385
  through absorbers, 131–147
  through back and side, 147–150
  from exterior, 151–154
  external, 386–398
  internal, 398–429
  in partially evacuated enclosures, 150–151
  in SHACOB program, 766–767, 770–771
  window system, 117–131
Heat transfer fluids
  durability of, 549–550
  materials for, 832, 856–861, 947–948
  R&D for, 1014–1021
HELIOS program, 91
Heliostats, 460–463
Helium, degradation from, 402–407
h forced convection heat transfer, 407–414
High temperature storage, 714, 1059

Hilbert integral, 331–332
Holography, 335–337
Honeycombs, 32–33, 851–854
Hottel-Whillier model, 8–9, 13, 29, 57, 135, 138, 142, 183, 385, 471–472
Hot water systems, 512–515, 517, 698–701
Houses, collector integration into, 569–570
Housings, durability of, 560, 603
Hull, John, 18

Ice for storage, 656, 719–720
Ignition temperatures, 542
IITRI testing program, 965
Image-forming systems
  aberrations of, 310–313
  characteristics of, 318–324
  mirror, 324–326
  principles of, 306–310
Image spreads, 81–86
Impermeable absorbers, 412
Incident angle modifier, 200–203, 291, 294
Inclination and convection, 123
Incongruent melting, 617
Indirect heat exchange, 661–664
Indium tin oxide for heat mirrors, 887–888, 986–987
Industrial process heat field collector tests, 299–303
Industrial storage applications, 817–821
Injection in solar ponds, 470–471
Inorganic compounds for storage, 661, 784–785
Instrumentation for solar ponds, 492–499
Insulation, 147–149
  durability of, 550–559
  materials for, 539, 832, 855–856
Integrated concentrators, 345–347
Intercept factor, 56–57, 67
Interference films, 919–923
Internal corrosion, 449
Internal flow, 375–384
Internal heat transfer, 398–429
Internal partial baffles for convection, 123–125
Interseasonal storage, 797
Intrinsic absorbers, 891–895, 994–995
Ion plating, 990
IPH testing program, 966–967
Irrigation pumping, 819–820
ISO testing for degradation, 958

Jet impingement heaters, 141–142

Knudsen number, 402
Kooi, C. F., 471–472
Kritchman, E. M., 95
Kutscher, Charles, 17

Langley, Samuel Pierpont, 7
Latent thermal storage, 617–618, 655
  for building applications, 806–807
  evaluation of, 747–760
  fusion, 656–671
  modeling of, 715–720
  R&D for, 782–785
  solid–solid transformation, 675–678
  vaporation, 671–675
Layered injection in solar ponds, 470–471
Leakage in air collectors, 213–215
Lenses, 45–47, 93–95, 308, 353, 451–452
LiBr chillers, 810–812
Limb darkening, 81
Line-focus concentrators, 347–348, 450–455
Line-focus Fresnel lenses, 47
Liners for solar ponds, 499–501
Lior, Noam, 14
Liquid–liquid heat exchangers, 766–767
Liquid phase reactions, 684
Liquids
  heat transfer through, 142–145
  for thermal storage, 616, 625–639, 648–655, 698–710
Literature on collectors, 4–6
Localized corrosion, 1018
Longitudinal collector load tests, 264
Long-term storage. See Seasonal thermal storage
Low concentration systems, 338–342
Low-cost collectors, degradation of, 20–21, 603–604
Loxsom, Fred M., 21
Lumped-parameter models, 697

Mach-Zehnder interferometers, 499
Macroencapsulation, 667, 671
Materials, 1040
  for absorbers, 242–243, 538–539, 831, 834–844, 891–899
  for collectors, 228–236, 240–243, 524–525, 933–937, 946–948
  for covers, 240–242
  for glazings, 831–832, 844–845, 874
  for heat mirrors, 884–891
  for heat transfer fluids, 832, 856–861, 947–948
  for heliostats, 460
  high temperature storage fluids, 1059
  initiatives and plans for, 833–834
  for insulation, 539, 832, 855–856
  modeling of (see Modeling)
  performance of, 832, 863–868
  for reflectors, 831–832, 844–855
  for sealants and gaskets, 832, 861–863
  for storage, 774–775, 1021–1028

Materials (cont.)
testing of, 14–16, 183–221, 245–246, 297–303
and thermal conduction, 401
Mathematical thermal research methods, 429–430
Matrix **M**, 919–923
Maxwell-Garnett theory, 912–914
Mechanical properties, characterization of, 953
Membrane liners for solar ponds, 499–501
Mesh coatings, 926–927
Metal-dielectric composites, 896–898, 996–998
Metal microgrid heat mirrors, 891, 989
Metals
for absorber substrates, 1001
corrosion of, 1015–1021
evaluation of, 933–937
for storage, 639, 648, 661
Microencapsulation, 671
Microgrid heat mirrors, 891, 989
Mining operations, solar ponds for, 504
MINSUN program, 709–710, 723
Mirrors, 324–326, 460–463, 884–891, 983–991
MIRVAL program, 91
MISR collector testing, 297–299
Modeling
of absorber surfaces, 904–919
of antireflection films, 927
of coatings, 919–927
of collectors, 12–14, 99
of thermal storage systems, 692–724
of reflectors, 927
Moisture damage, 595
Moisture in soils, 720
Molecular cavity desiccants, 1030
Molten salts for storage, 638–639, 822
Monitoring of solar ponds, 492–499
Monte Carlo techniques, 87, 91, 120
Moody diagram, 378
Multilayer absorbers, 898, 914–919, 998
Multiple reflections, 60–62, 65

National Bureau of Standards, 731, 867–868, 956–957, 965–966
Natural aquifers, 617, 651
Natural stratification, 616, 630
Negative wind loads, 261, 264
Nielsen, Carl, 18
Nonconcentrating collectors, 100–103, 442–449
Nonevacuated CPCs, 342–344
Nonimaging optics, 326–335
Nonisothermal collectors, 367

Nontracking collectors
compound parabolic concentrators, 69–78
flat plate, 68–69
low concentration, 338–342
and reflectors for evacuated tubes, 78–79
and side reflectors, 79–80
thermal performance of, 183–221
V-troughs, 79
Nontracking concentrators, 38–41, 67
Norrish reactions, 583
NTU model, 718
Nucleation, 617
Numerical aperture, 320–321
Numerical thermal research methods, 429–430

Oils for storage, 654–655
Optical efficiency, 56–65, 67, 362
Optical material evaluation, 960–963
Optical path length, 313–314
Optical properties, characterization of, 948–953
Optical-trapping absorber surfaces, 914
Optics
errors in, 81–86
geometric, 305–326
holography, 335–337
nonimaging, 326–335
R&D for, 338–354
Organic materials for storage, 635–638, 656, 748
Oriented polymers, 584
Overlapped glass plate heaters, 141
Ozone, 584–588, 594

Packed bed absorbers, 141
Paints, 32, 838–840, 897–898, 955–956, 997–998
Parabolic collectors and reflectors, 10, 14, 42, 86–89, 347. *See also* Compound parabolic concentrators
design of, 450–452
dirt accumulation on, 455
Paraffin storage, 656, 660, 768–769
Parallel channels, flow in, 380–384
Parallel plates, convection between, 126–128
Paraxial optics, 310
Parker, A. J., Jr., 1053
Partially evacuated enclosures, heat transfer in, 150–151
Particulate absorbers, 914
Passive heating, storage for, 626
PCM. *See* Phase change materials
Pebble-beds, 742–746
Peel strength, 953
Performance
of arrays, 163–164, 215–216

of collectors, 8–10, 163–164
of compound parabolic concentrators, 73
vs. cost, 413
and degradation, 955–973
of materials, 832, 863–868
of solar ponds, 471–475
Pettit, R. B., 85
Phase change materials (PCM), 617, 656, 660, 664
  evaluation of, 747–749
  modeling of, 715–718
  in paraffin, 768–769
  R&D for, 776–778, 783–784, 787–788, 1023
Phase segregation, 617
Phase space concept, 314–318
Phillips, W. F., 701
Photodegradation, 581–584, 875–881, 975–981
Photo-excitable isomers, 1023
Physical material evaluation, 963–964
Piping connections, 560–566
Pitting corrosion, 1019–1020
Planar reflectors, 447–448
Plane surfaces, convection from, 388–391
Plastic bubbles for collector protection, 458
Plastics, 597, 846–851, 1054–1056
  for absorbers, 33, 444–445
  for absorber substrates, 1005–1008
  evaluation of, 937
  vs. glass, 532, 598–599, 875
  photodegradation of, 581–584
  R&D for, 974–983
  ultraviolet stabilization of, 881–883
Plate absorbers, heat transfer through, 131–136, 140–142
Point-focus collectors, 455–459
Point-focus concentrators, 47–48, 348–353
Polished glass reflectors, 602
Polymers. See Plastics
Pond liners, 499–501
Porous desiccants, 1030
Porous matrix absorbers, 141
Positive wind loads, 260–261
Power generation storage applications, 492, 502, 504–505, 822
Pressure distribution on collectors, 376, 381
Processing variables and material degradation, 883

Quad-cavity receivers, 158

Rabl, Ari, 12
Radiation, 117–121, 125
  diffuse, 111, 146, 216–221
  external, 388
  internal, 399–400
Radiative energy input, 110

Rain
  effect of, on collectors, 246–256, 574–576
  and solar ponds, 485
Rankine chillers, 812–813
Ray tracing, 10, 120, 306
Receivers
  cavity, 103, 107, 157–158, 428
  central, 89–93, 374
  external, 107, 159–161
  thermal design of, 156–161
Reflections
  laws of, 57–58, 84
  multiple, 60–62, 65
Reflectors
  environmental degradation of, 601–603
  for evacuated tubes, 78–79
  Fresnel, 89–93
  materials for, 831–832, 844–855
  models for, 927
  parabolic, 86–89
  planar, 447–448
  for point-focus collectors, 456
  side, 79–80
  V-trough, 39, 78–79
Refraction, law of, 58–59, 305–306
Refractories for storage, 648
Reliability. See Durability of collectors
Research and development
  for absorber substrates, 1001–1008
  for adhesives, 1008–1014
  for collectors, 4–8, 16–17, 442–449, 455–459
  for degradation, 604–605
  for desiccants, 1028–1032
  for heat mirrors, 983–991
  for heat transfer fluids, 1014–1021
  optical, 338–354
  for polymer glazings, 974–983
  for selective absorbers, 991–1001
  SHACOB program, 764–787
  for solar ponds, 17–18
  STOR program, 787–789
  for thermal storage systems, 763, 789–791, 1021–1028
Response time, collector, 291
Reversible thermochemical reactions for storage, 678–684
Rhee, S. J., 411
Rock storage, 639–646, 651, 654, 710–714, 742–747, 775–776, 781
Roof structure integration, 566–569
Rupture collector tests, 270, 273–276

Salt gradients, 468–470, 501–502
  erosion in, 480–486
  monitoring, 494–495
  stability in, 475–479
Saltless solar ponds, 502–503

Salts for storage, 638–639, 660–661, 807, 822, 1027
Salt solution desiccants, 1031
Sand for storage, 646
Scanning injection in solar ponds, 470
Scattering effects in solar ponds, 496
Schladow, S. G., 486
SCRAM program, 351
Sealants
  durability of, 559–560
  materials for, 832, 861–863
Seasonal thermal storage, 468, 633–635, 797
  modeling of, 707–710
  R&D for, 782, 788, 790
  tank losses in, 721–722
Second law of thermodynamics, 66, 364–368
Selective absorbers, 891–899, 906–908, 991–1001
Selective coatings, 523
Semiconductors
  for coatings, 923–926
  for heat mirrors, 885–889, 986–988
  for selective absorbers, 895–896, 906–908, 995–996
SERI testing programs, 966–972
SHACOB storage system, 781–787
  initial projects, 764–770
  large solicitation for, 770–781
Shading effects, 145–147, 570
Shadis, W. J., 1053–1054
Side reflectors, 79–80
Side wall effects in solar ponds, 479
Slags for storage, 648, 667
Slats for convection suppression, 130
Snell's law of reflection, 57–58, 84
Snell's law of refraction, 58–59, 305–306
Soil liners for solar ponds, 500
Soil properties, measuring, 497–498
Soils, modeling of, 720–722
Solar Demonstration Program, 511
Solar fractions, 801
Solar One Plant, 395
Solar ponds, 34–35, 468–470
  alternative salts for, 501–502
  applications of, 503–505
  gradient zones for, 475–486
  heat extraction from, 486–492
  instrumentation for, 492–499
  liners for, 499–501
  R&D for, 17–18
  saltless, 502–503
  for storage, 633–635, 819, 825
  thermal performance of, 471–475
  as thin-film systems, 580
Solder joints, durability of, 566
Solid absorbents, 1056–1058
Solid–solid transformation, 675–678

Solid storage media, 616–617, 742–747, 775–776, 781
  earth and rock, 639–646
  with liquids, 648–655
  metals and slags, 648
  modeling of, 710–714
  refractories, 648
  sand, 646
SOLPOND program, 474–475
SOLRAY program, 162
SOLSYS program, 119, 704
Sorption chillers, 813–815
Sorption heat pumps, 815–816, 820–821
SRCC calculations, 218–219, 221
Stability, gradient zone, 475–479
Stagnation condition, 362
Steady-state solar pond model, 471–473
Storage fluids, high temperature, 1059
Storage technology. *See* Thermal storage systems
STOR program, 787–789
Stratification storage systems, 804–806, 825
  degradation of, 711
  modeling of, 701–707, 781
  natural, 616, 630
  R&D for, 785–786, 790
Structural tests for collectors, 260–276, 448
Substrates for absorbers, 1001–1008
Surface coatings, 598
Surface effects and gradient zones, 485–486
Swimming pool heating, 11, 504, 512
Switchable glazings, 1048–1054
SYMFIT method, 221

Tanks, heat loss from, 721–722
Tax credits, 517–518, 534
Technology for collectors, 10–11
Telecentric optical systems, 320
Temperature
  damage from, 594–595, 957
  monitoring of, for solar ponds, 493–494
Tempered glass for hail protection, 449, 595
Testing
  accelerated, 1027
  of collectors, 14–16, 183–221, 203–207, 289–303
  for degradation, 595–596, 956–958, 964–972
  of materials, 228–236, 240–243
  of selective absorbers, 1000
  of thermal storage systems, 731–735
Textured surface absorbers, 898, 999
Theoretical thermal research methods, 429–430
Theory for collectors, 12–14
Theory of uniform gradients, 476
Thermal boundary conditions, 424
Thermal conductivity of materials, 933

Thermal cycling, 256–260
Thermal energy balance, 100–117
Thermal energy transport, 823
Thermal flow behavior in collectors, 138–139
Thermal network models, 697
Thermal performance of collectors
  air collectors, 213–216
  all-day, 216–221
  basic equations for, 183–188
  under full-irradiance conditions, 188–207
  and material testing, 231–236
  under zero-irradiance conditions, 207–212
Thermal research, 358–359
  and fluid mechanics, 369–384
  for external heat transfer, 385, 386–398
  for internal heat transfer, 385, 398–429
  theoretical methods for, 429–430
  and thermodynamics, 360–368
Thermal storage systems, 611–612, 692–697,
  796
  agricultural and industrial, 817–821
  applications for, 618–619, 823–825
  for aquifers, 722–723
  in buildings, 797–817
  cooking applications, 822–823
  evaluation of, 735–760
  latent, 715–720 (see also Latent thermal
    storage)
  in liquids, 616, 625–639, 648–655, 698–710
  long-term (see Seasonal thermal storage)
  for power generation, 492, 502, 504–505, 822
  purposes and strategies of, 613–615
  R&D for, 763, 789–791, 1021–1028
  retrospective for, 619–623
  in reversible thermochemical reactions, 678–
    684
  salts for, 638–639, 660–661, 807, 822, 1027
  SHACOB program, 764–787
  in solids, 639–655, 710–714, 720–722, 742–
    747, 775–776, 781
  standards for, 731–735, 738, 741–742, 749–
    760
  STOR program, 787–789
  stratification (see Stratification storage
    systems)
  technical approaches to, 615–618
  thermochemical storage, 618, 678–684, 723–
    724, 789
Thermal theory of collectors, 99
Thermal traps, 32, 120–121
Thermic diode collectors, 33
Thermochemical storage, 618, 678–684
  modeling of, 723–724
  R&D for, 789
Thermoclines, 616, 638
Thermodynamics, 360–368
Thin-film plastic absorbers, 444–445

Thin-film systems, degradation of, 579–580
Thin-metal film heat mirrors, 889–890, 988–
  989
Tin oxide for heat mirrors, 887–888, 987–988
TOT method, 221
Tracking collectors, 80
  flexible hosing for, 455
  flow over, 372–375
  and Fresnel lenses, 93–95
  and Fresnel reflectors, 89–93
  and image spread, 81–86
  with line-focus, 347–348
  parabolic reflectors, 86–89
  with point-focus, 348–353
  testing, 15, 289–303
Tracking concentrators, 41–50
Transient effects, heat transfer from, 154–156
Transient response functions, 697
Transmittance-absorptance product, 185
Transmittance measurements for solar ponds,
  495–497
Transmittance spheres, 960–962
Transparent coatings, models for, 919–927
Transpired absorbers, 412, 446
Trickle collectors, 31–32
TRNSYS program, 696, 700, 709, 740–742,
  767, 772–773
Trombe wall glazing, 1041–1043
Tubular absorbers, loss in, 76
Turbulence and convection, 125–126
Two-dimensional concentrators, 328–331
Two-tank intermittent storage systems,
  678

U heat loss coefficient, 414–429
UH-XX programs, 91
Ultraviolet irradiance, 584–594, 602, 881–
  883, 975–981
Unconstrained storage systems, 797–799

Vaporation, 671–675
Vapor-deposited absorber coatings, 836–838
Visual material evaluation, 958–960
V-shaped cavities, 334
V-trough reflectors, 39, 78–79

Walton, I. C., 478
Warranties, 18
Water
  effect of, on collectors, 246–256, 574–576
  for thermal storage, 626–634, 735–742 767–
    768, 772–775
Water heating
  and annual storage, 803
  gas backup, 772–773
  storage for, 626
Weight and costs, 531

Wind loading
  and cavities, 398
  convective effects of, 152–153, 158
  damage from, 595
  on heliostats, 463
  on line-focus concentrators, 453–455
  on point-focus collectors, 458
  on stationary collectors, 370–372
  tests for, 260–261, 264
  on tracking collectors, 372–375
Window system, collector heat transfer in,
    117–131
Winston, Roland, 16
Wood deterioration, 539–543

Yearly averaging storage, 615
Yekutieli, G., 95

Zeopower-type chillers, 813–815
Zero-irradiance conditions, collector
    performance under, 207–212